P9-AGS-331

NMR Band Handbook

NMR
BAND HANDBOOK

BY HERMAN A. SZYMANSKI
AND ROBERT E. YELIN
Department of Chemistry
Canisius College, Buffalo, New York

IFI/PLENUM • NEW YORK–WASHINGTON • 1968

Library of Congress Catalog Card Number 66-20321

A Subsidiary of Plenum Publishing Corporation
227 West 17 Street, New York, N. Y. 10011

Printed in the United States of America

CONTENTS

INTRODUCTION

With the development in recent years of NMR spectrometers capable of recording with a minimum of effort data that are both accurate and reproducible, a need has arisen for spectra catalogs and data books to aid the researcher. The ability to correlate a spectrum to a precalibrated chart accurately (to within 0.02 ppm) has made chemical shift, along with the character of the spectral lines produced, an increasingly important tool in configurational analysis—a tool which this book is designed to complement. It is known that the chemical shift of a proton can be influenced by atoms as much as five or six carbons away, a fact of great importance to compound identification. The entries in this book, arranged according to proton environment, enable the researcher with a hypothetical structure already proposed for his unknown quickly to locate examples of compounds having similar structures. An index of molecular formulas and a shift index, which should prove useful in hypothesizing alternate structures for the unknown and in checking out the hypotheses, have also been provided.

The data collected represent some 4800 shifts from the spectra of about 1200 compounds that appear in the catalogs of Varian Associates, Sadtler Research Laboratories, and the American Petroleum Institute, whose cooperation is gratefully acknowledged. The alphanumerical code that is the basis of our presentation, and the explanation of its use, are borrowed intact from the NMR Spectra Catalog of Varian Associates, for whose permission to use this material we are especially grateful.

It now remains to discuss the various elements that comprise this volume and explain the symbolism used.

I. The Handbook Entries

Data are entered into the Handbook according to proton code designation. The method of coding protons is described in section II, and a tabulation of the structural groups to which the code numbers and letters refer will be found on page ix and, for convenience, on a foldout at the end of the book. The latter may be kept in view while using the Handbook and its indexes. Entries are arranged by the numerical order of main groups listed in the coding table, and then by the alphabetical sequence of, first, subgroups, and then, sub-subgroups.

▶ 1-Nav ❶

❷	Shift	Peak
▶ a	2.93	1A
b	~6.67	N
c	~7.15	O

(b) (c) (c) (b) (c) CH_3 N — CH_3(a) ❸

❹ ▶ $C_8H_{11}N$ ❺ ▶ 45.3 mg/0.50 ml $CDCl_3$ ❻ ▶ S 1

Each entry consists of the following elements:

1. *The proton code designation,* which determines the position of the entry in the collection. Where code designations are the same, entries are arranged according to increasing value of chemical shift. Parentheses have no effect on the ordering sequence.
2. *The spectral data.* This table contains a line for each proton group keyed on the structural formula. The line for the proton code designation of the entry in question is indicated by an arrowhead. Chemical shifts are reported to the nearest cycle per second. Centers of broad lines are estimated. In the case of multiplets with peaks of nearly first order, the center of the multiplet is recorded; for higher-order multiplets, the center of gravity is estimated if all the lines are observable. Where overlapping occurs, we list an approximate position if the assignment seems reasonable. The limitations of the peak code are discussed in section III, and the code designations are tabulated on page xiii and also on the foldout.
3. *The structural formula,* including letter designations that refer to the chemical shift table.
4. *The molecular formula,* for convenience in using the Molecular Formula Index.
5. *The solvent* used for the sample, and its concentration.
6. *The source* from which the data were obtained. V refers to *NMR Spectra Catalog,* Volumes 1 and 2, Varian Associates, Palo Alto, California; S to *NMR Spectra Catalog,* Sadtler Research Laboratories, Philadelphia, Pennsylvania; and API to *Nuclear Magnetic Resonance Spectral Data,* American Petroleum Institute, Research Project 44, Texas A & M University, College Station, Texas. The numbers following the letter designations are the spectrum numbers used in the source catalog.

II. The Proton Code

All proton chemical environments have been classified into twenty-one main groups, listed in the table on page ix and designated by the

MAIN GROUPS	FUNCTIONAL GROUPS
1. $R-CH_3$	a. $-CH_3$
2. $R-CH_2-R'$	b. $-CH_2-$
3. $R-CH(R')-R''$	c. $>CH-$
4. $R-CH=N-R'$	d. $>C<$
5. $R-CH=CH_2$	e. $-Cl$
6. $(H)(R)C=C(R')(H)$	f. $-F$
7. $(H)(R)C=C(H)(R')$	g. $-Br$
8. $(R')(R)C=C(H)(H)$	h. $-H$
9. $(R')(R)C=C(R'')(H)$	i. $-I$
10. $R-C\equiv CH$	j. $-C(=O)-$
11. $R-C(=O)H$	k. $-C(=O)O-$
12. $R-C(=O)OH$	l. $-C(-)=$
13. $R-OH$	m. $-N=$
14. Phenyl	n. $-N<$
15. Pyridyl	o. $-O-$ or $\rightarrow O$
16. Furanyl	p. $-OH$
17. Pyrryl	q. $-O-C(=O)-$
18. Thienyl	r. $-NO_2$
19. Misc. aromatic	s. $-S-$
20. $(R')(R)NH$	t. $-C\equiv C-$
21. H on other atoms	u. $-C\equiv N$
	v. Phenyl
	w. Aromatic ring junction
	x. Misc. aromatic
	y. Thienyl
	z. Misc. atom

Arabic numerals 1 to 21. Nearby arrangements of atoms are classified into functional subgroups and sub-subgroups and designated by capital and lower-case letters of the alphabet respectively. These are also listed in the table. Subgroups are bonded directly to the main group, and are coded by capital letters. Sub-subgroups are bonded directly to the subgroup, and are coded by lower-case letters.

In order to code a particular proton environment, write down the number of the main group in which it falls. If the main group is part of a ring, enclose the number in parentheses. This number is followed by a hyphen, and then by the capital letters corresponding to each subgroup. If the subgroup is part of a ring, the capital letter should be placed in parentheses. The capital letters should be arranged in alphabetical order except for groups 4-9, for which a special order applies. Each capital letter should be followed by the lower-case letters corresponding to the sub-subgroups, again in alphabetical order, before proceeding to the next capital letter. Lower-case letters are not put in parentheses, no distinction being made between cyclic and noncyclic sub-subgroups.

The procedure described above covers most straightforward situations.

Example: Code the underlined proton.

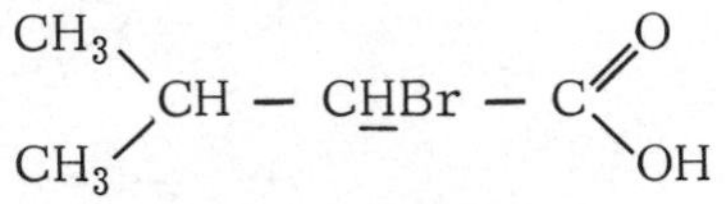

$$(CH_3)_2CH - C\underline{H}Br - C(=O)OH$$

The proton is in main group 3, and its immediate neighbors are CH, Br, and $C(=O)O$, which are coded C, G, and K in alphabetical order. The CH< group has two sub-subgroups, both methyl, coded aa, and the $C(=O)O$ group has one sub-subgroup, a proton, coded h. The entire code symbol describing the underlined proton is thus

3-CaaGKh

A few special rules must be observed in coding certain types of protons. These include the following:

A. It should be noted that olefinic protons are classified into six main groups, 4-9, depending upon whether the double bond is a C = C bond or a C = N bond, whether it is mono-, di- or tri- substituted, and upon the cis- or trans- nature of the substitution. In order to distinguish between the olefinic protons and/or locate the substituents in groups 4, 6, 7, 8, and 9, the subgroups are coded in the order gem-, cis-, trans- relative to the proton being described, rather than in the usual alphabetical order. Since the proton–proton relationship is defined by the main group number, however, the second olefinic proton is omitted in the subgroup coding. In the case of group 5, the two olefinic protons on the same carbon atom cannot be distinguished without coding the protons; therefore, they are not omitted in this case.

Example: Code protons (1) and (2).

$$\overset{(1)}{H}(CN)C{=}C\overset{(2)}{H}(CH_2OH)$$

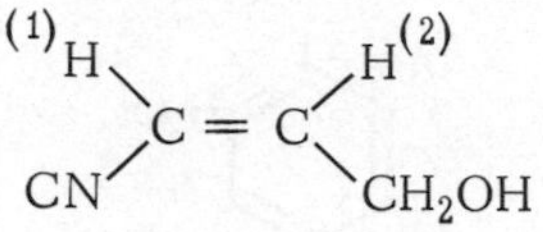

Both protons belong to group 7. Proton (1) is gem- to CN whose code is U. It is cis- to proton (2) whose code is omitted, and trans- to CH_2OH whose code is Bp. Proton (2) bears a reverse relationship to the substituents. Hence, proton (1) is coded 7-UBp and (2) is coded 7-BpU.

Example: Code protons (1), (2), and (3).

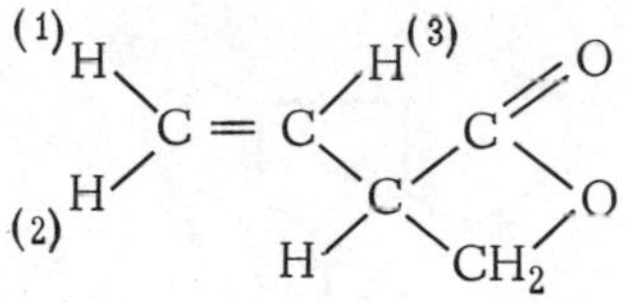

The code for the substituent group is (C)bk. Proton (1) is coded 5-HH(C)bk. Proton (2) is coded 5-H(C)bkH. Proton (3) is coded 5-(C)bkHH.

B. Conjugated double bonds result from bonding groups 4-9 to subgroups J, K, L, or M. Since subgroups L and M have an unused double bond, they will nearly always be followed by l or m. An exception is allene, $CH_2 = C = CH_2$, coded 8-Lhh.

C. Phenyl protons are coded with only the capital letters representing the two ortho-substituents in alphabetical order, since the rigid geometry of these molecules makes further coding meaningless.

Example: Code protons (1) and (2).

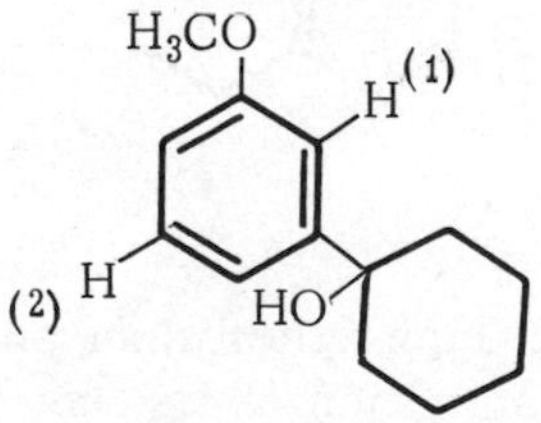

The ortho- substituents for proton (1) are, in alphabetical order, (D) and O. Thus, the code for proton (1) is 14-(D)O. The code for proton (2) is 14-HH.

D. When a benzene ring is a substituent, i.e., a subgroup, its code letter, V, is followed by lower-case letters denoting the two sub-subgroups ortho- to the point of attachment in alphabetical order.

Example: Code the carboxyl proton.

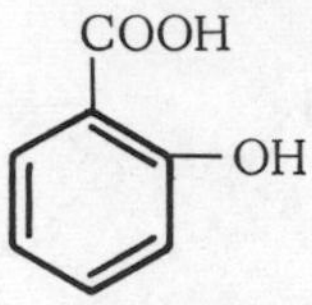

The proton belongs to group 12. The substituent is a benzene ring with a proton and hydroxyl ortho- to the point of attachment. Hence the code is 12-Vhp.

E. Indoles and other polynuclear heterocyclic compounds are coded as benzene derivatives, unless aromaticity arises exclusively from resonating double bonds.

Example: Code proton (1).

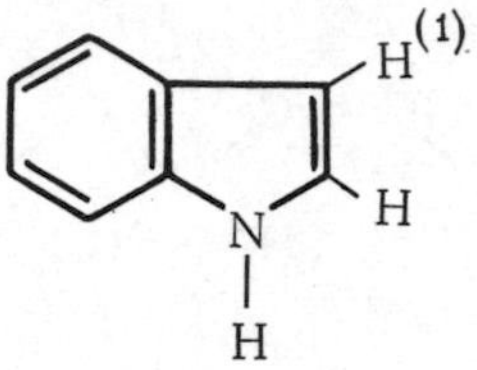

This proton belongs to cyclic group 7. The gem- substituent is phenyl, coded Vhn, the cis- substituent is a proton, omitted, and the trans- substituent is NH, coded (N)hv. The entire code is thus (7)-Vhn(N)hv.

F. In the case of groups 15, 16, 17, and 18, the position of the proton relative to the heterocyclic atom is denoted by the Greek letters α, β, γ, placed between the main group number and the first capital letter designating an adjacent substituent.

Example: Code the aromatic protons.

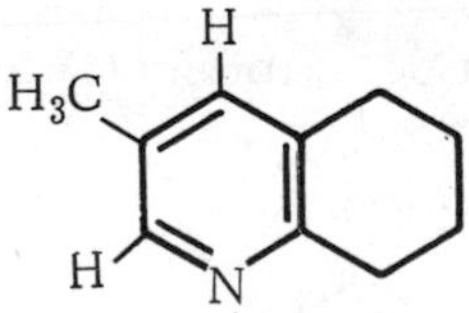

15 - α - A
15 - γ - A(B)

G. Note that if an aromatic ring other than phenyl is bonded to a main group, the subgroup designation X is not followed by lower-case designations for sub-subgroups in ortho- positions.

H. Miscellaneous aromatic protons are coded with a 19 followed by the chemical name of the ring type and the proton position in parentheses, and then by the code letters for the ortho- substituents.

I. Protons bonded to atoms which are not covered by groups 1-20 are coded 21 followed by the name of the atom in parentheses and then by the appropriate number of subgroups.

PEAK CODE

1A. Sharp singlet
1B. Medium-broad singlet
1C. Broad singlet
1D. Very broad singlet
1E. Singlet interfered with or overlapped by other bands
1F. Medium-broad singlet with fine line structure
1G. Singlet with shoulder
1H. Sharp singlet with fine line structure
1I. Singlet with fine line structure (upon expansion a more complex pattern is observed)

2A. Doublet, both lines sharp
2B. Doublet, both lines of medium width
2C. Doublet, half of an A_2B_2 or AB pattern
2D. Doublet with close spacing of lines
2E. Doublet which upon expansion shows a more complex pattern
2F. Doublet interfered with by other lines
2G. Distorted doublet
2H. Doublet overlapped by other bands

3A. Triplet, well formed and distinct
3B. Triplet, but with some distortion, not clear
3C. Triplet with fine line structure overlapping
3D. Triplet interfered with or overlapped by other signals
3E. Two or three sets of triplets

4A. Quartet– sharp, well-formed, distinct
4B. Quartet, poorly-formed but distinct
4C. Quartet overlapped by fine line structure
4D. Olefinic quartet of an ABX pattern
4E. Quartet overlapped by other signals

5A. Quintet
5B. Quintet with fine line structure

6A. Sextet
6B. Sextet with fine line structure
6C. Sextet overlapped by other signals

7A. Septet
7B. Septet with fine line structure
7X. Multiplicity greater than seven lines

M. Complex series of lines, small spacing of lines in the pattern
N. Complex series of lines over a medium ppm range
O. Complex series of lines, pattern
P. Complex series of lines covering an extensive region of the spectrum
Q. Pattern which is part of an A_2B_2, A_4B_4, etc.
R. Part of an ABX pattern
S. Pattern not easily identified or described, or pattern overlapped by other signals

III. The Peak Code

So that the reader may at least qualitatively visualize the line pattern for the proton, a description of the peak configuration is given in code. The code symbols are explained in the table on page xiii. It is advisable to be particularly cautions when using the pattern data for the following reasons:

The patterns are based on a so-called normal run, that is, one which is supposed to give the best resolution without blowing up. Since two spectroscopists may run a sample under different conditions (different amplitude, rf field, etc.), the spectrum might not appear quite the same in each case. The description of the pattern which is given is based on the available spectrum, and not necessarily on one that the authors consider optimum.

The coding categories are rather broad, and in places overlap, since, if they were too specific, there would be a need for a different classification for each spectrum. For example, one might be tempted to call a pattern N when we have called it O or S. Or, where does one draw the line between category 1I and the case when one faintly sees a doublet or triplet?

When a pattern is listed as a triplet or a quartet, this does not imply a first-order multiplet or an equally-spaced multiplet. It merely gives the number of lines observed.

It is hoped that the pattern data will prove helpful to the reader in spite of these hazards, but if a discrepancy arises between what is predicted and what is obtained, one may have to consult the original spectrum cited in the entry.

IV. The Indexes

To complement the data in the entries two indexes are provided. Of these it need perhaps only be noted here that the Shift Index makes no special distinction for "alternative" assignments (ones noted on the entries as either of two assignments). A word of caution is also in order concerning the Shift Index: although duplicate entries have been gleaned out, it should be kept in mind by the reader that an index entry of a single shift and a single proton code designation may represent many entries in the code listing. The Index of Molecular Formulas, however, reports every entry in the collection.

NMR BAND HANDBOOK

▶ 1-Bb

(c) (b) (a)
$H_3C\backslash C{=}C/CH_2CH_2CH_3$; H (d) on first C, CH_3 on second C

	Shift	Peak
▶a	.83	3C
b	1.33	S
c	1.87	3B
d	5.14	O

▶ C_7H_{14} ▶ liquid ▶ API 395

▶ 1-Bb

(a) (b) (d) (c)
$CH_3-(CH_2)_{10}-CH_2OH$

	Shift	Peak
▶a	0.85	3B
b	1.29	1A
c	1.83	1B
d	3.62	3B

▶ $C_{12}H_{26}O$ ▶ 44.4 mg/0.50 ml $CDCl_3$ ▶ S 103

▶ 1-Bb

(a) (b) (c) (b) (a)
$CH_3-CH_2-CH(CH_3)-(CH_2)_3-CH_3$; branch CH_3 = (a)

	Shift	Peak
▶a	.83	3B
b	1.23	1B
c	~1.27	S

▶ C_8H_{18} ▶ 50% CCl_4 ▶ API 341

▶ 1-Bb

(b) (e) (a) (d) (d) (c)
$CH_3-CH(CH_3)-C(CH_3)_2-CH_2-CH_2-CH_3$; CH-CH_3 = (b), C-$(CH_3)_2$ = (a)

	Shift	Peak
a	.80	1A
b	.86	2H
▶c	~ .86	3D
d	1.22	3A
e	~1.50	1D

▶ C_9H_{20} ▶ 50% CCl_4 ▶ API 213

▶ 1-Bb

(a) (b) (b) (b) (a)
$CH_3-CH_2-(CH_2)_4-CH_2-CH_3$

	Shift	Peak
▶a	.84	3B
b	1.24	1B

▶ C_8H_{18} ▶ 10% CCl_4 ▶ API 200

▶ 1-Bb

(b) (c) (a)
$C_6H_5-CH_2-(CH_2)_5-CH_3$

(d) = all ring H's

	Shift	Peak
▶a	.86	3B
b	1.28	1D
c	2.54	3B
d	7.09	1A

▶ $C_{13}H_{20}$ ▶ 50% CCl_4 ▶ API 357

▶ 1-Bb

(a) (b) (c) (d)
$H_3CH_2CH_2C(H)C{=}C(H)CH_2CH_2CH_3$ (c) (b) (a) ; H = (d)

	Shift	Peak
▶a	.85	3C
b	1.32	6B
c	1.91	N
d	5.32	N

▶ C_8H_{16} ▶ 50 vol.% CCl_4 ▶ API 432

▶ 1-Bb

(a) (b) (c)
$CH_3-CH_2-(CH_2)_{11}-CH_2-OH$

	Shift	Peak
▶a	0.86	3B
b	1.21	1A
c	~3.55	3B

▶ $C_{14}H_{30}O$ ▶ 44.9 mg/0.25 ml $CDCl_3$ ▶ S 294

▶ 1-Bb

(d) H, (e) H on C; (f) H on C
$H_2C{=}CH-CH_2(CH_2)_5CH_3$ (c) (b) (a)

	Shift	Peak
▶a	.85	3B
b	1.28	1B
c	1.99	3D
d	4.88	R
e	4.99	R
f	5.68	R

▶ C_9H_{18} ▶ 50% CCl_4 ▶ API 434

▶ 1-Bb

(b) (a)
$CH_3-CH_2-CH_2-CH_2-CH_3$

	Shift	Peak
▶a	.87	3B
b	1.28	1E

▶ C_5H_{12} ▶ 50% CCl_4 ▶ API 331

▶ 1-Bb

(a) (c) (e) (d) (b)
$CH_3-C(CH_3)_2-CH(CH_3)-CH_2-CH_2-CH_3$; C-CH_3 = (a), CH-CH_3 = (c)

	Shift	Peak
a	.85	1E
▶b	.85	1E
c	.85	1E
d	~1.03	S
e	~1.32	S

▶ C_9H_{20} ▶ 50% CCl_4 ▶ API 210

▶ 1-Bb

(a) (c) (d)
$CH_3-(CH_2)_3-CH(CH_3)-NH_2$; CH-CH_3 = (b)

	Shift	Peak
▶a	.87	3D
b	1.05	2H
c	1.25	1B
d	3.2	1C

▶ $C_6H_{15}N$ ▶ 34.4 mg/0.50 ml $CDCl_3$ ▶ S 28

▶ 1-Bb

$CH_3-C(CH_3)_2-CH_2-(CH_2)_2-CH_3$ (b) CH_3; (c) CH_2; (c) $(CH_2)_2$; (a) CH_3

	Shift	Peak
▶a	.87	1E
b	1.20	1E
c	1.20	1B

▶ C_8H_{18} ▶ 50% CCl_4 ▶ API 203

▶ 1-Bb

(a) (b) (c) $CH_3(CH_2)_{17}CH_2C(=O)-OH$

	Shift	Peak
▶a	0.87	3B
b	1.24	1A
c	~2.27	2B

▶ $C_{20}H_{40}O_2$ ▶ 45.2 mg/0.50 ml $CDCl_3$ ▶ S 260

▶ 1-Bb

$CH_3-C(CH_3)_2-CH_2-(CH_2)_3-CH_3$ (a) CH_3; (c) $(CH_2)_3$; (b) CH_3

	Shift	Peak
a	.87	1E
▶b	.87	1E
c	1.28	1B

▶ C_9H_{20} ▶ 50% CCl_4 ▶ API 209

▶ 1-Bb

(a) (b) (c) (e) (d) $CH_3-(CH_2)_2-CH_2-NH-C(=O)-NH_2$

	Shift	Peak
▶a	0.88	3D
b	~1.2-1.5	N
c	~3.08	4C
d	5.18	1B
e	5.95	3C

▶ $C_5H_{12}N_2O$ ▶ 45.3 mg/0.25 ml $CDCl_3$ ▶ S 282

▶ 1-Bb

Benzene ring bearing CH_3(c) and $CH_2CH_2CH_3$ (d)(b)(a); ring H: (e), (f), (g), (h)

	Shift	Peak
▶a	.87	3C
b	1.56	6B
c	2.17	1A
d	2.45	3B
e	6.82	S
f	6.82	S
g	6.82	S
h	6.82	S

▶ $C_{10}H_{14}$ ▶ liquid run, neat ▶ API 96

▶ 1-Bb

(c) H_3C(H(f))C=C(H(e))$CH_2-CH_2-CH_3$ (d) (b) (a)

	Shift	Peak
▶a	.88	3B
b	~1.33	O
c	1.69	N
d	1.87	1F
e	5.32	S
f	5.32	S

▶ C_6H_{12} ▶ 10% CCl_4 ▶ API 32

▶ 1-Bb

(a) (b) (c) $CH_3-(CH_2)_8-CH_2-C(H(f))=C(H(d))(H(e))$

	Shift	Peak
▶a	0.87	3B
b	1.28	1A
c	2.05	2G
d	4.80	R
e	~5.03	R
f	5.68	R

▶ $C_{12}H_{24}$ ▶ 44.4 mg/0.50 ml $CDCl_3$ ▶ S 202

▶ 1-Bb

H(e)H(f)C=C(H(g)) $CH_2CH_2CH_2CH_3$ (d) (c) (b) (a)

	Shift	Peak
▶a	.88	3C
b	~1.29 or 1.31	S
c	~1.31 or 1.29	S
d	2.00	3C
e	4.84	R
f	4.97	R
g	5.65	R

▶ C_6H_{12} ▶ liquid ▶ API 368

▶ 1-Bb

(a) (b) (c) $CH_3-(CH_2)_9-CH_2-CH(=O)$ (d)

	Shift	Peak
▶a	0.87	3A
b	1.29	1B
c	2.42	3B
d	9.77	3A

▶ $C_{12}H_{24}O$ ▶ 45.2 mg/0.50 ml $CDCl_3$ ▶ S 106

▶ 1-Bb

(c) H_3C(H(e))C=C(H(e)) $CH_2CH_2CH_3$ (d) (b) (a)

	Shift	Peak
▶a	.88	3C
b	1.28	S
c	1.60	2E
d	1.92	1F
e	5.35	O

▶ C_6H_{12} ▶ 50 vol.% CCl_4 ▶ API 370

▶ 1-Bb

(b) (a) $CH_3-(CH_2)_{13}-CH_3$

	Shift	Peak
▶a	.87	3B
b	1.30	1A

▶ $C_{15}H_{32}$ ▶ 50% CCl_4 ▶ API 221

▶ 1-Bb

(e)H (e)H C=C(CH_3 (c)) $CH_2CH_2CH_3$ (d) (b) (a)

	Shift	Peak
▶a	.88	3C
b	1.34	S
c	1.57	1H
d	1.90	S
e	4.60	N

▶ C_6H_{12} ▶ 50 vol.% CCl_4 ▶ API 373

▶ 1-Bb

	Shift	Peak
▶ a	.88	3B
b	1.30	1B

$\overset{(b)}{CH_3-(CH_2)_4}-\overset{(a)}{CH_3}$

▶ C_6H_{14} ▶ 50% CCl_4 ▶ API 333

▶ 1-Bb

	Shift	Peak
▶ a	0.88	3B
b	1.32	2E
c	1.82	3C
d	3.18	3A

(a) (b) (c) (d)
$CH_3-(CH_2)_4-CH_2-CH_2-I$

▶ $C_7H_{15}I$ ▶ 45.0 mg/0.50 ml $CDCl_3$ ▶ S 210

▶ 1-Bb

	Shift	Peak
▶ a	0.88	3B
b	1.36	1B
c	2.28	1A
d	3.62	3A

(a) (b) (d) (c)
$CH_3-(CH_2)_4-CH_2-OH$

▶ $C_6H_{14}O$ ▶ 45.2 mg/0.50 ml $CDCl_3$ ▶ S 198

▶ 1-Bb

	Shift	Peak
▶ a	.88	3B
b	1.30	1B

(a) (b)
$CH_3-(CH_2)_5-CH_3$

▶ C_7H_{16} ▶ 50% CCl_4 ▶ API 337

▶ 1-Bb

	Shift	Peak
▶ a	.88	3C
b	1.31	O
c	1.62	2E
d	1.92	1F
e	5.33 or 5.42	S
f	5.42 or 5.33	S

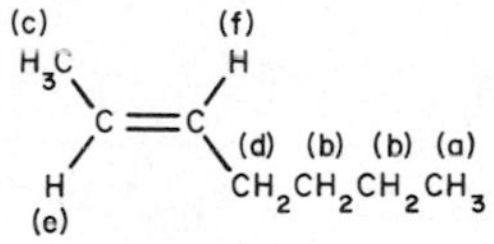

▶ C_7H_{14} ▶ liquid ▶ API 390

▶ 1-Bb

	Shift	Peak
▶ a	.88	3D
b	.88	S
c	1.24	S
d	~1.26	S

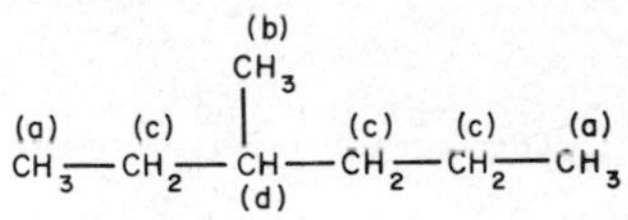

▶ C_7H_{16} ▶ 50% CCl_4 ▶ API 338

▶ 1-Bb

	Shift	Peak
▶ a	.88	3D
b	.98	2H
c	1.29	O
d	2.11	5A
e	4.86	R
f	4.97	R
g	~5.58	R

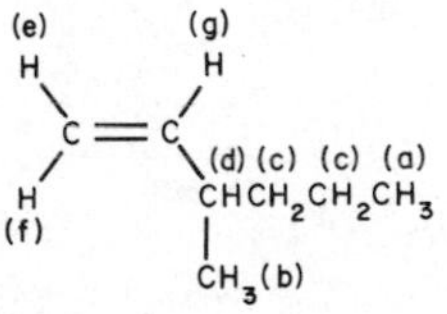

▶ C_7H_{14} ▶ liquid ▶ API 392

▶ 1-Db

	Shift	Peak
a	.88	1E
▶ b	.88	3D
c	1.20	N

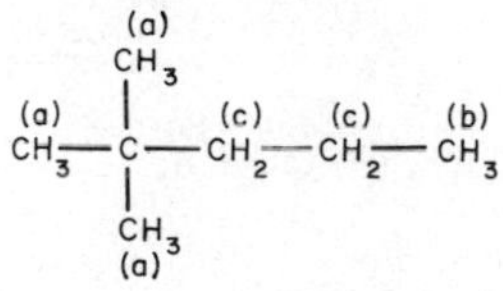

▶ C_7H_{16} ▶ 50% CCl_4 ▶ API 339

▶ 1-Bb

	Shift	Peak
▶ a	0.88	3B
b	1.32	1B
c	2.38	3B
d	11.47	1B

(a) (b) (c) (d)
$CH_3-(CH_2)_4-CH_2-C(=O)-OH$

▶ $C_7H_{14}O_2$ ▶ 45.6 mg/0.50 ml $CDCl_3$ ▶ S 49

▶ 1-Bb

	Shift	Peak
▶ a	0.88	3B
b	1.33	1B
c	1.95	1A
d	3.62	3B

(a) (b) (d) (c)
$CH_3-(CH_2)_5-CH_2-OH$

▶ $C_7H_{16}O$ ▶ 45.0 mg/0.50 ml $CDCl_3$ ▶ S 215

▶ 1-Bb

	Shift	Peak
▶ a	0.88	3B
b	1.31	1E
c	3.53	3A

(a) (b) (c)
$CH_3-(CH_2)_5-CH_2Cl$

▶ $C_7H_{15}Cl$ ▶ 44.8 mg/0.50 ml $CDCl_3$ ▶ S 209

▶ 1-Bb

	Shift	Peak
▶ a	0.88	3B
b	1.30	1B
c	2.43	3B
d	9.73	3A

(a) (b) (c) (d)
$CH_3-(CH_2)_5-CH_2-CH(=O)$

▶ $C_8H_{16}O$ ▶ 44.6 mg/0.50 ml $CDCl_3$ ▶ S 105

▶ 1-Bb

Shift		Peak
▶a	0.88	3B
b	1.30	1G
c	2.36	3B
d	11.47	1A

(a) (b) (c) (d)
$CH_3(CH_2)_5CH_2—C(=O)—OH$

▶ $C_8H_{16}O_2$ ▶ 45.6 mg/0.50 ml $CDCl_3$ ▶ S 157

▶ 1-Bb

Shift		Peak
▶a	0.88	3B
b	1.29	1B
c	2.30	1B
d	9.35	1A

(a) (b) (c) (d)
$CH_3—(CH_2)_6—CH_2—CH(=O)$

▶ $C_9H_{18}O$ ▶ 45.3 mg/0.50 ml $CDCl_3$ ▶ S 100

▶ 1-Bb

Shift		Peak
▶a	.88	3A
b	~1.25	S
c	~1.25	S
d	~1.25	S

$CH_3—CH_2—CH_2—CH(d)[(b)CH_3]—CH_2(c)—CH_2—CH_3(a)$

▶ C_8H_{18} ▶ 50% CCl_4 ▶ API 201

▶ 1-Bb

Shift		Peak
▶a	0.88	3B
b	1.28	1E
c	2.38	3B
d	10.53	1B

(a) (b) (c) (d)
$CH_3—(CH_2)_6—CH_2—C(=O)—OH$

▶ $C_9H_{18}O_2$ ▶ 35.7 mg/0.50 ml $CDCl_3$ ▶ S 9

▶ 1-Bb

Shift		Peak
▶a	~ .88	3D
b	~ .88	3D
c	~1.26	1F

$[(b)H_3C—H_2C(c)]_2CH—CH_2(c)—CH_2(c)—CH_3(a)$

▶ C_8H_{18} ▶ 50% CCl_4 ▶ API 202

▶ 1-Bb

Shift		Peak
▶a	0.88	3B
b	1.32	1E
c	4.16	3B
d	8.03	1B

(d) (c) (b) (a)
$HC(=O)—OCH_2—(CH_2)_6—CH_3$

▶ $C_9H_{18}O_2$ ▶ 44.4 mg/0.50 ml $CDCl_3$ ▶ S 114

▶ 1-Bb

Shift		Peak
▶a	0.88	3B
b	1.27 ± 0.03	1A
c	1.27 ± 0.03	1A

(a) (b) (c) (b) (a)
$CH_3—CH_2—(CH_2)_4—CH_2—CH_3$

▶ C_8H_{18} ▶ 7% $CDCl_3$ ▶ V 216

▶ 1-Bb

Shift		Peak
▶a	0.88	3B
b	1.26	S
c	1.33	S
d	2.28	3B
e	4.12	4A

(a) (c) (d) (e) (b)
$CH_3—(CH_2)_4—CH_2—C(=O)—OCH_2—CH_3$

▶ $C_9H_{18}O_2$ ▶ 44.5 mg/0.50 ml $CDCl_3$ ▶ S 283

▶ 1-Bb

Shift		Peak
▶a	.88	3B
b	1.12	2A
c	1.30	1B
d	3.70	4B
e	4.08	1A

(b) (d) (c) (c) (a)
$CH_3—CH[OH(e)]—CH_2—(CH_2)_4—CH_3$

▶ $C_8H_{18}O$ ▶ 20% CCl_4 ▶ API 170

▶ 1-Bb

Shift		Peak
▶a	.88	3B
b	1.28	1B

(a) (b)
$CH_3—(CH_2)_7—CH_3$

▶ C_9H_{20} ▶ 50% CCl_4 ▶ API 343

▶ 1-Bb

Shift		Peak
▶a	0.88	3B
b	1.28	1B
c	2.17	1A
d	2.46	3B

(c) (d) (b) (a)
$CH_3—C(=O)—CH_2—(CH_2)_5—CH_3$

▶ $C_9H_{18}O$ ▶ 33.7 mg/0.50 ml $CDCl_3$ ▶ S 48

▶ 1-Bb

Shift		Peak
▶a	0.88	3B
b	1.28	1A

(a) (b) (a)
$CH_3—(CH_2)_7—CH_3$

▶ C_9H_{20} ▶ 45.0 mg/0.50 ml $CDCl_3$ ▶ S 203

▶ 1-Bb

	Shift	Peak
▶ a	.88	3B
b	1.28	1B
c	3.44	3B
d	4.27	1A

$\overset{(a)}{CH_3}—\overset{(b)}{(CH_2)_7}—\overset{(c)}{CH_2}—\overset{(d)}{OH}$

▶ $C_9H_{20}O$ ▶ 20% CCl_4 ▶ API 171

▶ 1-Bb

	Shift	Peak
▶ a	.88	3B
b	1.28	1A

$CH_3—\overset{(b)}{(CH_2)_9}—\overset{(a)}{CH_3}$

▶ $C_{11}H_{24}$ ▶ 50% CCl_4 ▶ API 348

▶ 1-Bb

	Shift	Peak
▶ a	0.88	3B
b	1.28	3D
c	1.30	S
d	2.28	3B
e	4.12	4A

$\overset{(a)}{CH_3}—\overset{(c)}{(CH_2)_5}—\overset{(d)}{CH_2}—\overset{O}{\overset{||}{C}}—O\overset{(e)}{CH_2}\overset{(b)}{CH_3}$

▶ $C_{10}H_{20}O_2$ ▶ 45.2 mg/0.50 ml $CDCl_3$ ▶ S 244

▶ 1-Bb

	Shift	Peak
▶ a	.88	3B
b	1.28	1A

$\overset{(a)}{CH_3}—\overset{(b)}{(CH_2)_{10}}—CH_3$

▶ $C_{12}H_{26}$ ▶ 50% CCl_4 ▶ API 350

▶ 1-Bb

	Shift	Peak
▶ a	0.88	3B
b	1.28	1B
c	2.30	3B
d	3.68	1A

$\overset{(a)}{CH_3}—\overset{(b)}{(CH_2)_6}—\overset{(c)}{CH_2}\overset{O}{\overset{||}{C}}—\overset{(d)}{OCH_3}$

▶ $C_{10}H_{20}O_2$ ▶ 45.1 mg/0.50 ml $CDCl_3$ ▶ S 249

▶ 1-Bb

	Shift	Peak
▶ a	0.88	3B
b	1.28	1B
c	2.28	3B
d	3.67	1A

$\overset{(a)}{CH_3}—\overset{(b)}{(CH_2)_9}—\overset{(c)}{CH_2}—\overset{O}{\overset{||}{C}}—\overset{(d)}{OCH_3}$

▶ $C_{13}H_{26}O_2$ ▶ 45.7 mg/0.50 ml $CDCl_3$ ▶ S 125

▶ 1-Bb

	Shift	Peak
▶ a	0.88	3B
b	1.28	1E

$\overset{(a)}{CH_3}—\overset{(b)}{(CH_2)_8}—\overset{(a)}{CH_3}$

▶ $C_{10}H_{22}$ ▶ 44.8 mg/0.50 ml $CDCl_3$ ▶ S 201

▶ 1-Bb

	Shift	Peak
▶ a	.88	3B
b	1.27	1A
c	2.3	3B
d	4.88	7X
e	5.01	7X
f	5.72	R

$\overset{(d)}{H}(\underset{(e)}{H})C{=}C(\overset{(f)}{H})\overset{(c)}{CH_2}\overset{(b)}{(CH_2)_{10}}\overset{(a)}{CH_3}$

▶ $C_{14}H_{28}$ ▶ 50 vol.% CCl_4 ▶ API 436

▶ 1-Bb

	Shift	Peak
▶ a	.88	3B
b	1.27	1B
c	3.45	3B
d	4.28	1A

$\overset{(a)}{CH_3}—\overset{(b)}{(CH_2)_8}—\overset{(c)}{CH_2}—\overset{(d)}{OH}$

▶ $C_{10}H_{22}O$ ▶ 20% CCl_4 ▶ API 172

▶ 1-Bb

	Shift	Peak
▶ a	0.88	3B
b	1.26	1A
c	~2.35	1D
d	10.95	1B

$\overset{(a)}{CH_3}—\overset{(b)}{(CH_2)_{11}}—\overset{(c)}{CH_2}—\overset{O}{\overset{||}{C}}—\overset{(d)}{OH}$

▶ $C_{14}H_{28}O_2$ ▶ 44.8 mg/0.50 ml $CDCl_3$ ▶ S 167

▶ 1-Bb

	Shift	Peak
▶ a	0.88	3B
b	1.40	O
c	3.38	3A

$\overset{(a)}{CH_3}\overset{(b)}{(CH_2)_3}\overset{(c)}{CH_2}—O—\overset{(c)}{CH_2}\overset{(b)}{(CH_2)_3}\overset{(a)}{CH_3}$

▶ $C_{10}H_{22}O$ ▶ 44.6 mg/0.50 ml $CDCl_3$ ▶ S 151

▶ 1-Bb

	Shift	Peak
▶ a	0.88	3B
b	1.28	1E
c	2.28	3C
d	4.11	4A

$\overset{(a)}{CH_3}—\overset{(b)}{(CH_2)_9}—\overset{(c)}{CH_2}—\overset{O}{\overset{||}{C}}—O\overset{(d)}{CH_2}\overset{(a)}{CH_3}$

▶ $C_{14}H_{28}O_2$ ▶ 45.5 mg/0.50 ml $CDCl_3$ ▶ S 194

▶ 1-Bb

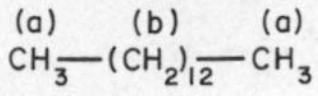

	Shift	Peak
▶ a	0.88	3B
b	1.28	1A

▶ $C_{14}H_{30}$ ▶ 44.8 mg/0.50 ml $CDCl_3$ ▶ S 204

▶ 1-Bb

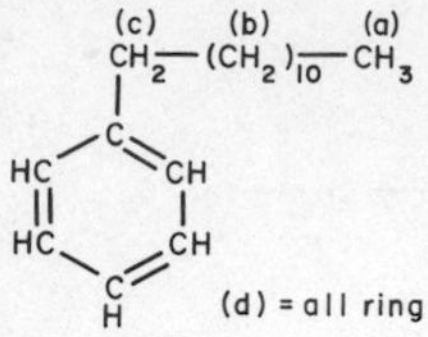

(d) = all ring H's

	Shift	Peak
▶ a	.88	3B
b	1.22	1A
c	2.54	3B
d	7.09	1G

▶ $C_{18}H_{30}$ ▶ 50% CCl_4 ▶ API 360

▶ 1-Bb

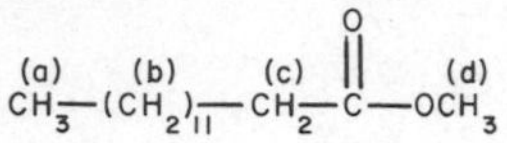

	Shift	Peak
▶ a	0.88	3B
b	1.29	1A
c	2.24	3B
d	3.68	1A

▶ $C_{15}H_{30}O_2$ ▶ 45.1 mg/0.50 ml $CDCl_3$ ▶ S 195

▶ 1-Bb

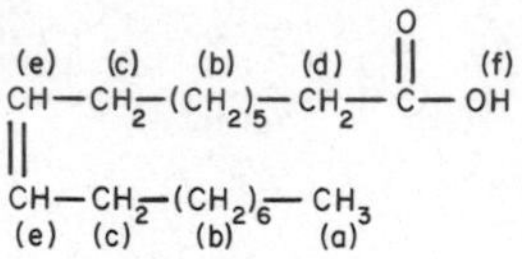

	Shift	Peak
▶ a	0.88	3B
b	1.29	1B
c	2.00	1F
d	2.25	3B
e	5.28	3D
f	11.45	1C

▶ $C_{18}H_{34}O_2$ ▶ 45.2 mg/0.50 ml $CDCl_3$ ▶ S 70

▶ 1-Bb

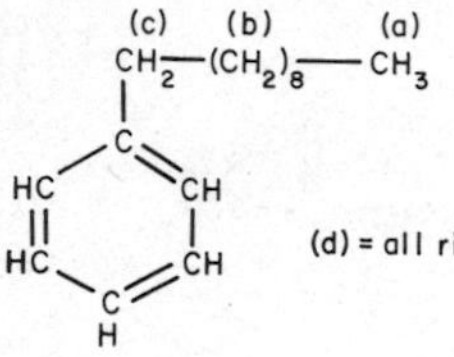

(d) = all ring H's

	Shift	Peak
▶ a	.88	3B
b	1.25	1B
c	2.50	3B
d	7.07	1A

▶ $C_{16}H_{26}$ ▶ 50% CCl_4 ▶ API 359

▶ 1-Bb

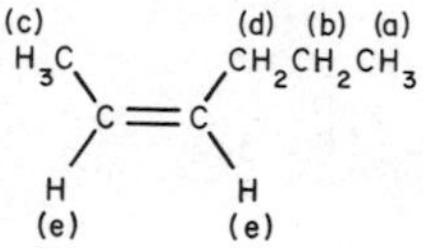

	Shift	Peak
▶ a	.89	3C
b	1.30	S
c	1.56	2E
d	2.00	4C
e	5.32	O

▶ C_6H_{12} ▶ liquid ▶ API 369

▶ 1-Bb

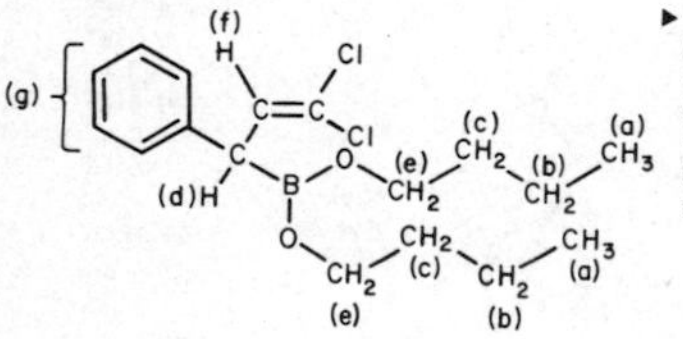

	Shift	Peak
▶ a	0.88	3C
b	~1.36	S
c	~1.50	S
d	3.58	2H
e	3.84	3B
f	6.24	2A
g	7.20	1A

▶ $C_{17}H_{25}BCl_2O_2$ ▶ 7% $CDCl_3$ ▶ V 667

▶ 1-Bb

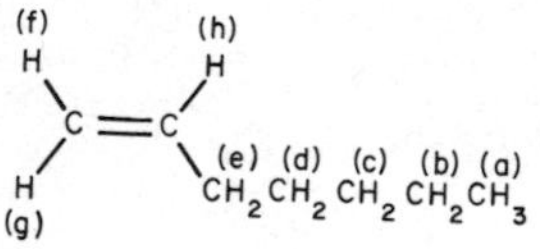

	Shift	Peak
▶ a	.89	3C
b	.90	S
c	~1.33 or 1.35	S
d	~1.35 or 1.33	S
e	1.97	4C
f	~4.84	R
g	~4.97	R
h	~5.70	R

▶ C_7H_{14} ▶ liquid ▶ API 384

▶ 1-Bb

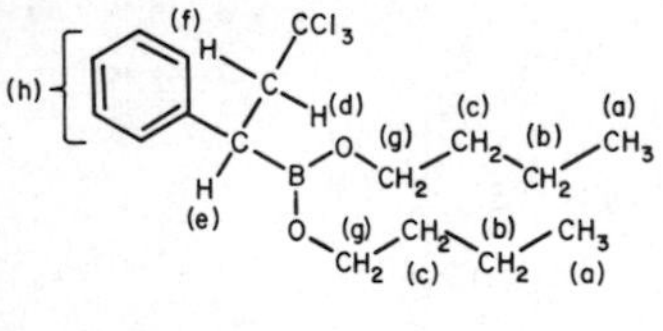

	Shift	Peak
▶ a	0.88	3C
b	~1.36	S
c	~1.50	S
d	2.86*	S
e	2.89*	S
f	3.65	S
g	3.88	3C
h	7.26	1A

*±0.04

▶ $C_{17}H_{26}BCl_3O_2$ ▶ 7% $CDCl_3$ ▶ V 668

▶ 1-Bb

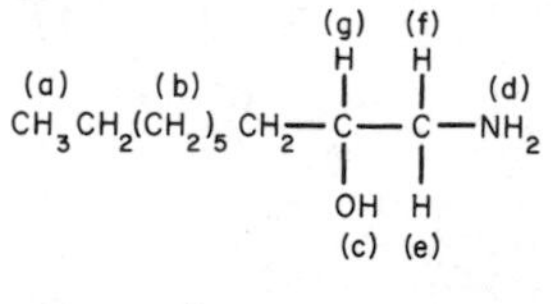

	Shift	Peak
▶ a	0.89	3B
b	1.28	1G
c	1.75	1E
d	1.75	1E
e	2.51	S
f	2.84	S
g	3.50	1D

▶ $C_{10}H_{23}NO$ ▶ 7% $CDCl_3$ ▶ V 575

▶ 1-Bb

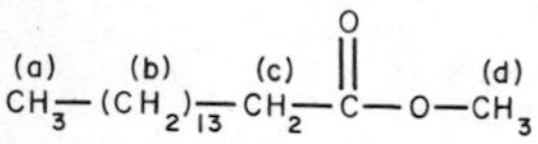

	Shift	Peak
▶ a	~0.88	1D
b	1.26	1A
c	~2.30	1D
d	3.63	1A

▶ $C_{17}H_{34}O_2$ ▶ 44.4 mg/0.50 ml $CDCl_3$ ▶ S 248

▶ 1-Bb

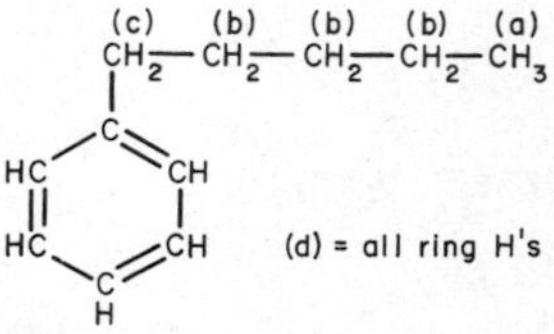

(d) = all ring H's

	Shift	Peak
▶ a	.89	3B
b	1.36	O
c	2.52	3B
d	7.04	1G

▶ $C_{11}H_{16}$ ▶ 50% CCl_4 ▶ API 355

▶ 1-Bb

$C_6H_5-CH_2-(CH_2)_4-CH_3$ — (b) CH_2, (c) $(CH_2)_4$, (a) CH_3; (d) = all ring H's

Shift	Peak
▶ a .89	3B
b 1.30	1F
c 2.54	3B
d 7.10	1A

▶ $C_{12}H_{18}$ ▶ 50% CCl_4 ▶ API 356

▶ 1-Bb

(a) CH_3- (b) $(CH_2)_3-$ (c) CH_2- (d) CH_2-I

Shift	Peak
▶ a 0.90	3B
b 1.32	2E
c 1.84	3C
d 3.20	3A

▶ $C_6H_{13}I$ ▶ 45.8 mg/0.50 ml $CDCl_3$ ▶ S 212

▶ 1-Bb

$C_6H_5-CH_2-(CH_2)_7-CH_3$ — (c) CH_2, (b) $(CH_2)_7$, (a) CH_3; (d) = all ring H's

Shift	Peak
▶ a .89	3B
b 1.25	1B
c 2.46	3B
d 7.07	1A

▶ $C_{15}H_{24}$ ▶ 50% CCl_4 ▶ API 358

▶ 1-Bb

(a) CH_3- (b) $(CH_2)_4-$ (d) CH_2- (c) NH_2

Shift	Peak
▶ a 0.90	3B
b 1.28	1B
c 1.58	S
d 2.70	1C

▶ $C_6H_{15}N$ ▶ 35.6 mg/0.50 ml $CDCl_3$ ▶ S 46

▶ 1-Bb

(f)(c)(b)(a) $CH_2CH_2(CH_2)_2CH_3$; H(j), H(h), (g) H, (e) CH_3, H (i), CH_3 (d), H(k)

Shift	Peak
▶ a 0.89	3B
b ~1.30	S
c ~1.40	S
d 1.72	1A
e 1.97	2A
f 2.94	3B
g 6.06	S
h 6.17	S
i 6.58	S
j 6.89	S
k 7.88	1G

▶ $C_{21}H_{22}O_5$ ▶ 7% $CDCl_3$ ▶ V 687

▶ 1-Bb

(c) $H_3C-CH=CH-$ (d) (b) (b) (a) $CH_2CH_2CH_2CH_3$; H (e), H (e)

Shift	Peak
▶ a .90	3C
b 1.31	O
c 1.57	2E
d 2.02	1F
e ~5.37	O

▶ C_7H_{14} ▶ liquid ▶ API 389

▶ 1-Bb

CH_3-CH_2- (b) CH_2- (a) CH_3

Shift	Peak
▶ a .9	3C
b 1.25	O

▶ C_4H_{10} ▶ 50% CCl_4 ▶ API 329

▶ 1-Bb

(b)(d) $H_3CH_2C-CH=CH-$ (d)(c)(a) $CH_2CH_2CH_3$; H (e), (e) H

Shift	Peak
▶ a .90	3D
b .97	3D
c 1.33	S
d 1.95	O
e 5.40	O

▶ C_7H_{14} ▶ liquid ▶ API 391

▶ 1 Bb

(e)H, (f)H $C=C$ H(g), (d) (c) (b) (a) $CH_2CH_2CH_2CH_3$

Shift	Peak
▶ a .90	3B
b ~1.34	S
c ~1.34	S
d 2.00	4C
e 4.85	R
f 4.97	R
g 5.68	R

▶ C_6H_{12} ▶ 10% CCl_4 ▶ API 155

▶ 1-Bb

(d) H_3C, (c) H_3C $C=C$ H (f), (e) (b) (a) $CH_2CH_2CH_3$

Shift	Peak
▶ a .90	3C
b 1.28	S
c ~1.58 or 1.68	
d ~1.68 or 1.58	
e 1.91	4B
f 5.08	3C

▶ C_7H_{14} ▶ liquid ▶ API 408

▶ 1-Bb

(a) CH_3- (b) $(CH_2)_3-$ (c) CH_2- (d) CH_2-Cl

Shift	Peak
▶ a 0.90	3B
b 1.33	2H
c 1.78	O
d 3.53	3A

▶ $C_6H_{13}Cl$ ▶ 44.6 mg/0.50 ml $CDCl_3$ ▶ S 211

▶ 1-Bb

(c) $CH_3-C(=O)-$ (d) CH_2- (b) $(CH_2)_3-$ (a) CH_3

Shift	Peak
▶ a 0.90	3B
b 1.38	O
c 2.15	1A
d 2.42	3B

▶ $C_7H_{14}O$ ▶ 35.4 mg/0.50 ml $CDCl_3$ ▶ S 41

▶ 1-Bb

(a) (b) (c) (d)
$CH_3-(CH_2)_4-CH_2-CH_2-Br$

	Shift	Peak
▶ a	0.90	3B
b	1.30	1B
c	1.85	1F
d	3.41	3A

▶ $C_7H_{15}Br$ ▶ 45.0 mg/0.50 ml $CDCl_3$ ▶ S 208

▶ 1-Bb

(a) (b) (a)
$CH_3-(CH_2)_9-CH_3$

	Shift	Peak
▶ a	0.90	3B
b	1.28	1G

▶ $C_{11}H_{24}$ ▶ 44.4 mg/0.50 ml $CDCl_3$ ▶ S 289

▶ 1-Bb

(a) (b) (c) (d)
$CH_3-(CH_2)_5-CH_2-OH$

	Shift	Peak
▶ a	.90	3B
b	1.32	1G
c	3.44	3B
d	4.42	1B

▶ $C_7H_{16}O$ ▶ 20% CCl_4 ▶ API 168

▶ 1-Bb

(a)CH_3–(b)CH_2–(b)CH_2–(d)CH_2–N(H(e))–C(=O)–N(H(h))–SO_2–C_6H_4[H(g), H(g), H(f), H(f)]–CH_3(c)

	Shift	Peak
▶ a	0.90	3B
b	~1.40	1F
c	2.46	1A
d	3.23	4C
e	6.53	3B
f	7.32	Q
g	7.79	Q
h	~8.60	1D

▶ $C_{12}H_{18}N_2O_3S$ ▶ 7% $CDCl_3$ ▶ V 602

▶ 1-Bb

(a) (b) (d) (c)
$CH_3-(CH_2)_5-CH_2SH$

	Shift	Peak
▶ a	0.90	3B
b	1.29	1E
c	2.09	1A
d	2.53	4B

▶ $C_7H_{16}S$ ▶ 46.9 mg/0.50 ml $CDCl_3$ ▶ S 297

▶ 1-Bb

(d)H, (e)H >C=C< H(f), (c)CH_2(b)$(CH_2)_{12}$(a)CH_3

	Shift	Peak
▶ a	.90	3B
b	1.24	1A
c	2.04	3B
d	4.88	R
e	5.02	R
f	5.72	R

▶ $C_{16}H_{32}$ ▶ 50 vol.% CCl_4 ▶ API 437

▶ 1-Bb

Ring: (e)H, (g)H, H(h); (d)CH_2; C(=O)–NH_2(f); N–(c)CH_2–(b)CH_2–(a)CH_3

	Shift	Peak
▶ a	0.90	3C
b	1.58	6B
c	3.09	2A
d	3.19	N
e	4.72	5B
f	5.49	1C
g	5.73	4A
h	7.05	2D

▶ $C_9H_{14}N_2O$ ▶ 7% $CDCl_3$ ▶ V 542

▶ 1-Bb

(b) (a)
$CH_3-(CH_2)_{14}-CH_3$

	Shift	Peak
▶ a	.90	3B
b	1.26	1A

▶ $C_{16}H_{34}$ ▶ 10% CCl_4 ▶ API 48

▶ 1-Bb

(a) (b) (d) (c)
$CH_3-(CH_2)_7-CH_2OH$

	Shift	Peak
▶ a	0.90	3B
b	1.32	1B
c	1.97	1A
d	3.62	3B

▶ $C_9H_{20}O$ ▶ 44.6 mg/0.50 ml $CDCl_3$ ▶ S 102

▶ 1-Bb

(f) (c) (b) (d) (e)
$CH-CH_2-(CH_2)_5-CH_2-C(=O)-OCH_3$
‖
$CH-CH_2-(CH_2)_6-CH_3$
(f) (c) (b) (a)

	Shift	Peak
▶ a	0.90	3B
b	1.29	1B
c	2.00	1C
d	2.32	S
e	3.70	1A
f	5.4	3B

▶ $C_{19}H_{36}O_2$ ▶ 33.7 mg/0.50 ml $CDCl_3$ ▶ S 71

▶ 1-Bb

Pyrrole: (d)H, (e)H, N–H(f); (c)$CH_2(CH_2)_3CH_3$(a); CH_3(b)

	Shift	Peak
▶ a	0.90	3B
b	2.15	1A
c	~2.38	S
d	5.98	3A
e	6.55	3A
f	7.67	1D

▶ $C_{10}H_{17}N$ ▶ 7% $CDCl_3$ ▶ V 278

▶ 1-Bb

H(d)
(e)H–C–O–C(=O)–(c)CH_2–(b)$(CH_2)_{15}$–(a)CH_3
(f)H–C–O–C(=O)–(c)CH_2–(b)$(CH_2)_{15}$–(a)CH_3
(e)H–C–O–C(=O)–(c)CH_2–(b)$(CH_2)_{15}$–(a)CH_3
(d)H

	Shift	Peak
▶ a	0.90	3B
b	1.25	1A
c	2.30	3B
d	4.18*	S
e	4.18*	S
f	5.28	S

*or 4.28

▶ $C_{57}H_{110}O_6$ ▶ 7% $CDCl_3$ ▶ V 368

▶ 1-Bb

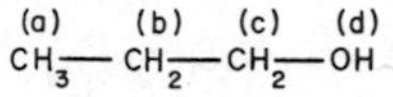

(a) (b) (c) (d)
$CH_3—CH_2—CH_2—OH$

	Shift	Peak
▶ a	.91	3C
b	1.43	6B
c	3.45	3A
d	4.68	1A

▶ C_3H_8O ▶ 20% CCl_4 ▶ API 165

▶ 1-Bb

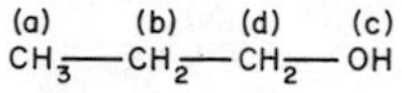

(a) (b) (d) (c)
$CH_3—CH_2—CH_2—OH$

	Shift	Peak
▶ a	0.92	3A
b	1.57	7A
c	~2.28	1C
d	3.58	3A

▶ C_3H_8O ▶ 7% $CDCl_3$ ▶ V 43

▶ 1-Bb

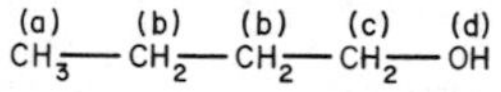

(a) (b) (b) (c) (d)
$CH_3—CH_2—CH_2—CH_2—OH$

	Shift	Peak
▶ a	.91	3B
b	~1.45	O
c	3.48	3C
d	4.58	1A

▶ $C_4H_{10}O$ ▶ 20% CCl_4 ▶ API 166

▶ 1-Bb

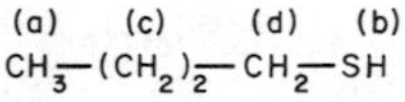

(a) (c) (d) (b)
$CH_3—(CH_2)_2—CH_2—SH$

	Shift	Peak
▶ a	0.92	3A
b	1.23	S
c	1.53	S
d	2.52	4B

▶ $C_4H_{10}S$ ▶ 46.0 mg/0.50 ml $CDCl_3$ ▶ S 274

▶ 1-Bb

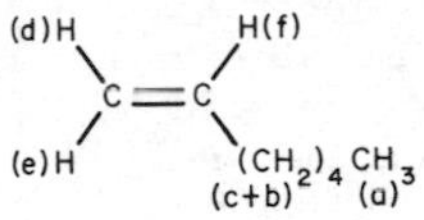

(d)H, (e)H $C=C$ H(f), $(CH_2)_4CH_3$ (c+b) (a)

	Shift	Peak
▶ a	.91	3B
b	1.31	S
c	2.02	3B
d	4.84	R
e	4.98	R
f	5.67	R

▶ C_7H_{14} ▶ 10% CCl_4 ▶ API 157

▶ 1-Bb

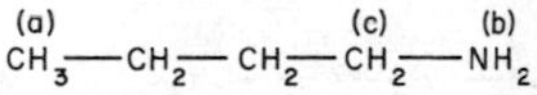

(a) (c) (b)
$CH_3—CH_2—CH_2—CH_2—NH_2$

	Shift	Peak
▶ a	0.92	3B
b	1.10	1A
c	2.70	3C

▶ $C_4H_{11}N$ ▶ 7% $CDCl_3$ ▶ V 89

▶ 1-Bb

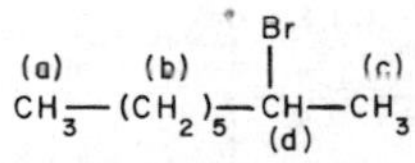

(a) (b) Br (c)
$CH_3—(CH_2)_5—CH(Br)—CH_3$
(d)

	Shift	Peak
▶ a	0.91	3B
b	1.28	1B
c	1.71	2A
d	4.12	5B

▶ $C_8H_{17}Br$ ▶ 45.2 mg/0.50 ml $CDCl_3$ ▶ S 227

▶ 1-Bb

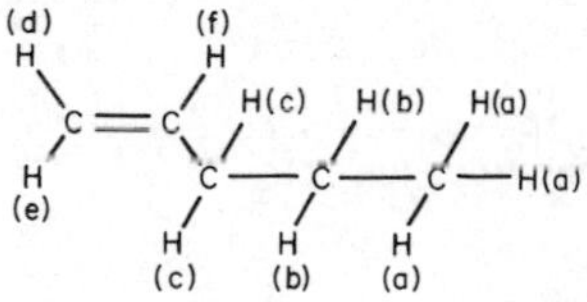

(d)H, (e)H $C=C$ H(f), C(H(c))(H(c)) — C(H(b))(H(b)) — C(H(a))(H(a))—H(a)

	Shift	Peak
▶ a	.92	3C
b	1.38	6B
c	1.93	4C
d	4.76	R
e	4.98	R
f	5.68	R

▶ C_5H_{10} ▶ liquid ▶ API 366

▶ 1-Bb

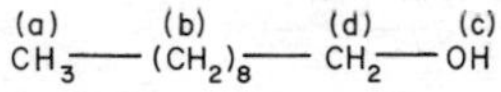

(a) (b) (d) (c)
$CH_3—(CH_2)_8—CH_2—OH$

	Shift	Peak
▶ a	0.91	3B
b	1.28	1B
c	2.03	1A
d	3.63	3A

▶ $C_{10}H_{22}O$ ▶ 7% $CDCl_3$ ▶ V 282

▶ 1-Bb

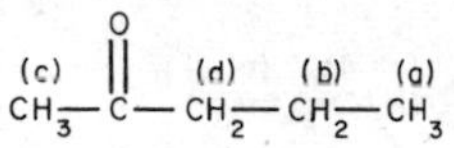

(c) (d) (b) (a)
$CH_3—C(=O)—CH_2—CH_2—CH_3$

	Shift	Peak
▶ a	0.92	3B
b	1.62	6B
c	2.14	1A
d	2.42	3A

▶ $C_5H_{10}O$ ▶ 45.4 mg/0.50 ml $CDCl_3$ ▶ S 177

▶ 1-Bb

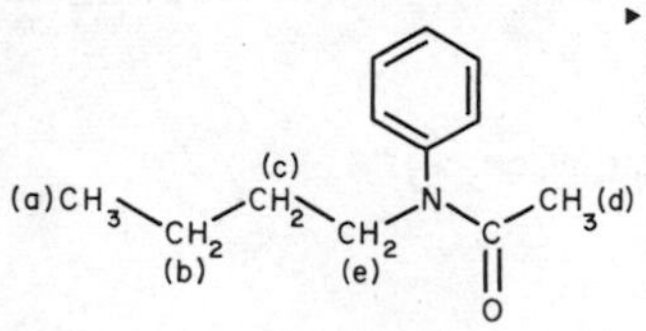

(a)CH_3—CH_2(b)—CH_2(c)—CH_2(e)—N(C_6H_5)—C(=O)—CH_3(d)

	Shift	Peak
▶ a	0.91	3B
b	1.44 ± 0.1	S
c	1.44 ± 0.1	S
d	1.87	1A
e	3.73	3A

▶ $C_{12}H_{17}NO$ ▶ 7% $CDCl_3$ ▶ V 601

▶ 1-Bb

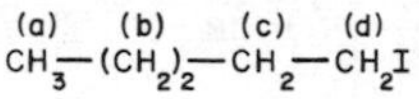

(a) (b) (c) (d)
$CH_3—(CH_2)_2—CH_2—CH_2I$

	Shift	Peak
▶ a	0.92	3B
b	1.38	2E
c	1.86	3C
d	3.20	3A

▶ $C_5H_{11}I$ ▶ 45.0 mg/0.50 ml $CDCl_3$ ▶ S 228

▶ 1-Bb

CH_3—CH(OH)—$(CH_2)_2$—CH_3; labels: (b) CH_3, (d) OH, (e) CH, (c) $(CH_2)_2$, (a) CH_3

	Shift	Peak
▶ a	0.92	3D
b	1.18	2H
c	1.33	N
d	2.22	1B
e	3.70	1F

▶ $C_5H_{12}O$ ▶ 44.6 mg/0.50 ml $CDCl_3$ ▶ S 60

▶ 1-Bb

(a) CH_3—(b) $(CH_2)_2$—(c) CH_2—S—CH_2—$(CH_2)_2$—CH_3

	Shift	Peak
▶ a	0.92	3B
b	1.49	O
c	2.49	3B

▶ $C_8H_{18}S$ ▶ 45.8 mg/0.50 ml $CDCl_3$ ▶ S 275

▶ 1-Bb

(a) CH_3—(b) $(CH_2)_3$—(c) CH_2—C(=O)—OH (d)

	Shift	Peak
▶ a	0.92	3B
b	1.45	O
c	2.36	3C
d	11.37	1B

▶ $C_6H_{12}O_2$ ▶ 44.7 mg/0.50 ml $CDCl_3$ ▶ S 37

▶ 1-Bb

C_6H_5 ring (H(d), (d)HC, (d)HC, CH(d), H(d))—(c) CH_2—(b) CH_2—(a) CH_3

	Shift	Peak
▶ a	.92	3C
b	1.58	6B
c	2.57	3C
d	7.10	1E

▶ C_9H_{12} ▶ 10% CCl_4 ▶ API 41

▶ 1-Bb

(a) CH_3—CH(d)((a) CH_3)—(c) CH_2—(c) CH_2—(c) CH_2—(b) CH_3

	Shift	Peak
a	.82	2A
▶ b	~.92	1E
c	~.93	1B
d	~.97	1D

▶ C_7H_{16} ▶ 50% CCl_4 ▶ API 195

▶ 1-Bb

(d) H, (e) H >C=C< H (f), (c) CH_2(b) $(CH_2)_8$(a) CH_3

	Shift	Peak
▶ a	.92	3B
b	1.32	1B
c	2.05	3B
d	4.86	R
e	5.00	R
f	5.68	R

▶ $C_{12}H_{24}$ ▶ 50 vol.% CCl_4 ▶ API 435

▶ 1-Bb

(b) CH_3(d) CH_2C(=O)—O(e) CH_2(c) $(CH_2)_3$(a) CH_3

	Shift	Peak
▶ a	0.92	3B
b	1.12	S
c	1.32	S
d	2.33	4A
e	4.05	3B

▶ $C_8H_{16}O_2$ ▶ 44.8 mg/0.50 ml $CDCl_3$ ▶ S 40

▶ 1-Bb

N[(c) CH_2—(b) $(CH_2)_2$—(a) CH_3]$_3$

	Shift	Peak
▶ a	0.92	3B
b	1.37	6B
c	2.40	3B

▶ $C_{12}H_{27}N$ ▶ 44.8 mg/0.50 ml $CDCl_3$ ▶ S 84

▶ 1-Bb

(c) CH_3C(=O)—O(d) CH_2(b) $(CH_2)_4$(a) CH_3

	Shift	Peak
▶ a	0.92	3B
b	1.37	1E
c	2.05	1A
d	4.05	3A

▶ $C_8H_{16}O_2$ ▶ 45.0 mg/0.50 ml $CDCl_3$ ▶ S 217

▶ 1-Bb

(a) CH_3—(b) CH_2—(d) CH_2—OH (c)

	Shift	Peak
▶ a	0.93	3A
b	1.56	6B
c	2.17	1B
d	3.62	3A

▶ C_3H_8O ▶ 34.7 mg/0.50 ml $CDCl_3$ ▶ S 10

▶ 1-Bb

(a) CH_3(b) $(CH_2)_3$—(d) CHCH_2OH (c); CH—(b) CH_2(a) CH_3

	Shift	Peak
▶ a	0.92	3B
b	1.30	1B
c	1.75	1B
d	3.52	1I

▶ $C_8H_{18}O$ ▶ 45.0 mg/0.50 ml $CDCl_3$ ▶ S 98

▶ 1-Bb

(a) $CH_3CH_2CH_2CH_2I$ (b)

	Shift	Peak
▶ a	0.93	3C
b	3.20	3A

▶ C_4H_9I ▶ 7% $CDCl_3$ ▶ V 81

▶ 1-Bb

Shift	Peak
▶ a 0.93	3A
b 1.64	5B
c 2.04	1A
d 4.02	3A

$CH_3C(=O)-OCH_2CH_2CH_3$ (c) (d) (b) (a)

▶ $C_5H_{10}O_2$ ▶ 44.9 mg/0.50 ml $CDCl_3$ ▶ S 115

▶ 1-Bb

Shift	Peak
▶ a 0.93	3B
b 4.13	3A

$O=C(-OCH_2CH_2CH_2CH_3)_2$ (b) (a)

▶ $C_9H_{18}O_3$ ▶ 7% $CDCl_3$ ▶ V 243

▶ 1-Bb

Shift	Peak
▶ a 0.93	3B
b 4.05	1A
c 4.20	3D

(a) $CH_3-CH_2-CH_2-$ (c) $CH_2-O-C(=O)-$ (b) CH_2-Cl

▶ $C_6H_{11}ClO_2$ ▶ 7% $CDCl_3$ ▶ V 138

▶ 1-Bb

Shift	Peak
▶ a .93	3C
b 1.47	S
c 1.47	S
d 2.57	3B
e 7.11	1A

C_6H_5- (d) CH_2- (c) CH_2- (b) CH_2- (a) CH_3

(e) = all ring H's

▶ $C_{10}H_{14}$ ▶ 10% CCl_4 ▶ API 42

▶ 1-Bb

Shift	Peak
▶ a 0.93	3B
b 1.49	O
c 2.33	3B

(a) CH_3- (b) $(CH_2)_3-$ (c) $CH_2-C\equiv N$

▶ $C_6H_{11}N$ ▶ 45.0 mg/0.50 ml $CDCl_3$ ▶ S 176

▶ 1-Bb

Shift	Peak
▶ a 0.94	3B
b 1.38	O
c 1.83	4C
d 3.42	3A

(a) CH_3 (b) CH_2CH_2 (c) CH_2 (d) CH_2Br

▶ $C_5H_{11}Br$ ▶ 44.5 mg/0.50 ml $CDCl_3$ ▶ S 120

▶ 1-Bb

Shift	Peak
▶ a 0.93	3B
b 2.08	1A
c 2.50	3B

(a) $CH_3-CH_2-CH_2-CH_2-$ (c) CH_2-S- (b) CH_3

▶ $C_6H_{14}S$ ▶ 7% $CDCl_3$ ▶ V 144

▶ 1-Bb

Shift	Peak
▶ a 0.94	3B
b 1.56	1F
c 4.05	1A
d 4.18	3A

$Cl-$ (c) $CH_2-C(=O)-O-$ (d) CH_2 (b) $(CH_2)_2$ (a) CH_3

▶ $C_6H_{11}ClO_2$ ▶ 44.5 mg/0.50 ml $CDCl_3$ ▶ S 92

▶ 1-Bb

Shift	Peak
▶ a .93	3C
b 1.38	O
c 1.60	1H
d 2.01	3B
e ~4.67	M
f ~4.67	M

(e) H, (f) H $C=C$ (c) CH_3, (d) CH_2 (b) CH_2 (b) CH_2 (a) CH_3

▶ C_7H_{14} ▶ liquid ▶ API 407

▶ 1-Bb

Shift	Peak
▶ a 0.94	3B
b 1.41	2A
c 1.46	S
d 3.02	1A
e 4.18	4C

(b) CH_3- (e) $CH(OH)$ (d) $-C(=O)-O$ (e) CH_2 (c) CH_2CH_2 (a) CH_3

▶ $C_7H_{14}O_3$ ▶ 44.6 mg/0.50 ml $CDCl_3$ ▶ S 246

▶ 1-Bb

Shift	Peak
▶ a 0.93	3B
b 3.22	1A

(a) $CH_3(CH_2)_3C(OCH_3)_3$ (b)

▶ $C_8H_{18}O_3$ ▶ 7% $CDCl_3$ ▶ V 218

▶ 1-Bb

Shift	Peak
a .83	2H
▶ b .94	S
c 1.30	1A

$CH_3-CH($ (a) $CH_3)-$ (c) $(CH_2)_7-$ (b) CH_3

▶ $C_{11}H_{24}$ ▶ 50% CCl_4 ▶ API 349

▶ 1-Bb

$CH_3CH_2-C(=O)-OCH_2CH_2CH_3$ — (b) (d) / (e) (c) (a)

	Shift	Peak
▶ a	0.95	3D
b	1.18	3D
c	1.60	6B
d	2.33	4A
e	4.04	3A

▶ $C_6H_{12}O_2$ ▶ 44.4 mg/0.50 ml $CDCl_3$ ▶ S 38

▶ 1-Bb

Syn– $CH_3(a)-CH_2(b)-CH_2-C(H(c))=N-OH(e)$

Anti– $CH_3(a)-CH_2(b)-CH_2-C(H(d))=N-OH(e)$

	Shift	Peak
▶ a	0.96	3C
b	1.55	6B
c	6.74	3A
d	7.45	3A
e	8.85	1D

▶ C_4H_9NO ▶ 7% $CDCl_3$ ▶ V 420

▶ 1-Bb

$CH_3(a)-CH_2-CH_2-CH_2(c)-O-C(=O)-CH_3(b)$

	Shift	Peak
▶ a	0.95	3B
b	2.03	1A
c	4.05	3A

▶ $C_6H_{12}O_2$ ▶ 7% $CDCl_3$ ▶ V 140

▶ 1-Bb

$CH_3CH_2CH_2C(=O)-OCH_2CH_3$ — (a) (c) (d) / (e) (b)

	Shift	Peak
▶ a	0.96	3A
b	1.28	3A
c	1.67	6B
d	2.28	3C
e	4.13	4A

▶ $C_6H_{12}O_2$ ▶ 45.0 mg/0.50 ml $CDCl_3$ ▶ S 42

▶ 1-Bb

$CH_3-CH(CH_3(a))(d)-CH_2(c)-CH_2(c)-CH_3(b)$

	Shift	Peak
a	.88	3D
▶ b	.95	2H
c	1.20	N
d	~1.50	1D

▶ C_6H_{14} ▶ 50% CCl_4 ▶ API 334

▶ 1-Bb

$CH_3-CH(CH_3(a))-(CH_2)_4(c)-CH_3(b)$

	Shift	Peak
a	.87	2H
▶ b	.96	S
c	1.27	1B

▶ C_8H_{18} ▶ 50% CCl_4 ▶ API 340

▶ 1-Bb

Pyrrole: (d) H, (e) H, H(f), N–H (g); $CH_2(c)-CH_2(b)-CH_3(a)$

	Shift	Peak
▶ a	0.95	3C
b	1.60	6B
c	2.48	3C
d	6.13	4B
e	6.70	S
f	6.55	S
g	~7.83	1A

▶ $C_7H_{11}N$ ▶ 7% $CDCl_3$ ▶ V 177

▶ 1-Bb

$C_6H_5-NH(d)-CH_2(c)-(CH_2)_2(b)-CH_3(a)$; ring (e), (e), (f), (f), (f)

	Shift	Peak
▶ a	0.96	3B
b	1.50	1F
c	3.08	3A
d	3.42	1A
e	6.52	3C
f	7.12	3C

▶ $C_{10}H_{15}N$ ▶ 45.4 mg/0.50 ml $CDCl_3$ ▶ S 280

▶ 1-Bb

$C_6H_5-NH(d)-C(=O)-CH_2(c)-CH_2(b)-CH_3(a)$

	Shift	Peak
▶ a	0.95	3A
b	1.71	6B
c	2.24	3C
d	~7.8	1C

▶ $C_{10}H_{13}NO$ ▶ 45.3 mg/0.50 ml $CDCl_3$ ▶ S 261

▶ 1-Bb

$C_6H_5-N[CH_2(c)-(CH_2)_2(b)-CH_3(a)]_2$

	Shift	Peak
▶ a	0.96	3B
b	1.48	1F
c	3.28	3C

▶ $C_{14}H_{23}N$ ▶ 45.2 mg/0.50 ml $CDCl_3$ ▶ S 147

▶ 1-Bb

$H_2N(d)-C_6H_4-C(=O)-O-CH_2(c)CH_2CH_2(b)CH_3(a)$; ring (f), (f), (e), (e)

	Shift	Peak
▶ a	0.95	3B
b	~1.60	1F
c	4.25	3D
d	~4.10	1E
e	6.59	Q
f	7.83	Q

▶ $C_{11}H_{15}NO_2$ ▶ 44.6 mg/0.50 ml $CDCl_3$ ▶ S 243

▶ 1-Bb

(d) CH_2–CH_2(c)–CH_3(a); (b) CH_3–CH(e); OCH_3(f); (h)HO; C–H(g)–O; (f)CH_3O

	Shift	Peak
▶ a	0.96	3B
b	1.37	2A
c	~1.65	S
d	2.58	3B
e	3.42	S
f	3.88	1A
g	5.08	2A
h	5.62	1A

▶ $C_{20}H_{24}O_4$ ▶ 7% $CDCl_3$ ▶ V 682

▶ 1-Bb

Shift	Peak
▶ a 0.97	3C
b 1.67	6B
c 2.42	3C
d 9.74	3A

(a) CH_3—(b) CH_2—(c) CH_2—C(=O)—H(d)

▶ C_4H_8O ▶ 7% $CDCl_3$ ▶ V 78

▶ 1-Bb

Shift	Peak
▶ a 1.08	3C
b 1.70	6B
c 2.30	3C

(a) CH_3—(b) CH_2—(c) CH_2—C≡N

▶ C_4H_7N ▶ 45.0 mg/0.50 ml $CDCl_3$ ▶ S 33

▶ 1-Bb

Shift	Peak
▶ a 0.98	3B
b 1.64	O
c 4.35	3A
d 7.48	N
e 8.06	N

C_6H_5—C(=O)—O(c) CH_2(b) $(CH_2)_2$(a) CH_3

▶ $C_{11}H_{14}O_2$ ▶ 44.8 mg/0.50 ml $CDCl_3$ ▶ S 158

▶ 1-Bb

Shift	Peak
▶ a 1.08	3A
b 1.73	6B
c 2.93	3A
d 7.10	Q
e 8.38	Q

4-(propylthio)pyridine: S—(c) CH_2—(b) CH_2—(a) CH_3; ring H(d), H(e)

▶ $C_8H_{11}NS$ ▶ 7% $CDCl_3$ ▶ V 509

▶ 1-Bb

Shift	Peak
▶ a .99	3B
b 1.25	1A
c 2.02	3B
d 4.90	R
e 5.08	R
f 5.68	R

H(e)H(d)C=C(f)H—(c) CH_2(b) $(CH_2)_{12}$(a) CH_3

▶ $C_{16}H_{32}$ ▶ 50% CCl_4 ▶ API 73

▶ 1-Bc

Shift	Peak
▶ a .77	3C
b 1.30	S

Ethylcyclobutane: ring CH—(b) CH_2—(a) CH_3

▶ C_6H_{12} ▶ liquid ▶ API 446

▶ 1-Bb

Shift	Peak
▶ a 1.01	3A
b 1.86	6B
c 3.18	3A

(a) CH_3—(b) CH_2—(c) CH_2—I

▶ C_3H_7I ▶ 45.0 mg/0.50 ml $CDCl_3$ ▶ S 230

▶ 1-Bc

Shift	Peak
▶ a ~ .79	3D
b ~ .82	2H
c ~1.22	1D
d ~1.58	S

(b) CH_3—(d) CH(CH_3(b))—(d) CH(CH_3(b))—(c) CH_2—(a) CH_3

▶ C_7H_{16} ▶ 10% CCl_4 ▶ API 197

▶ 1-Bb

Shift	Peak
▶ a 1.03	3A
b 2.07	6A
c 4.38	3A

(a) CH_3—(b) CH_2—(c) CH_2—NO_2

▶ $C_3H_7NO_2$ ▶ 7% $CDCl_3$ ▶ V 42

▶ 1-Bc

Shift	Peak
▶ a .80	3D
b .92	S
c ~.94	S

CH_3—CH_2—(c) CH((b) CH_2—(a) CH_3)—CH_2—CH_3

▶ C_7H_{16} ▶ 50% CCl_4 ▶ API 196

▶ 1-Bb

Shift	Peak
▶ a 1.03	3A
b 2.15	6A
c 4.74	3C
d 8.29	3C
e 8.97	R
f 9.12	R
g 9.37	1G

3-carbamoyl-1-propylpyridinium iodide: ring H(d), H(e), H(f), H(g); C(=O)NH_2; $N^{(+)}$—(c) CH_2—(b) CH_2—(a) CH_3; $I^{(-)}$

▶ $C_9H_{13}IN_2O$ ▶ D_2O ▶ V 539

▶ 1-Bc

Shift	Peak
▶ a 0.80	3A
b 1.22	2A
c 1.59	5B
d 2.53	6A
e 7.20	1A

(e) C_6H_5—(d) CH((b) CH_3)—(c) CH_2—(a) CH_3

▶ $C_{10}H_{14}$ ▶ 45.0 mg/0.50 ml $CDCl_3$ ▶ S 31

▶ 1-Bc

(b) (d)(c) (a)
$CH_3CHCH_2CH_3$ attached to C of benzene ring: (e)HC, CH(e), (f)HC, CH(f), H(f)

	Shift	Peak
▶ a	.82	3C
b	1.23	2A
c	1.60	5B
d	2.56	4C
e	7.11	1A
f	7.11	1A

▶ $C_{10}H_{14}$ ▶ 10% CCl_4 ▶ API 90

▶ 1-Bc

(a)
H–C(–CH_2CH_3) ring with H_2C, CH_2, H_2C, CH_2, $C H_2$

	Shift	Peak
a	.86	3D

▶ C_8H_{16} ▶ 10% CCl_4 ▶ API 149

▶ 1-Bc

(b) CH_3
CH_3—CH(d)—CH_2(c)—CH_3(a)

	Shift	Peak
▶ a	.83	3D
b	.90	2H
c	1.20	S
d	~1.67	1D

▶ C_5H_{12} ▶ 50% CCl_4 ▶ API 332

▶ 1-Bc

(b) (a)
CH_2CH_3
CH_3—CH_2—CH—CH_2(c)—OH(d)

	Shift	Peak
▶ a	.87	3C
b	1.29	3C
c	3.35	2G
d	4.18	1A

▶ $C_6H_{14}O$ ▶ 20% CCl_4 ▶ API 167

▶ 1-Bc

H, H(d)
C=C
H, (c) (b) (a) $CHCH_2CH_3$
CH_2CH_3

	Shift	Peak
▶ a	.83	3C
b	1.30	O
c	~1.72	S
d	~5.50	R

▶ C_7H_{14} ▶ liquid ▶ API 398

▶ 1-Bc

CH_3
CH_3—C(–CH_3(a))—CH_2(d)—CH(e)(–CH_3(c))—CH_2—CH_3(b)

	Shift	Peak
a	.87	1E
▶ b	.87	1E
c	.87	1E
d	~1.22	S
e	~1.32	1D

▶ C_9H_{20} ▶ 50% CCl_4 ▶ API 211

▶ 1-Bc

(a) CH_3
(a) CH_3—(b) CH_2—CH(c)—(b) $(CH_2)_3$—(a) CH_3

	Shift	Peak
▶ a	.83	3B
b	1.23	1B
c	~1.27	S

▶ C_8H_{18} ▶ 50% CCl_4 ▶ API 341

▶ 1-Bc

(b)
CH_3 CH_2—CH_3
CH_3—C(–CH_3(a))—CH(d)—CH_2(c)—CH_3

	Shift	Peak
a	.87	1E
▶ b	.87	1E
c	1.00	1B
d	~1.43	7X

▶ C_9H_{20} ▶ 50% CCl_4 ▶ API 216

▶ 1-Bc

(b) CH_3
CH_3—CH_2—CH(d)—CH_2(c)—CH_3(a)

	Shift	Peak
▶ a	.84	S
b	.84	S
c	1.20	O
d	~1.30	S

▶ C_6H_{14} ▶ 50% CCl_4 ▶ API 335

▶ 1-Bc

(f) H, H(g)
C=C, CH_3(b)
(e) H, $CHCH_2CH_3$
(d) (c) (a)

	Shift	Peak
▶ a	.88	3D
b	.99	2A
c	1.32	S
d	2.01	5B
e	4.83 or 4.96	R
f	4.96 or 4.83	R
g	5.53	R

▶ C_6H_{12} ▶ 10% CCl_4 ▶ API 156

▶ 1-Bc

(a)
CH_2CH_3
CH ring with H_2C, CH_2, H_2C, CH_2

	Shift	Peak
a	.86	3C

▶ C_7H_{14} ▶ 10% CCl_4 ▶ API 145

▶ 1-Bc

(e) H, (g) H
C=C
H (f), (d) (c) (a) $CHCH_2CH_3$
CH_3(b)

	Shift	Peak
▶ a	.88	3C
b	1.03	1E
c	1.24	S
d	1.96	7B
e	4.86	R
f	4.97	R
g	5.54	R

▶ C_6H_{12} ▶ liquid ▶ API 374

▶ 1-Bc

$CH_3-CH_2-CH(CH_3)-CH_2-CH_3$ (a: CH_3; b: branch CH_3; c: CH_2; d: CH)

	Shift	Peak
▶ a	.88	3D
b	.88	2H
c	1.29	S
d	~1.30	S

▶ C_6H_{14} ▶ 10% CCl_4 ▶ API 47

▶ 1-Bc

$(d)H_3C-CH=CH-CH(CH_3)CH_2CH_3$ (b: CH_3; e: CH; c: CH_2; a: CH_3; g, f: vinyl H)

	Shift	Peak
▶ a	.92	3C
b	.95	2H
c	1.25	O
d	1.58	2E
e	2.38	6B
f	~5.17 or 5.32	S
g	~5.32 or 5.17	S

▶ C_7H_{14} ▶ liquid run, neat ▶ API 409

▶ 1-Bc

$CH_3-CH_2-CH(CH_2CH_3)-CH_2-OH$ (a: CH_3; b: CH_2; d: CH_2O; c: OH)

	Shift	Peak
▶ a	0.88	3B
b	1.30	6B
c	1.56	S
d	3.54	2G

▶ $C_6H_{14}O$ ▶ 35.5 mg/0.50 ml $CDCl_3$ ▶ S 14

▶ 1-Bc

$CH_3(CH_2)_3-CH(CH_2CH_3)CH_2OH$ (a: CH_3; b: CH_2; d: CH_2O; c: OH)

	Shift	Peak
▶ a	0.92	3B
b	1.30	1B
c	1.75	1B
d	3.52	1I

▶ $C_8H_{18}O$ ▶ 45.0 mg/0.50 ml $CDCl_3$ ▶ S 98

▶ 1-Bc

$CH_3-CH_2-CH(CH_3)-CH_2-CH_2-CH_3$ (a: CH_3; b: branch CH_3; c: CH_2; d: CH)

	Shift	Peak
▶ a	.88	3D
b	.88	S
c	1.24	S
d	~1.26	S

▶ C_7H_{16} ▶ 50% CCl_4 ▶ API 338

▶ 1-Bc

$CH_3-CH_2-CH(NO_2)-CH_3$ (a, c, d, b)

	Shift	Peak
▶ a	0.95	3A
b	1.52	2A
c	~1.90	O
d	4.52	6A

▶ $C_4H_9NO_2$ ▶ 7% $CDCl_3$ ▶ V 84

▶ 1-Bc

$(H_3C-H_2C)_2CH-CH_2-CH_2-CH_3$ (b: CH_3; c: CH_2; d: CH; a: CH_3)

	Shift	Peak
a	~ .88	3D
▶ b	~ .88	3D
c	~1.26	1F

▶ C_8H_{18} ▶ 50% CCl_4 ▶ API 202

▶ 1-Bc

$H_2C=CH-CH_2CH(CH_3)CH_2CH_3$ (f, e: =CH₂; g: =CH; d: CH_2; c: CH; c: CH_2; b: CH_3; a: CH_3)

	Shift	Peak
a	~ .87	3D
▶ b	~ .95	2H
c	1.30	O
d	1.98	4C
e	~4.96	R
f	~4.96	R
g	5.67	R

▶ C_7H_{14} ▶ liquid run, neat ▶ API 393

▶ 1-Bc

$CH_3-CH_2-CH(NH_2)-CH_3$ (a, d, b; c: NH_2)

	Shift	Peak
▶ a	0.90	3D
b	1.05	2F
c	1.35	N
d	2.78	6A

▶ $C_4H_{11}N$ ▶ 7% $CDCl_3$ ▶ V 88

▶ 1-Bc

$CH_3-CH(CH_3)-CH(CH_2CH_3)-CH(CH_3)-CH_3$ (a: CH_3; d: CH; c: CH_2; b: CH_3)

	Shift	Peak
a	.86	2H
▶ b	.95	3D
c	~1.2	S
d	~1.60	S

▶ C_9H_{20} ▶ 50% CCl_4 ▶ API 217

▶ 1-Bc

$(a)CH_3-CH(CH_3)-CH(CH_2CH_3)-CH_2-CH_3(b)$ (a: CH_3; c: CH_2; b: CH_3; d: CH)

	Shift	Peak
a	~ .83	2H
▶ b	~ .91	3D
c	1.20	3B
d	~1.72	1D

▶ C_8H_{18} ▶ 50% CCl_4 ▶ API 206

▶ 1-Bc

$(b)CH_3-CH(e)(NH(d)C_6H_5)-CH_2-CH_3$ (c: CH_2; a: CH_3)

	Shift	Peak
▶ a	0.96	3D
b	1.17	2A
c	~1.50	O
d	~3.38	S
e	~3.41	S

▶ $C_{10}H_{15}N$ ▶ 7% $CDCl_3$ ▶ V 568

▶ 1-Bc

Shift	Peak
▶ a 0.97	3A
b 1.58	4A
c 2.22	5B
d 10.45	1C

▶ $C_6H_{12}O$ ▶ 36.9 mg/0.50 ml $CDCl_3$ ▶ S 74

▶ 1-Bc

Shift	Peak
▶ a 1.10	3A
b 2.32	5B
c 5.80	3A

▶ $C_3H_6ClNO_2$ ▶ 7% $CDCl_3$ ▶ V 385

▶ 1-Bc

Shift	Peak
▶ a 1.02	3C
b 1.70	4C
c 1.92	2A
d 4.17	4C

▶ C_4H_9I ▶ 7% $CDCl_3$ ▶ V 82

▶ 1-Bc

Shift	Peak
▶ a 1.13	3B
b 1.89	O
c 2.74	O
d 3.42	4A
e 3.77	1A
f 3.98	3E
g 4.59	2A
h 5.59	S
i 5.63*	S
j 5.63*	S
k 7.47	1A

*or 6.12

▶ $C_{15}H_{16}O_6$ ▶ 7% $CDCl_3$ ▶ V 641

▶ 1-Bc

Shift	Peak
▶ a 1.03	3A
b 1.30	3A
c 2.08	5B
d 4.20	S
e 4.27	4E

▶ $C_6H_{11}BrO_2$ ▶ 7% $CDCl_3$ ▶ V 137

▶ 1-Bd

Shift	Peak
▶ a 0.69	3A
b 1.22	1A
c 1.60	4A
d ~4.85	1D

▶ $C_{11}H_{16}O$ ▶ 44.6 mg/0.50 ml $CDCl_3$ ▶ S 256

▶ 1-Bc

Shift	Peak
▶ a 1.05	3A
b 3.62	2F
c 3.83	2F
d 4.13	O

▶ $C_4H_8Br_2$ ▶ 7% $CDCl_3$ ▶ V 74

▶ 1-Bd

Shift	Peak
▶ a .73	3C
b 1.18	4C

▶ C_9H_{20} ▶ 50% CCl_4 ▶ API 215

▶ 1-Bc

Shift	Peak
▶ a 1.08	3A
b 2.07	5B
c 4.23	3A
d 10.97	1A

▶ $C_4H_7BrO_2$ ▶ 7% $CDCl_3$ ▶ V 66

▶ 1-Bd

Shift	Peak
▶ a .78	3D
b .97	1E
c 1.28	O
d 5.69	4D

▶ C_7H_{14} ▶ liquid ▶ API 401

▶ 1-Bc

Shift	Peak
▶ a 1.08	3C
b 1.71	2H
c 1.79	6C
d 4.11	5A

▶ C_4H_9Br ▶ 7% $CDCl_3$ ▶ V 418

▶ 1-Bd

Shift	Peak
▶ a .78	3D
b ~ .79	1E
c 1.50	3C

▶ C_8H_{18} ▶ 50% CCl_4 ▶ API 207

▶ 1-Bd

	Shift	Peak
▶a	~ .82	3D
b	~ .82	1E
c	1.16	4C

$CH_3—CH_2—C(CH_3)(CH_3)—CH_2—CH_3$; (c) CH_2, (a) CH_3, (b) CH_3

▶ C_7H_{16} ▶ 50% CCl_4 ▶ API 199

▶ 1-Bd

	Shift	Peak
▶a	0.91	3A
b	1.06	1A
c	1.43	3B
d	2.43	1A
e	8.47	1D

(e) H–N; C=O; (d) CH_2; C; (c) CH_2–CH_3 (a); C=O; (d) CH_2; CH_3 (b)

▶ $C_8H_{13}NO_2$ ▶ 7% $CDCl_3$ ▶ V 514

▶ 1-Bd

	Shift	Peak
▶a	0.83	N
b	1.20	N
c	3.93	1A
d	5.52	1A

(b) CH_2—CH_3 (a); C; (c) CH_2, CH_2 (c), CH_2 (c); O, O, O; C; H(d)

▶ $C_7H_{12}O_3$ ▶ 7% $CDCl_3$ ▶ V 493

▶ 1-Bd

	Shift	Peak
▶a	0.92	3A
b	1.80	S
c	2.05	S
d	2.70	2C
e	2.97	2C
f	7.30	1A

(f); O; H(e); C–C; H—C—H; (c) (b); H(d); CH_3 (a)

▶ $C_{10}H_{12}O$ ▶ 7% $CDCl_3$ ▶ V 558

▶ 1-Bd

	Shift	Peak
▶a	0.84	3C
b	1.38	4C
c	2.83	3B
d	3.43	1E
e	3.59	2B
f	3.95	2E
g	5.14	R
h	5.22	R
i	5.87	R

(e) (c) CH_2—OH; (g) H, (h) H, (i) H, (f) CH_2, O, (d) CH_2, C, (d) CH_2, O, (f) CH_2, (i) H, (g) H, (h) H; CH_2—CH_3 (b) (a)

▶ $C_{12}H_{22}O_3$ ▶ 7% $CDCl_3$ ▶ V 603

▶ 1-Bd

	Shift	Peak
▶a	1.08	3B
b	1.58	1A
c	1.80	3D

$CH_3—CH_2—CCl(CH_3)—CH_3$ (c); (a) (b) (c)

▶ $C_5H_{11}Cl$ ▶ 7% $CDCl_3$ ▶ V 447

▶ 1-Bd

	Shift	Peak
a	.85	1E
▶b	.85	3D
c	1.19	3C

(a) CH_3; $CH_3—C—CH_2—CH_3$ (c) (b); CH_3

▶ C_6H_{14} ▶ 50% CCl_4 ▶ API 336

▶ 1-Bd

	Shift	Peak
▶a	1.24	3C
b	1.43	3D
c	1.78	1A
d	2.36	4C
e	3.45	4A

(a) CH_3—(d) CH_2; C; (c) CH_3; SO_2—CH_2—CH_3 (e) (b); SO_2—CH_2—CH_3 (e) (b)

▶ $C_8H_{18}O_4S_2$ ▶ 7% $CDCl_3$ ▶ V 519

▶ 1-Bd

	Shift	Peak
▶a	~ .85	3D
b	~ .85	1E
c	~ .92	2H
d	1.13	2A
e	1.63	7X

(c) CH_3, (b) CH_3; (c) (e) (c) (d) (a) $CH_3CH—CH_2CCH_2CH_3$; CH_3 (b)

▶ C_9H_{20} ▶ 50% CCl_4 ▶ API 426

▶ 1-De

	Shift	Peak
▶a	1.48	3A
b	3.57	4A

(a) (b) CH_3CH_2Cl

▶ C_2H_5Cl ▶ 7% $CDCl_3$ ▶ V 11

▶ 1-Bd

	Shift	Peak
a	.73	1A
b	.87 ± .03	1E
▶c	.87 ± .03	3D
d	1.31	4A

(b) CH_3, (a) CH_3; (b) $CH_3—C—C—CH_2—CH_3$ (d) (c); CH_3 (b), CH_3 (a)

▶ C_9H_{20} ▶ 50% CCl_4 ▶ API 218

▶ 1-Bg

	Shift	Peak
▶a	1.67	3A
b	3.43	4A

(a) (b) $CH_3—CH_2—Br$

▶ C_2H_5Br ▶ 7% $CDCl_3$ ▶ V 10

▶ 1-Bg

	Shift	Peak
▶ a	1.70	3A
b	3.40	4A

(a) (b)
CH_3CH_2Br

▶ C_2H_5Br ▶ 45.0 mg/0.50 ml $CDCl_3$ ▶ S 225

▶ 1-Bj

	Shift	Peak
▶ a	1.23	3A
b	2.93	4A
c	7.10	N
d	7.60	N
e	7.70	O

(c)H, H(e), (d)H, S, C, O, (b) (a) CH_2CH_3

▶ C_7H_8OS ▶ 7% $CDCl_3$ ▶ V 163

▶ 1-Bi

	Shift	Peak
▶ a	1.83	3A
b	3.20	4A

(a) (b)
CH_3CH_2I

▶ C_2H_5I ▶ 7% $CDCl_3$ ▶ V 13

▶ 1-Bk

	Shift	Peak
a	0.92	3B
▶ b	1.12	S
c	1.32	S
d	2.33	4A
e	4.05	3B

(b) (d) O (e) (c) (a)
$CH_3CH_2C{-}OCH_2(CH_2)_3CH_3$

▶ $C_8H_{16}O_2$ ▶ 44.8 mg/0.50 ml $CDCl_3$ ▶ S 40

▶ 1-Bj

	Shift	Peak
▶ a	1.05	3A
b	2.13	1A
c	2.47	4A

O=C, (b) CH_3, CH_2CH_3 (c) (a)

▶ C_4H_8O ▶ 7% $CDCl_3$ ▶ V 76

▶ 1-Bk

	Shift	Peak
a	0.92	2A
▶ b	1.13	3D
c	~1.85	S
d	2.31	4A
e	3.83	2A

(b) (d) O (e) (a) CH_3 (a)
$CH_3{-}CH_2{-}C{-}OCH_2{-}CH{-}CH_3$ (c)

▶ $C_7H_{14}O_2$ ▶ 4.5 mg/0.50 ml $CDCl_3$ ▶ S 55

▶ 1-Bj

	Shift	Peak
▶ a	1.13	3A
b	2.23	4C
c	~6.42	1D

(a) (b)
$CH_3{-}CH_2{-}C(=O){-}NH_2$ (c)

▶ C_3H_7NO ▶ 7% $CDCl_3$ ▶ V 40

▶ 1-Bk

	Shift	Peak
a	0.95	3D
▶ b	1.18	3D
c	1.60	6B
d	2.33	4A
e	4.04	3A

$CH_3CH_2{-}C(=O){-}OCH_2CH_2CH_3$
(b) (d) (e) (c) (a)

▶ $C_6H_{12}O_2$ ▶ 44.4 mg/0.50 ml $CDCl_3$ ▶ S 38

▶ 1-Bj

	Shift	Peak
▶ a	1.20	3A
b	2.38	1A
c	2.78	4A
d	6.13	2E
e	7.08	2A

(d)H, H(e), CH_3 (b), O, C(=O), CH_2 (c), CH_3 (a)

▶ $C_8H_{10}O_2$ ▶ 7% $CDCl_3$ ▶ V 206

▶ 1-Bk

	Shift	Peak
▶ a	1.30	3A
b	2.08	1A
c	4.28	4A
d	5.19	2A
e	6.60	1D

(b)CH_3C(=O)N(H)(e)–C(H)(d)[O–C(=O)–CH_2(c)–CH_3(a)]$_2$

▶ $C_9H_{15}NO_5$ ▶ 7% $CDCl_3$ ▶ V 547

▶ 1-Bj

	Shift	Peak
▶ a	1.22	3A
b	2.99	4A
c	7.50	N
d	7.98	N

(d), (c), (c), (d), (c); C(=O)–CH_2(b)–CH_3(a)

▶ $C_9H_{10}O$ ▶ 44.5 mg/0.50 ml $CDCl_3$ ▶ S 34

▶ 1-Bl

	Shift	Peak
▶ a	.90	3A
b	1.60	1A
c	2.05	4A

(b) H_3C, (b) CH_3, C=C, CH_3 (b), CH_2 (c)–CH_3 (a)

▶ C_7H_{14} ▶ 50% CCl_4 ▶ API 352

▶ 1-Bl

	Shift	Peak
▶ a	.91	3C
b	1.99	5B
c	5.25	O

▶ C_6H_{12} ▶ liquid ▶ API 371

▶ 1-Bl

	Shift	Peak
▶ a	.93	3C
b	1.58	2E
c	1.95	O
d	5.35 or 5.43	S
e	5.43 or 5.35	S

▶ C_5H_{10} ▶ liquid ▶ API 461

▶ 1-Bl

	Shift	Peak
▶ a	.91 or .98	3D
b	1.57	1I
c	1.94	5B
d	5.11	3C

▶ C_7H_{14} ▶ liquid ▶ API 411

▶ 1-Bl

	Shift	Peak
▶ a	.93	3C
b	1.92	1F
c	5.38	O

▶ C_6H_{12} ▶ liquid ▶ API 372

▶ 1-Bl

	Shift	Peak
▶ a	.91 or .96	3D
▶ b	.91 or .96	3D
c	1.61	1H
d	1.97	4A
e	5.07	3A

▶ C_7H_{14} ▶ liquid ▶ API 412

▶ 1-Bl

	Shift	Peak
▶ a	.93	3C
b	1.56	2E
c	2.04	4C
d	5.17	4A

▶ C_7H_{14} ▶ liquid ▶ API 414

▶ 1-Bl

	Shift	Peak
▶ a	.92	3A
b	1.57	1I
c	1.64	1I
d	1.97	4C
e	5.05	O

▶ C_6H_{12} ▶ 10% CCl_4 ▶ API 34

▶ 1-Bl

	Shift	Peak
▶ a	.94	3C
b	1.62	1B
c	2.03	4C

▶ C_7H_{14} ▶ liquid ▶ API 415

▶ 1-Bl

	Shift	Peak
▶ a	.92	3C
b	1.58	1I
c	1.65	1I
d	5.19	3C

▶ C_6H_{12} ▶ liquid ▶ API 376

▶ 1-Bl

	Shift	Peak
▶ a	.95	3C
b	1.46	1F
c	1.61	S
d	2.03	4A
e	~5.08	4C

▶ C_6H_{12} ▶ liquid ▶ API 377

▶ 1-Bl

	Shift	Peak
▶ a	.93	3C
b	1.52	2E
c	2.00	4C
d	5.30	N

▶ C_5H_{10} ▶ 50 vol.% CCl_4 ▶ API 429

▶ 1-Bl

	Shift	Peak
▶ a	.95	3A
b	1.60	1H
c	1.94	N
d	5.38	N

▶ C_5H_{10} ▶ 50 vol.% CCl_4 ▶ API 430

▶ 1-Bl

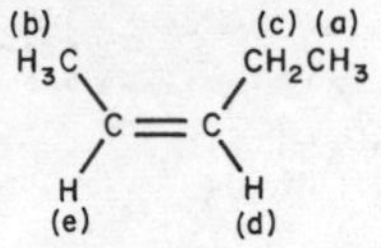

(b) H_3C (c) (a) CH_2CH_3 C=C H (e) H (d)

	Shift	Peak
▶ a	.95	3C
b	1.58	2E
c	2.05	5B
d	~5.28 or 5.38	S
e	~5.38 or 5.28	S

▶ C_5H_{10} ▶ liquid (neat) ▶ API 460

▶ 1-Bl

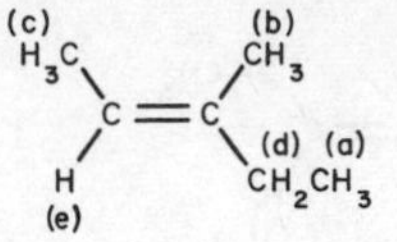

(c) H_3C (b) CH_3 C=C H (e) (d) (a) CH_2CH_3

	Shift	Peak
▶ a	.97	3C
b	1.49	S
c	1.56	S
d	1.95	4C
e	5.18	4C

▶ C_6H_{12} ▶ liquid ▶ API 378

▶ 1-Bl

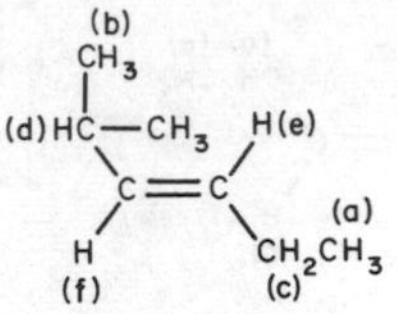

(b) CH_3 (d) HC—CH_3 H(e) C=C H (f) (a) CH_2CH_3 (c)

	Shift	Peak
▶ a	.96	3D
b	.97	2H
c	1.98	S
d	~2.28	S
e	~5.38	S
f	~5.38	S

▶ C_7H_{14} ▶ liquid ▶ API 397

▶ 1-Bl

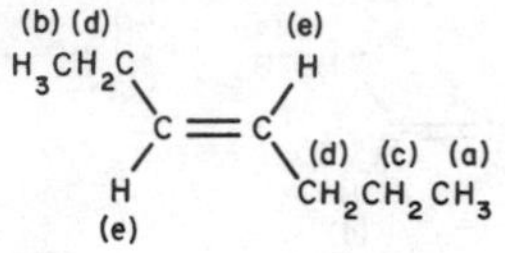

(b) (d) H_3CH_2C (e) H C=C H (e) (d) (c) (a) $CH_2CH_2CH_3$

	Shift	Peak
a	.90	3D
▶ b	.97	3D
c	1.33	S
d	1.95	O
e	5.40	O

▶ C_7H_{14} ▶ liquid ▶ API 391

▶ 1-Bl

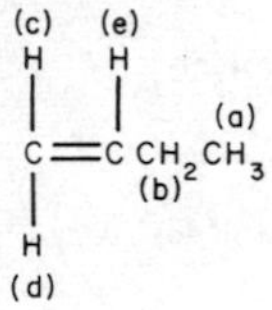

(c) H (e) H C=C CH_2CH_3 (b) (a) H (d)

	Shift	Peak
▶ a	.97	3C
b	2.02	5B
c	4.95	R
d	4.98	R
e	5.80	R

▶ C_4H_8 ▶ degassed liquid ▶ API 55

▶ 1-Bl

(a) CH_3–CH_2–CH(e)=CH(e)–(CH_2(c)–CH(e)=CH(e)$)_2$–CH_2–$(CH_2)_5$(b)–CH_2–C(=O)–OCH_3 (d)

	Shift	Peak
▶ a	0.97	3A
b	1.33	1B
c	2.80	3B
d	3.67	1A
e	~5.38	3C

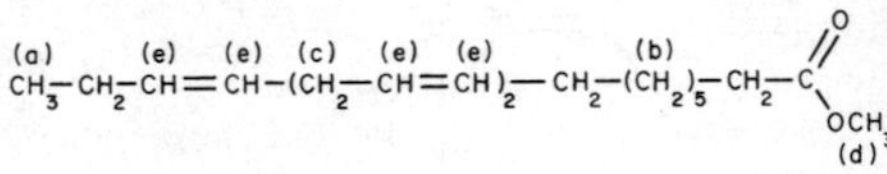

▶ $C_{19}H_{32}O_2$ ▶ 7% $CDCl_3$ ▶ V 337

▶ 1-Bl

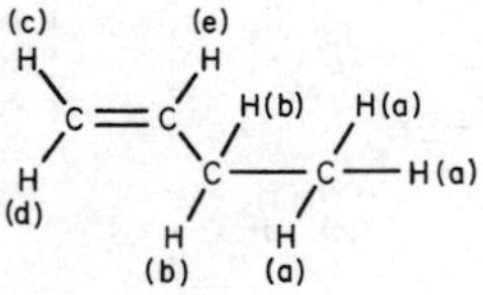

(c) H (e) H C=C H(b) H(a) H (d) C—C—H(a) H (b) H (a)

	Shift	Peak
▶ a	0.97	3C
b	1.98	5B
c	~4.88	R
d	~5.7	R
e	~5.9	R

▶ C_4H_8 ▶ liquid ▶ API 365

▶ 1-Bl

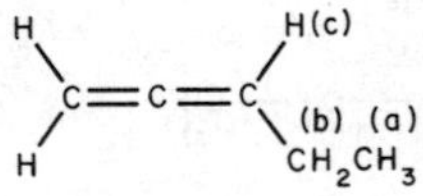

H H C=C=C H(c) (b) (a) CH_2CH_3

	Shift	Peak
▶ a	.98	3C
b	1.94	O
c	5.07	5B

▶ C_5H_8 ▶ liquid ▶ API 467

▶ 1-Bl

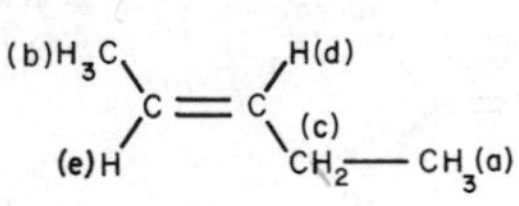

(b)H_3C H(d) C=C (e)H (c) CH_2—CH_3(a)

	Shift	Peak
▶ a	.97	3A
b	1.61	5A
c	1.97	1F
d	5.37	S
e	5.37	S

▶ C_5H_{10} ▶ 10% CCl_4 ▶ API 29

▶ 1-Bl

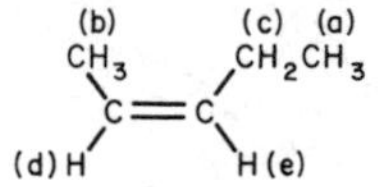

(b) CH_3 (c) (a) CH_2CH_3 C=C (d) H H (e)

	Shift	Peak
▶ a	.98	3A
b	1.58	2E
c	2.00	5B
d	5.23	S
e	5.30	S

▶ C_5H_{10} ▶ 10% CCl_4 ▶ API 154

▶ 1-Bl

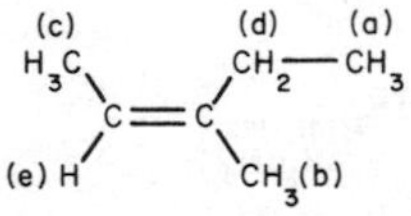

(c) H_3C (d) (a) CH_2—CH_3 C=C (e) H CH_3(b)

	Shift	Peak
▶ a	.97	3A
b	1.47	1E
c	1.57	1E
d	1.97	4A
e	5.13	2E

▶ C_6H_{12} ▶ 10% CCl_4 ▶ API 35

▶ 1-Bl

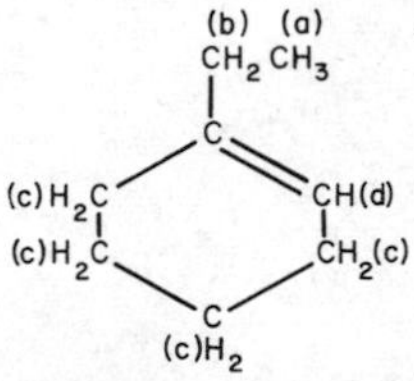

(b) (a) CH_2CH_3 C (c)H_2C CH(d) (c)H_2C CH_2(c) C (c)H_2

	Shift	Peak
▶ a	1.00	3A
b	~1.48 or 1.85	S
c	~1.85 or 1.48	S
d	5.32	1B

▶ C_8H_{14} ▶ 33% CCl_4 ▶ API 142

▶ 1-Bl

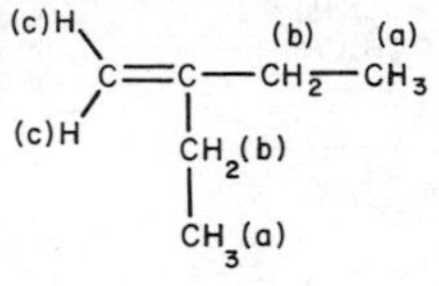

	Shift	Peak
▶ a	1.02	3C
b	2.08	4C
c	4.63	3A

▶ C_6H_{12} ▶ 10% CCl_4 ▶ API 36

▶ 1-Bn

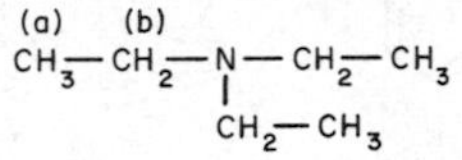

	Shift	Peak
▶ a	.98	3A
b	2.42	4A

▶ $C_6H_{15}N$ ▶ 34.8 mg/0.50 ml $CDCl_3$ ▶ S 29

▶ 1-Bl

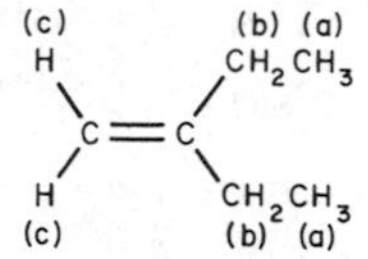

	Shift	Peak
▶ a	1.02	3C
b	2.00	4C
c	4.67	N

▶ C_6H_{12} ▶ liquid ▶ API 381

▶ 1-Bn

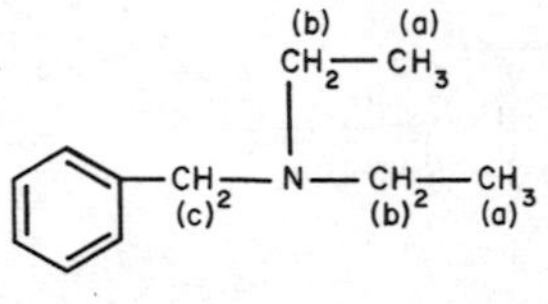

	Shift	Peak
▶ a	1.03	3A
b	2.52	4A
c	3.54	1A

▶ $C_{11}H_{17}N$ ▶ 44.8 mg/0.50 ml $CDCl_3$ ▶ S 152

▶ 1-Bl

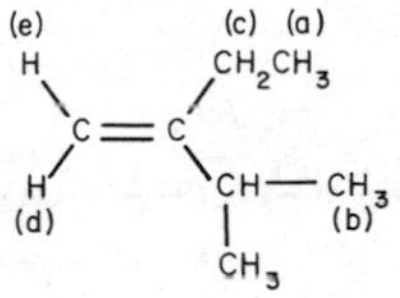

	Shift	Peak
▶ a	1.03	3D
b	1.03	2H
c	~2.13	S
d	4.70 ± 0.03	S
e	4.70 ± 0.03	S

▶ C_7H_{14} ▶ liquid ▶ API 416

▶ 1-Bn

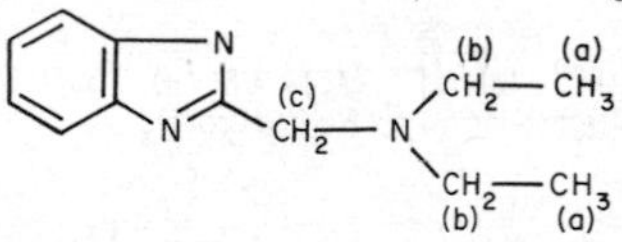

	Shift	Peak
▶ a	1.03	3A
b	2.63	4A
c	3.88	1A

▶ $C_{12}H_{17}N_3$ ▶ 7% $CDCl_3$ ▶ V 296

▶ 1-Bl

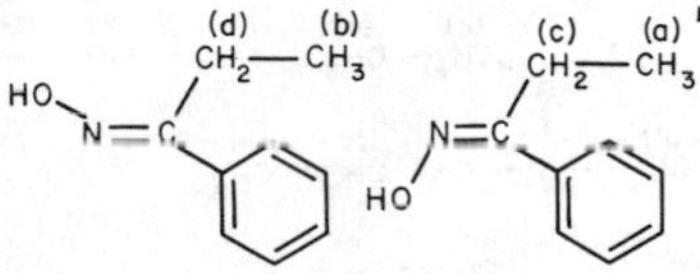

	Shift	Peak
▶ a	1.11	3A
▶ b	1.17	3D
c	2.58	4A
d	2.84	4E

mixture of anti- and syn- forms

▶ $C_9H_{11}NO$ ▶ 7% $CDCl_3$ ▶ V 533

▶ 1-Bn

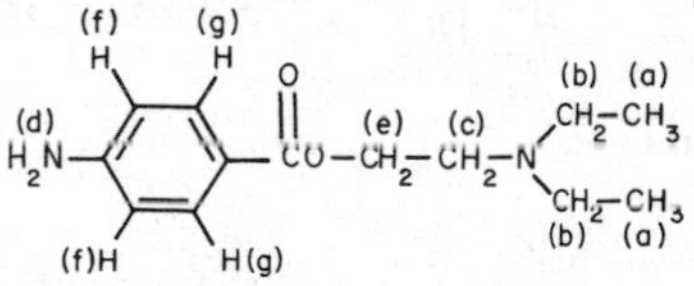

	Shift	Peak
▶ a	1.05	3A
b	2.62	4E
c	2.82	3D
d	4.13	1B
e	4.33	3D
f	0.00	Q
g	7.83	Q

▶ $C_{13}H_{20}N_2O_2$ ▶ 7% $CDCl_3$ ▶ V 302

▶ 1-Bl

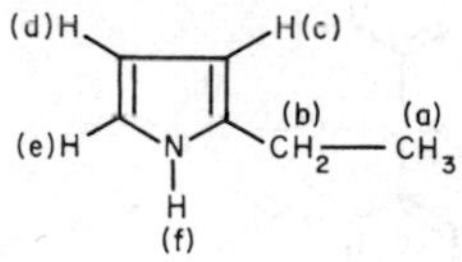

	Shift	Peak
▶ a	1.23	3A
b	2.63	6A
c	5.97	1I
d	6.17	4A
e	6.68	N
f	~7.85	1A

▶ C_6H_9N ▶ 7% $CDCl_3$ ▶ V 130

▶ 1-Bn

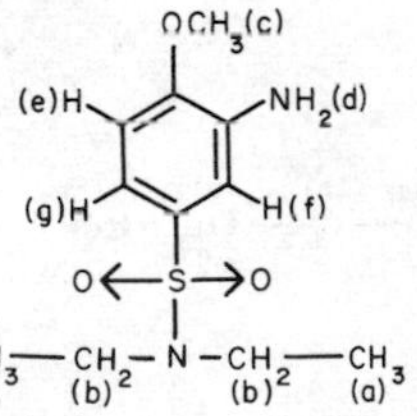

	Shift	Peak
▶ a	1.11	3A
b	3.20	4A
c	3.88	1E
d	3.88	1E
e	6.80	2E
f	7.13	1A
g	7.17	2A

▶ $C_{11}H_{18}N_2O_3S$ ▶ 7% $CDCl_3$ ▶ V 587

▶ 1-Bn

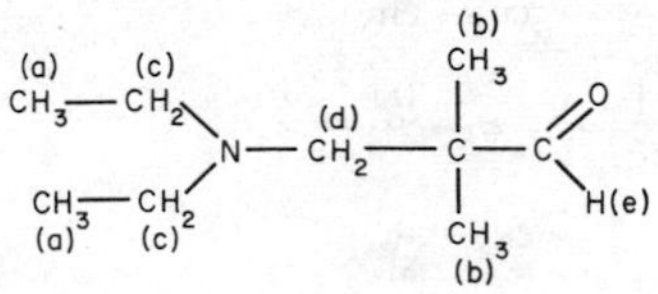

	Shift	Peak
▶ a	0.97	3D
b	1.07	1E
c	2.50	4E
d	2.55	1E
e	9.57	1A

▶ $C_9H_{19}NO$ ▶ 7% $CDCl_3$ ▶ V 548

▶ 1-Bn

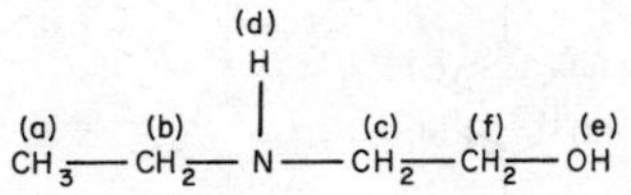

	Shift	Peak
▶ a	1.12	3A
b	2.68	4E
c	2.73	3D
d	3.53	1E
e	3.53	1E
f	3.65	3D

▶ $C_4H_{11}NO$ ▶ 7% $CDCl_3$ ▶ V 92

▶ 1-Bn

	Shift	Peak
▶ a	1.13	3A
b	2.16	1B
c	3.38	S
d	3.40	S
e	3.72	3D

(a) CH_3 (c) CH_2 N(phenyl) (d) CH_2 (e) CH_2 OH (b)

▶ $C_{10}H_{15}NO$ ▶ 7% $CDCl_3$ ▶ V 569

▶ 1-Bo

	Shift	Peak
▶ a	0.96	3A
b	4.08	4A
c	6.95	O
d	7.34	N

(b) (a) OCH_2CH_3, Cl, (c), (c), (c), (d)

▶ C_8H_9ClO ▶ 45.2 mg/0.50 ml $CDCl_3$ ▶ S 234

▶ 1-Bn

	Shift	Peak
▶ a	1.14	3A
b	2.02	1A
c	3.26	5B
d	8.08	1B

(b) $CH_3C(=O)—NHCH_2CH_3$ (d) (c) (a)

▶ C_4H_9NO ▶ 44.4 mg/0.50 ml $CDCl_3$ ▶ S 146

▶ 1-Bo

	Shift	Peak
▶ a	1.20	S
b	1.32	S
c	~3.50 or 3.70	2C
d	~3.50 or 3.70	2C
e	4.72	4A

(b) CH_3—(e) CH(O—CH(c)H(d)—CH_3(a))(O—CH(d)H(c)—CH_3(a))

▶ $C_6H_{14}O_2$ ▶ 7% $CDCl_3$ ▶ V 143

▶ 1-Bn

	Shift	Peak
▶ a	1.22	3A
b	3.49	5B
c	~6.21	1C

(a) CH_3—(b) CH_2—(c) NH—C(=S)—(c) NH—(b) CH_2—(a) CH_3

▶ $C_5H_{12}N_2S$ ▶ 45.2 mg/0.50 ml $CDCl_3$ ▶ S 190

▶ 1-Bo

	Shift	Peak
▶ a	1.20	3A
b	~3.50	4E
c	~3.63	1E

(a) (b) CH_3CH_2—O—CH_2CH_2(c)—O—CH_2CH_2(c)—O—CH_2CH_3 (b) (a)

▶ $C_8H_{18}O_3$ ▶ 45.3 mg/0.50 ml $CDCl_3$ ▶ S 285

▶ 1-Bn

	Shift	Peak
▶ a	1.42	3A
b	3.18	4A

(a) CH_3—(b) CH_2—N(—(b) CH_2—(a) CH_3)(—(b) CH_2—(a) CH_3) • HBr

▶ $C_6H_{15}N \cdot HBr$ ▶ 44.9 mg/0.50 ml $CDCl_3$ ▶ S 268

▶ 1-Bo

	Shift	Peak
a	0.12	1A
▶ b	1.22	3A
c	3.80	4A

(a) CH_3—Si(—O—(c) CH_2—(b) CH_3)₃

▶ $C_7H_{18}O_3Si$ ▶ 7% $CDCl_3$ ▶ V 184

▶ 1-Bn

	Shift	Peak
▶ a	1.45	3A
b	3.03	4A
c	9.1	1D

(a) CH_3—(b) CH_2—(c) NH—(b) CH_2—(a) CH_3 • HCl

▶ $C_4H_{11}N \cdot HCl$ ▶ 44.8 mg/0.50 ml $CDCl_3$ ▶ S 257

▶ 1-Bo

	Shift	Peak
▶ a	1.22	3A
b	3.87	4A

$(C_6H_5)_3$Si—O—(b) CH_2—(a) CH_3

▶ $C_{20}H_{20}OSi$ ▶ 7% $CDCl_3$ ▶ V 341

▶ 1-Bo

	Shift	Peak
▶ a	0.90	3A
b	3.99	4A
c	6.78	Q
d	7.21	Q

(b) (a) OCH_2CH_3, (c), (c), (d), (d), Cl

▶ C_8H_9ClO ▶ 45.4 mg/0.50 ml $CDCl_3$ ▶ S 235

▶ 1-Bo

	Shift	Peak
▶ a	1.23	3A
b	3.82	4A
c	5.98 ± 0.05	S
d	5.98 ± 0.05	S
e	5.98 ± 0.05	S

(c)H—C(=C(d)H—H(e))—Si(—O—(b) CH_2—(a) CH_3)₃

▶ $C_8H_{18}O_3Si$ ▶ 7% $CDCl_3$ ▶ V 219

▶ 1-Bo

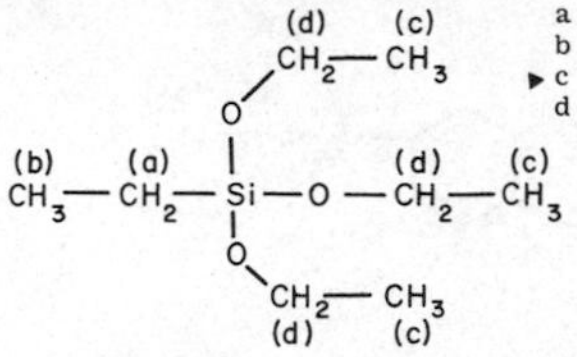

	Shift	Peak
a	~0.62	S
b	~1.00	S
▶ c	1.23	3A
d	3.82	4A

▶ $C_8H_{20}O_3Si$ ▶ 7% $CDCl_3$ ▶ V 222

▶ 1-Bo

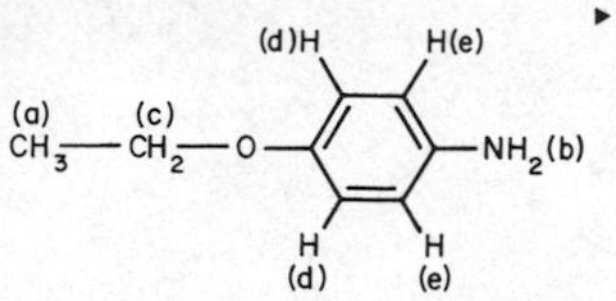

	Shift	Peak
▶ a	1.35	3A
b	3.30	1B
c	3.93	4A
d	6.63 or 6.70	Q
e	6.63 or 6.70	Q

▶ $C_8H_{11}NO$ ▶ 7% $CDCl_3$ ▶ V 208

▶ 1-Bo

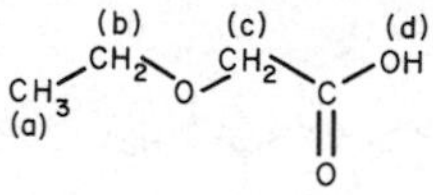

	Shift	Peak
▶ a	1.27	3A
b	3.66	4A
c	4.13	1A
d	10.95	1A

▶ $C_4H_8O_3$ ▶ 7% $CDCl_3$ ▶ V 417

▶ 1-Bo

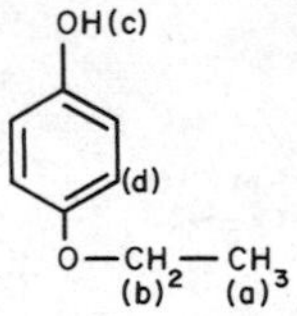

	Shift	Peak
▶ a	1.36	3A
b	3.93	4A
c	5.45	1C
d	6.67	1A

▶ $C_8H_{10}O_2$ ▶ 45.0 mg/0.50 ml $CDCl_3$ ▶ S 12

▶ 1-Bo

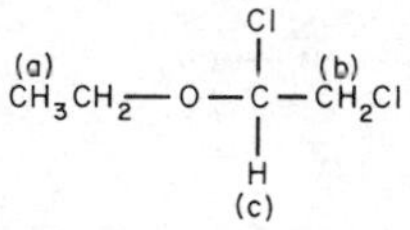

	Shift	Peak
▶ a	1.28	3A
b	3.70	O
c	5.63	3A

▶ $C_4H_8Cl_2O$ ▶ 7% $CDCl_3$ ▶ V 75

▶ 1-Bo

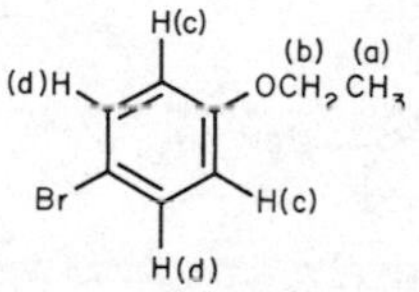

	Shift	Peak
▶ a	1.37	3A
b	3.93	4A
c	6.73	Q
d	7.28	Q

▶ C_8H_9BrO ▶ 7% $CDCl_3$ ▶ V 198

▶ 1-Bo

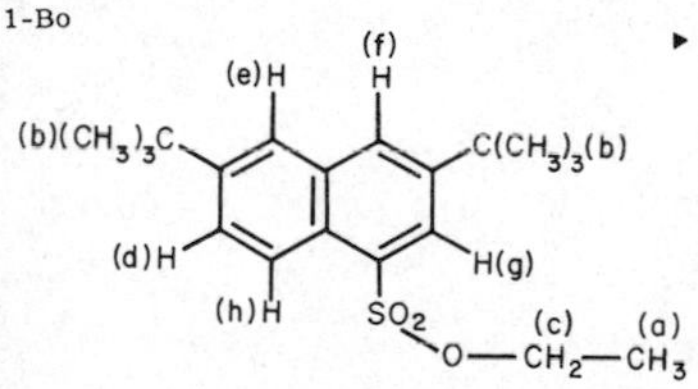

	Shift	Peak
▶ a	1.28	3D
b	1.43	1A
c	4.07	4A
d	7.73	2A
e	7.82	2G
f	8.01	2G
g	8.28	2A
h	8.45	2A

▶ $C_{20}H_{28}O_3S$ ▶ 7% $CDCl_3$ ▶ V 683

▶ 1-Bo

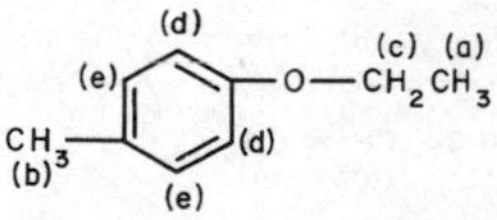

	Shift	Peak
▶ a	1.38	3A
b	2.28	1A
c	3.98	4A
d	6.75	Q
e	7.05	Q

▶ $C_9H_{12}O$ ▶ 45.0 mg/0.50 ml $CDCl_3$ ▶ S 284

▶ 1-Bo

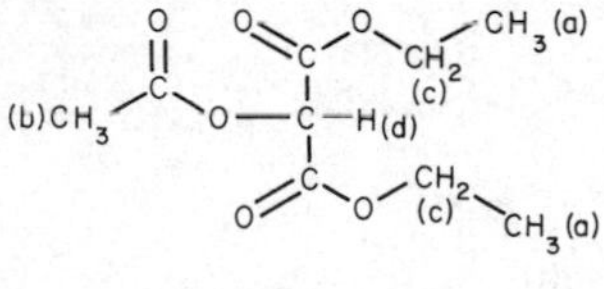

	Shift	Peak
▶ a	1.31	3A
b	2.22	1A
c	4.30	4A
d	5.52	1A

▶ $C_9H_{14}O_6$ ▶ 7% $CDCl_3$ ▶ V 546

▶ 1-Bo

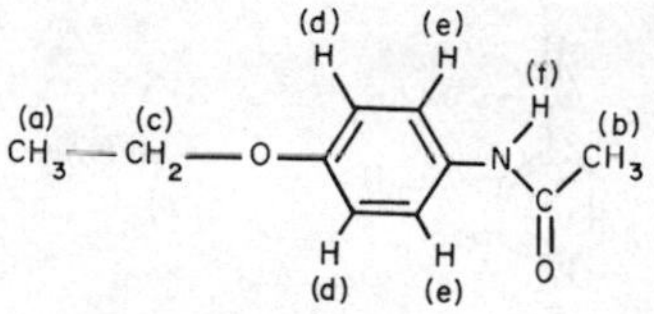

	Shift	Peak
▶ a	1.38	3A
b	2.12	1A
c	4.00	4A
d	6.83	Q
e	7.41	Q
f	7.91	1C

▶ $C_{10}H_{13}NO_2$ ▶ 7% $CDCl_3$ ▶ V 267

▶ 1-Bo

	Shift	Peak
▶ a	1.35	3C
b	4.11	5B

$(CH_3CH_2O)_3P \rightarrow O$

▶ $C_6H_{15}O_4P$ ▶ 7% $CDCl_3$ ▶ V 482

▶ 1-Bo

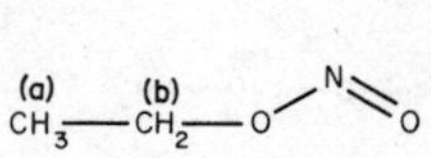

	Shift	Peak
▶ a	1.39	3A
b	4.78	4A

$CH_3—CH_2—O—N=O$

▶ $C_2H_5NO_2$ ▶ 7% $CDCl_3$ ▶ V 374

1-Bo

Shift	Peak
a 1.40	3A
b 4.04	4A

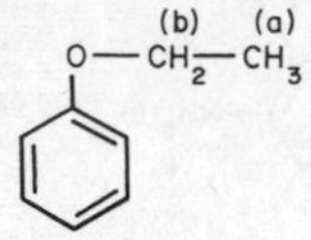

$C_8H_{10}O$ — 34.7 mg/0.50 ml $CDCl_3$ — S 26

1-Bo

Shift	Peak
a 1.45	3A
b 4.11	4A

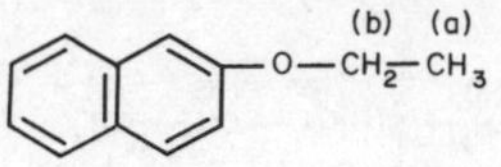

$C_{12}H_{12}O$ — 45.0 mg/0.50 ml $CDCl_3$ — S 72

1-Bo

Shift	Peak
a 1.40	3A
b 3.99	4A

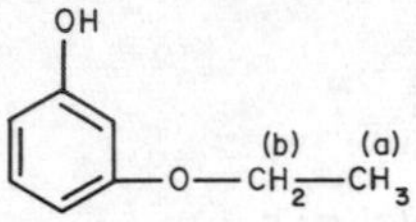

$C_8H_{10}O_2$ — 45.1 mg/0.50 ml $CDCl_3$ — S 267

1-Bo

Shift	Peak
a 1.45	3A
b 2.37	2D
c 4.08	4A
d 6.08	2E
e 6.73	1G
f ~6.78	2G
g 7.47	2E

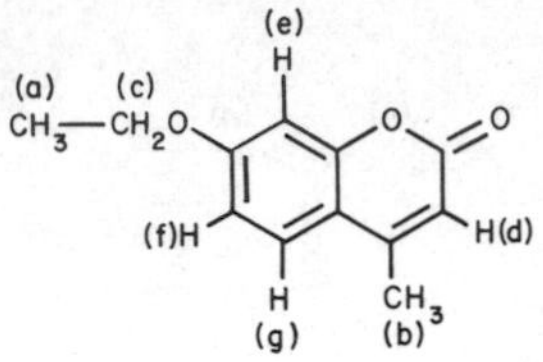

$C_{12}H_{12}O_3$ — 7% $CDCl_3$ — V 294

1-Bo

Shift	Peak
a 1.40	3A
b 3.72	1B
c 4.03	4A
d 6.72	1A

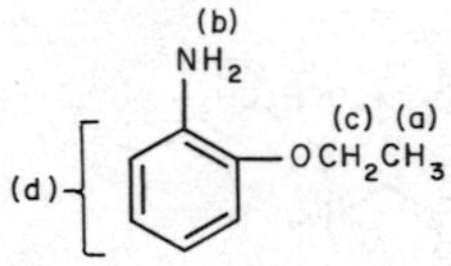

$C_8H_{11}NO$ — 45.1 mg/0.50 ml $CDCl_3$ — S 236

1-Bo

Shift	Peak
a 1.47	3A
b 4.42	4A
c 7.65	1A

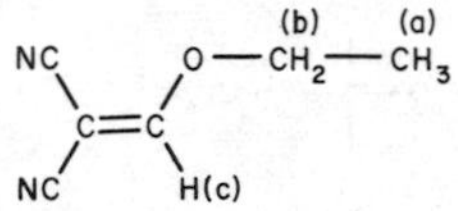

$C_6H_6N_2O$ — 7% $CDCl_3$ — V 452

1-Bo

Shift	Peak
a 1.40	3A
b 4.05	4A

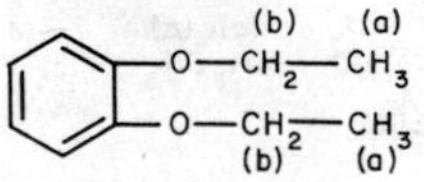

$C_{10}H_{14}O_2$ — 45.0 mg/0.50 ml $CDCl_3$ — S 171

1-Bp

Shift	Peak
a 1.22	3A
b 2.58	1B
c 3.70	4A

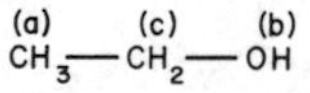

C_2H_6O — 7% $CDCl_3$ — V 14

1-Bo

Shift	Peak
a 1.41	3A
b 4.40	4A
c 6.92	O
d 7.37	O
e 7.83	2E
f 10.78	1A

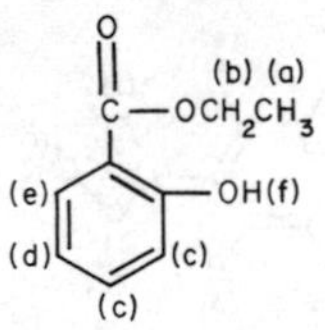

$C_9H_{10}O_3$ — 45.6 mg/0.50 ml $CDCl_3$ — S 196

1-Bq

Shift	Peak
a 0.83	3A
b 2.17	1A
c 2.37	1A
d 3.87	4A
e 5.13	1A
f 6.15	1C

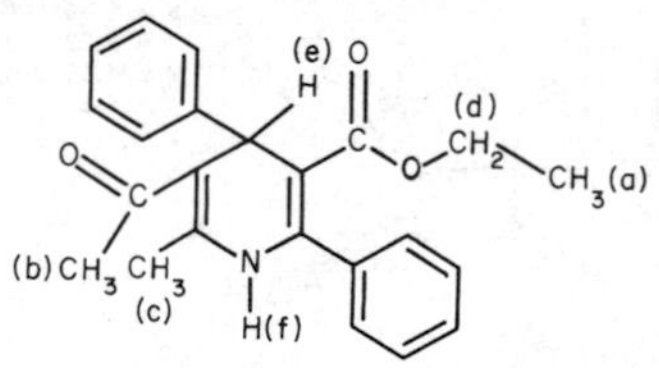

$C_{23}H_{23}NO_3$ — 7% $CDCl_3$ — V 695

1-Bo

Shift	Peak
a 1.44	3A
b 4.08	4A
c 5.70	1A

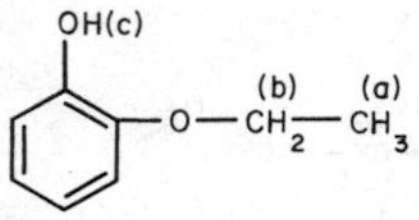

$C_8H_{10}O_2$ — 45.3 mg/0.50 ml $CDCl_3$ — S 170

1-Bq

Shift	Peak
a 0.88	3B
b 1.28	1E'
c 2.28	3C
d 4.11	4A

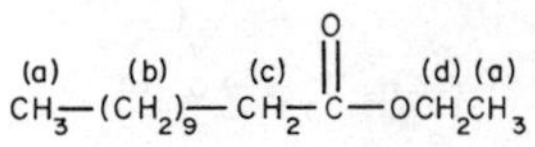

$C_{14}H_{28}O_2$ — 45.5 mg/0.50 ml $CDCl_3$ — S 194

▶ 1-Bq

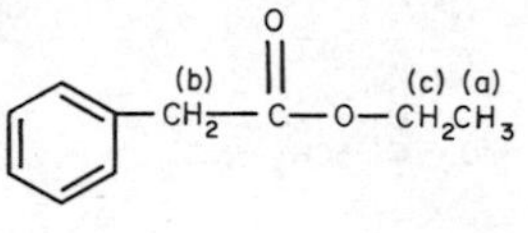

	Shift	Peak
▶ a	1.22	3A
b	3.55	1A
c	4.12	4A

▶ $C_{10}H_{12}O_2$ ▶ 44.9 mg/0.50 ml $CDCl_3$ ▶ S 66

▶ 1-Bq

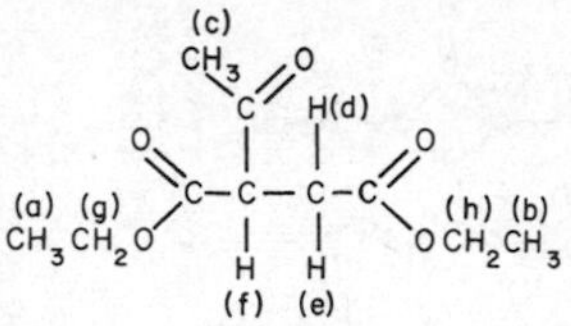

	Shift	Peak
▶ a	1.25*	3A
▶ b	1.25*	3A
c	2.35	1A
d	2.86	2A
e	2.89	2A
f	3.98	S
g	4.13†	4A
h	4.13†	4A

*or 1.27
†or 4.22

▶ $C_{10}H_{16}O_5$ ▶ 7% $CDCl_3$ ▶ V 277

▶ 1-Bq

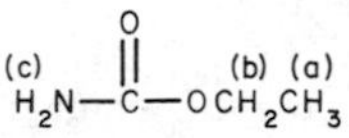

	Shift	Peak
▶ a	1.23	3A
b	4.10	4A
c	~5.17	1D

▶ $C_3H_7NO_2$ ▶ 44.4 mg/0.50 ml $CDCl_3$ ▶ S 245

▶ 1-Bq

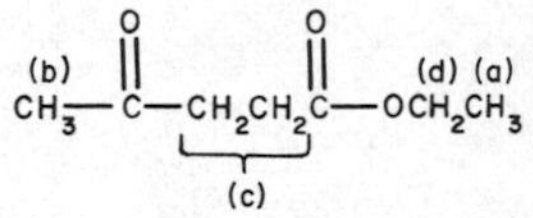

	Shift	Peak
▶ a	1.26	3A
b	2.20	1A
c	2.64	Q
d	4.12	4A

▶ $C_7H_{12}O_3$ ▶ 45.5 mg/0.50 ml $CDCl_3$ ▶ S 247

▶ 1-Bq

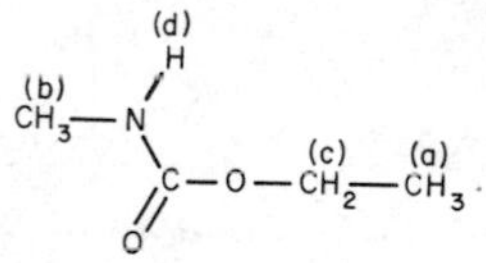

	Shift	Peak
▶ a	1.23	3A
b	2.78	2A
c	4.14	4A
d	5.16	1D

▶ $C_4H_9NO_2$ ▶ 7% $CDCl_3$ ▶ V 85

▶ 1-Bq

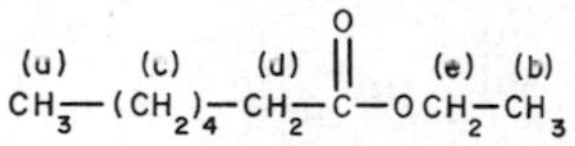

	Shift	Peak
a	0.88	3B
▶ b	1.26	S
c	1.33	S
d	2.28	3B
e	4.12	4A

▶ $C_9H_{18}O_2$ ▶ 44.5 mg/0.50 ml $CDCl_3$ ▶ S 283

▶ 1-Bq

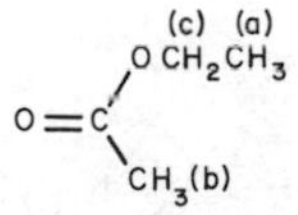

	Shift	Peak
▶ a	1.25	3A
b	2.03	1A
c	4.12	4A

▶ $C_4H_8O_2$ ▶ 7% $CDCl_3$ ▶ V 79

▶ 1-Bq

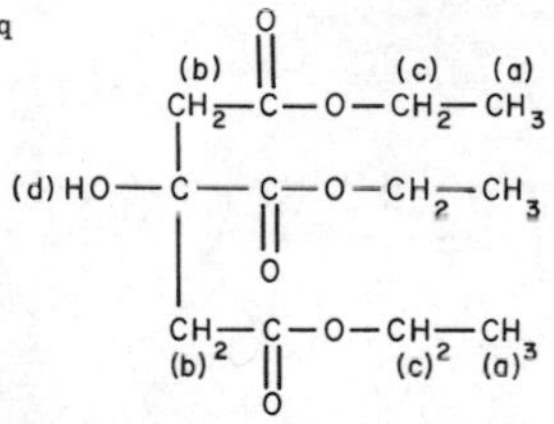

	Shift	Peak
▶ a	1.26	3D
b	2.86	1A
c	4.18	4A
d	4.47	1B

▶ $C_{12}H_{20}O_7$ ▶ 45.3 mg/0.50 ml $CDCl_3$ ▶ S 150

▶ 1-Bq

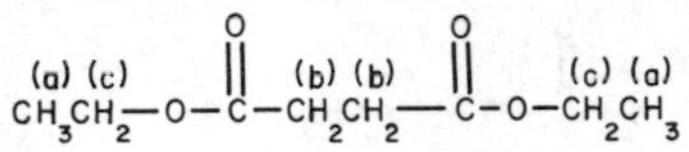

	Shift	Peak
▶ a	1.25	3A
b	2.62	1A
c	4.16	4A

▶ $C_8H_{14}O_4$ ▶ 45.5 mg/0.50 ml $CDCl_3$ ▶ S 183

▶ 1-Bq

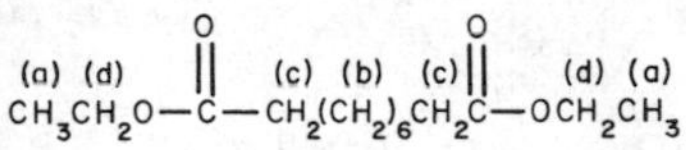

	Shift	Peak
▶ a	1.26	3D
b	1.35	1E
c	2.28	3B
d	4.13	4A

▶ $C_{14}H_{26}O_4$ ▶ 45.1 mg/0.50 ml $CDCl_3$ ▶ S 250

▶ 1-Bq

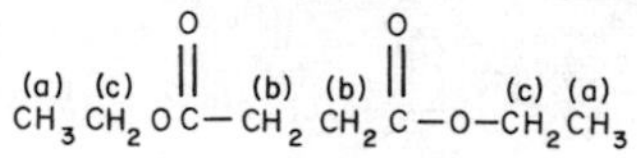

	Shift	Peak
▶ a	1.25	3A
b	2.62	1A
c	4.15	4A

▶ $C_8H_{14}O_4$ ▶ 7% $CDCl_3$ ▶ V 215

▶ 1-Bq

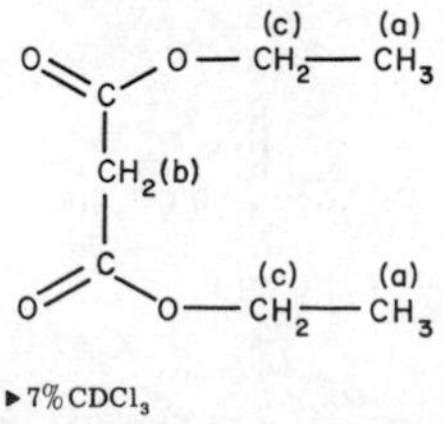

	Shift	Peak
▶ a	1.27	3A
b	3.37	1A
c	4.22	4A

▶ $C_7H_{12}O_4$ ▶ 7% $CDCl_3$ ▶ V 181

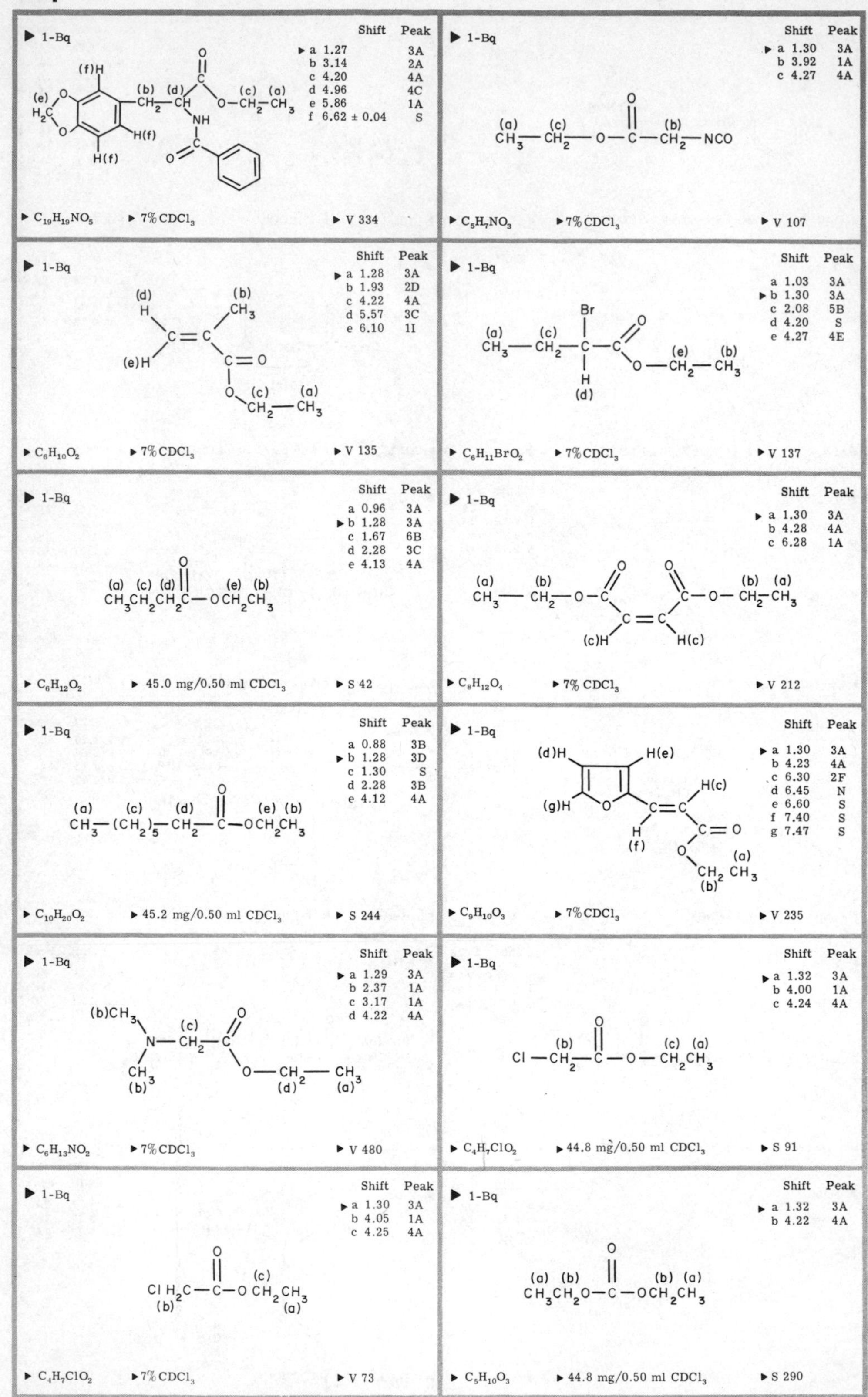
1-Bq
Shift Peak
a 1.27 3A
b 3.14 2A
c 4.20 4A
d 4.96 4C
e 5.86 1A
f 6.62 ± 0.04 S
C19H19NO5
7% CDCl3
V 334
1-Bq
Shift Peak
a 1.30 3A
b 3.92 1A
c 4.27 4A
C5H7NO3
7% CDCl3
V 107
1-Bq
Shift Peak
a 1.28 3A
b 1.93 2D
c 4.22 4A
d 5.57 3C
e 6.10 1I
C6H10O2
7% CDCl3
V 135
1-Bq
Shift Peak
a 1.03 3A
b 1.30 3A
c 2.08 5B
d 4.20 S
e 4.27 4E
C6H11BrO2
7% CDCl3
V 137
1-Bq
Shift Peak
a 0.96 3A
b 1.28 3A
c 1.67 6B
d 2.28 3C
e 4.13 4A
C6H12O2
45.0 mg/0.50 ml CDCl3
S 42
1-Bq
Shift Peak
a 1.30 3A
b 4.28 4A
c 6.28 1A
C8H12O4
7% CDCl3
V 212
1-Bq
Shift Peak
a 0.88 3B
b 1.28 3D
c 1.30 S
d 2.28 3B
e 4.12 4A
C10H20O2
45.2 mg/0.50 ml CDCl3
S 244
1-Bq
Shift Peak
a 1.30 3A
b 4.23 4A
c 6.30 2F
d 6.45 N
e 6.60 S
f 7.40 S
g 7.47 S
C9H10O3
7% CDCl3
V 235
1-Bq
Shift Peak
a 1.29 3A
b 2.37 1A
c 3.17 1A
d 4.22 4A
C6H13NO2
7% CDCl3
V 480
1-Bq
Shift Peak
a 1.32 3A
b 4.00 1A
c 4.24 4A
C4H7ClO2
44.8 mg/0.50 ml CDCl3
S 91
1-Bq
Shift Peak
a 1.30 3A
b 4.05 1A
c 4.25 4A
C4H7ClO2
7% CDCl3
V 73
1-Bq
Shift Peak
a 1.32 3A
b 4.22 4A
C5H10O3
44.8 mg/0.50 ml CDCl3
S 290

▶ 1-Bq

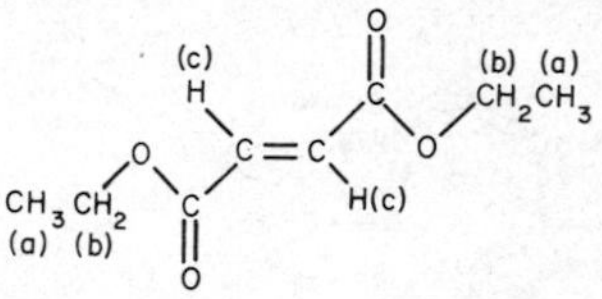

	Shift	Peak
▶ a	1.32	3A
b	4.27	4A
c	6.83	1A

▶ $C_8H_{12}O_4$ ▶ 7% $CDCl_3$ ▶ V 213

▶ 1-Bq

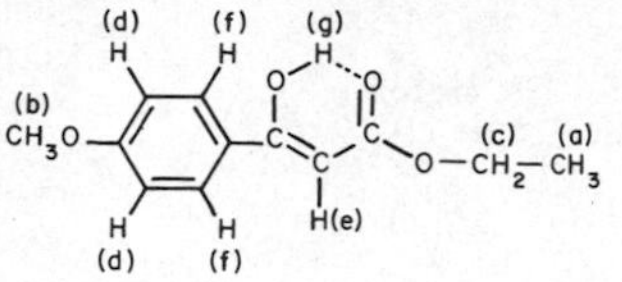

	Shift	Peak
▶ a	1.40	3A
b	3.87	1A
c	4.40	4A
d	6.97	2A
e	7.02	1A
f	7.98	2A
g	~15.33	1D

▶ $C_{12}H_{14}O_4$ ▶ 7% $CDCl_3$ ▶ V 295

▶ 1-Bq

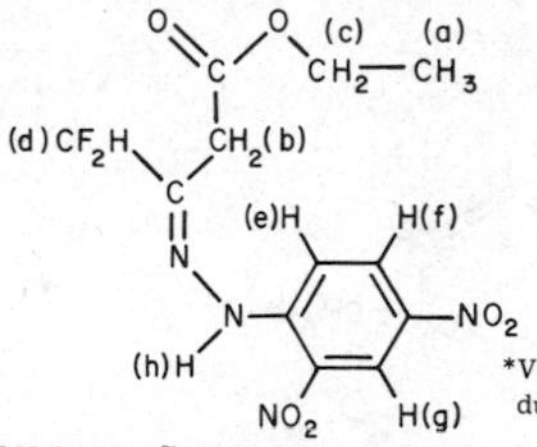

	Shift	Peak
▶ a	1.33	3A
b	3.67	1A
c	4.30	4A
d	6.30	3A*
e	8.03	2C
f	8.40	4A
g	9.08	2A
h	11.77	1B

*Very large splitting due to fluorine

▶ $C_{12}H_{12}F_2N_4O_6$ ▶ 7% $CDCl_3$ ▶ V 293

▶ 1-Br

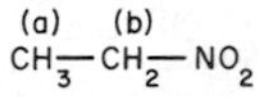

	Shift	Peak
▶ a	1.58	3A
b	4.33	4A

▶ $C_2H_5NO_2$ ▶ 35.8 mg/0.50 ml $CDCl_3$ ▶ S 2

▶ 1-Bq

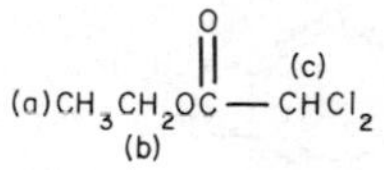

	Shift	Peak
▶ a	1.35	3A
b	4.33	4A
c	5.93	1A

▶ $C_4H_6Cl_2O_2$ ▶ 7% $CDCl_3$ ▶ V 59

▶ 1-Bs

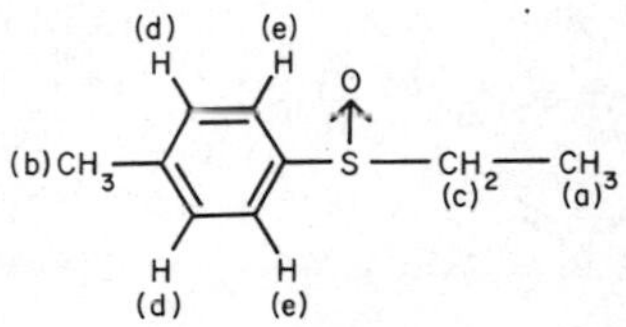

	Shift	Peak
▶ a	1.20	3A
b	2.43	1A
c	2.82	4C
d	7.33	Q
e	7.51	Q

▶ $C_9H_{12}OS$ ▶ 7% $CDCl_3$ ▶ V 538

▶ 1-Bq

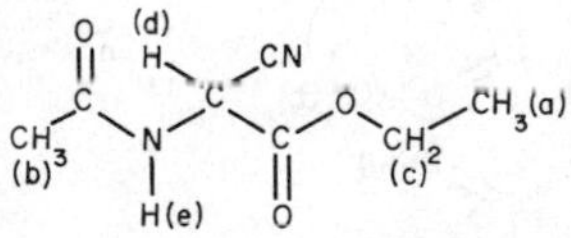

	Shift	Peak
▶ a	1.37	3A
b	2.12	1A
c	4.36	4A
d	5.50	2A
e	6.75	1D

▶ $C_7H_{10}N_2O_3$ ▶ 7% $CDCl_3$ ▶ V 491

▶ 1-Bs

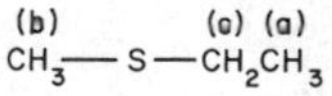

	Shift	Peak
▶ a	1.27	3A
b	2.10	1A
c	2.53	4A

▶ C_3H_8S ▶ 7% $CDCl_3$ ▶ V 46

▶ 1-Bq

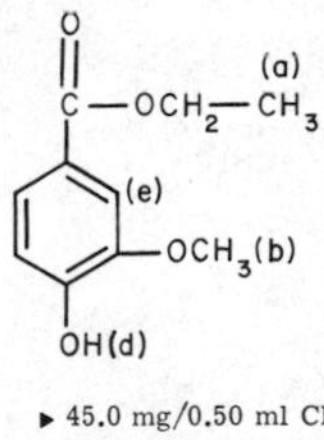

	Shift	Peak
▶ a	1.38	3A
b	3.93	1A
c	4.32	4A
d	6.80 or 6.94	1A
e	6.80 or 6.94	1A

▶ $C_{10}H_{12}O_4$ ▶ 45.0 mg/0.50 ml $CDCl_3$ ▶ S 127

▶ 1-Bs

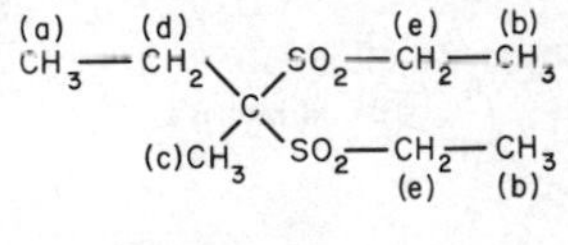

	Shift	Peak
a	1.24	3C
▶ b	1.43	3D
c	1.78	1A
d	2.36	4C
e	3.45	4A

▶ $C_8H_{18}O_4S_2$ ▶ 7% $CDCl_3$ ▶ V 519

▶ 1-Bq

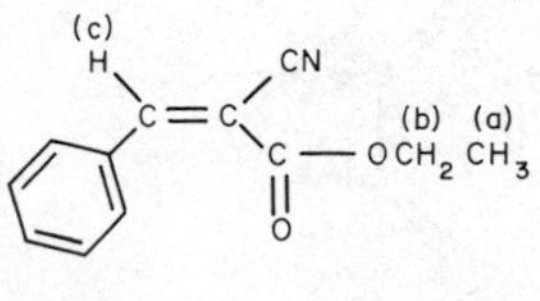

	Shift	Peak
▶ a	1.38	3A
b	4.38	4A
c	8.22	1A

▶ $C_{12}H_{11}NO_2$ ▶ 7% $CDCl_3$ ▶ V 290

▶ 1-Bs

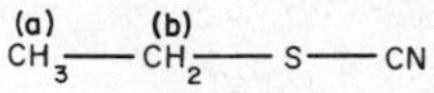

	Shift	Peak
▶ a	1.52	3A
b	2.99	4A

▶ C_3H_5NS ▶ 7% $CDCl_3$ ▶ V 384

▶ 1-Bt

(c) (a)
(b)HC≡CCH$_2$CH$_3$

	Shift	Peak
▶ a	1.12	3C
b	~1.90	N
c	~2.17	O

▶ C_4H_6 ▶ liquid ▶ API 464

▶ 1-Bv

	Shift	Peak
▶ a	1.14	3A
b	2.17	1A
c	2.40	4A
d	6.68	1B
e	6.68	1B

▶ $C_{10}H_{14}$ ▶ liquid ▶ API 12

▶ 1-Bt

(b) (c) (a)
HC≡C—CH$_2$—CH$_3$

	Shift	Peak
▶ a	1.15	3A
b	1.76	3A
c	~2.14	4C

▶ C_4H_6 ▶ 10% CCl_4 ▶ API 162

▶ 1-Bv

	Shift	Peak
▶ a	1.16	3A
b	2.58	4A

▶ $C_{10}H_{14}$ ▶ run neat ▶ API 180 A

▶ 1-Bv

(d) = all ring H's

	Shift	Peak
▶ a	1.09	3A
b	2.11	1A
c	2.45	4A
d	7.00	1A

▶ C_9H_{12} ▶ run neat ▶ API 177 A

▶ 1-Bv

	Shift	Peak
▶ a	1.16	3A
b	2.07	1A
c	2.47	4A
d	6.83	2D

▶ $C_{10}H_{14}$ ▶ run neat ▶ API 182 A

▶ 1-Bv

	Shift	Peak
▶ a	1.10	3A
b	2.47	4A
c	7.10	1A
d	7.10	1A

▶ C_8H_{10} ▶ liquid ▶ API 74

▶ 1-Bv

	Shift	Peak
▶ a	1.18	3A
b	2.20	1A
c	2.60	4A
d	6.84	1A

▶ $C_{10}H_{14}$ ▶ CCl_4 ▶ API 182 B

▶ 1-Bv

(c)= all ring H's

	Shift	Peak
▶ a	1.10	3A
b	2.5	4A
c	7.07	1A

▶ C_8H_{10} ▶ run neat ▶ API 173 A

▶ 1-Bv

(d) = all ring H's

	Shift	Peak
▶ a	1.21	3A
b	2.27	1A
c	2.63	4A
d	7.02	1A

▶ C_9H_{12} ▶ CCl_4 ▶ API 177B

▶ 1-Bv

	Shift	Peak
▶ a	1.10	3A
b	2.15	2A
c	2.21	4A
d	6.67	1E
e	7.08	1E

▶ $C_{10}H_{14}$ ▶ CCl_4 ▶ API 183 A

▶ 1-Bv

	Shift	Peak
▶ a	1.22	3A
b	2.54	4A
c	7.10	1E

▶ C_8H_{10} ▶ 10% CCl_4 ▶ API 38

▶ 1-Bv

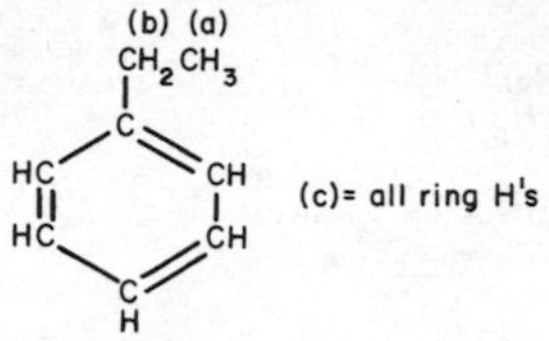

(c) = all ring H's

	Shift	Peak
▶ a	1.22	3A
b	2.63	4A
c	7.13	1A

▶ C_8H_{10} ▶ CCl_4 ▶ API 173 B

▶ 1-Bz

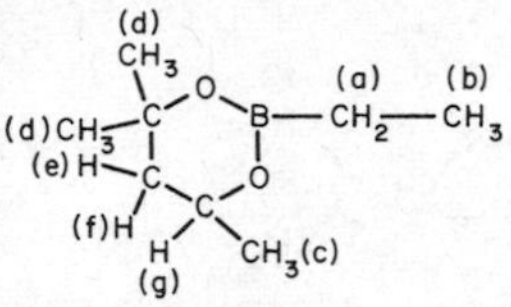

	Shift	Peak
a	0.61	S
▶ b	0.88	S
c	1.25	2H
d	1.29	1A
e	1.47	S
f	1.76	S
g	4.17	O

▶ $C_8H_{17}BO_2$ ▶ 7% $CDCl_3$ ▶ V 518

▶ 1-Bv

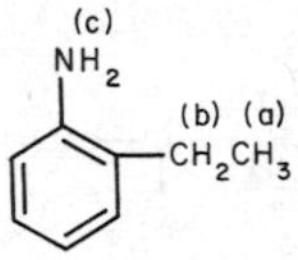

	Shift	Peak
▶ a	1.23	3A
b	2.46	4A
c	3.47	1A

▶ $C_8H_{11}N$ ▶ 45.3 mg/0.10 ml $CDCl_3$ ▶ S 111

▶ 1-Bz

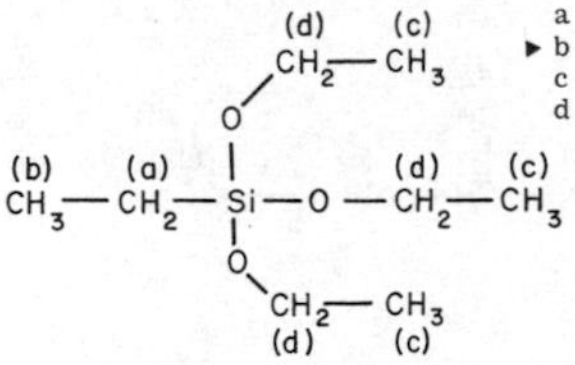

	Shift	Peak
a	~0.62	S
▶ b	~1.00	S
c	1.23	3A
d	3.82	4A

▶ $C_8H_{20}O_3Si$ ▶ 7% $CDCl_3$ ▶ V 222

▶ 1-Bv

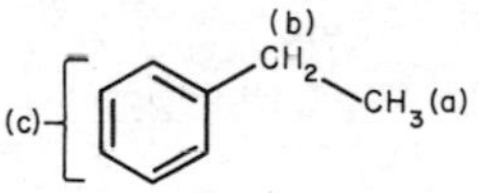

	Shift	Peak
▶ a	1.25	3A
b	2.68	4A
c	7.23	1A

▶ C_8H_{10} ▶ 7% $CDCl_3$ ▶ V 505

▶ 1-Caa

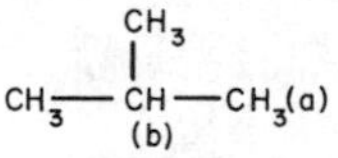

	Shift	Peak
▶ a	.88	2A
b	1.65	7X

▶ C_4H_{10} ▶ 50% CCl_4 ▶ API 330

▶ 1-Bv

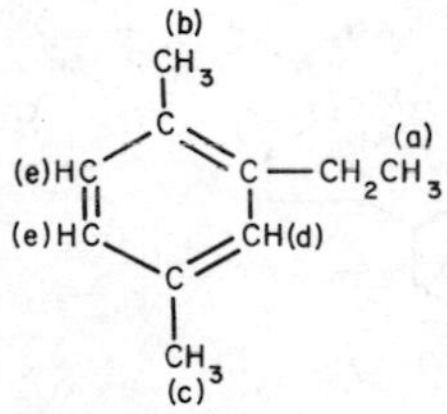

	Shift	Peak
▶ a	1.27	3A
b	2.27	2A
c	2.32	4A
d	6.71	1E
e	6.78	1E

▶ $C_{10}H_{14}$ ▶ run neat ▶ API 183 B

▶ 1-Cab

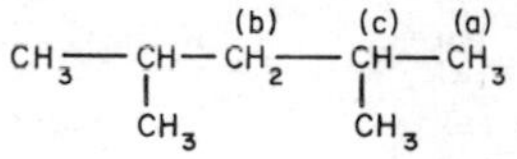

	Shift	Peak
▶ a	.80	2A
b	1.00	2A
c	1.58	1F

▶ C_7H_{16} ▶ 10% CCl_4 ▶ API 198

▶ 1-Bv

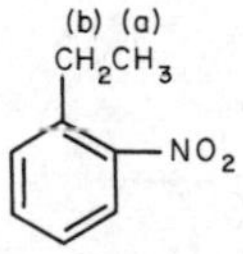

	Shift	Peak
▶ a	1.28	3A
b	2.92	4A

▶ $C_8H_9NO_2$ ▶ 45.7 mg/0.50 ml $CDCl_3$ ▶ S 109

▶ 1-Cab

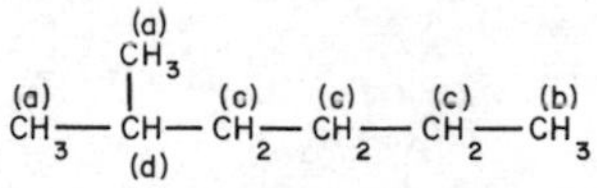

	Shift	Peak
▶ a	.82	2A
b	~.92	1E
c	~.93	1B
d	~.97	1D

▶ C_7H_{16} ▶ 50% CCl_4 ▶ API 195

▶ 1-Bv

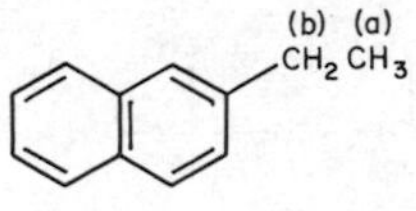

	Shift	Peak
▶ a	1.30	3A
b	2.77	4A

▶ $C_{12}H_{12}$ ▶ 7% $CDCl_3$ ▶ V 292

▶ 1-Cab

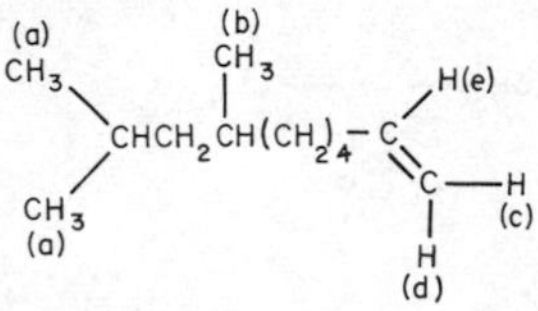

	Shift	Peak
▶ a	0.82 or 0.87	2A
b	0.82 or 0.87	2A
c	4.95	R
d	4.98	R
e	~5.80	R

▶ $C_{12}H_{24}$ ▶ 7% $CDCl_3$ ▶ V 298

▶ 1-Cab

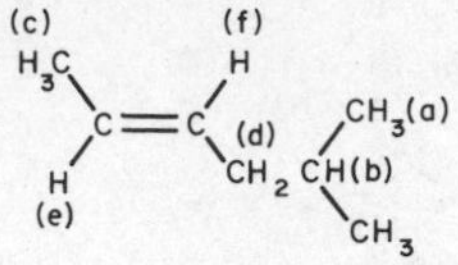

	Shift	Peak
▶ a	.83	2A
b	~1.40	S
c	1.62	2E
d	1.83	O
e	~5.38	O
f	~5.38	O

▶ C_7H_{14} ▶ liquid ▶ API 396

▶ 1-Cab

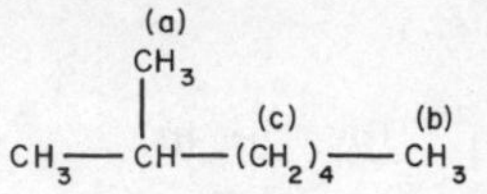

	Shift	Peak
▶ a	.87	2H
b	.96	S
c	1.27	1B

▶ C_8H_{18} ▶ 50% CCl_4 ▶ API 340

▶ 1-Cab

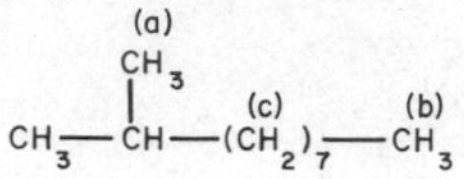

	Shift	Peak
▶ a	.83	2H
b	.94	S
c	1.30	1A

▶ $C_{11}H_{24}$ ▶ 50% CCl_4 ▶ API 349

▶ 1-Cab

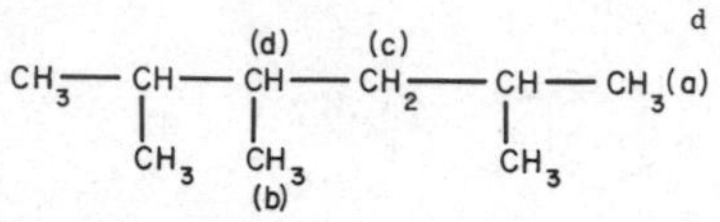

	Shift	Peak
▶ a	~ .87	2H
b	~ .87	2H
c	~1.15	S
d	~1.45	S

▶ C_9H_{20} ▶ 50% CCl_4 ▶ API 214

▶ 1-Cab

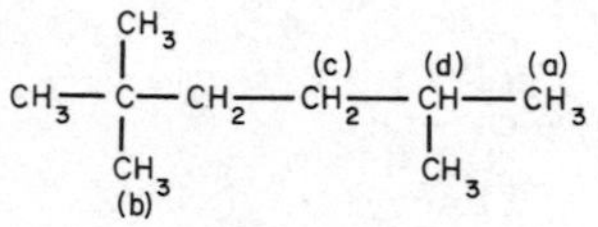

	Shift	Peak
▶ a	~ .84	1E
b	~ .84	1E
c	~1.13	1G
d	~1.37	1D

▶ C_9H_{20} ▶ 50% CCl_4 ▶ API 212

▶ 1-Cab

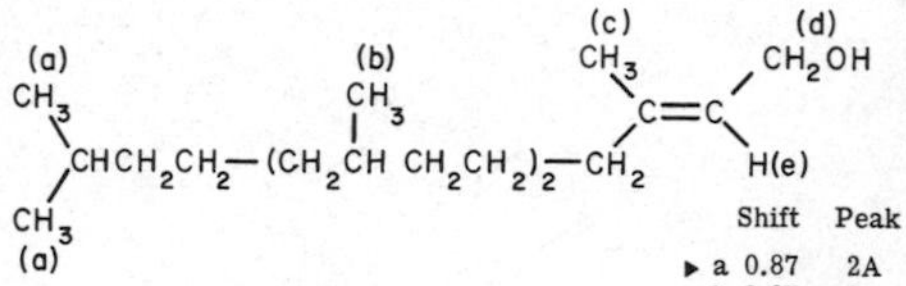

	Shift	Peak
▶ a	0.87	2A
b	0.87	2A
c	1.68	1A
d	4.13	2A
e	5.42	3C

▶ $C_{20}H_{40}O$ ▶ 7% $CDCl_3$ ▶ V 346

▶ 1-Cab

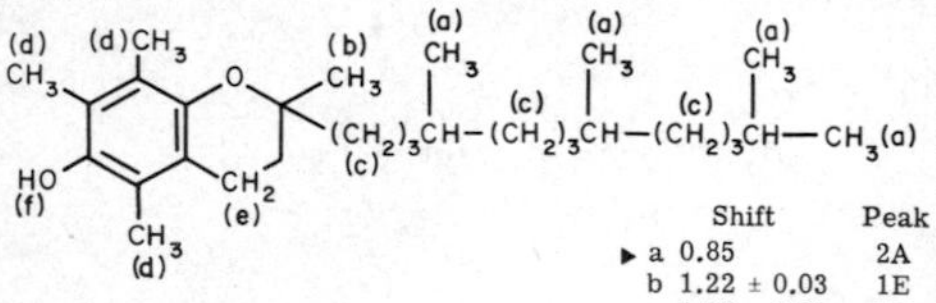

	Shift	Peak
▶ a	0.85	2A
b	1.22 ± 0.03	1E
c	1.22 ± 0.03	1E
d	2.12 ± 0.03	1B
e	2.62	3C
f	3.96	1D

▶ $C_{29}H_{50}O_2$ ▶ 7% $CDCl_3$ ▶ V 366

▶ 1-Cab

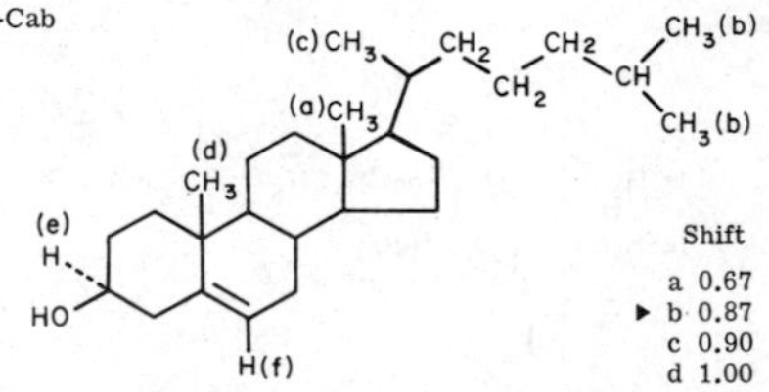

	Shift	Peak
a	0.67	1A
▶ b	0.87	2H
c	0.90	2H
d	1.00	1E
e	3.50	1D
f	5.36	2E

▶ $C_{27}H_{46}O$ ▶ 7% $CDCl_3$ ▶ V 363

▶ 1-Cab

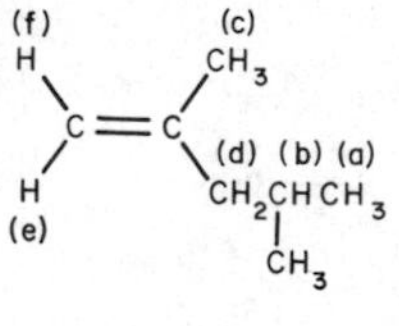

	Shift	Peak
▶ a	.87	2H
b	1.62	S
c	1.67	3C
d	1.82	2B
e	4.67 ± .03	N
f	4.67 ± .03	N

▶ C_7H_{14} ▶ liquid ▶ API 400

▶ 1-Cab

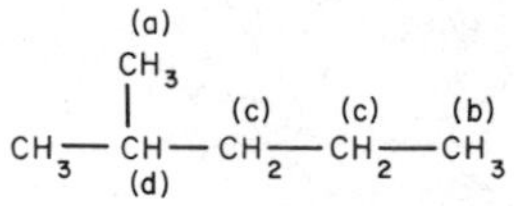

	Shift	Peak
▶ a	.88	3D
b	.95	2H
c	1.20	N
d	~1.50	1D

▶ C_6H_{14} ▶ 50% CCl_4 ▶ API 334

▶ 1-Cab

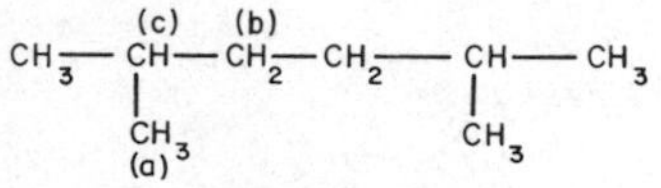

	Shift	Peak
▶ a	.87	2A
b	~1.17	1G
c	~1.40	1D

▶ C_8H_{18} ▶ 50% CCl_4 ▶ API 205

▶ 1-Cab

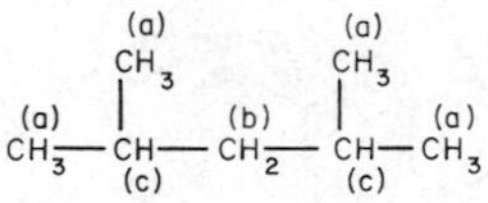

	Shift	Peak
▶ a	0.88	2A
b	1.03	3D
c	~1.55	7B

▶ C_7H_{16} ▶ 45.0 mg/0.50 ml $CDCl_3$ ▶ S 52

▶ 1-Cab

(a) CH_3—CH(c)(CH_3)—$(CH_2)_3$(b)—CH_2(d)—OH(e)

	Shift	Peak
▶ a	.88	2A
b	1.31	S
c	~1.63	S
d	3.48	3B
e	4.42	1A

▶ $C_7H_{16}O$ ▶ 20% CCl_4 ▶ API 169

▶ 1-Cab

H(e)H(d)C=C H(f) — CH_2(c) — CH(b)(CH_3(a))$_2$

	Shift	Peak
▶ a	.90	2A
b	1.48	S
c	1.88	S
d	4.77	R
e	4.77	R
f	5.65	R

▶ C_6H_{12} ▶ liquid ▶ API 375

▶ 1-Cab

CH_3—CH(CH_3)—$(CH_2)_4$(b)—CH(c)(CH_3(a))—CH_3

	Shift	Peak
▶ a	.88	2A
b	1.25	1B
c	~1.47	S

▶ $C_{10}H_{22}$ ▶ 50% CCl_4 ▶ API 346

▶ 1-Cab

(a)CH_3(a)CH_3 CH—CH_2(b)—CH_2(c)—C(=O)—N H(e) H(d)

	Shift	Peak
▶ a	0.90	2E
b	~1.53	3C
c	~2.20	4C
d	6.05 or 6.33	1D
e	6.05 or 6.33	1D

▶ $C_6H_{13}NO$ ▶ 7% $CDCl_3$ ▶ V 142

▶ 1-Cab

	Shift	Peak
a	0.66	1A
▶ b	0.88	S
c	0.95	S
d	1.48	S
e	2.53	S
f	3.90	S
g	4.03	S
h	4.12	S

▶ $C_{27}H_{44}O_2$ ▶ 7% $CDCl_3$ ▶ V 698

▶ 1-Cab

H(e)H(f)C=C H(g) — CH_2(d) CH_2(b) CH(c) CH_3(a) CH_3

	Shift	Peak
▶ a	.91	2C
b	1.32	S
c	1.60	S
d	2.01	4C
e	4.87	R
f	5.02	R
g	5.70	R

▶ C_7H_{14} ▶ liquid ▶ API 394

▶ 1-Cab

H(e)H(d)C=C H(f) — CH_2(c) — CH(b)(CH_3(a))—CH_3(a)

	Shift	Peak
▶ a	.89	2A
b	~1.60	S
c	1.87	S
d	4.83	R
e	5.03	R
f	~5.02	O

▶ C_6H_{12} ▶ 10% CCl_4 ▶ API 33

▶ 1-Cab

CH_3(a)—CH(b)(CH_3(a))—CH_2(d)—NH(c)—CH_2—CH(CH_3)—CH_3

	Shift	Peak
▶ a	0.91	2A
b	~1.6	S
c	1.77	1B
d	2.42	2A

▶ $C_8H_{19}N$ ▶ 44.7 mg/0.50 ml $CDCl_3$ ▶ S 136

▶ 1-Cab

CH_3—CH(d)(CH_3(b))—CH_2(c)—CH_3(a)

	Shift	Peak
a	.83	3D
▶ b	.90	2H
c	1.20	S
d	~1.67	1D

▶ C_5H_{12} ▶ 50% CCl_4 ▶ API 332

▶ 1-Cab

(a) CH_3—CH(CH_3)—CH_2 (b) —CH_2(c)—Br

	Shift	Peak
▶ a	0.92	2A
b	1.78	O
c	3.42	3A

▶ $C_5H_{11}Br$ ▶ 45.0 mg/0.50 ml $CDCl_3$ ▶ S 121

▶ 1-Cab

(a) CH_3—CH(CH_3(a))—CH_2(b)CH_2(c)C≡N

	Shift	Peak
▶ a	0.90	2A
b	1.60	3D
c	2.38	3A

▶ $C_6H_{11}N$ ▶ 44.8 mg/0.50 ml $CDCl_3$ ▶ S 165

▶ 1-Cab

(a) CH_3—CH(CH_3)—CH_2 (b) —CH_2(c)—Cl

	Shift	Peak
▶ a	0.92	2A
b	1.65	O
c	3.55	3B

▶ $C_5H_{11}Cl$ ▶ 45.0 mg/0.50 ml $CDCl_3$ ▶ S 122

▶ 1-Cab

$CH_3CH_2C(=O)OCH_2CH(CH_3)CH_3$ — (b) CH_3, (d) CH_2, (e) OCH_2, (c) CH, (a) CH_3

	Shift	Peak
▶ a	0.92	2A
b	1.13	3D
c	~1.85	S
d	2.31	4A
e	3.83	2A

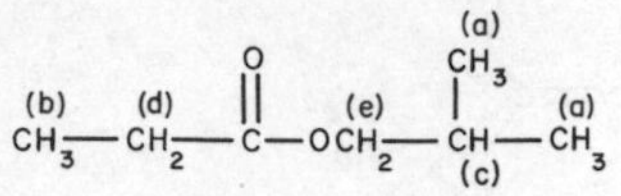

▶ $C_7H_{14}O_2$ ▶ 4.5 mg/0.50 ml $CDCl_3$ ▶ S 55

▶ 1-Cab

$CH_3CH(CH_3)C(=O)OCH_2CH(CH_3)CH_3$ — (b) CH_3, (d) CH, (e) CH_2, (c) CH, (a) CH_3

	Shift	Peak
▶ a	0.94	2A
b	1.18	2A
c	~1.82	7B
d	~2.50	5B
e	3.84	2A

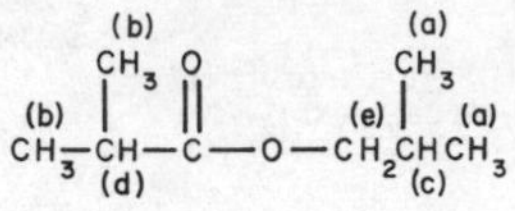

▶ $C_8H_{16}O_2$ ▶ 45.2 mg/0.50 ml $CDCl_3$ ▶ S 238

▶ 1-Cab

$CH_3C(=O)OCH_2CH_2CH(CH_3)CH_3$ — (c) CH_3, (d) CH_2, (b) CH_2–CH, (a) CH_3

	Shift	Peak
▶ a	0.92	2A
b	~1.50	O
c	2.03	1A
d	4.03	4B

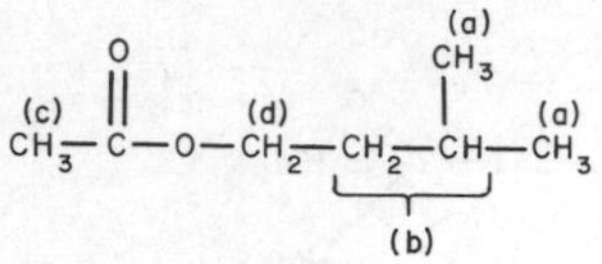

▶ $C_7H_{14}O_2$ ▶ 44.8 mg/0.50 ml $CDCl_3$ ▶ S 281

▶ 1-Cab

$CH_3CH(CH_3)CH_2C(=O)OCH_2CH(CH_3)CH_3$ — (a) CH_3, (b) CH, (c) CH_2, (d) OCH_2

	Shift	Peak
▶ a	0.94	2A
b	1.12	S
c	2.20	2D
d	3.88	2A

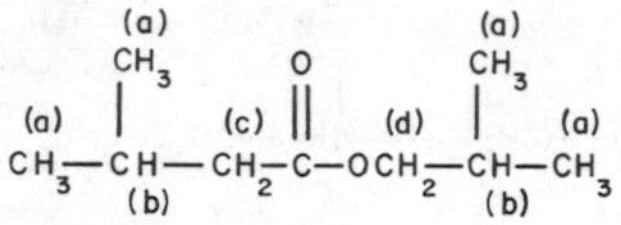

▶ $C_9H_{18}O_2$ ▶ 44.5 mg/0.50 ml $CDCl_3$ ▶ S 277

▶ 1-Cab

$CH_3CH(CH_3)CH_2C(CH_3)_2CH_2CH_3$ — (c) CH_3, (e) CH, (c) CH_2, (b) CH_3, (d) CH_2, (a) CH_3

	Shift	Peak
a	~ .85	3D
b	~ .85	1E
▶ c	~ .92	2H
d	1.13	2A
e	1.63	7X

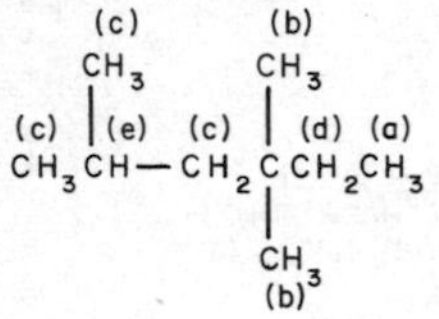

▶ C_9H_{20} ▶ 50% CCl_4 ▶ API 426

▶ 1-Cab

$(CH_3)_3CCH_2CH(CH_3)CH_3$ — (a) $C(CH_3)_3$, (c) CH_2, (d) CH, (b) CH_3

	Shift	Peak
a	~ .96	1E
▶ b	~ .96	2H
c	~1.12	2A
d	~1.67	7X

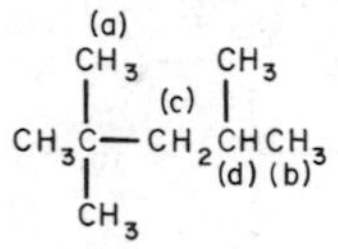

▶ C_8H_{18} ▶ 50% CCl_4 ▶ API 424

▶ 1-Cab

$(CH_3)_2CHCH_2C(=O)CH_3$ — (a) CH_3, (c) CH_2, (b) CH_3

	Shift	Peak
▶ a	0.93	2A
b	2.12	1A
c	~2.28	N

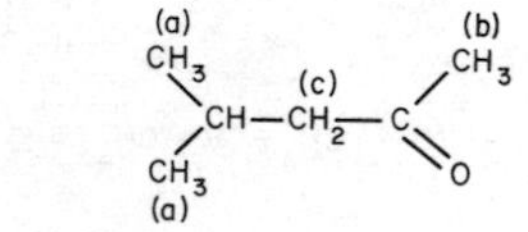

▶ $C_6H_{12}O$ ▶ 7% $CDCl_3$ ▶ V 139

▶ 1-Cab

$H_3C(H)C=C(H)CH_2CH(CH_3)_2$ — (c) H_3C, (e) H, (d) CH_2, (b) CH, (a) $(CH_3)_2$

	Shift	Peak
▶ a	.98	2A
b	1.38	S
c	1.60	2H
d	1.80	S
e	5.30	N

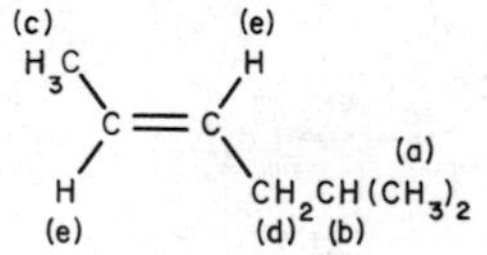

▶ C_7H_{14} ▶ 50 vol.% CCl_4 ▶ API 431

▶ 1-Cab

$CH_3CH(CH_3)CH_2CH_2C(=O)OH$ — (a) CH_3, (b) CH–CH_2, (c) CH_2, (d) OH

	Shift	Peak
▶ a	0.93	2A
b	~1.58	3C
c	2.36	3B
d	11.67	1A

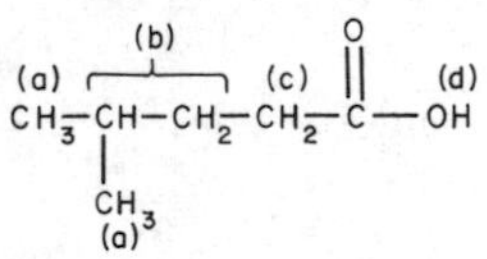

▶ $C_6H_{12}O_2$ ▶ 45.4 mg/0.50 ml $CDCl_3$ ▶ S 156

▶ 1-Cab

$C_6H_5C(=O)OCH_2CH(CH_3)CH_3$ — (c) CH_2, (b) CH, (a) CH_3, (d), (e) ring H

	Shift	Peak
▶ a	1.02	2A
b	2.05	5B
c	4.12	2A
d	7.5	N
e	~8.05	S

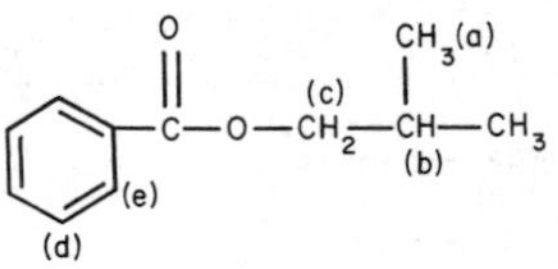

▶ $C_{11}H_{14}O_2$ ▶ 45.5 mg/0.50 ml $CDCl_3$ ▶ S 75

▶ 1-Cab

$(CH_3)_2CHCH_2C(C_6H_5)=NOH$ — (a) CH_3, (b) CH, (c) CH_2, (d) OH

	Shift	Peak
▶ a	0.93	2A
b	1.98	7B
c	2.77	2A
d	9.68	1D

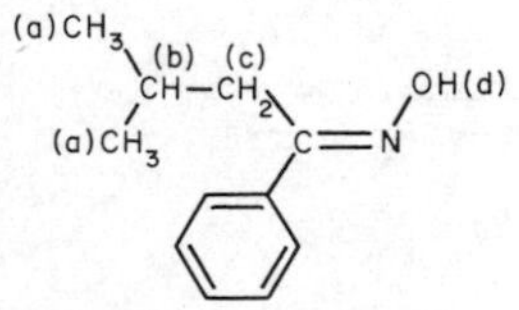

▶ $C_{11}H_{15}NO$ ▶ 7% $CDCl_3$ ▶ V 585

▶ 1-Cac

$CH_3CH(CH_3)CH(CH_3)CH_3$ — (b) CH, (a) CH_3

	Shift	Peak
▶ a	.82	2A
b	1.42	7B

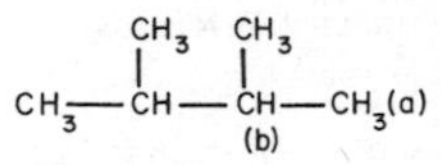

▶ C_6H_4 ▶ 50% CCl_4 ▶ API 194

▶ 1-Cac

(b) (d) (d) (c) (a)
CH_3—CH—CH—CH_2—CH_3
CH_3 CH_3
(b) (b)

	Shift	Peak
a	~ .79	3D
▶ b	~ .82	2H
c	~1.22	1D
d	~1.58	S

▶ C_7H_{16} ▶ 10% CCl_4 ▶ API 197

▶ 1-Cac

CH_3 CH_3
(d) (a)
CH_3—CH—CH—CH—CH_3
(d) (d)
CH_2—CH_3
(c) (b)

	Shift	Peak
▶ a	.86	2H
b	.95	3D
c	~1.2	S
d	~1.69	S

▶ C_9H_{20} ▶ 50% CCl_4 ▶ API 217

▶ 1-Cac

(a)
CH_3CHCH_3
H
C
H_2C CH_2
H_2C CH_2
C
H_2

	Shift	Peak
a	.82	2A

▶ C_9H_{18} ▶ 10% CCl_4 ▶ API 152

▶ 1-Cac

(d) (f)
H H (a)
CH_3
C=C (c)
H CH—CH—CH_3(a)
(e)
CH_3
(b)

	Shift	Peak
▶ a	.88	3A
b	.98	2H
c	1.73	7X
d	4.90	R
e	5.00	R
f	5.64	R

▶ C_7H_{14} ▶ liquid ▶ API 413

▶ 1-Cac

(b)
CH_3
H(e)
OH(d)
CH
CH_3 CH_3
(a) (c)

	Shift	Peak
▶ a	0.82 or 0.93	2A
b	0.90	1A
c	0.82 or 0.93	2A
d	1.75	1E
e	3.42	1D

▶ $C_{10}H_{20}O$ ▶ 7% $CDCl_3$ ▶ V 281

▶ 1-Cac

(a) (c) (b)
CH_3 CH_3 CH_3
(a) (b)
CH_3—C—CH—CH—CH_3
(d) (d)
CH_3
(a)

	Shift	Peak
a	.90	1E
▶ b	.90	1E
c	.90	1E
d	2.02	5A

▶ C_9H_{20} ▶ 50% CCl_4 ▶ API 219

▶ 1-Cac

(c)
CH_3
OH(d)
C—OCH_3(e)
O
CH
CH_3 CH_3
(a) (b)

	Shift	Peak
▶ a	0.82 or 0.87	2A
▶ b	0.82 or 0.87	2A
c	1.35	1E
d	1.67	1E
e	3.72	1A

▶ $C_{12}H_{22}O_3$ ▶ 7% $CDCl_3$ ▶ V 297

▶ 1-Cac

(c) (c) (c)
(b)CH_3—CH—CH—CH—CH_3
CH_3 CH_3 CH_3
(a)

	Shift	Peak
a	.80 ± .03 or .93 ± .03	S
▶ b	.93 ± .03 or .80 ± .03	S
c	1.67	6A

▶ C_8H_{18} ▶ 50% CCl_4 ▶ API 208

▶ 1-Cac

(a) (c) (b)
CH_3 CH_2—CH_3
(a)CH_3—CH—CH—CH_2—CH_3(b)
(d) (c)

	Shift	Peak
▶ a	~ .83	2H
b	~ .91	3D
c	1.20	3D
d	~1.72	1D

▶ C_8H_{18} ▶ 50% CCl_4 ▶ API 206

▶ 1-Cac

(e) (a) (f)
H CH_3 H
(c)CH_3 (d) (h)
CH—C—C—C—OH
CH_3 OH CH_3 H
(b) (h) (a) (g)

	Shift	Peak
a	0.92	1A
▶ b	0.95 or 1.00	2A
▶ c	0.95 or 1.00	2A
d	1.92	6B
e	3.38	2A
f	3.44 or 3.53	S
g	3.44 or 3.53	S
h	3.58	1A

▶ $C_8H_{18}O_2$ ▶ 7% $CDCl_3$ ▶ V 217

▶ 1-Cac

(c)
CH_3
H(d)
O
CH
CH_3 CH_3
(a) (b)

	Shift	Peak
▶ a	0.85 or 0.95	2A
b	0.85 or 0.95	2A
c	1.93	1B
d	5.87	N

▶ $C_{10}H_{16}O$ ▶ 7% $CDCl_3$ ▶ V 275

▶ 1-Cac

(h)H O (d)
H (d)
H CH_3(a)
CH
(c) CH_3(b)
N O H
CH_3O CH_3 (g)
(e) (f)

	Shift	Peak
▶ a	0.97*	2A
▶ b	0.97*	2A
c	2.00	6A
d	~3.20	S
e	3.87†	1A
f	3.87†	1A
g	4.72	4C
h	8.05	4A

*or 1.02
†or 3.90

▶ $C_{16}H_{19}NO_3$ ▶ 7% $CDCl_3$ ▶ V 325

▶ 1-Cac

(e) (a)
OH CH_3
(d)HC ≡ C—C—C—H(c)
H CH_3
(f) (b)

	Shift	Peak
▶ a	1.00	2D
▶ b	1.02	2D
c	1.85	7B
d	2.47	2E
e	2.82	1B
f	4.18	4A

▶ $C_6H_{10}O$ ▶ 7% $CDCl_3$ ▶ V 133

▶ 1-Cad

CH_3(b)
H(c)
O
O
H(d)
CH_3—C—H
(a)
CH_3(a)

	Shift	Peak
▶ a	1.00	2A
b	1.37	1A
c	6.42 or 6.47	2C
d	6.42 or 6.47	2C

▶ $C_{10}H_{16}O_2$ ▶ 7% $CDCl_3$ ▶ V 276

▶ 1-Cad

(a) (b) (a)
CH_3 CH_3 CH_3
(a) (a)
CH_3—CH—C—CH—CH_3
(c) (c)
CH_3
(b)

	Shift	Peak
▶ a	.70	2A
b	.87	1A
c	1.62	5B

▶ C_9H_{20} ▶ 50% CCl_4 ▶ API 345

▶ 1-Cag

Br
(a)CH_3—C—CH_3(a)
H(b)

	Shift	Peak
▶ a	1.71	2A
b	4.32	5A

▶ C_3H_7Br ▶ 7% $CDCl_3$ ▶ V 391

▶ 1-Cad

(g)
H O
O OH(h)
(d)H CH_3(c)
(f)H
CH(e)
CH_3 CH_3
(a) (b)

	Shift	Peak
▶ a	0.75	2A
▶ b	0.97	2A
c	1.45	1A
d	2.10	2C
e	2.40	S
f	2.60	2C
g	6.77	1A
h	12.00	1B

▶ $C_{10}H_{14}O_3$ ▶ 7% $CDCl_3$ ▶ V 567

▶ 1-Cai

(a)
CH_3
(b)H—C—I
CH_3
(a)

	Shift	Peak
▶ a	1.90	2A
b	4.34	5A

▶ C_3H_7I ▶ 7% $CDCl_3$ ▶ V 392

▶ 1-Cad

CH_3 CH_3
(a)H_3C—C—CH—CH_3(b)
(c)
CH_3

	Shift	Peak
a	.81	1E
▶ b	.81	S
c	~1.30	7X

▶ C_7H_{16} ▶ 50% CCl_4 ▶ API 419

▶ 1-Caj

(d) (c)
H CH_3
(e) O
H—C—C
(a)CH_3 (f)
CH C OH
N N
C H
(b)CH_3 O

	Shift	Peak
▶ a	1.10	2A
▶ b	1.12	2A
c	1.60	1A
d	2.81	2A
e	3.48	S
f	3.52	S

▶ $C_9H_{14}N_2O_3$ ▶ 7% $CDCl_3$ ▶ V 543

▶ 1-Cad

(f)
CH_3
Br
(e)H_2C
BrH_2C
(d) O
CH(c)
CH_3 CH_3
(a) (b)

	Shift	Peak
▶ a	0.84	2A
▶ b	0.92	2A
c	2.08	S
d	2.22	1H
e	2.61	1H
f	3.58	1A

▶ $C_{10}H_{14}Br_2O$ ▶ 7% $CDCl_3$ ▶ V 564

▶ 1-Caj

O
C CH_3(a)
C
H CH_3(a)
(b)

	Shift	Peak
▶ a	1.22	2A
b	3.58	5A

▶ $C_{10}H_{12}O$ ▶ 7% $CDCl_3$ ▶ V 559

▶ 1-Cad

(b) (a)
CH_3 CH_3
(b) (d) (d) (c)
CH_3—CH—C—CH_2—CH_2—CH_3
(e)
CH_3
(a)

	Shift	Peak
a	.80	1A
▶ b	.86	2H
c	~ .86	3D
d	1.22	3A
e	~1.50	1D

▶ C_9H_{20} ▶ 50% CCl_4 ▶ API 213

▶ 1-Cak

(b) (a)
CH_3 O CH_3
(b) (e) (a)
CH_3—CH—C—O—CH_2CHCH_3
(d) (c)

	Shift	Peak
a	0.94	2A
▶ b	1.18	2A
c	~1.82	7B
d	~2.50	5B
e	3.84	2A

▶ $C_8H_{16}O_2$ ▶ 45.2 mg/0.50 ml $CDCl_3$ ▶ S 238

▶ 1-Cak

	Shift	Peak
▶ a	1.22	2A
b	2.60	5B
c	11.55	1B

▶ $C_4H_8O_2$ ▶ 44.9 mg/0.50 ml $CDCl_3$ ▶ S 54

▶ 1-Cal

	Shift	Peak
▶ a	.93	2A
b	1.48	1E
c	1.52	1E
d	2.20	5A
e	5.20	4B

▶ C_7H_{14} ▶ 50% CCl_4 ▶ API 354

▶ 1-Cak

	Shift	Peak
▶ a	1.22	2A
b	2.68	7B

▶ $C_8H_{14}O_3$ ▶ 7% $CDCl_3$ ▶ V 516

▶ 1-Cal

	Shift	Peak
▶ a	.93	2A
b	1.61	N
c	2.42	6B
d	4.93	2E

▶ C_7H_{14} ▶ liquid ▶ API 402

▶ 1-Cal

	Shift	Peak
▶ a	.91	2A
b	~1.48 or 1.52	1E
c	~1.52 or 1.48	1E
d	2.74	5A
e	5.05	1D

▶ C_7H_{14} ▶ 50% CCl_4 ▶ API 353

▶ 1-Cal

	Shift	Peak
▶ a	.93	2A
b	1.47	1F
c	1.55	1F
d	2.84	5A
e	5.10	1F

▶ C_7H_{14} ▶ liquid ▶ API 403

▶ 1-Cal

	Shift	Peak
▶ a	.91	2A
b	1.50	1F
c	~1.62	S
d	~1.62	S
e	2.83	7A

▶ C_8H_{16} ▶ liquid ▶ API 463

▶ 1-Cal

	Shift	Peak
▶ a	.94	2A
b	1.60	2E
c	2.15	1D
d	5.36	S
e	5.45	S

▶ C_6H_{12} ▶ liquid run ▶ API 70

▶ 1-Cal

	Shift	Peak
▶ a	.93	2A
b	2.14	6B
c	4.60	R
d	4.75	R
e	5.48	R

▶ C_5H_{10} ▶ liquid run, neat ▶ API 387

▶ 1-Cal

	Shift	Peak
▶ a	0.94	2A
b	1.61	2A
c	2.63	6A
d	5.20 ± 0.10	O
e	5.30 ± 0.10	O

▶ C_6H_{12} ▶ 7% $CDCl_3$ ▶ V 471

▶ 1-Cal

	Shift	Peak
▶ a	.93	2A
b	1.57	2A
c	2.62	6B
d	5.12 or 5.23	O
e	5.23 or 5.12	O

▶ C_6H_{12} ▶ liquid ▶ API 379

▶ 1-Cal

	Shift	Peak
▶ a	.95	2A
b	1.63	2A
c	2.55	6B
d	5.11	N
e	5.25	N

▶ C_6H_{12} ▶ liquid run ▶ API 69

▶ 1-Cal

	Shift	Peak
a	.96	3D
▶ b	.97	2H
c	1.98	S
d	~2.28	S
e	~5.38	S
f	~5.38	S

▶ C_7H_{14} ▶ liquid ▶ API 397

▶ 1-Cal

	Shift	Peak
▶ a	1.27	2A
b	1.34	1E
c	1.69	O
d	2.78	O
e	3.28	S
f	5.27	1B
g	5.27	1B
h	7.19*	2F
i	7.19*	2F
j	7.39	1A

*or 7.31

▶ $C_{15}H_{22}O_3$ ▶ 7% $CDCl_3$ ▶ V 647

▶ 1-Cal

	Shift	Peak
▶ a	.97	2A
b	2.21	6A
c	5.22	O

▶ C_7H_{14} ▶ liquid ▶ API 404

▶ 1-Can

	Shift	Peak
▶ a	1.18	2A
b	4.17	6B
c	5.59	4A
d	~6.00	S
e	6.14	2A
f	6.24	1A

▶ $C_9H_{11}NO$ ▶ 7% $CDCl_3$ ▶ V 468

▶ 1-Cal

	Shift	Peak
▶ a	.98	2A
b	1.62	2E
c	2.22	1F
d	5.37 or 5.43	S
e	5.43 or 5.37	S

▶ C_6H_{12} ▶ 5 vol.% CCl_4 ▶ API 380

▶ 1-Cao

	Shift	Peak
▶ a	1.12	1A
b	1.18	1A
c	3.78	5A

▶ $C_7H_{16}O$ ▶ 7% $CDCl_3$ ▶ V 183

▶ 1-Cal

	Shift	Peak
▶ a	.99	2A
b	1.67	1H
c	2.23	5B
d	4.62	N

▶ C_6H_{12} ▶ liquid ▶ API 382

▶ 1-Cap

	Shift	Peak
▶ a	1.20	2A
b	1.60	1B
c	4.00	5A

▶ C_3H_8O ▶ 7% $CDCl_3$ ▶ V 44

▶ 1-Cal

	Shift	Peak
a	1.03	3D
▶ b	1.03	2H
c	~2.13	S
d	4.70 ± 0.03	S
e	4.70 ± 0.03	S

▶ C_7H_{14} ▶ liquid ▶ API 416

▶ 1-Caq

	Shift	Peak
▶ a	1.23	2A
b	2.01	1A
c	4.98	5A

▶ $C_5H_{10}O_2$ ▶ 45.0 mg/0.50 ml $CDCl_3$ ▶ S 99

▶ 1-Cal

	Shift	Peak
▶ a	1.27	2A
b	2.90	5B
c	8.32	1A
d	8.75	1B

▶ $C_{10}H_{12}O_2$ ▶ 7% $CDCl_3$ ▶ V 560

▶ 1-Caq

	Shift	Peak
▶ a	1.29	3A
b	5.13	5B
c	8.02	1B

impurity

▶ $C_4H_8O_2$ ▶ 7% $CDCl_3$ ▶ V 415

▶ 1-Car

(a) CH_3
(b)H — C — NO_2
CH_3 (a)

	Shift	Peak
▶ a	1.55	2A
b	4.67	M

▶ $C_3H_7NO_2$ ▶ 7% $CDCl_3$ ▶ V 41

▶ 1-Cav

CH_3(b)
C
HC CH(d)
HC CH(d)
C
CH_3 — CH — CH_3(a)
(c)

	Shift	Peak
▶ a	1.20	2A
b	2.24	1A
c	2.80	5A
d	6.99	1A

▶ $C_{10}H_{14}$ ▶ 50% CCl_4 ▶ API 225

▶ 1-Cas

(a)CH_3
CH — S — H(b)
(c)
(a)CH_3

	Shift	Peak
▶ a	1.34	2A
b	1.56	2A
c	3.16	6A

▶ C_3H_8S ▶ 7% $CDCl_3$ ▶ V 396

▶ 1-Cav

(b) CH_3
(d)H H(d)
(d)H H(d)
CH(c)
CH_3 CH_3
(a) (a)

	Shift	Peak
▶ a	1.22	2A
b	2.30	1A
c	2.87	5B
d	7.08	1A

▶ $C_{10}H_{14}$ ▶ 7% $CDCl_3$ ▶ V 268

▶ 1-Cau

(a) CH_3
(b)H — C — C ≡ N
CH_3 (a)

	Shift	Peak
▶ a	1.33	2A
b	2.72	6B

▶ C_4H_7N ▶ 7% $CDCl_3$ ▶ V 408

▶ 1-Cav

(f)H C=O···H(g)
O
(e)H
(d)H H(c)
C H(b)
(a)CH_3 CH_3(a)

	Shift	Peak
▶ a	1.23	2A
b	2.92	5A
c	6.82	N
d	6.88	N
e	7.42	2A
f	9.79	1A
g	11.00	1B

▶ $C_{10}H_{12}O_2$ ▶ 7% $CDCl_3$ ▶ V 259

▶ 1-Cav

H_3C CH_3(a)
CH(b)
C
HC CH
HC CH
C
H

(c) = all ring H's

	Shift	Peak
▶ a	1.15	2A
b	2.75	5A
c	7.09	1A

▶ C_9H_{12} ▶ run neat ▶ API 176

▶ 1-Cav

(b) CH_3
(f)H H(e)
(g)H OH(d)
(a)CH_3 — C — CH_3(a)
H
(c)

	Shift	Peak
▶ a	1.23	2A
b	2.25	1A
c	3.17	5B
d	4.75	1A
e	6.55	S
f	6.71	S
g	7.06	S

▶ $C_{10}H_{14}O$ ▶ 7% $CDCl_3$ ▶ V 270

▶ 1-Cav

CH_3(b)
C CH_3
HC C — CH(c)
HC CH CH_3(a)
C
H

(d) = all ring H's

	Shift	Peak
▶ a	1.15	2A
b	2.20	1A
c	3.01	5A
d	6.95	O

▶ $C_{10}H_{14}$ ▶ run neat ▶ API 181 A

▶ 1-Cav

(a) (a)
CH_3 CH_3
CH(b)
(c)

	Shift	Peak
▶ a	1.25	2A
b	2.90	7A
c	7.25	1A

▶ C_9H_{12} ▶ 7% $CDCl_3$ ▶ V 240

▶ 1-Cav

CH_3(b)
C CH_3(a)
HC C — CH(c)
HC CH CH_3(a)
C
H

(d) = all ring H's

	Shift	Peak
▶ a	1.17	2A
b	2.24	1A
c	3.04	5A
d	7.00	N

▶ $C_{10}H_{14}$ ▶ 50% CCl_4 ▶ API 224

▶ 1-Cav

CH_3(b)
C CH_3
HC C — CH(c)
HC CH CH_3(a)
C
H

(d) = all ring H's

	Shift	Peak
▶ a	1.25	2A
b	2.30	1A
c	3.05	5A
d	7.01	O

▶ $C_{10}H_{14}$ ▶ CCl_4 ▶ API 181 B

▶ 1-(C)bb

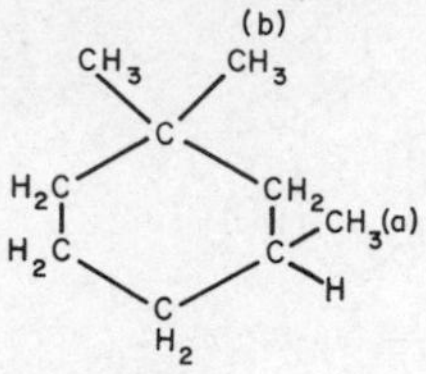

(b)
CH_3 CH_3
C
H_2C CH_2
H_2C C $CH_3(a)$
H
C
H_2

Shift		Peak
▶ a	.79	S
b	.89	S

▶ C_9H_{18} ▶ 10% CCl_4 ▶ API 153

▶ 1-Cbb

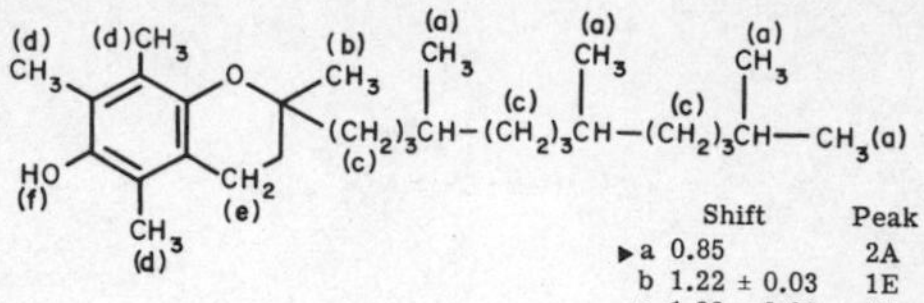

(d) (d)CH_3 (b) (a) (a) (a)
CH_3 O CH_3 CH_3 CH_3 CH_3
(c) (c)
$(CH_2)_3CH-(CH_2)_3CH-(CH_2)_3CH-CH_3(a)$
HO (c)
(f) CH_2
CH_3 (e)
(d)

Shift		Peak
▶ a	0.85	2A
b	1.22 ± 0.03	1E
c	1.22 ± 0.03	1E
d	2.12 ± 0.03	1B
e	2.62	3C
f	3.96	1D

▶ $C_{29}H_{50}O_2$ ▶ 7% $CDCl_3$ ▶ V 366

▶ 1-(C)bb

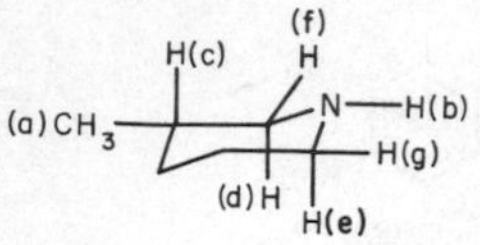

(f)
H(c) H
(a)CH_3 N—H(b)
H(g)
(d)H
H(e)

Shift		Peak
▶ a	0.82	2A
b	1.47	S
c	1.55 ± 0.10	S
d	2.21	S
e	2.52	S
f	3.00	S
g	3.00	S

▶ $C_6H_{13}N$ ▶ 7% $CDCl_3$ ▶ V 478

▶ 1-Cbb

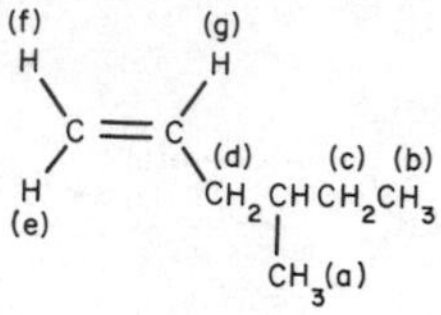

(f) (g)
H H
C=C
H (d) (c) (b)
(e) $CH_2CHCH_2CH_3$
$CH_3(a)$

Shift		Peak
▶ a	~ .87	3D
b	~ .95	2H
c	1.30	O
d	1.98	4C
e	~4.96	R
f	~4.96	R
g	5.67	R

▶ C_7H_{14} ▶ liquid run, neat ▶ API 393

▶ 1-(C)bb

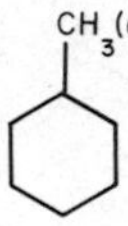

$CH_3(a)$

Shift		Peak
a	0.82	2H

▶ C_7H_{14} ▶ 35.3 mg/0.50 ml $CDCl_3$ ▶ S 27

▶ 1-Cbb

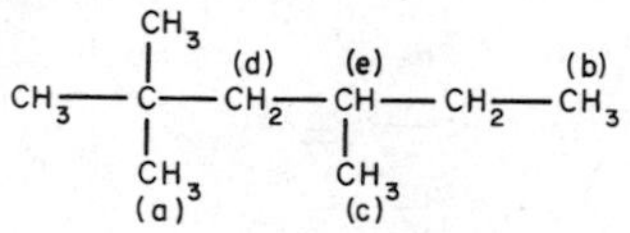

CH_3
(d) (e) (b)
CH_3—C—CH_2—CH—CH_2—CH_3
CH_3 CH_3
(a) (c)

Shift		Peak
a	.87	1E
b	.87	1E
▶ c	.87	1E
d	~1.22	S
e	~1.32	1D

▶ C_9H_{20} ▶ 50% CCl_4 ▶ API 211

▶ 1-Cbb

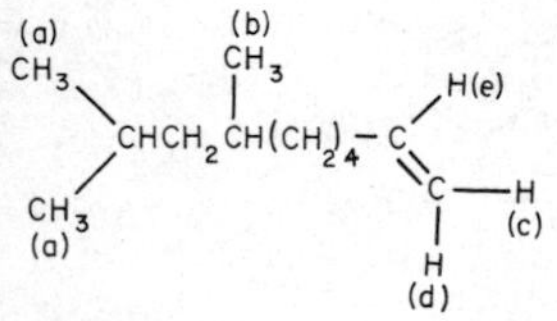

(a) (b)
CH_3 CH_3 H(e)
$CHCH_2CH(CH_2)_4$—C
CH_3 C—H
(a) H (c)
(d)

Shift		Peak
a	0.82 or 0.87	2A
▶ b	0.82 or 0.87	2A
c	4.95	R
d	4.98	R
e	~5.80	R

▶ $C_{12}H_{21}$ ▶ 7% $CDCl_3$ ▶ V 298

▶ 1-(C)bb

(b)
H CH_3
(b)H_3C C—C H
(a)
H H_3C C
C H H_3C
H (b) H
C—C
H H

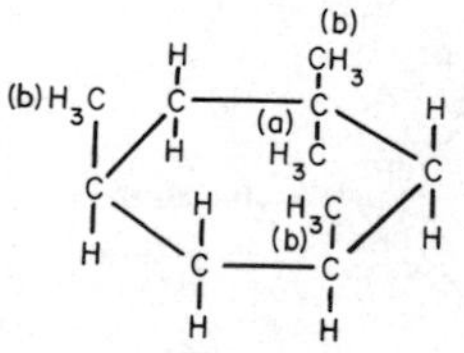

Shift		Peak
a	.78	1A
▶ b	.87	1A

▶ $C_{10}H_{20}$ ▶ 10% CCl_4 ▶ API 444 B

▶ 1-Cbb

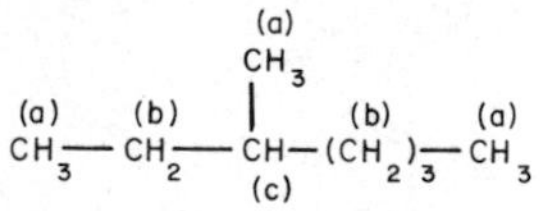

(a)
CH_3
(a) (b) (b) (a)
CH_3—CH_2—CH—$(CH_2)_3$—CH_3
(c)

Shift		Peak
▶ a	.83	3B
b	1.23	1B
c	~1.27	S

▶ C_8H_{18} ▶ 50% CCl_4 ▶ API 341

▶ 1-Cbb

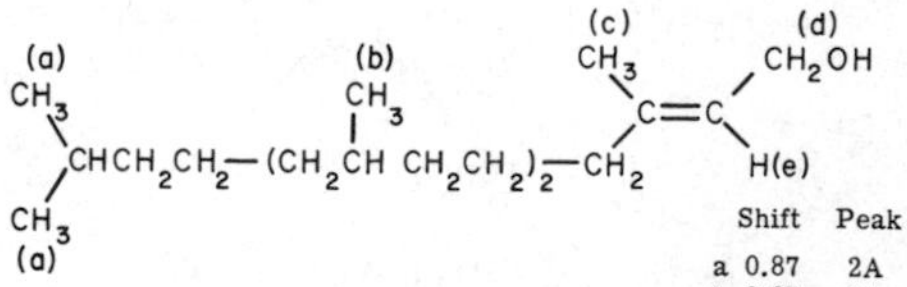

(c) (d)
(a) (b) CH_3 CH_2OH
CH_3 CH_3 C=C
$CHCH_2CH_2$—$(CH_2CH\ CH_2CH_2)_2$—CH_2 H(e)
CH_3
(a)

Shift		Peak
a	0.87	2A
▶ b	0.87	2A
c	1.68	1A
d	4.13	2A
e	5.42	3C

▶ $C_{20}H_{40}O$ ▶ 7% $CDCl_3$ ▶ V 346

▶ 1-Cbb

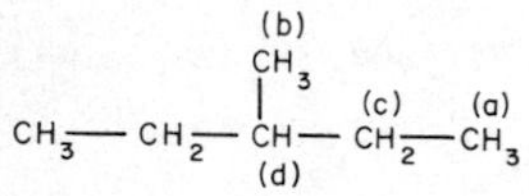

(b)
CH_3
(c) (a)
CH_3—CH_2—CH—CH_2—CH_3
(d)

Shift		Peak
a	.84	S
▶ b	.84	S
c	1.20	O
d	~1.30	S

▶ C_6H_{14} ▶ 50% CCl_4 ▶ API 335

▶ 1-Cbb

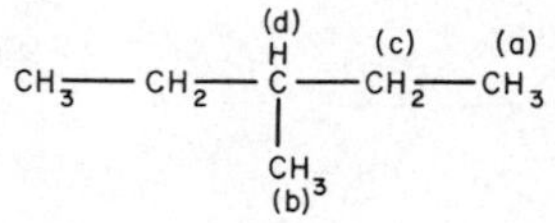

(d)
H (c) (a)
CH_3—CH_2—C—CH_2—CH_3
CH_3
(b)

Shift		Peak
a	.88	3D
▶ b	.88	2H
c	1.29	S
d	~1.30	S

▶ C_6H_{14} ▶ 10% CCl_4 ▶ API 47

1-Cbb

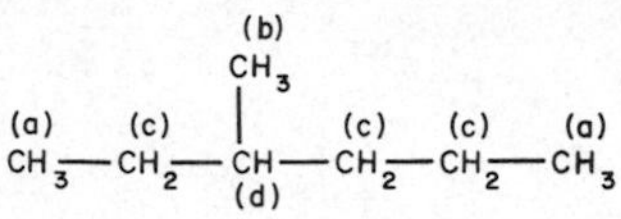

$CH_3(a)—CH_2(c)—CH(d)(CH_3(b))—CH_2(c)—CH_2(c)—CH_3(a)$

	Shift	Peak
a	.88	3D
▶ b	.88	S
c	1.24	S
d	~1.26	S

C_7H_{16} ▶ 50% CCl_4 ▶ API 338

1-Cbb

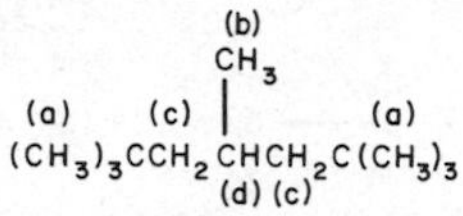

$(CH_3)_3CCH_2CH(CH_3)CH_2C(CH_3)_3$

	Shift	Peak
a	.95	1E
▶ b	~ .96	2H
c	1.17	S
d	1.57	7X

$C_{12}H_{26}$ ▶ 50% CCl_4 ▶ API 428

1-(C)bb

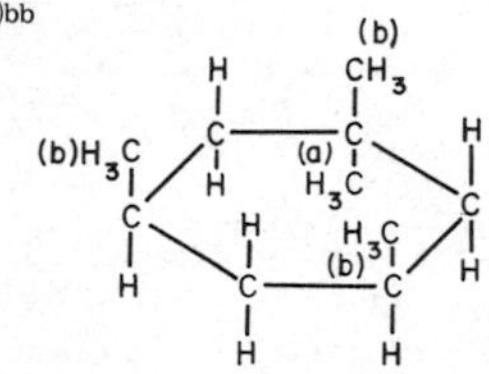

	Shift	Peak
a	.79	1A
▶ b	.89	1A

$C_{10}H_{20}$ ▶ run neat ▶ API 444 A

1-(C)bb

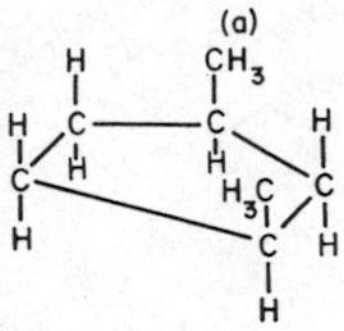

	Shift	Peak
a	.98	2A

C_7H_{14} ▶ liquid ▶ API 447

1-(C)bb

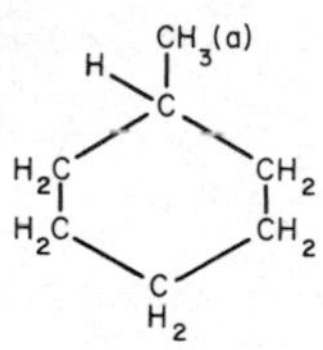

	Shift	Peak
a	0.90	2B

C_7H_{14} ▶ 10% CCl_4 ▶ API 148

1-(C)bb

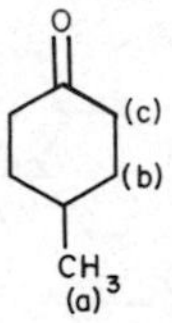

	Shift	Peak
▶ a	1.00	2A
b	~1.82	O
c	~2.20	3B

$C_7H_{12}O$ ▶ 37.0 mg/0.50 ml $CDCl_3$ ▶ S 8

1-(C)bb

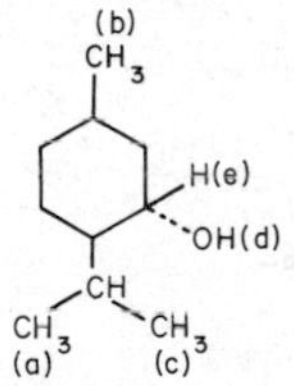

	Shift	Peak
a	0.82 or 0.93	2A
▶ b	0.90	1A
c	0.82 or 0.93	2A
d	1.75	1E
e	3.42	1D

$C_{10}H_{20}O$ ▶ 7% $CDCl_3$ ▶ V 281

1-Cbb

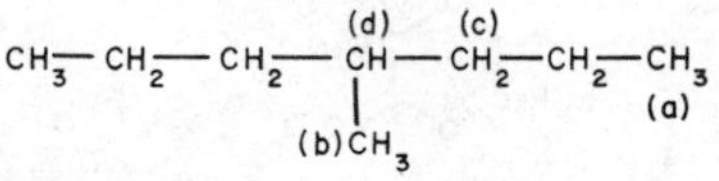

$CH_3—CH_2—CH_2—CH(d)(CH_3(b))—CH_2(c)—CH_2—CH_3(a)$

	Shift	Peak
a	.88	3A
▶ b	~1.25	S
c	~1.25	S
d	~1.25	S

C_8H_{18} ▶ 50% CCl_4 ▶ API 201

1-(C)bb

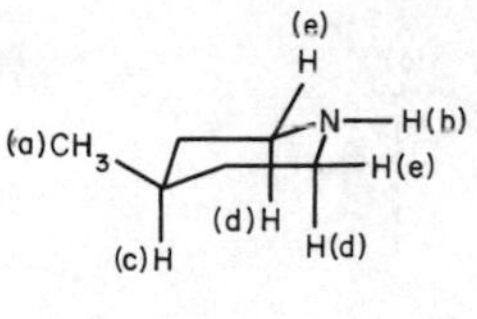

	Shift	Peak
▶ a	0.91	2A
b	1.49	S
c	1.53 ± .10	S
d	2.60	3C
e	3.09	2D

$C_6H_{13}N$ ▶ 7% $CDCl_3$ ▶ V 479

1-Cbc

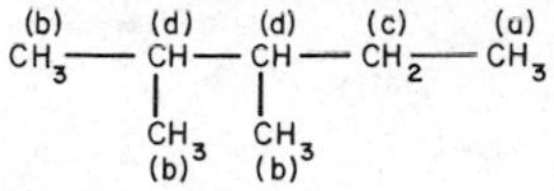

$CH_3(b)—CH(d)(CH_3(b))—CH(d)(CH_3(b))—CH_2(c)—CH_3(a)$

	Shift	Peak
a	~ .79	3D
▶ b	~ .82	2H
c	~1.22	1D
d	~1.58	S

C_7H_{16} ▶ 10% CCl_4 ▶ API 197

1-(C)bb

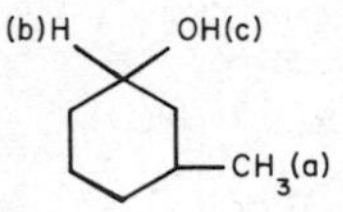

	Shift	Peak
▶ a	0.95	2A
b	~3.5	1D
c	4.02	1B

$C_7H_{14}O$ ▶ 45.1 mg/0.50 ml $CDCl_3$ ▶ S 79

1-(C)bc

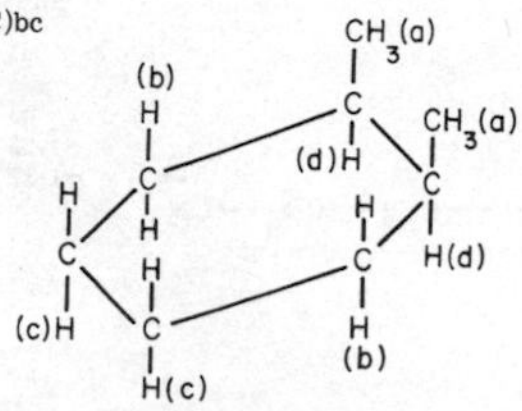

	Shift	Peak
▶ a	.83	2A
b	1.40	S
c	1.40	S
d	~1.67	S

C_8H_{16} ▶ liquid ▶ API 453

▶ 1-Cbc

	Shift	Peak
a	~ .87	2H
▶ b	~ .87	2H
c	~1.15	S
d	~1.45	S

(d) (c)
CH_3—CH—CH—CH_2—CH—CH_3(a)
CH_3 CH_3 CH_3
(b)

▶ C_9H_{20} ▶ 50% CCl_4 ▶ API 214

▶ 1-Cbe

	Shift	Peak
▶ a	1.60	2A
b	3.52	R
c	3.78	R
d	4.10	R

(b) (d)
H H
Cl—C—C—CH_3 (a)
H Cl
(c)

▶ $C_3H_6Cl_2$ ▶ 7% $CDCl_3$ ▶ V 30

▶ 1-Cbc

(c)CH_3 CH_2 CH_2 CH_2 CH CH_3(b) CH_3(b)
(a)CH_3
(d) CH_3
(e) H
HO
H(f)

	Shift	Peak
a	0.67	1A
b	0.87	2H
▶ c	0.90	2H
d	1.00	1E
e	3.50	1D
f	5.36	2E

▶ $C_{27}H_{46}O$ ▶ 7% $CDCl_3$ ▶ V 363

▶ 1-Cbg

	Shift	Peak
a	1.08	3C
▶ b	1.71	2H
c	1.79	6C
d	4.11	5A

(d)
H
(a)CH_3—CH_2—C—CH_3(b)
(c) Br

▶ C_4H_9Br ▶ 7% $CDCl_3$ ▶ V 418

▶ 1-Cbc

(c) CH_3
(f)H (g) H
O (d) (e) (a) CH_3
(h) H H H
(b) CH_3
CH_3 (b)
O
H

	Shift	Peak
a	0.66	1A
b	0.88	S
▶ c	0.95	S
d	1.48	S
e	2.53	S
f	3.90	S
g	4.03	S
h	4.12	S

▶ $C_{27}H_{44}O_2$ ▶ 7% $CDCl_3$ ▶ V 698

▶ 1-Cbg

	Shift	Peak
a	0.91	3B
b	1.28	1B
▶ c	1.71	2A
d	4.12	5B

(a) (b) Br (c)
CH_3—$(CH_2)_5$—CH—CH_3
(d)

▶ $C_8H_{17}Br$ ▶ 45.2 mg/0.50 ml $CDCl_3$ ▶ S 227

▶ 1-(C)bc

	Shift	Peak
▶ a	1.26	2A
b	3.28	1B
c	3.87	2A
d	4.34	3C

(a) CH_3
OH(b)
H(c)
(d)H_2C O O

▶ $C_6H_{10}O_3$ ▶ 7% $CDCl_3$ ▶ V 467

▶ 1-Cbg

	Shift	Peak
▶ a	1.83	2A
b	3.52	4E
c	3.75	S
d	4.18	S

(b) (d)
H H
Br—C—C—CH_3 (a)
H Br
(c)

▶ $C_3H_6Br_2$ ▶ 45.4 mg/0.50 ml $CDCl_3$ ▶ S 187

▶ 1-(C)bc

(n) (e) (b) (g)
(f)H CH_3 H(c)
H H H
(d) CH_3
(j) O (i) C O
(l)H— H (k) H H
O CH_3 (h)
(a) O C—H(m)
O

	Shift	Peak
a	1.22	1E
▶ b	1.29	2H
c	1.62	S
d	1.97	1A
e	2.06	S
f	2.60	3E
g	~3.10	S
h	~3.10	S
i	4.67	3C
j	5.63	2A
k	5.93	2A
l	6.14	2H
m	6.23	N
n	7.79	2F

▶ $C_{17}H_{20}O_5$ ▶ 7% $CDCl_3$ ▶ V 666

▶ 1-Cbi

	Shift	Peak
a	1.02	3C
b	1.70	4C
▶ c	1.92	2A
d	4.17	4C

(d)
H
(b)
(a) CH_3CH_2—C—CH_3(c)
I

▶ C_4H_9I ▶ 7% $CDCl_3$ ▶ V 82

▶ 1-Cbd

	Shift	Peak
a	.85	1E
b	.85	1E
▶ c	.85	1E
d	~1.03	S
e	~1.32	S

(a) (c)
CH_3 CH_3 (d) (b)
CH_3—C—CH—CH_2—CH_2—CH_3
CH_3 (e)

▶ C_9H_{20} ▶ 50% CCl_4 ▶ API 210

▶ 1-(C)bj

	Shift	Peak
▶ a	1.17	2A
b	2.55	6B
c	3.73	10A
d	3.84	10A
e	4.07	10A
f	4.49	3A

(a)
O CH_3
C—C
(d)H (b)H H(c)
C C
H O H
(e) (f)

▶ $C_5H_8O_2$ ▶ 7% $CDCl_3$ ▶ V 438

▶ 1-Cbl

	Shift	Peak
a	.92	3C
▶ b	.95	2H
c	1.25	O
d	1.58	2E
e	2.38	6B
f	~5.17 or 5.32	S
g	~5.32 or 5.17	S

▶ C_7H_{14} ▶ liquid run, neat ▶ API 409

▶ 1-(C)bn

	Shift	Peak
▶ a	1.05	2A
b	1.34	S
c	2.57 ± .10	S

▶ $C_6H_{13}N$ ▶ 7% $CDCl_3$ ▶ V 477

▶ 1-Cbl

	Shift	Peak
a	.88	3D
▶ b	.98	2H
c	1.29	O
d	2.11	5A
e	4.86	R
f	4.97	R
g	~5.58	R

▶ C_7H_{14} ▶ liquid ▶ API 392

▶ 1-Cbn

	Shift	Peak
a	.87	3D
▶ b	1.05	2H
c	1.25	1B
d	3.2	1C

▶ $C_6H_{15}N$ ▶ 34.4 mg/0.50 ml $CDCl_3$ ▶ S 28

▶ 1-Cbl

	Shift	Peak
a	.88	3D
▶ b	.99	2A
c	1.32	S
d	2.01	5B
e	4.83 or 4.96	R
f	4.96 or 4.83	R
g	5.53	R

▶ C_6H_{12} ▶ 10% CCl_4 ▶ API 156

▶ 1-Cbn

	Shift	Peak
a	0.96	3D
▶ b	1.17	2A
c	~1.50	O
d	~3.38	S
e	~3.41	S

▶ $C_{10}H_{15}N$ ▶ 7% $CDCl_3$ ▶ V 568

▶ 1-Cbl

	Shift	Peak
a	.88	3C
▶ b	1.03	1E
c	1.24	S
d	1.96	7B
e	4.86	R
f	4.97	R
g	5.54	R

▶ C_6H_{12} ▶ liquid ▶ API 374

▶ 1-(C)bn

	Shift	Peak
▶ a	1.54	1A
b	1.58	1E
c	1.62	1E
d	3.74	7X
e	5.33	6B
f	8.10	4A

▶ $C_{15}H_{22}ClNO_2$ ▶ D_2O ▶ V 646

▶ 1-Cbl

	Shift	Peak
a	.82	3C
▶ b	1.23	2A
c	1.60	5B
d	2.56	4C
e	7.11	1A
f	7.11	1A

▶ $C_{10}H_{14}$ ▶ 10% CCl_4 ▶ API 90

▶ 1-Cbn

	Shift	Peak
▶ a	1.68	2A
b	2.97	1A
c	3.27	S
d	3.56	S
e	4.62	1F

▶ $C_5H_{13}Cl_2N$ ▶ 7% $CDCl_3$ ▶ V 448

▶ 1-Cbn

	Shift	Peak
a	0.90	3D
▶ b	1.05	2F
c	1.25	N
d	2.78	6A

▶ $C_4H_{11}N$ ▶ 7% $CDCl_3$ ▶ V 88

▶ 1-(C)bn

	Shift	Peak
a	1.56	1A
▶ b	1.69	1A
c	1.75	1A
d	3.54	1D
e	5.27	1D
f	8.04	N

▶ $C_{15}H_{22}ClNO_2$ ▶ 7% $CDCl_3$ ▶ V 645

1-Cbo

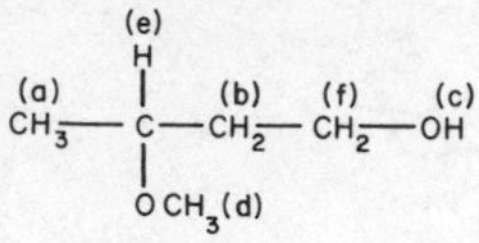

	Shift	Peak
▶ a	1.18	2A
b	1.72	4A
c	3.10	3A
d	3.33	1A
e	3.55	O
f	3.73	6B

▶ $C_5H_{12}O_2$ ▶ 7% $CDCl_3$ ▶ V 120

1-Cbp

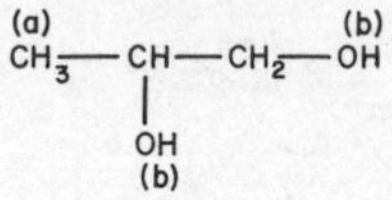

	Shift	Peak
▶ a	1.12	2A
b	4.28	1A

▶ $C_3H_8O_2$ ▶ 7% $CDCl_3$ ▶ V 45

1-(C)bo

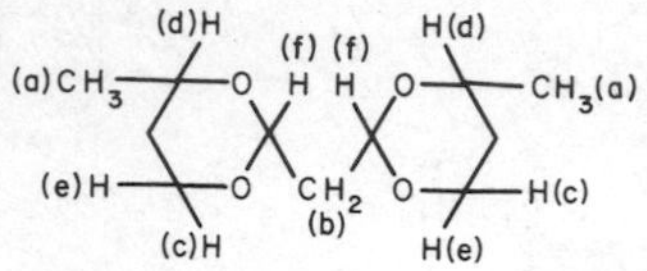

	Shift	Peak
▶ a	1.22	2A
b	1.92	3A
c	2.68±.02	S
d	2.80±.10	S
e	4.05	S
f	4.64	3A

▶ $C_{11}H_{20}O_4$ ▶ 7% $CDCl_3$ ▶ V 588

1-Cbp

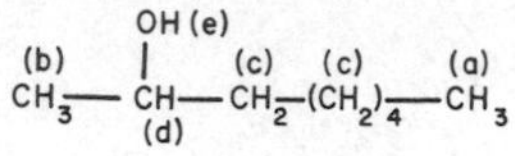

	Shift	Peak
a	.88	3B
▶ b	1.12	2A
c	1.30	1B
d	3.70	4B
e	4.08	1A

▶ $C_8H_{18}O$ ▶ 20% CCl_4 ▶ API 170

1-(C)bo

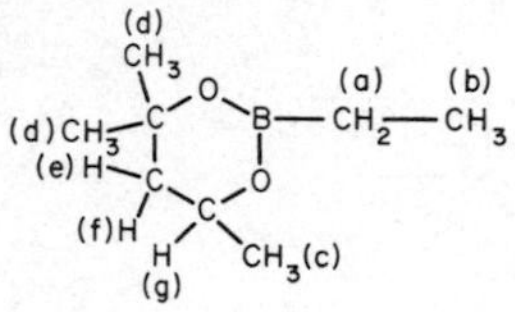

	Shift	Peak
a	0.61	S
b	0.88	S
▶ c	1.25	2H
d	1.29	1A
e	1.47	S
f	1.76	S
g	4.17	O

▶ $C_8H_{17}BO_2$ ▶ 7% $CDCl_3$ ▶ V 518

1-Cbp

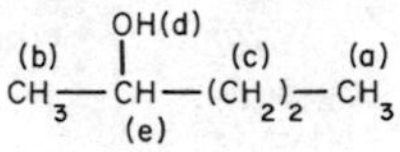

	Shift	Peak
a	0.92	3D
▶ b	1.18	2H
c	1.33	N
d	2.22	1B
e	3.70	1F

▶ $C_5H_{12}O$ ▶ 44.6 mg/0.50 ml $CDCl_3$ ▶ S 60

1-(C)bo

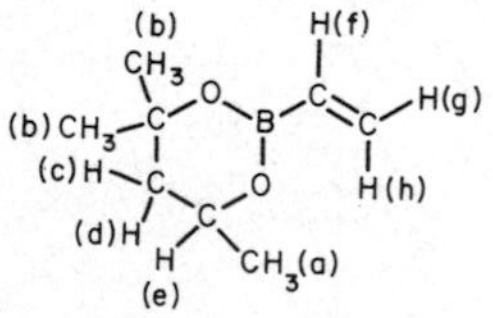

	Shift	Peak
▶ a	1.27	1A
b	1.31	1A
c	1.49	2G
d	1.81	2G
e	4.22	7X
f	~5.90	S
g	~5.90	S
h	~5.90	S

▶ $C_8H_{15}BO_2$ ▶ 7% $CDCl_3$ ▶ V 517

1-Cbp

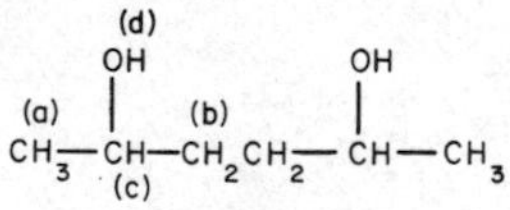

	Shift	Peak
▶ a	1.18	1A
b	1.52	3A
c	3.78	S
d	3.87	1A

▶ $C_6H_{14}O_2$ ▶ 44.6 mg/0.50 ml $CDCl_3$ ▶ S 216

1-(C)bo

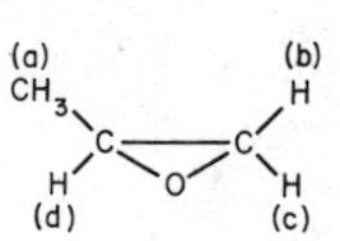

	Shift	Peak
▶ a	1.32	2A
b	2.42	N
c	2.72	N
d	~2.98	N

▶ C_3H_6O ▶ 7% $CDCl_3$ ▶ V 32

1-Cbp

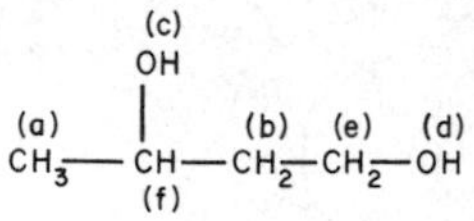

	Shift	Peak
▶ a	1.23	2A
b	1.68	4C
c	3.58 or 3.65	1A
d	3.58 or 3.65	1A
e	3.80	3A
f	4.03	4A

▶ $C_4H_{10}O_2$ ▶ 7% $CDCl_3$ ▶ V 86

1-(C)bo

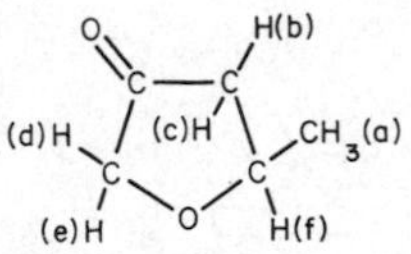

	Shift	Peak
▶ a	1.44	2A
b	2.15	10A
c	2.57	10A
d	3.83	10A
e	4.04	10A
f	4.35	7C

▶ $C_5H_8O_2$ ▶ 7% $CDCl_3$ ▶ V 439

1-Cbq

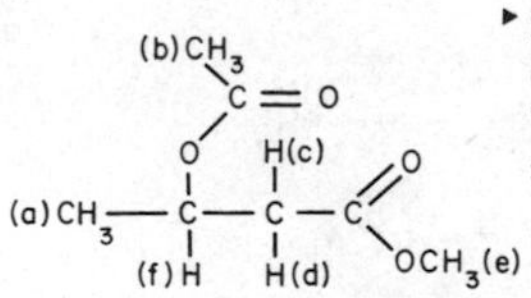

	Shift	Peak
▶ a	1.32	2A
b	2.03	1A
c	2.50 or 2.70	2A
d	2.50 or 2.70	2A
e	3.70	1A
f	5.30	4A

▶ $C_7H_{12}O_4$ ▶ 7% $CDCl_3$ ▶ V 182

▶ 1-Cbr

(a) (c) (d) (b)
CH_3—CH_2—CH—CH_3
NO_2

	Shift	Peak
a	0.95	3A
▶ b	1.52	2A
c	~1.90	O
d	4.52	6A

▶ $C_4H_9NO_2$ ▶ 7% $CDCl_3$ ▶ V 84

▶ 1-(C)cn

CH_3(c)
N
CH_2 CH—CH_3(b)
(a)CH_3
(d) CH_3
N
(d) CH_3
H(e)

	Shift	Peak
a	0.93	1A
▶ b	1.05	2A
c	2.22	1A
d	2.32	1A
e	5.38	1I

▶ $C_{24}H_{40}N_2$ ▶ 7% $CDCl_3$ ▶ V 359

▶ 1-Cbv

(b) CH_3
(e) CH—CH_2—CH_3 (a)
(d) (c)

	Shift	Peak
a	0.80	3A
▶ b	1.22	2A
c	1.59	5B
d	2.53	6A
e	7.20	1A

▶ $C_{10}H_{14}$ ▶ 45.0 mg/0.50 ml $CDCl_3$ ▶ S 31

▶ 1-Ccn

(c) (b)
H H
(d) C—C—CH_3(a)
OD ND_3 (+)
$Cl^{(-)}$

	Shift	Peak
▶ a	1.21	2A
b	3.72	O
c	5.00	2A
d	7.48	1A

▶ $C_9H_{14}ClNO$ ▶ D_2O ▶ V 565

▶ 1-Ccc

(c) (c) (c)
(b)CH_3—CH—CH—CH—CH_3
CH_3 CH_3 (a) CH_3

	Shift	Peak
▶ a	.80 ± .03 or .93 ± .03	S
b	.93 ± .03 or .80 ± .03	S
c	1.67	6A

▶ C_8H_{18} ▶ 50% CCl_4 ▶ API 208

▶ 1-Ccp

(b) (b)
OH OH
(a) (a)
CH_3—CH—CH—CH_3
(c) (c)

	Shift	Peak
▶ a	1.13	2A
b	3.05	1A
c	~3.80	O

▶ $C_4H_{10}O_2$ ▶ 7% $CDCl_3$ ▶ V 87

▶ 1-Ccd

(a) (c) (b)
CH_3 CH_3 CH_3
(a) (b)
CH_3—C—CH—CH—CH_3
(d) (d)
CH_3 (a)

	Shift	Peak
a	.90	1E
b	.90	1E
▶ c	.90	1E
d	2.02	5A

▶ C_9H_{20} ▶ 50% CCl_4 ▶ API 219

▶ 1-(C)cq

O
(f)H O C O
(g)H C C CH_3(a)
C H(e)
CH_3 C O OH(c) H(d)
(b) H (h)

	Shift	Peak
▶ a	1.65	2A
b	1.97	2E
c	3.86	1C
d	3.92	S
e	4.32	O
f	5.87	1I
g	6.06	2A
h	6.99	4A

▶ $C_{12}H_{12}O_5$ ▶ 7% $CDCl_3$ ▶ V 595

▶ 1-Ccg

O (c)H H(b)
C—C—C—CH_3(a)
(d)HO Br Br

	Shift	Peak
▶ a	1.93	N
b	4.42 or 4.43	S
c	4.42 or 4.43	S
d	11.58	1A

▶ $C_4H_6Br_2O_2$ ▶ 7% $CDCl_3$ ▶ V 403

▶ 1-(C)cv

(d) CH_2(c)
CH_2 CH_3(a)
(b) CH_3
CH(e) OCH_3(f)
(h)HO C O
H(g)
(f)CH_3O

	Shift	Peak
a	0.96	3B
▶ b	1.37	2A
c	~1.65	S
d	2.58	3B
e	3.42	S
f	3.88	1A
g	5.08	2A
h	5.62	1A

▶ $C_{20}H_{24}O_4$ ▶ 7% $CDCl_3$ ▶ V 682

▶ 1-Ccl

(d) (f)
H H (a)
CH_3
C=C (c)
H CH—CH—CH_3(a)
(e) CH_3 (b)

	Shift	Peak
a	.88	3A
▶ b	.98	2H
c	1.73	7X
d	4.90	R
e	5.00	R
f	5.64	R

▶ C_7H_{14} ▶ liquid ▶ API 413

▶ 1-Cek

(b)
H
(a) (c)
CH_3—C—COOH
Cl

	Shift	Peak
▶ a	1.73	2B
b	4.47	4A
c	11.22	1A

▶ $C_3H_5ClO_2$ ▶ 7% $CDCl_3$ ▶ V 25

1-Cgg

	Shift	Peak
▶ a	2.47	2A
b	5.86	4A

(a)CH$_3$—C(Br)(Br)—H(b)

▶ $C_2H_4Br_2$ ▶ 7% $CDCl_3$ ▶ V 370

1-Cnn

	Shift	Peak
▶ a	1.62	2A
b	2.12 ± .10	S
c	2.36 ± .10	S
d	3.52	3A

▶ $C_{10}H_{16}N_2O_2$ ▶ 7% $CDCl_3$ ▶ V 571

1-Cgk

	Shift	Peak
▶ a	1.86	2A
b	4.38	4A
c	9.95	1A

CH_3(a)—CH(Br)(b)—C(=O)—OH(c)

▶ $C_3H_5BrO_2$ ▶ 44.5 mg/0.50 ml $CDCl_3$ ▶ S 144

1-Cnv

	Shift	Peak
▶ a	1.38	2A
b	1.58	1A
c	4.10	4A
d	7.30 ± .03	1A

▶ $C_8H_{11}N$ ▶ 7% $CDCl_3$ ▶ V 207

1-Cgv

	Shift	Peak
▶ a	2.02	2A
b	5.20	4A

▶ C_8H_9Br ▶ 45.2 mg/0.50 ml $CDCl_3$ ▶ S 223

1-Coo

	Shift	Peak
a	1.20	S
▶ b	1.32	S
c	~3.50 or 3.70	2C
d	~3.50 or 3.70	2C
e	4.72	4A

▶ $C_6H_{14}O_2$ ▶ 7% $CDCl_3$ ▶ V 143

1-Ckn

	Shift	Peak
▶ a	1.48	2A
b	3.78	4A

(a)CH_3—CH(ND$_2$)(b)—C(=O)—OD

▶ $C_3H_7NO_2$ ▶ D_2O ▶ V 393

1-(C)oo

	Shift	Peak
▶ a	1.40	S
b	5.10	1C

▶ $C_6H_{12}O_3$ ▶ 7% $CDCl_3$ ▶ V 474

1-Ckp

	Shift	Peak
a	0.94	3B
▶ b	1.41	2A
c	1.46	S
d	3.02	1A
e	4.18	4C

(b)CH_3—CH(OH(d))(e)—C(=O)—OCH_2(e)CH_2CH_2(c)CH_3(a)

▶ $C_7H_{14}O_3$ ▶ 44.6 mg/0.50 ml $CDCl_3$ ▶ S 246

1-Cpv

	Shift	Peak
▶ a	1.38	2A
b	3.20	1B
c	4.75	4A
d	6.98	O
e	7.22	O

▶ C_8H_9FO ▶ 7% $CDCl_3$ ▶ V 200

1-Cks

	Shift	Peak
▶ a	1.56	2A
b	2.26	2A
c	3.58	5B
d	11.03	1B

(a)CH_3—C(SH(b))(H(c))—C(=O)—OH(d)

▶ $C_3H_6O_2S$ ▶ 7% $CDCl_3$ ▶ V 389

1-Cpv

	Shift	Peak
▶ a	1.42	2A
b	2.38	1A
c	4.78	4A
d	7.30	1A

▶ $C_8H_{10}O$ ▶ 45.2 mg/0.50 ml $CDCl_3$ ▶ S 133

▶ 1-Daaa

Shift		Peak
a	.96	1A

(a) CH_3 — $H_3C-C(CH_3)-CH_3$ with CH_3

▶ C_5H_{12} ▶ 50% CCl_4 ▶ API 418

▶ 1-Daab

Shift		Peak
▶ a	.87	1E
b	.87	1E
c	1.28	1B

$CH_3-C(CH_3)_2-CH_2-(CH_2)_3-CH_3$ — (a) CH_3, (c) $(CH_2)_3$, (b) CH_3

▶ C_9H_{20} ▶ 50% CCl_4 ▶ API 209

▶ 1-Daab

Shift		Peak
▶ a	0.72	1A
b	1.33	1A
c	1.68	1A
d	5.02	1C
e	6.75	Q
f	7.20	Q

OH(d), (e)H, H(e), (f)H, H(f); $CH_3-C(CH_3)(b)-CH_2(c)-C(CH_3)_3(a)$

▶ $C_{14}H_{22}O$ ▶ 7% $CDCl_3$ ▶ V 315

▶ 1-Daab

Shift		Peak
▶ a	.87	1E
b	.87	1E
c	.87	1E
d	~1.22	S
e	~1.32	1D

$CH_3-C(CH_3)_2-CH_2-CH(CH_3)-CH_2-CH_3$ — (a) CH_3, (d) CH_2, (e) CH, (c) CH_3, (b) CH_3

▶ C_9H_{20} ▶ 50% CCl_4 ▶ API 211

▶ 1-Daab

Shift		Peak
▶ a	.83	1A
b	1.92	2E
c	4.82	R
d	5.03	R
e	5.72	R

(c) H, (d) H, (e) H; $C=C$; (b) CH_2 (a) $C(CH_3)_3$

▶ C_7H_{14} ▶ liquid ▶ API 385

▶ 1-Daab

Shift		Peak
▶ a	.87	1A
b	1.13	1A

(a) (b) (b) (a)
$(CH_3)_3CCH_2CH_2C(CH_3)_3$

▶ $C_{10}H_{22}$ ▶ 10 vol.% CCl_4 ▶ API 439

▶ 1-Daab

Shift		Peak
a	~ .84	1E
▶ b	~ .84	1E
c	~1.13	1G
d	~1.37	1D

$CH_3-C(CH_3)_2-CH_2-CH_2-CH(CH_3)-CH_3$ — (b) CH_3, (c) CH_2, (d) CH, (a) CH_3

▶ C_9H_{20} ▶ 50% CCl_4 ▶ API 212

▶ 1-Daab

Shift		Peak
▶ a	.87	1A
b	1.18	1A

$CH_3-C(CH_3)_2-(CH_2)_4-C(CH_3)_2-CH_3$ — (b) $(CH_2)_4$, (a) CH_3

▶ $C_{12}H_{26}$ ▶ 50% CCl_4 ▶ API 351

▶ 1-Daab

Shift		Peak
▶ a	.85	1E
b	.85	3D
c	1.19	3C

$CH_3-C(CH_3)_2-CH_2-CH_3$ — (a) CH_3, (c) CH_2, (b) CH_3

▶ C_6H_{14} ▶ 50% CCl_4 ▶ API 336

▶ 1-Daab

Shift		Peak
▶ a	.88	1E
b	.88	3D
c	1.20	N

$CH_3-C(CH_3)_2-CH_2-CH_2-CH_3$ — (a) CH_3, (c) CH_2, (c) CH_2, (b) CH_3

▶ C_7H_{16} ▶ 50% CCl_4 ▶ API 339

▶ 1-Daab

Shift		Peak
a	.87	1E
▶ b	1.20	1E
c	1.20	1B

$CH_3-C(CH_3)_2-CH_2-(CH_2)_2-CH_3$ — (b) CH_3, (c) CH_2, (c) $(CH_2)_2$, (a) CH_3

▶ C_8H_{18} ▶ 50% CCl_4 ▶ API 203

▶ 1-Daab

Shift		Peak
▶ a	.88	1A
b	1.15	1A

$CH_3-C(CH_3)_2-CH_2-CH_2-C(CH_3)_2-CH_3$ — (b) CH_2, (a) CH_3

▶ $C_{10}H_{22}$ ▶ 10% CCl_4 ▶ API 220

▶ 1-Daab

(a)
CH_3
$CH_3CCH_2CHCH(CH_3)_2$
CH_3 CH_3

	Shift	Peak
a	.90	1E

▶ $C_{10}H_{22}$ ▶ 50% CCl_4 ▶ API 427

▶ 1-Daac

CH_3 CH_3
(a)H_3C—C—CH—CH_3(b)
(c)
CH_3

	Shift	Peak
▶ a	.81	1E
b	.81	S
c	~1.30	7X

▶ C_7H_{16} ▶ 50% CCl_4 ▶ API 419

▶ 1-Daab

(d) (b)
H CH_3
C=C
(a)
H $CH_2C(CH_3)_3$
(e) (c)

	Shift	Peak
▶ a	.92	1A
b	1.76	1A
c	1.94	1A
d	~4.70	1C
e	~4.80	1C

▶ C_8H_{16} ▶ 50 vol.% CCl_4 ▶ API 433

▶ 1-Daac

(a)
CH_3 CH_3
(a)CH_3C—$CHCH_2CH_3$
CH_3
(a)

	Shift	Peak
a	.85	1A

▶ C_8H_{18} ▶ 50% CCl_4 ▶ API 423

▶ 1-Daab

(b)
CH_3
(a) (c) (a)
$(CH_3)_3CCH_2CHCH_2C(CH_3)_3$
(d) (c)

	Shift	Peak
▶ a	.95	1E
b	~ .96	2H
c	1.17	S
d	1.57	7X

▶ $C_{12}H_{26}$ ▶ 50% CCl_4 ▶ API 428

▶ 1-Daac

(a) (c)
CH_3 CH_3
(d) (b)
CH_3—C—CH—CH_2—CH_2—CH_3
(e)
CH_3

	Shift	Peak
▶ a	.85	1E
b	.85	1E
c	.85	1E
d	~1.03	S
e	~1.32	S

▶ C_9H_{20} ▶ 50% CCl_4 ▶ API 210

▶ 1-Daab

(a)
CH_3 CH_3
(c)
CH_3C—CH_2CHCH_3
(d) (b)
CH_3

	Shift	Peak
▶ a	~ .96	1E
b	~ .96	2H
c	~1.12	2A
d	~1.67	7X

▶ C_8H_{18} ▶ 50% CCl_4 ▶ API 424

▶ 1-Daac

CH_3(a)
H_3C—C—CH_3
C
H H
C C
H H H H
C C
H H H
C
H

	Shift	Peak
a	.86	1E

▶ $C_{10}H_{20}$ ▶ liquid ▶ API 458

▶ 1-Daab

CH_3 CH_3
H_3C—C—CH_2—C—CH_3(a)
(b)
CH_3 CH_3

	Shift	Peak
▶ a	.98	1A
b	1.27	1A

▶ C_9H_{20} ▶ 10% CCl_4 ▶ API 28

▶ 1-Daac

(b)
CH_3 CH_2—CH_3
CH_3—C—CH—CH_2—CH_3
(d) (c)
CH_3
(a)

	Shift	Peak
▶ a	.87	1E
b	.87	1E
c	1.00	1B
d	~1.43	7X

▶ C_9H_{20} ▶ 50% CCl_4 ▶ API 216

▶ 1-Daab

(c) (a)
NH_2 CH_3
(b) (d) (a)
CH_3—C—CH_2—C—CH_3
CH_3 CH_3
(b) (a)

	Shift	Peak
▶ a	1.02	1A
b	1.18	1A
c	1.25	1A
d	1.45	1A

▶ $C_8H_{19}N$ ▶ 7% $CDCl_3$ ▶ V 220

▶ 1-Daac

(a) (c) (b)
CH_3 CH_3 CH_3
(a) (b)
CH_3—C—CH—CH—CH_3
(d) (d)
CH_3
(a)

	Shift	Peak
▶ a	.90	1E
b	.90	1E
c	.90	1E
d	2.02	5A

▶ C_9H_{20} ▶ 50% CCl_4 ▶ API 219

▶ 1-Daad

Shift	Peak
a .73	1A
▶b .87 ± .03	1E
c .87 ± .03	3D
d 1.31	4A

(b) (a)
CH_3 CH_3
(b) (d) (c)
CH_3—C——C——CH_2—CH_3
CH_3 CH_3
(b) (a)

▶ C_9H_{20} ▶ 50% CCl_4 ▶ API 218

▶ 1-Daal

Shift	Peak
▶a 1.08	1B
b 1.73	1H
c 4.62 or 4.72	N
d 4.72 or 4.62	N

(d) (b)
H CH_3
C=C CH_3
H C
(c) CH_3 CH_3(a)

▶ C_7H_{14} ▶ liquid ▶ API 417

▶ 1-Daag

Shift	Peak
a 1.82	1A

(a)
CH_3
(a) (a)
CH_3—C—CH_3
Br

▶ C_4H_9Br ▶ 44.7 mg/0.50 ml $CDCl_3$ ▶ S 229

▶ 1-Daal

Shift	Peak
▶a 1.10	1A
b 1.72	2A
c ~5.23	S
d ~5.23	S

CH_3
(a)CH_3—C CH_3(b)
CH_3 C=C
H H
(d) (c)

▶ C_7H_{14} ▶ 50% CCl_4 ▶ API 222

▶ 1-Daal

Shift	Peak
▶a .98	1A
b ~4.74	R
c ~4.92	R
d 5.78	4D

(b) (d)
H H
C=C
(a)
H C(CH_3)$_3$
(c)

▶ C_6H_{12} ▶ liquid ▶ API 383

▶ 1-Daal

Shift	Peak
▶a 1.11	1A
b 1.69	2E
c 5.23 ± .03	O
d 5.23 ± .03	O

CH_3(a)
(b)H_3C C—CH_3
C=C CH_3
H H
(d) (c)

▶ C_7H_{14} ▶ liquid ▶ API 405

▶ 1-Daal

Shift	Peak
▶a 1.00	1A
b 1.62	2E

(b)H_3C H
C=C CH_3(a)
H C—CH_3
CH_3

▶ C_7H_{14} ▶ liquid ▶ API 406

▶ 1-Daal

Shift	Peak
▶a 1.27	1A
b 6.47	1A

(b) O (a)
H C(CH_3)$_3$
(CH_3)$_3$C H(b)
(a) O

▶ $C_{14}H_{20}O_2$ ▶ 7% $CDCl_3$ ▶ V 314

▶ 1-Daal

Shift	Peak
▶a 1.03	1A
b 1.72	1A
c 4.60 or 4.67	1B
d 4.67 or 4.60	1B

(b)
CH_3 CH_3
(c+d)
CH_2=C——C——CH_3(a)
CH_3

▶ C_7H_{14} ▶ 50% CCl_4 ▶ API 223

▶ 1-Daan

Shift	Peak
▶a 1.10	1A
b 2.57	1E
c 2.70	3D
d 3.65	3C

OH(b)
CH_2(d)
(a)CH_3 CH_2(c)
(a)CH_3— C —N (c) (d)
CH_2— CH_2— OH(b)
(a)CH_3

▶ $C_8H_{19}NO_2$ ▶ 7% $CDCl_3$ ▶ V 221

▶ 1-Daal

Shift	Peak
▶a 1.05	1A
b 1.73	3A
c ~4.66	O

CH_3(b) CH_3(a)
(c)H_2C=C——C—CH_3(a)
CH_3(a)

▶ C_7H_{14} ▶ 10% CCl_4 ▶ API 158

▶ 1-Daan

Shift	Peak
▶a 1.15	1A
b 1.23	1A

(a)
CH_3
(a) (b)
CH_3— C —NH_2
CH_3
(a)

▶ $C_4H_{11}N$ ▶ 7% $CDCl_3$ ▶ V 90

▶ 1-Daao

Shift	Peak
a 1.12	1A
▶ b 1.18	1A
c 3.78	5A

▶ $C_7H_{16}O$ ▶ 7% $CDCl_3$ ▶ V 183

▶ 1-Daav

Shift	Peak
▶ a 1.29	1A
b 7.19	1A

▶ $C_{14}H_{22}$ ▶ CCl_4 ▶ API 186

▶ 1-Daap

Shift	Peak
▶ a 1.28	1A
b 1.35	1A

▶ $C_4H_{10}O$ ▶ 7% $CDCl_3$ ▶ V 423

▶ 1-Daav

Shift	Peak
▶ a 1.30	1A
b 3.80	1A
c 8.50	2C
d 8.89	2C

▶ $C_{20}H_{28}O_6S_2$ ▶ 7% $CDCl_3$ ▶ V 684

▶ 1-Daaq

Shift	Peak
▶ a 1.45	1A
b 1.97	1A

▶ $C_6H_{12}O_2$ ▶ 7% $CDCl_3$ ▶ V 141

▶ 1-Daav

Shift	Peak
▶ a 1.32	1A
b 2.33	1A
c 7.12	Q
d 7.28	Q

▶ $C_{11}H_{16}$ ▶ 7% $CDCl_3$ ▶ V 586

▶ 1-Daav

Shift	Peak
▶ a 1.23	1A
b 7.15	O

▶ $C_{10}H_{14}$ ▶ liquid run ▶ API 20

▶ 1-Daav

Shift	Peak
a 1.35	1A

▶ $C_{10}H_{14}$ ▶ 44.8 mg/0.50 ml $CDCl_3$ ▶ S 53

▶ 1-Daav

Shift	Peak
▶ a 1.25	1A
b 2.21	1A
c ~7.03	S
d 7.11	S

▶ $C_{11}H_{16}$ ▶ CCl_4 ▶ API 185 A

▶ 1-Daav

Shift	Peak
▶ a 1.40	1A
b 2.27	1A
c 4.75	1A
d 6.52	2A
e 6.84	2E or 4A
f 7.05	1I

▶ $C_{11}H_{16}O$ ▶ 7% $CDCl_3$ ▶ V 288

▶ 1-Daav

Shift	Peak
▶ a 1.26	1A
b ~5.5	1C

▶ $C_{10}H_{14}O$ ▶ 45.0 mg/0.25 ml $CDCl_3$ ▶ S 47

▶ 1-Daav

Shift	Peak
a 1.28	3D
▶ b 1.43	1A
c 4.07	4A
d 7.73	2A
e 7.82	2G
f 8.01	2G
g 8.28	2A
h 8.45	2A

▶ $C_{20}H_{28}O_3S$ ▶ 7% $CDCl_3$ ▶ V 683

▶ 1-(D)abb

(b) H CH_3 (a) H_3C C C H (a) H_3C C H C H H H_3C (b) C C H H H

	Shift	Peak
▶ a	.78	1A
▶ b	.87	1A

▶ $C_{10}H_{20}$ ▶ 10% CCl_4 ▶ API 444 B

▶ 1-(D)abb

(b) CH_3 CH_3 C H_2C CH_2 CH_3(a) H_2C C H C H_2

	Shift	Peak
a	.79	S
▶ b	.89	S

▶ C_9H_{18} ▶ 10% CCl_4 ▶ API 153

▶ 1-(D)abb

(b) H CH_3 (b)H_3C C C H (a) H_3C C H C H H H_3C (b) C C H H H

	Shift	Peak
▶ a	.79	1A
▶ b	.89	1A

▶ $C_{10}H_{20}$ ▶ run neat ▶ API 444 A

▶ 1-(D)abb

(a) CH_3 CH_3 C H_2C CH_2(c) H_2C CH_2(b)

	Shift	Peak
▶ a	.97	1G
b	1.32	Q
c	1.55	Q

▶ C_7H_{14} ▶ 10% CCl_4 ▶ API 146

▶ 1-Dabb

CH_3 (c) (a)
CH_3—CH_2—C—CH_2—CH_3
CH_3 (b)

	Shift	Peak
a	~ .82	3D
▶ b	~ .82	1E
c	1.16	4C

▶ C_7H_{16} ▶ 50% CCl_4 ▶ API 199

▶ 1-(D)abb

(c)H_2 CH_3(a) C C (c)H_2C (a)H_3C (a) CH_3 CH_2(b) C CH_3(a)

	Shift	Peak
▶ a	1.02	1A
b	1.30	1A
c	1.50	1A

▶ C_9H_{18} ▶ 10 vol.% CCl_4 ▶ API 440

▶ 1-Dabb

CH_3(a)
$CH_3CH_2CCH_2CH_2CH_3$
CH_3(a)

	Shift	Peak
a	.82	1E

▶ C_8H_{18} ▶ 50% CCl_4 ▶ API 421

▶ 1-(D)abb

(f) (g) (h) H O CH_2 O HO O (c)H_2C CH_2(c) (e)H_2C CH_2(d) C CH_3 CH_3 CH_3 CH_3 (a) (a) (b) (b)

keto-enol mixture

	Shift	Peak
▶ a	1.06	1E
▶ b	1.09	1E
c	2.29	1A
d	2.29	1A
e	2.56	1A
f	3.36	1B
g	5.51	1A
h	10.90	1A

▶ $C_8H_{12}O_2$ ▶ 7% $CDCl_3$ ▶ V 512

▶ 1-Dabb

(c) (b) CH_3 CH_3
(c) (e) (c) (d) (a)
$CH_3CH—CH_2CCH_2CH_3$
CH_3 (b)

	Shift	Peak
a	~ .85	3D
▶ b	~ .85	1E
c	~ .92	2H
d	1.13	2A
e	1.63	7X

▶ C_9H_{20} ▶ 50% CCl_4 ▶ API 426

▶ 1-(D)abb

(a) CH_3 (a)CH_3 CH_2(d) (e)H_2C N $C(CH_2CH_2CN)_3$ (b) (c)

	Shift	Peak
▶ a	1.10	1A
b	1.96	S
c	2.32	S
d	2.32	S
e	3.60	S

▶ $C_{16}H_{22}N_4$ ▶ 7% $CDCl_3$ ▶ V 660

▶ 1-(D)abb

(a) (a) CH_3 CH_3 C (b)H_2C CH_2(b) (b)H_2C CH_2(b) C H_2 (b)

	Shift	Peak
▶ a	.88	1A
b	~1.31	O

▶ C_8H_{16} ▶ 10% CCl_4 ▶ API 150

▶ 1-Dabc

(b) (a) CH_3 CH_3
(b) (d) (d) (c)
CH_3—CH—C—CH_2—CH_2—CH_3
(e) CH_3 (a)

	Shift	Peak
▶ a	.80	1A
b	.86	2H
c	~ .86	3D
d	1.22	3A
e	~1.50	1D

▶ C_9H_{20} ▶ 50% CCl_4 ▶ API 213

▶ 1-(D)abc

(a) CH_3 (b) CH_3 H(i) O
CH_3(e)
H
(d) H_2C (f) H(h)
CH_3 (c)
(g)H

Shift	Peak
▶a 0.86	1A
▶b 0.93	1A
c 1.57	2G
d 2.05	1C
e 2.26	1A
f 2.30	S
g 5.50	1C
h 6.06	2A
i 6.60	4A

▶ $C_{13}H_{20}O$ ▶ 7% $CDCl_3$ ▶ V 616

▶ 1-Dabg

(a) (b)
CH_3 H
Br—C—C—Br
CH_3 H
(a) (b)

Shift	Peak
▶a 1.90	1A
b 3.88	1A

▶ $C_4H_8Br_2$ ▶ 7% $CDCl_3$ ▶ V 412

▶ 1-Dabc

(e) (a) (f)
H CH_3 H
(c) CH_3 (d) (h)
CH— C — C — C— OH
CH_3 (b)
OH CH_3 H
(h) (a) (g)

Shift	Peak
▶a 0.92	1A
b 0.95 or 1.00	2A
c 0.95 or 1.00	2A
d 1.92	5B
e 3.38	2A
f 3.44 or 3.53	S
g 3.44 or 3.53	S
h 3.58	1A

▶ $C_8H_{18}O_2$ ▶ 7% $CDCl_3$ ▶ V 217

▶ 1-(D)abj

(b)
(a) (c) CH_3
CH_3— CH_2 O
(d)
N— CH_2 — C — C
CH_3— CH_2 H(e)
(a) (c)
CH_3
(b)

Shift	Peak
a 0.97	3D
▶b 1.07	1E
c 2.50	4E
d 2.55	1E
e 9.57	1A

▶ $C_9H_{19}NO$ ▶ 7% $CDCl_3$ ▶ V 548

▶ 1-Dabd

(b) (a)
CH_3 CH_3
(b) (d) (c)
CH_3— C — C — CH_2— CH_3
CH_3 CH_3
(b) (a)

Shift	Peak
▶a .73	1A
b .87 ± .03	1E
c .87 ± .03	3D
d 1.31	4A

▶ C_9H_{20} ▶ 50% CCl_4 ▶ API 218

▶ 1-Dabj

(a)
CH_3 O
(c)
NC CH_2 CH_2— C — C
(b)
H(d)
CH_3
(a)

Shift	Peak
▶a 1.12	1A
b 1.92	Q
c 2.27	Q
d 9.47	1A

▶ $C_7H_{11}NO$ ▶ 7% $CDCl_3$ ▶ V 178

▶ 1-(D)abd

CH_3(b) CH_3(a)
(b) CH_3 CH_3(a)
O CH_2
(c)
O

Shift	Peak
▶a 1.10	1A
b 1.33	1A
c 2.43	1A

▶ $C_8H_{14}O_2$ ▶ 7% $CDCl_3$ ▶ V 214

▶ 1-Dabl

(d)
H H
C = C CH_3(b)
H C CH_2CH_3
(c) (a)
CH_3

Shift	Peak
a .78	3D
▶b .97	1E
c 1.28	O
d 5.69	4D

▶ C_7H_{14} ▶ liquid ▶ API 401

▶ 1-Dabe

(c) (a)
CH_3
(b)
CH_2— C — CH_3(a)
Cl

Shift	Peak
▶a 1.57	1A
b 3.07	1A
c 7.27	1A

▶ $C_{10}H_{13}Cl$ ▶ 7% $CDCl_3$ ▶ V 263

▶ 1-(D)abl

(f)
(a) (a) H O
CH_3 CH_3
C CH_3(d)
H(e)
CH_2 CH_3
(c) (b)

Shift	Peak
▶a 1.07	1A
b 1.78	1A
c 2.09	1C
d 2.31	1A
e 6.11	2A
f 7.28	2B

▶ $C_{13}H_{20}O$ ▶ 7% $CDCl_3$ ▶ V 617

▶ 1-Dabe

Cl
CH_3— CH_2— C — CH_3(c)
(a) (b)
CH_3
(c)

Shift	Peak
a 1.08	3B
b 1.58	1A
▶c 1.80	3D

▶ $C_5H_{11}Cl$ ▶ 7% $CDCl_3$ ▶ V 447

▶ 1-(D)abl

(f) (d)
H H
(b) CH_3
O
O
(c) H_2 C (e) H
(a) CH_3 CH_3
(a)

Shift	Peak
▶a 1.20	1A
b 1.70	2D
c 2.73	1G
d 4.73	1B
e 6.14	S
f 6.16	S

▶ $C_{23}H_{22}O_2$ ▶ 7% $CDCl_3$ ▶ V 694

▶ 1-(D)abl

(a)
CH3
(b)
H2C — C — CH3(a)
(d)H C N C N
(e)H H2C — CH2
(c) (f)

Shift	Peak
▶a 1.29	1A
b 2.76	3A
c 3.40	3C
d 3.79	1F
e 3.79	1F
f 4.18	3C

▶ $C_9H_{14}N_2$ ▶ 7% $CDCl_3$ ▶ V 541

▶ 1-(D)abo

(b) H(f)
CH3
(b) CH3 — C — O — B — C — H(g)
(c) H — C O C
(d) H C H(h)
H CH3(a)
(e)

Shift	Peak
a 1.27	1A
▶b 1.31	1A
c 1.49	2G
d 1.81	2G
e 4.22	7X
f ~5.90	S
g ~5.90	S
h ~5.90	S

▶ $C_8H_{15}BO_2$ ▶ 7% $CDCl_3$ ▶ V 517

▶ 1-(D)abn

(a) (a)
CH3 CH3
CH2(d)
(c) CH3 — N = N — OH(e)
CH2(b)
(a) CH3 CH3
(a)

Shift	Peak
▶a 1.12	1A
b 2.24	1H
c 2.28	1A
d 2.56	1B
e 8.73	1B

▶ $C_{10}H_{20}N_2O$ ▶ 7% $CDCl_3$ ▶ V 574

▶ 1-Dabp

(g)H---O
O
H(i)
(h)
(j)H H (c) (b)
CH2 CH3
(a)CH3 — CH(e) CH2 C
(d) CH3(b) OH(f)
(a)CH3

Shift	Peak
a 1.27	2A
▶b 1.34	1E
c 1.69	O
d 2.78	O
e 3.28	S
f 5.27	1B
g 5.27	1B
h 7.19*	2F
i 7.19*	2F
j 7.39	1A

*or 7.31

▶ $C_{15}H_{22}O_3$ ▶ 7% $CDCl_3$ ▶ V 647

▶ 1-Dabn

(c) (a)
NH2 CH3
(b) (d) (a)
CH3 — C — CH2 — C — CH3
CH3 CH3
(b) (a)

Shift	Peak
a 1.02	1A
▶b 1.18	1A
c 1.25	1A
d 1.45	1A

▶ $C_8H_{19}N$ ▶ 7% $CDCl_3$ ▶ V 220

▶ 1-Dabr

(a)
CH3
(a) (b) (c)
CH3 — C — CH2 — OH
NO2

Shift	Peak
▶a 1.53	1A
b 3.81	2A
c 3.05	3B

▶ $C_4H_9NO_3$ ▶ 45.1 mg/0.25 ml $CDCl_3$ ▶ S 296

▶ 1-(D)abn

H(f) (a)
O CH3
C (c)
CH3
O (+)NH2 Cl(-)
H H H
(f) (e) CH3 (d)
(b)

Shift	Peak
▶a 1.56	1A
b 1.69	1A
▶c 1.75	1A
d 3.54	1D
e 5.27	1D
f 8.04	N

▶ $C_{15}H_{22}ClNO_2$ ▶ 7% $CDCl_3$ ▶ V 645

▶ 1-Dabr

NO2
(a)CH3 — C — CH2 — OH(b)
(c)
CH3
(a)

Shift	Peak
▶a 1.59	1A
b 2.52	3B
c 3.89	2B

▶ $C_4H_9NO_3$ ▶ 7% $CDCl_3$ ▶ V 422

▶ 1-(D)abn

H(f) (c)
O CH3
C (b)
CH3
O (+)ND2 Cl(-)
H H H(d)
(f) (e) CH3
(a)

Shift	Peak
a 1.54	1A
▶b 1.58	1E
▶c 1.62	1E
d 3.74	7X
e 5.33	6B
f 8.10	4A

▶ $C_{15}H_{22}ClNO_2$ ▶ D_2O ▶ V 646

▶ 1-Dabv

(d)
OH
(b) (c) (a)
CH3 — C — CH2 — CH3
CH3(b)

Shift	Peak
a 0.69	3A
▶b 1.22	1A
c 1.60	4A
d ~4.85	1D

▶ $C_{11}H_{16}O$ ▶ 44.6 mg/0.50 ml $CDCl_3$ ▶ S 256

▶ 1-(D)abo

(d)
CH3
(a) (b)
(d) CH3 — C — O — B — CH2 — CH3
(e) H — C O
(f) H C
H CH3(c)
(g)

Shift	Peak
a 0.61	S
b 0.88	S
c 1.25	2H
▶d 1.29	1A
e 1.47	S
f 1.76	S
g 4.17	O

▶ $C_8H_{17}BO_2$ ▶ 7% $CDCl_3$ ▶ V 518

▶ 1-Dabv

OH(d)
(e)H H(e)
(f)H H(f) (a)
CH3
(a)
CH3 — C — CH2 — C — CH3
(c)
CH3 CH3
(b) (a)

Shift	Peak
a 0.72	1A
▶b 1.33	1A
c 1.68	1A
d 5.02	1C
e 6.75	Q
f 7.20	Q

▶ $C_{14}H_{22}O$ ▶ 7% $CDCl_3$ ▶ V 315

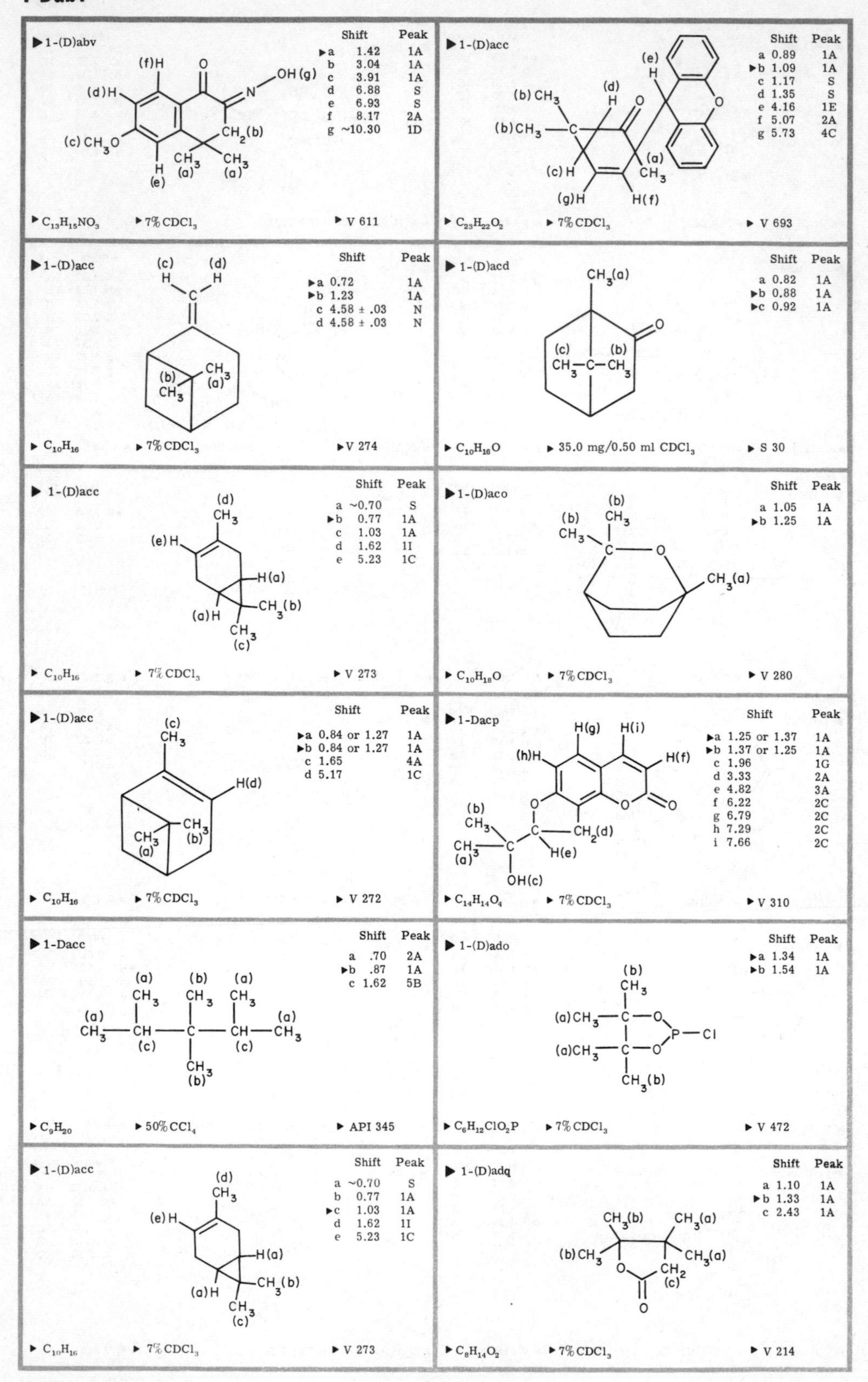

▶ 1-(D)abv
Shift Peak
▶a 1.42 1A
b 3.04 1A
c 3.91 1A
d 6.88 S
e 6.93 S
f 8.17 2A
g ~10.30 1D
▶ $C_{13}H_{15}NO_3$ ▶ 7% $CDCl_3$ ▶ V 611
▶ 1-(D)acc
Shift Peak
a 0.89 1A
▶b 1.09 1A
c 1.17 S
d 1.35 S
e 4.16 1E
f 5.07 2A
g 5.73 4C
▶ $C_{23}H_{22}O_2$ ▶ 7% $CDCl_3$ ▶ V 693
▶ 1-(D)acc
Shift Peak
▶a 0.72 1A
▶b 1.23 1A
c 4.58 ± .03 N
d 4.58 ± .03 N
▶ $C_{10}H_{16}$ ▶ 7% $CDCl_3$ ▶ V 274
▶ 1-(D)acd
Shift Peak
a 0.82 1A
▶b 0.88 1A
▶c 0.92 1A
▶ $C_{10}H_{16}O$ ▶ 35.0 mg/0.50 ml $CDCl_3$ ▶ S 30
▶ 1-(D)acc
Shift Peak
a ~0.70 S
▶b 0.77 1A
c 1.03 1A
d 1.62 1I
e 5.23 1C
▶ $C_{10}H_{16}$ ▶ 7% $CDCl_3$ ▶ V 273
▶ 1-(D)aco
Shift Peak
a 1.05 1A
▶b 1.25 1A
▶ $C_{10}H_{18}O$ ▶ 7% $CDCl_3$ ▶ V 280
▶ 1-(D)acc
Shift Peak
▶a 0.84 or 1.27 1A
▶b 0.84 or 1.27 1A
c 1.65 4A
d 5.17 1C
▶ $C_{10}H_{16}$ ▶ 7% $CDCl_3$ ▶ V 272
▶ 1-Dacp
Shift Peak
▶a 1.25 or 1.37 1A
▶b 1.37 or 1.25 1A
c 1.96 1G
d 3.33 2A
e 4.82 3A
f 6.22 2C
g 6.79 2C
h 7.29 2C
i 7.66 2C
▶ $C_{14}H_{14}O_4$ ▶ 7% $CDCl_3$ ▶ V 310
▶ 1-Dacc
Shift Peak
a .70 2A
▶b .87 1A
c 1.62 5B
▶ C_9H_{20} ▶ 50% CCl_4 ▶ API 345
▶ 1-(D)ado
Shift Peak
▶a 1.34 1A
▶b 1.54 1A
▶ $C_6H_{12}ClO_2P$ ▶ 7% $CDCl_3$ ▶ V 472
▶ 1-(D)acc
Shift Peak
a ~0.70 S
b 0.77 1A
▶c 1.03 1A
d 1.62 1I
e 5.23 1C
▶ $C_{10}H_{16}$ ▶ 7% $CDCl_3$ ▶ V 273
▶ 1-(D)adq
Shift Peak
a 1.10 1A
▶b 1.33 1A
c 2.43 1A
▶ $C_8H_{14}O_2$ ▶ 7% $CDCl_3$ ▶ V 214

▶ 1-(D)adv

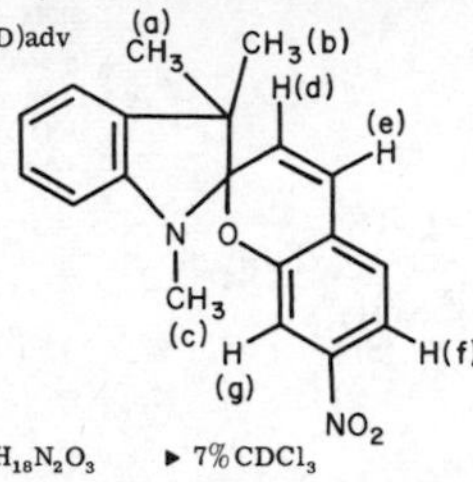

	Shift	Peak
▶a	1.19	1A
▶b	1.30	1A
c	2.74	1A
d	5.84	2A
e	6.93	S
f	8.01 ± .02	S
g	8.01 ± .02	S

▶ $C_{19}H_{18}N_2O_3$ ▶ 7% $CDCl_3$ ▶ V 677

▶ 1-(D)alo

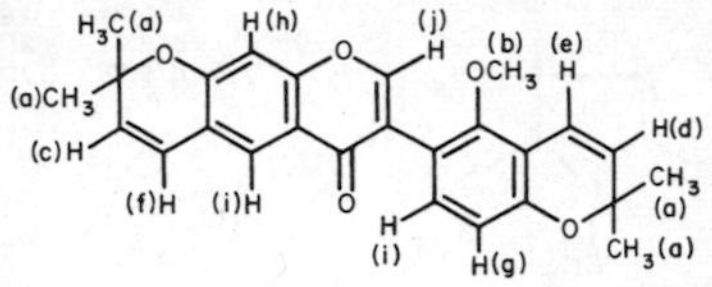

	Shift	Peak
▶a	1.46*	1A
b	3.58	1A
c	5.62†	2A
d	5.62†	2A
e	6.45‡	2A
f	6.45‡	2A
g	6.62	S
h	7.12	2A
i	7.88§	1A
j	7.88§	1A

*or 1.49 ‡or 6.60
†or 5.70 §or 7.90

▶ $C_{26}H_{24}O_5$ ▶ 7% $CDCl_3$ ▶ V 696

▶ 1-(D)ajl

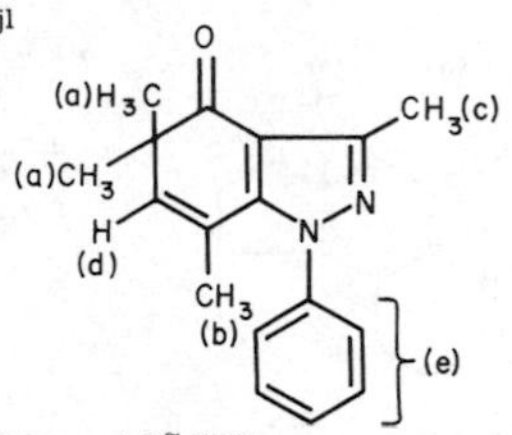

	Shift	Peak
▶a	1.28	1A
b	1.60	2A
c	2.57	1A
d	5.87	2A
e	7.46	1A

▶ $C_{17}H_{18}N_2O$ ▶ 7% $CDCl_3$ ▶ V 664

▶ 1-Dalp

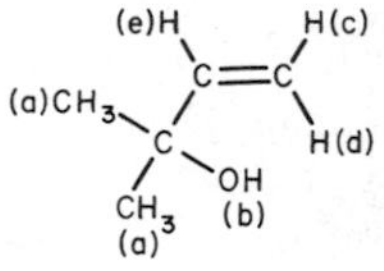

	Shift	Peak
▶a	1.32	1A
b	1.92	1B
c	4.98	R
d	5.17	R
e	6.00	4D

▶ $C_5H_{10}O$ ▶ 7% $CDCl_3$ ▶ V 444

▶ 1-(D)ajl

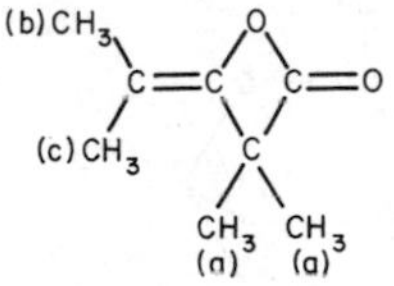

	Shift	Peak
▶a	1.47	1A
b	1.64 or 1.68	1A
c	1.64 or 1.68	1A

▶ $C_8H_{12}O_2$ ▶ 7% $CDCl_3$ ▶ V 513

▶ 1-Dalv

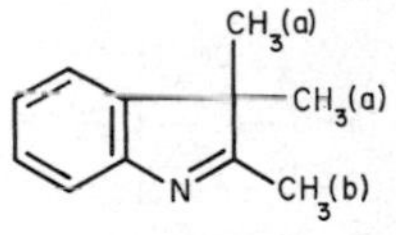

	Shift	Peak
▶a	1.28	1A
b	2.45	1A

▶ $C_{11}H_{13}N$ ▶ 7% $CDCl_3$ ▶ V 286

▶ 1-(D)akl

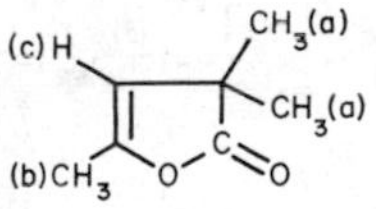

	Shift	Peak
▶a	1.28	1A
b	1.98	2A
c	5.16	2A

▶ $C_7H_{10}O_2$ ▶ 7% $CDCl_3$ ▶ V 175

▶ 1-(D)alv

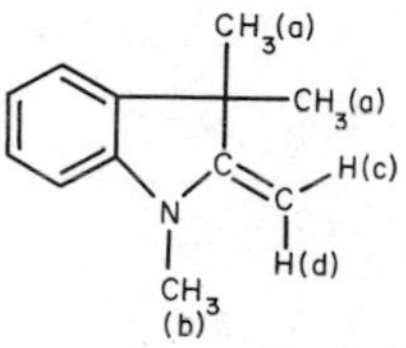

	Shift	Peak
▶a	1.35	1A
b	3.02	1A
c	3.82	1A
d	3.82	1A

▶ $C_{12}H_{15}N$ ▶ 7% $CDCl_3$ ▶ V 596

▶ 1-(D)all

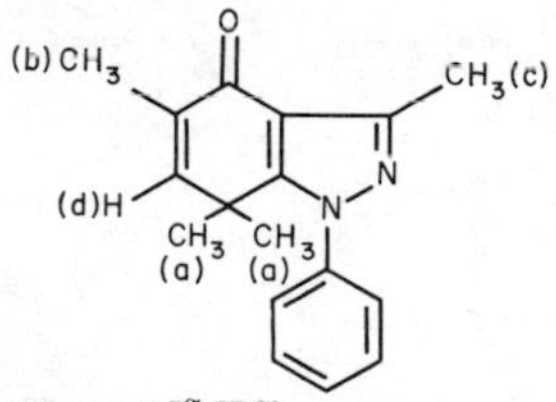

	Shift	Peak
▶a	1.26	1A
b	1.97	2D
c	2.59	1A
d	6.27	2A

▶ $C_{17}H_{18}N_2O$ ▶ 7% $CDCl_3$ ▶ V 665

▶ 1-Dalv

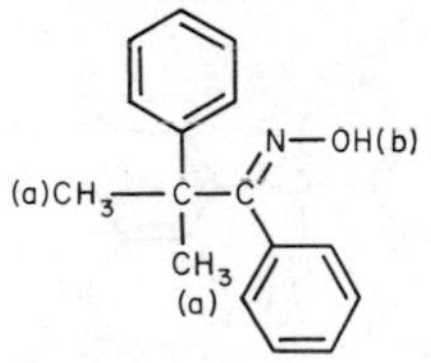

	Shift	Peak
▶a	1.53	1A
b	8.38	1A

▶ $C_{16}H_{17}NO$ ▶ 7% $CDCl_3$ ▶ V 658

▶ 1-(D)alo

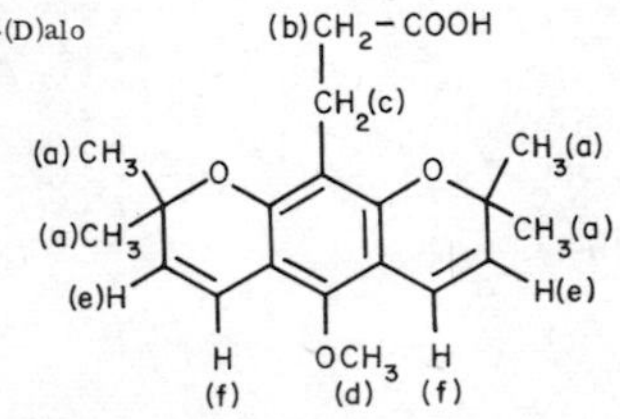

	Shift	Peak
▶a	1.42	1A
b	2.62	Q
c	2.86	Q
d	3.75	1A
e	5.50	2C
f	6.53	2C

▶ $C_{20}H_{24}O_5$ ▶ 7% $CDCl_3$ ▶ V 344

▶ 1-(D)alv

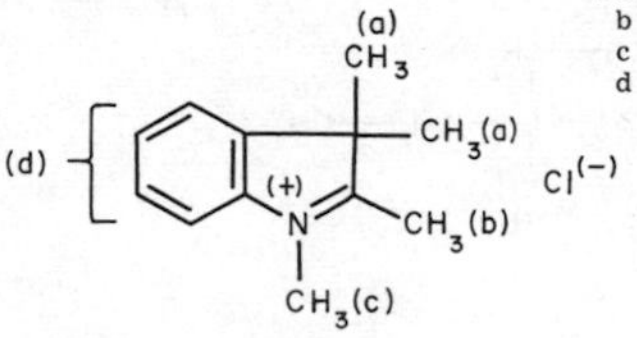

	Shift	Peak
▶a	1.56	1A
b	2.87	1B
c	4.10	1A
d	7.51 ± 0.05	1G

▶ $C_{12}H_{16}ClN$ ▶ 7% $CDCl_3$ ▶ V 599

1-(D)alv

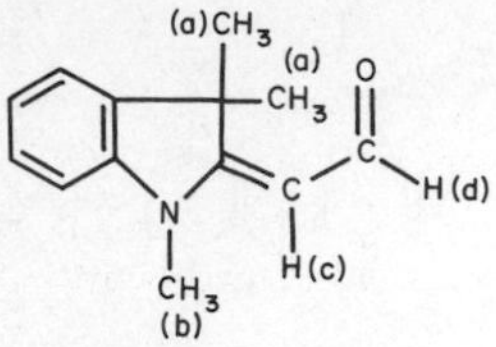

Shift	Peak
▸a 1.67	1A
b 3.22	1A
c 5.37	2C
d 9.98	2C

▸ $C_{13}H_{15}NO$ ▸ 7% $CDCl_3$ ▸ V 610

1-(D)bbb

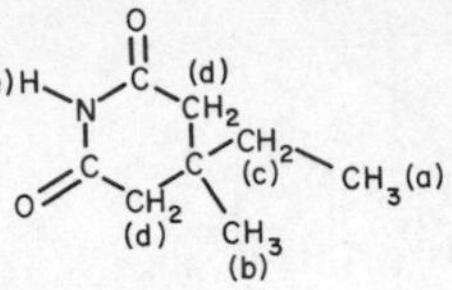

Shift	Peak
a 0.91	3A
▸b 1.06	1A
c 1.43	3B
d 2.43	1A
e 8.47	1D

▸ $C_8H_{13}NO_2$ ▸ 7% $CDCl_3$ ▸ V 514

1-Danp

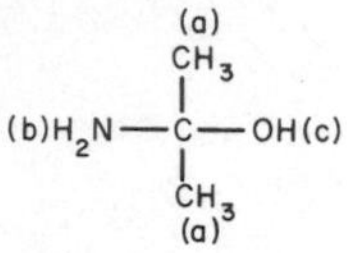

$(b)H_2N-C(CH_3)_2(a)-OH(c)$

Shift	Peak
▸a 1.12	1A
b 2.17	1B
c 3.30	1A

▸ C_3H_9NO ▸ 7% $CDCl_3$ ▸ V 397

1-(D)bbc

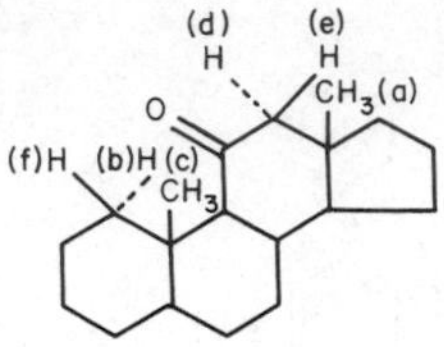

Shift	Peak
▸a 0.67	1E
b 0.78	S
c 1.03	1E
d 2.27	S
e 2.27	S
f 2.45	S

▸ $C_{19}H_{30}O$ ▸ 7% $CDCl_3$ ▸ V 680

1-Dapt

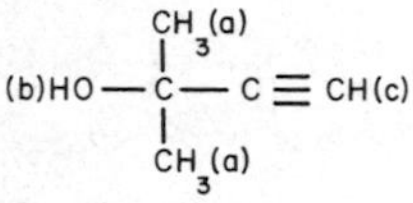

$(b)HO-C(CH_3)_2(a)-C\equiv CH(c)$

Shift	Peak
▸a 1.55	1A
b 2.27	1B
c 2.44	1A

▸ C_5H_8O ▸ 7% $CDCl_3$ ▸ V 436

1-(D)bbo

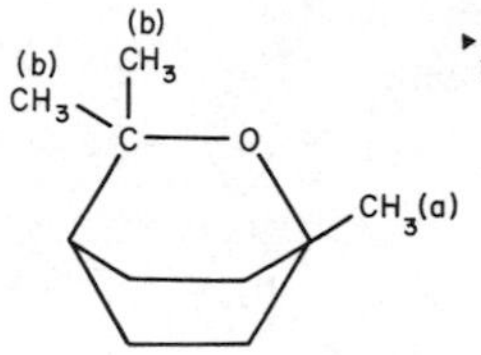

Shift	Peak
▸a 1.05	1A
b 1.25	1A

▸ $C_{10}H_{18}O$ ▸ 7% $CDCl_3$ ▸ V 280

1-Dapu

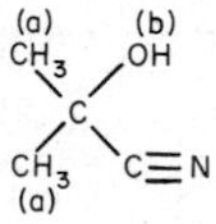

$(CH_3)_2(a)C(OH)(b)C\equiv N$

Shift	Peak
▸a 1.63	1A
b 3.25	1B

▸ C_4H_7NO ▸ 7% $CDCl_3$ ▸ V 70

1-(D)bbo

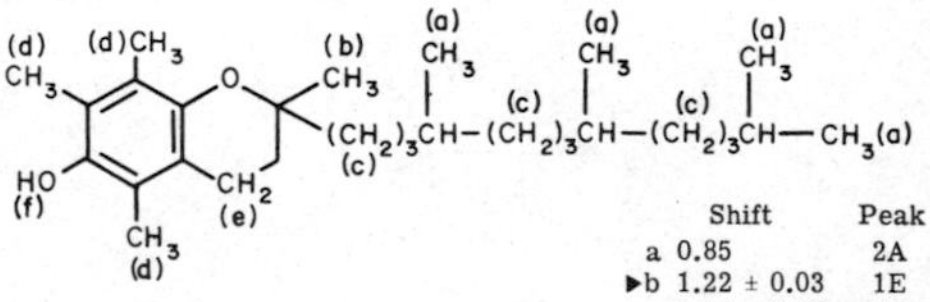

Shift	Peak
a 0.85	2A
▸b 1.22 ± 0.03	1E
c 1.22 ± 0.03	1E
d 2.12 ± 0.03	1B
e 2.62	3C
f 3.96	1D

▸ $C_{29}H_{50}O_2$ ▸ 7% $CDCl_3$ ▸ V 366

1-Davv

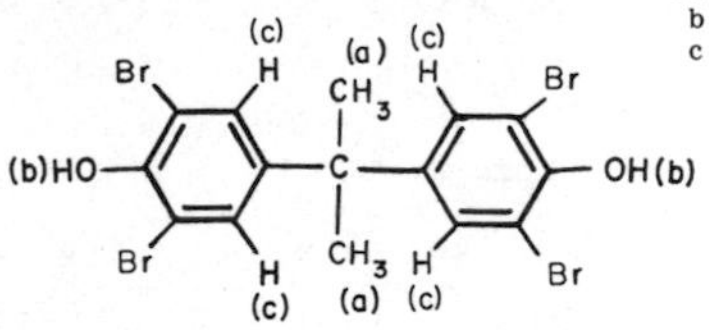

Shift	Peak
▸a 1.58	1A
b 5.66	1D
c 7.27	1A

▸ $C_{15}H_{12}Br_4O_2$ ▸ 7% $CDCl_3$ ▸ V 636

1-(D)bbp

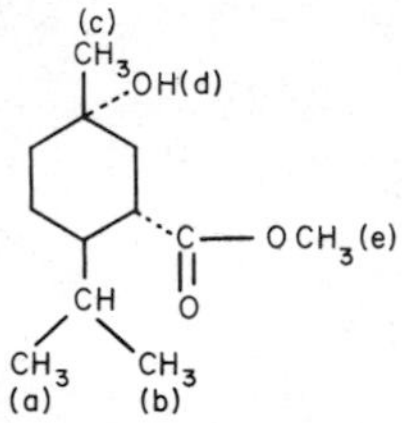

Shift	Peak
a 0.82 or 0.87	2A
b 0.82 or 0.87	2A
▸c 1.35	1E
d 1.67	1E
e 3.72	1A

▸ $C_{12}H_{22}O_3$ ▸ 7% $CDCl_3$ ▸ V 297

1-Dbbb

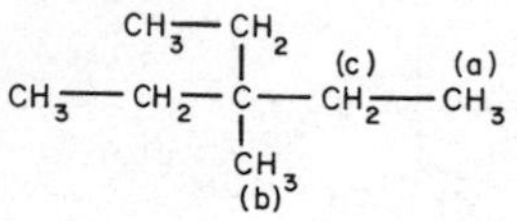

$CH_3CH_2-C(CH_3)(b)(CH_2CH_3)-CH_2(c)-CH_3(a)$

Shift	Peak
a .78	3D
▸b ~ .79	1E
c 1.50	3C

▸ C_8H_{18} ▸ 50% CCl_4 ▸ API 207

1-(D)bbp

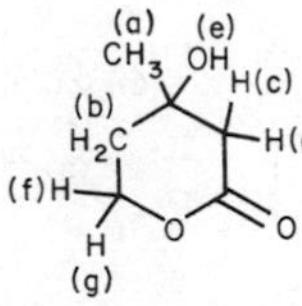

Shift	Peak
▸a 1.40	1A
b 1.92	4B
c 2.51	2G
d 2.68	2G
e 2.92	1A
f 4.33	N
g 4.60	N

▸ $C_6H_{10}O_3$ ▸ 7% $CDCl_3$ ▸ V 466

▶ 1-(D)bbv

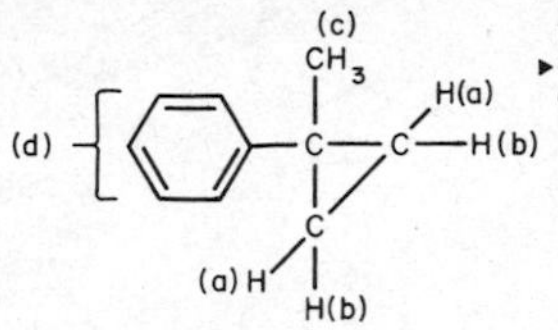

(c) CH_3, H(a), (d), C, C—H(b), C, (a) H, H(b)

	Shift	Peak
a	0.65 ± .03	N
b	0.81 ± .03	1I
▶c	1.37	1A
d	7.17	1A

▶ $C_{10}H_{12}$ ▶ 7% $CDCl_3$ ▶ V 556

▶ 1-(D)bcc

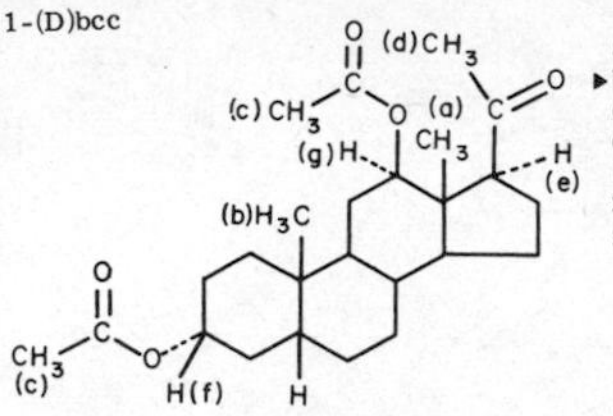

O, (d)CH_3, (c) CH_3, C, O (a), C=O, (g)H, CH_3, H (e), (b)H_3C, O, C, CH_3 (c), O, H(f), H

	Shift	Peak
a	0.70	1A
▶b	0.92	1A
c	2.03, 2.05	1A
d	2.18	1A
e	2.98	3C
f	~4.73	1D
g	5.17	1I

▶ $C_{25}H_{38}O_5$ ▶ 7% $CDCl_3$ ▶ V 361

▶ 1-(D)bcc

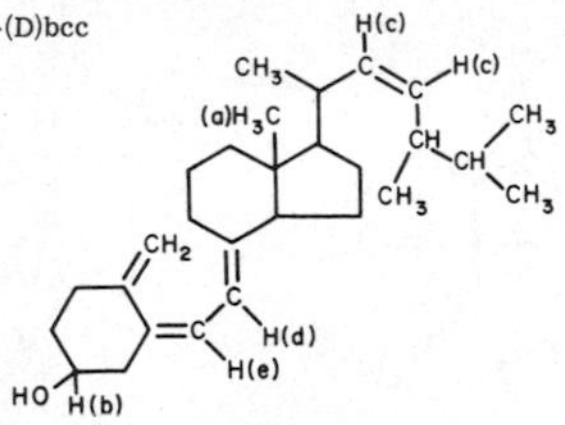

H(c), CH_3, C, C—H(c), (a)H_3C, CH_3, CH, CH, CH_3, CH_3, CH_2, C, C, H(d), H(e), HO, H(b)

	Shift	Peak
▶a	0.57	1A
b	3.92	1F
c	~5.20	N
d	6.05*	2C
e	6.05*	2C

*or 6.20

▶ $C_{28}H_{44}$ ▶ 7% $CDCl_3$ ▶ V 364

▶ 1-(D)bcc

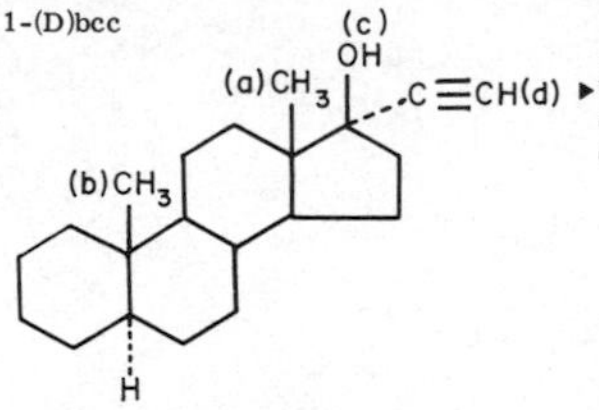

(c) OH, (a) CH_3, C≡CH(d), (b)CH_3, H

	Shift	Peak
a	0.80 or 0.83	1A
▶b	0.80 or 0.83	1A
c	1.88	1A
d	2.55	1A

▶ $C_{21}H_{32}O$ ▶ 7% $CDCl_3$ ▶ V 351

▶ 1-(D)bcc

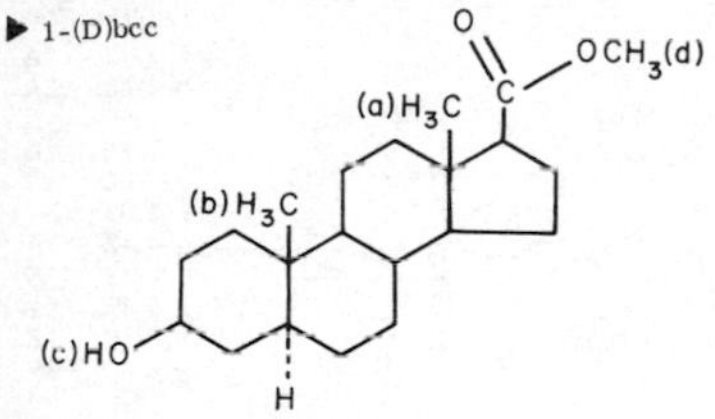

O, C, OCH_3(d), (a)H_3C, (b) H_3C, (c)HO, H

	Shift	Peak
▶a	0.63	1A
▶b	0.80	1A
c	1.67	1E
d	3.67	1A

▶ $C_{21}H_{34}O_3$ ▶ 7% $CDCl_3$ ▶ V 354

▶ 1-(D)bcc

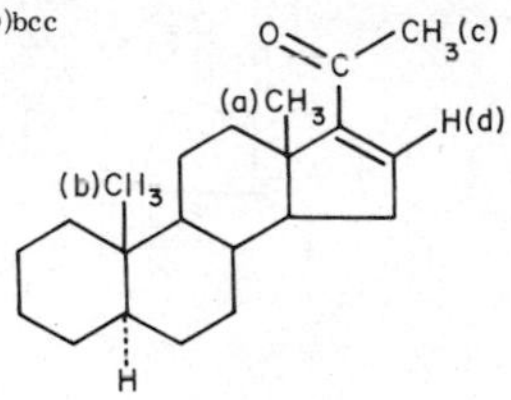

O, C, CH_3(c), (a) CH_3, H(d), (b)CH_3, H

	Shift	Peak
a	0.80 or 0.87	1A
▶b	0.80 or 0.87	1A
c	2.25	1A
d	6.72	N

▶ $C_{21}H_{35}O$ ▶ 7% $CDCl_3$ ▶ V 352

▶ 1-(D)bcc

(c) CH_3, (f)H (g), O, H, (d) (e), H H, (h) H, (a) CH_3, (b) CH_3, CH_3 (b), O, H

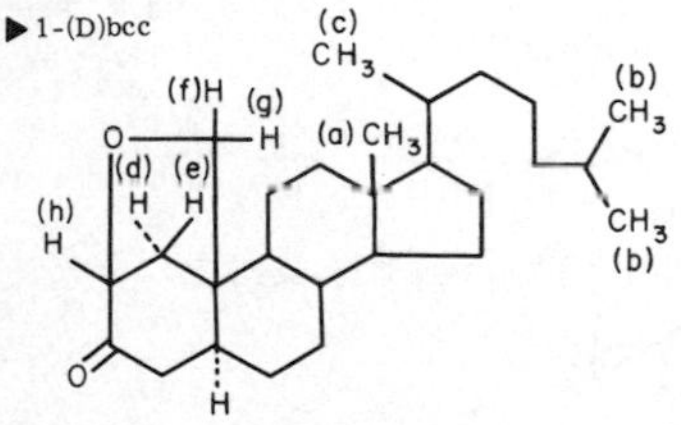

	Shift	Peak
▶a	0.66	1A
b	0.88	S
c	0.95	S
d	1.48	S
e	2.53	S
f	2.90	S
g	4.03	S
h	4.12	S

▶ $C_{27}H_{44}O_2$ ▶ 7% $CDCl_3$ ▶ V 698

▶ 1-(D)bcc

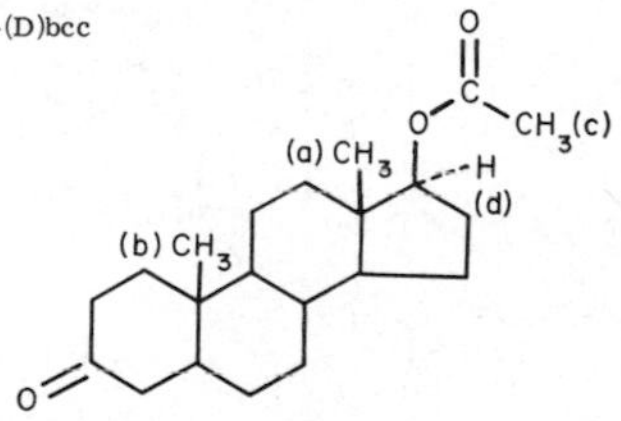

O, C, O, CH_3(c), (a) CH_3, H, (d), (b) CH_3, O

	Shift	Peak
▶a	0.80	1A
▶b	1.03	1A
c	2.05	1E
d	4.62	3D

▶ $C_{21}H_{32}O_3$ ▶ 7% $CDCl_3$ ▶ V 353

▶ 1-(D)bcc

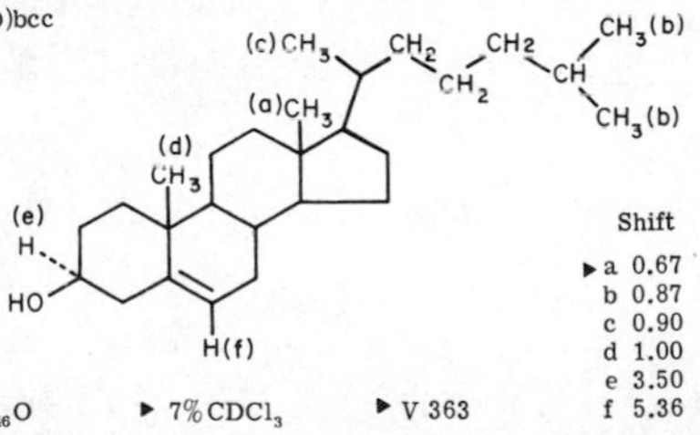

(c) CH_3, CH_2, CH_2, CH_2, CH, CH_3(b), CH_3(b), (a)CH_3, (d) CH_3, (e) H, HO, H(f)

	Shift	Peak
▶a	0.67	1A
b	0.87	2H
c	0.90	2H
d	1.00	1E
e	3.50	1D
f	5.36	2E

▶ $C_{27}H_{46}O$ ▶ 7% $CDCl_3$ ▶ V 363

▶ 1-(D)bcc

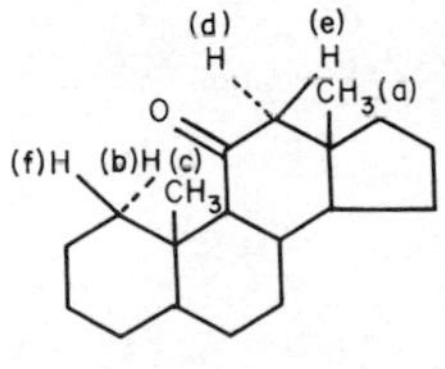

(d) H, (e) H, CH_3(a), O, (f)H, (b)H (c) CH_3

	Shift	Peak
a	0.67	1E
b	0.78	S
▶c	1.03	1E
d	2.27	S
e	2.27	S
f	2.45	S

▶ $C_{19}H_{30}O$ ▶ 7% $CDCl_3$ ▶ V 680

▶ 1-(D)bcc

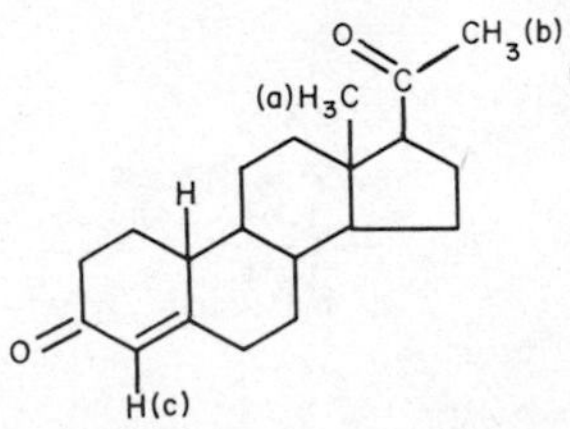

O, C, CH_3(b), (a)H_3C, H, O, H(c)

	Shift	Peak
▶a	0.70	1A
b	2.13	1E
c	5.87	1B

▶ $C_{20}H_{28}O_2$ ▶ 7% $CDCl_3$ ▶ V 345

▶ 1-(D)bcc

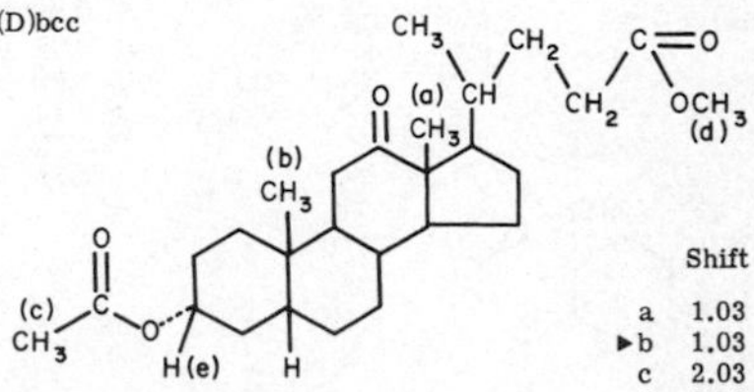

CH_3, CH_2, C=O, O (a) CH, CH_2, OCH_3 (d), CH_3, (b) CH_3, O, (c) C, CH_3, O, H(e), H

	Shift	Peak
a	1.03	1A
▶b	1.03	1A
c	2.03	1E
d	3.70	1A
e	~4.70	1D

▶ $C_{27}H_{42}O_5$ ▶ 7% $CDCl_3$ ▶ V 362

▶ 1-(D)bcd

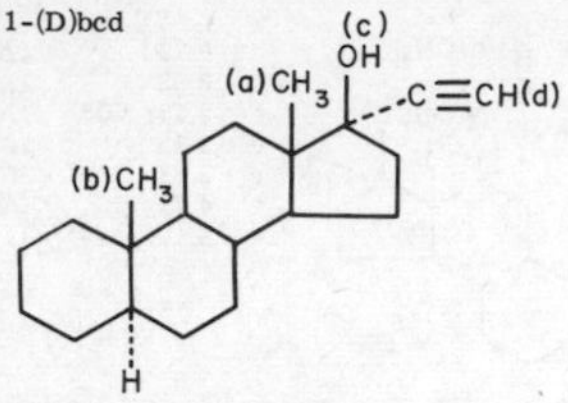

	Shift	Peak
▶a	0.80 or 0.83	1A
b	0.80 or 0.83	1A
c	1.88	1A
d	2.55	1A

▶ $C_{21}H_{32}O$ ▶ 7% $CDCl_3$ ▶ V 351

▶ 1-(D)bcl

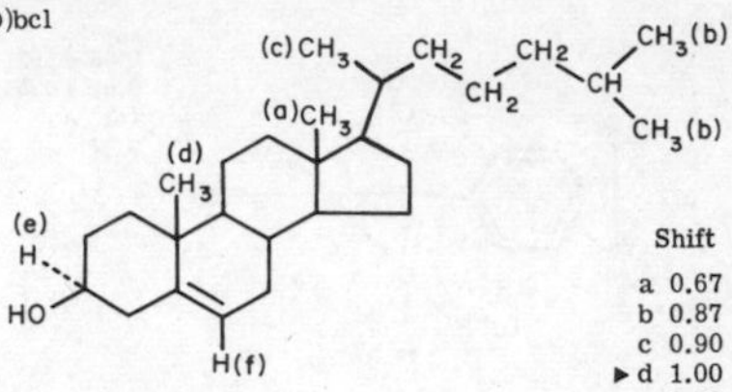

	Shift	Peak
a	0.67	1A
b	0.87	2H
c	0.90	2H
▶d	1.00	1E
e	3.50	1D
f	5.36	2E

▶ $C_{27}H_{46}O$ ▶ 7% $CDCl_3$ ▶ V 363

▶ 1-(D)bcd

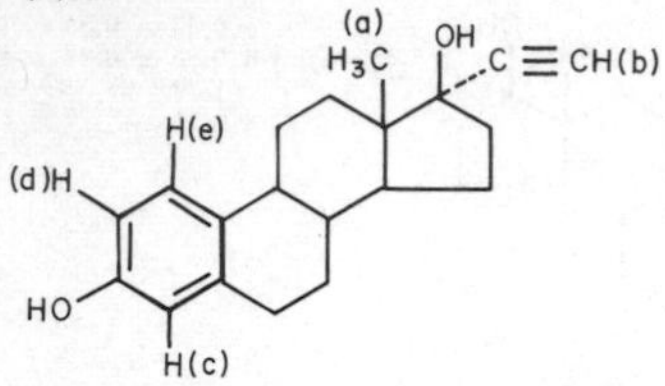

	Shift	Peak
▶a	0.88	1A
b	2.62	1A
c	6.60	1E
d	6.67	2H
e	7.18	2E

▶ $C_{20}H_{24}O_2$ ▶ 7% $CDCl_3$ ▶ V 343

▶ 1-(D)bcl

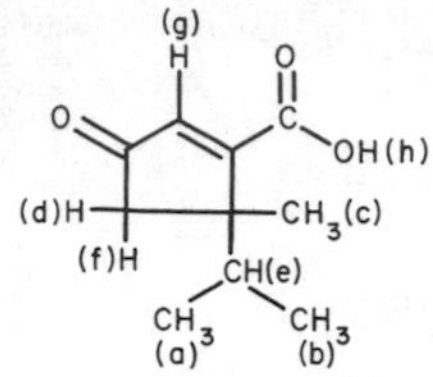

	Shift	Peak
a	0.75	2A
b	0.97	2A
▶c	1.45	1A
d	2.10	2C
e	2.40	S
f	2.60	2C
g	6.77	1A
h	12.00	1B

▶ $C_{10}H_{14}O_3$ ▶ 7% $CDCl_3$ ▶ V 567

▶ 1-(D)bcj

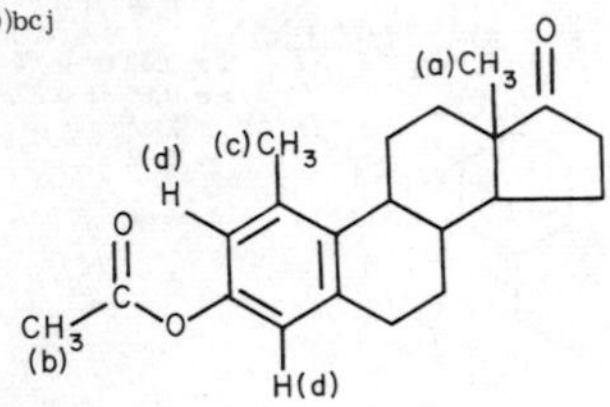

	Shift	Peak
▶a	0.95	1A
b	2.28	1A
c	2.33	1A
d	6.72	1B

▶ $C_{21}H_{26}O_3$ ▶ 7% $CDCl_3$ ▶ V 350

▶ 1-(D)bcv

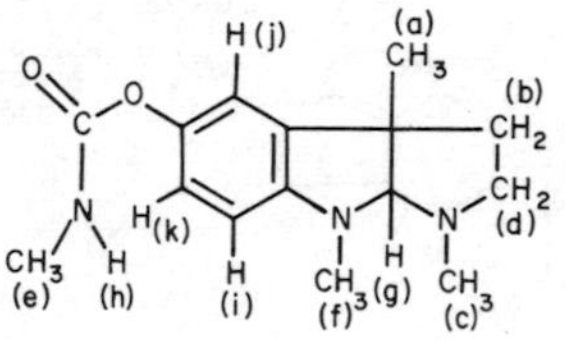

	Shift	Peak
▶a	1.42	1A
b	1.95	3A
c	2.55	1A
d	2.70	3D
e	2.82	1E
f	2.92	1E
g	4.12	1A
h	5.33	1D
i	6.37	2A
j	6.78	1E
k	6.87	2H

▶ $C_{15}H_{21}N_3O_2$ ▶ 7% $CDCl_3$ ▶ V 319

▶ 1-(D)bcj

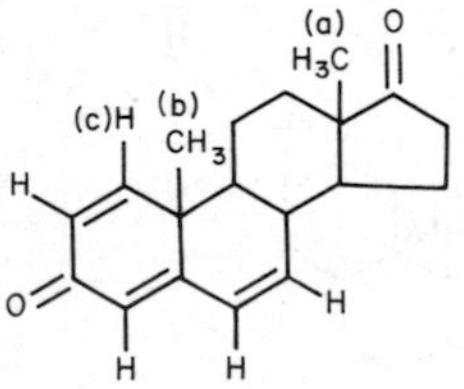

	Shift	Peak
▶a	1.00	1A
b	1.23	1A
c	7.12	2A

▶ $C_{19}H_{22}O_2$ ▶ 7% $CDCl_3$ ▶ V 335

▶ 1-(D)bdj

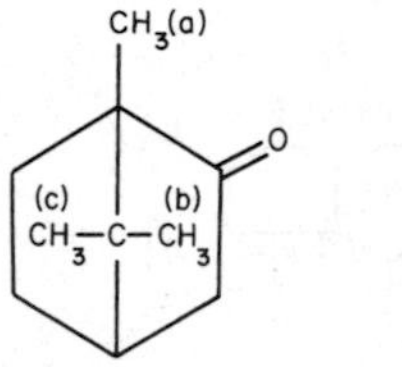

	Shift	Peak
▶a	0.82	1A
b	0.88	1A
c	0.92	1A

▶ $C_{10}H_{16}O$ ▶ 35.0 mg/0.50 ml $CDCl_3$ ▶ S 30

▶ 1-(D)bcl

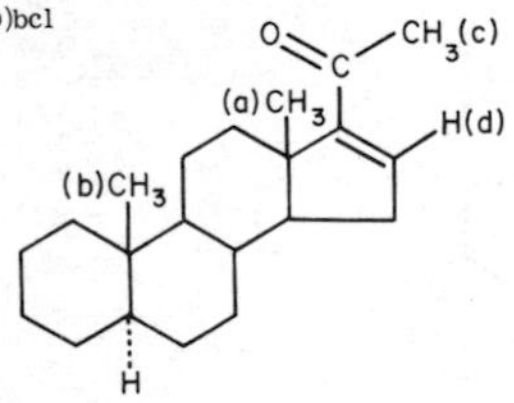

	Shift	Peak
▶a	0.80 or 0.87	1A
b	0.80 or 0.87	1A
c	2.25	1A
d	6.72	N

▶ $C_{21}H_{32}O$ ▶ 7% $CDCl_3$ ▶ V 352

▶ 1-Dbee

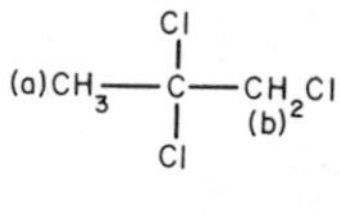

	Shift	Peak
▶a	2.20	1A
b	4.02	1A

▶ $C_3H_5Cl_3$ ▶ 7% $CDCl_3$ ▶ V 383

▶ 1-(D)bcl

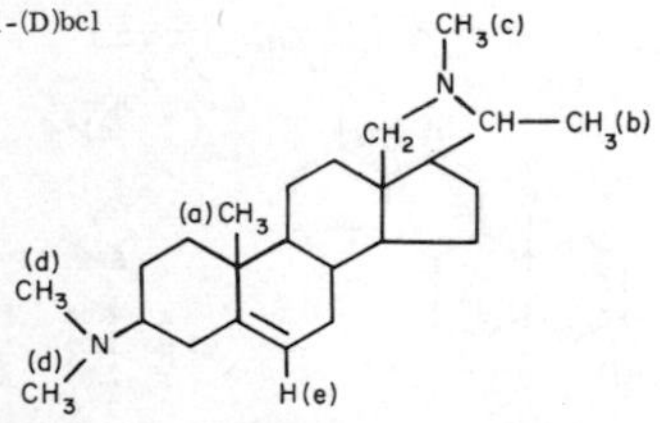

	Shift	Peak
▶a	0.93	1A
b	1.05	2A
c	2.22	1A
d	2.32	1A
e	5.38	1I

▶ $C_{24}H_{40}N_2$ ▶ 7% $CDCl_3$ ▶ V 359

▶ 1-Dbef

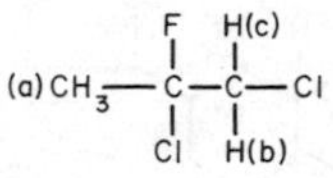

	Shift	Peak
▶a	2.03	2A
b	3.83	S
c	3.96	S

▶ $C_3H_5Cl_2F$ ▶ 7% $CDCl_3$ ▶ V 382

▶ 1-Dbff

F
(a)CH3—C—CH2Cl
(b)
F

	Shift	Peak
▶a	1.75	3A
b	3.63	3A

▶ $C_3H_5ClF_2$ ▶ 7% $CDCl_3$ ▶ V 381

▶ 1-(D)ccc

O (d)CH3
(c) CH3 C O (a) C=O
(g)H CH3 H
(e)
(b)H3C
O
CH3 C O
(c)
H(f) H

	Shift	Peak
▶a	0.70	1A
b	0.92	1A
c	2.03, 2.05	1A
d	2.18	1A
e	2.98	3C
f	~4.73	1D
g	5.17	1I

▶ $C_{25}H_{38}O_5$ ▶ 7% $CDCl_3$ ▶ V 361

▶ 1-(D)ccj

(h) NO2 (h)
H H (f)H H(e)
C
O2N C O C H(g)
H O (a) CH3 CH2
(h) CH3 CH2 H(c)
H(b) C
H(d)
CH3 CH3

	Shift	Peak
▶a	1.74	1E
b	2.40	O
c	4.49 or 4.82	1B
d	4.82 or 4.49	1B
e	5.22	R
f	5.24	R
g	6.08	4D
h	~8.01	N

▶ $C_{27}H_{36}N_2O_6$ ▶ 7% $CDCl_3$ ▶ V 697

▶ 1-(D)ccj

CH3 CH2 C=O
O (a) CH CH2 OCH3
CH3 (d)
(b)
CH3
O
(c) C O
CH3
H(e) H

	Shift	Peak
▶a	1.03	1A
b	1.03	1A
c	2.03	1E
d	3.70	1A
e	~4.70	1D

▶ $C_{27}H_{42}O_5$ ▶ 7% $CDCl_3$ ▶ V 362

▶ 1-(D)bkn

(d) (c)
H CH3 O
(e)
H—C C
(a) CH3 (f)
CH N OH
C N H
(b) CH3 O

	Shift	Peak
a	1.10	2A
b	1.12	2A
▶c	1.60	1A
d	2.81	2A
e	3.48	S
f	3.52	S

▶ $C_9H_{14}N_2O_3$ ▶ 7% $CDCl_3$ ▶ V 543

▶ 1-(D)ccj

(n) (e) (b) (g)
H (f)H CH3 H(c)
H H
(j) (d)
(l)H O C CH3
H (i) (k) O
CH3 (h) H
O (a) O H(m)
O

	Shift	Peak
▶a	1.22	1E
b	1.29	2H
c	1.62	S
d	1.97	1A
e	2.06	S
f	2.60	3E
g	~3.10	S
h	~3.10	S
i	4.67	3C
j	5.63	2A
k	5.93	2A
l	6.14	2H
m	6.23	N
n	7.79	2F

▶ $C_{17}H_{20}O_5$ ▶ 7% $CDCl_3$ ▶ V 666

▶ 1-(D)blo

CH3 (b)
H(c)
O
O
H(d)
CH3—C—H
(a)
CH3(a)

	Shift	Peak
a	1.00	2A
▶b	1.37	1A
c	6.42 or 6.47	2C
d	6.42 or 6.47	2C

▶ $C_{10}H_{16}O_2$ ▶ 7% $CDCl_3$ ▶ V 276

▶ 1-Dcee

CH3(a)
(a)CH3 Cl Cl—C—Cl
C O Cl
O O C—CH3(a)
Cl (b)H Cl
H(b)
H(b)

	Shift	Peak
▶a	2.15	1A
b	5.19	1A

▶ $C_9H_{12}Cl_6O_3$ ▶ 7% $CDCl_3$ ▶ V 536

▶ 1-Dblp

(f)H H(e)
C
HO C
(a)CH3 C H(g)
CH2
CH3 CH2 H(c)
C
H H(d)
(b)
CH3 CH3

	Shift	Peak
▶a	1.28	1E
b	2.40	S
c	4.52	1B
d	4.83	1B
e	5.08	R
f	5.22	R
g	5.92	4D

▶ $C_{20}H_{34}O$ ▶ 7% $CDCl_3$ ▶ V 685

▶ 1-(D)cjl

(e)
H
(d)
(b) CH3 H O O
(b)CH3—C
(a)
(c) H CH3
(g)H H(f)

	Shift	Peak
▶a	0.89	1A
b	1.09	1A
c	1.17	S
d	1.35	S
e	4.16	1E
f	5.07	2A
g	5.73	4C

▶ $C_{23}H_{22}O_2$ ▶ 7% $CDCl_3$ ▶ V 693

▶ 1-Dbss

(a) (d) (e) (b)
CH3—CH2 SO2—CH2—CH3
C
(c)CH3 SO2—CH2—CH3
(e) (b)

	Shift	Peak
a	1.24	3C
b	1.43	3D
▶c	1.78	1A
d	2.36	4C
e	3.45	4A

▶ $C_8H_{18}O_4S_2$ ▶ 7% $CDCl_3$ ▶ V 519

▶ 1-(D)cll

(a) O
H3C
(c)H (b)
CH3
H
O H
H H

	Shift	Peak
a	1.00	1A
▶b	1.23	1A
c	7.12	2A

▶ $C_{19}H_{22}O_2$ ▶ 7% $CDCl_3$ ▶ V 335

▶ 1-(D)jlq

	Shift	Peak
a	0.89	3B
b	~1.30	S
c	~1.40	S
▶d	1.72	1A
e	1.97	2A
f	2.94	3B
g	6.06	S
h	6.17	S
i	6.58	S
j	6.89	S
k	7.88	1G

▶ $C_{21}H_{22}O_5$ ▶ 7% $CDCl_3$ ▶ V 687

▶ 1-Jb

	Shift	Peak
a	0.90	3B
b	1.38	O
▶c	2.15	1A
d	2.42	3B

▶ $C_7H_{14}O$ ▶ 35.4 mg/0.50 ml $CDCl_3$ ▶ S 41

▶ 1-Ja

	Shift	Peak
a	2.06	5A

▶ C_3H_6O ▶ acetone D_6 ▶ V 388

▶ 1-Jb

	Shift	Peak
▶a	2.15	1E
b	3.13	2A
c	5.43	O

▶ $C_9H_{14}O$ ▶ 7% $CDCl_3$ ▶ V 545

▶ 1-Jb

	Shift	Peak
a	0.93	2A
▶b	2.12	1A
c	~2.28	N

▶ $C_6H_{12}O$ ▶ 7% $CDCl_3$ ▶ V 139

▶ 1-Jb

	Shift	Peak
a	0.88	3B
b	1.28	1B
▶c	2.17	1A
d	2.46	3B

▶ $C_9H_{18}O$ ▶ 33.7 mg/0.50 ml $CDCl_3$ ▶ S 48

▶ 1-Jb

	Shift	Peak
a	1.05	3A
▶b	2.13	1A
c	2.47	4A

▶ C_4H_8O ▶ 7% $CDCl_3$ ▶ V 76

▶ 1-Jb

	Shift	Peak
▶a	2.17	1A
b	3.52	1A
c	9.34	1C

Probably exists about 5% in enol form.

▶ $C_{10}H_{11}NO_2$ ▶ 7% $CDCl_3$ ▶ V 256

▶ 1-Jb

	Shift	Peak
a	0.92	3B
b	1.62	6B
▶c	2.14	1A
d	2.42	3A

▶ $C_5H_{10}O$ ▶ 45.4 mg/0.50 ml $CDCl_3$ ▶ S 177

▶ 1-Jb

	Shift	Peak
a	1.26	3A
▶b	2.20	1A
c	2.64	Q
d	4.12	4A

▶ $C_7H_{12}O_3$ ▶ 45.5 mg/0.50 ml $CDCl_3$ ▶ S 247

▶ 1-Jb

	Shift	Peak
▶a	2.14	1A
b	2.70	3A
c	2.94	1A
d	3.66	3A

▶ $C_{11}H_{15}NO$ ▶ 7% $CDCl_3$ ▶ V 584

▶ 1-Jc

	Shift	Peak
a	0.70	1A
▶b	2.13	1E
c	5.87	1B

▶ $C_{20}H_{28}O_2$ ▶ 7% $CDCl_3$ ▶ V 345

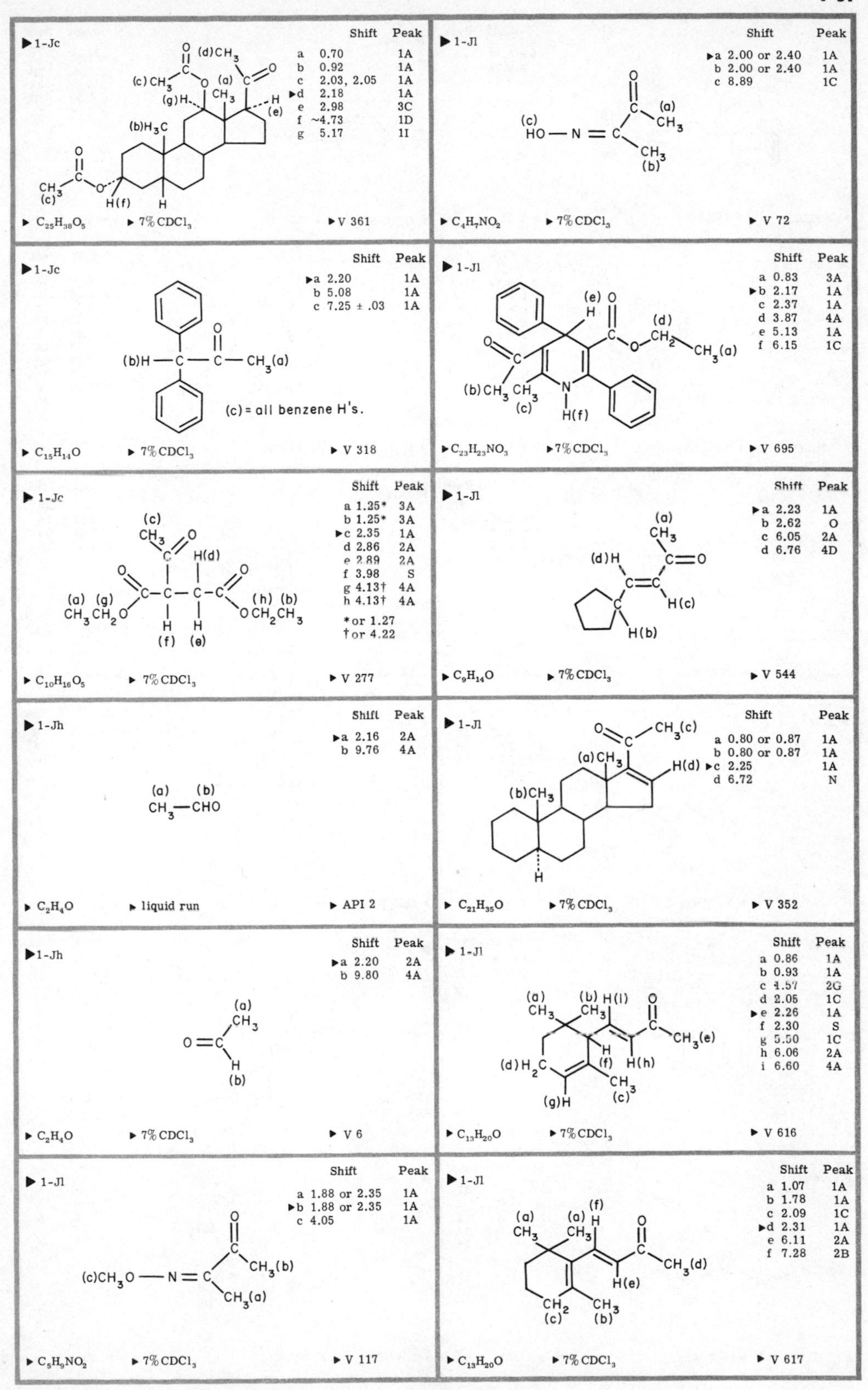

▶ 1-Jc

	Shift	Peak
a	0.70	1A
b	0.92	1A
c	2.03, 2.05	1A
▶d	2.18	1A
e	2.98	3C
f	~4.73	1D
g	5.17	1I

▶ $C_{25}H_{38}O_5$ ▶ 7% $CDCl_3$ ▶ V 361

▶ 1-Jl

	Shift	Peak
▶a	2.00 or 2.40	1A
b	2.00 or 2.40	1A
c	8.89	1C

▶ $C_4H_7NO_2$ ▶ 7% $CDCl_3$ ▶ V 72

▶ 1-Jc

	Shift	Peak
▶a	2.20	1A
b	5.08	1A
c	7.25 ± .03	1A

(c) = all benzene H's.

▶ $C_{15}H_{14}O$ ▶ 7% $CDCl_3$ ▶ V 318

▶ 1-Jl

	Shift	Peak
a	0.83	3A
▶b	2.17	1A
c	2.37	1A
d	3.87	4A
e	5.13	1A
f	6.15	1C

▶ $C_{23}H_{23}NO_3$ ▶ 7% $CDCl_3$ ▶ V 695

▶ 1-Jc

	Shift	Peak
a	1.25*	3A
b	1.25*	3A
▶c	2.35	1A
d	2.86	2A
e	2.89	2A
f	3.98	S
g	4.13†	4A
h	4.13†	4A

*or 1.27
†or 4.22

▶ $C_{10}H_{16}O_5$ ▶ 7% $CDCl_3$ ▶ V 277

▶ 1-Jl

	Shift	Peak
▶a	2.23	1A
b	2.62	O
c	6.05	2A
d	6.76	4D

▶ $C_9H_{14}O$ ▶ 7% $CDCl_3$ ▶ V 544

▶ 1-Jh

	Shift	Peak
▶a	2.16	2A
b	9.76	4A

▶ C_2H_4O ▶ liquid run ▶ API 2

▶ 1-Jl

	Shift	Peak
a	0.80 or 0.87	1A
b	0.80 or 0.87	1A
▶c	2.25	1A
d	6.72	N

▶ $C_{21}H_{35}O$ ▶ 7% $CDCl_3$ ▶ V 352

▶ 1-Jh

	Shift	Peak
▶a	2.20	2A
b	9.80	4A

▶ C_2H_4O ▶ 7% $CDCl_3$ ▶ V 6

▶ 1-Jl

	Shift	Peak
a	0.86	1A
b	0.93	1A
c	1.57	2G
d	2.05	1C
▶e	2.26	1A
f	2.30	S
g	5.50	1C
h	6.06	2A
i	6.60	4A

▶ $C_{13}H_{20}O$ ▶ 7% $CDCl_3$ ▶ V 616

▶ 1-Jl

	Shift	Peak
a	1.88 or 2.35	1A
▶b	1.88 or 2.35	1A
c	4.05	1A

▶ $C_5H_9NO_2$ ▶ 7% $CDCl_3$ ▶ V 117

▶ 1-Jl

	Shift	Peak
a	1.07	1A
b	1.78	1A
c	2.09	1C
▶d	2.31	1A
e	6.11	2A
f	7.28	2B

▶ $C_{13}H_{20}O$ ▶ 7% $CDCl_3$ ▶ V 617

▶ 1-Jl

(c)H, C=C, C(=O)CH$_3$(a), H(b)

	Shift	Peak
▶a	2.35	1A
b	6.70	2A
c	7.48	S

▶ $C_{10}H_{10}O$ ▶ 7% $CDCl_3$ ▶ V 251

▶ 1-Jn

(e) (b) (c) CH_2—CH_2—N(H(d))—C(=O)—CH_3(a)

	Shift	Peak
▶a	1.90	1A
b	2.80	3B
c	3.48	4A
d	6.50	1D
e	7.25	1G

▶ $C_{10}H_{13}NO$ ▶ 7% $CDCl_3$ ▶ V 265

▶ 1-Jl

(c)H, H(d), (e)H, N, CH$_3$ (b), C(=O)CH$_3$(a)

	Shift	Peak
▶a	2.40	1A
b	3.92	1A
c	6.10	4A
d	6.77	3A
e	6.92	4A

▶ C_7H_9NO ▶ 7% $CDCl_3$ ▶ V 174

▶ 1-Jn

O=C—CH$_3$(a), (f) H, (c) H_2C, (b) CH_2, N, H(j), H(e), H(i), CH_3O, CH_3O, CH_3O, H (g), (h)H, O, (d) OCH_3

	Shift	Peak
▶a	1.96	1A
b	2.43	S
c	2.43	S
d	3.67	1A
e	4.62	S
f	6.55	1A
g	6.92	2A
h	7.37	S
i	7.63	1B
j	8.38	2G

▶ $C_{22}H_{25}NO_6$ ▶ 7% $CDCl_3$ ▶ V 689

▶ 1-Jl

(c)H, H(d), (a)CH_3, O, C(=O)CH$_3$(b)

	Shift	Peak
a	2.38	2D
▶b	2.43	1A
c	6.17	2E
d	7.12	2E

▶ $C_7H_8O_2$ ▶ 7% $CDCl_3$ ▶ V 165

▶ 1-Jn

(b)CH_3C(=O)—NHCH_2CH_3 (d) (c) (a)

	Shift	Peak
a	1.14	3A
▶b	2.02	1A
c	3.26	5B
d	8.08	1B

▶ C_4H_9NO ▶ 44.4 mg/0.50 ml $CDCl_3$ ▶ S 146

▶ 1-Jl

(a)CH_3, O, O—H(d)···O, (c)H, O, C, CH_3(b)

	Shift	Peak
a	2.27	1A
▶b	2.68	1A
c	5.92	1A
d	16.71	1C

▶ $C_8H_8O_4$ ▶ 7% $CDCl_3$ ▶ V 504

▶ 1-Jn

(c) NH—C(=O)—CH_3 (b), CH_3(a)

	Shift	Peak
a	2.06 or 2.17	1A
▶b	2.06 or 2.17	1A
c	~7.5	1C

▶ $C_9H_{11}NO$ ▶ 45.2 mg/0.50 ml $CDCl_3$ ▶ S 175

▶ 1-Jn

(b) CH_3, (c) CH_3, N, C(=O)CH$_3$(a)

	Shift	Peak
▶a	1.78	1A
b	2.25	1A
c	3.20	1A

▶ $C_{10}H_{13}NO$ ▶ 7% $CDCl_3$ ▶ V 264

▶ 1-Jn

(f)H, (e)H, N—C(=O)CH$_3$(a), CH_3O, CH_3O, CH_3O, H(d), H(b), H (c), O, (g) H, OCH_3

	Shift	Peak
▶a	2.07	1G
b	3.63	2C
c	4.12	2C
d	4.84	1F
e	6.02	2E
f	6.49	1B
g	6.68	2A

▶ $C_{22}H_{25}NO_6$ ▶ 7% $CDCl_3$ ▶ V 690

▶ 1-Jn

(a)CH_3, (b) CH_2, (c) CH_2, (e) CH_2, N, C(=O)CH$_3$(d)

	Shift	Peak
a	0.91	3B
b	1.44 ± 0.1	S
c	1.44 ± 0.1	S
▶d	1.87	1A
e	3.73	3A

▶ $C_{12}H_{17}NO$ ▶ 7% $CDCl_3$ ▶ V 601

▶ 1-Jn

O=C—N, CH_3(c), CH_3 (a), CH_3 (b)

	Shift	Peak
▶a	2.08	1A
b	2.94	1A
c	3.02	1A

▶ C_4H_9NO ▶ 7% $CDCl_3$ ▶ V 421

▶ 1-Jn

	Shift	Peak
a	1.30	3A
▶b	2.08	1A
c	4.28	4A
d	5.19	2A
e	6.60	1D

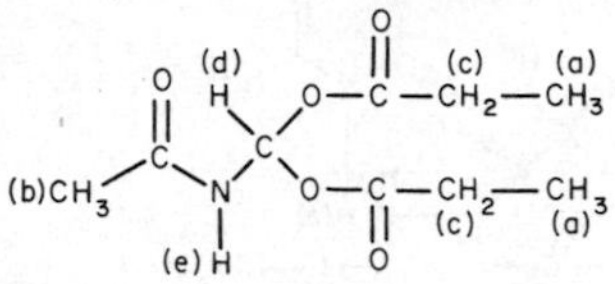

▶ $C_9H_{15}NO_5$ ▶ 7% $CDCl_3$ ▶ V 547

▶ 1-Js

	Shift	Peak
▶a	2.40	1A
b	4.73	1A

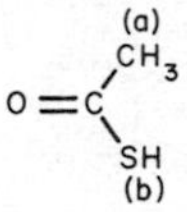

▶ C_2H_4OS ▶ 7% $CDCl_3$ ▶ V 7

▶ 1-Jn

	Shift	Peak
a	1.37	3A
▶b	2.12	1A
c	4.36	4A
d	5.50	2A
e	6.75	1D

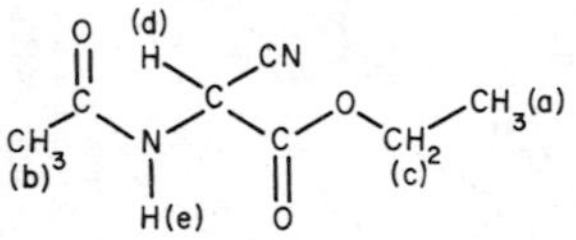

▶ $C_7H_{10}N_2O_3$ ▶ 7% $CDCl_3$ ▶ V 491

▶ 1-Jv

	Shift	Peak
▶a	2.47	1A
b	4.43	1C

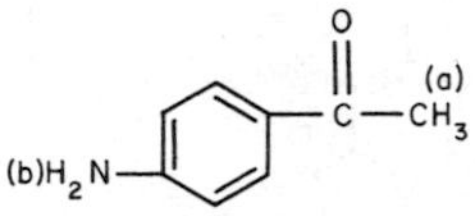

▶ C_8H_9NO ▶ 44.9 mg/0.25 ml $CDCl_3$ ▶ S 242

▶ 1-Jn

	Shift	Peak
▶a	2.12	1A
b	2.30	1A
c	7.12	Q
d	7.37	Q
e	7.88	1D

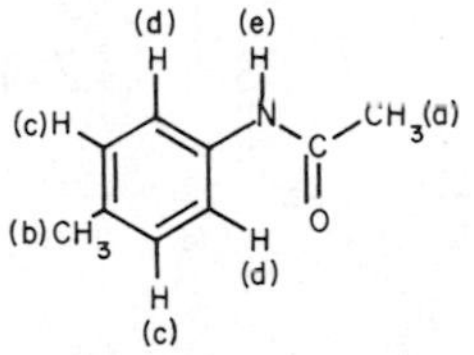

▶ $C_9H_{11}NO$ ▶ 7% $CDCl_3$ ▶ V 239

▶ 1-Jv

	Shift	Peak
a	2.47 or 2.58	1A
▶b	2.47 or 2.58	1A

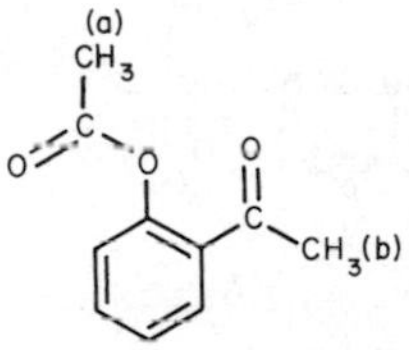

▶ $C_{10}H_{10}O_3$ ▶ 7% $CDCl_3$ ▶ V 254

▶ 1-Jn

	Shift	Peak
a	1.38	3A
▶b	2.12	1A
c	4.00	4A
d	6.83	Q
e	7.41	Q
f	7.91	1C

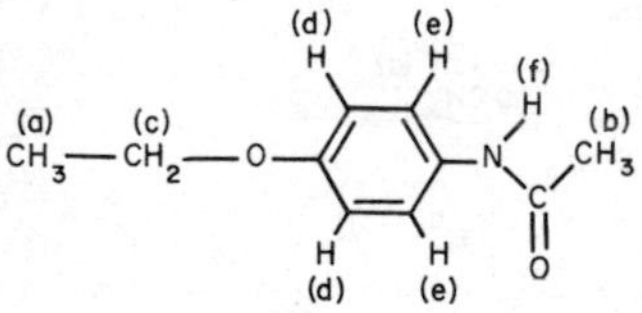

▶ $C_{10}H_{13}NO_2$ ▶ 7% $CDCl_3$ ▶ V 267

▶ 1-Jv

	Shift	Peak
▶a	2.53	1A
b	3.80	1B

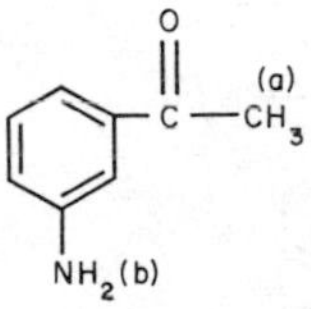

▶ C_8H_9NO ▶ 45.0 mg/0.50 ml $CDCl_3$ ▶ S 119

▶ 1-Jn

	Shift	Peak
▶a	2.16	1A
b	8.14	1F

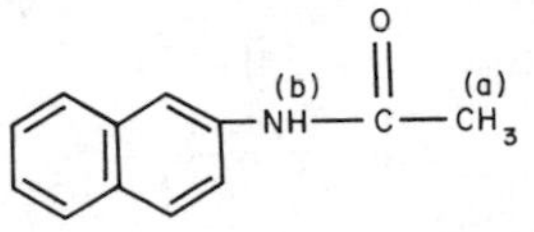

▶ $C_{12}H_{11}NO$ ▶ 44.8 mg/0.50 ml $CDCl_3$ ▶ S 131

▶ 1-Jv

	Shift	Peak
▶a	2.57	1A
b	3.93	1A
c	6.62	1C
d	6.97	2A
e	7.57	S
f	7.60	S

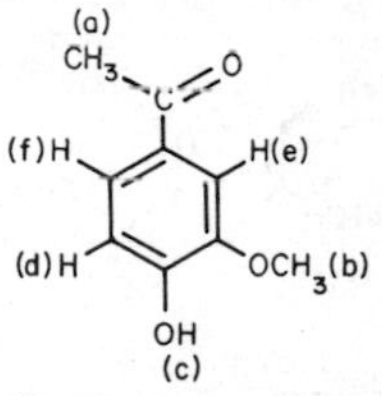

▶ $C_9H_{10}O_3$ ▶ 7% $CDCl_3$ ▶ V 234

▶ 1-Jn

	Shift	Peak
a	2.06	O
▶b	2.50	1A
c	2.62	S
d	3.83	3A

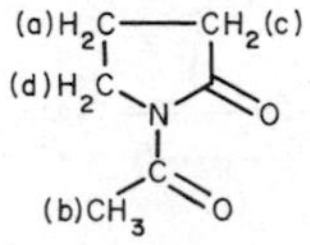

▶ $C_6H_9NO_2$ ▶ 7% $CDCl_3$ ▶ V 465

▶ 1-Jv

	Shift	Peak
a	2.58	1A

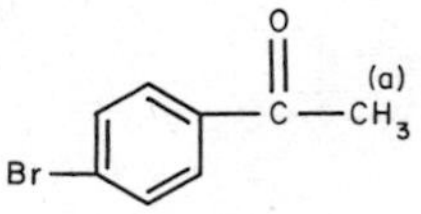

▶ C_8H_7BrO ▶ 44.4 mg/0.25 ml $CDCl_3$ ▶ S 219

▶ 1-Jv

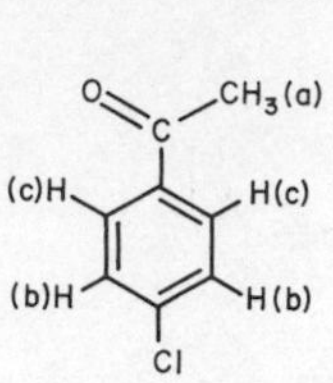

Shift		Peak
▶a	2.58	1A
b	7.45	Q
c	7.90	Q

▶ C_8H_7ClO ▶ 7% $CDCl_3$ ▶ V 188

▶ 1-Kb

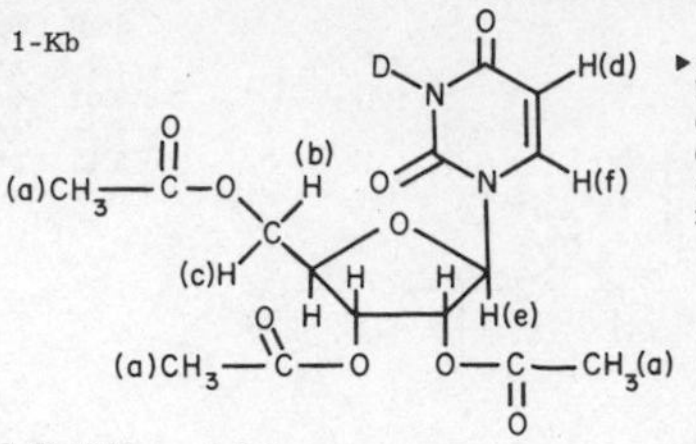

Shift		Peak
▶a	1.92	1A
b	3.80	S
c	3.87	S
d	5.81	S
e	5.83	S
f	7.65	2A

▶ $C_{15}H_{18}N_2O_9$ ▶ D_2O ▶ V 642

▶ 1-Jv

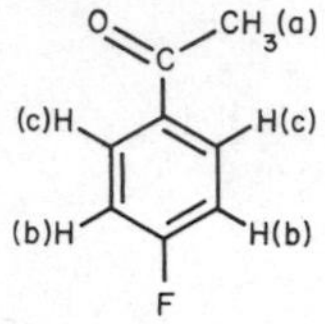

Shift		Peak
▶a	2.58	1A
b	7.12	3C
c	7.97	4C

▶ C_8H_7FO ▶ 7% $CDCl_3$ ▶ V 189

▶ 1-Kb

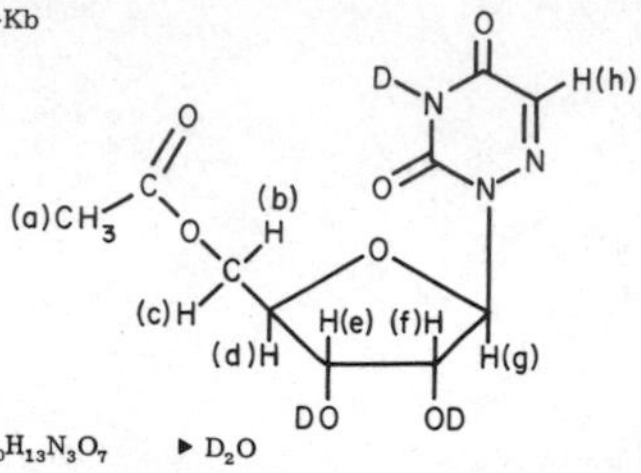

Shift		Peak
▶a	1.94	1A
b	3.64	S
c	3.78	S
d	3.92	S
e	4.10	3A
f	4.47	3A
g	6.00	2A
h	7.45	1A

▶ $C_{10}H_{13}N_3O_7$ ▶ D_2O ▶ V 562

▶ 1-Jv

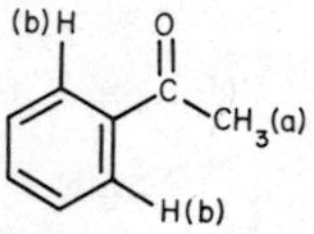

Shift		Peak
▶a	2.59	1A
b	7.91	N

▶ C_8H_8O ▶ 7% $CDCl_3$ ▶ V 192

▶ 1-Kb

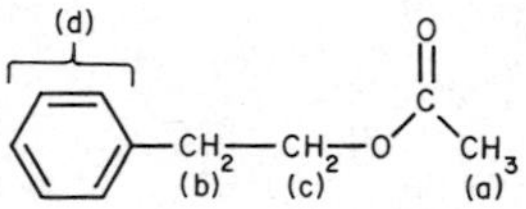

Shift		Peak
▶a	2.02	1A
b	2.93	3A
c	4.30	3A
d	7.29 ± 0.03	1A

▶ $C_{10}H_{12}O_2$ ▶ 7% $CDCl_3$ ▶ V 261

▶ 1-Jv

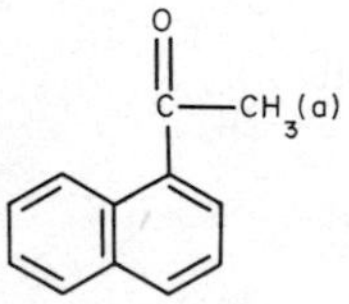

Shift		Peak
a	2.68	1A

▶ $C_{12}H_{10}O$ ▶ 46.0 mg/0.50 ml $CDCl_3$ ▶ S 298

▶ 1-Kb

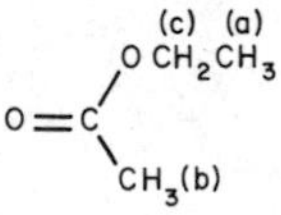

Shift		Peak
a	1.25	3A
▶b	2.03	1A
c	4.12	4A

▶ $C_4H_8O_2$ ▶ 7% $CDCl_3$ ▶ V 79

▶ 1-Jy

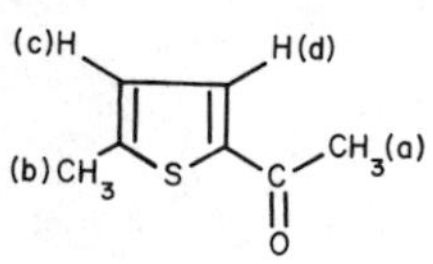

Shift		Peak
▶a	2.48	1E
b	2.52	1E
c	6.78	2E
d	7.50	2A

▶ C_7H_8OS ▶ 7% $CDCl_3$ ▶ V 164

▶ 1-Kb

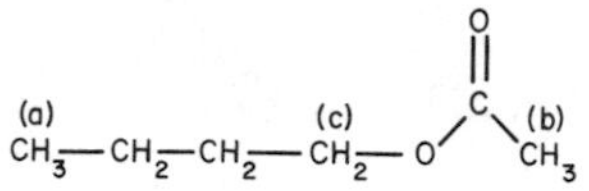

Shift		Peak
a	0.95	3B
▶b	2.03	1A
c	4.05	3A

▶ $C_6H_{12}O_2$ ▶ 7% $CDCl_3$ ▶ V 140

▶ 1-Jy

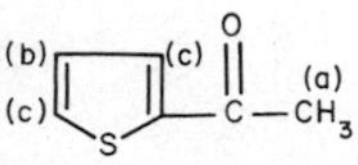

Shift		Peak
▶a	2.55	1A
b	7.10	4A
c	7.66	3C

▶ C_6H_6OS ▶ 44.8 mg/0.50 ml $CDCl_3$ ▶ S 123

▶ 1-Kb

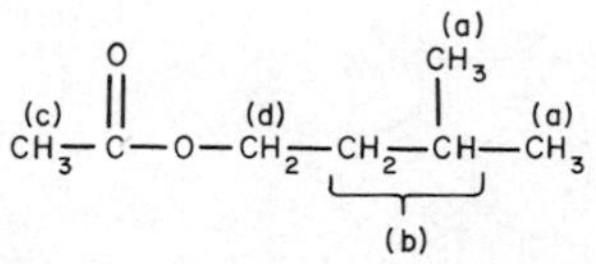

Shift		Peak
a	0.92	2A
b	~1.50	O
▶c	2.03	1A
d	4.03	4B

▶ $C_7H_{14}O_2$ ▶ 44.8 mg/0.50 ml $CDCl_3$ ▶ S 281

▶ 1-Kb

	Shift	Peak
a	0.93	3A
b	1.64	5B
▶c	2.04	1A
d	4.02	3A

(c) (d) (b) (a) $CH_3C(=O)—OCH_2CH_2CH_3$

▶ $C_5H_{10}O_2$ ▶ 44.9 mg/0.50 ml $CDCl_3$ ▶ S 115

▶ 1-Kb

	Shift	Peak
▶a	2.12	1A
b	3.13	1C
c	4.28	Q
d	4.71	Q

(a)CH_3—C(=O)—O—(d)CH_2—C≡C—(c)CH_2—OH(b)

▶ $C_6H_8O_3$ ▶ 7% $CDCl_3$ ▶ V 463

▶ 1-Kb

	Shift	Peak
a	0.92	3B
b	1.37	1E
▶c	2.05	1A
d	4.05	3A

(c) (d) (b) (a) $CH_3C(=O)—OCH_2(CH_2)_4CH_3$

▶ $C_8H_{16}O_2$ ▶ 45.0 mg/0.50 ml $CDCl_3$ ▶ S 217

▶ 1-Kb

	Shift	Peak
a	2.13	1A
▶b	2.13	1A
c	4.28	S
d	4.32	S
e	4.62	S
f	6.14	2A
g	7.67	1A

▶ $C_{14}H_{17}N_3O_9$ ▶ D_2O ▶ V 634

▶ 1-Kb

	Shift	Peak
▶a	2.06	1A
b	5.08	1A
c	7.31	1A

▶ $C_9H_{10}O_2$ ▶ 7% $CDCl_3$ ▶ V 530

▶ 1-Kc

	Shift	Peak
▶a	1.92	1A
b	3.80	S
c	3.87	S
d	5.81	S
e	5.83	S
f	7.65	2A

▶ $C_{15}H_{18}N_2O_9$ ▶ D_2O ▶ V 642

▶ 1-Kb

	Shift	Peak
▶a	2.07	1A
b	5.03	1A
c	6.35	1E
d	6.38	1E
e	7.40	N

▶ $C_7H_8O_3$ ▶ 7% $CDCl_3$ ▶ V 167

▶ 1-Kc

	Shift	Peak
▶a	1.92	1A
b	3.47	1A
c	5.07	2A
d	5.58	1E
e	~5.63	4E

▶ $C_{22}H_{24}N_2O_6$ ▶ 7% $CDCl_3$ ▶ V 356

▶ 1-Kb

	Shift	Peak
▶a	2.08	1A
b	~4.17	S
c	~4.28	S
d	5.25	5B

▶ $C_9H_{14}O_6$ ▶ 7% $CDCl_3$ ▶ V 242

▶ 1-Kc

	Shift	Peak
a	1.22	1E
b	1.29	2H
c	1.62	S
▶d	1.97	1A
e	2.06	S
f	2.60	3E
g	~3.10	S
h	~3.10	S
i	4.67	3C
j	5.63	2A
k	5.93	2A
l	6.14	2H
m	6.23	N
n	7.79	2F

▶ $C_{17}H_{20}O_5$ ▶ 7% $CDCl_3$ ▶ V 666

▶ 1-Kb

	Shift	Peak
▶a	2.10*	1A
b	4.42	1B
c	5.62	S
d	5.88	3A
e	6.17	2A
f	8.03†	1A
g	8.03†	1A
h	~13.1	S

*or 2.15
†or 8.35

▶ $C_{16}H_{18}N_4O_8$ ▶ 7% $CDCl_3$ ▶ V 324

▶ 1-Kc

	Shift	Peak
a	1.23	2A
▶b	2.01	1A
c	4.98	5A

(b) CH_3—C(=O)—O—(c)$CH(CH_3)_2$ (a)

▶ $C_5H_{10}O_2$ ▶ 45.0 mg/0.50 ml $CDCl_3$ ▶ S 99

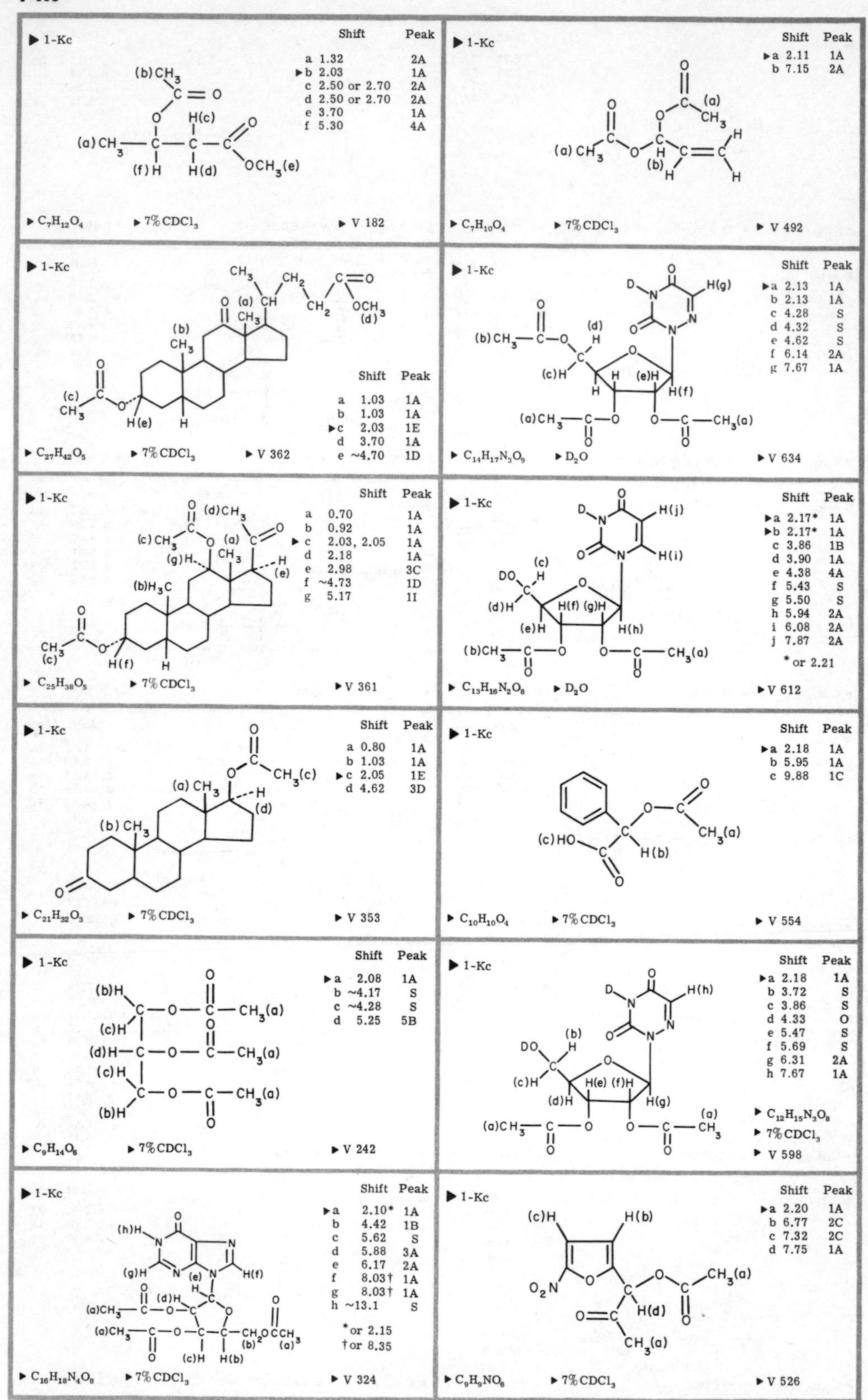

1-Kc
Shift Peak
a 1.32 2A
b 2.03 1A
c 2.50 or 2.70 2A
d 2.50 or 2.70 2A
e 3.70 1A
f 5.30 4A
C7H12O4 7% CDCl3 V 182
1-Kc
Shift Peak
a 2.11 1A
b 7.15 2A
C7H10O4 7% CDCl3 V 492
1-Kc
Shift Peak
a 1.03 1A
b 1.03 1A
c 2.03 1E
d 3.70 1A
e ~4.70 1D
C27H42O5 7% CDCl3 V 362
1-Kc
Shift Peak
a 2.13 1A
b 2.13 1A
c 4.28 S
d 4.32 S
e 4.62 S
f 6.14 2A
g 7.67 1A
C14H17N3O9 D2O V 634
1-Kc
Shift Peak
a 0.70 1A
b 0.92 1A
c 2.03, 2.05 1A
d 2.18 1A
e 2.98 3C
f ~4.73 1D
g 5.17 1I
C25H38O5 7% CDCl3 V 361
1-Kc
Shift Peak
a 2.17* 1A
b 2.17* 1A
c 3.86 1B
d 3.90 1A
e 4.38 4A
f 5.43 S
g 5.50 S
h 5.94 2A
i 6.08 2A
j 7.87 2A
* or 2.21
C13H16N2O8 D2O V 612
1-Kc
Shift Peak
a 0.80 1A
b 1.03 1A
c 2.05 1E
d 4.62 3D
C21H32O3 7% CDCl3 V 353
1-Kc
Shift Peak
a 2.18 1A
b 5.95 1A
c 9.88 1C
C10H10O4 7% CDCl3 V 554
1-Kc
Shift Peak
a 2.08 1A
b ~4.17 S
c ~4.28 S
d 5.25 5B
C9H14O6 7% CDCl3 V 242
1-Kc
Shift Peak
a 2.18 1A
b 3.72 S
c 3.86 S
d 4.33 O
e 5.47 S
f 5.69 S
g 6.31 2A
h 7.67 1A
C12H15N3O8 7% CDCl3 V 598
1-Kc
Shift Peak
a 2.10* 1A
b 4.42 1B
c 5.62 S
d 5.88 3A
e 6.17 2A
f 8.03† 1A
g 8.03† 1A
h ~13.1 S
*or 2.15
†or 8.35
C16H18N4O8 7% CDCl3 V 324
1-Kc
Shift Peak
a 2.20 1A
b 6.77 2C
c 7.32 2C
d 7.75 1A
C9H9NO6 7% CDCl3 V 526

▶ 1-Kc

OCH$_3$(c) O C H(d) O O Cl Cl C CH$_3$ (b) Cl H(e) O C O (a) CH$_3$

Shift		Peak
▶a	2.20	1A
b	2.25	1A
c	3.87	1A
d	6.05	1A
e	6.63	1A

▶ $C_{11}H_{11}Cl_3O_6$ ▶ 7% $CDCl_3$ ▶ V 284

▶ 1-Kl

O (c)H O C C=C CH$_3$(b) (d)H CH$_3$(a)

Shift		Peak
a	1.93	1A
▶b	2.12	1A
c	4.69	1G
d	4.69	1G

▶ $C_5H_8O_2$ ▶ 7% $CDCl_3$ ▶ V 440

▶ 1-Kc

O O O CH$_3$(a) C CH$_2$ (c) (b)CH$_3$ C O C H(d) C O CH$_2$ O (c) CH$_3$(a)

Shift		Peak
a	1.31	3A
▶b	2.22	1A
c	4.30	4A
d	5.52	1A

▶ $C_9H_{14}O_6$ ▶ 7% $CDCl_3$ ▶ V 546

▶ 1-Kl

OCH$_3$(c) O C H(d) O O Cl Cl C CH$_3$ (b) Cl H(e) O C = O (a) CH$_3$

Shift		Peak
a	2.20	1A
▶b	2.25	1A
c	3.87	1A
d	6.05	1A
e	6.63	1A

▶ $C_{11}H_{11}Cl_3O_6$ ▶ 7% $CDCl_3$ ▶ V 284

▶ 1 Kd

(a) CH$_3$ O (a) (b) CH$_3$—C—O—C—CH$_3$ CH$_3$ (a)

Shift		Peak
a	1.45	1A
▶b	1.97	1A

▶ $C_6H_{12}O_2$ ▶ 7% $CDCl_3$ ▶ V 141

▶ 1-Kv

O (d) H (g) H (j) H O (f) H (h) H O O (c) CH$_3$ C O H (i) O H (j) O C (b) CH$_3$ (e) H H(h) CH$_3$ C O (a)

Shift		Peak
▶a	2.06	1A
▶b	2.32	1A
▶c	2.32	1A
d	6.36	1G
e	6.83*	S
f	6.83*	S
g	6.89	S
h	7.15	S
i	7.33	S
j	7.53	2E

*or 7.08

▶ $C_{23}H_{18}O_8$ ▶ 7% $CDCl_3$ ▶ V 692

▶ 1-Kd

H(e) (d) (c)H H O C—OCH$_3$(b) (f)H N N CH$_3$(a)

Shift		Peak
a	2.25	1A
▶b	3.69	1A
c	7.14	S
d	7.34	S
e	7.64	S
f	8.54	2E

▶ $C_{13}H_{18}N_2O_2$ ▶ 7% $CDCl_3$ ▶ V 301

▶ 1-Kv

O (a) O—C—CH$_3$ (b) (b) (b) (b) O—C—CH$_3$ (a) O

Shift		Peak
▶a	2.25	1A
b	7.06	1A

▶ $C_{10}H_{10}O_4$ ▶ 44.8 mg/0.50 ml $CDCl_3$ ▶ S 252

▶ 1-Kh

(a) CH$_3$ O=C OH (b)

Shift		Peak
▶a	2.10	1A
b	11.37	1A

▶ $C_2H_4O_2$ ▶ 7% $CDCl_3$ ▶ V 8

▶ 1-Kv

(g) (f)H H (i) (c) C=C (d) H CH$_3$ (e)H CH$_2$ H(h) (b) CH$_3$—CO O O O CH$_3$ (a)

Shift		Peak
a	2.25 or 2.37	1A
▶b	2.25 or 2.37	1A
c	2.40	2D
d	3.33	2B
e	5.07	R
f	5.12	R
g	5.85	4D
h	6.22	1I
i	7.30	1I

▶ $C_{16}H_{16}O_4$ ▶ 7% $CDCl_3$ ▶ V 323

▶ 1-Kl

(b)H H(d) C=C (a) (c) H O—C—CH$_3$ O

Shift		Peak
▶a	2.12	1A
b	4.55	R
c	4.85	R
d	7.25	4D

▶ $C_4H_6O_2$ ▶ 7% $CDCl_3$ ▶ V 65

▶ 1-Kv

O (a)CH$_3$ (d) (c)CH$_3$ H O C CH$_3$ O (b) H(d)

Shift		Peak
a	0.95	1A
▶b	2.28	1A
c	2.33	1A
d	6.72	1B

▶ $C_{21}H_{26}O_3$ ▶ 7% $CDCl_3$ ▶ V 350

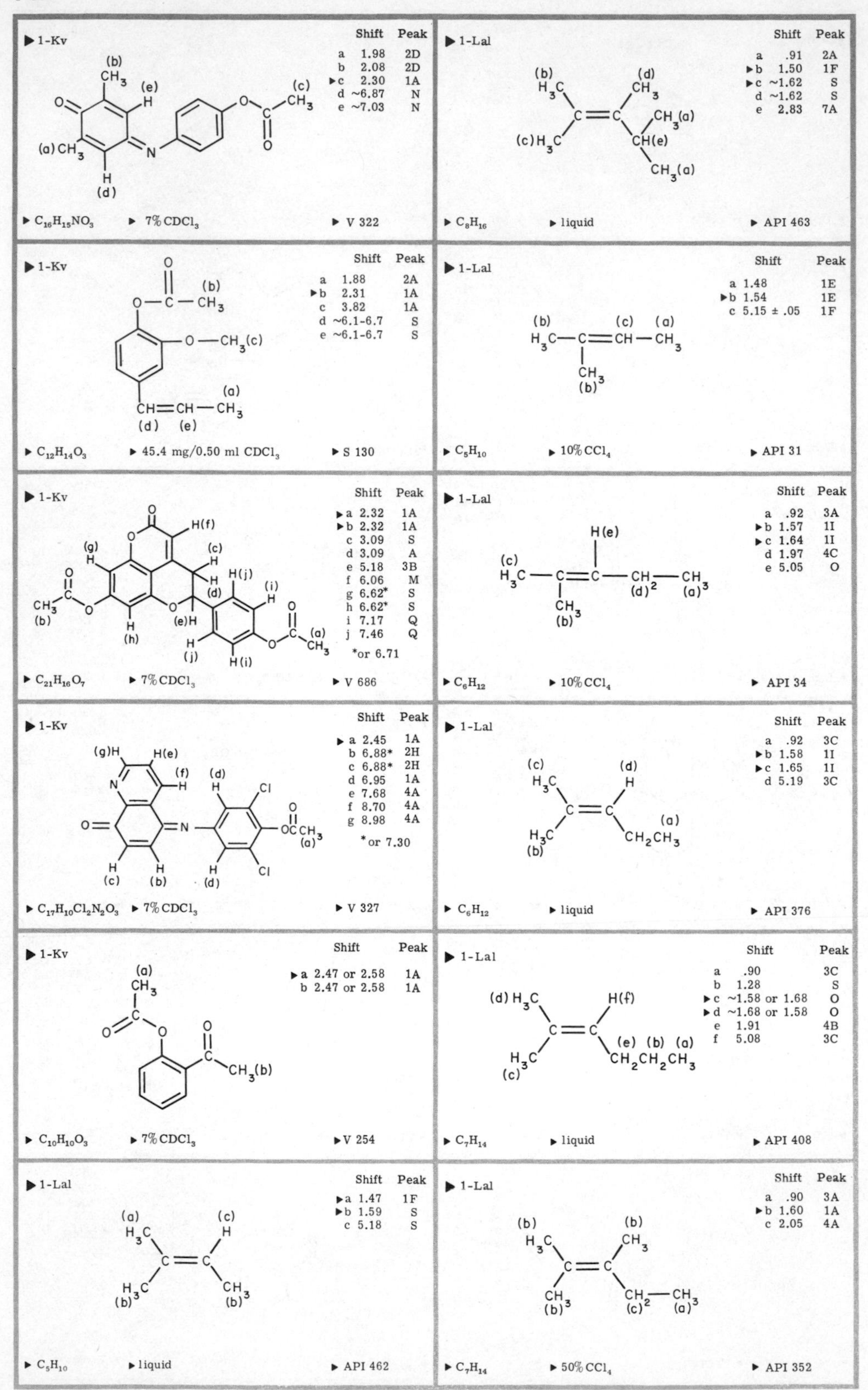

1-Kv
Shift Peak
a 1.98 2D
b 2.08 2D
c 2.30 1A
d ~6.87 N
e ~7.03 N
C16H15NO3 7% CDCl3 V 322
1-Lal
Shift Peak
a .91 2A
b 1.50 1F
c ~1.62 S
d ~1.62 S
e 2.83 7A
C8H16 liquid API 463
1-Kv
Shift Peak
a 1.88 2A
b 2.31 1A
c 3.82 1A
d ~6.1-6.7 S
e ~6.1-6.7 S
C12H14O3 45.4 mg/0.50 ml CDCl3 S 130
1-Lal
Shift Peak
a 1.48 1E
b 1.54 1E
c 5.15 ± .05 1F
C5H10 10% CCl4 API 31
1-Kv
Shift Peak
a 2.32 1A
b 2.32 1A
c 3.09 S
d 3.09 A
e 5.18 3B
f 6.06 M
g 6.62* S
h 6.62* S
i 7.17 Q
j 7.46 Q
*or 6.71
C21H16O7 7% CDCl3 V 686
1-Lal
Shift Peak
a .92 3A
b 1.57 1I
c 1.64 1I
d 1.97 4C
e 5.05 O
C6H12 10% CCl4 API 34
1-Kv
Shift Peak
a 2.45 1A
b 6.88* 2H
c 6.88* 2H
d 6.95 1A
e 7.68 4A
f 8.70 4A
g 8.98 4A
*or 7.30
C17H10Cl2N2O3 7% CDCl3 V 327
1-Lal
Shift Peak
a .92 3C
b 1.58 1I
c 1.65 1I
d 5.19 3C
C6H12 liquid API 376
1-Kv
Shift Peak
a 2.47 or 2.58 1A
b 2.47 or 2.58 1A
C10H10O3 7% CDCl3 V 254
1-Lal
Shift Peak
a .90 3C
b 1.28 S
c ~1.58 or 1.68 O
d ~1.68 or 1.58 O
e 1.91 4B
f 5.08 3C
C7H14 liquid API 408
1-Lal
Shift Peak
a 1.47 1F
b 1.59 S
c 5.18 S
C5H10 liquid API 462
1-Lal
Shift Peak
a .90 3A
b 1.60 1A
c 2.05 4A
C7H14 50% CCl4 API 352

▶ 1-Lal

	Shift	Peak
▶ a	1.60 or 1.68	1G
b	2.04 ± 0.05	1G
c	4.15	2G
d	5.12	1F

(a) CH_3(a)CH_3C=CH(d)–CH_2(b)–(CH_2(b)–C(CH_3(a))=CH(d)–CH_2(b))$_7$–CH_2(b)–C(CH_3(a))=CH–CH_2(c)–OH

▶ $C_{45}H_{74}O$ ▶ 7% $CDCl_3$ ▶ V 367

▶ 1-Lal

	Shift	Peak
▶ a	1.63	3A
b	4.43	7A

(b)H(b)HC=C=C(CH_3(a))CH_3(a)

▶ C_5H_8 ▶ liquid ▶ API 472

▶ 1-Lal

	Shift	Peak
a	1.61	1A

$(H_3C)_2C=C(CH_3)_2$ (a)

▶ C_6H_{12} ▶ 50% CCl_4 ▶ API 388

▶ 1-Lal

	Shift	Peak
a	1.47	1A
▶ b	1.64 or 1.68	1A
▶ c	1.64 or 1.68	1A

(b)CH_3, (c)CH_3, O, C=O, CH_3(a), CH_3(a)

▶ $C_8H_{12}O_2$ ▶ 7% $CDCl_3$ ▶ V 513

▶ 1-Lal

	Shift	Peak
a	.93	2A
▶ b	1.61	N
c	2.42	6B
d	4.93	2E

(b)H_3C, H_3C C=C H(d), CH(c)CH_3, CH_3(a)

▶ C_7H_{14} ▶ liquid ▶ API 402

▶ 1-Lal

	Shift	Peak
a	.90	3C
b	1.28	S
▶ c	~1.58 or 1.68	O
▶ d	~1.68 or 1.58	O
e	1.91	4B
f	5.08	3C

(d)H_3C, H_3C(c) C=C H(f), CH_2(e)CH_2(b)CH_3(a)

▶ C_7H_{14} ▶ liquid ▶ API 408

▶ 1-Lal

	Shift	Peak
a	.94	3C
▶ b	1.62	1B
c	2.03	4C

(b)H_3C, (b)H_3C C=C CH_3(b), CH_2(c)CH_3(a)

▶ C_7H_{14} ▶ liquid ▶ API 415

▶ 1-Lal

	Shift	Peak
▶ a	1.74	1A
▶ b	1.79	1A
c	5.98	1B

(a)CH_3, (b)CH_3 C=C H(c), C=C(H(c)) CH_3(b), CH_3(a)

▶ C_8H_{14} ▶ 7% $CDCl_3$ ▶ V 515

▶ 1-Lal

	Shift	Peak
a	.91	2A
▶ b	1.50	1F
▶ c	~1.62	S
d	~1.62	S
e	2.83	7A

(b)H_3C, (c)H_3C C=C CH_3(d), CH(e)(CH_3(a))CH_3(a)

▶ C_8H_{16} ▶ liquid ▶ API 463

▶ 1-Lal

	Shift	Peak
▶ a	~1.77	1A
▶ b	~1.77	1A
c	5.77	N

(a)CH_3, (b)CH_3 C=C Cl, H(c)

▶ C_4H_7Cl ▶ 7% $CDCl_3$ ▶ V 67

▶ 1-Lal

	Shift	Peak
▶ a	1.62 or 1.68	1A
b	2.05 ± 0.03	3A
c	4.15	2A
d	5.12	1C
e	5.45	3A

(a)CH_3(a)CH_3C=CH(d)–CH_2(b)–CH_2(b)–C(CH_3(a))=CH(e)–CH_2(c)–OH

▶ $C_{10}H_{18}O$ ▶ 7% $CDCl_3$ ▶ V 279

▶ 1-Lal

	Shift	Peak
▶ a	1.93	2D
▶ b	2.18	2D
c	5.72	M
d	11.95	1B

(a)CH_3, (b)CH_3 C=C H(c), C=O, OH(d)

▶ $C_5H_8O_2$ ▶ 7% $CDCl_3$ ▶ V 114

▶1-Lam

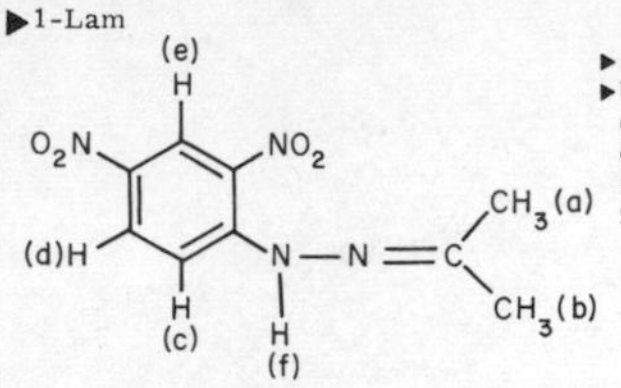

	Shift	Peak
▶a	2.10 or 2.18	1A
▶b	2.10 or 2.18	1A
c	7.90	2A
d	8.23	4A
e	9.04	2A
f	10.95	1B

▶ $C_9H_{10}N_4O_4$ ▶ 7% $CDCl_3$ ▶ V 233

▶1-Lbl

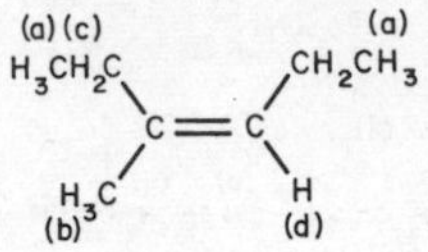

	Shift	Peak
a	.91 or .98	3D
▶b	1.57	1I
c	1.94	5B
d	5.11	3C

▶ C_7H_{14} ▶ liquid ▶ API 411

▶1-Lam

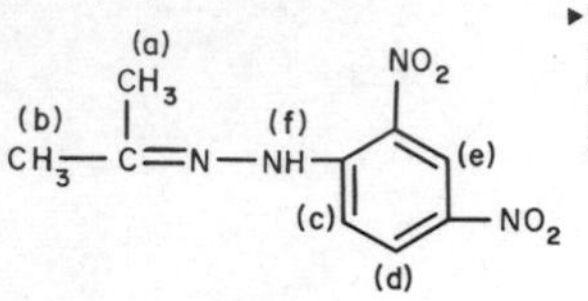

	Shift	Peak
▶a	2.10 or 2.20	1A
▶b	2.10 or 2.20	1A
c	7.85	2A
d	8.18	4A
e	8.99	2A
f	10.94	1C

▶ $C_9H_{10}N_4O_4$ ▶ 45.0 mg/0.25 ml $CDCl_3$ ▶ S 199

▶1-(L)bl

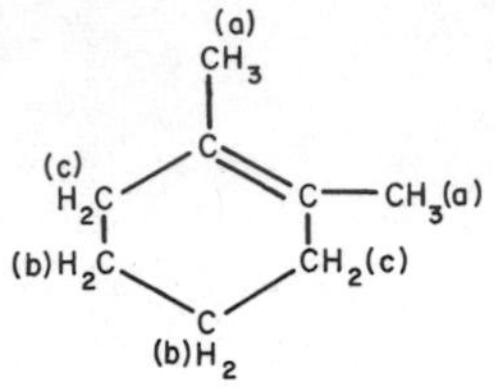

	Shift	Peak
▶a	1.58	1A
b	~1.87	1I
c	~1.94	S

▶ C_8H_{14} ▶ 33% CCl_4 ▶ API 143

▶1-Lbl

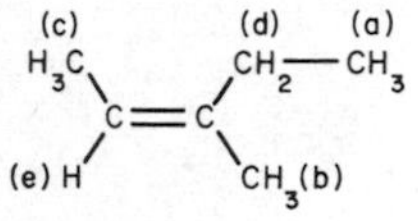

	Shift	Peak
a	.97	3A
▶b	1.47	1E
c	1.57	1E
d	1.97	4A
e	5.13	2E

▶ C_6H_{12} ▶ 10% CCl_4 ▶ API 35

▶1-Lbl

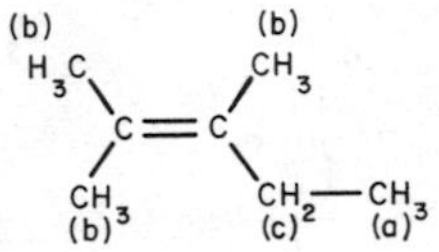

	Shift	Peak
a	.90	3A
▶b	1.60	1A
c	2.05	4A

▶ C_7H_{14} ▶ 50% CCl_4 ▶ API 352

▶1-Lbl

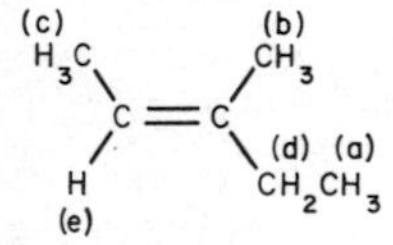

	Shift	Peak
a	.97	3C
▶b	1.49	S
c	1.56	S
d	1.95	4C
e	5.18	4C

▶ C_6H_{12} ▶ liquid ▶ API 378

▶1-Lbl

	Shift	Peak
▶a	1.60 or 1.68	1G
b	2.04 ± 0.05	1G
c	4.15	2G
d	5.12	1F

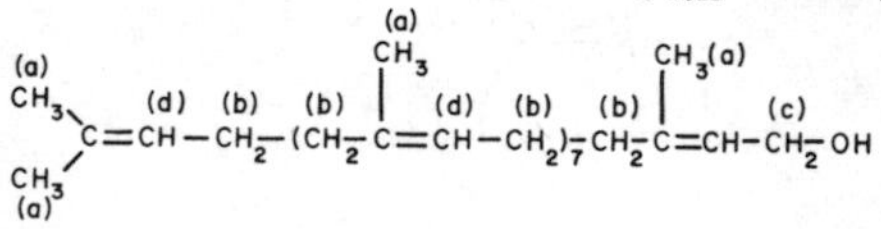

▶ $C_{45}H_{74}O$ ▶ 7% $CDCl_3$ ▶ V 367

▶1-(L)bl

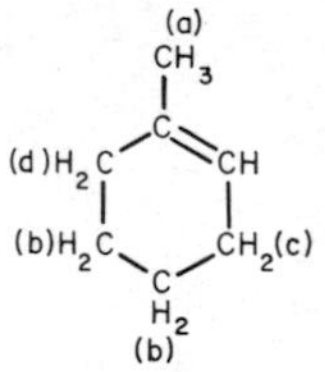

	Shift	Peak
▶a	~1.53	S
b	~1.53	S
c	~1.90	1F
d	5.34	1F

▶ C_7H_{12} ▶ 33 vol.% CCl_4 ▶ API 141

▶1-Lbl

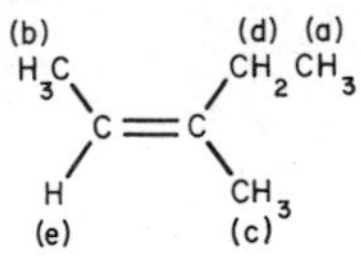

	Shift	Peak
a	.95	3C
b	1.46	1F
▶c	1.61	S
d	2.03	4A
e	~5.08	4C

▶ C_6H_{12} ▶ liquid ▶ API 377

▶1-Lbl

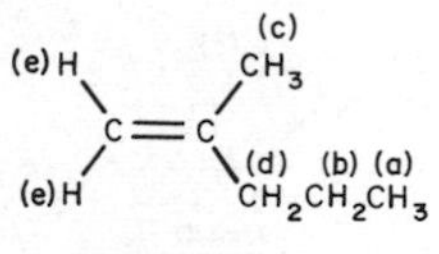

	Shift	Peak
a	.88	3C
b	1.34	S
▶c	1.57	1H
d	1.90	S
e	4.60	N

▶ C_6H_{12} ▶ 50 vol.% CCl_4 ▶ API 373

▶1-Lbl

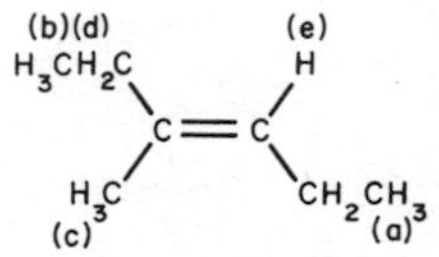

	Shift	Peak
a	.91 or .96	3D
b	.91 or .96	3D
▶c	1.61	1H
d	1.97	4A
e	5.07	3A

▶ C_7H_{14} ▶ liquid ▶ API 412

▶ 1-Lbl

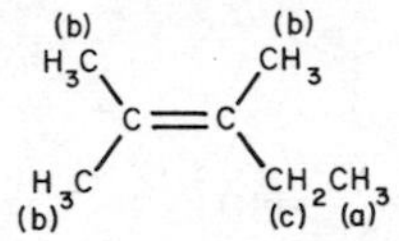

Shift		Peak
a	.94	3C
▶b	1.62	1B
c	2.03	4C

▶ C_7H_{14} ▶ liquid ▶ API 415

▶ 1-Lbl

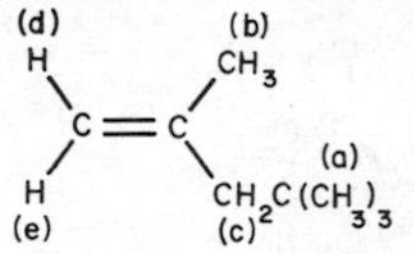

Shift		Peak
a	.92	1A
▶b	1.76	1A
c	1.94	1A
d	~4.70	1C
e	~4.80	1C

▶ C_8H_{16} ▶ 50 vol.% CCl_4 ▶ API 433

▶ 1-(L)bl

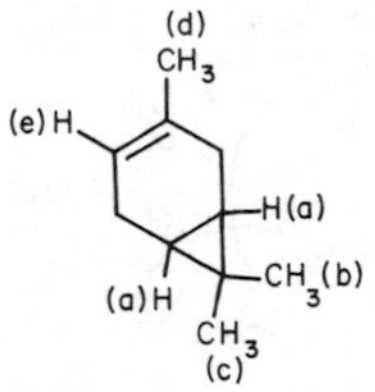

Shift		Peak
a	~0.70	S
b	0.77	1A
c	1.03	1A
▶d	1.62	1I
e	5.23	1C

▶ $C_{10}H_{16}$ ▶ 7% $CDCl_3$ ▶ V 273

▶ 1-(L)bl

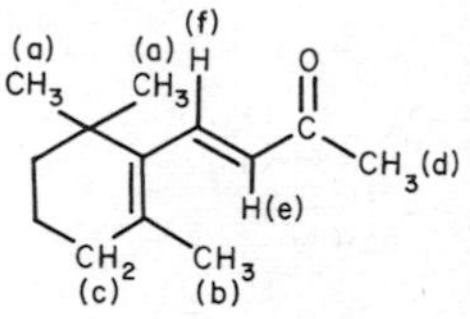

Shift		Peak
a	1.07	1A
▶b	1.78	1A
c	2.09	1C
d	2.31	1A
e	6.11	2A
f	7.28	2B

▶ $C_{13}H_{20}O$ ▶ 7% $CDCl_3$ ▶ V 617

▶ 1-Lbl

Shift		Peak
▶a	1.62 or 1.68	1A
b	2.05 ± 0.03	3A
c	4.15	2A
d	5.12	1C
e	5.45	3A

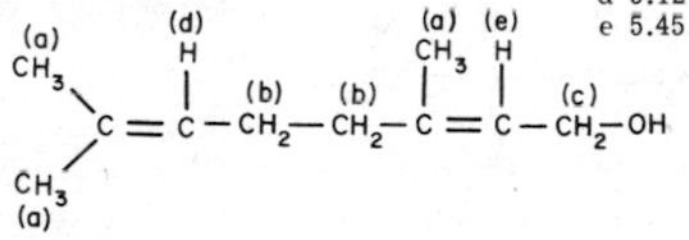

▶ $C_{10}H_{18}O$ ▶ 7% $CDCl_3$ ▶ V 279

▶ 1-(L)bl

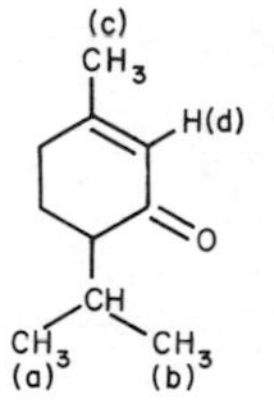

Shift		Peak
a	0.85 or 0.95	2A
b	0.85 or 0.95	2A
▶c	1.93	1B
d	5.87	N

▶ $C_{10}H_{16}O$ ▶ 7% $CDCl_3$ ▶ V 275

▶ 1-Lbl

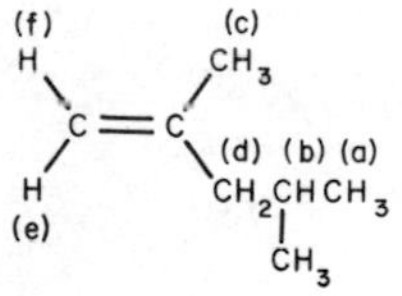

Shift		Peak
a	.87	2H
b	1.62	S
▶c	1.67	3C
d	1.82	2B
e	4.67 ± .03	N
f	4.67 ± .03	N

▶ C_7H_{14} ▶ liquid ▶ API 400

▶ 1-(L)bl

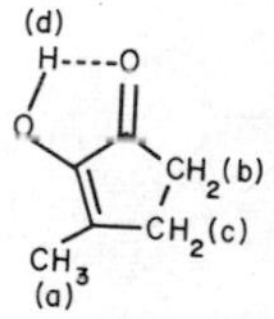

Shift		Peak
▶a	2.02	1A
b	2.43	1A
c	2.43	1A
d	6.77	1A

▶ $C_6H_8O_2$ ▶ 7% $CDCl_3$ ▶ V 129

▶ 1-Lbl

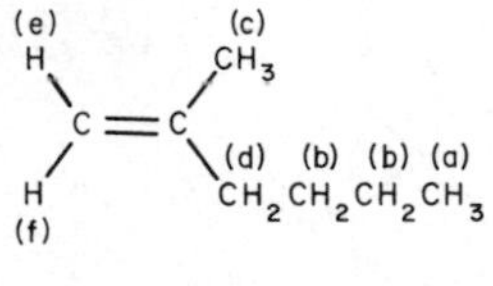

Shift		Peak
a	.93	3C
b	1.38	O
▶c	1.68	1H
d	2.01	3B
e	~4.07	M
f	~4.67	M

▶ C_7H_{14} ▶ liquid ▶ API 407

▶ 1-(L)bl

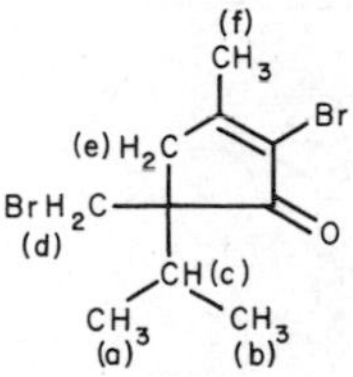

Shift		Peak
a	0.84	2A
b	0.92	2A
c	2.08	S
d	2.22	1H
e	2.61	1H
▶f	3.58	1A

▶ $C_{10}H_{14}Br_2O$ ▶ 7% $CDCl_3$ ▶ V 564

▶ 1-Lbl

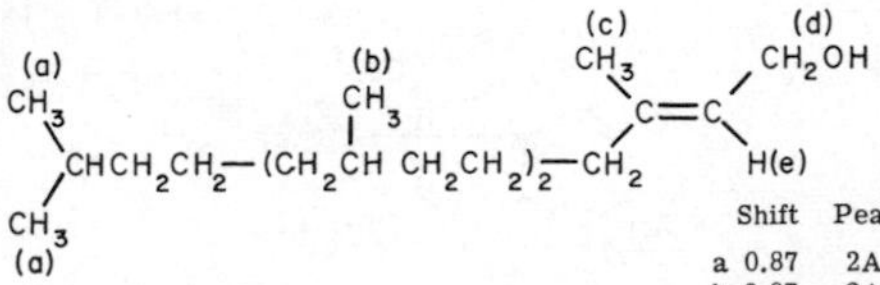

Shift		Peak
a	0.87	2A
b	0.87	2A
▶c	1.68	1A
d	4.13	2A
e	5.42	3C

▶ $C_{20}H_{40}O$ ▶ 7% $CDCl_3$ ▶ V 346

▶ 1-Lcl

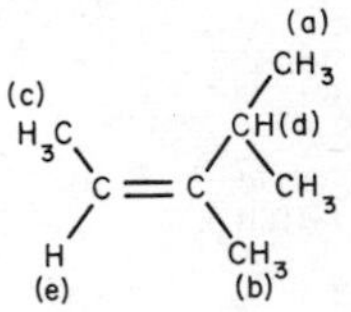

Shift		Peak
a	.93	2A
▶b	1.47	1F
c	1.55	1F
d	2.84	5A
e	5.10	1F

▶ C_7H_{14} ▶ liquid ▶ API 403

▶ 1-Lcl

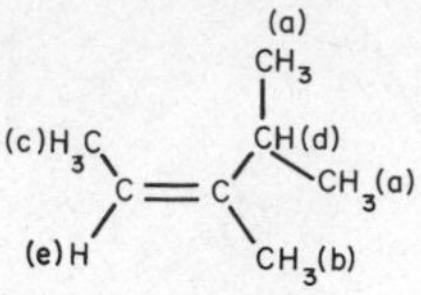

	Shift	Peak
a	.91	2A
▶b	~1.48 or 1.52	1E
c	~1.52 or 1.48	1E
d	2.74	5A
e	5.05	1D

▶ C_7H_{14} ▶ 50% CCl_4 ▶ API 353

▶ 1-Lcl

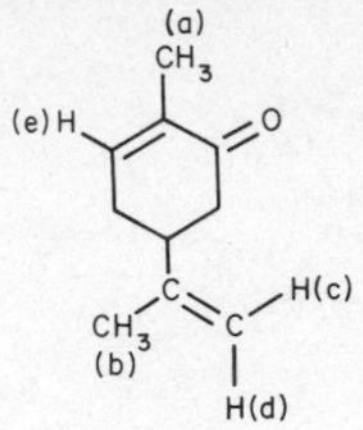

	Shift	Peak
a	1.75 ± .03	1E
▶b	1.75 ± .03	1E
c	4.78 ± .03	N
d	4.78 ± .03	N
e	6.75	1C

▶ $C_{10}H_{14}O$ ▶ 7% $CDCl_3$ ▶ V 271

▶ 1-Lcl

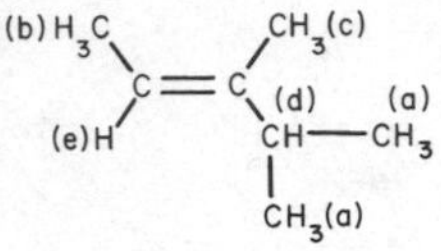

	Shift	Peak
a	.93	2A
b	1.48	1E
▶c	1.52	1E
d	2.20	5A
e	5.20	4B

▶ C_7H_{14} ▶ 50% CCl_4 ▶ API 354

▶ 1-(L)cl

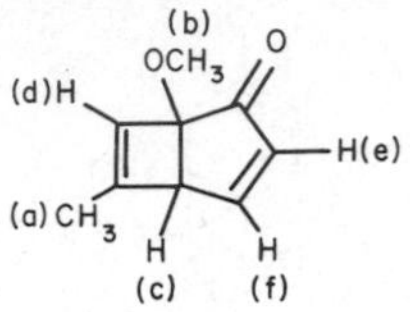

	Shift	Peak
▶a	2.13	2D
b	3.46	1A
c	3.69	1B
d	5.88	2A
e	6.43	2A
f	6.84	2G

▶ $C_9H_{10}O_2$ ▶ 7% $CDCl_3$ ▶ V 531

▶ 1-(L)cl

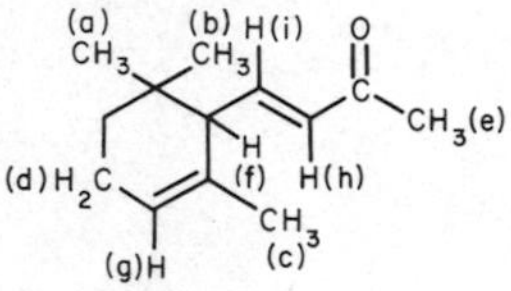

	Shift	Peak
a	0.86	1A
b	0.93	1A
▶c	1.57	2G
d	2.05	1C
e	2.26	1A
f	2.30	S
g	5.50	1C
h	6.06	2A
i	6.60	4A

▶ $C_{13}H_{20}O$ ▶ 7% $CDCl_3$ ▶ V 616

▶ 1-Ldl

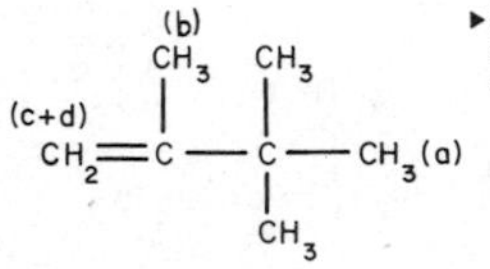

	Shift	Peak
a	1.03	1A
▶b	1.72	1A
c	4.60 or 4.67	1B
d	4.67 or 4.60	1B

▶ C_7H_{14} ▶ 50% CCl_4 ▶ API 223

▶ 1-Lcl

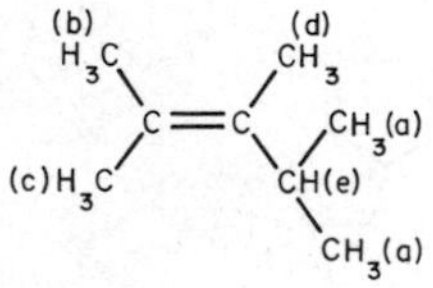

	Shift	Peak
a	.91	2A
b	1.50	1F
c	~1.62	S
▶d	~1.62	S
e	2.83	7A

▶ C_8H_{16} ▶ liquid ▶ API 463

▶ 1-Ldl

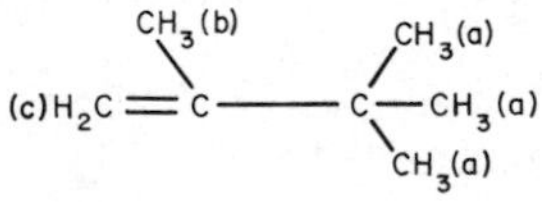

	Shift	Peak
a	1.05	1A
▶b	1.73	3A
c	~4.66	O

▶ C_7H_{14} ▶ 10% CCl_4 ▶ API 158

▶ 1-(L)cl

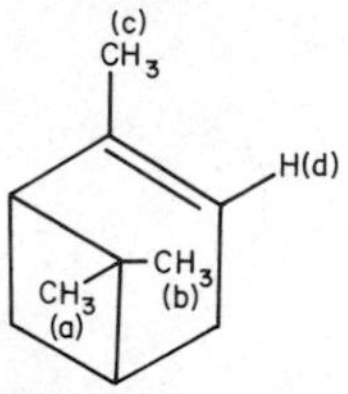

	Shift	Peak
a	0.84 or 1.27	1A
b	0.84 or 1.27	1A
▶c	1.65	4A
d	5.17	1C

▶ $C_{10}H_{16}$ ▶ 7% $CDCl_3$ ▶ V 272

▶ 1-Ldl

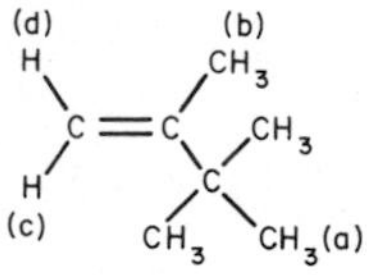

	Shift	Peak
a	1.08	1B
▶b	1.73	1H
c	4.62 or 4.72	N
d	4.72 or 4.62	N

▶ C_7H_{14} ▶ liquid ▶ API 417

▶ 1-Lcl

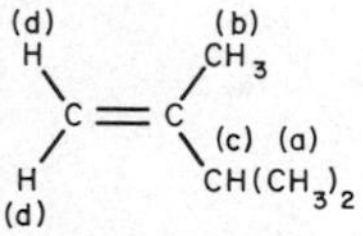

	Shift	Peak
a	.99	2A
▶b	1.67	1H
c	2.23	5B
d	4.62	N

▶ C_6H_{12} ▶ liquid ▶ API 382

▶ 1-Ldm

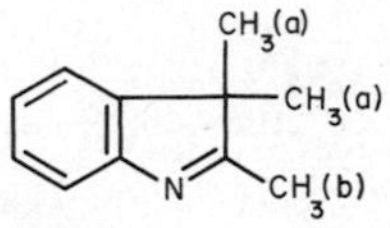

	Shift	Peak
a	1.28	1A
▶b	2.45	1A

▶ $C_{11}H_{13}N$ ▶ 7% $CDCl_3$ ▶ V 286

▶ 1-(L)dm

	Shift	Peak
a	1.56	1A
▶b	2.87	1B
c	4.10	1A
d	7.51 ± 0.05	1G

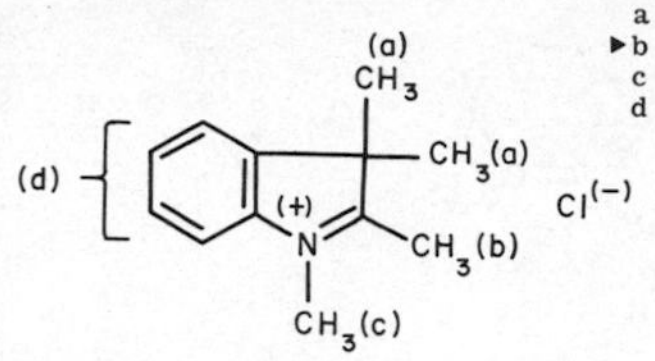

▶ $C_{12}H_{16}ClN$ ▶ 7% $CDCl_3$ ▶ V 599

▶ 1-Lhl

	Shift	Peak
a	.93	2A
▶b	1.48	1E
c	1.52	1E
d	2.20	5A
e	5.20	4B

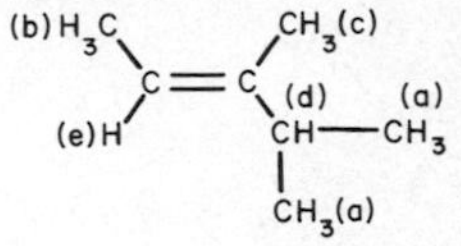

▶ C_7H_{14} ▶ 50% CCl_4 ▶ API 354

▶ 1-Lel

	Shift	Peak
▶a	2.30	2D
b	6.07	2E
c	11.73	1B

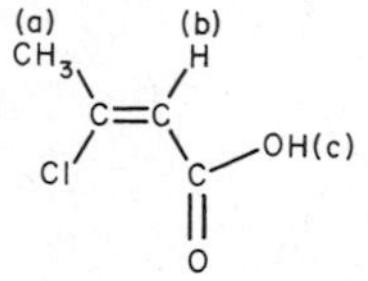

▶ $C_4H_5ClO_2$ ▶ 7% $CDCl_3$ ▶ V 53

▶ 1-Lhl

	Shift	Peak
a	.93	3C
▶b	1.52	2E
c	2.00	4C
d	5.30	N

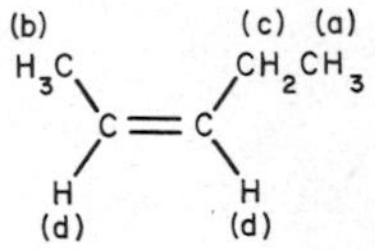

▶ C_5H_{10} ▶ 50 vol.% CCl_4 ▶ API 429

▶ 1-Lel

	Shift	Peak
▶a	2.58	2C
b	6.10	2E
c	11.82	1A

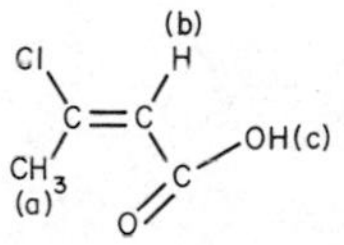

▶ $C_4H_5ClO_2$ ▶ 7% $CDCl_3$ ▶ V 54

▶ 1-Lhl

	Shift	Peak
a	.91	2A
b	~1.48 or 1.52	1E
▶c	~1.52 or 1.48	1E
d	2.74	5A
e	5.05	1D

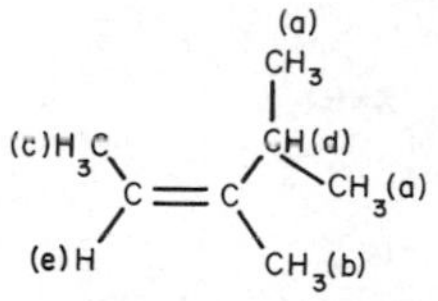

▶ C_7H_{14} ▶ 50% CCl_4 ▶ API 353

▶ 1-Lgl

	Shift	Peak
▶a	2.28	2E
b	5.33	M
c	5.52	M

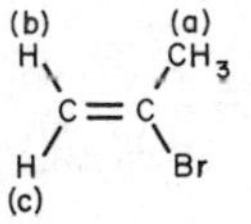

▶ C_3H_5Br ▶ 7% $CDCl_3$ ▶ V 23

▶ 1-Lhl

	Shift	Peak
a	.93	2A
b	1.47	1F
▶c	1.55	1F
d	2.84	5A
e	5.10	1F

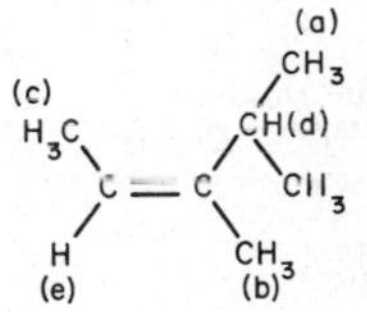

▶ C_7H_{14} ▶ liquid ▶ API 403

▶ 1-Lhl

	Shift	Peak
a	.95	3C
▶b	1.40	1F
c	1.61	S
d	2.03	4A
e	~5.08	4C

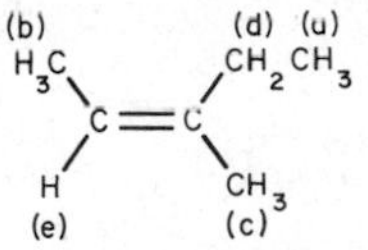

▶ C_6H_{12} ▶ liquid ▶ API 377

▶ 1-Lhl

	Shift	Peak
a	.89	3C
b	1.30	S
▶c	1.56	2E
d	2.00	4C
e	5.32	O

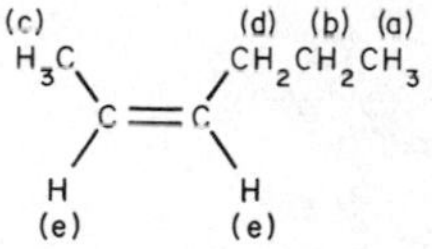

▶ C_6H_{12} ▶ liquid ▶ API 369

▶ 1-Lhl

	Shift	Peak
▶a	1.48	1E
b	1.54	1E
c	5.15 ± .05	1F

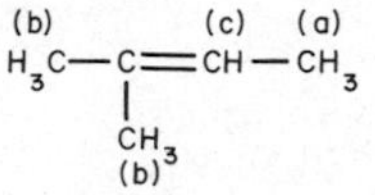

▶ C_5H_{10} ▶ 10% CCl_4 ▶ API 31

▶ 1-Lhl

	Shift	Peak
a	.97	3C
b	1.49	S
▶c	1.56	S
d	1.95	4C
e	5.18	4C

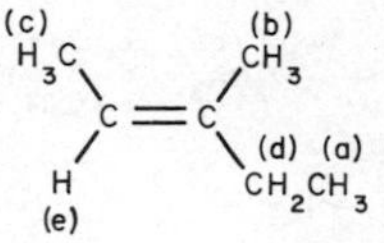

▶ C_6H_{12} ▶ liquid ▶ API 378

▶ 1-Lhl

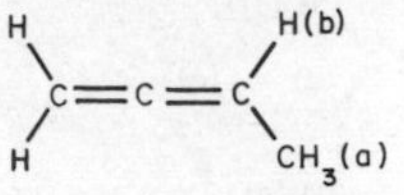

	Shift	Peak
▶a	1.57	5A
b	5.08	5B

▶ C_4H_6 ▶ liquid ▶ API 465

▶ 1-Lhl

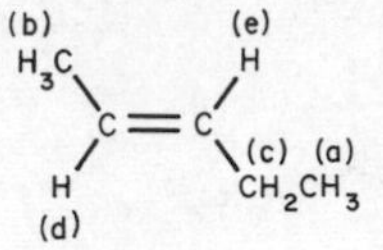

	Shift	Peak
a	.93	3C
▶b	1.58	2E
c	1.95	O
d	5.35 or 5.43	S
e	5.43 or 5.35	S

▶ C_5H_{10} ▶ liquid ▶ API 461

▶ 1-Lhl

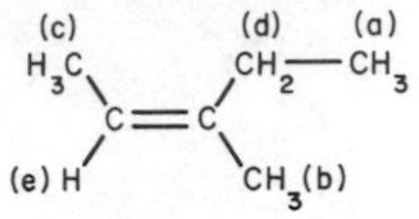

	Shift	Peak
a	.97	3A
b	1.47	1E
▶c	1.57	1E
d	1.97	4A
e	5.13	2E

▶ C_6H_{12} ▶ 10% CCl_4 ▶ API 35

▶ 1-Lhl

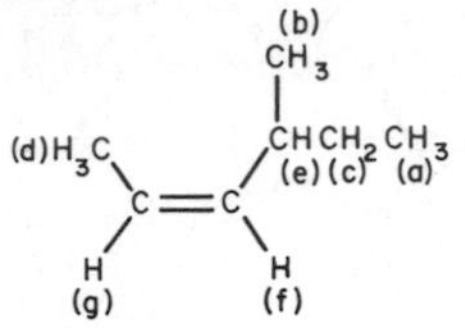

	Shift	Peak
a	.92	3C
b	.95	2H
c	1.25	O
▶d	1.58	2E
e	2.38	6B
f	~5.17 or 5.32	S
g	~5.32 or 5.17	S

▶ C_7H_{14} ▶ liquid run, neat ▶ API 409

▶ 1-Lhl

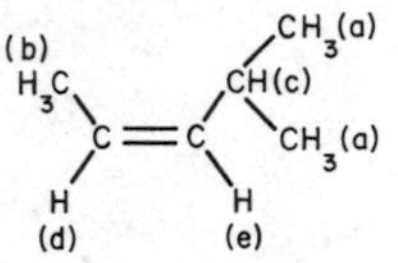

	Shift	Peak
a	.93	2A
▶b	1.57	2A
c	2.62	6B
d	5.12 or 5.23	O
e	5.23 or 5.12	O

▶ C_6H_{12} ▶ liquid ▶ API 379

▶ 1-Lhl

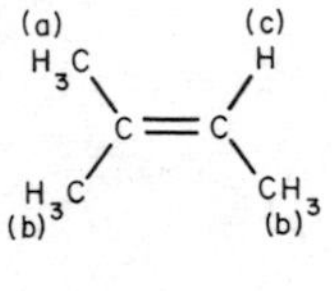

	Shift	Peak
a	1.47	1F
▶b	1.59	S
c	5.18	S

▶ C_5H_{10} ▶ liquid ▶ API 462

▶ 1-Lhl

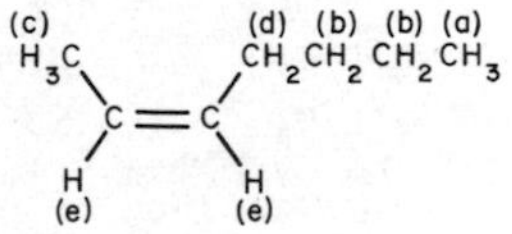

	Shift	Peak
a	.90	3C
b	1.31	O
▶c	1.57	2E
d	2.02	1F
e	~5.37	O

▶ C_7H_{14} ▶ liquid ▶ API 389

▶ 1-Lhl

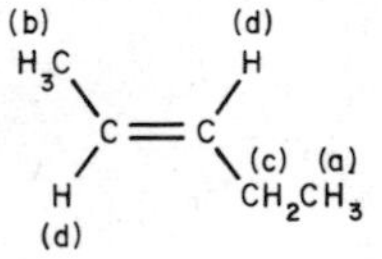

	Shift	Peak
a	.95	3A
▶b	1.60	1H
c	1.94	N
d	5.38	N

▶ C_5H_{10} ▶ 50 vol.% CCl_4 ▶ API 430

▶ 1-Lhl

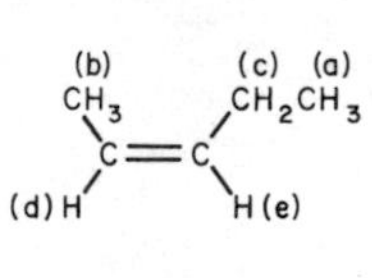

	Shift	Peak
a	.98	3A
▶b	1.58	2E
c	2.00	5B
d	5.23	S
e	5.30	S

▶ C_5H_{10} ▶ 10% CCl_4 ▶ API 154

▶ 1-Lhl

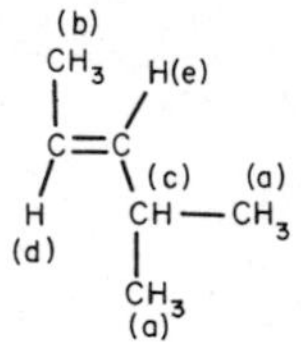

	Shift	Peak
a	.94	2A
▶b	1.60	2E
c	2.15	1D
d	5.36	S
e	5.45	S

▶ C_6H_{12} ▶ liquid run ▶ API 70

▶ 1-Lhl

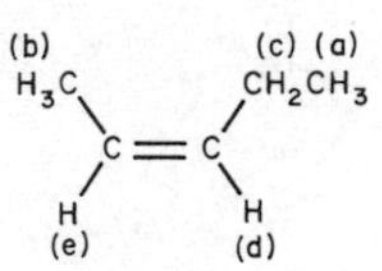

	Shift	Peak
a	.95	3C
▶b	1.58	2E
c	2.05	5B
d	~5.28 or 5.38	S
e	~5.38 or 5.28	S

▶ C_5H_{10} ▶ liquid (neat) ▶ API 460

▶ 1-Lhl

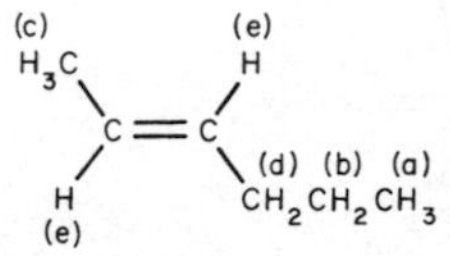

	Shift	Peak
a	.88	3C
b	1.28	S
▶c	1.60	2E
d	1.92	1F
e	5.35	O

▶ C_6H_{12} ▶ 50 vol.% CCl_4 ▶ API 370

▶ 1-Lhl

(c) H_3C, (e) H, C=C, H (e), $CH_2CH(CH_3)_2$ (d) (b) (a)

	Shift	Peak
a	.98	2A
b	1.38	S
▶c	1.60	2H
d	1.80	S
e	5.30	N

▶ C_7H_{14} ▶ 50 vol.% CCl_4 ▶ API 431

▶ 1-Lhl

(c) H_3C, (f) H, C=C, H (e), (d) CH_2, $CH(b)$, $CH_3(a)$, CH_3

	Shift	Peak
a	.83	2A
b	~1.40	S
▶c	1.62	2E
d	1.83	O
e	~5.38	O
f	~5.38	O

▶ C_7H_{14} ▶ liquid ▶ API 396

▶ 1-Lhl

D, C=C, C=C, $CH_3(a)$, H(c), $CH_3(b)$

	Shift	Peak
▶a	1.60	2A
b	2.08	1A
c	5.98	4A

▶ $C_{18}H_{17}D$ ▶ 7% $CDCl_3$ ▶ V 674

▶ 1-Lhl

(b) H_3C, H, C=C, H, $CH_3(a)$, C—CH_3, CH_3

	Shift	Peak
a	1.00	1A
▶b	1.62	2E

▶ C_7H_{14} ▶ liquid ▶ API 406

▶ 1-Lhl

(b) H_3C, H(d), C=C, (e) H, (c) CH_2—$CH_3(a)$

	Shift	Peak
a	.97	3A
▶b	1.61	5A
c	1.97	1F
d	5.37	S
e	5.37	S

▶ C_5H_{10} ▶ 10% CCl_4 ▶ API 29

▶ 1-Lhl

(d) H, H(e), C=C, CH_3 (b), (c) CH—$CH_3(a)$, CH_3 (a)

	Shift	Peak
a	.95	2A
▶b	1.63	2A
c	2.55	6B
d	5.11	N
e	5.25	N

▶ C_6H_{12} ▶ liquid run ▶ API 69

▶ 1-Lhl

(c) H, (a) CH_3, C, (a) CH_3, $CH_3(b)$, C=C, (d) H, H(e)

	Shift	Peak
a	0.94	2A
▶b	1.61	2A
c	2.63	6A
d	5.20 ± 0.10	O
e	5.30 ± 0.10	O

▶ C_6H_{12} ▶ 7% $CDCl_3$ ▶ V 471

▶ 1-Lhl

H, H, C=C, H, H, C=C, H, H, $CH_3(a)$

	Shift	Peak
a	1.67	2B

▶ C_5H_8 ▶ 50% CCl_4 ▶ API 469

▶ 1-Lhl

(b) H_3C, (c) H, C=C, H (d), $CH_3(a)$, CH(c), $CH_3(a)$

	Shift	Peak
a	.98	2A
▶b	1.62	2E
c	2.22	1F
d	5.37 or 5.43	S
e	5.43 or 5.37	S

▶ C_6H_{12} ▶ 5 vol.% CCl_4 ▶ API 380

▶ 1-Lhl

H, H, C=C, H, C=C, $CH_3(a)$, H, H

	Shift	Peak
a	1.69	2E

▶ C_5H_8 ▶ 50% CCl_4 ▶ API 468

▶ 1-Lhl

(c) H_3C, (f) H, C=C, H (e), (d) (b) (b) (a) $CH_2CH_2CH_2CH_3$

	Shift	Peak
a	.88	3C
b	1.31	O
▶c	1.62	2E
d	1.92	1F
e	5.33 or 5.42	S
f	5.42 or 5.33	S

▶ C_7H_{14} ▶ liquid ▶ API 390

▶ 1-Lhl

(c) H_3C, (e) H, C=C, H (f), CH_2—CH_2—CH_3 (d) (b) (a)

	Shift	Peak
a	.88	3B
b	~1.33	O
▶c	1.69	N
d	1.87	1F
e	5.32	S
f	5.32	S

▶ C_6H_{12} ▶ 10% CCl_4 ▶ API 32

▶ 1-Lhl

(b)H_3C — C=C — C(CH_3(a))(CH_3)(CH_3); H (d); H (c)

Shift	Peak
a 1.11	1A
▶b 1.69	2E
c 5.23 ± .03	O
d 5.23 ± .03	O

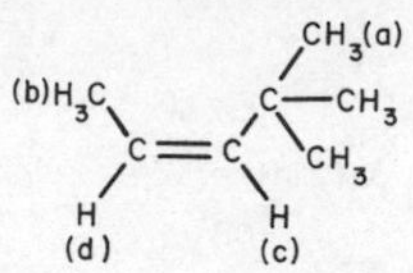

▶ C_7H_{14} ▶ liquid ▶ API 405

▶ 1-Lhl

OH(c); OCH_3(b); CH=CH—CH_3; (e) (d) (a)

Shift	Peak
▶a 1.82	2A
b 3.88	1A
c 5.48	1B
d 6.03	S
e 6.20	S

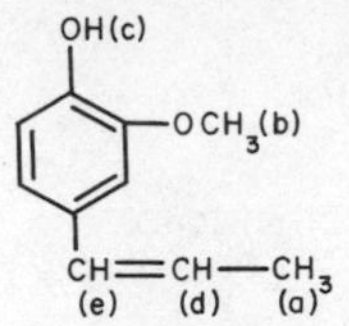

▶ $C_{10}H_{12}O_2$ ▶ 45.0 mg/0.5 ml $CDCl_3$ ▶ S 76

▶ 1-Lhl

(e)H; (c) (b) CH_2—OH; C=C; (a)CH_3; H(d)

Shift	Peak
▶a 1.72	N
b 2.07	1A
c 4.07	N
d ~5.68	S
e ~5.68	S

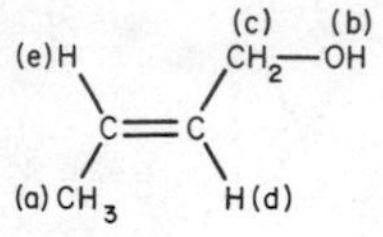

▶ C_4H_8O ▶ 7% $CDCl_3$ ▶ V 414

▶ 1-Lhl

(b) OCH_3; (e)H; H(e); (f)H; H(f); C=C; H(c); (d)H; CH_3(a)

Shift	Peak
▶a 1.83	2A
b 3.75	1A
c 6.08	N
d 6.28	N
e 6.80	Q
f 7.23	Q

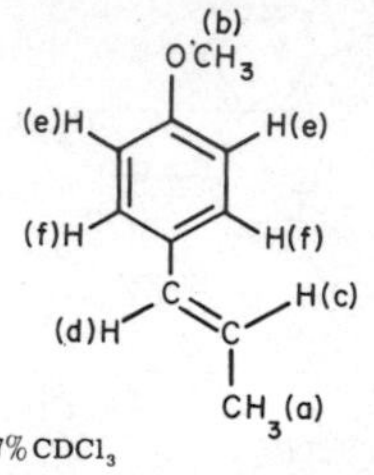

▶ $C_{10}H_{12}O$ ▶ 7% $CDCl_3$ ▶ V 258

▶ 1-Lhl

CH_3; (a)CH_3—C—; CH_3; C=C; CH_3(b); H (d); H (c)

Shift	Peak
a 1.10	1A
▶b 1.72	2A
c ~5.23	S
d ~5.23	S

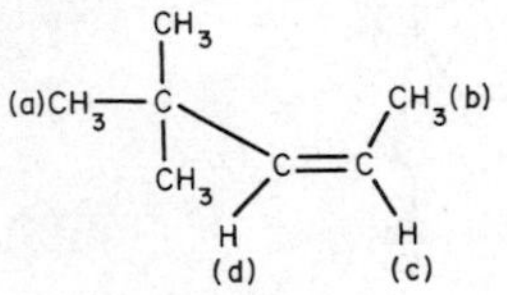

▶ C_7H_{14} ▶ 50% CCl_4 ▶ API 222

▶ 1-Lhl

H H; C=C; CH_3 (a); C=C; H(b); H (c); C—OH(d); O

Shift	Peak
▶a 1.87	4A
b 5.79	2A
c 7.36	O
d 12.03	1A

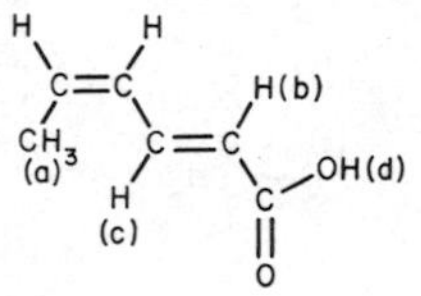

▶ $C_6H_8O_2$ ▶ 7% $CDCl_3$ ▶ V 462

▶ 1-Lhl

H H; C=C; H; C=C; CH_3(a); H H

Shift	Peak
a 1.74	2E

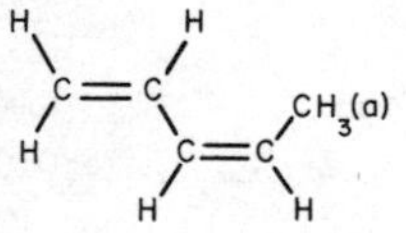

▶ C_5H_8 ▶ 10% CCl_4 ▶ API 159

▶ 1-Lhl

O; O—C—CH_3 (b); O—CH_3(c); (a); CH=CH—CH_3; (d) (e)

Shift	Peak
▶a 1.88	2A
b 2.31	1A
c 3.82	1A
d ~6.1-6.7	S
e ~6.1-6.7	S

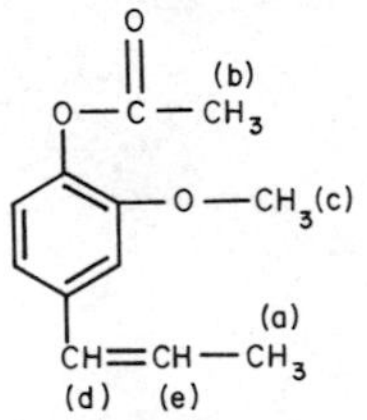

▶ $C_{12}H_{14}O_3$ ▶ 45.4 mg/0.50 ml $CDCl_3$ ▶ S 130

▶ 1-Lhl

H H; C=C; H; H; C=C; H; CH_3(a)

Shift	Peak
a 1.75	2A

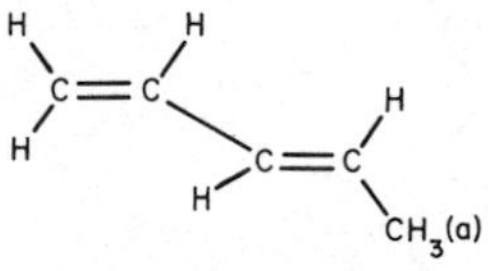

▶ C_5H_8 ▶ 10% CCl_4 ▶ API 160

▶ 1-Lhl

(a)CH_3; H(b); C=C; (c)H; C—OH(d); O

Shift	Peak
▶a 1.90	2E
b 5.83	2E
c 7.10	2E
d 12.18	O

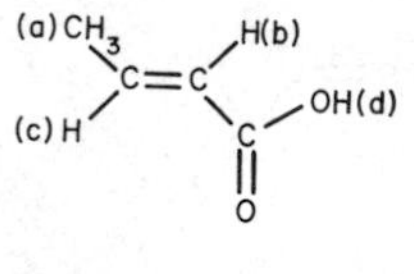

▶ $C_4H_6O_2$ ▶ 7% $CDCl_3$ ▶ V 61

▶ 1-Lhl

(a)CH_3; (d) H; H(g); C=C; (c)H; O; CH_2(b); O; (e)H; H (f)

Shift	Peak
▶a 1.80	2A
b 5.87	1A
c ~6.08	S
d 6.23	S
e 6.70	S
f 6.70	S
g 6.83	S

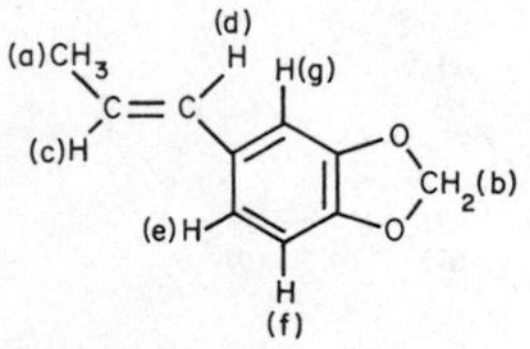

▶ $C_{10}H_{10}O_2$ ▶ 7% $CDCl_3$ ▶ V 252

▶ 1-Lhl

O; (f)H; O; C; O; CH_3(a); (g)H; C; H(e); O; OH(c); H(d); C; CH_3 (b); H (h)

Shift	Peak
a 1.65	2A
▶b 1.97	2E
c 3.86	1C
d 3.92	S
e 4.32	O
f 5.87	1I
g 6.06	2A
h 6.99	4A

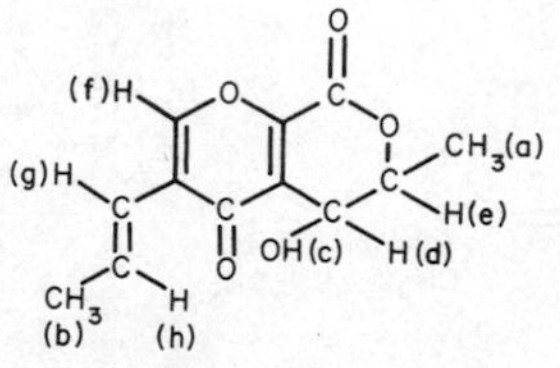

▶ $C_{12}H_{12}O_5$ ▶ 7% $CDCl_3$ ▶ V 595

▶ 1-Lhl

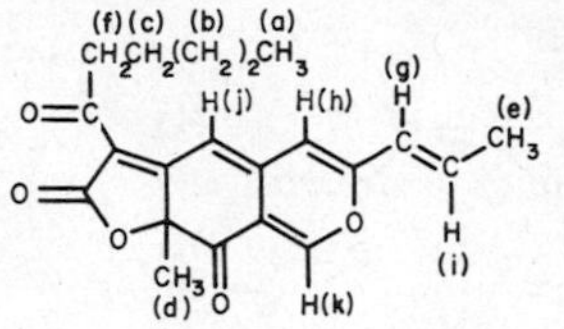

	Shift	Peak
a	0.89	3B
b	~1.30	S
c	~1.40	S
d	1.72	1A
▶e	1.97	2A
f	2.94	3B
g	6.06	S
h	6.17	S
i	6.58	S
j	6.89	S
k	7.88	1G

▶ $C_{21}H_{22}O_5$ ▶ 7% $CDCl_3$ ▶ V 687

▶ 1-(L)jl

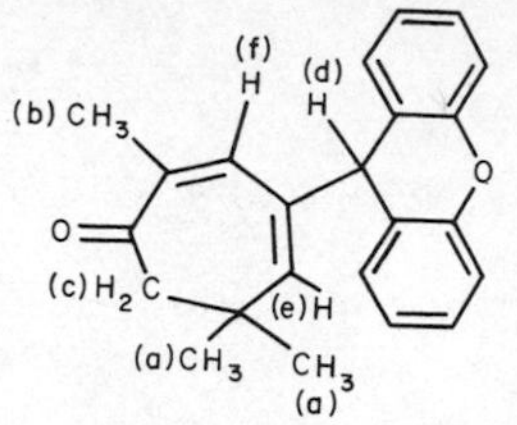

	Shift	Peak
a	1.20	1A
▶b	1.70	2D
c	2.73	1G
d	4.73	1B
e	6.14	S
f	6.16	S

▶ $C_{23}H_{22}O_2$ ▶ 7% $CDCl_3$ ▶ V 694

▶ 1-Lhl

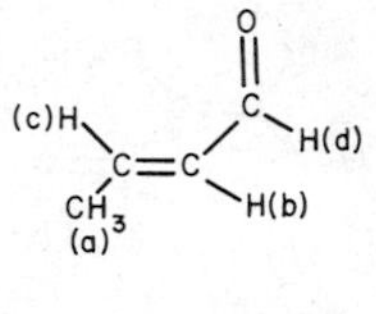

	Shift	Peak
▶a	2.03	2E
b	6.13	N
c	6.87	N
d	9.48	O

▶ C_4H_6O ▶ 7% $CDCl_3$ ▶ V 60

▶ 1-(L)jl

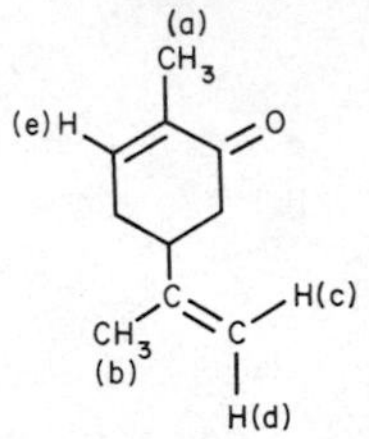

	Shift	Peak
▶a	1.75 ± .03	1E
b	1.75 ± .03	1E
c	4.78 ± .03	N
d	4.78 ± .03	N
e	6.75	1C

▶ $C_{10}H_{14}O$ ▶ 7% $CDCl_3$ ▶ V 271

▶ 1-Lhl

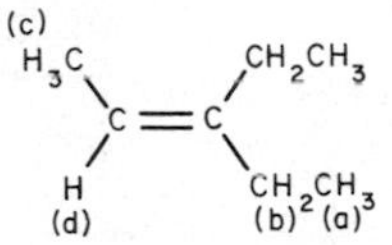

	Shift	Peak
a	.93	3C
b	1.56	2E
▶c	2.04	4C
d	5.17	4A

▶ C_7H_{14} ▶ liquid ▶ API 414

▶ 1-Ljl

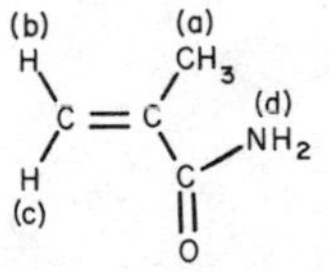

	Shift	Peak
▶a	1.95	1H
b	5.38	M
c	5.77	M
d	~6.38	1D

▶ C_4H_7NO ▶ 7% $CDCl_3$ ▶ V 71

▶ 1-Lhl

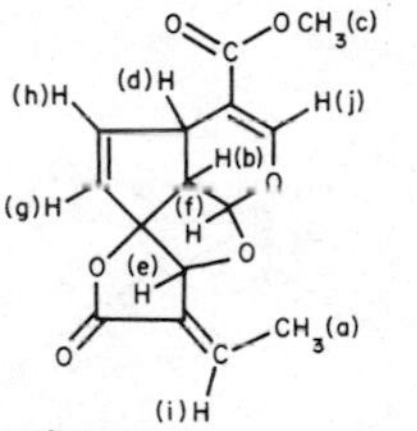

	Shift	Peak
▶a	2.11	2A
b	3.46	4A
c	3.79	1A
d	4.02	3E
e	5.12	1B
f	5.39	S
g	5.67*	S
h	5.67*	S
i	7.19	O
j	7.46	1A

*or 6.08

▶ $C_{15}H_{14}O_6$ ▶ 7% $CDCl_3$ ▶ V 640

▶ 1-(L)jl

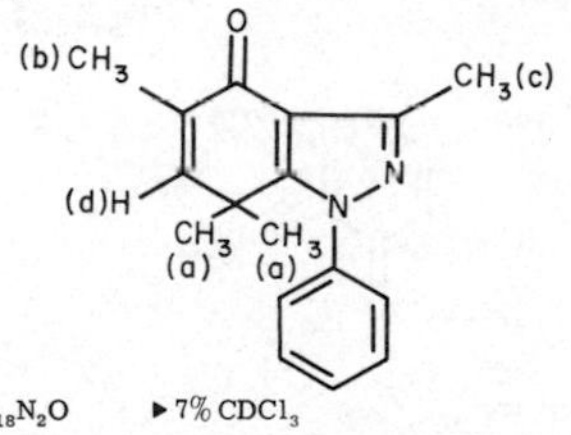

	Shift	Peak
a	1.26	1A
▶b	1.97	2D
c	2.59	1A
d	6.27	2A

▶ $C_{17}H_{18}N_2O$ ▶ 7% $CDCl_3$ ▶ V 665

▶ 1-Lhm

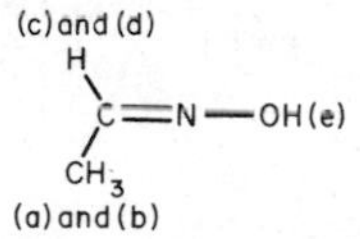

	Shift	Peak
▶a	1.86	1A
▶b	1.89	1A
c	6.84	4A
d	7.44	4A
e	~8.92	4D

syn-and anti-forms

▶ C_2H_5NO ▶ 7% $CDCl_3$ ▶ V 373

▶ 1-(L)jl

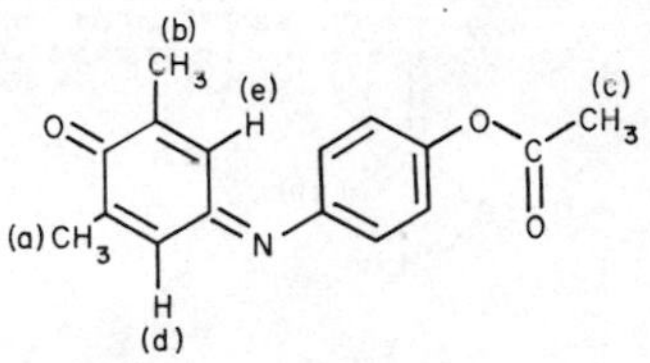

	Shift	Peak
▶a	1.98	2D
▶b	2.08	2D
c	2.30	1A
d	~6.87	N
e	~7.03	N

▶ $C_{16}H_{15}NO_3$ ▶ 7% $CDCl_3$ ▶ V 322

▶ 1-(L)jl

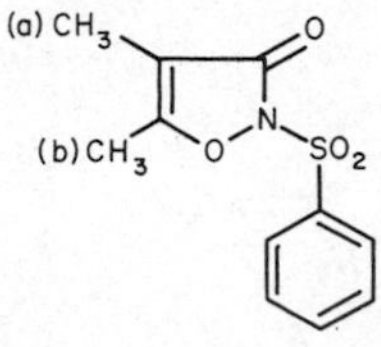

	Shift	Peak
▶a	1.61	1A
b	2.17	1A

▶ $C_{11}H_{11}NO_4S$ ▶ 7% $CDCl_3$ ▶ V 580

▶ 1-(L)jl

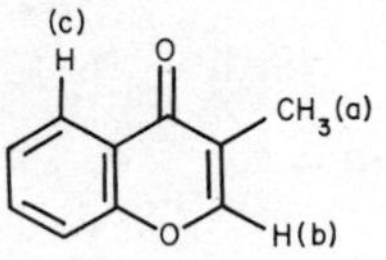

	Shift	Peak
▶a	2.02	2D
b	7.78	2D
c	8.24	2E

▶ $C_{10}H_8O_2$ ▶ 7% $CDCl_3$ ▶ V 552

▶ 1-Ljl

	Shift	Peak
▶a	2.04	1H
b	6.02	4A
c	6.51	1G

▶ C_4H_5ClO ▶ 7% $CDCl_3$ ▶ V 400

▶ 1-Lkl

	Shift	Peak
a	1.28	3A
▶b	1.93	2D
c	4.22	4A
d	5.57	3C
e	6.10	1I

▶ $C_6H_{10}O_2$ ▶ 7% $CDCl_3$ ▶ V 135

▶ 1-Ljl

	Shift	Peak
▶a	2.06	2D
b	5.43	1I
c	5.77	1I

▶ $C_{10}H_{11}NO$ ▶ 7% $CDCl_3$ ▶ V 555

▶ 1-Lkl

	Shift	Peak
▶a	1.95	2E
b	3.75	1A
c	5.57	M
d	6.10	M

▶ $C_5H_8O_2$ ▶ 7% $CDCl_3$ ▶ V 113

▶ 1-(L)jl

	Shift	Peak
▶a	2.22	1A
b	~7.45	1G

▶ $C_{10}H_8O_2$ ▶ 45.0 mg/0.50 ml $CDCl_3$ ▶ S 87

▶ 1-Lkl

	Shift	Peak
▶a	1.97	2E
b	5.72	M
c	6.30	2E
d	11.57	1B

▶ $C_4H_6O_2$ ▶ 7% $CDCl_3$ ▶ V 62

▶ 1-(L)jl

	Shift	Peak
▶a	2.23	1A
b	6.03	1A

▶ $C_7H_8O_2$ ▶ 7% $CDCl_3$ ▶ V 166

▶ 1-(L)kl

	Shift	Peak
▶a	2.19	2A
b	6.67	4A

▶ $C_5H_4O_3$ ▶ 7% $CDCl_3$ ▶ V 427

▶ 1-Ljm

	Shift	Peak
▶a	1.88 or 2.35	1A
b	1.88 or 2.35	1A
c	4.05	1A

▶ $C_5H_9NO_2$ ▶ 7% $CDCl_3$ ▶ V 117

▶ 1-(L)ll

	Shift	Peak
a	1.28	1A
▶b	1.60	2A
c	2.57	1A
d	5.87	2A
e	7.46	1A

▶ $C_{17}H_{18}N_2O$ ▶ 7% $CDCl_3$ ▶ V 664

▶ 1-Ljm

	Shift	Peak
a	2.00 or 2.40	1A
▶b	2.00 or 2.40	1A
c	8.89	1C

▶ $C_4H_7NO_2$ ▶ 7% $CDCl_3$ ▶ V 72

▶ 1-Lll

	Shift	Peak
a	1.79	1A

▶ C_5H_8 ▶ 50% CCl_4 ▶ API 473

▶ 1-Lll

(b) H, (a) CH_3, C=C, H(c), (c)H, C=C, CH_3 (a), H(b)

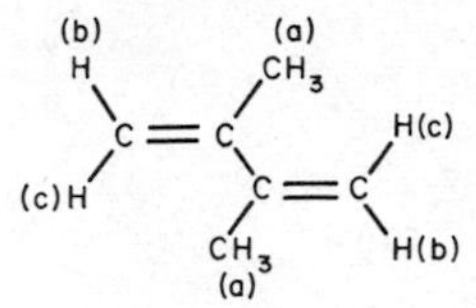

	Shift	Peak
▶a	1.88	1A
b	~4.93	S
c	~4.93	S

▶ C_6H_{10} ▶ liquid ▶ API 475

▶ 1-(L)lm

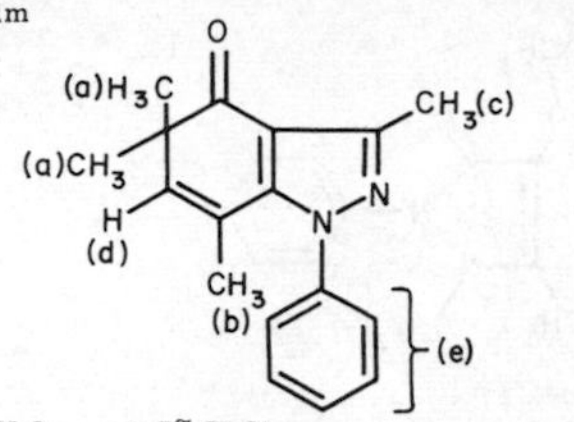

O, (a)H_3C, CH_3(c), (a)CH_3, N, N, H (d), N, CH_3 (b), (e)

	Shift	Peak
a	1.28	1A
b	1.60	2A
▶c	2.57	1A
d	5.87	2A
e	7.46	1A

▶ $C_{17}H_{18}N_2O$ ▶ 7% $CDCl_3$ ▶ V 664

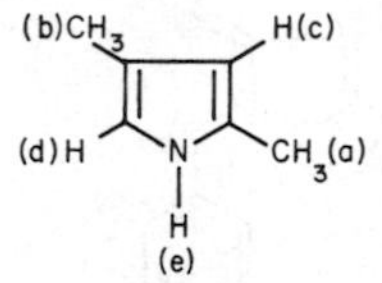

▶ 1-(L)ll

(b)CH_3, H(c), (d)H, N, CH_3(a), H (e)

	Shift	Peak
a	2.07 or 2.20	1A
▶b	2.07 or 2.20	1A
c	5.73	1B
d	6.37	1B
e	~7.50	1D

▶ C_6H_9N ▶ 7% $CDCl_3$ ▶ V 131

▶ 1-(L)lm

O, (b) CH_3, CH_3(c), N, (d)H, N, CH_3 CH_3 (a) (a)

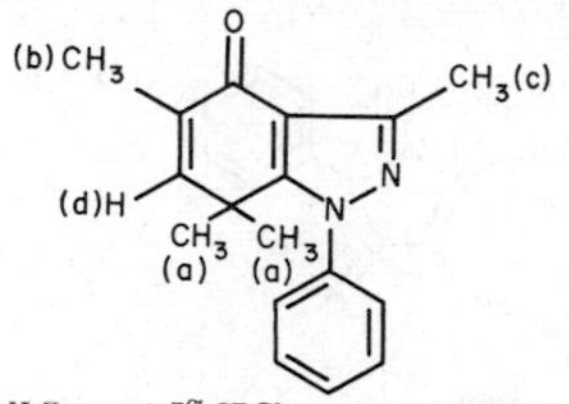

	Shift	Peak
a	1.26	1A
b	1.97	2D
▶c	2.59	1A
d	6.27	2A

▶ $C_{17}H_{18}N_2O$ ▶ 7% $CDCl_3$ ▶ V 665

▶ 1-Lll

CH_3(a), D, C=C, C=C, H(c), CH_3(b)

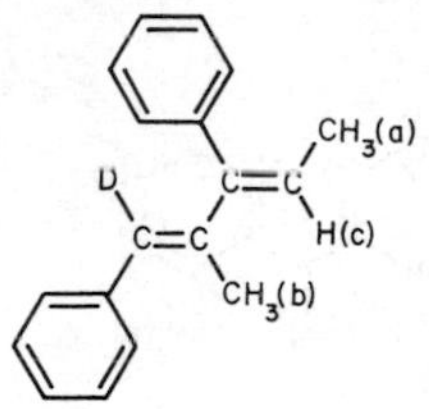

	Shift	Peak
a	1.60	2A
▶b	2.08	1A
c	5.98	4A

▶ $C_{18}H_{17}D$ ▶ 7% $CDCl_3$ ▶ V 674

▶ 1-Lln

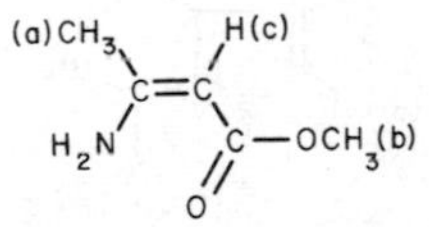

(a)CH_3, H(c), C=C, H_2N, C—OCH_3(b), O

	Shift	Peak
▶a	1.92	1A
b	3.62	1A
c	4.53	1B
d	~6.15	1D

▶ $C_5H_9NO_2$ ▶ 7% $CDCl_3$ ▶ V 442

▶ 1-(L)lm

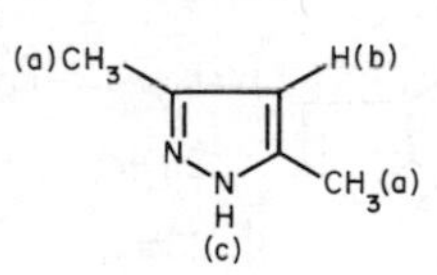

(a)CH_3, H(b), N, N, CH_3(a), H (c)

	Shift	Peak
▶a	2.27	1A
b	5.80	1B
c	11.92	1D

▶ $C_5H_8N_2$ ▶ 7% $CDCl_3$ ▶ V 441

▶ 1-(L)ln

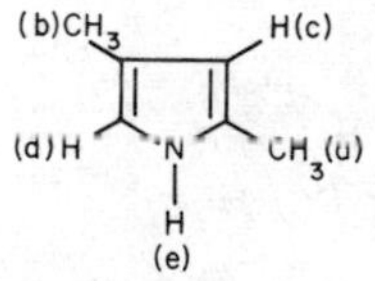

(b)CH_3, H(c), (d)H, N, CH_3(a), H (e)

	Shift	Peak
▶a	2.07 or 2.20	1A
b	2.07 or 2.20	1A
c	5.73	1B
d	6.37	1B
e	~7.50	1D

▶ C_6H_9N ▶ 7% $CDCl_3$ ▶ V 131

▶ 1-(L)lm

(a) CH_3, N, (c)H, S, CH_3 (b)

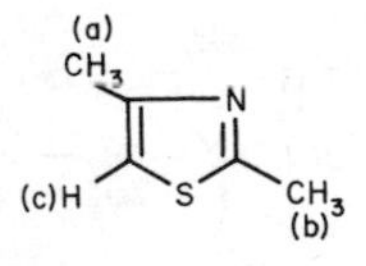

	Shift	Peak
▶a	2.38	2D
b	2.67	1A
c	6.66	2E

▶ C_5H_7NS ▶ 7% $CDCl_3$ ▶ V 108

▶ 1-(L)ln

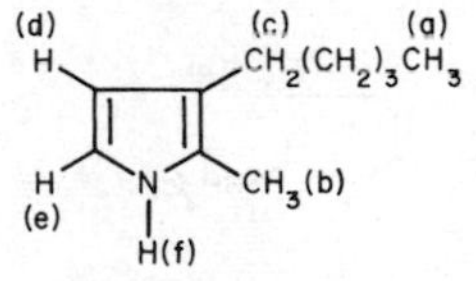

(d) H, (c) (a) $CH_2(CH_2)_3CH_3$, H (e), N, CH_3(b), H(f)

	Shift	Peak
a	0.90	3B
▶b	2.15	1A
c	~2.38	S
d	5.98	3A
e	6.55	3A
f	7.07	1D

▶ $C_{10}H_{17}N$ ▶ 7% $CDCl_3$ ▶ V 278

▶ 1-(L)lm

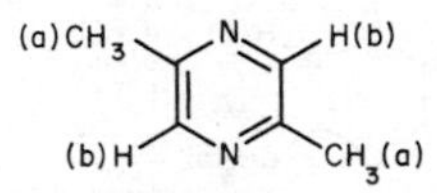

(a)CH_3, N, H(b), (b)H, N, CH_3(a)

	Shift	Peak
▶a	2.52	1A
b	8.35	1A

▶ $C_6H_8N_2$ ▶ 7% $CDCl_3$ ▶ V 459

▶ 1-(L)ln

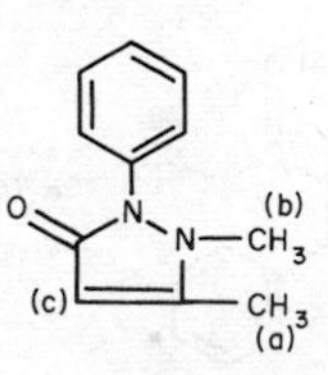

O, N, N—CH_3 (b), (c), CH_3 (a)

	Shift	Peak
▶a	2.19	1A
b	2.99	1A
c	5.31	1G

▶ $C_{11}H_{12}N_2O$ ▶ 44.7 mg/0.50 ml $CDCl_3$ ▶ S 258

▶ 1-(L)ln

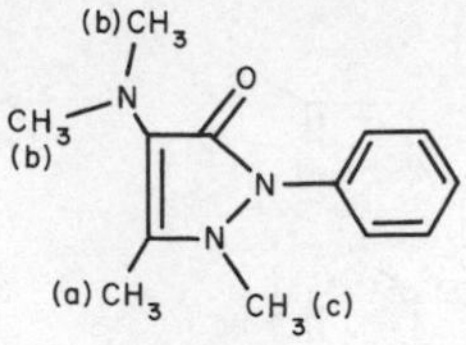

	Shift	Peak
▶a	2.21	1A
b	2.81	1A
c	2.96	1A

▶ $C_{13}H_{17}N_3O$ ▶ 7% $CDCl_3$ ▶ V 615

▶ 1-Llo

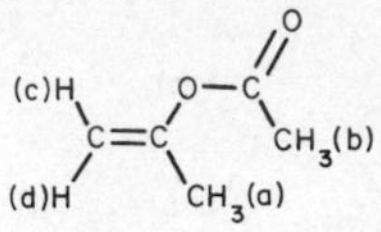

	Shift	Peak
▶a	1.93	1A
b	2.12	1A
c	4.69	1G
d	4.69	1G

▶ $C_5H_8O_2$ ▶ 7% $CDCl_3$ ▶ V 440

▶ 1-(L)ln

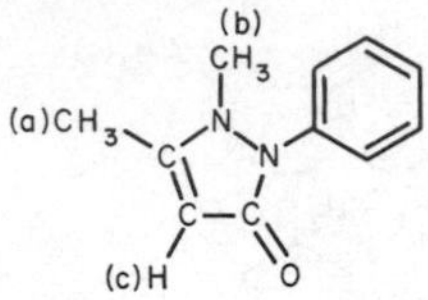

	Shift	Peak
▶a	2.23	1G
b	3.06	1A
c	5.38	2E

▶ $C_{11}H_{12}N_2O$ ▶ 7% $CDCl_3$ ▶ V 581

▶ 1-(L)lo

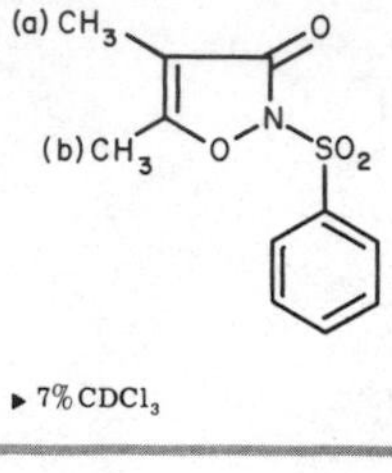

	Shift	Peak
a	1.61	1A
▶b	2.17	1A

▶ $C_{11}H_{11}NO_4S$ ▶ 7% $CDCl_3$ ▶ V 580

▶ 1-(L)ln

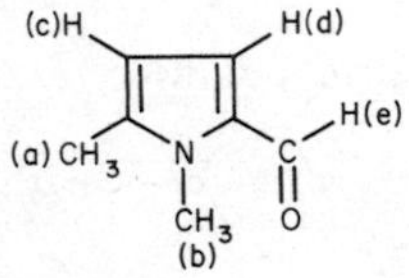

	Shift	Peak
▶a	2.27	1A
b	3.87	1A
c	6.02	2A
d	6.81	2A
e	9.38	1A

▶ C_7H_9NO ▶ 7% $CDCl_3$ ▶ V 170

▶ 1-(L)lo

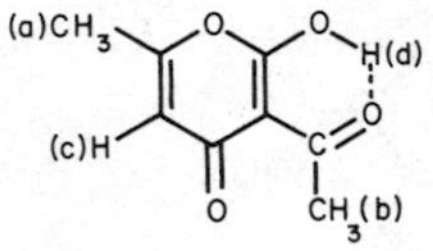

	Shift	Peak
▶a	2.27	1A
b	2.68	1A
c	5.92	1A
d	16.71	1C

▶ $C_8H_8O_4$ ▶ 7% $CDCl_3$ ▶ V 504

▶ 1-(L)ln

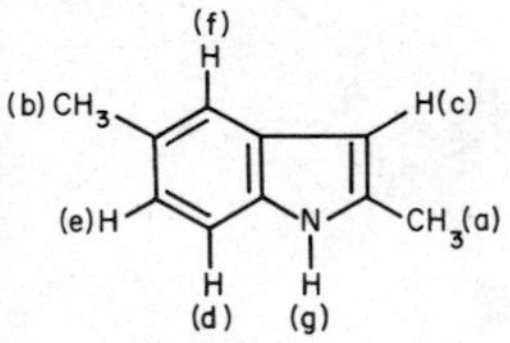

	Shift	Peak
▶a	2.28	1A
b	2.40	1A
c	6.10	1I
d	~6.90 or 7.02	N
e	~6.90 or 7.02	N
f	7.27	S
g	~7.30	S

▶ $C_{10}H_{11}N$ ▶ 7% $CDCl_3$ ▶ V 255

▶ 1-(L)lo

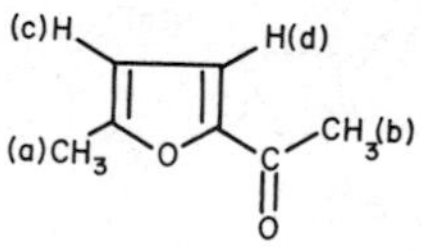

	Shift	Peak
▶a	2.38	2D
b	2.43	1A
c	6.17	2E
d	7.12	2E

▶ $C_7H_8O_2$ ▶ 7% $CDCl_3$ ▶ V 165

▶ 1-(L)ln

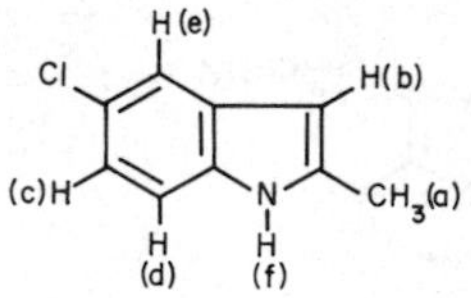

	Shift	Peak
▶a	2.33	2D
b	6.10	1B
c	7.02	2D
d	7.02	2D
e	7.44	1I
f	~7.65	1D

▶ C_9H_8ClN ▶ 7% $CDCl_3$ ▶ V 228

▶ 1-(L)lo

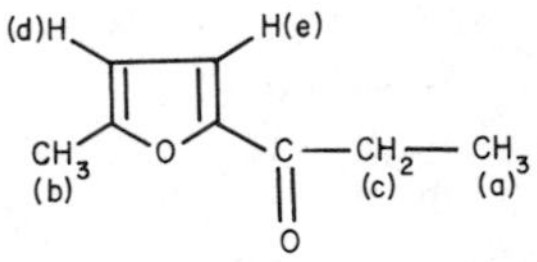

	Shift	Peak
a	1.20	3A
▶b	2.38	1A
c	2.78	4A
d	6.13	2E
e	7.08	2A

▶ $C_8H_{10}O_2$ ▶ 7% $CDCl_3$ ▶ V 206

▶ 1-(L)ln

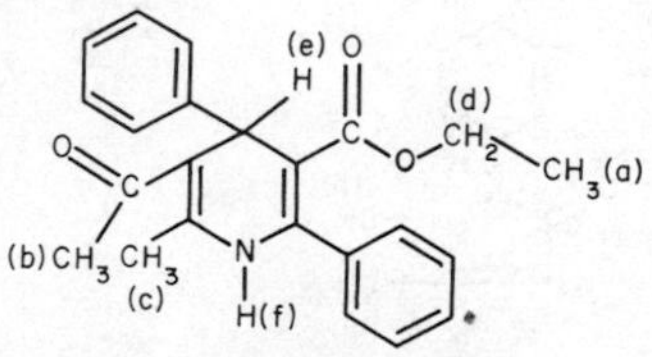

	Shift	Peak
a	0.83	3A
b	2.17	1A
▶c	2.37	1A
d	3.87	4A
e	5.13	1A
f	6.15	1C

▶ $C_{23}H_{23}NO_3$ ▶ 7% $CDCl_3$ ▶ V 695

▶ 1-(L)lo

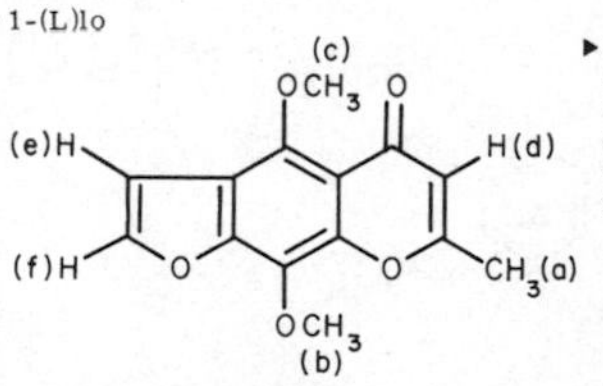

	Shift	Peak
▶a	2.40	1G
b	4.06 or 4.20	1A
c	4.06 or 4.20	1A
d	6.05	1G
e	7.01	2C
f	7.63	2C

▶ $C_{14}H_{12}O_5$ ▶ 7% $CDCl_3$ ▶ V 628

▶ 1-(L)lq

(c)H CH3(a) CH3(a) C=O (b)CH3 O

Shift	Peak
a 1.28	1A
▶ b 1.98	2A
c 5.16	2A

▶ $C_7H_{10}O_2$ ▶ 7% $CDCl_3$ ▶ V 175

▶ 1-(L)lv

(f)H (g)H (e)H C=C (d) CH2 (i) H (c) CH3 H(h) (b) CH3—CO O O=O CH3 (a) O

Shift	Peak
a 2.25 or 2.37	1A
b 2.25 or 2.37	1A
▶ c 2.40	2D
d 3.33	2B
e 5.07	R
f 5.12	R
g 5.85	4D
h 6.22	1I
i 7.30	1I

▶ $C_{16}H_{16}O_4$ ▶ 7% $CDCl_3$ ▶ V 323

▶ 1-Llt

(a) CH3 (b) HC≡C—C=C—H(d) H (c)

Shift	Peak
▶ a 1.90	2E
b 2.87	1A
c 5.27 or 5.37	N
d 5.27 or 5.37	N

▶ C_5H_6 ▶ 7% $CDCl_3$ ▶ V 99

▶ 1-(L)mm

CH3(a) C N N N N CH3 (b)

Shift	Peak
▶ a 2.52	1A
b 4.30	1A

▶ $C_3H_6N_4$ ▶ 7% $CDCl_3$ ▶ V 387

▶ 1-Llu

(b)H CH3(a) C=C (c)H CN

Shift	Peak
▶ a 2.00	1I
b 5.73	N
c 5.82	N

▶ C_4H_5N ▶ 7% $CDCl_3$ ▶ V 97

▶ 1-(L)ms

(a) CH3 N (c)H S CH3 (b)

Shift	Peak
a 2.38	2D
▶ b 2.67	1A
c 6.66	2E

▶ C_5H_7NS ▶ 7% $CDCl_3$ ▶ V 108

▶ 1-Llv

CH3(a) C=C—H(b) H(c)

Shift	Peak
▶ a 2.12	2D
b 5.05	N
c 5.36	N

▶ C_9H_{10} ▶ 7% $CDCl_3$ ▶ V 232

▶ 1-(L)ms

S C—CH3(a) N

Shift	Peak
▶ a 2.80	1A

▶ C_8H_7NS ▶ 7% $CDCl_3$ ▶ V 191

▶ 1-(L)lv

CH3(a) N H(b) H(c)

Shift	Peak
▶ a 2.28	2D
b 6.78	1I
c ~7.40	S

▶ C_9H_9N ▶ 7% $CDCl_3$ ▶ V 231

▶ 1-Lmv

(b) OCH3 (d)H H(d) (e)H H(e) C=N CH3 (a) OCH3(c)

Shift	Peak
▶ a 2.18	1A
b 3.86	1A
c 3.97	1A
d 6.87	Q
e 7.58	Q

▶ $C_{10}H_{13}NO_2$ ▶ 7% $CDCl_3$ ▶ V 561

▶ 1-(L)lv

(e) H (a) (c) CH3—CH2O O O (f)H H(d) H (g) CH3 (b)

Shift	Peak
a 1.45	3A
▶ b 2.37	2D
c 4.08	4A
d 6.08	2E
e 6.73	1G
f ~6.78	2G
g 7.47	2E

▶ $C_{12}H_{12}O_3$ ▶ 7% $CDCl_3$ ▶ V 294

▶ 1-(M)lv

(a) CH3 CH3(a) (d) (+) N CH3(b) Cl(−) CH3(c)

Shift	Peak
a 1.56	1A
b 2.87	1B
▶ c 4.10	1A
d 7.51 ± 0.05	1G

▶ $C_{12}H_{16}ClN$ ▶ 7% $CDCl_3$ ▶ V 599

▶ 1-Nab

(a) CH_3 (b) (c) (d)
N— CH_2— CH_2—OH
CH_3 (a)

	Shift	Peak
▶a	2.25	1A
b	2.45	3A
c	3.60	3D
d	3.60	1E

▶ $C_4H_{11}NO$ ▶ 7% $CDCl_3$ ▶ V 91

▶ 1-Nahx

CH_3 D CH_3 (c) N(+) (c)
N N H
2Cl(−) DO H H N N
(a) H (+)ND_3 H H(f) H (g) H(h)
(d) N OD D
CH_3O C—C—C (b)H H(e) O

	Shift	Peak
a	3.09	S
b	3.29	S
▶c	3.68	S
d	4.02*	S
e	4.37	S
f	4.60	S
g	4.82	S
h	6.12	S

*± 0.05

▶ $C_{22}H_{31}Cl_2N_7O_5$ ▶ D_2O ▶ V 691

▶ 1-Nab

H(d) CH_2—CH_2—N—CH_3(a)
(c)H_2C O O CH_3(a)
(f)H
(b) CH_3O H(e)
(b)CH_3O

	Shift	Peak
▶a	2.35	1A
b	3.93	1A
c	6.03	1A
d	6.65	1A
e	6.80	1A
f	7.77	1I

▶ $C_{21}H_{25}NO_4$ ▶ 7% $CDCl_3$ ▶ V 349

▶ 1-Naj

(a) CH_3
N—CH_3(a)
O═C—H
(b)

	Shift	Peak
▶a	2.88, 2.97	2A
b	8.02	1B

▶ C_3H_7NO ▶ 7% $CDCl_3$ ▶ V 39

▶ 1-Nab

(b)CH_3 (c) O
N—CH_2—C
CH_3 (b) O—CH_2—CH_3 (d) (a)

	Shift	Peak
a	1.29	3A
▶b	2.37	1A
c	3.17	1A
d	4.22	4A

▶ $C_6H_{13}NO_2$ ▶ 7% $CDCl_3$ ▶ V 480

▶ 1-Naj

O CH_3(c)
C—N
CH_3 (a) CH_3 (b)

	Shift	Peak
a	2.08	1A
▶b	2.94	1A
▶c	3.02	1A

▶ C_4H_9NO ▶ 7% $CDCl_3$ ▶ V 421

▶ 1-Nab

H(f) (c) (b) (a) CH_3
(e) H_2C O CH_2—CH_2—N CH_3 (a)
O H(i)
(j)H H(h)
(d) CH_3O H(g)
CH_3O(d)

	Shift	Peak
▶a	2.38	1A
b	~2.7	Q
c	~3.2	Q
d	4.05 and 4.08	1A
e	6.23	1A
f	7.13 or 7.23	1A
g	7.13 or 7.23	1A
h	7.53 or 7.81	2C
i	7.53 or 7.81	2C
j	8.52	1A

▶ $C_{21}H_{23}NO_4$ ▶ 7% $CDCl_3$ ▶ V 348

▶ 1-Naj

(d)H O (a) CH_3
C—C—N—CH_3(b)
O
N H H (c)
(e)

	Shift	Peak
▶a	2.99	1A
▶b	3.07	1A
c	7.69	1B
d	8.27	4C
e	10.76	1C

▶ $C_{12}H_{12}N_2O_2$ ▶ 7% $CDCl_3$ ▶ V 594

▶ 1-Nac

CH_3(c)
N
CH_2 CH—CH_3(b)
(a)CH_3
(d) CH_3
N
(d) CH_3 H(e)

	Shift	Peak
a	0.93	1A
b	1.05	2A
c	2.22	1A
▶d	2.32	1A
e	5.38	1I

▶ $C_{24}H_{40}N_2$ ▶ 7% $CDCl_3$ ▶ V 359

▶ 1-Nal

(b)CH_3 O
CH_3 (b) N
N
N
(a) CH_3 CH_3(c)

	Shift	Peak
a	2.21	1A
▶b	2.81	1A
c	2.96	1A

▶ $C_{13}H_{17}N_3O$ ▶ 7% $CDCl_3$ ▶ V 615

▶ 1-Nach

(c) (a) (b)
H CH_3 CH_3
Cl—C—C—N(+)—H Cl(−)
H H CH_3
(d) (e) (b)

	Shift	Peak
a	1.68	2A
▶b	2.97	1A
c	3.27	S
d	3.56	S
e	4.62	1F

▶ $C_5H_{13}Cl_2N$ ▶ 7% $CDCl_3$ ▶ V 448

▶ 1-Nam

(a) CH_3 O
N—N
CH_3 (b)

	Shift	Peak
▶a	3.09	1A
▶b	3.82	1A

▶ $C_2H_6N_2O$ ▶ 7% $CDCl_3$ ▶ V 375

▶ 1-Nav

	Shift	Peak
a	2.81	1A

▶ $C_8H_{11}NO$ ▶ 44.7 mg/0.50 ml $CDCl_3$ ▶ S 255

▶ 1-(N)bb

	Shift	Peak
▶a	1.92	1A
b	2.55	1F
c	2.88	1F
d	3.58	1A
e	3.78	1A
f	5.92, 5.96	1A
g	6.65	1A
h	6.69	1A
i	6.91	1A

▶ $C_{20}H_{19}NO_5$ ▶ 7% $CDCl_3$ ▶ V 339

▶ 1-Nav

	Shift	Peak
▶a	2.87	1A
b	5.29	1B
c	6.64	Q
d	6.95	Q

▶ $C_{25}H_{31}N_3$ ▶ 7% $CDCl_3$ ▶ V 360

▶ 1-(N)bb

	Shift	Peak
▶a	2.25	1A
b	3.69	1A
c	7.14	S
d	7.34	S
e	7.64	S
f	8.54	2E

▶ $C_{13}H_{18}N_2O_2$ ▶ 7% $CDCl_3$ ▶ V 301

▶ 1-Nav

	Shift	Peak
▶a	2.93	1A
b	~6.67	N
c	~7.15	O

▶ $C_8H_{11}N$ ▶ 45.3 mg/0.50 ml $CDCl_3$ ▶ S 1

▶ 1-(N)bb

	Shift	Peak
a	2.12	1A
▶b	2.27	1A
c	2.37	Q
d	2.88	Q

▶ $C_5H_{12}N_2$ ▶ 7% $CDCl_3$ ▶ V 119

▶ 1-Nav

	Shift	Peak
▶a	3.04	1A
b	6.70	Q
c	7.77	Q

▶ $C_{17}H_{20}N_2O$ ▶ 7% $CDCl_3$ ▶ V 329

▶ 1-(N)bc

	Shift	Peak
▶a	2.18	1A
b	7.30	4A
c	7.75	6A
d	8.55	2A
e	8.60	2A

▶ $C_{10}H_{14}N_2$ ▶ 7% $CDCl_3$ ▶ V 269

▶ 1-Nav

	Shift	Peak
▶a	3.05	1A
b	6.69	Q
c	7.71	Q
d	9.70	1A

▶ $C_9H_{11}NO$ ▶ 7% $CDCl_3$ ▶ V 238

▶ 1-(N)bc

	Shift	Peak
a	0.93	1A
b	1.05	2A
▶c	2.22	1A
d	2.32	1A
e	5.38	1I

▶ $C_{24}H_{40}N_2$ ▶ 7% $CDCl_3$ ▶ V 359

▶ 1-Naz

	Shift	Peak
a	2.64	2A

▶ $C_6H_{18}N_3PO$ ▶ 7% $CDCl_3$ ▶ V 145

▶ 1-(N)bc

	Shift	Peak
▶a	2.53	1E
b	3.88*	1A
c	3.88*	1A
d	3.97	S
e	5.47	2A
f	5.88	1A
g	6.38†	1E
h	6.38†	1E
i	6.52	2H
j	7.07	2C

*or 4.05
†or 6.57

▶ $C_{21}H_{21}NO_6$ ▶ 7% $CDCl_3$ ▶ V 347

▶ 1-(N)bc

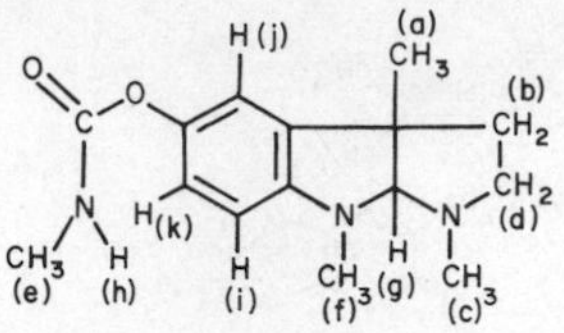

	Shift	Peak
a	1.42	1A
b	1.95	3A
▶c	2.55	1A
d	2.70	3D
e	2.82	1E
f	2.92	1E
g	4.12	1A
h	5.33	1D
i	6.37	2A
j	6.78	1E
k	6.87	2H

▶ $C_{15}H_{21}N_3O_2$ ▶ 7% $CDCl_3$ ▶ V 319

▶ 1-(N)cv

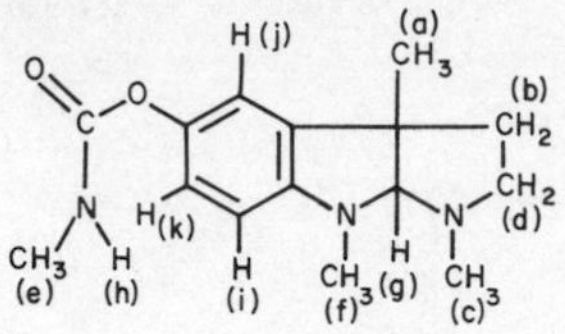

	Shift	Peak
a	1.42	1A
b	1.95	3A
c	2.55	1A
d	2.70	3D
e	2.82	1E
▶f	2.92	1E
g	4.12	1A
h	5.33	1D
i	6.37	2A
j	6.78	1E
k	6.87	2H

▶ $C_{15}H_{21}N_3O_2$ ▶ 7% $CDCl_3$ ▶ V 319

▶ 1-(N)bc

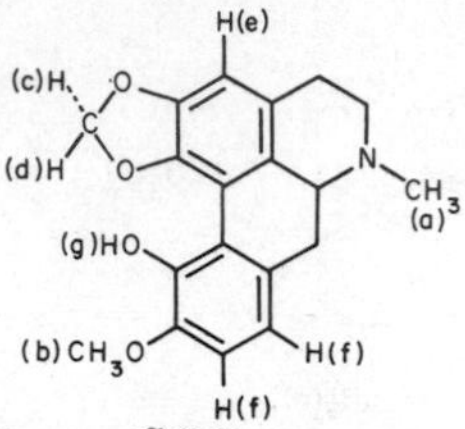

	Shift	Peak
▶a	2.57	1E
b	3.92	1A
c	5.98	2C
d	6.13	2C
e	6.68	1A
f	6.88	1E
g	6.88	1E

▶ $C_{19}H_{19}NO_4$ ▶ 7% $CDCl_3$ ▶ V 333

▶ 1-(N)dd

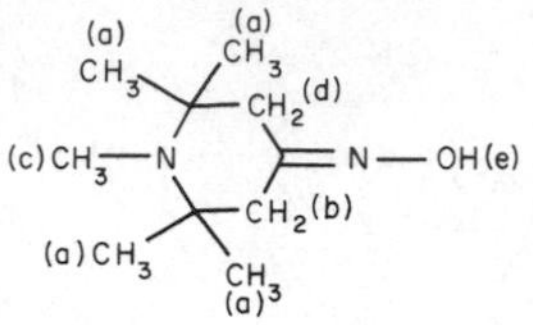

	Shift	Peak
a	1.12	1A
b	2.24	1H
▶c	2.28	1A
d	2.56	1B
e	8.73	1B

▶ $C_{10}H_{20}N_2O$ ▶ 7% $CDCl_3$ ▶ V 574

▶ 1-(N)bc

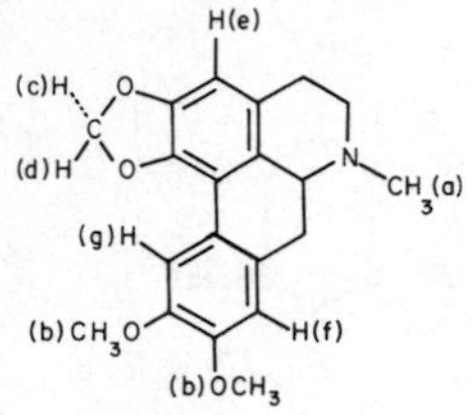

	Shift	Peak
▶a	2.57	1E
b	3.93	1A
c	5.95	2C
d	6.10	2C
e	6.55	1A
f	6.83	1A
g	7.72	1A

▶ $C_{20}H_{21}NO_4$ ▶ 7% $CDCl_3$ ▶ V 342

▶ 1-(N)dv

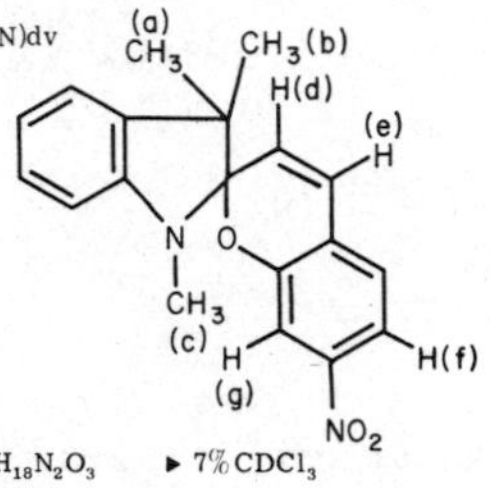

	Shift	Peak
a	1.19	1A
b	1.30	1A
▶c	2.74	1A
d	5.84	2A
e	6.93	S
f	8.01 ± .02	S
g	8.01 ± .02	S

▶ $C_{19}H_{18}N_2O_3$ ▶ 7% $CDCl_3$ ▶ V 677

▶ 1-(N)bj

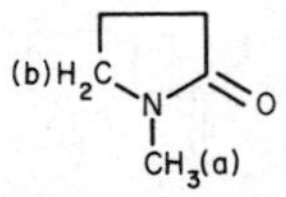

	Shift	Peak
▶a	2.83	1A
b	3.40	3A

▶ C_5H_9NO ▶ 7% $CDCl_3$ ▶ V 116

▶ 1-Nhj

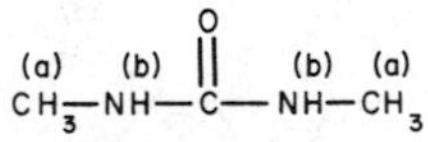

	Shift	Peak
▶a	2.72	2A
b	5.69	1C

▶ $C_3H_8N_2O$ ▶ 45.2 mg/0.50 ml $CDCl_3$ ▶ S 189

▶ 1-(N)bl

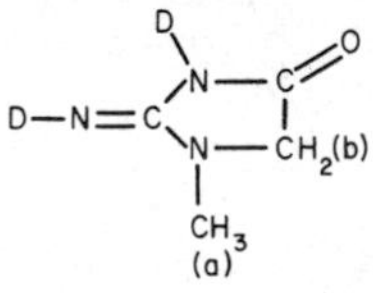

	Shift	Peak
▶a	3.06	1A
b	4.07	1A

▶ $C_4H_7N_3O$ ▶ D_2O ▶ V 411

▶ 1-Nhk

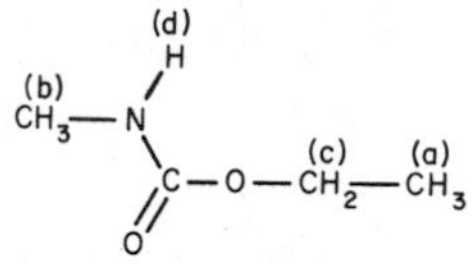

	Shift	Peak
a	1.23	3A
▶b	2.78	2A
c	4.14	4A
d	5.16	1D

▶ $C_4H_9NO_2$ ▶ 7% $CDCl_3$ ▶ V 85

▶ 1-Nbv

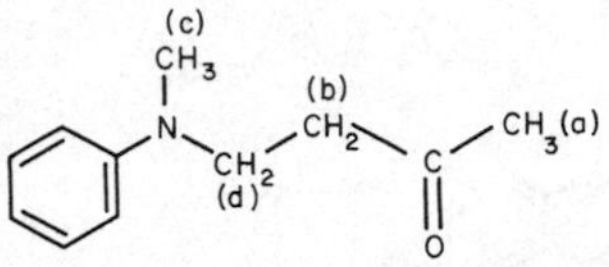

	Shift	Peak
a	2.14	1A
b	2.70	3A
▶c	2.94	1A
d	3.66	3A

▶ $C_{11}H_{15}NO$ ▶ 7% $CDCl_3$ ▶ V 584

▶ 1-Nhk

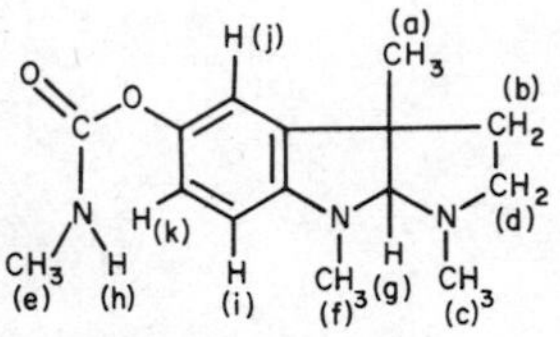

	Shift	Peak
a	1.42	1A
b	1.95	3A
c	2.55	1A
d	2.70	3D
▶e	2.82	1E
f	2.92	1E
g	4.12	1A
h	5.33	1D
i	6.37	2A
j	6.78	1E
k	6.87	2H

▶ $C_{15}H_{21}N_3O_2$ ▶ 7% $CDCl_3$ ▶ V 319

▶ 1-Nhv

H(c) H(b) (d)H N CH_3(a) O_2N H(c) H(d)

	Shift	Peak
▶ a	2.95	1A
b	4.63	1D
c	6.55	Q
d	8.10	Q

▶ $C_7H_8N_2O_2$ ▶ 7% $CDCl_3$ ▶ V 489

▶ 1-(N)jl

O (b) CH_3 N H(c) O N H(d) CH_3 (a)

	Shift	Peak
▶ a	3.37 or 3.43	1A
b	3.37 or 3.43	1A
c	5.73	2C
d	7.20	2C

▶ $C_6H_8N_2O_2$ ▶ 7% $CDCl_3$ ▶ V 460

▶ 1-Nhv

(d) H (e) (a) O H N CH_3 (c)H H(b) (c)H H(b) H (d) O H N CH_3 (e) (a)

	Shift	Peak
▶ a	3.04	2A
b	7.13	1G
c	7.67	Q
d	8.31	Q
e	10.46	1C

▶ $C_{16}H_{14}N_2O_2$ ▶ 7% $CDCl_3$ ▶ V 652

▶ 1-Njv

(b) (c) CH_3 CH_3 N C CH_3(a) O (d)

	Shift	Peak
a	1.78	1A
b	2.25	1A
▶ c	3.20	1A

▶ $C_{10}H_{13}NO$ ▶ 7% $CDCl_3$ ▶ V 264

▶ 1-(N)jj

O (b)CH_3 N H(c) O N H(d) CH_3(a)

	Shift	Peak
a	3.30 or 3.43	1A
▶ b	3.30 or 3.43	1A
c	5.84	2C
d	7.61	2C

▶ $C_6H_8N_2O_2$ ▶ D_2O ▶ V 461

▶ 1-(N)jv

(f) (a) H OCH_3 (e)H H(c) O N O (d)H_2C O CH_3(h)

	Shift	Peak
a	3.83 or 3.92	1A
▶ b	3.83 or 3.92	1A
c	5.92	1A
d	6.07	1A
e	6.80	2C
f	7.57	2C

▶ $C_{12}H_{11}NO_4$ ▶ 7% $CDCl_3$ ▶ V 291

▶ 1-(N)jj

O (b) CH_3 N H(c) O N H(d) CH_3 (a)

	Shift	Peak
a	3.37 or 3.43	1A
▶ b	3.37 or 3.43	1A
c	5.73	2C
d	7.20	2C

▶ $C_6H_8N_2O_2$ ▶ 7% $CDCl_3$ ▶ V 460

▶ 1-(N)jv

(e) CH_2 H(h) O CH_2(a) (g)H CH_2(b) (f)H N O CH_3O (c) CH_3 (d)

	Shift	Peak
a	2.03	5B
b	2.63	3A
c	3.87 or 3.92	1A
▶ d	3.87 or 3.92	1A
e	4.30	3A
f	~7.00	R
g	~7.10	R
h	7.52	4D

▶ $C_{14}H_{15}NO_3$ ▶ 7% $CDCl_3$ ▶ V 312

▶ 1-(N)jl

(b) (a) H_2C CH_2 S C N C O CH_3 (c)

	Shift	Peak
a	2.69 or 3.18	Q
b	2.69 or 3.18	Q
▶ c	3.28	1A

▶ C_5H_7NOS ▶ 7% $CDCl_3$ ▶ V 434

▶ 1-(N)jv

(d) OCH_3 (a) (c) (b) CH_2 CH_2OH N O OCH_3 CH_3 (e) (f)

	Shift	Peak
a	3.03	3A
b	~3.67	1D
c	3.92	3D
d	3.92 or 4.00	1E
e	3.92 or 4.00	1E
▶ f	3.92 or 4.00	1E

▶ $C_{14}H_{17}NO_4$ ▶ 7% $CDCl_3$ ▶ V 313

▶ 1-(N)jl

O (b)CH_3 N H(c) O N H(d) CH_3(a)

	Shift	Peak
▶ a	3.30 or 3.43	1A
b	3.30 or 3.43	1A
c	5.84	2C
d	7.61	2C

▶ $C_6H_8N_2O_2$ ▶ D_2O ▶ V 461

▶ 1-(N)ll

(c)H H(d) (a) CH_3 N C H(e) O CH_3 (b)

	Shift	Peak
a	2.27	1A
▶ b	3.87	1A
c	6.02	2A
d	6.81	2A
e	9.38	1A

▶ C_7H_9NO ▶ 7% $CDCl_3$ ▶ V 170

▶ 1-(N)ll

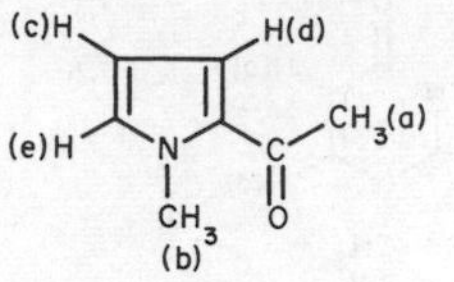

	Shift	Peak
a	2.40	1A
▶b	3.92	1A
c	6.10	4A
d	6.77	3A
e	6.92	4A

▶ C_7H_9NO ▶ 7% $CDCl_3$ ▶ V 174

▶ 1-(N)lv

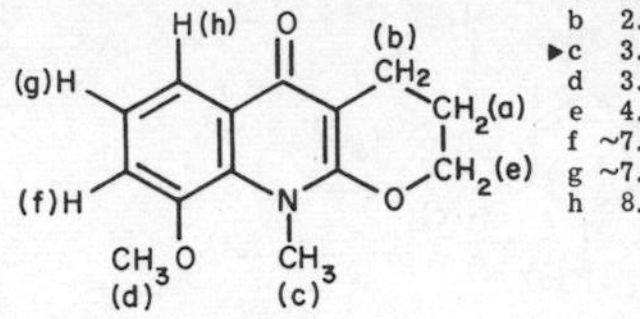

	Shift	Peak
a	2.05	5B
b	2.70	3A
▶c	3.77 or 3.87	1A
d	3.77 or 3.87	1A
e	4.33	3A
f	~7.00	S
g	~7.15	S
h	8.00	4D

▶ $C_{14}H_{15}NO_3$ ▶ 7% $CDCl_3$ ▶ V 311

▶ 1-(N)ln

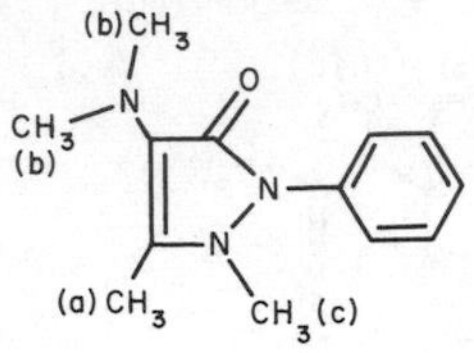

	Shift	Peak
a	2.21	1A
b	2.81	1A
▶c	2.96	1A

▶ $C_{13}H_{17}N_3O$ ▶ 7% $CDCl_3$ ▶ V 615

▶ 1-(N)lv

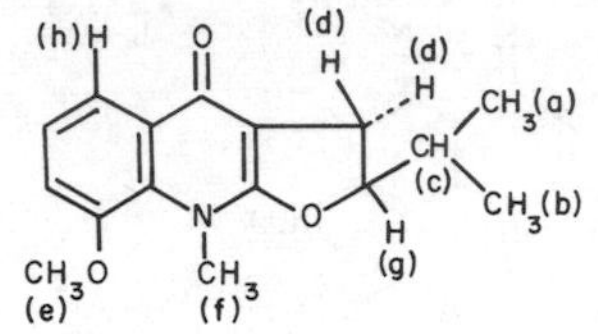

	Shift	Peak
a	0.97*	2A
b	0.97*	2A
c	2.00	6A
d	~3.20	S
e	3.87†	1A
▶f	3.87†	1A
g	4.72	4C
h	8.05	4A

*or 1.02
†or 3.90

▶ $C_{16}H_{19}NO_3$ ▶ 7% $CDCl_3$ ▶ V 325

▶ 1-(N)ln

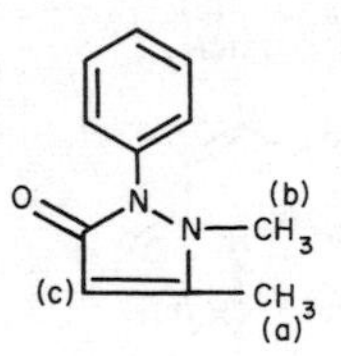

	Shift	Peak
a	2.19	1A
▶b	2.99	1A
c	5.31	1G

▶ $C_{11}H_{12}N_2O$ ▶ 44.7 mg/0.50 ml $CDCl_3$ ▶ S 258

▶ 1-Nlv

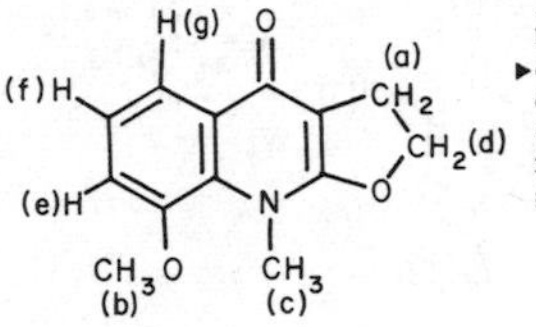

	Shift	Peak
a	3.27	Q
b	3.87 or 3.92	1A
▶c	3.87 or 3.92	1A
d	4.73	Q
e	~7.05	R
f	~7.23	R
g	8.05	4D

▶ $C_{13}H_{13}NO_3$ ▶ 7% $CDCl_3$ ▶ V 300

▶ 1-(N)ln

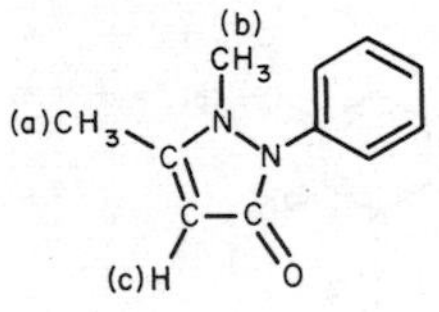

	Shift	Peak
a	2.23	1G
▶b	3.06	1A
c	5.38	2E

▶ $C_{11}H_{12}N_2O$ ▶ 7% $CDCl_3$ ▶ V 581

▶ 1-(N)mm

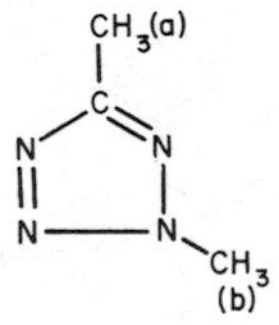

	Shift	Peak
a	2.52	1A
▶b	4.30	1A

▶ $C_3H_6N_4$ ▶ 7% $CDCl_3$ ▶ V 387

▶ 1-(N)lv

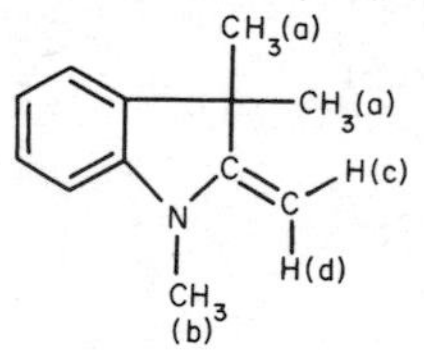

	Shift	Peak
a	1.35	1A
▶b	3.02	1A
c	3.82	1A
d	3.82	1A

▶ $C_{12}H_{15}N$ ▶ 7% $CDCl_3$ ▶ V 596

▶ 1-Nmv

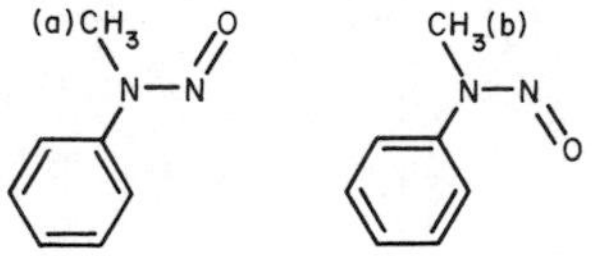

	Shift	Peak
▶a	3.46	1A
▶b	4.12	1A

▶ $C_7H_8N_2O$ ▶ 7% $CDCl_3$ ▶ V 488

▶ 1-(N)lv

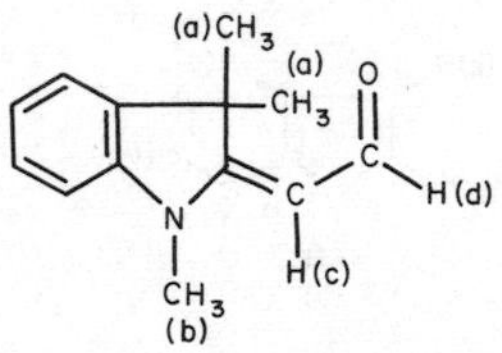

	Shift	Peak
a	1.67	1A
▶b	3.22	1A
c	5.37	2C
d	9.98	2C

▶ $C_{13}H_{15}NO$ ▶ 7% $CDCl_3$ ▶ V 610

▶ 1-Ob

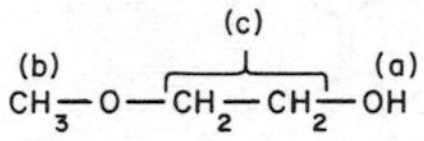

	Shift	Peak
a	2.72	1C
▶b	3.27	1A
c	~3.50	S

▶ $C_3H_8O_2$ ▶ 37.4 mg/0.50 ml $CDCl_3$ ▶ S 32

▶ 1-Ob

$CH_3OCH_2CH_2CN$ — (b) CH_3, (c) OCH_2, (a) CH_2CN

	Shift	Peak
a	2.62	3A
▶b	3.40	1A
c	3.62	3A

▶ C_4H_7NO ▶ 7% $CDCl_3$ ▶ V 69

▶ 1-Oc

	Shift	Peak
a	1.92	1A
▶b	3.47	1A
c	5.07	2A
d	5.58	1E
e	~5.63	4E

▶ $C_{22}H_{24}N_2O_6$ ▶ 7% $CDCl_3$ ▶ V 356

▶ 1-Ob

(a) (b) $CH_3O\,CH_2CN$

	Shift	Peak
▶a	3.47	1A
b	4.20	1A

▶ C_3H_5NO ▶ 7% $CDCl_3$ ▶ V 28

▶ 1-Od

(a) $CH_3(CH_2)_3\,C(OCH_3)_3$ (b)

	Shift	Peak
a	0.93	3B
▶b	3.22	1A

▶ $C_8H_{18}O_3$ ▶ 7% $CDCl_3$ ▶ V 218

▶ 1-Oc

(a) CH_3—C(H(e))(OCH_3(d))—(b) CH_2—(f) CH_2—(c) OH

	Shift	Peak
a	1.18	2A
b	1.72	4A
c	3.10	3A
▶d	3.33	1A
e	3.66	O
f	3.73	6B

▶ $C_5H_{12}O_2$ ▶ 7% $CDCl_3$ ▶ V 120

▶ 1-Od

	Shift	Peak
a	2.13	2D
▶b	3.46	1A
c	3.69	1B
d	5.88	2A
e	6.43	2A
f	6.84	2G

▶ $C_9H_{10}O_2$ ▶ 7% $CDCl_3$ ▶ V 531

▶ 1-Oc

Br—CH_2(a)—C(H(c))(OCH_3(b))(OCH_3(b))

	Shift	Peak
a	3.37	1A
▶b	3.40	1A
c	4.54	3A

▶ $C_4H_9BrO_2$ ▶ 7% $CDCl_3$ ▶ V 419

▶ 1-Oj

(a) CH_3(H_2N)C=C(H(c))—C(=O)—OCH_3(b)

	Shift	Peak
a	1.92	1A
▶b	3.62	1A
c	4.53	1B
d	~6.15	1D

▶ $C_5H_9NO_2$ ▶ 7% $CDCl_3$ ▶ V 442

▶ 1-Oc

	Shift	Peak
▶a	3.41	1A
b	3.67	S
c	3.82 ± 0.04	S
d	3.82 ± 0.04	S
e	3.89 ± 0.05	S
f	3.98 ± 0.05	1F
g	4.82	2G

▶ $C_6H_{12}O_5$ ▶ D_2O ▶ V 475

▶ 1-Ol

(a)H—C≡C—C(H(c))=C(H(d))—OCH_3(b)

	Shift	Peak
a	3.08	2E
▶b	3.80	1A
c	4.52	4D
d	6.35	2E

▶ C_5H_6O ▶ 7% $CDCl_3$ ▶ V 100

▶ 1-Oc

(c)H—C(OCH_3(a))(OCH_3(a))—C(=O)—OCH_3(b)

	Shift	Peak
▶a	3.44	1A
b	3.82	1A
c	4.82	1A

▶ $C_5H_{10}O_4$ ▶ 7% $CDCl_3$ ▶ V 446

▶ 1-Ol

	Shift	Peak
▶a	3.80	1A
b	3.92	1B
c	5.05	1I or 2D
d	~6.30	1D

▶ $C_5H_7NO_2$ ▶ 7% $CDCl_3$ ▶ V 105

▶ 1-Ol

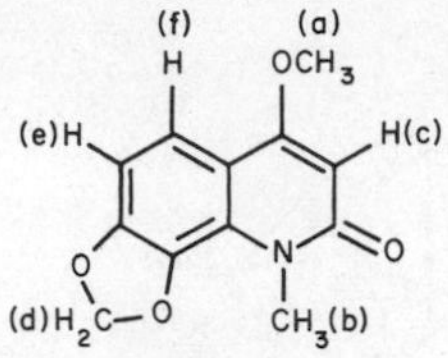

	Shift	Peak
▶a	3.83 or 3.92	1A
b	3.83 or 3.92	1A
c	5.92	1A
d	6.07	1A
e	6.80	2C
f	7.57	2C

▶ $C_{12}H_{11}NO_4$ ▶ 7% $CDCl_3$ ▶ V 291

▶ 1-Ov

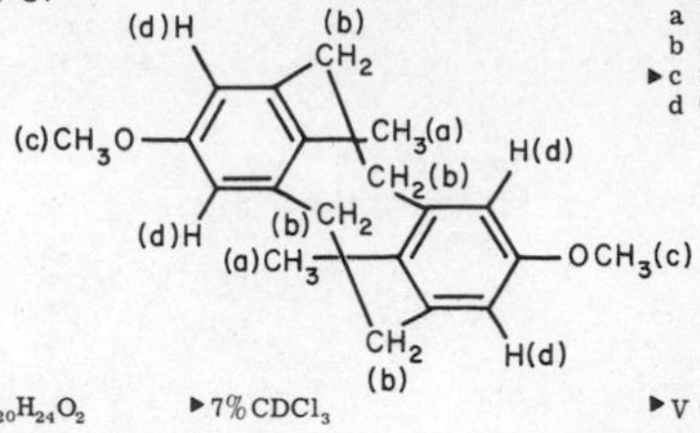

	Shift	Peak
a	1.53	1A
b	2.83	1A
▶c	3.67	1A
d	6.45	1A

▶ $C_{20}H_{24}O_2$ ▶ 7% $CDCl_3$ ▶ V 681

▶ 1-Ol

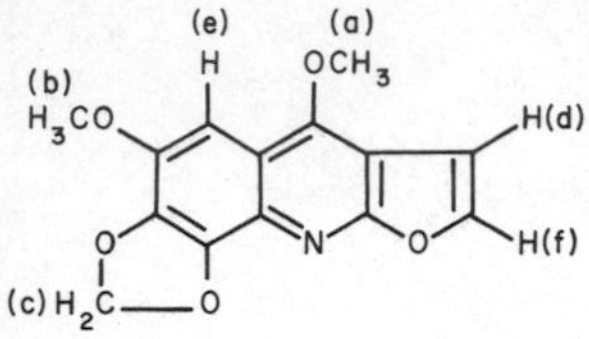

	Shift	Peak
a	4.28 or 4.40	1A
▶b	4.28 or 4.40	1A
c	6.08	1A
d	7.05	2C
e	7.28	1H
f	7.62	2C

▶ $C_{14}H_{11}NO_5$ ▶ 7% $CDCl_3$ ▶ V 304

▶ 1-Ov

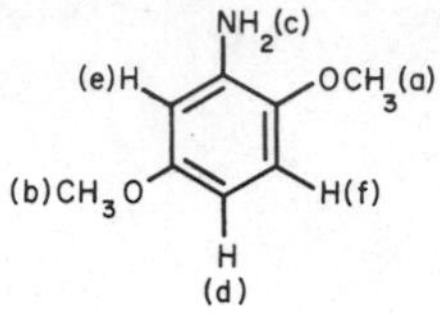

	Shift	Peak
▶a	3.72 or 3.79	1A
▶b	3.72 or 3.79	1A
c	~3.75	1E
d	6.25	S
e	6.32	S
f	6.71	2A

▶ $C_8H_{11}NO_2$ ▶ 7% $CDCl_3$ ▶ V 508

▶ 1-Om

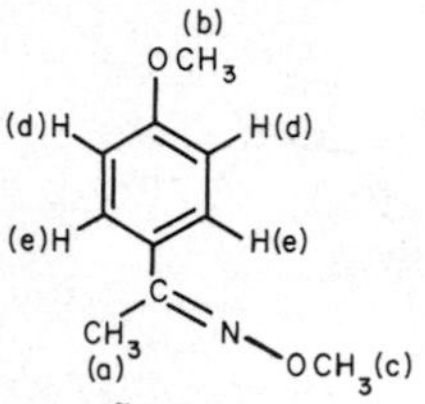

	Shift	Peak
a	2.18	1A
b	3.86	1A
▶c	3.97	1A
d	6.87	Q
e	7.58	Q

▶ $C_{10}H_{13}NO_2$ ▶ 7% $CDCl_3$ ▶ V 561

▶ 1-Ov

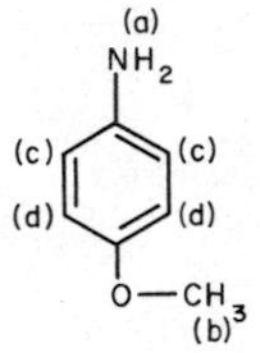

	Shift	Peak
a	3.38	1A
▶b	3.73	1A
c	~6.68 or 6.72	Q
d	~6.68 or 6.72	Q

▶ C_7H_9NO ▶ 45.2 mg/0.50 ml $CDCl_3$ ▶ S 94

▶ 1-Om

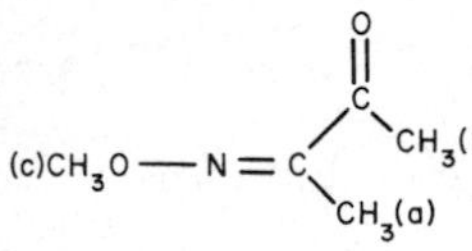

	Shift	Peak
a	1.88 or 2.35	1A
b	1.88 or 2.35	1A
▶c	4.05	1A

▶ $C_5H_9NO_2$ ▶ 7% $CDCl_3$ ▶ V 117

▶ 1-Ov

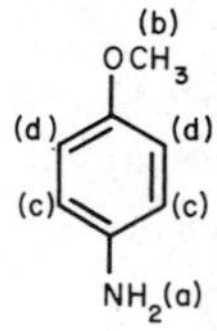

	Shift	Peak
a	3.40	1A
▶b	3.73	1A
c	~6.68 or 6.73	Q
d	~6.68 or 6.73	Q

▶ C_7H_9NO ▶ 7% $CDCl_3$ ▶ V 171

▶ 1-Os

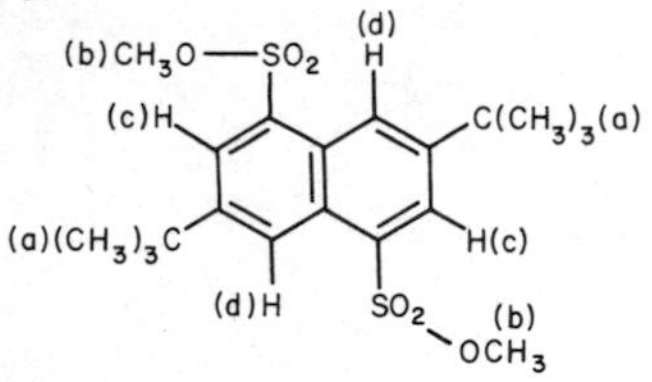

	Shift	Peak
a	1.30	1A
▶b	3.80	1A
c	8.50	2C
d	8.89	2C

▶ $C_{20}H_{28}O_6S_2$ ▶ 7% $CDCl_3$ ▶ V 684

▶ 1-Ov

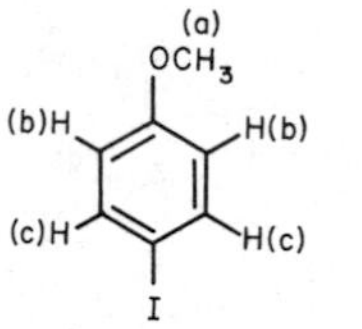

	Shift	Peak
▶a	3.75	1A
b	6.67	Q
c	7.53	Q

▶ C_7H_7IO ▶ 7% $CDCl_3$ ▶ V 153

▶ 1-Ov

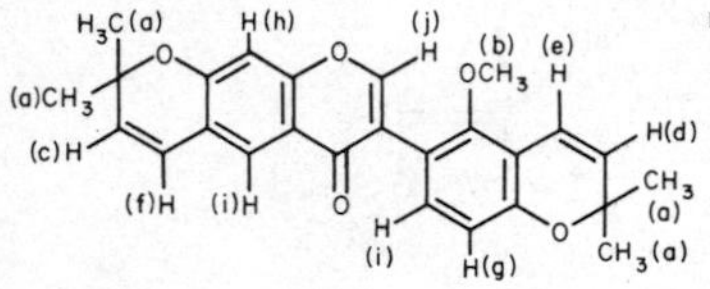

	Shift	Peak
a	1.46*	1A
▶b	3.58	1A
c	5.62†	2A
d	5.62†	2A
e	6.45‡	2A
f	6.45‡	2A
g	6.62	S
h	7.12	2A
i	7.88§	1A
j	7.88§	1A

*or 1.49 †or 5.70 ‡or 6.60 §or 7.90

▶ $C_{26}H_{24}O_5$ ▶ 7% $CDCl_3$ ▶ V 696

▶ 1-Ov

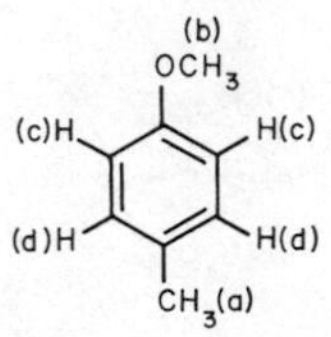

	Shift	Peak
a	2.28	1A
▶b	3.75	1A
c	6.80	Q
d	7.05	Q

▶ $C_8H_{10}O$ ▶ 7% $CDCl_3$ ▶ V 205

▶ 1-Ov

(b) OCH_3
(e)H H(e)
(f)H H(f)
(d)H C H(c)
CH_3(a)

	Shift	Peak
a	1.83	2A
▶b	3.75	1A
c	6.08	N
d	6.28	N
e	6.80	Q
f	7.23	Q

▶ $C_{10}H_{12}O$ ▶ 7% $CDCl_3$ ▶ V 258

▶ 1-Ov

H(h) O (b) CH_2
(g)H CH_2(a)
(f)H N O CH_2(e)
CH_3O (d) CH_3 (c)

	Shift	Peak
a	2.05	5B
b	2.70	3A
c	3.77 or 3.87	1A
▶d	3.77 or 3.87	1A
e	4.33	3A
f	~7.00	S
g	~7.15	S
h	8.00	4D

▶ $C_{14}H_{15}NO_3$ ▶ 7% $CDCl_3$ ▶ V 311

▶ 1-Ov

(e)H (g) H (d) H O
(b)CH_3O C C H(h)
(c) HO H H(f)
H (e) H (g) (h) H (f) OCH_3(a)

	Shift	Peak
▶a	3.75*	1A
▶b	3.75*	1A
c	4.54	1C
d	5.82	1B
e	6.83	S
f	6.87	S
g	7.24	2E
h	7.89	2E

*or 3.81

▶ $C_{16}H_{16}O_4$ ▶ 7% $CDCl_3$ ▶ V 655

▶ 1-Ov

(a) OCH_3
(b) (b)
(c) (c)
Br

	Shift	Peak
▶a	3.78	1A
b	6.75	Q
c	7.37	Q

▶ C_7H_7BrO ▶ 45.5 mg/0.50 ml $CDCl_3$ ▶ S 141

▶ 1-Ov

(b)CH_2 – COOH
CH_2(c)
(a) CH_3 O O CH_3(a)
(a)CH_3 CH_3(a)
(e)H H(e)
H (f) OCH_3 (d) H (f)

	Shift	Peak
a	1.42	1A
b	2.62	Q
c	2.86	Q
▶d	3.75	1A
e	5.50	2C
f	6.53	2C

▶ $C_{20}H_{24}O_5$ ▶ 7% $CDCl_3$ ▶ V 344

▶ 1-Ov

O CH_3(a)

	Shift	Peak
a	3.78	1A

▶ C_7H_8O ▶ 7% $CDCl_3$ ▶ V 162

▶ 1-Ov

(a) OCH_3
(b) (b)
(c) (c)
Cl

	Shift	Peak
▶a	3.77	1A
b	6.80	Q
c	7.20	Q

▶ C_7H_7ClO ▶ 44.8 mg/0.50 ml $CDCl_3$ ▶ S 164

▶ 1-Ov

OH
(a) O—CH_3

	Shift	Peak
a	3.78	1A

▶ $C_7H_8O_2$ ▶ 45.2 mg/0.50 ml $CDCl_3$ ▶ S 259

▶ 1-Ov

(a) OCH_3
(b) (b)
(c) (c)
F

	Shift	Peak
▶a	3.77	1A
b	6.82	S
c	6.95	S

▶ C_7H_7FO ▶ 45.2 mg/0.50 ml $CDCl_3$ ▶ S 233

▶ 1-Ov

(c) (b)
(d) O—CH_3
(a) H_3C (c)
(d)

	Shift	Peak
a	2.29	1A
▶b	3.78	1A
c	6.76	Q
d	7.08	Q

▶ $C_8H_{10}O$ ▶ 45.0 mg/0.50 ml $CDCl_3$ ▶ S 292

▶ 1-Ov

(e) CH_3O (b) CH_2
CH_2(a)
N CH_2(c)
H (d)

	Shift	Peak
a	1.93	5B
b	2.73	3A
c	3.22	3C
d	3.49	1B
▶e	3.77	1A

▶ $C_{10}H_{13}NO$ ▶ 7% $CDCl_3$ ▶ V 266

▶ 1-Ov

OCH_3
(b)
OCH_3(a)

	Shift	Peak
▶a	3.78	1A
b	6.85	1A

▶ $C_8H_{10}O_2$ ▶ 45.1 mg/0.50 ml $CDCl_3$ ▶ S 56

▶ 1-Ov

	Shift	Peak
a	2.22	3A
▶b	3.79	1A
c	4.17	5A
d	6.50	1A
e	6.83	3A

▶ $C_{13}H_{14}O_2$ ▶ 7% $CDCl_3$ ▶ V 608

▶ 1-Ov

	Shift	Peak
▶a	3.86	1A
b	6.82	Q
c	7.91	Q

▶ $C_9H_{10}O_3$ ▶ 45.0 mg/0.50 ml $CDCl_3$ ▶ S 124

▶ 1-Ov

	Shift	Peak
a	3.70	1B
▶b	3.82	1A
c	6.73	1A

▶ C_7H_9NO ▶ 45.2 mg/0.50 ml $CDCl_3$ ▶ S 192

▶ 1-Ov

	Shift	Peak
a	2.18	1A
▶b	3.86	1A
c	3.97	1A
d	6.87	Q
e	7.58	Q

▶ $C_{10}H_{13}NO_2$ ▶ 7% $CDCl_3$ ▶ V 561

▶ 1-Ov

	Shift	Peak
a	1.88	2A
b	2.31	1A
▶c	3.82	1A
d	~6.1-6.7	S
e	~6.1-6.7	S

▶ $C_{12}H_{14}O_3$ ▶ 45.4 mg/0.50 ml $CDCl_3$ ▶ S 130

▶ 1-Ov

	Shift	Peak
▶a	3.87	1A
b	7.02	Q
c	7.83	Q
d	9.87	1A

▶ $C_8H_8O_2$ ▶ 7% $CDCl_3$ ▶ V 194

▶ 1-Ov

	Shift	Peak
a	3.30	2E
▶b	3.83	1A
c	5.05	2E
d	5.05	2E
e	5.59	1B
f	~5.94	O
g	~6.64	S
h	~6.69	S
i	6.87	S

▶ $C_{10}H_{12}O_2$ ▶ 7% $CDCl_3$ ▶ V 260

▶ 1-Ov

	Shift	Peak
a	1.40	3A
▶b	3.87	1A
c	4.40	4A
d	6.97	2A
e	7.02	1A
f	7.98	2A
g	~15.33	1D

▶ $C_{12}H_{14}O_4$ ▶ 7% $CDCl_3$ ▶ V 295

▶ 1-Ov

	Shift	Peak
a	2.33	1A
b	3.13	3A
▶c	3.83	1A
d	4.28	4A
e	6.85 or 6.98	1A
f	6.85 or 6.98	1A
g	7.13	1B

▶ $C_{13}H_{13}NO_2$ ▶ 7% $CDCl_3$ ▶ V 299

▶ 1-Ov

	Shift	Peak
a	3.27	Q
▶b	3.87 or 3.92	1A
c	3.87 or 3.92	1A
d	4.73	Q
e	~7.05	R
f	~7.23	R
g	8.05	4D

▶ $C_{13}H_{13}NO_3$ ▶ 7% $CDCl_3$ ▶ V 300

▶ 1-Ov

	Shift	Peak
a	3.73	1A
▶b	3.85	1A
c	~6.80	1A

▶ C_7H_9NO ▶ 7% $CDCl_3$ ▶ V 172

▶ 1-Ov

	Shift	Peak
a	2.03	5B
b	2.63	3A
▶c	3.87 or 3.92	1A
d	3.87 or 3.92	1A
e	4.30	3A
f	~7.00	R
g	~7.10	R
h	7.52	4D

▶ $C_{14}H_{15}NO_3$ ▶ 7% $CDCl_3$ ▶ V 312

▶ 1-Ov

(h)H O (d) H (d) H CH(c) CH3(a) CH3(b) N O H (g) CH3O (e) CH3 (f)

	Shift	Peak
a	0.97*	2A
b	0.97*	2A
c	2.00	6A
d	~3.20	S
▶e	3.87†	1A
f	3.87†	1A
g	4.72	4C
h	8.05	4A

*or 1.02
†or 3.90

▶ $C_{16}H_{19}NO_3$ ▶ 7% $CDCl_3$ ▶ V 325

▶ 1-Ov

OCH3(c) O OCH3 (b) H(i) O (d) H H(j) (h)H (e) H N—CH3(a) (f)H2C O O H(g)

	Shift	Peak
a	2.53	1E
▶b	3.88*	1A
▶c	3.88*	1A
d	3.97	S
e	5.47	2A
f	5.88	1A
g	6.38†	1E
h	6.38†	1E
i	6.52	2H
j	7.07	2C

*or 4.05
†or 6.57

▶ $C_{21}H_{21}NO_6$ ▶ 7% $CDCl_3$ ▶ V 347

▶ 1-Ov

COOH (not shown)
(c)H H(c) (b)H H(b) OCH3(a)

	Shift	Peak
▶a	3.88	1A
b	6.98	Q
c	8.08	Q

▶ $C_8H_8O_3$ ▶ 7% $CDCl_3$ ▶ V 196

▶ 1-Ov

O C—Cl (c) (c) (b) (b) OCH3(a)

	Shift	Peak
▶a	3.90	1A
b	6.92	Q
c	8.04	Q

▶ $C_8H_7ClO_2$ ▶ 44.6 mg/0.50 ml $CDCl_3$ ▶ S 180

▶ 1-Ov

(e) H C=C NO2 H(d) (c) OCH3(b) OCH3 (a)

	Shift	Peak
▶a	3.88 or 3.92	1A
▶b	3.88 or 3.92	1A
c	7.07	1A
d	7.73	2A
e	8.20	3A

▶ $C_{10}H_{11}NO_4$ ▶ 7% $CDCl_3$ ▶ V 257

▶ 1-Ov

O C—H(b) OCH3(a)

	Shift	Peak
▶a	3.90	1A
b	9.87	1A

▶ $C_8H_8O_2$ ▶ 45.5 mg/0.50 ml $CDCl_3$ ▶ S 126

▶ 1-Ov

OH(c) OCH3(b) CH=CH—CH3 (e) (d) (a)

	Shift	Peak
a	1.82	2A
▶b	3.88	1A
c	5.48	1B
d	6.03	S
e	6.20	S

▶ $C_{10}H_{12}O_2$ ▶ 45.0 mg/0.5 ml $CDCl_3$ ▶ S 76

▶ 1-Ov

(f)H O N OH(g) (d)H CH2(b) (c) CH3O CH3 (a) CH3 (a) H (e)

	Shift	Peak
a	1.42	1A
b	3.04	1A
▶c	3.91	1A
d	6.88	S
e	6.93	S
f	8.17	2A
g	~10.30	1D

▶ $C_{13}H_{15}NO_3$ ▶ 7% $CDCl_3$ ▶ V 611

▶ 1-Ov

OCH3(c) (e)H NH2(d) (g)H H(f) O←S→O CH3—CH2—N—CH2—CH3 (a) (b) (b) (a)

	Shift	Peak
a	1.11	3A
b	3.20	4A
▶c	3.88	1E
d	3.88	1E
e	6.80	2E
f	7.13	1A
g	7.17	2A

▶ $C_{11}H_{18}N_2O_3S$ ▶ 7% $CDCl_3$ ▶ V 587

▶ 1-Ov

H(e) (c)H O C (d)H O N CH3 (a) (g)HO (b)CH3O H(f) H(f)

	Shift	Peak
a	2.57	1E
▶b	3.92	1A
c	5.98	2C
d	6.13	2C
e	6.68	1A
f	6.88	1E
g	6.88	1E

▶ $C_{19}H_{19}NO_4$ ▶ 7% $CDCl_3$ ▶ V 333

▶ 1-Ov

(d) CH2 CH2(c) CH3(a) (b)CH3 CH(e) OCH3(f) (h)HO C O H(g) (f)CH3O

	Shift	Peak
a	0.96	3B
b	1.37	2A
c	~1.65	S
d	2.58	3B
e	3.42	S
▶f	3.88	1A
g	5.08	2A
h	5.62	1A

▶ $C_{20}H_{24}O_4$ ▶ 7% $CDCl_3$ ▶ V 682

▶ 1-Ov

O C H(f) (d)H H(e) (c)H OCH3(a) OH(b)

	Shift	Peak
▶a	3.93	1A
b	6.47	1D
c	7.02	2A
d	7.40	N
e	7.40	N
f	9.78	1A

▶ $C_8H_8O_3$ ▶ 7% $CDCl_3$ ▶ V 197

▶ 1-Ov

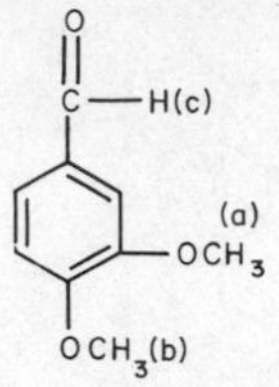

Shift	Peak
▶a 3.93 or 3.96	1A
▶b 3.93 or 3.96	1A
c 9.8	1A

▶ $C_9H_{10}O_3$ ▶ 45.2 mg/0.50 ml $CDCl_3$ ▶ S 197

▶ 1-Ov

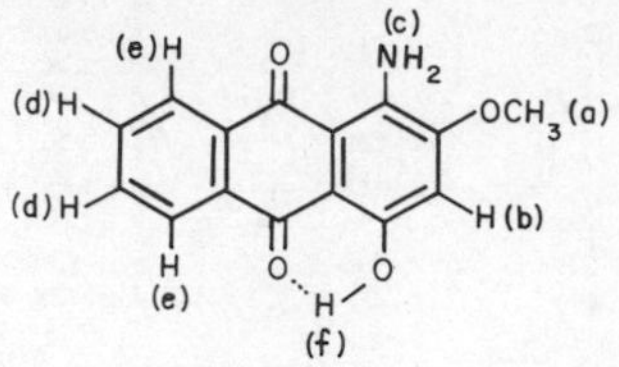

Shift	Peak
▶a 3.96	1A
b 6.52	1A
c ~ 7.30	1D
d 7.72	Q
e 8.34	Q
f 14.22	1A

▶ $C_{15}H_{11}NO_4$ ▶ 7% $CDCl_3$ ▶ V 635

▶ 1-Ov

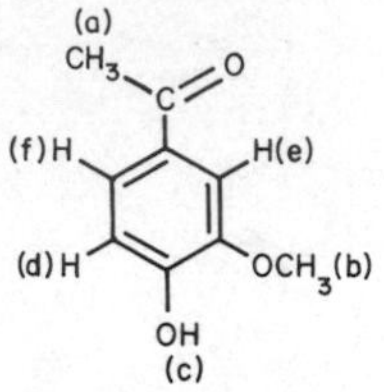

Shift	Peak
a 2.57	1A
▶b 3.93	1A
c 6.62	1C
d 6.97	2A
e 7.57	S
f 7.60	S

▶ $C_9H_{10}O_3$ ▶ 7% $CDCl_3$ ▶ V 234

▶ 1-Ov

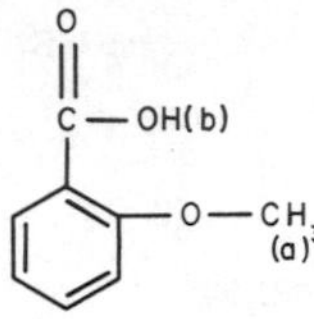

Shift	Peak
▶a 4.02	1A
b 10.30	1B

▶ $C_8H_8O_3$ ▶ 44.6 mg/0.50 ml $CDCl_3$ ▶ S 253

▶ 1-Ov

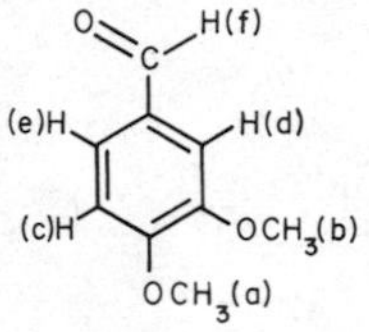

Shift	Peak
▶a 3.93 or 3.97	1A
▶b 3.93 or 3.97	1A
c 6.98	2E
d 7.40	S
e 7.45	S
f 9.98	1A

▶ $C_9H_{10}O_3$ ▶ 7% $CDCl_3$ ▶ V 236

▶ 1-Ov

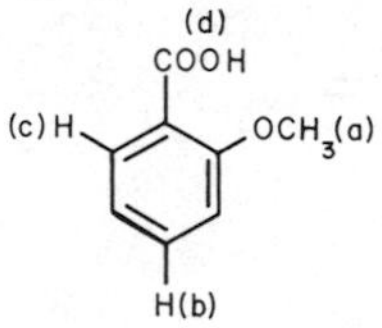

Shift	Peak
▶a 4.07	1A
b ~7.60	O
c 8.17	4A
d 11.00	1B

▶ $C_8H_8O_3$ ▶ 7% $CDCl_3$ ▶ V 195

▶ 1-Ov

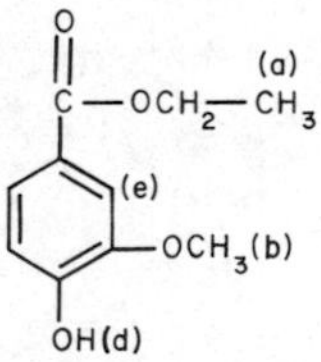

Shift	Peak
a 1.38	3A
▶b 3.93	1A
c 4.32	4A
d 6.80 or 6.94	1A
e 6.80 or 6.94	1A

▶ $C_{10}H_{12}O_4$ ▶ 45.0 mg/0.50 ml $CDCl_3$ ▶ S 127

▶ 1-Ov

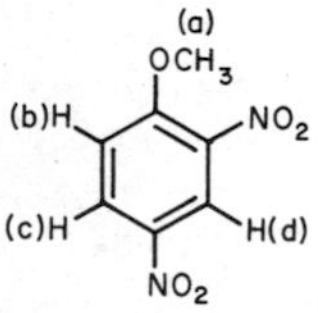

Shift	Peak
▶a 4.12	1A
b 7.28	3A
c 8.47	4A
d 8.72	2A

▶ $C_7H_6N_2O_5$ ▶ 7% $CDCl_3$ ▶ V 149

▶ 1-Ov

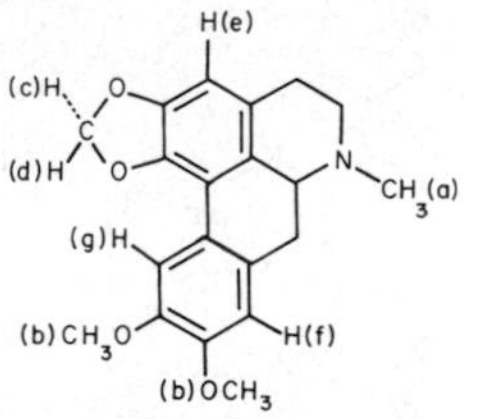

Shift	Peak
a 2.57	1E
▶b 3.93	1A
c 5.95	2C
d 6.10	2C
e 6.55	1A
f 6.83	1A
g 7.72	1A

▶ $C_{20}H_{21}NO_4$ ▶ 7% $CDCl_3$ ▶ V 342

▶ 1-Ox

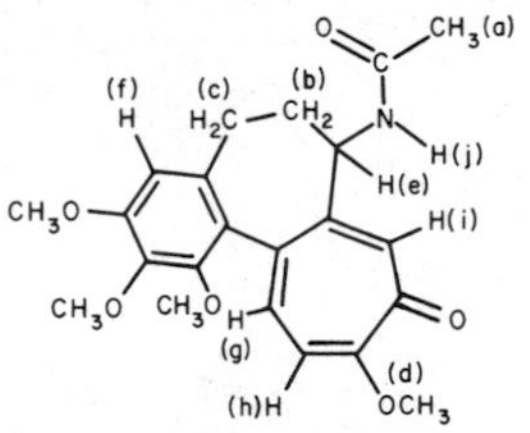

Shift	Peak
a 1.96	1A
b 2.43	S
c 2.43	S
▶d 3.67	1A
e 4.62	S
f 6.55	1A
g 6.92	2A
h 7.37	S
i 7.63	1B
j 8.38	2G

▶ $C_{22}H_{25}NO_6$ ▶ 7% $CDCl_3$ ▶ V 689

▶ 1-Ov

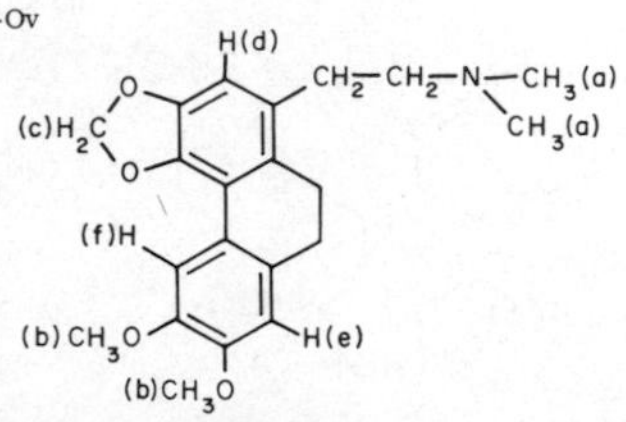

Shift	Peak
a 2.35	1A
▶b 3.93	1A
c 6.03	1A
d 6.65	1A
e 6.80	1A
f 7.77	1I

▶ $C_{21}H_{25}NO_4$ ▶ 7% $CDCl_3$ ▶ V 349

▶ 1-Ox

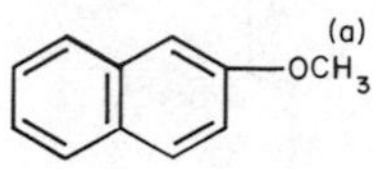

Shift	Peak
a 3.84	1A

▶ $C_{11}H_{10}O$ ▶ 44.8 mg/0.50 ml $CDCl_3$ ▶ S 77

▶ 1-Ox

	Shift	Peak
▶a	3.87	1A
b	7.00	2A
c	7.28	S
d	7.35	S
e	7.97	2E
f	8.71	4A

(b) H, (e) H, (a) CH_3O, H(c), (d)H, N, H(f), H (e)

▶ $C_{10}H_9NO$ ▶ 7% $CDCl_3$ ▶ V 249

▶ 1-P

	Shift	Peak
a	1.43	1A
▶b	3.47	1A

(b) CH_3 — OH (a)

▶ CH_4O ▶ 7% $CDCl_3$ ▶ V 1

▶ 1-Ox

	Shift	Peak
a	3.03	3A
b	~3.67	1D
c	3.92	3D
▶d	3.92 or 4.00	1E
▶e	3.92 or 4.00	1E
f	3.92 or 4.00	1E

(d) OCH_3, (a) (c) (b) CH_2CH_2OH, O, N, OCH_3 (e), CH_3 (f)

▶ $C_{14}H_{17}NO_4$ ▶ 7% $CDCl_3$ ▶ V 313

▶ 1-Qb

	Shift	Peak
a	~0.88	1D
b	1.26	1A
c	~2.30	1D
▶d	3.63	1A

(a) CH_3—(b) $(CH_2)_{13}$—(c) CH_2—C(=O)—O—(d) CH_3

▶ $C_{17}H_{34}O_2$ ▶ 44.4 mg/0.50 ml $CDCl_3$ ▶ S 248

▶ 1-Ox

	Shift	Peak
a	2.38	1A
b	~2.7	Q
c	~3.2	Q
▶d	4.05 and 4.08	1A
e	6.23	1A
f	7.13 or 7.23	1A
g	7.13 or 7.23	1A
h	7.53 or 7.81	2C
i	7.53 or 7.81	2C
j	8.52	1A

H(f), (c) (b) CH_2—CH_2—N, (a) CH_3, CH_3 (a), (e) H_2C, O, O, H(i), (j)H, H(h), (d) CH_3O, H(g), CH_3O(d)

▶ $C_{21}H_{23}NO_4$ ▶ 7% $CDCl_3$ ▶ V 348

▶ 1-Qb

	Shift	Peak
a	3.60	1A
▶b	3.66	1A
c	7.28	1A

(c) C_6H_5—(a) CH_2—C(=O)—(b) OCH_3

▶ $C_9H_{10}O_2$ ▶ 44.4 mg/0.50 ml $CDCl_3$ ▶ S 19

▶ 1-Ox

	Shift	Peak
a	2.40	1G
▶b	4.06 or 4.20	1A
▶c	4.06 or 4.20	1A
d	6.05	1G
e	7.01	2C
f	7.63	2C

(c) OCH_3, O, (e)H, H(d), (f)H, O, O, CH_3(a), OCH_3 (b)

▶ $C_{14}H_{12}O_5$ ▶ 7% $CDCl_3$ ▶ V 628

▶ 1-Qb

	Shift	Peak
a	0.88	3B
b	1.28	1B
c	2.28	3B
▶d	3.67	1A

(a) CH_3—(b) $(CH_2)_9$—(c) CH_2—C(=O)—(d) OCH_3

▶ $C_{13}H_{26}O_2$ ▶ 45.7 mg/0.50 ml $CDCl_3$ ▶ S 125

▶ 1-Ox

	Shift	Peak
▶a	4.28 or 4.40	1A
b	4.28 or 4.40	1A
c	6.08	1A
d	7.05	2C
e	7.28	1H
f	7.62	2C

(e) H, (a) OCH_3, (b) H_3CO, H(d), O, N, O, H(f), (c) H_2C, O

▶ $C_{14}H_{11}NO_5$ ▶ 7% $CDCl_3$ ▶ V 304

▶ 1-Qb

	Shift	Peak
a	2.28	3B
b	2.88	3B
▶c	3.67	1A
d	7.12	R
e	7.62	R
f	7.70	R
g	1.35	1F

(d) H, (f) H, H (e), S, —C(=O)—(h) CH_2—(g) $(CH_2)_6$—(a) CH_2—C(=O)—(c) OCH_3

▶ $C_{15}H_{22}O_3S$ ▶ 7% $CDCl_3$ ▶ V 320

▶ 1-Oz

	Shift	Peak
a	0.12	1H
▶b	3.55	1A

(a) CH_3—Si—(b) OCH_3, (b) OCH_3, OCH_3 (b)

▶ $C_4H_{12}O_3Si$ ▶ 7% $CDCl_3$ ▶ V 93

▶ 1-Qb

	Shift	Peak
a	0.97	3A
b	1.33	1B
c	2.80	3B
▶d	3.67	1A
e	~5.38	3C

(a) CH_3—(e) CH_2—(e) CH=(c) CH—$(CH_2$—(e) CH=(e) $CH)_2$—CH_2—(b) $(CH_2)_5$—CH_2—C(=O)—OCH_3 (d)

▶ $C_{19}H_{32}O_2$ ▶ 7% $CDCl_3$ ▶ V 337

▶ 1-Qb

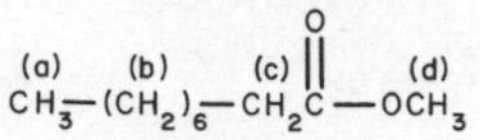

	Shift	Peak
a	0.88	3B
b	1.28	1B
c	2.30	3B
▶d	3.68	1A

(a) CH_3—(b) $(CH_2)_6$—(c) $CH_2C(=O)$—(d) OCH_3

▶ $C_{10}H_{20}O_2$ ▶ 45.1 mg/0.50 ml $CDCl_3$ ▶ S 249

▶ 1-Qb

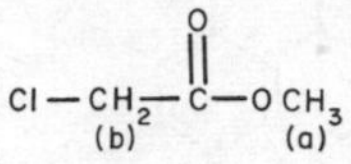

	Shift	Peak
▶a	3.82	1A
b	4.07	1A

Cl—(b) CH_2—C(=O)—O(a) CH_3

▶ $C_3H_5ClO_2$ ▶ 45.1 mg/0.50 ml $CDCl_3$ ▶ S 95

▶ 1-Qb

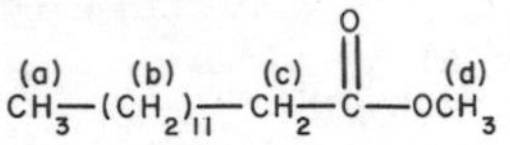

	Shift	Peak
a	0.88	3B
b	1.29	1A
c	2.24	3B
▶d	3.68	1A

(a) CH_3—(b) $(CH_2)_{11}$—(c) CH_2—C(=O)—(d) OCH_3

▶ $C_{15}H_{30}O_2$ ▶ 45.1 mg/0.50 ml $CDCl_3$ ▶ S 195

▶ 1-Qb

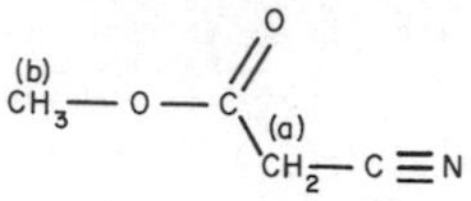

	Shift	Peak
a	3.48	1A
▶b	3.82	1A

(b) CH_3—O—C(=O)—(a) CH_2—C≡N

▶ $C_4H_5NO_2$ ▶ 7% $CDCl_3$ ▶ V 57

▶ 1-Qb

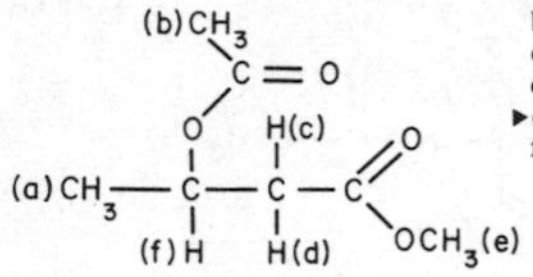

	Shift	Peak
a	1.32	2A
b	2.03	1A
c	2.50 or 2.70	2A
d	2.50 or 2.70	2A
▶e	3.70	1A
f	5.30	4A

(a) CH_3—CH(f)(O—C(=O)—(b) CH_3)—C H(c) H(d)—C(=O)—OCH_3(e)

▶ $C_7H_{12}O_4$ ▶ 7% $CDCl_3$ ▶ V 182

▶ 1-Qc

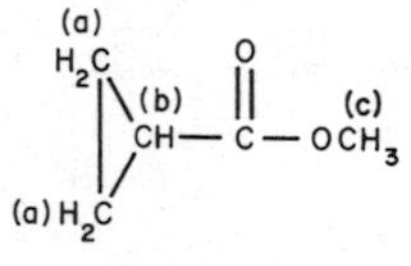

	Shift	Peak
a	~0.93	O
b	~1.63	O
▶c	3.67	1A

(a) H_2C, (a) H_2C (cyclopropane ring) (b) CH—C(=O)—(c) OCH_3

▶ $C_5H_8O_2$ ▶ 7% $CDCl_3$ ▶ V 112

▶ 1-Qb

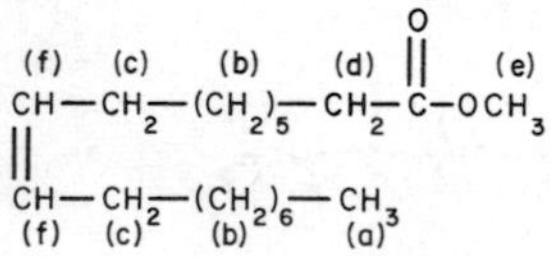

	Shift	Peak
a	0.90	3B
b	1.29	1B
c	2.00	1C
d	2.32	S
▶e	3.70	1A
f	5.4	3B

(f) CH—(c) CH_2—(b) $(CH_2)_5$—(d) CH_2—C(=O)—(e) OCH_3
‖
(f) CH—(c) CH_2—(b) $(CH_2)_6$—(a) CH_3

▶ $C_{19}H_{36}O_2$ ▶ 33.7 mg/0.50 ml $CDCl_3$ ▶ S 71

▶ 1-Qc

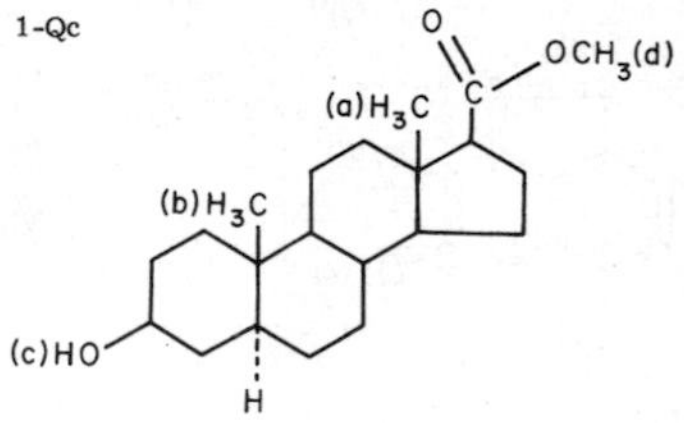

	Shift	Peak
a	0.63	1A
b	0.80	1A
c	1.67	1E
▶d	3.67	1A

(a) H_3C, (b) H_3C, (c) HO, H, C(=O)—OCH_3(d)

▶ $C_{21}H_{34}O_3$ ▶ 7% $CDCl_3$ ▶ V 354

▶ 1-Qb

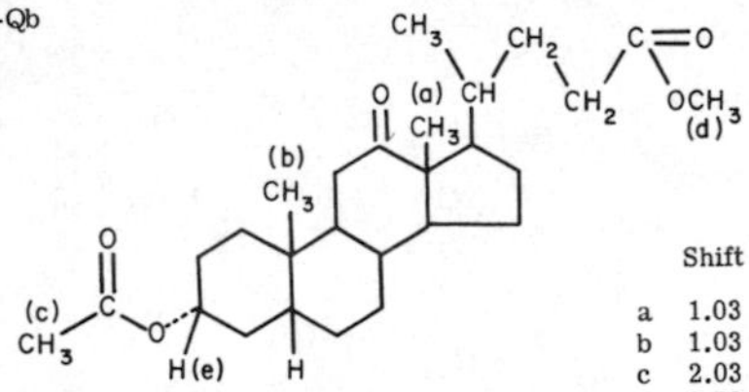

	Shift	Peak
a	1.03	1A
b	1.03	1A
c	2.03	1E
▶d	3.70	1A
e	~4.70	1D

CH_3, (a) CH, CH_2, CH_2, C=O, (d) OCH_3, O, (a) CH_3, (b) CH_3, (c) CH_3C(=O)O, H(e), H

▶ $C_{27}H_{42}O_5$ ▶ 7% $CDCl_3$ ▶ V 362

▶ 1-Qc

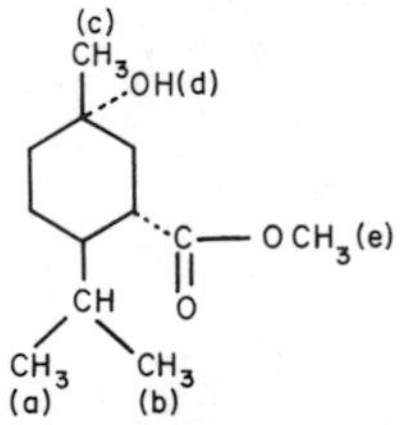

	Shift	Peak
a	0.82 or 0.87	2A
b	0.82 or 0.87	2A
c	1.35	1E
d	1.67	1E
▶e	3.72	1A

(c) CH_3, OH(d), C(=O)—OCH_3(e), CH, CH_3 (a), CH_3 (b)

▶ $C_{12}H_{22}O_3$ ▶ 7% $CDCl_3$ ▶ V 297

▶ 1-Qb

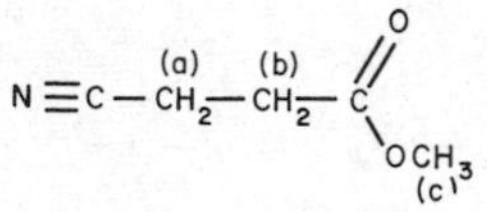

	Shift	Peak
a	2.68	1A
b	2.68	1A
▶c	3.75	1A

N≡C—(a) CH_2—(b) CH_2—C(=O)—OCH_3 (c)

▶ $C_5H_7NO_2$ ▶ 7% $CDCl_3$ ▶ V 106

▶ 1-Qc

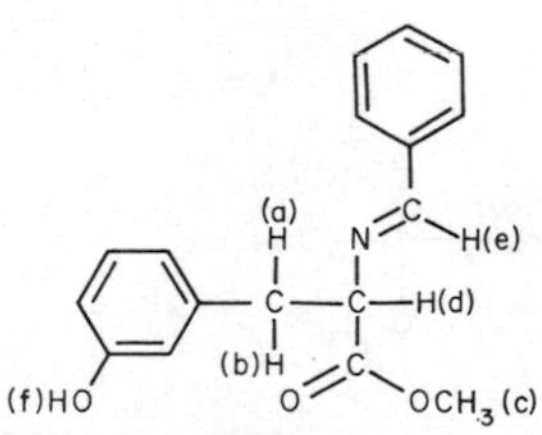

	Shift	Peak
a	3.23	N
b	3.52	N
▶c	3.72	1A
d	4.28	4A
e	7.89	1A
f	8.04	1F

(a) H, (b) H, N=C—H(e), C—H(d), C(=O)—OCH_3(c), (f) HO

▶ $C_{17}H_{17}NO_3$ ▶ 7% $CDCl_3$ ▶ V 663

▶ 1-Qc

	Shift	Peak
a	3.44	1A
▶b	3.82	1A
c	4.82	1A

▶ $C_5H_{10}O_4$ ▶ 7% $CDCl_3$ ▶ V 446

▶ 1-Ql

	Shift	Peak
▶a	3.73	1A
b	6.96	4C
c	12.68	1G

▶ $C_{12}H_{16}O_5$ ▶ 7% $CDCl_3$ ▶ V 600

▶ 1-Qc

	Shift	Peak
a	2.20	1A
b	2.25	1A
▶c	3.87	1A
d	6.05	1A
e	6.63	1A

▶ $C_{11}H_{11}Cl_3O_6$ ▶ 7% $CDCl_3$ ▶ V 284

▶ 1-Ql

	Shift	Peak
a	3.38	1A
▶b	3.73 or 3.84	1A
▶c	3.73 or 3.84	1A
d	7.82	1A
e	12.62	1A

▶ $C_{15}H_{14}O_5$ ▶ 7% $CDCl_3$ ▶ V 639

▶ 1-Qh

	Shift	Peak
▶a	3.77	2D
b	8.08	2E

▶ $C_2H_4O_2$ ▶ 7% $CDCl_3$ ▶ V 9

▶ 1-Ql

	Shift	Peak
▶a	3.75	1A
b	~5.82	4D
c	~6.20	R
d	~6.38	R

▶ $C_4H_6O_2$ ▶ 7% $CDCl_3$ ▶ V 64

▶ 1-Ql

	Shift	Peak
a	3.28	3C
▶b	3.62 or 3.70	1E
▶c	3.62 or 3.70	1E
d	3.74	S
e	6.97	S

▶ $C_{15}H_{21}NO_5$ ▶ 7% $CDCl_3$ ▶ V 643

▶ 1-Ql

	Shift	Peak
a	1.95	2E
▶b	3.75	1A
c	5.57	M
d	6.10	M

▶ $C_5H_8O_2$ ▶ 7% $CDCl_3$ ▶ V 113

▶ 1-Ql

	Shift	Peak
▶a	3.65 or 3.80	1A
▶b	3.80 or 3.65	1A
c	7.20	1A
d	7.83	1A
e	12.69	1A

▶ $C_{16}H_{16}O_5$ ▶ 7% $CDCl_3$ ▶ V 656

▶ 1-Ql

	Shift	Peak
a	1.13	3B
b	1.89	O
c	2.74	O
d	3.42	4A
▶e	3.77	1A
f	3.98	3E
g	4.59	2A
h	5.59	S
i	5.63*	S
j	5.63*	S
k	7.47	1A

*or 6.12

▶ $C_{15}H_{16}O_6$ ▶ 7% $CDCl_3$ ▶ V 641

▶ 1-Ql

	Shift	Peak
▶a	3.72	1A
b	7.10	O
c	~12.30	1D

▶ $C_{11}H_{14}O_5$ ▶ 7% $CDCl_3$ ▶ V 583

▶ 1-Ql

	Shift	Peak
▶a	3.78	1A
b	6.42 or 7.71	2C
c	6.42 or 7.71	2C

▶ $C_{10}H_{10}O_2$ ▶ 44.6 mg/0.25 ml $CDCl_3$ ▶ S 20

▶ 1-Ql

(a) CH_3—O—C(=O)—C—H (b)
(b) H—C—C(=O)—O—CH_3 (a)

	Shift	Peak
▶a	3.79	1A
b	6.88	1A

▶ $C_6H_8O_4$ ▶ 45.0 mg/0.10 ml $CDCl_3$ ▶ S 110

▶ 1-Qv

(c) (b) (b) (c) C(=O)—OCH_3 (a) C(=O)—OCH_3 (a)

	Shift	Peak
▶a	3.92	1A
b	7.58	Q
c	7.63	Q

▶ $C_{10}H_{10}O_4$ ▶ 46.0 mg/0.50 ml $CDCl_3$ ▶ S 89

▶ 1-Ql

O=C—OCH_3(c); (h)H; (d)H; H(j); H(b); (g)H; (f) H; (e) H; O; O; CH_3(a); (i)H

	Shift	Peak
a	2.11	2A
b	3.46	4A
▶c	3.79	1A
d	4.02	3E
e	5.12	1B
f	5.59	S
g	5.67*	S
h	5.67*	S
i	7.19	O
j	7.46	1A

*or 6.08

▶ $C_{15}H_{14}O_6$ ▶ 7% $CDCl_3$ ▶ V 640

▶ 1-Sb

(a) CH_3—CH_2—CH_2—CH_2—CH_2 (c)—S—CH_3 (b)

	Shift	Peak
a	0.93	3B
▶b	2.08	1A
c	2.50	3B

▶ $C_6H_{14}S$ ▶ 7% $CDCl_3$ ▶ V 144

▶ 1-Ql

(b)H; H(c); (d)H; O; C(=O)—O—CH_3(a)

	Shift	Peak
▶a	3.88	1A
b	6.51	4D
c	7.20	N
d	7.58	N

▶ $C_6H_6O_3$ ▶ 7% $CDCl_3$ ▶ V 125

▶ 1-Sb

(b) CH_3—S—$CH_2$$CH_3$ (c) (a)

	Shift	Peak
a	1.27	3A
▶b	2.10	1A
c	2.53	4A

▶ C_3H_8S ▶ 7% $CDCl_3$ ▶ V 46

▶ 1-Ql

(a) CH_3O; C=O; (c)H; C=C; CN; (b)H; H (b)

	Shift	Peak
▶a	3.96	1A
b	8.03	4A
c	8.27	1A

▶ $C_{11}H_9NO_2$ ▶ 7% $CDCl_3$ ▶ V 576

▶ 1-Sb

O=C—N—C(COOH(f))(H(d))—$CH_2$$CH_2SCH_3$ (b) (c) (a); (e)H

	Shift	Peak
▶a	2.10	1A
b	2.25	O
c	2.66	O
d	4.94	4B
e	7.14	3D
f	10.28	1B

▶ $C_{12}H_{15}NO_3S$ ▶ 7% $CDCl_3$ ▶ V 597

▶ 1-Qv

C(=O)—OCH_3 (a); (c) (c); (b) (b); OCH_3(a)

	Shift	Peak
▶a	3.86	1A
b	6.82	Q
c	7.91	Q

▶ $C_9H_{10}O_3$ ▶ 45.0 mg/0.50 ml $CDCl_3$ ▶ S 124

▶ 1-Sl

(c) H, (d) H; C=C; H (b); S—CH_3 (a)

	Shift	Peak
▶a	2.25	1A
b	4.95	2B
c	5.18	2B
d	6.43	4D

▶ C_3H_6S ▶ 7% $CDCl_3$ ▶ V 36

▶ 1-Qv

(c); (b); (b); (c); (b); C(=O)—O—CH_3 (a)

	Shift	Peak
▶a	3.92	1A
b	7.48	4A
c	~8.0	O

▶ $C_8H_8O_2$ ▶ 45.3 mg/0.50 ml $CDCl_3$ ▶ S 78

▶ 1-Sloo

(b) H, (d) H; C=C; H (c); SO_2; CH_3 (a)

	Shift	Peak
▶a	2.62	1A
b	5.95	2B
c	6.13	2B
d	6.70	4D

▶ $C_3H_6O_2S$ ▶ 7% $CDCl_3$ ▶ V 35

1-Sv

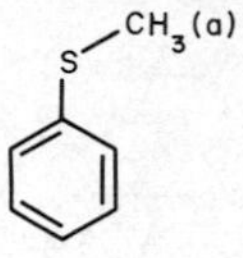

Shift	Peak
a 2.47	1A

S—CH3(a)

C_7H_8S · 7% $CDCl_3$ · V 490

1-Vah

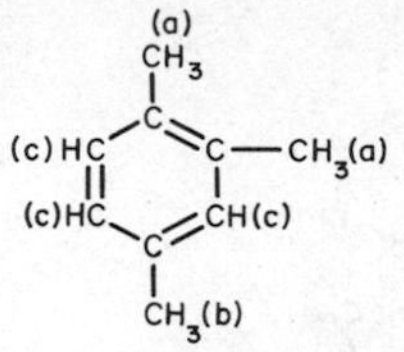

Shift	Peak
▸a 2.03	1A
b 2.15	1A
c 6.82	N

(a) CH3; (c)HC; C—CH3(a); (c)HC; CH(c); CH3(b)

C_9H_{12} · run neat · API 178 A

1-Sx

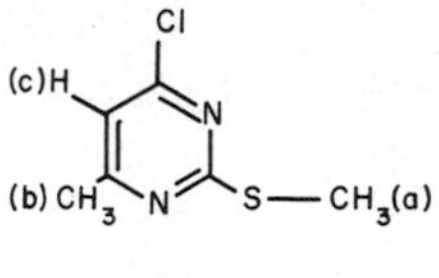

Shift	Peak
▸a 2.45 or 2.55	1A
b 2.45 or 2.55	1A
c 6.85	1A

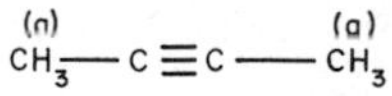

Cl; (c)H; N; (b) CH3; N; S—CH3(a)

$C_6H_7ClN_2S$ · 7% $CDCl_3$ · V 126

1-Vah

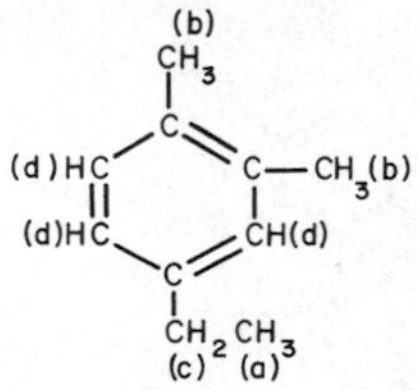

Shift	Peak
a 1.16	3A
▸b 2.07	1A
c 2.47	4A
d 6.83	2D

(b) CH3; (d)HC; C—CH3(b); (d)HC; CH(d); CH2CH3 (c) (a)

$C_{10}H_{14}$ · run neat · API 182 A

1-Ta

Shift	Peak
a 1.69	1A

(a) CH3—C≡C—CH3 (a)

C_4H_6 · 10% CCl_4 · API 163

1-Vah

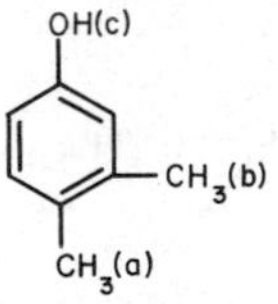

Shift	Peak
▸a 2.12	1A
▸b 2.12	1A
c 5.16	1B

OH(c); CH3(b); CH3(a)

$C_8H_{10}O$ · 45.5 mg/0.50 ml $CDCl_3$ · S 193

1-Th

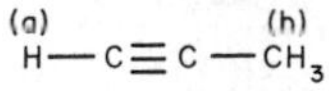

Shift	Peak
a 1.80	1A
▸b 1.80	1A

(a) H—C≡C—CH3 (b)

C_3H_4 · 7% $CDCl_3$ · V 16

1-Vah

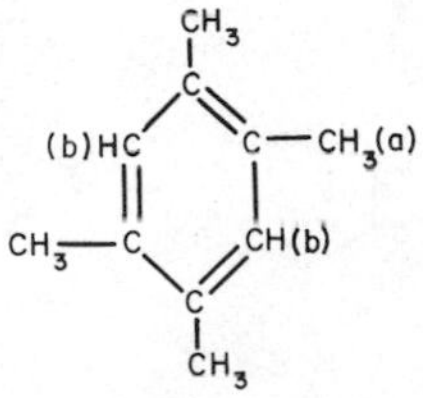

Shift	Peak
▸a 2.16	1A
b 6.74	1B

CH3; (b)HC; C—CH3(a); CH3—C; CH(b); CH3

$C_{10}H_{14}$ · CCl_4 · API 184

1-Vaa

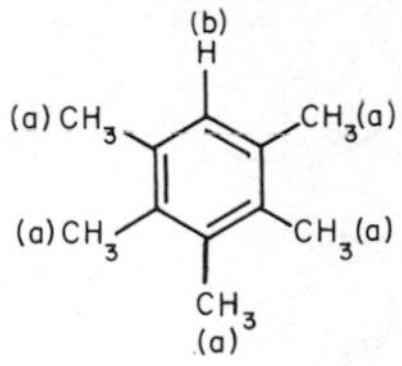

Shift	Peak
▸a 2.20 ± .03	1A
b 6.83	1B

(b) H; (a) CH3; CH3(a); (a) CH3; CH3(a); CH3 (a)

$C_{11}H_{16}$ · 7% $CDCl_3$ · V 287

1-Vah

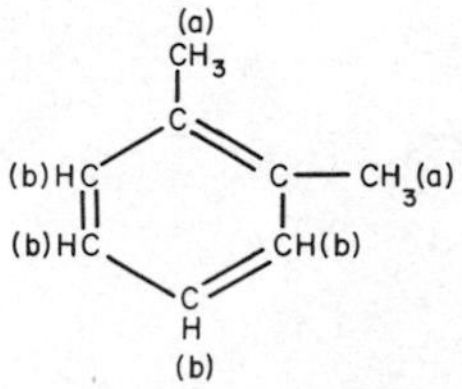

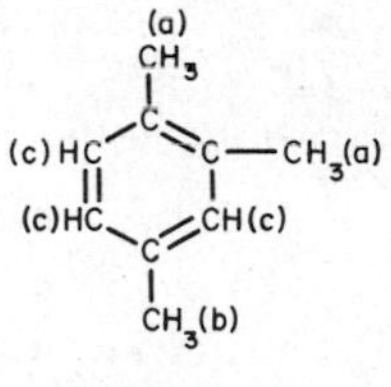

Shift	Peak
▸a 2.17	1A
b 2.23	1A
c 6.80	N

(a) CH3; (c)HC; C—CH3(a); (c)HC; CH(c); CH3(b)

C_9H_{12} · CCl_4 · API 178B

1-Vah

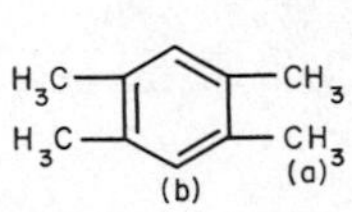

Shift	Peak
▸a 2.03	1A
b 6.93	1A

(a) CH3; (b)HC; C—CH3(a); (b)HC; CH(b); H (b)

C_8H_{10} · run neat · API 174 A

1-Vah

Shift	Peak
▸a 2.19	1A
b 6.87	1B

H3C; CH3; H3C; CH3 (a); (b)

$C_{10}H_{14}$ · 45.9 mg/0.50 ml $CDCl_3$ · S 85

▶ 1-Vah

(b) CH_3; (d)HC, (d)HC, C—CH_3(b), CH(d), CH_2CH_3 (c) (a)

	Shift	Peak
a	1.18	3A
▶b	2.20	1A
c	2.60	4A
d	6.84	1A

▶ $C_{10}H_{14}$ ▶ CCl_4 ▶ API 182 B

▶ 1-Vbb

(d)H, (b) CH_2, (c)CH_3O, CH_3(a), CH_2(b), H(d), (d)H, (b) CH_2, (a)CH_3, OCH_3(c), CH_2 (b), H(d)

	Shift	Peak
▶a	1.53	1A
b	2.83	1A
c	3.67	1A
d	6.45	1A

▶ $C_{20}H_{24}O_2$ ▶ 7% $CDCl_3$ ▶ V 681

▶ 1-Vah

(b) H; (a) CH_3, CH_3(a), (a) CH_3, CH_3(a), CH_3 (a)

	Shift	Peak
▶a	2.20 ± .03	1A
b	6.83	1B

▶ $C_{11}H_{16}$ ▶ 7% $CDCl_3$ ▶ V 287

▶ 1-Vbh

(b) CH_3; HC, HC, C—CH_2CH_3 (c) (a), CH, H

(d) = all ring H's

	Shift	Peak
a	1.09	3A
▶b	2.11	1A
c	2.45	4A
d	7.00	1A

▶ C_9H_{12} ▶ run neat ▶ API 177 A

▶ 1-Vah

(a) CH_3; (b)HC, (b)HC, C—CH_3(a), CH(b), H (b)

	Shift	Peak
▶a	2.24	1A
b	6.98	1A

▶ C_8H_{10} ▶ CCl_4 ▶ API 174 B

▶ 1-Vbh

(b) CH_3; (e)HC, (e)HC, C—CH_2CH_3 (a), CH(d), CH_3 (c)

	Shift	Peak
a	1.10	3A
▶b	2.15	2A
c	2.21	4A
d	6.67	1E
e	7.08	1E

▶ $C_{10}H_{14}$ ▶ CCl_4 ▶ API 183 A

▶ 1-Vah

(b) {ring}; CH_3(a), CH_3(a)

	Shift	Peak
▶a	2.25	1A
b	7.10	1A

▶ C_8H_{10} ▶ 7% $CDCl_3$ ▶ V 201

▶ 1-Vbh

(b) CH_3; HC, HC, C—CH_2CH_3 (c) (a), CH, H

(d) = all ring H's

	Shift	Peak
a	1.21	3A
▶b	2.27	1A
c	2.63	4A
d	7.02	1A

▶ C_9H_{12} ▶ CCl_4 ▶ API 177B

▶ 1-Vao

(d) CH_3, (d)CH_3, O, (b) CH_3, (a) CH_3, (a) CH_3, (a) CH_3, $(CH_2)_3CH-(CH_2)_3CH-(CH_2)_3CH-CH_3$(a), (c) (c) (c), HO (f), CH_2 (e), CH_3 (d)

	Shift	Peak
a	0.85	2A
b	1.22 ± 0.03	1E
c	1.22 ± 0.03	1E
▶d	2.12 ± 0.03	1B
e	2.62	3C
f	3.96	1D

▶ $C_{29}H_{50}O_2$ ▶ 7% $CDCl_3$ ▶ V 366

▶ 1-Vbh

(b) CH_3; (e)HC, (e)HC, C—CH_2CH_3 (a), CH(d), CH_3 (c)

	Shift	Peak
a	1.27	3A
▶b	2.27	2A
c	2.32	4A
d	6.71	1E
e	6.78	1E

▶ $C_{10}H_{14}$ ▶ run neat ▶ API 183 B

▶ 1-Vap

(d) CH_3, (d)CH_3, O, (b) CH_3, (a) CH_3, (a) CH_3, (a) CH_3, $(CH_2)_3CH-(CH_2)_3CH-(CH_2)_3CH-CH_3$(a), (c) (c) (c), HO (f), CH_2 (e), CH_3 (d)

	Shift	Peak
a	0.85	2A
b	1.22 ± 0.03	1E
c	1.22 ± 0.03	1E
▶d	2.12 ± 0.03	1B
e	2.62	3C
f	3.96	1D

▶ $C_{29}H_{50}O_2$ ▶ 7% $CDCl_3$ ▶ V 366

▶ 1-Vbh

(c) {ring}; (b) CH_2Br, CH_3 (a)

	Shift	Peak
▶a	2.41	1A
b	4.50	1A
c	7.19	1A

▶ C_8H_9Br ▶ 45.3 mg/0.50 ml $CDCl_3$ ▶ S 239

1-Vch

	Shift	Peak
a	1.15	2A
▶b	2.20	1A
c	3.01	5A
d	6.95	O

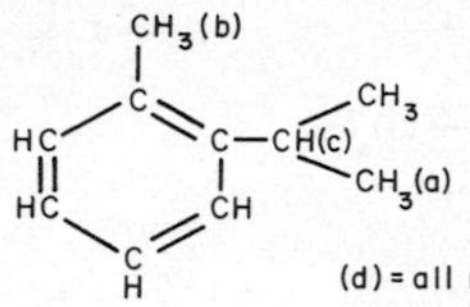

(d) = all ring H's

▶ $C_{10}H_{14}$ ▶ run neat ▶ API 181 A

1-Vhh

	Shift	Peak
▶a	1.97	1A
b	7.88	1A

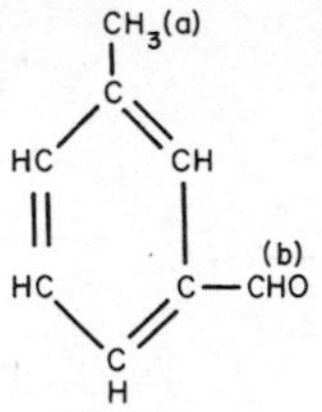

▶ C_8H_8O ▶ 10% CCl_4 ▶ API 56

1-Vch

	Shift	Peak
a	1.17	2A
▶b	2.24	1A
c	3.04	5A
d	7.00	N

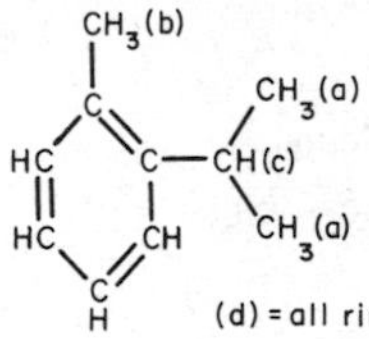

(d) = all ring H's

▶ $C_{10}H_{14}$ ▶ 50% CCl_4 ▶ API 224

1-Vhh

	Shift	Peak
▶a	2.12	1A
b	2.22	1A
c	3.34	1B
d	6.50	2A
e	6.76	2B, 2G

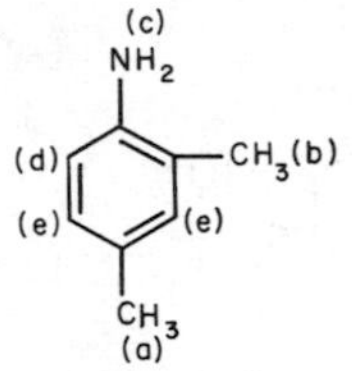

▶ $C_8H_{11}N$ ▶ 44.9 mg/0.50 ml $CDCl_3$ ▶ S 237

1-Vch

	Shift	Peak
a	1.25	2A
▶b	2.30	1A
c	3.05	5A
d	7.01	O

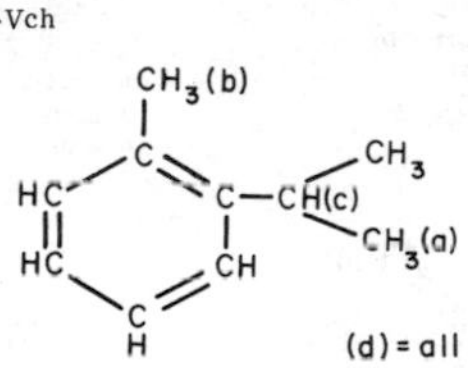

(d) = all ring H's

▶ $C_{10}H_{14}$ ▶ CCl_4 ▶ API 181 B

1-Vhh

	Shift	Peak
a	2.03	1A
▶b	2.15	1A
c	6.82	N

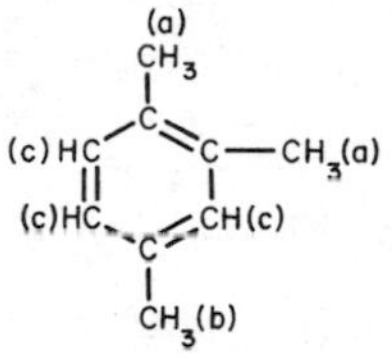

▶ C_9H_{12} ▶ run neat ▶ API 178 A

1-Vch

	Shift	Peak
a	0.95	1A
b	2.28	1A
▶c	2.33	1A
d	6.72	1B

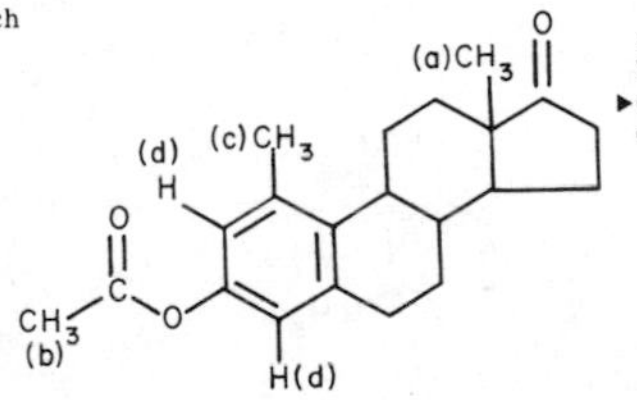

▶ $C_{21}H_{26}O_3$ ▶ 7% $CDCl_3$ ▶ V 350

1-Vhh

	Shift	Peak
▶a	2.15	1A
b	6.62	1F

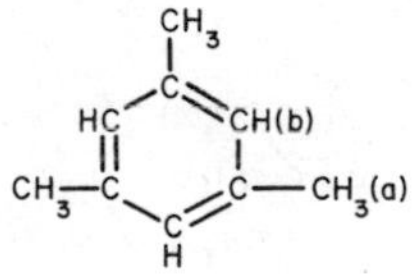

▶ C_9H_{12} ▶ run neat ▶ API 179 A

1-Vch

	Shift	Peak
a	2.37	1A

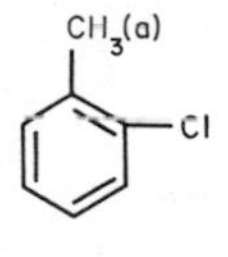

▶ C_7H_7Cl ▶ 35.0 mg/0.50 ml $CDCl_3$ ▶ S 13

1-Vhh

	Shift	Peak
a	1.14	3A
▶b	2.17	1A
c	2.40	4A
d	6.68	1B
e	6.68	1B

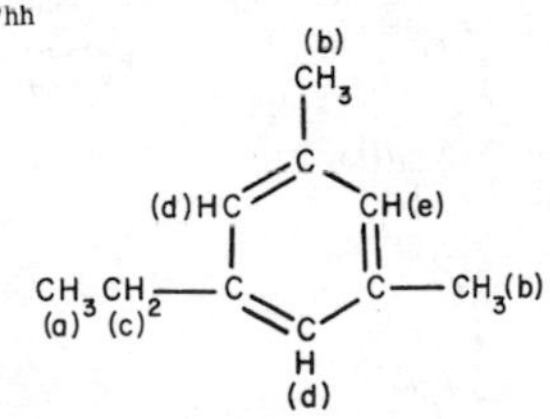

▶ $C_{10}H_{14}$ ▶ liquid ▶ API 12

1-Vgh

	Shift	Peak
▶a	2.40	1A
b	7.12	O
c	7.50	2E

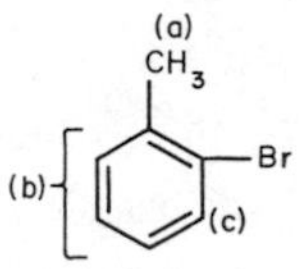

▶ C_7H_7Br ▶ 45.4 mg/0.50 ml $CDCl_3$ ▶ S 161

1-Vhh

	Shift	Peak
a	.87	3C
b	1.56	6B
▶c	2.17	1A
d	2.45	3B
e	6.82	S
f	6.82	S
g	6.82	S
h	6.82	S

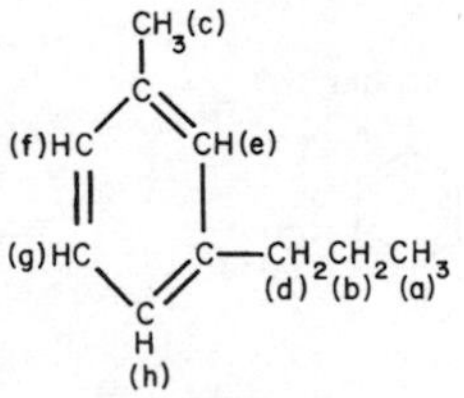

▶ $C_{10}H_{14}$ ▶ liquid run, neat ▶ API 96

▶ 1-Vhh

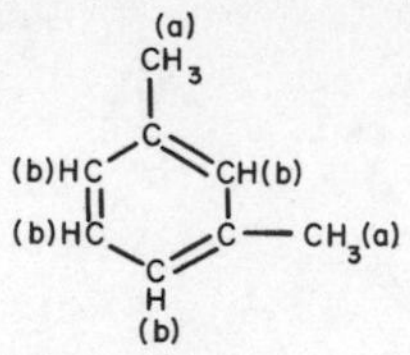

(a) CH3
(b)HC CH(b)
(b)HC C — CH3(a)
H (b)

Shift		Peak
▶a	2.18	1A
b	~6.93	N

▶ C_8H_{10} ▶ run neat ▶ API 175 A

▶ 1-Vhh

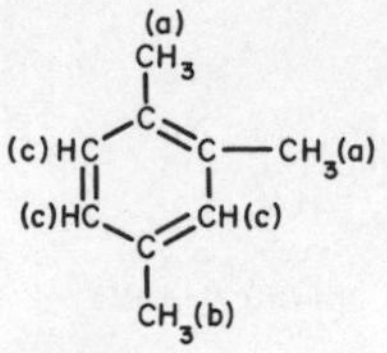

(a) CH3
(c)HC C — CH3(a)
(c)HC CH(c)
CH3(b)

Shift		Peak
a	2.17	1A
▶b	2.23	1A
c	6.80	N

▶ C_9H_{12} ▶ CCl_4 ▶ API 178B

▶ 1-Vhh

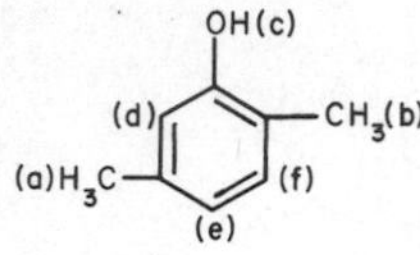

OH(c)
(d) — CH3(b)
(a)H3C- (f)
(e)

Shift		Peak
▶a	2.20	1A
b	2.25	1A
c	4.88	1A
d	6.55	1A
e	6.70	S
f	6.91	S

▶ $C_8H_{10}O$ ▶ 45.0 mg/0.50 ml $CDCl_3$ ▶ S 82

▶ 1-Vhh

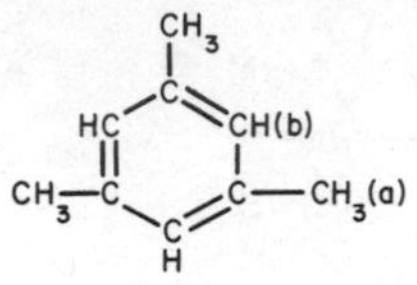

CH3
HC CH(b)
CH3—C C—CH3(a)
H

Shift		Peak
▶a	2.23	1A
b	6.69	1F

▶ C_9H_{12} ▶ CCl_4 ▶ API 179 B

▶ 1-Vhh

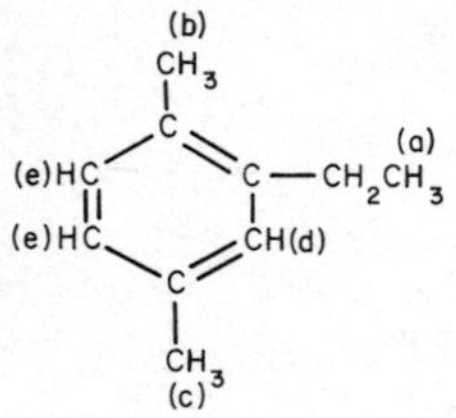

(b) CH3
(e)HC C — CH2CH3 (a)
(e)HC CH(d)
CH3 (c)

Shift		Peak
a	1.10	3A
b	2.15	2A
▶c	2.21	4A
d	6.67	1E
e	7.08	1E

▶ $C_{10}H_{14}$ ▶ CCl_4 ▶ API 183 A

▶ 1-Vhh

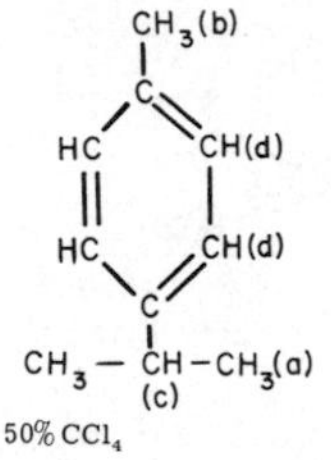

CH3(b)
HC CH(d)
HC CH(d)
CH3 — CH — CH3(a)
(c)

Shift		Peak
a	1.20	2A
▶b	2.24	1A
c	2.80	5A
d	6.99	1A

▶ $C_{10}H_{14}$ ▶ 50% CCl_4 ▶ API 225

▶ 1-Vhh

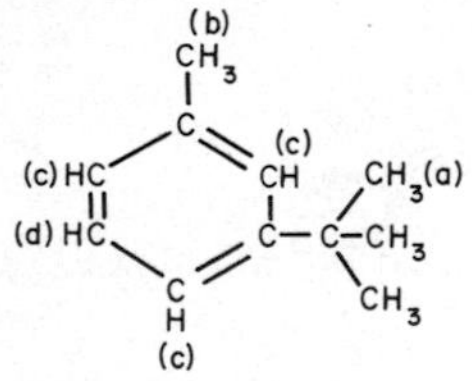

(b) CH3
(c) CH CH3(a)
(c)HC
(d)HC C— C—CH3
CH3
H (c)

Shift		Peak
a	1.25	1A
▶b	2.21	1A
c	~7.03	S
d	7.11	S

▶ $C_{11}H_{16}$ ▶ CCl_4 ▶ API 185 A

▶ 1-Vhh

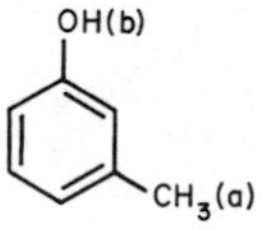

OH(b)
CH3(a)

Shift		Peak
▶a	2.25	1A
b	5.67	1A

▶ C_7H_8O ▶ 7% $CDCl_3$ ▶ V 160

▶ 1-Vhh

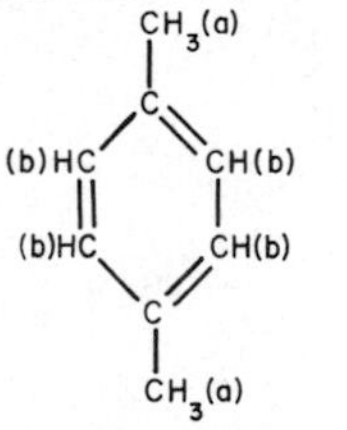

CH3(a)
(b)HC CH(b)
(b)HC CH(b)
CH3(a)

Shift		Peak
▶a	2.22	1A
b	6.95	1A

▶ C_8H_{10} ▶ liquid run, neat ▶ API 82

▶ 1-Vhh

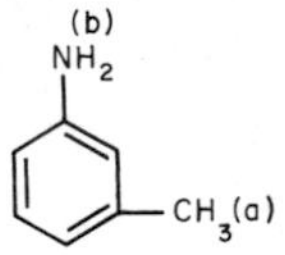

(b) NH2
CH3(a)

Shift		Peak
▶a	2.25	1A
b	3.46	1A

▶ C_7H_9N ▶ 44.6 mg/0.50 ml $CDCl_3$ ▶ S 108

▶ 1-Vhh

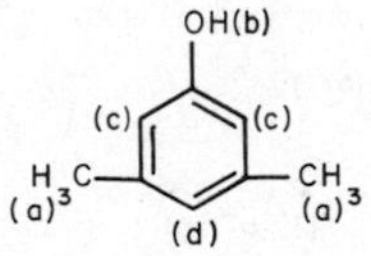

OH(b)
(c) (c)
H3C (a) CH3 (a)
(d)

Shift		Peak
▶a	2.22	1A
b	5.87	1A
c	6.34	1B
d	6.45	1B

▶ $C_8H_{10}O$ ▶ 44.8 mg/0.50 ml $CDCl_3$ ▶ S 83

▶ 1-Vhh

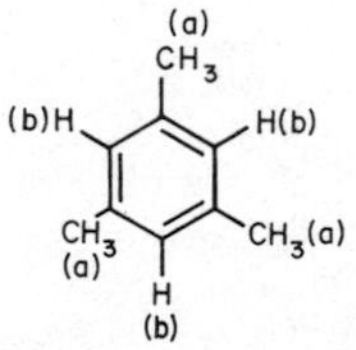

(a) CH3
(b)H H(b)
CH3 (a) CH3(a)
H (b)

Shift		Peak
▶a	2.25	1A
b	6.78	1B

▶ C_9H_{12} ▶ 7% $CDCl_3$ ▶ V 241

▶ 1-Vhh

(b) CH_3
(f)H H(e)
(g)H OH(d)
(a)CH_3—C—CH_3(a)
H
(c)

Shift	Peak
a 1.23	2A
▶b 2.25	1A
c 3.17	5B
d 4.75	1A
e 6.55	S
f 6.71	S
g 7.06	S

▶ $C_{10}H_{14}O$ ▶ 7% $CDCl_3$ ▶ V 270

▶ 1-Vhh

(b) OCH_3
(c)H H(c)
(d)H H(d)
CH_3(a)

Shift	Peak
▶a 2.28	1A
b 3.75	1A
c 6.80	Q
d 7.05	Q

▶ $C_8H_{10}O$ ▶ 7% $CDCl_3$ ▶ V 205

▶ 1-Vhh

CH_3(a)
C
(c)HC CH(b)
(d)HC C—CH_3(a)
C
H(c)

Shift	Peak
▶a 2.26	1A
b ~6.87	S
c ~6.87	S
d ~6.93	S

▶ C_8H_{10} ▶ 10% CCl_4 ▶ API 39

▶ 1-Vhh

(d) (c) (a)
(e) O—CH_2 CH_3
CH_3 (d)
(b) (e)

Shift	Peak
a 1.38	3A
▶b 2.28	1A
c 3.98	4A
d 6.75	Q
e 7.05	Q

▶ $C_9H_{12}O$ ▶ 45.0 mg/0.50 ml $CDCl_3$ ▶ S 284

▶ 1-Vhh

CH_3(a)
C
(b)HC CH(b)
(b)HC CH(b)
C
CH_3(a)

Shift	Peak
▶a 2.26	1A
b 6.93	1A

▶ C_8H_{10} ▶ 10% CCl_4 ▶ API 40

▶ 1-Vhh

(a)
CH_3
C
(b)HC CH(b)
(b)HC C—CH_3(a)
C
H
(b)

Shift	Peak
▶a 2.29	1A
b 6.86	N

▶ C_8H_{10} ▶ CCl_4 ▶ API 175 B

▶ 1-Vhh

(c)
OH
(d) (a)
H $C(CH_3)_3$
(e)H H(f)
CH_3
(b)

Shift	Peak
a 1.40	1A
▶b 2.27	1A
c 4.75	1A
d 6.52	2A
e 6.84	2E or 4A
f 7.05	11

▶ $C_{11}H_{16}O$ ▶ 7% $CDCl_3$ ▶ V 288

▶ 1-Vhh

(c) (b)
(d) O—CH_3
(a) H_3C (c)
(d)

Shift	Peak
▶a 2.29	1A
b 3.78	1A
c 6.76	Q
d 7.08	Q

▶ $C_8H_{10}O$ ▶ 45.0 mg/0.50 ml $CDCl_3$ ▶ S 292

▶ 1-Vhh

(a)
CH_3
(b)
(c)
Br

Shift	Peak
▶a 2.28	1A
b 7.05	Q
c 7.33	Q

▶ C_7H_7Br ▶ 44.8 mg/0.25 ml $CDCl_3$ ▶ S 163

▶ 1-Vhh

(a)
CH_3
SH(b)

Shift	Peak
▶a 2.30	1A
b 3.27	1A

▶ C_7H_8S ▶ 7% $CDCl_3$ ▶ V 168

▶ 1-Vhh

(a)
CH_3
I

Shift	Peak
a 2.28	1A

▶ C_7H_7I ▶ 45.0 mg/0.50 ml $CDCl_3$ ▶ S 173

▶ 1-Vhh

CH_3(a)
CH_3(a)

Shift	Peak
a 2.30	1A

▶ C_8H_{10} ▶ 7% $CDCl_3$ ▶ V 202

▶ 1-Vhh

	Shift	Peak
▶a	2.30	1A
b	7.05	1A

▶ C_8H_{10} ▶ 7% $CDCl_3$ ▶ V 203

▶ 1-Vhh

	Shift	Peak
▶a	2.32	1A
b	7.17 ± 0.03	1A

▶ C_7H_8 ▶ 7% $CDCl_3$ ▶ V 157

▶ 1-Vhh

	Shift	Peak
▶a	2.30	1A
b	3.55	1A
c	12.67	1A

▶ $C_9H_{10}O_2$ ▶ 44.9 mg/0.25 ml $CDCl_3$ ▶ S 272

▶ 1-Vhh

	Shift	Peak
▶a	2.32	2A
b	2.43	1B
c	4.47	1G

▶ $C_8H_{10}O$ ▶ 45.2 mg/0.50 ml $CDCl_3$ ▶ S 67

▶ 1-Vhh

	Shift	Peak
a	2.12	1A
▶b	2.30	1A
c	7.12	Q
d	7.37	Q
e	7.88	1D

▶ $C_9H_{11}NO$ ▶ 7% $CDCl_3$ ▶ V 239

▶ 1-Vhh

	Shift	Peak
a	1.27	3A
b	2.27	2A
▶c	2.32	4A
d	6.71	1E
e	6.78	1E

▶ $C_{10}H_{14}$ ▶ run neat ▶ API 183 B

▶ 1-Vhh

	Shift	Peak
a	1.22	2A
▶b	2.30	1A
c	2.87	5B
d	7.08	1A

▶ $C_{10}H_{14}$ ▶ 7% $CDCl_3$ ▶ V 268

▶ 1-Vhh

	Shift	Peak
a	1.32	1A
▶b	2.33	1A
c	7.12	Q
d	7.28	Q

▶ $C_{11}H_{16}$ ▶ 7% $CDCl_3$ ▶ V 586

▶ 1-Vhh

	Shift	Peak
a	2.31	1A

▶ C_7H_7Cl ▶ 45.0 mg/0.50 ml $CDCl_3$ ▶ S 51

▶ 1-Vhh

	Shift	Peak
▶a	2.35	1A
b	4.56	1A
c	7.24	1E

▶ C_8H_9Cl ▶ 45.2 mg/0.50 ml $CDCl_3$ ▶ S 68

▶ 1-Vhh

	Shift	Peak
▶a	2.32	1A
b	7.08	S
c	7.27	S

▶ C_7H_7Br ▶ 45.1 mg/0.50 ml $CDCl_3$ ▶ S 162

▶ 1-Vhh

	Shift	Peak
▶a	2.37	1A
b	7.07	1A
c	7.07	1A

▶ C_7H_8 ▶ liquid run, neat ▶ API 81

▶ 1-Vhh

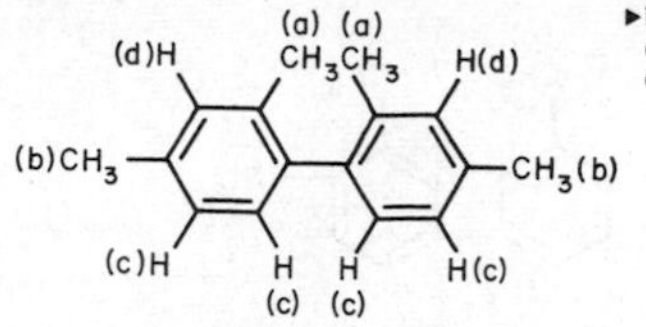

	Shift	Peak
a	2.03	1A
▶b	2.37	1A
c	7.02	S
d	7.09	S

▶ $C_{16}H_{18}$ ▶ 7% $CDCl_3$ ▶ V 659

▶ 1-Vhh

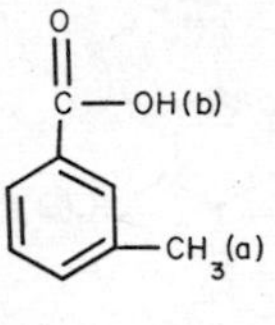

	Shift	Peak
▶a	2.40	1A
b	12.70	1A

▶ $C_8H_8O_2$ ▶ 45.0 mg/0.25 ml $CDCl_3$ ▶ S 18

▶ 1-Vhh

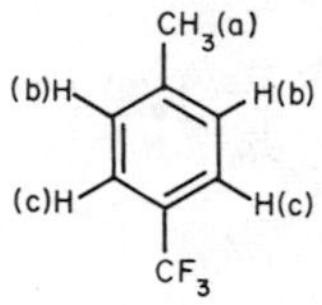

	Shift	Peak
▶a	2.38	1A
b	7.25	Q
c	7.48	Q

▶ $C_8H_7F_3$ ▶ 7% $CDCl_3$ ▶ V 190

▶ 1-Vhh

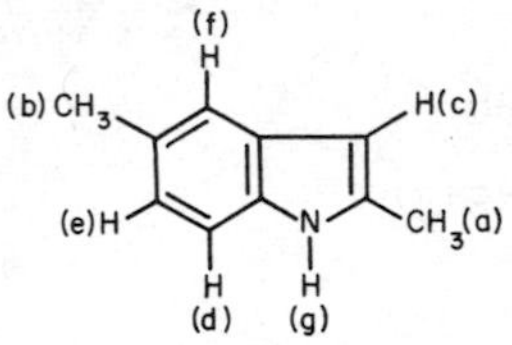

	Shift	Peak
a	2.28	1A
▶b	2.40	1A
c	6.10	1I
d	~6.90 or 7.02	N
e	~6.90 or 7.02	N
f	7.27	S
g	~7.30	S

▶ $C_{10}H_{11}N$ ▶ 7% $CDCl_3$ ▶ V 255

▶ 1-Vhh

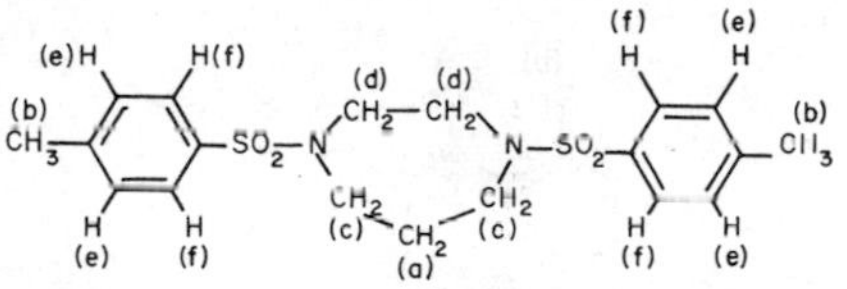

	Shift	Peak		Shift	Peak
a	1.95	5B	d	3.37	1E
▶ b	2.38	1A	e	7.30	Q
c	3.33	3A	f	7.63	Q

▶ $C_{19}H_{24}N_2O_4S_2$ ▶ 7% $CDCl_3$ ▶ V 336

▶ 1-Vhh

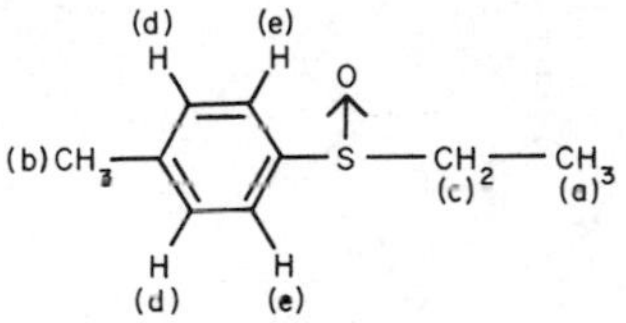

	Shift	Peak
a	1.20	3A
▶b	2.43	1A
c	2.82	4C
d	7.33	Q
e	7.51	Q

▶ $C_9H_{12}OS$ ▶ 7% $CDCl_3$ ▶ V 538

▶ 1-Vhh

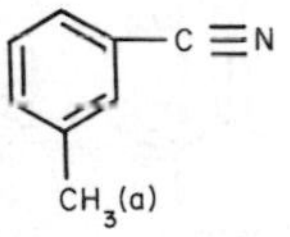

	Shift	Peak
a	2.39	1A

▶ C_8H_7N ▶ 45.2 mg/0.50 ml $CDCl_3$ ▶ S 168

▶ 1-Vhh

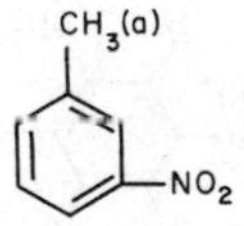

	Shift	Peak
a	2.45	1A

▶ $C_7H_7NO_2$ ▶ 36.3 mg/0.50 ml $CDCl_3$ ▶ S 22

▶ 1-Vhh

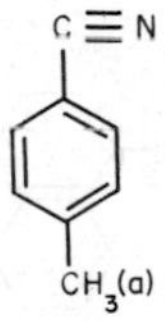

	Shift	Peak
a	2.4	1A

▶ C_8H_7N ▶ 44.7 mg/0.25 ml $CDCl_3$ ▶ S 169

▶ 1-Vhh

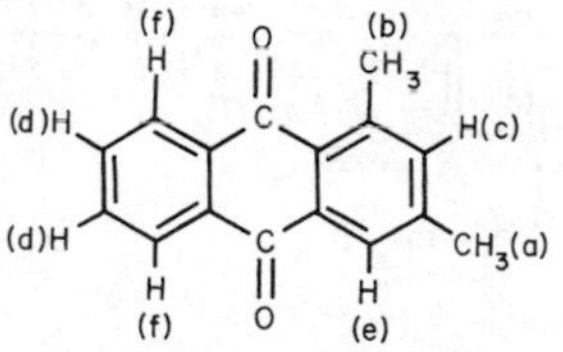

	Shift	Peak
▶a	2.45	1A
b	2.78	1A
c	7.33	2C
d	7.74	Q
e	7.99	2C
f	8.23	Q

▶ $C_{16}H_{12}O_2$ ▶ 7% $CDCl_3$ ▶ V 650

▶ 1-Vhh

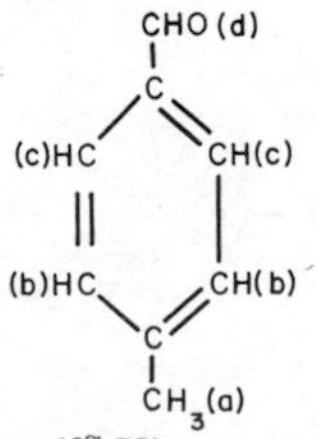

	Shift	Peak
▶a	2.40	1A
b	7.36	O
c	7.59	O
d	9.87	1A

▶ C_8H_8O ▶ 10% CCl_4 ▶ API 51

▶ 1-Vhh

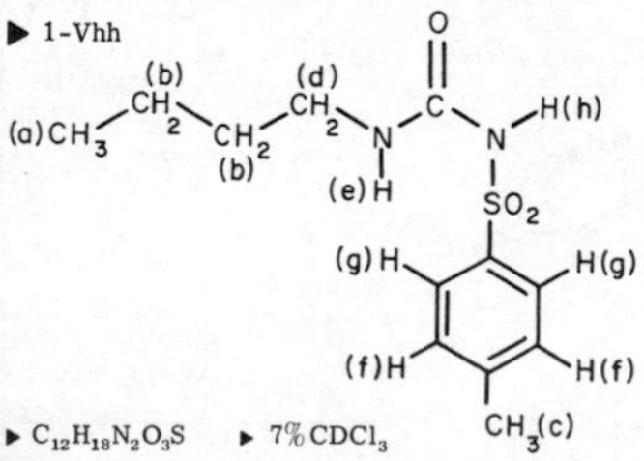

	Shift	Peak
a	0.90	3B
b	~1.40	1F
▶c	2.46	1A
d	3.23	4C
e	6.53	3B
f	7.32	Q
g	7.79	Q
h	~8.60	1D

▶ $C_{12}H_{18}N_2O_3S$ ▶ 7% $CDCl_3$ ▶ V 602

▶ 1-Vhj

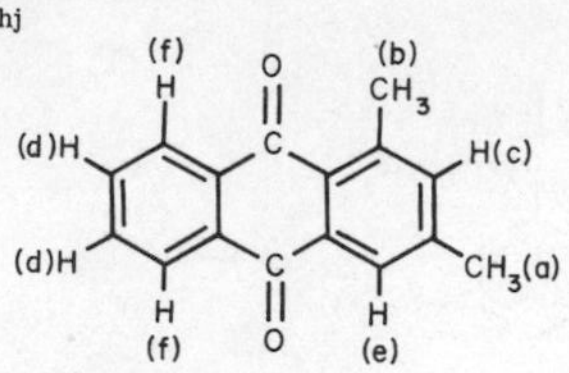

	Shift	Peak
a	2.45	1A
▶b	2.78	1A
c	7.33	2C
d	7.74	Q
e	7.99	2C
f	8.23	Q

▶ $C_{16}H_{12}O_2$ ▶ 7% $CDCl_3$ ▶ V 650

▶ 1-Vhn

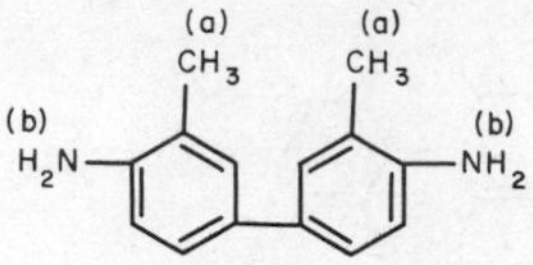

	Shift	Peak
▶a	2.13	1A
b	3.46	1E

▶ $C_{14}H_{16}N_2$ ▶ 44.7 mg/0.50 ml $CDCl_3$ ▶ S 269

▶ 1-Vhk

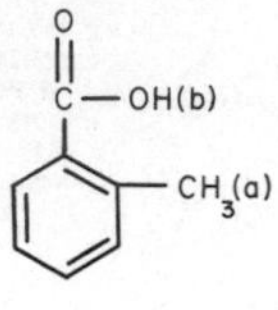

	Shift	Peak
▶a	2.64	M
b	12.63	1A

▶ $C_8H_8O_2$ ▶ 44.8 mg/0.25 ml $CDCl_3$ ▶ S 17

▶ 1-Vhn

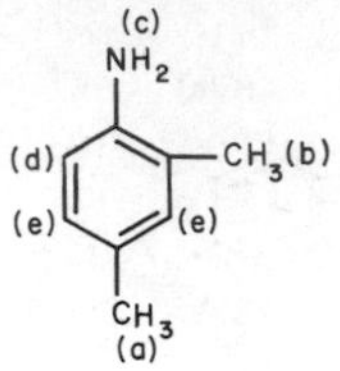

	Shift	Peak
a	2.12	1A
▶b	2.22	1A
c	3.34	1B
d	6.50	2A
e	6.76	2B, 2G

▶ $C_9H_{11}N$ ▶ 44.9 mg/0.50 ml $CDCl_3$ ▶ S 237

▶ 1-Vhl

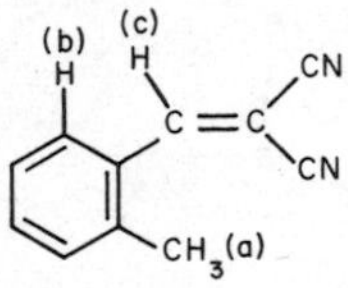

	Shift	Peak
▶a	2.45	1A
b	8.05	S
c	8.11	S

▶ $C_{11}H_8N_2$ ▶ 7% $CDCl_3$ ▶ V 283

▶ 1-Vhn

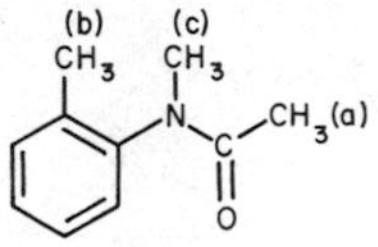

	Shift	Peak
a	1.78	1A
▶b	2.25	1A
c	3.20	1A

▶ $C_{10}H_{13}NO$ ▶ 7% $CDCl_3$ ▶ V 264

▶ 1-Vhm

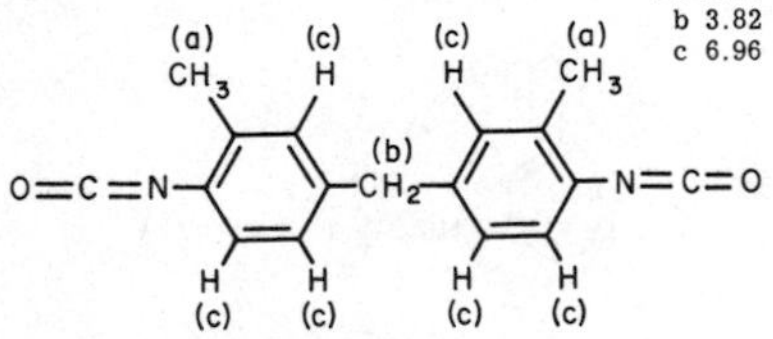

	Shift	Peak
▶a	2.26	1A
b	3.82	1A
c	6.96	1A

▶ $C_{17}H_{14}N_2O_2$ ▶ 7% $CDCl_3$ ▶ V 661

▶ 1-Vho

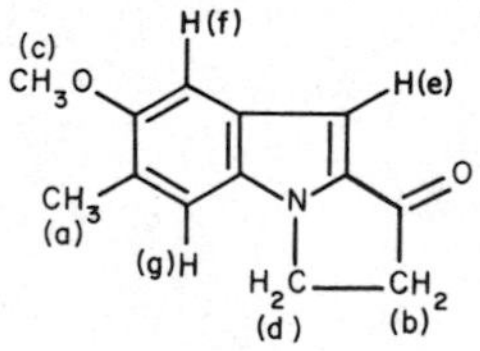

	Shift	Peak
▶a	2.33	1A
b	3.13	3A
c	3.83	1A
d	4.28	4A
e	6.85 or 6.98	1A
f	6.85 or 6.98	1A
g	7.13	1B

▶ $C_{13}H_{13}NO_2$ ▶ 7% $CDCl_3$ ▶ V 299

▶ 1-Vhn

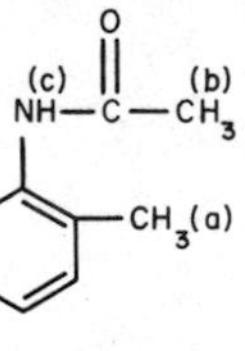

	Shift	Peak
▶a	2.06 or 2.17	1A
b	2.06 or 2.17	1A
c	~7.5	1C

▶ $C_9H_{11}NO$ ▶ 45.2 mg/0.50 ml $CDCl_3$ ▶ S 175

▶ 1-Vhp

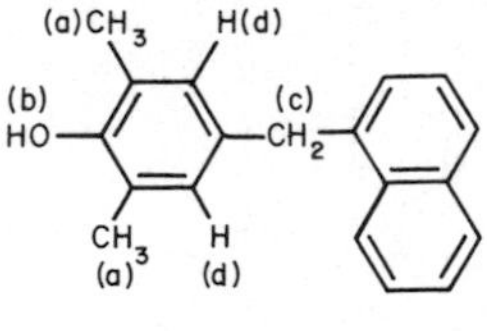

	Shift	Peak
▶a	2.12	1A
b	4.15	1D
c	4.27	1B
d	6.77	1B

▶ $C_{19}H_{18}O$ ▶ 7% $CDCl_3$ ▶ V 332

▶ 1-Vhn

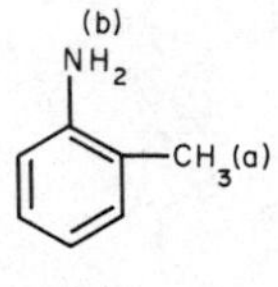

	Shift	Peak
▶a	2.13	1A
b	3.43	1B

▶ C_7H_9N ▶ 44.3 mg/0.50 ml $CDCl_3$ ▶ S 107

▶ 1-Vhp

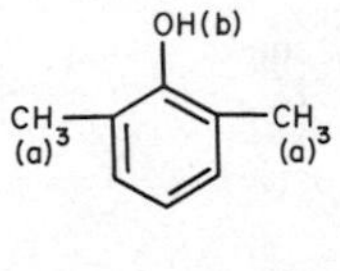

	Shift	Peak
▶a	2.20	1A
b	4.65	1A

▶ $C_8H_{10}O$ ▶ 45.2 mg/0.25 ml $CDCl_3$ ▶ S 36

▶ 1-Vhp

(a) (e) CH_3 H (d) HO (b) (c) CH_2—CH_2 CH_3 (a) H (e)

Shift	Peak
▶a 2.20	1A
b 2.90	Q
c 3.30	Q
d 4.18	1C
e 6.83	1B

▶ $C_{20}H_{20}O$ ▶ 7% $CDCl_3$ ▶ V 340

▶ 1-Vhv

(c)(c) H H (b) (b) H—C—C—H (d)H H(d) (d)H H(d) (d)H CH_3 CH_3 H(d) (a) (a)

Shift	Peak
▶a 2.29	1A
b 2.61	S
c 2.71	S
d 7.15	1A

▶ $C_{16}H_{16}$ ▶ 7% $CDCl_3$ ▶ V 654

▶ 1-Vhp

OH(c) (d) CH_3(b) (a)H_3C (f) (e)

Shift	Peak
a 2.20	1A
▶b 2.25	1A
c 4.88	1A
d 6.55	1A
e 6.70	S
f 6.91	S

▶ $C_8H_{10}O$ ▶ 45.0 mg/0.50 ml $CDCl_3$ ▶ S 82

▶ 1-Vqq

(g) (f)H H (i) (c) C=C (d) H CH_3 (e)H CH_2 H(h) (b) CH_3—CO O O O CH_3 (a)

Shift	Peak
▶a 2.25 or 2.37	1A
b 2.25 or 2.37	1A
c 2.40	2D
d 3.33	2B
e 5.07	R
f 5.12	R
g 5.85	4D
h 6.22	1I
i 7.30	1I

▶ $C_{16}H_{16}O_4$ ▶ 7% $CDCl_3$ ▶ V 323

▶ 1-Vhu

C≡N CH_3(a)

Shift	Peak
a 2.55	1A

▶ C_8H_7N ▶ 44.8 mg/0.50 ml $CDCl_3$ ▶ S 174

▶ 1-Vrr

(a) CH_3 NO_2 NO_2 (b)H H(b) NO_2

Shift	Peak
▶a 2.74	1A
b 8.84	1A

▶ $C_7H_5N_3O_6$ ▶ 7% $CDCl_3$ ▶ V 486

▶ 1-Vhv

(a) (a) (d)H CH_3CH_3 H(d) (b)CH_3 CH_3(b) (c)H H H H(c) (c) (c)

Shift	Peak
▶a 2.03	1A
b 2.37	1A
c 7.02	S
d 7.09	S

▶ $C_{16}H_{18}$ ▶ 7% $CDCl_3$ ▶ V 659

▶ 1-X

H(b) (c)H H(c) (a) CH_3 (d)H H(d) (d)H H(d) CH_3 (a) (c)H H(c) H(b)

Shift	Peak
▶a −4.23	1A
b 8.13	3B
c 8.63	1A
d 8.66	1A

▶ $C_{18}H_{16}$ ▶ 7% $CDCl_3$ ▶ V 672

▶ 1-Vhv

(b) H_2C S CH_2(b) CH_3 CH_3 (a) (a)

Shift	Peak
▶a 2.09	1A
b 3.27	1A

▶ $C_{16}H_{16}S$ ▶ 7% $CDCl_3$ ▶ V 657

▶ 1-X

OD N CH_3(a) DO N H(b)

Shift	Peak
▶a 1.83	2D
b 7.42	7B

▶ $C_5H_6N_2O_2$ ▶ D_2O ▶ V 432

▶ 1-Vhv

(b)H O H(b) (c)H—C C—H(c) CH_3 CH_3 (a) (a)

Shift	Peak
▶a 2.20	1A
b 3.31	2C
c 3.52	2C

▶ $C_{17}H_{16}O$ ▶ 7% $CDCl_3$ ▶ V 662

▶ 1-X

O D N CH_3(a) O N H(h) (c) DO H C O (d)H H(f) (b)H H H (e) (g) DO H(b)

Shift	Peak
▶a 1.91	2D
b 2.40	4A
c 3.79	2A
d 3.85	1A
e 4.03	N
f 4.49	4E
g 6.28	3A
h 7.65	2D

▶ $C_{10}H_{14}N_2O_5$ ▶ D_2O ▶ V 566

▶ 1-X

	Shift	Peak
▶a	2.17	1A
b	5.47	1A

▶ $C_{15}H_{21}O_6Rh$ ▶ 7% $CDCl_3$ ▶ V 644

▶ 1-X

	Shift	Peak
a	2.48	1A

▶ $C_{12}H_{12}$ ▶ 10% CCl_4 ▶ API 64

▶ 1-X

	Shift	Peak
▶a	2.33	1A
b	7.20 ± 0.10	S
c	7.20 ± 0.10	S
d	8.10 ± 0.05	S
e	8.10 ± 0.05	S

▶ C_6H_7NO ▶ 7% $CDCl_3$ ▶ V 457

▶ 1-X

	Shift	Peak
▶a	2.50	1A
▶b	2.72	1A
c	7.21	2C
d	7.90	2C
e	7.90	S

▶ $C_{11}H_{11}N$ ▶ 7% $CDCl_3$ ▶ V 579

▶ 1-X

	Shift	Peak
a	2.40	1A

▶ $C_{11}H_{10}$ ▶ CCl_4 ▶ API 190

▶ 1-X

	Shift	Peak
▶a	2.52	1A
b	6.95	2A
c	7.45	4A

▶ C_7H_9N ▶ 7% $CDCl_3$ ▶ V 169

▶ 1-X

	Shift	Peak
▶a	2.42	1A
▶b	2.62	1A

▶ $C_{12}H_{12}$ ▶ 10% CCl_4 ▶ API 58

▶ 1-X

	Shift	Peak
▶a	2.52	1E
▶b	2.70	1A

▶ $C_{12}H_{12}$ ▶ 10% CCl_4 ▶ API 60

▶ 1-X

	Shift	Peak
▶a	2.43	1A
b	7.31	2E
c	7.75	S
d	7.98	S

▶ $C_{14}H_{12}N_2O_2$ ▶ 7% $CDCl_3$ ▶ V 623

▶ 1-X

	Shift	Peak
a	2.57	1A

▶ $C_{12}H_{12}$ ▶ 10% CCl_4 ▶ API 59

▶ 1-X

	Shift	Peak
a	2.45 or 2.55	1A
▶b	2.45 or 2.55	1A
c	6.85	1A

▶ $C_6H_7ClN_2S$ ▶ 7% $CDCl_3$ ▶ V 126

▶ 1-X

	Shift	Peak
▶a	2.42	1A
▶b	2.62	1A

▶ $C_{12}H_{12}$ ▶ 10% CCl_4 ▶ API 58

► 1-X

Shift	Peak
►a 2.66	1A
►b 2.70	1A
c 7.14	1B

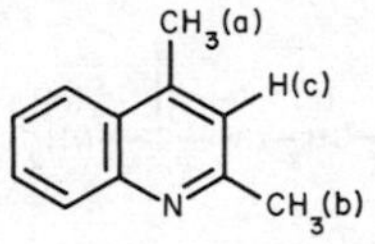

CH_3(a) H(c) N CH_3(b)

► $C_{11}H_{11}N$ ► 7% $CDCl_3$ ► V 578

► 1-Yh

Shift	Peak
a 2.48	1E
►b 2.52	1E
c 6.78	2E
d 7.50	2A

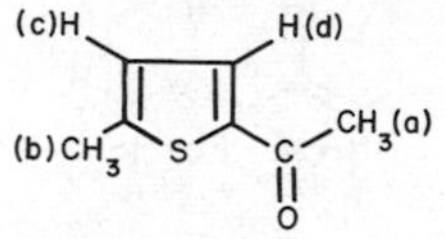

(c)H H(d) (b)CH_3 S C CH_3(a) O

► C_7H_8OS ► 7% $CDCl_3$ ► V 164

► 1-X

Shift	Peak
►a 2.52	1E
►b 2.70	1A

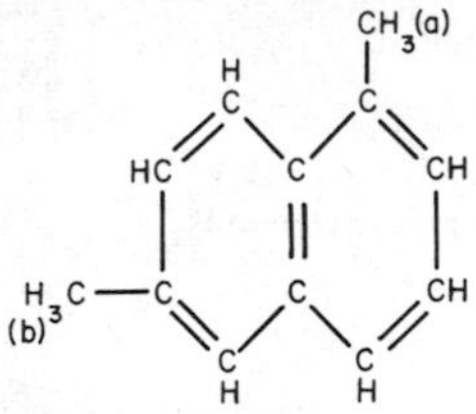

CH_3(a) H C C HC C CH H_3C (b) C C CH C C H H

► $C_{12}H_{12}$ ► 10% CCl_4 ► API 60

► 1-Z

Shift	Peak
►a 0.00	1A
b 0.60	N
c 1.78	O
d 2.82	3C

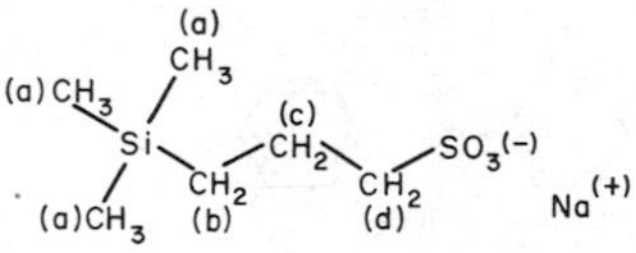

(a) CH_3 (a)CH_3 Si (c) CH_2 CH_2 CH_2 $SO_3^{(-)}$ (a)CH_3 (b) (d) $Na^{(+)}$

► $C_6H_{15}NaO_3SSi$ ► D_2O ► V 481

► 1-X

Shift	Peak
►a 2.72	1A
b ~7.00	1E
c 7.03	1E

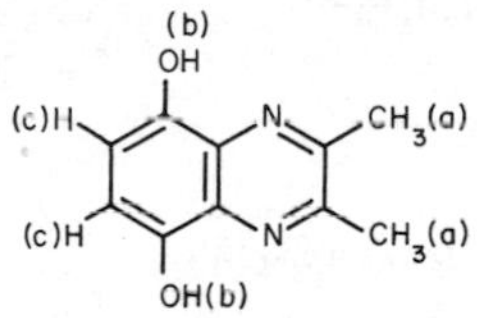

(b) OH (c)H N CH_3(a) (c)H N CH_3(a) OH(b)

► $C_{10}H_{10}N_2O_2$ ► 7% $CDCl_3$ ► V 553

► 1-Z

Shift	Peak
a 0.06	3A

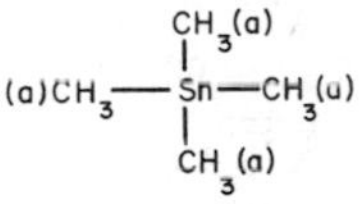

CH_3(a) (a)CH_3—Sn—CH_3(a) CH_3(a)

► $C_4H_{12}Sn$ ► 7% $CDCl_3$ ► V 424

► 1-X

Shift	Peak
a 2.80	1A

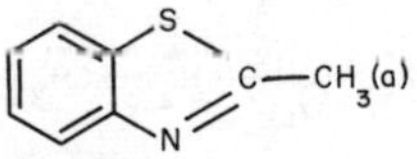

S C—CH_3(a) N

► C_8H_7NS ► 7% $CDCl_3$ ► V 191

► 1-Z

Shift	Peak
►a 0.10 or 0.12	1A
►b 0.10 or 0.12	1A
c 1.57	1A

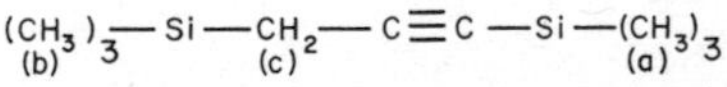

$(CH_3)_3$—Si—CH_2—C≡C—Si—$(CH_3)_3$ (b) (c) (a)

► $C_9H_{20}Si_2$ ► 7% $CDCl_3$ ► V 549

► 1-X

Shift	Peak
a 2.97	1A

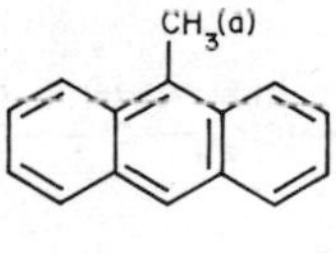

CH_3(a)

► $C_{15}H_{12}$ ► 7% $CDCl_3$ ► V 317

► 1-Z

Shift	Peak
►a 0.12	1H
b 3.55	1A

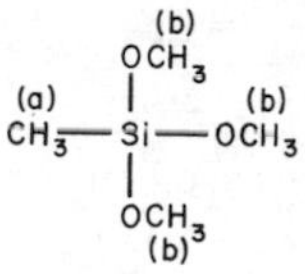

(b) OCH_3 (a) CH_3—Si—OCH_3 (b) OCH_3 (b)

► $C_4H_{12}O_3Si$ ► 7% $CDCl_3$ ► V 93

► 1-Yh

Shift	Peak
►a 2.48	2D
b 6.72	N
c 6.87	N
d 7.03	N

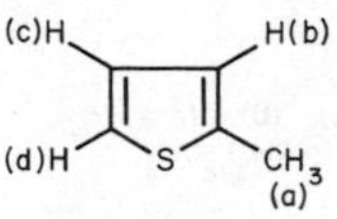

(c)H H(b) (d)H S CH_3 (a)

► C_5H_6S ► 7% $CDCl_3$ ► V 103

► 1-Z

Shift	Peak
►a 0.12	1A
b 1.22	3A
c 3.80	4A

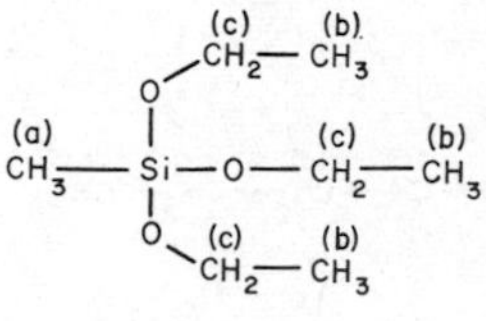

(c) (b) O CH_2—CH_3 (a) CH_3—Si—O—CH_2—CH_3 (c) (b) O CH_2—CH_3 (c) (b)

► $C_7H_{18}O_3Si$ ► 7% $CDCl_3$ ► V 184

▶ 1-Z

	Shift	Peak
a	0.14	1A

(divinyldimethylsilane: $H_2C{=}CH{-}Si(CH_3)_2{-}CH{=}CH_2$; (a) = CH_3)

▶ $C_6H_{12}Si$ ▶ 7% $CDCl_3$ ▶ V 476

▶ 2-ABb

	Shift	Peak
a	0.88	3D
▶b	~1.2-1.5	N
c	~3.08	4C
d	5.18	1B
e	5.95	3C

(a) (b) (c) (e) (d)
$CH_3{-}(CH_2)_2{-}CH_2{-}NH{-}C(=O){-}NH_2$

▶ $C_5H_{12}N_2O$ ▶ 45.3 mg/0.25 ml $CDCl_3$ ▶ S 282

▶ 1-Z

	Shift	Peak
▶a	0.62	2A
b	4.97	4A

$(C_6H_5)_2Si(CH_3)H$; (a) = CH_3, H(b)

▶ $C_{13}H_{14}Si$ ▶ 7% $CDCl_3$ ▶ V 609

▶ 2-ABb

	Shift	Peak
a	.87	1E
b	1.20	1E
▶c	1.20	1B

(b) (c) (c) (a)
$CH_3{-}C(CH_3)_2{-}CH_2{-}(CH_2)_2{-}CH_3$

▶ C_8H_{18} ▶ 50% CCl_4 ▶ API 203

▶ 1-Z

	Shift	Peak
a	2.52	N

$O{=}S(CD_2H)(CD_3)$; (a) = CD_2H

▶ C_2H_6OS ▶ DMS-D_6 ▶ V 376

▶ 2-ABb

	Shift	Peak
a	.88	3B
▶b	1.22	1A
c	2.54	3B
d	7.09	1G

(c) (b) (a)
$C_6H_5{-}CH_2{-}(CH_2)_{10}{-}CH_3$

(d) = all ring H's

▶ $C_{18}H_{30}$ ▶ 50% CCl_4 ▶ API 360

▶ **2** -ABa

	Shift	Peak
a	.9	3C
▶b	1.25	O

(b) (a)
$CH_3{-}CH_2{-}CH_2{-}CH_3$

▶ C_4H_{10} ▶ 50% CCl_4 ▶ API 329

▶ 2-ABb

	Shift	Peak
a	.83	3B
▶b	1.23	1B
c	~1.27	S

(a) (b) (c) (b) (a)
$CH_3{-}CH_2{-}CH(CH_3)_{(a)}{-}(CH_2)_3{-}CH_3$

▶ C_8H_{18} ▶ 50% CCl_4 ▶ API 341

▶ 2-ABb

	Shift	Peak
a	.89	3C
▶b	.90	S
c	~1.33 or 1.35	S
d	~1.35 or 1.33	S
e	1.97	4C
f	~4.84	R
g	~4.97	R
h	~5.70	R

H(f), H(g) $C{=}C$ H(h)
(e) (d) (c) (b) (a)
$CH_2{=}CH{-}CH_2CH_2CH_2CH_2CH_3$

▶ C_7H_{14} ▶ liquid ▶ API 384

▶ 2-ABb

	Shift	Peak
a	.84	3B
▶b	1.24	1B

(a) (b) (b) (b) (a)
$CH_3{-}CH_2{-}(CH_2)_4{-}CH_2{-}CH_3$

▶ C_8H_{18} ▶ 10% CCl_4 ▶ API 200

▶ 2-ABb

	Shift	Peak
a	.82	2A
b	~.92	1E
▶c	~.93	1B
d	~.97	1D

(a) (d) (c) (c) (c) (b)
$CH_3{-}CH(CH_3)_{(a)}{-}CH_2{-}CH_2{-}CH_2{-}CH_3$

▶ C_7H_{16} ▶ 50% CCl_4 ▶ API 195

▶ 2-ABb

	Shift	Peak
a	.90	3B
▶b	1.24	1A
c	2.04	3B
d	4.88	R
e	5.02	R
f	5.72	R

H(d), H(e) $C{=}C$ H(f)
(c) (b) (a)
$CH_2{=}CH{-}CH_2(CH_2)_{12}CH_3$

▶ $C_{16}H_{32}$ ▶ 50 vol.% CCl_4 ▶ API 437

▶ 2-ABb

	Shift	Peak
a	0.87	3B
▶b	1.24	1A
c	~2.27	2B

$CH_3(CH_2)_{17}CH_2C(=O)—OH$ (a) (b) (c)

▶ $C_{20}H_{40}O_2$ ▶ 45.2 mg/0.50 ml $CDCl_3$ ▶ S 260

▶ 2-ABb

	Shift	Peak
a	0.88	3B
▶b	1.26	1A
c	~2.35	1D
d	10.95	1B

$CH_3—(CH_2)_{11}—CH_2—C(=O)—OH$ (a) (b) (c) (d)

▶ $C_{14}H_{28}O_2$ ▶ 44.8 mg/0.50 ml $CDCl_3$ ▶ S 167

▶ 2-ABb

	Shift	Peak
a	.87	3D
b	1.05	2H
▶c	1.25	1B
d	3.2	1C

$CH_3—(CH_2)_3—CH(CH_3)—NH_2$ (a) (c) (b) (d)

▶ $C_6H_{15}N$ ▶ 34.4 mg/0.50 ml $CDCl_3$ ▶ S 28

▶ 2-ABb

	Shift	Peak
a	.90	3B
▶b	1.26	1A

$CH_3—(CH_2)_{14}—CH_3$ (b) (a)

▶ $C_{16}H_{34}$ ▶ 10% CCl_4 ▶ API 48

▶ 2-ABb

	Shift	Peak
a	.89	3B
▶b	1.25	1B
c	2.46	3B
d	7.07	1A

$C_6H_5—CH_2—(CH_2)_7—CH_3$ (c) (b) (a)

(d) = all ring H's

▶ $C_{15}H_{24}$ ▶ 50% CCl_4 ▶ API 358

▶ 2-ABb

	Shift	Peak
a	~0.88	1D
▶b	1.26	1A
c	~2.30	1D
d	3.63	1A

$CH_3—(CH_2)_{13}—CH_2—C(=O)—O—CH_3$ (a) (b) (c) (d)

▶ $C_{17}H_{34}O_2$ ▶ 44.4 mg/0.50 ml $CDCl_3$ ▶ S 248

▶ 2-ABb

	Shift	Peak
a	.88	3B
▶b	1.25	1B
c	2.50	3B
d	7.07	1A

$C_6H_5—CH_2—(CH_2)_8—CH_3$ (c) (b) (a)

(d) = all ring H's

▶ $C_{16}H_{26}$ ▶ 50% CCl_4 ▶ API 359

▶ 2-ABb

	Shift	Peak
a	.87	2H
b	.96	S
▶c	1.27	1B

$CH_3—CH(CH_3)—(CH_2)_4—CH_3$ (a) (c) (b)

▶ C_8H_{18} ▶ 50% CCl_4 ▶ API 340

▶ 2-ABb

	Shift	Peak
a	.99	3B
▶b	1.25	1A
c	2.02	3B
d	4.90	R
e	5.08	R
f	5.68	R

$H(e)H(d)C=CH(f)—CH_2(CH_2)_{12}CH_3$ (c) (b) (a)

▶ $C_{16}H_{32}$ ▶ 50% CCl_4 ▶ API 73

▶ 2-ABb

	Shift	Peak
a	0.88	3B
▶b	1.27 ± 0.03	1A
c	1.27 ± 0.03	1A

$CH_3—CH_2—(CH_2)_4—CH_2—CH_3$ (a) (b) (c) (b) (a)

▶ C_8H_{18} ▶ 7% $CDCl_3$ ▶ V 216

▶ 2-ABb

	Shift	Peak
a	0.90	3B
▶b	1.25	1A
c	2.30	3B
d	4.18*	S
e	4.18*	S
f	5.28	S

*or 4.28

(e)H—C(H(d))—O—C(=O)—CH₂—(CH₂)₁₅—CH₃ (c) (b) (a)
(f)H—C—O—C(=O)—CH₂—(CH₂)₁₅—CH₃ (c) (b) (a)
(e)H—C(H(d))—O—C(=O)—CH₂—(CH₂)₁₅—CH₃ (c) (b) (a)

▶ $C_{57}H_{110}O_6$ ▶ 7% $CDCl_3$ ▶ V 368

▶ 2-ABb

	Shift	Peak
a	.88	3B
▶b	1.27	1B
c	3.45	3B
d	4.28	1A

$CH_3—(CH_2)_8—CH_2—OH$ (a) (b) (c) (d)

▶ $C_{10}H_{22}O$ ▶ 20% CCl_4 ▶ API 172

▶ 2-ABb

$H_2C=CH-CH_2(CH_2)_{10}CH_3$ — (d) H, (e) H, (f) H, (c) CH_2, (b) $(CH_2)_{10}$, (a) CH_3

Shift		Peak
a	.88	3B
▶b	1.27	1A
c	2.3	3B
d	4.88	7X
e	5.01	7X
f	5.72	R

▶ $C_{14}H_{28}$ ▶ 50 vol.% CCl_4 ▶ API 436

▶ 2-ABb

(a) CH_3–(b) $(CH_2)_6$–(c) CH_2–C(=O)–(d) OH

Shift		Peak
a	0.88	3B
▶b	1.28	1E
c	2.38	3B
d	10.53	1B

▶ $C_9H_{18}O_2$ ▶ 35.7 mg/0.50 ml $CDCl_3$ ▶ S 9

▶ 2-ABb

CH_3—CH_2—CH_2—(b) CH_2—(a) CH_3

Shift		Peak
a	.87	3B
▶b	1.28	1E

▶ C_5H_{12} ▶ 50% CCl_4 ▶ API 331

▶ 2-ABb

CH_3—C(CH_3)(a CH_3)—CH_2—(c) $(CH_2)_3$—(b) CH_3

Shift		Peak
a	.87	1E
b	.87	1E
▶c	1.28	1B

▶ C_9H_{20} ▶ 50% CCl_4 ▶ API 209

▶ 2-ABb

(a) CH_3—(b) $(CH_2)_4$—(d) CH_2—(c) NH_2

Shift		Peak
a	0.90	3B
▶b	1.28	1B
c	1.58	S
d	2.70	1C

▶ $C_6H_{15}N$ ▶ 35.6 mg/0.50 ml $CDCl_3$ ▶ S 46

▶ 2-ABb

(a) CH_3—(b) $(CH_2)_7$—CH_3

Shift		Peak
a	.88	3B
▶b	1.28	1B

▶ C_9H_{20} ▶ 50% CCl_4 ▶ API 343

▶ 2-ABb

(a) CH_3—(b) $(CH_2)_5$—(d) CH(Br)—(c) CH_3

Shift		Peak
a	0.91	3B
▶b	1.28	1B
c	1.71	2A
d	4.12	5B

▶ $C_8H_{17}Br$ ▶ 45.2 mg/0.50 ml $CDCl_3$ ▶ S 227

▶ 2-ABb

(a) CH_3—(b) $(CH_2)_7$—(a) CH_3

Shift		Peak
a	0.88	3B
▶b	1.28	1A

▶ C_9H_{20} ▶ 45.0 mg/0.50 ml $CDCl_3$ ▶ S 203

▶ 2-ABb

$H_2C=CH-CH_2(CH_2)_5CH_3$ — (d) H, (e) H, (f) H, (c) CH_2, (b) $(CH_2)_5$, (a) CH_3

Shift		Peak
a	.85	3B
▶b	1.28	1B
c	1.99	3B
d	4.88	R
e	4.99	R
f	5.68	R

▶ C_9H_{18} ▶ 50% CCl_4 ▶ API 434

▶ 2-ABb

(a) CH_3—(b) $(CH_2)_7$—(c) CH_2—(d) OH

Shift		Peak
a	.88	3B
▶b	1.28	1B
c	3.44	3B
d	4.27	1A

▶ $C_9H_{20}O$ ▶ 20% CCl_4 ▶ API 171

▶ 2-ABb

(c) CH_3—C(=O)—(d) CH_2—(b) $(CH_2)_5$—(a) CH_3

Shift		Peak
a	0.88	3B
▶b	1.28	1B
c	2.17	1A
d	2.46	3B

▶ $C_9H_{18}O$ ▶ 33.7 mg/0.50 ml $CDCl_3$ ▶ S 48

▶ 2-ABb

(a) CH_3—(b) $(CH_2)_6$—(c) CH_2C(=O)—(d) OCH_3

Shift		Peak
a	0.88	3B
▶b	1.28	1B
c	2.30	3B
d	3.68	1A

▶ $C_{10}H_{20}O_2$ ▶ 45.1 mg/0.50 ml $CDCl_3$ ▶ S 249

▶ 2-ABb

	Shift	Peak
a	0.88	3B
▶b	1.28	1E

(a) (b) (a)
$CH_3-(CH_2)_8-CH_3$

▶ $C_{10}H_{22}$ ▶ 44.8 mg/0.50 ml $CDCl_3$ ▶ S 201

▶ 2-ABb

	Shift	Peak
a	0.88	3B
▶b	1.28	1E
c	2.28	3C
d	4.11	4A

(a) (b) (c) (d) (a)
$CH_3-(CH_2)_9-CH_2-C(=O)-OCH_2CH_3$

▶ $C_{14}H_{28}O_2$ ▶ 45.5 mg/0.50 ml $CDCl_3$ ▶ S 194

▶ 2-ABb

	Shift	Peak
a	.88	3B
▶b	1.28	1A

(b) (a)
$CH_3-(CH_2)_9-CH_3$

▶ $C_{11}H_{24}$ ▶ 50% CCl_4 ▶ API 348

▶ 2-ABb

	Shift	Peak
a	0.88	3B
▶b	1.28	1A

(a) (b) (a)
$CH_3-(CH_2)_{12}-CH_3$

▶ $C_{14}H_{30}$ ▶ 44.8 mg/0.50 ml $CDCl_3$ ▶ S 204

▶ 2-ABb

	Shift	Peak
a	0.90	3B
▶b	1.28	1G

(a) (b) (a)
$CH_3-(CH_2)_9-CH_3$

▶ $C_{11}H_{24}$ ▶ 44.4 mg/0.50 ml $CDCl_3$ ▶ S 289

▶ 2-ABb

	Shift	Peak
a	0.90	3B
▶b	1.29	1E
c	2.09	1A
d	2.53	4B

(a) (b) (d) (c)
$CH_3-(CH_2)_5-CH_2SH$

▶ $C_7H_{16}S$ ▶ 46.9 mg/0.50 ml $CDCl_3$ ▶ S 297

▶ 2-ABb

	Shift	Peak
a	0.87	3B
▶b	1.28	1A
c	2.05	2G
d	4.80	R
e	~5.03	R
f	5.68	R

(a) (b) (c) (f) (d) (e)
$CH_3-(CH_2)_8-CH_2-CH(f)=CH(d)H(e)$

▶ $C_{12}H_{24}$ ▶ 44.4 mg/0.50 ml $CDCl_3$ ▶ S 202

▶ 2-ABb

	Shift	Peak
a	0.88	3B
▶b	1.29	1B
c	2.30	1B
d	9.35	1A

(a) (b) (c) (d)
$CH_3-(CH_2)_6-CH_2-CH(=O)$

▶ $C_9H_{18}O$ ▶ 45.3 mg/0.50 ml $CDCl_3$ ▶ S 100

▶ 2-ABb

	Shift	Peak
a	.88	3B
▶b	1.28	1A

(a) (b)
$CH_3-(CH_2)_{10}-CH_3$

▶ $C_{12}H_{26}$ ▶ 50% CCl_4 ▶ API 350

▶ 2-ABb

	Shift	Peak
a	0.87	3A
▶b	1.29	1B
c	2.42	3B
d	0.77	3A

(a) (b) (c) (d)
$CH_3-(CH_2)_9-CH_2-CH(=O)$

▶ $C_{12}H_{24}O$ ▶ 45.2 mg/0.50 ml $CDCl_3$ ▶ S 106

▶ 2-ABb

	Shift	Peak
a	0.88	3B
▶b	1.28	1B
c	2.28	3B
d	3.67	1A

(a) (b) (c) (d)
$CH_3-(CH_2)_9-CH_2-C(=O)-OCH_3$

▶ $C_{13}H_{26}O_2$ ▶ 45.7 mg/0.50 ml $CDCl_3$ ▶ S 125

▶ 2-ABb

	Shift	Peak
a	0.85	3B
▶b	1.29	1A
c	1.83	1B
d	3.62	3B

(a) (b) (d) (c)
$CH_3-(CH_2)_{10}-CH_2OH$

▶ $C_{12}H_{26}O$ ▶ 44.4 mg/0.50 ml $CDCl_3$ ▶ S 103

▶ 2-ABb

	Shift	Peak
a	0.88	3B
▶b	1.29	1A
c	2.24	3B
d	3.68	1A

(a) (b) (c) (d)
$CH_3-(CH_2)_{11}-CH_2-C(=O)-OCH_3$

▶ $C_{15}H_{30}O_2$ ▶ 45.1 mg/0.50 ml $CDCl_3$ ▶ S 195

▶ 2-ABb

	Shift	Peak
a	.88	3B
▶b	1.30	1B

(a) (b)
$CH_3-(CH_2)_5-CH_3$

▶ C_7H_{16} ▶ 50% CCl_4 ▶ API 337

▶ 2-ABb

	Shift	Peak
a	0.88	3B
▶b	1.29	1B
c	2.00	1F
d	2.25	3B
e	5.28	3D
f	11.45	1C

(e) (c) (b) (d) (f)
$CH-CH_2-(CH_2)_5-CH_2-C(=O)-OH$
$\|$
$CH-CH_2-(CH_2)_6-CH_3$
(e) (c) (b) (a)

▶ $C_{18}H_{34}O_2$ ▶ 45.2 mg/0.50 ml $CDCl_3$ ▶ S 70

▶ 2-ABb

	Shift	Peak
a	0.88	3B
▶b	1.30	1B
c	2.43	3B
d	9.73	3A

(a) (b) (c)
$CH_3-(CH_2)_5-CH_2-CH(=O)$
(d)

▶ $C_8H_{16}O$ ▶ 44.6 mg/0.50 ml $CDCl_3$ ▶ S 105

▶ 2-ABb

	Shift	Peak
a	0.90	3B
▶b	1.29	1B
c	2.00	1C
d	2.32	S
e	3.70	1A
f	5.4	3B

(f) (c) (b) (d) (e)
$CH-CH_2-(CH_2)_5-CH_2-C(=O)-OCH_3$
$\|$
$CH-CH_2-(CH_2)_6-CH_3$
(f) (c) (b) (a)

▶ $C_{19}H_{36}O_2$ ▶ 33.7 mg/0.50 ml $CDCl_3$ ▶ S 71

▶ 2-ABb

	Shift	Peak
a	0.88	3B
▶b	1.30	1G
c	2.36	3B
d	11.47	1A

(a) (b) (c) (d)
$CH_3(CH_2)_5CH_2-C(=O)-OH$

▶ $C_8H_{16}O_2$ ▶ 45.6 mg/0.50 ml $CDCl_3$ ▶ S 157

▶ 2-ABb

	Shift	Peak
a	.88	3C
▶b	~1.29 or 1.31	S
c	~1.31 or 1.29	S
d	2.00	3C
e	4.84	R
f	4.97	R
g	5.65	R

(e) H, (f) H, (g) H
$H_2C=CH-CH_2CH_2CH_2CH_3$
(d) (c) (b) (a)

▶ C_6H_{12} ▶ liquid ▶ API 368

▶ 2-ABb

	Shift	Peak
a	.88	3B
b	1.12	2A
▶c	1.30	1B
d	3.70	4B
e	4.08	1A

OH (e)
(b) (c) (c) (a)
$CH_3-CH(OH)-CH_2-(CH_2)_4-CH_3$
(d)

▶ $C_8H_{18}O$ ▶ 20% CCl_4 ▶ API 170

▶ 2-ABb

	Shift	Peak
a	.88	3B
▶b	1.30	1B

(b) (a)
$CH_3-(CH_2)_4-CH_3$

▶ C_6H_{14} ▶ 50% CCl_4 ▶ API 333

▶ 2-ABb

	Shift	Peak
a	0.92	3B
▶b	1.30	1B
c	1.75	1B
d	3.52	1I

(b)
(a) (d) (c)
$CH_3(CH_2)_3-CH(CH_2CH_3)CH_2OH$
CH_2CH_3: (b) (a)

▶ $C_8H_{18}O$ ▶ 45.0 mg/0.50 ml $CDCl_3$ ▶ S 98

▶ 2-ABb

	Shift	Peak
a	0.90	3B
▶b	1.30	1B
c	1.85	1F
d	3.41	3A

(a) (b) (c) (d)
$CH_3-(CH_2)_4-CH_2-CH_2-Br$

▶ $C_7H_{15}Br$ ▶ 45.0 mg/0.50 ml $CDCl_3$ ▶ S 208

▶ 2-ABb

	Shift	Peak
a	0.88	3B
b	1.28	3D
▶c	1.30	S
d	2.28	3B
e	4.12	4A

(a) (c) (d) (e) (b)
$CH_3-(CH_2)_5-CH_2-C(=O)-OCH_2CH_3$

▶ $C_{10}H_{20}O_2$ ▶ 45.2 mg/0.50 ml $CDCl_3$ ▶ S 244

▶ 2-ABb

	Shift	Peak
a	.83	2H
b	.94	S
▶ c	1.30	1A

$CH_3-CH(CH_3)-(CH_2)_7-CH_3$ [(a) CH_3 on CH; (c) $(CH_2)_7$; (b) terminal CH_3]

▶ $C_{11}H_{24}$ ▶ 50% CCl_4 ▶ API 349

▶ 2-ABb

	Shift	Peak
a	0.88	3B
▶ b	1.31	1E
c	3.53	3A

(a) CH_3—(b) $(CH_2)_5$—(c) CH_2Cl

▶ $C_7H_{15}Cl$ ▶ 44.8 mg/0.50 ml $CDCl_3$ ▶ S 209

▶ 2-ABb

	Shift	Peak
a	.87	3B
▶ b	1.30	1A

CH_3—(b) $(CH_2)_{13}$—(a) CH_3

▶ $C_{15}H_{32}$ ▶ 50% CCl_4 ▶ API 221

▶ 2-ABb

	Shift	Peak
a	0.90	3B
▶ b	1.32	2E
c	1.84	3C
d	3.20	3A

(a) CH_3—(b) $(CH_2)_3$—(c) CH_2—(d) CH_2—I

▶ $C_6H_{13}I$ ▶ 45.8 mg/0.50 ml $CDCl_3$ ▶ S 212

▶ 2-ABb

	Shift	Peak
a	0.89	3B
▶ b	~1.30	S
c	~1.40	S
d	1.72	1A
e	1.97	2A
f	2.94	3B
g	6.06	S
h	6.17	S
i	6.58	S
j	6.89	S
k	7.88	1C

(f)(c)(b)(a) $CH_2CH_2(CH_2)_2CH_3$ acyl group on fused furanone–pyran ring system; H(j), H(h), H(k); (d) CH_3; (g) H—C=C(H (i))—(e) CH_3

▶ $C_{21}H_{22}O_5$ ▶ 7% $CDCl_3$ ▶ V 687

▶ 2-ABb

	Shift	Peak
a	0.88	3B
▶ b	1.32	1B
c	2.38	3B
d	11.47	1B

(a) CH_3—(b) $(CH_2)_4$—(c) CH_2—C(=O)—(d) OH

▶ $C_7H_{14}O_2$ ▶ 45.6 mg/0.50 ml $CDCl_3$ ▶ S 49

▶ 2-ABb

	Shift	Peak
a	.91	3B
▶ b	1.31	S
c	2.02	3B
d	4.84	R
e	4.98	R
f	5.67	R

(d)H, (e)H C=C H(f), $(CH_2)_4CH_3$ [(c+b) $(CH_2)_4$, (a) CH_3]

▶ C_7H_{14} ▶ 10% CCl_4 ▶ API 157

▶ 2-ABb

	Shift	Peak
a	0.88	3B
▶ b	1.32	2E
c	1.82	3C
d	3.18	3A

(a) CH_3 (b) $(CH_2)_4$—(c) CH_2—(d) CH_2—I

▶ $C_7H_{15}I$ ▶ 45.0 mg/0.50 ml $CDCl_3$ ▶ S 210

▶ 2-ABb

	Shift	Peak
a	.90	3C
▶ b	1.31	O
c	1.57	2E
d	2.02	1F
e	~5.37	O

(c) H_3C—CH(e)=CH(e)—(d)(b)(b)(a) $CH_2CH_2CH_2CH_3$

▶ C_7H_{14} ▶ liquid ▶ API 389

▶ 2-ABb

	Shift	Peak
a	.90	3B
▶ b	1.32	1G
c	3.44	3B
d	4.42	1B

(a) CH_3—(b) $(CH_2)_5$—(c) CH_2—(d) OH

▶ $C_7H_{16}O$ ▶ 20% CCl_4 ▶ API 168

▶ 2-ABb

	Shift	Peak
a	.88	3C
▶ b	1.31	O
c	1.62	2E
d	1.92	1F
e	5.33 or 5.42	S
f	5.42 or 5.33	S

(c) H_3C—CH(e)=CH(f)—(d)(b)(b)(a) $CH_2CH_2CH_2CH_3$

▶ C_7H_{14} ▶ liquid ▶ API 390

▶ 2-ABb

	Shift	Peak
a	0.92	3B
b	1.12	S
▶ c	1.32	S
d	2.33	4A
e	4.05	3B

(b)(d) $CH_3CH_2C(=O)$—O(e)CH_2(c)$(CH_2)_3$(a)CH_3

▶ $C_8H_{16}O_2$ ▶ 44.8 mg/0.50 ml $CDCl_3$ ▶ S 40

▶ 2-ABb

$HC(=O)-OCH_2-(CH_2)_6-CH_3$ (d) (c) (b) (a)

	Shift	Peak
a	0.88	3B
▶ b	1.32	1E
c	4.16	3B
d	8.03	1B

▶ $C_9H_{18}O_2$ ▶ 44.4 mg/0.50 ml $CDCl_3$ ▶ S 114

▶ 2-ABb

(e)H(f)H C=C H(g) $CH_2CH_2CH_2CH_3$ (d) (c) (b) (a)

	Shift	Peak
a	.90	3B
▶ b	~1.34	S
c	~1.34	S
d	2.00	4C
e	4.85	R
f	4.97	R
g	5.68	R

▶ C_6H_{12} ▶ 10% CCl_4 ▶ API 155

▶ 2-ABb

$CH_3-(CH_2)_7-CH_2OH$ (a) (b) (d) (c)

	Shift	Peak
a	0.90	3B
▶ b	1.32	1B
c	1.97	1A
d	3.62	3B

▶ $C_9H_{20}O$ ▶ 44.6 mg/0.50 ml $CDCl_3$ ▶ S 102

▶ 2-ABb

$CH_3-(CH_2)_4-CH_2-OH$ (a) (b) (d) (c)

	Shift	Peak
a	0.88	3B
▶ b	1.36	1B
c	2.28	1A
d	3.62	3A

▶ $C_6H_{14}O$ ▶ 45.2 mg/0.50 ml $CDCl_3$ ▶ S 198

▶ 2-ABb

(d)H (e)H C=C H(f) $CH_2(CH_2)_8CH_3$ (c) (b) (a)

	Shift	Peak
a	.92	3B
▶ b	1.32	1B
c	2.05	3B
d	4.86	R
e	5.00	R
f	5.68	R

▶ $C_{12}H_{24}$ ▶ 50 vol.% CCl_4 ▶ API 435

▶ 2-ABb

$C_6H_5-CH_2-CH_2-CH_2-CH_2-CH_3$ (c) (b) (b) (b) (a)

(d) = all ring H's

	Shift	Peak
a	.89	3B
▶ b	1.36	O
c	2.52	3B
d	7.04	1G

▶ $C_{11}H_{16}$ ▶ 50% CCl_4 ▶ API 355

▶ 2-ABb

$CH_3-(CH_2)_3-CH_2-CH_2-Cl$ (a) (b) (c) (d)

	Shift	Peak
a	0.90	3B
▶ b	1.33	2H
c	1.78	O
d	3.53	3A

▶ $C_6H_{13}Cl$ ▶ 44.6 mg/0.50 ml $CDCl_3$ ▶ S 211

▶ 2-ABb

	Shift	Peak
a	0.88	3C
▶ b	~1.36	S
c	~1.50	S
d	3.58	2H
e	3.84	3B
f	6.24	2A
g	7.20	1A

▶ $C_{17}H_{25}BCl_2O_2$ ▶ 7% $CDCl_3$ ▶ V 667

▶ 2-ABb

$CH_3-(CH_2)_5-CH_2-OH$ (a) (b) (d) (c)

	Shift	Peak
a	0.88	3B
▶ b	1.33	1B
c	1.95	1A
d	3.62	3B

▶ $C_7H_{16}O$ ▶ 45.0 mg/0.50 ml $CDCl_3$ ▶ S 215

▶ 2-ABb

	Shift	Peak
a	0.88	3C
▶ b	~1.36	S
c	~1.50	S
d	2.86*	S
e	2.89*	S
f	3.65	S
g	3.88	3C
h	7.26	1A

*±0.04

▶ $C_{17}H_{26}BCl_3O_2$ ▶ 7% $CDCl_3$ ▶ V 668

▶ 2-ABb

$CH_3-(CH_2)_4-CH_2-C(=O)-OCH_2-CH_3$ (a) (c) (d) (e) (b)

	Shift	Peak
a	0.88	3B
b	1.26	S
▶ c	1.33	S
d	2.28	3B
e	4.12	4A

▶ $C_9H_{18}O_2$ ▶ 44.5 mg/0.50 ml $CDCl_3$ ▶ S 283

▶ 2-ABb

$CH_3C(=O)-OCH_2(CH_2)_4CH_3$ (c) (d) (b) (a)

	Shift	Peak
a	0.92	3B
▶ b	1.37	1E
c	2.05	1A
d	4.05	3A

▶ $C_8H_{16}O_2$ ▶ 45.0 mg/0.50 ml $CDCl_3$ ▶ S 217

► 2-ABb

(a) CH_3—(b) $(CH_2)_2$—(c) CH_2—N [three identical butyl chains: CH_3—$(CH_2)_2$—CH_2—]

	Shift	Peak
a	0.92	3B
► b	1.37	6B
c	2.40	3B

► $C_{12}H_{27}N$ ► 44.8 mg/0.50 ml $CDCl_3$ ► S 84

► 2-ABb

(a) CH_3—(b) CH_2—(b) CH_2—(d) CH_2—N(e)H—C(=O)—N(h)H—SO_2—C_6H_4[(g)H, (g)H, (f)H, (f)H]—(c) CH_3

	Shift	Peak
a	0.90	3B
► b	~1.40	1F
c	2.46	1A
d	3.23	4C
e	6.53	3B
f	7.32	Q
g	7.79	Q
h	~8.60	1D

► $C_{12}H_{18}N_2O_3S$ ► 7% $CDCl_3$ ► V 602

► 2-ABb

(a) CH_3 (b) $CH_2CH_2CH_2$ (c) CH_2 (d) CH_2Br

	Shift	Peak
a	0.94	3B
► b	1.38	O
c	1.83	4C
d	3.42	3A

► $C_5H_{11}Br$ ► 44.5 mg/0.50 ml $CDCl_3$ ► S 120

► 2-ABb

(a) CH_3—(b) CH_2—(c) CH_2—(e) CH_2—N(C_6H_5)—C(=O)—(d) CH_3

	Shift	Peak
a	0.91	3B
► b	1.44 ± 0.1	S
c	1.44 ± 0.1	S
d	1.87	1A
e	3.73	3A

► $C_{12}H_{17}NO$ ► 7% $CDCl_3$ ► V 601

► 2-ABb

(a) CH_3—(b) $(CH_2)_2$—(c) CH_2—(d) CH_2I

	Shift	Peak
a	0.92	3B
► b	1.38	2E
c	1.86	3C
d	3.20	3A

► $C_5H_{11}I$ ► 45.0 mg/0.50 ml $CDCl_3$ ► S 228

► 2-ABb

(a) CH_3—(b) CH_2—(b) CH_2—(c) CH_2—(d) OH

	Shift	Peak
a	.91	3B
► b	~1.45	O
c	3.48	3C
d	4.58	1A

► $C_4H_{10}O$ ► 20% CCl_4 ► API 166

► 2-ABb

(e)H(f)H C=C (c) CH_3 / (d) CH_2 (b) CH_2 (b) CH_2 (a) CH_3

	Shift	Peak
a	.93	3C
► b	1.38	O
c	1.68	1H
d	2.01	3B
e	~4.67	M
f	~4.67	M

► C_7H_{14} ► liquid ► API 407

► 2-ABb

(a) CH_3—(b) $(CH_2)_3$—(c) CH_2—C(=O)—(d) OH

	Shift	Peak
a	0.92	3B
► b	1.45	O
c	2.36	3C
d	11.37	1B

► $C_6H_{12}O_2$ ► 44.7 mg/0.50 ml $CDCl_3$ ► S 37

► 2-ABb

(c) CH_3—C(=O)—(d) CH_2—(b) $(CH_2)_3$—(a) CH_3

	Shift	Peak
a	0.90	3B
► b	1.38	O
c	2.15	1A
d	2.42	3B

► $C_7H_{14}O$ ► 35.4 mg/0.50 ml $CDCl_3$ ► S 41

► 2-ABb

(b) CH_3—(e) CH((d) OH)—C(=O)—O(e) CH_2 (c) CH_2CH_2 (a) CH_3

	Shift	Peak
a	0.94	3B
b	1.41	2A
► c	1.46	S
d	3.02	1A
e	4.18	4C

► $C_7H_{14}O_3$ ► 44.6 mg/0.50 ml $CDCl_3$ ► S 246

► 2-ABb

(a) CH_3 (b) $(CH_2)_3$ (c) CH_2—O—(c) CH_2 (b) $(CH_2)_3$ (a) CH_3

	Shift	Peak
a	0.88	3B
► b	1.40	O
c	3.38	3A

► $C_{10}H_{22}O$ ► 44.6 mg/0.50 ml $CDCl_3$ ► S 151

► 2-ABb

C_6H_5—(d) CH_2—(c) CH_2—(b) CH_2—(a) CH_3

(e) = all ring H's

	Shift	Peak
a	.93	3C
► b	1.47	S
c	1.47	S
d	2.57	3B
e	7.11	1A

► $C_{10}H_{14}$ ► 10% CCl_4 ► API 42

▶ 2-ABb

(c) (b) (a)
$CH_2—(CH_2)_2—CH_3$
|
C_6H_5—N—CH_2—$(CH_2)_2$—CH_3
(c) (b) (c)

	Shift	Peak
a	0.96	3B
▶b	1.48	1F
c	3.28	3C

▶ $C_{14}H_{23}N$ ▶ 45.2 mg/0.50 ml $CDCl_3$ ▶ S 147

▶ 2-ABb

NH_2(d)–C_6H_4–C(=O)–O–$CH_2CH_2CH_2CH_3$
(c) (b) (a); ring H's (e), (e), (f), (f)

	Shift	Peak
a	0.95	3B
▶b	~1.60	1F
c	4.25	3D
d	~4.10	1E
e	6.59	Q
f	7.83	Q

▶ $C_{11}H_{15}NO_2$ ▶ 44.6 mg/0.50 ml $CDCl_3$ ▶ S 243

▶ 2-ABb

(a) (b) (c)
$CH_3—(CH_2)_3—CH_2—C≡N$

	Shift	Peak
a	0.93	3B
▶b	1.49	O
c	2.33	3B

▶ $C_6H_{11}N$ ▶ 45.0 mg/0.50 ml $CDCl_3$ ▶ S 176

▶ 2-ABb

C_6H_5–C(=O)–O $CH_2(CH_2)_2CH_3$
(c) (b) (a); ring H's (e), (e), (d), (d), (d)

	Shift	Peak
a	0.98	3B
▶b	1.64	O
c	4.35	3A
d	7.48	N
e	8.06	N

▶ $C_{11}H_{14}O_2$ ▶ 44.8 mg/0.50 ml $CDCl_3$ ▶ S 158

▶ 2-ABb

(a) (b) (c)
$CH_3—(CH_2)_2—CH_2—S—CH_2—(CH_2)_2—CH_3$

	Shift	Peak
a	0.92	3B
▶b	1.49	O
c	2.49	3B

▶ $C_8H_{18}S$ ▶ 45.8 mg/0.50 ml $CDCl_3$ ▶ S 275

▶ 2-ABb

(b) (c) (a)
C_6H_5–$CH_2—(CH_2)_4—CH_3$

(d) = all ring H's

	Shift	Peak
a	.89	3B
b	1.30	1F
▶c	2.54	3B
d	7.10	1A

▶ $C_{12}H_{18}$ ▶ 50% CCl_4 ▶ API 356

▶ 2-ABb

(d) (c) (b) (a)
C_6H_5–NH—$CH_2—(CH_2)_2—CH_3$
ring H's (e), (e), (f), (f), (f)

	Shift	Peak
a	0.96	3B
▶b	1.50	1F
c	3.08	3A
d	3.42	1A
e	6.52	3C
f	7.12	3C

▶ $C_{10}H_{15}N$ ▶ 45.4 mg/0.50 ml $CDCl_3$ ▶ S 280

▶ 2-ABb

(b) (c) (a)
C_6H_5–$CH_2—(CH_2)_5—CH_3$

(d) = all ring H's

	Shift	Peak
a	.86	3B
b	1.28	1B
▶c	2.54	3B
d	7.09	1A

▶ $C_{13}H_{20}$ ▶ 50% CCl_4 ▶ API 357

▶ 2-ABb

(a) (c) (d) (b)
$CH_3—(CH_2)_2—CH_2—SH$

	Shift	Peak
a	0.92	3A
b	1.23	S
▶c	1.53	S
d	2.52	4B

▶ $C_4H_{10}S$ ▶ 46.0 mg/0.50 ml $CDCl_3$ ▶ S 274

▶ 2-ABc

(a)
CH_3
|
CH_3—CH—CH_2—CH_2—CH_3
(d) (c) (c) (b)

	Shift	Peak
a	.88	3D
b	.95	2H
▶c	1.20	N
d	~1.50	1D

▶ C_6H_{14} ▶ 50% CCl_4 ▶ API 334

▶ 2-ABb

(c) (d) (b) (a)
Cl—CH_2—C(=O)—O—$CH_2(CH_2)_2CH_3$

	Shift	Peak
a	0.94	3B
▶b	1.56	1F
c	4.05	1A
d	4.18	3A

▶ $C_6H_{11}ClO_2$ ▶ 44.5 mg/0.50 ml $CDCl_3$ ▶ S 92

▶ 2-ABc

(b)
CH_3
|
$CH_3—CH_2$—CH—$CH_2—CH_2—CH_3$
(a) (c) (d) (c) (c) (a)

	Shift	Peak
a	.88	3D
b	.88	S
▶c	1.24	S
d	~1.26	S

▶ C_7H_{16} ▶ 50% CCl_4 ▶ API 338

▶ 2-ABc

$CH_3-CH_2-CH_2-CH(CH_3)-CH_2-CH_2-CH_3$ — (d) CH, (c) CH_2, (c) CH_2, (a) CH_3, (b) CH_3

	Shift	Peak
a	.88	3A
b	~1.25	S
▶ c	~1.25	S
d	~1.25	S

▶ C_8H_{18} ▶ 50% CCl_4 ▶ API 201

▶ 2-ABi

(a) (b) (c) $CH_3-CH_2-CH_2-I$

	Shift	Peak
a	1.01	3A
▶ b	1.86	6B
c	3.18	3A

▶ C_3H_7I ▶ 45.0 mg/0.50 ml $CDCl_3$ ▶ S 230

▶ 2-ABc

(b) H_3C-H_2C (c) / (b) H_3C-H_2C (c) $>CH-CH_2-CH_2-CH_3$ (c) (c) (a)

	Shift	Peak
a	~ .88	3D
b	~ .88	3D
▶ c	~1.26	1F

▶ C_8H_{18} ▶ 50% CCl_4 ▶ API 202

▶ 2-ABj

(c) (d) (b) (a) $CH_3-C(=O)-CH_2-CH_2-CH_3$

	Shift	Peak
a	0.92	3B
▶ b	1.62	6B
c	2.14	1A
d	2.42	3A

▶ $C_5H_{10}O$ ▶ 45.4 mg/0.50 ml $CDCl_3$ ▶ S 177

▶ 2-ABc

(e) H, (f) H $>C=C<$ H (g), $CHCH_2CH_2CH_3$ (d) (c) (c) (a), CH_3 (b)

	Shift	Peak
a	.88	3D
b	.98	2H
▶ c	1.29	O
d	2.11	5A
e	4.86	R
f	4.97	R
g	~5.58	R

▶ C_7H_{14} ▶ liquid ▶ API 392

▶ 2-ABj

(a) (b) (c) $CH_3-CH_2-CH_2-C(=O)-H$ (d)

	Shift	Peak
a	0.97	3C
▶ b	1.67	6B
c	2.42	3C
d	9.74	3A

▶ C_4H_8O ▶ 7% $CDCl_3$ ▶ V 78

▶ 2-ABc

(b) $CH_3-CH(OH)-(CH_2)_2-CH_3$ — OH (d), CH (e), $(CH_2)_2$ (c), CH_3 (a)

	Shift	Peak
a	0.92	3D
b	1.18	2H
▶ c	1.33	N
d	2.22	1B
e	3.70	1F

▶ $C_5H_{12}O$ ▶ 44.6 mg/0.50 ml $CDCl_3$ ▶ S 60

▶ 3-ABj

$C_6H_5-NH-C(=O)-CH_2-CH_2-CH_3$ — NH (d), CH_2 (c), CH_2 (b), CH_3 (a)

	Shift	Peak
a	0.95	3A
▶ b	1.71	6B
c	2.24	3C
d	~7.8	1C

▶ $C_{10}H_{13}NO$ ▶ 45.3 mg/0.50 ml $CDCl_3$ ▶ S 261

▶ 2-ABd

(a) $CH_3-C(CH_3)_2-CH_2-CH_2-CH_3$ — CH_3 (a) ×3, CH_2 (c), CH_2 (c), CH_3 (b)

	Shift	Peak
a	.88	1E
b	.88	3D
▶ c	1.20	N

▶ C_7H_{16} ▶ 50% CCl_4 ▶ API 339

▶ 2-ABk

(a) (c) (d) (e) (b) $CH_3CH_2CH_2C(=O)-OCH_2CH_3$

	Shift	Peak
a	0.96	3A
b	1.28	3A
▶ c	1.67	6B
d	2.28	3C
e	4.13	4A

▶ $C_6H_{12}O_2$ ▶ 45.0 mg/0.50 ml $CDCl_3$ ▶ S 42

▶ 2-ABd

(b) $CH_3-CH(CH_3)-C(CH_3)_2-CH_2-CH_2-CH_3$ — CH (e), CH_3 (b), CH_3 (a) ×2, CH_2 (d), CH_2 (d), CH_3 (c)

	Shift	Peak
a	.80	1A
b	.86	2H
c	~ .86	3D
▶ d	1.22	3A
e	~1.50	1D

▶ C_9H_{20} ▶ 50% CCl_4 ▶ API 213

▶ 2-ABl

(c) H_3C, (e) H $>C=C<$ H (e), $CH_2CH_2CH_3$ (d) (b) (a)

	Shift	Peak
a	.88	3C
▶ b	1.28	S
c	1.60	2E
d	1.92	1F
e	5.35	O

▶ C_6H_{12} ▶ 50 vol.% CCl_4 ▶ API 370

▶ 2-ABl

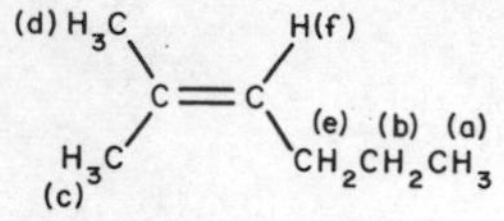

(d) H_3C, H(f), C=C, (c) H_3C, (e) (b) (a) $CH_2CH_2CH_3$

	Shift	Peak
a	.90	3C
▶ b	1.28	S
c	~1.58 or 1.68	O
d	~1.68 or 1.58	O
e	1.91	4B
f	5.08	3C

▶ C_7H_{14} ▶ liquid ▶ API 408

▶ 2-ABl

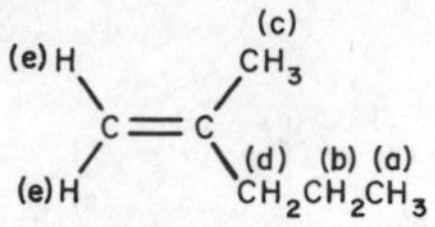

(e) H, (e) H, C=C, (c) CH_3, (d) (b) (a) $CH_2CH_2CH_3$

	Shift	Peak
a	.88	3C
▶ b	1.34	S
c	1.57	1H
d	1.90	S
e	4.60	N

▶ C_6H_{12} ▶ 50 vol.% CCl_4 ▶ API 373

▶ 2-ABl

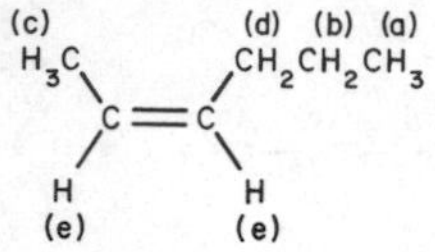

(c) H_3C, C=C, (d) (b) (a) $CH_2CH_2CH_3$, H (e), H (e)

	Shift	Peak
a	.89	3C
▶ b	1.30	S
c	1.56	2E
d	2.00	4C
e	5.32	O

▶ C_6H_{12} ▶ liquid ▶ API 369

▶ 2-ABl

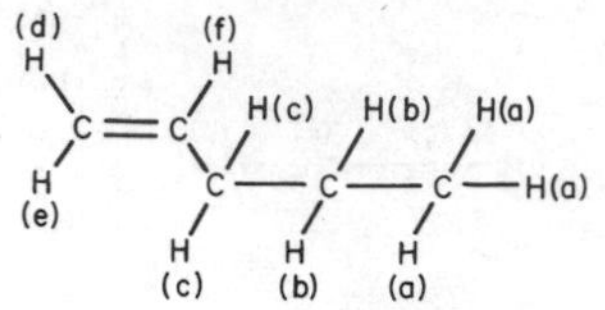

(d) H, (f) H, C=C, (e) H, H(c) H(b) H(a), C—C—C—H(a), H (c), H (b), H (a)

	Shift	Peak
a	.92	3C
▶ b	1.38	6B
c	1.93	4C
d	4.76	R
e	4.98	R
f	5.68	R

▶ C_5H_{10} ▶ liquid ▶ API 366

▶ 2-ABl

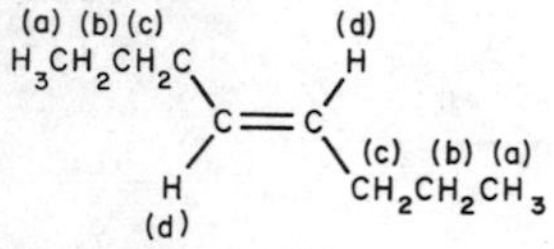

(a) (b) (c) $H_3CH_2CH_2C$, (d) H, C=C, H (d), (c) (b) (a) $CH_2CH_2CH_3$

	Shift	Peak
a	.85	3C
▶ b	1.32	6B
c	1.91	N
d	5.32	N

▶ C_8H_{16} ▶ 50 vol.% CCl_4 ▶ API 432

▶ 2-ABl

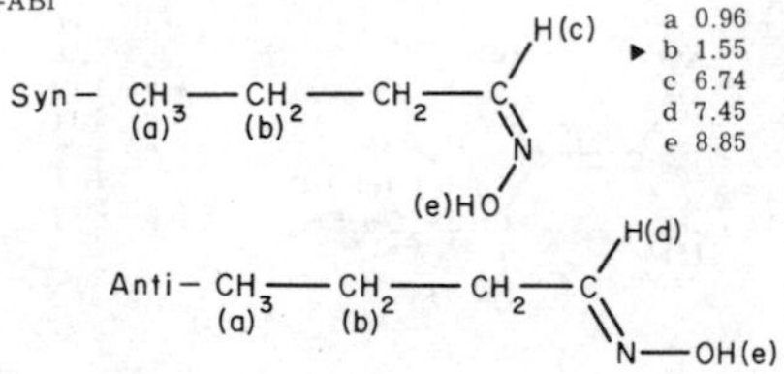

Syn– CH_3(a)—CH_2(b)—CH_2—C(H(c))=N—OH(e)

Anti– CH_3(a)—CH_2(b)—CH_2—C(H(d))=N—OH(e)

	Shift	Peak
a	0.96	3C
▶ b	1.55	6B
c	6.74	3A
d	7.45	3A
e	8.85	1D

▶ C_4H_9NO ▶ 7% $CDCl_3$ ▶ V 420

▶ 2-ABl

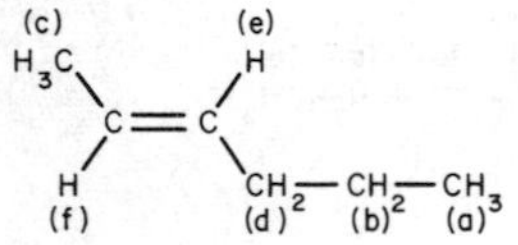

(c) H_3C, (e) H, C=C, H (f), CH_2(d)—CH_2(b)—CH_3(a)

	Shift	Peak
a	.88	3B
▶ b	~1.33	O
c	1.69	N
d	1.87	1F
e	5.32	S
f	5.32	S

▶ C_6H_{12} ▶ 10% CCl_4 ▶ API 32

▶ 2-ABl

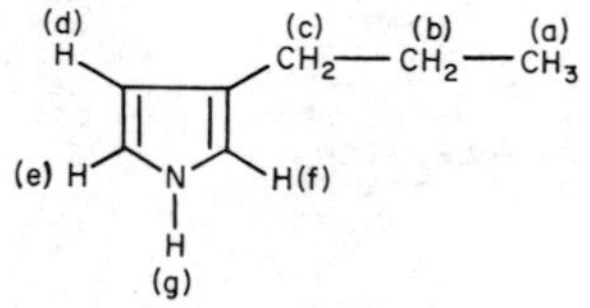

(d) H, (c) CH_2—(b) CH_2—(a) CH_3, (e) H, N, H(f), H (g)

	Shift	Peak
a	0.95	3C
▶ b	1.60	6B
c	2.48	3C
d	6.13	4B
e	6.70	S
f	6.55	S
g	~7.83	1A

▶ $C_7H_{11}N$ ▶ 7% $CDCl_3$ ▶ V 177

▶ 2-ABl

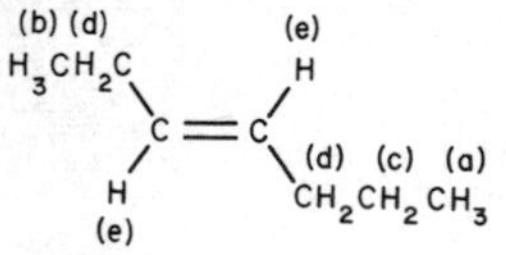

(b) (d) H_3CH_2C, (e) H, C=C, H (e), (d) (c) (a) $CH_2CH_2CH_3$

	Shift	Peak
a	.90	3D
b	.97	3D
▶ c	1.33	S
d	1.95	O
e	5.40	O

▶ C_7H_{14} ▶ liquid ▶ API 391

▶ 2-ABn

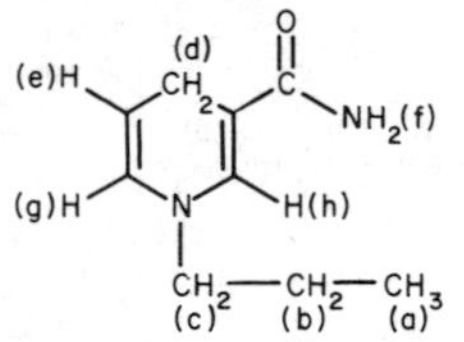

(e)H, (d) CH_2, O, C, NH_2(f), (g)H, N, H(h), CH_2(c)—CH_2(b)—CH_3(a)

	Shift	Peak
a	0.90	3C
▶ b	1.58	6B
c	3.09	2A
d	3.19	N
e	4.72	5B
f	5.49	1C
g	5.73	4A
h	7.05	2D

▶ $C_9H_{14}N_2O$ ▶ 7% $CDCl_3$ ▶ V 542

▶ 2-ABl

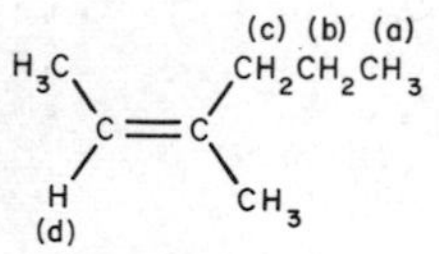

H_3C, C=C, H (d), (c) (b) (a) $CH_2CH_2CH_3$, CH_3

	Shift	Peak
a	.83	3C
▶ b	1.33	S
c	1.87	3B
d	5.14	O

▶ C_7H_{14} ▶ liquid ▶ API 395

▶ 2-ABp

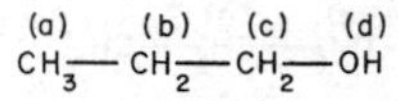

(a) CH_3—(b) CH_2—(c) CH_2—(d) OH

	Shift	Peak
a	.91	3C
▶ b	1.43	6B
c	3.45	3A
d	4.68	1A

▶ C_3H_8O ▶ 20% CCl_4 ▶ API 165

2-ABp

(a) (b) (d) (c)
$CH_3—CH_2—CH_2—OH$

	Shift	Peak
a	0.93	3A
▶ b	1.56	6B
c	2.17	1B
d	3.62	3A

▶ C_3H_8O ▶ 34.7 mg/0.50 ml $CDCl_3$ ▶ S 10

2-ABu

(a) (b) (c)
$CH_3—CH_2—CH_2—C≡N$

	Shift	Peak
a	1.08	3C
▶ b	1.70	6B
c	2.30	3C

▶ C_4H_7N ▶ 45.0 mg/0.50 ml $CDCl_3$ ▶ S 33

2-ABp

(a) (b) (d) (c)
$CH_3—CH_2—CH_2—OH$

	Shift	Peak
a	0.92	3A
▶ b	1.57	7A
c	~2.28	1C
d	3.58	3A

▶ C_3H_8O ▶ 7% $CDCl_3$ ▶ V 43

2-ABv

CH_3(c); (f)HC, CH(e), (g)HC, C—$CH_2CH_2CH_3$ (d)(b)(a); C—H (h)

	Shift	Peak
a	.87	3C
▶ b	1.56	6B
c	2.17	1A
d	2.45	3B
e	6.82	S
f	6.82	S
g	6.82	S
h	6.82	S

▶ $C_{10}H_{14}$ ▶ liquid run, neat ▶ API 96

2-ABq

$CH_3CH_2—C(=O)—O CH_2CH_2CH_3$
(b) (d) (e) (c) (a)

	Shift	Peak
a	0.95	3D
b	1.18	3D
▶ c	1.60	6B
d	2.33	4A
e	4.04	3A

▶ $C_6H_{12}O_2$ ▶ 44.4 mg/0.50 ml $CDCl_3$ ▶ S 38

2-ABv

H(d), (d)HC, C—$CH_2—CH_2—CH_3$ (c) (b) (a), (d)HC, CH(d), H(d)

	Shift	Peak
a	.92	3C
▶ b	1.58	6B
c	2.57	3C
d	7.10	1E

▶ C_9H_{12} ▶ 10% CCl_4 ▶ API 41

2-ABq

(c) (d) (b) (a)
$CH_3C(=O)—O CH_2CH_2CH_3$

	Shift	Peak
a	0.93	3A
▶ b	1.64	5B
c	2.04	1A
d	4.02	3A

▶ $C_5H_{10}O_2$ ▶ 44.9 mg/0.50 ml $CDCl_3$ ▶ S 115

2-ABv

(d) CH_2, CH_2(c), CH_3(a); (b) CH_3; CH(e); OCH_3(f); (h)HO; C—O; H(g); (f)CH_3O

	Shift	Peak
a	0.96	3B
b	1.37	2A
▶ c	~1.65	S
d	2.58	3B
e	3.42	S
f	3.88	1A
g	5.08	2A
h	5.62	1A

▶ $C_{20}H_{24}O_4$ ▶ 7% $CDCl_3$ ▶ V 682

2-ABr

(a) (b) (c)
$CH_3—CH_2—CH_2—NO_2$

	Shift	Peak
a	1.03	3A
▶ b	2.07	6A
c	4.38	3A

▶ $C_3H_7NO_2$ ▶ 7% $CDCl_3$ ▶ V 42

2-ABx

(e) H; C(=O)—NH_2; (d)H; (f)H; (+)N; H(g); $I^{(-)}$; $CH_2—CH_2—CH_3$ (c) (b) (a)

	Shift	Peak
a	1.03	3A
▶ b	2.15	6A
c	4.74	3C
d	8.29	3C
e	8.97	R
f	9.12	R
g	9.37	1G

▶ $C_9H_{13}IN_2O$ ▶ D_2O ▶ V 539

2-ABs

S—CH_2(c)—CH_2(b)—CH_3(a); (d)H, H(d); (e)H, N, H(e)

	Shift	Peak
a	1.08	3A
▶ b	1.73	6B
c	2.93	3A
d	7.10	Q
e	8.38	Q

▶ $C_8H_{11}NS$ ▶ 7% $CDCl_3$ ▶ V 509

2-ACaa

(b) CH_3
$CH_3—CH—CH_2—CH_3$
(d) (c) (a)

	Shift	Peak
a	.83	3D
b	.90	2H
▶ c	1.20	S
d	~1.67	1D

▶ C_5H_{12} ▶ 50% CCl_4 ▶ API 332

▶ 2-ACab

	Shift	Peak
a	.84	S
b	.84	S
▶ c	1.20	O
d	~1.30	S

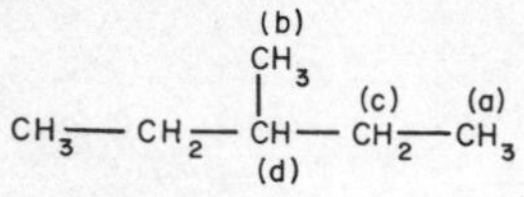

▶ C_6H_{14} ▶ 50% CCl_4 ▶ API 335

▶ 2-ACac

	Shift	Peak
a	1.28	1F

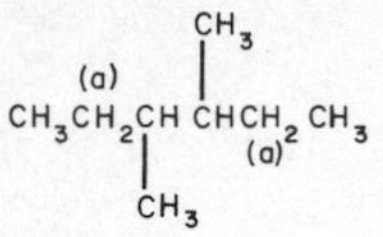

▶ C_8H_{18} ▶ 50% CCl_4 ▶ API 422

▶ 2-ACab

	Shift	Peak
a	.83	3B
▶ b	1.23	1B
c	~1.27	S

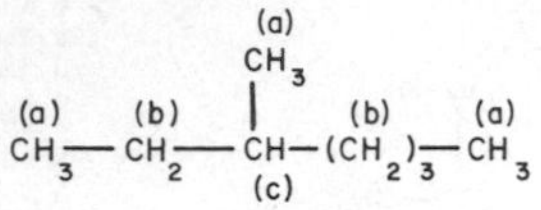

▶ C_8H_{18} ▶ 50% CCl_4 ▶ API 341

▶ 2-ACag

	Shift	Peak
a	1.08	3C
b	1.71	2H
▶ c	1.79	6C
d	4.11	5A

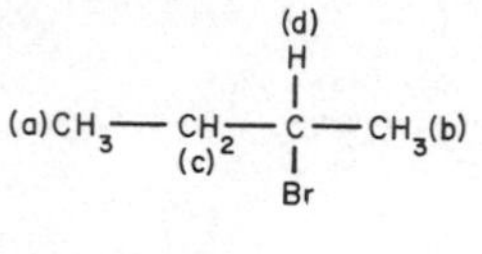

▶ C_4H_9Br ▶ 7% $CDCl_3$ ▶ V 418

▶ 2-ACab

	Shift	Peak
a	.88	3D
b	.88	S
▶ c	1.24	S
d	~1.26	S

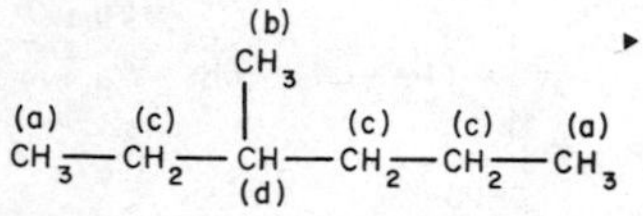

▶ C_7H_{16} ▶ 50% CCl_4 ▶ API 338

▶ 2-ACai

	Shift	Peak
a	1.02	3C
▶ b	1.70	4C
c	1.92	2A
d	4.17	4C

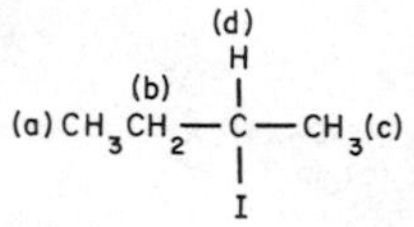

▶ C_4H_9I ▶ 7% $CDCl_3$ ▶ V 82

▶ 2-ACab

	Shift	Peak
a	.88	3D
b	.88	2H
▶ c	1.29	S
d	~1.30	S

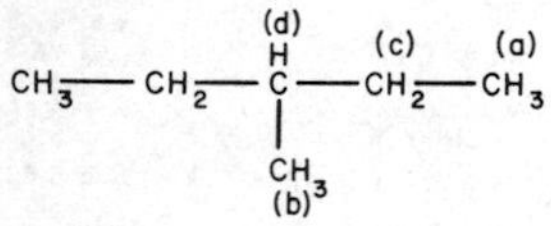

▶ C_6H_{14} ▶ 10% CCl_4 ▶ API 47

▶ 2-ACal

	Shift	Peak
a	.88	3C
b	1.03	1E
▶ c	1.24	S
d	1.96	7B
e	4.86	R
f	4.97	R
g	5.54	R

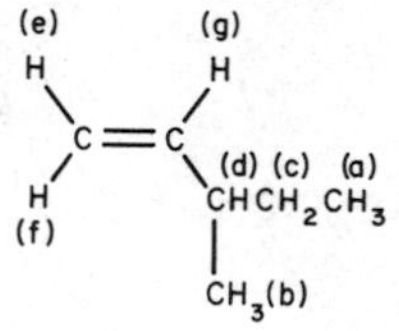

▶ C_6H_{12} ▶ liquid ▶ API 374

▶ 2-ACab

	Shift	Peak
a	~ .87	3D
b	~ .95	2H
▶ c	1.30	O
d	1.98	4C
e	~4.96	R
f	~4.96	R
g	5.67	R

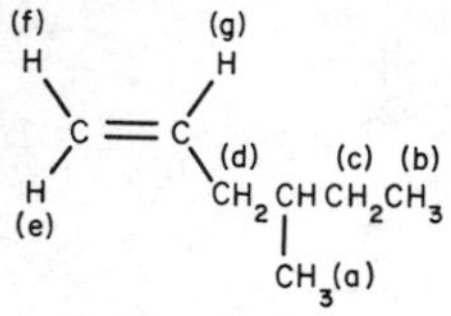

▶ C_7H_{14} ▶ liquid run, neat ▶ API 393

▶ 2-ACal

	Shift	Peak
a	.88	3D
b	.99	2A
▶ c	1.32	S
d	2.01	5B
e	4.83 or 4.96	R
f	4.96 or 4.83	R
g	5.53	R

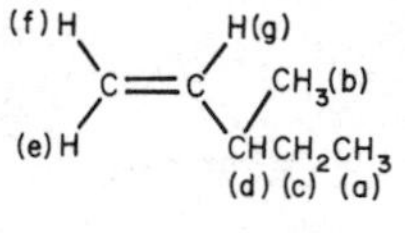

▶ C_6H_{12} ▶ 10% CCl_4 ▶ API 156

▶ 2-ACac

	Shift	Peak
a	~ .79	3D
b	~ .82	2H
▶ c	~1.22	1D
d	~1.58	S

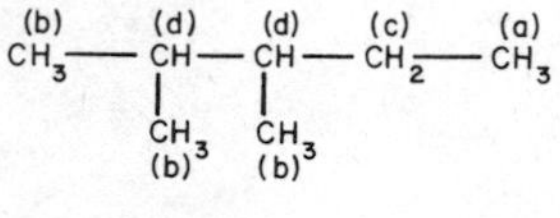

▶ C_7H_{16} ▶ 10% CCl_4 ▶ API 197

▶ 2-ACal

	Shift	Peak
a	.92	3C
b	.95	2H
c	1.25	O
▶ d	1.58	2E
e	2.38	6B
f	~5.17 or 5.32	S
g	~5.32 or 5.17	S

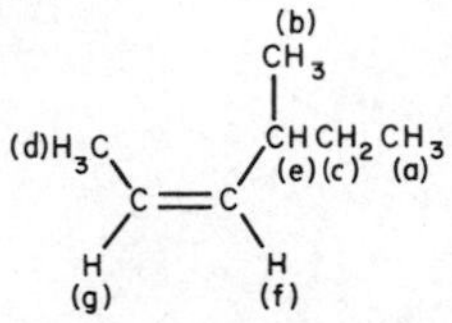

▶ C_7H_{14} ▶ liquid run, neat ▶ API 409

▶ 2-ACan

	Shift	Peak
a	0.96	3D
b	1.17	2A
▶ c	~1.50	O
d	~3.38	S
e	~3.41	S

▶ $C_{10}H_{15}N$ ▶ 7% $CDCl_3$ ▶ V 568

▶ 2-ACbb

	Shift	Peak
a	.87	3C
▶ b	1.29	3C
c	3.35	2G
d	4.18	1A

▶ $C_6H_{14}O$ ▶ 20% CCl_4 ▶ API 167

▶ 2-ACar

	Shift	Peak
a	0.95	3A
b	1.52	2A
▶ c	~1.90	O
d	4.52	6A

▶ $C_4H_9NO_2$ ▶ 7% $CDCl_3$ ▶ V 84

▶ 2-A(C)bb

	Shift	Peak
a	.77	3C
▶ b	1.30	S

▶ C_6H_{12} ▶ liquid ▶ API 446

▶ 2-ACav

	Shift	Peak
a	0.80	3A
b	1.22	2A
▶ c	1.59	5B
d	2.53	6A
e	7.20	1A

▶ $C_{10}H_{14}$ ▶ 45.0 mg/0.50 ml $CDCl_3$ ▶ S 31

▶ 2-ACbb

	Shift	Peak
a	0.88	3B
▶ b	1.30	6B
c	1.56	S
d	3.54	2G

▶ $C_6H_{14}O$ ▶ 35.5 mg/0.50 ml $CDCl_3$ ▶ S 14

▶ 2-ACav

	Shift	Peak
a	.82	3C
b	1.23	2A
▶ c	1.60	5B
d	2.56	4C
e	7.11	1A
f	7.11	1A

▶ $C_{10}H_{14}$ ▶ 10% CCl_4 ▶ API 90

▶ 2-ACbc

	Shift	Peak
a	~ .83	2H
b	~ .91	3D
▶ c	1.20	3B
d	~1.72	1D

▶ C_8H_{18} ▶ 50% CCl_4 ▶ API 206

▶ 2-ACbb

	Shift	Peak
a	.80	3B
▶ b	.92	3
c	~ .94	S

▶ C_7H_{16} ▶ 50% CCl_4 ▶ API 196

▶ 2-ACbd

	Shift	Peak
a	.87	1E
b	.87	1E
▶ c	1.00	1B
d	~1.43	7X

▶ C_9H_{20} ▶ 50% CCl_4 ▶ API 216

▶ 2-ACbb

	Shift	Peak
a	~ .88	3D
b	~ .88	3D
▶ c	~1.26	1F

▶ C_8H_{18} ▶ 50% CCl_4 ▶ API 202

▶ 2-ACbj

	Shift	Peak
a	0.97	3A
▶ b	1.58	4A
c	2.22	5B
d	10.45	1C

▶ $C_6H_{12}O$ ▶ 36.9 mg/0.50 ml $CDCl_3$ ▶ S 74

▶ 2-ACbl

$CH_2=CH-CH(CH_2CH_3)-CH_2CH_3$ [H(d) on =CH; (c) CH, (b) CH_2, (a) CH_3]

	Shift	Peak
a	.83	3C
▶ b	1.30	O
c	~1.72	S
d	~5.50	R

▶ C_7H_{14} ▶ liquid ▶ API 398

▶ 2-ADaaa

$CH_3-C(CH_3)_2-CH_2-CH_3$ [(a) CH_3, (c) CH_2, (b) CH_3]

	Shift	Peak
a	.85	1E
b	.85	3D
▶ c	1.19	3C

▶ C_6H_{14} ▶ 50% CCl_4 ▶ API 336

▶ 2-ACcc

$CH_3-CH(CH_3)-CH(CH_2CH_3)-CH(CH_3)-CH_3$ [(d) CH, (c) CH_2, (b) CH_3, (a) CH_3]

	Shift	Peak
a	.86	2H
b	.95	3D
▶ c	~1.2	S
d	~1.69	S

▶ C_9H_{20} ▶ 50% CCl_4 ▶ API 217

▶ 2-ADaab

$CH_3CH(CH_3)-CH_2C(CH_3)_2CH_2CH_3$ [(c) CH_3, (e) CH, (d) CH_2, (b) CH_3, (a) CH_3]

	Shift	Peak
a	~ .85	3D
b	~ .85	1E
c	~ .92	2H
▶ d	1.13	2A
e	1.63	7X

▶ C_9H_{20} ▶ 50% CCl_4 ▶ API 426

▶ 2-A(C)ck

[structure: OCH_3(e), (f)H, (i)H, (j)H, H(k), (d)H, (h)H, (g)H, CH_2(b), CH_3(a), (c)H]

	Shift	Peak
a	1.13	3B
▶ b	1.89	O
c	2.74	O
d	3.42	4A
e	3.77	1A
f	3.98	3E
g	4.59	2A
h	5.59	S
i	5.63*	S
j	5.63*	S
k	7.47	1A

*or 6.12

▶ $C_{15}H_{16}O_6$ ▶ 7% $CDCl_3$ ▶ V 641

▶ 2-ADaab

$CH_3-CH_2-C(CH_3)_2-CH_2-CH_3$ [(b) CH_3, (c) CH_2, (a) CH_3]

	Shift	Peak
a	~ .82	3D
b	~ .82	1E
▶ c	1.16	4C

▶ C_7H_{16} ▶ 50% CCl_4 ▶ API 199

▶ 2-ACer

(a)CH_3-CH_2(b)$-CH(Cl)(NO_2)$(c)

	Shift	Peak
a	1.10	3A
▶ b	2.32	5B
c	5.80	3A

▶ $C_3H_6ClNO_2$ ▶ $CDCl_3$ ▶ V 385

▶ 2-ADaad

$CH_3-C(CH_3)_2-C(CH_3)_2-CH_2-CH_3$ [(b) CH_3, (a) CH_3, (d) CH_2, (c) CH_3]

	Shift	Peak
a	.73	1A
b	.87 ± .03	1E
c	.87 ± .03	3D
▶ d	1.31	4A

▶ C_9H_{20} ▶ 50% CCl_4 ▶ API 218

▶ 2-ACgk

(a)CH_3-(b)CH_2-(c)$CH(Br)-$(d)$COOH$

	Shift	Peak
a	1.08	3A
▶ b	2.07	5B
c	4.23	3A
d	10.97	1A

▶ $C_4H_7BrO_2$ ▶ 7% $CDCl_3$ ▶ V 66

▶ 2-ADaae

CH_3(a)$-CH_2$(b)$-C(Cl)(CH_3)-CH_3$(c)

	Shift	Peak
a	1.08	3B
▶ b	1.58	1A
c	1.80	3D

▶ $C_5H_{11}Cl$ ▶ 7% $CDCl_3$ ▶ V 447

▶ 2-ACgk

(a)CH_3-(c)$CH_2-C(Br)(H)$(d)$-C(=O)-O-$(e)CH_2-(b)CH_3

	Shift	Peak
a	1.03	3A
b	1.30	3A
▶ c	2.08	5B
d	4.20	S
e	4.27	4E

▶ $C_6H_{11}BrO_2$ ▶ 7% $CDCl_3$ ▶ V 137

▶ 2-ADaal

$CH_2=CH-C(CH_3)(CH_3)CH_2CH_3$ [(d) H, (b) CH_3, (c) CH_2, (a) CH_3]

	Shift	Peak
a	.78	3D
b	.97	1E
▶ c	1.28	O
d	5.69	4D

▶ C_7H_{14} ▶ liquid ▶ API 401

▶ 2-ADaav

	Shift	Peak
a	0.69	3A
b	1.22	1A
▶ c	1.60	4A
d	~4.85	1D

▶ $C_{11}H_{16}O$ ▶ 44.6 mg/0.50 ml $CDCl_3$ ▶ S 256

▶ 2-ADbbb

	Shift	Peak
a	0.84	3C
▶ b	1.38	4C
c	2.83	3B
d	3.43	1E
e	3.59	2B
f	3.95	2E
g	5.14	R
h	5.22	R
i	5.87	R

▶ $C_{12}H_{22}O_3$ ▶ 7% $CDCl_3$ ▶ V 603

▶ 2-A(D)abb

	Shift	Peak
a	0.91	3A
b	1.06	1A
▶ c	1.43	3B
d	2.43	1A
e	8.47	1D

▶ $C_8H_{13}NO_2$ ▶ 7% $CDCl_3$ ▶ V 514

▶ 2-A(D)bov

	Shift	Peak
a	0.92	3A
▶ b	1.80	S
▶ c	2.05	S
d	2.70	2C
e	2.97	2C
f	7.30	1A

▶ $C_{10}H_{12}O$ ▶ 7% $CDCl_3$ ▶ V 558

▶ 2-ADabb

	Shift	Peak
a	.78	3D
b	~ .79	1E
▶ c	1.50	3C

▶ C_8H_{18} ▶ 50% CCl_4 ▶ API 207

▶ 2-AE

	Shift	Peak
a	1.48	3A
▶ b	3.57	4A

(a) (b)
CH_3CH_2Cl

▶ C_2H_5Cl ▶ 7% $CDCl_3$ ▶ V 11

▶ 2-ADass

	Shift	Peak
a	1.24	3C
b	1.43	3D
c	1.78	1A
▶ d	2.36	4C
e	3.45	4A

▶ $C_8H_{16}O_4S_2$ ▶ 7% $CDCl_3$ ▶ V 519

▶ 2-AG

	Shift	Peak
a	1.70	3A
▶ b	3.40	4A

(a) (b)
CH_3CH_2Br

▶ C_2H_5Br ▶ 45.0 mg/0.50 ml $CDCl_3$ ▶ S 225

▶ 2-ADbbb

	Shift	Peak
a	.73	3C
▶ b	1.18	4C

▶ C_9H_{20} ▶ 50% CCl_4 ▶ API 215

▶ 2-AG

	Shift	Peak
a	1.67	3A
▶ b	3.43	4A

(a) (b)
$CH_3—CH_2—Br$

▶ C_2H_5Br ▶ 7% $CDCl_3$ ▶ V 10

▶ 2-A(D)bbb

	Shift	Peak
a	0.83	N
▶ b	1.20	N
c	3.93	1A
d	5.52	1A

▶ $C_7H_{12}O_3$ ▶ 7% $CDCl_3$ ▶ V 493

▶ 2-AI

	Shift	Peak
a	1.83	3A
▶ b	3.20	4A

(a) (b)
CH_3CH_2I

▶ C_2H_5I ▶ 7% $CDCl_3$ ▶ V 13

▶ 2-AJa

	Shift	Peak
a	1.05	3A
b	2.13	1A
▶ c	2.47	4A

(b) CH_3 ; O=C ; CH_2CH_3 (c) (a)

▶ C_4H_8O ▶ 7% $CDCl_3$ ▶ V 76

▶ 2-AKb

	Shift	Peak
a	0.95	3D
b	1.18	3D
c	1.60	6B
▶ d	2.33	4A
e	4.04	3A

$CH_3CH_2—C(=O)—OCH_2CH_2CH_3$ (b) (d) (e) (c) (a)

▶ $C_6H_{12}O_2$ ▶ 44.4 mg/0.50 ml $CDCl_3$ ▶ S 38

▶ 2-AJl

	Shift	Peak
a	1.20	3A
b	2.38	1A
▶ c	2.78	4A
d	6.13	2E
e	7.08	2A

(d)H ; H(e) ; CH_3 (b) ; O ; C(=O)—CH_2—CH_3 (c) (a)

▶ $C_8H_{10}O_2$ ▶ 7% $CDCl_3$ ▶ V 206

▶ 2-AKb

	Shift	Peak
a	0.92	3B
b	1.12	S
c	1.32	S
▶ d	2.33	4A
e	4.05	3B

(b) (d) (e) (c) (a) $CH_3CH_2C(=O)—OCH_2(CH_2)_3CH_3$

▶ $C_8H_{16}O_2$ ▶ 44.8 mg/0.50 ml $CDCl_3$ ▶ S 40

▶ 2-AJn

	Shift	Peak
a	1.13	3A
▶ b	2.23	4C
c	~6.42	1D

(a) (b) $CH_3—CH_2—C(=O)—NH_2$ (c)

▶ C_3H_7NO ▶ 7% $CDCl_3$ ▶ V 40

▶ 2-AKc

	Shift	Peak
a	1.30	3A
b	2.08	1A
▶ c	4.28	4A
d	5.19	2A
e	6.60	1D

(b)CH_3—C(=O)—N(H (e))—C(H (d))(O—C(=O)—CH_2—CH_3)(O—C(=O)—CH_2—CH_3) (c) (a)

▶ $C_9H_{15}NO_5$ ▶ 7% $CDCl_3$ ▶ V 547

▶ 2-AJv

	Shift	Peak
a	1.22	3A
▶ b	2.99	4A
c	7.50	N
d	7.98	N

(c) (c) (c) (d) (d) ; C(=O)—CH_2—CH_3 (b) (a)

▶ $C_9H_{10}O$ ▶ 44.5 mg/0.50 ml $CDCl_3$ ▶ S 34

▶ 2-ALal

	Shift	Peak
a	.91 or .98	3D
b	1.57	1I
▶ c	1.94	5B
d	5.11	3C

(a) (c) H_3CH_2C ; CH_2CH_3 (a) ; C=C ; H_3C (b) ; H (d)

▶ C_7H_{14} ▶ liquid ▶ API 411

▶ 2-AJy

	Shift	Peak
a	1.23	3A
▶ b	2.93	4A
c	7.10	N
d	7.60	N
e	7.70	O

(c)H ; H(e) ; (d)H ; S ; C(=O)—CH_2CH_3 (b) (a)

▶ C_7H_8OS ▶ 7% $CDCl_3$ ▶ V 163

▶ 2-ALal

	Shift	Peak
a	.97	3C
b	1.49	S
c	1.56	S
▶ d	1.95	4C
e	5.18	4C

(c) H_3C ; (b) CH_3 ; C=C ; H (e) ; (d) (a) CH_2CH_3

▶ C_6H_{12} ▶ liquid ▶ API 378

▶ 2-AKb

	Shift	Peak
a	0.92	2A
b	1.13	3D
c	~1.85	S
▶ d	2.31	4A
e	3.83	2A

(b) (d) (e) (a) $CH_3—CH_2—C(=O)—OCH_2—CH(CH_3)—CH_3$ (c) (a)

▶ $C_7H_{14}O_2$ ▶ 4.5 mg/0.50 ml $CDCl_3$ ▶ S 55

▶ 2-ALal

	Shift	Peak
a	.97	3A
b	1.47	1E
c	1.57	1E
▶ d	1.97	4A
e	5.13	2E

(c) H_3C ; (d) (a) $CH_2—CH_3$; C=C ; (e) H ; CH_3(b)

▶ C_6H_{12} ▶ 10% CCl_4 ▶ API 35

▶ 2-ALal

	Shift	Peak
a	.91 or .96	3D
b	.91 or .96	3D
c	1.61	1H
▶d	1.97	4A
e	5.07	3A

▶ C_7H_{14} ▶ liquid ▶ API 412

▶ 2-ALbl

	Shift	Peak
a	1.02	3C
▶b	2.00	4C
c	4.67	N

▶ C_6H_{12} ▶ liquid ▶ API 381

▶ 2-ALal

	Shift	Peak
a	.95	3C
b	1.46	1F
c	1.61	S
▶d	2.03	4A
e	~5.08	4C

▶ C_6H_{12} ▶ liquid ▶ API 377

▶ 2-ALbl

	Shift	Peak
a	1.02	3C
▶ b	2.08	4C
c	4.63	3A

▶ C_6H_{12} ▶ 10% CCl_4 ▶ API 36

▶ 2-ALal

	Shift	Peak
a	.94	3C
b	1.62	1B
▶c	2.03	4C

▶ C_7H_{14} ▶ liquid ▶ API 415

▶ 2-ALcl

	Shift	Peak
a	1.03	3D
b	1.03	2H
▶c	~2.13	S
d	4.70 ± .03	S
e	4.70 ± .03	S

▶ C_7H_{14} ▶ liquid ▶ API 416

▶ 2-ALal

	Shift	Peak
a	.90	3A
b	1.60	1A
▶c	2.05	4A

▶ C_7H_{14} ▶ 50% CCl_4 ▶ API 352

▶ 2-ALhl

	Shift	Peak
a	.93	3C
▶b	1.92	1F
c	5.38	O

▶ C_6H_{12} ▶ liquid ▶ API 372

▶ 2-A(L)bl

	Shift	Peak
a	1.00	3A
▶b	~1.48 or 1.85	S
c	~1.85 or 1.48	S
d	5.32	1B

▶ C_8H_{14} ▶ 33% CCl_4 ▶ API 142

▶ 2-ALhl

	Shift	Peak
a	.98	3C
▶b	1.94	O
c	5.07	5B

▶ C_5H_8 ▶ liquid ▶ API 467

▶ 2-ALbl

	Shift	Peak
a	.93	3C
▶b	1.56	2E
c	2.04	4C
d	5.17	4A

▶ C_7H_{14} ▶ liquid ▶ API 414

▶ 2-ALhl

	Shift	Peak
a	.95	3A
b	1.60	1H
▶c	1.94	N
d	5.38	N

▶ C_5H_{10} ▶ 50 vol.% CCl_4 ▶ API 430

▶ 2-ALhl

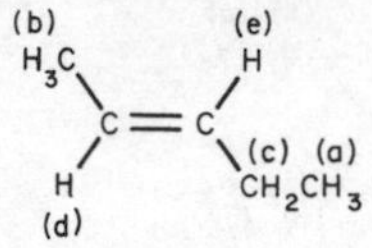

	Shift	Peak
a	.93	3C
b	1.58	2E
▶c	1.95	O
d	5.35 or 5.43	S
e	5.43 or 5.35	S

▶ C_5H_{10} ▶ liquid ▶ API 461

▶ 2-ALhl

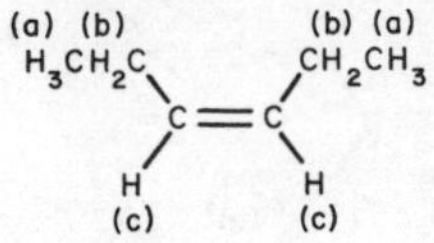

	Shift	Peak
a	.91	3C
▶b	1.99	5B
c	5.25	O

▶ C_6H_{12} ▶ liquid ▶ API 371

▶ 2-ALhl

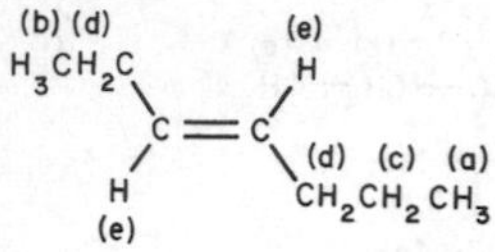

	Shift	Peak
a	.90	3D
b	.97	3D
c	1.33	S
▶d	1.95	O
e	5.40	O

▶ C_7H_{14} ▶ liquid ▶ API 391

▶ 2-ALhl

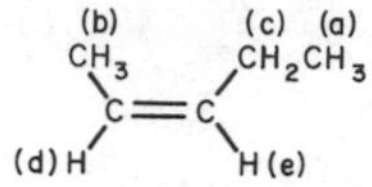

	Shift	Peak
a	.98	3A
b	1.58	2E
▶c	2.00	5B
d	5.23	S
e	5.30	S

▶ C_5H_{10} ▶ 10% CCl_4 ▶ API 154

▶ 2-ALhl

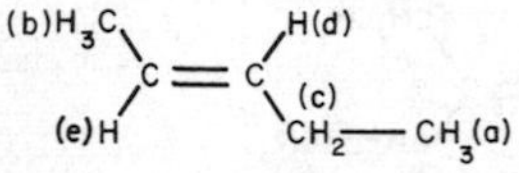

	Shift	Peak
a	.97	3A
b	1.61	5A
▶c	1.97	1F
d	5.37	S
e	5.37	S

▶ C_5H_{10} ▶ 10% CCl_4 ▶ API 29

▶ 2-ALhl

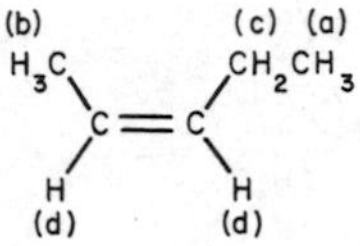

	Shift	Peak
a	.93	3C
b	1.52	2E
▶c	2.00	4C
d	5.30	N

▶ C_5H_{10} ▶ 50 vol.% CCl_4 ▶ API 429

▶ 2-ALhl

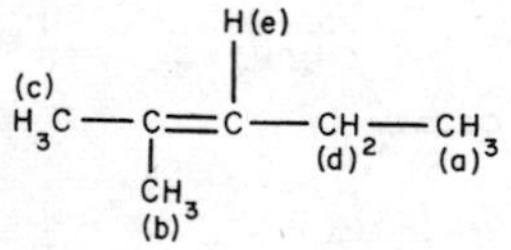

	Shift	Peak
a	.92	3A
b	1.57	1I
c	1.64	1I
▶d	1.97	4C
e	5.05	O

▶ C_6H_{12} ▶ 10% CCl_4 ▶ API 34

▶ 2-ALhl

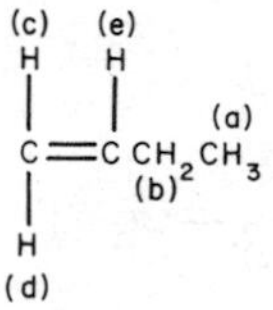

	Shift	Peak
a	.97	3C
▶b	2.02	5B
c	4.95	R
d	4.98	R
e	5.80	R

▶ C_4H_8 ▶ degassed liquid ▶ API 55

▶ 2-ALhl

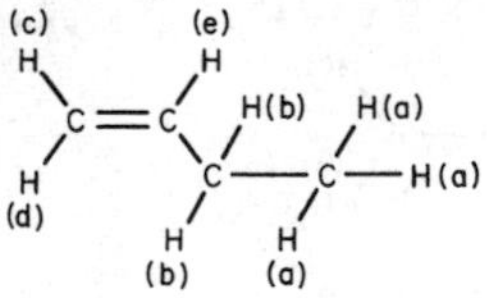

	Shift	Peak
a	0.97	3C
▶b	1.98	5B
c	~4.88	R
d	~5.7	R
e	~5.9	R

▶ C_4H_8 ▶ liquid ▶ API 365

▶ 2-ALhl

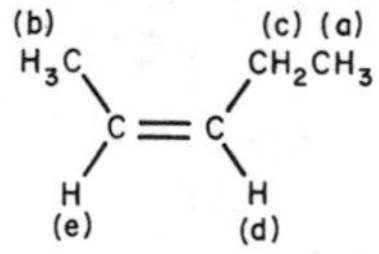

	Shift	Peak
a	.95	3C
b	1.58	2E
▶c	2.05	5B
d	~5.28 or 5.38	S
e	~5.38 or 5.28	S

▶ C_5H_{10} ▶ liquid (neat) ▶ API 460

▶ 2-ALhl

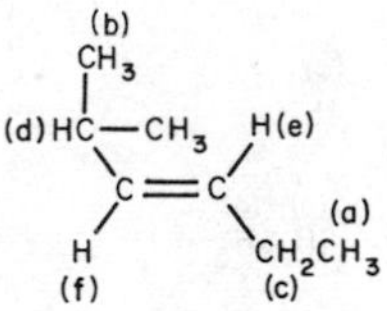

	Shift	Peak
a	.96	3D
b	.97	2H
▶c	1.98	S
d	~2.28	S
e	~5.38	S
f	~5.38	S

▶ C_7H_{14} ▶ liquid ▶ API 397

▶ 2-A(L)ln

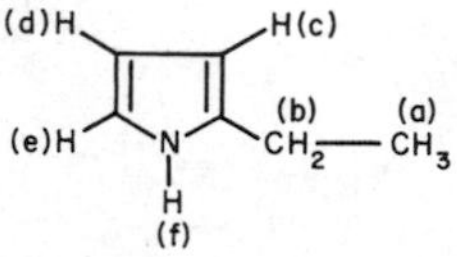

	Shift	Peak
a	1.23	3A
▶b	2.63	6A
c	5.97	1I
d	6.17	4A
e	6.68	N
f	~7.85	1A

▶ C_6H_9N ▶ 7% $CDCl_3$ ▶ V 130

▶ 2-ALmv

	Shift	Peak
a	1.11	3A
b	1.17	3D
▶c	2.58	4A
▶d	2.84	4E

(d) (b) $CH_2—CH_3$ (c) (a) $CH_2—CH_3$

HO N=C N=C HO

mixture of anti- and syn- forms

▶ $C_9H_{11}NO$ ▶ 7% $CDCl_3$ ▶ V 533

▶ 2-ANbb

	Shift	Peak
a	1.42	3A
▶b	3.18	4A

(b) (a) $CH_2—CH_3$

(a) (b) $CH_3—CH_2—N$ HBr

$CH_2—CH_3$ (b) (a)

▶ $C_6H_{15}N \cdot HBr$ ▶ 44.9 mg/0.50 ml $CDCl_3$ ▶ S 268

▶ 2-ANbb

	Shift	Peak
a	.98	3A
▶b	2.42	4A

(a) (b) $CH_3—CH_2—N—CH_2—CH_3$

$CH_2—CH_3$

▶ $C_6H_{15}N$ ▶ 34.8 mg/0.50 ml $CDCl_3$ ▶ S 29

▶ 2-ANbh

	Shift	Peak
a	1.12	3A
▶b	2.68	4E
c	2.73	3D
d	3.53	1E
e	3.53	1E
f	3.65	3D

(d) H

(a) (b) (c) (f) (e) $CH_3—CH_2—N—CH_2—CH_2—OH$

▶ $C_4H_{11}NO$ ▶ 7% $CDCl_3$ ▶ V 92

▶ 2-ANbb

	Shift	Peak
a	0.97	3D
b	1.07	1E
▶c	2.50	4E
d	2.55	1E
e	9.57	1A

(b) CH_3

(a) (c) $CH_3—CH_2$ (d) $N—CH_2—C—C$ O

$CH_3—CH_2$ (a) (c) CH_3 (b) H(e)

▶ $C_9H_{19}NO$ ▶ 7% $CDCl_3$ ▶ V 548

▶ 2-ANbh

	Shift	Peak
a	1.45	3A
▶b	3.03	4A
c	9.1	1D

(a) (b) (c) (b) (a) $CH_3—CH_2—NH—CH_2—CH_3$ · HCl

▶ $C_4H_{11}N \cdot HCl$ ▶ 44.8 mg/0.50 ml $CDCl_3$ ▶ S 257

▶ 2-ANbb

	Shift	Peak
a	1.03	3A
▶b	2.52	4A
c	3.54	1A

(b) (a) $CH_2—CH_3$

$CH_2—N—CH_2—CH_3$ (c) (b) (a)

▶ $C_{11}H_{17}N$ ▶ 44.8 mg/0.50 ml $CDCl_3$ ▶ S 152

▶ 2-ANbs

	Shift	Peak
a	1.11	3A
▶b	3.20	4A
c	3.88	1E
d	3.88	1E
e	6.80	2E
f	7.13	1A
g	7.17	2A

OCH_3(c)

(e)H NH_2(d)

(g)H H(f)

O←S→O

$CH_3—CH_2—N—CH_2—CH_3$ (a) (b) (b) (a)

▶ $C_{11}H_{18}N_2O_3S$ ▶ 7% $CDCl_3$ ▶ V 587

▶ 2-ANbb

	Shift	Peak
a	1.05	3A
▶b	2.62	4E
c	2.82	3D
d	4.13	1B
e	4.33	3D
f	6.63	Q
g	7.83	Q

(f) H (g) H O (b) (a) $CH_2—CH_3$

(d) H_2N (e) (c) $CO—CH_2—CH_2—N$

(f)H H(g) $CH_2—CH_3$ (b) (a)

▶ $C_{13}H_{20}N_2O_2$ ▶ 7% $CDCl_3$ ▶ V 302

▶ 2-ANbv

	Shift	Peak
a	1.13	3A
b	2.16	1B
▶c	3.38	S
d	3.40	S
e	3.72	3D

(a) CH_3 N (e) CH_2

CH_2 (c) CH_2 (d) OH (b)

▶ $C_{10}H_{15}NO$ ▶ 7% $CDCl_3$ ▶ V 569

▶ 2-ANbb

	Shift	Peak
a	1.03	3A
▶b	2.63	4A
c	3.88	1A

N (c) $CH_2—N$ (b) (a) $CH_2—CH_3$

$CH_2—CH_3$ (b) (a)

▶ $C_{12}H_{17}N_3$ ▶ 7% $CDCl_3$ ▶ V 296

▶ 2-ANhj

	Shift	Peak
a	1.14	3A
b	2.02	1A
▶c	3.26	5B
d	8.08	1B

O (d) (c) (a)

(b) $CH_3C—NHCH_2CH_3$

▶ C_4H_9NO ▶ 44.4 mg/0.50 ml $CDCl_3$ ▶ S 146

▶ 2-ANhl

	Shift	Peak
a	1.22	3A
▶b	3.49	5B
c	~6.21	1C

▶ $C_5H_{12}N_2S$ ▶ 45.2 mg/0.50 ml $CDCl_3$ ▶ S 190

▶ 2-AOs

	Shift	Peak
a	1.28	3D
b	1.43	1A
▶c	4.07	4A
d	7.73	2A
e	7.82	2G
f	8.01	2G
g	8.28	2A
h	8.45	2A

▶ $C_{20}H_{28}O_3S$ ▶ 7% $CDCl_3$ ▶ V 683

▶ 2-AOb

	Shift	Peak
a	1.20	3A
▶b	~3.50	4E
c	~3.63	1E

▶ $C_8H_{18}O_3$ ▶ 45.3 mg/0.50 ml $CDCl_3$ ▶ S 285

▶ 2-AOv

	Shift	Peak
a	1.38	3A
b	2.28	1A
▶c	3.98	4A
d	6.75	Q
e	7.05	Q

▶ $C_9H_{12}O$ ▶ 45.0 mg/0.50 ml $CDCl_3$ ▶ S 284

▶ 2-AOb

	Shift	Peak
a	1.27	3A
▶b	3.66	4A
c	4.13	1A
d	10.95	1A

▶ $C_4H_8O_3$ ▶ 7% $CDCl_3$ ▶ V 417

▶ 2-AOv

	Shift	Peak
a	1.37	3A
▶b	3.93	4A
c	6.73	Q
d	7.28	Q

▶ C_8H_9BrO ▶ 7% $CDCl_3$ ▶ V 198

▶ 2-AOc

	Shift	Peak
a	1.20	S
b	1.32	S
▶c	~3.50 or 3.70	2C
▶d	~3.50 or 3.70	2C
e	4.72	4A

▶ $C_6H_{14}O_2$ ▶ 7% $CDCl_3$ ▶ V 143

▶ 2-AOv

	Shift	Peak
a	1.36	3A
▶b	3.93	4A
c	5.45	1C
d	6.67	1A

▶ $C_8H_{10}O_2$ ▶ 45.0 mg/0.50 ml $CDCl_3$ ▶ S 12

▶ 2-AOl

	Shift	Peak
a	1.47	3A
▶b	4.42	4A
c	7.65	1A

▶ $C_6H_6N_2O$ ▶ 7% $CDCl_3$ ▶ V 452

▶ 2-AOv

	Shift	Peak
a	1.35	3A
b	3.30	1B
▶c	3.93	4A
d	6.63 or 6.70	Q
e	6.63 or 6.70	Q

▶ $C_8H_{11}NO$ ▶ 7% $CDCl_3$ ▶ V 208

▶ 2-AOm

	Shift	Peak
a	1.39	3A
▶b	4.78	4A

▶ $C_2H_5NO_2$ ▶ 7% $CDCl_3$ ▶ V 374

▶ 2-AOv

	Shift	Peak
a	0.90	3A
▶b	3.99	4A
c	6.78	Q
d	7.21	Q

▶ C_8H_9ClO ▶ 45.4 mg/0.50 ml $CDCl_3$ ▶ S 235

▶ 2-AOv

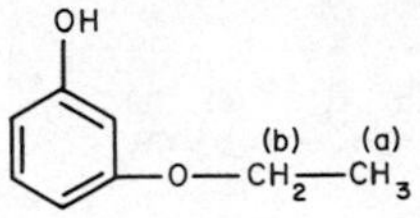

	Shift	Peak
a	1.40	3A
▶b	3.99	4A

▶ $C_8H_{10}O_2$ ▶ 45.1 mg/0.50 ml $CDCl_3$ ▶ S 267

▶ 2-AOv

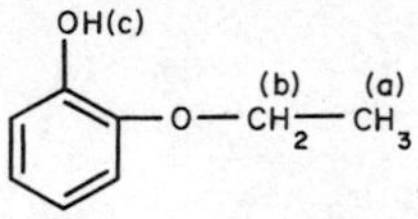

	Shift	Peak
a	1.44	3A
▶b	4.08	4A
c	5.70	1A

▶ $C_8H_{10}O_2$ ▶ 45.3 mg/0.50 ml $CDCl_3$ ▶ S 170

▶ 2-AOv

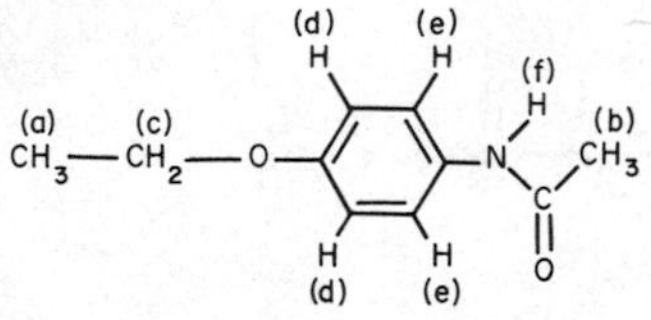

	Shift	Peak
a	1.38	3A
b	2.12	1A
▶c	4.00	4A
d	6.83	Q
e	7.41	Q
f	7.91	1C

▶ $C_{10}H_{13}NO_2$ ▶ 7% $CDCl_3$ ▶ V 267

▶ 2-AOv

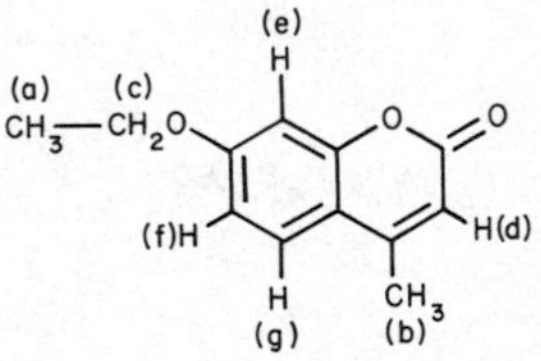

	Shift	Peak
a	1.45	3A
b	2.37	2D
▶c	4.08	4A
d	6.08	2E
e	6.73	1G
f	~6.78	2G
g	7.47	2E

▶ $C_{12}H_{12}O_3$ ▶ 7% $CDCl_3$ ▶ V 294

▶ 2-AOv

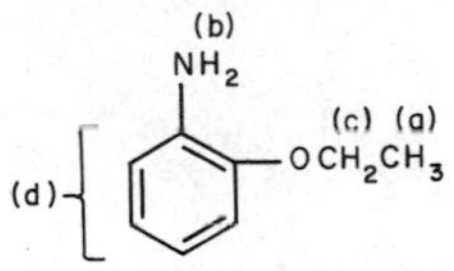

	Shift	Peak
a	1.40	3A
b	3.72	1B
▶c	4.03	4A
d	6.72	1A

▶ $C_8H_{11}NO$ ▶ 45.1 mg/0.50 ml $CDCl_3$ ▶ S 236

▶ 2-AOv

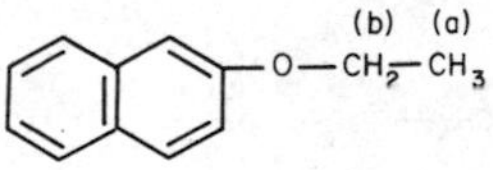

	Shift	Peak
a	1.45	3A
▶b	4.11	4A

▶ $C_{12}H_{12}O$ ▶ 45.0 mg/0.50 ml $CDCl_3$ ▶ S 72

▶ 2-AOv

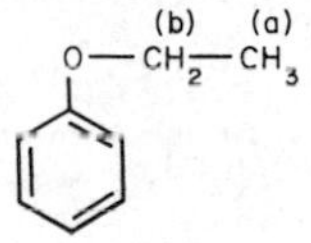

	Shift	Peak
a	1.40	3A
▶b	4.04	4A

▶ $C_8H_{10}O$ ▶ 34.7 mg/0.50 ml $CDCl_3$ ▶ S 26

▶ 2-AOz

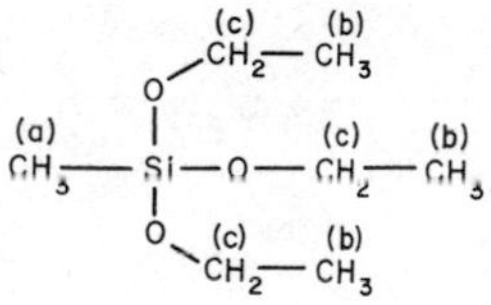

	Shift	Peak
a	0.12	1A
b	1.22	3A
▶c	3.80	4A

▶ $C_8H_{18}O_3Si$ ▶ 7% $CDCl_3$ ▶ V 184

▶ 2-AOv

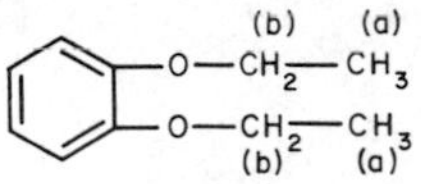

	Shift	Peak
a	1.40	3A
▶b	4.05	4A

▶ $C_{10}H_{14}O_2$ ▶ 45.0 mg/0.50 ml $CDCl_3$ ▶ S 171

▶ 2-AOz

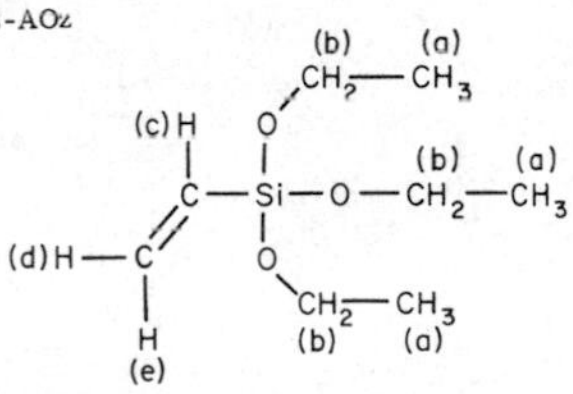

	Shift	Peak
a	1.23	3A
▶b	3.82	4A
c	5.98 ± 0.05	S
d	5.98 ± 0.05	S
e	5.98 ± 0.05	S

▶ $C_8H_{18}O_3Si$ ▶ 7% $CDCl_3$ ▶ V 219

▶ 2-AOv

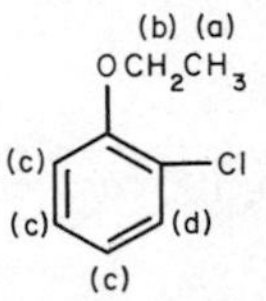

	Shift	Peak
a	0.96	3A
▶b	4.08	4A
c	6.95	O
d	7.34	N

▶ C_8H_9ClO ▶ 45.2 mg/0.50 ml $CDCl_3$ ▶ S 234

▶ 2-AOz

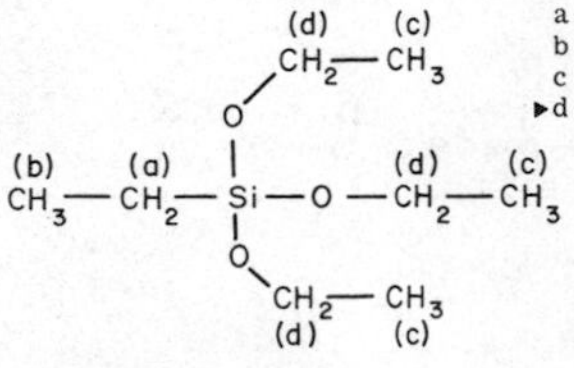

	Shift	Peak
a	~0.62	S
b	~1.00	S
c	1.23	3A
▶d	3.82	4A

▶ $C_8H_{20}O_3Si$ ▶ 7% $CDCl_3$ ▶ V 222

▶ 2-AOz

	Shift	Peak
a	1.22	3A
▶b	3.87	4A

$(C_6H_5)_3Si{-}O{-}CH_2{-}CH_3$ (b) = CH_2, (a) = CH_3

▶ $C_{20}H_{20}OSi$ ▶ 7% $CDCl_3$ ▶ V 341

▶ 2-AQb

	Shift	Peak
a	0.88	3B
b	1.26	S
c	1.33	S
d	2.28	3B
▶e	4.12	4A

$CH_3{-}(CH_2)_4{-}CH_2{-}C(=O){-}OCH_2{-}CH_3$ (a) (c) (d) (e) (b)

▶ $C_9H_{18}O_2$ ▶ 44.5 mg/0.50 ml $CDCl_3$ ▶ S 283

▶ 2-AOz

	Shift	Peak
a	1.35	3C
▶b	4.11	5B

$(CH_3CH_2O)_3P{\rightarrow}O$ (a) (b)

▶ $C_6H_{15}O_4P$ ▶ 7% $CDCl_3$ ▶ V 482

▶ 2-AQb

	Shift	Peak
a	1.22	3A
b	3.55	1A
▶c	4.12	4A

$C_6H_5{-}CH_2{-}C(=O){-}O{-}CH_2CH_3$ (b) (c) (a)

▶ $C_{10}H_{12}O_2$ ▶ 44.9 mg/0.50 ml $CDCl_3$ ▶ S 66

▶ 2-AP

	Shift	Peak
a	1.22	3A
b	2.58	1B
▶c	3.70	4A

$CH_3{-}CH_2{-}OH$ (a) (c) (b)

▶ C_2H_6O ▶ 7% $CDCl_3$ ▶ V 14

▶ 2-AQb

	Shift	Peak
a	0.88	3B
b	1.28	3D
c	1.30	S
d	2.28	3B
▶e	4.12	4A

$CH_3{-}(CH_2)_5{-}CH_2{-}C(=O){-}OCH_2CH_3$ (a) (c) (d) (e) (b)

▶ $C_{10}H_{20}O_2$ ▶ 45.2 mg/0.50 ml $CDCl_3$ ▶ S 244

▶ 2-AQa

	Shift	Peak
a	1.25	3A
b	2.03	1A
▶c	4.12	4A

$O{=}C(OCH_2CH_3)CH_3$ (c) (a); CH_3(b)

▶ $C_4H_8O_2$ ▶ 7% $CDCl_3$ ▶ V 79

▶ 2-AQb

	Shift	Peak
a	0.96	3A
b	1.28	3A
c	1.67	6B
d	2.28	3C
▶e	4.13	4A

$CH_3CH_2CH_2C(=O){-}OCH_2CH_3$ (a) (c) (d) (e) (b)

▶ $C_6H_{12}O_2$ ▶ 45.0 mg/0.50 ml $CDCl_3$ ▶ S 42

▶ 2-AQb

	Shift	Peak
a	0.88	3B
b	1.28	1E
c	2.28	3C
▶d	4.11	4A

$CH_3{-}(CH_2)_9{-}CH_2{-}C(=O){-}OCH_2CH_3$ (a) (b) (c) (d) (a)

▶ $C_{14}H_{28}O_2$ ▶ 45.5 mg/0.50 ml $CDCl_3$ ▶ S 194

▶ 2-AQb

	Shift	Peak
a	1.25*	3A
b	1.25*	3A
c	2.35	1A
d	2.86	2A
e	2.89	2A
f	3.98	S
g	4.13†	4A
▶h	4.13†	4A

*or 1.27
†or 4.22

$CH_3CH_2O{-}C(=O){-}CH(H_f)(C(=O)CH_3){-}CH(H_d)(H_e){-}C(=O){-}OCH_2CH_3$ — labels: (c) CH_3; (a) (g) CH_3CH_2O; (h) (b) OCH_2CH_3; H(d); H (f); H (e)

▶ $C_{10}H_{16}O_5$ ▶ 7% $CDCl_3$ ▶ V 277

▶ 2-AQb

	Shift	Peak
a	1.26	3A
b	2.20	1A
c	2.64	Q
▶d	4.12	4A

$CH_3{-}C(=O){-}CH_2CH_2C(=O){-}OCH_2CH_3$ (b) (c) (d) (a)

▶ $C_7H_{12}O_3$ ▶ 45.5 mg/0.50 ml $CDCl_3$ ▶ S 247

▶ 2-AQb

	Shift	Peak
a	1.26	3D
b	1.35	1E
c	2.28	3B
▶d	4.13	4A

$CH_3CH_2O{-}C(=O){-}CH_2(CH_2)_6CH_2C(=O){-}OCH_2CH_3$ (a) (d) (c) (b) (c) (d) (a)

▶ $C_{14}H_{26}O_4$ ▶ 45.1 mg/0.50 ml $CDCl_3$ ▶ S 250

▶ 2-AQb

Shift	Peak
a 1.25	3A
b 2.62	1A
▶c 4.15	4A

(a) (c) O (b) (b) O (c) (a)
$CH_3CH_2OC-CH_2CH_2C-O-CH_2CH_3$

▶ $C_8H_{14}O_4$ ▶ 7% $CDCl_3$ ▶ V 215

▶ 2-AQb

Shift	Peak
a 1.30	3A
b 4.05	1A
▶c 4.25	4A

$ClH_2C-C(=O)-OCH_2CH_3$ (b) (c) (a)

▶ $C_4H_7ClO_2$ ▶ 7% $CDCl_3$ ▶ V 73

▶ 2-AQb

Shift	Peak
a 1.25	3A
b 2.62	1A
▶c 4.16	4A

(a) (c) (b) (b) (c) (a)
$CH_3CH_2-O-C(=O)-CH_2CH_2-C(=O)-O-CH_2CH_3$

▶ $C_8H_{14}O_4$ ▶ 45.5 mg/0.50 ml $CDCl_3$ ▶ S 183

▶ 2-AQb

Shift	Peak
a 1.30	3A
b 3.92	1A
▶c 4.27	4A

(a) (c) (b)
$CH_3-CH_2-O-C(=O)-CH_2-NCO$

▶ $C_5H_7NO_3$ ▶ 7% $CDCl_3$ ▶ V 107

▶ 2-AQb

Shift	Peak
a 1.26	3D
b 2.86	1A
▶c 4.18	4A
d 4.47	1B

(b) $CH_2-C(=O)-O-CH_2-CH_3$ (c) (a)
(d) $HO-C-C(=O)-O-CH_2-CH_3$
(b) $CH_2-C(=O)-O-CH_2-CH_3$ (c) (a)

▶ $C_{12}H_{20}O_7$ ▶ 45.3 mg/0.50 ml $CDCl_3$ ▶ S 150

▶ 2-AQb

Shift	Peak
a 1.33	3A
b 3.67	1A
▶c 4.30	4A
d 6.30	3A*
e 8.03	2C
f 8.40	4A
g 9.08	2A
h 11.77	1B

*Very large splitting due to fluorine

(d) CF_2H, CH_2(b), O(c) CH_2-CH_3(a), (e)H, H(f), (h)H, NO_2, H(g)

▶ $C_{12}H_{12}F_2N_4O_6$ ▶ 7% $CDCl_3$ ▶ V 293

▶ 2-AQb

Shift	Peak
a 1.29	3A
b 2.37	1A
c 3.17	1A
▶d 4.22	4A

(b) CH_3 N CH_3 (b) — (c) $CH_2-C(=O)-O-CH_2-CH_3$ (d) (a)

▶ $C_6H_{13}NO_2$ ▶ 7% $CDCl_3$ ▶ V 480

▶ 2-AQc

Shift	Peak
a 1.25*	3A
b 1.25*	3A
c 2.35	1A
d 2.86	2A
e 2.89	2A
f 3.98	S
▶g 4.13†	4A
h 4.13†	4A

*or 1.27
†or 4.22

(c) CH_3, H(d), (a) (g) CH_3CH_2O, (h) (b) OCH_2CH_3, H (f), H (e)

▶ $C_{10}H_{16}O_5$ ▶ 7% $CDCl_3$ ▶ V 277

▶ 2-AQb

Shift	Peak
a 1.27	3A
b 3.37	1A
▶c 4.22	4A

$O=C(-O-CH_2-CH_3)-CH_2(b)-C(=O)-O-CH_2-CH_3$ (c) (a)

▶ $C_7H_{12}O_4$ ▶ 7% $CDCl_3$ ▶ V 181

▶ 2-AQc

Shift	Peak
a 1.27	3A
b 3.14	2A
▶c 4.20	4A
d 4.96	4C
e 5.86	1A
f 6.62 ± 0.04	S

(f)H, (e) H_2C, (b) CH_2, (d) CH, NH, (c) (a) $O-CH_2-CH_3$, H(f), H(f)

▶ $C_{19}H_{19}NO_5$ ▶ 7% $CDCl_3$ ▶ V 334

▶ 2-AQb

Shift	Peak
a 1.32	3A
b 4.00	1A
▶c 4.24	4A

(b) (c) (a)
$Cl-CH_2-C(=O)-O-CH_2CH_3$

▶ $C_4H_7ClO_2$ ▶ 44.8 mg/0.50 ml $CDCl_3$ ▶ S 91

▶ 2-AQc

Shift	Peak
a 1.03	3A
b 1.30	3A
c 2.08	5B
d 4.20	S
▶e 4.27	4E

(a) (c) Br (e) (b)
$CH_3-CH_2-C(Br)(H)-C(=O)-O-CH_2-CH_3$
(d)

▶ $C_6H_{11}BrO_2$ ▶ 7% $CDCl_3$ ▶ V 137

▶ 2-AQc

	Shift	Peak
a	1.31	3A
b	2.22	1A
▶c	4.30	4A
d	5.52	1A

(b)CH_3 C O O C H(d) C O O CH_2 (c) CH_3(a) C O O CH_2 (c) CH_3(a)

▶ $C_9H_{14}O_6$ ▶ 7% $CDCl_3$ ▶ V 546

▶ 2-AQl

	Shift	Peak
a	1.32	3A
▶b	4.27	4A
c	6.83	1A

(c) H C=C H(c) O C O (b) (a) CH_2CH_3 CH_3CH_2 (a) (b) O C O

▶ $C_8H_{12}O_4$ ▶ 7% $CDCl_3$ ▶ V 213

▶ 2-AQc

	Shift	Peak
a	1.35	3A
▶b	4.33	4A
c	5.93	1A

(a)CH_3CH_2OC — $CHCl_2$ (b) (c) O

▶ $C_4H_6Cl_2O_2$ ▶ 7% $CDCl_3$ ▶ V 59

▶ 2-AQl

	Shift	Peak
a	1.30	3A
▶b	4.28	4A
c	6.28	1A

(a) (b) CH_3— CH_2—O—C(=O) C=C (c)H H(c) C(=O)—O— CH_2—CH_3 (b) (a)

▶ $C_8H_{12}O_4$ ▶ 7% $CDCl_3$ ▶ V 212

▶ 2-AQc

	Shift	Peak
a	1.37	3A
b	2.12	1A
▶c	4.36	4A
d	5.50	2A
e	6.75	1D

O (d) H C CN CH_3 (b) C N H(e) C O O CH_2 (c) CH_3(a)

▶ $C_7H_{10}N_2O_3$ ▶ 7% $CDCl_3$ ▶ V 491

▶ 2-AQl

	Shift	Peak
a	1.38	3A
▶b	4.38	4A
c	8.22	1A

(c) H C=C CN C — OCH_2 CH_3 (b) (a) O

▶ $C_{12}H_{11}NO_2$ ▶ 7% $CDCl_3$ ▶ V 290

▶ 2-AQl

	Shift	Peak
a	0.83	3A
b	2.17	1A
c	2.37	1A
▶d	3.87	4A
e	5.13	1A
f	6.15	1C

(e) H O C O (d) CH_2 CH_3(a) O C (b) CH_3 CH_3 (c) N H(f)

▶ $C_{23}H_{23}NO_3$ ▶ 7% $CDCl_3$ ▶ V 695

▶ 2-AQl

	Shift	Peak
a	1.40	3A
b	3.87	1A
▶c	4.40	4A
d	6.97	2A
e	7.02	1A
f	7.98	2A
g	~15.33	1D

(d) H (f) H (g) O H O (b) CH_3O C C C H(e) O—CH_2—CH_3 (c) (a) H (d) H (f)

▶ $C_{12}H_{14}O_4$ ▶ 7% $CDCl_3$ ▶ V 295

▶ 2-AQl

	Shift	Peak
a	1.28	3A
b	1.93	2D
▶c	4.22	4A
d	5.57	3C
e	6.10	1I

(d) H (b) CH_3 C=C (e)H C=O O (c) CH_2—CH_3 (a)

▶ $C_6H_{10}O_2$ ▶ 7% $CDCl_3$ ▶ V 135

▶ 2-AQn

	Shift	Peak
a	1.23	3A
▶b	4.10	4A
c	~5.17	1D

(c) H_2N—C—OCH_2CH_3 (b) (a) O

▶ $C_3H_7NO_2$ ▶ 44.4 mg/0.50 ml $CDCl_3$ ▶ S 245

▶ 2-AQl

	Shift	Peak
a	1.30	3A
▶b	4.23	4A
c	6.30	2F
d	6.45	N
e	6.60	S
f	7.40	S
g	7.47	S

(d)H H(e) (g)H O C=C H(c) H (f) C=O O CH_2 CH_3 (b) (a)

▶ $C_9H_{10}O_3$ ▶ 7% $CDCl_3$ ▶ V 235

▶ 2-AQn

	Shift	Peak
a	1.23	3A
b	2.78	2A
▶c	4.14	4A
d	5.16	1D

(d) H (b) CH_3—N C—O—CH_2—CH_3 (c) (a) O

▶ $C_4H_9NO_2$ ▶ 7% $CDCl_3$ ▶ V 85

2-AQo

$CH_3CH_2O-C(=O)-OCH_2CH_3$ — labels: (a) CH_3, (b) CH_2

	Shift	Peak
a	1.32	3A
▸b	4.22	4A

▸ $C_5H_{10}O_3$ ▸ 44.8 mg/0.50 ml $CDCl_3$ ▸ S 290

2-ASov

(b)CH_3–C_6H_4–S(→O)–CH_2(c)–CH_3(a); ring H's (d), (e)

	Shift	Peak
a	1.20	3A
b	2.43	1A
▸c	2.82	4C
d	7.33	Q
e	7.51	Q

▸ $C_9H_{12}OS$ ▸ 7% $CDCl_3$ ▸ V 538

2-AQv

Ethyl 4-hydroxy-3-methoxybenzoate: C(=O)–OCH_2(c)–CH_3(a); –OCH_3(b); OH(d); ring H (e)

	Shift	Peak
a	1.38	3A
b	3.93	1A
▸c	4.32	4A
d	6.80 or 6.94	1A
e	6.80 or 6.94	1A

▸ $C_{10}H_{12}O_4$ ▸ 45.0 mg/0.50 ml $CDCl_3$ ▸ S 127

2-ASu

CH_3(a)–CH_2(b)–S–CN

	Shift	Peak
a	1.52	3A
▸b	2.99	4A

▸ C_3H_5NS ▸ 7% $CDCl_3$ ▸ V 384

2-AQv

Ethyl salicylate: C(=O)–OCH_2(b)CH_3(a); OH(f); ring H's (c), (c), (d), (e)

	Shift	Peak
a	1.41	3A
▸b	4.40	4A
c	6.92	O
d	7.37	O
e	7.83	2E
f	10.78	1A

▸ $C_9H_{10}O_3$ ▸ 45.6 mg/0.50 ml $CDCl_3$ ▸ S 196

2-ATh

(b) HC≡C–CH_2(c)–CH_3(a)

	Shift	Peak
a	1.15	3A
b	1.76	3A
▸c	~2.14	4C

▸ C_4H_6 ▸ 10% CCl_4 ▸ API 162

2-AR

CH_3(a)–CH_2(b)–NO_2

	Shift	Peak
a	1.58	3A
▸b	4.33	4A

▸ $C_2H_5NO_2$ ▸ 35.8 mg/0.50 ml $CDCl_3$ ▸ S 2

2-ATh

(b)HC≡CCH_2(c)CH_3(a)

	Shift	Peak
a	1.12	3C
b	~1.90	N
▸c	~2.17	O

▸ C_4H_6 ▸ liquid ▸ API 464

2-ASa

CH_3(b)–S–CH_2(c)CH_3(a)

	Shift	Peak
a	1.27	3A
b	2.10	1A
▸c	2.53	4A

▸ C_3H_8S ▸ 7% $CDCl_3$ ▸ V 46

2-AVah

Ring with CH_3(b) and CH_2(c)CH_3(a)

(d) = all ring H's

	Shift	Peak
a	1.09	3A
b	2.11	1A
▸c	2.46	4A
d	7.00	1A

▸ C_9H_{12} ▸ run neat ▸ API 177 A

2-ASdoo

CH_3(a)–CH_2(d)–C(CH_3(c))(SO_2–CH_2(e)–CH_3(b))$_2$

	Shift	Peak
a	1.24	3C
b	1.43	3D
c	1.78	1A
d	2.36	4C
▸e	3.45	4A

▸ $C_8H_{18}O_4S_2$ ▸ 7% $CDCl_3$ ▸ V 519

2-AVah

Ring with CH_3(b) and CH_2(c)CH_3(a)

(d) = all ring H's

	Shift	Peak
a	1.21	3A
b	2.27	1A
▸c	2.63	4A
d	7.02	1A

▸ C_9H_{12} ▸ CCl_4 ▸ API 177B

▶ 2-AVhh

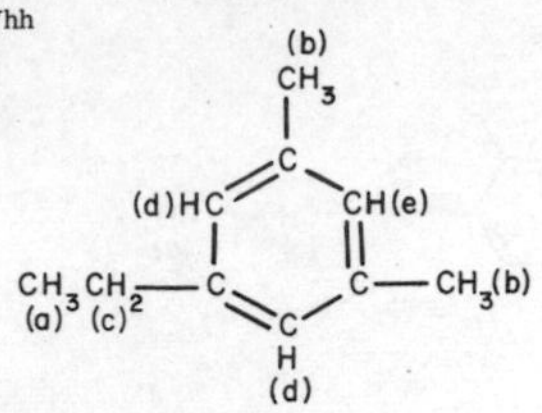

	Shift	Peak
a	1.14	3A
b	2.17	1A
▶c	2.40	4A
d	6.68	1B
e	6.68	1B

▶ $C_{10}H_{14}$ ▶ liquid ▶ API 12

▶ 2-AVhh

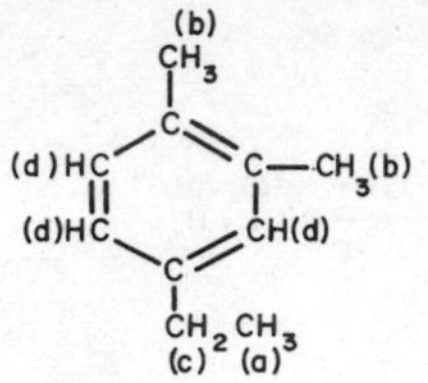

	Shift	Peak
a	1.18	3A
b	2.20	1A
▶c	2.60	4A
d	6.84	1A

▶ $C_{10}H_{14}$ ▶ CCl_4 ▶ API 182 B

▶ 2-AVhh

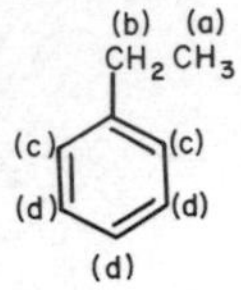

	Shift	Peak
a	1.10	3A
▶b	2.47	4A
c	7.10	1A
d	7.10	1A

▶ C_8H_{10} ▶ liquid ▶ API 74

▶ 2-AVhh

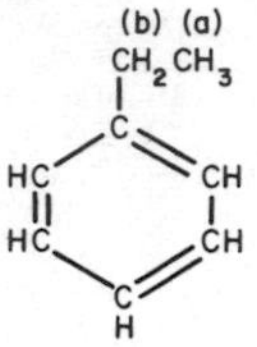

	Shift	Peak
a	1.22	3A
▶b	2.63	4A
c	7.13	1A

(c)= all ring H's

▶ C_8H_{10} ▶ CCl_4 ▶ API 173 B

▶ 2-AVhh

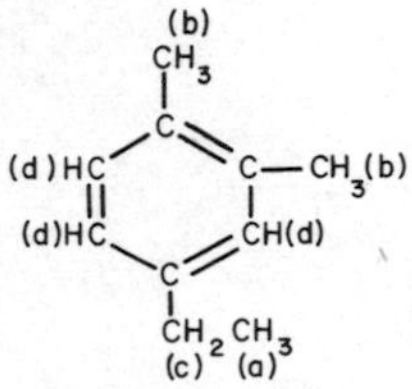

	Shift	Peak
a	1.16	3A
b	2.07	1A
▶c	2.47	4A
d	6.83	2D

▶ $C_{10}H_{14}$ ▶ run neat ▶ API 182 A

▶ 2-AVhh

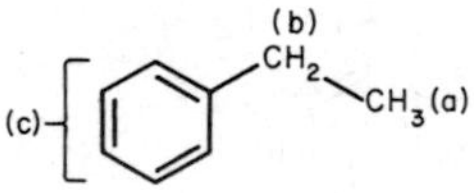

	Shift	Peak
a	1.25	3A
▶b	2.68	4A
c	7.23	1A

▶ C_8H_{10} ▶ 7% $CDCl_3$ ▶ V 505

▶ 2-AVhh

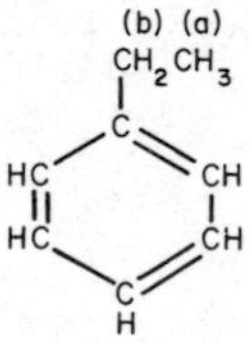

	Shift	Peak
a	1.10	3A
▶b	2.5	4A
c	7.07	1A

(c)= all ring H's

▶ C_8H_{10} ▶ run neat ▶ API 173 A

▶ 2-AVhn

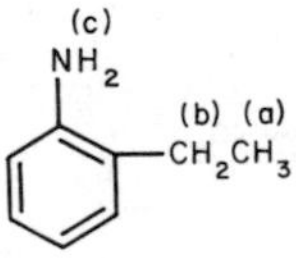

	Shift	Peak
a	1.23	3A
▶b	2.46	4A
c	3.47	1A

▶ $C_8H_{11}N$ ▶ 45.3 mg/0.10 ml $CDCl_3$ ▶ S 111

▶ 2-AVhh

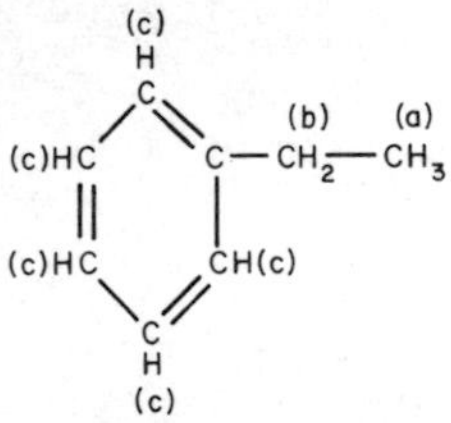

	Shift	Peak
a	1.22	3A
▶b	2.54	4A
c	7.10	1E

▶ C_8H_{10} ▶ 10% CCl_4 ▶ API 38

▶ 2-AVhr

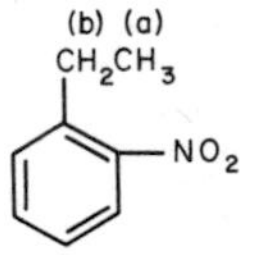

	Shift	Peak
a	1.28	3A
▶b	2.92	4A

▶ $C_8H_9NO_2$ ▶ 45.7 mg/0.50 ml $CDCl_3$ ▶ S 109

▶ 2-AVhh

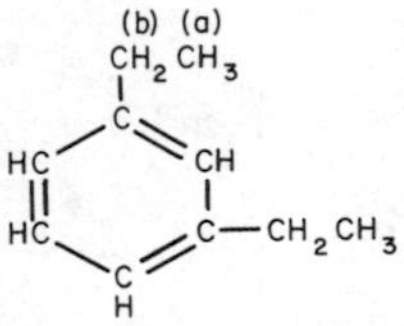

	Shift	Peak
a	1.16	3A
▶b	2.58	4A

▶ $C_{10}H_{14}$ ▶ run neat ▶ API 180 A

▶ 2-AX

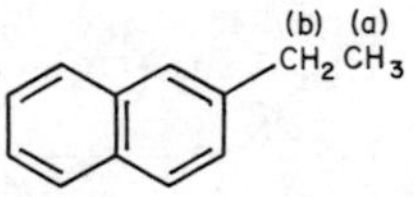

	Shift	Peak
a	1.30	3A
▶b	2.77	4A

▶ $C_{12}H_{12}$ ▶ 7% $CDCl_3$ ▶ V 292

▶ 2-A(Z)

(d) CH_3, (d) CH_3, (e) H, (f) H, (g) H — boronate ring with $B—CH_2—CH_3$ (a) (b), CH_3 (c)

	Shift	Peak
▶ a	0.61	S
b	0.88	S
c	1.25	2H
d	1.29	1A
e	1.47	S
f	1.76	S
g	4.17	O

▶ $C_8H_{17}BO_2$ ▶ 7% $CDCl_3$ ▶ V 518

▶ 2-BaBb

(a) (b) (c) $CH_3(CH_2)_{17}CH_2C(=O)—OH$

	Shift	Peak
a	0.87	3B
▶ b	1.24	1A
c	~2.27	2B

▶ $C_{20}H_{40}O_2$ ▶ 45.2 mg/0.50 ml $CDCl_3$ ▶ S 260

▶ 2-AZ

(b) (a) $CH_3—CH_2—Si(—O—CH_2—CH_3)_3$ (d) (c)

	Shift	Peak
▶ a	~0.62	S
b	~1.00	S
c	1.23	3A
d	3.82	4A

▶ $C_8H_{20}O_3Si$ ▶ 7% $CDCl_3$ ▶ V 222

▶ 2-BaBb

(c) (b) (a) $C_6H_5—CH_2—(CH_2)_7—CH_3$

(d) = all ring H's

	Shift	Peak
a	.89	3B
▶ b	1.25	1B
c	2.46	3B
d	7.07	1A

▶ $C_{15}H_{24}$ ▶ 50% CCl_4 ▶ API 358

▶ 2-BaBb

(a) (b) (c) $CH_3—CH_2—(CH_2)_{11}—CH_2—OH$

	Shift	Peak
a	0.86	3B
▶ b	1.21	1A
c	~3.55	3B

▶ $C_{14}H_{30}O$ ▶ 44.9 mg/0.25 ml $CDCl_3$ ▶ S 294

▶ 2-BaBb

(c) (b) (a) $C_6H_5—CH_2—(CH_2)_8—CH_3$

(d) = all ring H's

	Shift	Peak
a	.88	3B
▶ b	1.25	1B
c	2.50	3B
d	7.07	1A

▶ $C_{16}H_{26}$ ▶ 50% CCl_4 ▶ API 359

▶ 2-BaBb

(c) (b) (a) $C_6H_5—CH_2—(CH_2)_{10}—CH_3$

(d) = all ring H's

	Shift	Peak
a	.88	3B
▶ b	1.22	1A
c	2.54	3B
d	7.09	1G

▶ $C_{18}H_{30}$ ▶ 50% CCl_4 ▶ API 360

▶ 2-BaBb

(e) H, (d) H, (f) H: $H_2C=CH—CH_2(CH_2)_{12}CH_3$ (c) (b) (a)

	Shift	Peak
a	.99	3B
▶ b	1.25	1A
c	2.02	3B
d	4.90	R
e	5.08	R
f	5.68	R

▶ $C_{16}H_{32}$ ▶ 50% CCl_4 ▶ API 73

▶ 2-BaBb

(a) (b) (b) (b) (a) $CH_3—CH_2—(CH_2)_4—CH_2—CH_3$

	Shift	Peak
a	.84	3B
▶ b	1.24	1B

▶ C_8H_{18} ▶ 10% CCl_4 ▶ API 200

▶ 2-BaBb

(a) (b) (c) (d) $CH_3—(CH_2)_{11}—CH_2—C(=O)—OH$

	Shift	Peak
a	0.88	3B
▶ b	1.26	1A
c	~2.35	1D
d	10.95	1B

▶ $C_{14}H_{28}O_2$ ▶ 44.8 mg/0.50 ml $CDCl_3$ ▶ S 167

▶ 2-BaBb

(d) H, (e) H, (f) H: $H_2C=CH—CH_2(CH_2)_{12}CH_3$ (c) (b) (a)

	Shift	Peak
a	.90	3B
▶ b	1.24	1A
c	2.04	3B
d	4.88	R
e	5.02	R
f	5.72	R

▶ $C_{16}H_{32}$ ▶ 50 vol.% CCl_4 ▶ API 437

▶ 2-BaBb

(b) (a) $CH_3—(CH_2)_{14}—CH_3$

	Shift	Peak
a	.90	3B
▶ b	1.26	1A

▶ $C_{16}H_{34}$ ▶ 10% CCl_4 ▶ API 48

▶ 2-BaBb

$CH_3—(CH_2)_{13}—CH_2—C(=O)—O—CH_3$ — (a) CH_3, (b) $(CH_2)_{13}$, (c) CH_2, (d) OCH_3

	Shift	Peak
a	~0.88	1D
▶b	1.26	1A
c	~2.30	1D
d	3.63	1A

▶ $C_{17}H_{34}O_2$ ▶ 44.4 mg/0.50 ml $CDCl_3$ ▶ S 248

▶ 2-BaBb

$CH_3—(CH_2)_5—CH(Br)—CH_3$ — (a) CH_3, (b) $(CH_2)_5$, (c) CH_3, (d) CH

	Shift	Peak
a	0.91	3B
▶b	1.28	1B
c	1.71	2A
d	4.12	5B

▶ $C_8H_{17}Br$ ▶ 45.2 mg/0.50 ml $CDCl_3$ ▶ S 227

▶ 2-BaBb

$CH_3—CH(CH_3)—(CH_2)_4—CH_3$ — (a) CH_3 (branch), (c) $(CH_2)_4$, (b) CH_3

	Shift	Peak
a	.87	2H
b	.96	S
▶c	1.27	1B

▶ C_8H_{18} ▶ 50% CCl_4 ▶ API 340

▶ 2-BaBb

$H(d)H(e)C=CH(f)—CH_2(CH_2)_5CH_3$ — (c) CH_2, (b) $(CH_2)_5$, (a) CH_3

	Shift	Peak
a	.85	3B
▶b	1.28	1B
c	1.99	3B
d	4.88	R
e	4.99	R
f	5.68	R

▶ C_9H_{18} ▶ 50% CCl_4 ▶ API 434

▶ 2-BaBb

$CH_3—CH_2—(CH_2)_4—CH_2—CH_3$ — (a) CH_3, (b) CH_2, (c) $(CH_2)_4$, (b) CH_2, (a) CH_3

	Shift	Peak
a	0.88	3B
b	1.27 ± 0.03	1A
▶c	1.27 ± 0.03	1A

▶ C_8H_{18} ▶ 7% $CDCl_3$ ▶ V 216

▶ 2-BaBb

$CH_3—C(=O)—CH_2—(CH_2)_5—CH_3$ — (c) CH_3, (d) CH_2, (b) $(CH_2)_5$, (a) CH_3

	Shift	Peak
a	0.88	3B
▶b	1.28	1B
c	2.17	1A
d	2.46	3B

▶ $C_9H_{18}O$ ▶ 33.7 mg/0.50 ml $CDCl_3$ ▶ S 48

▶ 2-BaBb

$CH_3—(CH_2)_8—CH_2—OH$ — (a) CH_3, (b) $(CH_2)_8$, (c) CH_2, (d) OH

	Shift	Peak
a	.88	3B
▶b	1.27	1B
c	3.45	3B
d	4.28	1A

▶ $C_{10}H_{22}O$ ▶ 20% CCl_4 ▶ API 172

▶ 2-BaBb

$(CH_3)_3C—CH_2—(CH_2)_3—CH_3$ — (a) $C(CH_3)_3$, (c) $(CH_2)_3$, (b) CH_3

	Shift	Peak
a	.87	1E
b	.87	1E
▶c	1.28	1B

▶ C_9H_{20} ▶ 50% CCl_4 ▶ API 209

▶ 2-BaBb

$H(d)H(e)C=CH(f)—CH_2(CH_2)_{10}CH_3$ — (c) CH_2, (b) $(CH_2)_{10}$, (a) CH_3

	Shift	Peak
a	.88	3B
▶b	1.27	1A
c	2.3	3B
d	4.88	7X
e	5.01	7X
f	5.72	R

▶ $C_{14}H_{28}$ ▶ 50 vol.% CCl_4 ▶ API 436

▶ 2-BaBb

$CH_3—(CH_2)_7—CH_3$ — (a) CH_3, (b) $(CH_2)_7$

	Shift	Peak
a	.88	3B
▶b	1.28	1B

▶ C_9H_{20} ▶ 50% CCl_4 ▶ API 343

▶ 2-BaBb

$CH_3—(CH_2)_4—CH_2—NH_2$ — (a) CH_3, (b) $(CH_2)_4$, (d) CH_2, (c) NH_2

	Shift	Peak
a	0.90	3B
▶b	1.28	1B
c	1.58	S
d	2.70	1C

▶ $C_6H_{15}N$ ▶ 35.6 mg/0.50 ml $CDCl_3$ ▶ S 46

▶ 2-BaBb

$CH_3—(CH_2)_7—CH_3$ — (a) CH_3, (b) $(CH_2)_7$, (a) CH_3

	Shift	Peak
a	0.88	3B
▶b	1.28	1A

▶ C_9H_{20} ▶ 45.0 mg/0.50 ml $CDCl_3$ ▶ S 203

▶ 2-BaBb

(a) CH_3—(b) $(CH_2)_7$—(c) CH_2—(d) OH

	Shift	Peak
a	.88	3B
▶b	1.28	1B
c	3.44	3B
d	4.27	1A

▶ $C_9H_{20}O$ ▶ 20% CCl_4 ▶ API 171

▶ 2-BaBb

(a) CH_3—(b) $(CH_2)_8$—(c) CH_2—C(H (f))=C(H (d))(H (e))

	Shift	Peak
a	0.87	3B
▶b	1.28	1A
c	2.05	2G
d	4.80	R
e	~5.03	R
f	5.68	R

▶ $C_{12}H_{24}$ ▶ 44.4 mg/0.50 ml $CDCl_3$ ▶ S 202

▶ 2-BaBb

(a) CH_3—(b) $(CH_2)_6$—(c) CH_2C(=O)—(d) OCH_3

	Shift	Peak
a	0.88	3B
▶b	1.28	1B
c	2.30	3B
d	3.68	1A

▶ $C_{10}H_{20}O_2$ ▶ 45.1 mg/0.50 ml $CDCl_3$ ▶ S 249

▶ 2-BaBb

(a) CH_3—(b) $(CH_2)_{10}$—CH_3

	Shift	Peak
a	.88	3B
▶b	1.28	1A

▶ $C_{12}H_{26}$ ▶ 50% CCl_4 ▶ API 350

▶ 2-BaBb

(a) CH_3—(b) $(CH_2)_8$—(a) CH_3

	Shift	Peak
a	0.88	3B
▶b	1.28	1E

▶ $C_{10}H_{22}$ ▶ 44.8 mg/0.50 ml $CDCl_3$ ▶ S 201

▶ 2-BaBb

(a) CH_3—(b) $(CH_2)_9$—(c) CH_2—C(=O)—(d) OCH_3

	Shift	Peak
a	0.88	3B
▶b	1.28	1B
c	2.28	3B
d	3.67	1A

▶ $C_{13}H_{26}O_2$ ▶ 45.7 mg/0.50 ml $CDCl_3$ ▶ S 125

▶ 2-BaBb

(a) CH_3CH_2(b) $(CH_2)_5$ CH_2—C(H (g))(OH (c))—C(H (f))(H (e))—(d) NH_2

	Shift	Peak
a	0.89	3B
▶b	1.28	1G
c	1.75	1E
d	1.75	1E
e	2.51	S
f	2.84	S
g	3.50	1D

▶ $C_{10}H_{23}NO$ ▶ 7% $CDCl_3$ ▶ V 575

▶ 2-BaBb

(a) CH_3—(b) $(CH_2)_9$—(c) CH_2—C(=O)—O(d) CH_2(a) CH_3

	Shift	Peak
a	0.88	3B
▶b	1.28	1E
c	2.28	3C
d	4.11	4A

▶ $C_{14}H_{28}O_2$ ▶ 45.5 mg/0.50 ml $CDCl_3$ ▶ S 194

▶ 2-BaBb

CH_3—(b) $(CH_2)_9$—(a) CH_3

	Shift	Peak
a	.88	3B
▶b	1.28	1A

▶ $C_{11}H_{24}$ ▶ 50% CCl_4 ▶ API 348

▶ 2-DaBb

(a) CH_3—(b) $(CH_2)_{12}$—(a) CH_3

	Shift	Peak
a	0.88	3B
▶b	1.28	1A

▶ $C_{14}H_{30}$ ▶ 44.8 mg/0.50 ml $CDCl_3$ ▶ S 204

▶ 2-BaBb

(a) CH_3—(b) $(CH_2)_9$—(a) CH_3

	Shift	Peak
a	0.90	3B
▶b	1.28	1G

▶ $C_{11}H_{24}$ ▶ 44.4 mg/0.50 ml $CDCl_3$ ▶ S 289

▶ 2-BaBb

(a) CH_3—(b) $(CH_2)_5$—(d) CH_2(c) SH

	Shift	Peak
a	0.90	3B
▶b	1.29	1E
c	2.09	1A
d	2.53	4B

▶ $C_7H_{16}S$ ▶ 46.9 mg/0.50 ml $CDCl_3$ ▶ S 297

▶ 2-BaBb

	Shift	Peak
a	0.88	3B
▶b	1.29	1B
c	2.30	1B
d	9.35	1A

$\overset{(a)}{CH_3}-\overset{(b)}{(CH_2)_6}-\overset{(c)}{CH_2}-\underset{(d)}{\overset{O}{\overset{\|}{C}}H}$

▶ $C_9H_{18}O$ ▶ 45.3 mg/0.50 ml $CDCl_3$ ▶ S 100

▶ 2-BaBb

	Shift	Peak
a	.88	3B
▶b	1.30	1B

$CH_3-\overset{(b)}{(CH_2)_4}-\overset{(a)}{CH_3}$

▶ C_6H_{14} ▶ 50% CCl_4 ▶ API 333

▶ 2-BaBb

	Shift	Peak
a	0.87	3A
▶b	1.29	1B
c	2.42	3B
d	9.77	3A

$\overset{(a)}{CH_3}-\overset{(b)}{(CH_2)_9}-\overset{(c)}{CH_2}-\underset{(d)}{\overset{O}{\overset{\|}{C}}H}$

▶ $C_{12}H_{24}O$ ▶ 45.2 mg/0.50 ml $CDCl_3$ ▶ S 106

▶ 2-BaBb

	Shift	Peak
a	0.90	3B
▶b	1.30	1B
c	1.85	1F
d	3.41	3A

$\overset{(a)}{CH_3}-\overset{(b)}{(CH_2)_4}-\overset{(c)}{CH_2}-\overset{(d)}{CH_2}-Br$

▶ $C_7H_{15}Br$ ▶ 45.0 mg/0.50 ml $CDCl_3$ ▶ S 208

▶ 2-BaBb

	Shift	Peak
a	0.85	3B
▶b	1.29	1A
c	1.83	1B
d	3.62	3B

$\overset{(a)}{CH_3}-\overset{(b)}{(CH_2)_{10}}-\overset{(d)}{CH_2}\overset{(c)}{OH}$

▶ $C_{12}H_{26}O$ ▶ 44.4 mg/0.50 ml $CDCl_3$ ▶ S 103

▶ 2-BaBb

	Shift	Peak
a	.88	3B
▶b	1.30	1B

$\overset{(a)}{CH_3}-\overset{(b)}{(CH_2)_5}-CH_3$

▶ C_7H_{16} ▶ 50% CCl_4 ▶ API 337

▶ 2-BaBb

	Shift	Peak
a	0.88	3B
▶b	1.29	1A
c	2.24	3B
d	3.68	1A

$\overset{(a)}{CH_3}-\overset{(b)}{(CH_2)_{11}}-\overset{(c)}{CH_2}-\overset{O}{\overset{\|}{C}}-\overset{(d)}{OCH_3}$

▶ $C_{15}H_{30}O_2$ ▶ 45.1 mg/0.50 ml $CDCl_3$ ▶ S 195

▶ 2-BaBb

	Shift	Peak
a	0.88	3B
▶b	1.30	1B
c	2.43	3B
d	9.73	3A

$\overset{(a)}{CH_3}-\overset{(b)}{(CH_2)_5}-\overset{(c)}{CH_2}-\underset{(d)}{\overset{O}{\overset{\|}{C}}H}$

▶ $C_8H_{16}O$ ▶ 44.6 mg/0.50 ml $CDCl_3$ ▶ S 105

▶ 2-BaBb

	Shift	Peak
a	0.88	3B
▶b	1.29	1B
c	2.00	1F
d	2.25	3B
e	5.28	3D
f	11.45	1C

$\overset{(e)}{CH}-\overset{(c)}{CH_2}-\overset{(b)}{(CH_2)_5}-\overset{(d)}{CH_2}-\overset{O}{\overset{\|}{C}}-\overset{(f)}{OH}$
$\|$
$\underset{(e)}{CH}-\underset{(c)}{CH_2}-\underset{(b)}{(CH_2)_6}-\underset{(a)}{CH_3}$

▶ $C_{18}H_{34}O_2$ ▶ 45.2 mg/0.50 ml $CDCl_3$ ▶ S 70

▶ 2-BaBb

	Shift	Peak
a	0.88	3B
▶b	1.30	1G
c	2.36	3B
d	11.47	1A

$\overset{(a)}{CH_3}\overset{(b)}{(CH_2)_5}\overset{(c)}{CH_2}-\overset{O}{\overset{\|}{C}}-\overset{(d)}{OH}$

▶ $C_8H_{16}O_2$ ▶ 45.6 mg/0.50 ml $CDCl_3$ ▶ S 157

▶ 2-BaBb

	Shift	Peak
a	0.90	3B
▶b	1.29	1B
c	2.00	1C
d	2.32	S
e	3.70	1A
f	5.4	3B

$\overset{(f)}{CH}-\overset{(c)}{CH_2}-\overset{(b)}{(CH_2)_5}-\overset{(d)}{CH_2}-\overset{O}{\overset{\|}{C}}-\overset{(e)}{OCH_3}$
$\|$
$\underset{(f)}{CH}-\underset{(c)}{CH_2}-\underset{(b)}{(CH_2)_6}-\underset{(a)}{CH_3}$

▶ $C_{19}H_{36}O_2$ ▶ 33.7 mg/0.50 ml $CDCl_3$ ▶ S 71

▶ 2-BaBb

	Shift	Peak
a	.88	3B
b	1.12	2A
▶c	1.30	1B
d	3.70	4B
e	4.08	1A

$\overset{(b)}{CH_3}-\underset{(d)}{\overset{OH\,(e)}{\overset{|}{CH}}}-\overset{(c)}{CH_2}-\overset{(c)}{(CH_2)_4}-\overset{(a)}{CH_3}$

▶ $C_8H_{18}O$ ▶ 20% CCl_4 ▶ API 170

▶ 2-BaBb

$CH_3-(CH_2)_5-CH_2-C(=O)-OCH_2CH_3$ — (a) CH_3, (c) $(CH_2)_5$, (d) CH_2, (e) OCH_2, (b) CH_3

	Shift	Peak
a	0.88	3B
b	1.28	3D
▶c	1.30	S
d	2.28	3B
e	4.12	4A

▶ $C_{10}H_{20}O_2$ ▶ 45.2 mg/0.50 ml $CDCl_3$ ▶ S 244

▶ 2-BaBb

$CH_3-(CH_2)_4-CH_2-C(=O)-OH$ — (a) CH_3, (b) $(CH_2)_4$, (c) CH_2, (d) OH

	Shift	Peak
a	0.88	3B
▶b	1.32	1B
c	2.38	3B
d	11.47	1B

▶ $C_7H_{14}O_2$ ▶ 45.6 mg/0.50 ml $CDCl_3$ ▶ S 49

▶ 2-BaBb

$CH_3-CH(CH_3)-(CH_2)_7-CH_3$ — (a) CH_3 (both methyls on CH), (c) $(CH_2)_7$, (b) terminal CH_3

	Shift	Peak
a	.83	2H
b	.94	S
▶c	1.30	1A

▶ $C_{11}H_{24}$ ▶ 50% CCl_4 ▶ API 349

▶ 2-BaBb

$CH_3-(CH_2)_4-CH_2-CH_2-I$ — (a) CH_3, (b) $(CH_2)_4$, (c) CH_2, (d) CH_2

	Shift	Peak
a	0.88	3B
▶b	1.32	2E
c	1.82	3C
d	3.18	3A

▶ $C_7H_{15}I$ ▶ 45.0 mg/0.50 ml $CDCl_3$ ▶ S 210

▶ 2-BaBb

$CH_3-(CH_2)_{13}-CH_3$ — (b) $(CH_2)_{13}$, (a) CH_3

	Shift	Peak
a	.87	3B
▶b	1.30	1A

▶ $C_{15}H_{32}$ ▶ 50% CCl_4 ▶ API 221

▶ 2-BaBb

$CH_3-(CH_2)_5-CH_2-OH$ — (a) CH_3, (b) $(CH_2)_5$, (c) CH_2, (d) OH

	Shift	Peak
a	.90	3B
▶b	1.32	1G
c	3.44	3B
d	4.42	1B

▶ $C_7H_{16}O$ ▶ 20% CCl_4 ▶ API 168

▶ 2-BaBb

$(d)H(e)H\,C=C\,H(f)-(CH_2)_4CH_3$ — (d)H, (e)H, H(f), $(CH_2)_4$ (c+b), CH_3 (a)

	Shift	Peak
a	.91	3B
▶b	1.31	S
c	2.02	3B
d	4.84	R
e	4.98	R
f	5.67	R

▶ C_7H_{14} ▶ 10% CCl_4 ▶ API 157

▶ 2-BaBb

$CH_3CH_2C(=O)-OCH_2(CH_2)_3CH_3$ — (b) CH_3, (d) CH_2, (e) OCH_2, (c) $(CH_2)_3$, (a) CH_3

	Shift	Peak
a	0.92	3B
b	1.12	S
▶c	1.32	S
d	2.33	4A
e	4.05	3B

▶ $C_8H_{16}O_2$ ▶ 44.8 mg/0.50 ml $CDCl_3$ ▶ S 40

▶ 2-BaBb

$CH_3-(CH_2)_5-CH_2Cl$ — (a) CH_3, (b) $(CH_2)_5$, (c) CH_2Cl

	Shift	Peak
a	0.88	3B
▶b	1.31	1E
c	3.53	3A

▶ $C_7H_{15}Cl$ ▶ 44.8 mg/0.50 ml $CDCl_3$ ▶ S 209

▶ 2-BaBb

$HC(=O)-OCH_2-(CH_2)_6-CH_3$ — (d) HC, (c) OCH_2, (b) $(CH_2)_6$, (a) CH_3

	Shift	Peak
a	0.88	3B
▶b	1.32	1E
c	4.16	3B
d	8.03	1B

▶ $C_9H_{18}O_2$ ▶ 44.4 mg/0.50 ml $CDCl_3$ ▶ S 114

▶ 2-BaBb

$CH_3-(CH_2)_3-CH_2-CH_2-I$ — (a) CH_3, (b) $(CH_2)_3$, (c) CH_2, (d) CH_2

	Shift	Peak
a	0.90	3B
▶b	1.32	2E
c	1.84	3C
d	3.20	3A

▶ $C_6H_{13}I$ ▶ 45.8 mg/0.50 ml $CDCl_3$ ▶ S 212

▶ 2-BaBb

$CH_3-(CH_2)_7-CH_2OH$ — (a) CH_3, (b) $(CH_2)_7$, (d) CH_2, (c) OH

	Shift	Peak
a	0.90	3B
▶b	1.32	1B
c	1.97	1A
d	3.62	3B

▶ $C_9H_{20}O$ ▶ 44.6 mg/0.50 ml $CDCl_3$ ▶ S 102

▶ 2-BaBb

$CH_2=CH-CH_2(CH_2)_8CH_3$ — (d),(e) H on terminal C; (f) H; (c) (b) (a)

	Shift	Peak
a	.92	3B
▶b	1.32	1B
c	2.05	3B
d	4.86	R
e	5.00	R
f	5.68	R

▶ $C_{12}H_{24}$ ▶ 50 vol.% CCl_4 ▶ API 435

▶ 2-BaBb

$C_6H_5-CH_2-CH_2-CH_2-CH_2-CH_3$ — (c) (b) (b) (b) (a); (d) = all ring H's

	Shift	Peak
a	.89	3B
▶b	1.36	O
c	2.52	3B
d	7.04	1G

▶ $C_{11}H_{16}$ ▶ 50% CCl_4 ▶ API 355

▶ 2-BaBb

$CH_3-(CH_2)_3-CH_2-CH_2-Cl$ — (a) (b) (c) (d)

	Shift	Peak
a	0.90	3B
▶b	1.33	2H
c	1.78	O
d	3.53	3A

▶ $C_6H_{13}Cl$ ▶ 44.6 mg/0.50 ml $CDCl_3$ ▶ S 211

▶ 2-BaBb

$CH_3C(=O)-OCH_2(CH_2)_4CH_3$ — (c) (d) (b) (a)

	Shift	Peak
a	0.92	3B
▶b	1.37	1E
c	2.05	1A
d	4.05	3A

▶ $C_8H_{16}O_2$ ▶ 45.0 mg/0.50 ml $CDCl_3$ ▶ S 217

▶ 2-BaBb

$CH_3-(CH_2)_5-CH_2-OH$ — (a) (b) (d) (c)

	Shift	Peak
a	0.88	3B
▶b	1.33	1B
c	1.95	1A
d	3.62	3B

▶ $C_7H_{16}O$ ▶ 45.0 mg/0.50 ml $CDCl_3$ ▶ S 215

▶ 2-BaBb

$CH_3CH_2CH_2CH_2CH_2Br$ — (a) (b) (c) (d)

	Shift	Peak
a	0.94	3B
▶b	1.38	O
c	1.83	4C
d	3.42	3A

▶ $C_5H_{11}Br$ ▶ 44.5 mg/0.50 ml $CDCl_3$ ▶ S 120

▶ 2-BaBb

$CH_3-(CH_2)_4-CH_2-C(=O)-OCH_2-CH_3$ — (a) (c) (d) (e) (b)

	Shift	Peak
a	0.88	3B
b	1.26	S
▶c	1.33	S
d	2.28	3B
e	4.12	4A

▶ $C_9H_{18}O_2$ ▶ 44.5 mg/0.50 ml $CDCl_3$ ▶ S 283

▶ 2-BaBb

$CH_3-(CH_2)_2-CH_2-CH_2I$ — (a) (b) (c) (d)

	Shift	Peak
a	0.92	3B
▶b	1.38	2E
c	1.86	3C
d	3.20	3A

▶ $C_5H_{11}I$ ▶ 45.0 mg/0.50 ml $CDCl_3$ ▶ S 228

▶ 2-BaBb

$CH_2=CH-CH_2CH_2CH_2CH_2CH_3$ — (f),(g) H on terminal C; (h) H; (e) (d) (c) (b) (a)

	Shift	Peak
a	.89	3C
b	.90	S
▶c	~1.33 or 1.35	S
d	~1.35 or 1.33	S
e	1.97	4C
f	~4.84	R
g	~4.97	R
h	~5.70	R

▶ C_7H_{14} ▶ liquid ▶ API 384

▶ 2-BaBb

$CH_3-C(=O)-CH_2-(CH_2)_3-CH_3$ — (c) (d) (b) (a)

	Shift	Peak
a	0.90	3B
▶b	1.38	O
c	2.15	1A
d	2.42	3B

▶ $C_7H_{14}O$ ▶ 35.4 mg/0.50 ml $CDCl_3$ ▶ S 41

▶ 2-BaBb

$CH_3-(CH_2)_4-CH_2-OH$ — (a) (b) (d) (c)

	Shift	Peak
a	0.88	3B
▶b	1.36	1B
c	2.28	1A
d	3.62	3A

▶ $C_6H_{14}O$ ▶ 45.2 mg/0.50 ml $CDCl_3$ ▶ S 198

▶ 2-BaBb

$CH_3(CH_2)_3CH_2-O-CH_2(CH_2)_3CH_3$ — (a) (b) (c) (c) (b) (a)

	Shift	Peak
a	0.88	3B
▶b	1.40	O
c	3.38	3A

▶ $C_{10}H_{22}O$ ▶ 44.6 mg/0.50 ml $CDCl_3$ ▶ S 151

▶ 2-BaBb

	Shift	Peak
a	0.92	3B
▶b	1.45	O
c	2.36	3C
d	11.37	1B

(a) CH_3—(b) $(CH_2)_3$—(c) CH_2—C(=O)—(d) OH

▶ $C_6H_{12}O_2$ ▶ 44.7 mg/0.50 ml $CDCl_3$ ▶ S 37

▶ 2-BaBc

	Shift	Peak
a	0.92	3B
▶b	1.30	1B
c	1.75	1B
d	3.52	1I

(a) CH_3(b) $(CH_2)_3$—(d) CH(c) CH_2OH, with (b) CH_2(a) CH_3 on CH

▶ $C_8H_{18}O$ ▶ 45.0 mg/0.50 ml $CDCl_3$ ▶ S 98

▶ 2-BaBb

	Shift	Peak
a	0.93	3B
▶b	1.49	O
c	2.33	3B

(a) CH_3—(b) $(CH_2)_3$—(c) CH_2—C≡N

▶ $C_6H_{11}N$ ▶ 45.0 mg/0.50 ml $CDCl_3$ ▶ S 176

▶ 2-BaBd

	Shift	Peak
a	.87	1E
b	1.20	1E
▶c	1.20	1B

CH_3—C(b CH_3)(CH_3)—(c) CH_2—(c) $(CH_2)_2$—(a) CH_3

▶ C_8H_{18} ▶ 50% CCl_4 ▶ API 203

▶ 2-BaBb

	Shift	Peak
a	.86	3B
b	1.28	1B
▶c	2.54	3B
d	7.09	1A

C_6H_5—(c) CH_2—(b) $(CH_2)_5$—(a) CH_3

(d) = all ring H's

▶ $C_{13}H_{20}$ ▶ 50% CCl_4 ▶ API 357

▶ 2-BaBl

	Shift	Peak
a	.88	3C
b	~1.29 or 1.31	S
▶c	~1.31 or 1.29	S
d	2.00	3C
e	4.84	R
f	4.97	R
g	5.65	R

(e)H(f)H C=C H(g), (d) CH_2(c) CH_2(b) CH_2(a) CH_3

▶ C_6H_{12} ▶ liquid ▶ API 368

▶ 2-BaBc

	Shift	Peak
a	.82	2A
b	~.92	1E
▶c	~.93	1B
d	~.97	1D

(a) CH_3—(d) CH((a) CH_3)—(c) CH_2—(c) CH_2—(c) CH_2—(b) CH_3

▶ C_7H_{16} ▶ 50% CCl_4 ▶ API 195

▶ 2-BaBl

	Shift	Peak
a	.90	3C
▶b	1.31	O
c	1.57	2E
d	2.02	1F
e	~5.37	O

(c) H_3C(e)H C=C H(e), (d) CH_2(b) CH_2(b) CH_2(a) CH_3

▶ C_7H_{14} ▶ liquid ▶ API 389

▶ 2-BaBc

	Shift	Peak
a	.83	3B
▶b	1.23	1B
c	~1.27	S

(a) CH_3—(b) CH_2—(c) CH((a) CH_3)—(b) $(CH_2)_3$—(a) CH_3

▶ C_8H_{18} ▶ 50% CCl_4 ▶ API 341

▶ 2-BaBl

	Shift	Peak
a	.88	3C
▶b	1.31	O
c	1.62	2E
d	1.92	1F
e	5.33 or 5.42	S
f	5.42 or 5.33	S

(c) H_3C(e)H C=C H(f), (d) CH_2(b) CH_2(b) CH_2(a) CH_3

▶ C_7H_{14} ▶ liquid ▶ API 390

▶ 2-BaBc

	Shift	Peak
a	.87	3D
b	1.05	2H
▶c	1.25	1B
d	3.2	1C

(a) CH_3—(c) $(CH_2)_3$—CH((b) CH_3)—(d) NH_2

▶ $C_6H_{15}N$ ▶ 34.4 mg/0.50 ml $CDCl_3$ ▶ S 28

▶ 2-BaBl

	Shift	Peak
a	.90	3B
b	~1.34	S
▶c	~1.34	S
d	2.00	4C
e	4.85	R
f	4.97	R
g	5.68	R

(e)H(f)H C=C H(g), (d) CH_2(c) CH_2(b) CH_2(a) CH_3

▶ C_6H_{12} ▶ 10% CCl_4 ▶ API 155

▶ 2-BaBl

	Shift	Peak
a	.93	3C
▶b	1.38	O
c	1.68	1H
d	2.01	3B
e	~4.67	M
f	~4.67	M

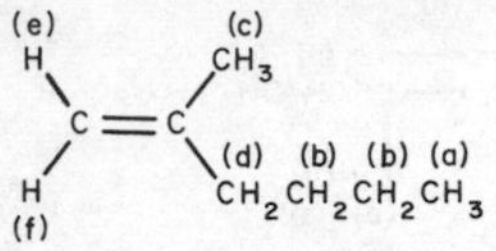

▶ C_7H_{14} ▶ liquid ▶ API 407

▶ 2-BaBn

	Shift	Peak
a	0.96	3B
▶b	1.50	1F
c	3.08	3A
d	3.42	1A
e	6.52	3C
f	7.12	3C

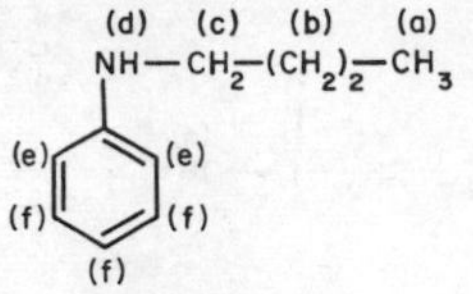

▶ $C_{10}H_{15}N$ ▶ 45.4 mg/0.50 ml $CDCl_3$ ▶ S 280

▶ 2-BaBn

	Shift	Peak
a	0.88	3D
▶b	~1.2-1.5	N
c	~3.08	4C
d	5.18	1B
e	5.95	3C

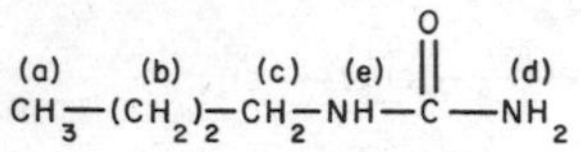

▶ $C_5H_{12}N_2O$ ▶ 45.3 mg/0.25 ml $CDCl_3$ ▶ S 282

▶ 2-BaBo

	Shift	Peak
a	0.88	3C
b	~1.36	S
▶c	~1.50	S
d	3.58	2H
e	3.84	3B
f	6.24	2A
g	7.20	1A

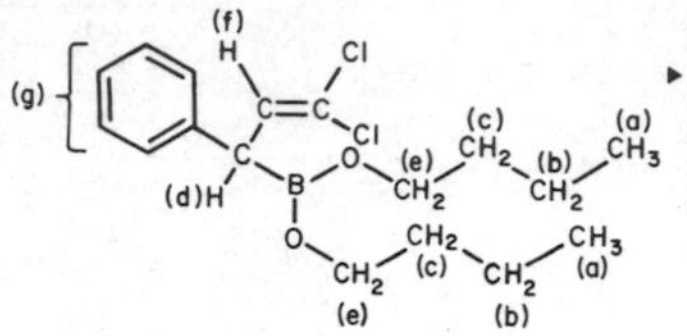

▶ $C_{17}H_{25}BCl_2O_2$ ▶ 7% $CDCl_3$ ▶ V 667

▶ 2-BaBn

	Shift	Peak
a	0.92	3B
▶b	1.37	6B
c	2.40	3B

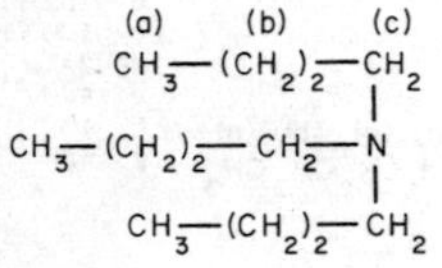

▶ $C_{12}H_{27}N$ ▶ 44.8 mg/0.50 ml $CDCl_3$ ▶ S 84

▶ 2-BaBo

	Shift	Peak
a	0.88	3C
b	~1.36	S
▶c	~1.50	S
d	2.86*	S
e	2.89*	S
f	3.65	S
g	3.88	3C
h	7.26	1A

*±0.04

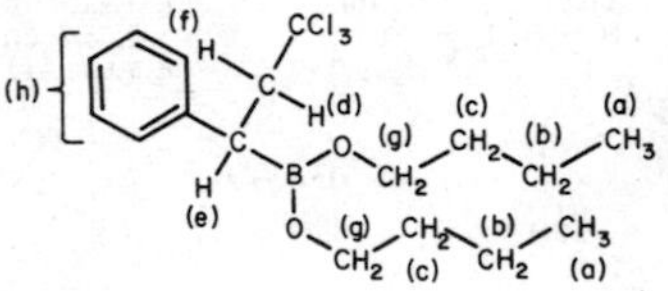

▶ $C_{17}H_{26}BCl_3O_2$ ▶ 7% $CDCl_3$ ▶ V 668

▶ 2-BaBn

	Shift	Peak
a	0.90	3B
▶b	~1.40	1F
c	2.46	1A
d	3.23	4C
e	6.53	3B
f	7.32	Q
g	7.79	Q
h	~8.60	1D

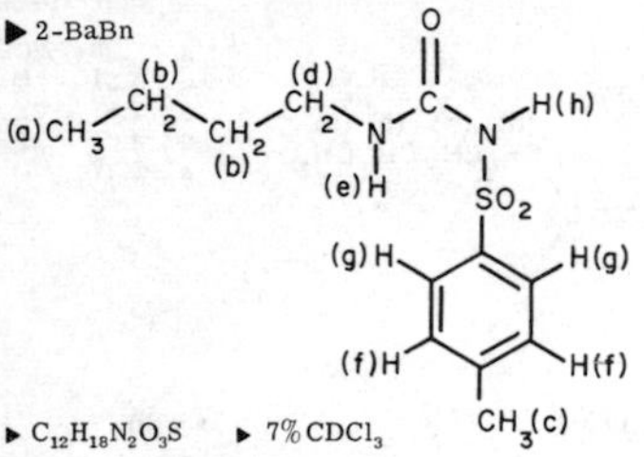

▶ $C_{12}H_{18}N_2O_3S$ ▶ 7% $CDCl_3$ ▶ V 602

▶ 2-BaBp

	Shift	Peak
a	.91	3B
▶b	~1.45	O
c	3.48	3C
d	4.58	1A

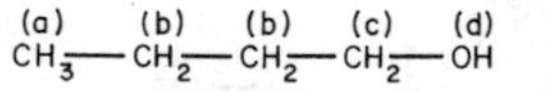

▶ $C_4H_{10}O$ ▶ 20% CCl_4 ▶ API 166

▶ 2-BaBn

	Shift	Peak
a	0.91	3B
b	1.44 ± 0.1	S
▶c	1.44 ± 0.1	S
d	1.87	1A
e	3.73	3A

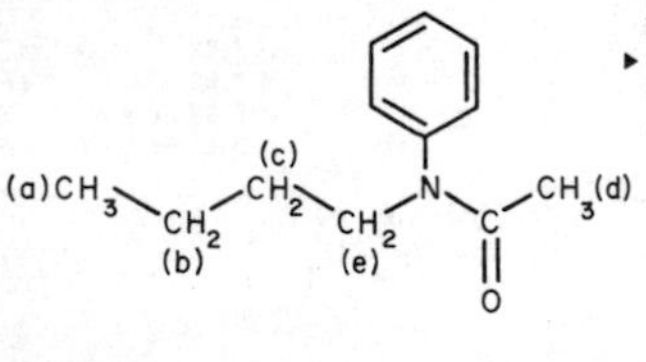

▶ $C_{12}H_{17}NO$ ▶ 7% $CDCl_3$ ▶ V 601

▶ 2-BaBq

	Shift	Peak
a	0.94	3B
b	1.41	2A
▶c	1.46	S
d	3.02	1A
e	4.18	4C

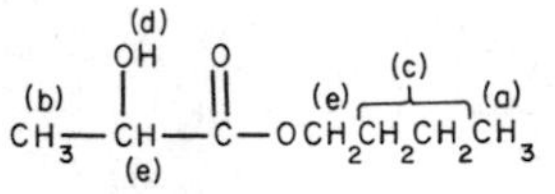

▶ $C_7H_{14}O_3$ ▶ 44.6 mg/0.50 ml $CDCl_3$ ▶ S 246

▶ 2-BaBn

	Shift	Peak
a	0.96	3B
▶b	1.48	1F
c	3.28	3C

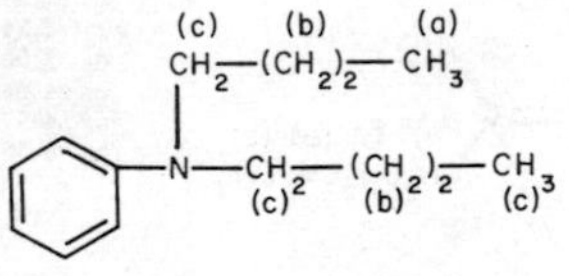

▶ $C_{14}H_{23}N$ ▶ 45.2 mg/0.50 ml $CDCl_3$ ▶ S 147

▶ 2-BaBq

	Shift	Peak
a	0.94	3B
▶b	1.56	1F
c	4.05	1A
d	4.18	3A

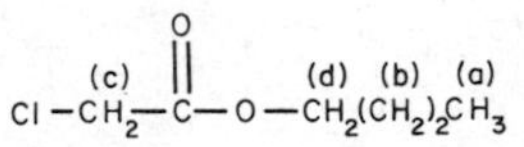

▶ $C_6H_{11}ClO_2$ ▶ 44.5 mg/0.50 ml $CDCl_3$ ▶ S 92

▶ 2-BaBq

$H_2N(d)$–C_6H_4(e, f)–C(=O)–O–CH_2(c)CH_2CH_2(b)CH_3(a)

	Shift	Peak
a	0.95	3B
▶b	~1.60	1F
c	4.25	3D
d	~4.10	1E
e	6.59	Q
f	7.83	Q

▶ $C_{11}H_{15}NO_2$ ▶ 44.6 mg/0.50 ml $CDCl_3$ ▶ S 243

▶ 2-BaBq

C_6H_5(d, e)–C(=O)–OCH_2(c)$(CH_2)_2$(b)CH_3(a)

	Shift	Peak
a	0.98	3B
▶b	1.64	O
c	4.35	3A
d	7.48	N
e	8.06	N

▶ $C_{11}H_{14}O_2$ ▶ 44.8 mg/0.50 ml $CDCl_3$ ▶ S 158

▶ 2-BaBs

(a) CH_3–(b) $(CH_2)_2$–(c) CH_2–S–CH_2–$(CH_2)_2$–CH_3

	Shift	Peak
a	0.92	3B
▶b	1.49	O
c	2.49	3B

▶ $C_8H_{18}S$ ▶ 45.8 mg/0.50 ml $CDCl_3$ ▶ S 275

▶ 2-BaBs

(a) CH_3–(c) $(CH_2)_2$–(d) CH_2–SH (b)

	Shift	Peak
a	0.92	3A
b	1.23	S
▶c	1.53	S
d	2.52	4B

▶ $C_4H_{10}S$ ▶ 46.0 mg/0.50 ml $CDCl_3$ ▶ S 274

▶ 2-BaBv

C_6H_5–(d) CH_2–(c) CH_2–(b) CH_2–(a) CH_3

(e) = all ring H's

	Shift	Peak
a	.93	3C
b	1.47	S
▶c	1.47	S
d	2.57	3B
e	7.11	1A

▶ $C_{10}H_{14}$ ▶ 10% CCl_4 ▶ API 42

▶ 2-BbBb

(a) CH_3–CH_2–(b) $(CH_2)_{11}$–(c) CH_2–OH

	Shift	Peak
a	0.86	3B
▶b	1.21	1A
c	~3.55	3B

▶ $C_{14}H_{30}O$ ▶ 44.9 mg/0.25 ml $CDCl_3$ ▶ S 294

▶ 2-BbBb

C_6H_5–(c) CH_2–(b) $(CH_2)_{10}$–(a) CH_3

(d) = all ring H's

	Shift	Peak
a	.88	3B
▶b	1.22	1A
c	2.54	3B
d	7.09	1G

▶ $C_{18}H_{30}$ ▶ 50% CCl_4 ▶ API 360

▶ 2-BbBb

(a) CH_3–(b) CH_2–(b) $(CH_2)_4$–(b) CH_2–(a) CH_3

	Shift	Peak
a	.84	3B
▶b	1.24	1B

▶ C_8H_{18} ▶ 10% CCl_4 ▶ API 200

▶ 2-BbBb

H(d)(H(e))C=C(H(f))–(c) CH_2(b) $(CH_2)_{12}$(a) CH_3

	Shift	Peak
a	.90	3B
▶b	1.24	1A
c	2.04	3B
d	4.88	R
e	5.02	R
f	5.72	R

▶ $C_{16}H_{32}$ ▶ 50 vol.% CCl_4 ▶ API 437

▶ 2-BbBb

(a) CH_3(b) $(CH_2)_{17}$(c) CH_2C(=O)–OH

	Shift	Peak
a	0.87	3B
▶b	1.24	1A
c	~2.27	2B

▶ $C_{20}H_{40}O_2$ ▶ 45.2 mg/0.50 ml $CDCl_3$ ▶ S 260

▶ 2-BbBb

C_6H_5–(c) CH_2–(b) $(CH_2)_7$–(a) CH_3

(d) = all ring H's

	Shift	Peak
a	.89	3B
▶b	1.26	1B
c	2.46	3B
d	7.07	1A

▶ $C_{15}H_{24}$ ▶ 50% CCl_4 ▶ API 358

▶ 2-BbBb

C_6H_5–(c) CH_2–(b) $(CH_2)_8$–(a) CH_3

(d) = all ring H's

	Shift	Peak
a	.88	3B
▶b	1.25	1B
c	2.50	3B
d	7.07	1A

▶ $C_{16}H_{26}$ ▶ 50% CCl_4 ▶ API 359

▶ 2-BbBb

	Shift	Peak
a	.99	3B
▶ b	1.25	1A
c	2.02	3B
d	4.90	R
e	5.08	R
f	5.68	R

$H^{(e)}H^{(d)}C{=}CH^{(f)}\text{—}\overset{(c)}{CH_2}\overset{(b)}{(CH_2)_{12}}\overset{(a)}{CH_3}$

▶ $C_{16}H_{32}$ ▶ 50% CCl_4 ▶ API 73

▶ 2-BbBb

	Shift	Peak
a	.88	3B
▶ b	1.27	1A
c	2.3	3B
d	4.88	7X
e	5.01	7X
f	5.72	R

$H^{(d)}H^{(e)}C{=}CH^{(f)}\text{—}\overset{(c)}{CH_2}\overset{(b)}{(CH_2)_{10}}\overset{(a)}{CH_3}$

▶ $C_{14}H_{28}$ ▶ 50 vol.% CCl_4 ▶ API 436

▶ 2-BbBb

	Shift	Peak
a	0.88	3B
▶ b	1.26	1A
c	~2.35	1D
d	10.95	1B

$\overset{(a)}{CH_3}\text{—}\overset{(b)}{(CH_2)_{11}}\text{—}\overset{(c)}{CH_2}\text{—}C(=O)\text{—}\overset{(d)}{OH}$

▶ $C_{14}H_{28}O_2$ ▶ 44.8 mg/0.50 ml $CDCl_3$ ▶ S 167

▶ 2-BbBb

	Shift	Peak
a	0.90	3B
▶ b	1.28	1B
c	1.58	S
d	2.70	1C

$\overset{(a)}{CH_3}\text{—}\overset{(b)}{(CH_2)_4}\text{—}\overset{(d)}{CH_2}\text{—}\overset{(c)}{NH_2}$

▶ $C_6H_{15}N$ ▶ 35.6 mg/0.50 ml $CDCl_3$ ▶ S 46

▶ 2-BbBb

	Shift	Peak
a	.90	3B
▶ b	1.26	1A

$CH_3\text{—}\overset{(b)}{(CH_2)_{14}}\text{—}\overset{(a)}{CH_3}$

▶ $C_{16}H_{34}$ ▶ 10% CCl_4 ▶ API 48

▶ 2-BbBb

	Shift	Peak
a	0.91	3B
▶ b	1.28	1B
c	1.71	2A
d	4.12	5B

$\overset{(a)}{CH_3}\text{—}\overset{(b)}{(CH_2)_5}\text{—}\underset{(d)}{CH}(Br)\text{—}\overset{(c)}{CH_3}$

▶ $C_8H_{17}Br$ ▶ 45.2 mg/0.50 ml $CDCl_3$ ▶ S 227

▶ 2-BbBb

	Shift	Peak
a	~0.88	1D
▶ b	1.26	1A
c	~2.30	1D
d	3.63	1A

$\overset{(a)}{CH_3}\text{—}\overset{(b)}{(CH_2)_{13}}\text{—}\overset{(c)}{CH_2}\text{—}C(=O)\text{—}O\text{—}\overset{(d)}{CH_3}$

▶ $C_{17}H_{34}O_2$ ▶ 44.4 mg/0.50 ml $CDCl_3$ ▶ S 248

▶ 2-BbBb

	Shift	Peak
a	.85	3B
▶ b	1.28	1B
c	1.99	3B
d	4.88	R
e	4.99	R
f	5.68	R

$H^{(d)}H^{(e)}C{=}CH^{(f)}\text{—}\overset{(c)}{CH_2}\overset{(b)}{(CH_2)_5}\overset{(a)}{CH_3}$

▶ C_9H_{18} ▶ 50% CCl_4 ▶ API 434

▶ 2-BbBb

	Shift	Peak
a	0.88	3B
b	1.27 ± 0.03	1A
▶ c	1.27 ± 0.03	1A

$\overset{(a)}{CH_3}\text{—}\overset{(b)}{CH_2}\text{—}\overset{(c)}{(CH_2)_4}\text{—}\overset{(b)}{CH_2}\text{—}\overset{(a)}{CH_3}$

▶ C_8H_{18} ▶ 7% $CDCl_3$ ▶ V 216

▶ 2-BbBb

	Shift	Peak
a	0.88	3B
▶ b	1.28	1B
c	2.17	1A
d	2.46	3B

$\overset{(c)}{CH_3}\text{—}C(=O)\text{—}\overset{(d)}{CH_2}\text{—}\overset{(b)}{(CH_2)_5}\text{—}\overset{(a)}{CH_3}$

▶ $C_9H_{18}O$ ▶ 33.7 mg/0.50 ml $CDCl_3$ ▶ S 48

▶ 2-BbBb

	Shift	Peak
a	.88	3B
▶ b	1.27	1B
c	3.45	3B
d	4.28	1A

$\overset{(a)}{CH_3}\text{—}\overset{(b)}{(CH_2)_8}\text{—}\overset{(c)}{CH_2}\text{—}\overset{(d)}{OH}$

▶ $C_{10}H_{22}O$ ▶ 20% CCl_4 ▶ API 172

▶ 2-BbBb

	Shift	Peak
a	.88	3B
▶ b	1.28	1B

$\overset{(a)}{CH_3}\text{—}\overset{(b)}{(CH_2)_7}\text{—}CH_3$

▶ C_9H_{20} ▶ 50% CCl_4 ▶ API 343

▶ 2-BbBb

	Shift	Peak
a	0.88	3B
▶b	1.28	1A

(a) (b) (a)
$CH_3-(CH_2)_7-CH_3$

▶ C_9H_{20} ▶ 45.0 mg/0.50 ml $CDCl_3$ ▶ S 203

▶ 2-BbBb

	Shift	Peak
a	0.90	3B
▶b	1.28	1G

(a) (b) (a)
$CH_3-(CH_2)_9-CH_3$

▶ $C_{11}H_{24}$ ▶ 44.4 mg/0.50 ml $CDCl_3$ ▶ S 289

▶ 2-BbBb

	Shift	Peak
a	.88	3B
▶b	1.28	1B
c	3.44	3B
d	4.27	1A

(a) (b) (c) (d)
$CH_3-(CH_2)_7-CH_2-OH$

▶ $C_9H_{20}O$ ▶ 20% CCl_4 ▶ API 171

▶ 2-BbBb

	Shift	Peak
a	.88	3B
▶b	1.28	1A

(a) (b)
$CH_3-(CH_2)_{10}-CH_3$

▶ $C_{12}H_{26}$ ▶ 50% CCl_4 ▶ API 350

▶ 2-BbBb

	Shift	Peak
a	0.88	3B
▶b	1.28	1B
c	2.30	3B
d	3.68	1A

(a) (b) (c) (d)
$CH_3-(CH_2)_6-CH_2C(=O)-OCH_3$

▶ $C_{10}H_{20}O_2$ ▶ 45.1 mg/0.50 ml $CDCl_3$ ▶ S 249

▶ 2-BbBb

	Shift	Peak
a	0.88	3B
▶b	1.28	1B
c	2.28	3B
d	3.67	1A

(a) (b) (c) (d)
$CH_3-(CH_2)_9-CH_2-C(=O)-OCH_3$

▶ $C_{13}H_{26}O_2$ ▶ 45.7 mg/0.50 ml $CDCl_3$ ▶ S 125

▶ 2-BbBb

	Shift	Peak
a	0.88	3B
▶b	1.28	1E

(a) (b) (a)
$CH_3-(CH_2)_8-CH_3$

▶ $C_{10}H_{22}$ ▶ 44.8 mg/0.50 ml $CDCl_3$ ▶ S 201

▶ 2-BbBb

	Shift	Peak
a	0.88	3B
▶b	1.28	1E
c	2.28	3C
d	4.11	4A

(a) (b) (c) (d) (a)
$CH_3-(CH_2)_9-CH_2-C(=O)-OCH_2CH_3$

▶ $C_{14}H_{28}O_2$ ▶ 45.5 mg/0.50 ml $CDCl_3$ ▶ S 194

▶ 2-BbBb

	Shift	Peak
a	0.91	3B
▶b	1.28	1B
c	2.03	1A
d	3.63	3A

(a) (b) (d) (c)
$CH_3-(CH_2)_8-CH_2-OH$

▶ $C_{10}H_{22}O$ ▶ 7% $CDCl_3$ ▶ V 282

▶ 2-BbBb

	Shift	Peak
a	0.88	3B
▶b	1.28	1A

(a) (b) (a)
$CH_3-(CH_2)_{12}-CH_3$

▶ $C_{14}H_{30}$ ▶ 44.8 mg/0.50 ml $CDCl_3$ ▶ S 204

▶ 2-BbBb

	Shift	Peak
a	.88	3B
▶b	1.28	1A

(b) (a)
$CH_3-(CH_2)_9-CH_3$

▶ $C_{11}H_{24}$ ▶ 50% CCl_4 ▶ API 348

▶ 2-BbBb

	Shift	Peak
a	0.90	3B
▶b	1.29	1E
c	2.09	1A
d	2.53	4B

(a) (b) (d) (c)
$CH_3-(CH_2)_5-CH_2SH$

▶ $C_7H_{16}S$ ▶ 46.9 mg/0.50 ml $CDCl_3$ ▶ S 297

▶ 2-BbBb

	Shift	Peak
a	0.88	3B
▶ b	1.29	1B
c	2.30	1B
d	9.35	1A

$CH_3-(CH_2)_6-CH_2-CH(=O)$ — (a) CH_3, (b) $(CH_2)_6$, (c) CH_2, (d) CH

▶ $C_9H_{18}O$ ▶ 45.3 mg/0.50 ml $CDCl_3$ ▶ S 100

▶ 2-BbBb

	Shift	Peak
a	0.90	3B
▶ b	1.30	1B
c	1.85	1F
d	3.41	3A

$CH_3-(CH_2)_4-CH_2-CH_2-Br$ — (a) CH_3, (b) $(CH_2)_4$, (c) CH_2, (d) CH_2

▶ $C_7H_{15}Br$ ▶ 45.0 mg/0.50 ml $CDCl_3$ ▶ S 208

▶ 2-BbBb

	Shift	Peak
a	0.87	3A
▶ b	1.29	1B
c	2.42	3B
d	9.77	3A

$CH_3-(CH_2)_9-CH_2-CH(=O)$ — (a) CH_3, (b) $(CH_2)_9$, (c) CH_2, (d) CH

▶ $C_{12}H_{24}O$ ▶ 45.2 mg/0.50 ml $CDCl_3$ ▶ S 106

▶ 2-BbBb

	Shift	Peak
a	.88	3B
▶ b	1.30	1B

$CH_3-(CH_2)_5-CH_3$ — (a) CH_3, (b) $(CH_2)_5$

▶ C_7H_{16} ▶ 50% CCl_4 ▶ API 337

▶ 2-BbBb

	Shift	Peak
a	0.85	3B
▶ b	1.29	1A
c	1.83	1B
d	3.62	3B

$CH_3-(CH_2)_{10}-CH_2OH$ — (a) CH_3, (b) $(CH_2)_{10}$, (d) CH_2, (c) OH

▶ $C_{12}H_{26}O$ ▶ 44.4 mg/0.50 ml $CDCl_3$ ▶ S 103

▶ 2-BbBb

	Shift	Peak
a	0.88	3B
▶ b	1.30	1B
c	2.43	3B
d	9.73	3A

$CH_3-(CH_2)_5-CH_2-CH(=O)$ — (a) CH_3, (b) $(CH_2)_5$, (c) CH_2, (d) CH

▶ $C_8H_{16}O$ ▶ 44.6 mg/0.50 ml $CDCl_3$ ▶ S 105

▶ 2-BbBb

	Shift	Peak
a	0.88	3B
▶ b	1.29	1A
c	2.24	3B
d	3.68	1A

$CH_3-(CH_2)_{11}-CH_2-C(=O)-OCH_3$ — (a) CH_3, (b) $(CH_2)_{11}$, (c) CH_2, (d) OCH_3

▶ $C_{15}H_{30}O_2$ ▶ 45.1 mg/0.50 ml $CDCl_3$ ▶ S 195

▶ 2-BbBb

	Shift	Peak
a	0.88	3B
▶ b	1.30	1G
c	2.36	3B
d	11.47	1A

$CH_3(CH_2)_5CH_2-C(=O)-OH$ — (a) CH_3, (b) $(CH_2)_5$, (c) CH_2, (d) OH

▶ $C_8H_{16}O_2$ ▶ 45.6 mg/0.50 ml $CDCl_3$ ▶ S 157

▶ 2-BbBb

	Shift	Peak
a	0.88	3B
▶ b	1.29	1B
c	2.00	1F
d	2.25	3B
e	5.28	3D
f	11.45	1C

$CH-CH_2-(CH_2)_5-CH_2-C(=O)-OH$ (e)(c)(b)(d)(f), double bond to $CH-CH_2-(CH_2)_6-CH_3$ (e)(c)(b)(a)

▶ $C_{18}H_{34}O_2$ ▶ 45.2 mg/0.50 ml $CDCl_3$ ▶ S 70

▶ 2-BbBb

	Shift	Peak
a	.88	3B
b	1.12	2A
▶ c	1.30	1B
d	3.70	4B
e	4.08	1A

$CH_3-CH(OH)-CH_2-(CH_2)_4-CH_3$ — (b) CH_3, (d) CH, OH (e), (c) CH_2, (c) $(CH_2)_4$, (a) CH_3

▶ $C_8H_{18}O$ ▶ 20% CCl_4 ▶ API 170

▶ 2-BbBb

	Shift	Peak
a	0.90	3B
▶ b	1.29	1B
c	2.00	1C
d	2.32	S
e	3.70	1A
f	5.4	3B

$CH-CH_2-(CH_2)_5-CH_2-C(=O)-OCH_3$ (f)(c)(b)(d)(e), double bond to $CH-CH_2-(CH_2)_6-CH_3$ (f)(c)(b)(a)

▶ $C_{19}H_{36}O_2$ ▶ 33.7 mg/0.50 ml $CDCl_3$ ▶ S 71

▶ 2-BbBb

	Shift	Peak
a	0.88	3B
b	1.28	3D
▶ c	1.30	S
d	2.28	3B
e	4.12	4A

$CH_3-(CH_2)_5-CH_2-C(=O)-OCH_2CH_3$ — (a) CH_3, (c) $(CH_2)_5$, (d) CH_2, (e) OCH_2, (b) CH_3

▶ $C_{10}H_{20}O_2$ ▶ 45.2 mg/0.50 ml $CDCl_3$ ▶ S 244

▶ 2-BbBb

	Shift	Peak
a	.83	2H
b	.94	S
▶c	1.30	1A

(a) CH_3 – CH_3—CH—$(CH_2)_7$—CH_3 (c) (b)

▶ $C_{11}H_{24}$ ▶ 50% CCl_4 ▶ API 349

▶ 2-BbBb

	Shift	Peak
a	0.88	3B
▶b	1.32	1B
c	2.38	3B
d	11.47	1B

(a) (b) (c) (d) CH_3—$(CH_2)_4$—CH_2—C(=O)—OH

▶ $C_7H_{14}O_2$ ▶ 45.6 mg/0.50 ml $CDCl_3$ ▶ S 49

▶ 2-BbBb

	Shift	Peak
a	.87	3B
▶b	1.30	1A

(a) (b) (a) CH_3—$(CH_2)_{13}$—CH_3

▶ $C_{15}H_{32}$ ▶ 50% CCl_4 ▶ API 221

▶ 2-BbBb

	Shift	Peak
a	0.88	3B
▶b	1.32	2E
c	1.82	3C
d	3.18	3A

(a) (b) (c) (d) CH_3—$(CH_2)_4$—CH_2—CH_2—I

▶ $C_7H_{15}I$ ▶ 45.0 mg/0.50 ml $CDCl_3$ ▶ S 210

▶ (2)-(B)b(B)b

	Shift	Peak
▶a	1.31	4E
b	1.78	3B
c	2.20	1E
d	~3.61	1C

(a) (a) (a) (b) (b) H(d) OH(c)

▶ $C_6H_{12}O$ ▶ 45.2 mg/0.50 ml $CDCl_3$ ▶ S 7

▶ 2-BbBb

	Shift	Peak
a	0.88	3B
▶b	1.32	1E
c	4.16	3B
d	8.03	1B

(d) HC(=O)—OCH_2—$(CH_2)_6$—CH_3 (c) (b) (a)

▶ $C_9H_{18}O_2$ ▶ 44.4 mg/0.50 ml $CDCl_3$ ▶ S 114

▶ 2-BbBb

	Shift	Peak
a	0.88	3B
▶b	1.31	1E
c	3.53	3A

(a) (b) (c) CH_3—$(CH_2)_5$—CH_2Cl

▶ $C_7H_{15}Cl$ ▶ 44.8 mg/0.50 ml $CDCl_3$ ▶ S 209

▶ 2-BbBb

	Shift	Peak
a	0.90	3B
▶b	1.32	1B
c	1.97	1A
d	3.62	3B

(a) (b) (d) (c) CH_3—$(CH_2)_7$—CH_2OH

▶ $C_9H_{20}O$ ▶ 44.6 mg/0.50 ml $CDCl_3$ ▶ S 102

▶ (2)-(B)b(B)b

	Shift	Peak
a	.88	1A
▶b	~1.31	O

(a) CH_3 (a) CH_3 C (b) H_2C CH_2(b) (b) H_2C CH_2(b) C H_2 (b)

▶ C_8H_{16} ▶ 10% CCl_4 ▶ API 150

▶ 2-BbBb

	Shift	Peak
a	.92	3B
▶b	1.32	1B
c	2.05	3B
d	4.86	R
e	5.00	R
f	5.68	R

(d) H (f) H C=C H (e) (c) (b) (a) $CH_2(CH_2)_8CH_3$

▶ $C_{12}H_{24}$ ▶ 50 vol.% CCl_4 ▶ API 435

▶ 2-BbBb

	Shift	Peak
a	0.90	3B
▶b	1.32	2E
c	1.84	3C
d	3.20	3A

(a) (b) (c) (d) CH_3—$(CH_2)_3$—CH_2—CH_2—I

▶ $C_6H_{13}I$ ▶ 45.8 mg/0.50 ml $CDCl_3$ ▶ S 212

▶ 2-BbBb

	Shift	Peak
a	0.90	3B
▶b	1.33	2H
c	1.78	O
d	3.53	3A

(a) (b) (c) (d) CH_3—$(CH_2)_3$—CH_2—CH_2—Cl

▶ $C_6H_{13}Cl$ ▶ 44.6 mg/0.50 ml $CDCl_3$ ▶ S 211

▶ 2-BbBb

Shift	Peak
a 0.88	3B
▶b 1.33	1B
c 1.95	1A
d 3.62	3B

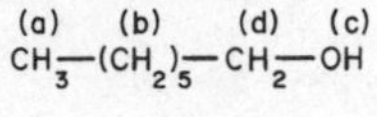

(a) (b) (d) (c)
$CH_3-(CH_2)_5-CH_2-OH$

▶ $C_7H_{16}O$ ▶ 45.0 mg/0.50 ml $CDCl_3$ ▶ S 215

▶ 2-BbBb

Shift	Peak
a 0.92	3B
▶b 1.37	1E
c 2.05	1A
d 4.05	3A

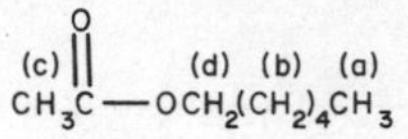

(c) (d) (b) (a)
$CH_3C(=O)-OCH_2(CH_2)_4CH_3$

▶ $C_8H_{16}O_2$ ▶ 45.0 mg/0.50 ml $CDCl_3$ ▶ S 217

▶ 2-BbBb

Shift	Peak
a 0.88	3B
b 1.26	S
▶c 1.33	S
d 2.28	3B
e 4.12	4A

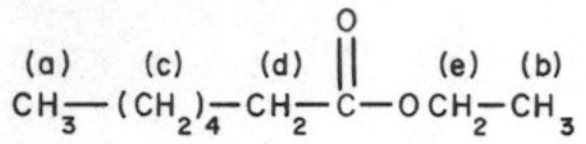

(a) (c) (d) (e) (b)
$CH_3-(CH_2)_4-CH_2-C(=O)-OCH_2-CH_3$

▶ $C_9H_{18}O_2$ ▶ 44.5 mg/0.50 ml $CDCl_3$ ▶ S 283

▶ (2)-(B)b(B)b

Shift	Peak
▶a ~1.4	S
b ~1.8	S
c 7.22	1A

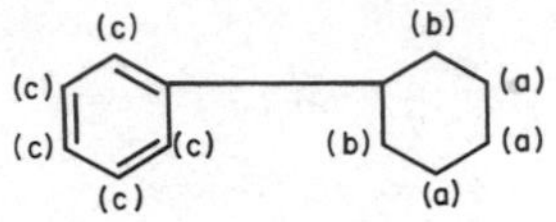

(c) (c) (c) (c) (c) (b) (a) (b) (a) (a)

▶ $C_{12}H_{16}$ ▶ 45.4 mg/0.50 ml $CDCl_3$ ▶ S 291

▶ 2-BbBb

Shift	Peak
a 0.97	3A
▶b 1.33	1B
c 2.80	3B
d 3.67	1A
e ~5.38	3C

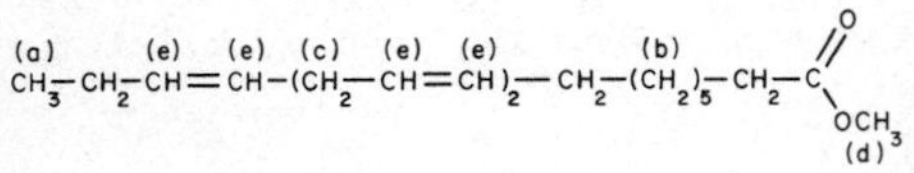

(a) (e) (e) (c) (e) (e) (b) (d)
$CH_3-CH_2-CH=CH-(CH_2-CH=CH)_2-CH_2-(CH_2)_5-CH_2-C(=O)OCH_3$

▶ $C_{19}H_{32}O_2$ ▶ 7% $CDCl_3$ ▶ V 337

▶ (2)-(B)b(B)b

Shift	Peak
a 1.43	1A

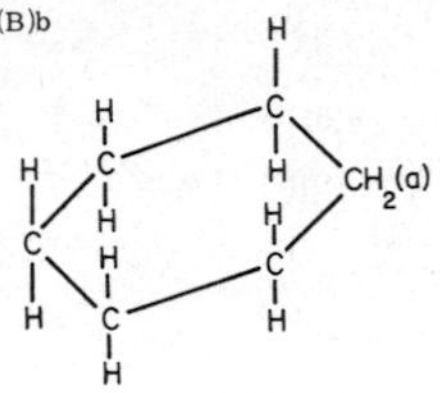

$CH_2(a)$

▶ C_6H_{12} ▶ liquid ▶ API 451

▶ 2-BbBb

Shift	Peak
a 1.26	3D
▶b 1.35	1E
c 2.28	3B
d 4.13	4A

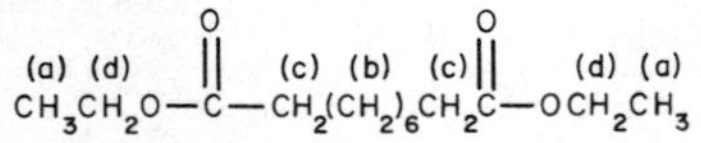

(a) (d) (c) (b) (c) (d) (a)
$CH_3CH_2O-C(=O)-CH_2(CH_2)_6CH_2C(=O)-OCH_2CH_3$

▶ $C_{14}H_{26}O_4$ ▶ 45.1 mg/0.50 ml $CDCl_3$ ▶ S 250

▶ (2)-(B)b(B)b

Shift	Peak
a 1.44	1A

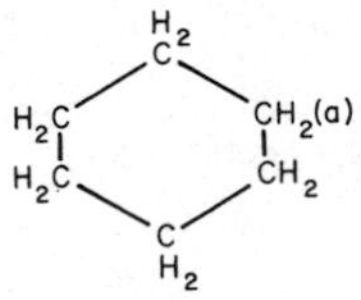

H_2C $CH_2(a)$ H_2C CH_2 H_2C CH_2

▶ C_6H_{12} ▶ 10% CCl_4 ▶ API 147

▶ 2-BbBb

Shift	Peak
a 2.28	3B
b 2.88	3B
c 3.67	1A
d 7.12	R
e 7.62	R
f 7.70	R
▶g 1.35	1F

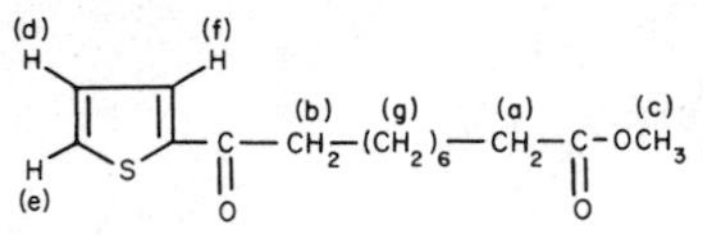

(d) (f) (e) (b) (g) (a) (c)
$C(=O)-CH_2-(CH_2)_6-CH_2-C(=O)-OCH_3$

▶ $C_{15}H_{22}O_3S$ ▶ 7% $CDCl_3$ ▶ V 320

▶ (2)-(B)b(B)b

Shift	Peak
a 1.32	1E
b 1.44	1E
▶c 1.44	1E

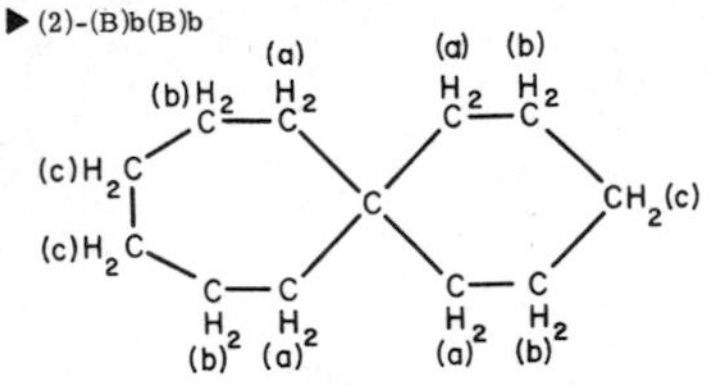

(a) (b) (a) (b) (c) (c) $CH_2(c)$ (b) (a) (a) (b)

▶ $C_{12}H_{22}$ ▶ 33 vol.% CCl_4 ▶ API 140

▶ 2-BbBb

Shift	Peak
a 0.88	3B
▶b 1.36	1B
c 2.28	1A
d 3.62	3A

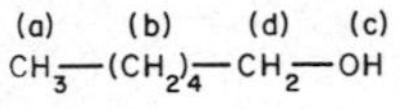

(a) (b) (d) (c)
$CH_3-(CH_2)_4-CH_2-OH$

▶ $C_6H_{14}O$ ▶ 45.2 mg/0.50 ml $CDCl_3$ ▶ S 198

▶ (2)-(B)b(B)b

Shift	Peak
▶a 1.51	N
b 1.82	O
c ~4.00	1F

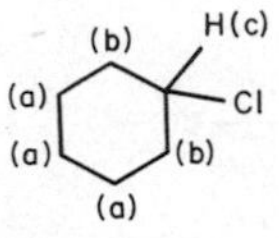

(b) H(c) (a) Cl (a) (b) (a)

▶ $C_6H_{11}Cl$ ▶ 45.0 mg/0.50 ml $CDCl_3$ ▶ S 25

▶ (2)-(B)b(B)b

	Shift	Peak
▶a	~1.62	1B
b	~1.62	1B
c	~3.57	1C
d	5.93	1A
e	6.43	2A
f	~7.40	S

▶ $C_{17}H_{19}NO_3$ ▶ 7% $CDCl_3$ ▶ V 328

▶ 2-BbBc

CH_3(a)—$(CH_2)_5$(b)—CH(d)(Br)—CH_3(c)

	Shift	Peak
a	0.91	3B
▶b	1.28	1B
c	1.71	2A
d	4.12	5B

▶ $C_8H_{17}Br$ ▶ 45.2 mg/0.50 ml $CDCl_3$ ▶ S 227

▶ 2-BbBb

C_6H_5—CH_2(b)—$(CH_2)_4$(c)—CH_3(a)

(d) = all ring H's

	Shift	Peak
a	.89	3B
b	1.30	1F
▶c	2.54	3B
d	7.10	1A

▶ $C_{12}H_{18}$ ▶ 50% CCl_4 ▶ API 356

▶ 2-BbBc

CH_3(b)—CH(d)(OH(e))—CH_2(c)—$(CH_2)_4$(c)—CH_3(a)

	Shift	Peak
a	.88	3B
b	1.12	2A
▶c	1.30	1B
d	3.70	4B
e	4.08	1A

▶ $C_8H_{18}O$ ▶ 20% CCl_4 ▶ API 170

▶ 2-BbBb

C_6H_5—CH_2(b)—$(CH_2)_5$(c)—CH_3(a)

(d) = all ring H's

	Shift	Peak
a	.86	3B
b	1.28	1B
▶c	2.54	3B
d	7.09	1A

▶ $C_{13}H_{20}$ ▶ 50% CCl_4 ▶ API 357

▶ 2-BbBc

CH_3—CH(CH_3(a))—$(CH_2)_7$(c)—CH_3(b)

	Shift	Peak
a	.83	2H
b	.94	S
▶c	1.30	1A

▶ $C_{11}H_{24}$ ▶ 50% CCl_4 ▶ API 349

▶ 2-BbBc

	Shift	Peak
a	0.85	2A
b	1.22 ± 0.03	1E
▶c	1.22 ± 0.03	1E
d	2.12 ± 0.03	1B
e	2.62	3C
f	3.96	1D

▶ $C_{29}H_{50}O_2$ ▶ 7% $CDCl_3$ ▶ V 366

▶ (2)-(B)b(B)c

	Shift	Peak
▶a	1.31	4E
b	1.78	3B
c	2.20	1E
d	~3.61	1C

▶ $C_6H_{12}O$ ▶ 45.2 mg/0.50 ml $CDCl_3$ ▶ S 7

▶ 2-BbBc

CH_3—CH(CH_3)—$(CH_2)_4$(b)—CH(c)(CH_3(a))—CH_3

	Shift	Peak
a	.88	2A
▶b	1.25	1B
c	~1.47	S

▶ $C_{10}H_{22}$ ▶ 50% CCl_4 ▶ API 346

▶ 2-BbBc

CH_3(a)—CH(c)(CH_3)—$(CH_2)_3$(b)—CH_2(d)—OH(e)

	Shift	Peak
a	.88	2A
▶b	1.31	S
c	~1.63	S
d	3.48	3B
e	4.42	1A

▶ $C_7H_{16}O$ ▶ 20% CCl_4 ▶ API 169

▶ 2-BbBc

CH_3—CH(CH_3(a))—$(CH_2)_4$(c)—CH_3(b)

	Shift	Peak
a	.87	2H
b	.96	S
▶c	1.27	1B

▶ C_9H_{18} ▶ 50% CCl_4 ▶ API 340

▶ (2)-(B)b(B)c

	Shift	Peak
▶a	~1.4	S
b	~1.8	S
c	7.22	1A

▶ $C_{12}H_{16}$ ▶ 45.4 mg/0.50 ml $CDCl_3$ ▶ S 291

▶ (2)-(B)b(B)c

CH$_3$(a) (b) H C CH$_3$(a) (d)H C H C H H C H(d) H C H (c)H C H (b) H(c)

	Shift	Peak
a	.83	2A
b	1.40	S
▶c	1.40	S
d	~1.67	S

▶ C_8H_{16} ▶ liquid ▶ API 453

▶ (2)-(B)b(B)d

(a) (a) CH$_3$ CH$_3$ C (b) H$_2$C CH$_2$(b) (b) H$_2$C CH$_2$(b) C H$_2$ (b)

	Shift	Peak
a	.88	1A
▶b	~1.31	O

▶ C_8H_{16} ▶ 10% CCl_4 ▶ API 150

▶ (2)-(B)b(B)c

(a) (a) H$_2$ H$_2$ C—C (a)H$_2$C H H CH$_2$(a) C—C (b) (b) (a)H$_2$C CH$_2$(a) C—C H$_2$ H$_2$ (a) (a)

	Shift	Peak
▶a	1.43	1B
b	1.63	1E

▶ $C_{10}H_{18}$ ▶ liquid run ▶ API 14

▶ (2)-(B)b(B)d

(a) CH$_3$ CH$_3$ C H$_2$C CH$_2$(c) H$_2$C —— CH$_2$(b)

	Shift	Peak
a	.97	1G
▶b	1.32	Q
c	1.55	Q

▶ C_7H_{14} ▶ 10% CCl_4 ▶ API 146

▶ (2)-(B)b(B)c

(b) H(c) (a) Cl (a) (b) (a)

	Shift	Peak
▶a	1.51	N
b	1.82	O
c	~4.00	1F

▶ $C_6H_{11}Cl$ ▶ 45.0 mg/0.50 ml $CDCl_3$ ▶ S 25

▶ (2)-(B)b(B)d

(a) (a) (b) (b)H$_2$ H$_2$ H$_2$ H$_2$ C—C C—C (c)H$_2$C CH$_2$(c) C (c)H$_2$C C—C C—C H$_2$ H$_2$ H$_2$ H$_2$ (b) (a) (a) (b)

	Shift	Peak
a	1.32	1E
▶b	1.44	1E
c	1.44	1E

▶ $C_{12}H_{22}$ ▶ 33 vol.% CCl_4 ▶ API 140

▶ (2)-(B)b(B)c

(b) (b) H$_2$ H$_2$ C (c) (c) C (a)H$_2$C CH—HC CH$_2$(a) C—C C—C H$_2$ H$_2$ H$_2$ H$_2$ (a) (b) (b) (a)

	Shift	Peak
▶a	~1.53	S
b	~1.53	S
c	~1.56	S

▶ $C_{10}H_{18}$ ▶ liquid ▶ API 13

▶ 2-BbBe

(a) (b) (c)
$CH_3—(CH_2)_5—CH_2Cl$

	Shift	Peak
a	0.88	3B
▶b	1.31	1E
c	3.53	3A

▶ $C_7H_{15}Cl$ ▶ 44.8 mg/0.50 ml $CDCl_3$ ▶ S 209

▶ 2-BbBd

(b)
$CH_3—C(CH_3)_2—(CH_2)_4—C(CH_3)_2—CH_3$
(a)

	Shift	Peak
a	.87	1A
▶b	1.18	1A

▶ $C_{12}H_{26}$ ▶ 50% CCl_4 ▶ API 351

▶ 2-BbBe

(a) (b) (c) (d)
$CH_3—(CH_2)_3—CH_2—CH_2—Cl$

	Shift	Peak
a	0.90	3B
b	1.33	2H
▶c	1.78	O
d	3.53	3A

▶ $C_6H_{13}Cl$ ▶ 44.6 mg/0.50 ml $CDCl_3$ ▶ S 211

▶ 2-BbBd

(c) (b)
$CH_3—C(CH_3)_2—CH_2—(CH_2)_3—CH_3$
(a)

	Shift	Peak
a	.87	1E
b	.87	1E
▶c	1.28	1B

▶ C_9H_{20} ▶ 50% CCl_4 ▶ API 209

▶ 2-BbBg

(b)
(a) (c) (d)
$CH_3CH_2CH_2CH_2CH_2Br$

	Shift	Peak
a	0.94	3B
b	1.38	O
▶c	1.83	4C
d	3.42	3A

▶ $C_5H_{11}Br$ ▶ 44.5 mg/0.50 ml $CDCl_3$ ▶ S 120

▶ 2-BbBg

(a) (b) (c) (d)
$CH_3—(CH_2)_4—CH_2—CH_2—Br$

	Shift	Peak
a	0.90	3B
b	1.30	1B
▶c	1.85	1F
d	3.41	3A

▶ $C_7H_{15}Br$ ▶ 45.0 mg/0.50 ml $CDCl_3$ ▶ S 208

▶ 2-BbBj

(a) (b) (c) (d)
$CH_3—(CH_2)_6—CH_2—CH(=O)$

	Shift	Peak
a	0.88	3B
▶b	1.29	1B
c	2.30	1B
d	9.35	1A

▶ $C_9H_{18}O$ ▶ 45.3 mg/0.50 ml $CDCl_3$ ▶ S 100

▶ 2-BbBi

(a) (b) (c) (d)
$CH_3—(CH_2)_4—CH_2—CH_2—I$

	Shift	Peak
a	0.88	3B
b	1.32	2E
▶c	1.82	3C
d	3.18	3A

▶ $C_7H_{15}I$ ▶ 45.0 mg/0.50 ml $CDCl_3$ ▶ S 210

▶ 2-BbBj

(a) (b) (c) (d)
$CH_3—(CH_2)_9—CH_2—CH(=O)$

	Shift	Peak
a	0.87	3A
▶b	1.29	1B
c	2.42	3B
d	9.77	3A

▶ $C_{12}H_{24}O$ ▶ 45.2 mg/0.50 ml $CDCl_3$ ▶ S 106

▶ 2-BbBi

(a) (b) (c) (d)
$CH_3—(CH_2)_3—CH_2—CH_2—I$

	Shift	Peak
a	0.90	3B
b	1.32	2E
▶c	1.84	3C
d	3.20	3A

▶ $C_6H_{13}I$ ▶ 45.8 mg/0.50 ml $CDCl_3$ ▶ S 212

▶ 2-BbBj

(a) (b) (c) (d)
$CH_3—(CH_2)_5—CH_2—CH(=O)$

	Shift	Peak
a	0.88	3B
▶b	1.30	1B
c	2.43	3B
d	9.73	3A

▶ $C_8H_{16}O$ ▶ 44.6 mg/0.50 ml $CDCl_3$ ▶ S 105

▶ 2-BbBi

(a) (b) (c) (d)
$CH_3—(CH_2)_2—CH_2—CH_2I$

	Shift	Peak
a	0.92	3B
b	1.38	2E
▶c	1.86	3C
d	3.20	3A

▶ $C_5H_{11}I$ ▶ 45.0 mg/0.50 ml $CDCl_3$ ▶ S 228

▶ 2-BbBj

(c) (d) (b) (a)
$CH_3—C(=O)—CH_2—(CH_2)_3—CH_3$

	Shift	Peak
a	0.90	3B
▶b	1.38	O
c	2.15	1A
d	2.42	3B

▶ $C_7H_{14}O$ ▶ 35.4 mg/0.50 ml $CDCl_3$ ▶ S 41

▶ 2-BbBj

(c) (d) (b) (a)
$CH_3—C(=O)—CH_2—(CH_2)_5—CH_3$

	Shift	Peak
a	0.88	3B
▶b	1.28	1D
c	2.17	1A
d	2.46	3B

▶ $C_9H_{18}O$ ▶ 33.7 mg/0.50 ml $CDCl_3$ ▶ S 48

▶ 2-BbBj

(f)(c) (b) (a)
$O=C—CH_2CH_2(CH_2)_2CH_3$; H(j), H(h), (g) H, (e) CH_3, H (i), CH_3 (d), H(k)

	Shift	Peak
a	0.89	3B
b	~1.30	S
▶c	~1.40	S
d	1.72	1A
e	1.97	2A
f	2.94	3B
g	6.06	S
h	6.17	S
i	6.58	S
j	6.89	S
k	7.88	1G

▶ $C_{21}H_{22}O_5$ ▶ 7% $CDCl_3$ ▶ V 687

▶ 2-BbBj

(a) (b) (c) (d)
$CH_3—(CH_2)_6—CH_2—C(=O)—OH$

	Shift	Peak
a	0.88	3B
▶b	1.28	1E
c	2.38	3B
d	10.53	1B

▶ $C_9H_{18}O_2$ ▶ 35.7 mg/0.50 ml $CDCl_3$ ▶ S 9

▶ 2-BbBk

(a) (b) (c)
$CH_3(CH_2)_{17}CH_2C(=O)—OH$

	Shift	Peak
a	0.87	3B
▶b	1.24	1A
c	~2.27	2B

▶ $C_{20}H_{40}O_2$ ▶ 45.2 mg/0.50 ml $CDCl_3$ ▶ S 260

▶ 2-BbBk

$\overset{(a)}{CH_3}$—$\overset{(b)}{(CH_2)_{11}}$—$\overset{(c)}{CH_2}$—C(=O)—$\overset{(d)}{OH}$

	Shift	Peak
a	0.88	3B
▶b	1.26	1A
c	~2.35	1D
d	10.95	1B

▶ $C_{14}H_{28}O_2$ ▶ 44.8 mg/0.50 ml $CDCl_3$ ▶ S 167

▶ 2-BbBk

$\overset{(e)}{CH}$—$\overset{(c)}{CH_2}$—$\overset{(b)}{(CH_2)_5}$—$\overset{(d)}{CH_2}$—C(=O)—$\overset{(f)}{OH}$
||
$\underset{(e)}{CH}$—$\underset{(c)}{CH_2}$—$\underset{(b)}{(CH_2)_6}$—$\underset{(a)}{CH_3}$

	Shift	Peak
a	0.88	3B
▶b	1.29	1B
c	2.00	1F
d	2.25	3B
e	5.28	3D
f	11.45	1C

▶ $C_{18}H_{34}O_2$ ▶ 45.2 mg/0.50 ml $CDCl_3$ ▶ S 70

▶ 2-BbBk

$\overset{(a)}{CH_3}$—$\overset{(b)}{(CH_2)_{13}}$—$\overset{(c)}{CH_2}$—C(=O)—O—$\overset{(d)}{CH_3}$

	Shift	Peak
a	~0.88	1D
▶b	1.26	1A
c	~2.30	1D
d	3.63	1A

▶ $C_{17}H_{34}O_2$ ▶ 44.4 mg/0.50 ml $CDCl_3$ ▶ S 248

▶ 2-BbBk

$\overset{(f)}{CH}$—$\overset{(c)}{CH_2}$—$\overset{(b)}{(CH_2)_5}$—$\overset{(d)}{CH_2}$—C(=O)—$\overset{(e)}{OCH_3}$
||
$\underset{(f)}{CH}$—$\underset{(c)}{CH_2}$—$\underset{(b)}{(CH_2)_6}$—$\underset{(a)}{CH_3}$

	Shift	Peak
a	0.90	3B
▶b	1.29	1B
c	2.00	1C
d	2.32	S
e	3.70	1A
f	5.4	3B

▶ $C_{19}H_{36}O_2$ ▶ 33.7 mg/0.50 ml $CDCl_3$ ▶ S 71

▶ 2-BbBk

$\overset{(a)}{CH_3}$—$\overset{(b)}{(CH_2)_6}$—$\overset{(c)}{CH_2}$C(=O)—$\overset{(d)}{OCH_3}$

	Shift	Peak
a	0.88	3B
▶b	1.28	1B
c	2.30	3B
d	3.68	1A

▶ $C_{10}H_{20}O_2$ ▶ 45.1 mg/0.50 ml $CDCl_3$ ▶ S 249

▶ 2-BbBk

$\overset{(a)}{CH_3}\overset{(b)}{(CH_2)_5}\overset{(c)}{CH_2}$—C(=O)—$\overset{(d)}{OH}$

	Shift	Peak
a	0.88	3B
▶b	1.30	1G
c	2.36	3B
d	11.47	1A

▶ $C_8H_{16}O_2$ ▶ 45.6 mg/0.50 ml $CDCl_3$ ▶ S 157

▶ 2-BbBk

$\overset{(a)}{CH_3}$—$\overset{(b)}{(CH_2)_9}$—$\overset{(c)}{CH_2}$—C(=O)—$\overset{(d)}{OCH_3}$

	Shift	Peak
a	0.88	3B
▶b	1.28	1B
c	2.28	3B
d	3.67	1A

▶ $C_{13}H_{26}O_2$ ▶ 45.7 mg/0.50 ml $CDCl_3$ ▶ S 125

▶ 2-BbBk

$\overset{(a)}{CH_3}$—$\overset{(c)}{(CH_2)_5}$—$\overset{(d)}{CH_2}$—C(=O)—$\overset{(e)}{OCH_2}\overset{(b)}{CH_3}$

	Shift	Peak
a	0.88	3B
b	1.28	3D
▶c	1.30	S
d	2.28	3B
e	4.12	4A

▶ $C_{10}H_{20}O_2$ ▶ 45.2 mg/0.50 ml $CDCl_3$ ▶ S 244

▶ 2-BbBk

$\overset{(a)}{CH_3}$—$\overset{(b)}{(CH_2)_9}$—$\overset{(c)}{CH_2}$—C(=O)—$\overset{(d)}{OCH_2}\overset{(a)}{CH_3}$

	Shift	Peak
a	0.88	3B
▶b	1.28	1E
c	2.28	3C
d	4.11	4A

▶ $C_{14}H_{28}O_2$ ▶ 45.5 mg/0.50 ml $CDCl_3$ ▶ S 194

▶ 2-BbBk

$\overset{(a)}{CH_3}$—$\overset{(b)}{(CH_2)_4}$—$\overset{(c)}{CH_2}$—C(=O)—$\overset{(d)}{OH}$

	Shift	Peak
a	0.88	3B
▶b	1.32	1B
c	2.38	3B
d	11.47	1B

▶ $C_7H_{14}O_2$ ▶ 45.6 mg/0.50 ml $CDCl_3$ ▶ S 49

▶ 2-BbBk

$\overset{(a)}{CH_3}$—$\overset{(b)}{(CH_2)_{11}}$—$\overset{(c)}{CH_2}$—C(=O)—$\overset{(d)}{OCH_3}$

	Shift	Peak
a	0.88	3B
▶b	1.29	1A
c	2.24	3B
d	3.68	1A

▶ $C_{15}H_{30}O_2$ ▶ 45.1 mg/0.50 ml $CDCl_3$ ▶ S 195

▶ 2-BbBk

$\overset{(a)}{CH_3}$—$\overset{(c)}{(CH_2)_4}$—$\overset{(d)}{CH_2}$—C(=O)—$\overset{(e)}{OCH_2}$—$\overset{(b)}{CH_3}$

	Shift	Peak
a	0.88	3D
b	1.26	S
▶c	1.33	S
d	2.28	3B
e	4.12	4A

▶ $C_9H_{18}O_2$ ▶ 44.5 mg/0.50 ml $CDCl_3$ ▶ S 283

▶ 2-BbBk

$CH_3CH_2O-C(=O)-CH_2(CH_2)_6CH_2C(=O)-OCH_2CH_3$ — (a) (d) ... (c) (b) (c) ... (d) (a)

	Shift	Peak
a	1.26	3D
▶b	1.35	1E
c	2.28	3B
d	4.13	4A

▶ $C_{14}H_{26}O_4$ ▶ 45.1 mg/0.50 ml $CDCl_3$ ▶ S 250

▶ 2-BbBl

$CH_3-(CH_2)_8-CH_2-C(H)=CH_2$ — (a) (b) (c); (f) H on C, (d) H and (e) H on terminal C

	Shift	Peak
a	0.87	3B
▶b	1.28	1A
c	2.05	2G
d	4.80	R
e	~5.03	R
f	5.68	R

▶ $C_{12}H_{24}$ ▶ 44.4 mg/0.50 ml $CDCl_3$ ▶ S 202

▶ 2-BbBk

$CH_3-(CH_2)_3-CH_2-C(=O)-OH$ — (a) (b) (c) (d)

	Shift	Peak
a	0.92	3B
▶b	1.45	O
c	2.36	3C
d	11.37	1B

▶ $C_6H_{12}O_2$ ▶ 44.7 mg/0.50 ml $CDCl_3$ ▶ S 37

▶ 2-BbBl

$CH-CH_2-(CH_2)_5-CH_2-C(=O)-OH$ / $\|$ / $CH-CH_2-(CH_2)_6-CH_3$ — (e) (c) (b) (d) (f); (e) (c) (b) (a)

	Shift	Peak
a	0.88	3B
▶b	1.29	1B
c	2.00	1F
d	2.25	3B
e	5.28	3D
f	11.45	1C

▶ $C_{18}H_{34}O_2$ ▶ 45.2 mg/0.50 ml $CDCl_3$ ▶ S 70

▶ 2-BbBl

(d)H, (e)H $C=C$ H(f), $CH_2(CH_2)_{12}CH_3$ — (c) (b) (a)

	Shift	Peak
a	.90	3B
▶b	1.24	1A
c	2.04	3B
d	4.88	R
e	5.02	R
f	5.72	R

▶ $C_{16}H_{32}$ ▶ 50 vol.% CCl_4 ▶ API 437

▶ 2-BbBl

$CH-CH_2-(CH_2)_5-CH_2-C(=O)-OCH_3$ / $\|$ / $CH-CH_2-(CH_2)_6-CH_3$ — (f) (c) (b) (d) (e); (f) (c) (b) (a)

	Shift	Peak
a	0.90	3B
▶b	1.29	1B
c	2.00	1C
d	2.32	S
e	3.70	1A
f	5.4	3B

▶ $C_{19}H_{36}O_2$ ▶ 33.7 mg/0.50 ml $CDCl_3$ ▶ S 71

▶ 2-BbBl

(e)H, (d)H $C=C$ H(f), $CH_2(CH_2)_{12}CH_3$ — (c) (b) (a)

	Shift	Peak
a	.99	3B
▶b	1.25	1A
c	2.02	3B
d	4.90	R
e	5.08	R
f	5.68	R

▶ $C_{16}H_{32}$ ▶ 50% CCl_4 ▶ API 73

▶ 2-BbBl

(d)H, (e)H $C=C$ H(f), $(CH_2)_4CH_3$ — (c+b) (a)

	Shift	Peak
a	.91	3B
▶b	1.31	S
c	2.02	3B
d	4.84	R
e	4.98	R
f	5.67	R

▶ C_7H_{14} ▶ 10% CCl_4 ▶ API 157

▶ 2-BbBl

(d)H, (e)H $C=C$ H(f), $CH_2(CH_2)_{10}CH_3$ — (c) (b) (a)

	Shift	Peak
a	.88	3B
▶b	1.27	1A
c	2.3	3B
d	4.88	7X
e	5.01	7X
f	5.72	R

▶ $C_{14}H_{28}$ ▶ 50 vol.% CCl_4 ▶ API 436

▶ 2-BbBl

(d)H, (e)H $C=C$ H(f), $CH_2(CH_2)_8CH_3$ — (c) (b) (a)

	Shift	Peak
a	.92	3B
▶b	1.32	1B
c	2.05	3B
d	4.86	R
e	5.00	R
f	5.68	R

▶ $C_{12}H_{24}$ ▶ 50 vol.% CCl_4 ▶ API 435

▶ 2-BbBl

(d)H, (e)H $C=C$ H(f), $CH_2(CH_2)_5CH_3$ — (c) (b) (a)

	Shift	Peak
a	.85	3B
▶b	1.28	1B
c	1.99	3B
d	4.88	R
e	4.99	R
f	5.68	R

▶ C_9H_{18} ▶ 50% CCl_4 ▶ API 434

▶ 2-BbBl

(f)H, (g)H $C=C$ H(h), $CH_2CH_2CH_2CH_2CH_3$ — (e) (d) (c) (b) (a)

	Shift	Peak
a	.89	3C
b	.90	S
c	~1.33 or 1.35	S
▶d	~1.35 or 1.33	S
e	1.97	4C
f	~4.84	R
g	~4.97	R
h	~5.70	R

▶ C_7H_{14} ▶ liquid ▶ API 384

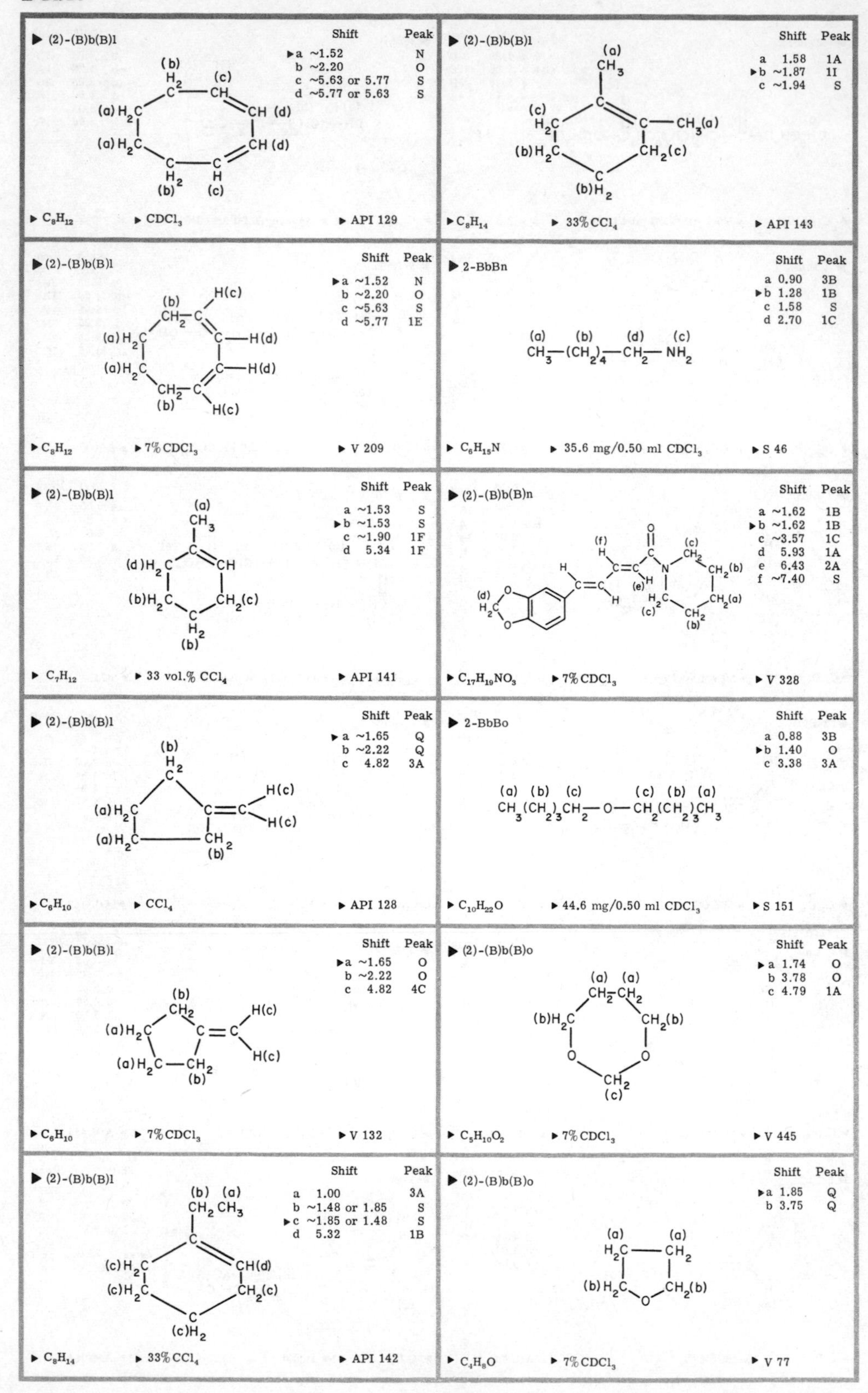
(2)-(B)b(B)l
Shift Peak
a ~1.52 N
b ~2.20 O
c ~5.63 or 5.77 S
d ~5.77 or 5.63 S
C8H12 CDCl3 API 129
(2)-(B)b(B)l
Shift Peak
a 1.58 1A
b ~1.87 1I
c ~1.94 S
C8H14 33% CCl4 API 143
(2)-(B)b(B)l
Shift Peak
a ~1.52 N
b ~2.20 O
c ~5.63 S
d ~5.77 1E
C8H12 7% CDCl3 V 209
2-BbBn
Shift Peak
a 0.90 3B
b 1.28 1B
c 1.58 S
d 2.70 1C
C6H15N 35.6 mg/0.50 ml CDCl3 S 46
(2)-(B)b(B)l
Shift Peak
a ~1.53 S
b ~1.53 S
c ~1.90 1F
d 5.34 1F
C7H12 33 vol.% CCl4 API 141
(2)-(B)b(B)n
Shift Peak
a ~1.62 1B
b ~1.62 1B
c ~3.57 1C
d 5.93 1A
e 6.43 2A
f ~7.40 S
C17H19NO3 7% CDCl3 V 328
(2)-(B)b(B)l
Shift Peak
a ~1.65 Q
b ~2.22 Q
c 4.82 3A
C6H10 CCl4 API 128
2-BbBo
Shift Peak
a 0.88 3B
b 1.40 O
c 3.38 3A
C10H22O 44.6 mg/0.50 ml CDCl3 S 151
(2)-(B)b(B)l
Shift Peak
a ~1.65 O
b ~2.22 O
c 4.82 4C
C6H10 7% CDCl3 V 132
(2)-(B)b(B)o
Shift Peak
a 1.74 O
b 3.78 O
c 4.79 1A
C5H10O2 7% CDCl3 V 445
(2)-(B)b(B)l
Shift Peak
a 1.00 3A
b ~1.48 or 1.85 S
c ~1.85 or 1.48 S
d 5.32 1B
C8H14 33% CCl4 API 142
(2)-(B)b(B)o
Shift Peak
a 1.85 Q
b 3.75 Q
C4H8O 7% CDCl3 V 77

▶ 2-BbBp

	Shift	Peak
a	0.86	3B
▶b	1.21	1A
c	~3.55	3B

$CH_3-CH_2-(CH_2)_{11}-CH_2-OH$ — (a) CH_3, (b) $(CH_2)_{11}$, (c) CH_2

▶ $C_{14}H_{30}O$ ▶ 44.9 mg/0.25 ml $CDCl_3$ ▶ S 294

▶ 2-BbBp

	Shift	Peak
a	0.90	3B
▶b	1.32	1B
c	1.97	1A
d	3.62	3B

$CH_3-(CH_2)_7-CH_2OH$ — (a) CH_3, (b) $(CH_2)_7$, (d) CH_2, (c) OH

▶ $C_9H_{20}O$ ▶ 44.6 mg/0.50 ml $CDCl_3$ ▶ S 102

▶ 2-BbBp

	Shift	Peak
a	.88	3B
▶b	1.27	1B
c	3.45	3B
d	4.28	1A

$CH_3-(CH_2)_8-CH_2-OH$ — (a) CH_3, (b) $(CH_2)_8$, (c) CH_2, (d) OH

▶ $C_{10}H_{22}O$ ▶ 20% CCl_4 ▶ API 172

▶ 2-BbBp

	Shift	Peak
a	0.88	3B
▶b	1.33	1B
c	1.95	1A
d	3.62	3B

$CH_3-(CH_2)_5-CH_2-OH$ — (a) CH_3, (b) $(CH_2)_5$, (d) CH_2, (c) OH

▶ $C_7H_{16}O$ ▶ 45.0 mg/0.50 ml $CDCl_3$ ▶ S 215

▶ 2-BbBp

	Shift	Peak
a	.88	3D
▶b	1.28	1B
c	3.44	3B
d	4.27	1A

$CH_3-(CH_2)_7-CH_2-OH$ — (a) CH_3, (b) $(CH_2)_7$, (c) CH_2, (d) OH

▶ $C_9H_{20}O$ ▶ 20% CCl_4 ▶ API 171

▶ 2-BbBp

	Shift	Peak
a	0.88	3B
▶b	1.36	1B
c	2.28	1A
d	3.62	3A

$CH_3-(CH_2)_4-CH_2-OH$ — (a) CH_3, (b) $(CH_2)_4$, (d) CH_2, (c) OH

▶ $C_6H_{14}O$ ▶ 45.2 mg/0.50 ml $CDCl_3$ ▶ S 198

▶ 2-BbBp

	Shift	Peak
a	0.85	3B
▶b	1.29	1A
c	1.83	1B
d	3.62	3B

$CH_3-(CH_2)_{10}-CH_2OH$ — (a) CH_3, (b) $(CH_2)_{10}$, (d) CH_2, (c) OH

▶ $C_{12}H_{26}O$ ▶ 44.4 mg/0.50 ml $CDCl_3$ ▶ S 103

▶ 2-BbBp

	Shift	Peak
▶a	~1.80	N
b	2.53	1A
c	3.60 or 3.70	S
d	3.70 or 3.60	S

$Cl-CH_2CH_2CH_2CH_2-OH$ — (c) $ClCH_2$, (a) central CH_2CH_2, (d) CH_2, (b) OH

▶ C_4H_9ClO ▶ 45.0 mg/0.50 ml $CDCl_3$ ▶ S 241

▶ 2-BbBp

	Shift	Peak
a	.88	2A
▶b	1.31	S
c	~1.63	S
d	3.48	3B
e	4.42	1A

$CH_3-CH(CH_3)-(CH_2)_3-CH_2-OH$ — (a) CH_3, (c) CH, (b) $(CH_2)_3$, (d) CH_2, (e) OH

▶ $C_7H_{16}O$ ▶ 20% CCl_4 ▶ API 169

▶ 2-BbBq

	Shift	Peak
a	0.92	3B
b	1.12	S
▶c	1.32	S
d	2.33	4A
e	4.05	3B

$CH_3CH_2C(=O)-OCH_2(CH_2)_3CH_3$ — (b) CH_3, (d) CH_2, (e) OCH_2, (c) $(CH_2)_3$, (a) CH_3

▶ $C_8H_{16}O_2$ ▶ 44.8 mg/0.50 ml $CDCl_3$ ▶ S 40

▶ 2-BbBp

	Shift	Peak
a	.90	3B
▶b	1.32	1G
c	3.44	3B
d	4.42	1B

$CH_3-(CH_2)_5-CH_2-OH$ — (a) CH_3, (b) $(CH_2)_5$, (c) CH_2, (d) OH

▶ $C_7H_{16}O$ ▶ 20% CCl_4 ▶ API 168

▶ 2-BbBq

	Shift	Peak
a	0.88	3B
▶b	1.32	1E
c	4.16	3B
d	8.03	1B

$HC(=O)-OCH_2-(CH_2)_6-CH_3$ — (d) HC, (c) OCH_2, (b) $(CH_2)_6$, (a) CH_3

▶ $C_9H_{18}O_2$ ▶ 44.4 mg/0.50 ml $CDCl_3$ ▶ S 114

▶ 2-BbBq

	Shift	Peak
a	0.92	3B
▶b	1.37	1E
c	2.05	1A
d	4.05	3A

(c) (d) (b) (a)
$CH_3C(=O)—OCH_2(CH_2)_4CH_3$

▶ $C_8H_{16}O_2$ ▶ 45.0 mg/0.50 ml $CDCl_3$ ▶ S 217

▶ 2-BbBv

	Shift	Peak
a	.89	3B
▶b	1.25	1B
c	2.46	3B
d	7.07	1A

(c) (b) (a)
$C_6H_5—CH_2—(CH_2)_7—CH_3$

(d) = all ring H's

▶ $C_{15}H_{24}$ ▶ 50% CCl_4 ▶ API 358

▶ 2-BbBs

	Shift	Peak
a	0.90	3B
▶b	1.29	1E
c	2.09	1A
d	2.53	4B

(a) (b) (d) (c)
$CH_3—(CH_2)_5—CH_2SH$

▶ $C_7H_{16}S$ ▶ 46.9 mg/0.50 ml $CDCl_3$ ▶ S 297

▶ 2-BbBv

	Shift	Peak
a	.88	3B
▶b	1.25	1B
c	2.50	3B
d	7.07	1A

(c) (b) (a)
$C_6H_5—CH_2—(CH_2)_8—CH_3$

(d) = all ring H's

▶ $C_{16}H_{26}$ ▶ 50% CCl_4 ▶ API 359

▶ (2)-(B)b(B)s

	Shift	Peak
▶a	1.93	Q
b	2.82	Q

Tetrahydrothiophene ring: (a) H_2C—CH_2 (a); (b) H_2C—S—CH_2 (b)

▶ C_4H_8S ▶ 7% $CDCl_3$ ▶ V 80

▶ 2-BbBv

	Shift	Peak
a	.89	3B
▶b	1.36	O
c	2.52	3B
d	7.04	1G

(c) (b) (b) (b) (a)
$C_6H_5—CH_2—CH_2—CH_2—CH_2—CH_3$

(d) = all ring H's

▶ $C_{11}H_{16}$ ▶ 50% CCl_4 ▶ API 355

▶ (2)-(B)b(B)s

	Shift	Peak
▶a	2.23	8A
b	3.00	8A

(a) H_2C—CH_2 (a); (b) H_2C—$S(O_2)$—CH_2 (b)

▶ $C_4H_8O_2S$ ▶ 7% $CDCl_3$ ▶ V 416

▶ (2)-(B)b(B)v

	Shift	Peak
▶a	1.79	O
b	2.76	O
c	7.07	1A

Tetralin: aromatic H (c); CH_2 (b), CH_2(a), CH_2(a), CH_2 (b)

▶ $C_{10}H_{12}$ ▶ 7% $CDCl_3$ ▶ V 557

▶ 2-BbBu

	Shift	Peak
a	0.93	3B
▶b	1.49	O
c	2.33	3B

(a) (b) (c)
$CH_3—(CH_2)_3—CH_2—C≡N$

▶ $C_6H_{11}N$ ▶ 45.0 mg/0.50 ml $CDCl_3$ ▶ S 176

▶ 2-BbBv

	Shift	Peak
a	.89	3B
b	1.30	1F
▶c	2.54	3B
d	7.10	1A

(b) (c) (a)
$C_6H_5—CH_2—(CH_2)_4—CH_3$

(d) = all ring H's

▶ $C_{12}H_{18}$ ▶ 50% CCl_4 ▶ API 356

▶ 2-BbBv

	Shift	Peak
a	.88	3B
▶b	1.22	1A
c	2.54	3B
d	7.09	1G

(c) (b) (a)
$C_6H_5—CH_2—(CH_2)_{10}—CH_3$

(d) = all ring H's

▶ $C_{18}H_{30}$ ▶ 50% CCl_4 ▶ API 360

▶ 2-BbBv

	Shift	Peak
a	.86	3B
b	1.28	1B
▶c	2.54	3B
d	7.09	1A

(b) (c) (a)
$C_6H_5—CH_2—(CH_2)_5—CH_3$

(d) = all ring H's

▶ $C_{13}H_{20}$ ▶ 50% CCl_4 ▶ API 357

▶ (2)-(B)c(B)c

	Shift	Peak
▶ a	2.12	Q
b	3.15	N
c	12.35	1A

▶ $C_5H_8O_2$ ▶ liquid run, neat ▶ API 191

▶ 2-BgBv

	Shift	Peak
▶ a	2.15	5B
b	2.75	3C
c	3.38	3A
d	7.22	1A

▶ $C_9H_{11}Br$ ▶ 7% $CDCl_3$ ▶ V 237

▶ (2)-(B)d(B)d

	Shift	Peak
▶ a	2.35	O
b	2.77	3B
c	10.83	1A

▶ $C_6H_8O_4$ ▶ D_2O ▶ API 192

▶ (2)-(B)j(B)j

	Shift	Peak
▶ a	1.98	5B
b	2.98	3B

▶ C_4H_6O ▶ liquid run, neat ▶ API 193

▶ 2-BeBe

	Shift	Peak
▶ a	2.20	5A
b	3.70	3A

▶ $C_3H_6Cl_2$ ▶ 7% $CDCl_3$ ▶ V 31

▶ (2)-(B)j(B)n

	Shift	Peak
▶ a	2.06	O
b	2.50	1A
c	2.62	S
d	3.83	3A

▶ $C_6H_9NO_2$ ▶ 7% $CDCl_3$ ▶ V 465

▶ 2-BeBg

	Shift	Peak
▶ a	2.28	5A
b	3.55	3D
c	3.70	3D

▶ C_3H_6BrCl ▶ 7% $CDCl_3$ ▶ V 29

▶ (2)-(B)j(B)n

	Shift	Peak
▶ a	2.08 ± .10	S
b	2.31 ± .10	S
c	3.47	1B
d	3.51	3D

▶ $C_{10}H_{16}N_2O_2$ ▶ 7% $CDCl_3$ ▶ V 572

▶ 2-BgBg

	Shift	Peak
▶ a	2.34	5B
b	3.58	3A

▶ $C_3H_6Br_2$ ▶ 44.6 mg/0.50 ml $CDCl_3$ ▶ S 188

▶ (2)-(B)j(B)n

	Shift	Peak
a	1.62	2A
▶ b	2.12 ± .10	S
c	2.36 ± .10	S
d	3.52	3A

▶ $C_{10}H_{16}N_2O_2$ ▶ 7% $CDCl_3$ ▶ V 571

▶ 2-BgBu

	Shift	Peak
▶ a	~2.23	O
b	2.58	O
c	3.53	3A

▶ C_4H_6BrN ▶ 7% $CDCl_3$ ▶ V 58

▶ (2)-(B)j(B)v

	Shift	Peak
▶ a	2.25	5B
b	2.77	3B
c	3.37	3B

▶ $C_{14}H_{12}O$ ▶ 7% $CDCl_3$ ▶ V 308

▶ (2)-(B)l(B)l

Shift		Peak
▶a	1.81	O
b	2.28	3C
c	5.65	1A

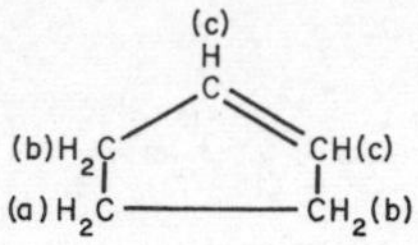

▶ C_5H_8 ▶ 10% CCl_4 ▶ API 164

▶ 2-BnBn

Shift		Peak
a	1.79	1E
▶b	2.46	4B
c	2.81	3A
d	3.74	3B

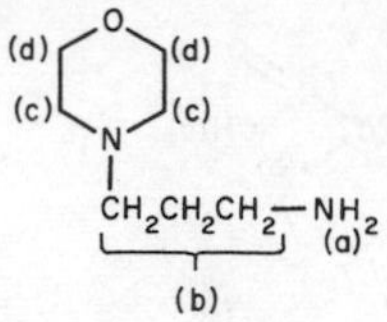

▶ $C_7H_{16}N_2O$ ▶ 33.2 mg/0.50 ml $CDCl_3$ ▶ S 69

▶ (2)-(B)l(B)l

Shift		Peak
▶a	1.92	6B
b	2.70	3C
c	4.70	3A

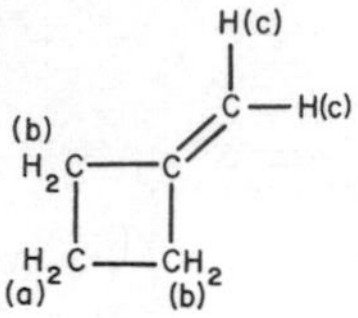

▶ C_5H_8 ▶ $CDCl_3$ ▶ API 127

▶ (2)-(B)n(B)v

Shift		Peak
▶a	1.93	5B
b	2.73	3A
c	3.22	3C
d	3.49	1B
e	3.77	1A

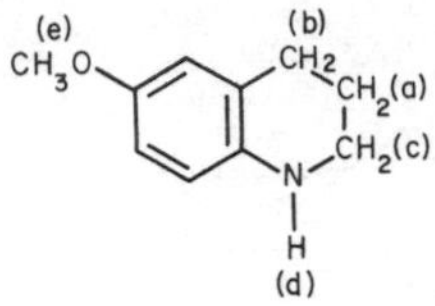

▶ $C_{10}H_{13}NO$ ▶ 7% $CDCl_3$ ▶ V 266

▶ (2)-(B)l(B)l

Shift		Peak
▶a	1.92	5B
b	2.70	3C
c	4.70	N

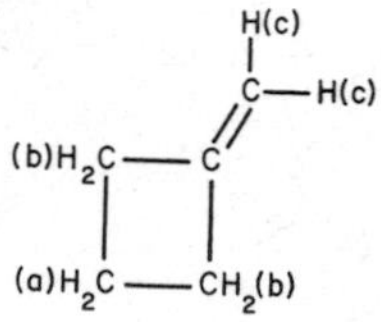

▶ C_5H_8 ▶ 7% $CDCl_3$ ▶ V 109

▶ (2)-(B)o(B)o

Shift		Peak
▶a	2.72	5A
b	4.73	3A

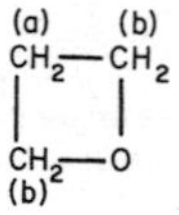

▶ C_3H_6O ▶ 7% $CDCl_3$ ▶ V 33

▶ (2)-(B)l(B)o

Shift		Peak
▶a	2.03	5B
b	2.63	3A
c	3.87 or 3.92	1A
d	3.87 or 3.92	1A
e	4.30	3A
f	~7.00	R
g	~7.10	R
h	7.52	4D

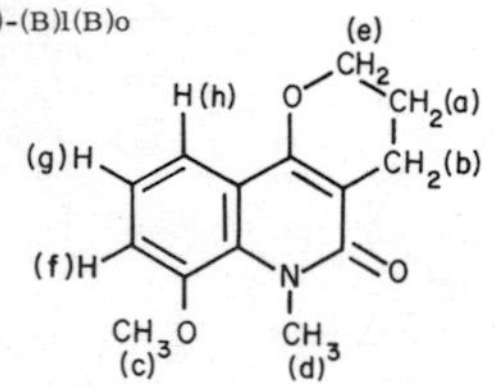

▶ $C_{14}H_{15}NO_3$ ▶ 7% $CDCl_3$ ▶ V 312

▶ (2)-(B)o(B)s

Shift		Peak
▶a	2.62	6B
b	3.28	3C
c	4.50	3A

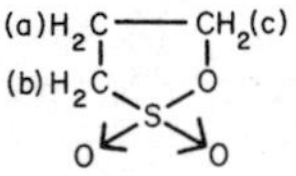

▶ $C_3H_6O_3S$ ▶ 7% $CDCl_3$ ▶ V 390

▶ (2)-(B)l(B)o

Shift		Peak
▶a	2.05	5B
b	2.70	3A
c	3.77 or 3.87	1A
d	3.77 or 3.87	1A
e	4.33	3A
f	~7.00	S
g	~7.15	S
h	8.00	4D

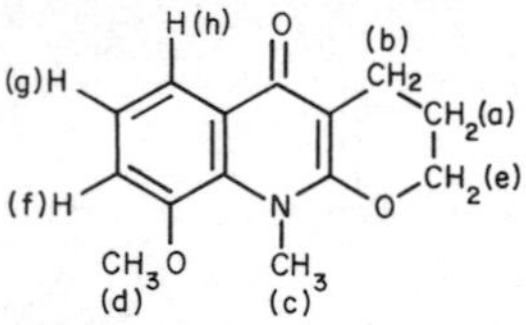

▶ $C_{14}H_{15}NO_3$ ▶ 7% $CDCl_3$ ▶ V 311

▶ 2-BpBv

Shift		Peak
▶a	1.89	5B
b	2.18	1A
c	2.69	3C
d	3.61	3A
e	7.18	1A

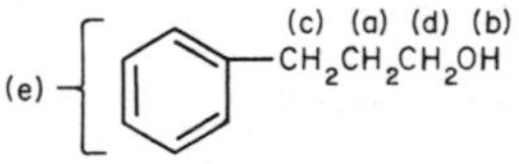

▶ $C_9H_{12}O$ ▶ 44.9 mg/0.50 ml $CDCl_3$ ▶ S 116

▶ (2)-(B)n(B)n

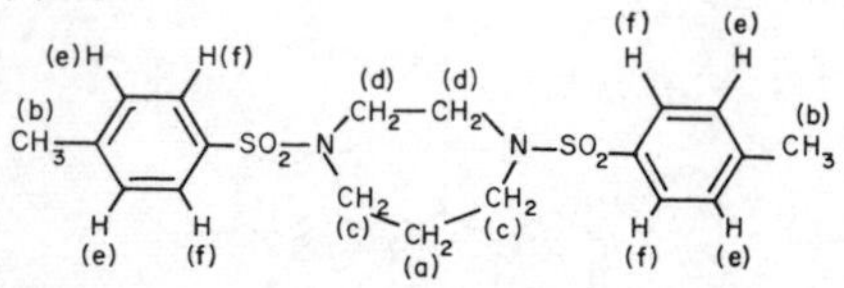

▶ $C_{19}H_{24}N_2O_4S_2$
▶ 7% $CDCl_3$
▶ V 336

Shift		Peak	Shift		Peak
▶a	1.95	5B	d	3.37	1E
b	2.38	1A	e	7.30	Q
c	3.33	3A	f	7.63	Q

▶ 2-BsBs

Shift		Peak
a	1.35	3A
▶b	1.88	5B
c	2.68	4C

▶ $C_3H_8S_2$ ▶ 7% $CDCl_3$ ▶ V 47

▶ 2-BsBz

	Shift	Peak
a	0.00	1A
b	0.60	N
▶c	1.78	O
d	2.82	3C

▶ $C_6H_{15}NaO_3SSi$ ▶ D_2O ▶ V 481

▶ 2-BaCab

	Shift	Peak
a	.88	3A
b	~1.25	S
▶c	~1.25	S
d	~1.25	S

▶ C_8H_{18} ▶ 50% CCl_4 ▶ API 201

▶ 2-BuBu

	Shift	Peak
▶a	2.03	5B
b	2.58	3C

▶ $C_5H_6N_2$ ▶ 45.3 mg/0.50 ml $CDCl_3$ ▶ S 251

▶ 2-BaCad

	Shift	Peak
a	.85	1E
b	.85	1E
c	.85	1E
▶d	~1.03	S
e	~1.32	S

▶ C_9H_{20} ▶ 50% CCl_4 ▶ API 210

▶ (2)-(B)v(B)v

	Shift	Peak
▶a	2.02	5B
b	2.86	3A
c	7.05	1I

▶ C_9H_{10} ▶ 10% CCl_4 ▶ API 188

▶ 2-BaCal

	Shift	Peak
a	.88	3D
b	.98	2H
▶c	1.29	O
d	2.11	5A
e	4.86	R
f	4.97	R
g	~5.58	R

▶ C_7H_{14} ▶ liquid ▶ API 392

▶ (2)-(B)v(B)v

	Shift	Peak
▶a	2.04	5B
b	2.91	3A
c	7.17	1A

▶ C_9H_{10} ▶ 7% $CDCl_3$ ▶ V 527

▶ 2-BaCap

	Shift	Peak
a	0.92	3D
b	1.18	2H
▶c	1.33	N
d	2.22	1B
e	3.70	1F

▶ $C_5H_{12}O$ ▶ 44.6 mg/0.50 ml $CDCl_3$ ▶ S 60

▶ 2-BaCaa

	Shift	Peak
a	.88	3D
b	.95	2H
▶c	1.20	N
d	~1.50	1D

▶ C_6H_{14} ▶ 50% CCl_4 ▶ API 334

▶ 2-BaCbb

	Shift	Peak
a	~ .88	3D
b	~ .88	3D
▶c	~1.26	1F

▶ C_8H_{18} ▶ 50% CCl_4 ▶ API 202

▶ 2-BaCab

	Shift	Peak
a	.88	3D
b	.88	S
▶c	1.24	S
d	~1.26	S

▶ C_7H_{16} ▶ 50% CCl_4 ▶ API 338

▶ 2-BbCaa

	Shift	Peak
a	.82	2A
b	~.92	1E
▶c	~.93	1B
d	~.97	1D

▶ C_7H_{16} ▶ 50% CCl_4 ▶ API 195

▶ 2-BbCaa

	Shift	Peak
a	.88	2A
▶b	1.25	1B
c	~1.47	S

(a)
CH_3 (b) CH_3
CH_3—CH—$(CH_2)_4$—CH—CH_3
(c)

▶ $C_{10}H_{22}$ ▶ 50% CCl_4 ▶ API 346

▶ 2-BbCag

	Shift	Peak
a	0.91	3B
▶b	1.28	1B
c	1.71	2A
d	4.12	5B

Br
(a) (b) (c)
CH_3—$(CH_2)_5$—CH—CH_3
(d)

▶ $C_8H_{17}Br$ ▶ 45.2 mg/0.50 ml $CDCl_3$ ▶ S 227

▶ 2-BbCaa

	Shift	Peak
a	.87	2H
b	.96	S
▶c	1.27	1B

(a)
CH_3
(c) (b)
CH_3—CH—$(CH_2)_4$—CH_3

▶ C_8H_{18} ▶ 50% CCl_4 ▶ API 340

▶ 2-BbCan

	Shift	Peak
a	.87	3D
b	1.05	2H
▶c	1.25	1B
d	3.2	1C

(a) (c) (d)
CH_3—$(CH_2)_3$—CH—NH_2
CH_3
(b)

▶ $C_6H_{15}N$ ▶ 34.4 mg/0.50 ml $CDCl_3$ ▶ S 28

▶ 2-BbCaa

	Shift	Peak
a	.83	2H
b	.94	S
▶c	1.30	1A

(a)
CH_3
(c) (b)
CH_3—CH—$(CH_2)_7$—CH_3

▶ $C_{11}H_{24}$ ▶ 50% CCl_4 ▶ API 349

▶ 2-BbCap

	Shift	Peak
a	.88	3B
b	1.12	2A
▶c	1.30	1B
d	3.70	4B
e	4.08	1A

OH (e)
(b) (c) (c) (a)
CH_3—CH—CH_2—$(CH_2)_4$—CH_3
(d)

▶ $C_8H_{18}O$ ▶ 20% CCl_4 ▶ API 170

▶ 2-BbCaa

	Shift	Peak
a	.88	2A
▶b	1.31	S
c	~1.63	S
d	3.48	3B
e	4.42	1A

CH_3
(a) (b) (d) (e)
CH_3—CH—$(CH_2)_3$—CH_2—OH
(c)

▶ $C_7H_{16}O$ ▶ 20% CCl_4 ▶ API 169

▶ 2-BbCbb

	Shift	Peak
a	0.92	3B
▶b	1.30	1B
c	1.75	1B
d	3.52	1I

(b)
(a) (d) (c)
$CH_3(CH_2)_3$—CH CH_2OH
CH_2CH_3
(b) (a)

▶ $C_8H_{18}O$ ▶ 45.0 mg/0.50 ml $CDCl_3$ ▶ S 98

▶ 2-BbCab

	Shift	Peak
a	.83	3B
▶b	1.23	1B
c	~1.27	S

(a)
CH_3
(a) (b) (b) (a)
CH_3—CH_2—CH—$(CH_2)_3$—CH_3
(c)

▶ C_8H_{18} ▶ 50% CCl_4 ▶ API 341

▶ (2)-(B)b(C)bc

	Shift	Peak
▶a	1.43	1B
b	1.63	1E

(a) (a)
H_2 H_2
C—C
(a)H_2C H H CH_2(a)
C—C
(b) (b)
(a)H_2C CH_2(a)
C—C
H_2 H_2
(a) (a)

▶ $C_{10}H_{18}$ ▶ liquid run ▶ API 14

▶ (2)-(B)b(C)ac

	Shift	Peak
a	.83	2A
▶b	1.40	S
c	1.40	S
d	~1.67	S

CH_3(a)
(b) C
H CH_3(a)
(d)H C
C
H H
C H H(d)
H C
(c)H C H
(b)
H(c)

▶ C_8H_{16} ▶ liquid ▶ API 453

▶ (2)-(B)b(C)bc

	Shift	Peak
a	~1.53	S
▶b	~1.53	S
c	~1.56	S

(b) (b)
H_2 H_2
C (c) (c) C
(a)H_2C CH—HC CH_2(a)
C—C C—C
H_2 H_2 H_2 H_2
(a) (b) (b) (a)

▶ $C_{10}H_{18}$ ▶ liquid ▶ API 13

▶ (2)-(B)b(C)be

	Shift	Peak
a	1.51	N
▶ b	1.82	O
c	~4.00	1F

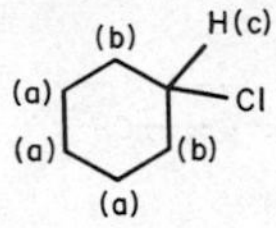

▶ $C_6H_{11}Cl$ ▶ 45.0 mg/0.50 ml $CDCl_3$ ▶ S 25

▶ (2)-(B)c(C)bk

	Shift	Peak
▶ a	~0.93	O
b	~1.63	O
c	3.67	1A

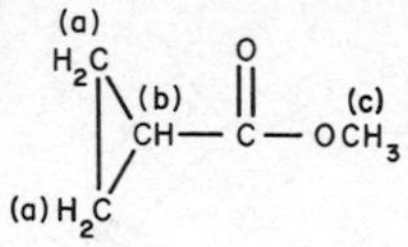

▶ $C_5H_8O_2$ ▶ 7% $CDCl_3$ ▶ V 112

▶ (2)-(B)b(C)bk

	Shift	Peak
▶ a	2.12	Q
b	3.15	N
c	12.35	1A

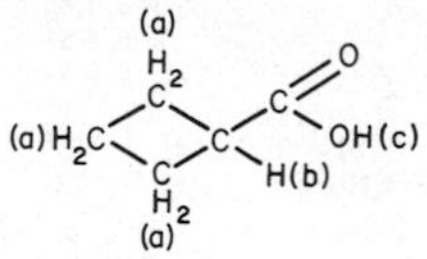

▶ $C_5H_8O_2$ ▶ liquid run, neat ▶ API 191

▶ (2)-(B)c(C)bn

	Shift	Peak
▶ a	~0.35	O
b	1.83	1A
c	2.30	O

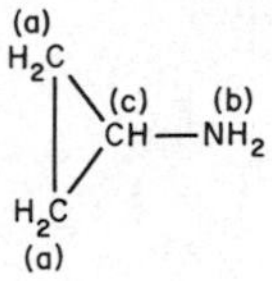

▶ C_3H_7N ▶ 7% $CDCl_3$ ▶ V 37

▶ (2)-(B)b(C)bp

	Shift	Peak
a	1.31	4E
▶ b	1.78	3B
c	2.20	1E
d	~3.61	1C

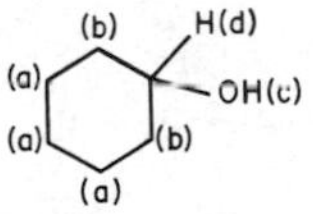

▶ $C_6H_{12}O$ ▶ 45.2 mg/0.50 ml $CDCl_3$ ▶ S 7

▶ (2)-(B)c(C)bv

	Shift	Peak
▶ a	0.68 ± .05	S
▶ b	0.74 ± .05	S
c	1.80	O
d	7.05	2A

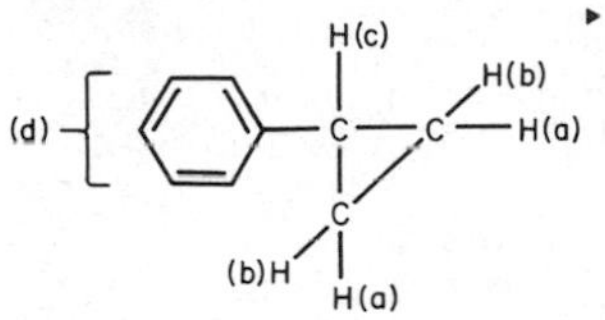

▶ C_9H_{10} ▶ 7% $CDCl_3$ ▶ V 528

▶ (2)-(B)b(C)bv

	Shift	Peak
a	~1.4	S
▶ b	~1.8	S
c	7.22	1A

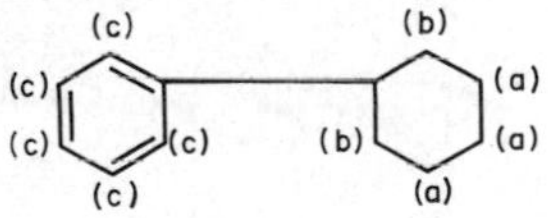

▶ $C_{12}H_{16}$ ▶ 45.4 mg/0.50 ml $CDCl_3$ ▶ S 291

▶ 2-BdCaa

	Shift	Peak
a	~ .84	1E
b	~ .84	1E
▶ c	~1.13	1G
d	~1.37	1D

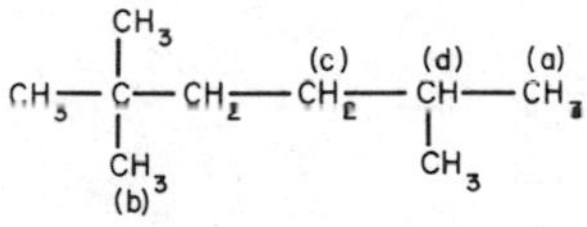

▶ C_9H_{20} ▶ 50% CCl_4 ▶ API 212

▶ 2-BcCaa

	Shift	Peak
a	.87	2A
▶ b	~1.17	1C
c	~1.40	1D

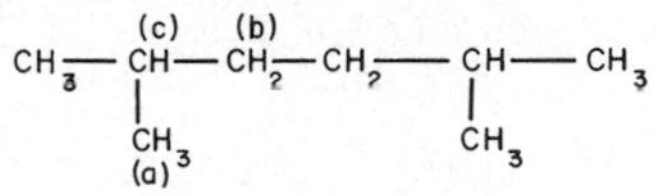

▶ C_8H_{18} ▶ 50% CCl_4 ▶ API 205

▶ 2-BeCaa

	Shift	Peak
a	0.92	2A
▶ b	1.65	O
c	3.55	3B

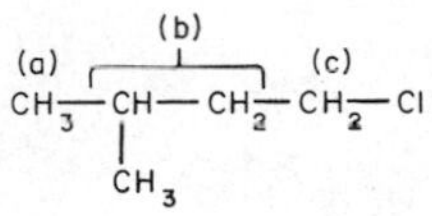

▶ $C_5H_{11}Cl$ ▶ 45.0 mg/0.50 ml $CDCl_3$ ▶ S 122

▶ 2-BcCap

	Shift	Peak
a	1.18	1A
▶ b	1.52	3A
c	3.78	S
d	3.87	1A

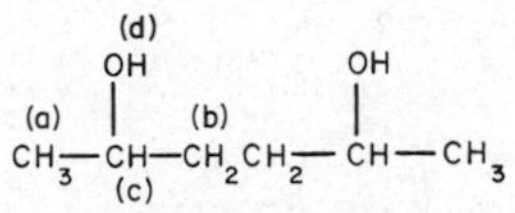

▶ $C_6H_{14}O_2$ ▶ 44.6 mg/0.50 ml $CDCl_3$ ▶ S 216

▶ 2-BgCaa

	Shift	Peak
a	0.92	2A
▶ b	1.78	O
c	3.42	3A

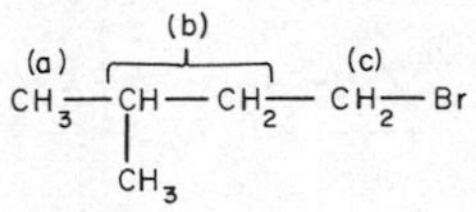

▶ $C_5H_{11}Br$ ▶ 45.0 mg/0.50 ml $CDCl_3$ ▶ S 121

▶ 2-BjCaa

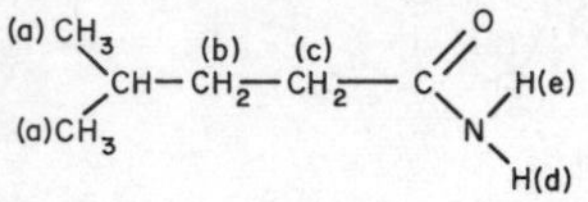

	Shift	Peak
a	0.90	2E
▶ b	~1.53	3C
c	~2.20	4C
d	6.05 or 6.33	1D
e	6.05 or 6.33	1D

▶ $C_6H_{13}NO$ ▶ 7% $CDCl_3$ ▶ V 142

▶ 2-BpCao

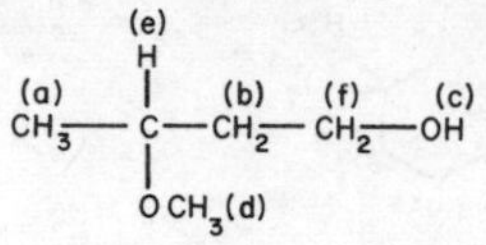

	Shift	Peak
a	1.18	2A
▶ b	1.72	4A
c	3.10	3A
d	3.33	1A
e	3.55	O
f	3.73	6B

▶ $C_5H_{12}O_2$ ▶ 7% $CDCl_3$ ▶ V 120

▶ (2)-(B)j(C)ab

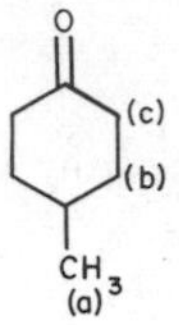

	Shift	Peak
a	1.00	2A
▶ b	~1.82	O
c	~2.20	3B

▶ $C_7H_{12}O$ ▶ 37.0 mg/0.50 ml $CDCl_3$ ▶ S 8

▶ 2-BpCap

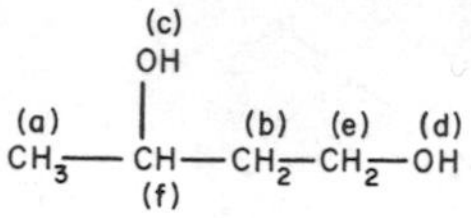

	Shift	Peak
a	1.23	2A
▶ b	1.68	4C
c	3.58 or 3.65	1A
d	3.58 or 3.65	1A
e	3.80	3A
f	4.03	4A

▶ $C_4H_{10}O_2$ ▶ 7% $CDCl_3$ ▶ V 86

▶ 2-BkCaa

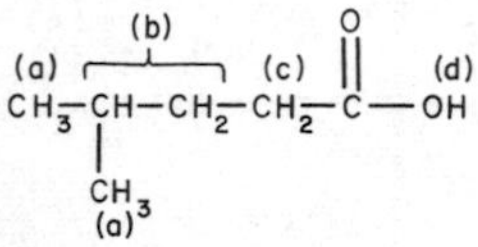

	Shift	Peak
a	0.93	2A
▶ b	~1.58	3C
c	2.36	3B
d	11.67	1A

▶ $C_6H_{12}O_2$ ▶ 45.4 mg/0.50 ml $CDCl_3$ ▶ S 156

▶ 2-BqCaa

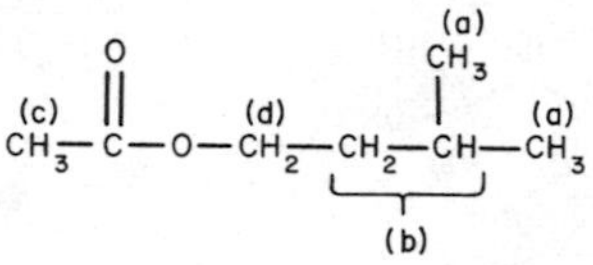

	Shift	Peak
a	0.92	2A
▶ b	~1.50	O
c	2.03	1A
d	4.03	4B

▶ $C_7H_{14}O_2$ ▶ 44.8 mg/0.50 ml $CDCl_3$ ▶ S 281

▶ 2-BkCkn

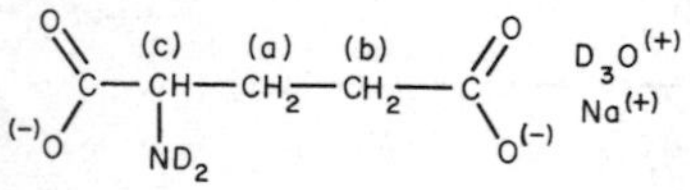

	Shift	Peak
▶ a	~2.12	S
b	~2.33	S
c	3.77	3B

▶ $C_5H_8NaNO_4$ ▶ D_2O ▶ V 435

▶ 2-BsCkn

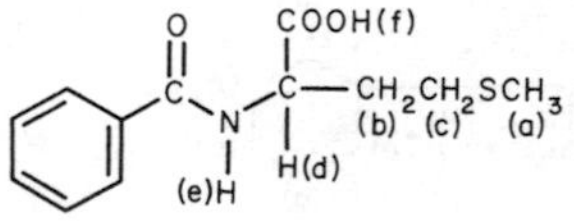

	Shift	Peak
a	2.10	1A
▶ b	2.25	O
c	2.66	O
d	4.94	4B
e	7.14	3D
f	10.28	1B

▶ $C_{12}H_{15}NO_3S$ ▶ 7% $CDCl_3$ ▶ V 597

▶ 2-BlCaa

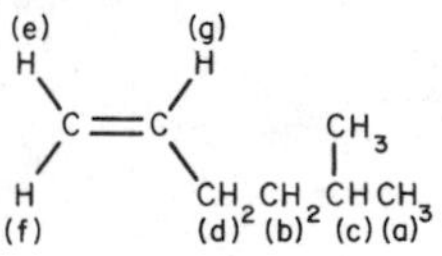

	Shift	Peak
a	.91	2C
▶ b	1.32	S
c	1.60	S
d	2.01	4C
e	4.87	R
f	5.02	R
g	5.70	R

▶ C_7H_{14} ▶ liquid ▶ API 394

▶ 2-BuCaa

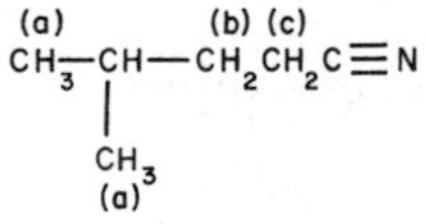

	Shift	Peak
a	0.90	2A
▶ b	1.60	3D
c	2.38	3A

▶ $C_6H_{11}N$ ▶ 44.8 mg/0.50 ml $CDCl_3$ ▶ S 165

▶ (2)-(B)n(C)bj

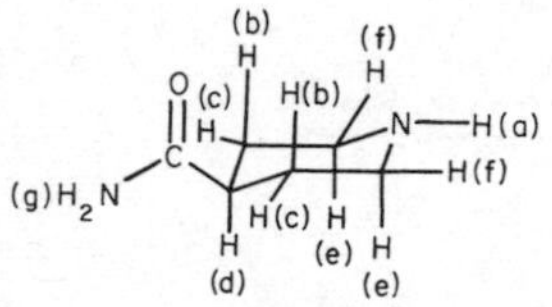

	Shift	Peak
a	1.53	1B
▶ b	1.60 ± 0.05	S
▶ c	1.82 ± 0.05	S
d	2.14	O
e	2.67	4A
f	3.17	6A
g	5.87	1D

▶ $C_6H_{12}N_2O$ ▶ 7% $CDCl_3$ ▶ V 473

▶ (2)-(B)v(C)ln

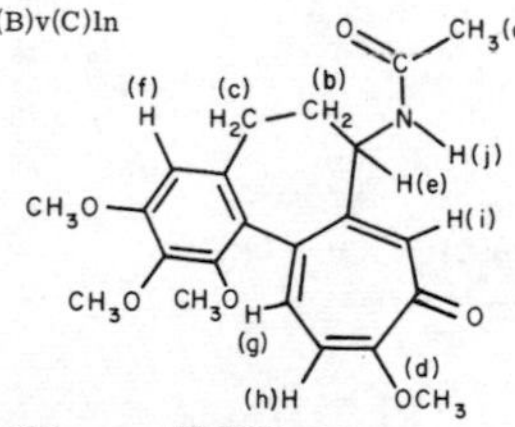

	Shift	Peak
a	1.96	1A
▶ b	2.43	S
c	2.43	S
d	3.67	1A
e	4.62	S
f	6.55	1A
g	6.92	2A
h	7.37	S
i	7.63	1B
j	8.38	2G

▶ $C_{22}H_{25}NO_6$ ▶ 7% $CDCl_3$ ▶ V 689

▶ 2-BaDaaa

	Shift	Peak
a	.88	1E
b	.88	3D
▶c	1.20	N

(a)
CH_3
(a) (c) (c) (b)
CH_3—C—CH_2—CH_2—CH_3
CH_3
(a)

▶ C_7H_{16} ▶ 50% CCl_4 ▶ API 339

▶ (2)-(B)b(D)aab

	Shift	Peak
a	.97	1G
b	1.32	Q
▶c	1.55	Q

(a)
CH_3 CH_3
C
H_2C CH_2(c)
H_2C CH_2(b)

▶ C_7H_{14} ▶ 10% CCl_4 ▶ API 146

▶ 2-BaDaac

	Shift	Peak
a	.80	1A
b	.86	2H
c	~ .86	3D
▶d	1.22	3A
e	~1.50	1D

(b) (a)
CH_3 CH_3
(b) (d) (d) (c)
CH_3—CH—C—CH_2—CH_2—CH_3
(e)
CH_3
(a)

▶ C_9H_{20} ▶ 50% CCl_4 ▶ API 213

▶ (2)-(B)b(D)acc

	Shift	Peak
a	0.67	1E
▶b	0.78	S
c	1.03	1E
d	2.27	S
e	2.27	S
▶f	2.45	S

(d) (e)
H H
O CH_3(a)
(f)H (b)H (c)
CH_3

▶ $C_{19}H_{30}O$ ▶ 7% $CDCl_3$ ▶ V 680

▶ 2-BbDaaa

	Shift	Peak
a	.87	1A
▶b	1.18	1A

CH_3 CH_3
(b)
CH_3—C—$(CH_2)_4$—C—CH_3
CH_3 CH_3
(a)

▶ $C_{12}H_{26}$ ▶ 50% CCl_4 ▶ API 351

▶ (2)-(B)b(D)bbb

	Shift	Peak
▶a	1.32	1E
b	1.44	1E
c	1.44	1E

(a) (a) (b)
(b)H_2 H_2 H_2 H_2
C—C C—C
(c)H_2C
C CH_2(c)
(c)H_2C
C—C C—C
H_2 H_2 H_2 H_2
(b) (a) (a) (b)

▶ $C_{12}H_{22}$ ▶ 33 vol.% CCl_4 ▶ API 140

▶ 2-BbDaaa

	Shift	Peak
a	.87	1E
b	1.20	1E
▶c	1.20	1B

(b)
CH_3
(c) (c) (a)
CH_3—C—CH_2—$(CH_2)_2$—CH_3
CH_3

▶ C_8H_{18} ▶ 50% CCl_4 ▶ API 203

▶ (2)-(B)b(D)bkk

	Shift	Peak
a	2.35	O
▶b	2.77	3B
c	10.83	1A

(b)
H_2
C O
C—OH(c)
(a)H_2C C
C—OH(c)
H_2 O
(b)

▶ $C_6H_8O_4$ ▶ D_2O ▶ API 192

▶ 2-BbDaaa

	Shift	Peak
a	.87	1E
b	.87	1E
▶c	1.28	1B

CH_3
(c) (b)
CH_3—C—CH_2—$(CH_2)_3$—CH_3
CH_3
(a)

▶ C_9H_{20} ▶ 50% CCl_4 ▶ API 209

▶ 2-BdDaaa

	Shift	Peak
a	.87	1A
▶b	1.13	1A

(a) (b) (b) (a)
$(CH_3)_3CCH_2CH_2C(CH_3)_3$

▶ $C_{10}H_{22}$ ▶ 10 vol.% CCl_4 ▶ API 439

▶ (2)-(B)b(D)aab

	Shift	Peak
a	.88	1A
▶b	~1.31	O

(a) (a)
CH_3 CH_3
C
(b)H_2C CH_2(b)
(b)H_2C CH_2(b)
C
H_2
(b)

▶ C_8H_{16} ▶ 10% CCl_4 ▶ API 150

▶ 2-BdDaaa

	Shift	Peak
a	.88	1A
▶b	1.15	1A

CH_3 CH_3
(b) (a)
CH_3—C—CH_2—CH_2—C—CH_3
CH_3 CH_3

▶ $C_{10}H_{22}$ ▶ 10% CCl_4 ▶ API 220

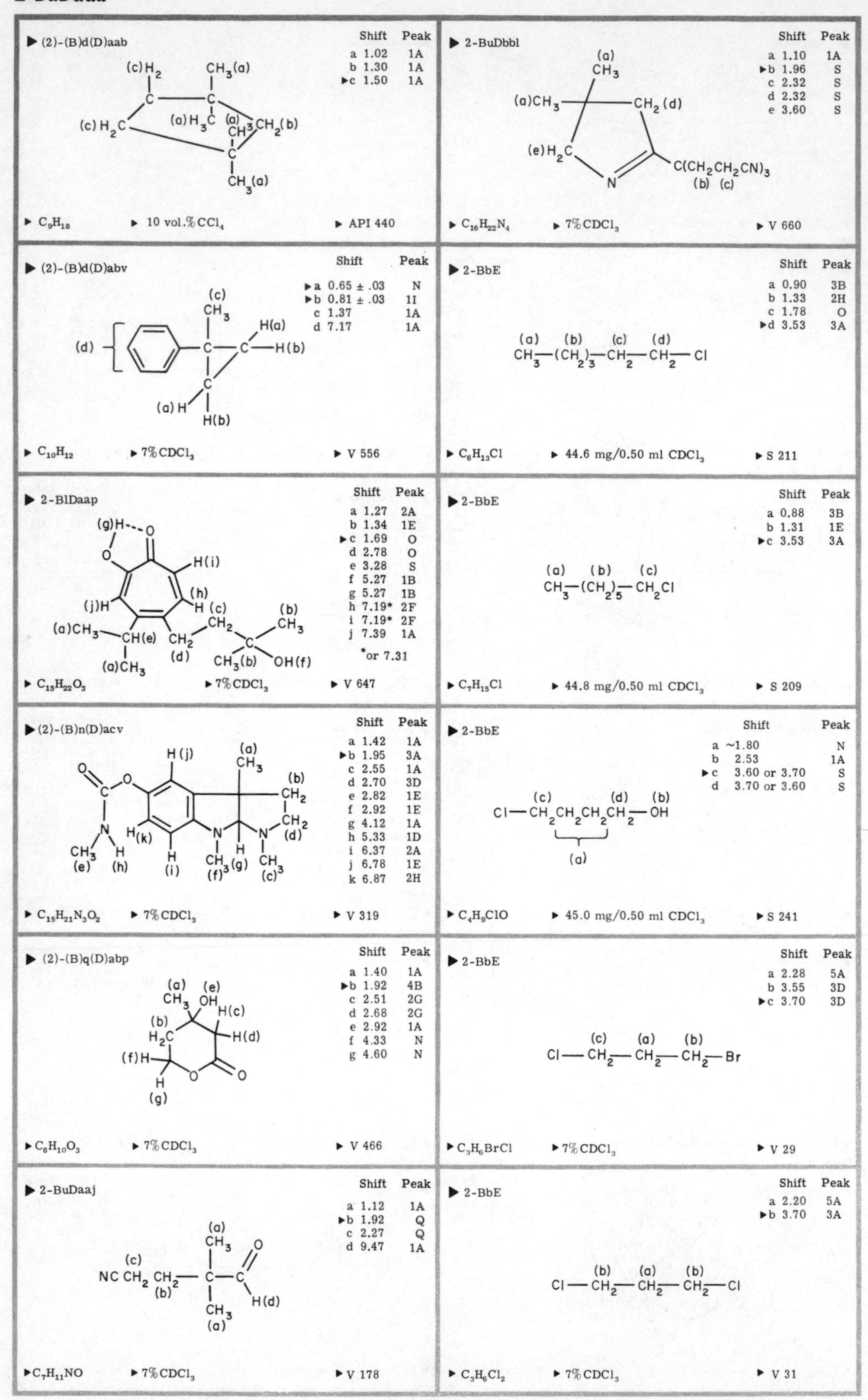

▶ (2)-(B)d(D)aab

	Shift	Peak
a	1.02	1A
b	1.30	1A
▶c	1.50	1A

▶ C_9H_{18} ▶ 10 vol.% CCl_4 ▶ API 440

▶ 2-BuDbbl

	Shift	Peak
a	1.10	1A
▶b	1.96	S
c	2.32	S
d	2.32	S
e	3.60	S

▶ $C_{16}H_{22}N_4$ ▶ 7% $CDCl_3$ ▶ V 660

▶ (2)-(B)d(D)abv

	Shift	Peak
▶a	0.65 ± .03	N
▶b	0.81 ± .03	1I
c	1.37	1A
d	7.17	1A

▶ $C_{10}H_{12}$ ▶ 7% $CDCl_3$ ▶ V 556

▶ 2-BbE

	Shift	Peak
a	0.90	3B
b	1.33	2H
c	1.78	O
▶d	3.53	3A

▶ $C_6H_{13}Cl$ ▶ 44.6 mg/0.50 ml $CDCl_3$ ▶ S 211

▶ 2-BlDaap

	Shift	Peak
a	1.27	2A
b	1.34	1E
▶c	1.69	O
d	2.78	O
e	3.28	S
f	5.27	1B
g	5.27	1B
h	7.19*	2F
i	7.19*	2F
j	7.39	1A

*or 7.31

▶ $C_{15}H_{22}O_3$ ▶ 7% $CDCl_3$ ▶ V 647

▶ 2-BbE

	Shift	Peak
a	0.88	3B
b	1.31	1E
▶c	3.53	3A

▶ $C_7H_{15}Cl$ ▶ 44.8 mg/0.50 ml $CDCl_3$ ▶ S 209

▶ (2)-(B)n(D)acv

	Shift	Peak
a	1.42	1A
▶b	1.95	3A
c	2.55	1A
d	2.70	3D
e	2.82	1E
f	2.92	1E
g	4.12	1A
h	5.33	1D
i	6.37	2A
j	6.78	1E
k	6.87	2H

▶ $C_{15}H_{21}N_3O_2$ ▶ 7% $CDCl_3$ ▶ V 319

▶ 2-BbE

	Shift	Peak
a	~1.80	N
b	2.53	1A
▶c	3.60 or 3.70	S
d	3.70 or 3.60	S

▶ C_4H_9ClO ▶ 45.0 mg/0.50 ml $CDCl_3$ ▶ S 241

▶ (2)-(B)q(D)abp

	Shift	Peak
a	1.40	1A
▶b	1.92	4B
c	2.51	2G
d	2.68	2G
e	2.92	1A
f	4.33	N
g	4.60	N

▶ $C_6H_{10}O_3$ ▶ 7% $CDCl_3$ ▶ V 466

▶ 2-BbE

	Shift	Peak
a	2.28	5A
b	3.55	3D
▶c	3.70	3D

▶ C_3H_6BrCl ▶ 7% $CDCl_3$ ▶ V 29

▶ 2-BuDaaj

	Shift	Peak
a	1.12	1A
▶b	1.92	Q
c	2.27	Q
d	9.47	1A

▶ $C_7H_{11}NO$ ▶ 7% $CDCl_3$ ▶ V 178

▶ 2-BbE

	Shift	Peak
a	2.20	5A
▶b	3.70	3A

▶ $C_3H_6Cl_2$ ▶ 7% $CDCl_3$ ▶ V 31

▶ 2-BcE

	Shift	Peak
a	0.92	2A
b	1.65	O
▶ c	3.55	3B

(a) CH_3–(b) CH(–CH_3)–(b) CH_2–(c) CH_2–Cl

▶ $C_5H_{11}Cl$ ▶ 45.0 mg/0.50 ml $CDCl_3$ ▶ S 122

▶ 2-BbG

	Shift	Peak
a	~2.23	O
b	2.58	O
▶ c	3.53	3A

Br–(c) CH_2–(a) CH_2–(b) CH_2–CN

▶ C_4H_6BrN ▶ 7% $CDCl_3$ ▶ V 58

▶ 2-BoE

	Shift	Peak
▶ a	3.78	Q
b	4.22	Q

C_6H_5–O–(b) CH_2–(a) CH_2–Cl

▶ C_8H_9ClO ▶ 7% $CDCl_3$ ▶ V 199

▶ 2-BbG

	Shift	Peak
a	2.28	5A
▶ b	3.55	3D
c	3.70	3D

Cl–(c) CH_2–(a) CH_2–(b) CH_2–Br

▶ C_3H_6BrCl ▶ 7% $CDCl_3$ ▶ V 29

▶ 2-BpE

	Shift	Peak
a	2.80	1A
b	3.68 or 3.83	Q
▶ c	3.68 or 3.83	Q

Cl–(c) CH_2–(b) CH_2–(a) OH

▶ C_2H_5ClO ▶ 7% $CDCl_3$ ▶ V 12

▶ 2-BbG

	Shift	Peak
a	2.34	5B
▶ b	3.58	3A

Br–(b) CH_2–(a) CH_2–(b) CH_2–Br

▶ $C_3H_6Br_2$ ▶ 44.6 mg/0.50 ml $CDCl_3$ ▶ S 188

▶ 2-BbG

	Shift	Peak
a	2.15	5B
b	2.75	3C
▶ c	3.38	3A
d	7.22	1A

(d) C_6H_5–(b) CH_2–(a) CH_2–(c) CH_2–Br

▶ $C_9H_{11}Br$ ▶ 7% $CDCl_3$ ▶ V 237

▶ 2-BcG

	Shift	Peak
a	0.92	2A
b	1.78	O
▶ c	3.42	3A

(a) CH_3–(b) CH(–CH_3)–(b) CH_2–(c) CH_2–Br

▶ $C_5H_{11}Br$ ▶ 45.0 mg/0.50 ml $CDCl_3$ ▶ S 121

▶ 2-BbG

	Shift	Peak
a	0.90	3B
b	1.30	1B
c	1.85	1F
▶ d	3.41	3A

(a) CH_3–(b) $(CH_2)_4$–(c) CH_2–(d) CH_2–Br

▶ $C_7H_{15}Br$ ▶ 45.0 mg/0.50 ml $CDCl_3$ ▶ S 208

▶ 2-BnG

	Shift	Peak
▶ a	3.63	3C
b	4.13	3C

Phthalimido N–(b) CH_2–(a) CH_2–Br

▶ $C_{10}H_8BrNO_2$ ▶ 7% $CDCl_3$ ▶ V 246

▶ 2-BbG

	Shift	Peak
a	0.94	3B
b	1.38	O
c	1.83	4C
▶ d	3.42	3A

(a) CH_3 (b) CH_2CH_2 (c) CH_2 (d) CH_2Br

▶ $C_5H_{11}Br$ ▶ 44.5 mg/0.50 ml $CDCl_3$ ▶ S 120

▶ 2-BuG

	Shift	Peak
a	2.98	Q
▶ b	3.52	Q

Br–(b) CH_2–(a) CH_2–C≡N

▶ C_3H_4BrN ▶ 45.2 mg/0.50 ml $CDCl_3$ ▶ S 145

▶ 2-BaI

	Shift	Peak
a	1.01	3A
b	1.86	6B
▶c	3.18	3A

(a) (b) (c) $CH_3-CH_2-CH_2-I$

▶ C_3H_7I ▶ 45.0 mg/0.50 ml $CDCl_3$ ▶ S 230

▶ 2-BaJa

	Shift	Peak
a	0.92	3B
b	1.62	6B
c	2.14	1A
▶d	2.42	3A

(c) (d) (b) (a) $CH_3-C(=O)-CH_2-CH_2-CH_3$

▶ $C_5H_{10}O$ ▶ 45.4 mg/0.50 ml $CDCl_3$ ▶ S 177

▶ 2-BbI

	Shift	Peak
a	0.88	3B
b	1.32	2E
c	1.82	3C
▶d	3.18	3A

(a) (b) (c) (d) $CH_3-(CH_2)_4-CH_2-CH_2-I$

▶ $C_7H_{15}I$ ▶ 45.0 mg/0.50 ml $CDCl_3$ ▶ S 210

▶ 2-BaJh

	Shift	Peak
a	0.97	3C
b	1.67	6B
▶c	2.42	3C
d	9.74	3A

(a) (b) (c) $CH_3-CH_2-CH_2-C(=O)-H$(d)

▶ C_4H_8O ▶ 7% $CDCl_3$ ▶ V 78

▶ 2-BbI

	Shift	Peak
a	0.93	3C
▶b	3.20	3A

(a) $CH_3CH_2CH_2CH_2I$ (b)

▶ C_4H_9I ▶ 7% $CDCl_3$ ▶ V 81

▶ 2-BaJn

	Shift	Peak
a	0.95	3A
b	1.71	6B
▶c	2.24	3C
d	~7.8	1C

(d) (c) (b) (a) $C_6H_5-NH-C(=O)-CH_2-CH_2-CH_3$

▶ $C_{10}H_{13}NO$ ▶ 45.3 mg/0.50 ml $CDCl_3$ ▶ S 261

▶ 2-BbI

	Shift	Peak
a	0.92	3B
b	1.38	2E
c	1.86	3C
▶d	3.20	3A

(a) (b) (c) (d) $CH_3-(CH_2)_2-CH_2-CH_2I$

▶ $C_5H_{11}I$ ▶ 45.0 mg/0.50 ml $CDCl_3$ ▶ S 228

▶ 2-BbJa

	Shift	Peak
a	0.90	3B
b	1.38	O
c	2.15	1A
▶d	2.42	3B

(c) (d) (b) (a) $CH_3-C(=O)-CH_2-(CH_2)_3-CH_3$

▶ $C_7H_{14}O$ ▶ 35.4 mg/0.50 ml $CDCl_3$ ▶ S 41

▶ 2-BbI

	Shift	Peak
a	0.90	3B
b	1.32	2E
c	1.84	3C
▶d	3.20	3A

(a) (b) (c) (d) $CH_3-(CH_2)_3-CH_2-CH_2-I$

▶ $C_6H_{13}I$ ▶ 45.8 mg/0.50 ml $CDCl_3$ ▶ S 212

▶ 2-BbJa

	Shift	Peak
a	0.88	3B
b	1.28	1B
c	2.17	1A
▶d	2.46	3B

(c) (d) (b) (a) $CH_3-C(=O)-CH_2-(CH_2)_5-CH_3$

▶ $C_9H_{18}O$ ▶ 33.7 mg/0.50 ml $CDCl_3$ ▶ S 48

▶ 2-BkI

	Shift	Peak
a	3.07	Q
▶b	3.30	Q
c	11.52	1A

(b) (a) (c) ICH_2CH_2COOH

▶ $C_3H_5IO_2$ ▶ 7% $CDCl_3$ ▶ V 27

▶ (2)-(B)b(J)b

	Shift	Peak
a	1.98	5B
▶b	2.98	3B

Cyclobutanone: (a)H_2C with two (b) CH_2 groups, C=O

▶ C_4H_6O ▶ liquid run, neat ▶ API 193

▶ 2-BbJh

	Shift	Peak
a	0.88	3B
b	1.29	1B
▶c	2.30	1B
d	9.35	1A

$CH_3-(CH_2)_6-CH_2-CHO$

▶ $C_9H_{18}O$ ▶ 45.3 mg/0.50 ml $CDCl_3$ ▶ S 100

▶ (2)-(B)b(J)n

	Shift	Peak
a	2.06	O
b	2.50	1A
▶c	2.62	S
d	3.83	3A

▶ $C_6H_9NO_2$ ▶ 7% $CDCl_3$ ▶ V 465

▶ 2-BbJh

	Shift	Peak
a	0.87	3A
b	1.29	1B
▶c	2.42	3B
d	9.77	3A

$CH_3-(CH_2)_9-CH_2-CHO$

▶ $C_{12}H_{24}O$ ▶ 45.2 mg/0.50 ml $CDCl_3$ ▶ S 106

▶ 2-BbJp

	Shift	Peak
a	0.88	3B
b	1.28	1E
▶c	2.38	3B
d	10.53	1B

$CH_3-(CH_2)_6-CH_2-C(=O)-OH$

▶ $C_9H_{18}O_2$ ▶ 35.7 mg/0.50 ml $CDCl_3$ ▶ S 9

▶ 2-BbJh

	Shift	Peak
a	0.88	3B
b	1.30	1B
▶c	2.43	3B
d	9.73	3A

$CH_3-(CH_2)_5-CH_2-CHO$

▶ $C_8H_{16}O$ ▶ 44.6 mg/0.50 ml $CDCl_3$ ▶ S 105

▶ (2)-(B)b(J)v

	Shift	Peak
a	2.25	5B
▶b	2.77	3B
c	3.37	3B

▶ $C_{14}H_{12}O$ ▶ 7% $CDCl_3$ ▶ V 308

▶ 2-BbJl

	Shift	Peak
a	0.89	3B
b	~1.30	S
c	~1.40	S
d	1.72	1A
e	1.97	2A
▶f	2.94	3B
g	6.06	S
h	6.17	S
i	6.58	S
j	6.89	S
k	7.88	1G

▶ $C_{21}H_{22}O_5$ ▶ 7% $CDCl_3$ ▶ V 687

▶ 2-BbJy

	Shift	Peak
a	2.28	3B
▶b	2.88	3B
c	3.67	1A
d	7.12	R
e	7.62	R
f	7.70	R
g	1.35	1F

▶ $C_{15}H_{22}O_3S$ ▶ 7% $CDCl_3$ ▶ V 320

▶ (2)-(B)b(J)n

	Shift	Peak
a	2.08 ± .10	S
▶b	2.31 ± .10	S
c	3.47	1B
d	3.51	3D

▶ $C_{10}H_{16}N_2O_2$ ▶ 7% $CDCl_3$ ▶ V 572

▶ (2)-(B)c(J)b

	Shift	Peak
a	1.00	2A
b	~1.82	O
▶c	~2.20	3B

▶ $C_7H_{12}O$ ▶ 37.0 mg/0.50 ml $CDCl_3$ ▶ S 8

▶ (2)-(B)b(J)n

	Shift	Peak
a	1.62	2A
b	2.12 ± .10	S
▶c	2.36 ± .10	S
d	3.52	3A

▶ $C_{10}H_{16}N_2O_2$ ▶ 7% $CDCl_3$ ▶ V 571

▶ 2-BcJn

	Shift	Peak
a	0.90	2E
b	~1.53	3C
▶c	~2.20	4C
d	6.05 or 6.33	1D
e	6.05 or 6.33	1D

▶ $C_6H_{13}NO$ ▶ 7% $CDCl_3$ ▶ V 142

▶ 2-BkJa

CH_3(b)—C(=O)—CH_2CH_2(c)—C(=O)—OCH_2(d)CH_3(a)

	Shift	Peak
a	1.26	3A
b	2.20	1A
▶c	2.64	Q
d	4.12	4A

▶ $C_7H_{12}O_3$ ▶ 45.5 mg/0.50 ml $CDCl_3$ ▶ S 247

▶ (2)-(B)v(J)v

	Shift	Peak
▶a	2.60	N
b	3.08	3C

▶ C_9H_8O ▶ 7% $CDCl_3$ ▶ V 229

▶ (2)-(B)l(J)l

	Shift	Peak
a	2.02	1A
▶b	2.43	1A
c	2.43	1A
d	6.77	1A

▶ $C_6H_8O_2$ ▶ 7% $CDCl_3$ ▶ V 129

▶ 2-BaKb

CH_3(a)CH_2(c)CH_2(d)C(=O)—OCH_2(e)CH_3(b)

	Shift	Peak
a	0.96	3A
b	1.28	3A
c	1.67	6B
▶d	2.28	3C
e	4.13	4A

▶ $C_6H_{12}O_2$ ▶ 45.0 mg/0.50 ml $CDCl_3$ ▶ S 42

▶ (2)-(B)l(J)n

	Shift	Peak
▶a	2.69 or 3.18	Q
b	2.69 or 3.18	Q
c	3.28	1A

▶ C_5H_7NOS ▶ 7% $CDCl_3$ ▶ V 434

▶ 2-BbKa

CH_3(a)—$(CH_2)_{11}$(b)—CH_2(c)—C(=O)—OCH_3(d)

	Shift	Peak
a	0.88	3B
b	1.29	1A
▶c	2.24	3B
d	3.68	1A

▶ $C_{15}H_{30}O_2$ ▶ 45.1 mg/0.50 ml $CDCl_3$ ▶ S 195

▶ 2-BnJa

	Shift	Peak
a	2.14	1A
▶b	2.70	3A
c	2.94	1A
d	3.66	3A

▶ $C_{11}H_{15}NO$ ▶ 7% $CDCl_3$ ▶ V 584

▶ 2-BbKa

CH_3(a)—$(CH_2)_9$(b)—CH_2(c)—C(=O)—OCH_3(d)

	Shift	Peak
a	0.88	3B
b	1.28	1B
▶c	2.28	3B
d	3.67	1A

▶ $C_{13}H_{26}O_2$ ▶ 45.7 mg/0.50 ml $CDCl_3$ ▶ S 125

▶ (2)-(B)n(J)l

	Shift	Peak
a	2.33	1A
▶b	3.13	3A
c	3.83	1A
d	4.28	4A
e	6.85 or 6.98	1A
f	6.85 or 6.98	1A
g	7.13	1B

▶ $C_{13}H_{13}NO_2$ ▶ 7% $CDCl_3$ ▶ V 299

▶ 2-BbKa

	Shift	Peak
▶a	2.28	3B
b	2.88	3B
c	3.67	1A
d	7.12	R
e	7.62	R
f	7.70	R
g	1.35	1F

▶ $C_{15}H_{22}O_3S$ ▶ 7% $CDCl_3$ ▶ V 320

▶ 2-BvJh

	Shift	Peak
▶a	2.74	Q
b	2.97	Q
c	7.22	1A
d	9.81	3A

▶ $C_9H_{10}O$ ▶ 7% $CDCl_3$ ▶ V 529

▶ 2-BbKa

CH_3(a)—$(CH_2)_6$(b)—CH_2(c)C(=O)—OCH_3(d)

	Shift	Peak
a	0.88	3B
b	1.28	1B
▶c	2.30	3B
d	3.68	1A

▶ $C_{10}H_{20}O_2$ ▶ 45.1 mg/0.50 ml $CDCl_3$ ▶ S 249

▶ 2-BbKa

(a) (b) (c) (d)

$CH_3-(CH_2)_{13}-CH_2-C(=O)-O-CH_3$

	Shift	Peak
a	~0.88	1D
b	1.26	1A
▶c	~2.30	1D
d	3.63	1A

▶ $C_{17}H_{34}O_2$ ▶ 44.4 mg/0.50 ml $CDCl_3$ ▶ S 248

▶ 2-BbKb

$(e)H-C(H(d))-O-C(=O)-CH_2-(CH_2)_{15}-CH_3$ (c) (b) (a)

$(f)H-C-O-C(=O)-CH_2-(CH_2)_{15}-CH_3$ (c) (b) (a)

$(e)H-C(H(d))-O-C(=O)-CH_2-(CH_2)_{15}-CH_3$ (c) (b) (a)

	Shift	Peak
a	0.90	3B
b	1.25	1A
▶c	2.30	3B
d	4.18*	S
e	4.18*	S
f	5.28	S

*or 4.28

▶ $C_{57}H_{110}O_6$ ▶ 7% $CDCl_3$ ▶ V 368

▶ 2-BbKa

(f) (c) (b) (d) (e)

$CH-CH_2-(CH_2)_5-CH_2-C(=O)-OCH_3$

$||$

$CH-CH_2-(CH_2)_6-CH_3$

(f) (c) (b) (a)

	Shift	Peak
a	0.90	3B
b	1.29	1B
c	2.00	1C
▶d	2.32	S
e	3.70	1A
f	5.4	3B

▶ $C_{19}H_{36}O_2$ ▶ 33.7 mg/0.50 ml $CDCl_3$ ▶ S 71

▶ 2-BbKc

$(e)H-C(H(d))-O-C(=O)-CH_2-(CH_2)_{15}-CH_3$ (c) (b) (a)

$(f)H-C-O-C(=O)-CH_2-(CH_2)_{15}-CH_3$ (c) (b) (a)

$(e)H-C(H(d))-O-C(=O)-CH_2-(CH_2)_{15}-CH_3$ (c) (b) (a)

	Shift	Peak
a	0.90	3B
b	1.25	1A
▶c	2.30	3B
d	4.18*	S
e	4.18*	S
f	5.28	S

*or 4.28

▶ $C_{57}H_{110}O_6$ ▶ 7% $CDCl_3$ ▶ V 368

▶ 2-BbKb

(a) (c) (d) (e) (b)

$CH_3-(CH_2)_4-CH_2-C(=O)-OCH_2-CH_3$

	Shift	Peak
a	0.88	3B
b	1.26	S
c	1.33	S
▶d	2.28	3B
e	4.12	4A

▶ $C_9H_{18}O_2$ ▶ 44.5 mg/0.50 ml $CDCl_3$ ▶ S 283

▶ 2-BbKh

(e) (c) (b) (d) (f)

$CH-CH_2-(CH_2)_6-CH_2-C(=O)-OH$

$||$

$CH-CH_2-(CH_2)_6-CH_3$

(e) (c) (b) (a)

	Shift	Peak
a	0.88	3B
b	1.29	1B
c	2.00	1F
▶d	2.25	3B
e	5.28	3D
f	11.45	1C

▶ $C_{18}H_{34}O_2$ ▶ 45.2 mg/0.50 ml $CDCl_3$ ▶ S 70

▶ 2-BbKb

(a) (c) (d) (e) (b)

$CH_3-(CH_2)_5-CH_2-C(=O)-OCH_2CH_3$

	Shift	Peak
a	0.88	3B
b	1.28	3D
c	1.30	S
▶d	2.28	3B
e	4.12	4A

▶ $C_{10}H_{20}O_2$ ▶ 45.2 mg/0.50 ml $CDCl_3$ ▶ S 244

▶ 2-BbKh

(a) (b) (c)

$CH_3(CH_2)_{17}CH_2C(=O)-OH$

	Shift	Peak
a	0.87	3B
b	1.24	1A
▶c	~2.27	2B

▶ $C_{20}H_{40}O_2$ ▶ 45.2 mg/0.50 ml $CDCl_3$ ▶ S 260

▶ 2-BbKb

(a) (d) (c) (b) (c) (d) (a)

$CH_3CH_2O-C(=O)-CH_2(CH_2)_6CH_2C(=O)-OCH_2CH_3$

	Shift	Peak
a	1.26	3D
b	1.35	1E
▶c	2.28	3B
d	4.13	4A

▶ $C_{14}H_{26}O_4$ ▶ 45.1 mg/0.50 ml $CDCl_3$ ▶ S 250

▶ 2-BbKh

(a) (b) (c) (d)

$CH_3-(CH_2)_{11}-CH_2-C(=O)-OH$

	Shift	Peak
a	0.88	3B
b	1.26	1A
▶c	~2.35	1D
d	10.95	1B

▶ $C_{14}H_{28}O_2$ ▶ 44.8 mg/0.50 ml $CDCl_3$ ▶ S 167

▶ 2-BbKb

(a) (b) (c) (d) (a)

$CH_3-(CH_2)_9-CH_2-C(=O)-OCH_2CH_3$

	Shift	Peak
a	0.88	3B
b	1.28	1E
▶c	2.28	3C
d	4.11	4A

▶ $C_{14}H_{28}O_2$ ▶ 45.5 mg/0.50 ml $CDCl_3$ ▶ S 194

▶ 2-BbKh

(a) (b) (c) (d)

$CH_3-(CH_2)_3-CH_2-C(=O)-OH$

	Shift	Peak
a	0.92	3B
b	1.45	O
▶c	2.36	3C
d	11.37	1B

▶ $C_6H_{12}O_2$ ▶ 44.7 mg/0.50 ml $CDCl_3$ ▶ S 37

▶ 2-BbKh

	Shift	Peak
a	0.88	3B
b	1.30	1G
▶c	2.36	3B
d	11.47	1A

(a) (b) (c) (d)
$CH_3(CH_2)_5CH_2—C(=O)—OH$

▶ $C_8H_{16}O_2$ ▶ 45.6 mg/0.50 ml $CDCl_3$ ▶ S 157

▶ 2-BkKb

	Shift	Peak
a	1.25	3A
▶b	2.62	1A
c	4.16	4A

(a) (c) (b) (b) (c) (a)
$CH_3CH_2—O—C(=O)—CH_2CH_2—C(=O)—O—CH_2CH_3$

▶ $C_8H_{14}O_4$ ▶ 45.5 mg/0.50 ml $CDCl_3$ ▶ S 183

▶ 2-BbKh

	Shift	Peak
a	0.88	3B
b	1.32	1B
▶c	2.38	3B
d	11.47	1B

(a) (b) (c) (d)
$CH_3—(CH_2)_4—CH_2—C(=O)—OH$

▶ $C_7H_{14}O_2$ ▶ 45.6 mg/0.50 ml $CDCl_3$ ▶ S 49

▶ 2-BkKb

	Shift	Peak
a	1.25	3A
▶b	2.62	1A
c	4.15	4A

(a) (c) (b) (b) (c) (a)
$CH_3CH_2OC(=O)—CH_2CH_2C(=O)—O—CH_2CH_3$

▶ $C_8H_{14}O_4$ ▶ 7% $CDCl_3$ ▶ V 215

▶ 2-BcK

	Shift	Peak
a	~2.12	S
▶b	~2.33	S
c	3.77	3B

(c) (a) (b)
$^{(-)}O(O=)C—CH(ND_2)—CH_2—CH_2—C(=O)O^{(-)}$ $D_3O^{(+)}$ $Na^{(+)}$

▶ $C_5H_8NaNO_4$ ▶ D_2O ▶ V 435

▶ 2-BkKb

	Shift	Peak
▶a	2.70	1A
b	5.10	1A

(b) (a) (a) (b)
$C_6H_5—CH_2—O—C(=O)—CH_2—CH_2—C(=O)—O—CH_2—C_6H_5$

▶ $C_{18}H_{18}O_4$ ▶ 45.0 mg/0.50 ml $CDCl_3$ ▶ S 140

▶ 2-BcKh

	Shift	Peak
a	0.93	2A
b	~1.58	3C
▶c	2.36	3B
d	11.67	1A

(a) (b) (c) (d)
$CH_3—CH(CH_3)—CH_2—CH_2—C(=O)—OH$
(a)

▶ $C_6H_{12}O_2$ ▶ 45.4 mg/0.50 ml $CDCl_3$ ▶ S 156

▶ 2-BnKh

	Shift	Peak
a	2.54	Q
▶b	3.19	Q

(a) (b)
$D_2N—CH_2—CH_2—C(=O)OD$

▶ $C_3H_7NO_2$ ▶ D_2O ▶ V 394

▶ 2-BiKh

	Shift	Peak
▶a	3.07	Q
b	3.30	Q
c	11.52	1A

(b) (a) (c)
ICH_2CH_2COOH

▶ $C_3H_5IO_2$ ▶ 7% $CDCl_3$ ▶ V 27

▶ (2)-(B)q(K)b

	Shift	Peak
▶a	3.56	8A
b	4.29	8A

O=C—CH_2(a), O—CH_2(b) (four-membered ring)

▶ $C_3H_4O_2$ ▶ 7% $CDCl_3$ ▶ V 409

▶ 2-BjKb

	Shift	Peak
a	1.26	3A
b	2.20	1A
▶c	2.64	Q
d	4.12	4A

(b) (d) (a)
$CH_3—C(=O)—CH_2CH_2C(=O)—OCH_2CH_3$
(c)

▶ $C_7H_{12}O_3$ ▶ 45.5 mg/0.50 ml $CDCl_3$ ▶ S 247

▶ 2-BuKa

	Shift	Peak
a	2.68	1A
▶b	2.68	1A
c	3.75	1A

(a) (b)
$N≡C—CH_2—CH_2—C(=O)OCH_3$
(c)

▶ $C_5H_7NO_2$ ▶ 7% $CDCl_3$ ▶ V 106

▶ 2-BvKh

(b) CH_2 – COOH
CH_2(c)
(a) CH_3 O O CH_3(a)
(a) CH_3 CH_3(a)
(e)H H(e)
H OCH_3 H
(f) (d) (f)

	Shift	Peak
a	1.42	1A
▶b	2.62	Q
c	2.86	Q
d	3.75	1A
e	5.50	2C
f	6.53	2C

▶ $C_{20}H_{24}O_5$ ▶ 7% $CDCl_3$ ▶ V 344

▶ 2-BaLhl

(a) (b) (c) (d)
$H_3CH_2CH_2C$ H
C = C
H (c) (b) (a)
(d) $CH_2CH_2CH_3$

	Shift	Peak
a	.85	3C
b	1.32	6B
▶c	1.91	N
d	5.32	N

▶ C_8H_{16} ▶ 50 vol.% CCl_4 ▶ API 432

▶ 2-BvKh

O
(a) (b) ‖ (c)
CH_2CH_2C — OH

	Shift	Peak
a	~2.79	Q
▶b	~2.79	Q
c	11.71	1A

▶ $C_9H_{10}O_2$ ▶ 45.7 mg/0.25 ml $CDCl_3$ ▶ S 299

▶ 2-BaLhl

(c) (e)
H_3C H
C = C
(d) (b) (a)
H $CH_2CH_2CH_3$
(e)

	Shift	Peak
a	.88	3C
b	1.28	S
c	1.60	2E
▶d	1.92	1F
e	5.35	O

▶ C_6H_{12} ▶ 50 vol.% CCl_4 ▶ API 370

▶ 2-BaLal

(c) (b) (a)
H_3C $CH_2CH_2CH_3$
C = C
H CH_3
(d)

	Shift	Peak
a	.83	3C
b	1.33	S
▶c	1.87	3B
d	5.14	O

▶ C_7H_{14} ▶ liquid ▶ API 395

▶ 2-BaLhl

(d) (f)
H H
C = C H(c) H(b) H(a)
H C — C — C — H(a)
(e) H H H
(c) (b) (a)

	Shift	Peak
a	.92	3C
b	1.38	6B
▶c	1.93	4C
d	4.76	R
e	4.98	R
f	5.68	R

▶ C_5H_{10} ▶ liquid ▶ API 366

▶ 2-BaLal

(c)
(e)H CH_3
C = C
(d) (b) (a)
(e)H $CH_2CH_2CH_3$

	Shift	Peak
a	.88	3C
b	1.34	S
c	1.57	1H
▶d	1.90	S
e	4.60	N

▶ C_6H_{12} ▶ 50 vol.% CCl_4 ▶ API 373

▶ 2-BaLhl

(b) (d) (e)
H_3CH_2C H
C = C
(d) (c) (a)
H $CH_2CH_2CH_3$
(e)

	Shift	Peak
a	.90	3D
b	.97	3D
c	1.33	S
▶d	1.95	O
e	5.40	O

▶ C_7H_{14} ▶ liquid ▶ API 391

▶ 2-BaLhl

(c) (e)
H_3C H
C = C
H CH_2 — CH_2 — CH_3
(f) (d) (b) (a)

	Shift	Peak
a	.88	3B
b	~1.33	O
c	1.69	N
▶d	1.87	1F
e	5.32	S
f	5.32	S

▶ C_6H_{12} ▶ 10% CCl_4 ▶ API 32

▶ 2-BaLhl

(c) (d) (b) (a)
H_3C $CH_2CH_2CH_3$
C = C
H H
(e) (e)

	Shift	Peak
a	.89	3C
b	1.30	S
c	1.56	2E
▶d	2.00	4C
e	5.32	O

▶ C_6H_{12} ▶ liquid ▶ API 369

▶ 2-BaLhl

(d) H_3C H(f)
C = C
(e) (b) (a)
H_3C $CH_2CH_2CH_3$
(c)

	Shift	Peak
a	.90	3C
b	1.28	S
c	~1.58 or 1.68	O
d	~1.68 or 1.58	O
▶e	1.91	4B
f	5.08	3C

▶ C_7H_{14} ▶ liquid ▶ API 408

▶ (2)-(B)b(L)al

(a)
CH_3
C
(c) H_2C C — CH_3(a)
(b) H_2C CH_2(c)
C
(b) H_2

	Shift	Peak
a	1.58	1A
b	~1.87	1I
▶c	~1.94	S

▶ C_8H_{14} ▶ 33% CCl_4 ▶ API 143

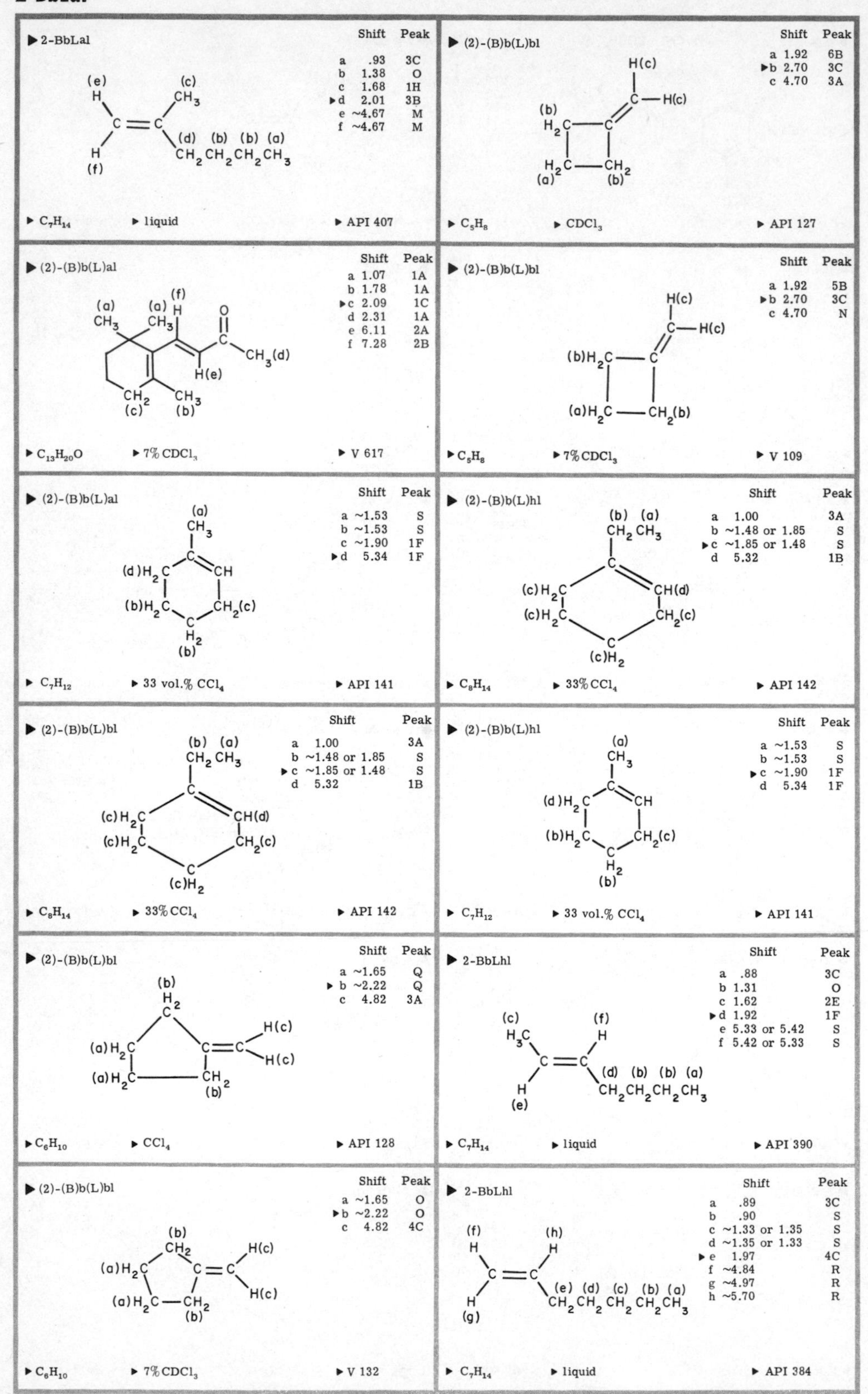
2-BbLal
Shift Peak
a .93 3C
b 1.38 O
c 1.68 1H
d 2.01 3B
e ~4.67 M
f ~4.67 M
C7H14 liquid API 407
(2)-(B)b(L)bl
Shift Peak
a 1.92 6B
b 2.70 3C
c 4.70 3A
C5H8 CDCl3 API 127
(2)-(B)b(L)al
Shift Peak
a 1.07 1A
b 1.78 1A
c 2.09 1C
d 2.31 1A
e 6.11 2A
f 7.28 2B
C13H20O 7% CDCl3 V 617
(2)-(B)b(L)bl
Shift Peak
a 1.92 5B
b 2.70 3C
c 4.70 N
C5H8 7% CDCl3 V 109
(2)-(B)b(L)al
Shift Peak
a ~1.53 S
b ~1.53 S
c ~1.90 1F
d 5.34 1F
C7H12 33 vol.% CCl4 API 141
(2)-(B)b(L)hl
Shift Peak
a 1.00 3A
b ~1.48 or 1.85 S
c ~1.85 or 1.48 S
d 5.32 1B
C8H14 33% CCl4 API 142
(2)-(B)b(L)bl
Shift Peak
a 1.00 3A
b ~1.48 or 1.85 S
c ~1.85 or 1.48 S
d 5.32 1B
C8H14 33% CCl4 API 142
(2)-(B)b(L)hl
Shift Peak
a ~1.53 S
b ~1.53 S
c ~1.90 1F
d 5.34 1F
C7H12 33 vol.% CCl4 API 141
(2)-(B)b(L)bl
Shift Peak
a ~1.65 Q
b ~2.22 Q
c 4.82 3A
C6H10 CCl4 API 128
2-BbLhl
Shift Peak
a .88 3C
b 1.31 O
c 1.62 2E
d 1.92 1F
e 5.33 or 5.42 S
f 5.42 or 5.33 S
C7H14 liquid API 390
(2)-(B)b(L)bl
Shift Peak
a ~1.65 O
b ~2.22 O
c 4.82 4C
C6H10 7% CDCl3 V 132
2-BbLhl
Shift Peak
a .89 3C
b .90 S
c ~1.33 or 1.35 S
d ~1.35 or 1.33 S
e 1.97 4C
f ~4.84 R
g ~4.97 R
h ~5.70 R
C7H14 liquid API 384

▶ 2-BbLhl

(d) H, (f) H, C=C, H (e), $CH_2(CH_2)_5CH_3$ (c) (b) (a)

	Shift	Peak
a	.85	3B
b	1.28	1B
▶c	1.99	3B
d	4.88	R
e	4.99	R
f	5.68	R

▶ C_9H_{18} ▶ 50% CCl_4 ▶ API 434

▶ 2-BbLhl

(c) H_3C, C=C, $CH_2CH_2CH_2CH_3$ (d) (b) (b) (a), H (e), H (e)

	Shift	Peak
a	.90	3C
b	1.31	O
c	1.57	2E
▶d	2.02	1F
e	~5.37	O

▶ C_7H_{14} ▶ liquid ▶ API 389

▶ 2-BbLhl

(e)H, H(g), C=C, (f)H, $CH_2CH_2CH_2CH_3$ (d) (c) (b) (a)

	Shift	Peak
a	.90	3B
b	~1.34	S
c	~1.34	S
▶d	2.00	4C
e	4.85	R
f	4.97	R
g	5.68	R

▶ C_6H_{12} ▶ 10% CCl_4 ▶ API 155

▶ 2-BbLhl

(e) H, (f) H, C=C, H (d), $CH_2(CH_2)_{12}CH_3$ (c) (b) (a)

	Shift	Peak
a	.99	3B
b	1.25	1A
▶c	2.02	3B
d	4.90	R
e	5.08	R
f	5.68	R

▶ $C_{16}H_{32}$ ▶ 50% CCl_4 ▶ API 73

▶ 2-BbLhl

(e) H, (g) H, C=C, H (f), $CH_2CH_2CH_2CH_3$ (d) (c) (b) (a)

	Shift	Peak
a	.88	3C
b	~1.29 or 1.31	S
c	~1.31 or 1.29	S
▶d	2.00	3C
e	4.84	R
f	4.97	R
g	5.65	R

▶ C_6H_{12} ▶ liquid ▶ API 368

▶ 2-BbLhi

(d) H, (f) H, C=C, H (e), $CH_2(CH_2)_{12}CH_3$ (c) (b) (a)

	Shift	Peak
a	.90	3B
b	1.24	1A
▶c	2.04	3B
d	4.88	R
e	5.02	R
f	5.72	R

▶ $C_{16}H_{32}$ ▶ 50 vol.% CCl_4 ▶ API 437

▶ 2-BbLhl

(e) (c) (b) (d) (f): $CH-CH_2-(CH_2)_5-CH_2-C(=O)-OH$ ‖ $CH-CH_2-(CH_2)_6-CH_3$ (e) (c) (b) (a)

	Shift	Peak
a	0.88	3B
b	1.29	1B
▶c	2.00	1F
d	2.25	3B
e	5.28	3D
f	11.45	1C

▶ $C_{18}H_{34}O_2$ ▶ 45.2 mg/0.50 ml $CDCl_3$ ▶ S 70

▶ 2-BbLhl

(d) H, (f) H, C=C, H (e), $CH_2(CH_2)_8CH_3$ (c) (b) (a)

	Shift	Peak
a	.92	3B
b	1.32	1B
▶c	2.05	3B
d	4.86	R
e	5.00	R
f	5.68	R

▶ $C_{12}H_{24}$ ▶ 50 vol.% CCl_4 ▶ API 435

▶ 2-BbLhl

(f) (c) (b) (d) (e): $CH-CH_2-(CH_2)_5-CH_2-C(=O)-OCH_3$ ‖ $CH-CH_2-(CH_2)_6-CH_3$ (f) (c) (b) (a)

	Shift	Peak
a	0.90	3B
b	1.29	1B
▶c	2.00	1C
d	2.32	S
e	3.70	1A
f	5.4	3B

▶ $C_{19}H_{36}O_2$ ▶ 33.7 mg/0.50 ml $CDCl_3$ ▶ S 71

▶ 2-BbLhl

(a) (b) (c): $CH_3-(CH_2)_8-CH_2-C(H)(f)=C(H)(d)H(e)$

	Shift	Peak
a	0.87	3B
b	1.28	1A
▶c	2.05	2G
d	4.80	R
e	~5.03	R
f	5.68	R

▶ $C_{12}H_{24}$ ▶ 44.4 mg/0.50 ml $CDCl_3$ ▶ S 202

▶ 2-BbLhl

(d)H, H(f), C=C, (e)H, $(CH_2)_4CH_3$ (c+b) (a)

	Shift	Peak
a	.91	3B
b	1.31	S
▶c	2.02	3B
d	4.84	R
e	4.98	R
f	5.67	R

▶ C_7H_{14} ▶ 10% CCl_4 ▶ API 157

▶ (2)-(B)b(L)hl

(b) H_2C, (c) CH, CH (d), (a) H_2C, (a) H_2C, CH (d), H_2C (b), CH (c)

	Shift	Peak
a	~1.52	N
▶b	~2.20	O
c	~5.63 or 5.77	S
d	~5.77 or 5.63	S

▶ C_8H_{12} ▶ $CDCl_3$ ▶ API 129

▶ (2)-(B)b(L)hl

(b) H(c)
CH2—C
(a)H2C C—H(d)
(a)H2C C—H(d)
CH2—C
(b) H(c)

	Shift	Peak
a	~1.52	N
▶b	~2.20	O
c	~5.63	S
d	~5.77	1E

▶ C_8H_{12} ▶ 7% $CDCl_3$ ▶ V 209

▶ 2-BcLhl

(e) (g)
H H
C=C CH3
H CH2CH2CHCH3
(f) (d) (b) (c) (a)

	Shift	Peak
a	.91	2C
b	1.32	S
c	1.60	S
▶d	2.01	4C
e	4.87	R
f	5.02	R
g	5.70	R

▶ C_7H_{14} ▶ liquid ▶ API 394

▶ (2)-(B)b(L)hl

(c)
H
C
(b)H2C CH(c)
(a)H2C CH2(b)

	Shift	Peak
a	1.81	O
▶b	2.28	3C
c	5.65	1A

▶ C_5H_8 ▶ 10% CCl_4 ▶ API 164

▶ (2)-(B)d(L)hl

(a) (b) H(i) O
CH3 CH3
C—CH3(e)
H
(d)H2C (f) H(h)
CH3
(g)H (c)

	Shift	Peak
a	0.86	1A
b	0.93	1A
c	1.57	2G
▶d	2.05	1C
e	2.26	1A
f	2.30	S
g	5.50	1C
h	6.06	2A
i	6.60	4A

▶ $C_{13}H_{20}O$ ▶ 7% $CDCl_3$ ▶ V 616

▶ 2-BbLhl

(d) (f)
H H
C=C
H (c) (b) (a)
(e) CH2(CH2)10CH3

	Shift	Peak
a	.88	3B
b	1.27	1A
▶c	2.3	3B
d	4.88	7X
e	5.01	7X
f	5.72	R

▶ $C_{14}H_{28}$ ▶ 50 vol.% CCl_4 ▶ API 436

▶ 2-Bd(L)ll

(g)H---O
O
H(i)
(h)
(j)H H (c) (b)
CH2 CH3
(a)CH3—CH(e) CH2 C
(d) CH3(b) OH(f)
(a)CH3

	Shift	Peak
a	1.27	2A
b	1.34	1E
c	1.69	O
▶d	2.78	O
e	3.28	S
f	5.27	1B
g	5.27	1B
h	7.19*	2F
i	7.19*	2F
j	7.39	1A

*or 7.31

▶ $C_{15}H_{22}O_3$ ▶ 7% $CDCl_3$ ▶ V 647

▶ (2)-(B)b(L)jl

(e)
CH2
H(h) O CH2(a)
(g)H CH2(b)
(f)H N O
CH3O CH3
(c) (d)

	Shift	Peak
a	2.03	5B
▶b	2.63	3A
c	3.87 or 3.92	1A
d	3.87 or 3.92	1A
e	4.30	3A
f	~7.00	R
g	~7.10	R
h	7.52	4D

▶ $C_{14}H_{15}NO_3$ ▶ 7% $CDCl_3$ ▶ V 312

▶ (2)-(B)j(L)al

(d)
H---O
O
CH2(b)
CH2(c)
CH3
(a)

	Shift	Peak
a	2.02	1A
b	2.43	1A
▶c	2.43	1A
d	6.77	1A

▶ $C_6H_8O_2$ ▶ 7% $CDCl_3$ ▶ V 129

▶ (2)-(B)b(L)jl

H(h) O (b)
CH2
(g)H CH2(a)
CH2(e)
(f)H N O
CH3O CH3
(d) (c)

	Shift	Peak
a	2.05	5B
▶b	2.70	3A
c	3.77 or 3.87	1A
d	3.77 or 3.87	1A
e	4.33	3A
f	~7.00	S
g	~7.15	S
h	8.00	4D

▶ $C_{14}H_{15}NO_3$ ▶ 7% $CDCl_3$ ▶ V 311

▶ (2)-(B)j(L)ns

(b) (a)
H2C—CH2
S C N C O
CH3
(c)

	Shift	Peak
a	2.69 or 3.18	Q
▶b	2.69 or 3.18	Q
c	3.28	1A

▶ C_5H_7NOS ▶ 7% $CDCl_3$ ▶ V 434

▶ 2-Bb(L)ll

(d) (c) (a)
H CH2(CH2)3CH3
H N CH3(b)
(e) H(f)

	Shift	Peak
a	0.90	3B
b	2.15	1A
▶c	~2.38	S
d	5.98	3A
e	6.55	3A
f	7.67	1D

▶ $C_{10}H_{17}N$ ▶ 7% $CDCl_3$ ▶ V 278

▶ 2-BlLal

(a) (a)
CH3 CH3(a)
CH3 (d) (b) (b) (d) (b) (b) (c)
C=CH—CH2—(CH2—C=CH—CH2)7—CH2—C=CH—CH2—OH
CH3
(a)

	Shift	Peak
a	1.60 or 1.68	1G
▶b	2.04 ± 0.05	1G
c	4.15	2G
d	5.12	1F

▶ $C_{45}H_{74}O$ ▶ 7% $CDCl_3$ ▶ V 367

▶ 2-BlLal

	Shift	Peak
	a 1.62 or 1.68	1A
▶	b 2.05 ± 0.03	3A
	c 4.15	2A
	d 5.12	1C
	e 5.45	3A

▶ $C_{10}H_{18}O$ ▶ 7% $CDCl_3$ ▶ V 279

▶ (2)-(B)n(L)hl

	Shift	Peak
	a 1.63	1A
▶	b 2.07	O
	c 2.95	3A
	d 3.33	N
	e 5.72 or 5.77	1E
	f 5.72 or 5.77	1E

▶ C_5H_9N ▶ 7% $CDCl_3$ ▶ V 115

▶ 2-BlLhl

	Shift	Peak
	a 1.60 or 1.68	1G
▶	b 2.04 ± 0.05	1G
	c 4.15	2G
	d 5.12	1F

▶ $C_{45}H_{74}O$ ▶ 7% $CDCl_3$ ▶ V 367

▶ (2)-Bn(L)ln

	Shift	Peak
▶	a 2.75	S
	b 3.02	S
	c 4.96	1E
	d 4.96	1E
	e 6.82	1A
	f 7.55	1A

▶ $C_5H_9N_3$ ▶ 7% $CDCl_3$ ▶ V 443

▶ 2-BlLhl

	Shift	Peak
	a 1.62 or 1.68	1A
▶	b 2.05 ± 0.03	3A
	c 4.15	2A
	d 5.12	1C
	e 5.45	3A

▶ $C_{10}H_{18}O$ ▶ 7% $CDCl_3$ ▶ V 279

▶ (2)-(B)o(L)jl

	Shift	Peak
▶	a 3.27	Q
	b 3.87 or 3.92	1A
	c 3.87 or 3.92	1A
	d 4.73	Q
	e ~7.05	R
	f ~7.23	R
	g 8.05	4D

▶ $C_{13}H_{13}NO_3$ ▶ 7% $CDCl_3$ ▶ V 300

▶ 2-BlLhl

	Shift	Peak
▶	a 2.08	2E
	b ~4.88 or 4.99	R
	c ~4.99 or 4.88	R
	d ~5.67	R

▶ C_6H_{10} ▶ liquid ▶ API 474

▶ 2-Bp(L)jl

	Shift	Peak
▶	a 3.03	3A
	b ~3.67	1D
	c 3.92	3D
	d 3.92 or 4.00	1E
	e 3.92 or 4.00	1E
	f 3.92 or 4.00	1E

▶ $C_{14}H_{17}NO_4$ ▶ 7% $CDCl_3$ ▶ V 313

▶ 2-BlLhl

	Shift	Peak
▶	a 2.13	3A
	b 4.93	R
	c ~5.00	R
	d ~5.70	R

▶ C_6H_{10} ▶ 10% CCl_4 ▶ API 161

▶ 2-BmMl

	Shift	Peak
▶	a 3.92	1A
	b 5.97	1A
	c 6.79	2A
	d 7.07	S
	e 7.32	4A
	f 8.17	2A

▶ $C_{18}H_{16}N_2O_4$ ▶ 7% $CDCl_3$ ▶ V 673

▶ (2)-(B)l(L)hl

	Shift	Peak
▶	a 2.38	1B
	b 5.58	1C

▶ C_8H_{12} ▶ 7% $CDCl_3$ ▶ V 511

▶ (2)-(B)n(M)l

	Shift	Peak
	a 1.29	1A
	b 2.76	3A
	c 3.40	3C
	d 3.79	1F
	e 3.79	1F
▶	f 4.18	3C

▶ $C_9H_{14}N_2$ ▶ 7% $CDCl_3$ ▶ V 541

▶ 2-Ba(N)ll

	Shift	Peak
	a 0.90	3C
	b 1.58	6B
▶	c 3.09	2A
	d 3.19	N
	e 4.72	5B
	f 5.49	1C
	g 5.73	4A
	h 7.05	2D

▶ $C_9H_{14}N_2O$ ▶ 7% $CDCl_3$ ▶ V 542

▶ (2)-(B)b(N)bj

	Shift	Peak
	a 2.08 ± .10	S
	b 2.31 ± .10	S
	c 3.47	1B
▶	d 3.51	3D

▶ $C_{10}H_{16}N_2O_2$ ▶ 7% $CDCl_3$ ▶ V 572

▶ (2)-(B)b(N)aj

	Shift	Peak
	a 2.83	1A
▶	b 3.40	3A

▶ C_5H_9NO ▶ 7% $CDCl_3$ ▶ V 116

▶ (2)-(B)b(N)bj

	Shift	Peak
	a ~1.62	1B
	b ~1.62	1B
▶	c ~3.57	1C
	d 5.93	1A
	e 6.43	2A
	f ~7.40	S

▶ $C_{17}H_{19}NO_3$ ▶ 7% $CDCl_3$ ▶ V 328

▶ 2-BbNbb

	Shift	Peak
	a 0.92	3B
	b 1.37	6B
▶	c 2.40	3B

▶ $C_{12}H_{27}N$ ▶ 44.8 mg/0.50 ml $CDCl_3$ ▶ S 84

▶ (2)-(B)b(N)bs

	Shift	Peak	Shift	Peak
	a 1.95	5B	d 3.37	1E
	b 2.38	1A	e 7.30	Q
▶	c 3.33	3A	f 7.63	Q

▶ $C_{19}H_{24}N_2O_4S_2$
▶ 7% $CDCl_3$
▶ V 336

▶ 2-Bb(N)bb

	Shift	Peak
	a 1.79	1E
▶	b 2.46	4B
	c 2.81	3A
	d 3.74	3B

▶ $C_7H_{16}N_2O$ ▶ 33.2 mg/0.50 ml $CDCl_3$ ▶ S 69

▶ 2-BbNbv

	Shift	Peak
	a 0.96	3B
	b 1.48	1F
▶	c 3.28	3C

▶ $C_{14}H_{23}N$ ▶ 45.2 mg/0.50 ml $CDCl_3$ ▶ S 147

▶ (2)-(B)b(N)bh

	Shift	Peak
	a 0.82	2A
	b 1.47	S
	c 1.55 ± 0.10	S
	d 2.21	S
▶	e 2.52	S
	f 3.00	S
	g 3.00	S

▶ $C_6H_{13}N$ ▶ 7% $CDCl_3$ ▶ V 478

▶ (2)-(B)b(N)cj

	Shift	Peak
	a 1.62	2A
	b 2.12 ± .10	S
	c 2.36 ± .10	S
▶	d 3.52	3A

▶ $C_{10}H_{16}N_2O_2$ ▶ 7% $CDCl_3$ ▶ V 571

▶ (2)-(B)b(N)bh

	Shift	Peak
	a 0.82	2A
	b 1.47	S
	c 1.55 ± 0.10	S
	d 2.21	S
	e 2.52	S
	f 3.00	S
▶	g 3.00	S

▶ $C_6H_{13}N$ ▶ 7% $CDCl_3$ ▶ V 478

▶ 2-BbNhh

	Shift	Peak
	a 1.79	1E
▶	b 2.46	4B
	c 2.81	3A
	d 3.74	3B

▶ $C_7H_{16}N_2O$ ▶ 33.2 mg/0.50 ml $CDCl_3$ ▶ S 69

▶ 2-BbNhh

(a) (c) (b)
$CH_3—CH_2—CH_2—CH_2—NH_2$

	Shift	Peak
a	0.92	3B
b	1.10	1A
▶ c	2.70	3C

▶ $C_4H_{11}N$ ▶ 7% $CDCl_3$ ▶ V 89

▶ (2)-(B)b(N)hv

(e) CH_3O, (b) CH_2, CH_2(a), CH_2(c), N, H, (d)

	Shift	Peak
a	1.93	5B
b	2.73	3A
▶ c	3.22	3C
d	3.49	1B
e	3.77	1A

▶ $C_{10}H_{13}NO$ ▶ 7% $CDCl_3$ ▶ V 266

▶ 2-BbNhh

(a) (b) (d) (c)
$CH_3—(CH_2)_4—CH_2—NH_2$

	Shift	Peak
a	0.90	3B
b	1.28	1B
c	1.58	S
▶ d	2.70	1C

▶ $C_6H_{15}N$ ▶ 35.6 mg/0.50 ml $CDCl_3$ ▶ S 46

▶ (2)-(B)b(N)jj

(a)H_2C—CH_2(c), (d)H_2C, N, O, C, (b)CH_3, O

	Shift	Peak
a	2.06	O
b	2.50	1A
c	2.62	S
▶ d	3.83	3A

▶ $C_6H_9NO_2$ ▶ 7% $CDCl_3$ ▶ V 465

▶ 2-BbNhj

(a) (b) (c) (e) O (d)
$CH_3—(CH_2)_2—CH_2—NH—C(=O)—NH_2$

	Shift	Peak
a	0.88	3D
b	~1.2-1.5	N
▶ c	~3.08	4C
d	5.18	1B
e	5.95	3C

▶ $C_5H_{12}N_2O$ ▶ 45.3 mg/0.25 ml $CDCl_3$ ▶ S 282

▶ 2-BbNjv

(a)CH_3, CH_2 (b), (c) CH_2, CH_2 (e), N, C, CH_3(d), O

	Shift	Peak
a	0.91	3B
b	1.44 ± 0.1	S
c	1.44 ± 0.1	S
d	1.87	1A
▶ e	3.73	3A

▶ $C_{12}H_{17}NO$ ▶ 7% $CDCl_3$ ▶ V 601

▶ 2-BbNhj

(a)CH_3, (b) CH_2, CH_2 (b), (d) CH_2, N, (e) H, O, C, N, H(h), SO_2, (g) H, H(g), (f)H, H(f), CH_3(c)

	Shift	Peak
a	0.90	3B
b	~1.40	1F
c	2.46	1A
▶ d	3.23	4C
e	6.53	3B
f	7.32	Q
g	7.79	Q
h	~8.60	1D

▶ $C_{12}H_{18}N_2O_3S$ ▶ 7% $CDCl_3$ ▶ V 602

▶ (2)-(B)c(N)bh

(e) H, N—H(b), (a)CH_3, H(e), (d) H, H(d), (c) H

	Shift	Peak
a	0.91	2A
b	1.49	S
c	1.53 ± .10	S
▶ d	2.60	3C
▶ e	3.09	2D

▶ $C_6H_{13}N$ ▶ 7% $CDCl_3$ ▶ V 479

▶ (2)-(B)b(N)hj

H_2C—CH_2, H_2C (a), N, O, H, (b)

	Shift	Peak
▶ a	3.40	3B
b	7.67	1C

▶ C_4H_7NO ▶ 7% $CDCl_3$ ▶ V 68

▶ (2)-(B)c(N)bh

(b) H, (f) H, O, (c) H, H(b), N—H(a), C, H(f), (g)H_2N, H(c) H, H (d), (e) H (e)

	Shift	Peak
a	1.53	1B
b	1.60 ± 0.05	S
c	1.82 ± 0.05	S
d	2.14	O
▶ e	2.67	4A
▶ f	3.17	6A
g	5.87	1D

▶ $C_6H_{12}N_2O$ ▶ 7% $CDCl_3$ ▶ V 473

▶ 2-BbNhv

(d) (c) (b) (a)
$NH—CH_2—(CH_2)_2—CH_3$
(e) (e) (f) (f) (f)

	Shift	Peak
a	0.96	3B
b	1.50	1F
▶ c	3.08	3A
d	3.42	1A
e	6.52	3C
f	7.12	3C

▶ $C_{10}H_{15}N$ ▶ 45.4 mg/0.50 ml $CDCl_3$ ▶ S 280

▶ (2)-(B)c(N)ch

OD, H(c), (d) H, H(a), H, C, OD, H, N, O, H(b) H, D, H

	Shift	Peak
▶ a	3.32	3A
▶ b	3.32	3A
c	3.93	4A
d	4.22	5A

▶ $C_6H_{11}NO_3$ ▶ 7% $CDCl_3$ ▶ V 469

▶ (2)-(B)d(N)ac

O C O N H(j) (a) CH_3 (b) CH_2 CH_2 N (d) N $H(k)$ CH_3 (e) H (h) H (i) CH_3 (f) (g) H CH_3 (c)

	Shift	Peak
a	1.42	1A
b	1.95	3A
c	2.55	1A
▶ d	2.70	3D
e	2.82	1E
f	2.92	1E
g	4.12	1A
h	5.33	1D
i	6.37	2A
j	6.78	1E
k	6.87	2H

▶ $C_{15}H_{21}N_3O_2$ ▶ 7% $CDCl_3$ ▶ V 319

▶ 2-BlNhh

(e)H N (b) CH_2 NH_2 (c) CH_2 (a) N H(f) H (d)

	Shift	Peak
a	2.75	S
▶ b	3.02	S
c	4.96	1E
d	4.96	1E
e	6.82	1A
f	7.55	1A

▶ $C_5H_9N_3$ ▶ 7% $CDCl_3$ ▶ V 443

▶ 2-Bg(N)jj

O C N—(b) CH_2—(a) CH_2—Br C O

	Shift	Peak
a	3.63	3C
▶ b	4.13	3C

▶ $C_{10}H_8BrNO_2$ ▶ 7% $CDCl_3$ ▶ V 246

▶ (2)-(B)m(N)ll

(a) CH_3 (b) H_2C C CH_3(a) (d)H C C C N N (e)H H_2C CH_2 (c) (f)

	Shift	Peak
a	1.29	1A
b	2.76	3A
▶ c	3.40	3C
d	3.79	1F
e	3.79	1F
f	4.18	3C

▶ $C_9H_{14}N_2$ ▶ 7% $CDCl_3$ ▶ V 541

▶ 2-BjNav

(c) CH_3 N (b) CH_2 CH_2 (d) C CH_3(a) O

	Shift	Peak
a	2.14	1A
b	2.70	3A
c	2.94	1A
▶ d	3.66	3A

▶ $C_{11}H_{15}NO$ ▶ 7% $CDCl_3$ ▶ V 584

▶ (2)-(B)n(N)ab

(a) H N (d)H_2C CH_2(d) (c)H_2C CH_2(c) N CH_3(b)

	Shift	Peak
a	2.12	1A
b	2.27	1A
▶ c	2.37	Q
d	2.88	Q

▶ $C_5H_{12}N_2$ ▶ 7% $CDCl_3$ ▶ V 119

▶ (2)-(B)j(N)lv

H(f) (c) CH_3O H(e) O CH_3 (a) N (g)H H_2C CH_2 (d) (b)

	Shift	Peak
a	2.33	1A
b	3.13	3A
c	3.83	1A
▶ d	4.28	4A
e	6.85 or 6.98	1A
f	6.85 or 6.98	1A
g	7.13	1B

▶ $C_{13}H_{13}NO_2$ ▶ 7% $CDCl_3$ ▶ V 299

▶ (2)-(B)n(N)bb

(d) (c) CH_2 N (a) CH_2 (b) CH_2 N SO_2 CH_2 CH_2 (a) (b)

	Shift	Peak
▶ a	2.51	Q
b	3.02	Q
c	3.48	1A
d	7.23	1A

▶ $C_{17}H_{20}N_2O_2S$ ▶ 7% $CDCl_3$ ▶ V 330

▶ 2-BkNhh

D N D (a) CH_2 (b) CH_2 C O OD

	Shift	Peak
▶ a	2.54	Q
b	3.19	Q

▶ $C_2H_7NO_2$ ▶ D_2O ▶ V 394

▶ (2)-(B)n(N)bh

(b)H_2C CH_2(b) N H (a)

	Shift	Peak
a	0.03	1D
▶ b	1.62	1B

▶ C_2H_5N ▶ 7% $CDCl_3$ ▶ V 372

▶ (2)-(B)l(N)bh

(f) H C (b)H_2C CH(e) (c)H_2C CH_2(d) N H(a)

	Shift	Peak
a	1.63	1A
b	2.07	O
▶ c	2.95	3A
d	3.33	N
e	5.72 or 5.77	1E
f	5.72 or 5.77	1E

▶ C_5H_9N ▶ 7% $CDCl_3$ ▶ V 115

▶ (2)-(B)n(N)bh

(a) H N (d)H_2C CH_2(d) (c)H_2C CH_2(c) N CH_3(b)

	Shift	Peak
a	2.12	1A
b	2.27	1A
c	2.37	Q
▶ d	2.88	Q

▶ $C_5H_{12}N_2$ ▶ 7% $CDCl_3$ ▶ V 119

▶ (2)-(B)n(N)bh

(a) H N (b) (b) (b) (b) N (c) (c) (e) (e) (d)

	Shift	Peak
a	1.78	1A
▶ b	3.08	1A
c	6.88	2H
d	7.05	S
e	7.18	2H

▶ $C_{10}H_{14}N_2$ ▶ 45.2 mg/0.50 ml $CDCl_3$ ▶ S 300

▶ (2)-(B)o(N)bb

O (d) (d) (c) (c) N $CH_2CH_2CH_2—NH_2$ (a) (b)

	Shift	Peak
a	1.79	1E
b	2.46	4B
▶ c	2.81	3A
d	3.74	3B

▶ $C_7H_{16}N_2O$ ▶ 33.2 mg/0.50 ml $CDCl_3$ ▶ S 69

▶ 2-BnNbj

O O (b)H_2C (c) (c) CH_2(b) N—CH_2— CH_2—N (a)$H_2C—CH_2$ $H_2C—CH_2$(a) (d) (d)

	Shift	Peak
a	2.08 ± .10	S
b	2.31 ± .10	S
▶ c	3.47	1B
d	3.51	3D

▶ $C_{10}H_{16}N_2O_2$ ▶ 7% $CDCl_3$ ▶ V 572

▶ (2)-(B)o(N)bh

O (c)H_2C CH_2(c) (b)H_2C CH_2(b) N H (a)

	Shift	Peak
a	1.92	1A
▶ b	2.87	Q
c	3.67	Q

▶ C_4H_9NO ▶ 7% $CDCl_3$ ▶ V 83

▶ (2)-(B)n(N)bs

(d) (c) CH_2—N (a) (b) $CH_2—CH_2$ N—SO_2 $CH_2—CH_2$ (a) (b)

	Shift	Peak
a	2.51	Q
▶ b	3.02	Q
c	3.48	1A
d	7.23	1A

▶ $C_{17}H_{20}N_2O_2S$ ▶ 7% $CDCl_3$ ▶ V 330

▶ (2)-(B)o(N)bl

(c)CH_3O C=O O (e)H C—OCH_3(b) (a)$H_2C—CH_2$(d) N O (a)$H_2C—CH_2$(d)

	Shift	Peak
▶ a	3.28	3C
b	3.62 or 3.70	1E
c	3.62 or 3.70	1E
d	3.74	S
e	6.97	S

▶ $C_{15}H_{21}NO_5$ ▶ 7% $CDCl_3$ ▶ V 643

▶ (2)-(B)n(N)bs

(e)H H(f) (f) (e) H H (b) CH_3 SO_2—N (d) (d) $CH_2—CH_2$ N—SO_2 CH_3 (b) CH_2 CH_2 CH_2 (c) (a) (c) H H H H (e) (f) (f) (e)

	Shift	Peak
a	1.95	5B
b	2.38	1A
c	3.33	3A
▶ d	3.37	1E
e	7.30	Q
f	7.63	Q

▶ $C_{19}H_{24}N_2O_4S_2$
▶ 7% $CDCl_3$
▶ V 336

▶ 2-BpNaa

(a) CH_3 (b) (c) (d) N—CH_2—CH_2—OH CH_3 (a)

	Shift	Peak
a	2.25	1A
▶ b	2.45	3A
c	3.60	3D
d	3.60	1E

▶ $C_4H_{11}NO$ ▶ 7% $CDCl_3$ ▶ V 91

▶ (2)-(B)n(N)bv

(a) H N (b) (b) (b) (b) N (c) (c) (e) (e) (d)

	Shift	Peak
a	1.78	1A
▶ b	3.08	1A
c	6.88	2H
d	7.05	S
e	7.18	2H

▶ $C_{10}H_{14}N_2$ ▶ 45.2 mg/0.50 ml $CDCl_3$ ▶ S 300

▶ 2-Bp(N)bb

(b) (d) (c) $CH_2—CH_2$—OH N (a) (a) (e) (e) O

	Shift	Peak
a	2.51 or 2.56	S
▶ b	2.56 or 2.51	S
c	3.08	1A
d	3.64	1E
e	3.72	3D

▶ $C_6H_{13}NO_2$ ▶ 45.8 mg/0.50 ml $CDCl_3$ ▶ S 185

▶ (2)-(B)o(N)bb

(b) (d) (c) $CH_2—CH_2$—OH N (a) (a) (e) (e) O

	Shift	Peak
▶ a	2.51 or 2.56	S
b	2.56 or 2.51	S
c	3.08	1A
d	3.64	1E
e	3.72	3D

▶ $C_6H_{13}NO_2$ ▶ 45.8 mg/0.50 ml $CDCl_3$ ▶ S 185

▶ 2-BpNbd

OH(b) CH_2(d) (a)CH_3 CH_2(c) (a)CH_3—C—N (c) (d) CH_2— CH_2— OH(b) (a)CH_3

	Shift	Peak
a	1.10	1A
b	2.57	1E
▶ c	2.70	3D
d	3.65	3C

▶ $C_8H_{19}NO_2$ ▶ 7% $CDCl_3$ ▶ V 221

▶ 2-BpNbh

(d) H
(a) CH_3—(b) CH_2—N—(c) CH_2—(f) CH_2—(e) OH

	Shift	Peak
a	1.12	3A
b	2.68	4E
▶ c	2.73	3D
d	3.53	1E
e	3.53	1E
f	3.65	3D

▶ $C_4H_{11}NO$ ▶ 7% $CDCl_3$ ▶ V 92

▶ 2-BvNhj

(e) C_6H_5—(b) CH_2—(c) CH_2—N(H (d))—C(=O)—CH_3(a)

	Shift	Peak
a	1.90	1A
b	2.80	3B
▶ c	3.48	4A
d	6.50	1D
e	7.25	1G

▶ $C_{10}H_{13}NO$ ▶ 7% $CDCl_3$ ▶ V 265

▶ 2-BpNbv

(a) CH_3—CH_2(c)—N(C_6H_5)—CH_2(d)—CH_2(e)—OH (b)

	Shift	Peak
a	1.13	3A
b	2.16	1B
c	3.38	S
▶ d	3.40	S
e	3.72	3D

▶ $C_{10}H_{15}NO$ ▶ 7% $CDCl_3$ ▶ V 569

▶ 2-BxNaa

(e) H_2C(O)(O); H(f); (c) CH_2—(b) CH_2—N(CH_3 (a))(CH_3 (a)); H(i); H(h); (j)H; (d) CH_3O; H(g); CH_3O(d)

	Shift	Peak
a	2.38	1A
▶ b	~2.7	Q
c	~3.2	Q
d	4.05 and 4.08	1A
e	6.23	1A
f	7.13 or 7.23	1A
g	7.13 or 7.23	1A
h	7.53 or 7.81	2C
i	7.53 or 7.81	2C
j	8.52	1A

▶ $C_{21}H_{23}NO_4$ ▶ 7% $CDCl_3$ ▶ V 348

▶ 2-BpNhv

(d), (f), (e), (d), (f) ring positions; —NH(a) CH_2(b) CH_2(c) OH(a)

	Shift	Peak
a	3.12	1E
▶ b	3.23	3A
c	3.78	3A
d	6.61	S
e	6.78	S
f	7.17	3B

▶ $C_8H_{11}NO$ ▶ 45.4 mg/0.50 ml $CDCl_3$ ▶ S 295

▶ 2-BxNbp

H(e), (c) H, H(d), (g) H; (f) OH; (b) CH_2(a) CH_2N CH_2(a) CH_2(b); (d)H, H(e), (c) H, H (g)

	Shift	Peak
▶ a	3.13	1A
b	3.13	1A
c	7.07	S
d	7.07	S
e	7.53	S
f	7.90	S
g	8.48	S

▶ $C_{14}H_{17}N_3O$ ▶ 7% $CDCl_3$ ▶ V 633

▶ 2-BqNbb

(d) H_2N—C_6H_4(H (f), H (g))—CO—O—(e) CH_2—(c) CH_2—N[(b) CH_2—(a) CH_3]$_2$

	Shift	Peak
a	1.05	3A
b	2.62	4E
▶ c	2.82	3D
d	4.13	1B
e	4.33	3D
f	6.63	Q
g	7.83	Q

▶ $C_{13}H_{20}N_2O_2$ ▶ 7% $CDCl_3$ ▶ V 302

▶ 2-BbOb

(a) CH_3(b) $(CH_2)_3$(c) CH_2—O—(c) CH_2(b) $(CH_2)_3$(a) CH_3

	Shift	Peak
a	0.88	3B
b	1.40	O
▶ c	3.38	3A

▶ $C_{10}H_{22}O$ ▶ 44.6 mg/0.50 ml $CDCl_3$ ▶ S 151

▶ (2)-(B)v(N)ab

(g)H; (f) H_2C(O)(O); (c) CH_2 (b) CH_2—N—CH_3(a); CH_2(d); (i)H; C(=O)—CH_2 (e); (f) CH_2(O)(O); (h)H; H(h)

	Shift	Peak
a	1.92	1A
▶ b	2.55	1F
c	2.88	1F
d	3.58	1A
e	3.78	1A
f	5.92, 5.96	1A
g	6.65	1A
h	6.69	1A
i	6.91	1A

▶ $C_{20}H_{19}NO_5$ ▶ 7% $CDCl_3$ ▶ V 339

▶ (2)-(B)b(O)b

(a) H_2C—CH_2 (a); (b) H_2C—O—CH_2(b)

	Shift	Peak
a	1.85	Q
▶ b	3.75	Q

▶ C_4H_8O ▶ 7% $CDCl_3$ ▶ V 77

▶ 2-BvNhh

(d) C_6H_5—(b) CH_2(c) CH_2(a) NH_2

	Shift	Peak
a	1.34	1B
b	2.76	Q
▶ c	2.97	Q
d	7.24	1A

▶ $C_8H_{11}N$ ▶ 45.0 mg/0.50 ml $CDCl_3$ ▶ S 35

▶ (2)-(B)b(O)b

(a) CH_2—CH_2 (a); (b)H_2C, CH_2(b); O, O; CH_2 (c)

	Shift	Peak
a	1.74	O
▶ b	3.78	O
c	4.79	1A

▶ $C_5H_{10}O_2$ ▶ 7% $CDCl_3$ ▶ V 445

▶ (2)-(B)b(O)b

	Shift	Peak
a	2.72	5A
▶ b	4.73	3A

▶ C_3H_6O ▶ 7% $CDCl_3$ ▶ V 33

▶ 2-BbOz

	Shift	Peak
a	0.88	3C
b	~1.36	S
c	~1.50	S
d	3.58	2H
▶ e	3.84	3B
f	6.24	2A
g	7.20	1A

▶ $C_{17}H_{25}BCl_2O_2$ ▶ 7% $CDCl_3$ ▶ V 667

▶ (2)-(B)b(O)l

	Shift	Peak
▶ a	3.97	N
b	4.65	N
c	6.37	N

▶ C_5H_8O ▶ 7% $CDCl_3$ ▶ V 111

▶ (2)-(B)c(O)c

	Shift	Peak
a	1.22	2A
b	1.92	3A
▶ c	2.68±.02	S
d	2.80±.10	S
▶ e	4.05	S
f	4.64	3A

▶ $C_{11}H_{20}O_4$ ▶ 7% $CDCl_3$ ▶ V 588

▶ (2)-(B)b(O)l

	Shift	Peak
a	2.03	5B
b	2.63	3A
c	3.87 or 3.92	1A
d	3.87 or 3.92	1A
▶ e	4.30	3A
f	~7.00	R
g	~7.10	R
h	7.52	4D

▶ $C_{14}H_{15}NO_3$ ▶ 7% $CDCl_3$ ▶ V 312

▶ 2-BeOv

	Shift	Peak
a	3.78	Q
▶ b	4.22	Q

▶ C_8H_9ClO ▶ 7% $CDCl_3$ ▶ V 199

▶ (2)-(B)b(O)l

	Shift	Peak
a	2.05	5B
b	2.70	3A
c	3.77 or 3.87	1A
d	3.77 or 3.87	1A
▶ e	4.33	3A
f	~7.00	3
g	~7.15	S
h	8.00	4D

▶ $C_{14}H_{15}NO_3$ ▶ 7% $CDCl_3$ ▶ V 311

▶ (2)-(B)l(O)l

	Shift	Peak
a	3.27	Q
b	3.87 or 3.92	1A
c	3.87 or 3.92	1A
▶ d	4.73	Q
e	~7.05	R
f	~7.23	R
g	8.05	4D

▶ $C_{13}H_{13}NO_3$ ▶ 7% $CDCl_3$ ▶ V 300

▶ (2)-(B)b(O)s

	Shift	Peak
a	2.62	6B
b	3.28	3C
▶ c	4.50	3A

▶ $C_3H_6O_3S$ ▶ 7% $CDCl_3$ ▶ V 390

▶ (2)-(B)n(O)b

	Shift	Peak
a	1.92	1A
b	2.87	Q
▶ c	3.67	Q

▶ C_4H_9NO ▶ 7% $CDCl_3$ ▶ V 83

▶ 2-BbOz

	Shift	Peak
a	0.88	3C
b	~1.36	S
c	~1.50	S
d	2.86*	S
e	2.89*	S
f	3.65	S
▶ g	3.88	3C
h	7.26	1A

*±0.04

▶ $C_{17}H_{26}BCl_3O_2$ ▶ 7% $CDCl_3$ ▶ V 668

▶ (2)-(B)n(O)b

	Shift	Peak
a	2.51 or 2.56	S
b	2.56 or 2.51	S
c	3.08	1A
d	3.64	1E
▶ e	3.72	3D

▶ $C_6H_{13}NO_2$ ▶ 45.8 mg/0.50 ml $CDCl_3$ ▶ S 185

▶ (2)-(B)n(O)b

	Shift	Peak
a	1.79	1E
b	2.46	4B
c	2.81	3A
▶d	3.74	3B

(d) (d) (c) (c) N — $CH_2CH_2CH_2$— NH_2 (a) (b)

▶ $C_7H_{16}N_2O$ ▶ 33.2 mg/0.50 ml $CDCl_3$ ▶ S 69

▶ 2-BpOv

	Shift	Peak
a	2.65	1B
b	3.97 ± 0.10	Q
▶c	3.97 ± 0.10	Q

O—CH_2 (c) —CH_2 (b) —OH (a)

▶ $C_8H_{10}O_2$ ▶ 7% $CDCl_3$ ▶ V 506

▶ (2)-(B)n(O)b

	Shift	Peak
a	3.28	3C
b	3.62 or 3.70	1E
c	3.62 or 3.70	1E
▶d	3.74	S
e	6.97	S

(c) CH_3O—C=O; C(=O)—OCH_3 (b); (e)H; (a)H_2C—CH_2(d); N; O; (a)H_2C—CH_2(d)

▶ $C_{15}H_{21}NO_5$ ▶ 7% $CDCl_3$ ▶ V 643

▶ 2-BuOa

	Shift	Peak
a	2.62	3A
b	3.40	1A
▶c	3.62	3A

(b) (c) (a)
$CH_3OCH_2CH_2CN$

▶ C_4H_7NO ▶ 7% $CDCl_3$ ▶ V 69

▶ 2-BoOb

	Shift	Peak
a	1.20	3A
b	~3.50	4E
▶c	~3.63	1E

(a) (b) (c) (c) (b) (a)
CH_3CH_2—O—CH_2CH_2—O—CH_2CH_2—O—CH_2CH_3

▶ $C_8H_{18}O_3$ ▶ 45.3 mg/0.50 ml $CDCl_3$ ▶ S 285

▶ 2-BaP

	Shift	Peak
a	.91	3C
b	1.43	6B
▶c	3.45	3A
d	4.68	1A

(a) (b) (c) (d)
CH_3—CH_2—CH_2—OH

▶ C_3H_8O ▶ 20% CCl_4 ▶ API 165

▶ (2)-(B)o(O)s

	Shift	Peak
▶a	~4.35	Q
▶b	~4.67	Q

(a)H H(a); (b) H—C—C—H(b); O O; S; O

▶ $C_2H_4O_3S$ ▶ 7% $CDCl_3$ ▶ V 371

▶ 2-BaP

	Shift	Peak
a	0.92	3A
b	1.57	7A
c	~2.28	1C
▶d	3.58	3A

(a) (b) (d) (c)
CH_3—CH_2—CH_2—OH

▶ C_3H_8O ▶ 7% $CDCl_3$ ▶ V 43

▶ 2-BpOa

	Shift	Peak
a	2.72	1C
b	3.27	1A
▶c	~3.50	S

(b) (c) (a)
CH_3—O—CH_2—CH_2—OH

▶ $C_3H_8O_2$ ▶ 37.4 mg/0.50 ml $CDCl_3$ ▶ S 32

▶ 2-BaP

	Shift	Peak
a	0.93	3A
b	1.56	6B
c	2.17	1B
▶d	3.62	3A

(a) (b) (d) (c)
CH_3—CH_2—CH_2—OH

▶ C_3H_8O ▶ 34.7 mg/0.50 ml $CDCl_3$ ▶ S 10

▶ 2-BpOb

	Shift	Peak
▶a	3.68	2E
b	4.45	1C

(b) (a) (a) (b)
HO—CH_2—CH_2—O—CH_2—CH_2—OH

▶ $C_4H_{10}O_3$ ▶ 45.6 mg/0.50 ml $CDCl_3$ ▶ S 179

▶ 2-BbP

	Shift	Peak
a	.90	3B
b	1.32	1G
▶c	3.44	3B
d	4.42	1B

(a) (b) (c) (d)
CH_3—$(CH_2)_5$—CH_2—OH

▶ $C_7H_{16}O$ ▶ 20% CCl_4 ▶ API 168

▶ 2-BbP

Shift	Peak
a .88	3B
b 1.28	1B
▶c 3.44	3B
d 4.27	1A

(a) CH_3—(b) $(CH_2)_7$—(c) CH_2—(d) OH

▶ $C_9H_{20}O$ ▶ 20% CCl_4 ▶ API 171

▶ 2-BbP

Shift	Peak
a 0.88	3B
b 1.36	1B
c 2.28	1A
▶d 3.62	3A

(a) CH_3—(b) $(CH_2)_4$—(d) CH_2—(c) OH

▶ $C_6H_{14}O$ ▶ 45.2 mg/0.50 ml $CDCl_3$ ▶ S 198

▶ 2-BbP

Shift	Peak
a .88	3B
b 1.27	1B
▶c 3.45	3B
d 4.28	1A

(a) CH_3—(b) $(CH_2)_8$—(c) CH_2—(d) OH

▶ $C_{10}H_{22}O$ ▶ 20% CCl_4 ▶ API 172

▶ 2-BbP

Shift	Peak
a 0.88	3B
b 1.33	1B
c 1.95	1A
▶d 3.62	3B

(a) CH_3—(b) $(CH_2)_5$—(d) CH_2—(c) OH

▶ $C_7H_{16}O$ ▶ 45.0 mg/0.50 ml $CDCl_3$ ▶ S 215

▶ 2-BbP

Shift	Peak
a .91	3B
b ~1.45	O
▶c 3.48	3C
d 4.58	1A

(a) CH_3—(b) CH_2—(b) CH_2—(c) CH_2—(d) OH

▶ $C_4H_{10}O$ ▶ 20% CCl_4 ▶ API 166

▶ 2-BbP

Shift	Peak
a 0.90	3B
b 1.32	1B
c 1.97	1A
▶d 3.62	3B

(a) CH_3—(b) $(CH_2)_7$—(d) CH_2(c) OH

▶ $C_9H_{20}O$ ▶ 44.6 mg/0.50 ml $CDCl_3$ ▶ S 102

▶ 2-BbP

Shift	Peak
a .88	2A
b 1.31	S
c ~1.63	S
▶d 3.48	3B
e 4.42	1A

(a) CH_3—(c) $CH(CH_3)$—(b) $(CH_2)_3$—(d) CH_2—(e) OH

▶ $C_7H_{16}O$ ▶ 20% CCl_4 ▶ API 169

▶ 2-BbP

Shift	Peak
a 0.85	3B
b 1.29	1A
c 1.83	1B
▶d 3.62	3B

(a) CH_3—(b) $(CH_2)_{10}$—(d) CH_2(c) OH

▶ $C_{12}H_{26}O$ ▶ 44.4 mg/0.50 ml $CDCl_3$ ▶ S 103

▶ 2-BbP

Shift	Peak
a 0.86	3B
b 1.21	1A
▶c ~3.55	3B

(a) CH_3—CH_2—(b) $(CH_2)_{11}$—(c) CH_2—OH

▶ $C_{14}H_{30}O$ ▶ 44.9 mg/0.25 ml $CDCl_3$ ▶ S 294

▶ 2-BbP

Shift	Peak
a 0.91	3B
b 1.28	1B
c 2.03	1A
▶d 3.63	3A

(a) CH_3—(b) $(CH_2)_8$—(d) CH_2—(c) OH

▶ $C_{10}H_{22}O$ ▶ 7% $CDCl_3$ ▶ V 282

▶ 2-BbP

Shift	Peak
a 1.89	5B
b 2.18	1A
c 2.69	3C
▶d 3.61	3A
e 7.18	1A

(e) C_6H_5—(c) CH_2(a) CH_2(d) CH_2(b) OH

▶ $C_9H_{12}O$ ▶ 44.9 mg/0.50 ml $CDCl_3$ ▶ S 116

▶ 2-BbP

Shift	Peak
a ~1.80	N
b 2.53	1A
c 3.60 or 3.70	S
▶d 3.70 or 3.60	S

Cl—(c) CH_2(a) CH_2CH_2(d) CH_2—(b) OH

▶ C_4H_9ClO ▶ 45.0 mg/0.50 ml $CDCl_3$ ▶ S 241

▶ 2-BcP

CH_3(a)—C(e)H[OCH_3(d)]—CH_2(b)—CH_2(f)—OH(c)

	Shift	Peak
a	1.18	2A
b	1.72	4A
c	3.10	3A
d	3.33	1A
e	3.55	O
▶f	3.73	6B

▶ $C_5H_{12}O_2$ ▶ 7% $CDCl_3$ ▶ V 120

▶ 2-BnP

CH_3(a)—CH_2(b)—N[H(d)]—CH_2(c)—CH_2(f)—OH(e)

	Shift	Peak
a	1.12	3A
b	2.68	4E
c	2.73	3D
d	3.53	1E
e	3.53	1E
▶f	3.65	3D

▶ $C_4H_{11}NO$ ▶ 7% $CDCl_3$ ▶ V 92

▶ 2-BcP

CH_3(a)—CH(f)[OH(c)]—CH_2(b)—CH_2(e)—OH(d)

	Shift	Peak
a	1.23	2A
b	1.68	4C
c	3.58 or 3.65	1A
d	3.58 or 3.65	1A
▶e	3.80	3A
f	4.03	4A

▶ $C_4H_{10}O_2$ ▶ 7% $CDCl_3$ ▶ V 86

▶ 2-BnP

(a)CH_3—C[(a)CH_3][(a)CH_3]—N[CH_2(c)—CH_2(d)—OH(b)]—CH_2(c)—CH_2(d)—OH(b)

	Shift	Peak
a	1.10	1A
b	2.57	1E
c	2.70	3D
▶d	3.65	3C

▶ $C_8H_{19}NO_2$ ▶ 7% $CDCl_3$ ▶ V 221

▶ 2-BeP

Cl—CH_2(c)—CH_2(b)—OH(a)

	Shift	Peak
a	2.80	1A
▶b	3.68 or 3.83	Q
c	3.68 or 3.83	Q

▶ C_2H_5ClO ▶ 7% $CDCl_3$ ▶ V 12

▶ 2-BnP

(a) CH_3—CH_2(c)—N(C_6H_5)—CH_2(d)—CH_2(e)—OH(b)

	Shift	Peak
a	1.13	3A
b	2.16	1B
c	3.38	S
d	3.40	S
▶e	3.72	3D

▶ $C_{10}H_{15}NO$ ▶ 7% $CDCl_3$ ▶ V 569

▶ 2-BlP

(d) OCH_3; CH_2(a) CH_2(c) OH(b); N; O; OCH_3 (e); CH_3 (f)

	Shift	Peak
a	3.03	3A
b	~3.67	1D
▶c	3.92	3D
d	3.92 or 4.00	1E
e	3.92 or 4.00	1E
f	3.92 or 4.00	1E

▶ $C_{14}H_{17}NO_4$ ▶ 7% $CDCl_3$ ▶ V 313

▶ 2-BnP

C_6H_5[(d), (e), (f)]—NH(a) CH_2(b) CH_2(c) OH(a)

	Shift	Peak
a	3.12	1E
b	3.23	3A
▶c	3.78	3A
d	6.61	S
e	6.78	S
f	7.17	3B

▶ $C_8H_{11}NO$ ▶ 45.4 mg/0.50 ml $CDCl_3$ ▶ S 295

▶ 2-BnP

(a)CH_3—N[CH_3(a)]—CH_2(b)—CH_2(c)—OH(d)

	Shift	Peak
a	2.25	1A
b	2.45	3A
▶c	3.60	3D
d	3.60	1E

▶ $C_4H_{11}NO$ ▶ 7% $CDCl_3$ ▶ V 91

▶ 2-BoP

CH_3(b)—O—CH_2—CH_2(c)—OH(a)

	Shift	Peak
a	2.72	1C
b	3.27	1A
▶c	~3.50	S

▶ $C_3H_8O_2$ ▶ 37.4 mg/0.50 ml $CDCl_3$ ▶ S 32

▶ 2-BnP

(morpholine: (a), (a), (e), (e), O)N—CH_2(b)—CH_2(d)—OH(c)

	Shift	Peak
a	2.51 or 2.56	S
b	2.56 or 2.51	S
c	3.08	1A
▶d	3.64	1E
e	3.72	3D

▶ $C_6H_{13}NO_2$ ▶ 45.8 mg/0.50 ml $CDCl_3$ ▶ S 185

▶ 2-BoP

(b) HO—CH_2—CH_2(a)—O—CH_2—CH_2(a)—OH (b)

	Shift	Peak
▶a	3.68	2E
b	4.45	1C

▶ $C_4H_{10}O_3$ ▶ 45.6 mg/0.50 ml $CDCl_3$ ▶ S 179

▶ 2-BoP

Shift	Peak
a 2.65	1B
▶b 3.97 ± 0.10	Q
c 3.97 ± 0.10	Q

C_6H_5–O–CH_2(c)–CH_2(b)–OH(a)

▶ $C_8H_{10}O_2$ ▶ 7% $CDCl_3$ ▶ V 506

▶ 2-BbQb

Shift	Peak
a 0.92	3B
b 1.12	S
c 1.32	S
d 2.33	4A
▶e 4.05	3B

CH_3(b)CH_2(d)C(=O)–OCH_2(e)$(CH_2)_3$(c)CH_3(a)

▶ $C_8H_{16}O_2$ ▶ 44.8 mg/0.50 ml $CDCl_3$ ▶ S 40

▶ 2-BvP

Shift	Peak
a 2.08	1A
b 2.82	Q
▶c 3.78	Q
d 7.24	1A

(d) C_6H_5–CH_2(b)CH_2(c)OH(a)

▶ $C_8H_{10}O$ ▶ 45.4 mg/0.50 ml $CDCl_3$ ▶ S 113

▶ 2-BbQb

Shift	Peak
a 0.94	3B
b 1.56	1F
c 4.05	1A
▶d 4.18	3A

Cl–CH_2(c)–C(=O)–O–CH_2(d)$(CH_2)_2$(b)CH_3(a)

▶ $C_6H_{11}ClO_2$ ▶ 44.5 mg/0.50 ml $CDCl_3$ ▶ S 92

▶ 2-BaQa

Shift	Peak
a 0.93	3A
b 1.64	5B
c 2.04	1A
▶d 4.02	3A

CH_3(c)C(=O)–OCH_2(d)CH_2(b)CH_3(a)

▶ $C_5H_{10}O_2$ ▶ 44.9 mg/0.50 ml $CDCl_3$ ▶ S 115

▶ 2-BbQb

Shift	Peak
a 0.93	3B
b 4.05	1A
▶c 4.20	3D

CH_3(a)–CH_2–CH_2–CH_2(c)–O–C(=O)–CH_2(b)–Cl

▶ $C_6H_{11}ClO_2$ ▶ 7% $CDCl_3$ ▶ V 138

▶ 2-BaQb

Shift	Peak
a 0.95	3D
b 1.18	3D
c 1.60	6B
d 2.33	4A
▶e 4.04	3A

CH_3(b)CH_2(d)–C(=O)–OCH_2(e)CH_2(c)CH_3(a)

▶ $C_6H_{12}O_2$ ▶ 44.4 mg/0.50 ml $CDCl_3$ ▶ S 38

▶ (2)-(B)b(Q)b

Shift	Peak
a 4.37	3B

H_2C–C(=O)–O–CH_2(a)–CH_2 (ring)

▶ $C_4H_6O_2$ ▶ 7% $CDCl_3$ ▶ V 63

▶ 2-BbQa

Shift	Peak
a 0.95	3B
b 2.03	1A
▶c 4.05	3A

CH_3(a)–CH_2–CH_2–CH_2(c)–O–C(=O)–CH_3(b)

▶ $C_6H_{12}O_2$ ▶ 7% $CDCl_3$ ▶ V 140

▶ 2-BbQc

Shift	Peak
a 0.94	3B
b 1.41	2A
c 1.46	S
d 3.02	1A
▶e 4.18	4C

CH_3(b)–CH(e)(OH)(d)–C(=O)–OCH_2(e)CH_2CH_2(c)CH_3(a)

▶ $C_7H_{14}O_3$ ▶ 44.6 mg/0.50 ml $CDCl_3$ ▶ S 246

▶ 2-BbQa

Shift	Peak
a 0.92	3B
b 1.37	1E
c 2.05	1A
▶d 4.05	3A

CH_3(c)C(=O)–OCH_2(d)$(CH_2)_4$(b)CH_3(a)

▶ $C_8H_{16}O_2$ ▶ 45.0 mg/0.50 ml $CDCl_3$ ▶ S 217

▶ 2-BbQh

Shift	Peak
a 0.88	3B
b 1.32	1E
▶c 4.16	3B
d 8.03	1B

HC(d)(=O)–OCH_2(c)–$(CH_2)_6$(b)–CH_3(a)

▶ $C_9H_{18}O_2$ ▶ 44.4 mg/0.50 ml $CDCl_3$ ▶ S 114

▶ 2-BbQo

	Shift	Peak
a	0.93	3B
▶b	4.13	3A

$O{=}C(-O\,CH_2CH_2CH_2CH_3)_2$ — (b) CH_2, (a) CH_3

▶ $C_9H_{18}O_3$ ▶ 7% $CDCl_3$ ▶ V 243

▶ (2)-(B)k(Q)b

	Shift	Peak
a	3.56	8A
▶b	4.29	8A

▶ $C_3H_4O_2$ ▶ 7% $CDCl_3$ ▶ V 409

▶ 2-BbQv

	Shift	Peak
a	0.95	3B
b	~1.60	1F
▶ c	4.25	3D
d	~4.10	1E
e	6.59	Q
f	7.83	Q

▶ $C_{11}H_{15}NO_2$ ▶ 44.6 mg/0.50 ml $CDCl_3$ ▶ S 243

▶ 2-BnQv

	Shift	Peak
a	1.05	3A
b	2.62	4E
c	2.82	3D
d	4.13	1B
▶e	4.33	3D
f	6.63	Q
g	7.83	Q

▶ $C_{13}H_{20}N_2O_2$ ▶ 7% $CDCl_3$ ▶ V 302

▶ 2-BbQv

	Shift	Peak
a	0.98	3B
b	1.64	O
▶c	4.35	3A
d	7.48	N
e	8.06	N

▶ $C_{11}H_{14}O_2$ ▶ 44.8 mg/0.50 ml $CDCl_3$ ▶ S 158

▶ 2-BvQa

	Shift	Peak
a	2.02	1A
b	2.93	3A
▶c	4.30	3A
d	7.29 ± 0.03	1A

▶ $C_{10}H_{12}O_2$ ▶ 7% $CDCl_3$ ▶ V 261

▶ 2-BcQa

	Shift	Peak
a	0.92	2A
b	~1.50	O
c	2.03	1A
▶ d	4.03	4B

▶ $C_7H_{14}O_2$ ▶ 44.8 mg/0.50 ml $CDCl_3$ ▶ S 281

▶ 2-BaR

	Shift	Peak
a	1.03	3A
b	2.07	6A
▶c	4.38	3A

(a) CH_3 — (b) CH_2 — (c) CH_2 — NO_2

▶ $C_3H_7NO_2$ ▶ 7% $CDCl_3$ ▶ V 42

▶ (2)-(B)c(Q)c

	Shift	Peak
a	1.26	2A
b	3.28	1B
c	3.87	2A
▶d	4.34	3C

▶ $C_6H_{10}O_3$ ▶ 7% $CDCl_3$ ▶ V 467

▶ 2-BaSx

	Shift	Peak
a	1.08	3A
b	1.73	6B
▶c	2.93	3A
d	7.10	Q
e	8.38	Q

▶ $C_8H_{11}NS$ ▶ 7% $CDCl_3$ ▶ V 509

▶ (2)-(B)d(Q)b

	Shift	Peak
a	1.40	1A
b	1.92	4B
c	2.51	2G
d	2.68	2G
e	2.92	1A
▶ f	4.33	N
▶ g	4.60	N

▶ $C_6H_{10}O_3$ ▶ 7% $CDCl_3$ ▶ V 466

▶ 2-BbSa

	Shift	Peak
a	0.93	3B
b	2.08	1A
▶ c	2.50	3B

(a) CH_3 — CH_2 — CH_2 — CH_2 — (c) CH_2 — S — (b) CH_3

▶ $C_6H_{14}S$ ▶ 7% $CDCl_3$ ▶ V 144

▶ 2-BbSb

	Shift	Peak
a	0.92	3B
b	1.49	O
▶c	2.49	3B

(a) (b) (c)
$CH_3-(CH_2)_2-CH_2-S-CH_2-(CH_2)_2-CH_3$

▶ $C_8H_{18}S$ ▶ 45.8 mg/0.50 ml $CDCl_3$ ▶ S 275

▶ 2-BbSh

	Shift	Peak
a	1.35	3A
b	1.88	5B
▶c	2.68	4C

(a) (c) (b) (c) (a)
$HSCH_2CH_2CH_2SH$

▶ $C_3H_8S_2$ ▶ 7% $CDCl_3$ ▶ V 47

▶ (2)-(B)b(S)b

	Shift	Peak
a	~2.57	2E

(a)H_2C — S — CH_2(a) (six-membered ring)

▶ $C_5H_{10}S$ ▶ 7% $CDCl_3$ ▶ V 118

▶ 2-BbSooo

	Shift	Peak
a	0.00	1A
b	0.60	N
c	1.78	O
▶d	2.82	3C

(a)CH_3, (a)CH_3, (a)CH_3 — Si — CH_2(b) — CH_2(c) — CH_2(d) — $SO_3^{(-)}$ $Na^{(+)}$

▶ $C_6H_{15}NaO_3SSi$ ▶ D_2O ▶ V 481

▶ (2)-(B)b(S)b

	Shift	Peak
a	1.93	Q
▶b	2.82	Q

(a) H_2C — CH_2 (a)
(b) H_2C — S — CH_2(b)

▶ C_4H_8S ▶ 7% $CDCl_3$ ▶ V 80

▶ (2)-(B)b(S)ooo

	Shift	Peak
a	2.62	6B
▶b	3.28	3C
c	4.50	3A

(a)H_2C — CH_2(c)
(b)H_2C — S(→O)(→O) — O

▶ $C_3H_6O_3S$ ▶ 7% $CDCl_3$ ▶ V 390

▶ (2)-(B)b(S)boo

	Shift	Peak
a	2.23	8A
▶b	3.00	8A

(a)H_2C — CH_2(a)
(b)H_2C — S(→O)(→O) — CH_2(b)

▶ $C_4H_8O_2S$ ▶ 7% $CDCl_3$ ▶ V 416

▶ 2-BcSa

	Shift	Peak
a	2.10	1A
b	2.25	O
▶c	2.66	O
d	4.94	4B
e	7.14	3D
f	10.28	1B

C_6H_5–C(=O)–N(H(e))–C(H(d))(COOH(f))–$CH_2CH_2SCH_3$
(b) (c) (a)

▶ $C_{12}H_{15}NO_3S$ ▶ 7% $CDCl_3$ ▶ V 597

▶ 2-BbSh

	Shift	Peak
a	0.92	3A
b	1.23	S
c	1.53	S
▶d	2.52	4B

(a) (c) (d) (b)
$CH_3-(CH_2)_2-CH_2-SH$

▶ $C_4H_{10}S$ ▶ 46.0 mg/0.50 ml $CDCl_3$ ▶ S 274

▶ 2-BuSb

	Shift	Peak
a	2.72	Q
▶b	2.88	Q

(a) (b) (b) (a)
$NC-CH_2-CH_2-S-CH_2-CH_2-CN$

▶ $C_6H_8N_2S$ ▶ 7% $CDCl_3$ ▶ V 127

▶ 2-BbSh

	Shift	Peak
a	0.90	3B
b	1.29	1E
c	2.09	1A
▶d	2.53	4B

(a) (b) (d) (c)
$CH_3-(CH_2)_5-CH_2SH$

▶ $C_7H_{16}S$ ▶ 46.9 mg/0.50 ml $CDCl_3$ ▶ S 297

▶ 2-BaU

	Shift	Peak
a	1.08	3C
b	1.70	6B
▶c	2.30	3C

(a) (b) (c)
$CH_3-CH_2-CH_2-C\equiv N$

▶ C_4H_7N ▶ 45.0 mg/0.50 ml $CDCl_3$ ▶ S 33

▶ 2-BbU

Shift	Peak
a 0.93	3B
b 1.49	O
▶c 2.33	3B

(a) (b) (c)
$CH_3-(CH_2)_3-CH_2-C\equiv N$

▶ $C_6H_{11}N$ ▶ 45.0 mg/0.50 ml $CDCl_3$ ▶ S 176

▶ 2-BdU

Shift	Peak
a 1.10	1A
b 1.96	S
▶c 2.32	S
d 2.32	S
e 3.60	S

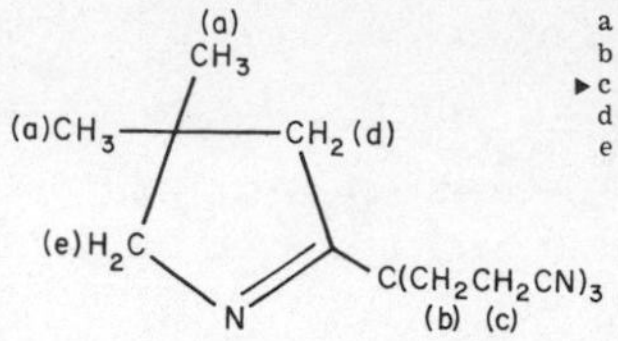

(a) CH_3, (a) CH_3, CH_2 (d), (e) H_2C, N, $C(CH_2CH_2CN)_3$ (b) (c)

▶ $C_{16}H_{22}N_4$ ▶ 7% $CDCl_3$ ▶ V 660

▶ 2-BbU

Shift	Peak
a 2.36	3B

(a) (a)
$N\equiv C-CH_2-(CH_2)_6-CH_2-C\equiv N$

▶ $C_{10}H_{16}N_2$ ▶ 7% $CDCl_3$ ▶ V 570

▶ 2-BgU

Shift	Peak
▶a 2.98	Q
b 3.52	Q

(b) (a)
$Br-CH_2-CH_2-C\equiv N$

▶ C_3H_4BrN ▶ 45.2 mg/0.50 ml $CDCl_3$ ▶ S 145

▶ 2-BbU

Shift	Peak
a ~2.23	O
▶b 2.58	O
c 3.53	3A

(c) (a) (b)
$Br-CH_2-CH_2-CH_2-CN$

▶ C_4H_6BrN ▶ 7% $CDCl_3$ ▶ V 58

▶ 2-BkU

Shift	Peak
▶a 2.68	1A
b 2.68	1A
c 3.75	1A

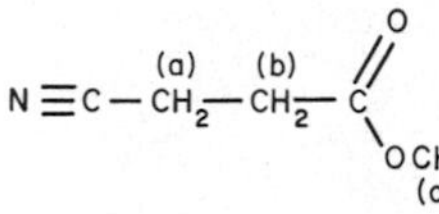

(a) (b)
$N\equiv C-CH_2-CH_2-C(=O)OCH_3$ (c)

▶ $C_5H_7NO_2$ ▶ 7% $CDCl_3$ ▶ V 106

▶ 2-BbU

Shift	Peak
a 2.03	5B
▶b 2.58	3C

(b) (a) (b)
$N\equiv C-CH_2CH_2CH_2-C\equiv N$

▶ $C_5H_6N_2$ ▶ 45.3 mg/0.50 ml $CDCl_3$ ▶ S 251

▶ 2-BoU

Shift	Peak
▶a 2.62	3A
b 3.40	1A
c 3.62	3A

(b) (c) (a)
$CH_3OCH_2CH_2CN$

▶ C_4H_7NO ▶ 7% $CDCl_3$ ▶ V 69

▶ 2-BcU

Shift	Peak
a 0.90	2A
b 1.60	3D
▶c 2.38	3A

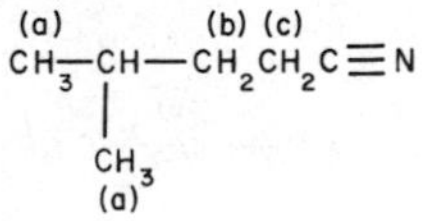

(a) (b) (c)
$CH_3-CH(CH_3)-CH_2CH_2C\equiv N$ (CH_3 (a))

▶ $C_6H_{11}N$ ▶ 44.8 mg/0.50 ml $CDCl_3$ ▶ S 165

▶ 2-BsU

Shift	Peak
▶a 2.72	Q
b 2.88	Q

(a) (b) (b) (a)
$NC-CH_2-CH_2-S-CH_2-CH_2-CN$

▶ $C_6H_8N_2S$ ▶ 7% $CDCl_3$ ▶ V 127

▶ 2-BdU

Shift	Peak
a 1.12	1A
b 1.92	Q
▶c 2.27	Q
d 9.47	1A

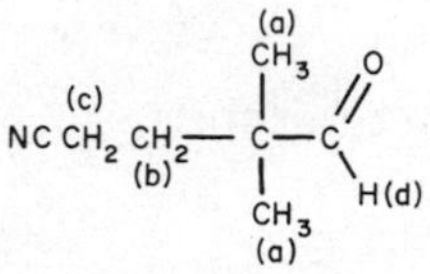

(c) NC CH_2 CH_2 (b) — C(CH_3 (a))(CH_3 (a)) — C(=O)H(d)

▶ $C_7H_{11}NO$ ▶ 7% $CDCl_3$ ▶ V 178

▶ 2-BaVhh

Shift	Peak
a .87	3C
b 1.56	6B
c 2.17	1A
▶d 2.45	3B
e 6.82	S
f 6.82	S
g 6.82	S
h 6.82	S

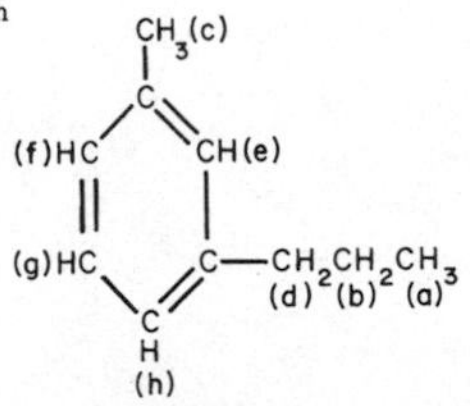

CH_3(c), (f)HC, CH(e), (g)HC, C—$CH_2CH_2CH_3$ (d) (b) (a), C H (h)

▶ $C_{10}H_{14}$ ▶ liquid run, neat ▶ API 96

▶ 2-BaVhh

	Shift	Peak
a	.92	3C
b	1.58	6B
▶c	2.57	3C
d	7.10	1E

▶ C_9H_{12} ▶ 10% CCl_4 ▶ API 41

▶ 2-BaVhh

	Shift	Peak
a	0.96	3B
b	1.37	2A
c	~1.65	S
▶d	2.58	3B
e	3.42	S
f	3.88	1A
g	5.08	2A
h	5.62	1A

▶ $C_{20}H_{24}O_4$ ▶ 7% $CDCl_3$ ▶ V 682

▶ (2)-(B)bVbh

	Shift	Peak
a	1.79	O
▶b	2.76	O
c	7.07	1A

▶ $C_{10}H_{12}$ ▶ 7% $CDCl_3$ ▶ V 557

▶ (2)-(B)bVbh

	Shift	Peak
a	2.02	5B
▶b	2.86	3A
c	7.05	1I

▶ C_9H_{10} ▶ 10% CCl_4 ▶ API 188

▶ (2)-(B)bVbh

	Shift	Peak
a	2.04	5B
▶b	2.91	3A
c	7.17	1A

▶ C_9H_{10} ▶ 7% $CDCl_3$ ▶ V 527

▶ 2-BbVhh

	Shift	Peak
a	.86	3B
▶b	1.28	1B
c	2.54	3B
d	7.09	1A

(d) = all ring H's

▶ $C_{13}H_{20}$ ▶ 50% CCl_4 ▶ API 357

▶ 2-BbVhh

	Shift	Peak
a	.89	3B
▶b	1.30	1F
c	2.54	3B
d	7.10	1A

(d) = all ring H's

▶ $C_{12}H_{18}$ ▶ 50% CCl_4 ▶ API 356

▶ 2-BbVhh

	Shift	Peak
a	.89	3B
b	1.25	1B
▶c	2.46	3B
d	7.07	1A

(d) = all ring H's

▶ $C_{15}H_{24}$ ▶ 50% CCl_4 ▶ API 358

▶ 2-BbVhh

	Shift	Peak
a	.88	3B
b	1.25	1B
▶c	2.50	3B
d	7.07	1A

(d) = all ring H's

▶ $C_{16}H_{26}$ ▶ 50% CCl_4 ▶ API 359

▶ 2-BbVhh

	Shift	Peak
a	.89	3B
b	1.36	O
▶c	2.52	3B
d	7.04	1G

(d) = all ring H's

▶ $C_{11}H_{16}$ ▶ 50% CCl_4 ▶ API 355

▶ 2-BbVhh

	Shift	Peak
a	.88	3B
b	1.22	1A
▶c	2.54	3B
d	7.09	1G

(d) = all ring H's

▶ $C_{18}H_{30}$ ▶ 50% CCl_4 ▶ API 360

▶ 2-BbVhh

	Shift	Peak
a	.93	3C
b	1.47	S
c	1.47	S
▶d	2.57	3B
e	7.11	1A

(e) = all ring H's

▶ $C_{10}H_{14}$ ▶ 10% CCl_4 ▶ API 42

▶ 2-BbVhh

	Shift	Peak
a	1.89	5B
b	2.18	1A
▶c	2.69	3C
d	3.61	3A
e	7.18	1A

(c) (a) (d) (b) $CH_2CH_2CH_2OH$ (e)

▶ $C_9H_{12}O$ ▶ 44.9 mg/0.50 ml $CDCl_3$ ▶ S 116

▶ 2-BjVhh

	Shift	Peak
a	2.74	Q
▶b	2.97	Q
c	7.22	1A
d	9.81	3A

▶ $C_9H_{10}O$ ▶ 7% $CDCl_3$ ▶ V 529

▶ 2-BbVhh

	Shift	Peak
a	2.15	5B
▶b	2.75	3C
c	3.38	3A
d	7.22	1A

(d) (b) (a) (c) $CH_2-CH_2-CH_2-Br$

▶ $C_9H_{11}Br$ ▶ 7% $CDCl_3$ ▶ V 237

▶ (2)-(B)jVhj

	Shift	Peak
a	2.60	N
▶b	3.08	3C

▶ C_9H_8O ▶ 7% $CDCl_3$ ▶ V 229

▶ (2)-(B)bVhn

	Shift	Peak
a	1.93	5B
▶b	2.73	3A
c	3.22	3C
d	3.49	1B
e	3.77	1A

▶ $C_{10}H_{13}NO$ ▶ 7% $CDCl_3$ ▶ V 266

▶ 2-BkVhh

	Shift	Peak
▶a	~2.79	Q
b	~2.79	Q
c	11.71	1A

(a) (b) (c) CH_2CH_2C-OH

▶ $C_9H_{10}O_2$ ▶ 45.7 mg/0.25 ml $CDCl_3$ ▶ S 299

▶ (2)-(B)bVjw

	Shift	Peak
a	2.25	5B
b	2.77	3B
▶c	3.37	3B

▶ $C_{14}H_{12}O$ ▶ 7% $CDCl_3$ ▶ V 308

▶ 2-BkVoo

	Shift	Peak
a	1.42	1A
b	2.62	Q
▶c	2.86	Q
d	3.75	1A
e	5.50	2C
f	6.53	2C

▶ $C_{20}H_{24}O_5$ ▶ 7% $CDCl_3$ ▶ V 344

▶ (2)-(B)cVhl

	Shift	Peak
a	1.96	1A
b	2.43	S
▶c	2.43	S
d	3.67	1A
e	4.62	S
f	6.55	1A
g	6.92	2A
h	7.37	S
i	7.63	1B
j	8.38	2G

▶ $C_{22}H_{25}NO_6$ ▶ 7% $CDCl_3$ ▶ V 689

▶ 2-BnVhh

	Shift	Peak
a	1.34	1B
▶b	2.76	Q
c	2.97	Q
d	7.24	1A

(d) (b) (c) (a) $CH_2CH_2NH_2$

▶ $C_8H_{11}N$ ▶ 45.0 mg/0.50 ml $CDCl_3$ ▶ S 35

▶ (2)-(B)dVao

	Shift	Peak
a	0.85	2A
b	1.22 ± 0.03	1E
c	1.22 ± 0.03	1E
d	2.12 ± 0.03	1B
▶e	2.62	3C
f	3.96	1D

▶ $C_{29}H_{50}O_2$ ▶ 7% $CDCl_3$ ▶ V 366

▶ 2-BnVhh

	Shift	Peak
a	1.90	1A
▶b	2.80	3B
c	3.48	4A
d	6.50	1D
e	7.25	1G

▶ $C_{10}H_{13}NO$ ▶ 7% $CDCl_3$ ▶ V 265

▶ (2)-(B)nVhj

(g)H (c) CH_2 (b) CH_2 N CH_3(a) CH_2(d) (f) H_2C O O C CH_2 (e) (i)H O O (f) CH_2 O (h)H H(h)

	Shift	Peak
a	1.92	1A
b	2.55	1F
▶c	2.88	1F
d	3.58	1A
e	3.78	1A
f	5.92, 5.96	1A
g	6.65	1A
h	6.69	1A
i	6.91	1A

▶ $C_{20}H_{19}NO_5$ ▶ 7% $CDCl_3$ ▶ V 339

▶ (2)-(B)vVhv

(c)(c) H H (b) H—C—C—H (b) (d)H H(d) (d)H H(d) (d)H CH_3 CH_3 H(d) (a) (a)

	Shift	Peak
a	2.29	1A
▶b	2.61	S
▶c	2.71	S
d	7.15	1A

▶ $C_{16}H_{16}$ ▶ 7% $CDCl_3$ ▶ V 654

▶ 2-BpVhh

(d) (b) (a) CH_2CH_2OH (c)

	Shift	Peak
a	2.08	1A
▶b	2.82	Q
c	3.78	Q
d	7.24	1A

▶ $C_8H_{10}O$ ▶ 45.4 mg/0.50 ml $CDCl_3$ ▶ S 113

▶ (2)-(B)vVhv

(a) H_2C CH_2 (a)

	Shift	Peak
a	2.86	1A

▶ $C_{14}H_{12}$ ▶ 7% $CDCl_3$ ▶ V 622

▶ 2-BqVhh

(d) CH_2—CH_2—O—C(=O)—CH_3 (b) (c) (a)

	Shift	Peak
a	2.02	1A
▶b	2.93	3A
c	4.30	3A
d	7.29 ± 0.03	1A

▶ $C_{10}H_{12}O_2$ ▶ 7% $CDCl_3$ ▶ V 261

▶ 2-BxVhh

(a) CH_3 (e) H (d) HO (b) CH_2—CH_2 (c) CH_3 (a) H (e)

	Shift	Peak
a	2.20	1A
▶b	2.90	Q
c	3.30	Q
d	4.18	1C
e	6.83	1B

▶ $C_{20}H_{20}O$ ▶ 7% $CDCl_3$ ▶ V 340

▶ (2)-(B)vVah

(d)H (b) CH_2 (c)CH_3O CH_3(a) CH_2(b) H(d) (d)H (b)CH_2 (a)CH_3 OCH_3(c) CH_2 (b) H(d)

	Shift	Peak
a	1.53	1A
▶b	2.83	1A
c	3.67	1A
d	6.45	1A

▶ $C_{20}H_{24}O_2$ ▶ 7% $CDCl_3$ ▶ V 681

▶ 2-BaX

(d) H (c) CH_2—(b) CH_2—(a) CH_3 (e) H N H(f) H (g)

	Shift	Peak
a	0.95	3C
b	1.60	6B
▶c	2.48	3C
d	6.13	4B
e	6.70	S
f	6.55	S
g	~7.83	1A

▶ $C_7H_{11}N$ ▶ 7% $CDCl_3$ ▶ V 177

▶ 2-BvVhh

(a) CH_2—CH_2

	Shift	Peak
a	2.89	1A

▶ $C_{14}H_{14}$ ▶ 45.4 mg/0.50 ml $CDCl_3$ ▶ S 184

▶ 2-BaX

(e) H O C NH_2 (d)H (+) (f)H N H(g) $I^{(-)}$ CH_2—CH_2—CH_3 (c) (b) (a)

	Shift	Peak
a	1.03	3A
b	2.15	6A
▶c	4.74	3C
d	8.29	3C
e	8.97	R
f	9.12	R
g	9.37	1C

▶ $C_9H_{13}IN_2O$ ▶ D_2O ▶ V 539

▶ (2)-(B)vVhn

(a) (a) H_2C—CH_2 N H(b)

	Shift	Peak
▶a	3.07	1A
b	5.90	1C

▶ $C_{14}H_{13}N$ ▶ 7% $CDCl_3$ ▶ V 631

▶ 2-BnX

H(e) (c) H H(d) (f) OH (d)H H(e) (c) H H N (b) $CH_2CH_2NCH_2CH_2$ (a) (a) (b) N H (g) (g)

	Shift	Peak
a	3.13	1A
▶b	3.13	1A
c	7.07	S
d	7.07	S
e	7.53	S
f	7.90	S
g	8.48	S

▶ $C_{14}H_{17}N_3O$ ▶ 7% $CDCl_3$ ▶ V 633

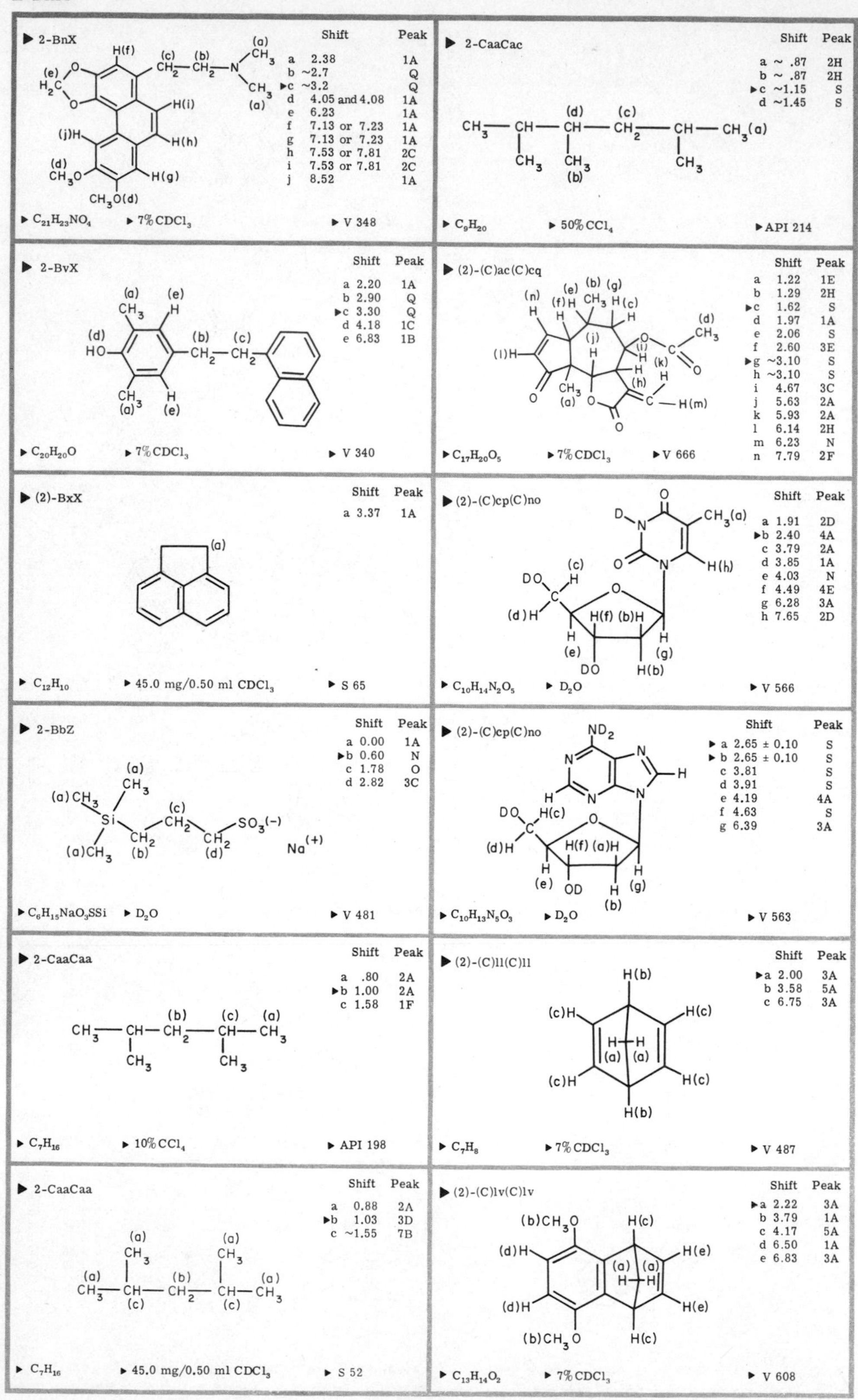
▶ 2-BnX
Shift Peak
a 2.38 1A
b ~2.7 Q
▶c ~3.2 Q
d 4.05 and 4.08 1A
e 6.23 1A
f 7.13 or 7.23 1A
g 7.13 or 7.23 1A
h 7.53 or 7.81 2C
i 7.53 or 7.81 2C
j 8.52 1A
▶ C21H23NO4 ▶ 7% CDCl3 ▶ V 348
▶ 2-CaaCac
Shift Peak
a ~ .87 2H
b ~ .87 2H
▶c ~1.15 S
d ~1.45 S
▶ C9H20 ▶ 50% CCl4 ▶ API 214
▶ 2-BvX
Shift Peak
a 2.20 1A
b 2.90 Q
▶c 3.30 Q
d 4.18 1C
e 6.83 1B
▶ C20H20O ▶ 7% CDCl3 ▶ V 340
▶ (2)-(C)ac(C)cq
Shift Peak
a 1.22 1E
b 1.29 2H
▶c 1.62 S
d 1.97 1A
e 2.06 S
f 2.60 3E
▶g ~3.10 S
h ~3.10 S
i 4.67 3C
j 5.63 2A
k 5.93 2A
l 6.14 2H
m 6.23 N
n 7.79 2F
▶ C17H20O5 ▶ 7% CDCl3 ▶ V 666
▶ (2)-BxX
Shift Peak
a 3.37 1A
▶ C12H10 ▶ 45.0 mg/0.50 ml CDCl3 ▶ S 65
▶ (2)-(C)cp(C)no
Shift Peak
a 1.91 2D
▶b 2.40 4A
c 3.79 2A
d 3.85 1A
e 4.03 N
f 4.49 4E
g 6.28 3A
h 7.65 2D
▶ C10H14N2O5 ▶ D2O ▶ V 566
▶ 2-BbZ
Shift Peak
a 0.00 1A
▶b 0.60 N
c 1.78 O
d 2.82 3C
▶ C6H15NaO3SSi ▶ D2O ▶ V 481
▶ (2)-(C)cp(C)no
Shift Peak
▶ a 2.65 ± 0.10 S
▶ b 2.65 ± 0.10 S
c 3.81 S
d 3.91 S
e 4.19 4A
f 4.63 S
g 6.39 3A
▶ C10H13N5O3 ▶ D2O ▶ V 563
▶ 2-CaaCaa
Shift Peak
a .80 2A
▶b 1.00 2A
c 1.58 1F
▶ C7H16 ▶ 10% CCl4 ▶ API 198
▶ (2)-(C)ll(C)ll
Shift Peak
▶a 2.00 3A
b 3.58 5A
c 6.75 3A
▶ C7H8 ▶ 7% CDCl3 ▶ V 487
▶ 2-CaaCaa
Shift Peak
a 0.88 2A
▶b 1.03 3D
c ~1.55 7B
▶ C7H16 ▶ 45.0 mg/0.50 ml CDCl3 ▶ S 52
▶ (2)-(C)lv(C)lv
Shift Peak
▶a 2.22 3A
b 3.79 1A
c 4.17 5A
d 6.50 1A
e 6.83 3A
▶ C13H14O2 ▶ 7% CDCl3 ▶ V 608

▶ 2-(C)oo(C)oo

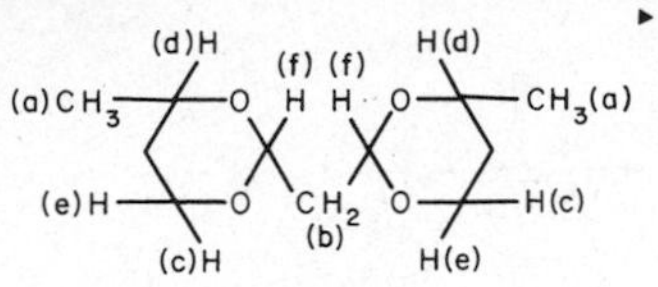

	Shift	Peak
a	1.22	2A
▶b	1.92	3A
c	2.68±.02	S
d	2.80±.10	S
e	4.05	S
f	4.64	3A

▶ $C_{11}H_{20}O_4$ ▶ 7% $CDCl_3$ ▶ V 588

▶ (2)-(C)ao(D)aao

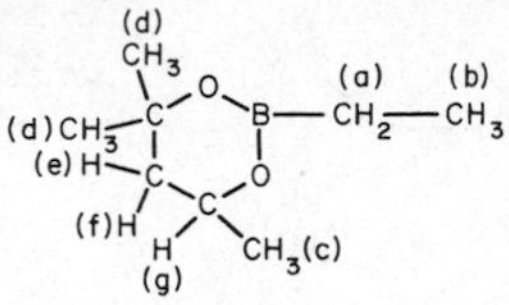

	Shift	Peak
a	0.61	S
b	0.88	S
c	1.25	2H
d	1.29	1A
▶e	1.47	S
▶f	1.76	S
g	4.17	O

▶ $C_8H_{17}BO_2$ ▶ 7% $CDCl_3$ ▶ V 518

▶ 2-CaaDaaa

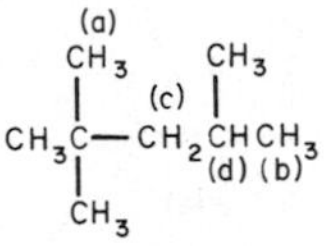

	Shift	Peak
a	~ .96	1E
b	~ .96	2H
▶c	~1.12	2A
d	~1.67	7X

▶ C_8H_{18} ▶ 50% CCl_4 ▶ API 424

▶ (2)-(C)ao(D)aao

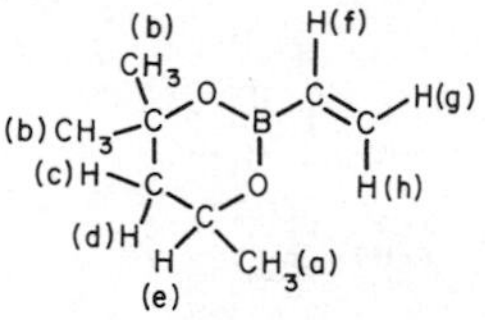

	Shift	Peak
a	1.27	1A
b	1.31	1A
▶c	1.49	2G
▶d	1.81	2G
e	4.22	7X
f	~5.90	S
g	~5.90	S
h	~5.90	S

▶ $C_8H_{15}BO_2$ ▶ 7% $CDCl_3$ ▶ V 517

▶ 2-CaaDaab

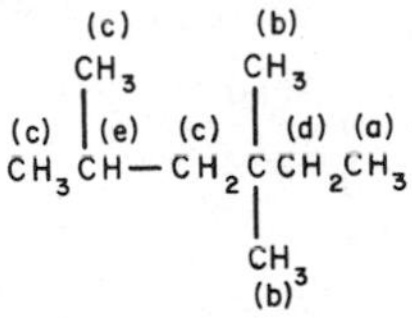

	Shift	Peak
a	~ .85	3D
b	~ .85	1E
▶c	~ .92	2H
d	1.13	2A
e	1.63	7X

▶ C_9H_{20} ▶ 50% CCl_4 ▶ API 426

▶ (2)-(C)jo(D)bcc

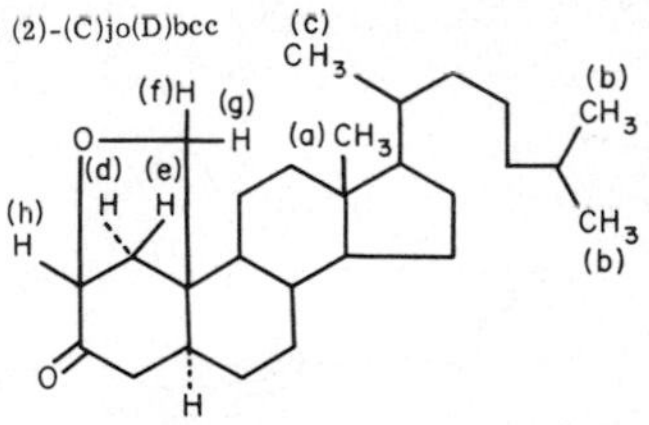

	Shift	Peak
a	0.66	1A
b	0.88	S
c	0.95	S
▶d	1.48	S
▶e	2.53	S
f	3.90	S
g	4.03	S
h	4.12	S

▶ $C_{27}H_{44}O_2$ ▶ 7% $CDCl_3$ ▶ V 698

▶ 2-CaaDaab

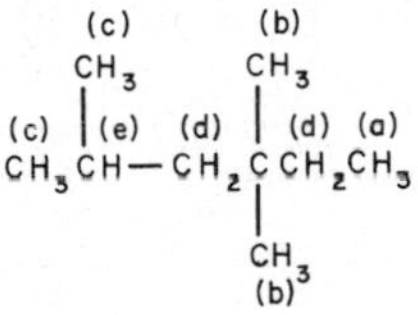

	Shift	Peak
a	~ .85	3D
b	~ .85	1E
c	~ .92	2H
▶d	1.13	2A
e	1.63	7X

▶ C_9H_{20} ▶ 50% CCl_4 ▶ API 426

▶ 2-CvzDeee

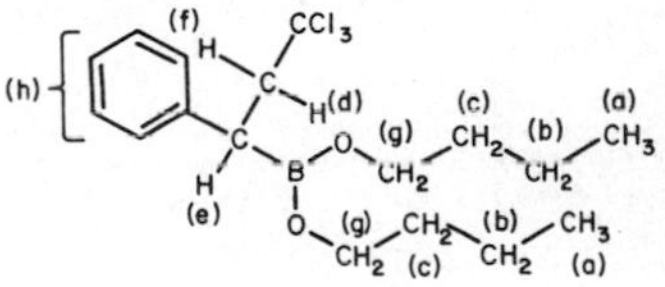

	Shift	Peak
a	0.88	3C
b	~1.36	S
c	~1.50	S
▶d	2.86*	S
e	2.89*	S
f	3.65	S
g	3.88	3C
h	7.26	1A

*±0.04

▶ $C_{17}H_{26}BCl_3O_2$ ▶ 7% $CDCl_3$ ▶ V 668

▶ 2-CabDaaa

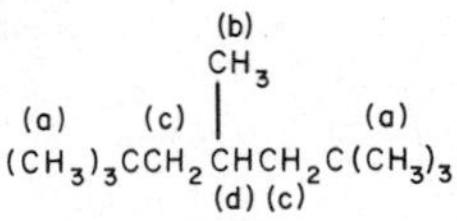

	Shift	Peak
a	.95	1E
b	~ .96	2H
▶c	1.17	S
d	1.57	7X

▶ $C_{12}H_{26}$ ▶ 50% CCl_4 ▶ API 428

▶ 2-CaeE

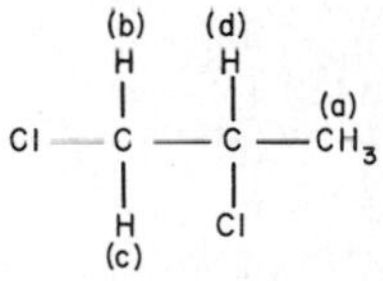

	Shift	Peak
a	1.60	2A
▶b	3.52	R
▶c	3.78	R
d	4.10	R

▶ $C_3H_6Cl_2$ ▶ 7% $CDCl_3$ ▶ V 30

▶ 2-CabDaaa

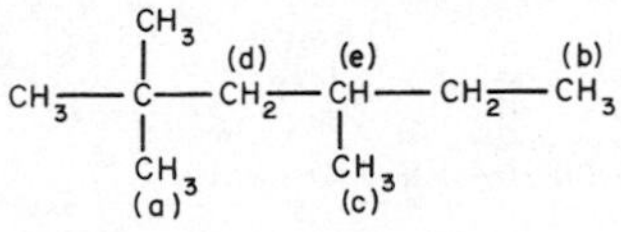

	Shift	Peak
a	.87	1E
b	.87	1E
c	.87	1E
▶d	~1.22	S
e	~1.32	1D

▶ C_9H_{20} ▶ 50% CCl_4 ▶ API 211

▶ 2-CanE

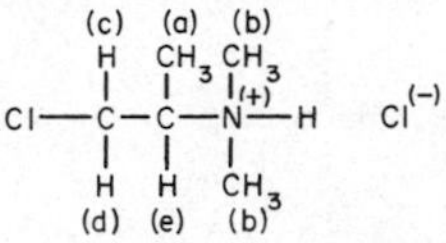

	Shift	Peak
a	1.68	2A
b	2.97	1A
▶c	3.27	S
▶d	3.56	S
e	4.62	1F

▶ $C_5H_{13}Cl_2N$ ▶ 7% $CDCl_3$ ▶ V 448

▶ 2-(C)boE

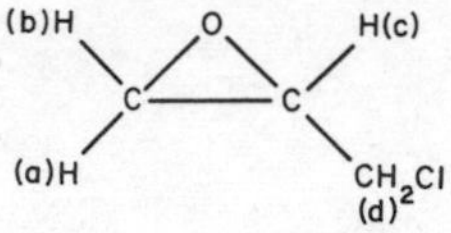

Shift	Peak
a 2.70	4A
b 2.88	3A
c 3.28	O
▶d 3.58	2A

▶ C_3H_5ClO ▶ 45.0 mg/0.50 ml $CDCl_3$ ▶ S 16

▶ 2-CbgG

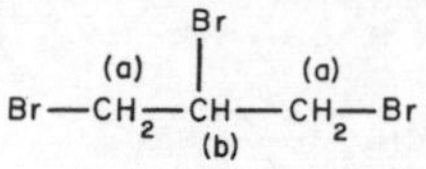

Shift	Peak
▶ a 3.88	2A
b 4.38	6B

Br—CH2—CH(Br)—CH2—Br

▶ $C_3H_5Br_3$ ▶ 44.5 mg/0.50 ml $CDCl_3$ ▶ S 231

▶ 2-CbpE

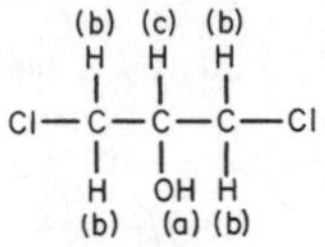

Shift	Peak
a 2.68	1B
▶ b 3.72	2A
c 4.07	6B

▶ $C_3H_6Cl_2O$ ▶ 7% $CDCl_3$ ▶ V 386

▶ 2-CgvG

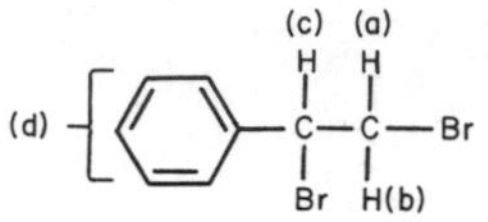

Shift	Peak
▶ a 4.02	S
▶ b 4.07	S
c 5.15	4A
d 7.38	1A

▶ $C_8H_8Br_2$ ▶ 7% $CDCl_3$ ▶ V 503

▶ 2-CeeE

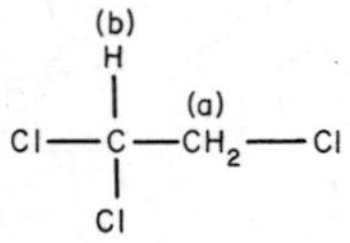

Shift	Peak
▶ a 3.95	2A
b 5.77	3A

▶ $C_2H_3Cl_3$ ▶ 7% $CDCl_3$ ▶ V 2

▶ 2-CooG

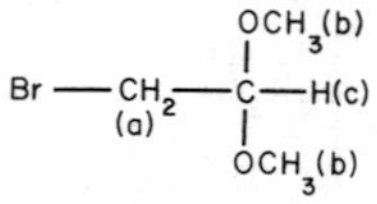

Shift	Peak
▶ a 3.37	1A
b 3.40	1A
c 4.54	3A

▶ $C_4H_9BrO_2$ ▶ 7% $CDCl_3$ ▶ V 419

▶ 2-CeoE

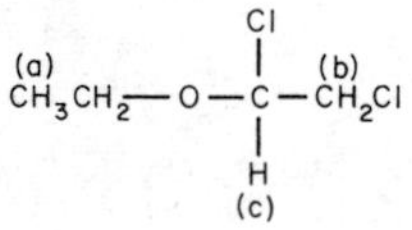

Shift	Peak
a 1.28	3A
▶ b 3.70	O
c 5.63	3A

▶ $C_4H_8Cl_2O$ ▶ 7% $CDCl_3$ ▶ V 75

▶ 2-CaaJa

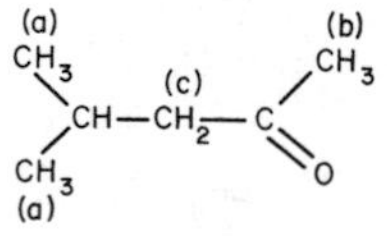

Shift	Peak
a 0.93	2A
b 2.12	1A
▶ c ~2.28	N

▶ $C_6H_{12}O$ ▶ 7% $CDCl_3$ ▶ V 139

▶ 2-CagG

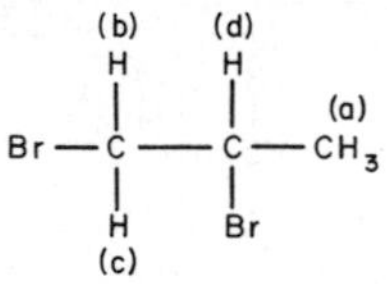

Shift	Peak
a 1.83	2A
▶ b 3.52	4E
▶ c 3.75	S
d 4.18	S

▶ $C_3H_6Br_2$ ▶ 45.4 mg/0.50 ml $CDCl_3$ ▶ S 187

▶ (2)-(C)ao(J)b

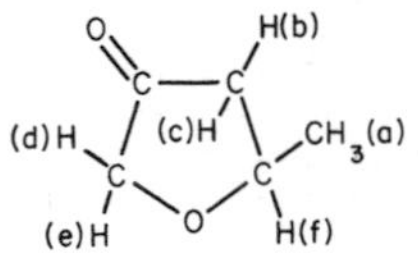

Shift	Peak
a 1.44	2A
▶ b 2.15	10A
▶ c 2.57	10A
d 3.83	10A
e 4.04	10A
f 4.35	7C

▶ $C_5H_8O_2$ ▶ 7% $CDCl_3$ ▶ V 439

▶ 2-CbgG

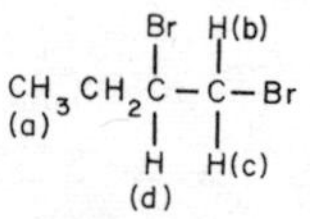

Shift	Peak
a 1.05	3A
▶ b 3.62	2F
▶ c 3.83	2F
d 4.13	O

▶ $C_4H_8Br_2$ ▶ 7% $CDCl_3$ ▶ V 74

▶ 2-CaaKb

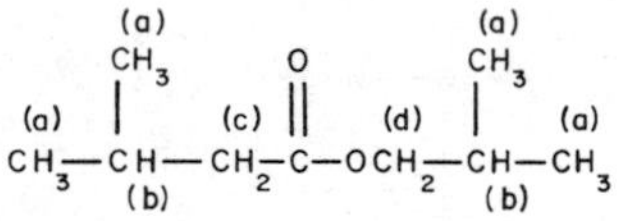

Shift	Peak
a 0.94	2A
b 1.12	S
▶ c 2.20	2D
d 3.88	2A

▶ $C_9H_{18}O_2$ ▶ 44.5 mg/0.50 ml $CDCl_3$ ▶ S 277

▶ 2-CaqKa

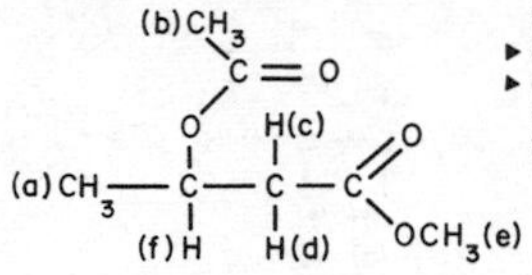

(b)CH_3 C=O O H(c) O (a)CH_3—C—C—C (f)H H(d) OCH_3(e)

	Shift	Peak
a	1.32	2A
b	2.03	1A
▶ c	2.50 or 2.70	2A
▶ d	2.50 or 2.70	2A
e	3.70	1A
f	5.30	4A

▶ $C_7H_{12}O_4$ ▶ 7% $CDCl_3$ ▶ V 182

▶ 2-CaaLhl

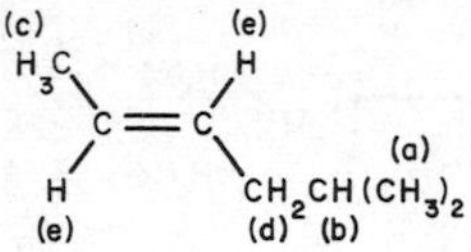

(c) H_3C (e) H C=C H (e) (a) $CH_2CH(CH_3)_2$ (d) (b)

	Shift	Peak
a	.98	2A
b	1.38	S
c	1.60	2H
▶ d	1.80	S
e	5.30	N

▶ C_7H_{14} ▶ 50 vol.% CCl_4 ▶ API 431

▶ (2)-(C)ck(K)c

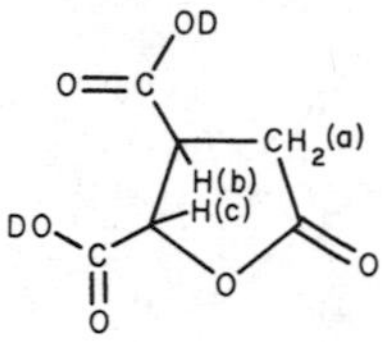

OD O=C CH_2(a) H(b) H(c) DO C O O O

	Shift	Peak
▶ a	2.98	2E
b	3.86	4A
c	5.24	2A

▶ $C_6H_6O_6$ ▶ D_2O ▶ V 456

▶ 2-CaaLhl

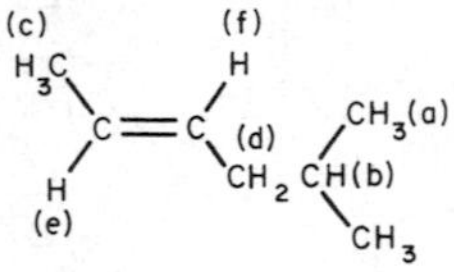

(c) H_3C (f) H C=C H (e) (d) CH_2 CH(b) CH_3(a) CH_3

	Shift	Peak
a	.83	2A
b	~1.40	S
c	1.62	2E
▶ d	1.83	O
e	~5.38	O
f	~5.38	O

▶ C_7H_{14} ▶ liquid ▶ API 396

▶ 2-CjkKb

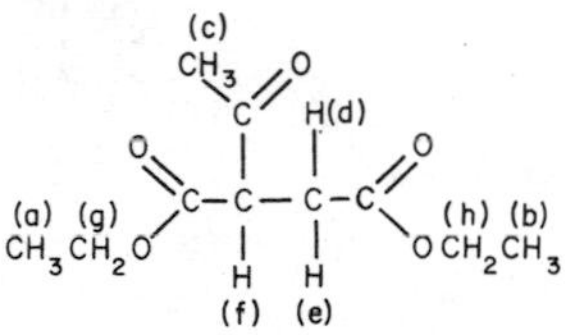

(c) CH_3 O H(d) O O (a) (g) CH_3CH_2O—C—C—C—C—OCH_2CH_3 (h) (b) H H (f) (e)

	Shift	Peak
a	1.25*	3A
b	1.25*	3A
c	2.35	1A
▶ d	2.86	2A
▶ e	2.89	2A
f	3.98	S
g	4.13†	4A
h	4.13†	4A

*or 1.27
†or 4.22

▶ $C_{10}H_{16}O_5$ ▶ 7% $CDCl_3$ ▶ V 277

▶ 2-CaaLhl

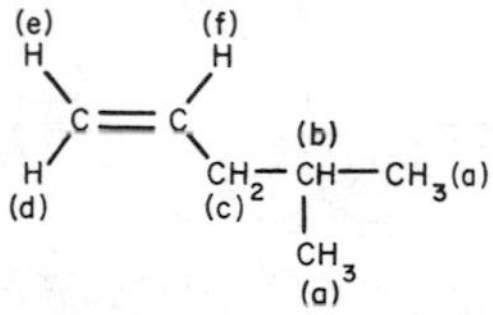

(e) H (f) H C=C H (d) CH_2 (c) (b) CH—CH_3(a) CH_3 (a)

	Shift	Peak
a	.89	2A
b	~1.60	S
▶ c	1.87	S
d	4.83	R
e	5.03	R
f	~5.62	O

▶ C_6H_{12} ▶ 10% CCl_4 ▶ API 33

▶ 2-CknKh

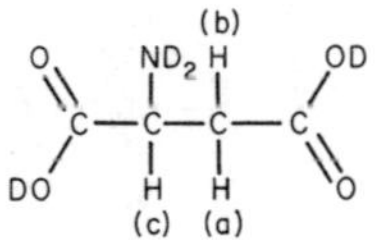

(b) O ND_2 H OD C—C—C—C DO H H O (c) (a)

	Shift	Peak
▶ a	2.83	2A
▶ b	2.90	1A
c	3.98	4A

▶ $C_4H_7NO_4$ ▶ D_2O ▶ V 410

▶ 2-CaaLhl

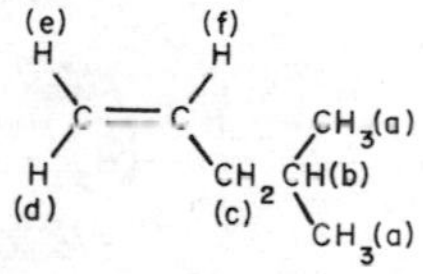

(e) H (f) H C=C H (d) CH_2 (c) CH(b) CH_3(a) CH_3(a)

	Shift	Peak
a	.90	2A
b	1.48	S
▶ c	1.88	S
d	4.77	R
e	4.77	R
f	5.05	R

▶ C_6H_{12} ▶ liquid ▶ API 375

▶ 2-CksKh

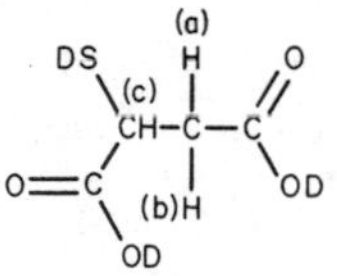

(a) DS H O (c) CH-C-C O=C (b)H OD OD

	Shift	Peak
▶ a	2.92	2D
▶ b	3.02	2D
c	3.85	3A

▶ $C_4H_6O_4S$ ▶ D_2O ▶ V 407

▶ 2-CaaLmv

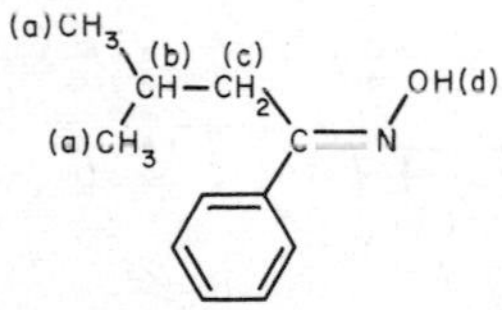

(a)CH_3 (b) (c) CH—CH_2 OH(d) (a)CH_3 C=N

	Shift	Peak
a	0.93	2A
b	1.98	7B
▶ c	2.77	2A
d	9.68	1D

▶ $C_{11}H_{15}NO$ ▶ 7% $CDCl_3$ ▶ V 585

▶ 2-CaaLal

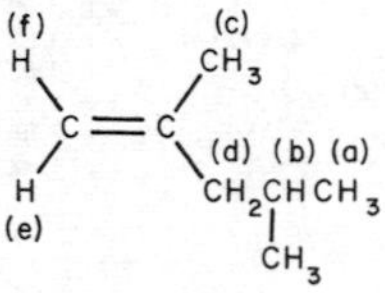

(f) H (c) CH_3 C=C H (e) (d) (b) (a) CH_2CHCH_3 CH_3

	Shift	Peak
a	.87	2H
b	1.62	S
c	1.67	3C
▶ d	1.82	2B
e	4.67 ± .03	N
f	4.67 ± .03	N

▶ C_7H_{14} ▶ liquid ▶ API 400

▶ 2-CabLhl

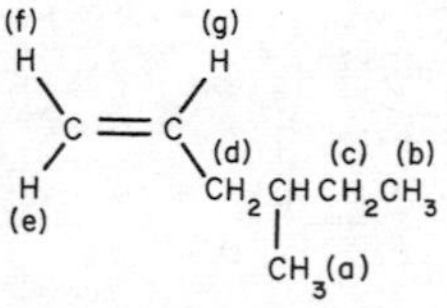

(f) H (g) H C=C H (e) (d) (c) (b) $CH_2CHCH_2CH_3$ CH_3(a)

	Shift	Peak
a	~ .87	3D
b	~ .95	2H
c	1.30	O
▶ d	1.98	4C
e	~4.96	R
f	~4.96	R
g	5.67	R

▶ C_7H_{14} ▶ liquid run, neat ▶ API 393

▶ (2)-(C)co(L)jl

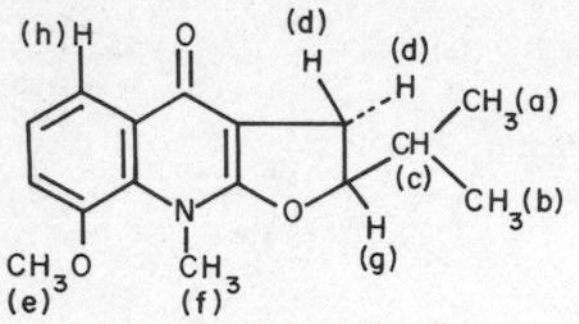

	Shift	Peak
a	0.97*	2A
b	0.97*	2A
c	2.00	6A
▶ d	~3.20	S
e	3.87†	1A
f	3.87†	1A
g	4.72	4C
h	8.05	4A

*or 1.02
†or 3.90

▶ $C_{16}H_{19}NO_3$ ▶ 7% $CDCl_3$ ▶ V 325

▶ 2-CbpNhh

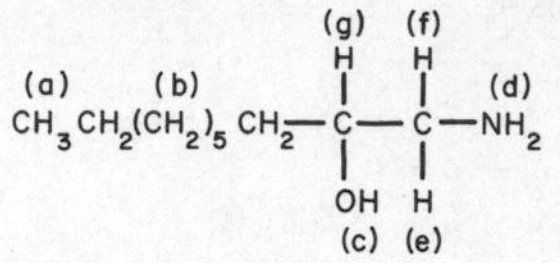

	Shift	Peak
a	0.89	3B
b	1.28	1G
c	1.75	1E
d	1.75	1E
▶ e	2.51	S
▶ f	2.84	S
g	3.50	1D

▶ $C_{10}H_{23}NO$ ▶ 7% $CDCl_3$ ▶ V 575

▶ 2-Ckn(L)lv

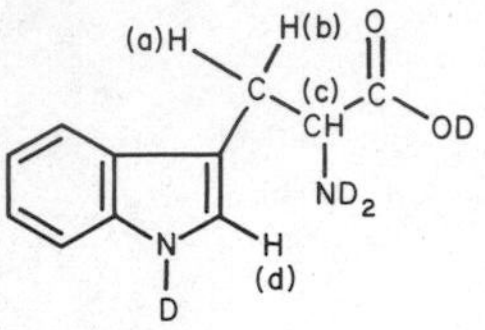

	Shift	Peak
▶ a	3.10	S
▶ b	3.33	S
c	3.75	O
d	7.28	S

▶ $C_{11}H_{12}N_2O_2$ ▶ D_2O ▶ V 582

▶ (2)-(C)aj(O)b

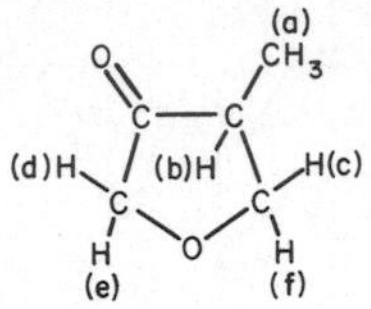

	Shift	Peak
a	1.17	2A
b	2.55	6B
▶ c	3.73	10A
d	3.84	10A
e	4.07	10A
▶ f	4.49	3A

▶ $C_5H_8O_2$ ▶ 7% $CDCl_3$ ▶ V 438

▶ (2)-(L)lvVhv

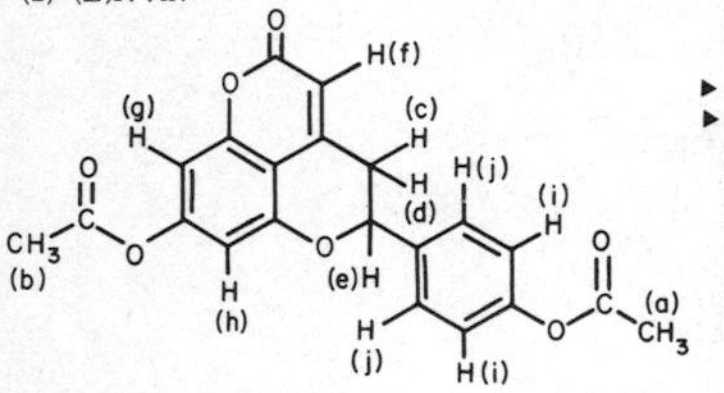

	Shift	Peak
a	2.32	1A
b	2.32	1A
▶ c	3.09	S
▶ d	3.09	A
e	5.18	3B
f	6.06	M
g	6.62*	S
h	6.62*	S
i	7.17	Q
j	7.46	Q

*or 6.71

▶ $C_{21}H_{16}O_7$ ▶ 7% $CDCl_3$ ▶ V 686

▶ (2)-(C)ao(O)c

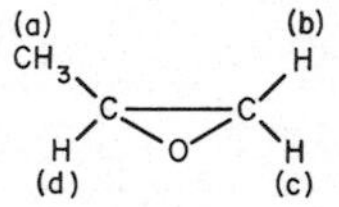

	Shift	Peak
a	1.32	2A
▶ b	2.42	N
▶ c	2.72	N
d	~2.98	N

▶ C_3H_6O ▶ 7% $CDCl_3$ ▶ V 32

▶ 2-CaaNbh

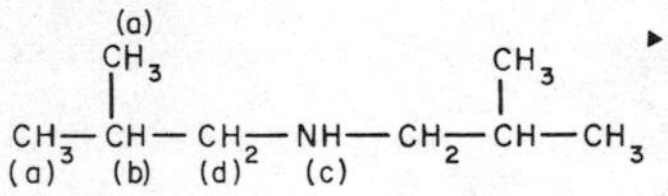

	Shift	Peak
a	0.91	2A
b	~1.6	S
c	1.77	1B
▶ d	2.42	2A

▶ $C_8H_{19}N$ ▶ 44.7 mg/0.50 ml $CDCl_3$ ▶ S 136

▶ (2)-(C)bo(O)c

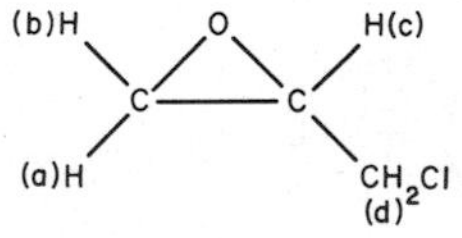

	Shift	Peak
▶ a	2.70	4A
b	2.88	3A
c	3.28	O
d	3.58	2A

▶ C_3H_5ClO ▶ 45.0 mg/0.50 ml $CDCl_3$ ▶ S 16

▶ (2)-(C)ab(N)bh

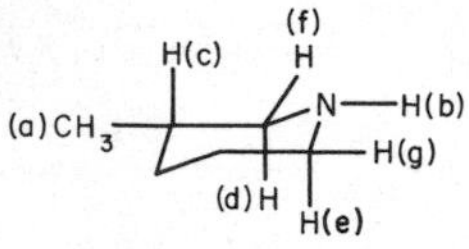

	Shift	Peak
a	0.82	2A
b	1.47	S
c	1.55 ± 0.10	S
▶ d	2.21	S
e	2.52	S
▶ f	3.00	S
g	3.00	S

▶ $C_6H_{13}N$ ▶ 7% $CDCl_3$ ▶ V 478

▶ (2)-(C)co(O)b

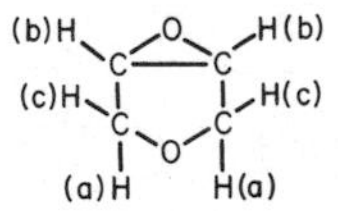

	Shift	Peak
▶ a	3.64	2H
b	3.79	2H
▶ c	4.01	4E

▶ $C_4H_6O_2$ ▶ 7% $CDCl_3$ ▶ V 405

▶ (2)-(C)bp(N)ch

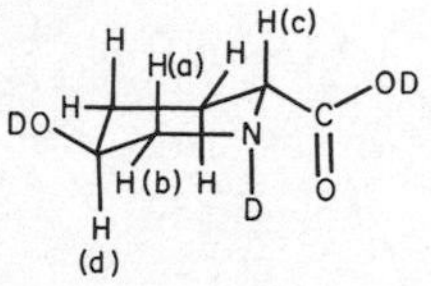

	Shift	Peak
▶ a	2.88	4A
▶ b	3.50	S
c	3.75	S
d	4.01	S

▶ $C_6H_{11}NO_3$ ▶ D_2O ▶ V 470

▶ (2)-(C)cp(O)c

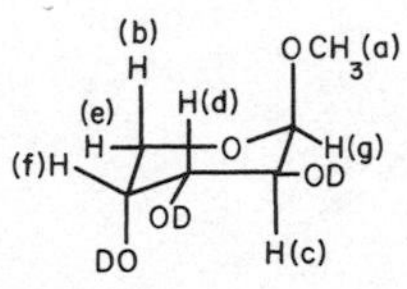

	Shift	Peak
a	3.41	1A
▶ b	3.67	S
c	3.82 ± 0.04	S
d	3.82 ± 0.04	S
▶ e	3.89 ± 0.05	S
f	3.98 ± 0.05	1F
g	4.82	2G

▶ $C_6H_{12}O_5$ ▶ D_2O ▶ V 475

▶ (2)-(C)ov(O)c

(b)H H(c)
(a)H O
(d)

	Shift	Peak
▶ a	2.77	4A
▶ b	3.12	4A
c	3.83	4A
d	7.28	1A

▶ C_8H_8O ▶ 7% $CDCl_3$ ▶ V 193

▶ 2-(C)coP

O
D N H(h)
O N N
(b)
DO H
C O
(c)H H(e) (f)H
(d)H H(g)
(a)CH$_3$—C—O O—C—CH$_3$ (a)
O O

	Shift	Peak
a	2.18	1A
▶ b	3.72	S
▶ c	3.86	S
d	4.33	O
e	5.47	S
f	5.69	S
g	6.31	2A
h	7.67	1A

▶ $C_{12}H_{15}N_3O_8$ ▶ 7% $CDCl_3$ ▶ V 598

▶ 2-CbbP

(b) (a)
CH$_2$CH$_3$
CH$_3$—CH$_2$—CH—CH$_2$—OH
(c) (d)

	Shift	Peak
a	.87	3C
b	1.29	3C
▶ c	3.35	2G
d	4.18	1A

▶ $C_6H_{14}O$ ▶ 20% CCl_4 ▶ API 167

▶ 2-(C)coP

O
D N CH$_3$(a)
O N H(h)
(c)
DO H
C O
(d)H H(f) (b)H
H H
(e) (g)
DO H(b)

	Shift	Peak
a	1.91	2D
b	2.40	4A
▶ c	3.79	2A
▶ d	3.85	1A
e	4.03	N
f	4.49	4E
g	6.28	3A
h	7.65	2D

▶ $C_{10}H_{14}N_2O_5$ ▶ D_2O ▶ V 566

▶ 2-CbbP

(b)
(a) (d) (c)
CH$_3$(CH$_2$)$_3$—CHCH$_2$OH
CH$_2$CH$_3$
(b) (a)

	Shift	Peak
a	0.92	3B
b	1.30	1B
c	1.75	1B
▶ d	3.52	1I

▶ $C_8H_{18}O$ ▶ 45.0 mg/0.50 ml $CDCl_3$ ▶ S 98

▶ 2-(C)coP

ND$_2$
N N
H
H N N
DO H(c)
C O
(d)H H(f) (a)H
H H
(e) OD H (g)
(b)

	Shift	Peak
a	2.65 ± 0.10	S
b	2.65 ± 0.10	S
▶ c	3.81	S
▶ d	3.91	S
e	4.19	4A
f	4.63	S
g	6.39	3A

▶ $C_{10}H_{13}N_5O_3$ ▶ D_2O ▶ V 563

▶ 2-CbbP

(a) (b)
CH$_3$CH$_2$
(c)
CH$_3$—CH$_2$—CH—CH$_2$—OH
(d)

	Shift	Peak
a	0.88	3B
b	1.30	6B
c	1.56	S
▶ d	3.54	2G

▶ $C_6H_{14}O$ ▶ 35.5 mg/0.50 ml $CDCl_3$ ▶ S 14

▶ 2-(C)coP

O
D N H(g)
O N H(h)
(a)
DO H
C O
(b)H H(d) (e)H
(c)H H(f)
DO OD

	Shift	Peak
▶ a	3.82	S
▶ b	3.92	S
c	4.18 ± 0.03	S
d	4.25	S
e	4.35	S
f	5.90	S
g	5.90	S
h	7.87	2A

▶ $C_9H_{11}N_2O_6$ ▶ D_2O ▶ V 535

▶ 2-CbpP

(b)
H
(a) (a)
DO—CH$_2$—C—CH$_2$—OD
OD

	Shift	Peak
▶ a	3.61	1E
b	3.79	S

▶ $C_3H_8O_3$ ▶ D_2O ▶ V 395

▶ 2-(C)coP

NH$_2$
N H(e)
O N H(f)
DO H(a)
C O
(b)H H (c)H
H(d)
DO OD

	Shift	Peak
▶ a	3.82	S
▶ b	3.96	S
c	4.32	3C
d	5.91	3A
e	6.04	1A

▶ $C_9H_{13}N_3O_5$ ▶ D_2O ▶ V 540

▶ 2-(C)coP

O
D N H(g)
O N N
(b)
DO H
C O
(a)H (d) (e)
H H
(c)H H(f)
DO OD

	Shift	Peak
▶ a	3.71	S
▶ b	3.83	S
c	4.13	S
d	4.38	S
e	4.60	S
f	6.12	2A
g	7.64	1A

▶ $C_8H_{11}N_3O_6$ ▶ D_2O ▶ V 510

▶ 2-(C)coP

O
D N H(j)
O N H(i)
(c)
DO H
C O
(d)H H(f) (g)H
(e)H H(h)
(b)CH$_3$—C—O O—C—CH$_3$(a)
O O

	Shift	Peak
a	2.17*	1A
b	2.17*	1A
▶ c	3.86	1B
▶ d	3.90	1A
e	4.38	4A
f	5.43	S
g	5.50	S
h	5.94	2A
i	6.08	2A
j	7.87	2A

* or 2.21

▶ $C_{13}H_{16}N_2O_8$ ▶ D_2O ▶ V 612

▶ 2-(C)coP

O D N H(f) O N H(h) DO H(a) C O (b)H OD (e) H H H (c) H(d) OD(g)

	Shift	Peak
▶a	3.95	S
▶b	3.99	S
c	4.18	S
d	4.40	3A
e	4.61	S
f	5.84	2C
g	6.13	2C
h	7.94	2A

▶ $C_9H_{12}N_2O_6$ ▶ D_2O ▶ V 537

▶ 2-CaaQv

O CH_3(a) (c) C—O—CH_2—CH—CH_3 (b) (e) (d)

	Shift	Peak
a	1.02	2A
b	2.05	5B
▶c	4.12	2A
d	7.5	N
e	~8.05	S

▶ $C_{11}H_{14}O_2$ ▶ 45.5 mg/0.50 ml $CDCl_3$ ▶ S 75

▶ 2-CcpP

(c) H O O C OD H(a) C—C—OD DO OD H H (b) (a)

	Shift	Peak
▶a	3.78	1A
b	4.10	O
c	4.97	2D

▶ $C_6H_8O_6$ ▶ D_2O ▶ V 464

▶ 2-CbqQa

(b)H C—O—C—CH_3(a) (c)H O (d)H—C—O—C—CH_3(a) (c)H O C—O—C—CH_3(a) (b)H O

	Shift	Peak
a	2.08	1A
▶b	~4.17	S
▶c	~4.28	S
d	5.25	5B

▶ $C_9H_{14}O_6$ ▶ 7% $CDCl_3$ ▶ V 242

▶ 2-CjoP

CH_3O CH_3O (e) H H(c) H (d) C—C—O OCH_3 O CH_2OH (b) (a)

	Shift	Peak
a	3.35	3B
▶b	4.10	S
c	5.41	3A
d	7.63	N
e	7.77	2A

▶ $C_{18}H_{20}O_6$ ▶ 7% $CDCl_3$ ▶ V 675

▶ 2-CbqQb

H(d) O (e)H—C—O—C—CH_2—$(CH_2)_{15}$—CH_3 (c) (b) (a)
(f)H—C—O—C—CH_2—$(CH_2)_{15}$—CH_3 (c) (b) (a)
(e)H—C—O—C—CH_2—$(CH_2)_{15}$—CH_3 (c) (b) (a) (d)H O

	Shift	Peak
a	0.90	3B
b	1.25	1A
c	2.30	3B
▶d	4.18*	S
▶e	4.18*	S
f	5.28	S

*or 4.28

▶ $C_{57}H_{110}O_6$ ▶ 7% $CDCl_3$ ▶ V 368

▶ 2-CaaQb

(b) (d) O (e) (a) CH_3 (a) CH_3—CH_2—C—OCH_2—CH—CH_3 (c)

	Shift	Peak
a	0.92	2A
b	1.13	3D
c	~1.85	S
d	2.31	4A
▶e	3.83	2A

▶ $C_7H_{14}O_2$ ▶ 4.5 mg/0.50 ml $CDCl_3$ ▶ S 55

▶ 2-(C)coQa

O D N H(h) O N N O (a)CH_3 C O (b) H (c)H O (d)H H(e) (f)H H(g) DO OD

	Shift	Peak
a	1.94	1A
▶b	3.64	S
▶c	3.78	S
d	3.92	S
e	4.10	3A
f	4.47	3A
g	6.00	2A
h	7.45	1A

▶ $C_{10}H_{13}N_3O_7$ ▶ D_2O ▶ V 562

▶ 2-CaaQb

(a) CH_3 O (a) CH_3 (a) (c) (d) (a) CH_3—CH—CH_2—C—OCH_2—CH—CH_3 (b) (b)

	Shift	Peak
a	0.94	2A
b	1.12	S
c	2.20	2D
▶d	3.88	2A

▶ $C_9H_{18}O_2$ ▶ 44.5 mg/0.50 ml $CDCl_3$ ▶ S 277

▶ 2-(C)coQa

O D N H(d) O (a)CH_3—C—O (b) H O N H(f) (c)H O H H H H(e) O (a)CH_3—C—O O—C—CH_3(a) O

	Shift	Peak
a	1.92	1A
▶b	3.80	S
▶c	3.87	S
d	5.81	S
e	5.83	S
f	7.65	2A

▶ $C_{15}H_{18}N_2O_9$ ▶ D_2O ▶ V 642

▶ 2-CaaQc

(b) CH_3 O (a) CH_3 (b) (e) (a) CH_3—CH—C—O—CH_2CHCH_3 (d) (c)

	Shift	Peak
a	0.94	2A
b	1.18	2A
c	~1.82	7B
d	~2.50	5B
▶e	3.84	2A

▶ $C_8H_{16}O_2$ ▶ 45.2 mg/0.50 ml $CDCl_3$ ▶ S 238

▶ 2-(C)coQa

O D N H(g) O N N O (b)CH_3—C—O (d) H O N (c)H H (e)H H H(f) (a)CH_3—C—O O—C—CH_3(a) O O

	Shift	Peak
a	2.13	1A
b	2.13	1A
▶c	4.28	S
▶d	4.32	S
e	4.62	S
f	6.14	2A
g	7.67	1A

▶ $C_{14}H_{17}N_3O_9$ ▶ D_2O ▶ V 634

▶ 2-(C)coQa

	Shift	Peak
a	2.10*	1A
▶b	4.42	1B
c	5.62	S
d	5.88	3A
e	6.17	2A
f	8.03†	1A
g	8.03†	1A
h	~13.1	S

*or 2.15
†or 8.35

▶ $C_{16}H_{18}N_4O_8$ ▶ 7% $CDCl_3$ ▶ V 324

▶ 2-CjnVhh

	Shift	Peak
▶a	3.09	S
▶b	3.29	S
c	3.68	S
d	4.02*	S
e	4.37	S
f	4.60	S
g	4.82	S
h	6.12	S

*± 0.05

▶ $C_{22}H_{31}Cl_2N_7O_5$ ▶ D_2O ▶ V 691

▶ 2-(C)coQa

	Shift	Peak
▶a	4.43	1E
b	4.43	1E
c	5.70	1C
d	5.95	3A
e	6.20	2A
f	6.37	1B
g	7.98 or 8.35	1A
h	7.98 or 8.35	1A

▶ $C_{16}H_{19}N_5O_8$ ▶ 7% $CDCl_3$ ▶ V 326

▶ 2-CkmVhh

	Shift	Peak
▶a	3.23	N
▶b	3.52	N
c	3.72	1A
d	4.28	4A
e	7.89	1A
f	8.04	1F

▶ $C_{17}H_{17}NO_3$ ▶ 7% $CDCl_3$ ▶ V 663

▶ 2-CcvR

	Shift	Peak
a	2.66	S
b	3.75	6A
▶c	4.63	S
▶d	4.91	S

▶ $C_{14}H_{17}NO_3$ ▶ 7% $CDCl_3$ ▶ V 632

▶ 2-CknVhh

	Shift	Peak
▶a	2.86	S
▶b	3.07	S
c	3.59	3C
d	7.35	1A

▶ $C_9H_{11}NO_2$ ▶ D_2O ▶ V 534

▶ (2)-(C)coVch

	Shift	Peak
▶a	2.96	S
▶b	3.20	S
c	4.08	3A
d	4.22	2A

▶ C_9H_8O ▶ 7% $CDCl_3$ ▶ V 524

▶ 2-CknVhh

	Shift	Peak
a	1.27	3A
▶b	3.14	2A
c	4.20	4A
d	4.96	4C
e	5.86	1A
f	6.62 ± 0.04	S

▶ $C_{19}H_{19}NO_5$ ▶ 7% $CDCl_3$ ▶ V 334

▶ (2)-(C)cpVoo

	Shift	Peak
a	1.73	1B
▶b	2.95	2A
c	4.28	1B
d	4.96	1I
e	6.12 or 6.20	2C
f	6.12 or 6.20	2C

▶ $C_{19}H_{22}O_6$ ▶ 7% $CDCl_3$ ▶ V 679

▶ 2-DaaaDaaa

	Shift	Peak
a	.98	1A
▶b	1.27	1A

▶ C_9H_{20} ▶ 10% CCl_4 ▶ API 28

▶ (2)-(C)doVoo

	Shift	Peak
a	1.25 or 1.37	1A
b	1.37 or 1.25	1A
c	1.96	1G
▶d	3.33	2A
e	4.82	3A
f	6.22	2C
g	6.79	2C
h	7.29	2C
i	7.66	2C

▶ $C_{14}H_{14}O_4$ ▶ 7% $CDCl_3$ ▶ V 310

▶ 2-DaaaDaan

	Shift	Peak
a	1.02	1A
b	1.18	1A
c	1.25	1A
▶d	1.45	1A

▶ $C_8H_{19}N$ ▶ 7% $CDCl_3$ ▶ V 220

2-DaaaDaav

	Shift	Peak
a	0.72	1A
b	1.33	1A
▸c	1.68	1A
d	5.02	1C
e	6.75	Q
f	7.20	Q

▸ $C_{14}H_{22}O$ ▸ 7% $CDCl_3$ ▸ V 315

2-DaagG

	Shift	Peak
a	1.90	1A
▸b	3.88	1A

▸ $C_4H_8Br_2$ ▸ 7% $CDCl_3$ ▸ V 412

(2)-(D)aab(D)aab

	Shift	Peak
a	1.02	1A
▸b	1.30	1A
c	1.50	1A

▸ C_9H_{18} ▸ 10 vol.% CCl_4 ▸ API 440

2-(D)bcjG

	Shift	Peak
a	0.84	2A
b	0.92	2A
c	2.08	S
▸d	2.22	1H
e	2.61	1H
f	3.58	1A

▸ $C_{10}H_{14}Br_2O$ ▸ 7% $CDCl_3$ ▸ V 564

2-DaeeE

	Shift	Peak
a	2.20	1A
▸b	4.02	1A

▸ $C_3H_5Cl_3$ ▸ 7% $CDCl_3$ ▸ V 383

(2)-(D)aab(J)b

	Shift	Peak
a	1.06	1E
b	1.09	1E
▸c	2.29	1A
d	2.29	1A
e	2.56	1A
f	3.36	1B
g	5.51	1A
h	10.90	1A

keto-enol mixture

▸ $C_8H_{12}O_2$ ▸ 7% $CDCl_3$ ▸ V 512

2-DaefE

	Shift	Peak
a	2.03	2A
▸b	3.83	S
▸c	3.96	S

▸ $C_3H_5Cl_2F$ ▸ 7% $CDCl_3$ ▸ V 382

(2)-(D)aab(J)l

	Shift	Peak
a	1.06	1E
b	1.09	1E
c	2.29	1A
▸d	2.29	1A
e	2.56	1A
f	3.36	1B
g	5.51	1A
h	10.90	1A

keto-enol mixture

▸ $C_8H_{12}O_2$ ▸ 7% $CDCl_3$ ▸ V 512

2-DaffE

	Shift	Peak
a	1.75	3A
▸b	3.63	3A

▸ $C_3H_5ClF_2$ ▸ 7% $CDCl_3$ ▸ V 381

(2)-(D)aal(J)l

	Shift	Peak
a	1.20	1A
b	1.70	2D
▸c	2.73	1G
d	4.73	1B
e	6.14	S
f	6.16	S

▸ $C_{23}H_{22}O_2$ ▸ 7% $CDCl_3$ ▸ V 694

2-DeffE

	Shift	Peak
a	4.00	3A

▸ $C_2H_2Cl_2F_2$ ▸ 7% $CDCl_3$ ▸ V 369

(2)-(D)abb(J)n

	Shift	Peak
a	0.91	3A
b	1.06	1A
c	1.43	3B
▸d	2.43	1A
e	8.47	1D

▸ $C_8H_{13}NO_2$ ▸ 7% $CDCl_3$ ▸ V 514

▶ (2)-(D)abc(J)c

(d) (e)
H H
CH_3(a)
O
(f)H (b)H (c)
CH_3

	Shift	Peak
a	0.67	1E
b	0.78	S
c	1.03	1E
▶ d	2.27	S
▶ e	2.27	S
f	2.45	S

▶ $C_{19}H_{30}O$ ▶ 7% $CDCl_3$ ▶ V 680

▶ 2-DaaaLhl

(c) (e)
H H
C=C (b) (a)
H $CH_2C(CH_3)_3$
(d)

	Shift	Peak
a	.83	1A
▶ b	1.92	2E
c	4.82	R
d	5.03	R
e	5.72	R

▶ C_7H_{14} ▶ liquid ▶ API 385

▶ (2)-(D)acl(J)l

(g)
H O
O C OH(h)
(d)H CH_3(c)
(f)H CH(e)
CH_3 CH_3
(a) (b)

	Shift	Peak
a	0.75	2A
b	0.97	2A
c	1.45	1A
▶ d	2.10	2C
e	2.40	S
▶ f	2.60	2C
g	6.77	1A
h	12.00	1B

▶ $C_{10}H_{14}O_3$ ▶ 7% $CDCl_3$ ▶ V 567

▶ (2)-(D)aab(L)dm

(a)
CH_3
(a)CH_3 CH_2(d)
(e)H_2C
N $C(CH_2CH_2CN)_3$
(b) (c)

	Shift	Peak
a	1.10	1A
b	1.96	S
c	2.32	S
▶ d	2.32	S
e	3.60	S

▶ $C_{16}H_{22}N_4$ ▶ 7% $CDCl_3$ ▶ V 660

▶ (2)-(D)aad(K)d

CH_3(b) CH_3(a)
(b)CH_3 CH_3(a)
O CH_2
(c)
O

	Shift	Peak
a	1.10	1A
b	1.33	1A
▶ c	2.43	1A

▶ $C_8H_{14}O_2$ ▶ 7% $CDCl_3$ ▶ V 214

▶ (2)-(D)aab(L)lp

(g)
(f) (h) H
O CH_2 O HO O
(c)H_2C CH_2(c) (e)H_2C CH_2(d)
C
CH_3 CH_3 CH_3 CH_3
(a) (a) (b) (b)

	Shift	Peak
a	1.06	1E
b	1.09	1E
c	2.29	1A
d	2.29	1A
▶ e	2.56	1A
f	3.36	1B
g	5.51	1A
h	10.90	1A

keto-enol mixture

▶ $C_8H_{12}O_2$ ▶ 7% $CDCl_3$ ▶ V 512

▶ (2)-(D)abp(K)b

(a) (e)
CH_3 OH
H(c)
(b)
H_2C H(d)
(f)H
O O
H
(g)

	Shift	Peak
a	1.40	1A
b	1.92	4B
▶ c	2.51	2G
▶ d	2.68	2G
e	2.92	1A
f	4.33	N
g	4.00	N

▶ $C_6H_{10}O_3$ ▶ 7% $CDCl_3$ ▶ V 466

▶ (2)-(D)aal(L)ln

(a)
CH_3
(b)
H_2C — C — CH_3(a)
(d)H C C
C N N
(e)H H_2C — CH_2
(c) (f)

	Shift	Peak
a	1.29	1A
▶ b	2.76	3A
c	3.40	3C
d	3.79	1F
e	3.79	1F
f	4.18	3C

▶ $C_9H_{14}N_2$ ▶ 7% $CDCl_3$ ▶ V 541

▶ 2-DbkpKb

O
(b) (c) (a)
CH_2—C—O—CH_2—CH_3
(d) HO—C—C—O—CH_2—CH_3
O
CH_2—C—O—CH_2—CH_3
(b) (c) (a)
O

	Shift	Peak
a	1.26	3D
▶ b	2.86	1A
c	4.18	4A
d	4.47	1B

▶ $C_{12}H_{20}O_7$ ▶ 45.3 mg/0.50 ml $CDCl_3$ ▶ S 150

▶ (2)-(D)aan(L)bm

(a) (a)
CH_3 CH_3
CH_2(d)
(c)CH_3—N N—OH(e)
CH_2(b)
(a)CH_3 CH_3
(a)

	Shift	Peak
a	1.12	1A
▶ b	2.24	1H
c	2.28	1A
▶ d	2.56	1B
e	8.73	1B

▶ $C_{10}H_{20}N_2O$ ▶ 7% $CDCl_3$ ▶ V 574

▶ 2-DaaaLal

(d) (b)
H CH_3
C=C
(a)
H $CH_2C(CH_3)_3$
(e) (c)

	Shift	Peak
a	.92	1A
b	1.76	1A
▶ c	1.94	1A
d	~4.70	1C
e	~4.80	1C

▶ C_8H_{16} ▶ 50 vol.% CCl_4 ▶ API 433

▶ (2)-(D)aav(L)jm

(f)H O
OH(g)
N
(d)H
CH_2(b)
(c) CH_3O
CH_3 CH_3
H (a) (a)
(e)

	Shift	Peak
a	1.42	1A
▶ b	3.04	1A
c	3.91	1A
d	6.88	S
e	6.93	S
f	8.17	2A
g	~10.30	1D

▶ $C_{13}H_{15}NO_3$ ▶ 7% $CDCl_3$ ▶ V 611

▶ (2)-(D)akn(L)jm

	Shift	Peak
a	1.10	2A
b	1.12	2A
c	1.60	1A
▶ d	2.81	2A
▶ e	3.48	S
f	3.52	S

▶ $C_9H_{14}N_2O_3$ ▶ 7% $CDCl_3$ ▶ V 543

▶ (2)-(D)bcc(O)c

	Shift	Peak
a	0.66	1A
b	0.88	S
c	0.95	S
d	1.48	S
e	2.53	S
▶ f	3.90	S
▶ g	4.03	S
h	4.12	S

▶ $C_{27}H_{44}O_2$ ▶ 7% $CDCl_3$ ▶ V 698

▶ (2)-(D)bcj(L)al

	Shift	Peak
a	0.84	2A
b	0.92	2A
c	2.08	S
d	2.22	1H
▶ e	2.61	1H
f	3.58	1A

▶ $C_{10}H_{14}Br_2O$ ▶ 7% $CDCl_3$ ▶ V 564

▶ (2)-(D)bov(O)d

	Shift	Peak
a	0.92	3A
b	1.80	S
c	2.05	S
▶ d	2.70	2C
▶ e	2.97	2C
f	7.30	1A

▶ $C_{10}H_{12}O$ ▶ 7% $CDCl_3$ ▶ V 558

▶ (2)-(D)aab(M)l

	Shift	Peak
a	1.10	1A
b	1.96	S
c	2.32	S
d	2.32	S
▶ e	3.60	S

▶ $C_{16}H_{22}N_4$ ▶ 7% $CDCl_3$ ▶ V 660

▶ 2-DaacP

	Shift	Peak
a	0.92	1A
b	0.95 or 1.00	2A
c	0.95 or 1.00	2A
d	1.92	5B
e	3.38	2A
▶ f	3.44 or 3.53	S
▶ g	3.44 or 3.53	S
h	3.58	1A

▶ $C_8H_{18}O_2$ ▶ 7% $CDCl_3$ ▶ V 217

▶ 2-DaajNbb

	Shift	Peak
a	0.97	3D
b	1.07	1E
c	2.50	4E
▶ d	2.55	1E
e	9.57	1A

▶ $C_9H_{19}NO$ ▶ 7% $CDCl_3$ ▶ V 548

▶ 2-DaarP

	Shift	Peak
a	1.53	1A
▶ b	3.81	2A
c	3.05	3B

▶ $C_4H_9NO_3$ ▶ 45.1 mg/0.25 ml $CDCl_3$ ▶ S 296

▶ 2-DbbbOb

	Shift	Peak
a	0.84	3C
b	1.38	4C
c	2.83	3B
▶ d	3.43	1E
e	3.59	2B
f	3.95	2E
g	5.14	R
h	5.22	R
i	5.87	R

▶ $C_{12}H_{22}O_3$ ▶ 7% $CDCl_3$ ▶ V 603

▶ 2-DaarP

	Shift	Peak
a	1.59	1A
b	2.52	3B
▶ c	3.89	2B

▶ $C_4H_9NO_3$ ▶ 7% $CDCl_3$ ▶ V 422

▶ (2)-(D)bbb(O)c

	Shift	Peak
a	0.83	N
b	1.20	N
▶ c	3.93	1A
d	5.52	1A

▶ $C_7H_{12}O_3$ ▶ 7% $CDCl_3$ ▶ V 493

▶ 2-DbbbP

	Shift	Peak
a	0.84	3C
b	1.38	4C
c	2.83	3B
d	3.43	1E
▶ e	3.59	2B
f	3.95	2E
g	5.14	R
h	5.22	R
i	5.87	R

▶ $C_{12}H_{22}O_3$ ▶ 7% $CDCl_3$ ▶ V 603

▶ 2-DcffP

	Shift	Peak
a	2.72	1B
▶b	3.97	3C
c	5.93	3E

(c) H—CF₂—CF₂—CH₂—OH (b) (a)

▶ $C_3H_4F_4O$ ▶ 7% $CDCl_3$ ▶ V 19

▶ 2-EKb

	Shift	Peak
a	1.30	3A
▶b	4.05	1A
c	4.25	4A

$ClH_2C-C(=O)-OCH_2CH_3$ (b) (c) (a)

▶ $C_4H_7ClO_2$ ▶ 7% $CDCl_3$ ▶ V 73

▶ 2-DfffP

	Shift	Peak
a	3.38	1B
▶b	3.93	4A

CF_3-CH_2-OH (b) (a)

▶ $C_2H_3F_3O$ ▶ 7% $CDCl_3$ ▶ V 4

▶ 2-EKb

	Shift	Peak
a	0.94	3B
b	1.56	1F
▶c	4.05	1A
d	4.18	3A

$Cl-CH_2-C(=O)-O-CH_2(CH_2)_2CH_3$ (c) (d) (b) (a)

▶ $C_6H_{11}ClO_2$ ▶ 44.5 mg/0.50 ml $CDCl_3$ ▶ S 92

▶ 2-DaaeVhh

	Shift	Peak
a	1.57	1A
▶b	3.07	1A
c	7.27	1A

$C_6H_5(c)-CH_2(b)-C(Cl)(CH_3(a))-CH_3(a)$

▶ $C_{10}H_{13}Cl$ ▶ 7% $CDCl_3$ ▶ V 263

▶ 2-EKb

	Shift	Peak
a	0.93	3B
▶b	4.05	1A
c	4.20	3D

$CH_3-CH_2-CH_2-CH_2-O-C(=O)-CH_2-Cl$ (a) (c) (b)

▶ $C_6H_{11}ClO_2$ ▶ 7% $CDCl_3$ ▶ V 138

▶ 2-EJn

	Shift	Peak
a	3.65	1A
▶b	4.17	1A
c	7.31 ± .05	1H
d	8.32	1C

$C_6H_4(c)(CH_2CN(a))-NH(d)-C(=O)-CH_2Cl(b)$

▶ $C_{10}H_9ClN_2O$ ▶ 7% $CDCl_3$ ▶ V 248

▶ 2-EKh

	Shift	Peak
▶a	4.10	1A
b	10.58	1B

$Cl-CH_2-C(=O)-OH$ (a) (b)

▶ $C_2H_3ClO_2$ ▶ 45.0 mg/0.50 ml $CDCl_3$ ▶ S 128

▶ 2-EKa

	Shift	Peak
a	3.82	1A
▶b	4.07	1A

$Cl-CH_2-C(=O)-OCH_3$ (b) (a)

▶ $C_3H_5ClO_2$ ▶ 45.1 mg/0.50 ml $CDCl_3$ ▶ S 95

▶ 2-ELel

	Shift	Peak
▶a	4.15	2D
b	5.42	2A
c	5.59	M

$H(b)(H(c))C=C(CH_2Cl(a))Cl$

▶ $C_3H_4Cl_2$ ▶ 7% $CDCl_3$ ▶ V 18

▶ 2-EKb

	Shift	Peak
a	1.32	3A
▶b	4.00	1A
c	4.24	4A

$Cl-CH_2-C(=O)-O-CH_2CH_3$ (b) (c) (a)

▶ $C_4H_7ClO_2$ ▶ 44.8 mg/0.50 ml $CDCl_3$ ▶ S 91

▶ 2-ELhl

	Shift	Peak
▶a	4.08	7B
b	5.93	7C

$H(b)(ClCH_2(a))C=C(CH_2Cl(a))H(b)$

▶ $C_4H_6Cl_2$ ▶ 7% $CDCl_3$ ▶ V 404

▶ 2-EVhh

Shift	Peak
▶a 4.55	1A
b 7.32	1A

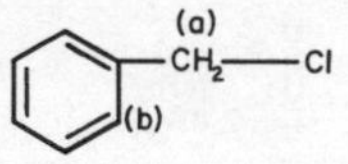

▶ C_7H_7Cl ▶ 45.2 mg/0.50 ml $CDCl_3$ ▶ S 3

▶ 2-GLhl

Shift	Peak
▶a 3.93	2B
b ~5.12	R
c ~5.32	R
d ~6.05	R

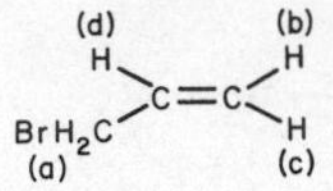

▶ C_3H_5Br ▶ 7% $CDCl_3$ ▶ V 24

▶ 2-EVhh

Shift	Peak
a 2.35	1A
▶b 4.56	1A
c 7.24	1E

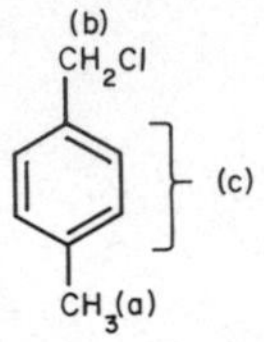

▶ C_8H_9Cl ▶ 45.2 mg/0.50 ml $CDCl_3$ ▶ S 68

▶ 2-GTh

Shift	Peak
a 2.46	1H
▶b 3.86	2A

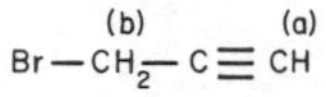

▶ C_3H_3Br ▶ 45.2 mg/0.50 ml $CDCl_3$ ▶ S 273

▶ 2-GJv

Shift	Peak
a 4.38	1A

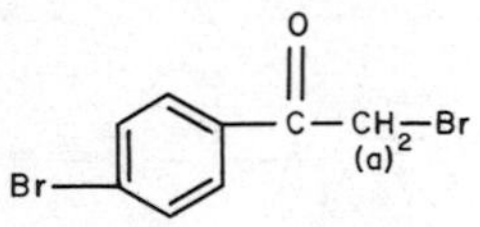

▶ $C_8H_6Br_2O$ ▶ 44.4 mg/0.25 ml $CDCl_3$ ▶ S 279

▶ 2-GVah

Shift	Peak
a 2.41	1A
▶b 4.50	1A
c 7.19	1A

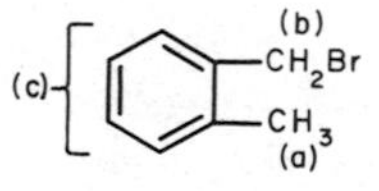

▶ C_8H_9Br ▶ 45.3 mg/0.50 ml $CDCl_3$ ▶ S 239

▶ 2-GJv

Shift	Peak
a 4.43	1A

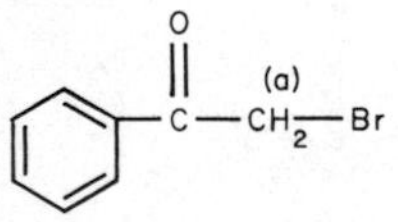

▶ C_8H_7BrO ▶ 45.3 mg/0.50 ml $CDCl_3$ ▶ S 240

▶ 2-GVhh

Shift	Peak
▶a 4.50	1A
b 7.37	1A

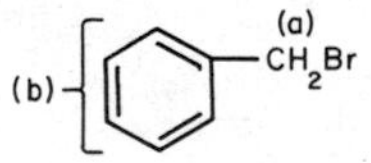

▶ C_7H_7Br ▶ 45.2 mg/0.50 ml $CDCl_3$ ▶ S 154

▶ 2-GJv

Shift	Peak
▶a 4.45	1A
b 7.70	Q
c 8.07	Q

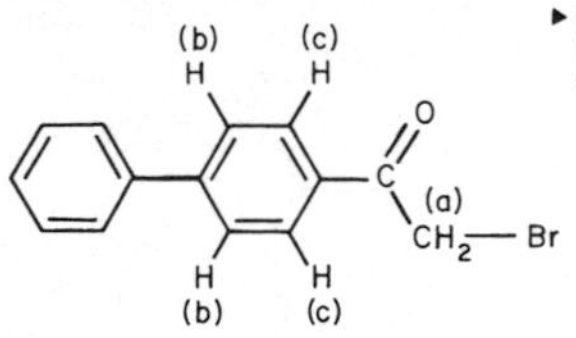

▶ $C_{14}H_{11}BrO$ ▶ 7% $CDCl_3$ ▶ V 303

▶ 2-II

Shift	Peak
a 3.89	1A

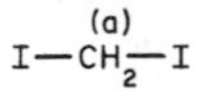

▶ CH_2I_2 ▶ 45.2 mg/0.50 ml $CDCl_3$ ▶ S 226

▶ 2-GLgl

Shift	Peak
▶a 4.17	2D
b 5.62	2A
c 6.02	M

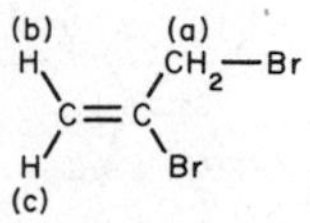

▶ $C_3H_4Br_2$ ▶ 7% $CDCl_3$ ▶ V 17

▶ 2-IKh

Shift	Peak
▶a 3.72	1A
b 10.25	1A

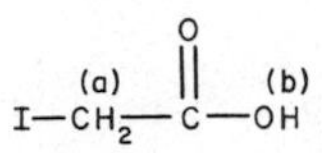

▶ $C_2H_3IO_2$ ▶ 44.8 mg/0.05 ml $CDCl_3$ ▶ S 214

▶ 2-ILhl

Shift	Peak
▶a 3.87	2E
b 5.05	R
c 5.15	R
d 5.93	R

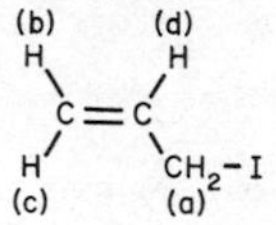

▶ C_3H_5I ▶ 7% $CDCl_3$ ▶ V 26

▶ 2-JnNhh

Shift	Peak
▶a 3.83	1A
b 3.87	1A

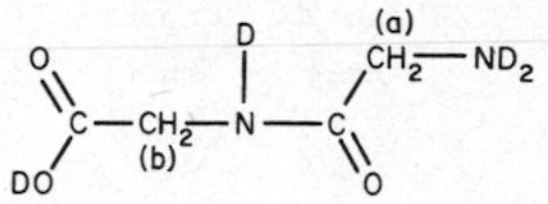

▶ $C_4H_8N_2O_3$ ▶ D_2O ▶ V 413

▶ 2-JaJn

Shift	Peak
a 2.17	1A
▶b 3.52	1A
c 9.34	1C

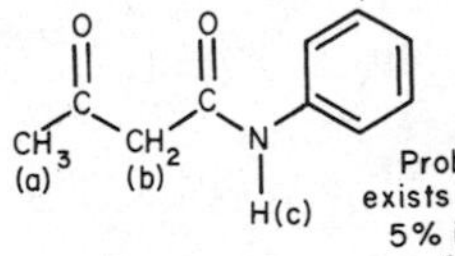

Probably exists about 5% in enol form.

▶ $C_{10}H_{11}NO_2$ ▶ 7% $CDCl_3$ ▶ V 256

▶ (2)-(J)b(O)c

Shift	Peak
a 1.44	2A
b 2.15	10A
c 2.57	10A
▶d 3.83	10A
▶e 4.04	10A
f 4.35	7C

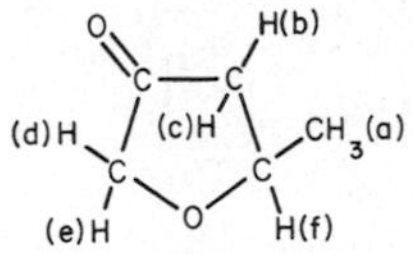

▶ $C_5H_8O_2$ ▶ 7% $CDCl_3$ ▶ V 439

▶ (2)-(J)b(J)b

Shift	Peak
a 1.06	1E
b 1.09	1E
c 2.29	1A
d 2.29	1A
e 2.56	1A
▶f 3.36	1B
g 5.51	1A
h 10.90	1A

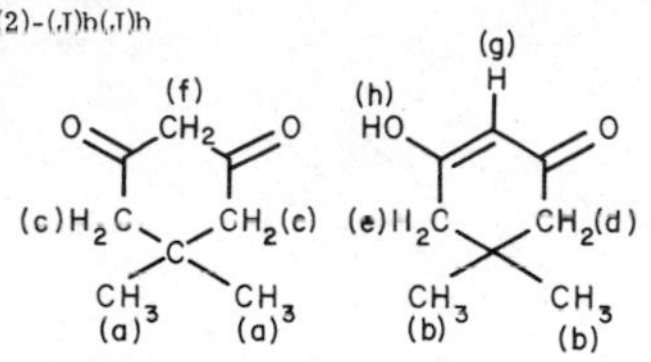

keto-enol mixture

▶ $C_8H_{12}O_2$ ▶ 7% $CDCl_3$ ▶ V 512

▶ (2)-(J)c(O)b

Shift	Peak
a 1.17	2A
b 2.55	6B
c 3.73	10A
▶d 3.84	10A
▶e 4.07	10A
f 4.49	3A

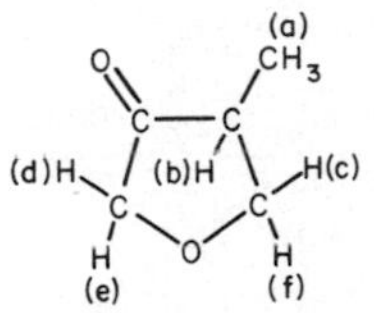

▶ $C_5H_8O_2$ ▶ 7% $CDCl_3$ ▶ V 438

▶ (2)-(J)v(J)v

Shift	Peak
a 3.23	1A

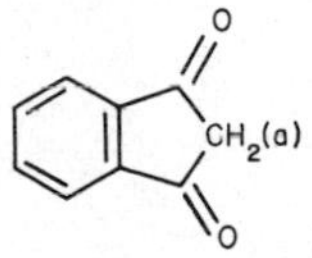

▶ $C_9H_6O_2$ ▶ 7% $CDCl_3$ ▶ V 224

▶ (2)-(J)bVbh

Shift	Peak
▶a 3.50	1A
b 7.25	1A

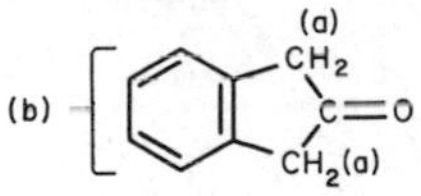

▶ C_9H_8O ▶ 7% $CDCl_3$ ▶ V 523

▶ 2-JaLhl

Shift	Peak
a 2.15	1E
▶b 3.13	2A
c 5.43	O

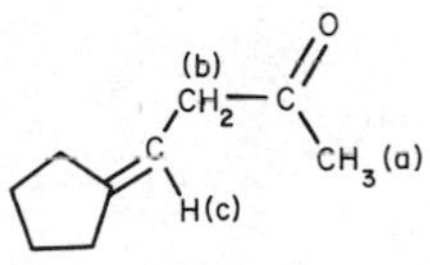

▶ $C_9H_{14}O$ ▶ 7% $CDCl_3$ ▶ V 545

▶ 2-JbVhh

Shift	Peak
a 3.70	1A

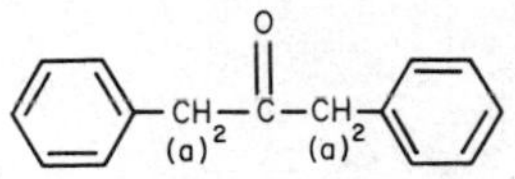

▶ $C_{15}H_{14}O$ ▶ 7% $CDCl_3$ ▶ V 638

▶ (2)-(J)n(N)al

Shift	Peak
a 3.06	1A
▶b 4.07	1A

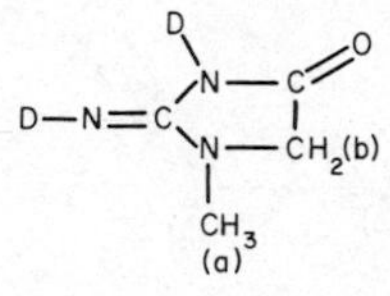

▶ $C_4H_7N_3O$ ▶ D_2O ▶ V 411

▶ (2)-(J)bVhv

Shift	Peak
a 2.20	1A
▶b 3.31	2C
c 3.52	2C

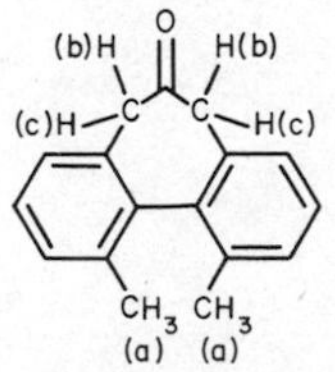

▶ $C_{17}H_{16}O$ ▶ 7% $CDCl_3$ ▶ V 662

▶ (2)-(J)bVhv

	Shift	Peak
a	3.54	1A

▶ $C_{15}H_{12}O$ ▶ 7% $CDCl_3$ ▶ V 637

▶ 2-KbLcm

	Shift	Peak
a	1.33	3A
▶b	3.67	1A
c	4.30	4A
d	6.30	3A*
e	8.03	2C
f	8.40	4A
g	9.08	2A
h	11.77	1B

*Very large splitting due to fluorine

▶ $C_{12}H_{12}F_2N_4O_6$ ▶ 7% $CDCl_3$ ▶ V 293

▶ 2-JeVhh

	Shift	Peak
▶a	4.10	1A
b	7.30	1A

▶ C_8H_7ClO ▶ 44.8 mg/0.50 ml $CDCl_3$ ▶ S 218

▶ 2-KhLkl

	Shift	Peak
▶a	3.46	1G
b	5.92	1G
c	6.40	1A

▶ $C_5H_6O_4$ ▶ D_2O ▶ V 433

▶ 2-JkVer

	Shift	Peak
▶a	4.78	1A
b	7.48	4A
c	7.75	4A
d	8.00	S
e	9.63	1B

▶ $C_9H_6ClNO_5$ ▶ 7% $CDCl_3$ ▶ V 223

▶ (2)-(K)m(L)mv

	Shift	Peak
a	3.82	1A

▶ $C_9H_7NO_2$ ▶ 7% $CDCl_3$ ▶ V 521

▶ (2)-(J)vVbh

	Shift	Peak
a	1.92	1A
b	2.55	1F
c	2.88	1F
d	3.58	1A
▶e	3.78	1A
f	5.92, 5.96	1A
g	6.65	1A
h	6.69	1A
i	6.91	1A

▶ $C_{20}H_{19}NO_5$ ▶ 7% $CDCl_3$ ▶ V 339

▶ 2-KbMj

	Shift	Peak
a	1.30	3A
▶b	3.92	1A
c	4.27	4A

▶ $C_5H_7NO_3$ ▶ 7% $CDCl_3$ ▶ V 107

▶ 2-JnX

	Shift	Peak
▶a	3.72	1A
b	6.46	1D
c	~7.17	N
d	~7.25	M
e	7.65	6A
f	8.52	2E

▶ $C_7H_8N_2O$ ▶ 7% $CDCl_3$ ▶ V 159

▶ 2-KbNaa

	Shift	Peak
a	1.29	3A
b	2.37	1A
▶c	3.17	1A
d	4.22	4A

▶ $C_6H_{13}NO_2$ ▶ 7% $CDCl_3$ ▶ V 480

▶ 2-KbKb

	Shift	Peak
a	1.27	3A
▶b	3.37	1A
c	4.22	4A

▶ $C_7H_{12}O_4$ ▶ 7% $CDCl_3$ ▶ V 181

▶ 2-KhNhj

	Shift	Peak
a	3.83	1A
▶b	3.87	1A

▶ $C_4H_8N_2O_3$ ▶ D_2O ▶ V 413

▶ 2-KhOb

	Shift	Peak
a	1.27	3A
b	3.66	4A
▶ c	4.13	1A
d	10.95	1A

▶ $C_4H_8O_3$ ▶ 7% $CDCl_3$ ▶ V 417

▶ 2-KbVhh

	Shift	Peak
a	1.22	3A
▶ b	3.55	1A
c	4.12	4A

▶ $C_{10}H_{12}O_2$ ▶ 44.9 mg/0.50 ml $CDCl_3$ ▶ S 66

▶ 2-KhOv

	Shift	Peak
▶ a	4.60	1A
b	11.10	1A

▶ $C_8H_8O_3$ ▶ 45.4 mg/0.50 ml $CDCl_3$ ▶ S 264

▶ 2-KhVhh

	Shift	Peak
a	2.30	1A
▶ b	3.55	1A
c	12.67	1A

▶ $C_9H_{10}O_2$ ▶ 44.9 mg/0.25 ml $CDCl_3$ ▶ S 272

▶ 2-KhOv

	Shift	Peak
▶ a	4.67	1A
b	6.88	Q
c	7.27	Q
d	8.86	1C

▶ $C_8H_7ClO_3$ ▶ 7% $CDCl_3$ ▶ V 500

▶ 2-KhVhh

	Shift	Peak
▶ a	3.58	1A
b	11.28	1A

▶ $C_8H_8O_2$ ▶ 44.9 mg/0.50 ml $CDCl_3$ ▶ S 117

▶ 2-KhSb

	Shift	Peak
▶ a	3.10	1A
b	3.87	1A
c	7.32	1A
d	10.97	1B

▶ $C_9H_{10}O_2S$ ▶ 7% $CDCl_3$ ▶ V 532

▶ (2)-(L)hl(L)hl

	Shift	Peak
▶ a	2.20	3A
b	5.28	4C
c	6.12	O
d	6.55	3C

▶ C_7H_8 ▶ 7% $CDCl_3$ ▶ V 158

▶ 2-KaU

	Shift	Peak
▶ a	3.48	1A
b	3.82	1A

▶ $C_4H_5NO_2$ ▶ 7% $CDCl_3$ ▶ V 57

▶ 2-LhlLhl

	Shift	Peak
▶ a	2.73	3C
b	~4.90	R
c	~5.07	R
d	~5.69	R

▶ C_5H_8 ▶ 50% CCl_4 ▶ API 470

▶ 2-KaVhh

	Shift	Peak
▶ a	3.60	1A
b	3.66	1A
c	7.28	1A

▶ $C_9H_{10}O_2$ ▶ 44.4 mg/0.50 ml $CDCl_3$ ▶ S 19

▶ 2-LhlLhl

	Shift	Peak
a	0.97	3A
b	1.33	1B
▶ c	2.80	3B
d	3.67	1A
e	~5.38	3C

▶ $C_{19}H_{32}O_2$ ▶ 7% $CDCl_3$ ▶ V 337

▶ (2)-(L)hl(L)jl

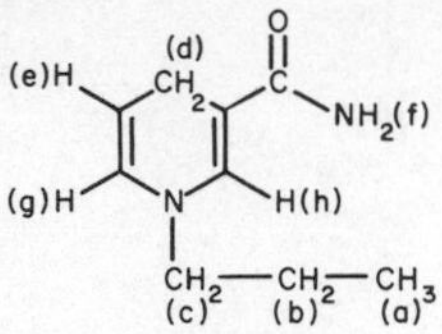

Shift	Peak
a 0.90	3C
b 1.58	6B
c 3.09	2A
▶d 3.19	N
e 4.72	5B
f 5.49	1C
g 5.73	4A
h 7.05	2D

▶ $C_9H_{14}N_2O$ ▶ 7% $CDCl_3$ ▶ V 542

▶ 2-LhlOb

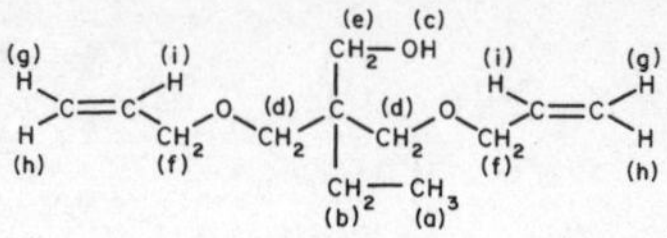

Shift	Peak
a 0.84	3C
b 1.38	4C
c 2.83	3B
d 3.43	1E
e 3.59	2B
▶f 3.95	2E
g 5.14	R
h 5.22	R
i 5.87	R

▶ $C_{12}H_{22}O_3$ ▶ 7% $CDCl_3$ ▶ V 603

▶ (2)-(L)hl(N)bh

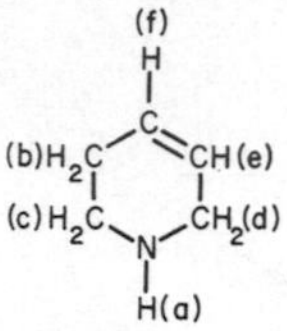

Shift	Peak
a 1.63	1A
b 2.07	O
c 2.95	3A
▶ d 3.33	N
e 5.72 or 5.77	1E
f 5.72 or 5.77	1E

▶ C_5H_9N ▶ 7% $CDCl_3$ ▶ V 115

▶ 2-LhlOb

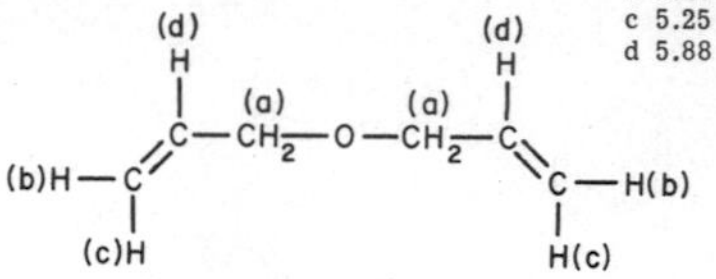

Shift	Peak
▶a 3.97	2E
b 5.17	R
c 5.25	R
d 5.88	R

▶ $C_6H_{10}O$ ▶ 7% $CDCl_3$ ▶ V 134

▶ 2-LhlNhh

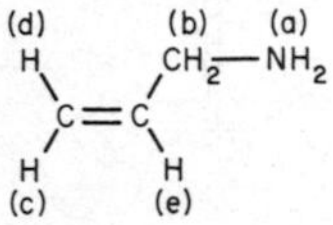

Shift	Peak
a 1.53	1B
▶ b 3.30	2E
c 5.03	R
d 5.13	R
e 5.92	R

▶ C_3H_7N ▶ 7% $CDCl_3$ ▶ V 38

▶ (2)-(L)hl(O)b

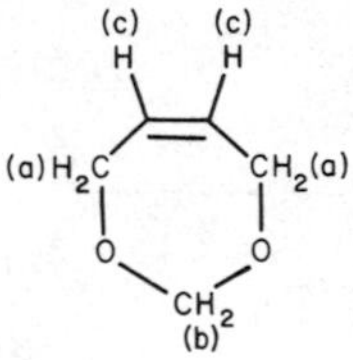

Shift	Peak
▶a 4.31	2D
b 4.88	1A
c 5.74	3A

▶ $C_5H_8O_2$ ▶ 7% $CDCl_3$ ▶ V 437

▶ 2-(L)loNhh

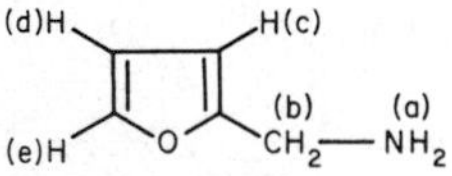

Shift	Peak
a 1.53	1A
▶ b 3.80	1A
c 6.13	N
d 6.30	N
e 7.33	N

▶ C_5H_7NO ▶ 7% $CDCl_3$ ▶ V 104

▶ 2-LhlOl

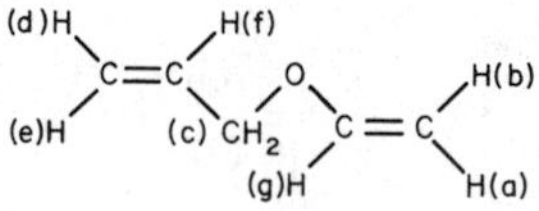

Shift	Peak
a 4.05	R
b 4.20	R
▶ c 4.20	R
d 5.20	R
e 5.32	R
f 5.90	R
g 6.45	4D

▶ C_5H_8O ▶ 7% $CDCl_3$ ▶ V 110

▶ (2)-(L)lo(N)hj

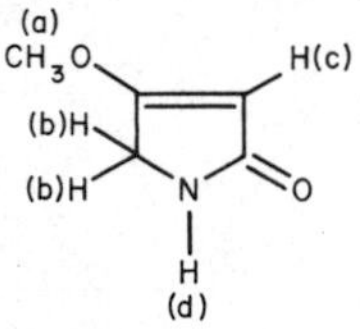

Shift	Peak
a 3.80	1A
▶b 3.92	1B
c 5.05	1I or 2D
d ~6.30	1D

▶ $C_5H_7NO_2$ ▶ 7% $CDCl_3$ ▶ V 105

▶ 2-LhlP

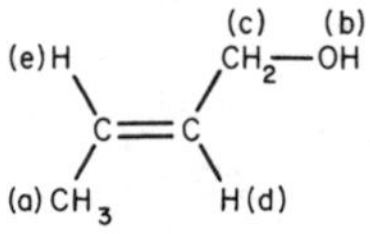

Shift	Peak
a 1.72	N
b 2.07	1A
▶ c 4.07	N
d ~5.68	S
e ~5.68	S

▶ C_4H_8O ▶ 7% $CDCl_3$ ▶ V 414

▶ 2-(L)mnNbb

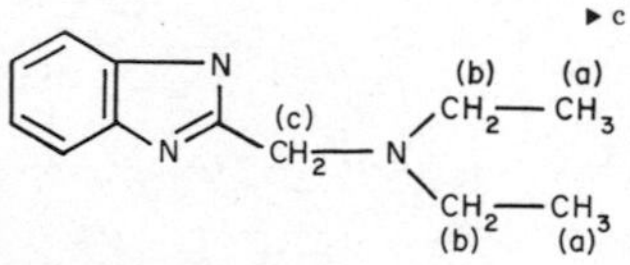

Shift	Peak
a 1.03	3A
b 2.63	4A
▶ c 3.88	1A

▶ $C_{12}H_{17}N_3$ ▶ 7% $CDCl_3$ ▶ V 296

▶ 2-LhlP

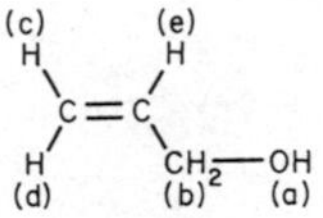

Shift	Peak
a 3.58	1A
▶b 4.13	2E
c 5.13	R
d 5.25	R
e ~6.00	R

▶ C_3H_6O ▶ 7% $CDCl_3$ ▶ V 34

▶ 2-LhlP

(c) CH_3 (d) CH_2OH
(a) CH_3 (b) CH_3
$CHCH_2CH_2-(CH_2CH\ CH_2CH_2)_2-CH_2$ C=C H(e)
CH_3 (a)

Shift	Peak
a 0.87	2A
b 0.87	2A
c 1.68	1A
▶d 4.13	2A
e 5.42	3C

▶ $C_{20}H_{40}O$ ▶ 7% $CDCl_3$ ▶ V 346

▶ 2-LhlQl

(c) H (b) H—C=C(H)(d)—CH_2(a)—O—C(=O)—C(H)(e)=C(H)(e)—C(=O)—O—CH_2(a)—C(H)(d)=C—H(b) H(c)

Shift	Peak
▶a 4.68	2D
b 5.25	R
c 5.32	R
d 5.88	R
e 6.27	1A

▶ $C_{10}H_{12}O_4$ ▶ 7% $CDCl_3$ ▶ V 262

▶ 2-LhlP

(a) CH_3 (d) H (a) (e) CH_3 H
C = C — CH_2(b) — CH_2(b) — C = C — CH_2(c) — OH
CH_3 (a)

Shift	Peak
a 1.62 or 1.68	1A
b 2.05 ± 0.03	3A
▶c 4.15	2A
d 5.12	1C
e 5.45	3A

▶ $C_{10}H_{18}O$ ▶ 7% $CDCl_3$ ▶ V 279

▶ (2)-Lhl(Q)l

(c)H H(b)
(a) H_2C O O

Shift	Peak
▶a 4.92	3C
b 6.15	N
c 7.63	N

▶ $C_4H_4O_2$ ▶ 7% $CDCl_3$ ▶ V 51

▶ 2-LhlP

(a) CH_3 (a) CH_3(a)
(a) CH_3 C=CH(d)—CH_2(b)—(CH_2(b)—C=CH(d)—CH_2(b))$_7$—CH_2(b)—C=CH—CH_2(c)—OH
CH_3 (a)

Shift	Peak
a 1.60 or 1.68	1G
b 2.04 ± 0.05	1G
▶c 4.15	2G
d 5.12	1F

▶ $C_{45}H_{74}O$ ▶ 7% $CDCl_3$ ▶ V 367

▶ 2-(L)loQa

(c)H H(d)
(e)H O CH_2(b) — O — C(=O) — CH_3(a)

Shift	Peak
a 2.07	1A
▶b 5.03	1A
c 6.35	1E
d 6.38	1E
e 7.40	N

▶ $C_7H_8O_3$ ▶ 7% $CDCl_3$ ▶ V 167

▶ 2-LhlP

(c) H — C — CH_2(b) — OH(a)
C(c) H

Shift	Peak
a 2.05	1B
▶b 4.27	4A
c 6.5	S

▶ $C_9H_{10}O$ ▶ 35.6 mg/0.50 ml $CDCl_3$ ▶ S 23

▶ 2-LhlSb

(d) H (d) H
(c) H—C=C—CH_2(a)—S—CH_2(a)—C=C—H(c)
(b)H H(b)

Shift	Peak
▶a 3.08	2E
b 5.07	R
c 5.08	R
d 5.72	R

▶ $C_6H_{10}S$ ▶ 7% $CDCl_3$ ▶ V 136

▶ 2-(L)loP

(c)H O (a) CH_2 OD
DO O H(b)

Shift	Peak
▶a 4.54	1A
b 6.59	1A
c 8.10	1A

▶ $C_6H_6O_4$ ▶ D_2O ▶ V 455

▶ (2)-(L)hl(S)boo

(b) H (b) H
(a) H_2C CH_2(a)
S
O O

Shift	Peak
▶a 3.74	1A
b 6.08	1A

▶ $C_4H_6O_2S$ ▶ 7% $CDCl_3$ ▶ V 406

▶ 2-(L)loP

(c)H H(c)
(d)H O CH_2—OH
(b) (a)

Shift	Peak
a 2.83	1B
▶b 4.57	1A
c 6.33 ± 0.03	M
d 7.44	M

▶ $C_5H_6O_2$ ▶ 7% $CDCl_3$ ▶ V 102

▶ 2-LhlSd

C — S — CH_2(a) — C(H(d)) = C — H(c)
H(b)

Shift	Peak
▶a 2.82	2A
b 5.00 ± 0.05	R
c 5.00 ± 0.05	R
d ~5.60	R

▶ $C_{22}H_{20}S$ ▶ 7% $CDCl_3$ ▶ V 355

▶ 2-LhlSl

N, C—S—CH_2—C, S, (a), H(d), (b), C—H, H(c)

Shift		Peak
▶a	3.95	2A
b	5.17	R
c	5.33	R
d	~6.00	R

▶ $C_{10}H_9NS_2$ ▶ 7% $CDCl_3$ ▶ V 250

▶ (2)-(L)hl Vhl

H(c), CH_2, (a), H(b)

Shift		Peak
▶a	3.33	1I
b	6.50	N
c	6.82	N

▶ C_9H_8 ▶ 7% $CDCl_3$ ▶ V 227

▶ (2)-(L)hl(S)loo

(c) H, (b) H, H_2C—SO_2, (a)

Shift		Peak
▶a	4.58	2E
b	6.80	2E
c	7.22	M

▶ $C_3H_4O_2S$ ▶ 7% $CDCl_3$ ▶ V 22

▶ 2-LhlVhq

(f)H, (g) H, (i) H, (c) CH_3, C=C, (d) CH_2, (e)H, H(h), (b) CH_3—CO, O, O, O, CH_3 (a)

Shift		Peak
a	2.25 or 2.37	1A
b	2.25 or 2.37	1A
c	2.40	2D
▶d	3.33	2B
e	5.07	R
f	5.12	R
g	5.85	4D
h	6.22	1I
i	7.30	1I

▶ $C_{16}H_{16}O_4$ ▶ 7% $CDCl_3$ ▶ V 323

▶ 2-(L)loSh

(d)H, H(c), (e)H, O, CH_2—SH, (b), (a)

Shift		Peak
a	1.90	3A
▶b	3.73	2A
c	6.17	N
d	6.28	N
e	7.33	M

▶ C_5H_6OS ▶ 7% $CDCl_3$ ▶ V 101

▶ (2)-(L)jlVhj

(a) CH_2, C, H (b), H (c), O

Shift		Peak
▶a	4.04	2D
b	7.55	S
c	7.92	S

▶ $C_{16}H_{12}O$ ▶ 7% $CDCl_3$ ▶ V 649

▶ 2-LhlU

(a)

$CH_2{=}CH{-}CH_2CN$

Shift		Peak
a	3.15	N

▶ C_4H_5N ▶ 7% $CDCl_3$ ▶ V 56

▶ (2)-LjlVhl

(e) HO, C=O, (a) CH_2, H(c), H(b), H (d)

Shift		Peak
▶a	3.38	1A
b	6.56	4A
c	7.27	S
d	7.30	S
e	11.05	1C

▶ $C_{12}H_{10}O_2$ ▶ 7% $CDCl_3$ ▶ V 591

▶ 2-LhlVhh

(e) O—CH_2, O, (g)H, (h)H, H(f), (a)CH_2, H(b), C=C, (d)H, H(c)

Shift		Peak
▶a	3.30	2E
b	5.03	2E
c	5.03	2E
d	~5.88	O
e	5.88	1A
f	6.67 ± .03	1H
g	6.67 ± .03	1H
h	6.67 ± .03	1H

▶ $C_{10}H_{10}O_2$ ▶ 7% $CDCl_3$ ▶ V 253

▶ (2)-(L)lpVhl

(e) OH, (a) CH_2, OCH_3(b), C=O, C=O, (d)H, OCH_3(c)

Shift		Peak
▶a	3.38	1A
b	3.73 or 3.84	1A
c	3.73 or 3.84	1A
d	7.82	1A
e	12.62	1A

▶ $C_{15}H_{14}O_5$ ▶ 7% $CDCl_3$ ▶ V 639

▶ 2-LhlVhh

(g) H, (f) H, (c) H, (i)H, (a) CH_2, C=C, H(d), (e)HO, H(h), OCH_3 (b)

Shift		Peak
▶a	3.30	2E
b	3.83	1A
c	5.05	2E
d	5.05	2E
e	5.59	1B
f	~5.94	O
g	~6.64	S
h	~6.69	S
i	6.87	S

▶ $C_{10}H_{12}O_2$ ▶ 7% $CDCl_3$ ▶ V 260

▶ 2-NbbNbb

(b) CH_2, (a), N—CH_2—N, (b) CH_2, CH_2 (b), CH_2 (b)

Shift		Peak
▶a	3.10	1A
b	3.62	1A
c	7.25 ± .03	1A

(c) = all benzene H's.

▶ $C_{29}H_{30}N_2$ ▶ 7% $CDCl_3$ ▶ V 365

▶ 2-NbbU

	Shift	Peak
▶a	3.58	1A
b	3.78	1A
c	7.25	1A

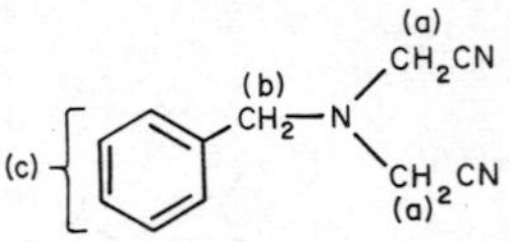

▶ $C_{11}H_{11}N_3$ ▶ 7% $CDCl_3$ ▶ V 285

▶ 2-NhhVhh

	Shift	Peak
a	1.62	1B
▶b	3.87	1A

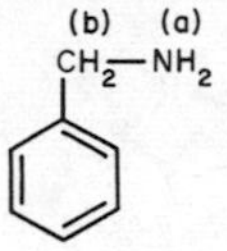

▶ C_7H_9N ▶ 44.9 mg/0.50 ml $CDCl_3$ ▶ S 63

▶ (2)-(N)abVbo

	Shift	Peak
a	1.92	1A
b	2.55	1F
c	2.88	1F
▶d	3.58	1A
e	3.78	1A
f	5.92, 5.96	1A
g	6.65	1A
h	6.69	1A
i	6.91	1A

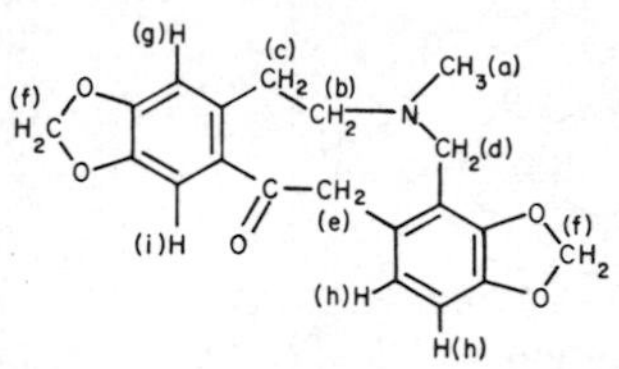

▶ $C_{20}H_{19}NO_5$ ▶ 7% $CDCl_3$ ▶ V 339

▶ 2-NhvVhh

	Shift	Peak
a	~3.82	1C
▶b	4.29	1A

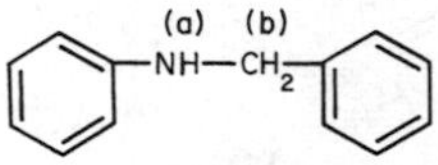

▶ $C_{13}H_{13}N$ ▶ 44.4 mg/0.50 ml $CDCl_3$ ▶ S 148

▶ 2-(N)bbVhh

	Shift	Peak
a	2.51	Q
b	3.02	Q
▶c	3.48	1A
d	7.23	1A

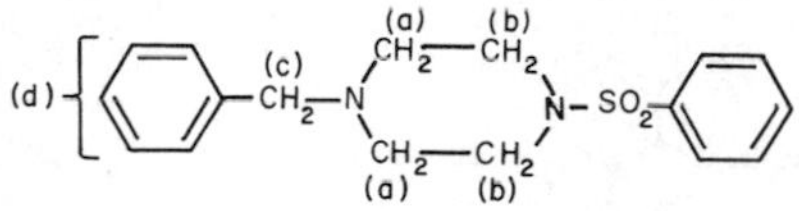

▶ $C_{17}H_{20}N_2O_2S$ ▶ 7% $CDCl_3$ ▶ V 330

▶ 2-(N)1lX

	Shift	Peak
▶a	4.96	1A
b	6.13	S
c	6.20 ± 0.10	S
d	6.20 ± 0.10	S
e	6.67	3A
f	7.32	3A

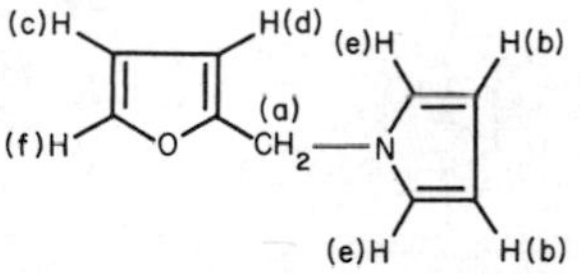

▶ C_9H_9NO ▶ 7% $CDCl_3$ ▶ V 525

▶ 2-NbbVhh

	Shift	Peak
a	1.03	3A
b	2.52	4A
▶c	3.54	1A

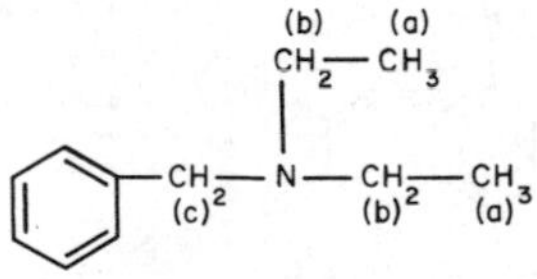

▶ $C_{11}H_{17}N$ ▶ 44.8 mg/0.50 ml $CDCl_3$ ▶ S 152

▶ (2)-(O)b(O)b

	Shift	Peak
a	1.74	O
b	3.78	O
▶c	4.79	1A

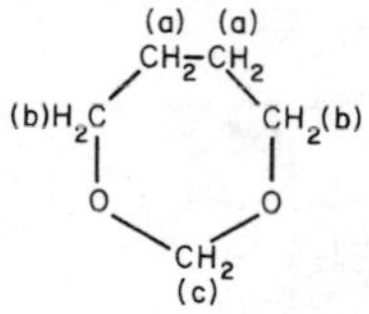

▶ $C_5H_{10}O_2$ ▶ 7% $CDCl_3$ ▶ V 445

▶ 2-NbbVhh

	Shift	Peak
a	3.10	1A
▶b	3.62	1A
c	7.25 ± .03	1A

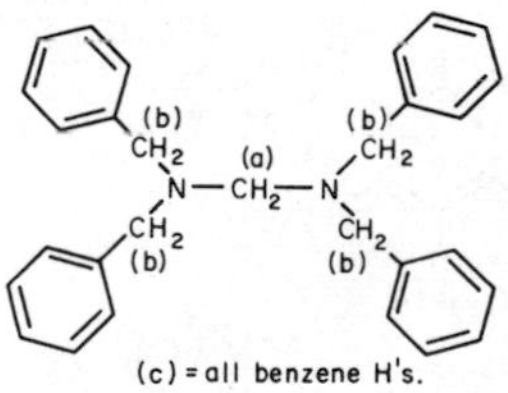

(c) = all benzene H's.

▶ $C_{29}H_{30}N_2$ ▶ 7% $CDCl_3$ ▶ V 365

▶ (2)-(O)b(O)b

	Shift	Peak
a	4.31	2D
▶b	4.88	1A
c	5.74	3A

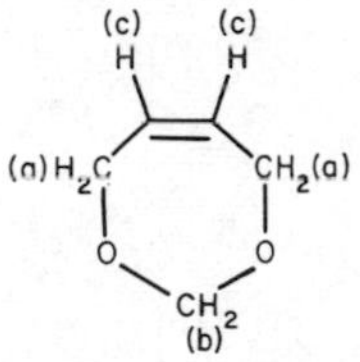

▶ $C_5H_8O_2$ ▶ 7% $CDCl_3$ ▶ V 437

▶ 2-NbbVhh

	Shift	Peak
a	3.58	1A
▶b	3.78	1A
c	7.25	1A

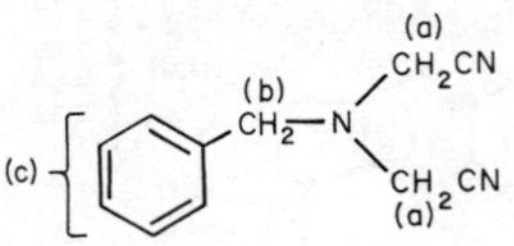

▶ $C_{11}H_{11}N_3$ ▶ 7% $CDCl_3$ ▶ V 285

▶ (2)-(O)v(O)v

	Shift	Peak
a	1.27	3A
b	3.14	2A
c	4.20	4A
d	4.96	4C
▶e	5.86	1A
f	6.62 ± 0.04	S

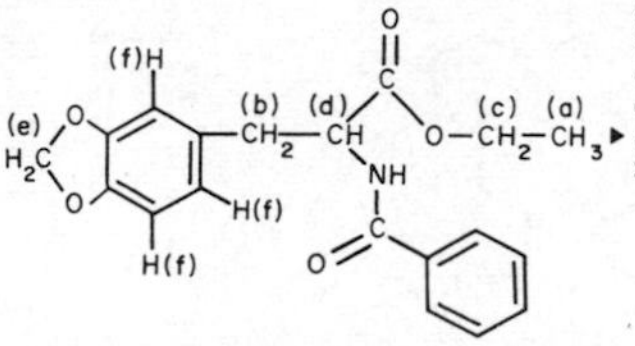

▶ $C_{19}H_{19}NO_5$ ▶ 7% $CDCl_3$ ▶ V 334

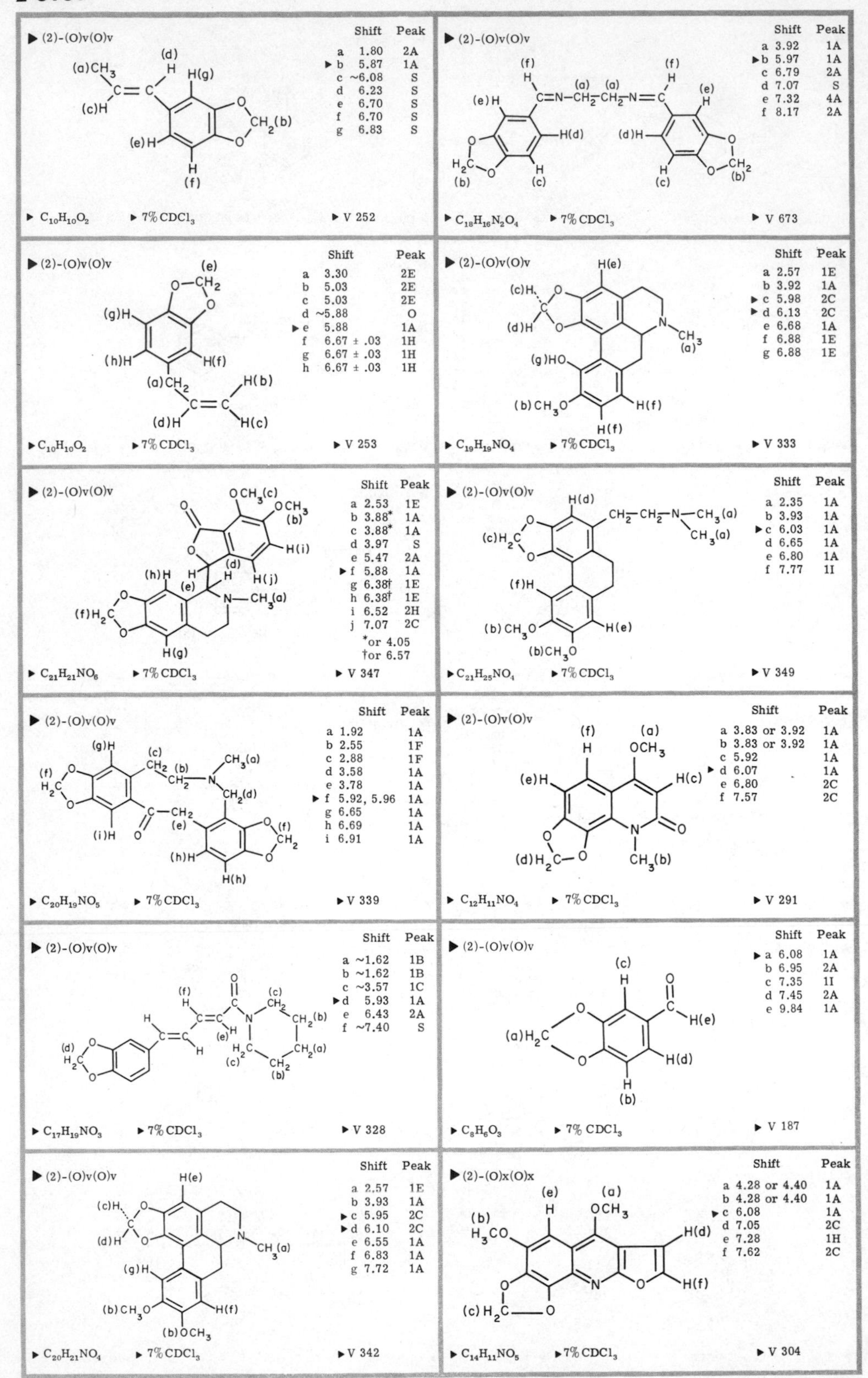

▶ (2)-(O)v(O)v

	Shift	Peak
a	1.80	2A
▶ b	5.87	1A
c	~6.08	S
d	6.23	S
e	6.70	S
f	6.70	S
g	6.83	S

▶ $C_{10}H_{10}O_2$ ▶ 7% $CDCl_3$ ▶ V 252

▶ (2)-(O)v(O)v

	Shift	Peak
a	3.92	1A
▶ b	5.97	1A
c	6.79	2A
d	7.07	S
e	7.32	4A
f	8.17	2A

▶ $C_{18}H_{16}N_2O_4$ ▶ 7% $CDCl_3$ ▶ V 673

▶ (2)-(O)v(O)v

	Shift	Peak
a	3.30	2E
b	5.03	2E
c	5.03	2E
d	~5.88	O
▶ e	5.88	1A
f	6.67 ± .03	1H
g	6.67 ± .03	1H
h	6.67 ± .03	1H

▶ $C_{10}H_{10}O_2$ ▶ 7% $CDCl_3$ ▶ V 253

▶ (2)-(O)v(O)v

	Shift	Peak
a	2.57	1E
b	3.92	1A
▶ c	5.98	2C
▶ d	6.13	2C
e	6.68	1A
f	6.88	1E
g	6.88	1E

▶ $C_{19}H_{19}NO_4$ ▶ 7% $CDCl_3$ ▶ V 333

▶ (2)-(O)v(O)v

	Shift	Peak
a	2.53	1E
b	3.88*	1A
c	3.88*	1A
d	3.97	S
e	5.47	2A
▶ f	5.88	1A
g	6.38†	1E
h	6.38†	1E
i	6.52	2H
j	7.07	2C

*or 4.05
†or 6.57

▶ $C_{21}H_{21}NO_6$ ▶ 7% $CDCl_3$ ▶ V 347

▶ (2)-(O)v(O)v

	Shift	Peak
a	2.35	1A
b	3.93	1A
▶ c	6.03	1A
d	6.65	1A
e	6.80	1A
f	7.77	1I

▶ $C_{21}H_{25}NO_4$ ▶ 7% $CDCl_3$ ▶ V 349

▶ (2)-(O)v(O)v

	Shift	Peak
a	1.92	1A
b	2.55	1F
c	2.88	1F
d	3.58	1A
e	3.78	1A
▶ f	5.92, 5.96	1A
g	6.65	1A
h	6.69	1A
i	6.91	1A

▶ $C_{20}H_{19}NO_5$ ▶ 7% $CDCl_3$ ▶ V 339

▶ (2)-(O)v(O)v

	Shift	Peak
a	3.83 or 3.92	1A
b	3.83 or 3.92	1A
c	5.92	1A
▶ d	6.07	1A
e	6.80	2C
f	7.57	2C

▶ $C_{12}H_{11}NO_4$ ▶ 7% $CDCl_3$ ▶ V 291

▶ (2)-(O)v(O)v

	Shift	Peak
a	~1.62	1B
b	~1.62	1B
c	~3.57	1C
▶ d	5.93	1A
e	6.43	2A
f	~7.40	S

▶ $C_{17}H_{19}NO_3$ ▶ 7% $CDCl_3$ ▶ V 328

▶ (2)-(O)v(O)v

	Shift	Peak
▶ a	6.08	1A
b	6.95	2A
c	7.35	1I
d	7.45	2A
e	9.84	1A

▶ $C_8H_6O_3$ ▶ 7% $CDCl_3$ ▶ V 187

▶ (2)-(O)v(O)v

	Shift	Peak
a	2.57	1E
b	3.93	1A
▶ c	5.95	2C
▶ d	6.10	2C
e	6.55	1A
f	6.83	1A
g	7.72	1A

▶ $C_{20}H_{21}NO_4$ ▶ 7% $CDCl_3$ ▶ V 342

▶ (2)-(O)x(O)x

	Shift	Peak
a	4.28 or 4.40	1A
b	4.28 or 4.40	1A
▶ c	6.08	1A
d	7.05	2C
e	7.28	1H
f	7.62	2C

▶ $C_{14}H_{11}NO_5$ ▶ 7% $CDCl_3$ ▶ V 304

▶ (2)-(O)x(O)x

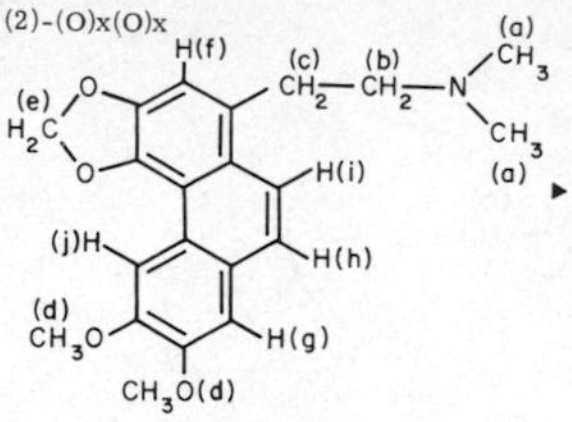

	Shift	Peak
a	2.38	1A
b	~2.7	Q
c	~3.2	Q
d	4.05 and 4.08	1A
▶ e	6.23	1A
f	7.13 or 7.23	1A
g	7.13 or 7.23	1A
h	7.53 or 7.81	2C
i	7.53 or 7.81	2C
j	8.52	1A

▶ $C_{21}H_{23}NO_4$ ▶ 7% $CDCl_3$ ▶ V 348

▶ 2-PVhh

	Shift	Peak
a	2.32	2A
b	2.43	1B
▶ c	4.47	1G

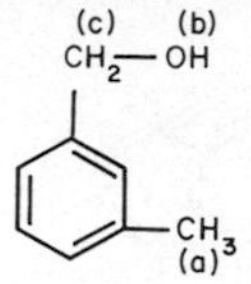

▶ $C_8H_{10}O$ ▶ 45.2 mg/0.50 ml $CDCl_3$ ▶ S 67

▶ 2-OaU

	Shift	Peak
a	3.47	1A
▶ b	4.20	1A

(a) (b)
$CH_3O\,CH_2CN$

▶ C_3H_5NO ▶ 7% $CDCl_3$ ▶ V 28

▶ 2-PVhh

	Shift	Peak
a	2.43	1A
▶ b	4.58	1A
c	7.28	1A

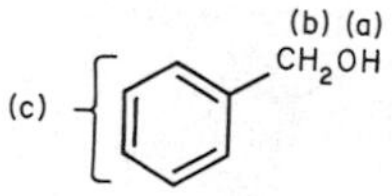

▶ C_7H_8O ▶ 7% $CDCl_3$ ▶ V 161

▶ 2-ObVhh

	Shift	Peak
▶ a	4.52	1A
b	7.33	1A

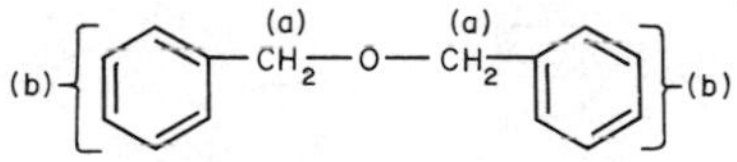

▶ $C_{14}H_{14}O$ ▶ 45.4 mg/0.50 ml $CDCl_3$ ▶ S 118

▶ 2-QaTb

	Shift	Peak
a	2.12	1A
b	3.13	1C
c	4.28	Q
▶ d	4.71	Q

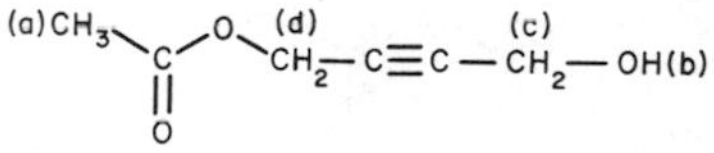

▶ $C_6H_8O_3$ ▶ 7% $CDCl_3$ ▶ V 463

▶ (2)-(O)bVhv

	Shift	Peak
a	4.37	1A

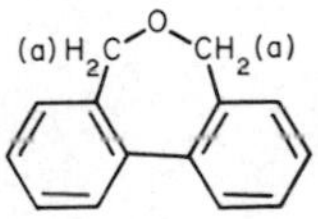

▶ $C_{14}H_{12}O$ ▶ 7% $CDCl_3$ ▶ V 624

▶ 2-QaVhh

	Shift	Peak
a	2.06	1A
▶ b	5.08	1A
c	7.31	1A

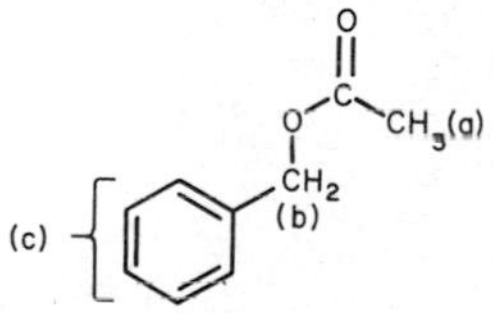

▶ $C_9H_{10}O_2$ ▶ 7% $CDCl_3$ ▶ V 530

▶ 2-PTb

	Shift	Peak
a	2.12	1A
b	3.13	1C
▶ c	4.28	Q
d	4.71	Q

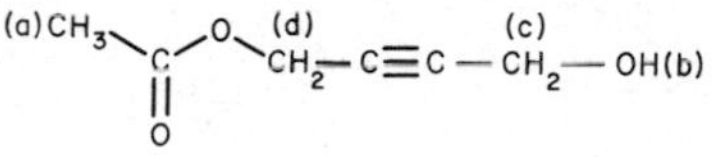

▶ $C_6H_8O_3$ ▶ 7% $CDCl_3$ ▶ V 463

▶ 2-QbVhh

	Shift	Peak
a	2.70	1A
▶ b	5.10	1A

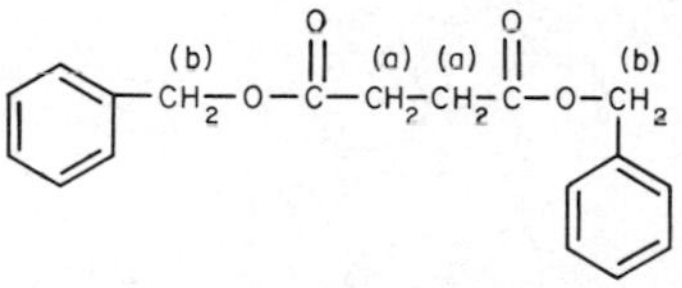

▶ $C_{18}H_{18}O_4$ ▶ 45.0 mg/0.50 ml $CDCl_3$ ▶ S 140

▶ 2-PTh

	Shift	Peak
a	2.50	3C
b	2.82	1A
▶ c	4.28	2A

(a) (c) (b)
H— C≡C—CH_2—OH

▶ C_3H_4O ▶ 7% $CDCl_3$ ▶ V 21

▶ 2-QlVhh

	Shift	Peak
▶ a	5.25	1A
b	6.47	2C
c	7.77	2C

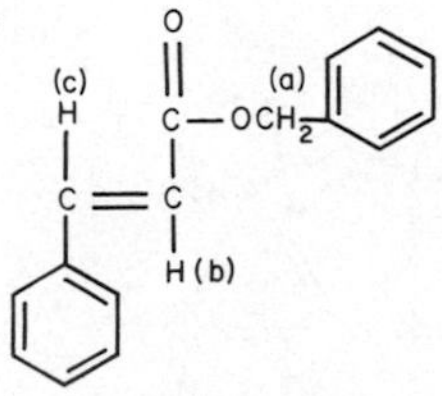

▶ $C_{16}H_{14}O_2$ ▶ 44.9 mg/0.50 ml $CDCl_3$ ▶ S 104

▶ 2-QvVhh

	Shift	Peak
▶	a 5.34	1A
	b 8.07	S

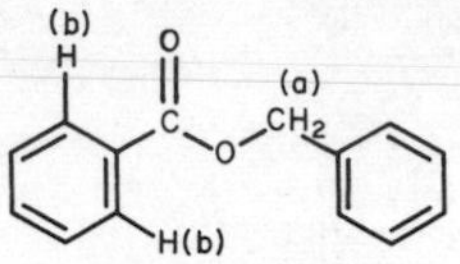

▶ $C_{14}H_{12}O_2$ ▶ 7% $CDCl_3$ ▶ V 627

▶ 2-ScVhh

	Shift	Peak
▶	a 3.68	1A
	b 4.20	1A

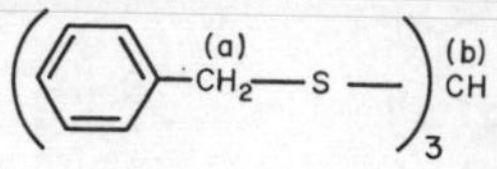

▶ $C_{22}H_{22}S_3$ ▶ 7% $CDCl_3$ ▶ V 688

▶ (2)-(Q)vVhk

	Shift	Peak
▶	a 5.32	1A
	b 7.92	S

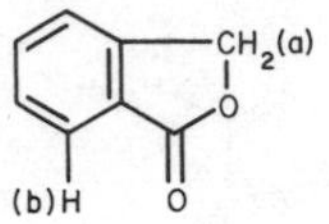

▶ $C_8H_6O_2$ ▶ 7% $CDCl_3$ ▶ V 496

▶ 2-ShVhh

	Shift	Peak
	a 1.62	3A
▶	b 3.70	2A
	c 7.28	1A

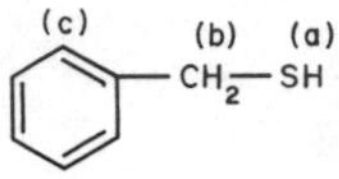

▶ C_7H_8S ▶ 43.6 mg/0.50 ml $CDCl_3$ ▶ S 276

▶ 2-SbVhh

	Shift	Peak
	a 3.10	1A
▶	b 3.87	1A
	c 7.32	1A
	d 10.97	1B

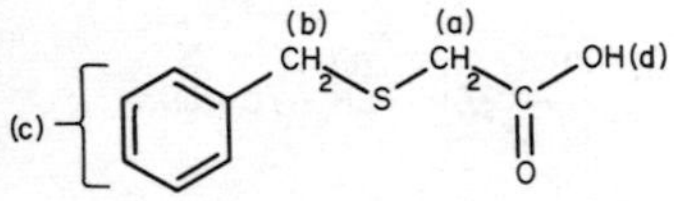

▶ $C_9H_{10}O_2S$ ▶ 7% $CDCl_3$ ▶ V 532

▶ 2-SlVhh

	Shift	Peak
▶	a 4.46	1A
	b 7.50	1A

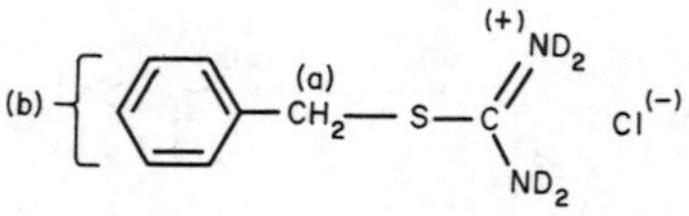

▶ $C_8H_{11}ClN_2S$ ▶ D_2O ▶ V 507

▶ (2)-(S)bVhv

	Shift	Peak
	a 2.09	1A
▶	b 3.27	1A

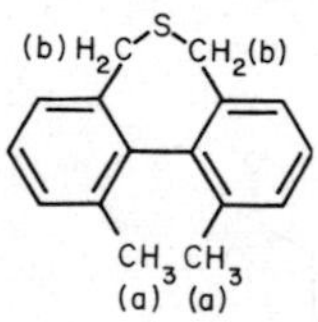

▶ $C_{16}H_{16}S$ ▶ 7% $CDCl_3$ ▶ V 657

▶ (2)-(S)lX

	Shift	Peak
▶	a 4.30	1A
	b 9.70	1I
	c 9.80	2A

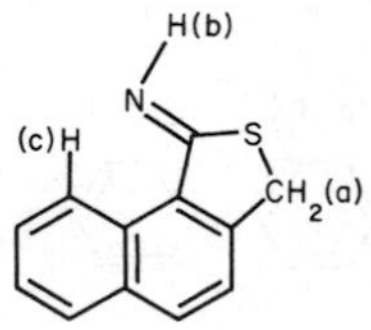

▶ $C_{12}H_9NS$ ▶ 7% $CDCl_3$ ▶ V 590

▶ (2)-(S)bVhv

	Shift	Peak
▶	a 3.32	2C
▶	b 3.43	2C

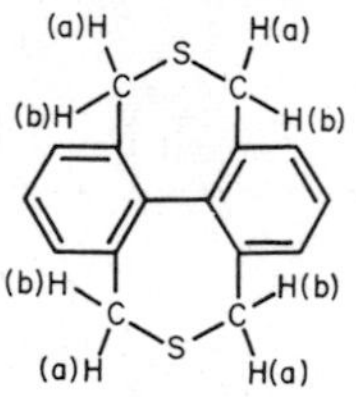

▶ $C_{16}H_{14}S_2$ ▶ 7% $CDCl_3$ ▶ V 653

▶ 2-TzZ

	Shift	Peak
	a 0.10 or 0.12	1A
	b 0.10 or 0.12	1A
▶	c 1.57	1A

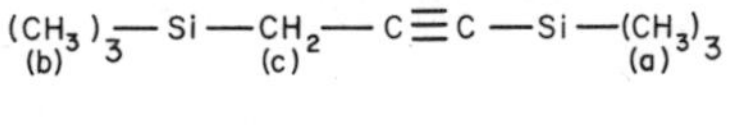

▶ $C_9H_{20}Si_2$ ▶ 7% $CDCl_3$ ▶ V 549

▶ (2)-(S)bVhv

	Shift	Peak
▶	a 3.36	2C
▶	b 3.47	2C

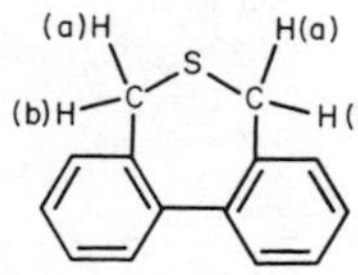

▶ $C_{14}H_{12}S$ ▶ 7% $CDCl_3$ ▶ V 630

▶ 2-UVhh

	Shift	Peak
▶	a 3.68	1A
	b 7.34	1A

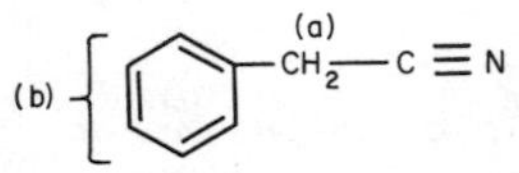

▶ C_8H_7N ▶ 44.5 mg/0.50 ml $CDCl_3$ ▶ S 135

▶ 2-UVhh

	Shift	Peak
▶ a	3.92	1A
b	7.57	Q
c	8.25	Q

(c)H H(b)
NO_2 (a) CH_2—CN
(c)H H(b)

▶ $C_8H_6N_2O_2$ ▶ 7% $CDCl_3$ ▶ V 495

▶ 2-VhhVhh

	Shift	Peak
▶ a	3.97	1A
b	7.05 or 7.13	Q
c	7.05 or 7.13	Q
d	7.22	1A

(b) (c)
H H
O
(d) (a) CH_2 C
NH_2
H H
(b) (c)

▶ $C_{14}H_{13}NO$ ▶ 7% $CDCl_3$ ▶ V 309

▶ 2-UVhn

	Shift	Peak
▶ a	3.65	1A
b	4.17	1A
c	7.31 ± .05	1H
d	8.32	1C

(a)
CH_2CN
(c)
(b)
N—C—CH_2Cl
H
(d) O

▶ $C_{10}H_9ClN_2O$ ▶ 7% $CDCl_3$ ▶ V 248

▶ (2)-VhjVhj

	Shift	Peak
a	4.23	1A

O
(a)

▶ $C_{14}H_{10}O$ ▶ 45.3 mg/0.50 ml $CDCl_3$ ▶ S 254

▶ 2-VaaVah

	Shift	Peak
▶ a	3.95	1B
b	6.28	1B

CH_3 CH_3 CH_3 CH_3
(a)
CH_3 CH_2 CH_3
CH_3 CH_3 H CH_3
(b)

▶ $C_{22}H_{30}$ ▶ 7% $CDCl_3$ ▶ V 357

▶ (2)-VhjVhj

	Shift	Peak
▶ a	4.32	1A
b	8.36	O

(b) (b)
H O H
CH_2
(a)

▶ $C_{14}H_{10}O$ ▶ 7% $CDCl_3$ ▶ V 621

▶ (2)-VbhVbh

	Shift	Peak
▶ a	3.87	1A
b	7.18	1A

(b)
(a)
CH_2
CH_2
(a)

▶ $C_{14}H_{12}$ ▶ 7% $CDCl_3$ ▶ V 307

▶ (2)-VhvVhv

	Shift	Peak
a	3.76	1A

(a)
H H

▶ $C_{13}H_{10}$ ▶ 45.0 mg/0.25 ml $CDCl_3$ ▶ S 288

▶ 2-VepVep

	Shift	Peak
▶ a	4.47	1A
b	5.75	1B
c	7.39	1A

(b) (b)
OH OH
(a)
Cl CH_2 Cl
(c)H Cl Cl H(c)
Cl Cl

▶ $C_{13}H_6Cl_6O_2$ ▶ 7% $CDCl_3$ ▶ V 604

▶ 2-VhhX

	Shift	Peak
▶ a	3.75	1A
b	5.79	1A
c	7.33	1A

(a)
DS N CH_2
(c)
N H
(b)
OD

▶ $C_{11}H_{10}N_2OS$ ▶ D_2O ▶ V 577

▶ 2-VhhVhh

	Shift	Peak
a	2.26	1A
▶ b	3.82	1A
c	6.96	1A

(a) (c) (c) (a)
CH_3 H H CH_3
(b)
O=C=N CH_2 N=C=O
H H H H
(c) (c) (c) (c)

▶ $C_{17}H_{14}N_2O_2$ ▶ 7% $CDCl_3$ ▶ V 661

▶ 2-VhhX

	Shift	Peak
▶ a	4.18	1A
b	7.27	1E
c	7.54	S
d	8.56	3C

H(c)
(d)H N CH_2 (b)
(a)

▶ $C_{12}H_{11}N$ ▶ 7% $CDCl_3$ ▶ V 593

▶ 2-VhhX

	Shift	Peak
a	2.12	1A
b	4.15	1D
▶ c	4.27	1B
d	6.77	1B

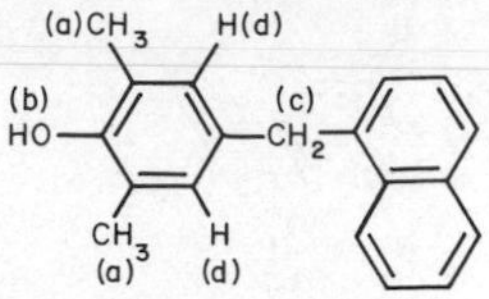

▶ $C_{19}H_{18}O$ ▶ 7% $CDCl_3$ ▶ V 332

▶ 3-AABb

	Shift	Peak
a	.88	2A
b	1.25	1B
▶ c	~1.47	S

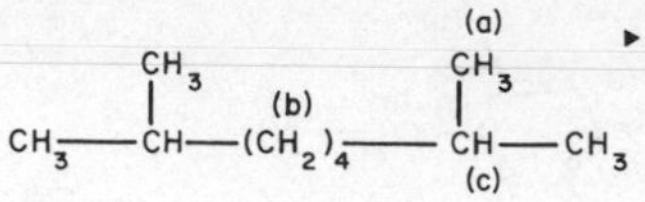

▶ $C_{10}H_{22}$ ▶ 50% CCl_4 ▶ API 346

▶ **3**-AAA

	Shift	Peak
a	.88	2A
▶ b	1.65	7X

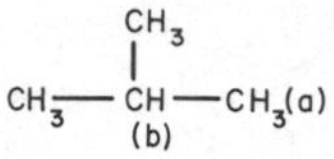

▶ C_4H_{10} ▶ 50% CCl_4 ▶ API 330

▶ 3-AABb

	Shift	Peak
a	.88	3D
b	.95	2H
c	1.20	N
▶ d	~1.50	1D

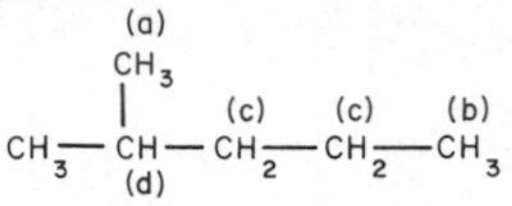

▶ C_6H_{14} ▶ 50% CCl_4 ▶ API 334

▶ 3-AABa

	Shift	Peak
a	.83	3D
b	.90	2H
c	1.20	S
▶ d	~1.67	1D

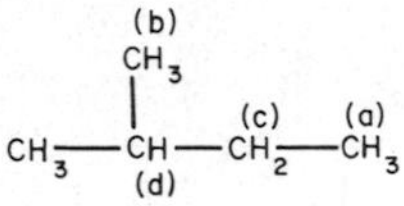

▶ C_5H_{12} ▶ 50% CCl_4 ▶ API 332

▶ 3-AABb

	Shift	Peak
a	0.92	2A
▶ b	~1.50	O
c	2.03	1A
d	4.03	4B

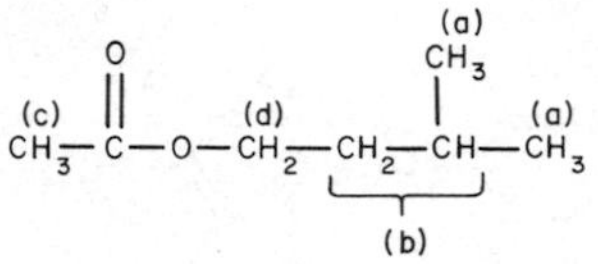

▶ $C_7H_{14}O_2$ ▶ 44.8 mg/0.50 ml $CDCl_3$ ▶ S 281

▶ 3-AABb

	Shift	Peak
a	.82	2A
b	~.92	1E
c	~.93	1B
▶ d	~.97	1D

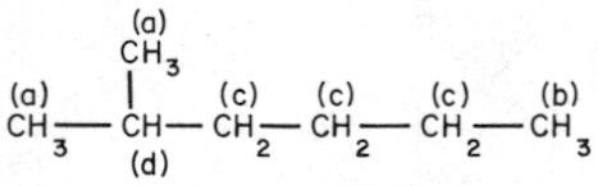

▶ C_7H_{16} ▶ 50% CCl_4 ▶ API 195

▶ 3-AABb

	Shift	Peak
a	0.93	2A
▶ b	~1.58	3C
c	2.36	3B
d	11.67	1A

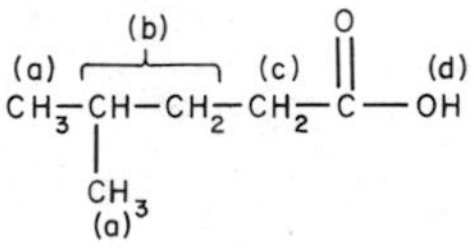

▶ $C_6H_{12}O_2$ ▶ 45.4 mg/0.50 ml $CDCl_3$ ▶ S 156

▶ 3-AABb

	Shift	Peak
a	~.84	1E
b	~.84	1E
c	~1.13	1G
▶ d	~1.37	1D

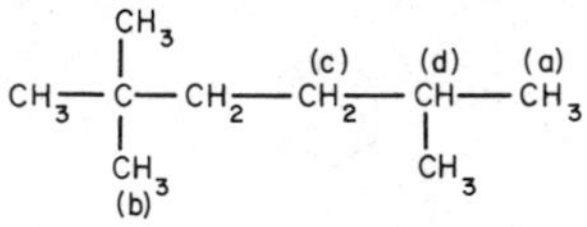

▶ C_9H_{20} ▶ 50% CCl_4 ▶ API 212

▶ 3-AABb

	Shift	Peak
a	.91	2C
b	1.32	S
▶ c	1.60	S
d	2.01	4C
e	4.87	R
f	5.02	R
g	5.70	R

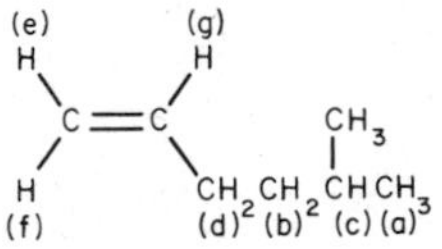

▶ C_7H_{14} ▶ liquid ▶ API 394

▶ 3-AABb

	Shift	Peak
a	.87	2A
b	~1.17	1G
▶ c	~1.40	1D

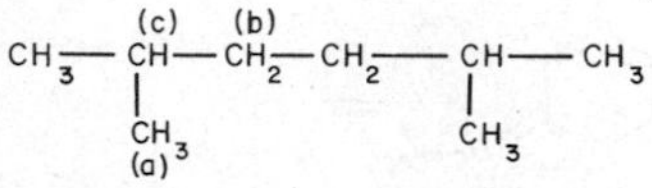

▶ C_8H_{18} ▶ 50% CCl_4 ▶ API 205

▶ 3-AABb

	Shift	Peak
a	.88	2A
b	1.31	S
▶ c	~1.63	S
d	3.48	3B
e	4.42	1A

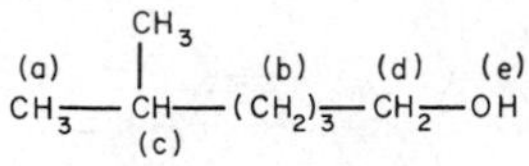

▶ $C_7H_{16}O$ ▶ 20% CCl_4 ▶ API 169

▶ 3-AABb

$CH_3-CH(CH_3)-CH_2-CH_2-Cl$
(a) CH₃; (b) CH–CH₂; (c) CH₂Cl

	Shift	Peak
a	0.92	2A
▶ b	1.65	O
c	3.55	3B

▶ $C_5H_{11}Cl$ ▶ 45.0 mg/0.50 ml $CDCl_3$ ▶ S 122

▶ 3-AABk

$(CH_3)_2CH-CH_2-C(=O)-OCH_2-CH(CH_3)-CH_3$
(a) CH_3; (b) CH; (c) CH_2; (d) OCH_2

	Shift	Peak
a	0.94	2A
▶ b	1.12	S
c	2.20	2D
d	3.88	2A

▶ $C_9H_{18}O_2$ ▶ 44.5 mg/0.50 ml $CDCl_3$ ▶ S 277

▶ 3-AABb

$CH_3-CH(CH_3)-CH_2-CH_2-Br$
(a) CH₃; (b) CH–CH₂; (c) CH₂Br

	Shift	Peak
a	0.92	2A
▶ b	1.78	O
c	3.42	3A

▶ $C_5H_{11}Br$ ▶ 45.0 mg/0.50 ml $CDCl_3$ ▶ S 121

▶ 3-AABl

$H_3C(c)-CH(e)=CH(e)-CH_2CH(CH_3)_2$
(a) $(CH_3)_2$; (b) CH; (c) H_3C; (d) CH_2; (e) H

	Shift	Peak
a	.98	2A
▶ b	1.38	S
c	1.60	2H
d	1.80	S
e	5.30	N

▶ C_7H_{14} ▶ 50 vol.% CCl_4 ▶ API 431

▶ 3-AABc

$CH_3-CH(CH_3)-CH_2-CH(CH_3)-CH_3$
(a) CH_3; (b) CH_2; (c) CH

	Shift	Peak
a	0.88	2A
b	1.03	3D
▶ c	~1.55	7B

▶ C_7H_{16} ▶ 45.0 mg/0.50 ml $CDCl_3$ ▶ S 52

▶ 3-AABl

$H_3C(c)-CH(e)=CH(f)-CH_2(d)-CH(b)(CH_3(a))-CH_3$

	Shift	Peak
a	.83	2A
▶ b	~1.40	S
c	1.62	2E
d	1.83	O
e	~5.38	O
f	~5.38	O

▶ C_7H_{14} ▶ liquid ▶ API 396

▶ 3-AABc

$CH_3-CH(CH_3)-CH_2(b)-CH(c)(CH_3)-CH_3(a)$

	Shift	Peak
a	.80	2A
b	1.00	2A
▶ c	1.58	1F

▶ C_7H_{16} ▶ 10% CCl_4 ▶ API 198

▶ 3-AABl

$H(d)H(e)C=CH(f)-CH_2(c)-CH(b)(CH_3(a))_2$

	Shift	Peak
a	.90	2A
▶ b	1.48	S
c	1.88	S
d	4.77	R
e	4.77	R
f	5.05	R

▶ C_6H_{12} ▶ liquid ▶ API 375

▶ 3-AABd

$CH_3(c)CH(e)(CH_3(c))-CH_2(c)C(CH_3(b))_2CH_2(d)CH_3(a)$

	Shift	Peak
a	~ .85	3D
b	~ .85	1E
c	~ .92	2H
d	1.13	2A
▶ e	1.63	7X

▶ C_9H_{20} ▶ 50% CCl_4 ▶ API 426

▶ 3-AABl

$H(d)H(e)C=CH(f)-CH_2(c)-CH(b)(CH_3(a))-CH_3(a)$

	Shift	Peak
a	.89	2A
▶ b	~1.60	S
c	1.87	S
d	4.83	R
e	5.03	R
f	~5.62	O

▶ C_6H_{12} ▶ 10% CCl_4 ▶ API 33

▶ 3-AABd

$(CH_3(a))_3C-CH_2(c)CH(d)(CH_3)CH_3(b)$

	Shift	Peak
a	~ .96	1E
b	~ .96	2H
c	~1.12	2A
▶ d	~1.67	7X

▶ C_8H_{18} ▶ 50% CCl_4 ▶ API 424

▶ 3-AABl

$H(f)H(e)C=C(CH_3(c))-CH_2(d)CH(b)(CH_3)CH_3(a)$

	Shift	Peak
a	.87	2H
▶ b	1.62	S
c	1.67	3C
d	1.82	2B
e	4.67 ± .03	N
f	4.67 ± .03	N

▶ C_7H_{14} ▶ liquid ▶ API 400

▶ 3-AABl

	Shift	Peak
a	0.93	2A
▶ b	1.98	7B
c	2.77	2A
d	9.68	1D

▶ $C_{11}H_{15}NO$ ▶ 7% $CDCl_3$ ▶ V 585

▶ 3-AACaa

	Shift	Peak
a	.82	2A
▶ b	1.42	7B

▶ C_6H_4 ▶ 50% CCl_4 ▶ API 194

▶ 3-AABn

	Shift	Peak
a	0.91	2A
▶ b	~1.6	S
c	1.77	1B
d	2.42	2A

▶ $C_8H_{19}N$ ▶ 44.7 mg/0.50 ml $CDCl_3$ ▶ S 136

▶ 3-AACac

	Shift	Peak
a	.80 ± .03 or .93 ± .03	S
b	.93 ± .03 or .80 ± .03	S
▶ c	1.67	6A

▶ C_8H_{18} ▶ 50% CCl_4 ▶ API 208

▶ 3-AABq

	Shift	Peak
a	0.94	2A
▶ b	1.12	S
c	2.20	2D
d	3.88	2A

▶ $C_9H_{18}O_2$ ▶ 44.5 mg/0.50 ml $CDCl_3$ ▶ S 277

▶ 3-AACad

	Shift	Peak
a	.90	1E
b	.90	1E
c	.90	1E
▶ d	2.02	5A

▶ C_9H_{20} ▶ 50% CCl_4 ▶ API 219

▶ 3-AABq

	Shift	Peak
a	0.94	2A
b	1.18	2A
▶ c	~1.82	7B
d	~2.50	5B
e	3.84	2A

▶ $C_8H_{16}O_2$ ▶ 45.2 mg/0.50 ml $CDCl_3$ ▶ S 238

▶ 3-AACbb

	Shift	Peak
a	~ .83	2H
b	~ .91	3D
c	1.20	3B
▶ d	~1.72	1D

▶ C_8H_{18} ▶ 50% CCl_4 ▶ API 206

▶ 3-AABq

	Shift	Peak
a	0.92	2A
b	1.13	3D
▶ c	~1.85	S
d	2.31	4A
e	3.83	2A

▶ $C_7H_{14}O_2$ ▶ 4.5 mg/0.50 ml $CDCl_3$ ▶ S 55

▶ 3-AACbc

	Shift	Peak
a	.86	2H
b	.95	3D
c	~1.2	S
▶ d	~1.69	S

▶ C_9H_{20} ▶ 50% CCl_4 ▶ API 217

▶ 3-AABq

	Shift	Peak
a	1.02	2A
▶ b	2.05	5B
c	4.12	2A
d	7.5	N
e	~8.05	S

▶ $C_{11}H_{14}O_2$ ▶ 45.5 mg/0.50 ml $CDCl_3$ ▶ S 75

▶ 3-AA(C)bo

	Shift	Peak
a	0.97*	2A
b	0.97*	2A
▶ c	2.00	6A
d	~3.20	S
e	3.87†	1A
f	3.87†	1A
g	4.72	4C
h	8.05	4A

*or 1.02
†or 3.90

▶ $C_{16}H_{19}NO_3$ ▶ 7% $CDCl_3$ ▶ V 325

▶ 3-AACdp

(e) (a) (f)
H CH3 H
(c) CH3 (d) (h)
CH—C—C—C—OH
CH3 (b) OH CH3 H
(h) (a) (g)

	Shift	Peak
a	0.92	1A
b	0.95 or 1.00	2A
c	0.95 or 1.00	2A
▶ d	1.92	5B
e	3.38	2A
f	3.44 or 3.53	S
g	3.44 or 3.53	S
h	3.58	1A

▶ $C_8H_{18}O_2$ ▶ 7% $CDCl_3$ ▶ V 217

▶ 3-AA(D)bbj

(f)
CH3
Br
(e) H2C
BrH2C
(d)
O
CH (c)
CH3 CH3
(a) (b)

	Shift	Peak
a	0.84	2A
b	0.92	2A
▶ c	2.08	S
d	2.22	1H
e	2.61	1H
f	3.58	1A

▶ $C_{10}H_{14}Br_2O$ ▶ 7% $CDCl_3$ ▶ V 564

▶ 3-AACpt

(e) (a)
OH CH3
(d)HC ≡ C—C—C—H(c)
H CH3
(f) (b)

	Shift	Peak
a	1.00	2D
b	1.02	2D
▶ c	1.85	7B
d	2.47	2E
e	2.82	1B
f	4.18	4A

▶ $C_6H_{10}O$ ▶ 7% $CDCl_3$ ▶ V 133

▶ 3-AAG

Br
(a)CH3—C—CH3(a)
H(b)

	Shift	Peak
a	1.71	2A
▶ b	4.32	5A

▶ C_3H_7Br ▶ 7% $CDCl_3$ ▶ V 391

▶ 3-AADaaa

CH3 CH3
(a)H3C—C—CH—CH3(b)
(c)
CH3

	Shift	Peak
a	.81	1E
b	.81	S
▶ c	~1.30	7X

▶ C_7H_{16} ▶ 50% CCl_4 ▶ API 419

▶ 3-AAI

(a)
CH3
(b)H—C—I
CH3
(a)

	Shift	Peak
a	1.90	2A
▶ b	4.34	5A

▶ C_3H_7I ▶ 7% $CDCl_3$ ▶ V 392

▶ 3-AADaab

(b) (a)
CH3 CH3
(b) (d) (d) (c)
CH3—CH—C—CH2—CH2—CH3
(c)
CH3
(a)

	Shift	Peak
a	.80	1A
b	.86	2H
c	~ .86	3D
d	1.22	3A
▶ e	~1.50	1D

▶ C_9H_{20} ▶ 50% CCl_4 ▶ API 213

▶ 3-AAJl

(d) (c)
H CH3
(e) O
H—C—C—C
(a) CH3 (f)
CH
N—N OH
C H
(b) CH3 O

	Shift	Peak
a	1.10	2A
b	1.12	2A
c	1.60	1A
d	2.81	2A
e	3.48	S
▶ f	3.52	S

▶ $C_9H_{14}N_2O_3$ ▶ 7% $CDCl_3$ ▶ V 543

▶ 3-AADaac

(a) (b) (a)
CH3 CH3 CH3
(a) (a)
CH3—CH—C—CH—CH3
(c) (c)
CH3
(b)

	Shift	Peak
a	.70	2A
b	.87	1A
▶ c	1.62	5B

▶ C_9H_{20} ▶ 50% CCl_4 ▶ API 345

▶ 3-AAJv

O
C CH3(a)
C
H CH3(a)
(b)

	Shift	Peak
a	1.22	2A
▶ b	3.50	5A

▶ $C_{10}H_{12}O$ ▶ 7% $CDCl_3$ ▶ V 559

▶ (3)-AA(D)abl

(g)
H O
O C
OH (h)
(d)H CH3(c)
(f)H CH(e)
CH3 CH3
(a) (b)

	Shift	Peak
a	0.75	2A
b	0.97	2A
c	1.45	1A
d	2.10	2C
▶ e	2.40	S
f	2.60	2C
g	6.77	1A
h	12.00	1B

▶ $C_{10}H_{14}O_3$ ▶ 7% $CDCl_3$ ▶ V 567

▶ 3-AAKb

(b) (a)
CH3 O CH3
(b) (e) (a)
CH3—CH—C—O—CH2CHCH3
(d) (c)

	Shift	Peak
a	0.94	2A
b	1.18	2A
c	~1.82	7B
▶ d	~2.50	5B
e	3.84	2A

▶ $C_8H_{16}O_2$ ▶ 45.2 mg/0.50 ml $CDCl_3$ ▶ S 238

▶ 3-AAKh

	Shift	Peak
a	1.22	2A
▶ b	2.60	5B
c	11.55	1B

▶ $C_4H_8O_2$ ▶ 44.9 mg/0.50 ml $CDCl_3$ ▶ S 54

▶ 3-AALal

	Shift	Peak
a	.91	2A
b	1.50	1F
c	~1.62	S
d	~1.62	S
▶ e	2.83	7A

▶ C_8H_{16} ▶ liquid ▶ API 463

▶ 3-AAKj

	Shift	Peak
a	1.22	2A
▶ b	2.68	7B

▶ $C_8H_{14}O_3$ ▶ 7% $CDCl_3$ ▶ V 516

▶ 3-AALal

	Shift	Peak
a	.93	2A
b	1.47	1F
c	1.55	1F
▶ d	2.84	5A
e	5.10	1F

▶ C_7H_{14} ▶ liquid ▶ API 403

▶ 3-AALal

	Shift	Peak
a	.93	2A
b	1.48	1E
c	1.52	1E
▶ d	2.20	5A
e	5.20	4B

▶ C_7H_{14} ▶ 50% CCl_4 ▶ API 354

▶ 3-AALhl

	Shift	Peak
a	.93	2A
▶ b	2.14	6B
c	4.60	R
d	4.75	R
e	5.48	R

▶ C_5H_{10} ▶ liquid run, neat ▶ API 387

▶ 3-AALal

	Shift	Peak
a	.97	2A
▶ b	2.21	6A
c	5.22	O

▶ C_7H_{14} ▶ liquid ▶ API 404

▶ 3-AALhl

	Shift	Peak
a	.94	2A
b	1.60	2E
▶ c	2.15	1D
d	5.36	S
e	5.45	S

▶ C_6H_{12} ▶ liquid run ▶ API 70

▶ 3-AALal

	Shift	Peak
a	.99	2A
b	1.67	1H
▶ c	2.23	5B
d	4.62	N

▶ C_6H_{12} ▶ liquid ▶ API 382

▶ 3-AALhl

	Shift	Peak
a	.98	2A
b	1.62	2E
▶ c	2.22	1F
d	5.37 or 5.43	S
e	5.43 or 5.37	S

▶ C_6H_{12} ▶ 5 vol.% CCl_4 ▶ API 380

▶ 3-AALal

	Shift	Peak
a	.91	2A
b	~1.48 or 1.52	1E
c	~1.52 or 1.48	1E
▶ d	2.74	5A
e	5.05	1D

▶ C_7H_{14} ▶ 50% CCl_4 ▶ API 353

▶ 3-AALhl

	Shift	Peak
a	.96	3D
b	.97	2H
c	1.98	S
▶ d	~2.28	S
e	~5.38	S
f	~5.38	S

▶ C_7H_{14} ▶ liquid ▶ API 397

▶ 3-AALhl

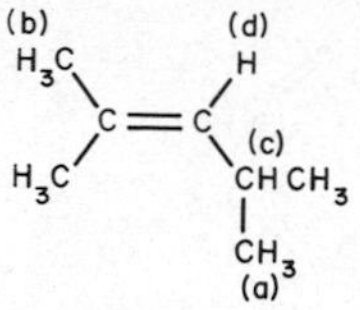

	Shift	Peak
a	.93	2A
b	1.61	N
▶ c	2.42	6B
d	4.93	2E

▶ C_7H_{14} ▶ liquid ▶ API 402

▶ 3-AANhj

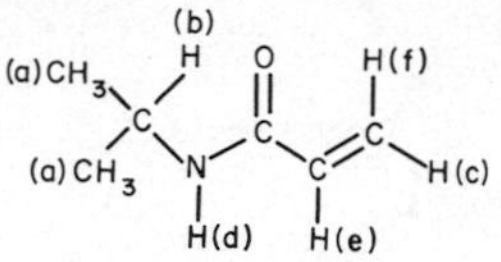

	Shift	Peak
a	1.18	2A
▶ b	4.17	6B
c	5.59	4A
d	~6.00	S
e	6.14	2A
f	6.24	1A

▶ $C_6H_{11}NO$ ▶ 7% $CDCl_3$ ▶ V 468

▶ 3-AALhl

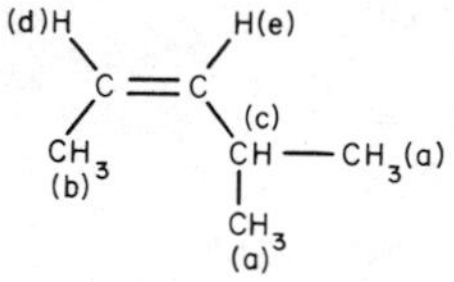

	Shift	Peak
a	.95	2A
b	1.63	2A
▶ c	2.55	6B
d	5.11	N
e	5.25	N

▶ C_6H_{12} ▶ liquid run ▶ API 69

▶ 3-AAOd

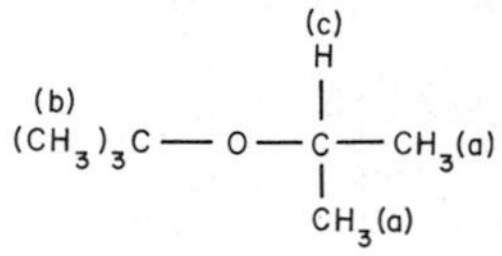

	Shift	Peak
a	1.12	1A
b	1.18	1A
▶ c	3.78	5A

▶ $C_7H_{16}O$ ▶ 7% $CDCl_3$ ▶ V 183

▶ 3-AALhl

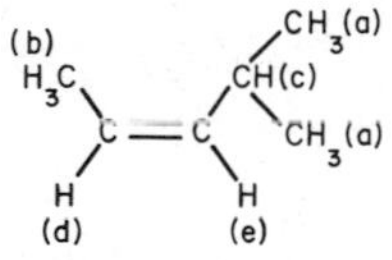

	Shift	Peak
a	.93	2A
b	1.57	2A
▶ c	2.62	6B
d	5.12 or 5.23	O
e	5.23 or 5.12	O

▶ C_6H_{12} ▶ liquid ▶ API 379

▶ 3-AAP

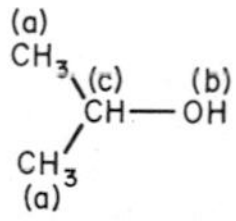

	Shift	Peak
a	1.20	2A
b	1.60	1B
▶ c	4.00	5A

▶ C_3H_8O ▶ 7% $CDCl_3$ ▶ V 44

▶ 3-AALhl

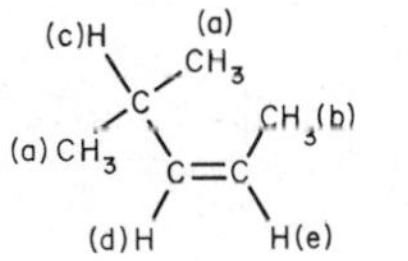

	Shift	Peak
a	0.94	2A
b	1.61	2A
▶ c	2.63	6A
d	5.20 ± 0.10	O
e	5.30 ± 0.10	O

▶ C_6H_{12} ▶ 7% $CDCl_3$ ▶ V 471

▶ 3-AAQa

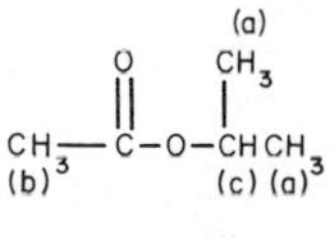

	Shift	Peak
a	1.23	2A
b	2.01	1A
▶ c	4.98	5A

▶ $C_5H_{10}O_2$ ▶ 45.0 mg/0.50 ml $CDCl_3$ ▶ S 99

▶ 3-AA(L)ll

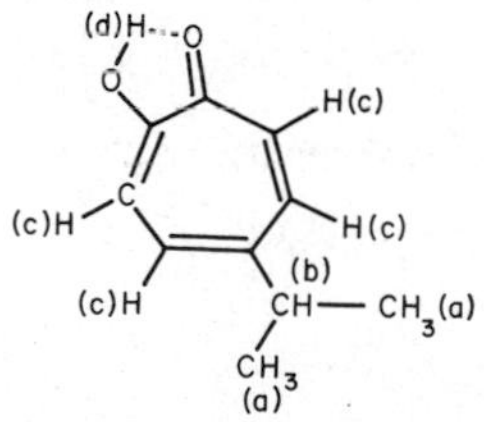

	Shift	Peak
a	1.27	2A
▶ b	2.90	5B
c	8.32	1A
d	8.75	1B

▶ $C_{10}H_{12}O_2$ ▶ 7% $CDCl_3$ ▶ V 560

▶ 3-AAQh

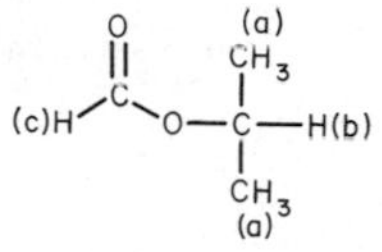

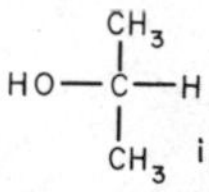

	Shift	Peak
a	1.29	3A
▶ b	5.13	5B
c	8.02	1B

impurity

▶ $C_4H_8O_2$ ▶ 7% $CDCl_3$ ▶ V 415

▶ 3-AA(L)ll

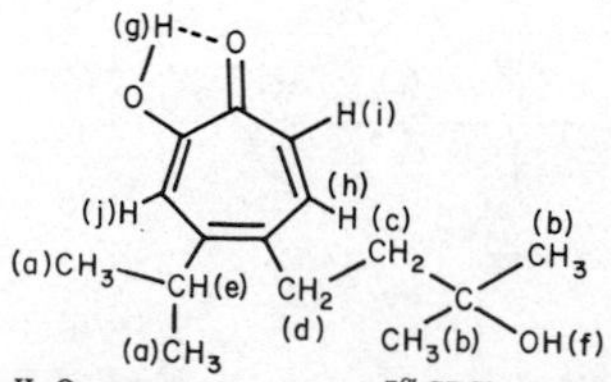

	Shift	Peak
a	1.27	2A
b	1.34	1E
c	1.69	O
d	2.78	O
▶ e	3.28	S
f	5.27	1B
g	5.27	1B
h	7.19*	2F
i	7.19*	2F
j	7.39	1A

*or 7.31

▶ $C_{15}H_{22}O_3$ ▶ 7% $CDCl_3$ ▶ V 647

▶ 3-AAR

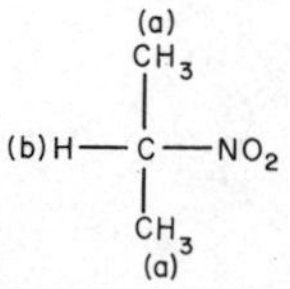

	Shift	Peak
a	1.55	2A
▶ b	4.67	M

▶ $C_3H_7NO_2$ ▶ 7% $CDCl_3$ ▶ V 41

▶ 3-AASh

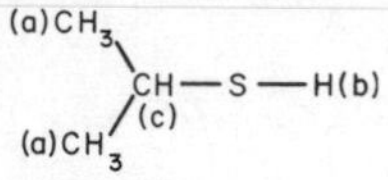

	Shift	Peak
a	1.34	2A
b	1.56	2A
▶ c	3.16	6A

▶ C_3H_8S ▶ 7% $CDCl_3$ ▶ V 396

▶ 3-AAVhh

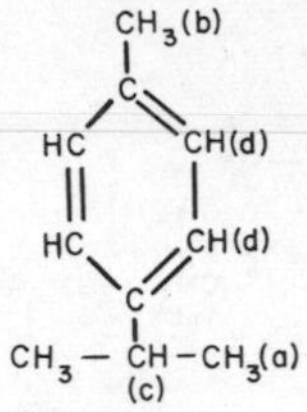

	Shift	Peak
a	1.20	2A
b	2.24	1A
▶ c	2.80	5A
d	6.99	1A

▶ $C_{10}H_{14}$ ▶ 50% CCl_4 ▶ API 225

▶ 3-AAU

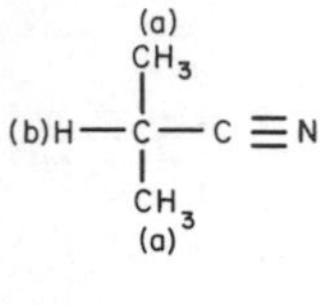

	Shift	Peak
a	1.33	2A
▶ b	2.72	6B

▶ C_4H_7N ▶ 7% $CDCl_3$ ▶ V 408

▶ 3-AAVhh

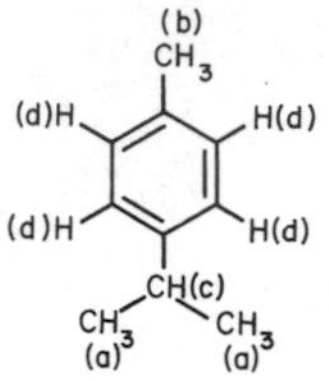

	Shift	Peak
a	1.22	2A
b	2.30	1A
▶ c	2.87	5B
d	7.08	1A

▶ $C_{10}H_{14}$ ▶ 7% $CDCl_3$ ▶ V 268

▶ 3-AAVah

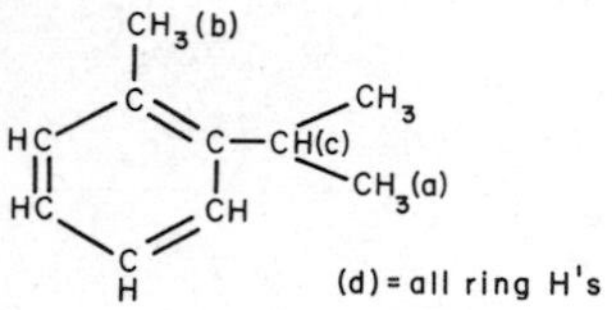

(d) = all ring H's

	Shift	Peak
a	1.15	2A
b	2.20	1A
▶ c	3.01	5A
d	6.95	O

▶ $C_{10}H_{14}$ ▶ run neat ▶ API 181 A

▶ 3-AAVhh

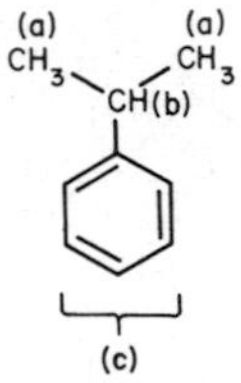

	Shift	Peak
a	1.25	2A
▶ b	2.90	7A
c	7.25	1A

▶ C_9H_{12} ▶ 7% $CDCl_3$ ▶ V 240

▶ 3-AAVah

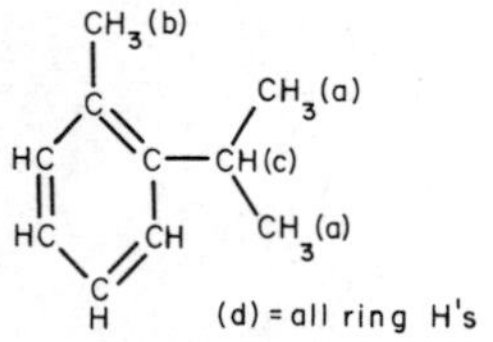

(d) = all ring H's

	Shift	Peak
a	1.17	2A
b	2.24	1A
▶ c	3.04	5A
d	7.00	N

▶ $C_{10}H_{14}$ ▶ 50% CCl_4 ▶ API 224

▶ 3-AAVhh

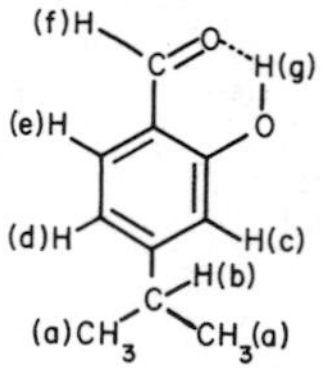

	Shift	Peak
a	1.23	2A
▶ b	2.92	5A
c	6.82	N
d	6.88	N
e	7.42	2A
f	9.79	1A
g	11.00	1B

▶ $C_{10}H_{12}O_2$ ▶ 7% $CDCl_3$ ▶ V 259

▶ 3-AAVah

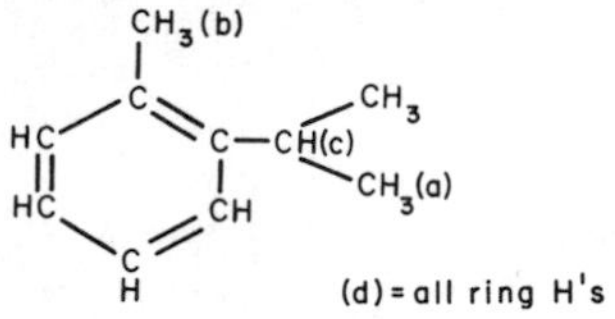

(d) = all ring H's

	Shift	Peak
a	1.25	2A
b	2.30	1A
▶ c	3.05	5A
d	7.01	O

▶ $C_{10}H_{14}$ ▶ CCl_4 ▶ API 181 B

▶ 3-AAVhp

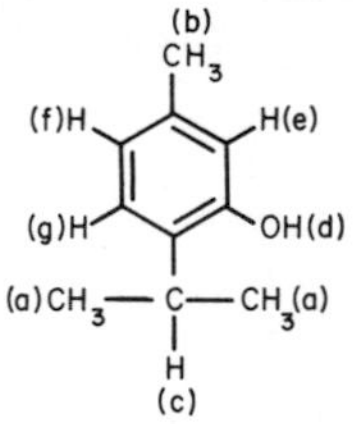

	Shift	Peak
a	1.23	2A
b	2.25	1A
▶ c	3.17	5B
d	4.75	1A
e	6.55	S
f	6.71	S
g	7.06	S

▶ $C_{10}H_{14}O$ ▶ 7% $CDCl_3$ ▶ V 270

▶ 3-AAVhh

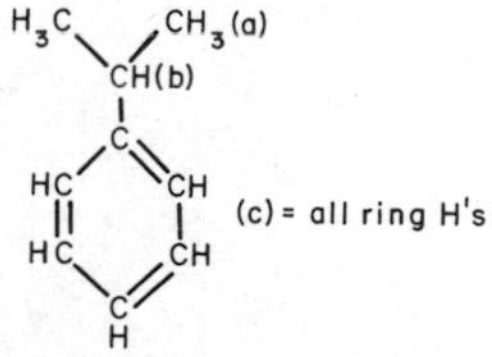

(c) = all ring H's

	Shift	Peak
a	1.15	2A
▶ b	2.75	5A
c	7.09	1A

▶ C_9H_{12} ▶ run neat ▶ API 176

▶ 3-ABaBa

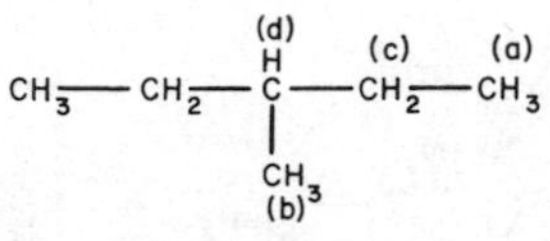

	Shift	Peak
a	.88	3D
b	.88	2H
c	1.29	S
▶ d	~1.30	S

▶ C_6H_{14} ▶ 10% CCl_4 ▶ API 47

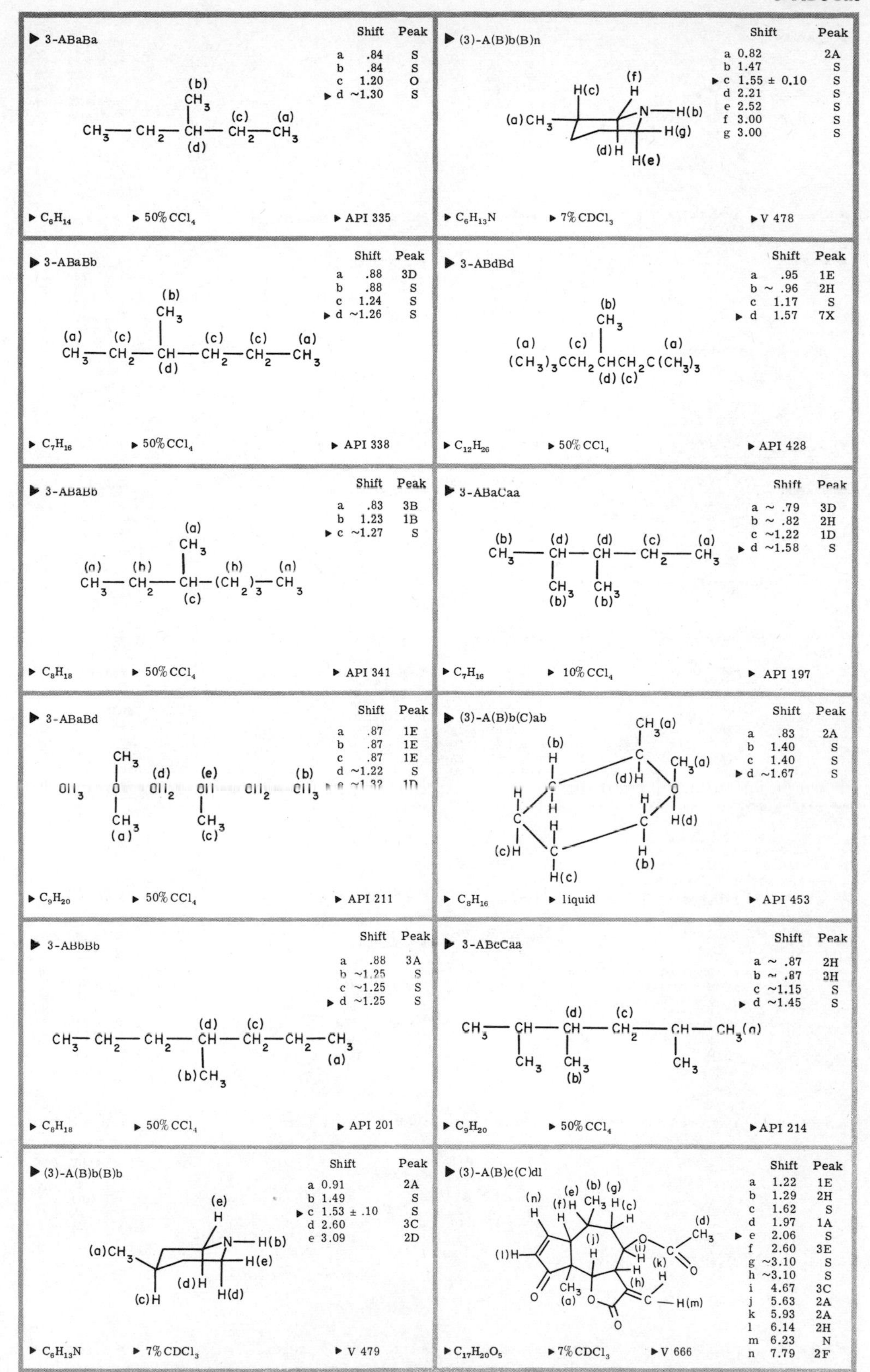
3-ABaBa
Shift Peak
a .84 S
b .84 S
c 1.20 O
d ~1.30 S
C6H14
50% CCl4
API 335
(3)-A(B)b(B)n
Shift Peak
a 0.82 2A
b 1.47 S
c 1.55 ± 0.10 S
d 2.21 S
e 2.52 S
f 3.00 S
g 3.00 S
C6H13N
7% CDCl3
V 478
3-ABaBb
Shift Peak
a .88 3D
b .88 S
c 1.24 S
d ~1.26 S
C7H16
50% CCl4
API 338
3-ABdBd
Shift Peak
a .95 1E
b ~ .96 2H
c 1.17 S
d 1.57 7X
C12H26
50% CCl4
API 428
3-ABaBb
Shift Peak
a .83 3B
b 1.23 1B
c ~1.27 S
C8H18
50% CCl4
API 341
3-ABaCaa
Shift Peak
a ~ .79 3D
b ~ .82 2H
c ~1.22 1D
d ~1.58 S
C7H16
10% CCl4
API 197
3-ABaBd
Shift Peak
a .87 1E
b .87 1E
c .87 1E
d ~1.22 S
C9H20
50% CCl4
API 211
(3)-A(B)b(C)ab
Shift Peak
a .83 2A
b 1.40 S
c 1.40 S
d ~1.67 S
C8H16
liquid
API 453
3-ABbBb
Shift Peak
a .88 3A
b ~1.25 S
c ~1.25 S
d ~1.25 S
C8H18
50% CCl4
API 201
3-ABcCaa
Shift Peak
a ~ .87 2H
b ~ .87 3H
c ~1.15 S
d ~1.45 S
C9H20
50% CCl4
API 214
(3)-A(B)b(B)b
Shift Peak
a 0.91 2A
b 1.49 S
c 1.53 ± .10 S
d 2.60 3C
e 3.09 2D
C6H13N
7% CDCl3
V 479
(3)-A(B)c(C)dl
Shift Peak
a 1.22 1E
b 1.29 2H
c 1.62 S
d 1.97 1A
e 2.06 S
f 2.60 3E
g ~3.10 S
h ~3.10 S
i 4.67 3C
j 5.63 2A
k 5.93 2A
l 6.14 2H
m 6.23 N
n 7.79 2F
C17H20O5
7% CDCl3
V 666

3-ABbDaaa

	Shift	Peak
a	.85	1E
b	.85	1E
c	.85	1E
d	~1.03	S
▶ e	~1.32	S

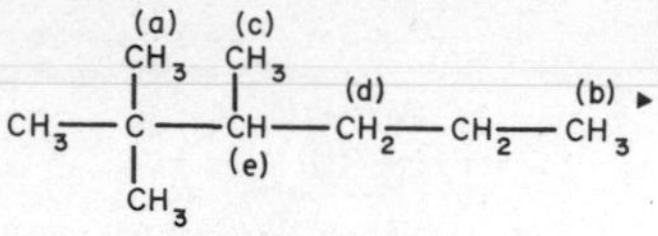

▶ C_9H_{20} ▶ 50% CCl_4 ▶ API 210

(3)-A(B)o(J)b

	Shift	Peak
a	1.17	2A
▶ b	2.55	6B
c	3.73	10A
d	3.84	10A
e	4.07	10A
f	4.49	3A

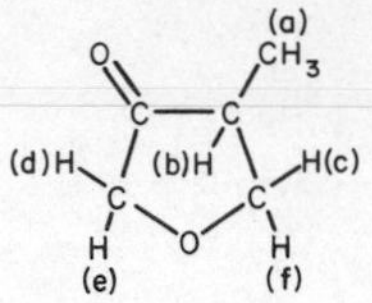

▶ $C_5H_8O_2$ ▶ 7% $CDCl_3$ ▶ V 438

3-ABeE

	Shift	Peak
a	1.60	2A
b	3.52	R
c	3.78	R
▶ d	4.10	R

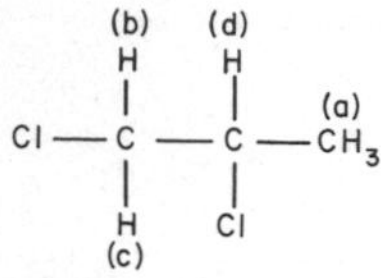

▶ $C_3H_6Cl_2$ ▶ 7% $CDCl_3$ ▶ V 30

3-ABaLhl

	Shift	Peak
a	.88	3C
b	1.03	1E
c	1.24	S
▶ d	1.96	7B
e	4.86	R
f	4.97	R
g	5.54	R

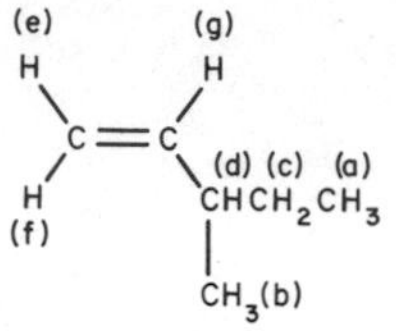

▶ C_6H_{12} ▶ liquid ▶ API 374

3-ABaG

	Shift	Peak
a	1.08	3C
b	1.71	2H
c	1.79	6C
▶ d	4.11	5A

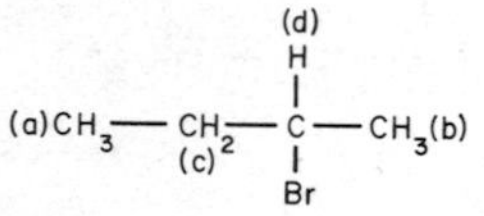

▶ C_4H_9Br ▶ 7% $CDCl_3$ ▶ V 418

3-ABaLhl

	Shift	Peak
a	.88	3D
b	.99	2A
c	1.32	S
▶ d	2.01	5B
e	4.83 or 4.96	R
f	4.96 or 4.83	R
g	5.53	R

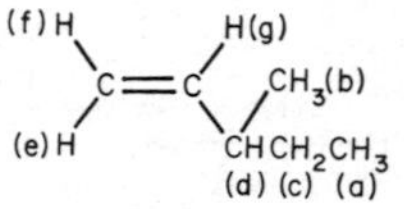

▶ C_6H_{12} ▶ 10% CCl_4 ▶ API 156

3-ABbG

	Shift	Peak
a	0.91	3B
b	1.28	1B
c	1.71	2A
▶ d	4.12	5B

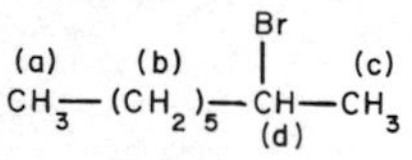

▶ $C_8H_{17}Br$ ▶ 45.2 mg/0.50 ml $CDCl_3$ ▶ S 227

3-ABaLhl

	Shift	Peak
a	.92	3C
b	.95	2H
c	1.25	O
d	1.58	2E
▶ e	2.38	6B
f	~5.17 or 5.32	S
g	~5.32 or 5.17	S

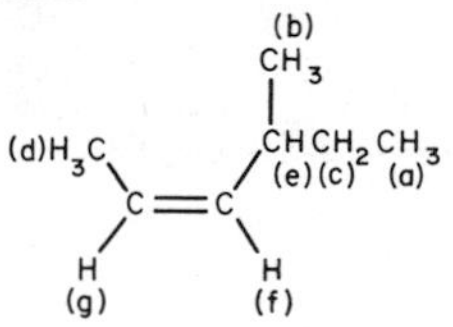

▶ C_7H_{14} ▶ liquid run, neat ▶ API 409

3-ABgG

	Shift	Peak
a	1.83	2A
b	3.52	4E
c	3.75	S
▶ d	4.18	S

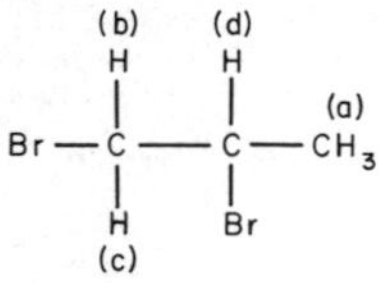

▶ $C_3H_6Br_2$ ▶ 45.4 mg/0.50 ml $CDCl_3$ ▶ S 187

3-ABbLhl

	Shift	Peak
a	.88	3D
b	.98	2H
c	1.29	O
▶ d	2.11	5A
e	4.86	R
f	4.97	R
g	~5.58	R

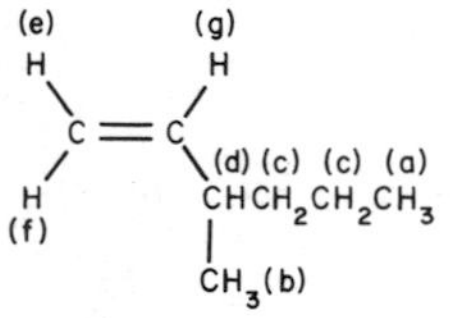

▶ C_7H_{14} ▶ liquid ▶ API 392

3-ABaI

	Shift	Peak
a	1.02	3C
b	1.70	4C
c	1.92	2A
▶ d	4.17	4C

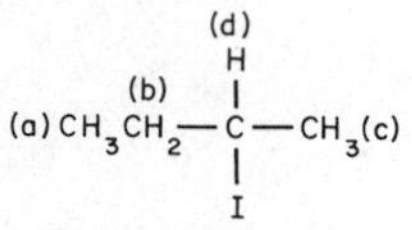

▶ C_4H_9I ▶ 7% $CDCl_3$ ▶ V 82

3-ABaNhh

	Shift	Peak
a	0.90	3D
b	1.05	2F
c	1.25	N
▶ d	2.78	6A

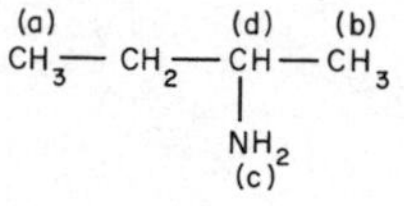

▶ $C_4H_{11}N$ ▶ 7% $CDCl_3$ ▶ V 88

3-ABaNhv

	Shift	Peak
a	0.96	3D
b	1.17	2A
c	~1.50	O
d	~3.38	S
▶ e	~3.41	S

▶ $C_{10}H_{15}N$ ▶ 7% $CDCl_3$ ▶ V 568

(3)-A(B)b(O)c

	Shift	Peak
a	1.22	2A
b	1.92	3A
c	2.68±.02	S
▶ d	2.80±.10	S
e	4.05	S
f	4.64	3A

▶ $C_{11}H_{20}O_4$ ▶ 7% $CDCl_3$ ▶ V 588

(3)-A(B)b(N)bh

	Shift	Peak
a	1.05	2A
b	1.34	S
▶ c	2.57 ± .10	S

▶ $C_6H_{13}N$ ▶ 7% $CDCl_3$ ▶ V 477

(3)-A(B)d(O)z

	Shift	Peak
a	0.61	S
b	0.88	S
c	1.25	2H
d	1.29	1A
e	1.47	S
f	1.76	S
▶ g	4.17	O

▶ $C_8H_{17}BO_2$ ▶ 7% $CDCl_3$ ▶ V 518

(3)-A(B)c(N)dhh

	Shift	Peak
a	1.56	1A
b	1.69	1A
c	1.75	1A
▶ d	3.54	1D
e	5.27	1D
f	8.04	N

▶ $C_{15}H_{22}ClNO_2$ ▶ 7% $CDCl_3$ ▶ V 645

(3)-A(B)d(O)z

	Shift	Peak
a	1.27	1A
b	1.31	1A
c	1.49	2G
d	1.81	2G
▶ e	4.22	7X
f	~5.90	S
g	~5.90	S
h	~5.90	S

▶ $C_8H_{15}BO_2$ ▶ 7% $CDCl_3$ ▶ V 517

(3)-A(B)c(N)dhh

	Shift	Peak
a	1.54	1A
b	1.58	1E
c	1.62	1E
▶ d	3.74	7X
e	5.33	6B
f	8.10	4A

▶ $C_{15}H_{22}ClNO_2$ ▶ D_2O ▶ V 646

(3)-A(B)j(O)b

	Shift	Peak
a	1.44	2A
b	2.15	10A
c	2.57	10A
d	3.83	10A
e	4.04	10A
▶ f	4.35	7C

▶ $C_5H_8O_2$ ▶ 7% $CDCl_3$ ▶ V 439

3-ABeNaah

	Shift	Peak
a	1.68	2A
b	2.97	1A
c	3.27	S
d	3.56	S
▶ e	4.62	1F

▶ $C_5H_{13}Cl_2N$ ▶ 7% $CDCl_3$ ▶ V 448

(3)-A(B)o(O)b

	Shift	Peak
a	1.32	2A
b	2.42	N
c	2.72	N
▶ d	~2.98	N

▶ C_3H_6O ▶ 7% $CDCl_3$ ▶ V 32

3-ABbOa

	Shift	Peak
a	1.18	2A
b	1.72	4A
c	3.10	3A
d	3.33	1A
▶ e	3.55	O
f	3.73	6B

▶ $C_5H_{12}O_2$ ▶ 7% $CDCl_3$ ▶ V 120

3-ABbP

	Shift	Peak
a	0.92	3D
b	1.18	2H
c	1.33	N
d	2.22	1B
▶ e	3.70	1F

▶ $C_5H_{12}O$ ▶ 44.6 mg/0.50 ml $CDCl_3$ ▶ S 60

▶ 3-ABbP

	Shift	Peak
a	.88	3B
b	1.12	2A
c	1.30	1B
▶ d	3.70	4B
e	4.08	1A

(b) CH_3—CH(OH (e))(d)—(c) CH_2—(c) $(CH_2)_4$—(a) CH_3

▶ $C_8H_{18}O$ ▶ 20% CCl_4 ▶ API 170

▶ 3-ABaVhh

	Shift	Peak
a	.82	3C
b	1.23	2A
c	1.60	5B
▶ d	2.56	4C
e	7.11	1A
f	7.11	1A

(b) (d) (c) (a) $CH_3CHCH_2CH_3$ attached to C of ring: (e)HC, CH(e), (f)HC, CH(f), C H(f)

▶ $C_{10}H_{14}$ ▶ 10% CCl_4 ▶ API 90

▶ 3-ABbP

	Shift	Peak
a	1.18	1A
b	1.52	3A
▶ c	3.78	S
d	3.87	1A

(a) CH_3—CH(OH (d))(c)—(b) CH_2CH_2—CH(OH)—CH_3

▶ $C_6H_{14}O_2$ ▶ 44.6 mg/0.50 ml $CDCl_3$ ▶ S 216

▶ 3-ACaaCaa

	Shift	Peak
a	.80 ± .03 or .93 ± .03	S
b	.93 ± .03 or .80 ± .03	S
▶ c	1.67	6A

(b) CH_3—CH(c)(CH_3)—CH(c)(CH_3 (a))—CH(c)(CH_3)—CH_3

▶ C_8H_{18} ▶ 50% CCl_4 ▶ API 208

▶ 3-ABbP

	Shift	Peak
a	1.23	2A
b	1.68	4C
c	3.58 or 3.65	1A
d	3.58 or 3.65	1A
e	3.80	3A
▶ f	4.03	4A

(a) CH_3—CH(OH (c))(f)—(b) CH_2—(e) CH_2—OH (d)

▶ $C_4H_{10}O_2$ ▶ 7% $CDCl_3$ ▶ V 86

▶ 3-ACgkG

	Shift	Peak
a	1.93	N
▶ b	4.42 or 4.43	S
c	4.42 or 4.43	S
d	11.58	1A

O=C((d)HO)—C((c)H)(Br)—C(H(b))(Br)—CH_3(a)

▶ $C_4H_6Br_2O_2$ ▶ 7% $CDCl_3$ ▶ V 403

▶ 3-ABkQa

	Shift	Peak
a	1.32	2A
b	2.03	1A
c	2.50 or 2.70	2A
d	2.50 or 2.70	2A
e	3.70	1A
▶ f	5.30	4A

(b) CH_3—C(=O)—O—C((a) CH_3)((f) H)—C(H(c))(H(d))—C(=O)—OCH_3(e)

▶ $C_7H_{12}O_4$ ▶ 7% $CDCl_3$ ▶ V 182

▶ 3-ACaaLhl

	Shift	Peak
a	.88	3A
b	.98	2H
▶ c	1.73	7X
d	4.90	R
e	5.00	R
f	5.64	R

(d) H, (e) H C=C (f) H—CH(CH_3 (b))—(c) CH(CH_3 (a))—CH_3(a)

▶ C_7H_{14} ▶ liquid ▶ API 413

▶ 3-ABaR

	Shift	Peak
a	0.95	3A
b	1.52	2A
c	~1.90	O
▶ d	4.52	6A

(a) CH_3—(c) CH_2—(d) CH(NO_2)—(b) CH_3

▶ $C_4H_9NO_2$ ▶ 7% $CDCl_3$ ▶ V 84

▶ 3-ACpvNhhh

	Shift	Peak
a	1.21	2A
▶ b	3.72	O
c	5.00	2A
d	7.48	1A

(d) C_6H_5—C((c) H)(OD)—C((b) H)($ND_3^{(+)}$)—CH_3(a) $Cl^{(-)}$

▶ $C_9H_{14}ClNO$ ▶ D_2O ▶ V 565

▶ 3-ABaVhh

	Shift	Peak
a	0.80	3A
b	1.22	2A
c	1.59	5B
▶ d	2.53	6A
e	7.20	1A

(e) C_6H_5—CH((b) CH_3)(d)—CH_2(c)—CH_3(a)

▶ $C_{10}H_{14}$ ▶ 45.0 mg/0.50 ml $CDCl_3$ ▶ S 31

▶ 3-ACapP

	Shift	Peak
a	1.13	2A
b	3.05	1A
▶ c	~3.80	O

(a) CH_3—CH(OH (b))(c)—CH(OH (b))(c)—CH_3(a)

▶ $C_4H_{10}O_2$ ▶ 7% $CDCl_3$ ▶ V 87

▶ (3)-A(C)lp(Q)l

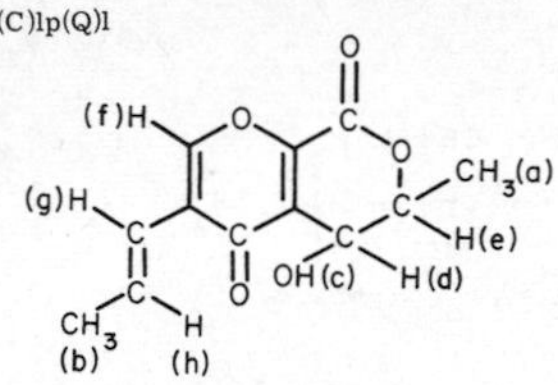

	Shift	Peak
a	1.65	2A
b	1.97	2E
c	3.86	1C
d	3.92	S
▶ e	4.32	O
f	5.87	1I
g	6.06	2A
h	6.99	4A

▶ $C_{12}H_{12}O_5$ ▶ 7% $CDCl_3$ ▶ V 595

▶ 3-AKhNhh

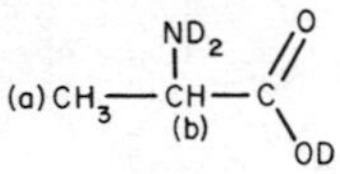

	Shift	Peak
a	1.48	2A
▶ b	3.78	4A

▶ $C_3H_7NO_2$ ▶ D_2O ▶ V 393

▶ (3)-A(C)ovVho

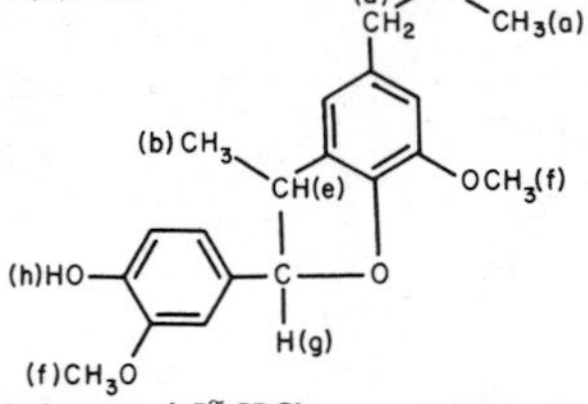

	Shift	Peak
a	0.96	3B
b	1.37	2A
c	~1.65	S
d	2.58	3B
▶ e	3.42	S
f	3.88	1A
g	5.08	2A
h	5.62	1A

▶ $C_{20}H_{24}O_4$ ▶ 7% $CDCl_3$ ▶ V 682

▶ 3-AKhSh

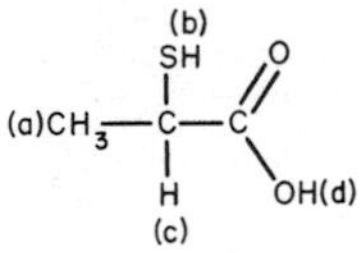

	Shift	Peak
a	1.56	2A
b	2.26	2A
▶ c	3.58	5B
d	11.03	1B

▶ $C_3H_6O_2S$ ▶ 7% $CDCl_3$ ▶ V 389

▶ 3-AEKh

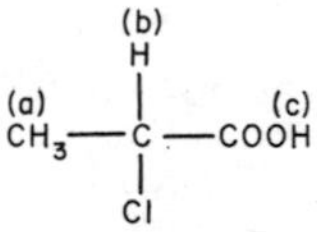

	Shift	Peak
a	1.73	2B
▶ b	4.47	4A
c	11.22	1A

▶ $C_3H_5ClO_2$ ▶ 7% $CDCl_3$ ▶ V 25

▶ 3-ANhhVhh

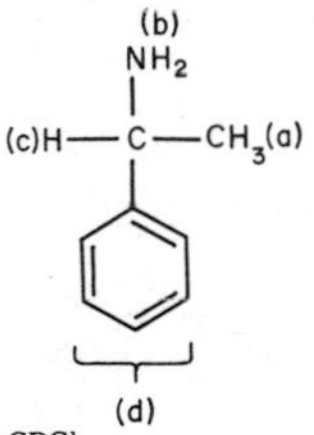

	Shift	Peak
a	1.38	2A
b	1.58	1A
▶ c	4.10	4A
d	7.30 ± .03	1A

▶ $C_8H_{11}N$ ▶ 7% $CDCl_3$ ▶ V 207

▶ 3-AGG

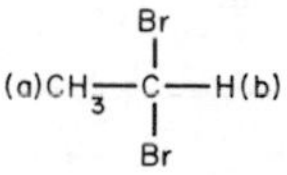

	Shift	Peak
a	2.47	2A
▶ b	5.86	4A

▶ $C_2H_4Br_2$ ▶ 7% $CDCl_3$ ▶ V 370

▶ 3-AObOb

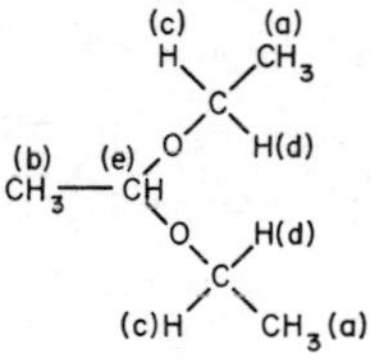

	Shift	Peak
a	1.20	S
b	1.32	S
c	~3.50 or 3.70	2C
d	~3.50 or 3.70	2C
▶ e	4.72	4A

▶ $C_6H_{14}O_2$ ▶ 7% $CDCl_3$ ▶ V 143

▶ 3-AGKh

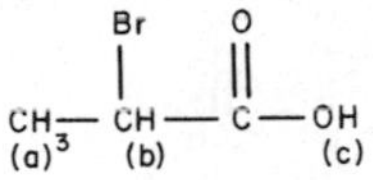

	Shift	Peak
a	1.86	2A
▶ b	4.38	4A
c	9.95	1A

▶ $C_3H_5BrO_2$ ▶ 44.5 mg/0.50 ml $CDCl_3$ ▶ S 144

▶ (3)-A(O)c(O)c

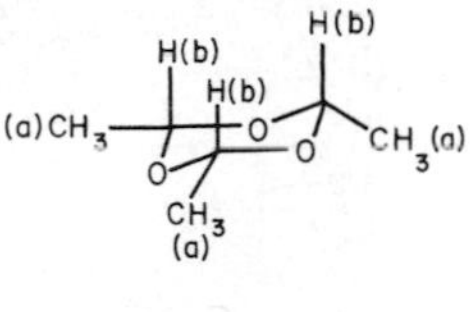

	Shift	Peak
a	1.40	S
▶ b	5.10	1C

▶ $C_6H_{12}O_3$ ▶ 7% $CDCl_3$ ▶ V 474

▶ 3-AGVhh

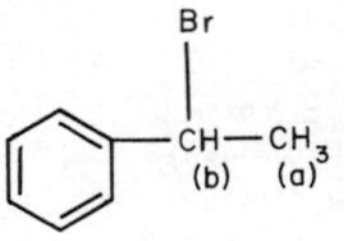

	Shift	Peak
a	2.02	2A
▶ b	5.20	4A

▶ C_8H_9Br ▶ 45.2 mg/0.50 ml $CDCl_3$ ▶ S 223

▶ 3-APVhh

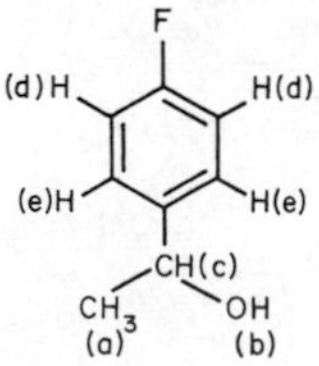

	Shift	Peak
a	1.38	2A
b	3.20	1B
▶ c	4.75	4A
d	6.98	O
e	7.22	O

▶ C_8H_9FO ▶ 7% $CDCl_3$ ▶ V 200

▶ 3-APVhh

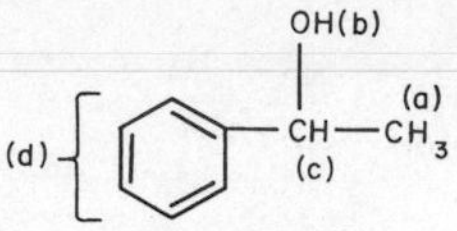

	Shift	Peak
a	1.42	2A
b	2.38	1A
▶ c	4.78	4A
d	7.30	1A

▶ $C_8H_{10}O$ ▶ 45.2 mg/0.50 ml $CDCl_3$ ▶ S 133

▶ 3-BaBaDaaa

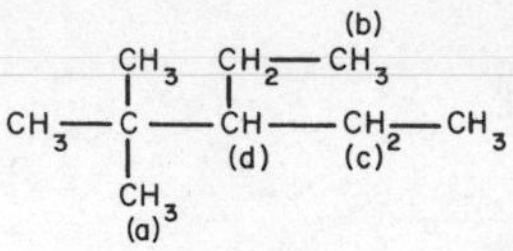

	Shift	Peak
a	.87	1E
b	.87	1E
c	1.00	1B
▶ d	~1.43	7X

▶ C_9H_{20} ▶ 50% CCl_4 ▶ API 216

▶ 3-BaBaBa

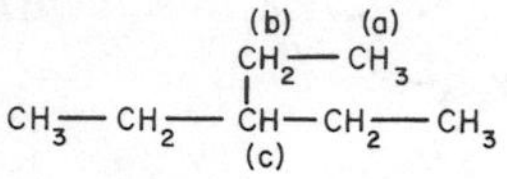

	Shift	Peak
a	.80	3B
b	.92	S
▶ c	~.94	S

▶ C_7H_{16} ▶ 50% CCl_4 ▶ API 196

▶ (3)-(B)b(B)bE

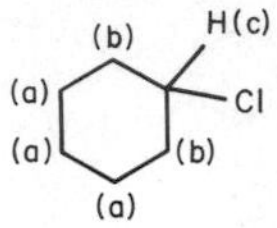

	Shift	Peak
a	1.51	N
b	1.82	O
▶ c	~4.00	1F

▶ $C_6H_{11}Cl$ ▶ 45.0 mg/0.50 ml $CDCl_3$ ▶ S 25

▶ 3-BaBbBp

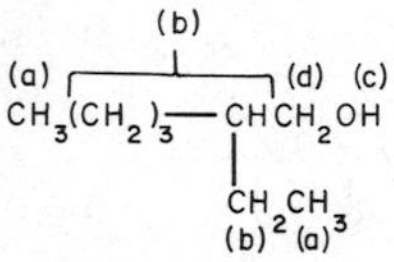

	Shift	Peak
a	0.92	3B
▶ b	1.30	1B
c	1.75	1B
d	3.52	1I

▶ $C_8H_{18}O$ ▶ 45.0 mg/0.50 ml $CDCl_3$ ▶ S 98

▶ 3-BaBgG

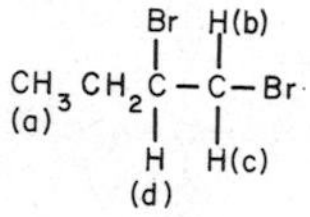

	Shift	Peak
a	1.05	3A
b	3.62	2F
c	3.83	2F
▶ d	4.13	O

▶ $C_4H_8Br_2$ ▶ 7% $CDCl_3$ ▶ V 74

▶ 3-BaBaCaa

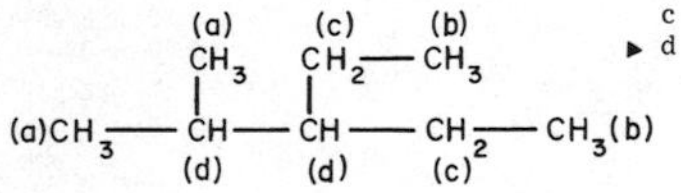

	Shift	Peak
a	~ .83	2H
b	~ .91	3D
c	1.20	3B
▶ d	~1.72	1D

▶ C_8H_{18} ▶ 50% CCl_4 ▶ API 206

▶ 3-BgBgG

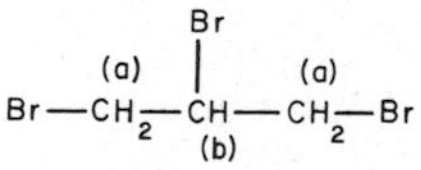

	Shift	Peak
a	3.88	2A
▶ b	4.38	6B

▶ $C_3H_5Br_3$ ▶ 44.5 mg/0.50 ml $CDCl_3$ ▶ S 231

▶ (3)-(B)b(B)b(C)bb

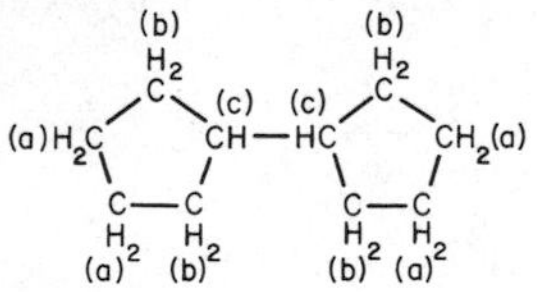

	Shift	Peak
a	~1.53	S
b	~1.53	S
▶ c	~1.56	S

▶ $C_{10}H_{18}$ ▶ liquid ▶ API 13

▶ 3-BaBaJh

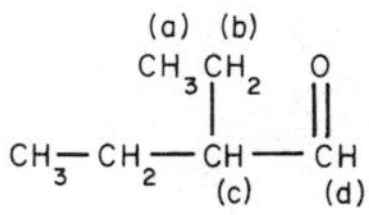

	Shift	Peak
a	0.97	3A
b	1.58	4A
▶ c	2.22	5B
d	10.45	1C

▶ $C_6H_{12}O$ ▶ 36.9 mg/0.50 ml $CDCl_3$ ▶ S 74

▶ (3)-(B)b(B)b(C)bb

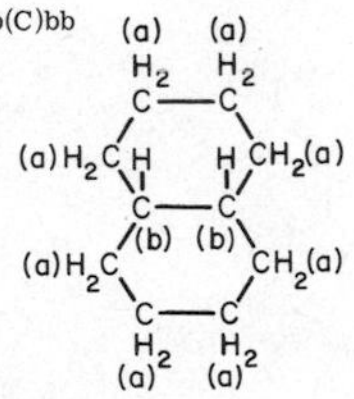

	Shift	Peak
a	1.43	1B
▶ b	1.63	1E

▶ $C_{10}H_{18}$ ▶ liquid run ▶ API 14

▶ (3)-(B)b(B)bJn

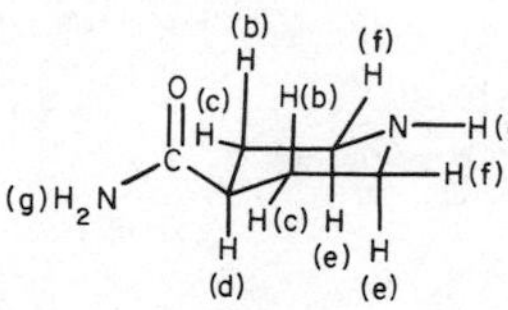

	Shift	Peak
a	1.53	1B
b	1.60 ± 0.05	S
c	1.82 ± 0.05	S
▶ d	2.14	O
e	2.67	4A
f	3.17	6A
g	5.87	1D

▶ $C_6H_{12}N_2O$ ▶ 7% $CDCl_3$ ▶ V 473

▶ (3)-(B)b(B)bJv

	Shift	Peak
a	3.27	1C

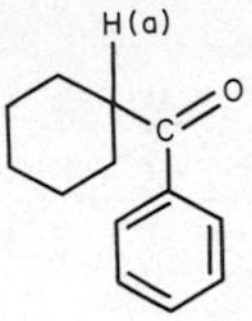

▶ $C_{13}H_{16}O$ ▶ 7% $CDCl_3$ ▶ V 613

▶ (3)-(B)b(B)bLmv

	Shift	Peak
▶ a	2.48±0.05	1C
b	9.02	1B

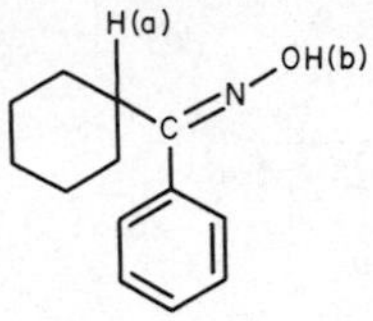

▶ $C_{13}H_{17}NO$ ▶ 7% $CDCl_3$ ▶ V 614

▶ (3)-(B)b(B)bKa

	Shift	Peak
a	~0.93	O
▶ b	~1.63	O
c	3.67	1A

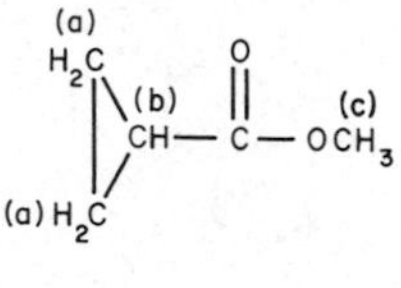

▶ $C_5H_8O_2$ ▶ 7% $CDCl_3$ ▶ V 112

▶ (3)-(B)b(B)bNhh

	Shift	Peak
a	~0.35	O
b	1.83	1A
▶ c	2.30	O

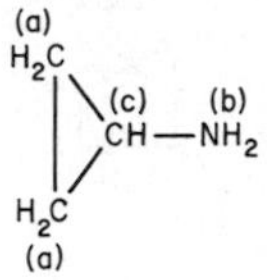

▶ C_3H_7N ▶ 7% $CDCl_3$ ▶ V 37

▶ (3)-(B)b(B)bKh

	Shift	Peak
a	2.12	Q
▶ b	3.15	N
c	12.35	1A

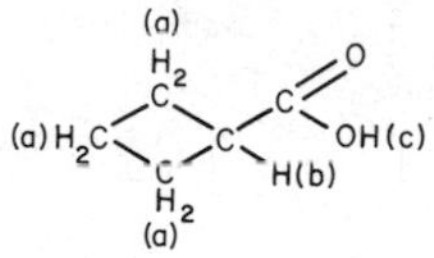

▶ $C_5H_8O_2$ ▶ liquid run, neat ▶ API 191

▶ (3)-Be(B)o(O)b

	Shift	Peak
a	2.70	4A
b	2.88	3A
▶ c	3.28	O
d	3.58	2A

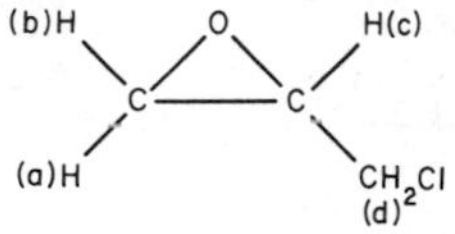

▶ C_3H_5ClO ▶ 45.0 mg/0.50 ml $CDCl_3$ ▶ S 16

▶ 3-BaBaLhl

	Shift	Peak
a	.83	3C
b	1.30	O
▶ c	~1.72	S
d	~5.50	R

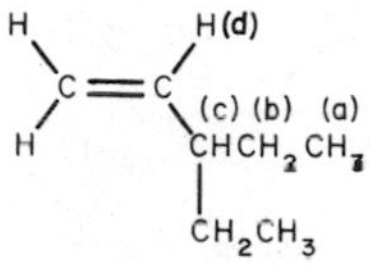

▶ C_7H_{14} ▶ liquid ▶ API 398

▶ (3)-(B)a(B)bP

	Shift	Peak
a	0.95	2A
▶ b	~3.5	1D
c	4.02	1B

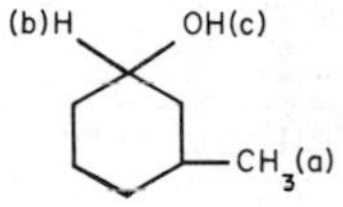

▶ $C_7H_{14}O$ ▶ 45.1 mg/0.50 ml $CDCl_3$ ▶ S 79

▶ (3)-(B)b(B)bLcm

	Shift	Peak
▶ a	2.46	O
b	9.40	1C

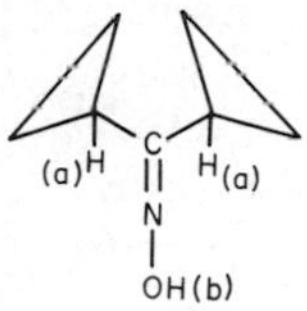

▶ $C_7H_{11}NO$ ▶ 7% $CDCl_3$ ▶ V 179

▶ (3)-(B)b(B)bP

	Shift	Peak
a	1.31	4E
b	1.78	3B
c	2.20	1E
▶ d	~3.61	1C

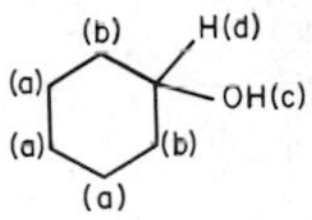

▶ $C_6H_{12}O$ ▶ 45.2 mg/0.50 ml $CDCl_3$ ▶ S 7

▶ (3)-(B)b(B)bLhl

	Shift	Peak
a	2.23	1A
▶ b	2.62	O
c	6.05	2A
d	6.76	4D

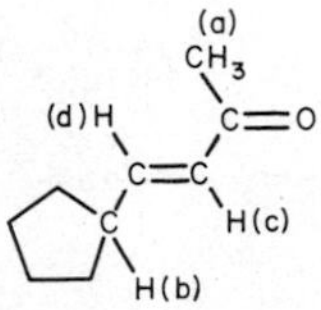

▶ $C_9H_{14}O$ ▶ 7% $CDCl_3$ ▶ V 544

▶ (3)-BbBbP

	Shift	Peak
a	1.88	1A
▶ b	4.33	1C

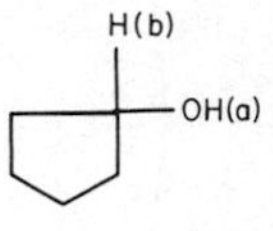

▶ $C_5H_{10}O$ ▶ 34.5 mg/0.50 ml $CDCl_3$ ▶ S 44

▶ (3)-(B)b(B)cP

	Shift	Peak
a	3.32	3A
b	3.32	3A
c	3.93	4A
▶ d	4.22	5A

▶ $C_6H_{11}NO_3$ ▶ 7% $CDCl_3$ ▶ V 469

▶ 3-BpBpP

	Shift	Peak
a	3.61	1E
▶ b	3.79	S

▶ $C_3H_8O_3$ ▶ D_2O ▶ V 395

▶ (3)-(B)b(B)lP

	Shift	Peak
a	0.67	1A
b	0.87	2H
c	0.90	2H
d	1.00	1E
▶ e	3.50	1D
f	5.36	2E

▶ $C_{27}H_{46}O$ ▶ 7% $CDCl_3$ ▶ V 363

▶ (3)-(B)b(B)cQa

	Shift	Peak
a	1.03	1A
b	1.03	1A
c	2.03	1E
d	3.70	1A
▶ e	~4.70	1D

▶ $C_{27}H_{42}O_5$ ▶ 7% $CDCl_3$ ▶ V 362

▶ (3)-(B)b(B)lP

	Shift	Peak
a	0.57	1A
▶ b	3.92	1F
c	~5.20	N
d	6.05*	2C
e	6.05*	2C

*or 6.20

▶ $C_{28}H_{44}$ ▶ 7% $CDCl_3$ ▶ V 364

▶ (3)-(B)b(B)cQa

	Shift	Peak
a	0.70	1A
b	0.92	1A
c	2.03, 2.05	1A
d	2.18	1A
e	2.98	3C
▶ f	~4.73	1D
g	5.17	1I

▶ $C_{25}H_{38}O_5$ ▶ 7% $CDCl_3$ ▶ V 361

▶ 3-BbBnP

	Shift	Peak
a	0.89	3B
b	1.28	1G
c	1.75	1E
d	1.75	1E
e	2.51	S
f	2.84	S
▶ g	3.50	1D

▶ $C_{10}H_{23}NO$ ▶ 7% $CDCl_3$ ▶ V 575

▶ (3)-(B)c(B)dQv

	Shift	Peak
a	1.56	1A
b	1.69	1A
c	1.75	1A
d	3.54	1D
▶ e	5.27	1D
f	8.04	N

▶ $C_{15}H_{22}ClNO_2$ ▶ 7% $CDCl_3$ ▶ V 645

▶ (3)-(B)b(B)nP

	Shift	Peak
a	2.88	4A
b	3.50	S
c	3.75	S
▶ d	4.01	S

▶ $C_6H_{11}NO_3$ ▶ D_2O ▶ V 470

▶ (3)-(B)c(B)dQv

	Shift	Peak
a	1.54	1A
b	1.58	1E
c	1.62	1E
d	3.74	7X
▶ e	5.33	6B
f	8.10	4A

▶ $C_{15}H_{22}ClNO_2$ ▶ D_2O ▶ V 646

▶ 3-BeBeP

	Shift	Peak
a	2.68	1B
b	3.72	2A
▶ c	4.07	6B

▶ $C_3H_6Cl_2O$ ▶ 7% $CDCl_3$ ▶ V 386

▶ 3-BqBqQa

	Shift	Peak
a	2.08	1A
b	~4.17	S
c	~4.28	S
▶ d	5.25	5B

▶ $C_9H_{14}O_6$ ▶ 7% $CDCl_3$ ▶ V 242

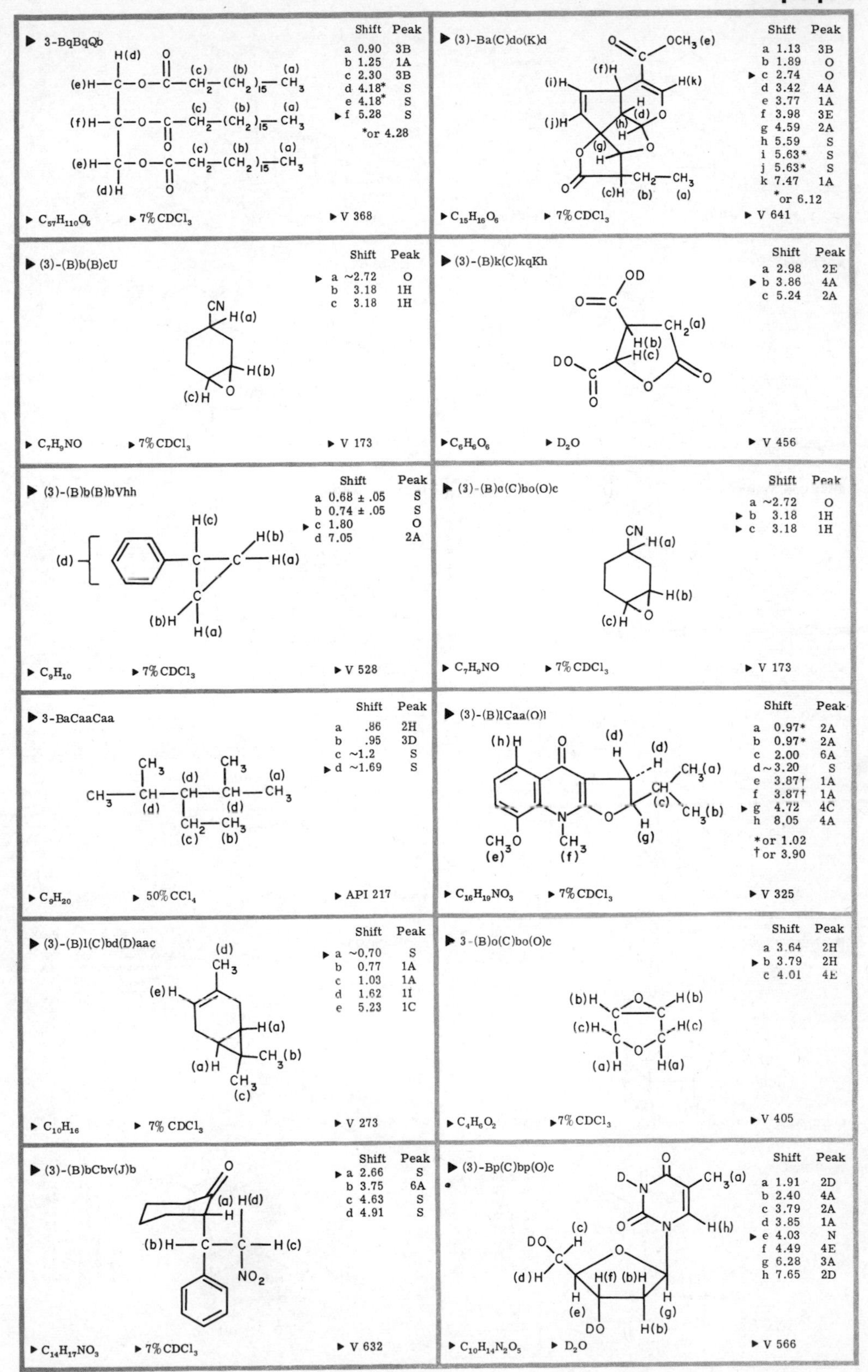

▶ 3-BqBqQb

	Shift	Peak
a	0.90	3B
b	1.25	1A
c	2.30	3B
d	4.18*	S
e	4.18*	S
▶ f	5.28	S

*or 4.28

▶ $C_{57}H_{110}O_6$ ▶ 7% $CDCl_3$ ▶ V 368

▶ (3)-Ba(C)do(K)d

	Shift	Peak
a	1.13	3B
b	1.89	O
▶ c	2.74	O
d	3.42	4A
e	3.77	1A
f	3.98	3E
g	4.59	2A
h	5.59	S
i	5.63*	S
j	5.63*	S
k	7.47	1A

*or 6.12

▶ $C_{15}H_{16}O_6$ ▶ 7% $CDCl_3$ ▶ V 641

▶ (3)-(B)b(B)cU

	Shift	Peak
▶ a	~2.72	O
b	3.18	1H
c	3.18	1H

▶ C_7H_9NO ▶ 7% $CDCl_3$ ▶ V 173

▶ (3)-(B)k(C)kqKh

	Shift	Peak
a	2.98	2E
▶ b	3.86	4A
c	5.24	2A

▶ $C_6H_6O_6$ ▶ D_2O ▶ V 456

▶ (3)-(B)b(B)bVhh

	Shift	Peak
a	0.68 ± .05	S
b	0.74 ± .05	S
▶ c	1.80	O
d	7.05	2A

▶ C_9H_{10} ▶ 7% $CDCl_3$ ▶ V 528

▶ (3)-(B)o(C)bo(O)c

	Shift	Peak
a	~2.72	O
▶ b	3.18	1H
▶ c	3.18	1H

▶ C_7H_9NO ▶ 7% $CDCl_3$ ▶ V 173

▶ 3-BaCaaCaa

	Shift	Peak
a	.86	2H
b	.95	3D
c	~1.2	S
▶ d	~1.69	S

▶ C_9H_{20} ▶ 50% CCl_4 ▶ API 217

▶ (3)-(B)lCaa(O)l

	Shift	Peak
a	0.97*	2A
b	0.97*	2A
c	2.00	6A
d	~3.20	S
e	3.87†	1A
f	3.87†	1A
▶ g	4.72	4C
h	8.05	4A

*or 1.02
†or 3.90

▶ $C_{16}H_{19}NO_3$ ▶ 7% $CDCl_3$ ▶ V 325

▶ (3)-(B)l(C)bd(D)aac

	Shift	Peak
▶ a	~0.70	S
b	0.77	1A
c	1.03	1A
d	1.62	1I
e	5.23	1C

▶ $C_{10}H_{16}$ ▶ 7% $CDCl_3$ ▶ V 273

▶ 3-(B)o(C)bo(O)c

	Shift	Peak
a	3.64	2H
▶ b	3.79	2H
c	4.01	4E

▶ $C_4H_6O_2$ ▶ 7% $CDCl_3$ ▶ V 405

▶ (3)-(B)bCbv(J)b

	Shift	Peak
▶ a	2.66	S
b	3.75	6A
c	4.63	S
d	4.91	S

▶ $C_{14}H_{17}NO_3$ ▶ 7% $CDCl_3$ ▶ V 632

▶ (3)-Bp(C)bp(O)c

	Shift	Peak
a	1.91	2D
b	2.40	4A
c	3.79	2A
d	3.85	1A
▶ e	4.03	N
f	4.49	4E
g	6.28	3A
h	7.65	2D

▶ $C_{10}H_{14}N_2O_5$ ▶ D_2O ▶ V 566

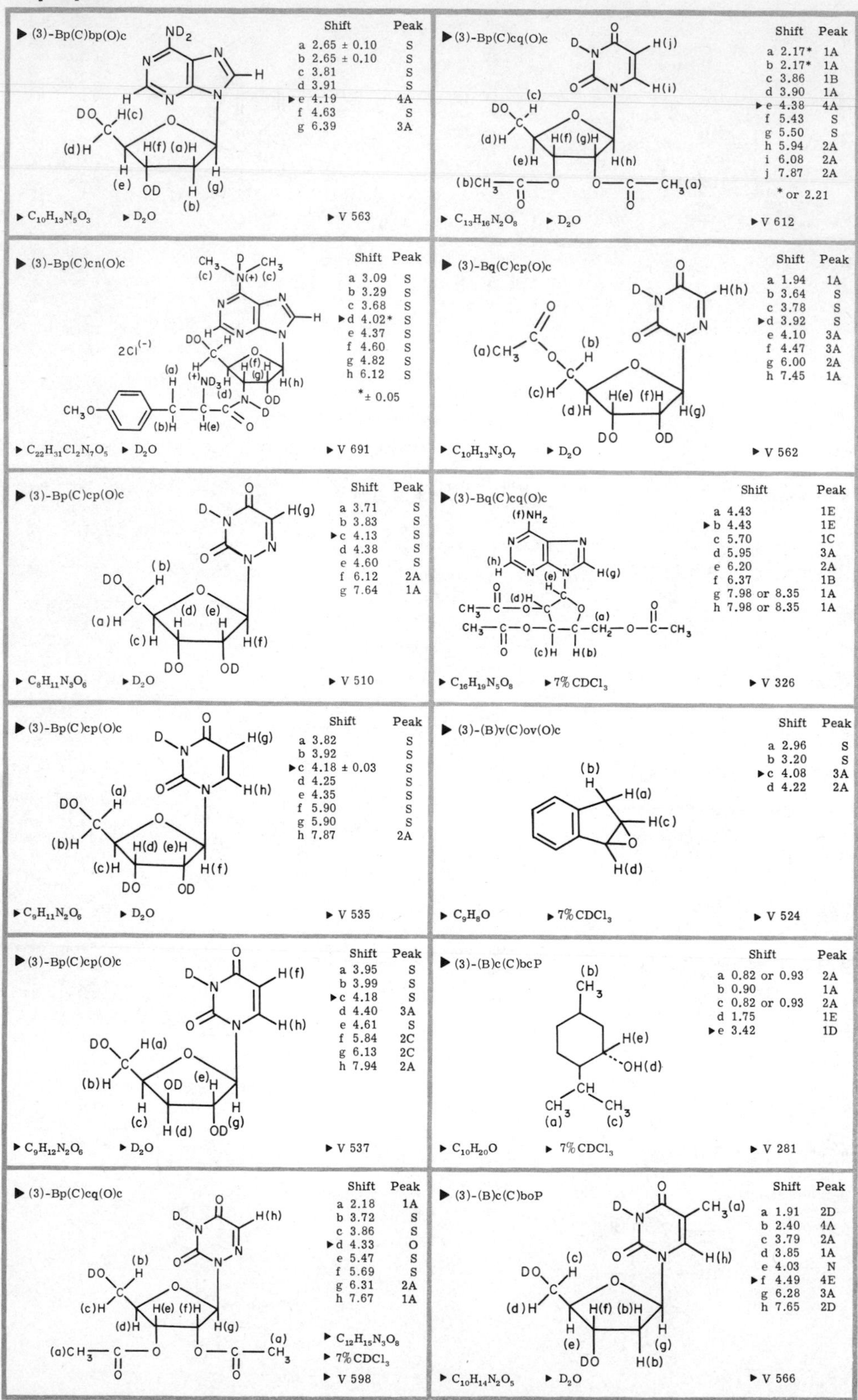
(3)-Bp(C)bp(O)c
Shift Peak
a 2.65 ± 0.10 S
b 2.65 ± 0.10 S
c 3.81 S
d 3.91 S
e 4.19 4A
f 4.63 S
g 6.39 3A
C10H13N5O3
D2O
V 563
(3)-Bp(C)cq(O)c
Shift Peak
a 2.17* 1A
b 2.17* 1A
c 3.86 1B
d 3.90 1A
e 4.38 4A
f 5.43 S
g 5.50 S
h 5.94 2A
i 6.08 2A
j 7.87 2A
* or 2.21
C13H16N2O8
D2O
V 612
(3)-Bp(C)cn(O)c
Shift Peak
a 3.09 S
b 3.29 S
c 3.68 S
d 4.02* S
e 4.37 S
f 4.60 S
g 4.82 S
h 6.12 S
*± 0.05
2Cl(-)
C22H31Cl2N7O5
D2O
V 691
(3)-Bq(C)cp(O)c
Shift Peak
a 1.94 1A
b 3.64 S
c 3.78 S
d 3.92 S
e 4.10 3A
f 4.47 3A
g 6.00 2A
h 7.45 1A
C10H13N3O7
D2O
V 562
(3)-Bp(C)cp(O)c
Shift Peak
a 3.71 S
b 3.83 S
c 4.13 S
d 4.38 S
e 4.60 S
f 6.12 2A
g 7.64 1A
C8H11N3O6
D2O
V 510
(3)-Bq(C)cq(O)c
Shift Peak
a 4.43 1E
b 4.43 1E
c 5.70 1C
d 5.95 3A
e 6.20 2A
f 6.37 1B
g 7.98 or 8.35 1A
h 7.98 or 8.35 1A
C16H19N5O8
7% CDCl3
V 326
(3)-Bp(C)cp(O)c
Shift Peak
a 3.82 S
b 3.92 S
c 4.18 ± 0.03 S
d 4.25 S
e 4.35 S
f 5.90 S
g 5.90 S
h 7.87 2A
C9H11N2O6
D2O
V 535
(3)-(B)v(C)ov(O)c
Shift Peak
a 2.96 S
b 3.20 S
c 4.08 3A
d 4.22 2A
C9H8O
7% CDCl3
V 524
(3)-Bp(C)cp(O)c
Shift Peak
a 3.95 S
b 3.99 S
c 4.18 S
d 4.40 3A
e 4.61 S
f 5.84 2C
g 6.13 2C
h 7.94 2A
C9H12N2O6
D2O
V 537
(3)-(B)c(C)bcP
Shift Peak
a 0.82 or 0.93 2A
b 0.90 1A
c 0.82 or 0.93 2A
d 1.75 1E
e 3.42 1D
C10H20O
7% CDCl3
V 281
(3)-Bp(C)cq(O)c
Shift Peak
a 2.18 1A
b 3.72 S
c 3.86 S
d 4.33 O
e 5.47 S
f 5.69 S
g 6.31 2A
h 7.67 1A
C12H15N3O8
7% CDCl3
V 598
(3)-(B)c(C)boP
Shift Peak
a 1.91 2D
b 2.40 4A
c 3.79 2A
d 3.85 1A
e 4.03 N
f 4.49 4E
g 6.28 3A
h 7.65 2D
C10H14N2O5
D2O
V 566

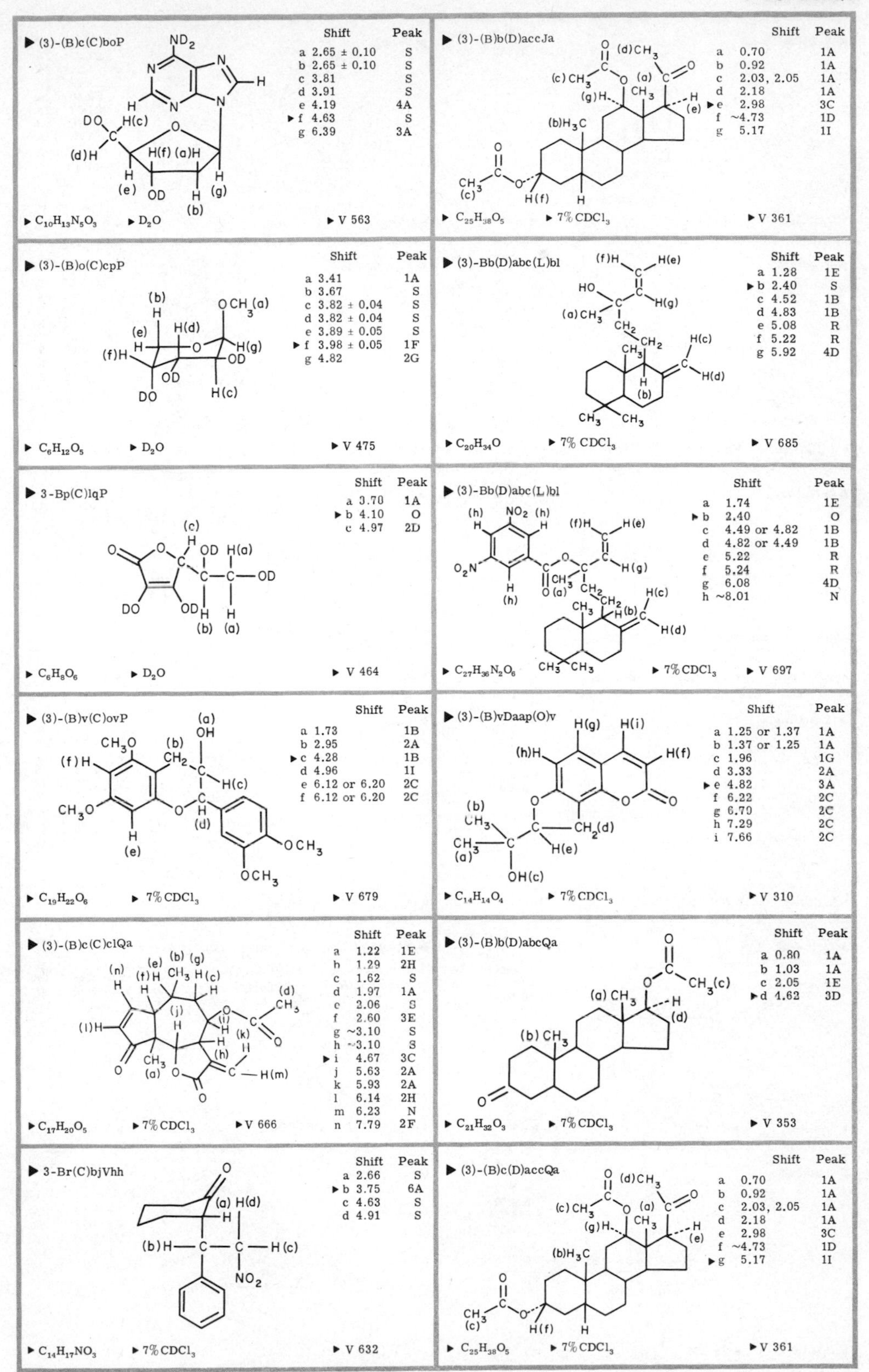
(3)-(B)c(C)boP
Shift Peak
a 2.65 ± 0.10 S
b 2.65 ± 0.10 S
c 3.81 S
d 3.91 S
e 4.19 4A
f 4.63 S
g 6.39 3A
C10H13N5O5 D2O V 563
(3)-(B)b(D)accJa
Shift Peak
a 0.70 1A
b 0.92 1A
c 2.03, 2.05 1A
d 2.18 1A
e 2.98 3C
f ~4.73 1D
g 5.17 1I
C25H38O5 7% CDCl3 V 361
(3)-(B)o(C)cpP
Shift Peak
a 3.41 1A
b 3.67 S
c 3.82 ± 0.04 S
d 3.82 ± 0.04 S
e 3.89 ± 0.05 S
f 3.98 ± 0.05 1F
g 4.82 2G
C6H12O5 D2O V 475
(3)-Bb(D)abc(L)bl
Shift Peak
a 1.28 1E
b 2.40 S
c 4.52 1B
d 4.83 1B
e 5.08 R
f 5.22 R
g 5.92 4D
C20H34O 7% CDCl3 V 685
3-Bp(C)lqP
Shift Peak
a 3.78 1A
b 4.10 O
c 4.97 2D
C6H8O6 D2O V 464
(3)-Bb(D)abc(L)bl
Shift Peak
a 1.74 1E
b 2.40 O
c 4.49 or 4.82 1B
d 4.82 or 4.49 1B
e 5.22 R
f 5.24 R
g 6.08 4D
h ~8.01 N
C27H36N2O6 7% CDCl3 V 697
(3)-(B)v(C)ovP
Shift Peak
a 1.73 1B
b 2.95 2A
c 4.28 1B
d 4.96 1I
e 6.12 or 6.20 2C
f 6.12 or 6.20 2C
C19H22O6 7% CDCl3 V 679
(3)-(B)vDaap(O)v
Shift Peak
a 1.25 or 1.37 1A
b 1.37 or 1.25 1A
c 1.96 1G
d 3.33 2A
e 4.82 3A
f 6.22 2C
g 6.70 2C
h 7.29 2C
i 7.66 2C
C14H14O4 7% CDCl3 V 310
(3)-(B)c(C)clQa
Shift Peak
a 1.22 1E
b 1.29 2H
c 1.62 S
d 1.97 1A
e 2.06 S
f 2.60 3E
g ~3.10 S
h ~3.10 S
i 4.67 3C
j 5.63 2A
k 5.93 2A
l 6.14 2H
m 6.23 N
n 7.79 2F
C17H20O5 7% CDCl3 V 666
(3)-(B)b(D)abcQa
Shift Peak
a 0.80 1A
b 1.03 1A
c 2.05 1E
d 4.62 3D
C21H32O3 7% CDCl3 V 353
3-Br(C)bjVhh
Shift Peak
a 2.66 S
b 3.75 6A
c 4.63 S
d 4.91 S
C14H17NO3 7% CDCl3 V 632
(3)-(B)c(D)accQa
Shift Peak
a 0.70 1A
b 0.92 1A
c 2.03, 2.05 1A
d 2.18 1A
e 2.98 3C
f ~4.73 1D
g 5.17 1I
C25H38O5 7% CDCl3 V 361

▶ 3-BeEE

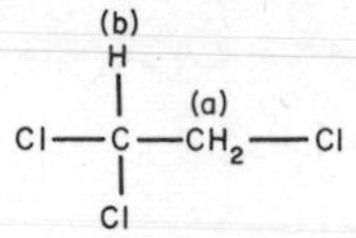

Shift	Peak
a 3.95	2A
▶b 5.77	3A

▶ $C_2H_3Cl_3$ ▶ 7% $CDCl_3$ ▶ V 2

▶ 3-BkJaKb

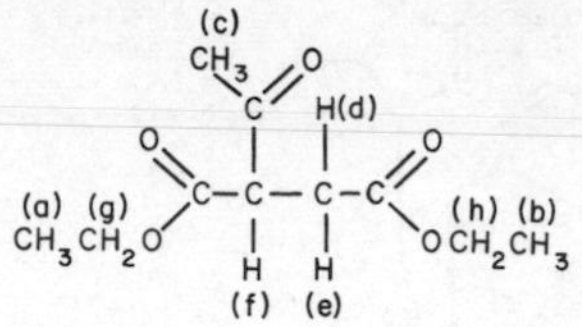

Shift	Peak
a 1.25*	3A
b 1.25*	3A
c 2.35	1A
d 2.86	2A
e 2.89	2A
▶f 3.98	S
g 4.13†	4A
h 4.13†	4A

*or 1.27
†or 4.22

▶ $C_{10}H_{16}O_5$ ▶ 7% $CDCl_3$ ▶ V 277

▶ 3-BeEOb

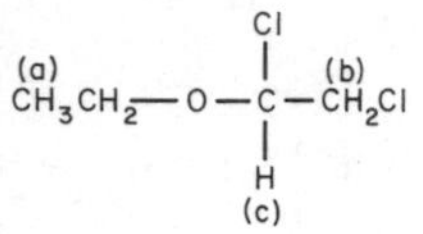

Shift	Peak
a 1.28	3A
b 3.70	O
▶c 5.63	3A

▶ $C_4H_8Cl_2O$ ▶ 7% $CDCl_3$ ▶ V 75

▶ 3-BvJnNhhh

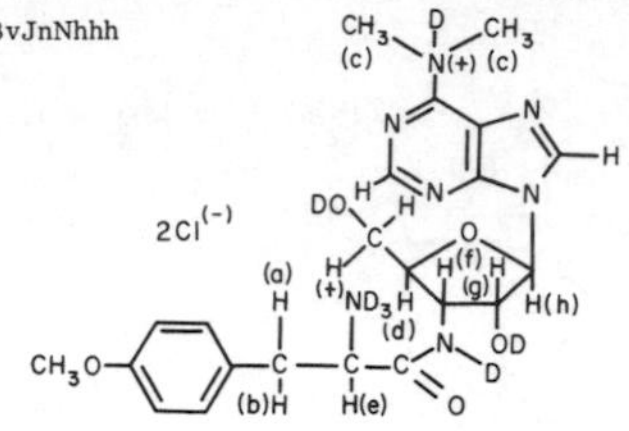

Shift	Peak
a 3.09	S
b 3.29	S
c 3.68	S
d 4.02*	S
▶e 4.37	S
f 4.60	S
g 4.82	S
h 6.12	S

*± 0.05

▶ $C_{22}H_{31}Cl_2N_7O_5$ ▶ D_2O ▶ V 691

▶ 3-BaER

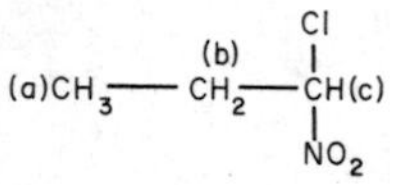

Shift	Peak
a 1.10	3A
b 2.32	5B
▶c 5.80	3A

▶ $C_3H_6ClNO_2$ ▶ 7% $CDCl_3$ ▶ V 385

▶ (3)-(B)d(J)b(O)b

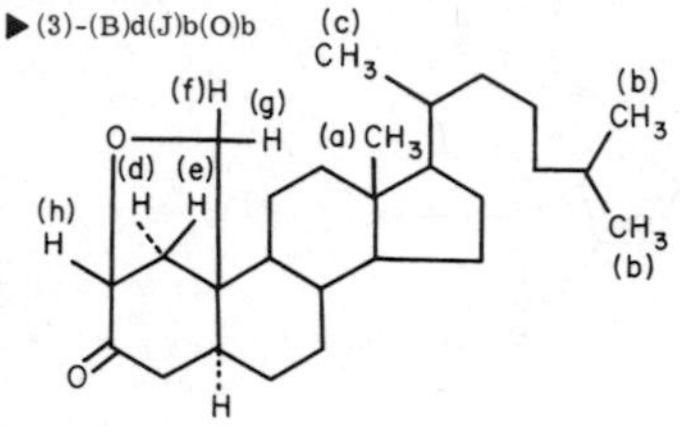

Shift	Peak
a 0.66	1A
b 0.88	S
c 0.95	S
d 1.48	S
e 2.53	S
f 3.90	S
g 4.03	S
▶h 4.12	S

▶ $C_{27}H_{44}O_2$ ▶ 7% $CDCl_3$ ▶ V 698

▶ 3-BaGKb

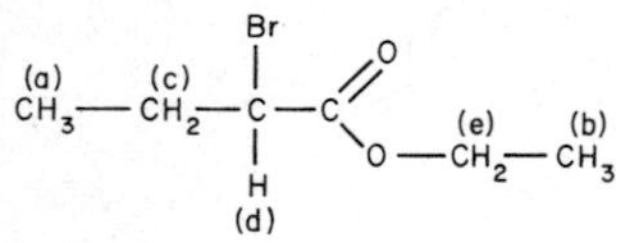

Shift	Peak
a 1.03	3A
b 1.30	3A
c 2.08	5B
▶d 4.20	S
e 4.27	4E

▶ $C_6H_{11}BrO_2$ ▶ 7% $CDCl_3$ ▶ V 137

▶ 3-BpJvOv

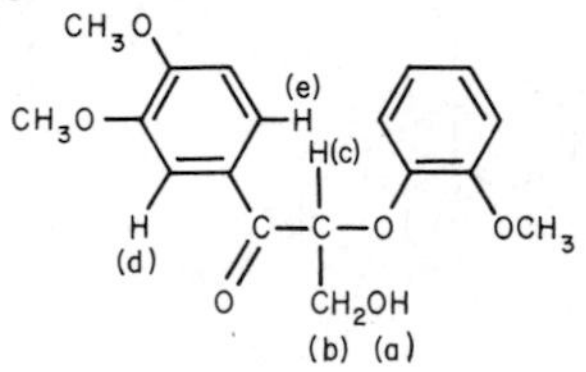

Shift	Peak
a 3.35	3B
b 4.10	S
▶c 5.41	3A
d 7.63	N
e 7.77	2A

▶ $C_{18}H_{20}O_6$ ▶ 7% $CDCl_3$ ▶ V 675

▶ 3-BaGKh

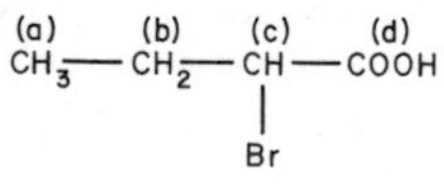

Shift	Peak
a 1.08	3A
b 2.07	5B
▶c 4.23	3A
d 10.97	1A

▶ $C_4H_7BrO_2$ ▶ 7% $CDCl_3$ ▶ V 66

▶ (3)-(B)bKh(L)bl

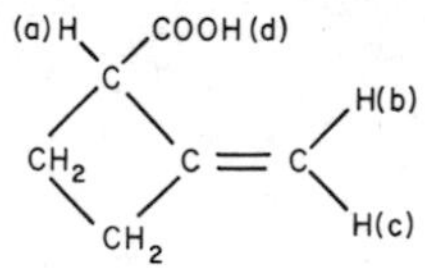

Shift	Peak
▶a 3.80	O
b 4.89 or 5.08	N
c 4.89 or 5.08	N
d 12.12	1A

▶ $C_6H_8O_2$ ▶ 7% $CDCl_3$ ▶ V 128

▶ 3-BgGVhh

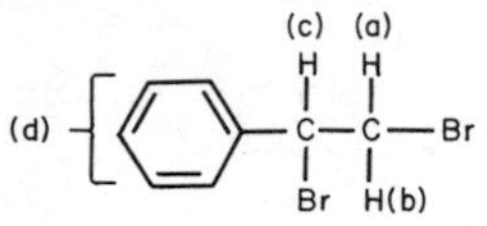

Shift	Peak
a 4.02	S
b 4.07	S
▶c 5.15	4A
d 7.38	1A

▶ $C_8H_8Br_2$ ▶ 7% $CDCl_3$ ▶ V 503

▶ 3-BvKaMl

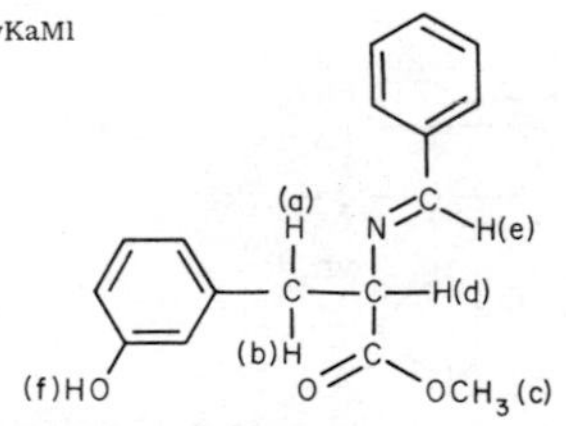

Shift	Peak
a 3.23	N
b 3.52	N
c 3.72	1A
▶d 4.28	4A
e 7.89	1A
f 8.04	1F

▶ $C_{17}H_{17}NO_3$ ▶ 7% $CDCl_3$ ▶ V 663

▶ 3-BbKNhh

	Shift	Peak
a	~2.12	S
b	~2.33	S
▶ c	3.77	3B

O (c) (a) (b) O
C—CH—CH2—CH2—C
(-)O ND2 O(-)
D3O(+) Na(+)

▶ $C_5H_8NNaO_4$ ▶ D_2O ▶ V 435

▶ 3-BvKbNhj

	Shift	Peak
a	1.27	3A
b	3.14	2A
c	4.20	4A
▶ d	4.96	4C
e	5.86	1A
f	6.62 ± 0.04	S

(f)H O
(e) O (b) (d) C (c) (a)
H2C CH2—CH O—CH2—CH3
O NH
H(f) C
H(f) O

▶ $C_{19}H_{19}NO_5$ ▶ 7% $CDCl_3$ ▶ V 334

▶ (3)-(B)bKh(N)bh

	Shift	Peak
a	2.88	4A
b	3.50	S
▶ c	3.75	S
d	4.01	S

H H(c)
H H(a) H OD
DO H N C
H(b) H D O
H
(d)

▶ $C_6H_{11}NO_3$ ▶ D_2O ▶ V 470

▶ 3-BvKhNhh

	Shift	Peak
a	2.86	S
b	3.07	S
▶ c	3.59	3C
d	7.35	1A

(b)H H(c) O
(d) C—C—C
H ND2 OD
(a)

▶ $C_9H_{11}NO_2$ ▶ D_2O ▶ V 534

▶ 3-BbKhNhj

	Shift	Peak
a	2.10	1A
b	2.25	O
c	2.66	O
▶ d	4.94	4B
e	7.14	3D
f	10.28	1B

O COOH(f)
C C—CH2CH2SCH3
N (b) (c) (a)
(e)H H(d)

▶ $C_{12}H_{15}NO_3S$ ▶ 7% $CDCl_3$ ▶ V 597

▶ 3-BkKhSh

	Shift	Peak
a	2.92	2D
b	3.02	2D
▶ c	3.85	3A

(a)
DS H O
(c)
CH—C—C
O=C OD
(b)H
OD

▶ $C_4H_6O_4S$ ▶ D_2O ▶ V 407

▶ (3)-(B)cKh(N)bh

	Shift	Peak
a	3.32	3A
b	3.32	3A
▶ c	3.93	4A
d	4.22	5A

OD H(c)
(d) H(a) H OD
H N C
H(b) H O
H D

▶ $C_6H_{11}NO_3$ ▶ 7% $CDCl_3$ ▶ V 469

▶ (3)-(B)c(L)hl(L)hl

	Shift	Peak
a	2.00	3A
▶ b	3.58	5A
c	6.75	3A

H(b)
(c)H H(c)
H—H
(a) (a)
(c)H H(c)
H(b)

▶ C_7H_8 ▶ 7% $CDCl_3$ ▶ V 487

▶ 3-BkKhNhh

	Shift	Peak
a	2.83	2A
b	2.90	1A
▶ c	3.98	4A

(b)
O ND2 H OD
C—C—C—C
DO H H O
(c) (a)

▶ $C_4H_7NO_4$ ▶ D_2O ▶ V 410

▶ (3)-(B)b(L)clNhj

	Shift	Peak
a	2.07	1G
b	3.63	2C
c	4.12	2C
▶ d	4.84	1F
e	6.02	2E
f	6.49	1B
g	6.68	2A

(e)H O
(f)H N—C
CH3O CH3(a)
CH3O H(d)
CH3O H(b)
H
(c)
(g)H O
OCH3

▶ $C_{22}H_{25}NO_6$ ▶ 7% $CDCl_3$ ▶ V 690

▶ 3-BlKhNhh

	Shift	Peak
a	3.10	S
b	3.33	S
▶ c	3.75	O
d	7.28	S

(a)H H(b) O
C (c) C
CH OD
ND2
N H
D (d)

▶ $C_{11}H_{12}N_2O_2$ ▶ D_2O ▶ V 582

▶ (3)-(B)b(L)llNhj

	Shift	Peak
a	1.96	1A
b	2.43	S
c	2.43	S
d	3.67	1A
▶ e	4.62	S
f	6.55	1A
g	6.92	2A
h	7.37	S
i	7.63	1B
j	8.38	2G

O CH3(a)
(f) (c) (b) C
H H2C CH2 N
H(j)
H(e)
CH3O H(i)
CH3O CH3O H O
(g)
(h)H (d)
OCH3

▶ $C_{22}H_{25}NO_6$ ▶ 7% $CDCl_3$ ▶ V 689

▶ (3)-(B)c(L)hlVco

	Shift	Peak
a	2.22	3A
b	3.79	1A
▶ c	4.17	5A
d	6.50	1A
e	6.83	3A

▶ $C_{13}H_{14}O_2$ ▶ 7% $CDCl_3$ ▶ V 608

▶ (3)-(B)o(O)bVhh

	Shift	Peak
a	2.77	4A
b	3.12	4A
▶ c	3.83	4A
d	7.28	1A

▶ C_8H_8O ▶ 7% $CDCl_3$ ▶ V 193

▶ (3)-(B)c(N)jl(O)c

	Shift	Peak
a	1.91	2D
b	2.40	4A
c	3.79	2A
d	3.85	1A
e	4.03	N
f	4.49	4E
▶ g	6.28	3A
h	7.65	2D

▶ $C_{10}H_{14}N_2O_5$ ▶ D_2O ▶ V 566

▶ 3-BdVhhZ

	Shift	Peak
a	0.88	3C
b	~1.36	S
c	~1.50	S
d	2.86*	S
▶ e	2.89*	S
f	3.65	S
g	3.88	3C
h	7.26	1A

*±0.04

▶ $C_{17}H_{26}BCl_3O_2$ ▶ 7% $CDCl_3$ ▶ V 668

▶ (3)-(B)c(N)lx(O)c

	Shift	Peak
a	2.65 ± 0.10	S
b	2.65 ± 0.10	S
c	3.81	S
d	3.91	S
e	4.19	4A
f	4.63	S
▶ g	6.39	3A

▶ $C_{10}H_{13}N_5O_3$ ▶ D_2O ▶ V 563

▶ (3)-(C)cd(C)dv(D)del

	Shift	Peak
▶ a	3.53	2A
b	3.97	2A

▶ $C_{20}H_8Cl_{12}$ ▶ 7% $CDCl_3$ ▶ V 338

▶ (3)-Bc(O)b(O)c

	Shift	Peak
a	1.22	2A
b	1.92	3A
c	2.68±.02	S
d	2.80±.10	S
e	4.05	S
▶ f	4.64	3A

▶ $C_{11}H_{20}O_4$ ▶ 7% $CDCl_3$ ▶ V 588

▶ (3)-(C)ll(C)oo(D)clq

	Shift	Peak
a	1.13	3B
b	1.89	O
c	2.74	O
▶ d	3.42	4A
e	3.77	1A
f	3.98	3E
g	4.59	2A
h	5.59	S
i	5.63*	S
j	5.63*	S
k	7.47	1A

*or 6.12

▶ $C_{15}H_{16}O_6$ ▶ 7% $CDCl_3$ ▶ V 641

▶ 3-BgOaOa

	Shift	Peak
a	3.37	1A
b	3.40	1A
▶ c	4.54	3A

▶ $C_4H_9BrO_2$ ▶ 7% $CDCl_3$ ▶ V 419

▶ (3)-(C)ll(C)oo(D)clq

	Shift	Peak
a	2.11	2A
▶ b	3.46	4A
c	3.79	1A
d	4.02	3E
e	5.12	1B
f	5.59	S
g	5.67*	S
h	5.67*	S
i	7.19	O
j	7.46	1A

*or 6.08

▶ $C_{15}H_{14}O_6$ ▶ 7% $CDCl_3$ ▶ V 640

▶ (3)-(B)l(O)vVhh

	Shift	Peak
a	2.32	1A
b	2.32	1A
c	3.09	S
d	3.09	A
▶ e	5.18	3B
f	6.06	M
g	6.62*	S
h	6.62*	S
i	7.17	Q
j	7.46	Q

*or 6.71

▶ $C_{21}H_{16}O_7$ ▶ 7% $CDCl_3$ ▶ V 686

▶ 3-CeeCeeE

	Shift	Peak
▶ a	4.52	3A
b	6.07	2A

▶ $C_3H_3Cl_5$ ▶ 7% $CDCl_3$ ▶ V 377

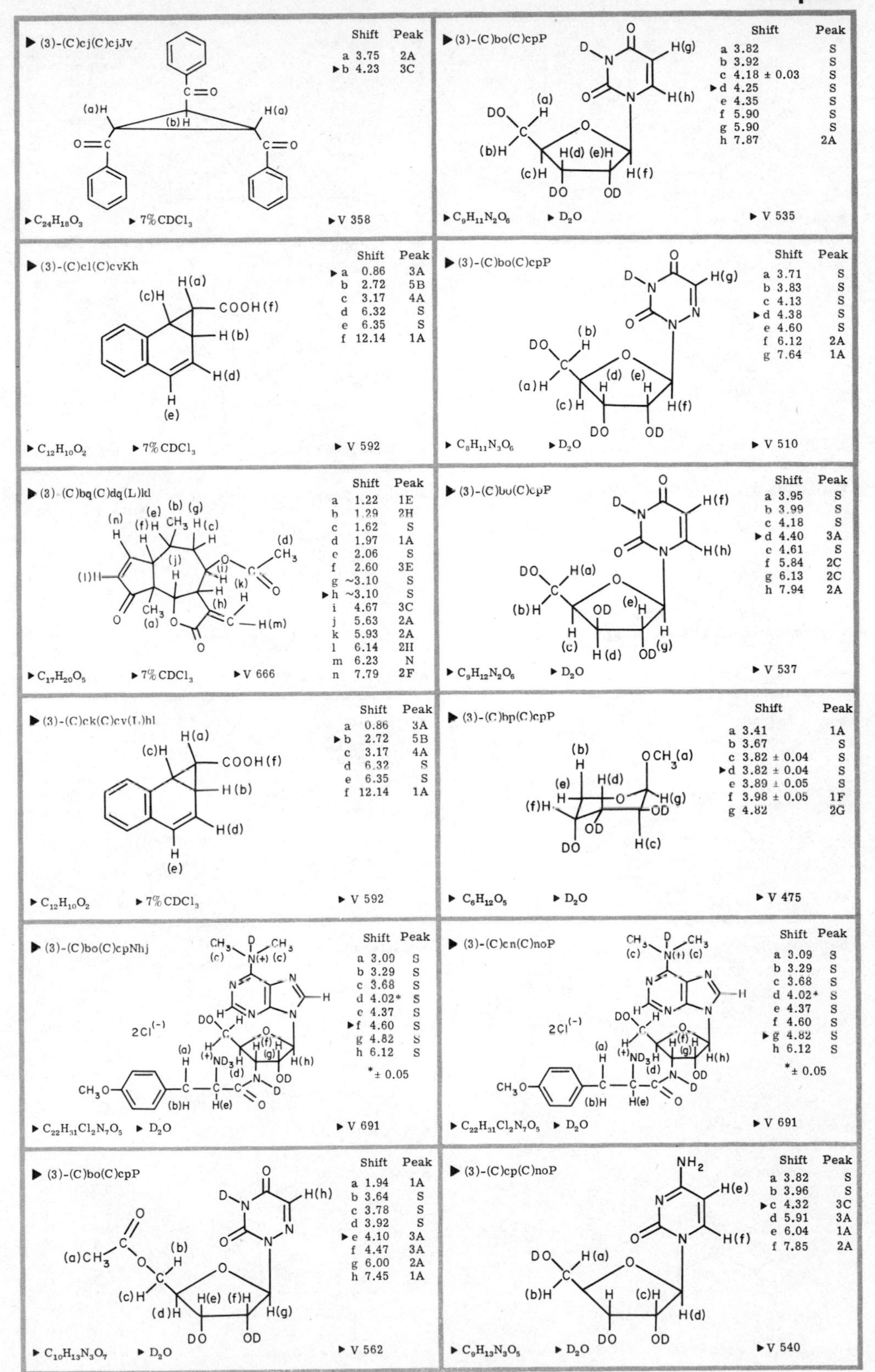

▶ (3)-(C)cj(C)cjJv

	Shift	Peak
a	3.75	2A
▶b	4.23	3C

▶ $C_{24}H_{18}O_3$ ▶ 7% $CDCl_3$ ▶ V 358

▶ (3)-(C)bo(C)cpP

	Shift	Peak
a	3.82	S
b	3.92	S
c	4.18 ± 0.03	S
▶d	4.25	S
e	4.35	S
f	5.90	S
g	5.90	S
h	7.87	2A

▶ $C_9H_{11}N_2O_6$ ▶ D_2O ▶ V 535

▶ (3)-(C)cl(C)cvKh

	Shift	Peak
▶a	0.86	3A
b	2.72	5B
c	3.17	4A
d	6.32	S
e	6.35	S
f	12.14	1A

▶ $C_{12}H_{10}O_2$ ▶ 7% $CDCl_3$ ▶ V 592

▶ (3)-(C)bo(C)cpP

	Shift	Peak
a	3.71	S
b	3.83	S
c	4.13	S
▶d	4.38	S
e	4.60	S
f	6.12	2A
g	7.64	1A

▶ $C_8H_{11}N_3O_6$ ▶ D_2O ▶ V 510

▶ (3)-(C)bq(C)dq(L)kl

	Shift	Peak
a	1.22	1E
b	1.29	2H
c	1.62	S
d	1.97	1A
e	2.06	S
f	2.60	3E
g	~3.10	S
▶h	~3.10	S
i	4.67	3C
j	5.63	2A
k	5.93	2A
l	6.14	2H
m	6.23	N
n	7.79	2F

▶ $C_{17}H_{20}O_5$ ▶ 7% $CDCl_3$ ▶ V 666

▶ (3)-(C)bo(C)cpP

	Shift	Peak
a	3.95	S
b	3.99	S
c	4.18	S
▶d	4.40	3A
e	4.61	S
f	5.84	2C
g	6.13	2C
h	7.94	2A

▶ $C_9H_{12}N_2O_6$ ▶ D_2O ▶ V 537

▶ (3)-(C)ck(C)cv(L)hl

	Shift	Peak
a	0.86	3A
▶b	2.72	5B
c	3.17	4A
d	6.32	S
e	6.35	S
f	12.14	1A

▶ $C_{12}H_{10}O_2$ ▶ 7% $CDCl_3$ ▶ V 592

▶ (3)-(C)bp(C)cpP

	Shift	Peak
a	3.41	1A
b	3.67	S
c	3.82 ± 0.04	S
▶d	3.82 ± 0.04	S
e	3.89 ± 0.05	S
f	3.98 ± 0.05	1F
g	4.82	2G

▶ $C_6H_{12}O_5$ ▶ D_2O ▶ V 475

▶ (3)-(C)bo(C)cpNhj

	Shift	Peak
a	3.09	S
b	3.29	S
c	3.68	S
d	4.02*	S
e	4.37	S
▶f	4.60	S
g	4.82	S
h	6.12	S

*± 0.05

▶ $C_{22}H_{31}Cl_2N_7O_5$ ▶ D_2O ▶ V 691

▶ (3)-(C)cn(C)noP

	Shift	Peak
a	3.09	S
b	3.29	S
c	3.68	S
d	4.02*	S
e	4.37	S
f	4.60	S
▶g	4.82	S
h	6.12	S

*± 0.05

▶ $C_{22}H_{31}Cl_2N_7O_5$ ▶ D_2O ▶ V 691

▶ (3)-(C)bo(C)cpP

	Shift	Peak
a	1.94	1A
b	3.64	S
c	3.78	S
d	3.92	S
▶e	4.10	3A
f	4.47	3A
g	6.00	2A
h	7.45	1A

▶ $C_{10}H_{13}N_3O_7$ ▶ D_2O ▶ V 562

▶ (3)-(C)cp(C)noP

	Shift	Peak
a	3.82	S
b	3.96	S
▶c	4.32	3C
d	5.91	3A
e	6.04	1A
f	7.85	2A

▶ $C_9H_{13}N_3O_5$ ▶ D_2O ▶ V 540

3-CcpCnoP

▶ (3)-(C)cp(C)noP

	Shift	Peak
a	3.82	S
b	3.92	S
c	4.18 ± 0.03	S
d	4.25	S
▶e	4.35	S
f	5.90	S
g	5.90	S
h	7.87	2A

▶ $C_9H_{11}N_2O_6$ ▶ D_2O ▶ V 535

▶ (3)-(C)bo(C)cqQa

	Shift	Peak
a	2.18	1A
b	3.72	S
c	3.86	S
d	4.33	O
▶e	5.47	S
f	5.69	S
g	6.31	2A
h	7.67	1A

▶ $C_{12}H_{15}N_3O_8$ ▶ 7% $CDCl_3$ ▶ V 598

▶ (3)-(C)cp(C)noP

	Shift	Peak
a	1.94	1A
b	3.64	S
c	3.78	S
d	3.92	S
e	4.10	3A
▶f	4.47	3A
g	6.00	2A
h	7.45	1A

▶ $C_{10}H_{13}N_3O_7$ ▶ D_2O ▶ V 562

▶ (3)-(C)bo(C)cqQa

	Shift	Peak
a	2.10*	1A
b	4.42	1B
▶c	5.62	S
d	5.88	3A
e	6.17	2A
f	8.03†	1A
g	8.03†	1A
h	~13.1	S

*or 2.15
†or 8.35

▶ $C_{16}H_{18}N_4O_8$ ▶ 7% $CDCl_3$ ▶ V 324

▶ (3)-(C)cp(C)noP

	Shift	Peak
a	3.71	S
b	3.83	S
c	4.13	S
d	4.38	S
▶e	4.60	S
f	6.12	2A
g	7.64	1A

▶ $C_8H_{11}N_3O_6$ ▶ D_2O ▶ V 510

▶ (3)-(C)bo(C)cqQa

	Shift	Peak
a	4.43	1E
b	4.43	1E
▶c	5.70	1C
d	5.95	3A
e	6.20	2A
f	6.37	1B
g	7.98 or 8.35	1A
h	7.98 or 8.35	1A

▶ $C_{16}H_{19}N_5O_8$ ▶ 7% $CDCl_3$ ▶ V 326

▶ (3)-(C)cp(C)noP

	Shift	Peak
a	3.95	S
b	3.99	S
c	4.18	S
d	4.40	3A
▶e	4.61	S
f	5.84	2C
g	6.13	2C
h	7.94	2A

▶ $C_9H_{12}N_2O_6$ ▶ D_2O ▶ V 537

▶ (3)-(C)cm(C)ooQa

	Shift	Peak
a	1.92	1A
b	3.47	1A
c	5.07	2A
d	5.58	1E
▶e	~5.63	4E

▶ $C_{22}H_{24}N_2O_6$ ▶ 7% $CDCl_3$ ▶ V 356

▶ (3)-(C)cp(C)ooP

	Shift	Peak
a	3.41	1A
b	3.67	S
▶c	3.82 ± 0.04	S
d	3.82 ± 0.04	S
e	3.89 ± 0.05	S
f	3.98 ± 0.05	1F
g	4.82	2G

▶ $C_6H_{12}O_5$ ▶ D_2O ▶ V 475

▶ (3)-(C)cq(C)noQa

	Shift	Peak
a	2.13	1A
b	2.13	1A
c	4.28	S
d	4.32	S
▶e	4.62	S
f	6.14	2A
g	7.67	1A

▶ $C_{14}H_{17}N_3O_9$ ▶ D_2O ▶ V 634

▶ (3)-(C)bo(C)cqQa

	Shift	Peak
a	2.17*	1A
b	2.17*	1A
c	3.86	1B
d	3.90	1A
e	4.38	4A
▶f	5.43	S
g	5.50	S
h	5.94	2A
i	6.08	2A
j	7.87	2A

*or 2.21

▶ $C_{13}H_{16}N_2O_8$ ▶ D_2O ▶ V 612

▶ (3)-(C)cq(C)noQa

	Shift	Peak
a	2.17*	1A
b	2.17*	1A
c	3.86	1B
d	3.90	1A
e	4.38	4A
f	5.43	S
▶g	5.50	S
h	5.94	2A
i	6.08	2A
j	7.87	2A

*or 2.21

▶ $C_{13}H_{16}N_2O_8$ ▶ D_2O ▶ V 612

▶ (3)-(C)cq(C)noQa

D—N, O, H(h), N, N, O, (b), DO, H, C, O, (c)H, H(e) (f)H, (d)H, H(g), (a), (a)CH_3—C—O, O—C—CH_3, O, O

	Shift	Peak
a	2.18	1A
b	3.72	S
c	3.86	S
d	4.33	O
e	5.47	S
▸f	5.69	S
g	6.31	2A
h	7.67	1A

▸ $C_{12}H_{15}N_3O_8$ ▸ 7% $CDCl_3$ ▸ V 598

▶ (3)-(C)dj(D)aac(L)hl

(e) H, (d) H, O, O, (b) CH_3, (b) CH_3—C, (a) CH_3, (c) H, (g)H, H(f)

	Shift	Peak
a	0.89	1A
b	1.09	1A
▸c	1.17	S
d	1.35	S
e	4.16	1E
f	5.07	2A
g	5.73	4C

▸ $C_{23}H_{22}O_2$ ▸ 7% $CDCl_3$ ▸ V 693

▶ (3)-(C)cq(C)noQa

O, (h)H—N, N, (g)H, N, (e), N, H(f), H, O, (d)H, O, (a)CH_3—C—O—C, O, (a)CH_3—C—O—C—C—CH_2O C CH_3, (b), (a), O, (c)H, H(b)

	Shift	Peak
a	2.10*	1A
b	4.42	1B
c	5.62	S
▸d	5.88	3A
e	6.17	2A
f	8.03†	1A
g	8.03†	1A
h	~13.1	S

*or 2.15
†or 8.35

▸ $C_{16}H_{18}N_4O_8$ ▸ 7% $CDCl_3$ ▸ V 324

▶ (3)-(C)bk(D)clq(O)c

O, C, OCH_3 (e), (f)H, (i)H, H(k), H, (d), O, (j)H, (h), H, O, (g), H, O, O, CH_2—CH_3, (c)H, (b), (a)

	Shift	Peak
a	1.13	3B
b	1.89	O
c	2.74	O
d	3.42	4A
e	3.77	1A
f	3.98	3E
▸g	4.59	2A
h	5.59	S
i	5.63*	S
j	5.63*	S
k	7.47	1A

*or 6.12

▸ $C_{15}H_{16}O_6$ ▸ 7% $CDCl_3$ ▸ V 641

▶ (3)-(C)cq(C)noQa

(f)NH_2, N, N, (h) H, N, N, H(g), (e), H, O, (d)H, O, CH_3—C—O—C, (a), O, CH_3—C—O—C—C—CH_2—O—C—CH_3, O, (c)H, H(b)

	Shift	Peak
a	4.43	1E
b	4.43	1E
c	5.70	1C
▸d	5.95	3A
e	6.20	2A
f	6.37	1B
g	7.98 or 8.35	1A
h	7.98 or 8.35	1A

▸ $C_{16}H_{19}N_5O_8$ ▸ 7% $CDCl_3$ ▸ V 326

▶ 3-CaaDaabP

(e) H, (a) CH_3, (f) H, (c) CH_3, (d) CH—C—C—C—OH (h), CH_3 (b), OH (h), CH_3 (a), H (g)

	Shift	Peak
a	0.92	1A
b	0.95 or 1.00	2A
c	0.95 or 1.00	2A
d	1.92	5B
▸e	3.38	2A
f	3.44 or 3.53	S
g	3.44 or 3.53	S
h	3.58	1A

▸ $C_8H_{18}O_2$ ▸ 7% $CDCl_3$ ▸ V 217

▶ (3)-(C)ck(C)clVhl

H(a), (c)H, COOH(f), H(b), H(d), H, (e)

	Shift	Peak
a	0.86	3A
b	2.72	5B
▸c	3.17	4A
d	6.33	S
e	6.35	S
f	12.14	1A

▸ $C_{12}H_{10}O_2$ ▸ 7% $CDCl_3$ ▸ V 592

▶ (3)-(C)cl(D)acj(Q)l

(n) H, (e) (f)H, (b) CH_3 (g) H(c), H, H, (d) CH_3, (j), O, (i), H, C, (k), (l)H, H, O, H, H, (h), CH_3, (a), O, O, C—H(m), O, O

	Shift	Peak
a	1.22	1E
b	1.29	2H
c	1.62	S
d	1.97	1A
e	2.06	S
f	2.60	3E
g	~3.10	S
h	~3.10	S
i	4.67	3C
▸j	5.63	2A
k	5.93	2A
l	6.14	2H
m	6.23	N
n	7.79	2F

▸ $C_{17}H_{20}O_5$ ▸ 7% $CDCl_3$ ▸ V 666

▶ (3)-(C)dl(D)aac(J)d

(e) H, (d) H, O, O, (b) CH_3, (b) CH_3—C, (a) CH_3, (c) H, (g)H, H(f)

	Shift	Peak
a	0.89	1A
b	1.09	1A
c	1.17	S
▸d	1.35	S
e	4.16	1E
f	5.07	2A
g	5.73	4C

▸ $C_{23}H_{22}O_2$ ▸ 7% $CDCl_3$ ▸ V 693

▶ (3)-(C)cd(D)delVch

Cl, Cl, Cl, Cl, Cl, H(b), Cl, (a)H, (a)H, Cl, Cl, Cl, H(b), Cl, Cl, Cl

	Shift	Peak
a	3.53	2A
▸b	3.97	2A

▸ $C_{20}H_8Cl_{12}$ ▸ 7% $CDCl_3$ ▸ V 338

▶ (3)-(C)ab(D)acj(L)hl

(n) H, (e) (f)H, (b) CH_3 (g) H(c), H, H, (d) CH_3, (j), O, (i), H, C, (k), (l)H, H, O, H, H, (h), CH_3, (a), O, O, C—H(m), O

	Shift	Peak
a	1.22	1E
b	1.29	2H
c	1.62	S
d	1.97	1A
e	2.06	S
▸f	2.60	3E
g	~3.10	S
h	~3.10	S
i	4.67	3C
j	5.63	2A
k	5.93	2A
l	6.14	2H
m	6.23	N
n	7.79	2F

▸ $C_{17}H_{20}O_5$ ▸ 7% $CDCl_3$ ▸ V 666

▶ 3-CceEE

(a), Cl H Cl, (b)H—C—C—C—H(b), Cl Cl Cl

	Shift	Peak
a	4.52	3A
▸b	6.07	2A

▸ $C_3H_3Cl_5$ ▸ 7% $CDCl_3$ ▸ V 377

▶ 3-CgjGJv

(b) meso, (a) dl

	Shift	Peak
▶ a	5.74	1A
b	5.93	1A

▶ $C_{16}H_{12}Br_2O_2$ ▶ 7% $CDCl_3$ ▶ V 648

▶ (3)-(C)cd(L)hl(L)kl

	Shift	Peak
a	1.13	3B
b	1.89	O
c	2.74	O
d	3.42	4A
e	3.77	1A
▶ f	3.98	3E
g	4.59	2A
h	5.59	S
i	5.63*	S
j	5.63*	S
k	7.47	1A

*or 6.12

▶ $C_{15}H_{16}O_6$ ▶ 7% $CDCl_3$ ▶ V 641

▶ 3-CgjGJv

(b) meso, (a) dl

	Shift	Peak
a	5.74	1A
▶ b	5.93	1A

▶ $C_{16}H_{12}Br_2O_2$ ▶ 7% $CDCl_3$ ▶ V 648

▶ (3)-(C)cd(L)hl(L)kl

	Shift	Peak
a	2.11	2A
b	3.46	4A
c	3.79	1A
▶ d	4.02	3E
e	5.12	1B
f	5.59	S
g	5.67*	S
h	5.67*	S
i	7.19	O
j	7.46	1A

*or 6.08

▶ $C_{15}H_{14}O_6$ ▶ 7% $CDCl_3$ ▶ V 640

▶ 3-CagGKh

	Shift	Peak
a	1.93	N
b	4.42 or 4.43	S
▶ c	4.42 or 4.43	S
d	11.58	1A

▶ $C_4H_6Br_2O_2$ ▶ 7% $CDCl_3$ ▶ V 403

▶ (3)-(C)jl(L)hl(L)lv

	Shift	Peak
a	2.07	1G
b	3.63	2C
▶ c	4.12	2C
d	4.84	1F
e	6.02	2E
f	6.49	1B
g	6.68	2A

▶ $C_{22}H_{25}NO_6$ ▶ 7% $CDCl_3$ ▶ V 690

▶ (3)-(C)ll(J)l(L)cl

	Shift	Peak
a	2.07	1G
▶ b	3.63	2C
c	4.12	2C
d	4.84	1F
e	6.02	2E
f	6.49	1B
g	6.68	2A

▶ $C_{22}H_{25}NO_6$ ▶ 7% $CDCl_3$ ▶ V 690

▶ 3-(C)aq(L)jlP

	Shift	Peak
a	1.65	2A
b	1.97	2E
c	3.86	1C
▶ d	3.92	S
e	4.32	O
f	5.87	1I
g	6.06	2A
h	6.99	4A

▶ $C_{12}H_{12}O_5$ ▶ 7% $CDCl_3$ ▶ V 595

▶ (3)-(C)ab(K)bP

	Shift	Peak
a	1.26	2A
b	3.28	1B
▶ c	3.87	2A
d	4.34	3C

▶ $C_6H_{10}O_3$ ▶ 7% $CDCl_3$ ▶ V 467

▶ (3)-Cbp(L)lp(Q)l

	Shift	Peak
a	3.78	1A
b	4.10	O
▶ c	4.97	2D

▶ $C_6H_8O_6$ ▶ D_2O ▶ V 464

▶ (3)-(C)bkKh(Q)b

	Shift	Peak
a	2.98	2E
b	3.86	4A
▶ c	5.24	2A

▶ $C_6H_6O_6$ ▶ D_2O ▶ V 456

▶ (3)-(C)cp(N)jl(O)c

	Shift	Peak
a	3.82	S
b	3.92	S
c	4.18 ± 0.03	S
d	4.25	S
e	4.35	S
▶ f	5.90	S
g	5.90	S
h	7.87	2A

▶ $C_9H_{11}N_2O_6$ ▶ D_2O ▶ V 535

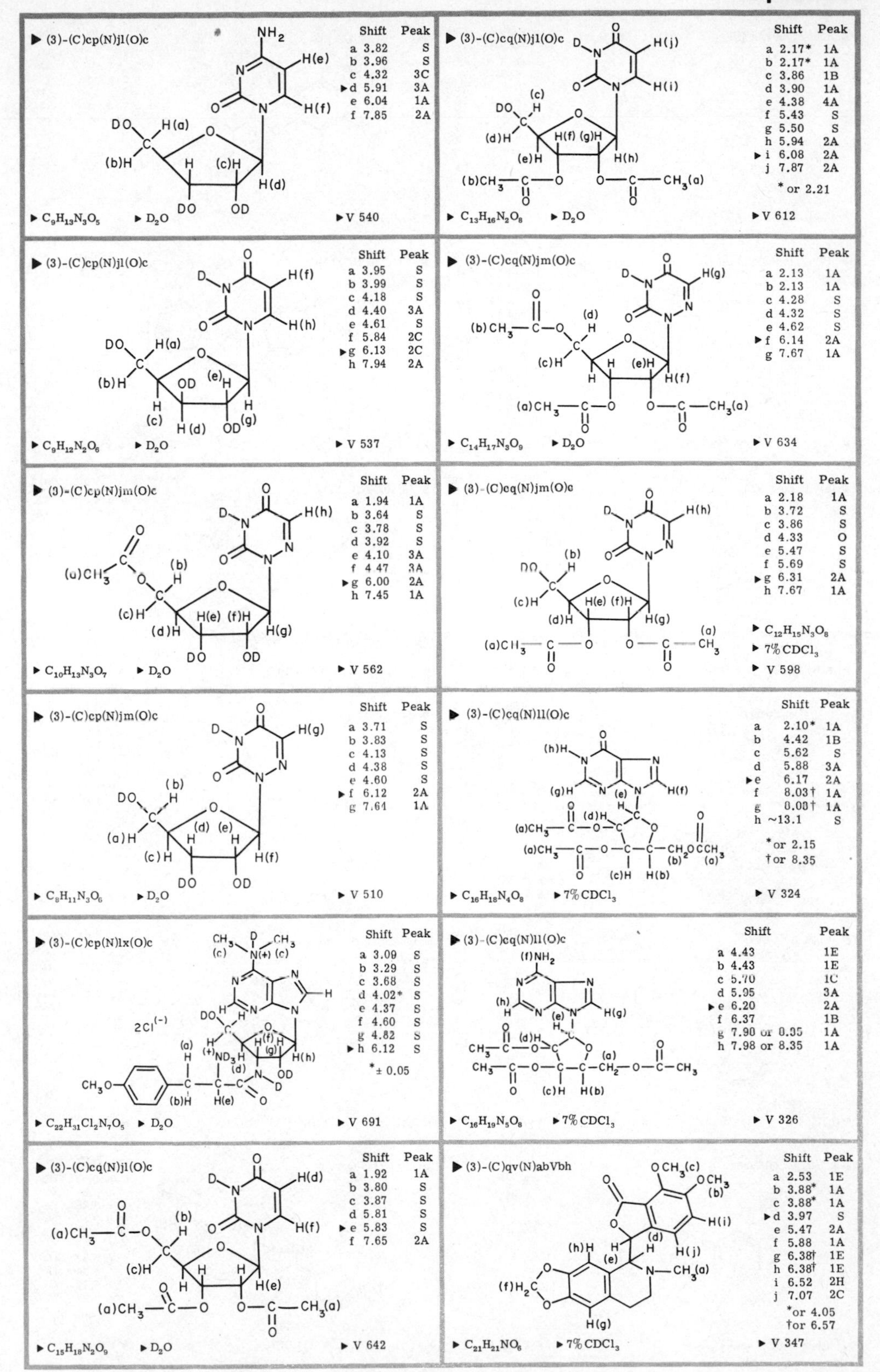

▶ (3)-(C)cp(N)jl(O)c

	Shift	Peak
a	3.82	S
b	3.96	S
c	4.32	3C
▶d	5.91	3A
e	6.04	1A
f	7.85	2A

▶ $C_9H_{13}N_3O_5$ ▶ D_2O ▶ V 540

▶ (3)-(C)cq(N)jl(O)c

	Shift	Peak
a	2.17*	1A
b	2.17*	1A
c	3.86	1B
d	3.90	1A
e	4.38	4A
f	5.43	S
g	5.50	S
h	5.94	2A
▶i	6.08	2A
j	7.87	2A

* or 2.21

▶ $C_{13}H_{16}N_2O_8$ ▶ D_2O ▶ V 612

▶ (3)-(C)cp(N)jl(O)c

	Shift	Peak
a	3.95	S
b	3.99	S
c	4.18	S
d	4.40	3A
e	4.61	S
f	5.84	2C
▶g	6.13	2C
h	7.94	2A

▶ $C_9H_{12}N_2O_6$ ▶ D_2O ▶ V 537

▶ (3)-(C)cq(N)jm(O)c

	Shift	Peak
a	2.13	1A
b	2.13	1A
c	4.28	S
d	4.32	S
e	4.62	S
▶f	6.14	2A
g	7.67	1A

▶ $C_{14}H_{17}N_3O_9$ ▶ D_2O ▶ V 634

▶ (3)-(C)cp(N)jm(O)c

	Shift	Peak
a	1.94	1A
b	3.64	S
c	3.78	S
d	3.92	S
e	4.10	3A
f	4.47	3A
▶g	6.00	2A
h	7.45	1A

▶ $C_{10}H_{13}N_3O_7$ ▶ D_2O ▶ V 562

▶ (3)-(C)cq(N)jm(O)c

	Shift	Peak
a	2.18	1A
b	3.72	S
c	3.86	S
d	4.33	O
e	5.47	S
f	5.69	S
▶g	6.31	2A
h	7.67	1A

▶ $C_{12}H_{15}N_3O_8$ ▶ 7% $CDCl_3$ ▶ V 598

▶ (3)-(C)cp(N)jm(O)c

	Shift	Peak
a	3.71	S
b	3.83	S
c	4.13	S
d	4.38	S
e	4.60	S
▶f	6.12	2A
g	7.64	1A

▶ $C_8H_{11}N_3O_6$ ▶ D_2O ▶ V 510

▶ (3)-(C)cq(N)ll(O)c

	Shift	Peak
a	2.10*	1A
b	4.42	1B
c	5.62	S
d	5.88	3A
▶e	6.17	2A
f	8.03†	1A
g	8.03†	1A
h	~13.1	S

*or 2.15
†or 8.35

▶ $C_{16}H_{18}N_4O_8$ ▶ 7% $CDCl_3$ ▶ V 324

▶ (3)-(C)cp(N)lx(O)c

	Shift	Peak
a	3.09	S
b	3.29	S
c	3.68	S
d	4.02*	S
e	4.37	S
f	4.60	S
g	4.82	S
▶h	6.12	S

*± 0.05

▶ $C_{22}H_{31}Cl_2N_7O_5$ ▶ D_2O ▶ V 691

▶ (3)-(C)cq(N)ll(O)c

	Shift	Peak
a	4.43	1E
b	4.43	1E
c	5.70	1C
d	5.95	3A
▶e	6.20	2A
f	6.37	1B
g	7.98 or 8.35	1A
h	7.98 or 8.35	1A

▶ $C_{16}H_{19}N_5O_8$ ▶ 7% $CDCl_3$ ▶ V 326

▶ (3)-(C)cq(N)jl(O)c

	Shift	Peak
a	1.92	1A
b	3.80	S
c	3.87	S
d	5.81	S
▶e	5.83	S
f	7.65	2A

▶ $C_{15}H_{18}N_2O_9$ ▶ D_2O ▶ V 642

▶ (3)-(C)qv(N)abVbh

	Shift	Peak
a	2.53	1E
b	3.88*	1A
c	3.88*	1A
▶d	3.97	S
e	5.47	2A
f	5.88	1A
g	6.38†	1E
h	6.38†	1E
i	6.52	2H
j	7.07	2C

*or 4.05
†or 6.57

▶ $C_{21}H_{21}NO_6$ ▶ 7% $CDCl_3$ ▶ V 347

▶ (3)-(C)cd(O)c(O)l

O OCH3(c)
(h)H (d)H H(j)
H(b) O
(g)H (f) H O
O (e) H
O CH3(a)
(i)H

	Shift	Peak
a	2.11	2A
b	3.46	4A
c	3.79	1A
d	4.02	3E
e	5.12	1B
▶ f	5.59	S
g	5.67*	S
h	5.67*	S
i	7.19	O
j	7.46	1A

*or 6.08

▶ $C_{15}H_{14}O_6$ ▶ 7% $CDCl_3$ ▶ V 640

▶ (3)-(C)bp(O)vVhh

(a) OH
CH3O (b) CH2 C
(f)H H(c)
CH3O O C H
(d)
H (e) OCH3
OCH3

	Shift	Peak
a	1.73	1B
b	2.95	2A
c	4.28	1B
▶ d	4.96	1I
e	6.12 or 6.20	2C
f	6.12 or 6.20	2C

▶ $C_{19}H_{22}O_6$ ▶ 7% $CDCl_3$ ▶ V 679

▶ (3)-(C)cd(O)c(O)l

O OCH3(e)
(f)H C
(i)H H(k)
H (d) O
(j)H (h) H
O (g) H O
O (c)H CH2 CH3
(b) (a)

	Shift	Peak
a	1.13	3B
b	1.89	O
c	2.74	O
d	3.42	4A
e	3.77	1A
f	3.98	3E
g	4.59	2A
▶ h	5.59	S
i	5.63*	S
j	5.63*	S
k	7.47	1A

*or 6.12

▶ $C_{15}H_{16}O_6$ ▶ 7% $CDCl_3$ ▶ V 641

▶ (3)-(C)ov(O)cVhh

(b) (b)
C C
H O H
(a) (a)

	Shift	Peak
▶ a	3.88	1A
b	7.38	1A

▶ $C_{14}H_{12}O$ ▶ 7% $CDCl_3$ ▶ V 625

▶ (3)-(C)cpOa(O)b

(b) H OCH3(a)
(e) H(d)
H O H(g)
(f)H OD
OD
DO H(c)

	Shift	Peak
a	3.41	1A
b	3.67	S
c	3.82 ± 0.04	S
d	3.82 ± 0.04	S
e	3.89 ± 0.05	S
f	3.98 ± 0.05	1F
▶ g	4.82	2G

▶ $C_6H_{12}O_5$ ▶ D_2O ▶ V 475

▶ (3)-(C)ov(O)cVhh

(b) H(a)
C C
(a)H O
(b)

	Shift	Peak
▶ a	4.37	1A
b	7.20	1A

▶ $C_{14}H_{12}O$ ▶ 7% $CDCl_3$ ▶ V 626

▶ (3)-(C)cqOa(O)c

H(d)
O
O O
H(e)
H(c)
N
N H O OCH3(b)
C
O CH3(a)

	Shift	Peak
a	1.92	1A
b	3.47	1A
▶ c	5.07	2A
d	5.58	1E
e	~5.63	4E

▶ $C_{22}H_{24}N_2O_8$ ▶ 7% $CDCl_3$ ▶ V 356

▶ 3-CaaPTh

(e) (a)
OH CH3
(d)HC ≡ C — C — C — H(c)
H CH3
(f) (b)

	Shift	Peak
a	1.00	2D
b	1.02	2D
c	1.85	7B
d	2.47	2E
e	2.82	1B
▶ f	4.18	4A

▶ $C_6H_{10}O$ ▶ 7% $CDCl_3$ ▶ V 133

▶ (3)-(C)av(O)vVhh

(d) CH2(c)
CH2 CH3(a)
(b)CH3
CH(e) OCH3(f)
(h)HO C O
H(g)
(f)CH3O

	Shift	Peak
a	0.96	3B
b	1.37	2A
c	~1.65	S
d	2.58	3B
e	3.42	S
f	3.88	1A
▶ g	5.08	2A
h	5.62	1A

▶ $C_{20}H_{24}O_4$ ▶ 7% $CDCl_3$ ▶ V 682

▶ 3-CanPVhh

(c) (b)
H H
(d) C — C — CH3(a)
(+)
OD ND3
Cl(−)

	Shift	Peak
a	1.21	2A
b	3.72	O
▶ c	5.00	2A
d	7.48	1A

▶ $C_9H_{14}ClNO$ ▶ D_2O ▶ V 565

▶ (3)-(C)bo(O)cVbh

(b)
H H(a)
H(c)
O
H(d)

	Shift	Peak
a	2.96	S
b	3.20	S
c	4.08	3A
▶ d	4.22	2A

▶ C_9H_8O ▶ 7% $CDCl_3$ ▶ V 524

▶ (3)-(C)nv(Q)vVhk

O OCH3(c)
OCH3
(b)
O H(i)
(d)
(h)H H H H(j)
(e)
O N — CH3(a)
(f)H2C
O
H(g)

	Shift	Peak
a	2.53	1E
b	3.88*	1A
c	3.88*	1A
d	3.97	S
▶ e	5.47	2A
f	5.88	1A
g	6.38†	1E
h	6.38†	1E
i	6.52	2H
j	7.07	2C

*or 4.05
†or 6.57

▶ $C_{21}H_{21}NO_6$ ▶ 7% $CDCl_3$ ▶ V 347

▶ (3)-(C)sv(S)cVhh

	Shift	Peak
▶a	3.96	1A
b	7.37	1A

▶ $C_{14}H_{12}S$ ▶ 7% $CDCl_3$ ▶ V 629

▶ (3)-(D)clq(L)kl(O)c

	Shift	Peak
a	2.11	2A
b	3.46	4A
c	3.79	1A
d	4.02	3E
▶e	5.12	1B
f	5.59	S
g	5.67*	S
h	5.67*	S
i	7.19	O
j	7.46	1A

*or 6.08

▶ $C_{15}H_{14}O_6$ ▶ 7% $CDCl_3$ ▶ V 640

▶ 3-DbffFF

	Shift	Peak
a	2.72	1B
b	3.97	3C
▶c	5.93	3E

▶ $C_3H_4F_4O$ ▶ 7% $CDCl_3$ ▶ V 19

▶ (3)-(D)cee(L)elQa

	Shift	Peak
a	2.20	1A
b	2.25	1A
c	3.87	1A
d	6.05	1A
▶e	6.63	1A

▶ $C_{11}H_{11}Cl_3O_6$ ▶ 7% $CDCl_3$ ▶ V 284

▶ (3)-(D)ceeKa(L)lq

	Shift	Peak
a	2.20	1A
b	2.25	1A
c	3.87	1A
▶d	6.05	1A
e	6.63	1A

▶ $C_{11}H_{11}Cl_3O_6$ ▶ 7% $CDCl_3$ ▶ V 284

▶ (3)-(D)abv(N)ab(N)av

	Shift	Peak
a	1.42	1A
b	1.95	3A
c	2.55	1A
d	2.70	3D
e	2.82	1E
f	2.92	1E
▶g	4.12	1A
h	5.33	1D
i	6.37	2A
j	6.78	1E
k	6.87	2H

▶ $C_{15}H_{21}N_3O_2$ ▶ 7% $CDCl_3$ ▶ V 319

▶ (3)-Deee(K)c(O)c

	Shift	Peak
▶a	5.12	2D
b	6.09	2D

▶ $C_5H_2Cl_6O_3$ ▶ 7% $CDCl_3$ ▶ V 425

▶ (3)-Daee(O)c(O)c

	Shift	Peak
a	2.15	1A
▶b	5.19	1A

▶ $C_9H_{12}Cl_6O_3$ ▶ 7% $CDCl_3$ ▶ V 536

▶ (3)-(D)aab(L)alLhl

	Shift	Peak
a	0.86	1A
b	0.93	1A
c	1.57	2G
d	2.05	1C
e	2.26	1A
▶f	2.30	S
g	5.50	1C
h	6.06	2A
i	6.60	4A

▶ $C_{13}H_{20}O$ ▶ 7% $CDCl_3$ ▶ V 616

▶ (3)-Deee(O)c(O)c

	Shift	Peak
a	5.46	1A

▶ $C_6H_3Cl_9O_3$ ▶ 7% $CDCl_3$ ▶ V 450

▶ (3)-(D)jlo(L)al(L)hl

	Shift	Peak
a	2.13	2D
b	3.46	1A
▶c	3.69	1B
d	5.88	2A
e	6.43	2A
f	6.84	2G

▶ $C_9H_{10}O_2$ ▶ 7% $CDCl_3$ ▶ V 531

▶ (3)-Deee(O)c(O)c

	Shift	Peak
▶a	5.67	1A
▶b	6.15	1A

▶ $C_6H_3Cl_9O_3$ ▶ 7% $CDCl_3$ ▶ V 449

▶ (3)-Deee(O)c(Q)c

	Shift	Peak
a	5.12	2D
▶b	6.09	2D

▶ $C_5H_2Cl_6O_3$ ▶ 7% $CDCl_3$ ▶ V 425

▶ 3-EVhhVhh

	Shift	Peak
▶a	6.12	1A
b	7.33 ± .05	1I

(b) = all benzene H's.

▶ $C_{13}H_{11}Cl$ ▶ 7% $CDCl_3$ ▶ V 176

▶ (3)-(D)ajlVhoVho

	Shift	Peak
a	0.89	1A
b	1.09	1A
c	1.17	S
d	1.35	S
▶e	4.16	1E
f	5.07	2A
g	5.73	4C

▶ $C_{23}H_{22}O_2$ ▶ 7% $CDCl_3$ ▶ V 693

▶ 3-FFLbm

	Shift	Peak
a	1.33	3A
b	3.67	1A
c	4.30	4A
▶d	6.30	3A*
e	8.03	2C
f	8.40	4A
g	9.08	2A
h	11.77	1B

*Very large splitting due to fluorine

▶ $C_{12}H_{12}F_2N_4O_6$ ▶ 7% $CDCl_3$ ▶ V 293

▶ 3-DeeeVhhVhh

	Shift	Peak
a	5.02	1A

▶ $C_{14}H_9Cl_5$ ▶ 45.4 mg/0.25 ml $CDCl_3$ ▶ S 15

▶ 3-GVhhVhh

	Shift	Peak
a	6.27	1A

▶ $C_{13}H_{11}Br$ ▶ 7% $CDCl_3$ ▶ V 606

▶ 3-DeeeVhhVhh

	Shift	Peak
a	5.02	1A

▶ $C_{14}H_9Cl_5$ ▶ 7% $CDCl_3$ ▶ V 620

▶ 3-Jl(L)loP

	Shift	Peak
a	4.02	1C
▶b	5.81	1A
c	6.37	N
d	6.40	N
e	6.55	N
f	7.27	N
g	7.38	N
h	7.62	N

▶ $C_{10}H_8O_4$ ▶ 7% $CDCl_3$ ▶ V 247

▶ 3-EEKb

	Shift	Peak
a	1.35	3A
b	4.33	4A
▶c	5.93	1A

▶ $C_4H_6Cl_2O_2$ ▶ 7% $CDCl_3$ ▶ V 59

▶ 3-JvPVhh

	Shift	Peak
a	3.75*	1A
b	3.75*	1A
c	4.54	1C
▶d	5.82	1B
e	6.83	S
f	6.87	S
g	7.24	2E
h	7.89	2E

*or 3.81

▶ $C_{16}H_{16}O_4$ ▶ 7% $CDCl_3$ ▶ V 655

▶ 3-EEKh

	Shift	Peak
▶a	5.98	1A
b	11.08	1A

▶ $C_2H_2Cl_2O_2$ ▶ 45.7 mg/0.50 ml $CDCl_3$ ▶ S 166

▶ 3-JaVhhVhh

	Shift	Peak
a	2.20	1A
▶b	5.08	1A
c	7.25 ± .03	1A

(c) = all benzene H's.

▶ $C_{15}H_{14}O$ ▶ 7% $CDCl_3$ ▶ V 318

▶ 3-KbKbQa

(b)CH_3 O C O C H(d) C O O CH_2 (c) CH_3(a) C O O CH_2 (c) CH_3(a)

Shift	Peak
a 1.31	3A
b 2.22	1A
c 4.30	4A
▶d 5.52	1A

▶ $C_9H_{14}O_6$ ▶ 7% $CDCl_3$ ▶ V 546

▶ (3)-(L)jl(L)klVhh

(e) H O C O CH_2 (d) CH_3(a) O C (b)CH_3 CH_3 (c) N H(f)

Shift	Peak
a 0.83	3A
b 2.17	1A
c 2.37	1A
d 3.87	4A
▶e 5.13	1A
f 6.15	1C

▶ $C_{23}H_{23}NO_3$ ▶ 7% $CDCl_3$ ▶ V 695

▶ 3-KbNhjU

O C CH_3 (b) N H(e) (d) H C CN C O O CH_2 (c) CH_3(a)

Shift	Peak
a 1.37	3A
b 2.12	1A
c 4.36	4A
▶d 5.50	2A
e 6.75	1D

▶ $C_7H_{10}N_2O_3$ ▶ 7% $CDCl_3$ ▶ V 491

▶ 3-LhlQaQa

(a) CH_3 C O O C H (b) O C O CH_3 (a) C=C H H H

Shift	Peak
a 2.11	1A
▶b 7.15	2A

▶ $C_7H_{10}O_4$ ▶ 7% $CDCl_3$ ▶ V 492

▶ 3-KaOaOa

(a)CH_3O (c)H — C — C O OCH_3(b) (a)CH_3O

Shift	Peak
a 3.44	1A
b 3.82	1A
▶c 4.82	1A

▶ $C_5H_{10}O_4$ ▶ 7% $CDCl_3$ ▶ V 446

▶ 3-(L)loQaQa

(c)H H(b) O_2N O C O C CH_3(a) O H(d) O=C CH_3(a)

Shift	Peak
a 2.20	1A
b 6.77	2C
c 7.32	2C
▶d 7.75	1A

▶ $C_9H_9NO_6$ ▶ 7% $CDCl_3$ ▶ V 526

▶ 3-KhPVhh

(c) (a) (b) H H OH (c)H C O H(b) O Cl H (c) H (c)

Shift	Peak
▶a 5.23	1A
b 6.63	1B
c 7.38	1A

▶ $C_8H_7ClO_3$ ▶ 7% $CDCl_3$ ▶ V 499

▶ 3-LhlVhoVho

H(d) H(a) C=C (f)H H(b) C OCH_3 CH_3O (c)H CH_3O OCH_3 H(e)

Shift	Peak
a 4.77	R
b 5.15	R
▶c 5.47	2E
d 6.27	R
e 6.53	2A
f 6.68	O

▶ $C_{19}H_{22}O_4$ ▶ 7% $CDCl_3$ ▶ V 678

▶ 3-KhQaVhh

O C O CH_3(a) C (c)HO C O H(b)

Shift	Peak
a 2.18	1A
▶b 5.95	1A
c 9.88	1C

▶ $C_{10}H_{10}O_4$ ▶ 7% $CDCl_3$ ▶ V 554

▶ (3)-(L)llVhoVho

(f) H (d) H (b) CH_3 O (c) H_2 C (e)H (a)CH_3 CH_3 (a) O

Shift	Peak
a 1.20	1A
b 1.70	2D
c 2.73	1G
▶d 4.73	1B
e 6.14	S
f 6.16	S

▶ $C_{23}H_{22}O_2$ ▶ 7% $CDCl_3$ ▶ V 694

▶ 3-KhVhhVhh

(a) CH — C — OH (b) O

Shift	Peak
▶a 5.05	1A
b 10.92	1A

▶ $C_{14}H_{12}O_2$ ▶ 45.0 mg/0.50 ml $CDCl_3$ ▶ S 129

▶ 3-LhlVhhZ

(g) (f) H Cl C=C Cl (d)H C B O CH_2 (e) CH_2 (c) CH_2 (b) CH_3 (a) O CH_2 (e) CH_2 (c) CH_2 (b) CH_3 (a)

Shift	Peak
a 0.88	3C
b ~1.36	S
c ~1.50	S
▶d 3.58	2H
e 3.84	3B
f 6.24	2A
g 7.20	1A

▶ $C_{17}H_{25}BCl_2O_2$ ▶ 7% $CDCl_3$ ▶ V 667

▶ 3-NhjQbQb

	Shift	Peak
a	1.30	3A
b	2.08	1A
c	4.28	4A
▶d	5.19	2A
e	6.60	1D

▶ $C_9H_{15}NO_5$ ▶ 7% $CDCl_3$ ▶ V 547

▶ 3-VhhVhhVhh

	Shift	Peak
a	2.87	1A
▶b	5.29	1B
c	6.64	Q
d	6.95	Q

▶ $C_{25}H_{31}N_3$ ▶ 7% $CDCl_3$ ▶ V 360

▶ (3)-(O)b(O)b(O)b

	Shift	Peak
a	0.83	N
b	1.20	N
c	3.93	1A
▶d	5.52	1A

▶ $C_7H_{12}O_3$ ▶ 7% $CDCl_3$ ▶ V 493

▶ 3-VhhVhhVhh

	Shift	Peak
a	5.54	1B

▶ $C_{19}H_{16}$ ▶ 45.7 mg/0.25 ml $CDCl_3$ ▶ S 220

▶ (3)-(O)b(O)c Vhh

	Shift	Peak
a	1.92	1A
b	3.47	1A
c	5.07	2A
▶d	5.58	1E
e	~5.63	4E

▶ $C_{22}H_{24}N_2O_6$ ▶ 7% $CDCl_3$ ▶ V 356

▶ **4**-AP

	Shift	Peak
a	1.86	1A
b	1.89	1A
▶c	6.84	4A
▶d	7.44	4A
e	~8.92	4D

syn- and anti-forms

▶ C_2H_5NO ▶ 7% $CDCl_3$ ▶ V 373

▶ 3-PVhhVhh

	Shift	Peak
a	2.37	2G
▶b	5.75	2G

▶ $C_{13}H_{12}O$ ▶ 45.4 mg/0.50 ml $CDCl_3$ ▶ S 186

▶ 4-BbP

	Shift	Peak
a	0.96	3C
b	1.55	6B
▶c	6.74	3A
▶d	7.45	3A
e	8.85	1D

Syn- / Anti-

▶ C_4H_9NO ▶ 7% $CDCl_3$ ▶ V 420

▶ 3-PVhhVhh

	Shift	Peak
a	2.29	1B
▶b	5.80	1G
c	7.32	1E

▶ $C_{13}H_{12}O$ ▶ 7% $CDCl_3$ ▶ V 607

▶ 4-BbVhh

	Shift	Peak
a	3.92	1A
b	5.97	1A
c	6.79	2A
d	7.07	S
e	7.32	4A
▶ f	8.17	2A

▶ $C_{18}H_{16}N_2O_4$ ▶ 7% $CDCl_3$ ▶ V 673

▶ 3-SbSbSb

	Shift	Peak
a	3.68	1A
▶b	4.20	1A

▶ $C_{22}H_{22}S_3$ ▶ 7% $CDCl_3$ ▶ V 688

▶ (4)-(J)n(N)cj

	Shift	Peak
a	1.94	1A
b	3.64	S
c	3.78	S
d	3.92	S
e	4.10	3A
f	4.47	3A
g	6.00	2A
▶h	7.45	1A

▶ $C_{10}H_{13}N_3O_7$ ▶ D_2O ▶ V 562

▶ (4)-(J)n(N)cj

O, D, N, H(g), O, N, N, (b), DO, H, O, (a)H, (d), (e), (c)H, H, H, H(f), DO, OD

	Shift	Peak
a	3.71	S
b	3.83	S
c	4.13	S
d	4.38	S
e	4.60	S
f	6.12	2A
▶ g	7.64	1A

▶ $C_8H_{11}N_3O_6$ ▶ D_2O ▶ V 510

▶ (4)-(N)hj(L)ln

O, (h)H, N, N, (g)H, N, (e), N, H(f), O, (d)H, H, O, (a)CH_3—C—O—C, O, (a)CH_3—C—O—C—C—CH_2OCCH$_3$, (b), (a), O, (c)H, H(b)

	Shift	Peak
a	2.10*	1A
b	4.42	1B
c	5.62	S
d	5.88	3A
e	6.17	2A
f	8.03†	1A
▶ g	8.03†	1A
h	~13.1	S

*or 2.15
†or 8.35

▶ $C_{16}H_{18}N_4O_8$ ▶ 7% $CDCl_3$ ▶ V 324

▶ (4)-(J)n(N)cj

O, D, N, H(h), O, N, N, (b), DO, H, O, (c)H, H(e), (f)H, (d)H, H(g), (a), (a)CH_3—C—O, O—C—CH_3, O, O

	Shift	Peak
a	2.18	1A
b	3.72	S
c	3.86	S
d	4.33	O
e	5.47	S
f	5.69	S
g	6.31	2A
▶ h	7.67	1A

▶ $C_{12}H_{15}N_3O_8$ ▶ 7% $CDCl_3$ ▶ V 598

▶ (4)-(N)hl(L)hl

(e)H, N, (b), CH_2, NH_2, (c), CH_2, (a), N, H(f), H, (d)

	Shift	Peak
a	2.75	S
b	3.02	S
c	4.96	1E
d	4.96	1E
e	6.82	1A
▶ f	7.55	1A

▶ $C_5H_9N_3$ ▶ 7% $CDCl_3$ ▶ V 443

▶ (4)-(J)n(N)cj

O, D, N, H(g), O, N, N, O, (d), (b)CH_3—C—O, H, O, (c)H, H, (e)H, H, H(f), (a)CH_3—C—O, O—C—CH_3(a), O, O

	Shift	Peak
a	2.13	1A
b	2.13	1A
c	4.28	S
d	4.32	S
e	4.62	S
f	6.14	2A
▶ g	7.67	1A

▶ $C_{14}H_{17}N_3O_9$ ▶ D_2O ▶ V 634

▶ (4)-(N)hl(L)hl

H, N, (a), H, N, H(b), H, (c)

	Shift	Peak
a	7.14	1C
▶ b	7.71	1C
c	13.38	1C

▶ $C_3H_4N_2$ ▶ 7% $CDCl_3$ ▶ V 20

▶ 4-JvP

(a), H, O, H(b), N, OH(c), H, (a)

	Shift	Peak
a	8.04	O
▶ b	8.13	1E
c	9.06	1B

▶ $C_8H_7NO_2$ ▶ 7% $CDCl_3$ ▶ V 501

▶ (4)-(N)cwW

(f)NH_2, N, N, (h), H, N, (e), N, H(g), O, (d)H, H, O, CH_3—C—O—C, (a), O, CH_3—C—O—C—C—CH_2—O—C—CH_3, O, (c)H, H(b)

	Shift	Peak
a	4.43	1E
b	4.43	1E
c	5.70	1C
d	5.95	3A
e	6.20	2A
f	6.37	1B
▶ g	7.98 or 8.35	1A
h	7.98 or 8.35	1A

▶ $C_{16}H_{19}N_5O_8$ ▶ 7% $CDCl_3$ ▶ V 326

▶ (4)-(N)al(L)ln

O, CH_3, CH_3, N, N, H(a), O, N, N, CH_3

	Shift	Peak
a	7.60	1A

▶ $C_8H_{10}N_4O_2$ ▶ 7% $CDCl_3$ ▶ V 204

▶ (4)-(N)hwW

OD, N, N, H(a), D_2N, N, N, D

	Shift	Peak
a	7.68	1A

▶ $C_5H_5N_5O$ ▶ 7% $CDCl_3$ ▶ V 430

▶ (4)-(N)cl(L)jl

O, (h)H, N, N, (g)H, N, (e), N, H(f), O, (d)H, H, O, (a)CH_3—C—O—C, O, (a)CH_3—C—O—C—C—CH_2OCCH$_3$, (b), (a), O, (c)H, H(b)

	Shift	Peak
a	2.10*	1A
b	4.42	1B
c	5.62	S
d	5.88	3A
e	6.17	2A
▶ f	8.03†	1A
g	8.03†	1A
h	~13.1	S

*or 2.15
†or 8.35

▶ $C_{16}H_{18}N_4O_8$ ▶ 7% $CDCl_3$ ▶ V 324

▶ (4)-(N)hwW

OD, N, N, H(b), (a)H, N, N, D

	Shift	Peak
a	7.87	1A
▶ b	8.08	1A

▶ $C_5H_4N_4O$ ▶ D_2O ▶ V 426

▶ (4)-(N)hwW

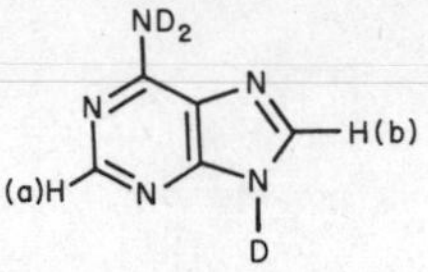

	Shift	Peak
a	8.20 or 8.32	1A
▶b	8.20 or 8.32	1A

▶ $C_5H_5N_5$ ▶ 7% $CDCl_3$ ▶ V 429

▶ 5-BaHH

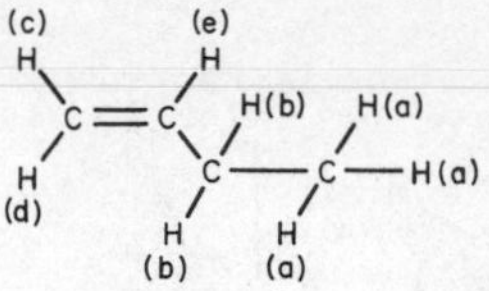

	Shift	Peak
a	0.97	3C
b	1.98	5B
c	~4.88	R
d	~5.7	R
▶e	~5.9	R

▶ C_4H_8 ▶ liquid ▶ API 365

▶ (4)-NhwW

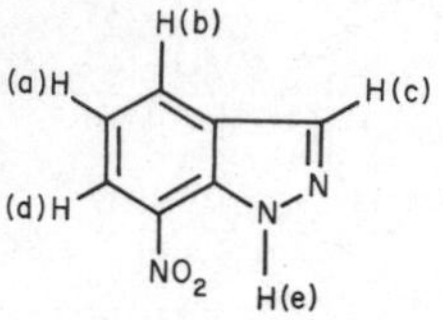

	Shift	Peak
a	7.34	4A
b	8.18	2D
▶c	8.32	2E
d	8.48	2D
e	11.63	1D

▶ $C_7H_5N_3O_2$ ▶ 7% $CDCl_3$ ▶ V 485

▶ 5-BbHH

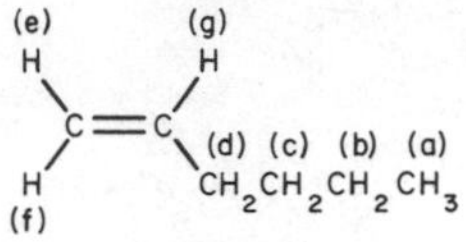

	Shift	Peak
a	.88	3C
b	~1.29 or 1.31	S
c	~1.31 or 1.29	S
d	2.00	3C
e	4.84	R
f	4.97	R
▶g	5.65	R

▶ C_6H_{12} ▶ liquid ▶ API 368

▶ (4)-(O)vVho

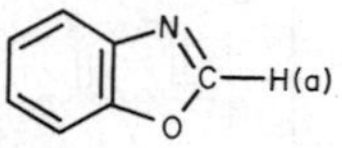

	Shift	Peak
a	8.07	1A

▶ C_7H_5NO ▶ 7% $CDCl_3$ ▶ V 147

▶ 5-BbHH

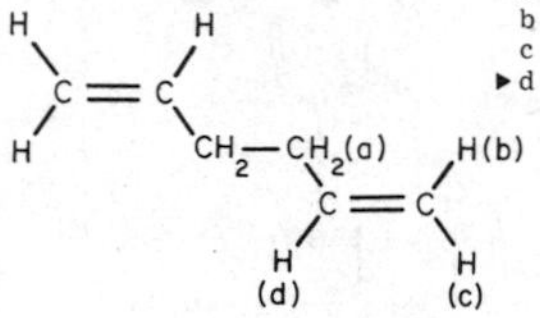

	Shift	Peak
a	2.08	2E
b	~4.88 or 4.99	R
c	~4.99 or 4.88	R
▶d	~5.67	R

▶ C_6H_{10} ▶ liquid ▶ API 474

▶ 4-VhhCbk

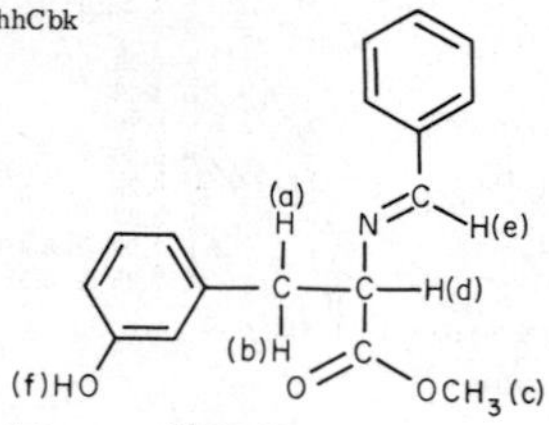

	Shift	Peak
a	3.23	N
b	3.52	N
c	3.72	1A
d	4.28	4A
▶e	7.89	1A
f	8.04	1F

▶ $C_{17}H_{17}NO_3$ ▶ 7% $CDCl_3$ ▶ V 663

▶ 5-BbHH

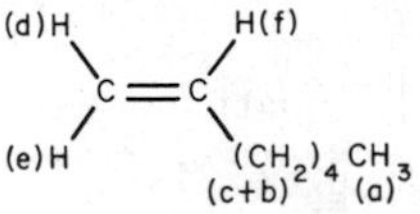

	Shift	Peak
a	.91	3B
b	1.31	S
c	2.02	3B
d	4.84	R
e	4.98	R
▶f	5.67	R

▶ C_7H_{14} ▶ 10% CCl_4 ▶ API 157

▶ 4-VhpP

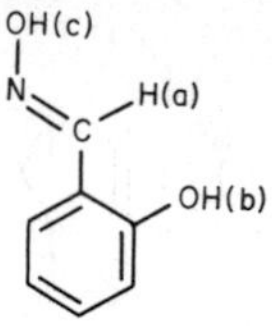

	Shift	Peak
▶a	8.18	1A
b	~8.18	1A
c	10.05	1D

▶ $C_7H_7NO_2$ ▶ 7% $CDCl_3$ ▶ V 156

▶ 5-BbHH

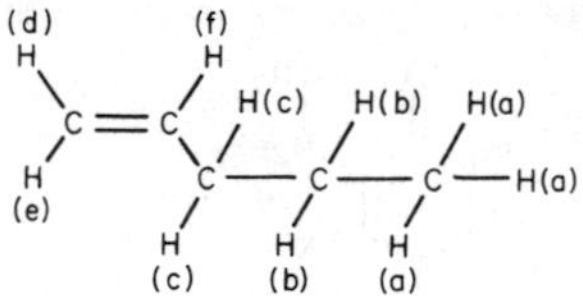

	Shift	Peak
a	.92	3C
b	1.38	6B
c	1.93	4C
d	4.76	R
e	4.98	R
▶f	5.68	R

▶ C_5H_{10} ▶ liquid ▶ API 366

▶**5**-BaHH

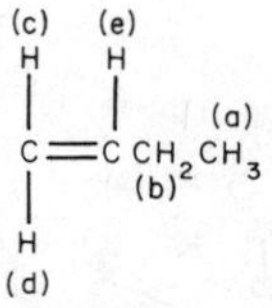

	Shift	Peak
a	.97	3C
b	2.02	5B
c	4.95	R
d	4.98	R
▶e	5.80	R

▶ C_4H_8 ▶ degassed liquid ▶ API 55

▶ 5-BbHH

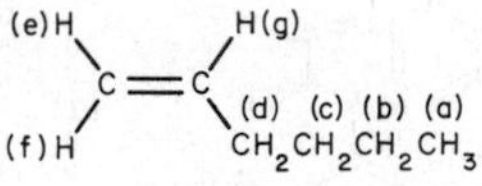

	Shift	Peak
a	.90	3B
b	~1.34	S
c	~1.34	S
d	2.00	4C
e	4.85	R
f	4.97	R
▶g	5.68	R

▶ C_6H_{12} ▶ 10% CCl_4 ▶ API 155

▶ 5-BbHH

(d) H, (f) H, C=C, H (e); (c) (b) (a) $CH_2(CH_2)_5CH_3$

	Shift	Peak
a	.85	3B
b	1.28	1B
c	1.99	3B
d	4.88	R
e	4.99	R
▶f	5.68	R

▶ C_9H_{18} ▶ 50% CCl_4 ▶ API 434

▶ 5-BbHH

(e) H, (g) H, C=C, H (f); CH_3; $CH_2CH_2CHCH_3$ (d) (b) (c) (a)

	Shift	Peak
a	.91	2C
b	1.32	S
c	1.60	S
d	2.01	4C
e	4.87	R
f	5.02	R
▶g	5.70	R

▶ C_7H_{14} ▶ liquid ▶ API 394

▶ 5-BbHH

(d) H, (f) H, C=C, H (e); (c) (b) (a) $CH_2(CH_2)_8CH_3$

	Shift	Peak
a	.92	3B
b	1.32	1B
c	2.05	3B
d	4.86	R
e	5.00	R
▶f	5.68	R

▶ $C_{12}H_{24}$ ▶ 50 vol.% CCl_4 ▶ API 435

▶ 5-BbHH

(d) H, (f) H, C=C, H (e); (c) (b) (a) $CH_2(CH_2)_{10}CH_3$

	Shift	Peak
a	.88	3B
b	1.27	1A
c	2.3	3B
d	4.88	7X
e	5.01	7X
▶f	5.72	R

▶ $C_{14}H_{28}$ ▶ 50 vol.% CCl_4 ▶ API 436

▶ 5-BbHH

(a) (b) (c) (f) (d); $CH_3-(CH_2)_8-CH_2-C=C$; H H; H (e)

	Shift	Peak
a	0.87	3B
b	1.28	1A
c	2.05	2G
d	4.80	R
e	~5.03	R
▶f	5.68	R

▶ $C_{12}H_{24}$ ▶ 44.4 mg/0.50 ml $CDCl_3$ ▶ S 202

▶ 5-BbHH

(d) H, (f) H, C=C, H (e); (c) (b) (a) $CH_2(CH_2)_{12}CH_3$

	Shift	Peak
a	.90	3B
b	1.24	1A
c	2.04	3B
d	4.88	R
e	5.02	R
▶f	5.72	R

▶ $C_{16}H_{32}$ ▶ 50 vol.% CCl_4 ▶ API 437

▶ 5-DbIIII

(e) H, (f) H, C=C, H (d); (c) (b) (a) $CH_2(CH_2)_{12}CH_3$

	Shift	Peak
a	.99	3B
b	1.25	1A
c	2.02	3B
d	4.90	R
e	5.08	R
▶f	5.68	R

▶ $C_{16}H_{32}$ ▶ 50% CCl_4 ▶ API 73

▶ 5-DbIIII

(a) CH_3, (b) CH_3, H(e); $CHCH_2CH(CH_2)_4-C=C-H$ (c); CH_3 (a); H (d)

	Shift	Peak
a	0.82 or 0.87	2A
b	0.82 or 0.87	2A
c	4.95	R
d	4.98	R
▶e	~5.80	R

▶ $C_{12}H_{24}$ ▶ 7% $CDCl_3$ ▶ V 298

▶ 5-BbHH

(b,c) (d) (a) (a) (d) (b,c); $H_2C=CH-CH_2-CH_2-CH=CH_2$

	Shift	Peak
a	2.13	3A
b	4.93	R
c	~5.00	R
▶d	~5.70	R

▶ C_6H_{10} ▶ 10% CCl_4 ▶ API 161

▶ 5-BcHH

(e) H, (f) H, C=C, H (d); (b); $CH_2-CH-CH_3$(a); (c); CH_3 (a)

	Shift	Peak
a	.89	2A
b	~1.60	S
c	1.87	S
d	4.83	R
e	5.03	R
▶f	~5.62	O

▶ C_6H_{12} ▶ 10% CCl_4 ▶ API 33

▶ 5-BbHH

(f) H, (h) H, C=C, H (g); (e) (d) (c) (b) (a) $CH_2CH_2CH_2CH_2CH_3$

	Shift	Peak
a	.89	3C
b	.90	S
c	~1.33 or 1.35	S
d	~1.35 or 1.33	S
e	1.97	4C
f	~4.84	R
g	~4.97	R
▶h	~5.70	R

▶ C_7H_{14} ▶ liquid ▶ API 384

▶ 5-BcHH

(e) H, (f) H, C=C, H (d); CH_3(a); CH_2CH(b); (c); CH_3(a)

	Shift	Peak
a	.90	2A
b	1.48	S
c	1.88	S
d	4.77	R
e	4.77	R
▶f	5.65	R

▶ C_6H_{12} ▶ liquid ▶ API 375

▶ 5-BcHH

(f) H (g) H
C=C
H (e) (d) (c) (b) $CH_2CHCH_2CH_3$
CH_3(a)

	Shift	Peak
a	~ .87	3D
b	~ .95	2H
c	1.30	O
d	1.98	4C
e	~4.96	R
f	~4.96	R
▶g	5.67	R

▶ C_7H_{14} ▶ liquid run, neat ▶ API 393

▶ 5-BoHH

(g) H (i) H (e) CH_2–OH (c) (i) H (g) H
H C=C CH_2 (f) O (d) CH_2 C (d) CH_2 O CH_2 (f) C=C H
(h) CH_2–CH_3 (b) (a) (h)

	Shift	Peak
a	0.84	3C
b	1.38	4C
c	2.83	3B
d	3.43	1E
e	3.59	2B
f	3.95	2E
g	5.14	R
h	5.22	R
▶i	5.87	R

▶ $C_{12}H_{22}O_3$ ▶ 7% $CDCl_3$ ▶ V 603

▶ 5-BdHH

(c) H (e) H
C=C
H (d) (b) (a) $CH_2C(CH_3)_3$

	Shift	Peak
a	.83	1A
b	1.92	2E
c	4.82	R
d	5.03	R
▶e	5.72	R

▶ C_7H_{14} ▶ liquid ▶ API 385

▶ 5-BoHH

(d) H (d) H
(a) (a)
(b)H–C=C–CH_2–O–CH_2–C=C–H(b)
(c)H H(c)

	Shift	Peak
a	3.97	2E
b	5.17	R
c	5.25	R
▶d	5.88	R

▶ $C_6H_{10}O$ ▶ 7% $CDCl_3$ ▶ V 134

▶ 5-BgHH

(d) H (b) H
C=C
BrH_2C (a) H (c)

	Shift	Peak
a	3.93	2B
b	~5.12	R
c	~5.32	R
▶d	~6.05	R

▶ C_3H_5Br ▶ 7% $CDCl_3$ ▶ V 24

▶ 5-BoHH

(d)H H(f)
C=C O H(b)
(e)H (c) CH_2 C=C
(g)H H(a)

	Shift	Peak
a	4.05	R
b	4.20	R
c	4.20	R
d	5.20	R
e	5.32	R
▶f	5.90	R
g	6.45	4D

▶ C_5H_8O ▶ 7% $CDCl_3$ ▶ V 110

▶ 5-BiHH

(b) H (d) H
C=C
H (c) CH_2–I (a)

	Shift	Peak
a	3.87	2E
b	5.05	R
c	5.15	R
▶d	5.93	R

▶ C_3H_5I ▶ 7% $CDCl_3$ ▶ V 26

▶ 5-BpHH

(c) H (e) H
C=C
H (d) CH_2–OH (b) (a)

	Shift	Peak
a	3.58	1A
b	4.13	2E
c	5.13	R
d	5.25	R
▶e	~6.00	R

▶ C_3H_6O ▶ 7% $CDCl_3$ ▶ V 34

▶ 5-BlHH

(b) H (d) H
C=C H(a)
(c)H C H
H C=C
H H

	Shift	Peak
a	2.73	3C
b	~4.90	R
c	~5.07	R
▶d	~5.69	R

▶ C_5H_8 ▶ 50% CCl_4 ▶ API 470

▶ 5-BqHH

(c) H O O (c) H
(b) H–C (a) C C (a) C–H (b)
C–CH_2 O C=C O CH_2–C
H (d) H (e) H (e) H (d)

	Shift	Peak
a	4.68	2D
b	5.25	R
c	5.32	R
▶d	5.88	R
e	6.27	1A

▶ $C_{10}H_{12}O_4$ ▶ 7% $CDCl_3$ ▶ V 262

▶ 5-BnHH

(d) H (b) (a) CH_2–NH_2
C=C
H (c) H (e)

	Shift	Peak
a	1.53	1B
b	3.30	2E
c	5.03	R
d	5.13	R
▶e	5.92	R

▶ C_3H_7N ▶ 7% $CDCl_3$ ▶ V 38

▶ 5-BsHH

(a) H(d)
C–S–CH_2–C
C–H(c)
H(b)

	Shift	Peak
a	2.82	2A
b	5.00 ± 0.05	R
c	5.00 ± 0.05	R
▶d	~5.60	R

▶ $C_{22}H_{20}S$ ▶ 7% $CDCl_3$ ▶ V 355

▶ 5-BsHH

(d) H, (a), (a), (d) H, (c) H—C, C—CH₂—S—CH₂—C, C—H (c), (b)H, H(b)

	Shift	Peak
a	3.08	2E
b	5.07	R
c	5.08	R
▶d	5.72	R

▶ $C_6H_{10}S$ ▶ 7% $CDCl_3$ ▶ V 136

▶ 5-CabHH

(f) H, H(g), C=C, CH_3(b), (e) H, $CHCH_2CH_3$, (d) (c) (a)

	Shift	Peak
a	.88	3D
b	.99	2A
c	1.32	S
d	2.01	5B
e	4.83 or 4.96	R
f	4.96 or 4.83	R
▶g	5.53	R

▶ C_6H_{12} ▶ 10% CCl_4 ▶ API 156

▶ 5-BsHH

N, S, C—S—CH_2—C, (a), H(d), (b), C—H, H(c)

	Shift	Peak
a	3.95	2A
b	5.17	R
c	5.33	R
▶d	~6.00	R

▶ $C_{10}H_9NS_2$ ▶ 7% $CDCl_3$ ▶ V 250

▶ 5-CabHH

(e) H, (g) H, C=C, H (f), (d) (c) (a), $CHCH_2CH_3$, CH_3(b)

	Shift	Peak
a	.88	3C
b	1.03	1E
c	1.24	S
d	1.96	7B
e	4.86	R
f	4.97	R
▶g	5.54	R

▶ C_6H_{12} ▶ liquid ▶ API 374

▶ 5-BvHH

(f)H, (g) H, (i) H, (c) CH_3, C=C, (d) CH_2, (e)H, H(h), (b) CH_3—CO, O, O, O, CH_3 (a)

	Shift	Peak
a	2.25 or 2.37	1A
b	2.25 or 2.37	1A
c	2.40	2D
d	3.33	2B
e	5.07	R
f	5.12	R
▶g	5.85	4D
h	6.22	1I
i	7.30	1I

▶ $C_{16}H_{16}O_4$ ▶ 7% $CDCl_3$ ▶ V 323

▶ 5-CabHH

(e) H, (g) H, C=C, H (f), (d) (c) (c) (a), $CHCH_2CH_2CH_3$, CH_3(b)

	Shift	Peak
a	.88	3D
b	.98	2H
c	1.29	O
d	2.11	5A
e	4.86	R
f	4.97	R
▶g	~5.58	R

▶ C_7H_{14} ▶ liquid ▶ API 392

▶ 5-BvHH

(e) O—CH_2—O, (g)H, (h)H, H(f), (a)CH_2, H(b), C=C, (d)H, H(c)

	Shift	Peak
a	3.30	2E
b	5.03	2E
c	5.03	2E
▶d	~5.88	O
e	5.88	1A
f	6.67 ± .03	1H
g	6.67 ± .03	1H
h	6.67 ± .03	1H

▶ $C_{10}H_{10}O_2$ ▶ 7% $CDCl_3$ ▶ V 253

▶ 5-CacHH

(d) H, (f) H, (a) CH_3, C=C, (c), CH—CH—CH_3(a), H (e), CH_3 (b)

	Shift	Peak
a	.88	3A
b	.98	2H
c	1.73	7X
d	4.90	R
e	5.00	R
▶f	5.64	R

▶ C_7H_{14} ▶ liquid ▶ API 413

▶ 5-BvHH

(f) H, (c) H, (g) H, (a) CH_2, C=C, H(d), (i)H, (e)HO, H(h), OCH_3 (b)

	Shift	Peak
a	3.30	2E
b	3.83	1A
c	5.05	2E
d	5.05	2E
e	5.59	1B
▶f	~5.94	O
g	~6.64	S
h	~6.69	S
i	6.87	S

▶ $C_{10}H_{12}O_2$ ▶ 7% $CDCl_3$ ▶ V 260

▶ 5-CbbHH

H, H(d), C=C, H, (c) (b) (a), $CHCH_2CH_3$, CH_2CH_3

	Shift	Peak
a	.83	3C
b	1.30	O
c	~1.72	S
▶d	~5.50	R

▶ C_7H_{14} ▶ liquid ▶ API 398

▶ 5-CaaHH

(c) H, (e) H, C=C, H (d), (b) (a), $CH(CH_3)_2$

	Shift	Peak
a	.93	2A
b	2.14	6B
c	4.60	R
d	4.75	R
▶e	5.48	R

▶ C_5H_{10} ▶ liquid run, neat ▶ API 387

▶ 5-(C)bbHH

H(c), (d)H, H(a), C=C, (e)H, H(b)

	Shift	Peak
a	4.93	R
b	5.02	R
c	~5.68	1E
d	~5.68	1E
▶e	5.83	4D

▶ C_8H_{12} ▶ 7% $CDCl_3$ ▶ V 210

▶ 5-CvvHH

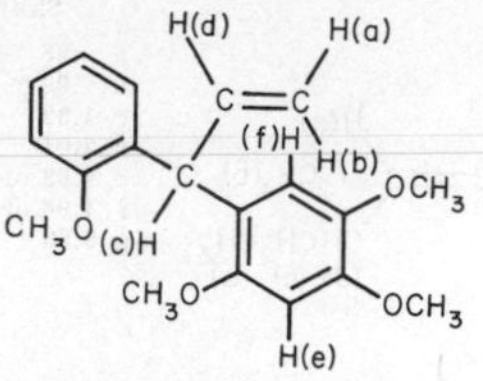

	Shift	Peak
a	4.77	R
b	5.15	R
c	5.47	2E
▶d	6.27	R
e	6.53	2A
f	6.68	O

▶ $C_{19}H_{22}O_4$ ▶ 7% $CDCl_3$ ▶ V 678

▶ 5-HBaH

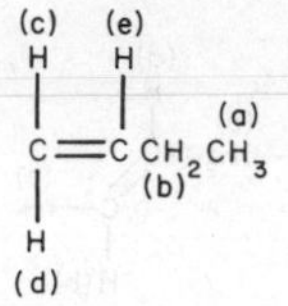

	Shift	Peak
a	.97	3C
b	2.02	5B
c	4.95	R
▶d	4.98	R
e	5.80	R

▶ C_4H_8 ▶ degassed liquid ▶ API 55

▶ 5-DaaaHH

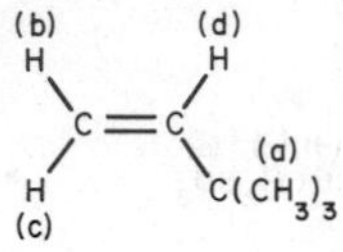

	Shift	Peak
a	.98	1A
b	~4.74	R
c	~4.92	R
▶d	5.78	4D

▶ C_6H_{12} ▶ liquid ▶ API 383

▶ 5-HBaH

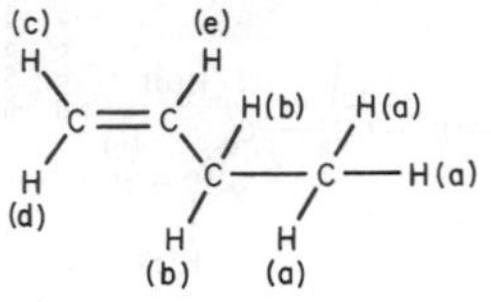

	Shift	Peak
a	0.97	3C
b	1.98	5B
c	~4.88	R
▶d	~5.7	R
e	~5.9	R

▶ C_4H_8 ▶ liquid ▶ API 365

▶ 5-DaabHH

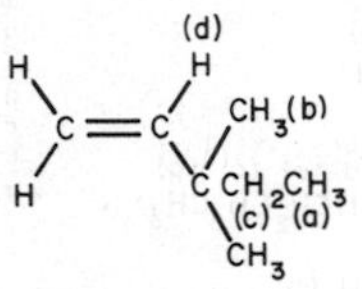

	Shift	Peak
a	.78	3D
b	.97	1E
c	1.28	O
▶d	5.69	4D

▶ C_7H_{14} ▶ liquid ▶ API 401

▶ 5-HBbH

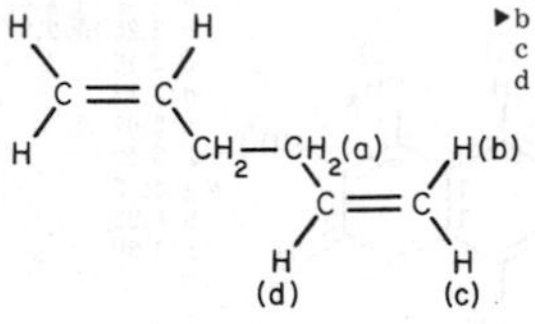

	Shift	Peak
a	2.08	2E
▶b	~4.88 or 4.99	R
c	~4.99 or 4.88	R
d	~5.67	R

▶ C_6H_{10} ▶ liquid ▶ API 474

▶ 5-DaapHH

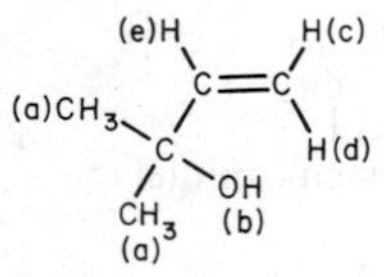

	Shift	Peak
a	1.32	1A
b	1.92	1B
c	4.98	R
d	5.17	R
▶e	6.00	4D

▶ $C_5H_{10}O$ ▶ 7% $CDCl_3$ ▶ V 444

▶ 5-HBbH

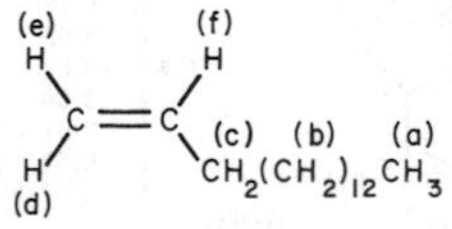

	Shift	Peak
a	.99	3B
b	1.25	1A
c	2.02	3B
▶d	4.90	R
e	5.08	R
f	5.68	R

▶ $C_{16}H_{32}$ ▶ 50% CCl_4 ▶ API 73

▶ 5-DabpHH

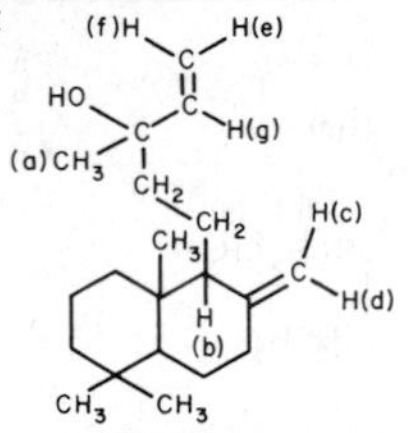

	Shift	Peak
a	1.28	1E
b	2.40	S
c	4.52	1B
d	4.83	1B
e	5.08	R
f	5.22	R
▶g	5.92	4D

▶ $C_{20}H_{34}O$ ▶ 7% $CDCl_3$ ▶ V 685

▶ 5-HBbH

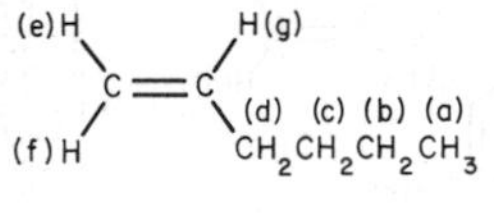

	Shift	Peak
a	.90	3B
b	~1.34	S
c	~1.34	S
d	2.00	4C
e	4.85	R
▶f	4.97	R
g	5.68	R

▶ C_6H_{12} ▶ 10% CCl_4 ▶ API 155

▶ 5-DabqHH

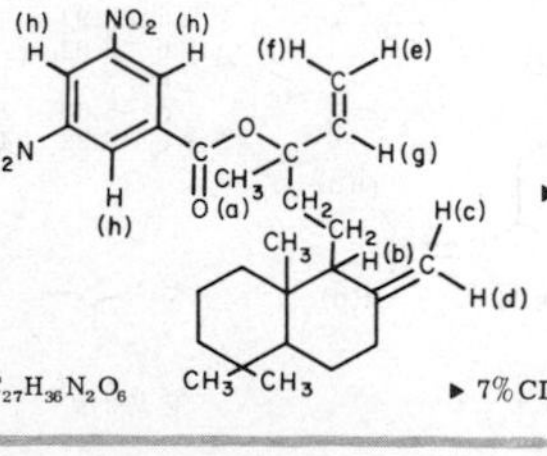

	Shift	Peak
a	1.74	1E
b	2.40	O
c	4.49 or 4.82	1B
d	4.82 or 4.49	1B
e	5.22	R
f	5.24	R
▶g	6.08	4D
h	~8.01	N

▶ $C_{27}H_{36}N_2O_6$ ▶ 7% $CDCl_3$ ▶ V 697

▶ 5-HBbH

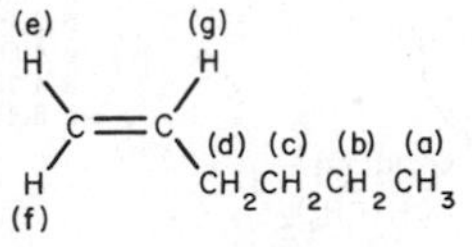

	Shift	Peak
a	.88	3C
b	~1.29 or 1.31	S
c	~1.31 or 1.29	S
d	2.00	3C
e	4.84	R
▶f	4.97	R
g	5.65	R

▶ C_6H_{12} ▶ liquid ▶ API 368

▶ 5-HBbH

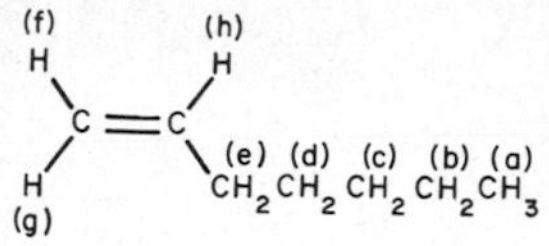

(f) H, (g) H, (h) H — C=C — $CH_2CH_2CH_2CH_2CH_3$ (e) (d) (c) (b) (a)

	Shift	Peak
a	.89	3C
b	.90	S
c	~1.33 or 1.35	S
d	~1.35 or 1.33	S
e	1.97	4C
f	~4.84	R
▶g	~4.97	R
h	~5.70	R

▶ C_7H_{14} ▶ liquid ▶ API 384

▶ 5-HBbH

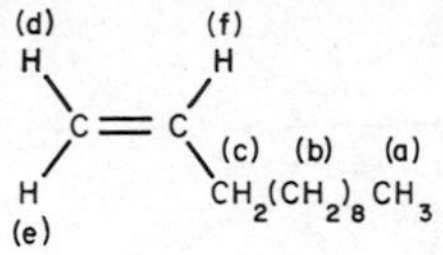

(d) H, (e) H, (f) H — C=C — $CH_2(CH_2)_8CH_3$ (c) (b) (a)

	Shift	Peak
a	.92	3B
b	1.32	1B
c	2.05	3B
d	4.86	R
▶e	5.00	R
f	5.68	R

▶ $C_{12}H_{24}$ ▶ 50 vol.% CCl_4 ▶ API 435

▶ 5-HBbH

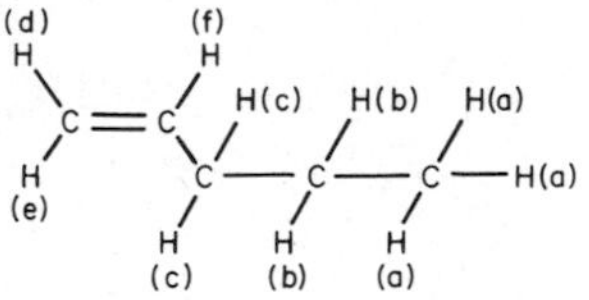

(d) H, (e) H, (f) H — C=C — C(H(c))(H(c)) — C(H(b))(H(b)) — C(H(a))(H(a))—H(a)

	Shift	Peak
a	.92	3C
b	1.38	6B
c	1.93	4C
d	4.76	R
▶e	4.98	R
f	5.68	R

▶ C_5H_{10} ▶ liquid ▶ API 366

▶ 5-HBbH

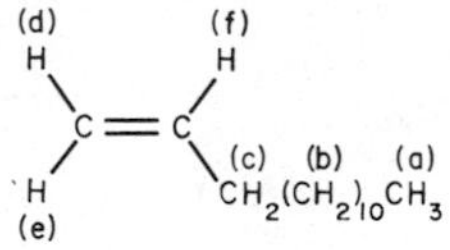

(d) H, (e) H, (f) H — C=C — $CH_2(CH_2)_{10}CH_3$ (c) (b) (a)

	Shift	Peak
a	.88	3B
b	1.27	1A
c	2.3	3B
d	4.88	7X
▶e	5.01	7X
f	5.72	R

▶ $C_{14}H_{28}$ ▶ 50 vol.% CCl_4 ▶ API 436

▶ 5-HBbH

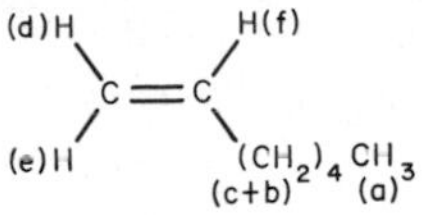

(d)H, (e)H, H(f) — C=C — $(CH_2)_4CH_3$ (c+b) (a)

	Shift	Peak
a	.91	3B
b	1.31	S
c	2.02	3B
d	4.84	R
▶e	4.98	R
f	5.67	R

▶ C_7H_{14} ▶ 10% CCl_4 ▶ API 157

▶ 5-HBbH

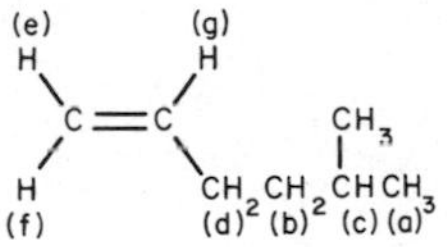

(e) H, (f) H, (g) H — C=C — $CH_2CH_2CH(CH_3)CH_3$ (d) (b) (c) (a)

	Shift	Peak
a	.91	2C
b	1.32	S
c	1.60	S
d	2.01	4C
e	4.87	R
▶f	5.02	R
g	5.70	R

▶ C_7H_{14} ▶ liquid ▶ API 394

▶ 5-HBbH

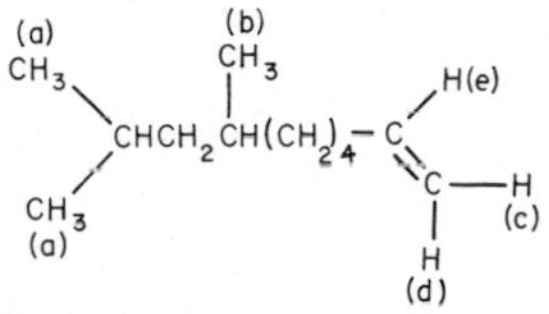

(a) CH_3 / (a) CH_3 >CHCH$_2$CH(CH$_3$ (b))(CH$_2$)$_4$—C(H(e))=C(H(c))H(d)

	Shift	Peak
a	0.82 or 0.87	2A
b	0.82 or 0.87	2A
c	4.95	R
▶d	4.98	R
e	~5.80	R

▶ $C_{12}H_{24}$ ▶ 7% $CDCl_3$ ▶ V 298

▶ 5-HBbH

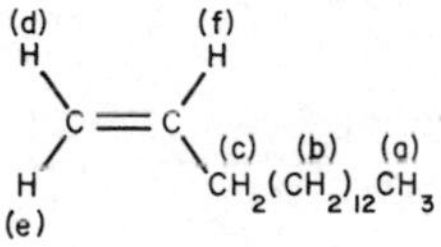

(d) H, (e) H, (f) H — C=C — $CH_2(CH_2)_{12}CH_3$ (c) (b) (a)

	Shift	Peak
a	.90	3B
b	1.24	1A
c	2.04	3B
d	4.88	R
▶e	5.02	R
f	5.72	R

▶ $C_{16}H_{32}$ ▶ 50 vol.% CCl_4 ▶ API 437

▶ 5-HBbH

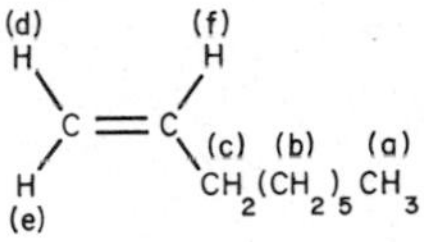

(d) H, (e) H, (f) H — C=C — $CH_2(CH_2)_5CH_3$ (c) (b) (a)

	Shift	Peak
a	.85	3B
b	1.28	1B
c	1.99	3B
d	4.88	R
▶e	4.99	R
f	5.68	R

▶ C_9H_{18} ▶ 50% CCl_4 ▶ API 434

▶ 5-HBbH

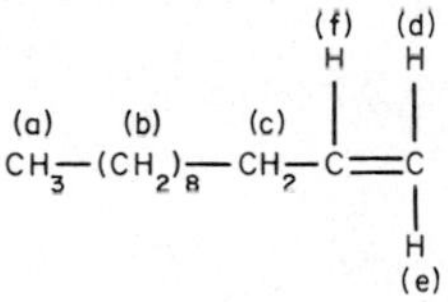

$CH_3—(CH_2)_8—CH_2—C(H(f))=C(H(d))H(e)$ (a) (b) (c)

	Shift	Peak
a	0.87	3D
b	1.28	1A
c	2.05	2G
d	4.80	R
▶e	~5.03	R
f	5.68	R

▶ $C_{12}H_{24}$ ▶ 44.4 mg/0.50 ml $CDCl_3$ ▶ S 202

▶ 5-HBbH

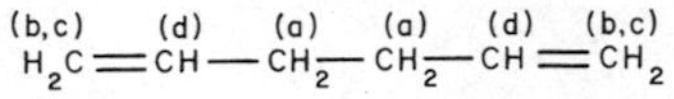

$H_2C=CH—CH_2—CH_2—CH=CH_2$ (b,c) (d) (a) (a) (d) (b,c)

	Shift	Peak
a	2.13	3A
b	4.93	R
▶c	~5.00	R
d	~5.70	R

▶ C_6H_{10} ▶ 10% CCl_4 ▶ API 161

▶ 5-HBcH

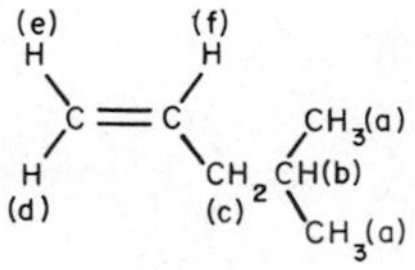

(d) H, (e) H, (f) H — C=C — CH_2 (c) CH(b)(CH_3(a))CH_3(a)

	Shift	Peak
a	.90	2A
b	1.48	S
c	1.88	S
▶d	4.77	R
e	4.77	R
f	5.65	R

▶ C_6H_{12} ▶ liquid ▶ API 375

▶ 5-HBcH

	Shift	Peak
a	.89	2A
b	~1.60	S
c	1.87	S
▶d	4.83	R
e	5.03	R
f	~5.62	O

▶ C_6H_{12} ▶ 10% CCl_4 ▶ API 33

▶ 5-HBnH

	Shift	Peak
a	1.53	1B
b	3.30	2E
c	5.03	R
▶d	5.13	R
e	5.92	R

▶ C_3H_7N ▶ 7% $CDCl_3$ ▶ V 38

▶ 5-HBcH

	Shift	Peak
a	~ .87	3D
b	~ .95	2H
c	1.30	O
d	1.98	4C
▶e	~4.96	R
f	~4.96	R
g	5.67	R

▶ C_7H_{14} ▶ liquid run, neat ▶ API 393

▶ 5-HBoH

	Shift	Peak
a	0.84	3C
b	1.38	4C
c	2.83	3B
d	3.43	1E
e	3.59	2B
f	3.95	2E
g	5.14	R
▶h	5.22	R
i	5.87	R

▶ $C_{12}H_{22}O_3$ ▶ 7% $CDCl_3$ ▶ V 603

▶ 5-HBdH

	Shift	Peak
a	.83	1A
b	1.92	2E
c	4.82	R
▶d	5.03	R
e	5.72	R

▶ C_7H_{14} ▶ liquid ▶ API 385

▶ 5-HBoH

	Shift	Peak
a	3.97	2E
b	5.17	R
▶c	5.25	R
d	5.88	R

▶ $C_6H_{10}O$ ▶ 7% $CDCl_3$ ▶ V 134

▶ 5-HBgH

	Shift	Peak
a	3.93	2B
b	~5.12	R
▶c	~5.32	R
d	~6.05	R

▶ C_3H_5Br ▶ 7% $CDCl_3$ ▶ V 24

▶ 5-HBoH

	Shift	Peak
a	4.05	R
b	4.20	R
c	4.20	R
d	5.20	R
▶e	5.32	R
f	5.90	R
g	6.45	4D

▶ C_5H_8O ▶ 7% $CDCl_3$ ▶ V 110

▶ 5-HBiH

	Shift	Peak
a	3.87	2E
b	5.05	R
▶c	5.15	R
d	5.93	R

▶ C_3H_5I ▶ 7% $CDCl_3$ ▶ V 26

▶ 5-HBpH

	Shift	Peak
a	3.58	1A
b	4.13	2E
c	5.13	R
▶d	5.25	R
e	~6.00	R

▶ C_3H_6O ▶ 7% $CDCl_3$ ▶ V 34

▶ 5-HBlH

	Shift	Peak
a	2.73	3C
b	~4.90	R
▶c	~5.07	R
d	~5.69	R

▶ C_5H_8 ▶ 50% CCl_4 ▶ API 470

▶ 5-HBqH

	Shift	Peak
a	4.68	2D
b	5.25	R
▶c	5.32	R
d	5.88	R
e	6.27	1A

▶ $C_{10}H_{12}O_4$ ▶ 7% $CDCl_3$ ▶ V 262

▶ 5-HBsH

	Shift	Peak
a	2.82	2A
▶b	5.00 ± 0.05	R
c	5.00 ± 0.05	R
d	~5.60	R

▶ $C_{22}H_{20}S$ ▶ 7% $CDCl_3$ ▶ V 355

▶ 5-HCaaH

	Shift	Peak
a	.93	2A
b	2.14	6B
c	4.60	R
▶d	4.75	R
e	5.48	R

▶ C_5H_{10} ▶ liquid run, neat ▶ API 387

▶ 5-HBsH

	Shift	Peak
a	3.08	2E
▶b	5.07	R
c	5.08	R
d	5.72	R

▶ $C_6H_{10}S$ ▶ 7% $CDCl_3$ ▶ V 136

▶ 5-HCabH

	Shift	Peak
a	.88	3D
b	.99	2A
c	1.32	S
d	2.01	5B
▶e	4.83 or 4.96	R
f	4.96 or 4.83	R
g	5.53	R

▶ C_6H_{12} ▶ 10% CCl_4 ▶ API 156

▶ 5-HBsH

	Shift	Peak
a	3.95	2A
b	5.17	R
▶c	5.33	R
d	~6.00	R

▶ $C_{10}H_9NS_2$ ▶ 7% $CDCl_3$ ▶ V 250

▶ 5-HCabH

	Shift	Peak
a	.88	3C
b	1.03	1E
c	1.24	S
d	1.96	7B
e	4.86	R
▶f	4.97	R
g	5.54	R

▶ C_6H_{12} ▶ liquid ▶ API 374

▶ 5-HBvH

	Shift	Peak
a	3.30	2E
▶b	5.03	2E
c	5.03	2E
d	~5.88	O
e	5.88	1A
f	6.67 ± .03	1H
g	6.67 ± .03	1H
h	6.67 ± .03	1H

▶ $C_{10}H_{10}O_2$ ▶ 7% $CDCl_3$ ▶ V 253

▶ 5-HCabH

	Shift	Peak
a	.88	3D
b	.98	2H
c	1.29	O
d	2.11	5A
e	4.86	R
▶f	4.97	R
g	~5.58	R

▶ C_7H_{14} ▶ liquid ▶ API 392

▶ 5-HBvH

	Shift	Peak
a	3.30	2E
b	3.83	1A
c	5.05	2E
▶d	5.05	2E
e	5.59	1B
f	~5.94	O
g	~6.64	S
h	~6.69	S
i	6.87	S

▶ $C_{10}H_{12}O_2$ ▶ 7% $CDCl_3$ ▶ V 260

▶ 5-HCacH

	Shift	Peak
a	.88	3A
b	.98	2H
c	1.73	7X
d	4.90	R
▶e	5.00	R
f	5.64	R

▶ C_7H_{14} ▶ liquid ▶ API 413

▶ 5-HBvH

	Shift	Peak
a	2.25 or 2.37	1A
b	2.25 or 2.37	1A
c	2.40	2D
d	3.33	2B
▶e	5.07	R
f	5.12	R
g	5.85	4D
h	6.22	1I
i	7.30	1I

▶ $C_{16}H_{16}O_4$ ▶ 7% $CDCl_3$ ▶ V 323

▶ 5-H(C)bbH

	Shift	Peak
▶a	4.93	R
b	5.02	R
c	~5.68	1E
d	~5.68	1E
e	5.83	4D

▶ C_8H_{12} ▶ 7% $CDCl_3$ ▶ V 210

▶ 5-HDaaaH

	Shift	Peak
a	.98	1A
b	~4.74	R
▶c	~4.92	R
d	5.78	4D

▶ C_6H_{12} ▶ liquid ▶ API 383

▶ 5-HHBb

	Shift	Peak
a	.92	3C
b	1.38	6B
c	1.93	4C
▶d	4.76	R
e	4.98	R
f	5.68	R

▶ C_5H_{10} ▶ liquid ▶ API 366

▶ 5-HDaapH

	Shift	Peak
a	1.32	1A
b	1.92	1B
c	4.98	R
▶d	5.17	R
e	6.00	4D

▶ $C_5H_{10}O$ ▶ 7% $CDCl_3$ ▶ V 444

▶ 5-HHBb

	Shift	Peak
a	0.87	3B
b	1.28	1A
c	2.05	2G
▶d	4.80	R
e	~5.03	R
f	5.68	R

▶ $C_{12}H_{24}$ ▶ 44.4 mg/0.50 ml $CDCl_3$ ▶ S 202

▶ 5-HDabpH

	Shift	Peak
a	1.28	1E
b	2.40	S
c	4.52	1B
d	4.83	1B
e	5.08	R
▶f	5.22	R
g	5.92	4D

▶ $C_{20}H_{34}O$ ▶ 7% $CDCl_3$ ▶ V 685

▶ 5-HHBb

	Shift	Peak
a	.88	3C
b	~1.29 or 1.31	S
c	~1.31 or 1.29	S
d	2.00	3C
▶e	4.84	R
f	4.97	R
g	5.65	R

▶ C_6H_{12} ▶ liquid ▶ API 368

▶ 5-HDabqH

	Shift	Peak
a	1.74	1E
b	2.40	O
c	4.49 or 4.82	1B
d	4.82 or 4.49	1B
e	5.22	R
▶f	5.24	R
g	6.08	4D
h	~8.01	N

▶ $C_{27}H_{36}N_2O_6$ ▶ 7% $CDCl_3$ ▶ V 697

▶ 5-HHBb

	Shift	Peak
a	.91	3B
b	1.31	S
c	2.02	3B
▶d	4.84	R
e	4.98	R
f	5.67	R

▶ C_7H_{14} ▶ 10% CCl_4 ▶ API 157

▶ 5-HHBa

	Shift	Peak
a	0.97	3C
b	1.98	5B
▶c	~4.88	R
d	~5.7	R
e	~5.9	R

▶ C_4H_8 ▶ liquid ▶ API 365

▶ 5-HHBb

	Shift	Peak
a	.89	3C
b	.90	S
c	~1.33 or 1.35	S
d	~1.35 or 1.33	S
e	1.97	4C
▶f	~4.84	R
g	~4.97	R
h	~5.70	R

▶ C_7H_{14} ▶ liquid ▶ API 384

▶ 5-HHBa

	Shift	Peak
a	.97	3C
b	2.02	5D
▶c	4.95	R
d	4.98	R
e	5.80	R

▶ C_4H_8 ▶ degassed liquid ▶ API 55

▶ 5-HHBb

	Shift	Peak
a	.90	3B
b	~1.34	S
c	~1.34	S
d	2.00	4C
▶e	4.85	R
f	4.97	R
g	5.68	R

▶ C_6H_{12} ▶ 10% CCl_4 ▶ API 155

▶ 5-HHBb

	Shift	Peak
a	.92	3B
b	1.32	1B
c	2.05	3B
▶ d	4.86	R
e	5.00	R
f	5.68	R

(d)H, (e)H C=C H(f), $CH_2(CH_2)_8CH_3$ (c) (b) (a)

▶ $C_{12}H_{24}$ ▶ 50 vol.% CCl_4 ▶ API 435

▶ 5-HHBb

	Shift	Peak
a	0.82 or 0.87	2A
b	0.82 or 0.87	2A
▶ c	4.95	R
d	4.98	R
e	~5.80	R

$(CH_3)_2CHCH_2CH(CH_3)(CH_2)_4-C(H)(e)=C(H)(H)$ (a) (b) (c) (d)

▶ $C_{12}H_{24}$ ▶ 7% $CDCl_3$ ▶ V 298

▶ 5-HHBb

	Shift	Peak
a	.91	2C
b	1.32	S
c	1.60	S
d	2.01	4C
▶ e	4.87	R
f	5.02	R
g	5.70	R

(e)H, (f)H C=C H(g), $CH_2CH_2CH(CH_3)CH_3$ (d) (b) (c) (a)

▶ C_7H_{14} ▶ liquid ▶ API 394

▶ 5-HHBb

	Shift	Peak
a	2.08	2E
b	~4.88 or 4.99	R
▶ c	~4.99 or 4.88	R
d	~5.67	R

$H_2C=CH-CH_2-CH_2(a)-CH(d)=CH_2$ (b)(c)

▶ C_6H_{10} ▶ liquid ▶ API 474

▶ 5-HHBb

	Shift	Peak
a	.85	3B
b	1.28	1B
c	1.99	3B
▶ d	4.88	R
e	4.99	R
f	5.68	R

(d)H, (e)H C=C H(f), $CH_2(CH_2)_5CH_3$ (c) (b) (a)

▶ C_9H_{18} ▶ 50% CCl_4 ▶ API 434

▶ 5-HHBb

	Shift	Peak
a	.99	3B
b	1.25	1A
c	2.02	3B
d	4.90	R
▶ e	5.08	R
f	5.68	R

(e)H, (d)H C=C H(f), $CH_2(CH_2)_{12}CH_3$ (c) (b) (a)

▶ $C_{16}H_{32}$ ▶ 50% CCl_4 ▶ API 73

▶ 5-HHBb

	Shift	Peak
a	.88	3D
b	1.27	1A
c	2.3	3B
▶ d	4.88	7X
e	5.01	7X
f	5.72	R

(d)H, (e)H C=C H(f), $CH_2(CH_2)_{10}CH_3$ (c) (b) (a)

▶ $C_{14}H_{28}$ ▶ 50 vol.% CCl_4 ▶ API 436

▶ 5-HHBc

	Shift	Peak
a	.90	2A
b	1.48	S
c	1.88	S
d	4.77	R
▶ e	4.77	R
f	5.65	R

(e)H, (d)H C=C H(f), $CH_2CH(CH_3)_2$ (c) (b) (a)

▶ C_6H_{12} ▶ liquid ▶ API 375

▶ 5-HHBb

	Shift	Peak
a	.90	3B
b	1.24	1A
c	2.04	3B
▶ d	4.88	R
e	5.02	R
f	5.72	R

(d)H, (e)H C=C H(f), $CH_2(CH_2)_{12}CH_3$ (c) (b) (a)

▶ $C_{16}H_{32}$ ▶ 50 vol.% CCl_4 ▶ API 437

▶ 5-HHBc

	Shift	Peak
a	~ .87	3D
b	~ .95	2H
c	1.30	O
d	1.90	4C
e	~4.96	R
▶ f	~4.96	R
g	5.07	R

(f)H, (e)H C=C H(g), $CH_2CH(CH_3)CH_2CH_3$ (d) (c) (a) (b)

▶ C_7H_{14} ▶ liquid run, neat ▶ API 393

▶ 5-HHBb

	Shift	Peak
a	2.13	3A
▶ b	4.93	R
c	~5.00	R
d	~5.70	R

$H_2C=CH-CH_2-CH_2-CH=CH_2$ (b,c) (d) (a) (a) (d) (b,c)

▶ C_6H_{10} ▶ 10% CCl_4 ▶ API 161

▶ 5-HHBc

	Shift	Peak
a	.89	2A
b	~1.60	S
c	1.87	S
d	4.83	R
▶ e	5.03	R
f	~5.62	O

(e)H, (d)H C=C H(f), $CH_2-CH(CH_3)-CH_3$ (c) (b) (a) (a)

▶ C_6H_{12} ▶ 10% CCl_4 ▶ API 33

▶ 5-HHBd

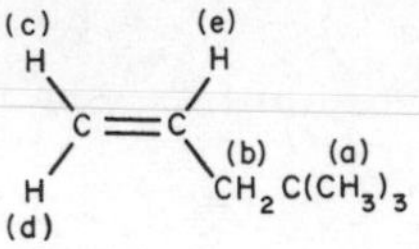

Shift	Peak
a .83	1A
b 1.92	2E
▶c 4.82	R
d 5.03	R
e 5.72	R

▶ C_7H_{14} ▶ liquid ▶ API 385

▶ 5-HHBo

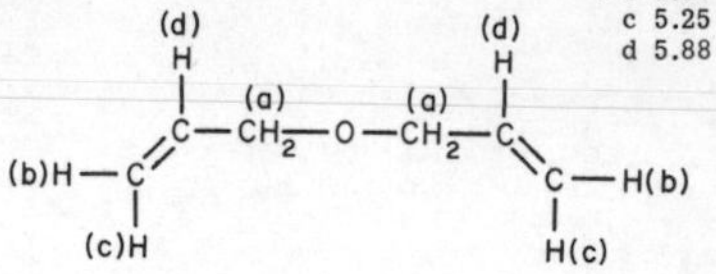

Shift	Peak
a 3.97	2E
▶b 5.17	R
c 5.25	R
d 5.88	R

▶ $C_6H_{10}O$ ▶ 7% $CDCl_3$ ▶ V 134

▶ 5-HHBg

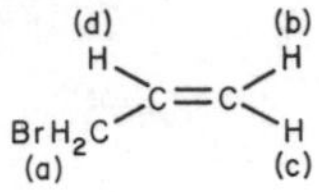

Shift	Peak
a 3.93	2B
▶b ~5.12	R
c ~5.32	R
d ~6.05	R

▶ C_3H_5Br ▶ 7% $CDCl_3$ ▶ V 24

▶ 5-HHBo

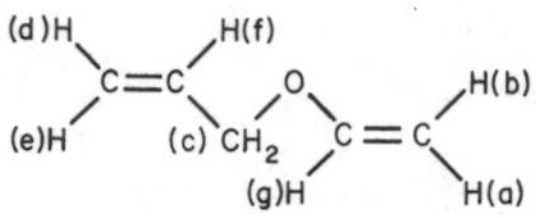

Shift	Peak
a 4.05	R
b 4.20	R
c 4.20	R
▶d 5.20	R
e 5.32	R
f 5.90	R
g 6.45	4D

▶ C_5H_8O ▶ 7% $CDCl_3$ ▶ V 110

▶ 5-HHBi

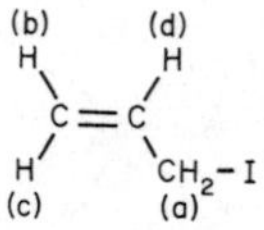

Shift	Peak
a 3.87	2E
▶b 5.05	R
c 5.15	R
d 5.93	R

▶ C_3H_5I ▶ 7% $CDCl_3$ ▶ V 26

▶ 5-HHBp

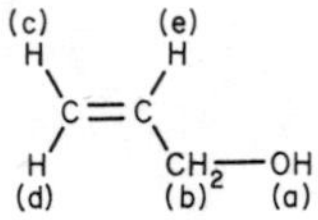

Shift	Peak
a 3.58	1A
b 4.13	2E
▶c 5.13	R
d 5.25	R
e ~6.00	R

▶ C_3H_6O ▶ 7% $CDCl_3$ ▶ V 34

▶ 5-HHBl

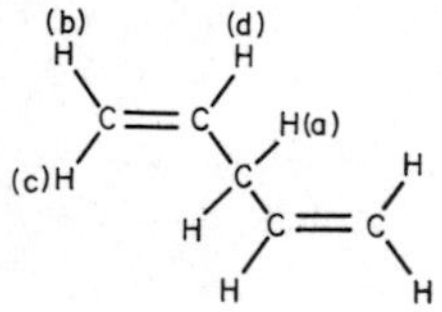

Shift	Peak
a 2.73	3C
▶b ~4.90	R
c ~5.07	R
d ~5.69	R

▶ C_5H_8 ▶ 50% CCl_4 ▶ API 470

▶ 5-HHBq

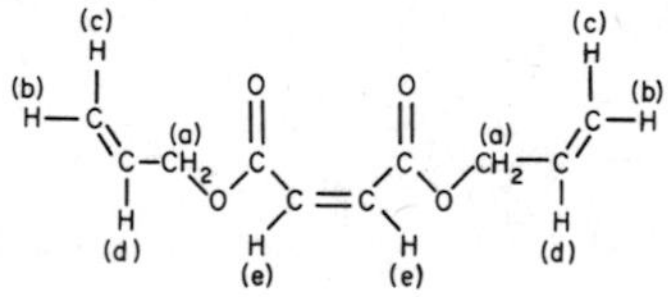

Shift	Peak
a 4.68	2D
▶b 5.25	R
c 5.32	R
d 5.88	R
e 6.27	1A

▶ $C_{10}H_{12}O_4$ ▶ 7% $CDCl_3$ ▶ V 262

▶ 5-HHBn

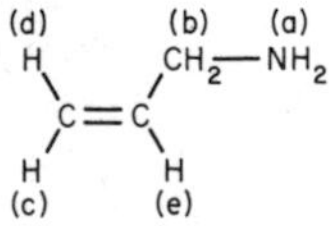

Shift	Peak
a 1.53	1B
b 3.30	2E
▶c 5.03	R
d 5.13	R
e 5.92	R

▶ C_3H_7N ▶ 7% $CDCl_3$ ▶ V 38

▶ 5-HHBs

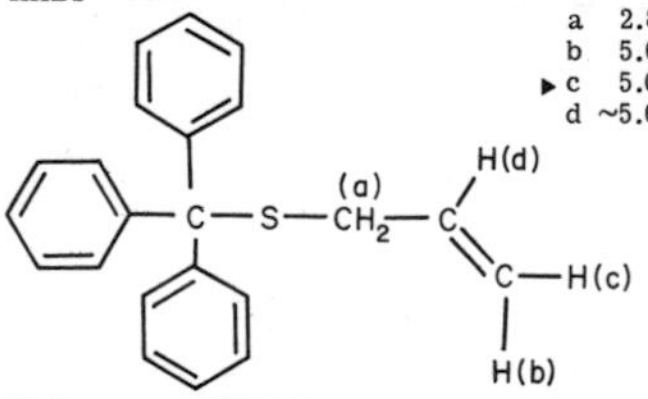

Shift	Peak
a 2.82	2A
b 5.00 ± 0.05	R
▶c 5.00 ± 0.05	R
d ~5.60	R

▶ $C_{22}H_{20}S$ ▶ 7% $CDCl_3$ ▶ V 355

▶ 5-HHBo

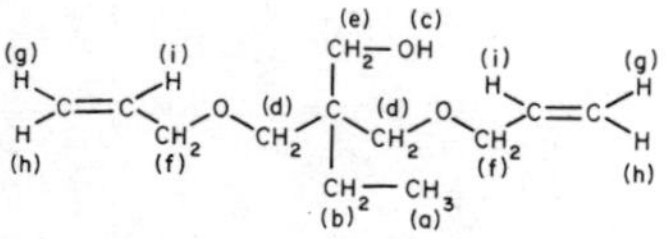

Shift	Peak
a 0.84	3C
b 1.38	4C
c 2.83	3B
d 3.43	1E
e 3.59	2B
f 3.95	2E
▶g 5.14	R
h 5.22	R
i 5.87	R

▶ $C_{12}H_{22}O_3$ ▶ 7% $CDCl_3$ ▶ V 603

▶ 5-HHBs

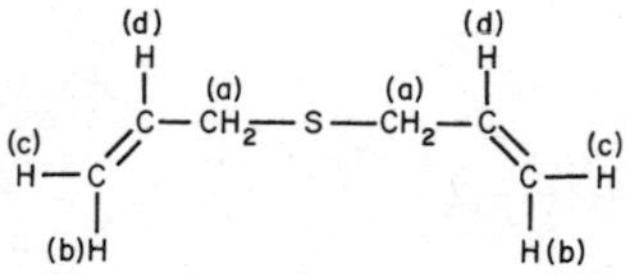

Shift	Peak
a 3.08	2E
b 5.07	R
▶c 5.08	R
d 5.72	R

▶ $C_6H_{10}S$ ▶ 7% $CDCl_3$ ▶ V 136

▶ 5-HHBs

N, S, C—S—CH_2—C, (a), H(d), (b), C—H, H(c)

	Shift	Peak
a	3.95	2A
▶ b	5.17	R
c	5.33	R
d	~6.00	R

▶ $C_{10}H_9NS_2$ ▶ 7% $CDCl_3$ ▶ V 250

▶ 5-HHCab

(e) H, (g) H, C=C, H (f), (d) (c) (c) (a), $CHCH_2CH_2CH_3$, CH_3(b)

	Shift	Peak
a	.88	3D
b	.98	2H
c	1.29	O
d	2.11	5A
▶ e	4.86	R
f	4.97	R
g	~5.58	R

▶ C_7H_{14} ▶ liquid ▶ API 392

▶ 5-HHBv

(e), O—CH_2, O, (g)H, (h)H, H(f), (a)CH_2, H(b), C=C, (d)H, H(c)

	Shift	Peak
a	3.30	2E
b	5.03	2E
▶ c	5.03	2E
d	~5.88	O
e	5.88	1A
f	6.67 ± .03	1H
g	6.67 ± .03	1H
h	6.67 ± .03	1H

▶ $C_{10}H_{10}O_2$ ▶ 7% $CDCl_3$ ▶ V 253

▶ 5-HHCab

(f) H, H(g), C=C, CH_3(b), (e) H, $CHCH_2CH_3$, (d) (c) (a)

	Shift	Peak
a	.88	3D
b	.99	2A
c	1.32	S
d	2.01	5B
e	4.83 or 4.96	R
▶ f	4.96 or 4.83	R
g	5.53	R

▶ C_6H_{12} ▶ 10% CCl_4 ▶ API 156

▶ 5-HHBv

(f) H, (c) H, (g) H, (a), C=C, CH_2, H(d), (i)H, (e)HO, H(h), OCH_3 (b)

	Shift	Peak
a	3.30	2E
b	3.83	1A
▶ c	5.05	2E
d	5.05	2E
e	5.59	1B
f	~5.94	O
g	~6.64	S
h	~6.69	S
i	6.87	S

▶ $C_{10}H_{12}O_2$ ▶ 7% $CDCl_3$ ▶ V 260

▶ 5-HHCac

(d) H, (f) H, (a) CH_3, C=C, (c), CH—CH—CH_3(a), H (e), CH_3 (b)

	Shift	Peak
a	.88	3A
b	.98	2H
c	1.73	7X
▶ d	4.90	R
e	5.00	R
f	5.64	R

▶ C_7H_{14} ▶ liquid ▶ API 413

▶ 5-HHBv

(g) H, (f)H, (i) H, (c) CH_3, C=C (d), (e)H, CH_2, H(h), (b) CH_3—CO, O, O, O, CH_3 (a)

	Shift	Peak
a	2.25 or 2.37	1A
b	2.25 or 2.37	1A
c	2.40	2D
d	3.33	2B
e	5.07	R
▶ f	5.12	R
g	5.85	4D
h	6.22	1I
i	7.30	1I

▶ $C_{16}H_{16}O_4$ ▶ 7% $CDCl_3$ ▶ V 323

▶ 5-HH(C)bb

H(c), (d)H, H(a), C=C, (e)H, H(b)

	Shift	Peak
a	4.93	R
▶ b	5.02	R
c	~5.68	1E
d	~5.68	1E
e	5.83	4D

▶ C_8H_{12} ▶ 7% $CDCl_3$ ▶ V 210

▶ 5-HHCaa

(c) H, (e) H, C=C, (b) (a), H (d), $CH(CH_3)_2$

	Shift	Peak
a	.93	2A
b	2.14	6B
▶ c	4.60	R
d	4.75	R
e	5.48	R

▶ C_5H_{10} ▶ liquid run, neat ▶ API 387

▶ 5-HHCvv

H(d), H(a), C=C, (f)H, H(b), C, OCH_3, CH_3O (c)H, CH_3O, OCH_3, H(e)

	Shift	Peak
▶ a	4.77	R
b	5.15	R
c	5.47	2E
d	6.27	R
e	6.53	2A
f	6.68	O

▶ $C_{19}H_{22}O_4$ ▶ 7% $CDCl_3$ ▶ V 678

▶ 5-HHCab

(e) H, (g) H, C=C, (d) (c) (a), H (f), $CHCH_2CH_3$, CH_3(b)

	Shift	Peak
a	.88	3C
b	1.03	1E
c	1.24	S
d	1.96	7B
▶ e	4.86	R
f	4.97	R
g	5.54	R

▶ C_6H_{12} ▶ liquid ▶ API 374

▶ 5-HHCvv

H(d), H(a), C=C, (f)H, H(b), C, OCH_3, CH_3O (c)H, CH_3O, OCH_3, H(e)

	Shift	Peak
a	4.77	R
▶ b	5.15	R
c	5.47	2E
d	6.27	R
e	6.53	2A
f	6.68	O

▶ $C_{19}H_{22}O_4$ ▶ 7% $CDCl_3$ ▶ V 678

▶ 5-HHDaaa

	Shift	Peak
a	.98	1A
▶ b	~4.74	R
c	~4.92	R
d	5.78	4D

▶ C_6H_{12} ▶ liquid ▶ API 383

▶ 5-HHOb

	Shift	Peak
▶ a	4.05	R
b	4.20	R
c	4.20	R
d	5.20	R
e	5.32	R
f	5.90	R
g	6.45	4D

▶ C_5H_8O ▶ 7% $CDCl_3$ ▶ V 110

▶ 5-HHDaap

	Shift	Peak
a	1.32	1A
b	1.92	1B
▶ c	4.98	R
d	5.17	R
e	6.00	4D

▶ $C_5H_{10}O$ ▶ 7% $CDCl_3$ ▶ V 444

▶ 5-HHQa

	Shift	Peak
a	2.12	1A
▶ b	4.55	R
c	4.85	R
d	7.25	4D

▶ $C_4H_6O_2$ ▶ 7% $CDCl_3$ ▶ V 65

▶ 5-HHDabp

	Shift	Peak
a	1.28	1E
b	2.40	S
c	4.52	1B
d	4.83	1B
▶ e	5.08	R
f	5.22	R
g	5.92	4D

▶ $C_{20}H_{34}O$ ▶ 7% $CDCl_3$ ▶ V 685

▶ 5-HHSa

	Shift	Peak
a	2.25	1A
b	4.95	2B
▶ c	5.18	2B
d	6.43	4D

▶ C_3H_6S ▶ 7% $CDCl_3$ ▶ V 36

▶ 5-HHDabq

	Shift	Peak
a	1.74	1E
b	2.40	O
c	4.49 or 4.82	1B
d	4.82 or 4.49	1B
▶ e	5.22	R
f	5.24	R
g	6.08	4D
h	~8.01	N

▶ $C_{27}H_{36}N_2O_6$ ▶ 7% $CDCl_3$ ▶ V 697

▶ 5-HHSaoo

	Shift	Peak
a	2.62	1A
▶ b	5.95	2B
c	6.13	2B
d	6.70	4D

▶ $C_3H_6O_2S$ ▶ 7% $CDCl_3$ ▶ V 35

▶ 5-HHJn

	Shift	Peak
a	1.18	2A
b	4.17	6B
▶ c	5.59	4A
d	~6.00	S
e	6.14	2A
f	6.24	1A

▶ $C_6H_{11}NO$ ▶ 7% $CDCl_3$ ▶ V 468

▶ 5-HHSl

	Shift	Peak
▶ a	5.67	S
b	5.72	S
c	7.05	4D

▶ $C_9H_7NS_2$ ▶ 7% $CDCl_3$ ▶ V 226

▶ 5-HHKa

	Shift	Peak
a	3.75	1A
▶ b	~5.82	4D
c	~6.20	R
d	~6.38	R

▶ $C_4H_6O_2$ ▶ 7% $CDCl_3$ ▶ V 64

▶ 5-HHVhh

	Shift	Peak
▶ a	5.28	R
b	5.73	R
c	6.69	4D
d	7.32	1A
e	7.32	1A

▶ C_8H_7Cl ▶ 7% $CDCl_3$ ▶ V 498

▶ 5-HHX

	Shift	Peak
▶ a	5.43	3C
b	5.90	2D
c	6.62	4D
d	7.22	Q
e	8.52	Q

▶ C_7H_7N ▶ 7% $CDCl_3$ ▶ V 155

▶ 5-HKaH

	Shift	Peak
a	3.75	1A
b	~5.82	4D
c	~6.20	R
▶ d	~6.38	R

▶ $C_4H_6O_2$ ▶ 7% $CDCl_3$ ▶ V 64

▶ 5-HHX

	Shift	Peak
▶ a	5.45	S
b	6.22	S
c	6.75	S
d	7.08	S
e	7.27	S
f	7.55	S
g	8.52	S

▶ C_7H_7N ▶ 7% $CDCl_3$ ▶ V 154

▶ 5-HObH

	Shift	Peak
a	4.05	R
▶ b	4.20	R
c	4.20	R
d	5.20	R
e	5.32	R
f	5.90	R
g	6.45	4D

▶ C_5H_8O ▶ 7% $CDCl_3$ ▶ V 110

▶ 5-HHZ

	Shift	Peak
a	1.27	1A
b	1.31	1A
c	1.49	2G
d	1.81	2G
e	4.22	7X
f	~5.90	S
▶ g	~5.90	S
h	~5.90	S

▶ $C_8H_{15}BO_2$ ▶ 7% $CDCl_3$ ▶ V 517

▶ 5-HQaH

	Shift	Peak
a	2.12	1A
b	4.55	R
▶ c	4.85	R
d	7.25	4D

▶ $C_4H_6O_2$ ▶ 7% $CDCl_3$ ▶ V 65

▶ 5-HHZ

	Shift	Peak
a	1.23	3A
b	3.82	4A
c	5.98 ± 0.05	S
▶ d	5.98 ± 0.05	S
e	5.98 ± 0.05	S

▶ $C_8H_{18}O_3Si$ ▶ 7% $CDCl_3$ ▶ V 219

▶ 5-HSaH

	Shift	Peak
a	2.25	1A
▶ b	4.95	2B
c	5.18	2B
d	6.43	4D

▶ C_3H_6S ▶ 7% $CDCl_3$ ▶ V 36

▶ 5-HHZ

	Shift	Peak
a	6.25 ± 0.03	1H
▶ b	6.25 ± 0.03	1H
c	6.25 ± 0.03	1H

▶ $C_2H_3Cl_3Si$ ▶ 7% $CDCl_3$ ▶ V 3

▶ 5-HSaooH

	Shift	Peak
a	2.62	1A
b	5.95	2B
▶ c	6.13	2B
d	6.70	4D

▶ $C_3H_6O_2S$ ▶ 7% $CDCl_3$ ▶ V 35

▶ 5-HJnH

	Shift	Peak
a	1.18	2A
b	4.17	6B
c	5.59	4A
d	~6.00	S
e	6.14	2A
▶ f	6.24	1A

▶ $C_6H_{11}NO$ ▶ 7% $CDCl_3$ ▶ V 468

▶ 5-HSlH

	Shift	Peak
a	5.67	S
▶ b	5.72	S
c	7.05	4D

▶ $C_9H_7NS_2$ ▶ 7% $CDCl_3$ ▶ V 226

▶ 5-HVhhH

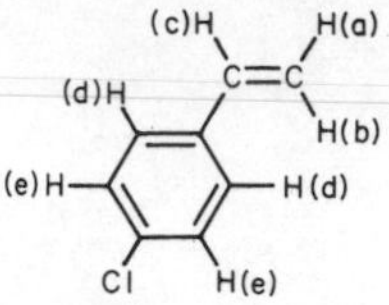

	Shift	Peak
a	5.28	R
▶ b	5.73	R
c	6.69	4D
d	7.32	1A
e	7.32	1A

▶ C_8H_7Cl ▶ 7% $CDCl_3$ ▶ V 498

▶ 5-JnHH

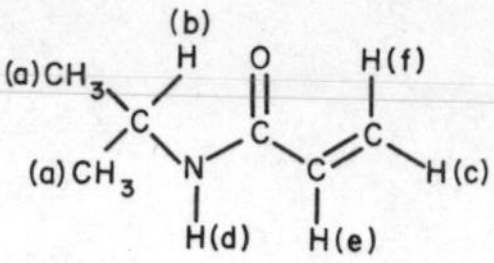

	Shift	Peak
a	1.18	2A
b	4.17	6B
c	5.59	4A
d	~6.00	S
▶ e	6.14	2A
f	6.24	1A

▶ $C_6H_{11}NO$ ▶ 7% $CDCl_3$ ▶ V 468

▶ 5-HXH

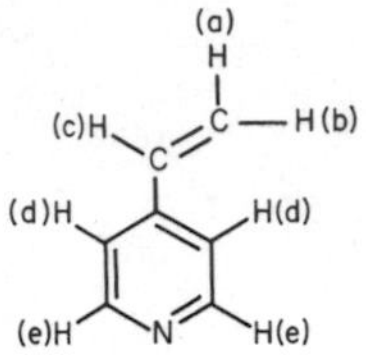

	Shift	Peak
a	5.43	3C
▶ b	5.90	2D
c	6.62	4D
d	7.22	Q
e	8.52	Q

▶ C_7H_7N ▶ 7% $CDCl_3$ ▶ V 155

▶ 5-KaHH

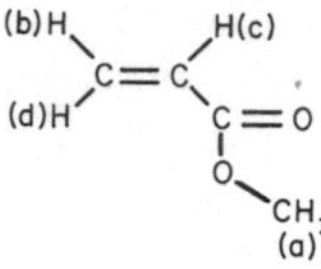

	Shift	Peak
a	3.75	1A
b	~5.82	4D
▶ c	~6.20	R
d	~6.38	R

▶ $C_4H_6O_2$ ▶ 7% $CDCl_3$ ▶ V 64

▶ 5-HXH

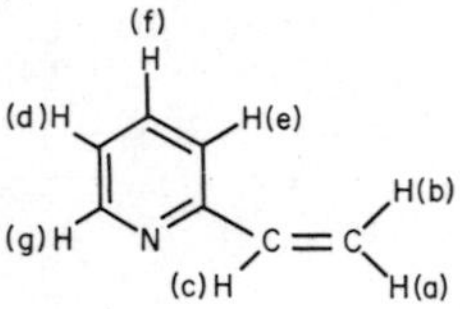

	Shift	Peak
a	5.45	S
▶ b	6.22	S
c	6.75	S
d	7.08	S
e	7.27	S
f	7.55	S
g	8.52	S

▶ C_7H_7N ▶ 7% $CDCl_3$ ▶ V 154

▶ 5-ObHH

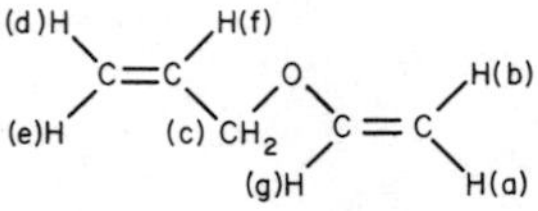

	Shift	Peak
a	4.05	R
b	4.20	R
c	4.20	R
d	5.20	R
e	5.32	R
f	5.90	R
▶ g	6.45	4D

▶ C_5H_8O ▶ 7% $CDCl_3$ ▶ V 110

▶ 5-HZH

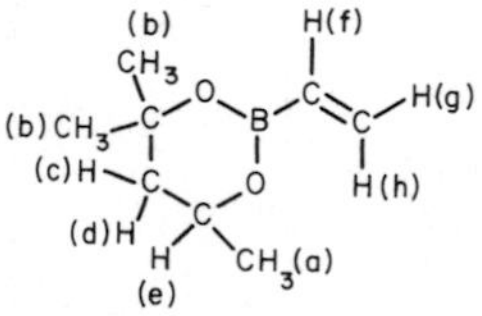

	Shift	Peak
a	1.27	1A
b	1.31	1A
c	1.49	2G
d	1.81	2G
e	4.22	7X
f	~5.90	S
g	~5.90	S
▶ h	~5.90	S

▶ $C_8H_{15}BO_2$ ▶ 7% $CDCl_3$ ▶ V 517

▶ 5-QaHH

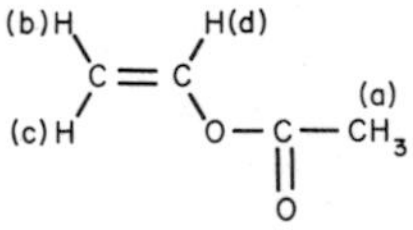

	Shift	Peak
a	2.12	1A
b	4.55	R
c	4.85	R
▶ d	7.25	4D

▶ $C_4H_6O_2$ ▶ 7% $CDCl_3$ ▶ V 65

▶ 5-HZH

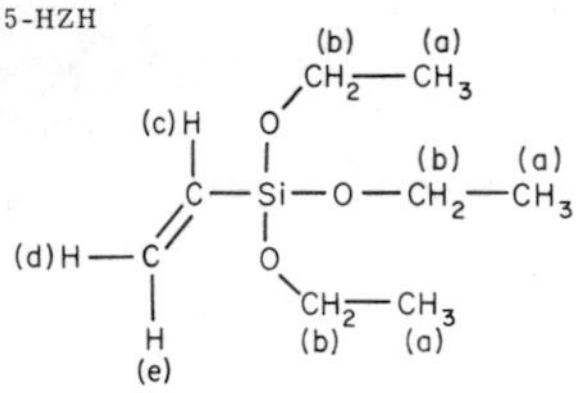

	Shift	Peak
a	1.23	3A
b	3.82	4A
c	5.98 ± 0.05	S
d	5.98 ± 0.05	S
▶ e	5.98 ± 0.05	S

▶ $C_8H_{18}O_3Si$ ▶ 7% $CDCl_3$ ▶ V 219

▶ 5-SaHH

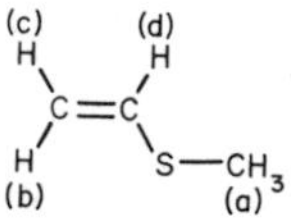

	Shift	Peak
a	2.25	1A
b	4.95	2B
c	5.18	2B
▶ d	6.43	4D

▶ C_3H_6S ▶ 7% $CDCl_3$ ▶ V 36

▶ 5-HZH

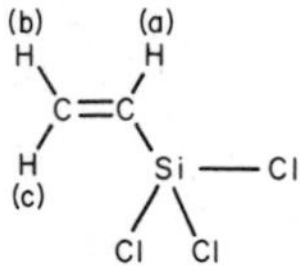

	Shift	Peak
a	6.25 ± 0.03	1H
b	6.25 ± 0.03	1H
▶ c	6.25 ± 0.03	1H

▶ $C_2H_3Cl_3Si$ ▶ 7% $CDCl_3$ ▶ V 3

▶ 5-SaooHH

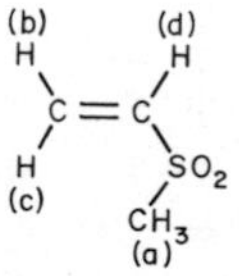

	Shift	Peak
a	2.62	1A
b	5.95	2B
c	6.13	2B
▶ d	6.70	4D

▶ $C_3H_6O_2S$ ▶ 7% $CDCl_3$ ▶ V 35

▶ 5-SlHH

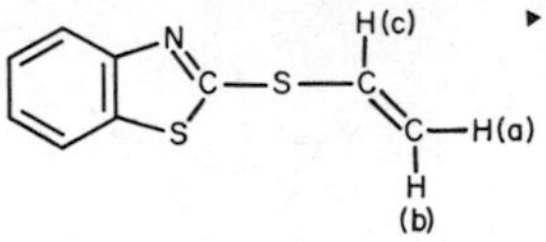

Shift	Peak
a 5.67	S
b 5.72	S
▶ c 7.05	4D

▶ $C_9H_7NS_2$ ▶ 7% $CDCl_3$ ▶ V 226

▶ 5-ZHH

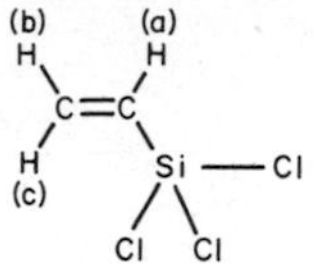

Shift	Peak
▶ a 6.25 ± 0.03	1H
b 6.25 ± 0.03	1H
c 6.25 ± 0.03	1H

▶ $C_2H_3Cl_3Si$ ▶ 7% $CDCl_3$ ▶ V 3

▶ 5-VhhHH

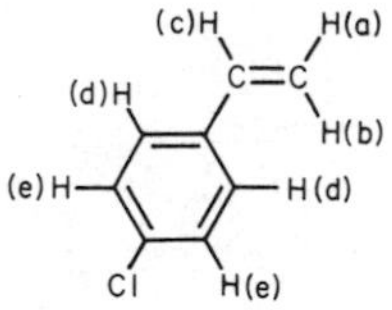

Shift	Peak
a 5.28	R
b 5.73	R
▶ c 6.69	4D
d 7.32	1A
e 7.32	1A

▶ C_8H_7Cl ▶ 7% $CDCl_3$ ▶ V 498

▶ 6-ABa

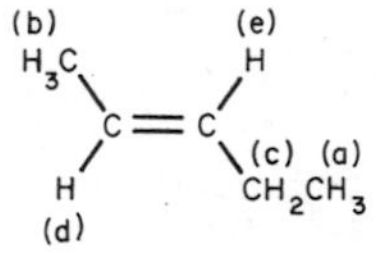

Shift	Peak
a .93	3C
b 1.58	2E
c 1.95	O
▶ d 5.35 or 5.43	S
e 5.43 or 5.35	S

▶ C_5H_{10} ▶ liquid ▶ API 461

▶ 5-XHH

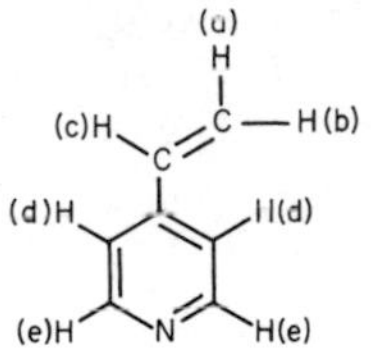

Shift	Peak
a 5.43	3C
b 5.90	2D
▶ c 6.62	4D
d 7.22	Q
e 8.52	Q

▶ C_7H_7N ▶ 7% $CDCl_3$ ▶ V 155

▶ 6-ABa

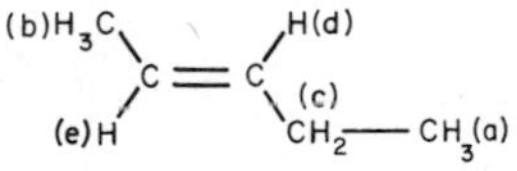

Shift	Peak
a .97	3A
b 1.61	5A
c 1.97	1F
d 5.37	S
▶ e 5.37	S

▶ C_5H_{10} ▶ 10% CCl_4 ▶ API 29

▶ 5-XHH

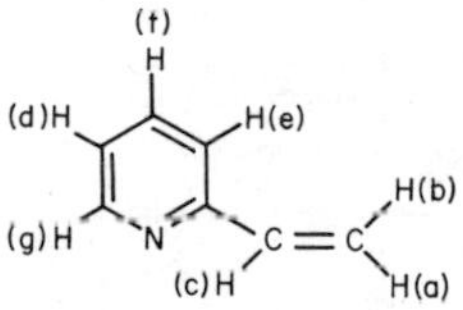

Shift	Peak
a 5.45	S
b 6.22	S
▶ c 6.75	S
d 7.08	S
e 7.27	S
f 7.55	S
g 8.52	S

▶ C_7H_7N ▶ 7% $CDCl_3$ ▶ V 154

▶ 6-ABa

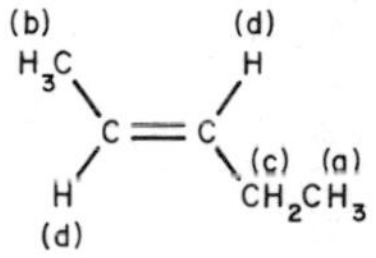

Shift	Peak
a .95	3A
b 1.60	1H
c 1.94	N
▶ d 5.38	N

▶ C_5H_{10} ▶ 50 vol.% CCl_4 ▶ API 430

▶ 5-ZHH

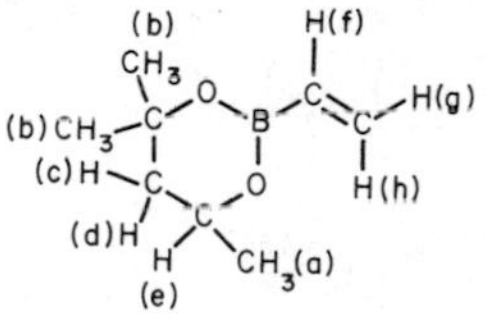

Shift	Peak
a 1.27	1A
b 1.31	1A
c 1.49	2G
d 1.81	2G
e 4.22	7X
▶ f ~5.90	S
g ~5.90	S
h ~5.90	S

▶ $C_8H_{15}BO_2$ ▶ 7% $CDCl_3$ ▶ V 517

▶ 6-ABb

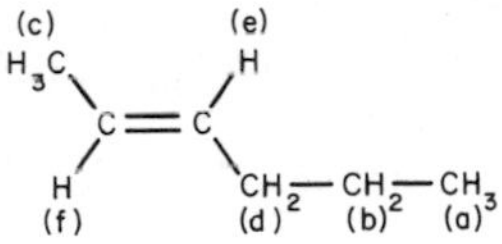

Shift	Peak
a .88	3B
b ~1.33	O
c 1.69	N
d 1.87	1F
e 5.32	S
▶ f 5.32	S

▶ C_6H_{12} ▶ 10% CCl_4 ▶ API 32

▶ 5-ZHH

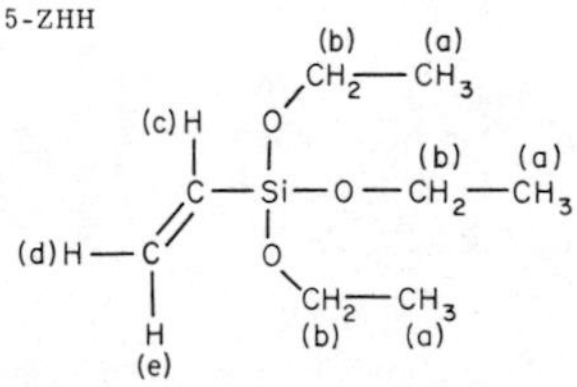

Shift	Peak
a 1.23	3A
b 3.82	4A
▶ c 5.98 ± 0.05	S
d 5.98 ± 0.05	S
e 5.98 ± 0.05	S

▶ $C_8H_{18}O_3Si$ ▶ 7% $CDCl_3$ ▶ V 219

▶ 6-ABb

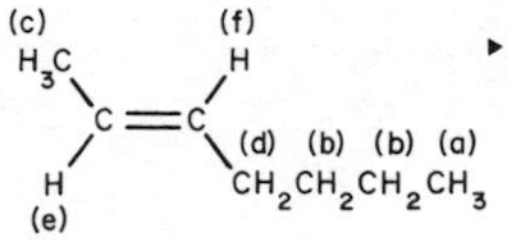

Shift	Peak
a .88	3C
b 1.31	O
c 1.62	2E
d 1.92	1F
▶ e 5.33 or 5.42	S
f 5.42 or 5.33	S

▶ C_7H_{14} ▶ liquid ▶ API 390

▶ 6-ABb

	Shift	Peak
a	.88	3C
b	1.28	S
c	1.60	2E
d	1.92	1F
▶ e	5.35	O

▶ C_6H_{12} ▶ 50 vol.% CCl_4 ▶ API 370

▶ 6-AJh

	Shift	Peak
a	2.03	2E
b	6.13	N
▶ c	6.87	N
d	9.48	O

▶ C_4H_6O ▶ 7% $CDCl_3$ ▶ V 60

▶ 6-ABc

	Shift	Peak
a	.98	2A
b	1.38	S
c	1.60	2H
d	1.80	S
▶ e	5.30	N

▶ C_7H_{14} ▶ 50 vol.% CCl_4 ▶ API 431

▶ 6-AKh

	Shift	Peak
a	1.90	2E
b	5.83	2E
▶ c	7.10	2E
d	12.18	O

▶ $C_4H_6O_2$ ▶ 7% $CDCl_3$ ▶ V 61

▶ 6-ABc

	Shift	Peak
a	.83	2A
b	~1.40	S
c	1.62	2E
d	1.83	O
▶ e	~5.38	O
f	~5.38	O

▶ C_7H_{14} ▶ liquid ▶ API 396

▶ 6-A(L)jl

	Shift	Peak
a	1.65	2A
b	1.97	2E
c	3.86	1C
d	3.92	S
e	4.32	O
f	5.87	1I
g	6.06	2A
▶ h	6.99	4A

▶ $C_{12}H_{12}O_5$ ▶ 7% $CDCl_3$ ▶ V 595

▶ 6-ABp

	Shift	Peak
a	1.72	N
b	2.07	1A
c	4.07	N
d	~5.68	S
▶ e	~5.68	S

▶ C_4H_8O ▶ 7% $CDCl_3$ ▶ V 414

▶ 6-A(L)lo

	Shift	Peak
a	0.89	3B
b	~1.30	S
c	~1.40	S
d	1.72	1A
e	1.97	2A
f	2.94	3B
g	6.06	S
h	6.17	S
▶ i	6.58	S
j	6.89	S
k	7.88	1G

▶ $C_{21}H_{22}O_5$ ▶ 7% $CDCl_3$ ▶ V 687

▶ 6-ACaa

	Shift	Peak
a	.94	2A
b	1.60	2E
c	2.15	1D
▶ d	5.36	S
e	5.45	S

▶ C_6H_{12} ▶ liquid run ▶ API 70

▶ 6-AVhh

	Shift	Peak
a	1.82	2A
b	3.88	1A
c	5.48	1B
▶ d	6.03	S
e	6.20	S

▶ $C_{10}H_{12}O_2$ ▶ 45.0 mg/0.5 ml $CDCl_3$ ▶ S 76

▶ 6-ACaa

	Shift	Peak
a	.98	2A
b	1.62	2E
c	2.22	1F
▶ d	5.37 or 5.43	S
e	5.43 or 5.37	S

▶ C_6H_{12} ▶ 5 vol.% CCl_4 ▶ API 380

▶ 6-AVhh

	Shift	Peak
a	1.80	2A
b	5.87	1A
▶ c	~6.08	S
d	6.23	S
e	6.70	S
f	6.70	S
g	6.83	S

▶ $C_{10}H_{10}O_2$ ▶ 7% $CDCl_3$ ▶ V 252

▶ 6-AVhh

(b) OCH_3
(e)H H(e)
(f)H H(f)
C H(c)
(d)H C
CH_3(a)

	Shift	Peak
a	1.83	2A
b	3.75	1A
▶ c	6.08	N
d	6.28	N
e	6.80	Q
f	7.23	Q

▶ $C_{10}H_{12}O$ ▶ 7% $CDCl_3$ ▶ V 258

▶ 6-BbA

(c) H_3C (e) H
C=C
H (e) (d) (b) (a) $CH_2CH_2CH_3$

	Shift	Peak
a	.88	3C
b	1.28	S
c	1.60	2E
d	1.92	1F
▶ e	5.35	O

▶ C_6H_{12} ▶ 50 vol.% CCl_4 ▶ API 370

▶ 6-AVhh

O
O—C—CH_3 (b)
O—CH_3(c)
(a) CH=CH—CH_3
(d) (e)

	Shift	Peak
a	1.88	2A
b	2.31	1A
c	3.82	1A
d	~6.1-6.7	S
▶ e	~6.1-6.7	S

▶ $C_{12}H_{14}O_3$ ▶ 45.4 mg/0.50 ml $CDCl_3$ ▶ S 130

▶ 6-BbA

(c) H_3C (f) H
C=C
H (e) (d) (b) (b) (a) $CH_2CH_2CH_2CH_3$

	Shift	Peak
a	.88	3C
b	1.31	O
c	1.62	2E
d	1.92	1F
e	5.33 or 5.42	S
▶ f	5.42 or 5.33	S

▶ C_7H_{14} ▶ liquid ▶ API 390

▶ 6-BaA

(b)H_3C H(d)
C=C
(e)H (c) CH_2—CH_3(a)

	Shift	Peak
a	.97	3A
b	1.61	5A
c	1.97	1F
▶ d	5.37	S
e	5.37	S

▶ C_5H_{10} ▶ 10% CCl_4 ▶ API 29

▶ 6-BcA

(c) H_3C (e) H
C=C
H (e) (a) $CH_2CH(CH_3)_2$ (d) (b)

	Shift	Peak
a	.98	2A
b	1.38	S
c	1.60	2H
d	1.80	S
▶ e	5.30	N

▶ C_7H_{14} ▶ 50 vol.% CCl_4 ▶ API 431

▶ 6-BaA

(b) H_3C (d) H
C=C
H (d) (c) (a) CH_2CH_3

	Shift	Peak
a	.95	3A
b	1.60	1H
c	1.94	N
▶ d	5.38	N

▶ C_5H_{10} ▶ 50 vol.% CCl_4 ▶ API 430

▶ 6-BcA

(c) H_3C (f) H
C=C
H (e) (d) CH_2 CH(b) CH_3(a) CH_3

	Shift	Peak
a	.83	2A
b	~1.40	S
c	1.62	2E
d	1.83	O
e	~5.38	O
▶ f	~5.38	O

▶ C_7H_{14} ▶ liquid ▶ API 396

▶ 6-BaA

(b) H_3C (e) H
C=C
H (d) (c) (a) CH_2CH_3

	Shift	Peak
a	.93	3C
b	1.58	2E
c	1.95	O
d	5.35 or 5.43	S
▶ e	5.43 or 5.35	S

▶ C_5H_{10} ▶ liquid ▶ API 461

▶ 6-BpA

(e)H (c) (b) CH_2—OH
C=C
(a) CH_3 H(d)

	Shift	Peak
a	1.72	N
b	2.07	1A
c	4.07	N
▶ d	~5.68	S
e	~5.68	S

▶ C_4H_8O ▶ 7% $CDCl_3$ ▶ V 414

▶ 6-BbA

(c) H_3C (e) H
C=C
H (f) CH_2—CH_2—CH_3 (d) (b) (a)

	Shift	Peak
a	.88	3B
b	~1.33	O
c	1.69	N
d	1.87	1F
▶ e	5.32	S
f	5.32	S

▶ C_6H_{12} ▶ 10% CCl_4 ▶ API 32

▶ 6-BaBa

(c) H (b) (a) CH_2CH_3
C=C
(a) (b) CH_3H_2C H (c)

	Shift	Peak
a	.93	3C
b	1.92	1F
▶ c	5.38	O

▶ C_6H_{12} ▶ liquid ▶ API 372

▶ 6-BaBb

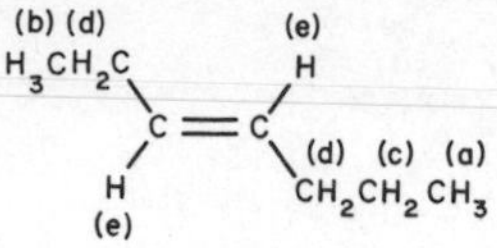

	Shift	Peak
a	.90	3D
b	.97	3D
c	1.33	S
d	1.95	O
▶ e	5.40	O

▶ C_7H_{14} ▶ liquid ▶ API 391

▶ 6-CaaA

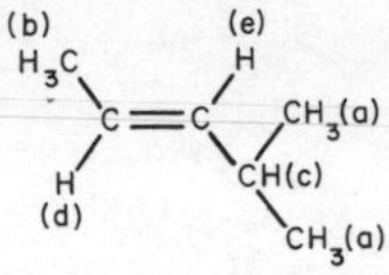

	Shift	Peak
a	.98	2A
b	1.62	2E
c	2.22	1F
d	5.37 or 5.43	S
▶ e	5.43 or 5.37	S

▶ C_6H_{12} ▶ 5 vol.% CCl_4 ▶ API 380

▶ 6-BbBa

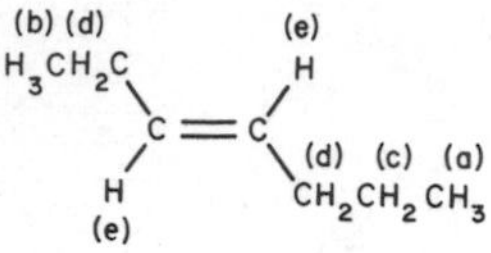

	Shift	Peak
a	.90	3D
b	.97	3D
c	1.33	S
d	1.95	O
▶ e	5.40	O

▶ C_7H_{14} ▶ liquid ▶ API 391

▶ 6-CaaA

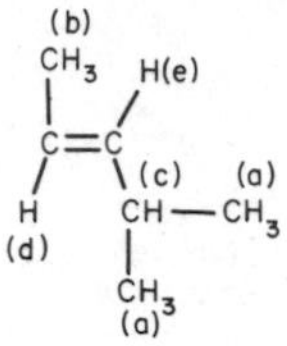

	Shift	Peak
a	.94	2A
b	1.60	2E
c	2.15	1D
d	5.36	S
▶ e	5.45	S

▶ C_6H_{12} ▶ liquid run ▶ API 70

▶ 6-BbBb

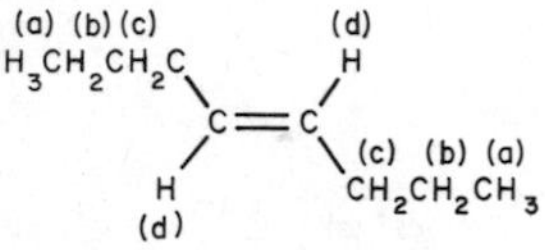

	Shift	Peak
a	.85	3C
b	1.32	6B
c	1.91	N
▶ d	5.32	N

▶ C_8H_{16} ▶ 50 vol.% CCl_4 ▶ API 432

▶ 6-CaaBa

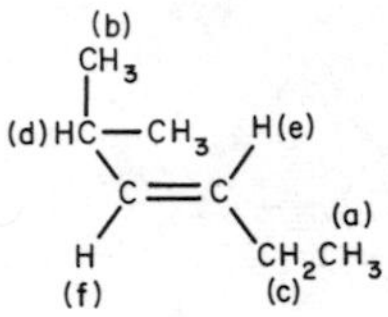

	Shift	Peak
a	.96	3D
b	.97	2H
c	1.98	S
d	~2.28	S
e	~5.38	S
▶ f	~5.38	S

▶ C_7H_{14} ▶ liquid ▶ API 397

▶ 6-BeBe

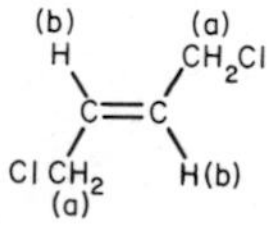

	Shift	Peak
a	4.08	7B
▶ b	5.93	7C

▶ $C_4H_6Cl_2$ ▶ 7% $CDCl_3$ ▶ V 404

▶ 6-(C)bbJa

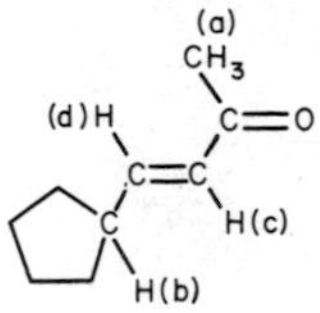

	Shift	Peak
a	2.23	1A
b	2.62	O
c	6.05	2A
▶ d	6.76	4D

▶ $C_9H_{14}O$ ▶ 7% $CDCl_3$ ▶ V 544

▶ 6-BaCaa

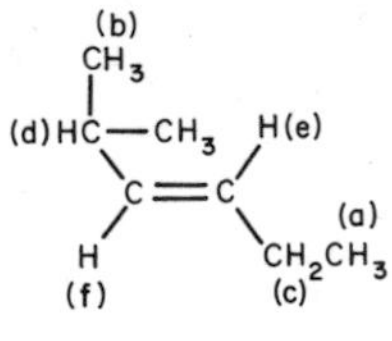

	Shift	Peak
a	.96	3D
b	.97	2H
c	1.98	S
d	~2.28	S
▶ e	~5.38	S
f	~5.38	S

▶ C_7H_{14} ▶ liquid ▶ API 397

▶ 6-(C)dlJa

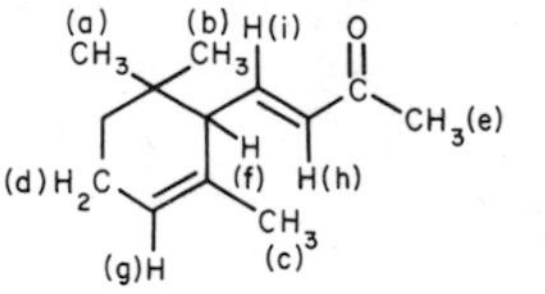

	Shift	Peak
a	0.86	1A
b	0.93	1A
c	1.57	2G
d	2.05	1C
e	2.26	1A
f	2.30	S
g	5.50	1C
h	6.06	2A
▶ i	6.60	4A

▶ $C_{13}H_{20}O$ ▶ 7% $CDCl_3$ ▶ V 616

▶ 6-BpVhh

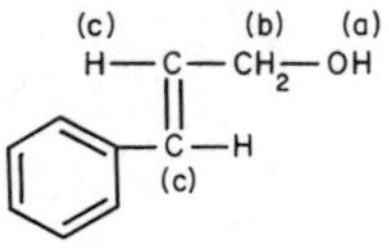

	Shift	Peak
a	2.05	1B
b	4.27	4A
▶ c	6.5	S

▶ $C_9H_{10}O$ ▶ 35.6 mg/0.50 ml $CDCl_3$ ▶ S 23

▶ 6-GVhh

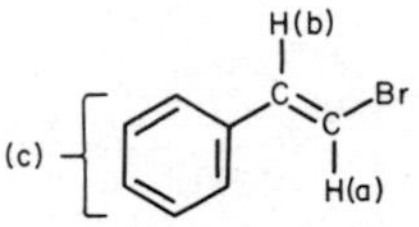

	Shift	Peak
▶ a	6.75	2C
b	7.10	2C
c	7.29	1A

▶ C_8H_7Br ▶ 7% $CDCl_3$ ▶ V 497

▶ 6-JhA

	Shift	Peak
a	2.03	2E
▶b	6.13	N
c	6.87	N
d	9.48	O

O
(c)H C H(d)
C=C
CH3 H(b)
(a)

▶ C_4H_6O ▶ 7% $CDCl_3$ ▶ V 60

▶ 6-JhVhh

	Shift	Peak
▶a	6.56	3A
b	6.80	2A
c	9.67	2A

O
(a)
H—C—CH
(c)
C—H
(b)

▶ C_9H_8O ▶ 45.0 mg/0.50 ml $CDCl_3$ ▶ S 101

▶ 6-Ja(C)bb

	Shift	Peak
a	2.23	1A
b	2.62	O
▶c	6.05	2A
d	6.76	4D

(a)
CH3
(d)H C=O
C=C
C H(c)
H(b)

▶ $C_9H_{14}O$ ▶ 7% $CDCl_3$ ▶ V 544

▶ 6-JaX

	Shift	Peak
a	6.55	S
▶b	6.57	S
c	6.77	S
d	7.20	S
e	7.55	S
f	9.57	2A

(a)H H(c)
H(b)
(e)H O C=C
H C=O
(d) H
(f)

▶ $C_7H_6O_2$ ▶ 7% $CDCl_3$ ▶ V 152

▶ (6)-Ja(C)dl

	Shift	Peak
a	0.86	1A
b	0.93	1A
c	1.57	2G
d	2.05	1C
e	2.26	1A
f	2.30	S
g	5.50	1C
▶h	6.06	2A
i	6.60	4A

(a) (b) H(i) O
CH3 CH3
C CH3(e)
H
(d)H2C (f) H(h)
CH3
(g)H (c)

▶ $C_{13}H_{20}O$ ▶ 7% $CDCl_3$ ▶ V 616

▶ 6-KhA

	Shift	Peak
a	1.90	2E
▶b	5.83	2E
c	7.10	2E
d	12.18	O

(a)CH3 H(b)
C=C
(c)H C OH(d)
O

▶ $C_4H_6O_2$ ▶ 7% $CDCl_3$ ▶ V 61

▶ 6-Ja(L)dl

	Shift	Peak
a	1.07	1A
b	1.78	1A
c	2.09	1C
d	2.31	1A
▶e	6.11	2A
f	7.28	2B

(f)
(a) (a) H O
CH3 CH3
C CH3(d)
H(e)
CH2 CH3
(c) (b)

▶ $C_{13}H_{20}O$ ▶ 7% $CDCl_3$ ▶ V 617

▶ 6-KaKa

	Shift	Peak
a	3.79	1A
▶b	6.88	1A

O
(a) (b)
CH3—O—C—C—H
(a)
H—C—C—O—CH3
(b) O

▶ $C_6H_8O_4$ ▶ 45.0 mg/0.10 ml $CDCl_3$ ▶ S 110

▶ 6-JnLhl

	Shift	Peak
a	~1.62	1B
b	~1.62	1B
c	~3.57	1C
d	5.93	1A
▶e	6.43	2A
f	~7.40	S

(f) O (c)
H C CH2
H C=C N CH2(b)
(a) O C=C (e)H
H2C H H2C CH2(a)
O (c) CH2
(b)

▶ $C_{17}H_{19}NO_3$ ▶ 7% $CDCl_3$ ▶ V 328

▶ 6-KbKb

	Shift	Peak
a	1.32	3A
b	4.27	4A
▶c	6.83	1A

O
(c) (b) (a)
H C CH2CH3
O C=C O
CH3CH2 C H(c)
(a) (b) O

▶ $C_8H_{12}O_4$ ▶ 7% $CDCl_3$ ▶ V 213

▶ 6-JaVhh

	Shift	Peak
a	2.35	1A
▶b	6.70	2A
c	7.48	S

O
(c)H C CH3(a)
C=C
H
(b)

▶ $C_{10}H_{10}O$ ▶ 7% $CDCl_3$ ▶ V 251

▶ 6-Kb(L)lo

	Shift	Peak
a	1.30	3A
b	4.23	4A
▶c	6.30	2F
d	6.45	N
e	6.60	S
f	7.40	S
g	7.47	S

(d)H H(e)
H(c)
(g)H O C=C
H C=O
(f) O (a)
CH2 CH3
(b)

▶ $C_9H_{10}O_3$ ▶ 7% $CDCl_3$ ▶ V 235

▶ 6-KhLhl

	Shift	Peak
a	1.87	4A
▶b	5.79	2A
c	7.36	O
d	12.03	1A

▶ $C_6H_8O_2$ ▶ 7% $CDCl_3$ ▶ V 462

▶ 6-(L)dlJa

	Shift	Peak
a	1.07	1A
b	1.78	1A
c	2.09	1C
d	2.31	1A
e	6.11	2A
▶ f	7.28	2B

▶ $C_{13}H_{20}O$ ▶ 7% $CDCl_3$ ▶ V 617

▶ 6-KaVhh

	Shift	Peak
a	3.78	1A
▶ b	6.42 or 7.71	2C
c	6.42 or 7.71	2C

▶ $C_{10}H_{10}O_2$ ▶ 44.6 mg/0.25 ml $CDCl_3$ ▶ S 20

▶ 6-LhlJn

	Shift	Peak
a	~1.62	1B
b	~1.62	1B
c	~3.57	1C
d	5.93	1A
e	6.43	2A
▶ f	~7.40	S

▶ $C_{17}H_{19}NO_3$ ▶ 7% $CDCl_3$ ▶ V 328

▶ 6-KbVhh

	Shift	Peak
a	5.25	1A
▶ b	6.47	2C
c	7.77	2C

▶ $C_{16}H_{14}O_2$ ▶ 44.9 mg/0.50 ml $CDCl_3$ ▶ S 104

▶ 6-(L)loJh

	Shift	Peak
a	6.55	S
b	6.57	S
c	6.77	S
▶d	7.20	S
e	7.55	S
f	9.57	2A

▶ $C_7H_6O_2$ ▶ 7% $CDCl_3$ ▶ V 152

▶ 6-KhVhh

	Shift	Peak
▶ a	6.46	1A
b	7.83	1A
c	13.21	1A

▶ $C_9H_8O_2$ ▶ 7% $CDCl_3$ ▶ V 230

▶ 6-LhlKh

	Shift	Peak
a	1.87	4A
b	5.79	2A
▶ c	7.36	O
d	12.03	1A

▶ $C_6H_8O_2$ ▶ 7% $CDCl_3$ ▶ V 462

▶ 6-(L)jlA

	Shift	Peak
a	1.65	2A
b	1.97	2E
c	3.86	1C
d	3.92	S
e	4.32	O
f	5.87	1I
▶ g	6.06	2A
h	6.99	4A

▶ $C_{12}H_{12}O_5$ ▶ 7% $CDCl_3$ ▶ V 595

▶ 6-(L)loKb

	Shift	Peak
a	1.30	3A
b	4.23	4A
c	6.30	2F
d	6.45	N
e	6.60	S
▶ f	7.40	S
g	7.47	S

▶ $C_9H_{10}O_3$ ▶ 7% $CDCl_3$ ▶ V 235

▶ 6-(L)loA

	Shift	Peak
a	0.89	3B
b	~1.30	S
c	~1.40	S
d	1.72	1A
e	1.97	2A
f	2.94	3B
▶ g	6.06	S
h	6.17	S
i	6.58	S
j	6.89	S
k	7.88	1G

▶ $C_{21}H_{22}O_5$ ▶ 7% $CDCl_3$ ▶ V 687

▶ 6-(L)lvVhh

	Shift	Peak
a	2.06	1A
b	2.32	1A
c	2.32	1A
d	6.36	1G
e	6.83*	S
f	6.83*	S
g	6.89	S
h	7.15	S
▶ i	7.33	S
j	7.53	2E

*or 7.08

▶ $C_{23}H_{18}O_8$ ▶ 7% $CDCl_3$ ▶ V 692

▶ 6-RVho

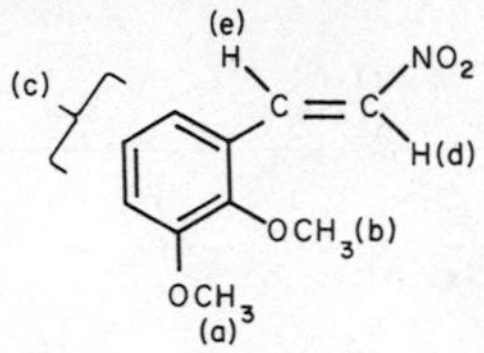

	Shift	Peak
a	3.88 or 3.92	1A
b	3.88 or 3.92	1A
c	7.07	1A
▶d	7.73	2A
e	8.20	2A

▶ $C_{10}H_{11}NO_4$ ▶ 7% $CDCl_3$ ▶ V 257

▶ 6-VhhG

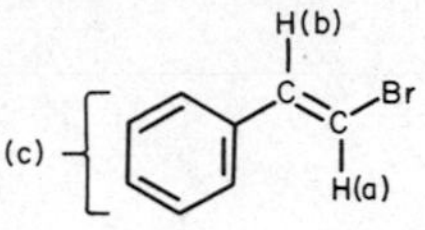

	Shift	Peak
a	6.75	2C
▶b	7.10	2C
c	7.29	1A

▶ C_8H_7Br ▶ 7% $CDCl_3$ ▶ V 497

▶ 6-VhhA

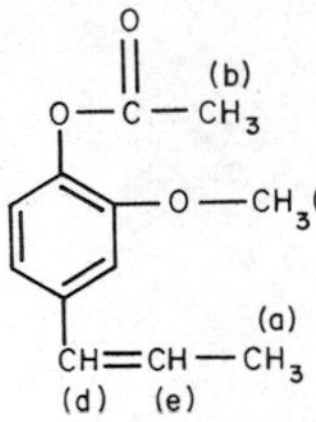

	Shift	Peak
a	1.88	2A
b	2.31	1A
c	3.82	1A
▶d	~6.1-6.7	S
e	~6.1-6.7	S

▶ $C_{12}H_{14}O_3$ ▶ 45.4 mg/0.50 ml $CDCl_3$ ▶ S 130

▶ 6-VhhJa

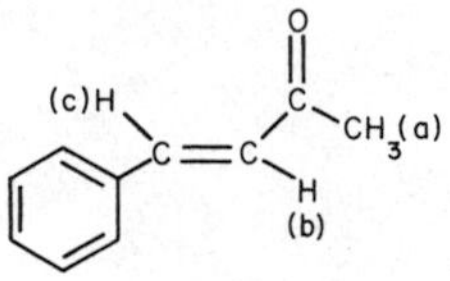

	Shift	Peak
a	2.35	1A
b	6.70	2A
▶c	7.48	S

▶ $C_{10}H_{10}O$ ▶ 7% $CDCl_3$ ▶ V 251

▶ 6-VhhA

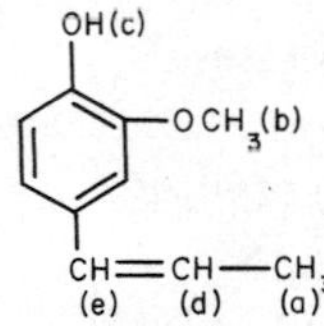

	Shift	Peak
a	1.82	2A
b	3.88	1A
c	5.48	1B
d	6.03	S
▶e	6.20	S

▶ $C_{10}H_{12}O_2$ ▶ 45.0 mg/0.5 ml $CDCl_3$ ▶ S 76

▶ 6-VhhJh

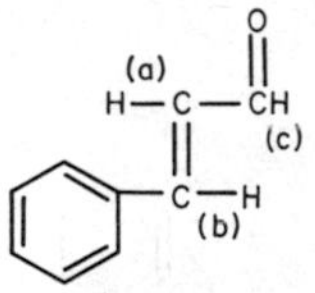

	Shift	Peak
a	6.56	3A
▶b	6.80	2A
c	9.67	2A

▶ C_9H_8O ▶ 45.0 mg/0.50 ml $CDCl_3$ ▶ S 101

▶ 6-VhhA

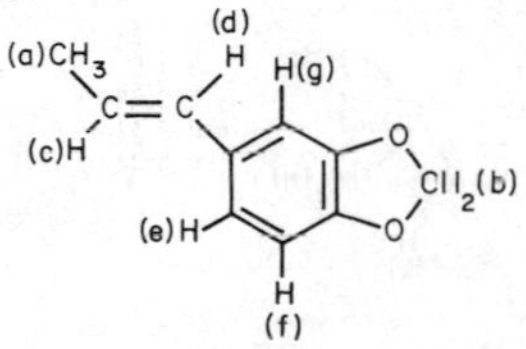

	Shift	Peak
a	1.80	2A
b	5.87	1A
c	~6.08	S
▶d	6.23	S
e	6.70	S
f	6.70	S
g	6.83	S

▶ $C_{10}H_{10}O_2$ ▶ 7% $CDCl_3$ ▶ V 252

▶ 6-VhhKa

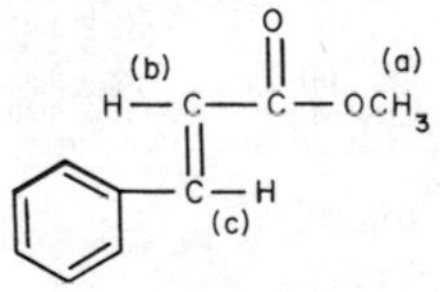

	Shift	Peak
a	3.78	1A
b	6.42 or 7.71	2C
▶c	6.42 or 7.71	2C

▶ $C_{10}H_{10}O_2$ ▶ 44.6 mg/0.25 ml $CDCl_3$ ▶ S 20

▶ 6-VhhA

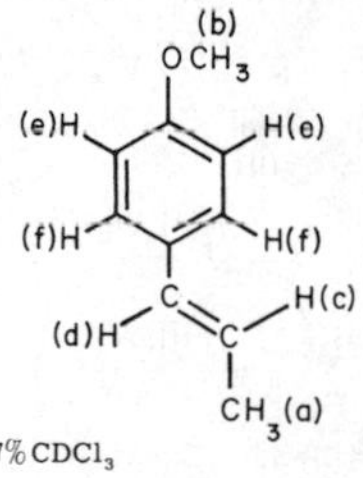

	Shift	Peak
a	1.83	2A
b	3.75	1A
c	6.08	N
▶d	6.28	N
e	6.80	Q
f	7.23	Q

▶ $C_{10}H_{12}O$ ▶ 7% $CDCl_3$ ▶ V 258

▶ 6-VhhKb

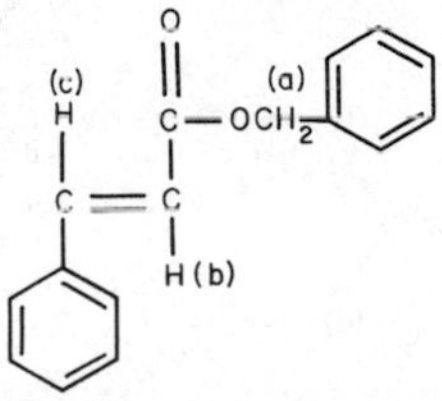

	Shift	Peak
a	5.25	1A
b	6.47	2C
▶c	7.77	2C

▶ $C_{16}H_{14}O_2$ ▶ 44.9 mg/0.50 ml $CDCl_3$ ▶ S 104

▶ 6-VhhBp

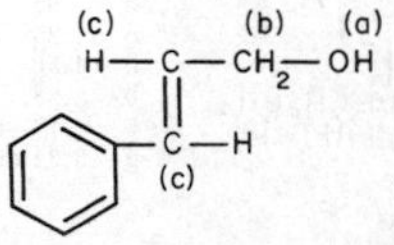

	Shift	Peak
a	2.05	1B
b	4.27	4A
▶c	6.5	S

▶ $C_9H_{10}O$ ▶ 35.6 mg/0.50 ml $CDCl_3$ ▶ S 23

▶ 6-VhhKh

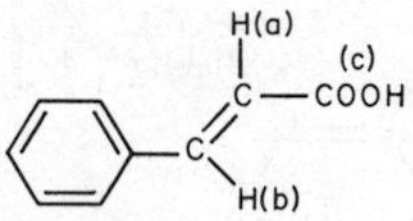

	Shift	Peak
a	6.46	1A
▶b	7.83	1A
c	13.21	1A

▶ $C_9H_8O_2$ ▶ 7% $CDCl_3$ ▶ V 230

▶ 6-Vhh(L)lv

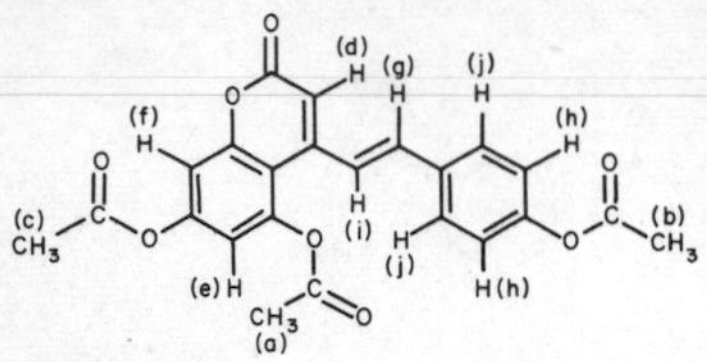

	Shift	Peak
a	2.06	1A
b	2.32	1A
c	2.32	1A
d	6.36	1G
e	6.83*	S
f	6.83*	S
▶ g	6.89	S
h	7.15	S
i	7.33	S
j	7.53	2E

*or 7.08

▶ $C_{23}H_{18}O_8$ ▶ 7% $CDCl_3$ ▶ V 692

▶ 7-ABb

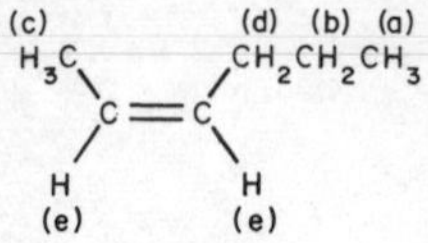

	Shift	Peak
a	.89	3C
b	1.30	S
c	1.56	2E
d	2.00	4C
▶ e	5.32	O

▶ C_6H_{12} ▶ liquid ▶ API 369

▶ 6-VhoR

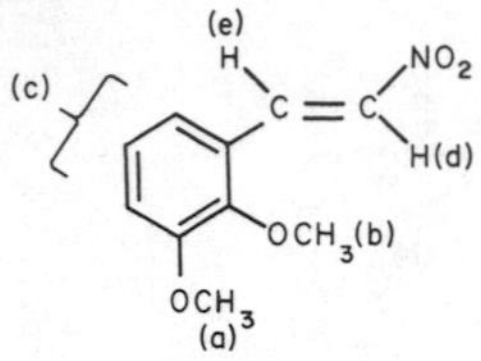

	Shift	Peak
a	3.88 or 3.92	1A
b	3.88 or 3.92	1A
c	7.07	1A
d	7.73	2A
▶ e	8.20	2A

▶ $C_{10}H_{11}NO_4$ ▶ 7% $CDCl_3$ ▶ V 257

▶ 7-ABb

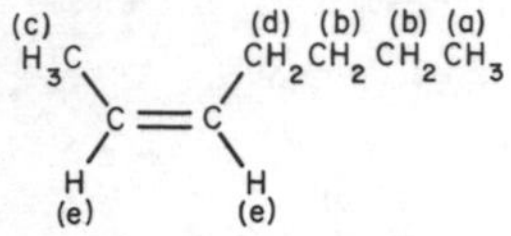

	Shift	Peak
a	.90	3C
b	1.31	O
c	1.57	2E
d	2.02	1F
▶ e	~5.37	O

▶ C_7H_{14} ▶ liquid ▶ API 389

▶ 6-VhhVhh

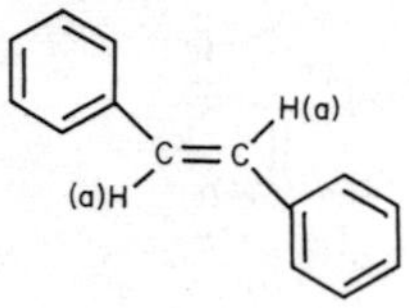

	Shift	Peak
a	7.10	1A

▶ $C_{14}H_{12}$ ▶ 7% $CDCl_3$ ▶ V 306

▶ 7-ACaa

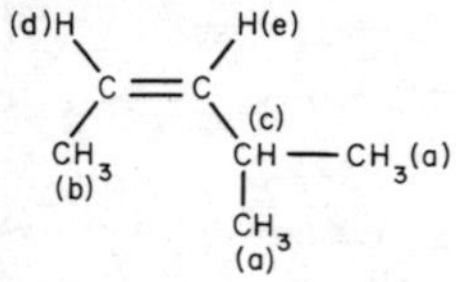

	Shift	Peak
a	.95	2A
b	1.63	2A
c	2.55	6B
▶ d	5.11	N
e	5.25	N

▶ C_6H_{12} ▶ liquid run ▶ API 69

▶ **7** -ABa

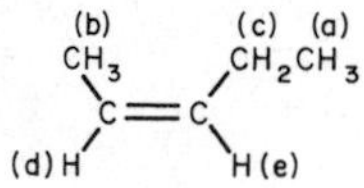

	Shift	Peak
a	.98	3A
b	1.58	2E
c	2.00	5B
▶ d	5.23	S
e	5.30	S

▶ C_5H_{10} ▶ 10% CCl_4 ▶ API 154

▶ 7-ACaa

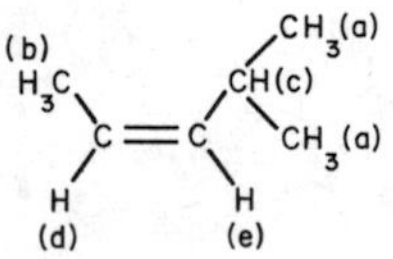

	Shift	Peak
a	.93	2A
b	1.57	2A
c	2.62	6B
▶ d	5.12 or 5.23	O
e	5.23 or 5.12	O

▶ C_6H_{12} ▶ liquid ▶ API 379

▶ 7-ABa

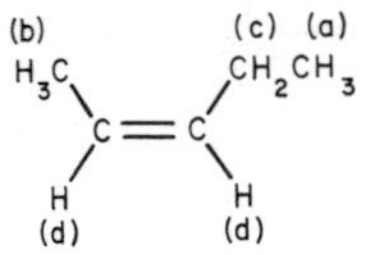

	Shift	Peak
a	.93	3C
b	1.52	2E
c	2.00	4C
▶ d	5.30	N

▶ C_5H_{10} ▶ 50 vol.% CCl_4 ▶ API 429

▶ 7-ACaa

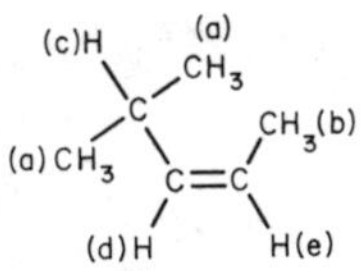

	Shift	Peak
a	0.94	2A
b	1.61	2A
c	2.63	6A
d	5.20 ± 0.10	O
▶ e	5.30 ± 0.10	O

▶ C_6H_{12} ▶ 7% $CDCl_3$ ▶ V 471

▶ 7-ABa

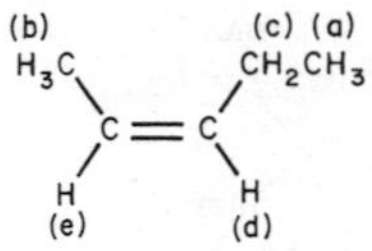

	Shift	Peak
a	.95	3C
b	1.58	2E
c	2.05	5B
d	~5.28 or 5.38	S
▶ e	~5.38 or 5.28	S

▶ C_5H_{10} ▶ liquid (neat) ▶ API 460

▶ 7-ACab

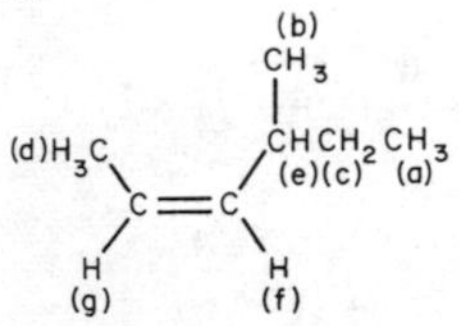

	Shift	Peak
a	.92	3C
b	.95	2H
c	1.25	O
d	1.58	2E
e	2.38	6B
f	~5.17 or 5.32	S
▶ g	~5.32 or 5.17	S

▶ C_7H_{14} ▶ liquid run, neat ▶ API 409

▶ 7-ADaaa

(a)CH_3—C(CH_3)(CH_3)—CH(d)=CH(c)—CH_3(b)

	Shift	Peak
a	1.10	1A
b	1.72	2A
▶ c	~5.23	S
d	~5.23	S

▶ C_7H_{14} ▶ 50% CCl_4 ▶ API 222

▶ 7-BbA

(c)H_3C—CH(e)=CH(e)—$CH_2CH_2CH_2CH_3$ (d) (b) (b) (a)

	Shift	Peak
a	.90	3C
b	1.31	O
c	1.57	2E
d	2.02	1F
▶ e	~5.37	O

▶ C_7H_{14} ▶ liquid ▶ API 389

▶ 7-ADaaa

(b)H_3C—CH(d)=CH(c)—C(CH_3(a))(CH_3)(CH_3)

	Shift	Peak
a	1.11	1A
b	1.69	2E
c	5.23 ± .03	O
▶ d	5.23 ± .03	O

▶ C_7H_{14} ▶ liquid ▶ API 405

▶ 7-BaBa

(a) (b) H_3CH_2C—CH(c)=CH(c)—CH_2CH_3 (b) (a)

	Shift	Peak
a	.91	3C
b	1.99	5B
▶ c	5.25	O

▶ C_6H_{12} ▶ liquid ▶ API 371

▶ 7-BaA

(b)H_3C—CH(e)=CH(d)—CH_2CH_3 (c) (a)

	Shift	Peak
a	.95	3C
b	1.58	2E
c	2.05	5B
▶ d	~5.28 or 5.38	S
e	~5.38 or 5.28	S

▶ C_5H_{10} ▶ liquid (neat) ▶ API 460

▶ 7-BaBl

(a) CH_3—(e) CH_2—(e) CH=(e) CH—(c) $(CH_2$—(e) CH=(e) $CH)_2$—(b) CH_2—$(CH_2)_5$—CH_2—C(=O)—OCH_3 (d)

	Shift	Peak
a	0.97	3A
b	1.33	1B
c	2.80	3B
d	3.67	1A
▶ e	~5.38	3C

▶ $C_{19}H_{32}O_2$ ▶ 7% $CDCl_3$ ▶ V 337

▶ 7-BaA

(b)CH_3—CH(d)=CH(e)—CH_2CH_3 (c) (a)

	Shift	Peak
a	.98	3A
b	1.58	2E
c	2.00	5B
d	5.23	S
▶ e	5.30	S

▶ C_5H_{10} ▶ 10% CCl_4 ▶ API 154

▶ 7-BbBb

(e) CH—(c) CH_2—(b) $(CH_2)_5$—(d) CH_2—C(=O)—OH (f)
‖
(e) CH—(c) CH_2—(b) $(CH_2)_6$—(a) CH_3

	Shift	Peak
a	0.88	3B
b	1.29	1B
c	2.00	1F
d	2.25	3B
▶ e	5.28	3D
f	11.45	1C

▶ $C_{18}H_{34}O_2$ ▶ 45.2 mg/0.50 ml $CDCl_3$ ▶ S 70

▶ 7-BaA

(b)H_3C—CH(d)=CH(d)—CH_2CH_3 (c) (a)

	Shift	Peak
a	.93	3C
b	1.52	2E
c	2.00	4C
▶ d	5.30	N

▶ C_5H_{10} ▶ 50 vol.% CCl_4 ▶ API 429

▶ 7-BbBb

(f) CH—(c) CH_2—(b) $(CH_2)_5$—(d) CH_2—C(=O)—OCH_3 (e)
‖
(f) CH—(c) CH_2—(b) $(CH_2)_6$—(a) CH_3

	Shift	Peak
a	0.90	3B
b	1.29	1B
c	2.00	1C
d	2.32	S
e	3.70	1A
▶ f	5.4	3B

▶ $C_{19}H_{36}O_2$ ▶ 33.7 mg/0.50 ml $CDCl_3$ ▶ S 71

▶ 7-BbA

(c)H_3C—CH(e)=CH(e)—$CH_2CH_2CH_3$ (d) (b) (a)

	Shift	Peak
a	.89	3C
b	1.30	S
c	1.56	2E
d	2.00	4C
▶ e	5.32	O

▶ C_6H_{12} ▶ liquid ▶ API 369

▶ (7)-(B)b(B)b

Cycloocta-1,5-diene: (b)H—C=C—H(b); (a)H_2C, CH_2(a); (a)H_2C, CH_2(a); (b)H—C=C—H(b)

	Shift	Peak
a	2.38	1B
▶ b	5.58	1C

▶ C_8H_{12} ▶ 7% $CDCl_3$ ▶ V 511

▶ (7)-(B)b(B)b

Shift		Peak
a	1.81	O
b	2.28	3C
▶ c	5.65	1A

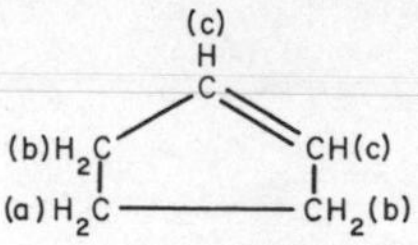
(c)
H
C
(b) H2C
CH(c)
(a) H2C
CH2(b)

▶ C_5H_8 ▶ 10% CCl_4 ▶ API 164

▶ 7-BlBb

Shift		Peak
a	0.97	3A
b	1.33	1B
c	2.80	3B
d	3.67	1A
▶ e	~5.38	3C

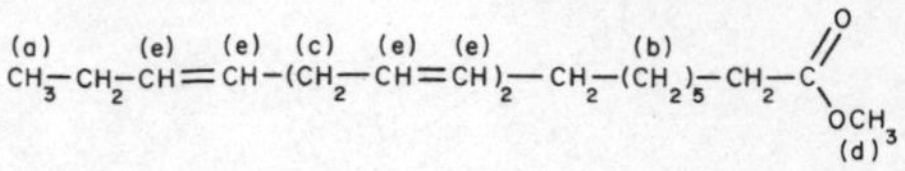
(a) (e) (e) (c) (e) (e) (b)
CH3–CH2–CH=CH–(CH2–CH=CH)2–CH2–(CH2)5–CH2–C
O
OCH3
(d)

▶ $C_{19}H_{32}O_2$ ▶ 7% $CDCl_3$ ▶ V 337

▶ (7)-(B)b(B)c

Shift		Peak
a	4.93	R
b	5.02	R
c	~5.68	1E
▶ d	~5.68	1E
e	5.83	4D

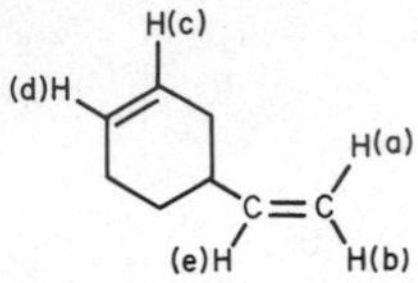
H(c)
(d)H
H(a)
C=C
(e)H
H(b)

▶ C_8H_{12} ▶ 7% $CDCl_3$ ▶ V 210

▶ 7-BlBl

Shift		Peak
a	0.97	3A
b	1.33	1B
c	2.80	3B
d	3.67	1A
▶ e	~5.38	3C

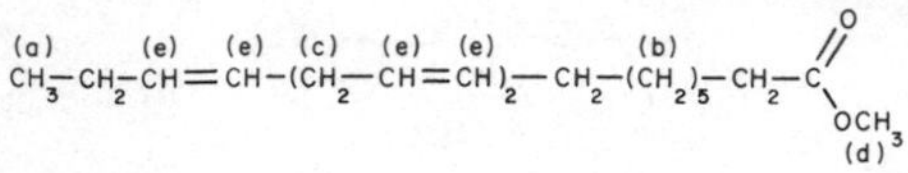
(a) (e) (e) (c) (e) (e) (b)
CH3–CH2–CH=CH–(CH2–CH=CH)2–CH2–(CH2)5–CH2–C
O
OCH3
(d)

▶ $C_{19}H_{32}O_2$ ▶ 7% $CDCl_3$ ▶ V 337

▶ 7-BbBl

Shift		Peak
a	0.97	3A
b	1.33	1B
c	2.80	3B
d	3.67	1A
▶ e	~5.38	3C

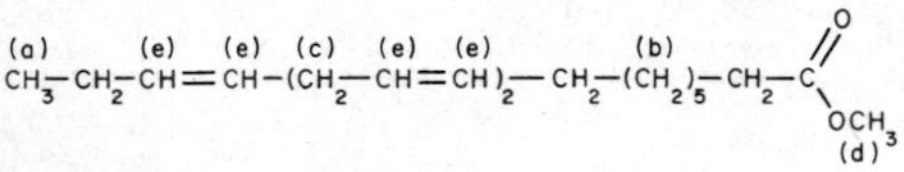
(a) (e) (e) (c) (e) (e) (b)
CH3–CH2–CH=CH–(CH2–CH=CH)2–CH2–(CH2)5–CH2–C
O
OCH3
(d)

▶ $C_{19}H_{32}O_2$ ▶ 7% $CDCl_3$ ▶ V 337

▶ 7-(B)n(B)b

Shift		Peak
a	1.63	1A
b	2.07	O
c	2.95	3A
d	3.33	N
▶ e	5.72 or 5.77	1E
f	5.72 or 5.77	1E

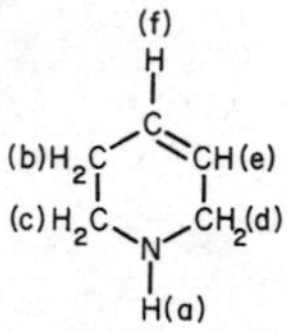
(f)
H
C
(b)H2C
CH(e)
(c)H2C
CH2(d)
N
H(a)

▶ C_5H_9N ▶ 7% $CDCl_3$ ▶ V 115

▶ 7-(B)b(B)n

Shift		Peak
a	1.63	1A
b	2.07	O
c	2.95	3A
d	3.33	N
e	5.72 or 5.77	1E
▶ f	5.72 or 5.77	1E

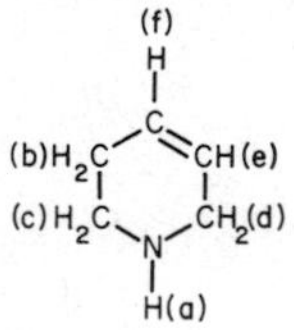
(f)
H
C
(b)H2C
CH(e)
(c)H2C
CH2(d)
N
H(a)

▶ C_5H_9N ▶ 7% $CDCl_3$ ▶ V 115

▶ (7)-(B)o(B)o

Shift		Peak
a	4.31	2D
b	4.88	1A
▶ c	5.74	3A

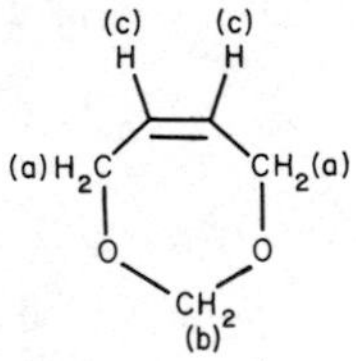
(c) (c)
H H
(a) H2C
CH2(a)
O
O
CH2
(b)

▶ $C_5H_8O_2$ ▶ 7% $CDCl_3$ ▶ V 437

▶ (7)-(B)c(B)b

Shift		Peak
a	4.93	R
b	5.02	R
▶ c	~5.68	1E
d	~5.68	1E
e	5.83	4D

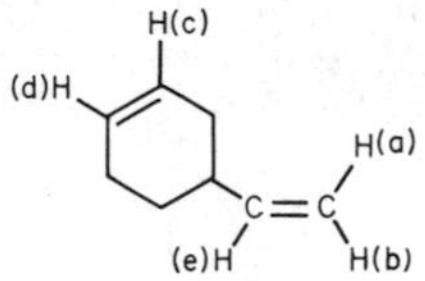
H(c)
(d)H
H(a)
C=C
(e)H
H(b)

▶ C_8H_{12} ▶ 7% $CDCl_3$ ▶ V 210

▶ (7)-(B)s(B)s

Shift		Peak
a	3.74	1A
▶ b	6.08	1A

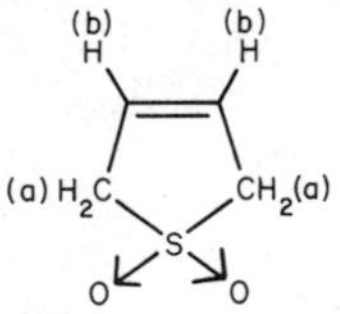
(b) (b)
H H
(a) H2C
CH2(a)
S
O O

▶ $C_4H_6O_2S$ ▶ 7% $CDCl_3$ ▶ V 406

▶ 7-BlBa

Shift		Peak
a	0.97	3A
b	1.33	1B
c	2.80	3B
d	3.67	1A
▶ e	~5.38	3C

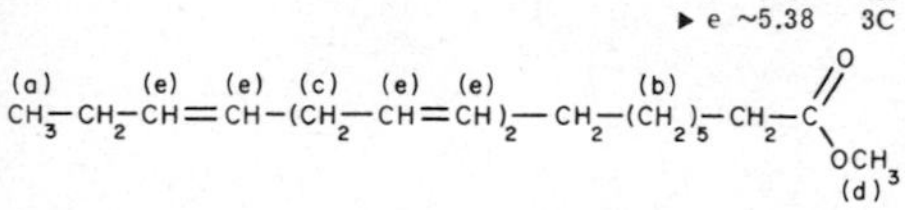
(a) (e) (e) (c) (e) (e) (b)
CH3–CH2–CH=CH–(CH2–CH=CH)2–CH2–(CH2)5–CH2–C
O
OCH3
(d)

▶ $C_{19}H_{32}O_2$ ▶ 7% $CDCl_3$ ▶ V 337

▶ (7)-(B)q(K)b

Shift		Peak
a	4.92	3C
b	6.15	N
▶ c	7.63	N

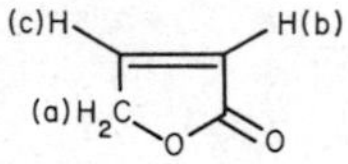
(c)H
H(b)
(a)H2C
O
O

▶ $C_4H_4O_2$ ▶ 7% $CDCl_3$ ▶ V 51

▶ (7)-(B)b(L)hl

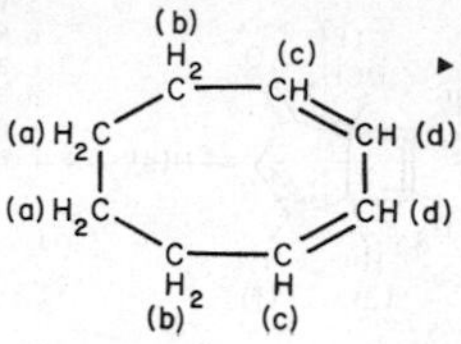

	Shift	Peak
a	~1.52	N
b	~2.20	O
▶ c	~5.63 or 5.77	S
d	~5.77 or 5.63	S

▶ C_8H_{12} ▶ $CDCl_3$ ▶ API 129

▶ (7)-(B)vVbh

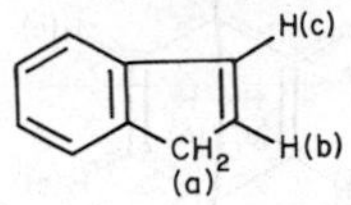

	Shift	Peak
a	3.33	1I
▶ b	6.50	N
c	6.82	N

▶ C_9H_8 ▶ 7% $CDCl_3$ ▶ V 227

▶ (7)-(B)b(L)hl

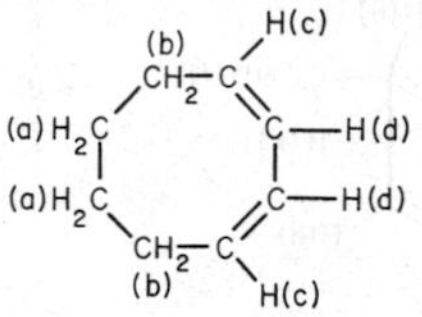

	Shift	Peak
a	~1.52	N
b	~2.20	O
▶ c	~5.63	S
d	~5.77	1E

▶ C_8H_{12} ▶ 7% $CDCl_3$ ▶ V 209

▶ 7-CaaA

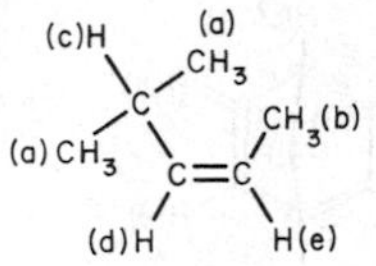

	Shift	Peak
a	0.94	2A
b	1.61	2A
c	2.63	6A
▶ d	5.20 ± 0.10	O
e	5.30 ± 0.10	O

▶ C_6H_{12} ▶ 7% $CDCl_3$ ▶ V 471

▶ (7)-(B)l(L)hl

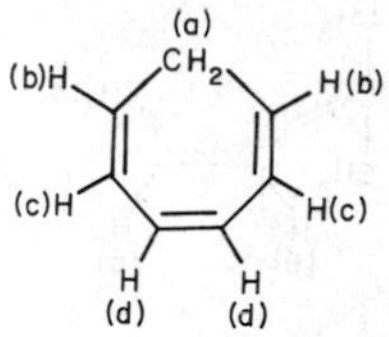

	Shift	Peak
a	2.20	3A
▶ b	5.28	4C
c	6.12	O
d	6.55	3C

▶ C_7H_8 ▶ 7% $CDCl_3$ ▶ V 158

▶ 7-CaaA

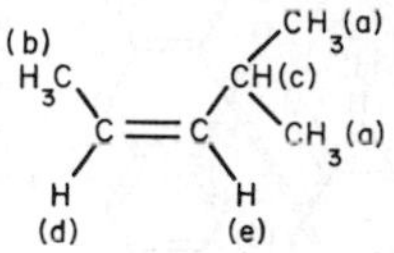

	Shift	Peak
a	.93	2A
b	1.57	2A
c	2.62	6B
d	5.12 or 5.23	O
▶ e	5.23 or 5.12	O

▶ C_6H_{12} ▶ liquid ▶ API 379

▶ (7)-(B)l(N)bl

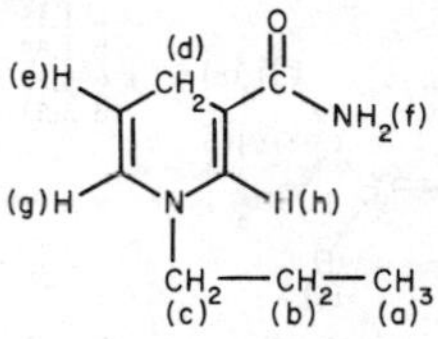

	Shift	Peak
a	0.90	3C
b	1.58	6B
c	3.09	2A
d	3.19	N
▶ e	4.72	5B
f	5.49	1C
g	5.73	4A
h	7.05	2D

▶ $C_9H_{14}N_2O$ ▶ 7% $CDCl_3$ ▶ V 542

▶ 7-CaaA

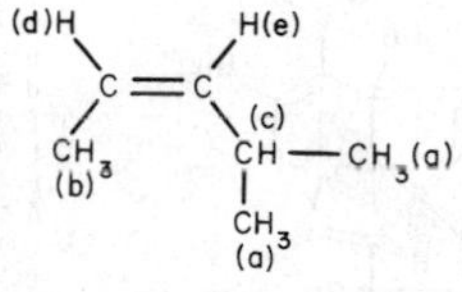

	Shift	Peak
a	.95	2A
b	1.63	2A
c	2.55	6B
d	5.11	N
▶ e	5.25	N

▶ C_6H_{12} ▶ liquid run ▶ API 69

▶ (7)-(B)b(O)b

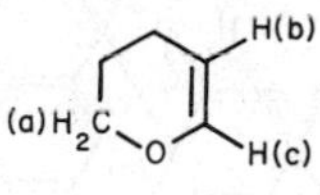

	Shift	Peak
a	3.97	N
▶ b	4.65	N
c	6.37	N

▶ C_5H_8O ▶ 7% $CDCl_3$ ▶ V 111

▶ 7-CabA

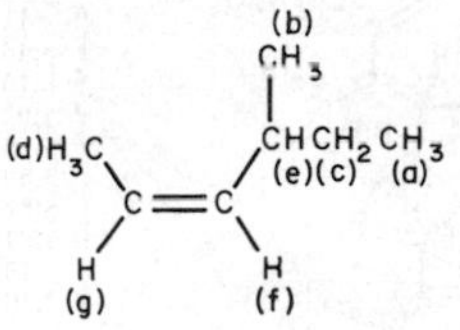

	Shift	Peak
a	.92	3C
b	.95	2H
c	1.25	O
d	1.58	2E
e	2.38	6B
▶ f	~5.17 or 5.32	S
g	~5.32 or 5.17	S

▶ C_7H_{14} ▶ liquid run, neat ▶ API 409

▶ (7)-(B)s(S)boo

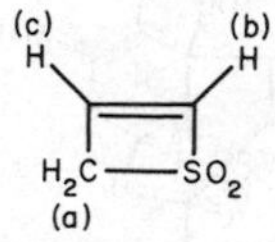

	Shift	Peak
a	4.58	2E
b	6.80	2E
▶ c	7.22	M

▶ $C_3H_4O_2S$ ▶ 7% $CDCl_3$ ▶ V 22

▶ (7)-CacCac

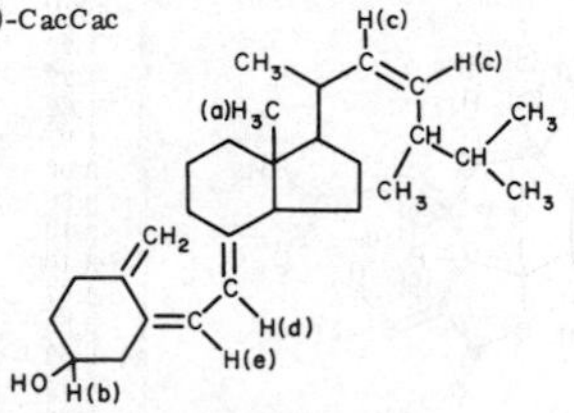

	Shift	Peak
a	0.57	1A
b	3.92	1F
▶ c	~5.20	N
d	6.05*	2C
e	6.05*	2C

*or 6.20

▶ $C_{28}H_{44}$ ▶ 7% $CDCl_3$ ▶ V 364

▶ (7)-(C)bl(C)bl

H(b)
(c)H H(c)
H—H
(a) (a)
(c)H H(c)
H(b)

	Shift	Peak
a	2.00	3A
b	3.58	5A
▶ c	6.75	3A

▶ C_7H_8 ▶ 7% $CDCl_3$ ▶ V 487

▶ (7)-(C)dl(J)d

(b)
OCH_3 O
(d)H H(e)
(a) CH_3
H H
(c) (f)

	Shift	Peak
a	2.13	2D
b	3.46	1A
c	3.69	1B
d	5.88	2A
e	6.43	2A
▶ f	6.84	2G

▶ $C_9H_{10}O_2$ ▶ 7% $CDCl_3$ ▶ V 531

▶ (7)-(C)bv(C)bv

(b) CH_3O H(c)
(d)H H(e)
(a) (a)
H—H
(d)H H(e)
(b) CH_3O H(c)

	Shift	Peak
a	2.22	3A
b	3.79	1A
c	4.17	5A
d	6.50	1A
▶ e	6.83	3A

▶ $C_{13}H_{14}O_2$ ▶ 7% $CDCl_3$ ▶ V 608

▶ (7)-(C)ccV

H(a)
(c)H COOH(f)
H(b)
H(d)
H
(e)

	Shift	Peak
a	0.86	3A
b	2.72	5B
c	3.17	4A
▶ d	6.32	S
e	6.35	S
f	12.14	1A

▶ $C_{12}H_{10}O_2$ ▶ 7% $CDCl_3$ ▶ V 592

▶ (7)-(C)cd(D)acj

(e)
H
(d)
(b) CH_3 H O O
(b) CH_3 —C
(a)
(c) H CH_3
(g)H H(f)

	Shift	Peak
a	0.89	1A
b	1.09	1A
c	1.17	S
d	1.35	S
e	4.16	1E
f	5.07	2A
▶ g	5.73	4C

▶ $C_{23}H_{22}O_2$ ▶ 7% $CDCl_3$ ▶ V 693

▶ 7-DaaaA

CH_3
(a) CH_3 — C CH_3(b)
CH_3 C=C
H H
(d) (c)

	Shift	Peak
a	1.10	1A
b	1.72	2A
c	~5.23	S
▶ d	~5.23	S

▶ C_7H_{14} ▶ 50% CCl_4 ▶ API 222

▶ (7)-(C)cl(D)ccq

O OCH_3(e)
C
(f)H
(i)H H(k)
H
(d) O
(j)H (h)
H
(g) O
O H
CH_2—CH_3
O (c)H (b) (a)

	Shift	Peak
a	1.13	3B
b	1.89	O
c	2.74	O
d	3.42	4A
e	3.77	1A
f	3.98	3E
g	4.59	2A
h	5.59	S
▶ i	5.63*	S
j	5.63*	S
k	7.47	1A

*or 6.12

▶ $C_{15}H_{16}O_6$ ▶ 7% $CDCl_3$ ▶ V 641

▶ 7-DaaaA

CH_3(a)
(b) H_3C C—CH_3
C=C CH_3
H H
(d) (c)

	Shift	Peak
a	1.11	1A
b	1.69	2E
▶ c	5.23 ± .03	O
d	5.23 ± .03	O

▶ C_7H_{14} ▶ liquid ▶ API 405

▶ (7)-(C)cl(D)ccq

O OCH_3(c)
C
(h)H (d)H H(j)
H(b)
(g)H O
(f)
H O
O (e)
H
O CH_3(a)
(i)H

	Shift	Peak
a	2.11	2A
b	3.46	4A
c	3.79	1A
d	4.02	3E
e	5.12	1B
f	5.59	S
g	5.67*	S
▶ h	5.67*	S
i	7.19	O
j	7.46	1A

*or 6.08

▶ $C_{15}H_{14}O_6$ ▶ 7% $CDCl_3$ ▶ V 640

▶ (7)-(D)acj(C)cd

(e)
H
(d)
(b) CH_3 H O O
(b) CH_3 —C
(a)
(c) H CH_3
(g)H H(f)

	Shift	Peak
a	0.89	1A
b	1.09	1A
c	1.17	S
d	1.35	S
e	4.16	1E
▶ f	5.07	2A
g	5.73	4C

▶ $C_{23}H_{22}O_2$ ▶ 7% $CDCl_3$ ▶ V 693

▶ (7)-(C)cd(J)d

(e) (b) (g)
(n) (f)H CH_3 H(c)
H H H (d)
(j) O CH_3
(l)H (i) C
H H (k) O
H H
O CH_3 (h)
(a) O C — H(m)
O

	Shift	Peak
a	1.22	1E
b	1.29	2H
c	1.62	S
d	1.97	1A
e	2.06	S
f	2.60	3E
g	~3.10	S
h	~3.10	S
i	4.67	3C
j	5.63	2A
k	5.93	2A
l	6.14	2H
m	6.23	N
▶ n	7.79	2F

▶ $C_{17}H_{20}O_5$ ▶ 7% $CDCl_3$ ▶ V 666

▶ (7)-(D)ccq(C)cl

O OCH_3(e)
C
(f)H
(i)H H(k)
H
(d) O
(j)H (h)
H
(g) O
O H
CH_2—CH_3
O (c)H (b) (a)

	Shift	Peak
a	1.13	3B
b	1.89	O
c	2.74	O
d	3.42	4A
e	3.77	1A
f	3.98	3E
g	4.59	2A
h	5.59	S
i	5.63*	S
▶ j	5.63*	S
k	7.47	1A

*or 6.12

▶ $C_{15}H_{16}O_6$ ▶ 7% $CDCl_3$ ▶ V 641

▶ (7)-(D)ccq(C)cl

O=C–O CH$_3$(c)
(h)H (d)H H(j)
H(b)
(g)H (f) H O
O (e) H O
O CH$_3$(a)
(i)H

	Shift	Peak
a	2.11	2A
b	3.46	4A
c	3.79	1A
d	4.02	3E
e	5.12	1B
f	5.59	S
▶ g	5.67*	S
h	5.67*	S
i	7.19	O
j	7.46	1A

*or 6.08

▶ $C_{15}H_{14}O_6$ ▶ 7% $CDCl_3$ ▶ V 640

▶ (7)-(D)aaoVoo

H$_3$C(a) H(h) (j)
O O H (b) (e)
(a)CH$_3$ OCH$_3$ H
(c)H H(d)
CH$_3$ (a)
(f)H (i)H O H
(i) O CH$_3$(a)
H(g)

	Shift	Peak
a	1.46*	1A
b	3.58	1A
c	5.62†	2A
▶ d	5.62†	2A
e	6.45‡	2A
f	6.45‡	2A
g	6.62	S
h	7.12	2A
i	7.88§	1A
j	7.88§	1A

*or 1.49 ‡or 6.60
†or 5.70 §or 7.90

▶ $C_{26}H_{24}O_5$ ▶ 7% $CDCl_3$ ▶ V 696

▶ (7)-(D)abo(D)bco

CH$_3$(b)
H(c)
O
O
H(d)
CH$_3$(a) — C — H
CH$_3$(a)

	Shift	Peak
a	1.00	2A
b	1.37	1A
▶ c	6.42 or 6.47	2C
d	6.42 or 6.47	2C

▶ $C_{10}H_{16}O_2$ ▶ 7% $CDCl_3$ ▶ V 276

▶ (7)-(D)dnoVho

CH$_3$(b)
(a) CH$_3$ H(d)
(e)
H
N O
CH$_3$
(c) H H(f)
(g)
NO$_2$

	Shift	Peak
a	1.19	1A
b	1.30	1A
c	2.74	1A
▶ d	5.84	2A
e	6.93	S
f	8.01 ± .02	S
g	8.01 ± .02	S

▶ $C_{19}H_{18}N_2O_3$ ▶ 7% $CDCl_3$ ▶ V 677

▶ (7)-(D)bco(D)abo

CH$_3$(b)
H(c)
O
O
H(d)
CH$_3$(a) — C — H
CH$_3$(a)

	Shift	Peak
a	1.00	2A
b	1.37	1A
c	6.42 or 6.47	2C
▶ d	6.42 or 6.47	2C

▶ $C_{10}H_{16}O_2$ ▶ 7% $CDCl_3$ ▶ V 276

▶ (7)-(J)d(C)cd

(e) (b) (g)
(n) (f)H CH$_3$ H(c)
H H H
(d)
(j) O CH$_3$
(l)H H (i) C
H (k) O
H H
O CH$_3$ (h)
(a) O C — H(m)
O

	Shift	Peak
a	1.22	1E
b	1.29	2H
c	1.62	S
d	1.97	1A
e	2.06	S
f	2.60	3E
g	~3.10	S
h	~3.10	S
i	4.67	3C
j	5.63	2A
k	5.93	2A
▶ l	6.14	2H
m	6.23	N
n	7.79	2F

▶ $C_{17}H_{20}O_5$ ▶ 7% $CDCl_3$ ▶ V 666

▶ (7)-(D)acl(J)l

(a) O
H$_3$C
(c)H (b)
CH$_3$
H
O H
H H

	Shift	Peak
a	1.00	1A
b	1.23	1A
▶ c	7.12	2A

▶ $C_{19}H_{22}O_2$ ▶ 7% $CDCl_3$ ▶ V 335

▶ (7)-(J)d(C)dl

(b)
OCH$_3$ O
(d)H
H(e)
(a) CH$_3$
H H
(c) (f)

	Shift	Peak
a	2.13	2D
b	3.46	1A
c	3.69	1B
d	5.88	2A
▶ e	6.43	2A
f	6.84	2G

▶ $C_9H_{10}O_2$ ▶ 7% $CDCl_3$ ▶ V 531

▶ (7)-(D)aaoVho

H$_3$C(a) H(h) (j)
O O H (b) (e)
(a)CH$_3$ OCH$_3$ H
(c)H H(d)
CH$_3$ (a)
(f)H (i)H O H
(i) O CH$_3$(a)
H(g)

	Shift	Peak
a	1.46*	1A
b	3.58	1A
▶ c	5.62†	2A
d	5.62†	2A
e	6.45‡	2A
f	6.45‡	2A
g	6.62	S
h	7.12	2A
i	7.88§	1A
j	7.88§	1A

*or 1.49 ‡or 6.60
†or 5.70 §or 7.90

▶ $C_{26}H_{24}O_5$ ▶ 7% $CDCl_3$ ▶ V 696

▶ (7)-(J)v(J)v

(c)
H O
(b)H H(a)
(b)H H(a)
H O
(c)

	Shift	Peak
▶ a	6.97	1A
b	7.77	Q
c	8.07	Q

▶ $C_{10}H_6O_2$ ▶ 7% $CDCl_3$ ▶ V 550

▶ (7)-(D)aaoVoo

(b)CH$_2$ — COOH
CH$_2$(c)
(a) CH$_3$ O O CH$_3$(a)
(a)CH$_3$ CH$_3$(a)
(e)H H(e)
H OCH$_3$ H
(f) (d) (f)

	Shift	Peak
a	1.42	1A
b	2.62	Q
c	2.86	Q
d	3.75	1A
▶ e	5.50	2C
f	6.53	2C

▶ $C_{20}H_{24}O_5$ ▶ 7% $CDCl_3$ ▶ V 344

▶ (7)-(J)l(L)cl

(d)H---O
O
H(c)
(c)H C
H(c)
(c)H (b)
CH — CH$_3$(a)
CH$_3$
(a)

	Shift	Peak
a	1.27	2A
b	2.90	5B
▶ c	8.32	1A
d	8.75	1B

▶ $C_{10}H_{12}O_2$ ▶ 7% $CDCl_3$ ▶ V 560

▶ (7)-(J)n(N)aj

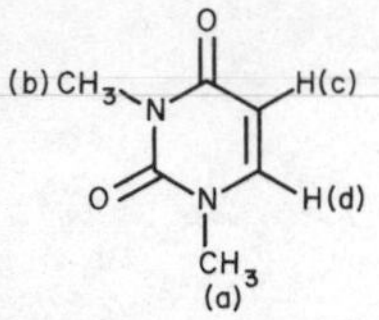

	Shift	Peak
a	3.37 or 3.43	1A
b	3.37 or 3.43	1A
▶ c	5.73	2C
d	7.20	2C

▶ $C_6H_8N_2O_2$ ▶ 7% $CDCl_3$ ▶ V 460

▶ (7)-(K)b(B)q

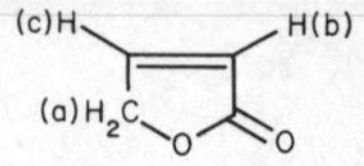

	Shift	Peak
a	4.92	3C
▶ b	6.15	N
c	7.63	N

▶ $C_4H_4O_2$ ▶ 7% $CDCl_3$ ▶ V 51

▶ (7)-(J)n(N)aj

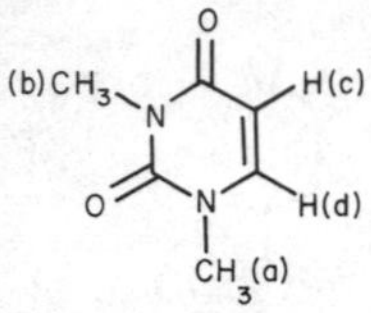

	Shift	Peak
a	3.30 or 3.43	1A
b	3.30 or 3.43	1A
▶ c	5.84	2C
d	7.61	2C

▶ $C_6H_8N_2O_2$ ▶ D_2O ▶ V 461

▶ 7-KbKb

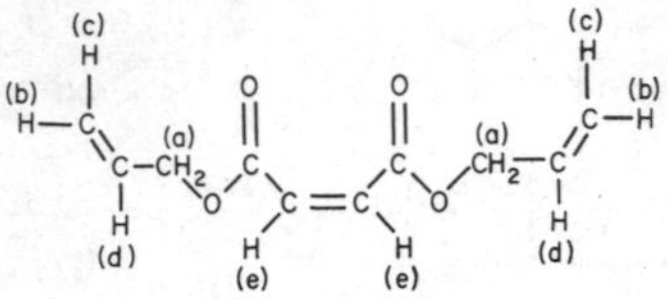

	Shift	Peak
a	4.68	2D
b	5.25	R
c	5.32	R
d	5.88	R
▶ e	6.27	1A

▶ $C_{10}H_{12}O_4$ ▶ 7% $CDCl_3$ ▶ V 262

▶ (7)-(J)n(N)cj

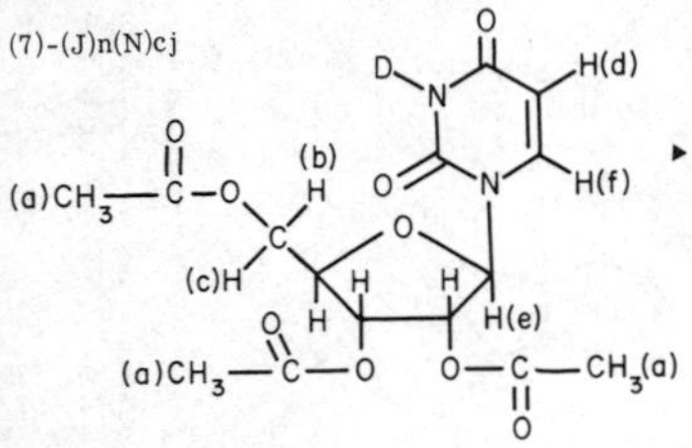

	Shift	Peak
a	1.92	1A
b	3.80	S
c	3.87	S
▶ d	5.81	S
e	5.83	S
f	7.65	2A

▶ $C_{15}H_{18}N_2O_9$ ▶ D_2O ▶ V 642

▶ 7-KbKb

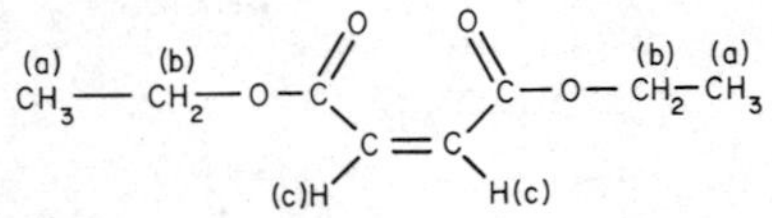

	Shift	Peak
a	1.30	3A
b	4.28	4A
▶ c	6.28	1A

▶ $C_8H_{12}O_4$ ▶ 7% $CDCl_3$ ▶ V 212

▶ (7)-(J)n(N)cj

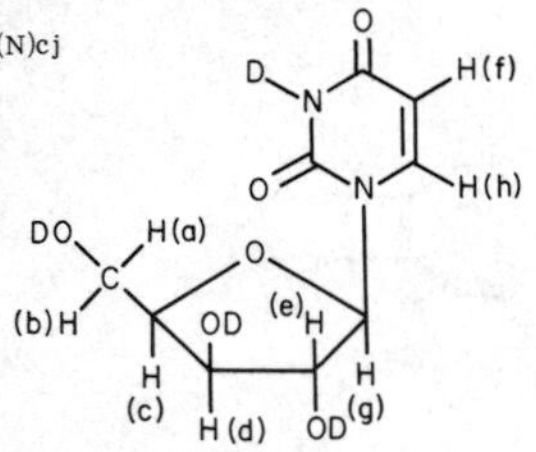

	Shift	Peak
a	3.95	S
b	3.99	S
c	4.18	S
d	4.40	3A
e	4.61	S
▶ f	5.84	2C
g	6.13	2C
h	7.94	2A

▶ $C_9H_{12}N_2O_6$ ▶ D_2O ▶ V 537

▶ (7)-(K)j(K)j

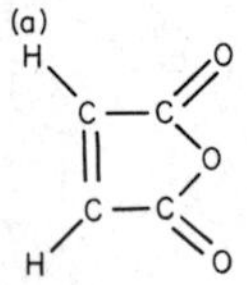

	Shift	Peak
a	7.10	1A

▶ $C_4H_2O_3$ ▶ 7% $CDCl_3$ ▶ V 48

▶ (7)-(J)n(N)cj

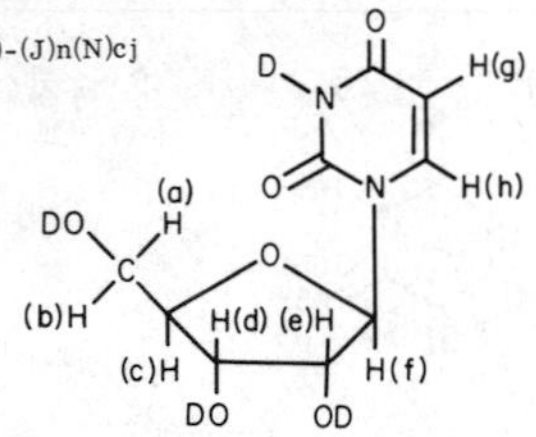

	Shift	Peak
a	3.82	S
b	3.92	S
c	4.18 ± 0.03	S
d	4.25	S
e	4.35	S
f	5.90	S
▶ g	5.90	S
h	7.87	2A

▶ $C_9H_{11}N_2O_6$ ▶ D_2O ▶ V 535

▶ (7)-(K)x(L)n

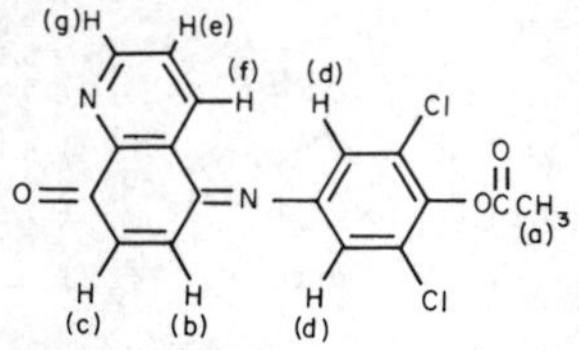

	Shift	Peak
a	2.45	1A
b	6.88*	2H
▶ c	6.88*	2H
d	6.95	1A
e	7.68	4A
f	8.70	4A
g	8.98	4A

*or 7.30

▶ $C_{17}H_{10}Cl_2N_2O_3$ ▶ 7% $CDCl_3$ ▶ V 327

▶ (7)-(J)n(N)cj

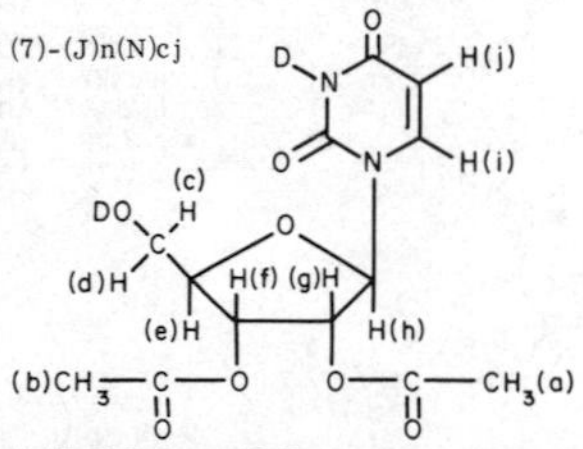

	Shift	Peak
a	2.17*	1A
b	2.17*	1A
c	3.86	1B
d	3.90	1A
e	4.38	4A
f	5.43	S
g	5.50	S
▶ h	5.94	2A
i	6.08	2A
j	7.87	2A

*or 2.21

▶ $C_{13}H_{16}N_2O_8$ ▶ D_2O ▶ V 612

▶ (7)-(K)vVho

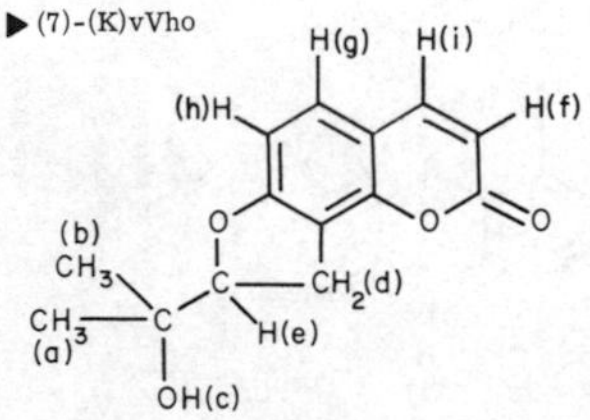

	Shift	Peak
a	1.25 or 1.37	1A
b	1.37 or 1.25	1A
c	1.96	1G
d	3.33	2A
e	4.82	3A
▶ f	6.22	2C
g	6.79	2C
h	7.29	2C
i	7.66	2C

▶ $C_{14}H_{14}O_4$ ▶ 7% $CDCl_3$ ▶ V 310

▶ (7)-(K)vVhq

	Shift	Peak
▶ a	6.42	2A
b	7.72	2A

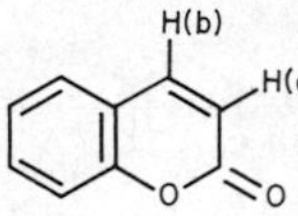

▶ $C_9H_6O_2$ ▶ 7% $CDCl_3$ ▶ V 225

▶ (7)-(L)hl(L)hl

	Shift	Peak
a	2.20	3A
b	5.28	4C
c	6.12	O
▶ d	6.55	3C

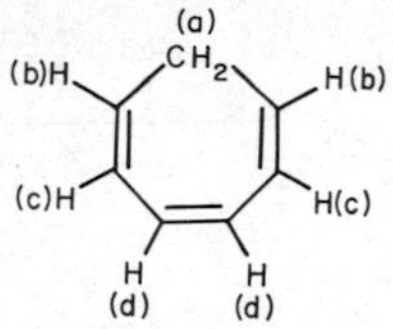

▶ C_7H_8 ▶ 7% $CDCl_3$ ▶ V 158

▶ (7)-(L)hl(B)b

	Shift	Peak
a	~1.52	N
b	~2.20	O
c	~5.63 or 5.77	S
▶ d	~5.77 or 5.63	S

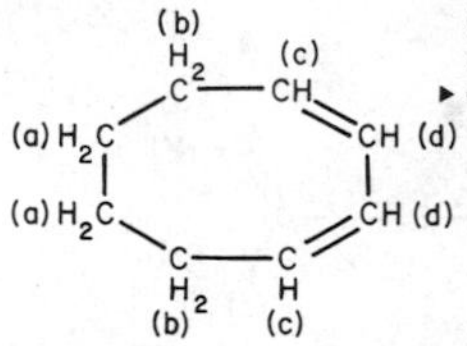

▶ C_8H_{12} ▶ $CDCl_3$ ▶ API 129

▶ (7)-(L)mn(N)cj

	Shift	Peak
a	3.82	S
b	3.96	S
c	4.32	3C
d	5.91	3A
▶ e	6.04	1A
f	7.85	2A

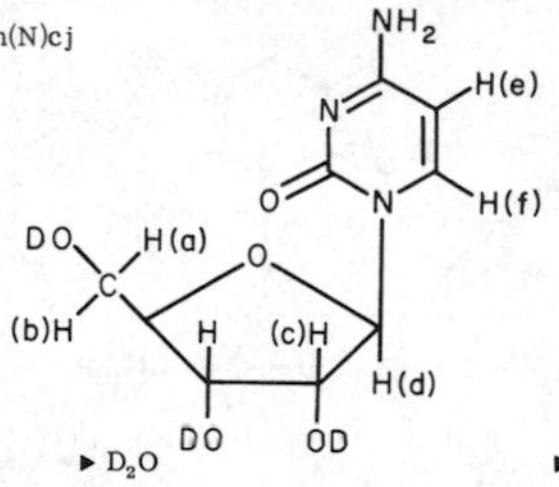

▶ $C_9H_{13}N_3O_5$ ▶ D_2O ▶ V 540

▶ (7)-(L)hl(B)b

	Shift	Peak
a	~1.52	N
b	~2.20	O
c	~5.63	S
▶ d	~5.77	1E

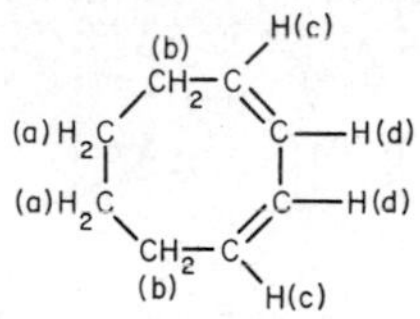

▶ C_8H_{12} ▶ 7% $CDCl_3$ ▶ V 209

▶ (7)-(L)hlVbh

	Shift	Peak
a	3.38	1A
▶ b	6.56	4A
c	7.27	S
d	7.30	S
e	11.05	1C

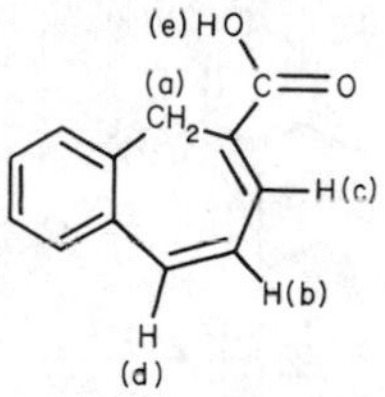

▶ $C_{12}H_{10}O_2$ ▶ 7% $CDCl_3$ ▶ V 591

▶ (7)-(L)hl(B)l

	Shift	Peak
a	2.20	3A
b	5.28	4C
▶ c	6.12	O
d	6.55	3C

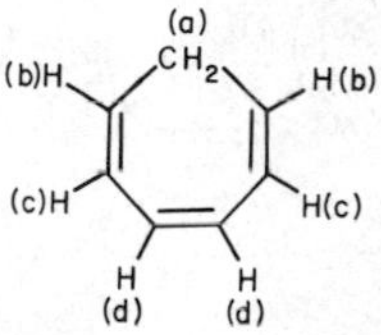

▶ C_7H_8 ▶ 7% $CDCl_3$ ▶ V 158

▶ (7)-(M)l(N)hl

	Shift	Peak
▶ a	7.14	1C
b	7.71	1C
c	13.38	1C

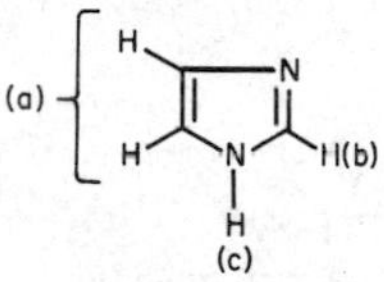

▶ $C_3H_4N_2$ ▶ 7% $CDCl_3$ ▶ V 20

▶ (7)-(L)cl(J)l

	Shift	Peak
a	1.27	2A
b	2.90	5B
▶ c	8.32	1A
d	8.75	1B

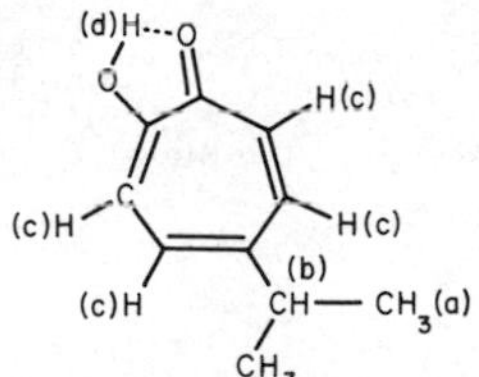

▶ $C_{10}H_{12}O_2$ ▶ 7% $CDCl_3$ ▶ V 560

▶ (7)-(N)bl(B)l

	Shift	Peak
a	0.90	3C
b	1.58	6B
c	3.09	2A
d	3.19	N
e	4.72	5B
f	5.49	1C
▶ g	5.73	4A
h	7.05	2D

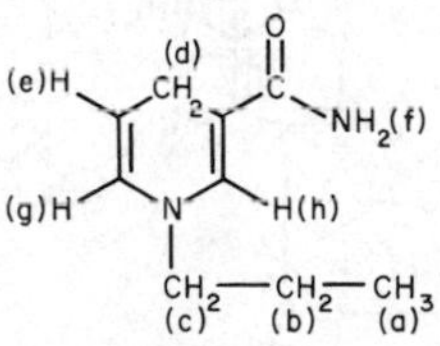

▶ $C_9H_{14}N_2O$ ▶ 7% $CDCl_3$ ▶ V 542

▶ (7)-(L)n(K)x

	Shift	Peak
a	2.45	1A
▶ b	6.88*	2H
c	6.88*	2H
d	6.95	1A
e	7.68	4A
f	8.70	4A
g	8.98	4A

*or 7.30

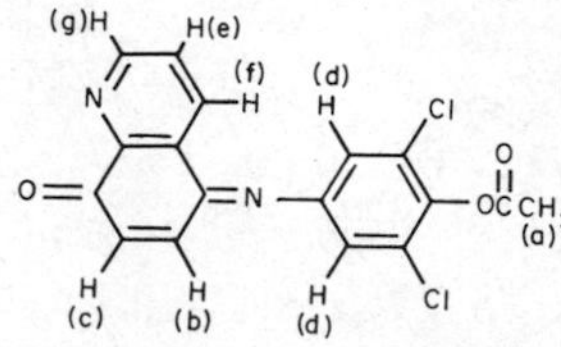

▶ $C_{17}H_{10}Cl_2N_2O_3$ ▶ 7% $CDCl_3$ ▶ V 327

▶ (7)-(N)aj(J)n

	Shift	Peak
a	3.37 or 3.43	1A
b	3.37 or 3.43	1A
c	5.73	2C
▶ d	7.20	2C

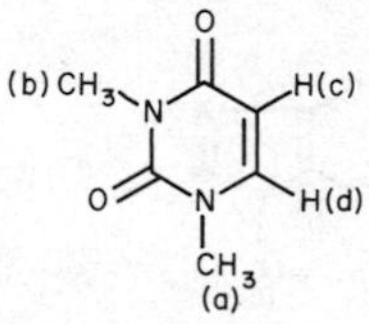

▶ $C_6H_8N_2O_2$ ▶ 7% $CDCl_3$ ▶ V 460

▶ (7)-(N)aj(J)n

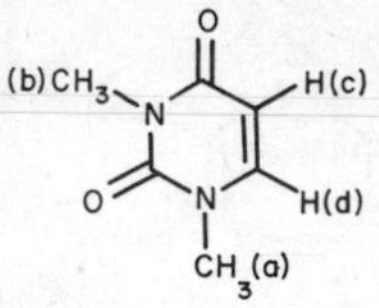

	Shift	Peak
a	3.30 or 3.43	1A
b	3.30 or 3.43	1A
c	5.84	2C
▶ d	7.61	2C

▶ $C_6H_8N_2O_2$ ▶ D_2O ▶ V 461

▶ (7)-(O)b(B)b

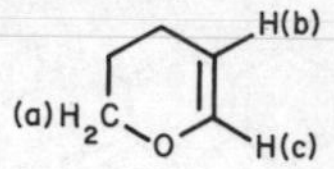

	Shift	Peak
a	3.97	N
b	4.65	N
▶ c	6.37	N

▶ C_5H_8O ▶ 7% $CDCl_3$ ▶ V 111

▶ (7)-(N)cj(J)n

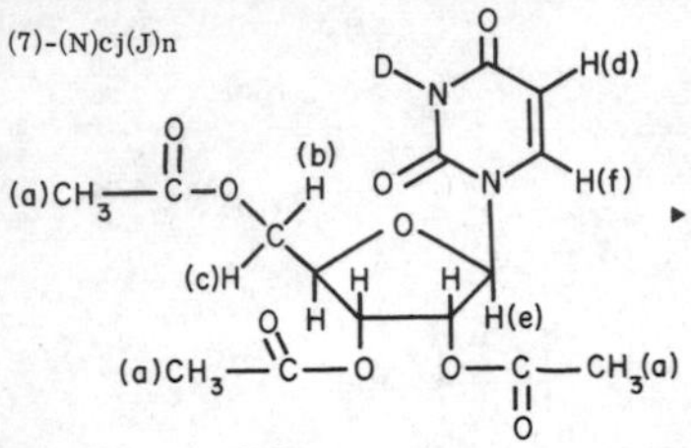

	Shift	Peak
a	1.92	1A
b	3.80	S
c	3.87	S
d	5.81	S
e	5.83	S
▶ f	7.65	2A

▶ $C_{15}H_{18}N_2O_9$ ▶ D_2O ▶ V 642

▶ 7-OaTh

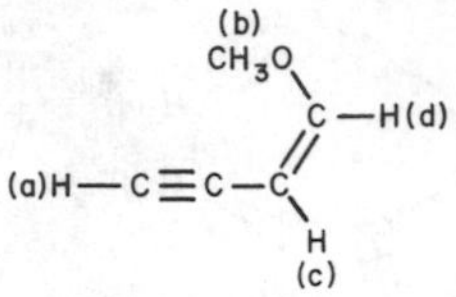

	Shift	Peak
a	3.08	2E
b	3.80	1A
c	4.52	4D
▶ d	6.35	2E

▶ C_5H_6O ▶ 7% $CDCl_3$ ▶ V 100

▶ 7-(N)cj(J)n

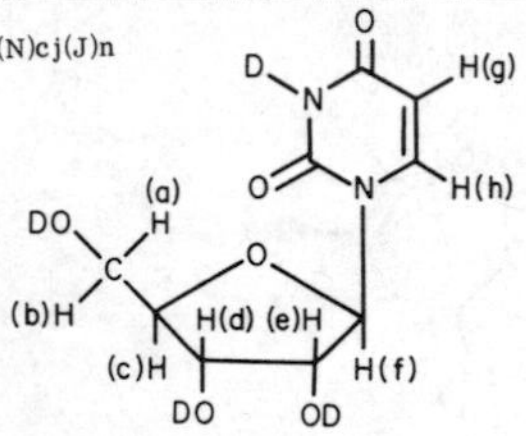

	Shift	Peak
a	3.82	S
b	3.92	S
c	4.18 ± 0.03	S
d	4.25	S
e	4.35	S
f	5.90	S
g	5.90	S
▶ h	7.87	2A

▶ $C_9H_{11}N_2O_6$ ▶ D_2O ▶ V 535

▶ (7)-(O)xX

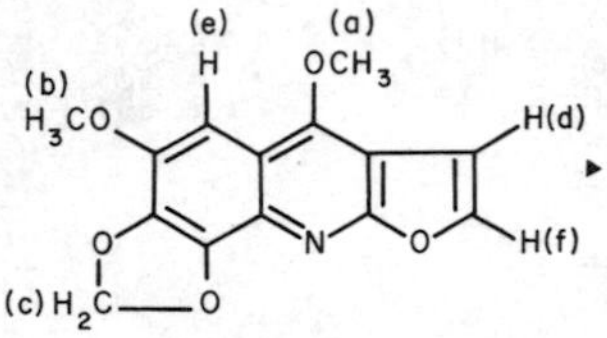

	Shift	Peak
a	4.28 or 4.40	1A
b	4.28 or 4.40	1A
c	6.08	1A
d	7.05	2C
e	7.28	1H
▶ f	7.62	2C

▶ $C_{14}H_{11}NO_5$ ▶ 7% $CDCl_3$ ▶ V 304

▶ (7)-(N)cj(J)n

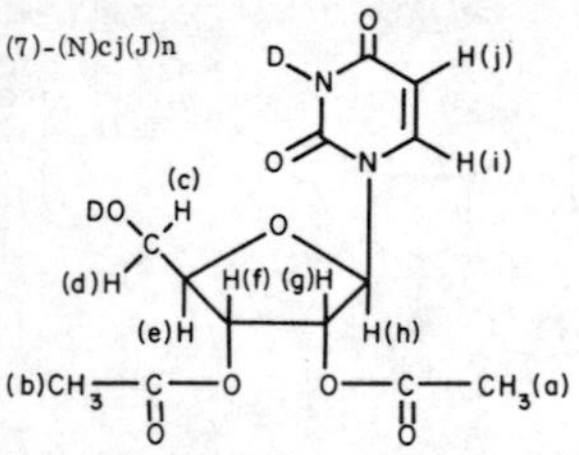

	Shift	Peak
a	2.17*	1A
b	2.17*	1A
c	3.86	1B
d	3.90	1A
e	4.38	4A
f	5.43	S
g	5.50	S
h	5.94	2A
i	6.08	2A
▶ j	7.87	2A

* or 2.21

▶ $C_{13}H_{16}N_2O_8$ ▶ D_2O ▶ V 612

▶ (7)-(S)boo(B)s

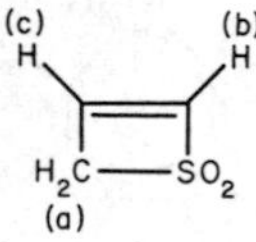

	Shift	Peak
a	4.58	2E
▶ b	6.80	2E
c	7.22	M

▶ $C_3H_4O_2S$ ▶ 7% $CDCl_3$ ▶ V 22

▶ (7)-(N)cj(J)n

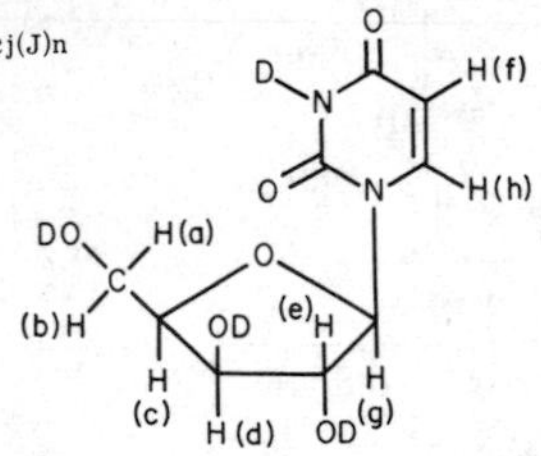

	Shift	Peak
a	3.95	S
b	3.99	S
c	4.18	S
d	4.40	3A
e	4.61	S
f	5.84	2C
g	6.13	2C
▶ h	7.94	2A

▶ $C_9H_{12}N_2O_6$ ▶ D_2O ▶ V 537

▶ 7-ThOa

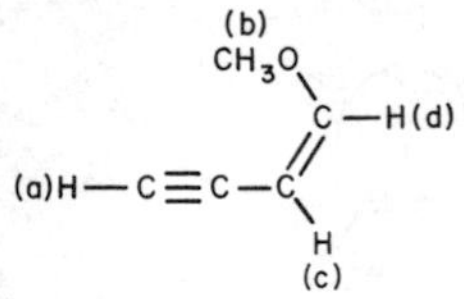

	Shift	Peak
a	3.08	2E
b	3.80	1A
▶ c	4.52	4D
d	6.35	2E

▶ C_5H_6O ▶ 7% $CDCl_3$ ▶ V 100

▶ (7)-(N)hl(M)l

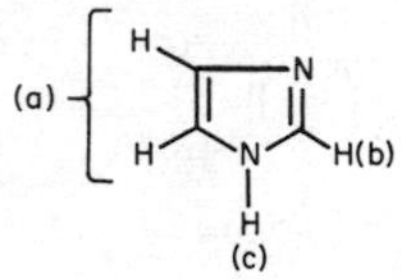

	Shift	Peak
▶ a	7.14	1C
b	7.71	1C
c	13.38	1C

▶ $C_3H_4N_2$ ▶ 7% $CDCl_3$ ▶ V 20

▶ (7)-Vbh(B)v

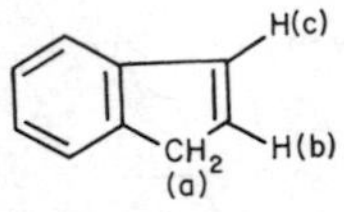

	Shift	Peak
a	3.33	1I
b	6.50	N
▶ c	6.82	N

▶ C_9H_8 ▶ 7% $CDCl_3$ ▶ V 227

▶ (7)-Vch(C)cc

(c)H H(a) COOH(f) H(b) H(d) H (e)

	Shift	Peak
a	0.86	3A
b	2.72	5B
c	3.17	4A
d	6.32	S
▶ e	6.35	S
f	12.14	1A

▶ $C_{12}H_{10}O_2$ ▶ 7% $CDCl_3$ ▶ V 592

▶ (7)-Vbh(L)hl

(e) HO (a) CH_2 C=O H(c) H(b) H (d)

	Shift	Peak
a	3.38	1A
b	6.56	4A
c	7.27	S
▶ d	7.30	S
e	11.05	1C

▶ $C_{12}H_{10}O_2$ ▶ 7% $CDCl_3$ ▶ V 591

▶ (7)-Vho(D)aoo

H_3C(a) H(h) (j) H (b) (e) OCH_3 H (a)CH_3 (c)H H(d) CH_3 (a) (f)H (i)H O H (i) H(g) CH_3(a)

	Shift	Peak
a	1.46*	1A
b	3.58	1A
c	5.62†	2A
d	5.62†	2A
▶ e	6.45‡	2A
▶ f	6.45‡	2A
g	6.62	S
h	7.12	2A
i	7.88§	1A
j	7.88§	1A

*or 1.49 ‡or 6.60
†or 5.70 §or 7.90

▶ $C_{26}H_{21}O_5$ ▶ 7% $CDCl_3$ ▶ V 696

▶ 7-VhhVhh

(a) H (a) H C=C

	Shift	Peak
▶ a	6.55	1A
b	7.18 ± .03	1A

(b) = all benzene H's.

▶ $C_{14}H_{12}$ ▶ 7% $CDCl_3$ ▶ V 305

▶ (7)-Vho(D)dno

(a) CH_3 CH_3(b) H(d) (e) H N O CH_3 (c) H (g) H(f) NO_2

	Shift	Peak
a	1.19	1A
b	1.30	1A
c	2.74	1A
d	5.84	2A
▶ e	6.93	S
f	8.01 ± .02	S
g	8.01 ± .02	S

▶ $C_{19}H_{18}N_2O_3$ ▶ 7% $CDCl_3$ ▶ V 677

▶ (7)-X(O)x

(e) H (a) OCH_3 (b) H_3CO H(d) N O H(f) O O (c) H_2C

	Shift	Peak
a	4.28 or 4.40	1A
b	4.28 or 4.40	1A
c	6.08	1A
▶ d	7.05	2C
e	7.28	1H
f	7.62	2C

▶ $C_{14}H_{11}NO_5$ ▶ 7% $CDCl_3$ ▶ V 304

▶ (7)-Voo(D)aao

(b)CH_2–COOH CH_2(c) (a) CH_3 O O CH_3(a) (a)CH_3 CH_3(a) (e)H H(e) H OCH_3 H (f) (d) (f)

	Shift	Peak
a	1.42	1A
b	2.62	Q
c	2.86	Q
d	3.75	1A
e	5.50	2C
▶ f	6.53	2C

▶ $C_{20}H_{24}O_5$ ▶ 7% $CDCl_3$ ▶ V 344

▶ **8**-ABb

(c) CH_3 (e) H C=C (e) H (d) (b) (a) $CH_2CH_2CH_3$

	Shift	Peak
a	.88	3C
b	1.34	S
c	1.57	1H
d	1.90	S
▶ e	4.60	N

▶ C_6H_{12} ▶ 50 vol.% CCl_4 ▶ API 373

▶ (7)-Vho(K)v

H(g) H(i) (h)H H(f) O O O (b) CH_3 C CH_2(d) CH_3 (a) C H(e) OH(c)

	Shift	Peak
a	1.25 or 1.37	1A
b	1.37 or 1.25	1A
c	1.96	1G
d	3.33	2A
e	4.82	3A
f	6.22	2C
g	6.79	2C
h	7.29	2C
▶ i	7.66	2C

▶ $C_{14}H_{14}O_4$ ▶ 7% $CDCl_3$ ▶ V 310

▶ 8-ABb

(e) H (c) CH_3 C=C H (f) (d) (b) (b) (a) $CH_2CH_2CH_2CH_3$

	Shift	Peak
a	.93	3C
b	1.38	O
c	1.68	1H
d	2.01	3B
▶ e	~4.67	M
f	~4.67	M

▶ C_7H_{14} ▶ liquid ▶ API 407

▶ (7)-Vhq(K)v

H(b) H(a) O O

	Shift	Peak
a	6.42	2A
▶ b	7.72	2A

▶ $C_9H_6O_2$ ▶ 7% $CDCl_3$ ▶ V 225

▶ 8-ABc

(f) H (c) CH_3 C=C H (e) (d) (b) (a) CH_2CHCH_3 CH_3

	Shift	Peak
a	.87	2H
b	1.62	S
c	1.67	3C
d	1.82	2B
e	4.67 ± .03	N
▶ f	4.67 ± .03	N

▶ C_7H_{14} ▶ liquid ▶ API 400

▶ 8-ABd

(d) H, (b) CH_3, C=C, H (e), (a) $CH_2C(CH_3)_3$ (c)

Shift	Peak
a .92	1A
b 1.76	1A
c 1.94	1A
▶ d ~4.70	1C
e ~4.80	1C

▶ C_8H_{16} ▶ 50 vol.% CCl_4 ▶ API 433

▶ 8-AG

(b) H, (a) CH_3, C=C, H (c), Br

Shift	Peak
a 2.28	2E
▶ b 5.33	M
c 5.52	M

▶ C_3H_5Br ▶ 7% $CDCl_3$ ▶ V 23

▶ 8-ACaa

(d) H, (b) CH_3, C=C, H (d), (c) (a) $CH(CH_3)_2$

Shift	Peak
a .99	2A
b 1.67	1H
c 2.23	5B
▶ d 4.62	N

▶ C_6H_{12} ▶ liquid ▶ API 382

▶ 8-AJe

Cl, O=C, H(c), C=C, CH_3 (a), H(b)

Shift	Peak
a 2.04	1H
▶ b 6.02	4A
c 6.51	1G

▶ C_4H_5ClO ▶ 7% $CDCl_3$ ▶ V 400

▶ 8-A(C)bb

(a) CH_3, (e) H, O, CH_3 (b), C=C, H(c), H(d)

Shift	Peak
a 1.75 ± .03	1E
b 1.75 ± .03	1E
c 4.78 ± .03	N
▶ d 4.78 ± .03	N
e 6.75	1C

▶ $C_{10}H_{14}O$ ▶ 7% $CDCl_3$ ▶ V 271

▶ 8-AJn

(b) H, (a) CH_3, C=C, (d) NH_2, H (c), C, O

Shift	Peak
a 1.95	1H
▶ b 5.38	M
c 5.77	M
d ~6.38	1D

▶ C_4H_7NO ▶ 7% $CDCl_3$ ▶ V 71

▶ 8-ADaaa

(b) CH_3, CH_3, (c+d) CH_2=C—C—CH_3 (a), CH_3

Shift	Peak
a 1.03	1A
b 1.72	1A
▶ c 4.60 or 4.67	1B
d 4.67 or 4.60	1B

▶ C_7H_{14} ▶ 50% CCl_4 ▶ API 223

▶ 8-AJn

O, N, H, C, C=C, H(c), CH_3 (a), H(b)

Shift	Peak
a 2.06	2D
▶ b 5.43	1I
c 5.77	1I

▶ $C_{10}H_{11}NO$ ▶ 7% $CDCl_3$ ▶ V 555

▶ 8-ADaaa

CH_3(b), CH_3(a), (c)H_2C=C—C—CH_3(a), CH_3(a)

Shift	Peak
a 1.05	1A
b 1.73	3A
▶ c ~4.66	O

▶ C_7H_{14} ▶ 10% CCl_4 ▶ API 158

▶ 8-AKa

(c)H, CH_3(a), C=C, (d)H, C=O, OCH_3(b)

Shift	Peak
a 1.95	2E
b 3.75	1A
▶ c 5.57	M
d 6.10	M

▶ $C_5H_8O_2$ ▶ 7% $CDCl_3$ ▶ V 113

▶ 8-ADaaa

(d) H, (b) CH_3, C=C, CH_3, H (c), C, CH_3, CH_3(a)

Shift	Peak
a 1.08	1B
b 1.73	1H
c 4.62 or 4.72	N
▶ d 4.72 or 4.62	N

▶ C_7H_{14} ▶ liquid ▶ API 417

▶ 8-AKb

(d) H, (b) CH_3, C=C, (e)H, C=O, O, (c) CH_2—CH_3 (a)

Shift	Peak
a 1.28	3A
b 1.93	2D
c 4.22	4A
▶ d 5.57	3C
e 6.10	1I

▶ $C_6H_{10}O_2$ ▶ 7% $CDCl_3$ ▶ V 135

▶ 8-AKh

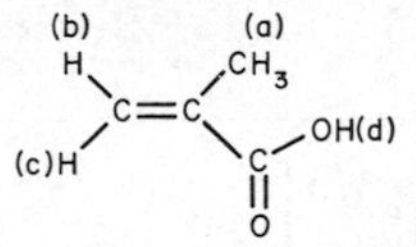

	Shift	Peak
a	1.97	2E
▶ b	5.72	M
c	6.30	2E
d	11.57	1B

▶ $C_4H_6O_2$ ▶ 7% $CDCl_3$ ▶ V 62

▶ 8-BbA

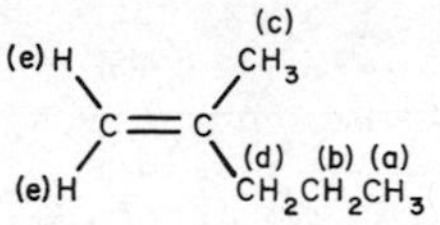

	Shift	Peak
a	.88	3C
b	1.34	S
c	1.57	1H
d	1.90	S
▶ e	4.60	N

▶ C_6H_{12} ▶ 50 vol.% CCl_4 ▶ API 373

▶ 8-ALal

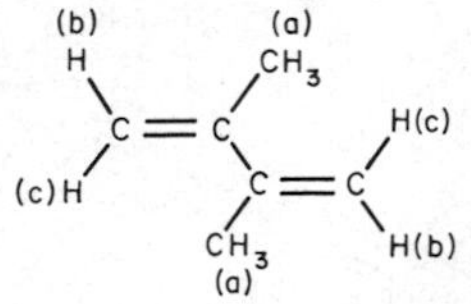

	Shift	Peak
a	1.88	1A
▶ b	~4.93	S
c	~4.93	S

▶ C_6H_{10} ▶ liquid ▶ API 475

▶ 8-BbA

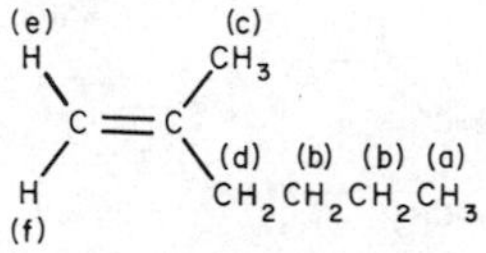

	Shift	Peak
a	.93	3C
b	1.38	O
c	1.68	1H
d	2.01	3B
e	~4.67	M
▶ f	~4.67	M

▶ C_7H_{14} ▶ liquid ▶ API 407

▶ 8-AQa

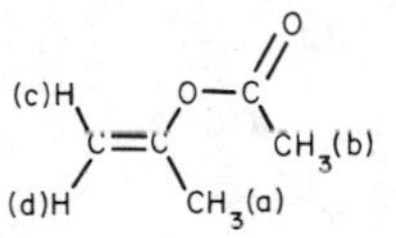

	Shift	Peak
a	1.93	1A
b	2.12	1A
c	4.69	1G
▶ d	4.69	1G

▶ $C_5H_8O_2$ ▶ 7% $CDCl_3$ ▶ V 440

▶ 8-BcA

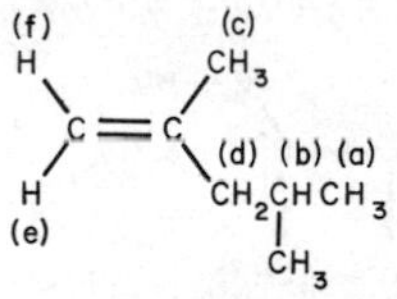

	Shift	Peak
a	.87	2H
b	1.62	S
c	1.67	3C
d	1.82	2B
▶ e	4.67 ± .03	N
f	4.67 ± .03	N

▶ C_7H_{14} ▶ liquid ▶ API 400

▶ 8-ATh

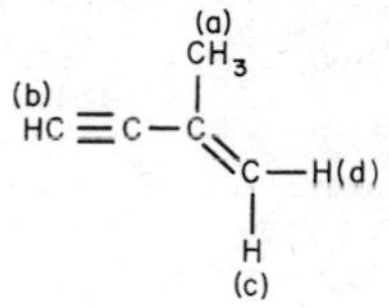

	Shift	Peak
a	1.90	2E
b	2.87	1A
c	5.27 or 5.37	N
▶ d	5.27 or 5.37	N

▶ C_5H_6 ▶ 7% $CDCl_3$ ▶ V 99

▶ 8-BdA

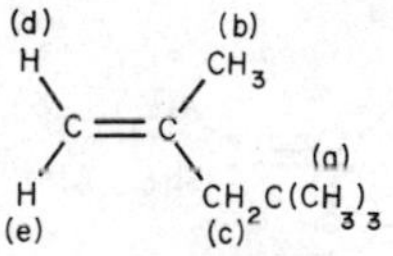

	Shift	Peak
a	.92	1A
b	1.76	1A
c	1.94	1A
d	~4.70	1C
▶ e	~4.80	1C

▶ C_8H_{16} ▶ 50 vol.% CCl_4 ▶ API 433

▶ 8-AU

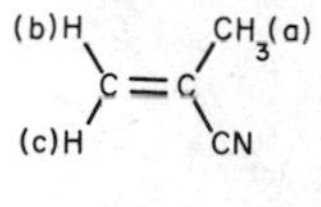

	Shift	Peak
a	2.00	1I
▶ b	5.73	N
c	5.82	N

▶ C_4H_5N ▶ 7% $CDCl_3$ ▶ V 97

▶ 8-BaBa

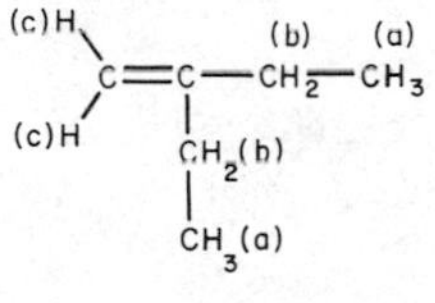

	Shift	Peak
a	1.02	3C
b	2.08	4C
▶ c	4.63	3A

▶ C_6H_{12} ▶ 10% CCl_4 ▶ API 36

▶ 8-AVhh

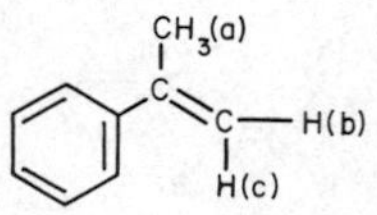

	Shift	Peak
a	2.12	2D
▶ b	5.05	N
c	5.36	N

▶ C_9H_{10} ▶ 7% $CDCl_3$ ▶ V 232

▶ 8-BaBa

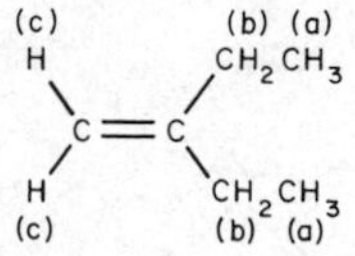

	Shift	Peak
a	1.02	3C
b	2.00	4C
▶ c	4.67	N

▶ C_6H_{12} ▶ liquid ▶ API 381

▶ (8)-(B)b(B)b

	Shift	Peak
a	4.55	1A

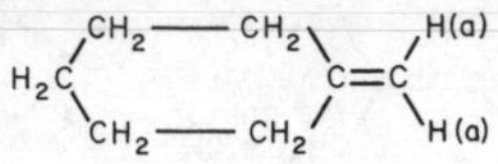

CH_2—CH_2 H(a)
H_2C C=C
CH_2—CH_2 H(a)

▶ C_7H_{12} ▶ 7% $CDCl_3$ ▶ V 180

▶ 8-(B)b(C)bd

	Shift	Peak
a	0.72	1A
b	1.23	1A
c	4.58 ± .03	N
▶ d	4.58 ± .03	N

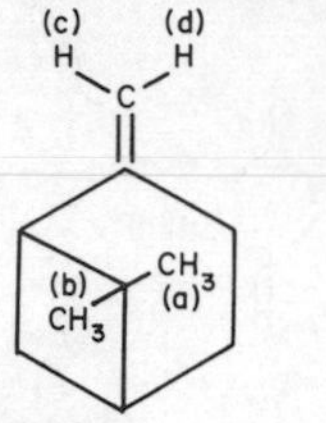

(c) (d)
H H
C
(b) CH_3 (a)
CH_3

▶ $C_{10}H_{16}$ ▶ 7% $CDCl_3$ ▶ V 274

▶ (8)-(B)b(B)b

	Shift	Peak
a	1.92	6B
b	2.70	3C
▶ c	4.70	3A

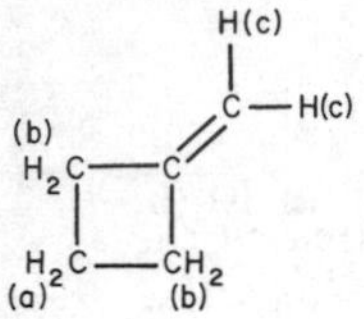

H(c)
C—H(c)
(b)
H_2C—C
H_2C—CH_2
(a) (b)

▶ C_5H_8 ▶ $CDCl_3$ ▶ API 127

▶ (8)-(B)b(C)bd

	Shift	Peak
a	1.74	1E
b	2.40	O
c	4.49 or 4.82	1B
▶ d	4.82 or 4.49	1B
e	5.22	R
f	5.24	R
g	6.08	4D
h	~8.01	N

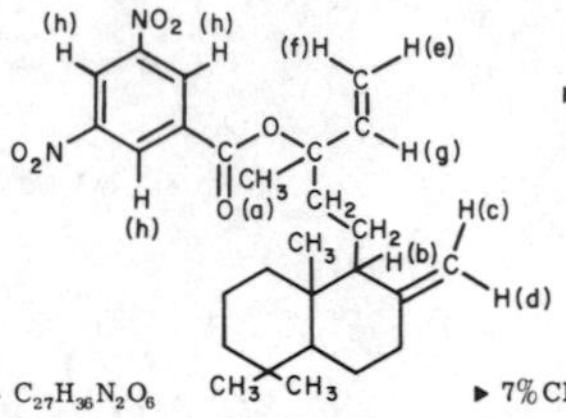

(h) NO_2 (h)
H H (f)H H(e)
O_2N C O C C H(g)
H O (a) CH_3 CH_2
(h) CH_3 CH_2 H(c)
H (b) C
H(d)
CH_3 CH_3

▶ $C_{27}H_{36}N_2O_6$ ▶ 7% $CDCl_3$ ▶ V 697

▶ (8)-(B)b(B)b

	Shift	Peak
a	1.92	5B
b	2.70	3C
▶ c	4.70	N

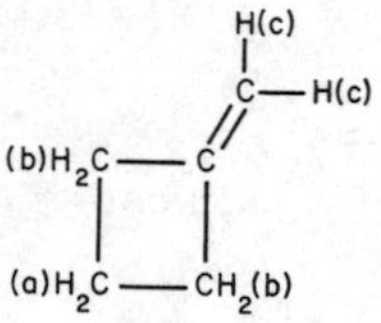

H(c)
C—H(c)
(b)H_2C—C
(a)H_2C—CH_2(b)

▶ C_5H_8 ▶ 7% $CDCl_3$ ▶ V 109

▶ (8)-(B)b(C)bd

	Shift	Peak
a	1.28	1E
b	2.40	S
c	4.52	1B
▶ d	4.83	1B
e	5.08	R
f	5.22	R
g	5.92	4D

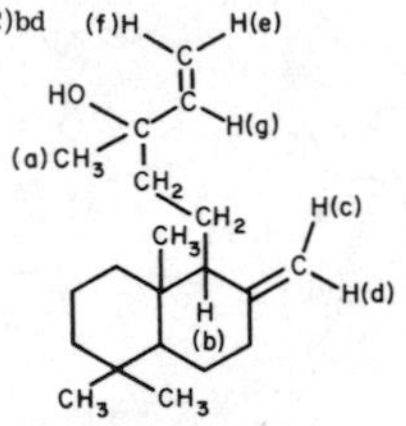

(f)H H(e)
C
HO C
C H(g)
(a)CH_3
CH_2
CH_2 H(c)
CH_3 C
H H(d)
(b)
CH_3 CH_3

▶ $C_{20}H_{34}O$ ▶ 7% $CDCl_3$ ▶ V 685

▶ (8)-(B)b(B)b

	Shift	Peak
a	~1.65	Q
b	~2.22	Q
▶ c	4.82	3A

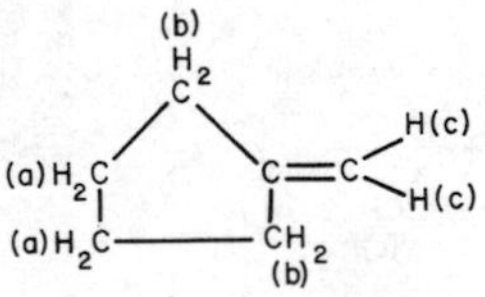

(b)
H_2
C H(c)
(a)H_2C C=C
(a)H_2C—CH_2 H(c)
(b)

▶ C_6H_{10} ▶ CCl_4 ▶ API 128

▶ (8)-(B)b(C)bk

	Shift	Peak
a	3.80	O
b	4.89 or 5.08	N
▶ c	4.89 or 5.08	N
d	12.12	1A

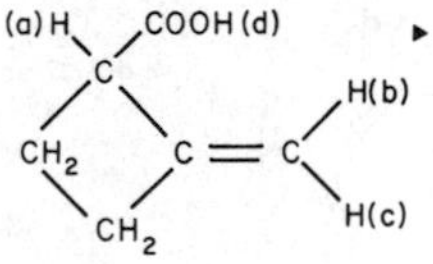

(a)H COOH(d)
C H(b)
CH_2 C=C
CH_2 H(c)

▶ $C_6H_8O_2$ ▶ 7% $CDCl_3$ ▶ V 128

▶ (8)-(B)b(B)b

	Shift	Peak
a	~1.65	O
b	~2.22	O
▶ c	4.82	4C

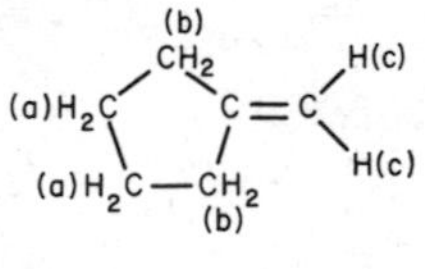

(b)
CH_2 H(c)
(a)H_2C C=C
(a)H_2C—CH_2 H(c)
(b)

▶ C_6H_{10} ▶ 7% $CDCl_3$ ▶ V 132

▶ 8-BeE

	Shift	Peak
a	4.15	2D
▶ b	5.42	2A
c	5.59	M

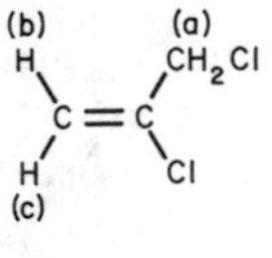

(b) (a)
H CH_2Cl
C=C
H Cl
(c)

▶ $C_3H_4Cl_2$ ▶ 7% $CDCl_3$ ▶ V 18

▶ 8-BaCaa

	Shift	Peak
a	1.03	3D
b	1.03	2H
c	~2.13	S
d	4.70 ± .03	S
▶ e	4.70 ± .03	S

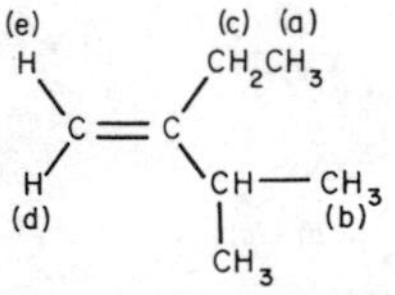

(e) (c) (a)
H CH_2CH_3
C=C
H CH—CH_3
(d) (b)
CH_3

▶ C_7H_{14} ▶ liquid ▶ API 416

▶ 8-BgG

	Shift	Peak
a	4.17	2D
▶ b	5.62	2A
c	6.02	M

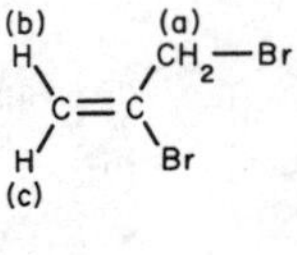

(b) (a)
H CH_2—Br
C=C
H Br
(c)

▶ $C_3H_4Br_2$ ▶ 7% $CDCl_3$ ▶ V 17

▶ 8-BkKh

	Shift	Peak
a	3.46	1G
▶ b	5.92	1G
c	6.40	1A

▶ $C_5H_6O_4$ ▶ D_2O ▶ V 433

▶ (8)-(C)bd(B)b

	Shift	Peak
a	1.28	1E
b	2.40	S
▶ c	4.52	1B
d	4.83	1B
e	5.08	R
f	5.22	R
g	5.92	4D

▶ $C_{20}H_{34}O$ ▶ 7% $CDCl_3$ ▶ V 685

▶ 8-(B)d(N)bl

	Shift	Peak
a	1.29	1A
b	2.76	3A
c	3.40	3C
▶ d	3.79	1F
e	3.79	1F
f	4.18	3C

▶ $C_9H_{14}N_2$ ▶ 7% $CDCl_3$ ▶ V 541

▶ 8-(C)bd(B)b

	Shift	Peak
a	0.72	1A
b	1.23	1A
▶ c	4.58 ± .03	N
d	4.58 ± .03	N

▶ $C_{10}H_{16}$ ▶ 7% $CDCl_3$ ▶ V 274

▶ 8-CaaA

	Shift	Peak
a	.99	2A
b	1.67	1H
c	2.23	5B
▶ d	4.62	N

▶ C_6H_{12} ▶ liquid ▶ API 382

▶ (8)-(C)bk(D)b

	Shift	Peak
a	3.80	O
▶ b	4.89 or 5.08	N
c	4.89 or 5.08	N
d	12.12	1A

▶ $C_6H_8O_2$ ▶ 7% $CDCl_3$ ▶ V 128

▶ 8-(C)bbA

	Shift	Peak
a	1.75 ± .03	1E
b	1.75 ± .03	1E
▶ c	4.78 ± .03	N
d	4.78 ± .03	N
e	6.75	1C

▶ $C_{10}H_{14}O$ ▶ 7% $CDCl_3$ ▶ V 271

▶ (8)-(C)cc(K)c

	Shift	Peak
a	1.22	1E
b	1.29	2H
c	1.62	S
d	1.97	1A
e	2.06	S
f	2.60	3E
g	~3.10	S
h	~3.10	S
i	4.67	3C
j	5.63	2A
▶ k	5.93	2A
l	6.14	2H
m	6.23	N
n	7.79	2F

▶ $C_{17}H_{20}O_5$ ▶ 7% $CDCl_3$ ▶ V 666

▶ 8-CaaBa

	Shift	Peak
a	1.03	3D
b	1.03	2H
c	~2.13	S
▶ d	4.70 ± .03	S
e	4.70 ± .03	S

▶ C_7H_{14} ▶ liquid ▶ API 416

▶ 8-DaaaA

	Shift	Peak
a	1.08	1B
b	1.73	1H
▶ c	4.62 or 4.72	N
d	4.72 or 4.62	N

▶ C_7H_{14} ▶ liquid ▶ API 417

▶ (8)-(C)bd(B)b

	Shift	Peak
a	1.74	1E
b	2.40	O
▶ c	4.49 or 4.82	1B
d	4.82 or 4.49	1B
e	5.22	R
f	5.24	R
g	6.08	4D
h	~8.01	N

▶ $C_{27}H_{36}N_2O_6$ ▶ 7% $CDCl_3$ ▶ V 697

▶ 8-DaaaA

	Shift	Peak
a	1.05	1A
b	1.73	3A
▶ c	~4.66	O

▶ C_7H_{14} ▶ 10% CCl_4 ▶ API 158

▶ 8-DaaaA

(b) CH_3 CH_3
(c+d) CH_2=C—C—CH_3(a)
CH_3

	Shift	Peak
a	1.03	1A
b	1.72	1A
c	4.60 or 4.67	1B
▶ d	4.67 or 4.60	1B

▶ C_7H_{14} ▶ 50% CCl_4 ▶ API 223

▶ 8-JnA

(b) H (a) CH_3
C=C
(d) NH_2
H (c)
C=O

	Shift	Peak
a	1.95	1H
b	5.38	M
▶ c	5.77	M
d	~6.38	1D

▶ C_4H_7NO ▶ 7% $CDCl_3$ ▶ V 71

▶ 8-(D)aav(N)av

CH_3(a)
CH_3(a)
H(c)
N
H(d)
CH_3 (b)

	Shift	Peak
a	1.35	1A
b	3.02	1A
▶ c	3.82	1A
d	3.82	1A

▶ $C_{12}H_{15}N$ ▶ 7% $CDCl_3$ ▶ V 596

▶ 8-JnA

O
N—C—C=C H(c)
H CH_3 (a) H(b)

	Shift	Peak
a	2.06	2D
b	5.43	1I
▶ c	5.77	1I

▶ $C_{10}H_{11}NO$ ▶ 7% $CDCl_3$ ▶ V 555

▶ 8-EBe

(b) H (a) CH_2Cl
C=C
H (c) Cl

	Shift	Peak
a	4.15	2D
b	5.42	2A
▶ c	5.59	M

▶ $C_3H_4Cl_2$ ▶ 7% $CDCl_3$ ▶ V 18

▶ 8-KaA

(c)H CH_3(a)
C=C
(d)H C=O
OCH_3(b)

	Shift	Peak
a	1.95	2E
b	3.75	1A
c	5.57	M
▶ d	6.10	M

▶ $C_5H_8O_2$ ▶ 7% $CDCl_3$ ▶ V 113

▶ 8-GA

(b) H (a) CH_3
C=C
H (c) Br

	Shift	Peak
a	2.28	2E
b	5.33	M
▶ c	5.52	M

▶ C_3H_5Br ▶ 7% $CDCl_3$ ▶ V 23

▶ 8-KbA

(d) H (b) CH_3
C=C
(e)H C=O
O (c) CH_2—CH_3 (a)

	Shift	Peak
a	1.28	3A
b	1.93	2D
c	4.22	4A
d	5.57	3C
▶ e	6.10	1I

▶ $C_6H_{10}O_2$ ▶ 7% $CDCl_3$ ▶ V 135

▶ 8-GBg

(b) H (a) CH_2—Br
C=C
H (c) Br

	Shift	Peak
a	4.17	2D
b	5.62	2A
▶ c	6.02	M

▶ $C_3H_4Br_2$ ▶ 7% $CDCl_3$ ▶ V 17

▶ 8-KhA

(b) H (a) CH_3
C=C
(c)H C—OH(d)
O

	Shift	Peak
a	1.97	2E
b	5.72	M
▶ c	6.30	2E
d	11.57	1B

▶ $C_4H_6O_2$ ▶ 7% $CDCl_3$ ▶ V 62

▶ 8-JeA

Cl
O=C H(c)
C=C
CH_3 (a) H(b)

	Shift	Peak
a	2.04	1H
b	6.02	4A
▶ c	6.51	1G

▶ C_4H_5ClO ▶ 7% $CDCl_3$ ▶ V 400

▶ 8-KhBk

O
(c)H C—OD
C=C
(b)H CH_2(a)
O=C
OD

	Shift	Peak
a	3.46	1G
b	5.92	1G
▶ c	6.40	1A

▶ $C_5H_6O_4$ ▶ D_2O ▶ V 433

▶ (8)-(K)c(C)cc

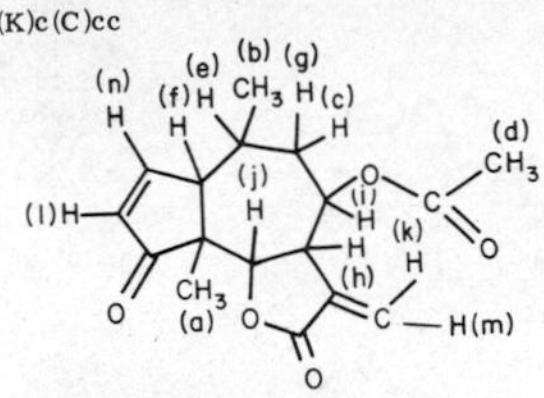

	Shift	Peak
a	1.22	1E
b	1.29	2H
c	1.62	S
d	1.97	1A
e	2.06	S
f	2.60	3E
g	~3.10	S
h	~3.10	S
i	4.67	3C
j	5.63	2A
k	5.93	2A
l	6.14	2H
▶ m	6.23	N
n	7.79	2F

▶ $C_{17}H_{20}O_5$ ▶ 7% $CDCl_3$ ▶ V 666

▶ 8-ThA

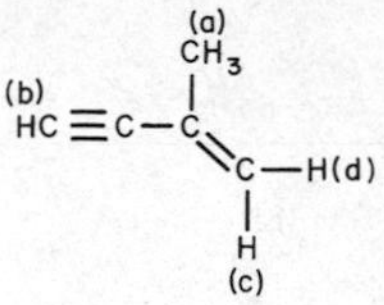

	Shift	Peak
a	1.90	2E
b	2.87	1A
▶ c	5.27 or 5.37	N
d	5.27 or 5.37	N

▶ C_5H_6 ▶ 7% $CDCl_3$ ▶ V 99

▶ 8-Laa

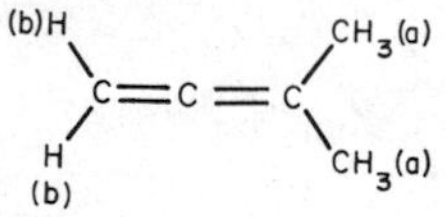

	Shift	Peak
a	1.63	3A
▶ b	4.43	7A

▶ C_5H_8 ▶ liquid ▶ API 472

▶ 8-UA

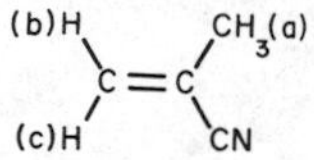

	Shift	Peak
a	2.00	1I
b	5.73	N
▶ c	5.82	N

▶ C_4H_5N ▶ 7% $CDCl_3$ ▶ V 97

▶ 8-LalA

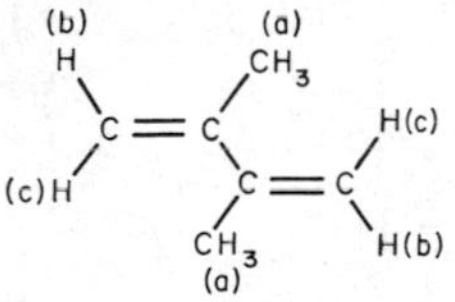

	Shift	Peak
a	1.88	1A
b	~4.93	S
▶ c	~4.93	S

▶ C_6H_{10} ▶ liquid ▶ API 475

▶ 8-VhhA

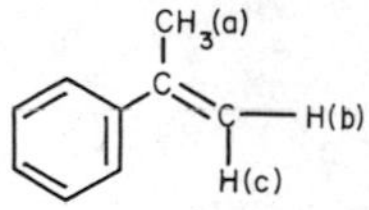

	Shift	Peak
a	2.12	2D
b	5.05	N
▶ c	5.36	N

▶ C_9H_{10} ▶ 7% $CDCl_3$ ▶ V 232

▶ 8-(N)bl(B)d

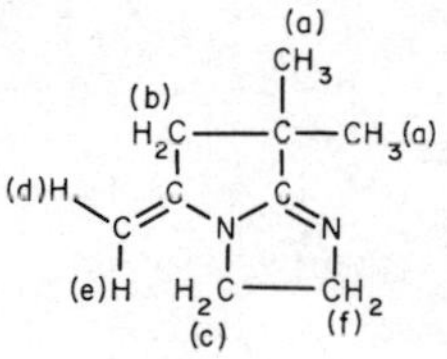

	Shift	Peak
a	1.29	1A
b	2.76	3A
c	3.40	3C
d	3.79	1F
▶ e	3.79	1F
f	4.18	3C

▶ $C_9H_{14}N_2$ ▶ 7% $CDCl_3$ ▶ V 541

▶ **9**-AAA

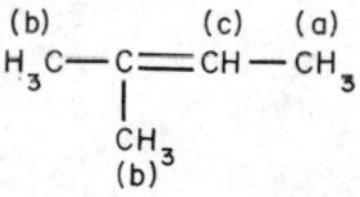

	Shift	Peak
a	1.48	1E
b	1.54	1E
▶ c	5.15 ± .05	1F

▶ C_5H_{10} ▶ 10% CCl_4 ▶ API 31

▶ 8-(N)av(D)aav

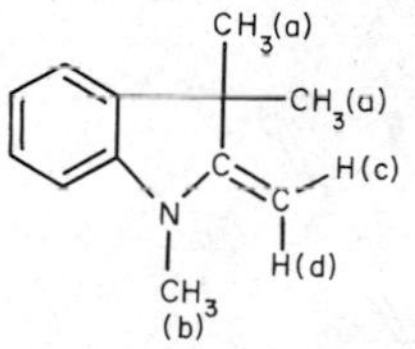

	Shift	Peak
a	1.35	1A
b	3.02	1A
c	3.82	1A
▶ d	3.82	1A

▶ $C_{12}H_{15}N$ ▶ 7% $CDCl_3$ ▶ V 596

▶ 9-AAA

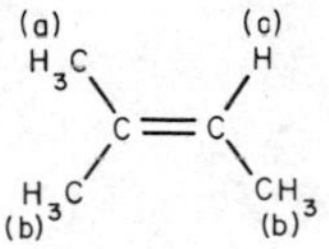

	Shift	Peak
a	1.47	1F
b	1.59	S
▶ c	5.18	S

▶ C_5H_{10} ▶ liquid ▶ API 462

▶ 8-QaA

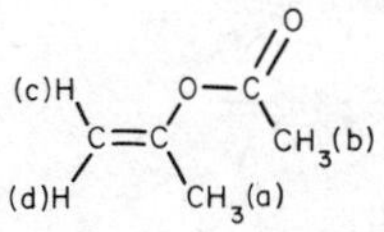

	Shift	Peak
a	1.93	1A
b	2.12	1A
▶ c	4.69	1G
d	4.69	1G

▶ $C_5H_8O_2$ ▶ 7% $CDCl_3$ ▶ V 440

▶ 9-AABa

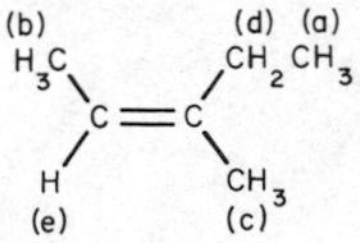

	Shift	Peak
a	.95	3C
b	1.46	1F
c	1.61	S
d	2.03	4A
▶ e	~5.08	4C

▶ C_6H_{12} ▶ liquid ▶ API 377

▶ 9-AABa

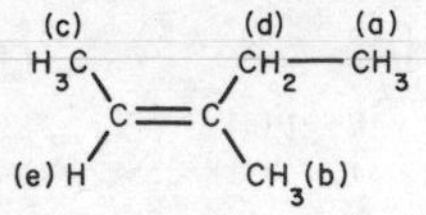

	Shift	Peak
a	.97	3A
b	1.47	1E
c	1.57	1E
d	1.97	4A
▶ e	5.13	2E

▶ C_6H_{12} ▶ 10% CCl_4 ▶ API 35

▶ 9-ACaaA

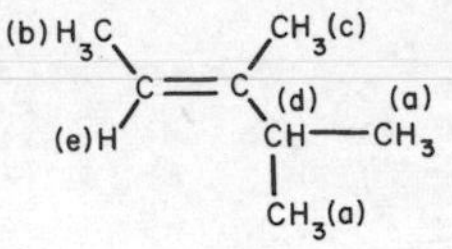

	Shift	Peak
a	.93	2A
b	1.48	1E
c	1.52	1E
d	2.20	5A
▶ e	5.20	4B

▶ C_7H_{14} ▶ 50% CCl_4 ▶ API 354

▶ 9-AABb

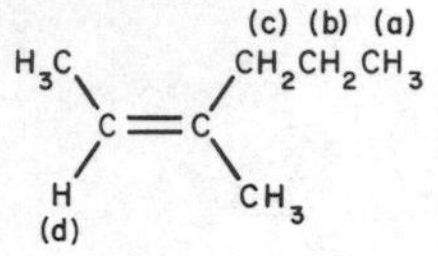

	Shift	Peak
a	.83	3C
b	1.33	S
c	1.87	3B
▶ d	5.14	O

▶ C_7H_{14} ▶ liquid ▶ API 395

▶ 9-ACaaA

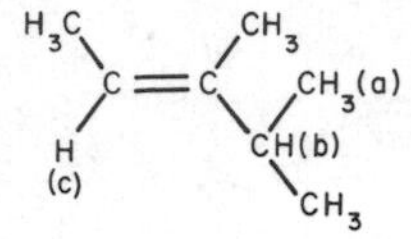

	Shift	Peak
a	.97	2A
b	2.21	6A
▶ c	5.22	O

▶ C_7H_{14} ▶ liquid ▶ API 404

▶ 9-AACaa

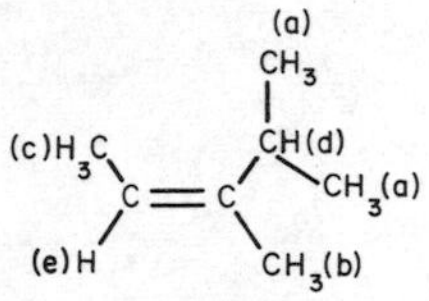

	Shift	Peak
a	.91	2A
b	~1.48 or 1.52	1E
c	~1.52 or 1.48	1E
d	2.74	5A
▶ e	5.05	1D

▶ C_7H_{14} ▶ 50% CCl_4 ▶ API 353

▶ (9)-A(K)d(C)do

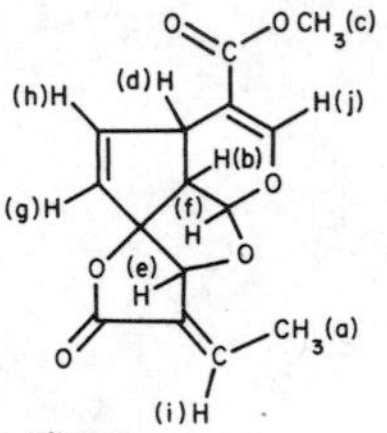

	Shift	Peak
a	2.11	2A
b	3.46	4A
c	3.79	1A
d	4.02	3E
e	5.12	1B
f	5.59	S
g	5.67*	S
h	5.67*	S
▶ i	7.19	O
j	7.46	1A

*or 6.08

▶ $C_{15}H_{14}O_6$ ▶ 7% $CDCl_3$ ▶ V 640

▶ 9-AACaa

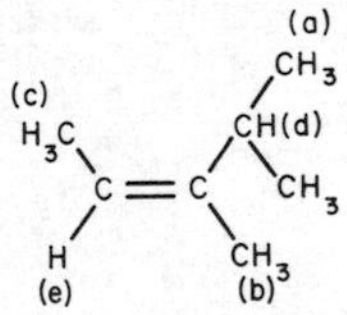

	Shift	Peak
a	.93	2A
b	1.47	1F
c	1.55	1F
d	2.84	5A
▶ e	5.10	1F

▶ C_7H_{14} ▶ liquid ▶ API 403

▶ 9-ALhh

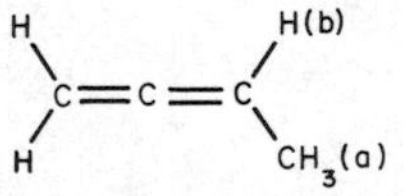

	Shift	Peak
a	1.57	5A
▶ b	5.08	5B

▶ C_4H_6 ▶ liquid ▶ API 465

▶ 9-ABaA

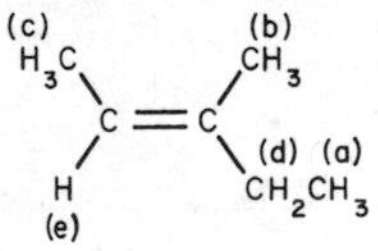

	Shift	Peak
a	.97	3C
b	1.49	S
c	1.56	S
d	1.95	4C
▶ e	5.18	4C

▶ C_6H_{12} ▶ liquid ▶ API 378

▶ 9-ALalVhh

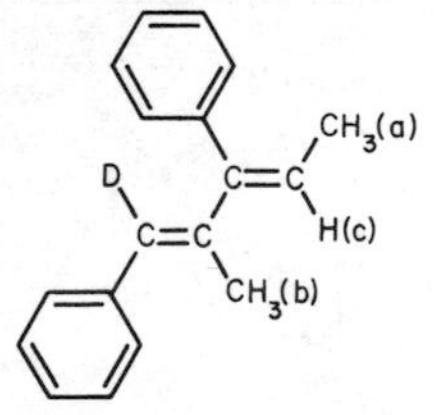

	Shift	Peak
a	1.60	2A
b	2.08	1A
▶ c	5.98	4A

▶ $C_{18}H_{17}D$ ▶ 7% $CDCl_3$ ▶ V 674

▶ 9-ABaBa

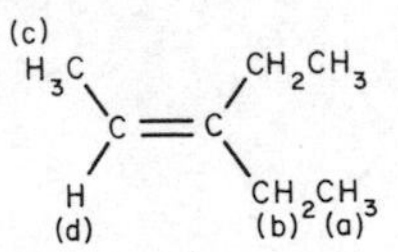

	Shift	Peak
a	.93	3C
b	1.56	2E
c	2.04	4C
▶ d	5.17	4A

▶ C_7H_{14} ▶ liquid ▶ API 414

▶ 9-BaAA

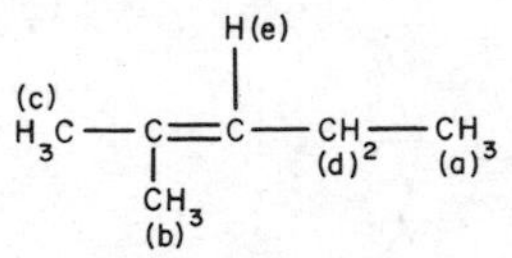

	Shift	Peak
a	.92	3A
b	1.57	1I
c	1.64	1I
d	1.97	4C
▶ e	5.05	O

▶ C_6H_{12} ▶ 10% CCl_4 ▶ API 34

▶ 9-BaAA

	Shift	Peak
a	.92	3C
b	1.58	1I
c	1.65	1I
▶ d	5.19	3C

▶ C_6H_{12} ▶ liquid ▶ API 376

▶ (9)-(B)cA(B)c

	Shift	Peak
a	~0.70	S
b	0.77	1A
c	1.03	1A
d	1.62	1I
▶ e	5.23	1C

▶ $C_{10}H_{16}$ ▶ 7% $CDCl_3$ ▶ V 273

▶ 9-BbAA

	Shift	Peak
a	.90	3C
b	1.28	S
c	~1.58 or 1.68	O
d	~1.68 or 1.58	O
e	1.91	4B
▶ f	5.08	3C

▶ C_7H_{14} ▶ liquid ▶ API 408

▶ 9-BpABb

	Shift	Peak
a	1.62 or 1.68	1A
b	2.05 ± 0.03	3A
c	4.15	2A
d	5.12	1C
▶ e	5.45	3A

▶ $C_{10}H_{18}O$ ▶ 7% $CDCl_3$ ▶ V 279

▶ 9-DbAA

	Shift	Peak
a	1.62 or 1.68	1A
b	2.05 ± 0.03	3A
c	4.15	2A
▶ d	5.12	1C
e	5.45	3A

▶ $C_{10}H_{18}O$ ▶ 7% $CDCl_3$ ▶ V 279

▶ (9)-(B)bA(C)dl

	Shift	Peak
a	0.86	1A
b	0.93	1A
c	1.57	2G
d	2.05	1C
e	2.26	1A
f	2.30	S
▶ g	5.50	1C
h	6.06	2A
i	6.60	4A

▶ $C_{13}H_{20}O$ ▶ 7% $CDCl_3$ ▶ V 616

▶ 9-BbAA

	Shift	Peak
a	1.60 or 1.68	1G
b	2.04 ± 0.05	1G
c	4.15	2G
▶ d	5.12	1F

▶ $C_{45}H_{74}O$ ▶ 7% $CDCl_3$ ▶ V 367

▶ (9)-(B)cA(C)bd

	Shift	Peak
a	0.84 or 1.27	1A
b	0.84 or 1.27	1A
c	1.65	4A
▶ d	5.17	1C

▶ $C_{10}H_{16}$ ▶ 7% $CDCl_3$ ▶ V 272

▶ 9-BaABa

	Shift	Peak
a	.91 or .98	3D
b	1.57	1I
c	1.94	5B
▶ d	5.11	3C

▶ C_7H_{14} ▶ liquid ▶ API 411

▶ (9)-(B)cA(J)b

	Shift	Peak
a	1.75 ± .03	1E
b	1.75 ± .03	1E
c	4.78 ± .03	N
d	4.78 ± .03	N
▶ e	6.75	1C

▶ $C_{10}H_{14}O$ ▶ 7% $CDCl_3$ ▶ V 271

▶ 9-BbABb

	Shift	Peak
a	1.60 or 1.68	1G
b	2.04 ± 0.05	1G
c	4.15	2G
▶ d	5.12	1F

▶ $C_{45}H_{74}O$ ▶ 7% $CDCl_3$ ▶ V 367

▶ 9-BaBaA

	Shift	Peak
a	.91 or .96	3D
b	.91 or .96	3D
c	1.61	1H
d	1.97	4A
▶ e	5.07	3A

▶ C_7H_{14} ▶ liquid ▶ API 412

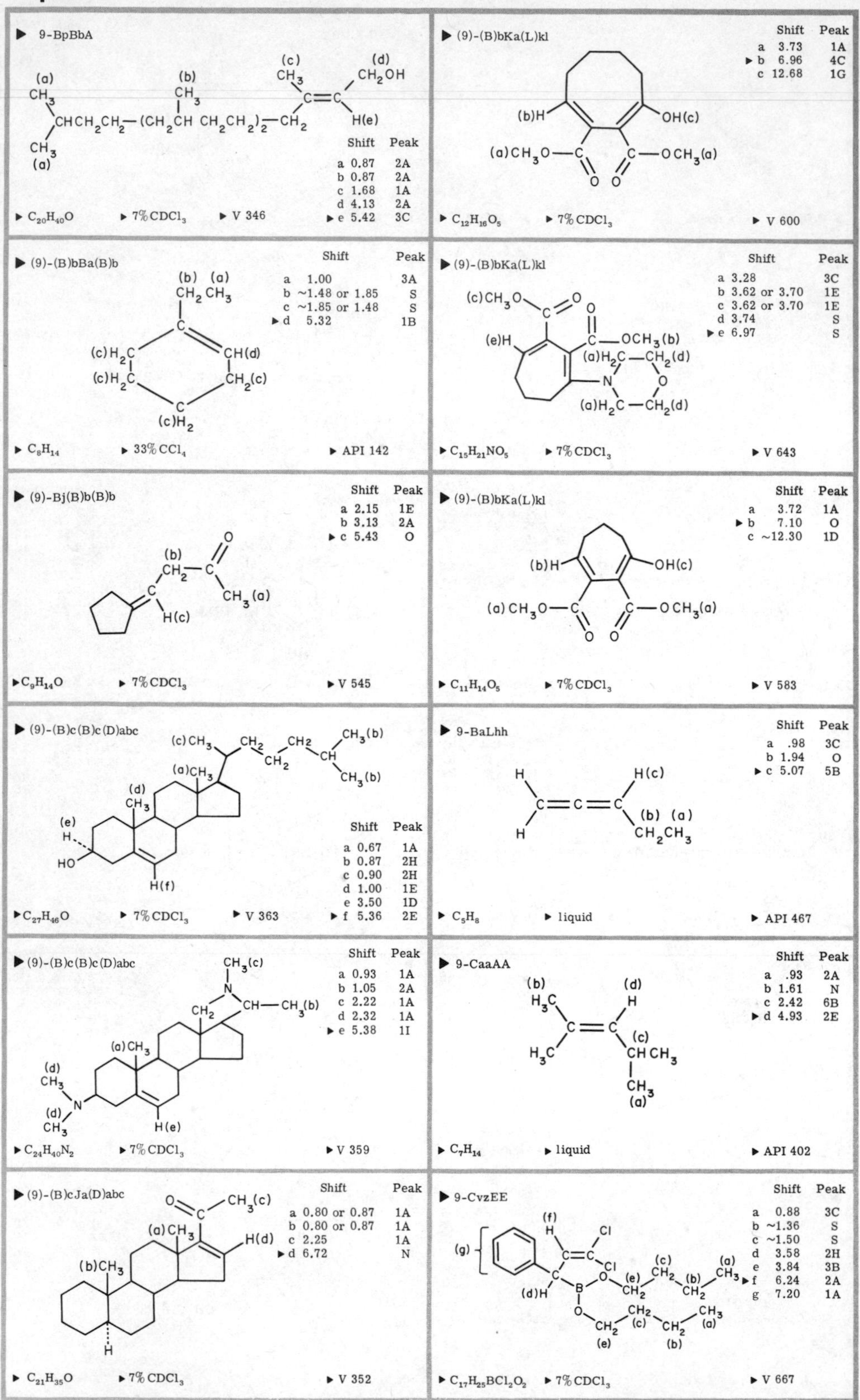
▶ 9-BpBbA
Shift Peak
a 0.87 2A
b 0.87 2A
c 1.68 1A
d 4.13 2A
▶ e 5.42 3C
▶ $C_{20}H_{40}O$ ▶ 7% $CDCl_3$ ▶ V 346
▶ (9)-(B)bKa(L)kl
Shift Peak
a 3.73 1A
▶ b 6.96 4C
c 12.68 1G
▶ $C_{12}H_{16}O_5$ ▶ 7% $CDCl_3$ ▶ V 600
▶ (9)-(B)bBa(B)b
Shift Peak
a 1.00 3A
b ~1.48 or 1.85 S
c ~1.85 or 1.48 S
▶ d 5.32 1B
▶ C_8H_{14} ▶ 33% CCl_4 ▶ API 142
▶ (9)-(B)bKa(L)kl
Shift Peak
a 3.28 3C
b 3.62 or 3.70 1E
c 3.62 or 3.70 1E
d 3.74 S
▶ e 6.97 S
▶ $C_{15}H_{21}NO_5$ ▶ 7% $CDCl_3$ ▶ V 643
▶ (9)-Bj(B)b(B)b
Shift Peak
a 2.15 1E
b 3.13 2A
▶ c 5.43 O
▶ $C_9H_{14}O$ ▶ 7% $CDCl_3$ ▶ V 545
▶ (9)-(B)bKa(L)kl
Shift Peak
a 3.72 1A
▶ b 7.10 O
c ~12.30 1D
▶ $C_{11}H_{14}O_5$ ▶ 7% $CDCl_3$ ▶ V 583
▶ (9)-(B)c(B)c(D)abc
Shift Peak
a 0.67 1A
b 0.87 2H
c 0.90 2H
d 1.00 1E
e 3.50 1D
▶ f 5.36 2E
▶ $C_{27}H_{46}O$ ▶ 7% $CDCl_3$ ▶ V 363
▶ 9-BaLhh
Shift Peak
a .98 3C
b 1.94 O
▶ c 5.07 5B
▶ C_5H_8 ▶ liquid ▶ API 467
▶ (9)-(B)c(B)c(D)abc
Shift Peak
a 0.93 1A
b 1.05 2A
c 2.22 1A
d 2.32 1A
▶ e 5.38 1I
▶ $C_{24}H_{40}N_2$ ▶ 7% $CDCl_3$ ▶ V 359
▶ 9-CaaAA
Shift Peak
a .93 2A
b 1.61 N
c 2.42 6B
▶ d 4.93 2E
▶ C_7H_{14} ▶ liquid ▶ API 402
▶ (9)-(B)cJa(D)abc
Shift Peak
a 0.80 or 0.87 1A
b 0.80 or 0.87 1A
c 2.25 1A
▶ d 6.72 N
▶ $C_{21}H_{35}O$ ▶ 7% $CDCl_3$ ▶ V 352
▶ 9-CvzEE
Shift Peak
a 0.88 3C
b ~1.36 S
c ~1.50 S
d 3.58 2H
e 3.84 3B
▶ f 6.24 2A
g 7.20 1A
▶ $C_{17}H_{25}BCl_2O_2$ ▶ 7% $CDCl_3$ ▶ V 667

▶ (9)-(C)c1Oa(J)c

(f)H (e)H O
CH_3O N—C
CH_3(a)
CH_3O H(d)
CH_3O H(b)
H
(c)
O
(g)H
OCH_3

	Shift	Peak
a	2.07	1G
b	3.63	2C
c	4.12	2C
d	4.84	1F
e	6.02	2E
f	6.49	1B
▶ g	6.68	2A

▶ $C_{22}H_{25}NO_6$ ▶ 7% $CDCl_3$ ▶ V 690

▶ 9-EAA

(a)
CH_3 Cl
C=C
CH_3 H
(b) (c)

	Shift	Peak
a	~1.77	1A
b	~1.77	1A
▶ c	5.77	N

▶ C_4H_7Cl ▶ 7% $CDCl_3$ ▶ V 67

▶ (9)-(D)cjoA(C)dl

(b)
OCH_3 O
(d)H
H(e)
(a)CH_3
H H
(c) (f)

	Shift	Peak
a	2.13	2D
b	3.46	1A
c	3.69	1B
▶ d	5.88	2A
e	6.43	2A
f	6.84	2G

▶ $C_9H_{10}O_2$ ▶ 7% $CDCl_3$ ▶ V 531

▶ (9)-(J)cA(B)b

(c)
CH_3
H(d)
O
CH
CH_3 CH_3
(a) (b)

	Shift	Peak
a	0.85 or 0.95	2A
b	0.85 or 0.95	2A
c	1.93	1B
▶ d	5.87	N

▶ $C_{10}H_{16}O$ ▶ 7% $CDCl_3$ ▶ V 275

▶ (9)-(D)aalA(J)l

O
(b)CH_3 CH_3(c)
N
(d)H N
CH_3 CH_3
(a) (a)

	Shift	Peak
a	1.26	1A
b	1.97	2D
c	2.59	1A
▶ d	6.27	2A

▶ $C_{17}H_{18}N_2O$ ▶ 7% $CDCl_3$ ▶ V 665

▶ (9)-(J)nA(N)an

O N (b)
N—CH_3
(c) CH_3
(a)

	Shift	Peak
a	2.19	1A
b	2.99	1A
▶ c	5.31	1G

▶ $C_{11}H_{12}N_2O$ ▶ 44.7 mg/0.50 ml $CDCl_3$ ▶ S 258

▶ (9)-(D)aajA(L)ln

O
(a)H_3C CH_3(c)
(a)CH_3
N
H N
(d)
CH_3
(b) (e)

	Shift	Peak
a	1.28	1A
b	1.60	2A
c	2.57	1A
▶ d	5.87	2A
e	7.46	1A

▶ $C_{17}H_{18}N_2O$ ▶ 7% $CDCl_3$ ▶ V 664

▶ (9)-(J)nA(N)an

(b)
CH_3
(a)CH_3 C N
N
C—C
(c)H O

	Shift	Peak
a	2.23	1G
b	3.06	1A
▶ c	5.38	2E

▶ $C_{11}H_{12}N_2O$ ▶ 7% $CDCl_3$ ▶ V 581

▶ (9)-(D)aakA(Q)d

CH_3(a)
(c)H
CH_3(a)
C
(b)CH_3 O O

	Shift	Peak
a	1.28	1A
b	1.98	2A
▶ c	5.16	2A

▶ $C_7H_{10}O_2$ ▶ 7% $CDCl_3$ ▶ V 175

▶ (9)-(J)1A(O)1

(a)CH_3 O O
H(d)
O
(c)H C
O CH_3(b)

	Shift	Peak
a	2.27	1A
b	2.68	1A
▶ c	5.92	1A
d	16.71	1C

▶ $C_8H_8O_4$ ▶ 7% $CDCl_3$ ▶ V 504

▶ (9)-(D)aab(C)vv(L)hl

(f) (d)
H H
(b) CH_3
O
O
(c)H_2C
(e)H
(a)CH_3 CH_3
(a)

	Shift	Peak
a	1.20	1A
b	1.70	2D
c	2.73	1G
d	4.73	1B
▶ e	6.14	S
f	6.16	S

▶ $C_{23}H_{22}O_2$ ▶ 7% $CDCl_3$ ▶ V 694

▶ (9)-(J)vA(O)v

(c)
OCH_3 O
(e)H H(d)
(f)H O O CH_3(a)
OCH_3
(b)

	Shift	Peak
a	2.40	1G
b	4.06 or 4.20	1A
c	4.06 or 4.20	1A
▶ d	6.05	1G
e	7.01	2C
f	7.63	2C

▶ $C_{14}H_{12}O_5$ ▶ 7% $CDCl_3$ ▶ V 628

▶ (9)-(J)b(B)b(C)bc

O, CH_3(b), (a)H_3C, H, O, H(c)

Shift		Peak
a	0.70	1A
b	2.13	1E
▶ c	5.87	1B

▶ $C_{20}H_{28}O_2$ ▶ 7% $CDCl_3$ ▶ V 345

▶ (9)-(J)nOa(B)n

(a) CH_3O, H(c), (b)H, (b)H, N, O, H (d)

Shift		Peak
a	3.80	1A
b	3.92	1B
▶ c	5.05	1I or 2D
d	~6.30	1D

▶ $C_5H_7NO_2$ ▶ 7% $CDCl_3$ ▶ V 105

▶ (9)-(J)1Bp(O)1

(c)H, O, (a) CH_2, OD, DO, H(b), O

Shift		Peak
a	4.54	1A
▶ b	6.59	1A
c	8.10	1A

▶ $C_6H_6O_4$ ▶ D_2O ▶ V 455

▶ (9)-(J)nOaVhn

(f) H, (a) OCH_3, (e)H, H(c), O, O, N, O, (d)H_2C, CH_3(b)

Shift		Peak
a	3.83 or 3.92	1A
b	3.83 or 3.92	1A
▶ c	5.92	1A
d	6.07	1A
e	6.80	2C
f	7.57	2C

▶ $C_{12}H_{11}NO_4$ ▶ 7% $CDCl_3$ ▶ V 291

▶ (9)-(J)1(C)bn(L)1v

O, CH_3(a), (f) H, (c) H_2C, (b) CH_2, N, H(j), H(e), H(i), CH_3O, CH_3O, CH_3O, H (g), O, (h)H, (d) OCH_3

Shift		Peak
a	1.96	1A
b	2.43	S
c	2.43	S
d	3.67	1A
e	4.62	S
f	6.55	1A
g	6.92	2A
h	7.37	S
▶ i	7.63	1B
j	8.38	2G

▶ $C_{22}H_{25}NO_6$ ▶ 7% $CDCl_3$ ▶ V 689

▶ (9)-(J)bP(B)d

(f) CH_2, O, O, (c)H_2C, CH_2(c), CH_3 (a), CH_3 (a); (g) H, (h) HO, O, (e)H_2C, CH_2(d), CH_3 (b), CH_3 (b)

keto-enol mixture

Shift		Peak
a	1.06	1E
b	1.09	1E
c	2.29	1A
d	2.29	1A
e	2.56	1A
f	3.36	1B
▶ g	5.51	1A
h	10.90	1A

▶ $C_8H_{12}O_2$ ▶ 7% $CDCl_3$ ▶ V 512

▶ (9)-(L)1Daaa(J)1

(b) H, O, (a) $C(CH_3)_3$, $(CH_3)_3C$ (a), H(b), O

Shift		Peak
a	1.27	1A
▶ b	6.47	1A

▶ $C_{14}H_{20}O_2$ ▶ 7% $CDCl_3$ ▶ V 314

▶ 9-JvVhhP

(b) H, O, O, C, C, C, H (a), Br, Br

Shift		Peak
▶ a	6.70	1A
b	16.61	1B

▶ $C_{15}H_{10}Br_2O_2$ ▶ 7% $CDCl_3$ ▶ V 316

▶ (9)-(J)bKh(D)abc

(g) H, O, O, OH(h), (d)H, CH_3(c), (f)H, CH(e), CH_3 (a), CH_3 (b)

Shift		Peak
a	0.75	2A
b	0.97	2A
c	1.45	1A
d	2.10	2C
e	2.40	S
f	2.60	2C
▶ g	6.77	1A
h	12.00	1B

▶ $C_{10}H_{14}O_3$ ▶ 7% $CDCl_3$ ▶ V 567

▶ (9)-JdYP

(b) H, (d) H, (a) H, CF_3, (c)H, S, O, O, H (e)

(f) = CH_2 in Keto form.

Shift		Peak
▶ a	6.42	1A
b	7.18	3C
c	7.73	N
d	7.82	N
e	14.17	1D
f	3.33	1B

▶ $C_8H_5F_3O_2S$ ▶ 7% $CDCl_3$ ▶ V 185

▶ (9)-Jh(N)av(D)aav

(a)CH_3, (a) CH_3, O, H(d), N, C, H(c), CH_3 (b)

Shift		Peak
a	1.67	1A
b	3.22	1A
▶ c	5.37	2C
d	9.98	2C

▶ $C_{13}H_{15}NO$ ▶ 7% $CDCl_3$ ▶ V 610

▶ 9-KhAA

(a) CH_3, H(c), C=C, CH_3 (b), C=O, OH(d)

Shift		Peak
a	1.93	2D
b	2.18	2D
▶ c	5.72	M
d	11.95	1B

▶ $C_5H_8O_2$ ▶ 7% $CDCl_3$ ▶ V 114

▶ 9-KhAE

	Shift	Peak
a	2.30	2D
▶ b	6.07	2E
c	11.73	1B

(a) CH_3 (b) H C=C Cl C OH(c) O

▶ $C_4H_5ClO_2$ ▶ 7% $CDCl_3$ ▶ V 53

▶ 9-KhEA

	Shift	Peak
a	2.58	2C
▶ b	6.10	2E
c	11.82	1A

Cl (b) H C=C CH_3 (a) C OH(c) O

▶ $C_4H_5ClO_2$ ▶ 7% $CDCl_3$ ▶ V 54

▶ (9)-(K)jA(K)j

	Shift	Peak
a	2.19	2A
▶ b	6.67	4A

(a) CH_3 H(b) O O O

▶ $C_5H_4O_3$ ▶ 7% $CDCl_3$ ▶ V 427

▶ 9-(K)jE(K)j

	Shift	Peak
a	6.98	1A

O Cl—C—C O (a)H—C—C O

▶ C_4HClO_3 ▶ 45.0 mg/0.50 ml $CDCl_3$ ▶ S 181

▶ 9-KaANhh

	Shift	Peak
a	1.92	1A
b	3.62	1A
▶ c	4.53	1B
d	~6.15	1D

(a) CH_3 H(c) C=C H_2N C—OCH_3(b) O

▶ $C_5H_9NO_2$ ▶ 7% $CDCl_3$ ▶ V 442

▶ (9)-(K)vLhlVqq

	Shift	Peak
a	2.06	1A
b	2.32	1A
c	2.32	1A
▶ d	6.36	1G
e	6.83*	S
f	6.83*	S
g	6.89	S
h	7.15	S
i	7.33	S
j	7.53	2E

*or 7.08

O (d) H (g) H (j) H (f) H (h) H O (c) CH_3 O H (i) H (j) (b) CH_3 (e) H H(h) CH_3 (a) O

▶ $C_{23}H_{18}O_8$ ▶ 7% $CDCl_3$ ▶ V 692

▶ (9)-KvAVho

	Shift	Peak
a	2.25 or 2.37	1A
b	2.25 or 2.37	1A
c	2.40	2D
d	3.33	2B
e	5.07	R
f	5.12	R
g	5.85	4D
▶ h	6.22	1I
i	7.30	1I

(f)H (g) H (i) H (c) CH_3 C=C (d) CH_2 (e)H H(h) (b) CH_3—C O O O O CH_3 (a)

▶ $C_{16}H_{16}O_4$ ▶ 7% $CDCl_3$ ▶ V 323

▶ 9-KbVhhP

	Shift	Peak
a	1.40	3A
b	3.87	1A
c	4.40	4A
d	6.97	2A
▶ e	7.02	1A
f	7.98	2A
g	~15.33	1D

(d) H (f) H (g) O H O (b) CH_3O C=C C O (c) CH_2 (a) CH_3 H H H(e) (d) (f)

▶ $C_{12}H_{14}O_4$ ▶ 7% $CDCl_3$ ▶ V 295

▶ (9)-(K)vAVhq

	Shift	Peak
a	1.45	3A
b	2.37	2D
c	4.08	4A
▶ d	6.08	2E
e	6.73	1G
f	~6.78	2G
g	7.47	2E

(e) H (a) (c) CH_3—CH_2O O O (f)H H(d) H (g) CH_3 (b)

▶ $C_{12}H_{12}O_3$ ▶ 7% $CDCl_3$ ▶ V 294

▶ 9-LhlAA

	Shift	Peak
a	1.74	1A
b	1.79	1A
▶ c	5.98	1B

(a) CH_3 H(c) C=C CH_3(b) (b) CH_3 C=C (c)H CH_3(a)

▶ C_8H_{14} ▶ 7% $CDCl_3$ ▶ V 515

▶ (9)-(K)v(B)cVoq

	Shift	Peak
a	2.32	1A
b	2.32	1A
c	3.09	S
d	3.09	A
e	5.18	3B
▶ f	6.06	M
g	6.62*	S
h	6.62*	S
i	7.17	Q
j	7.46	Q

*or 6.71

O O H(f) (g) H (c) H H H(j) (i) H O (d) CH_3 (b) O O (e)H O H (h) H (j) H(i) C (a) CH_3

▶ $C_{21}H_{16}O_7$ ▶ 7% $CDCl_3$ ▶ V 686

▶ (9)-(L)clA(J)b

	Shift	Peak
a	1.20	1A
b	1.70	2D
c	2.73	1G
d	4.73	1B
e	6.14	S
▶ f	6.16	S

(f) H (d) H (b) CH_3 O O (c) H_2C (e) H (a) CH_3 CH_3 (a)

▶ $C_{23}H_{22}O_2$ ▶ 7% $CDCl_3$ ▶ V 694

▶ (9)-(L)lmA(J)l

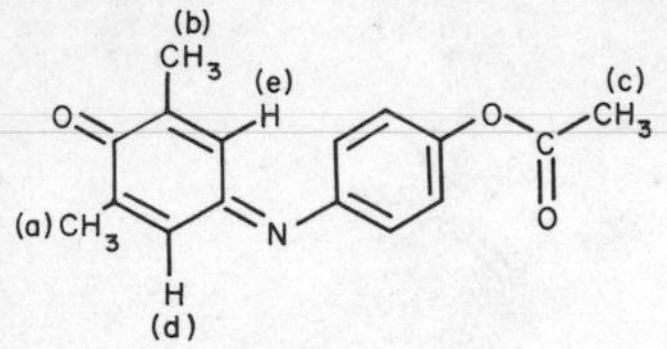

	Shift	Peak
a	1.98	2D
b	2.08	2D
c	2.30	1A
▶ d	~6.87	N
e	~7.03	N

▶ $C_{16}H_{15}NO_3$ ▶ 7% $CDCl_3$ ▶ V 322

▶ (9)-(L)lLhl(O)l

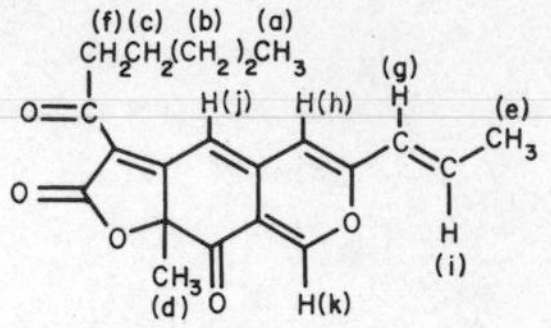

	Shift	Peak
a	0.89	3B
b	~1.30	S
c	~1.40	S
d	1.72	1A
e	1.97	2A
f	2.94	3B
g	6.06	S
▶ h	6.17	S
i	6.58	S
j	6.89	S
k	7.88	1G

▶ $C_{21}H_{22}O_5$ ▶ 7% $CDCl_3$ ▶ V 687

▶ 9-Lhl(B)c(L)bb

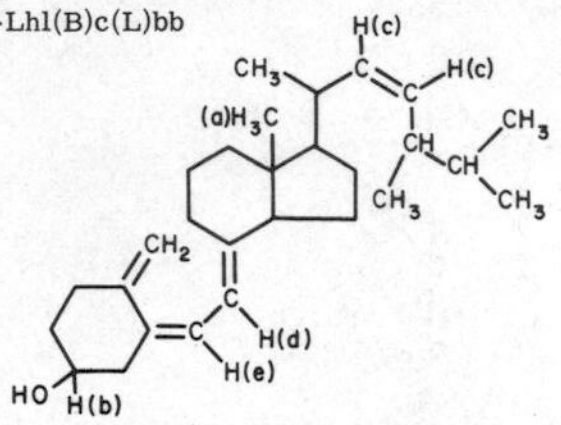

	Shift	Peak
a	0.57	1A
b	3.92	1F
c	~5.20	N
d	6.05*	2C
▶ e	6.05*	2C

*or 6.20

▶ $C_{28}H_{44}$ ▶ 7% $CDCl_3$ ▶ V 364

▶ (9)-(L)hlOa(J)l

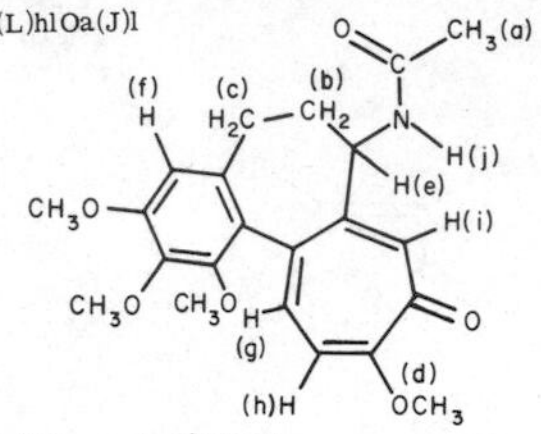

	Shift	Peak
a	1.96	1A
b	2.43	S
c	2.43	S
d	3.67	1A
e	4.62	S
f	6.55	1A
g	6.92	2A
▶ h	7.37	S
i	7.63	1B
j	8.38	2G

▶ $C_{22}H_{25}NO_6$ ▶ 7% $CDCl_3$ ▶ V 689

▶ 9-Lhl(C)bd(B)b

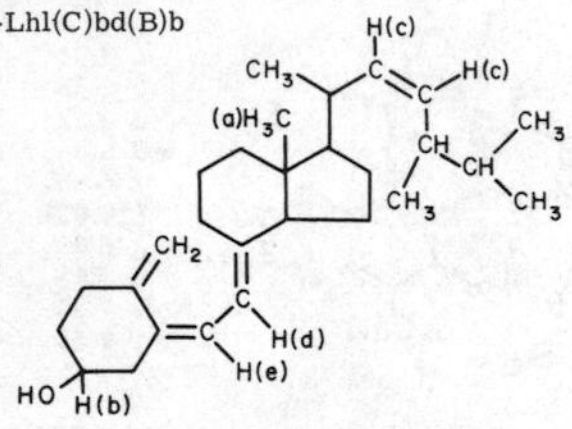

	Shift	Peak
a	0.57	1A
b	3.92	1F
c	~5.20	N
▶ d	6.05*	2C
e	6.05*	2C

*or 6.20

▶ $C_{28}H_{44}$ ▶ 7% $CDCl_3$ ▶ V 364

▶ (9)-(L)hlP(J)l

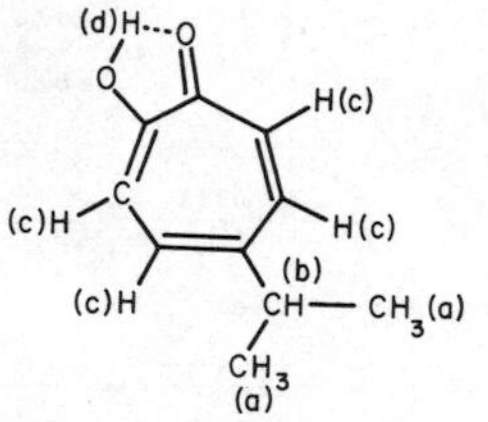

	Shift	Peak
a	1.27	2A
b	2.90	5B
▶ c	8.32	1A
d	8.75	1B

▶ $C_{10}H_{12}O_2$ ▶ 7% $CDCl_3$ ▶ V 560

▶ (9)-(L)hlCaa(L)hl

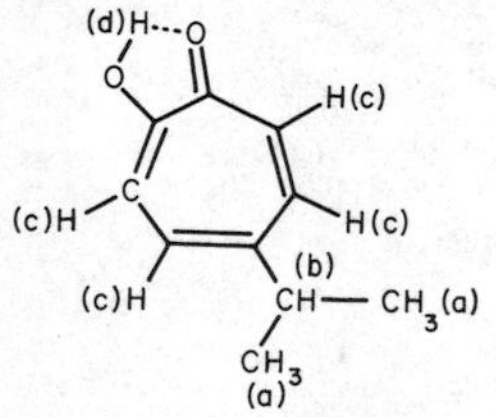

	Shift	Peak
a	1.27	2A
b	2.90	5B
▶ c	8.32	1A
d	8.75	1B

▶ $C_{10}H_{12}O_2$ ▶ 7% $CDCl_3$ ▶ V 560

▶ (9)-(L)hlVbo(L)cl

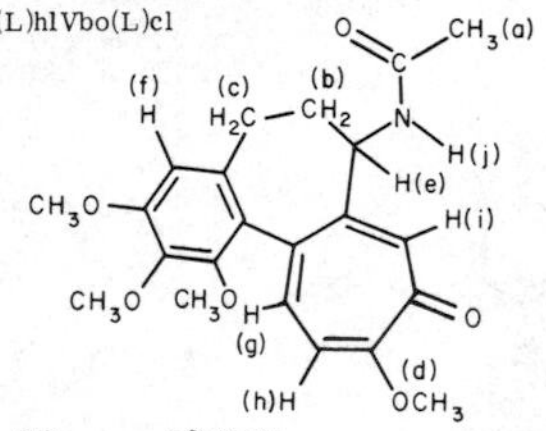

	Shift	Peak
a	1.96	1A
b	2.43	S
c	2.43	S
d	3.67	1A
e	4.62	S
f	6.55	1A
▶ g	6.92	2A
h	7.37	S
i	7.63	1B
j	8.38	2G

▶ $C_{22}H_{25}NO_6$ ▶ 7% $CDCl_3$ ▶ V 689

▶ (9)-(L)hlKh(B)v

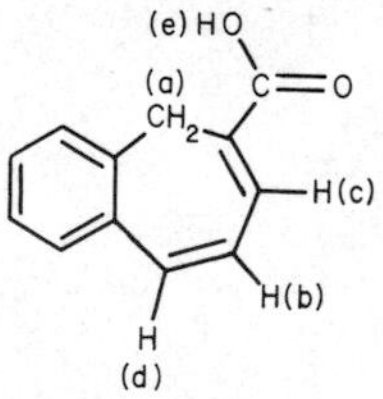

	Shift	Peak
a	3.38	1A
b	6.56	4A
▶ c	7.27	S
d	7.30	S
e	11.05	1C

▶ $C_{12}H_{10}O_2$ ▶ 7% $CDCl_3$ ▶ V 591

▶ 9-Lhl VhhVhh

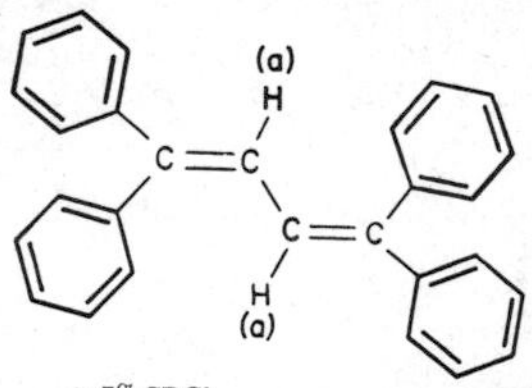

	Shift	Peak
a	6.80	1A

▶ $C_{28}H_{22}$ ▶ 7% $CDCl_3$ ▶ V 699

▶ (9)-(L)dl(L)hl(L)jl

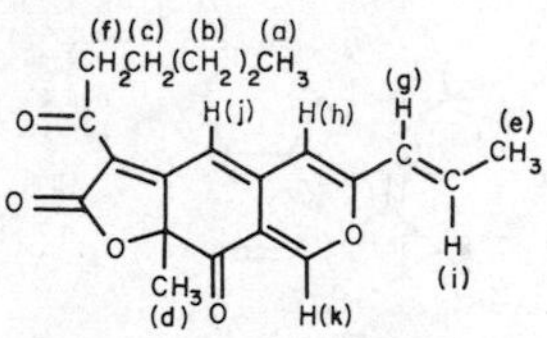

	Shift	Peak
a	0.89	3B
b	~1.30	S
c	~1.40	S
d	1.72	1A
e	1.97	2A
f	2.94	3B
g	6.06	S
h	6.17	S
i	6.58	S
▶ j	6.89	S
k	7.88	1G

▶ $C_{21}H_{22}O_5$ ▶ 7% $CDCl_3$ ▶ V 687

▶ (9)-(M)lBb(N)hl

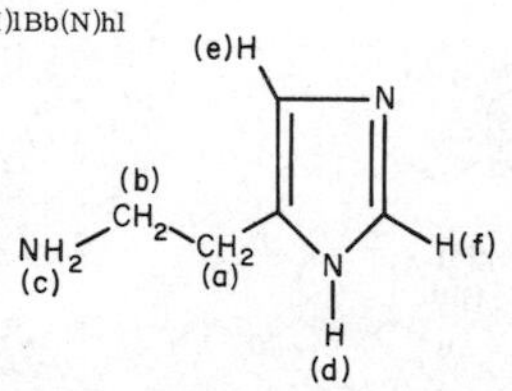

	Shift	Peak
a	2.75	S
b	3.02	S
c	4.96	1E
d	4.96	1E
▶ e	6.82	1A
f	7.55	1A

▶ $C_5H_9N_3$ ▶ 7% $CDCl_3$ ▶ V 443

▶ (9)-(N)cjA(J)n

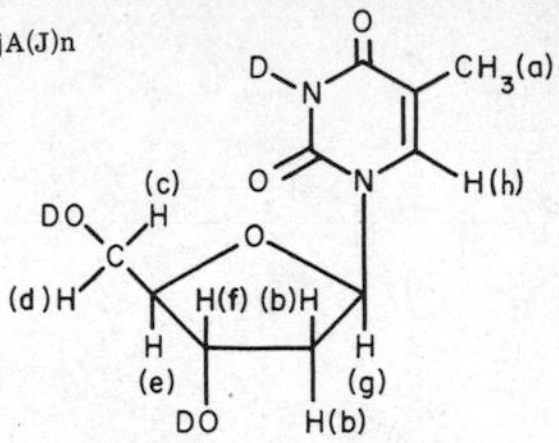

	Shift	Peak
a	1.91	2D
b	2.40	4A
c	3.79	2A
d	3.85	1A
e	4.03	N
f	4.49	4E
g	6.28	3A
▶ h	7.65	2D

▶ $C_{10}H_{14}N_2O_5$ ▶ D_2O ▶ V 566

▶ (9)-(O)l(J)d(L)ll

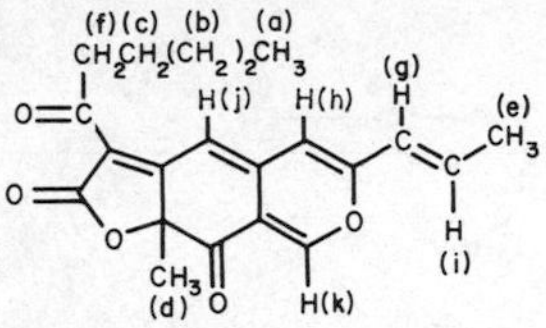

	Shift	Peak
a	0.89	3B
b	~1.30	S
c	~1.40	S
d	1.72	1A
e	1.97	2A
f	2.94	3B
g	6.06	S
h	6.17	S
i	6.58	S
j	6.89	S
▶ k	7.88	1G

▶ $C_{21}H_{22}O_5$ ▶ 7% $CDCl_3$ ▶ V 687

▶ (9)-(N)hvAVhn

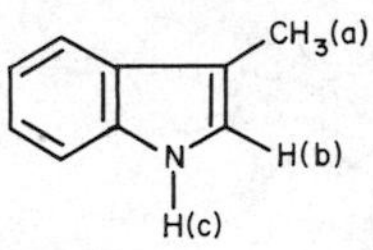

	Shift	Peak
a	2.28	2D
▶ b	6.78	1I
c	~7.40	S

▶ C_9H_9N ▶ 7% $CDCl_3$ ▶ V 231

▶ (9)-(O)cKa(C)cl

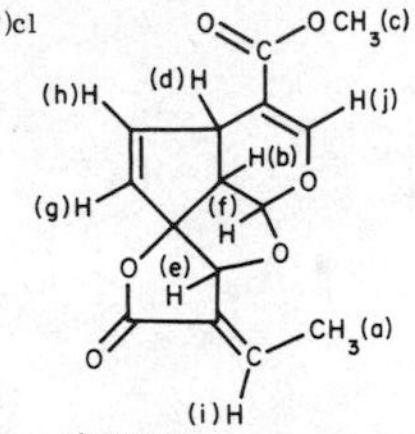

	Shift	Peak
a	2.11	2A
b	3.46	4A
c	3.79	1A
d	4.02	3E
e	5.12	1B
f	5.59	S
g	5.67*	S
h	5.67*	S
i	7.19	O
▶ j	7.46	1A

*or 6.08

▶ $C_{15}H_{14}O_6$ ▶ 7% $CDCl_3$ ▶ V 640

▶ (9)-(N)hvBcVhn

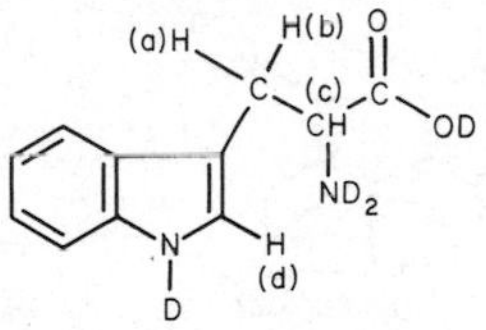

	Shift	Peak
a	3.10	S
b	3.33	S
c	3.75	O
▶ d	7.28	S

▶ $C_{11}H_{12}N_2O_2$ ▶ D_2O ▶ V 582

▶ (9)-(O)cKa(C)cl

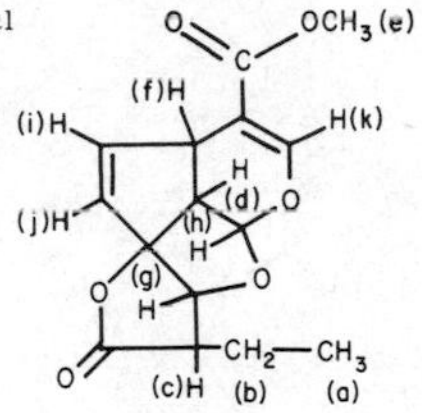

	Shift	Peak
a	1.13	3B
b	1.89	O
c	2.74	O
d	3.42	4A
e	3.77	1A
f	3.98	3E
g	4.59	2A
h	5.59	S
i	5.63*	S
j	5.63*	S
▶ k	7.47	1A

*or 6.12

▶ $C_{15}H_{16}O_6$ ▶ 7% $CDCl_3$ ▶ V 641

▶ (9)-(N)hvJjVhn

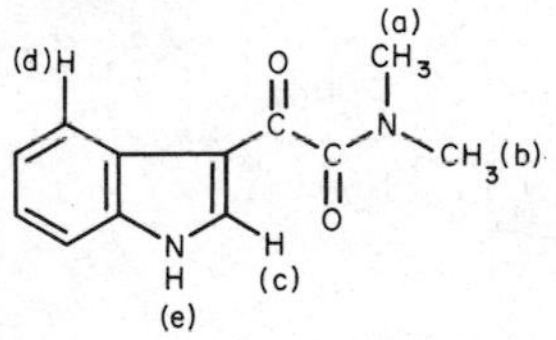

	Shift	Peak
a	2.99	1A
b	3.07	1A
▶ c	7.69	1B
d	8.27	4C
e	10.76	1C

▶ $C_{12}H_{12}N_2O_2$ ▶ 7% $CDCl_3$ ▶ V 594

▶ 9-ObKbKb

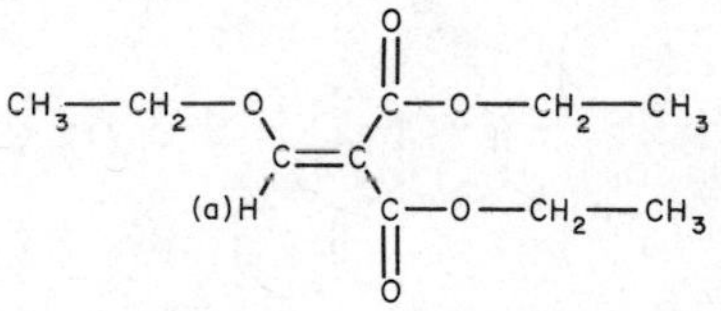

	Shift	Peak
a	7.56	1A

▶ $C_{10}H_{16}O_5$ ▶ 7% $CDCl_3$ ▶ V 573

▶ (9)-(O)lA(J)l

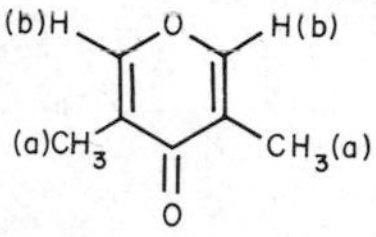

	Shift	Peak
a	2.23	1A
▶ b	6.03	1A

▶ $C_7H_8O_2$ ▶ 7% $CDCl_3$ ▶ V 166

▶ (9)-(O)lLhl(J)l

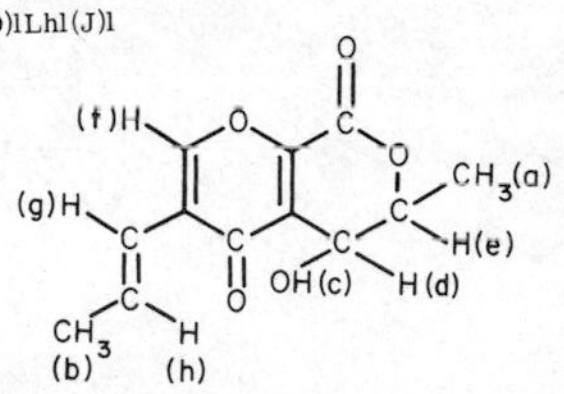

	Shift	Peak
a	1.65	2A
b	1.97	2E
c	3.86	1C
d	3.92	S
e	4.32	O
▶ f	5.87	1I
g	6.06	2A
h	6.99	4A

▶ $C_{12}H_{12}O_5$ ▶ 7% $CDCl_3$ ▶ V 595

▶ (9)-(O)vA(J)v

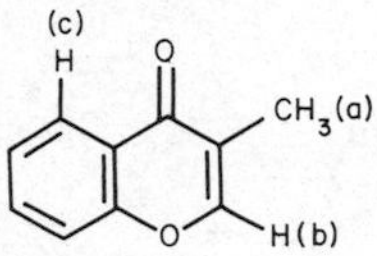

	Shift	Peak
a	2.02	2D
▶ b	7.78	2D
c	8.24	2E

▶ $C_{10}H_8O_2$ ▶ 7% $CDCl_3$ ▶ V 552

▶ (9)-(O)lP(J)l

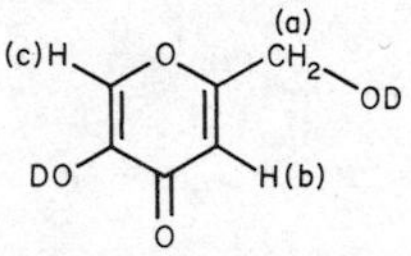

	Shift	Peak
a	4.54	1A
b	6.59	1A
▶ c	8.10	1A

▶ $C_6H_6O_4$ ▶ D_2O ▶ V 455

▶ 9-ObUU

NC, O—CH2—CH3 (b) (a), C=C, NC, H(c)

	Shift	Peak
a	1.47	3A
b	4.42	4A
▶ c	7.65	1A

▶ $C_6H_6N_2O$ ▶ 7% $CDCl_3$ ▶ V 452

▶ (9)-VhnA(N)hv

(f) H, (b) CH3, H(c), (e)H, N, CH3(a), H (d), H (g)

	Shift	Peak
a	2.28	1A
b	2.40	1A
▶ c	6.10	1I
d	~6.90 or 7.02	N
e	~6.90 or 7.02	N
f	7.27	S
g	~7.30	S

▶ $C_{10}H_{11}N$ ▶ 7% $CDCl_3$ ▶ V 255

▶ (9)-(O)vVho(J)w

H3C(a), H(h), (j), (b) (e), (a)CH3, OCH3, H, (c)H, H(d), (f)H, (i)H, O, CH3 (a), H (i), H(g), CH3(a)

	Shift	Peak
a	1.46*	1A
b	3.58	1A
c	5.62†	2A
d	5.62†	2A
e	6.45‡	2A
f	6.45‡	2A
g	6.62	S
h	7.12	2A
▶ i	7.88§	1A
j	7.88§	1A

*or 1.49 ‡or 6.60
†or 5.70 §or 7.90

▶ $C_{26}H_{24}O_5$ ▶ 7% $CDCl_3$ ▶ V 696

▶ (9)-Vhh(J)v(B)v

(a) CH2, C, H (b), H (c), O

	Shift	Peak
a	4.04	2D
▶ b	7.55	S
c	7.92	S

▶ $C_{16}H_{12}O$ ▶ 7% $CDCl_3$ ▶ V 649

▶ (9)-(S)1A(M)1

(a) CH3, N, (c)H, S, CH3 (b)

	Shift	Peak
a	2.38	2D
b	2.67	1A
▶ c	6.66	2E

▶ C_5H_7NS ▶ 7% $CDCl_3$ ▶ V 108

▶ 9-Vhn(J)b(N)bv

H(f), (c) CH3O, H(e), O, CH3 (a), N, (g)H, H2C—CH2 (d) (b)

	Shift	Peak
a	2.33	1A
b	3.13	3A
c	3.83	1A
d	4.28	4A
▶ e	6.85 or 6.98	1A
f	6.85 or 6.98	1A
g	7.13	1B

▶ $C_{13}H_{13}NO_2$ ▶ 7% $CDCl_3$ ▶ V 299

▶ (9)-(S)1Vhh(S)1

S, H(a), (a)H, S

	Shift	Peak
a	6.53	1A

▶ $C_{16}H_{12}S_2$ ▶ 7% $CDCl_3$ ▶ V 651

▶ (9)-VbhKa(L)kl

(e) OH, (a) CH2, OCH3(b), C, O, C=O, (d)H, OCH3(c)

	Shift	Peak
a	3.38	1A
b	3.73 or 3.84	1A
c	3.73 or 3.84	1A
▶ d	7.82	1A
e	12.62	1A

▶ $C_{15}H_{14}O_5$ ▶ 7% $CDCl_3$ ▶ V 639

▶ (9)-VhoA(K)v

O, O, CH3 (a), (b)

	Shift	Peak
a	2.22	1A
▶ b	~7.45	1G

▶ $C_{10}H_8O_2$ ▶ 45.0 mg/0.50 ml $CDCl_3$ ▶ S 87

▶ (9)-VbhKa(L)kl

(e) OH, O, C, OCH3(a), OCH3(b), C, O, H (d), O, (c)

	Shift	Peak
a	3.65 or 3.80	1A
b	3.80 or 3.65	1A
c	7.20	1A
▶ d	7.83	1A
e	12.69	1A

▶ $C_{16}H_{16}O_5$ ▶ 7% $CDCl_3$ ▶ V 656

▶ (9)-VhnA(N)hv

H (e), Cl, H(b), (c)H, N, CH3(a), H (d), H (f)

	Shift	Peak
a	2.33	2D
▶ b	6.10	1B
c	7.02	2D
d	7.02	2D
e	7.44	1I
f	~7.65	1D

▶ C_9H_8ClN ▶ 7% $CDCl_3$ ▶ V 228

▶ 9-VhhKaU

(a) CH3O, (c)H, C=O, (b)H, C=C, CN, H (b)

	Shift	Peak
a	3.96	1A
b	8.03	4A
▶ c	8.27	1A

▶ $C_{11}H_9NO_2$ ▶ 7% $CDCl_3$ ▶ V 576

▶ (9)-VhsNhh(S)v

Shift	Peak
a 3.89	1C
▶b 6.19	1A

▶ C_8H_7NS ▶ 7% $CDCl_3$ ▶ V 502

▶ 10-Bg

Shift	Peak
▶a 2.46	1H
b 3.86	2A

Br—CH_2(b)—C≡CH(a)

▶ C_3H_3Br ▶ 45.2 mg/0.50 ml $CDCl_3$ ▶ S 273

▶ 9-VhhUKb

Shift	Peak
a 1.38	3A
b 4.38	4A
▶c 8.22	1A

▶ $C_{12}H_{11}NO_2$ ▶ 7% $CDCl_3$ ▶ V 290

▶ 10-Bp

Shift	Peak
▶a 2.50	3C
b 2.82	1A
c 4.28	2A

(a)H—C≡C—CH_2(c)—OH(b)

▶ C_3H_4O ▶ 7% $CDCl_3$ ▶ V 21

▶ 9-VahIII

Shift	Peak
a 2.45	1A
b 8.05	S
▶c 8.11	S

▶ $C_{11}H_8N_2$ ▶ 7% $CDCl_3$ ▶ V 283

▶ 10-Ccp

Shift	Peak
a 1.00	2D
b 1.02	2D
c 1.85	7B
▶d 2.47	2E
e 2.82	1B
f 4.18	4A

▶ $C_6H_{10}O$ ▶ 7% $CDCl_3$ ▶ V 133

▶10-A

Shift	Peak
▶a 1.80	1A
b 1.80	1A

(a)H—C≡C—CH_3(b)

▶ C_3H_4 ▶ 7% $CDCl_3$ ▶ V 16

▶ 10-Daap

Shift	Peak
a 1.55	1A
b 2.27	1B
▶c 2.44	1A

▶ C_5H_8O ▶ 7% $CDCl_3$ ▶ V 436

▶ 10-Ba

Shift	Peak
a 1.15	3A
▶b 1.76	3A
c ~2.14	4C

(b)HC≡C—CH_2(c)—CH_3(a)

▶ C_4H_6 ▶ 10% CCl_4 ▶ API 162

▶ 10-(D)bbp

Shift	Peak
▶a 2.48	1A
b 2.78	1A

▶ $C_8H_{12}O$ ▶ 7% $CDCl_3$ ▶ V 211

▶ 10-Ba

Shift	Peak
a 1.12	3C
▶b ~1.90	N
c ~2.17	O

(b)HC≡CCH_2(c)CH_3(a)

▶ C_4H_6 ▶ liquid ▶ API 464

▶ 10-(D)bdp

Shift	Peak
a 0.80 or 0.83	1A
b 0.80 or 0.83	1A
c 1.88	1A
▶d 2.55	1A

▶ $C_{21}H_{32}O$ ▶ 7% $CDCl_3$ ▶ V 351

▶ 10-(D)bdp

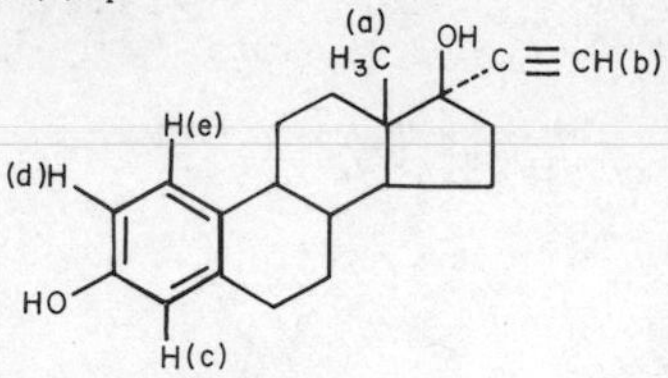

Shift	Peak
a 0.88	1A
▶b 2.62	1A
c 6.60	1E
d 6.67	2H
e 7.18	2E

▶ $C_{20}H_{24}O_2$ ▶ 7% $CDCl_3$ ▶ V 343

▶ 11-A

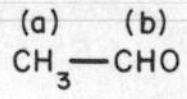

Shift	Peak
a 2.16	2A
▶b 9.76	4A

(a) (b) CH_3—CHO

▶ C_2H_4O ▶ liquid run ▶ API 2

▶ 10-Lal

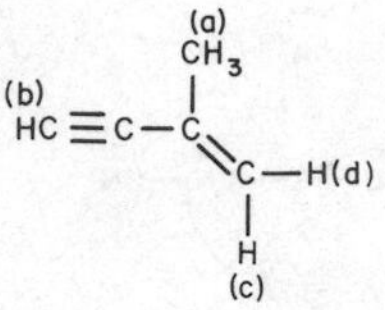

Shift	Peak
a 1.90	2E
▶b 2.87	1A
c 5.27 or 5.37	N
d 5.27 or 5.37	N

▶ C_5H_6 ▶ 7% $CDCl_3$ ▶ V 99

▶ 11-A

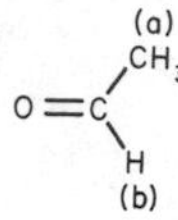

Shift	Peak
a 2.20	2A
▶b 9.80	4A

▶ C_2H_4O ▶ 7% $CDCl_3$ ▶ V 6

▶ 10-Lhl

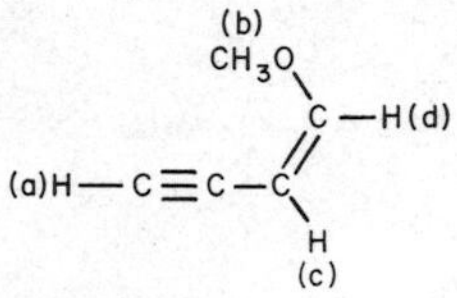

Shift	Peak
▶a 3.08	2E
b 3.80	1A
c 4.52	4D
d 6.35	2E

▶ C_5H_6O ▶ 7% $CDCl_3$ ▶ V 100

▶ 11-Bb

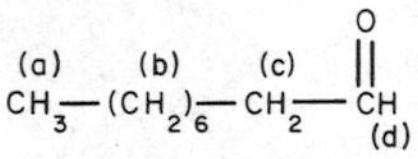

Shift	Peak
a 0.88	3B
b 1.29	1B
c 2.30	1B
▶d 9.35	1A

(a) (b) (c) (d) CH_3—$(CH_2)_6$—CH_2—CHO

▶ $C_9H_{18}O$ ▶ 45.3 mg/0.50 ml $CDCl_3$ ▶ S 100

▶ 10-Vhh

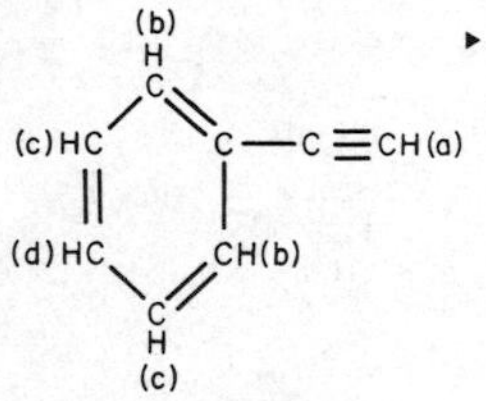

Shift	Peak
▶ a 3.05	1A
b 7.24 or 7.45	S
c 7.24 or 7.45	S
d 7.32	S

▶ C_8H_6 ▶ $CDCl_3$ ▶ API 136

▶ 11-Bb

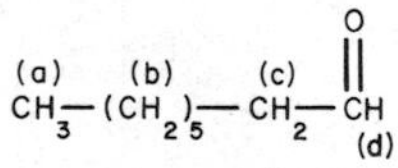

Shift	Peak
a 0.88	3B
b 1.30	1B
c 2.43	3B
▶d 9.73	3A

(a) (b) (c) (d) CH_3—$(CH_2)_5$—CH_2—CHO

▶ $C_8H_{16}O$ ▶ 44.6 mg/0.50 ml $CDCl_3$ ▶ S 105

▶ 10-Vhh

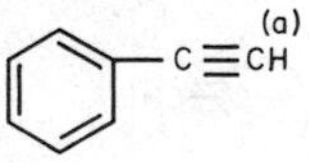

Shift	Peak
a 3.08	1A

▶ C_8H_6 ▶ 45.3 mg/0.50 ml $CDCl_3$ ▶ S 222

▶ 11-Bb

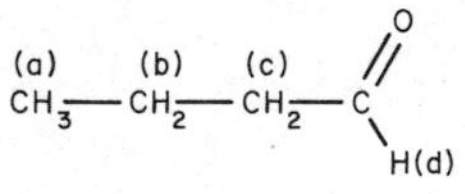

Shift	Peak
a 0.97	3C
b 1.67	6B
c 2.42	3C
▶d 9.74	3A

(a) (b) (c) CH_3—CH_2—CH_2—C(=O)—H(d)

▶ C_4H_8O ▶ 7% $CDCl_3$ ▶ V 78

▶ 10-Vhh

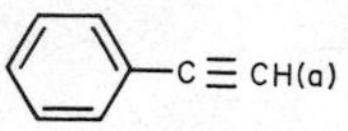

Shift	Peak
a 3.05	1A

▶ C_8H_6 ▶ 7% $CDCl_3$ ▶ V 186

▶ 11-Bb

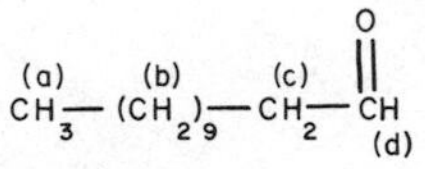

Shift	Peak
a 0.87	3A
b 1.29	1B
c 2.42	3B
▶d 9.77	3A

(a) (b) (c) (d) CH_3—$(CH_2)_9$—CH_2—CHO

▶ $C_{12}H_{24}O$ ▶ 45.2 mg/0.50 ml $CDCl_3$ ▶ S 106

▶ 11-Bb

	Shift	Peak
a	2.74	Q
b	2.97	Q
c	7.22	1A
▶d	9.81	3A

▶ $C_9H_{10}O$ ▶ 7% $CDCl_3$ ▶ V 529

▶ 11-Lhl

	Shift	Peak
a	6.56	3A
b	6.80	2A
▶c	9.67	2A

▶ C_9H_8O ▶ 45.0 mg/0.50 ml $CDCl_3$ ▶ S 101

▶ 11-Cbb

	Shift	Peak
a	0.97	3A
b	1.58	4A
c	2.22	5B
▶d	10.45	1C

▶ $C_6H_{12}O$ ▶ 36.9 mg/0.50 ml $CDCl_3$ ▶ S 74

▶ 11-Lhl

	Shift	Peak
a	1.67	1A
b	3.22	1A
c	5.37	2C
▶d	9.98	2C

▶ $C_{13}H_{15}NO$ ▶ 7% $CDCl_3$ ▶ V 610

▶ 11-Daab

	Shift	Peak
a	1.12	1A
b	1.92	Q
c	2.27	Q
▶d	9.47	1A

▶ $C_7H_{11}NO$ ▶ 7% $CDCl_3$ ▶ V 178

▶ 11 (L)ln

	Shift	Peak
a	2.27	1A
b	3.87	1A
c	6.02	2A
d	6.81	2A
▶e	9.38	1A

▶ C_7H_9NO ▶ 7% $CDCl_3$ ▶ V 170

▶ 11-Daab

	Shift	Peak
a	0.97	3D
b	1.07	1E
c	2.50	4E
d	2.55	1E
▶e	9.57	1A

▶ $C_9H_{19}NO$ ▶ 7% $CDCl_3$ ▶ V 548

▶ 11-(L)ln

	Shift	Peak
a	6.30	N
b	6.98	N
c	7.17	N
▶d	9.45	2D
e	~11.08	1D

▶ C_5H_5NO ▶ 7% $CDCl_3$ ▶ V 98

▶ 11-Lhl

	Shift	Peak
a	2.03	2E
b	6.13	N
c	6.87	N
▶d	9.48	O

▶ C_4H_6O ▶ 7% $CDCl_3$ ▶ V 60

▶ 11-(L)lo

	Shift	Peak
a	6.63	4D
b	7.28	N
c	7.72	M
▶d	9.67	1A

▶ $C_5H_4O_2$ ▶ 7% $CDCl_3$ ▶ V 95

▶ 11-Lhl

	Shift	Peak
a	6.55	S
b	6.57	S
c	6.77	S
d	7.20	S
e	7.55	S
▶f	9.57	2A

▶ $C_7H_6O_2$ ▶ 7% $CDCl_3$ ▶ V 152

▶ 11-Naa

	Shift	Peak
a	2.88, 2.97	2A
▶b	8.02	1B

▶ C_3H_7NO ▶ 7% $CDCl_3$ ▶ V 39

▶ 11-Oa

	Shift	Peak
a	3.77	2D
▶b	8.08	2E

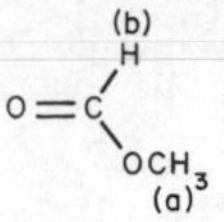

▶ $C_2H_4O_2$ ▶ 7% $CDCl_3$ ▶ V 9

▶ 11-Vhh

	Shift	Peak
a	3.93 or 3.96	1A
b	3.93 or 3.96	1A
▶c	9.8	1A

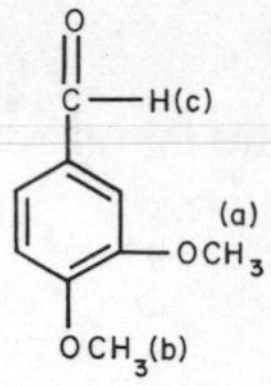

▶ $C_9H_{10}O_3$ ▶ 45.2 mg/0.50 ml $CDCl_3$ ▶ S 197

▶ 11-Ob

	Shift	Peak
a	0.88	3B
b	1.32	1E
c	4.16	3B
▶d	8.03	1B

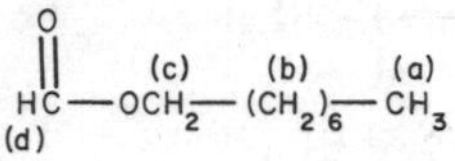

▶ $C_9H_{18}O_2$ ▶ 44.4 mg/0.50 ml $CDCl_3$ ▶ S 114

▶ 11-Vhh

	Shift	Peak
a	6.08	1A
b	6.95	2A
c	7.35	1I
d	7.45	2A
▶e	9.84	1A

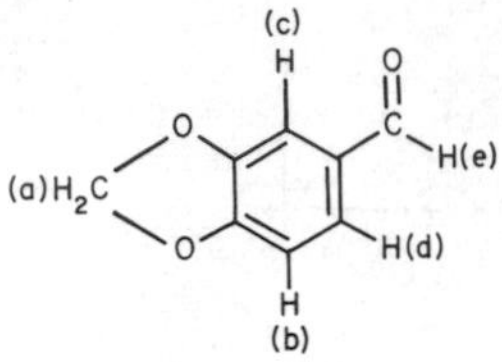

▶ $C_8H_6O_3$ ▶ 7% $CDCl_3$ ▶ V 187

▶ 11-Oc

	Shift	Peak
a	1.29	3A
b	5.13	5B
▶c	8.02	1B

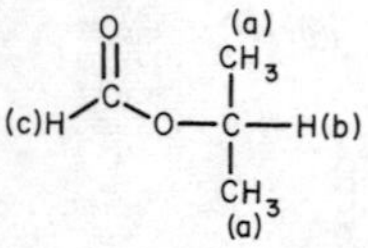

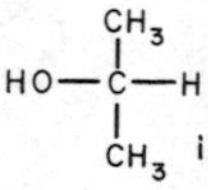

impurity

▶ $C_4H_8O_2$ ▶ 7% $CDCl_3$ ▶ V 415

▶ 11-Vhh

	Shift	Peak
a	2.40	1A
b	7.36	O
c	7.59	O
▶d	9.87	1A

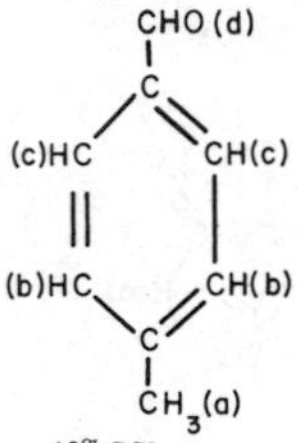

▶ C_8H_8O ▶ 10% CCl_4 ▶ API 51

▶ 11-Veh

	Shift	Peak
a	7.95	S
▶b	10.52	1A

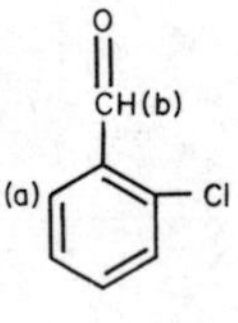

▶ C_7H_5ClO ▶ 35.7 mg/0.50 ml $CDCl_3$ ▶ S 50

▶ 11-Vhh

	Shift	Peak
a	3.90	1A
▶b	9.87	1A

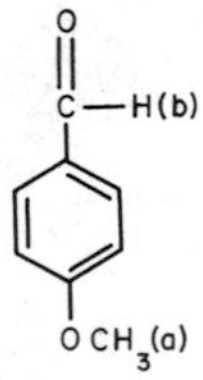

▶ $C_8H_8O_2$ ▶ 45.5 mg/0.50 ml $CDCl_3$ ▶ S 126

▶ 11-Vhh

	Shift	Peak
a	3.05	1A
b	6.69	Q
c	7.71	Q
▶d	9.70	1A

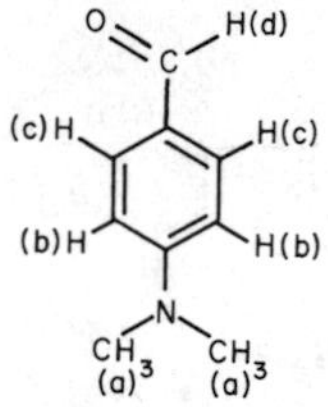

▶ $C_9H_{11}NO$ ▶ 7% $CDCl_3$ ▶ V 238

▶ 11-Vhh

	Shift	Peak
a	3.87	1A
b	7.02	Q
c	7.83	Q
▶d	9.87	1A

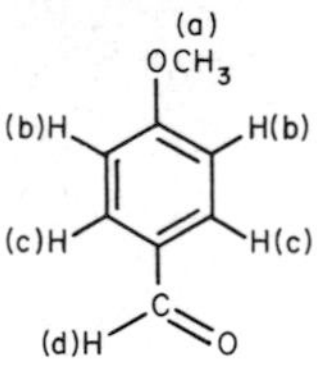

▶ $C_8H_8O_2$ ▶ 7% $CDCl_3$ ▶ V 194

▶ 11-Vhh

	Shift	Peak
a	3.93	1A
b	6.47	1D
c	7.02	2A
d	7.40	N
e	7.40	N
▶f	9.78	1A

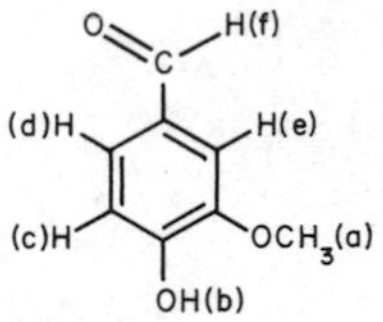

▶ $C_8H_8O_3$ ▶ 7% $CDCl_3$ ▶ V 197

▶ 11-Vhh

	Shift	Peak
a	1.97	1A
▶b	7.88	1A

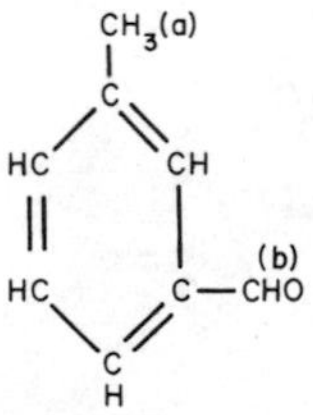

▶ C_8H_8O ▶ 10% CCl_4 ▶ API 56

▶11-Vhh

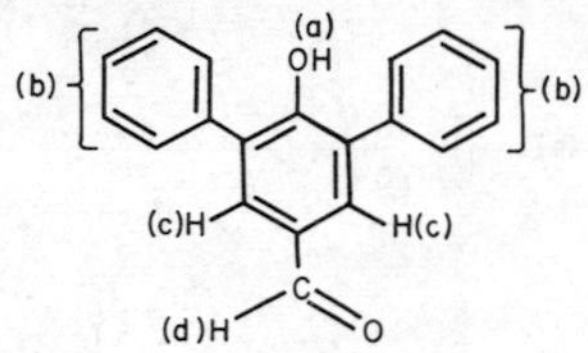

	Shift	Peak
a	6.11	1B
b	7.52±0.05	1E
c	7.83	1A
▶d	9.92	1A

▶$C_{19}H_{14}O_2$ ▶7% $CDCl_3$ ▶V 676

▶11-Vhp

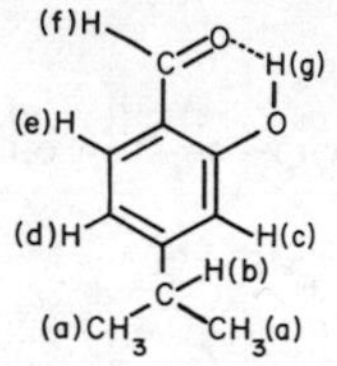

	Shift	Peak
a	1.23	2A
b	2.92	5A
c	6.82	N
d	6.88	N
e	7.42	2A
▶f	9.79	1A
g	11.00	1B

▶$C_{10}H_{12}O_2$ ▶7% $CDCl_3$ ▶V 259

▶11-Vhh

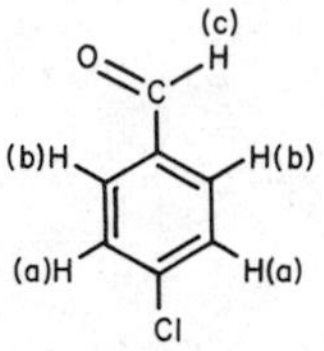

	Shift	Peak
a	7.50	Q
b	7.75	Q
▶c	9.97	1A

▶C_7H_5ClO ▶7% $CDCl_3$ ▶V 146

▶11-Vhr

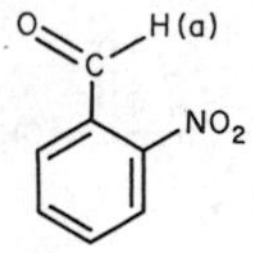

	Shift	Peak
a	10.43	1A

▶$C_7H_5NO_3$ ▶7% $CDCl_3$ ▶V 148

▶11 Vhh

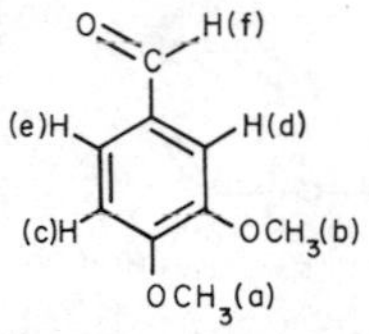

	Shift	Peak
a	3.93 or 3.97	1A
b	3.93 or 3.97	1A
c	6.98	2E
d	7.40	S
e	7.45	S
▶f	9.98	1A

▶$C_9H_{10}O_3$ ▶7% $CDCl_3$ ▶V 236

▶11-Y

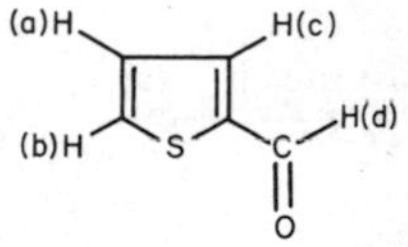

	Shift	Peak
a	7.22	N
b	7.78	M
c	7.78	M
▶d	9.92	2D

▶C_5H_4OS ▶7% $CDCl_3$ ▶V 94

▶11-Vhh

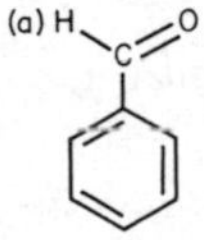

	Shift	Peak
a	10.00	1A

▶C_7H_6O ▶7% $CDCl_3$ ▶V 151

▶12-A

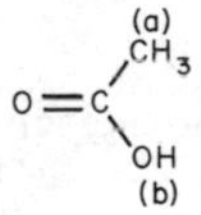

	Shift	Peak
a	2.10	1A
▶b	11.37	1A

▶$C_2H_4O_2$ ▶7% $CDCl_3$ ▶V 8

▶11-Vhj

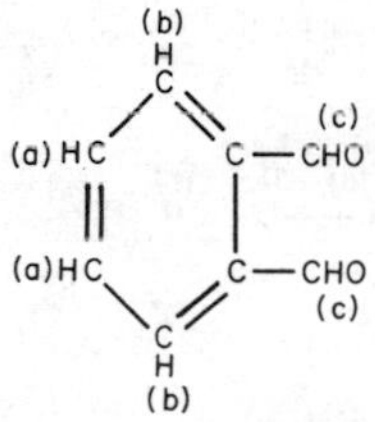

	Shift	Peak
a	7.78	Q
b	7.84	Q
▶c	10.38	1A

▶$C_8H_6O_2$ ▶10% CCl_4 ▶API 52

▶12-Bb

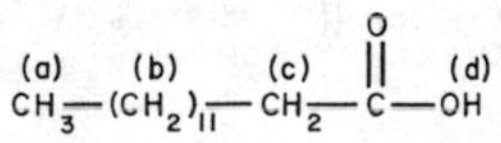

	Shift	Peak
a	0.88	3B
b	1.26	1A
c	~2.35	1D
▶d	10.95	1B

CH_3—$(CH_2)_{11}$—CH_2—C(=O)—OH

▶$C_{14}H_{28}O_2$ ▶44.8 mg/0.50 ml $CDCl_3$ ▶S 167

▶11-Vhp

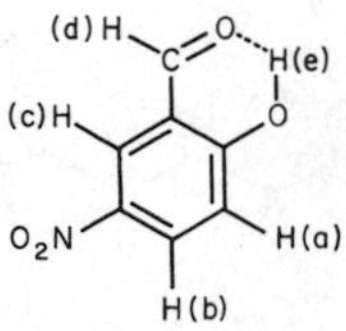

	Shift	Peak
a	7.15	2A
b	8.46	2A
c	8.61	3A
▶d	10.06	1A
e	11.63	1A

▶$C_7H_5NO_4$ ▶7% $CDCl_3$ ▶V 484

▶12-Bb

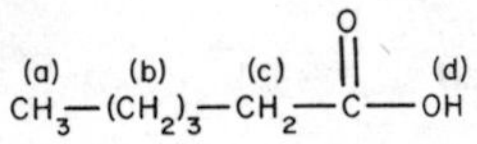

	Shift	Peak
a	0.92	3B
b	1.45	O
c	2.36	3C
▶d	11.37	1B

CH_3—$(CH_2)_3$—CH_2—C(=O)—OH

▶$C_6H_{12}O_2$ ▶44.7 mg/0.50 ml $CDCl_3$ ▶S 37

▶ 12-Bb

	Shift	Peak
a	0.88	3B
b	1.29	1B
c	2.00	1F
d	2.25	3B
e	5.28	3D
▶f	11.45	1C

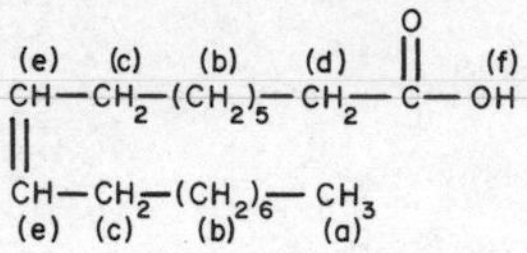

▶ $C_{18}H_{34}O_2$ ▶ 45.2 mg/0.50 ml $CDCl_3$ ▶ S 70

▶ 12-Be

	Shift	Peak
a	4.10	1A
▶b	10.58	1B

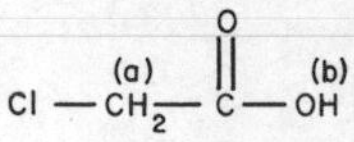

▶ $C_2H_3ClO_2$ ▶ 45.0 mg/0.50 ml $CDCl_3$ ▶ S 128

▶ 12-Bb

	Shift	Peak
a	0.88	3B
b	1.32	1B
c	2.38	3B
▶d	11.47	1B

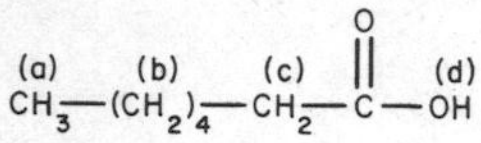

▶ $C_7H_{14}O_2$ ▶ 45.6 mg/0.50 ml $CDCl_3$ ▶ S 49

▶ 12-Bi

	Shift	Peak
a	3.72	1A
▶b	10.25	1A

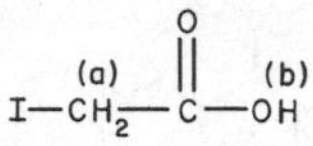

▶ $C_2H_3IO_2$ ▶ 44.8 mg/0.05 ml $CDCl_3$ ▶ S 214

▶ 12-Bb

	Shift	Peak
a	0.88	3B
b	1.30	1G
c	2.36	3B
▶d	11.47	1A

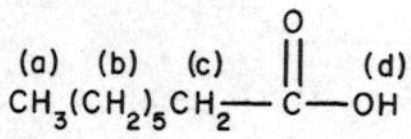

▶ $C_8H_{16}O_2$ ▶ 45.6 mg/0.50 ml $CDCl_3$ ▶ S 157

▶ 12-Bo

	Shift	Peak
a	4.67	1A
b	6.88	Q
c	7.27	Q
▶d	8.86	1C

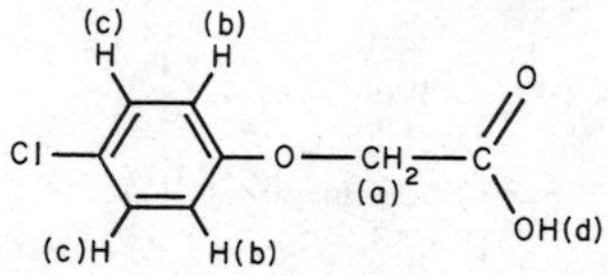

▶ $C_8H_7ClO_3$ ▶ 7% $CDCl_3$ ▶ V 500

▶ 12-Bb

	Shift	Peak
a	3.07	Q
b	3.30	Q
▶c	11.52	1A

▶ $C_3H_5IO_2$ ▶ 7% $CDCl_3$ ▶ V 27

▶ 12-Bo

	Shift	Peak
a	1.27	3A
b	3.66	4A
c	4.13	1A
▶d	10.95	1A

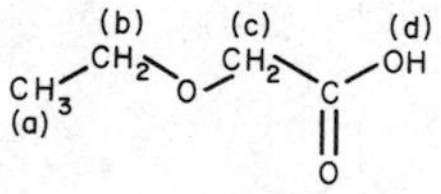

▶ $C_4H_8O_3$ ▶ 7% $CDCl_3$ ▶ V 417

▶ 12-Bb

	Shift	Peak
a	0.93	2A
b	~1.58	3C
c	2.36	3B
▶d	11.67	1A

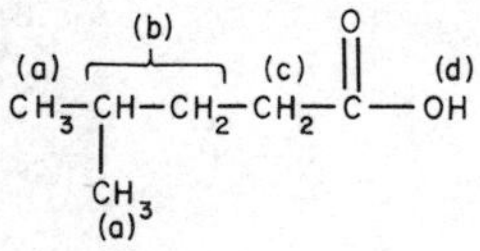

▶ $C_6H_{12}O_2$ ▶ 45.4 mg/0.50 ml $CDCl_3$ ▶ S 156

▶ 12-Bo

	Shift	Peak
a	4.60	1A
▶b	11.10	1A

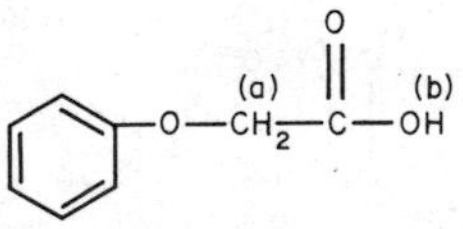

▶ $C_8H_8O_3$ ▶ 45.4 mg/0.50 ml $CDCl_3$ ▶ S 264

▶ 12-Bb

	Shift	Peak
a	~2.79	Q
b	~2.79	Q
▶c	11.71	1A

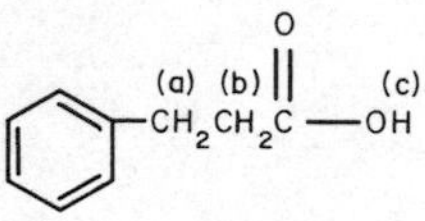

▶ $C_9H_{10}O_2$ ▶ 45.7 mg/0.25 ml $CDCl_3$ ▶ S 299

▶ 12-Bs

	Shift	Peak
a	3.10	1A
b	3.87	1A
c	7.32	1A
▶d	10.97	1B

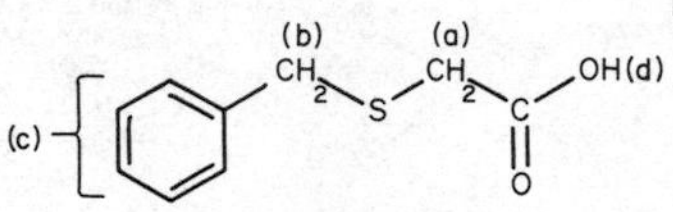

▶ $C_9H_{10}O_2S$ ▶ 7% $CDCl_3$ ▶ V 532

▶ 12-Bv

	Shift	Peak
a	3.58	1A
▶b	11.28	1A

▶ $C_8H_8O_2$ ▶ 44.9 mg/0.50 ml $CDCl_3$ ▶ S 117

▶ 12-(C)bb

	Shift	Peak
a	2.12	Q
b	3.15	N
▶c	12.35	1A

▶ $C_5H_8O_2$ ▶ liquid run, neat ▶ API 191

▶ 12-Bv

	Shift	Peak
a	2.30	1A
b	3.55	1A
▶c	12.67	1A

▶ $C_9H_{10}O_2$ ▶ 44.9 mg/0.25 ml $CDCl_3$ ▶ S 272

▶ 12-Cbg

	Shift	Peak
a	1.08	3A
b	2.07	5B
c	4.23	3A
▶d	10.97	1A

▶ $C_4H_7BrO_2$ ▶ 7% $CDCl_3$ ▶ V 66

▶ 12-Caa

	Shift	Peak
a	1.22	2A
b	2.60	5B
▶c	11.55	1B

▶ $C_4H_8O_2$ ▶ 44.9 mg/0.50 ml $CDCl_3$ ▶ S 54

▶ 12-(C)bl

	Shift	Peak
a	3.80	O
b	4.89 or 5.08	N
c	4.89 or 5.08	N
▶d	12.12	1A

▶ $C_6H_8O_2$ ▶ 7% $CDCl_3$ ▶ V 128

▶ 12-Cae

	Shift	Peak
a	1.73	2B
b	4.47	4A
▶c	11.22	1A

▶ $C_3H_5ClO_2$ ▶ 7% $CDCl_3$ ▶ V 25

▶ 12-Cbn

	Shift	Peak
a	2.10	1A
b	2.25	O
c	2.66	O
d	4.94	4B
e	7.14	3D
▶f	10.28	1B

▶ $C_{12}H_{15}NO_3S$ ▶ 7% $CDCl_3$ ▶ V 597

▶ 12-Cag

	Shift	Peak
a	1.86	2A
b	4.38	4A
▶c	9.95	1A

▶ $C_3H_5BrO_2$ ▶ 44.5 mg/0.50 ml $CDCl_3$ ▶ S 144

▶ 12 (C)cc

	Shift	Peak
a	0.86	3A
b	2.72	5B
c	3.17	4A
d	6.32	S
e	6.35	S
▶f	12.14	1A

▶ $C_{12}H_{10}O_2$ ▶ 7% $CDCl_3$ ▶ V 592

▶ 12-Cas

	Shift	Peak
a	1.56	2A
b	2.26	2A
c	3.58	5B
▶d	11.03	1B

▶ $C_3H_6O_2S$ ▶ 7% $CDCl_3$ ▶ V 389

▶ 12-Ccg

	Shift	Peak
a	1.93	N
b	4.42 or 4.43	S
c	4.42 or 4.43	S
▶d	11.58	1A

▶ $C_4H_6Br_2O_2$ ▶ 7% $CDCl_3$ ▶ V 403

▶ 12-Cee

	Shift	Peak
a	5.98	1A
▶b	11.08	1A

▶ $C_2H_2Cl_2O_2$ ▶ 45.7 mg/0.50 ml $CDCl_3$ ▶ S 166

▶ 12-Lal

	Shift	Peak
a	1.97	2E
b	5.72	M
c	6.30	2E
▶d	11.57	1B

▶ $C_4H_6O_2$ ▶ 7% $CDCl_3$ ▶ V 62

▶ 12-Cov

	Shift	Peak
a	2.18	1A
b	5.95	1A
▶c	9.88	1C

▶ $C_{10}H_{10}O_4$ ▶ 7% $CDCl_3$ ▶ V 554

▶ 12-(L)bl

	Shift	Peak
a	3.38	1A
b	6.56	4A
c	7.27	S
d	7.30	S
▶e	11.05	1C

▶ $C_{12}H_{10}O_2$ ▶ 7% $CDCl_3$ ▶ V 591

▶ 12-Cpv

	Shift	Peak
a	5.23	1A
▶b	6.63	1B
c	7.38	1A

▶ $C_8H_7ClO_3$ ▶ 7% $CDCl_3$ ▶ V 499

▶ 12-(L)dl

	Shift	Peak
a	0.75	2A
b	0.97	2A
c	1.45	1A
d	2.10	2C
e	2.40	S
f	2.60	2C
g	6.77	1A
▶h	12.00	1B

▶ $C_{10}H_{14}O_3$ ▶ 7% $CDCl_3$ ▶ V 567

▶ 12-Cvv

	Shift	Peak
a	5.05	1A
▶b	10.92	1A

▶ $C_{14}H_{12}O_2$ ▶ 45.0 mg/0.50 ml $CDCl_3$ ▶ S 129

▶ 12-Lhl

	Shift	Peak
a	2.30	2D
b	6.07	2E
▶c	11.73	1B

▶ $C_4H_5ClO_2$ ▶ 7% $CDCl_3$ ▶ V 53

▶ 12-(D)bbk

	Shift	Peak
a	2.35	O
b	2.77	3B
▶c	10.83	1A

▶ $C_6H_8O_4$ ▶ D_2O ▶ API 192

▶ 12-Lhl

	Shift	Peak
a	2.58	2C
b	6.10	2E
▶c	11.82	1A

▶ $C_4H_5ClO_2$ ▶ 7% $CDCl_3$ ▶ V 54

▶ 12-Jb

	Shift	Peak
a	4.78	1A
b	7.48	4A
c	7.75	4A
d	8.00	S
▶e	9.63	1B

▶ $C_9H_6ClNO_5$ ▶ 7% $CDCl_3$ ▶ V 223

▶ 12-Lhl

	Shift	Peak
a	1.93	2D
b	2.18	2D
c	5.72	M
▶d	11.95	1B

▶ $C_5H_8O_2$ ▶ 7% $CDCl_3$ ▶ V 114

▶ 12-Lhl

	Shift	Peak
a	1.87	4A
b	5.79	2A
c	7.36	O
▶d	12.03	1A

H, H, C=C, CH_3 (a), C=C, H(b), H (c), C, OH(d), O

▶ $C_6H_8O_2$ ▶ 7% $CDCl_3$ ▶ V 462

▶ 12-Vho

	Shift	Peak
a	4.02	1A
▶b	10.30	1B

O, C—OH(b), O—CH_3 (a)

▶ $C_8H_8O_3$ ▶ 44.6 mg/0.50 ml $CDCl_3$ ▶ S 253

▶ 12-Lhl

	Shift	Peak
a	1.90	2E
b	5.83	2E
c	7.10	2E
▶d	12.18	O

(a) CH_3, H(b), C=C, (c) H, C, OH(d), O

▶ $C_4H_6O_2$ ▶ 7% $CDCl_3$ ▶ V 61

▶ 12-Vho

	Shift	Peak
a	4.07	1A
b	~7.60	O
c	8.17	4A
▶ d	11.00	1B

(d) COOH, (c) H, OCH_3(a), H(b)

▶ $C_8H_8O_3$ ▶ 7% $CDCl_3$ ▶ V 195

▶ 12-Lhl

	Shift	Peak
a	6.46	1A
b	7.83	1A
▶c	13.21	1A

H(a), C—COOH (c), C, H(b)

▶ $C_9H_8O_2$ ▶ 7% $CDCl_3$ ▶ V 230

▶ **13**-A

	Shift	Peak
▶a	1.43	1A
b	3.47	1A

(b) CH_3 — (a) OH

▶ CH_4O ▶ 7% $CDCl_3$ ▶ V 1

▶ 12-Vah

	Shift	Peak
a	2.64	M
▶b	12.63	1A

O, C—OH(b), CH_3(a)

▶ $C_8H_8O_2$ ▶ 44.8 mg/0.25 ml $CDCl_3$ ▶ S 17

▶ 13-Ba

	Shift	Peak
a	1.22	3A
▶b	2.58	1B
c	3.70	4A

(a) CH_3 — (c) CH_2 — (b) OH

▶ C_2H_6O ▶ 7% $CDCl_3$ ▶ V 14

▶ 12-Vhh

	Shift	Peak
a	7.48	N
b	8.15	N
▶c	12.28	1G

O, (b), C—OH(c), (a), (a), (b), (a)

▶ $C_7H_6O_2$ ▶ 45.1 mg/0.50 ml $CDCl_3$ ▶ S 57

▶ 13-Bb

	Shift	Peak
a	0.85	3B
b	1.29	1A
▶c	1.83	1B
d	3.62	3B

(a) CH_3—(b) $(CH_2)_{10}$—(d) CH_2(c) OH

▶ $C_{12}H_{26}O$ ▶ 44.4 mg/0.50 ml $CDCl_3$ ▶ S 103

▶ 12-Vhh

	Shift	Peak
a	2.40	1A
▶b	12.70	1A

O, C—OH(b), CH_3(a)

▶ $C_8H_8O_2$ ▶ 45.0 mg/0.25 ml $CDCl_3$ ▶ S 18

▶ 13-Bb

	Shift	Peak
a	0.88	3B
b	1.33	1B
▶c	1.95	1A
d	3.62	3B

(a) CH_3—(b) $(CH_2)_5$—(d) CH_2—(c) OH

▶ $C_7H_{16}O$ ▶ 45.0 mg/0.50 ml $CDCl_3$ ▶ S 215

▶ 13-Bb

	Shift	Peak
a	0.90	3B
b	1.32	1B
▶c	1.97	1A
d	3.62	3B

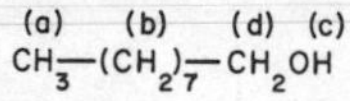

(a) (b) (d) (c)
$CH_3-(CH_2)_7-CH_2OH$

▶ $C_9H_{20}O$ ▶ 44.6 mg/0.50 ml $CDCl_3$ ▶ S 102

▶ 13-Bb

	Shift	Peak
a	0.92	3A
b	1.57	7A
▶c	~2.28	1C
d	3.58	3A

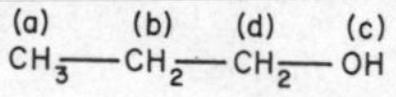

(a) (b) (d) (c)
$CH_3-CH_2-CH_2-OH$

▶ C_3H_8O ▶ 7% $CDCl_3$ ▶ V 43

▶ 13-Bb

	Shift	Peak
a	0.91	3B
b	1.28	1B
▶c	2.03	1A
d	3.63	3A

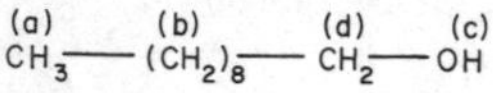

(a) (b) (d) (c)
$CH_3-(CH_2)_8-CH_2-OH$

▶ $C_{10}H_{22}O$ ▶ 7% $CDCl_3$ ▶ V 282

▶ 13-Bb

	Shift	Peak
a	0.88	3B
b	1.36	1B
▶c	2.28	1A
d	3.62	3A

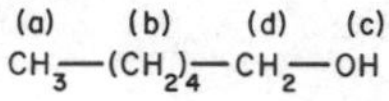

(a) (b) (d) (c)
$CH_3-(CH_2)_4-CH_2-OH$

▶ $C_6H_{14}O$ ▶ 45.2 mg/0.50 ml $CDCl_3$ ▶ S 198

▶ 13-Bb

	Shift	Peak
▶a	2.08	1A
b	2.82	Q
c	3.78	Q
d	7.24	1A

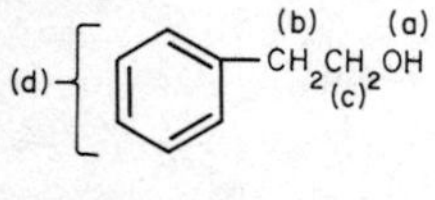

(d) $C_6H_5-CH_2CH_2OH$ (b) (c) (a)

▶ $C_8H_{10}O$ ▶ 45.4 mg/0.50 ml $CDCl_3$ ▶ S 113

▶ 13-Bb

	Shift	Peak
a	~1.80	N
▶b	2.53	1A
c	3.60 or 3.70	S
d	3.70 or 3.60	S

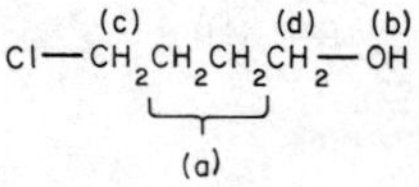

(c) (d) (b)
$Cl-CH_2CH_2CH_2CH_2-OH$
(a)

▶ C_4H_9ClO ▶ 45.0 mg/0.50 ml $CDCl_3$ ▶ S 241

▶ 13-Bb

	Shift	Peak
a	1.13	3A
▶b	2.16	1B
c	3.38	S
d	3.40	S
e	3.72	3D

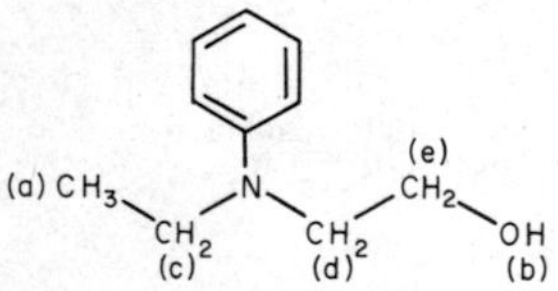

(a) CH_3CH_2(c)$-N(C_6H_5)-CH_2$(d)CH_2(e)OH(b)

▶ $C_{10}H_{15}NO$ ▶ 7% $CDCl_3$ ▶ V 569

▶ 13-Bb

	Shift	Peak
a	1.10	1A
▶b	2.57	1E
c	2.70	3D
d	3.65	3C

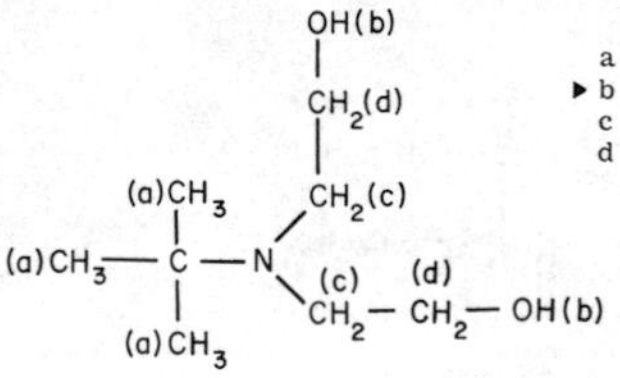

(a)$(CH_3)_3C-N[CH_2$(c)CH_2(d)OH(b)$]_2$

▶ $C_8H_{19}NO_2$ ▶ 7% $CDCl_3$ ▶ V 221

▶ 13-Bb

	Shift	Peak
a	0.93	3A
b	1.56	6B
▶c	2.17	1B
d	3.62	3A

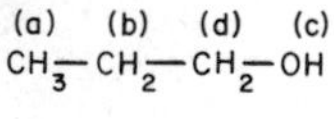

(a) (b) (d) (c)
$CH_3-CH_2-CH_2-OH$

▶ C_3H_8O ▶ 34.7 mg/0.50 ml $CDCl_3$ ▶ S 10

▶ 13-Bb

	Shift	Peak
▶a	2.65	1B
b	3.97 ± 0.10	Q
c	3.97 ± 0.10	Q

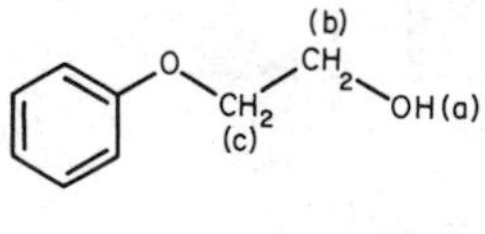

$C_6H_5-O-CH_2$(c)CH_2(b)OH(a)

▶ $C_8H_{10}O_2$ ▶ 7% $CDCl_3$ ▶ V 506

▶ 13-Bb

	Shift	Peak
a	1.89	5B
▶b	2.18	1A
c	2.69	3C
d	3.61	3A
e	7.18	1A

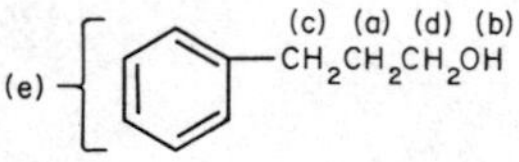

(e) $C_6H_5-CH_2CH_2CH_2OH$ (c) (a) (d) (b)

▶ $C_9H_{12}O$ ▶ 44.9 mg/0.50 ml $CDCl_3$ ▶ S 116

▶ 13-Bb

	Shift	Peak
▶a	2.72	1C
b	3.27	1A
c	~3.50	S

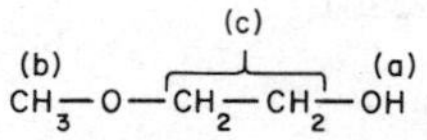

(b) (c) (a)
$CH_3-O-CH_2-CH_2-OH$

▶ $C_3H_8O_2$ ▶ 37.4 mg/0.50 ml $CDCl_3$ ▶ S 32

▶ 13-Bb

Cl—CH_2(c)—CH_2(b)—OH(a)

	Shift	Peak
▶ a	2.80	1A
b	3.68 or 3.83	Q
c	3.68 or 3.83	Q

▶ C_2H_5ClO ▶ 7% $CDCl_3$ ▶ V 12

▶ 13-Bb

$(CH_3)_2$(a)N—CH_2(b)—CH_2(c)—OH(d)

	Shift	Peak
a	2.25	1A
b	2.45	3A
c	3.60	3D
▶ d	3.60	1E

▶ $C_4H_{11}NO$ ▶ 7% $CDCl_3$ ▶ V 91

▶ 13-Bb

4-(CH_2(b)—CH_2(d)—OH(c))morpholine; ring CH_2 next to N (a), ring CH_2 next to O (e)

	Shift	Peak
a	2.51 or 2.56	S
b	2.56 or 2.51	S
▶ c	3.08	1A
d	3.64	1E
e	3.72	3D

▶ $C_6H_{13}NO_2$ ▶ 45.8 mg/0.50 ml $CDCl_3$ ▶ S 185

▶ 13-Bb

Quinolinone: 4-OCH_3(d), 3-CH_2(a)CH_2(c)OH(b), 2=O, N-CH_3(f), 8-OCH_3(e)

	Shift	Peak
a	3.03	3A
▶ b	~3.67	1D
c	3.92	3D
d	3.92 or 4.00	1E
e	3.92 or 4.00	1E
f	3.92 or 4.00	1E

▶ $C_{14}H_{17}NO_4$ ▶ 7% $CDCl_3$ ▶ V 313

▶ 13-Bb

CH_3(a)—CH(H (e))(OCH_3(d))—CH_2(b)—CH_2(f)—OH(c)

	Shift	Peak
a	1.18	2A
b	1.72	4A
▶ c	3.10	3A
d	3.33	1A
e	3.55	O
f	3.73	6B

▶ $C_5H_{12}O_2$ ▶ 7% $CDCl_3$ ▶ V 120

▶ 13-Bb

CH_3(a)—$(CH_2)_7$(b)—CH_2(c)—OH(d)

	Shift	Peak
a	.88	3B
b	1.28	1B
c	3.44	3B
▶ d	4.27	1A

▶ $C_9H_{20}O$ ▶ 20% CCl_4 ▶ API 171

▶ 13-Bb

C_6H_5(d, e, f)—NH(a)CH_2(b)CH_2(c)OH(a)

	Shift	Peak
▶ a	3.12	1E
b	3.23	3A
c	3.78	3A
d	6.61	S
e	6.78	S
f	7.17	3B

▶ $C_8H_{11}NO$ ▶ 45.4 mg/0.50 ml $CDCl_3$ ▶ S 295

▶ 13-Bb

CH_3(a)—$(CH_2)_8$(b)—CH_2(c)—OH(d)

	Shift	Peak
a	.88	3B
b	1.27	1B
c	3.45	3B
▶ d	4.28	1A

▶ $C_{10}H_{22}O$ ▶ 20% CCl_4 ▶ API 172

▶ 13-Bb

CH_3(a)—CH_2(b)—N(H (d))—CH_2(c)—CH_2(f)—OH(e)

	Shift	Peak
a	1.12	3A
b	2.68	4E
c	2.73	3D
d	3.53	1E
▶ e	3.53	1E
f	3.65	3D

▶ $C_4H_{11}NO$ ▶ 7% $CDCl_3$ ▶ V 92

▶ 13-Bb

CH_3(a)—$(CH_2)_5$(b)—CH_2(c)—OH(d)

	Shift	Peak
a	.90	3B
b	1.32	1G
c	3.44	3B
▶ d	4.42	1B

▶ $C_7H_{16}O$ ▶ 20% CCl_4 ▶ API 168

▶ 13-Bb

CH_3(a)—CH(OH (c))(f)—CH_2(b)—CH_2(e)—OH(d)

	Shift	Peak
a	1.23	2A
b	1.68	4C
c	3.58 or 3.65	1A
▶ d	3.58 or 3.65	1A
e	3.80	3A
f	4.03	4A

▶ $C_4H_{10}O_2$ ▶ 7% $CDCl_3$ ▶ V 86

▶ 13-Bb

CH_3(a)—CH(CH_3)(c)—$(CH_2)_3$(b)—CH_2(d)—OH(e)

	Shift	Peak
a	.88	2A
b	1.31	S
c	~1.63	S
d	3.48	3B
▶ e	4.42	1A

▶ $C_7H_{16}O$ ▶ 20% CCl_4 ▶ API 169

▶ 13-Bb

$HO-CH_2-CH_2-O-CH_2-CH_2-OH$ (b) HO; (a) CH_2CH_2; (a) CH_2CH_2; (b) OH

	Shift	Peak
a	3.68	2E
▶b	4.45	1C

▶ $C_4H_{10}O_3$ ▶ 45.6 mg/0.50 ml $CDCl_3$ ▶ S 179

▶ 13-Bc

$CH_3-CH_2-CH(CH_2CH_3)-CH_2-OH$ (b) CH_2 (a) CH_3; (c) CH_2; (d) OH

	Shift	Peak
a	.87	3C
b	1.29	3C
c	3.35	2G
▶d	4.18	1A

▶ $C_6H_{14}O$ ▶ 20% CCl_4 ▶ API 167

▶ 13-Bb

$CH_3-CH_2-CH_2-CH_2-OH$ (a) (b) (b) (c) (d)

	Shift	Peak
a	.91	3B
b	~1.45	O
c	3.48	3C
▶d	4.58	1A

▶ $C_4H_{10}O$ ▶ 20% CCl_4 ▶ API 166

▶ 13-Bc

$CH_3-CH(OH)-CH_2-OH$ (a) CH_3; (b) OH; (b) OH

	Shift	Peak
a	1.12	2A
▶b	4.28	1A

▶ $C_3H_8O_2$ ▶ 7% $CDCl_3$ ▶ V 45

▶ 13-Bb

$CH_3-CH_2-CH_2-OH$ (a) (b) (c) (d)

	Shift	Peak
a	.91	3C
b	1.43	6B
c	3.45	3A
▶d	4.68	1A

▶ C_3H_8O ▶ 20% CCl_4 ▶ API 165

▶ 13-Bd

(a) $CH_3-C(NO_2)(CH_3)-CH_2-OH$ (b); (a) CH_3; (c) CH_2

	Shift	Peak
a	1.59	1A
▶b	2.52	3B
c	3.89	2B

▶ $C_4H_9NO_3$ ▶ 7% $CDCl_3$ ▶ V 422

▶ 13-Bc

$CH_3-CH_2-CH(CH_2CH_3)-CH_2-OH$ (a) CH_3 (b) CH_2; (d) CH_2; (c) OH

	Shift	Peak
a	0.88	3B
b	1.30	6B
▶c	1.56	S
d	3.54	2G

▶ $C_6H_{14}O$ ▶ 35.5 mg/0.50 ml $CDCl_3$ ▶ S 14

▶ 13-Bd

$H-CF_2-CF_2-CH_2-OH$ (c) H; (b) CH_2; (a) OH

	Shift	Peak
▶a	2.72	1B
b	3.97	3C
c	5.93	3E

▶ $C_3H_4F_4O$ ▶ 7% $CDCl_3$ ▶ V 19

▶ 13-Bc

$CH_3(CH_2)_3-CH(CH_2CH_3)CH_2OH$ (a) CH_3; (b) $(CH_2)_3$, CH; (d) CH_2; (c) OH; (b) CH_2 (a) CH_3

	Shift	Peak
a	0.92	3B
b	1.30	1B
▶c	1.75	1B
d	3.52	1I

▶ $C_8H_{18}O$ ▶ 45.0 mg/0.50 ml $CDCl_3$ ▶ S 98

▶ 13-Bd

$(CH_2=CH-CH_2-O-CH_2)_2C(CH_2OH)(CH_2CH_3)$ (g)/(h) =CH_2; (i) =CH; (f) CH_2; (d) CH_2; (e) CH_2 (c) OH; (b) CH_2 (a) CH_3

	Shift	Peak
a	0.84	3C
b	1.38	4C
▶c	2.83	3B
d	3.43	1E
e	3.59	2B
f	3.95	2E
g	5.14	R
h	5.22	R
i	5.87	R

▶ $C_{12}H_{22}O_3$ ▶ 7% $CDCl_3$ ▶ V 603

▶ 13-Bc

3,4-$(CH_3O)_2C_6H_3$-CO-CH(c)(CH_2OH)-O-C_6H_4-2-OCH_3; (e) H; (d) H; (b) CH_2 (a) OH

	Shift	Peak
▶a	3.35	3B
b	4.10	S
c	5.41	3A
d	7.63	N
e	7.77	2A

▶ $C_{18}H_{20}O_6$ ▶ 7% $CDCl_3$ ▶ V 675

▶ 13-Bd

$CH_3-C(CH_3)(NO_2)-CH_2-OH$ (a) CH_3; (a) CH_3; (b) CH_2; (c) OH

	Shift	Peak
a	1.53	1A
b	3.81	2A
▶c	3.05	3B

▶ $C_4H_9NO_3$ ▶ 45.1 mg/0.25 ml $CDCl_3$ ▶ S 296

▶ 13-Bd

(b) (a): $CF_3—CH_2—OH$

	Shift	Peak
▶a	3.38	1B
b	3.93	4A

▶ $C_2H_3F_3O$ ▶ 7% $CDCl_3$ ▶ V 4

▶ 13-Bt

(a) (c) (b): $H—C{\equiv}C—CH_2—OH$

	Shift	Peak
a	2.50	3C
▶b	2.82	1A
c	4.28	2A

▶ C_3H_4O ▶ 7% $CDCl_3$ ▶ V 21

▶ 13-Bd

(c) CH_3 and (b) CH_3 on (d) CH—C(H(e))(OH(h))—C((a) CH_3)((a) CH_3)—C(H(f))(H(g))—OH(h)

	Shift	Peak
a	0.92	1A
b	0.95 or 1.00	2A
c	0.95 or 1.00	2A
d	1.92	5B
e	3.38	2A
f	3.44 or 3.53	S
g	3.44 or 3.53	S
▶h	3.58	1A

▶ $C_8H_{18}O_2$ ▶ 7% $CDCl_3$ ▶ V 217

▶ 13-Bt

(a) CH_3—C(=O)—O—(d) CH_2—C≡C—(c) CH_2—OH(b)

	Shift	Peak
a	2.12	1A
▶b	3.13	1C
c	4.28	Q
d	4.71	Q

▶ $C_6H_8O_3$ ▶ 7% $CDCl_3$ ▶ V 463

▶ 13-Bl

(c) H—C(—(b) CH_2—OH (a))=C(c)—H, phenyl

	Shift	Peak
▶a	2.05	1B
b	4.27	4A
c	6.5	S

▶ $C_9H_{10}O$ ▶ 35.6 mg/0.50 ml $CDCl_3$ ▶ S 23

▶ 13-Bv

(c) phenyl—CH_2OH (b) (a)

	Shift	Peak
▶a	2.43	1A
b	4.58	1A
c	7.28	1A

▶ C_7H_8O ▶ 7% $CDCl_3$ ▶ V 161

▶ 13-Bl

(e)H and (a) CH_3 on C=C bearing (c) CH_2—OH (b) and H(d)

	Shift	Peak
a	1.72	N
▶b	2.07	1A
c	4.07	N
d	~5.68	S
e	~5.68	S

▶ C_4H_8O ▶ 7% $CDCl_3$ ▶ V 414

▶ 13-Bv

(c) CH_2—OH (b) on benzene ring with CH_3 (a) in meta position

	Shift	Peak
a	2.32	2A
▶b	2.43	1B
c	4.47	1G

▶ $C_8H_{10}O$ ▶ 45.2 mg/0.50 ml $CDCl_3$ ▶ S 67

▶ 13-Bl

Furan ring with (c)H, H(c), (d)H and CH_2—OH (b) (a)

	Shift	Peak
▶a	2.83	1B
b	4.57	1A
c	6.33 ± 0.03	M
d	7.44	M

▶ $C_5H_6O_2$ ▶ 7% $CDCl_3$ ▶ V 102

▶ 13-Caa

(a) CH_3, (a) CH_3 on (c) CH—OH (b)

	Shift	Peak
a	1.20	2A
▶b	1.60	1B
c	4.00	5A

▶ C_3H_8O ▶ 7% $CDCl_3$ ▶ V 44

▶ 13-Bl

(c) H and (d) H on C=C bearing (e) H and CH_2—OH (b) (a)

	Shift	Peak
▶a	3.58	1A
b	4.13	2E
c	5.13	R
d	5.25	R
e	~6.00	R

▶ C_3H_6O ▶ 7% $CDCl_3$ ▶ V 34

▶ 13-Cab

(b) CH_3—CH(OH(d))(e)—(c) $(CH_2)_2$—(a) CH_3

	Shift	Peak
a	0.92	3D
b	1.18	2H
c	1.33	N
▶d	2.22	1B
e	3.70	1F

▶ $C_5H_{12}O$ ▶ 44.6 mg/0.50 ml $CDCl_3$ ▶ S 60

▶ 13-Cab

CH_3(a)–CH(f)(OH(c))–CH_2(b)–CH_2(e)–OH(d)

	Shift	Peak
a	1.23	2A
b	1.68	4C
▶ c	3.58 or 3.65	1A
d	3.58 or 3.65	1A
e	3.80	3A
f	4.03	4A

▶ $C_4H_{10}O_2$ ▶ 7% $CDCl_3$ ▶ V 86

▶ 13-Cav

(d) C_6H_5–CH(c)(OH(b))–CH_3(a)

	Shift	Peak
a	1.42	2A
▶ b	2.38	1A
c	4.78	4A
d	7.30	1A

▶ $C_8H_{10}O$ ▶ 45.2 mg/0.50 ml $CDCl_3$ ▶ S 133

▶ 13-Cab

CH_3(a)–CH(c)(OH(d))–CH_2CH_2(b)–CH(OH)–CH_3

	Shift	Peak
a	1.18	1A
b	1.52	3A
c	3.78	S
▶ d	3.87	1A

▶ $C_6H_{14}O_2$ ▶ 44.6 mg/0.50 ml $CDCl_3$ ▶ S 216

▶ 13-Cav

F, (d)H, H(d), (e)H, H(e), CH(c), CH_3 (a), OH (b)

	Shift	Peak
a	1.38	2A
▶ b	3.20	1B
c	4.75	4A
d	6.98	O
e	7.22	O

▶ C_8H_9FO ▶ 7% $CDCl_3$ ▶ V 200

▶ 13-Cab

CH_3(b)–CH(d)(OH(e))–CH_2(c)–$(CH_2)_4$(c)–CH_3(a)

	Shift	Peak
a	.88	3B
b	1.12	2A
c	1.30	1B
d	3.70	4B
▶ e	4.08	1A

▶ $C_8H_{18}O$ ▶ 20% CCl_4 ▶ API 170

▶ 13-(C)bb

(a)H_3C, (b)H_3C, (c)HO, H, O, C, OCH_3(d)

	Shift	Peak
a	0.63	1A
b	0.80	1A
▶ c	1.67	1E
d	3.67	1A

▶ $C_{21}H_{34}O_3$ ▶ 7% $CDCl_3$ ▶ V 354

▶ 13-Cab

CH_3(a)–CH(OH(b))–CH_2–OH(b)

	Shift	Peak
a	1.12	2A
▶ b	4.28	1A

▶ $C_3H_8O_2$ ▶ 7% $CDCl_3$ ▶ V 45

▶ 13-Cbb

$CH_3CH_2(CH_2)_5CH_2$–C(H(g))(OH(c))–C(H(f))(H(e))–NH_2(d); (a), (b)

	Shift	Peak
a	0.89	3B
b	1.28	1G
▶ c	1.75	1E
d	1.75	1E
e	2.51	S
f	2.84	S
g	3.50	1D

▶ $C_{10}H_{23}NO$ ▶ 7% $CDCl_3$ ▶ V 575

▶ 13-Cac

CH_3(a)–CH(c)(OH(b))–CH(c)(OH(b))–CH_3(a)

	Shift	Peak
a	1.13	2A
▶ b	3.05	1A
c	~3.80	O

▶ $C_4H_{10}O_2$ ▶ 7% $CDCl_3$ ▶ V 87

▶ 13-(C)bb

H(b), OH(a)

	Shift	Peak
▶ a	1.88	1A
b	4.33	1C

▶ $C_5H_{10}O$ ▶ 34.5 mg/0.50 ml $CDCl_3$ ▶ S 44

▶ 13-Cak

CH_3(b)–CH(e)(OH(d))–C(=O)–OCH_2(e)CH_2CH_2(c)CH_3(a)

	Shift	Peak
a	0.94	3B
b	1.41	2A
c	1.46	S
▶ d	3.02	1A
e	4.18	4C

▶ $C_7H_{14}O_3$ ▶ 44.6 mg/0.50 ml $CDCl_3$ ▶ S 246

▶ 13-(C)bb

H(d), OH(c), (a), (b)

	Shift	Peak
a	1.31	4E
b	1.78	3B
▶ c	2.20	1E
d	~3.61	1C

▶ $C_6H_{12}O$ ▶ 45.2 mg/0.50 ml $CDCl_3$ ▶ S 7

▶ 13-Cbb

	Shift	Peak
▶ a	2.68	1B
b	3.72	2A
c	4.07	6B

▶ $C_3H_6Cl_2O$ ▶ 7% $CDCl_3$ ▶ V 386

▶ 13-Cct

	Shift	Peak
a	1.00	2D
b	1.02	2D
c	1.85	7B
d	2.47	2E
▶ e	2.82	1B
f	4.18	4A

▶ $C_6H_{10}O$ ▶ 7% $CDCl_3$ ▶ V 133

▶ 13-(C)bb

	Shift	Peak
a	0.95	2A
b	~3.5	1D
▶ c	4.02	1B

▶ $C_7H_{14}O$ ▶ 45.1 mg/0.50 ml $CDCl_3$ ▶ S 79

▶ 13-Cjl

	Shift	Peak
▶ a	4.02	1C
b	5.81	1A
c	6.37	N
d	6.40	N
e	6.55	N
f	7.27	N
g	7.38	N
h	7.62	N

▶ $C_{10}H_8O_4$ ▶ 7% $CDCl_3$ ▶ V 247

▶ 13-(C)bc

	Shift	Peak
▶ a	1.73	1B
b	2.95	2A
c	4.28	1B
d	4.96	1I
e	6.12 or 6.20	2C
f	6.12 or 6.20	2C

▶ $C_{19}H_{22}O_6$ ▶ 7% $CDCl_3$ ▶ V 679

▶ 13-Cjv

	Shift	Peak
a	3.75*	1A
b	3.75*	1A
▶ c	4.54	1C
d	5.82	1B
e	6.83	S
f	6.87	S
g	7.24	2E
h	7.89	2E

*or 3.81

▶ $C_{16}H_{16}O_4$ ▶ 7% $CDCl_3$ ▶ V 655

▶ 13-(C)bc

	Shift	Peak
a	0.82 or 0.93	2A
b	0.90	1A
c	0.82 or 0.93	2A
▶ d	1.75	1E
e	3.42	1D

▶ $C_{10}H_{20}O$ ▶ 7% $CDCl_3$ ▶ V 281

▶ 13-Ckv

	Shift	Peak
a	5.23	1A
▶ b	6.63	1B
c	7.38	1A

▶ $C_8H_7ClO_3$ ▶ 7% $CDCl_3$ ▶ V 499

▶ 13-(C)ck

	Shift	Peak
a	1.26	2A
▶ b	3.28	1B
c	3.87	2A
d	4.34	3C

▶ $C_6H_{10}O_3$ ▶ 7% $CDCl_3$ ▶ V 467

▶ 13-Cvv

	Shift	Peak
▶ a	2.29	1B
b	5.80	1G
c	7.32	1E

▶ $C_{13}H_{12}O$ ▶ 7% $CDCl_3$ ▶ V 607

▶ 13-(C)cl

	Shift	Peak
a	1.65	2A
b	1.97	2E
▶ c	3.86	1C
d	3.92	S
e	4.32	O
f	5.87	1I
g	6.06	2A
h	6.99	4A

▶ $C_{12}H_{12}O_5$ ▶ 7% $CDCl_3$ ▶ V 595

▶ 13-Cvv

	Shift	Peak
a	2.37	2G
▶ b	5.75	2G

▶ $C_{13}H_{12}O$ ▶ 45.4 mg/0.50 ml $CDCl_3$ ▶ S 186

▶ 13-Daaa

Shift	Peak
a 1.28	1A
▶b 1.35	1A

▶ $C_4H_{10}O$ ▶ 7% $CDCl_3$ ▶ V 423

▶ 13-Daau

Shift	Peak
a 1.63	1A
▶b 3.25	1B

▶ C_4H_7NO ▶ 7% $CDCl_3$ ▶ V 70

▶ 13-Daab

Shift	Peak
a 1.27	2A
b 1.34	1E
c 1.69	O
d 2.78	O
e 3.28	S
▶f 5.27	1B
g 5.27	1B
h 7.19*	2F
i 7.19*	2F
j 7.39	1A

*or 7.31

▶ $C_{15}H_{22}O_3$ ▶ 7% $CDCl_3$ ▶ V 647

▶ 13-(D)abb

Shift	Peak
a 0.82 or 0.87	2A
b 0.82 or 0.87	2A
c 1.35	1E
▶d 1.67	1E
e 3.72	1A

▶ $C_{12}H_{22}O_3$ ▶ 7% $CDCl_3$ ▶ V 297

▶ 13-Daac

Shift	Peak
a 1.25 or 1.37	1A
b 1.37 or 1.25	1A
▶c 1.96	1G
d 3.33	2A
e 4.82	3A
f 6.22	2C
g 6.79	2C
h 7.29	2C
i 7.66	2C

▶ $C_{14}H_{14}O_4$ ▶ 7% $CDCl_3$ ▶ V 310

▶ 13-(D)abb

Shift	Peak
a 1.40	1A
b 1.92	4B
c 2.51	2G
d 2.68	2G
▶e 2.92	1A
f 4.33	N
g 4.60	N

▶ $C_6H_{10}O_3$ ▶ 7% $CDCl_3$ ▶ V 466

▶ 13-Daal

Shift	Peak
a 1.32	1A
▶b 1.92	1B
c 4.98	R
d 5.17	R
e 6.00	4D

▶ $C_5H_{10}O$ ▶ 7% $CDCl_3$ ▶ V 444

▶ 13-Dbbk

Shift	Peak
a 1.26	3D
b 2.86	1A
c 4.18	4A
▶d 4.47	1B

▶ $C_{12}H_{20}O_7$ ▶ 45.3 mg/0.50 ml $CDCl_3$ ▶ S 150

▶ 13-Daan

Shift	Peak
a 1.12	1A
b 2.17	1B
▶c 3.30	1A

▶ C_3H_9NO ▶ 7% $CDCl_3$ ▶ V 397

▶ 13-(D)bbt

Shift	Peak
a 2.48	1A
▶b 2.78	1A

▶ $C_8H_{12}O$ ▶ 7% $CDCl_3$ ▶ V 211

▶ 13-Daat

Shift	Peak
a 1.55	1A
▶b 2.27	1B
c 2.44	1A

▶ C_5H_8O ▶ 7% $CDCl_3$ ▶ V 436

▶ 13-(D)bdt

Shift	Peak
a 0.80 or 0.83	1A
b 0.80 or 0.83	1A
▶c 1.88	1A
d 2.55	1A

▶ $C_{21}H_{32}O$ ▶ 7% $CDCl_3$ ▶ V 351

▶ 13-Jb

	Shift	Peak
a	0.88	3B
b	1.28	1E
c	2.38	3B
▶d	10.53	1B

(a) $CH_3-(CH_2)_6-CH_2-C(=O)-OH$ (b) (c) (d)

▶ $C_9H_{18}O_2$ ▶ 35.7 mg/0.50 ml $CDCl_3$ ▶ S 9

▶ 13-Lbl

	Shift	Peak
a	3.65 or 3.80	1A
b	3.80 or 3.65	1A
c	7.20	1A
d	7.83	1A
▶e	12.69	1A

▶ $C_{16}H_{16}O_5$ ▶ 7% $CDCl_3$ ▶ V 656

▶ 13-Jd

	Shift	Peak
a	8.38	1B

$Cl_3C-C(=O)-OH$ (a)

▶ $C_2HCl_3O_2$ ▶ 44.8 mg/0.50 ml $CDCl_3$ ▶ S 6

▶ 13-(L)jl

	Shift	Peak
a	1.27	2A
b	1.34	1E
c	1.69	O
d	2.78	O
e	3.28	S
f	5.27	1B
▶g	5.27	1B
h	7.19*	2F
i	7.19*	2F
j	7.39	1A

*or 7.31

▶ $C_{15}H_{22}O_3$ ▶ 7% $CDCl_3$ ▶ V 647

▶ 13-(L)bl

	Shift	Peak
a	1.06	1E
b	1.09	1E
c	2.29	1A
d	2.29	1A
e	2.56	1A
f	3.36	1B
g	5.51	1A
▶h	10.90	1A

keto-enol mixture

▶ $C_8H_{12}O_2$ ▶ 7% $CDCl_3$ ▶ V 512

▶ 13-(L)jl

	Shift	Peak
a	2.02	1A
b	2.43	1A
c	2.43	1A
▶d	6.77	1A

▶ $C_6H_8O_2$ ▶ 7% $CDCl_3$ ▶ V 129

▶ 13-(L)bl

	Shift	Peak
a	3.72	1A
b	7.10	O
▶c	~12.30	1D

▶ $C_{11}H_{14}O_5$ ▶ 7% $CDCl_3$ ▶ V 583

▶ 13-(L)jl

	Shift	Peak
a	1.27	2A
b	2.90	5B
c	8.32	1A
▶d	8.75	1B

▶ $C_{10}H_{12}O_2$
▶ 7% $CDCl_3$
▶ V 560

▶ 13-(L)bl

	Shift	Peak
a	3.38	1A
b	3.73 or 3.84	1A
c	3.73 or 3.84	1A
d	7.82	1A
▶e	12.62	1A

▶ $C_{15}H_{14}O_5$ ▶ 7% $CDCl_3$ ▶ V 639

▶ 13-(L)lo

	Shift	Peak
a	2.27	1A
b	2.68	1A
c	5.92	1A
▶d	16.71	1C

▶ $C_8H_8O_4$ ▶ 7% $CDCl_3$ ▶ V 504

▶ 13-(L)bl

	Shift	Peak
a	3.73	1A
b	6.96	4C
▶c	12.68	1G

▶ $C_{12}H_{16}O_5$ ▶ 7% $CDCl_3$ ▶ V 600

▶ 13-Llv

	Shift	Peak
a	1.40	3A
b	3.87	1A
c	4.40	4A
d	6.97	2A
e	7.02	1A
f	7.98	2A
▶g	~15.33	1D

▶ $C_{12}H_{14}O_4$ ▶ 7% $CDCl_3$ ▶ V 295

▶ 13-(L)lv

	Shift	Peak
a	6.70	1A
▶b	16.61	1B

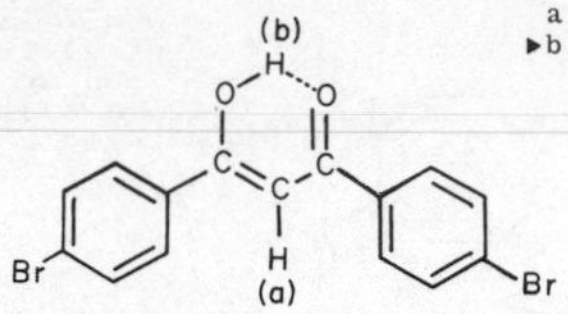

▶ $C_{15}H_{10}Br_2O_2$ ▶ 7% $CDCl_3$ ▶ V 316

▶ 13-Ml

	Shift	Peak
a	1.86	1A
b	1.89	1A
c	6.84	4A
d	7.44	4A
▶e	~8.92	4D

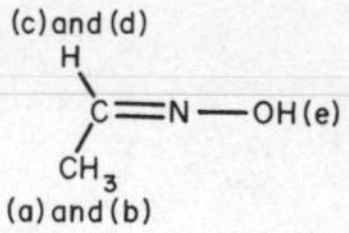

syn- and anti-forms

▶ C_2H_5NO ▶ 7% $CDCl_3$ ▶ V 373

▶ 13-Lly

	Shift	Peak
a	6.42	1A
b	7.18	3C
c	7.73	N
d	7.82	N
▶e	14.17	1D

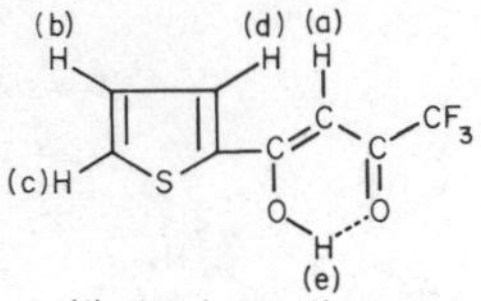

(f) = CH_2 in Keto form.

▶ $C_8H_5F_3O_2S$ ▶ 7% $CDCl_3$ ▶ V 185

▶ 13-Ml

	Shift	Peak
a	2.48±0.05	1C
▶b	9.02	1B

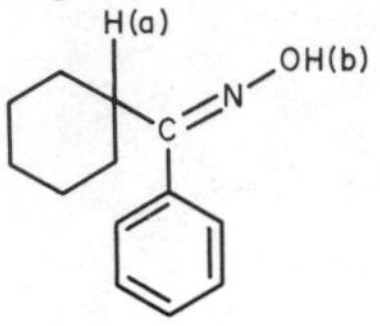

▶ $C_{13}H_{17}NO$ ▶ 7% $CDCl_3$ ▶ V 614

▶ 13-Ml

	Shift	Peak
a	1.53	1A
▶b	8.38	1A

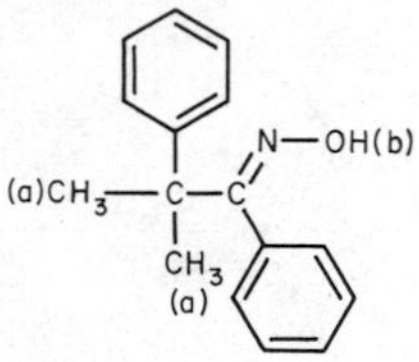

▶ $C_{16}H_{17}NO$ ▶ 7% $CDCl_3$ ▶ V 658

▶ 13-Ml

	Shift	Peak
a	8.04	O
b	8.13	1E
▶c	9.06	1B

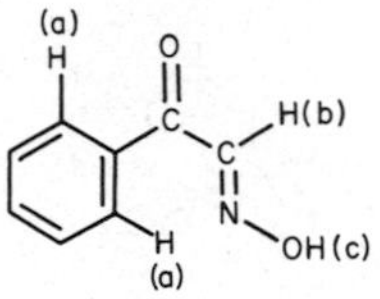

▶ $C_8H_7NO_2$ ▶ 7% $CDCl_3$ ▶ V 501

▶ 13-Ml

	Shift	Peak
a	1.12	1A
b	2.24	1H
c	2.28	1A
d	2.56	1B
▶e	8.73	1B

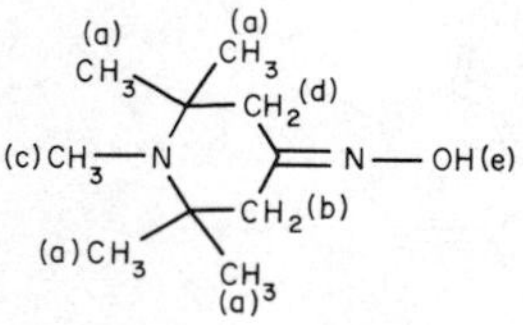

▶ $C_{10}H_{20}N_2O$ ▶ 7% $CDCl_3$ ▶ V 574

▶ 13-Ml

	Shift	Peak
a	2.46	O
▶b	9.40	1C

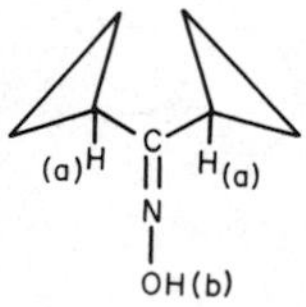

▶ $C_7H_{11}NO$ ▶ 7% $CDCl_3$ ▶ V 179

▶ 13-Ml

	Shift	Peak
a	0.96	3C
b	1.55	6B
c	6.74	3A
d	7.45	3A
▶e	8.85	1D

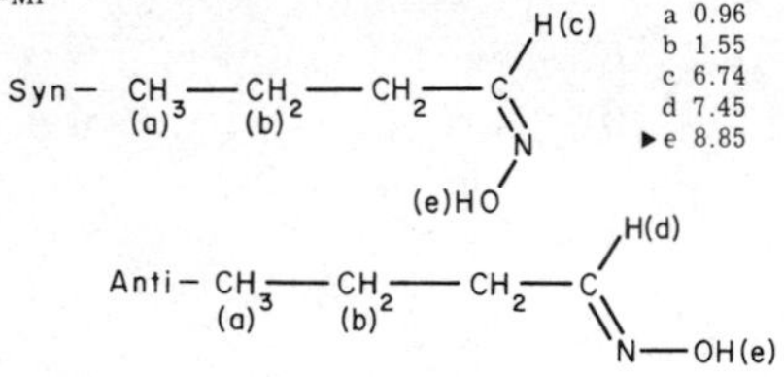

Syn – CH_3(a) — CH_2(b) — CH_2 — C(H(c))=N—OH(e)

Anti – CH_3(a) — CH_2(b) — CH_2 — C(H(d))=N—OH(e)

▶ C_4H_9NO ▶ 7% $CDCl_3$ ▶ V 420

▶ 13-Ml

	Shift	Peak
a	0.93	2A
b	1.98	7B
c	2.77	2A
▶d	9.68	1D

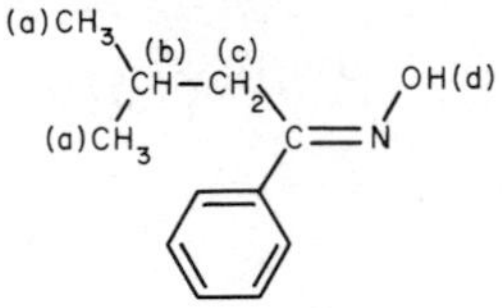

▶ $C_{11}H_{15}NO$ ▶ 7% $CDCl_3$ ▶ V 585

▶ 13-Ml

	Shift	Peak
a	2.00 or 2.40	1A
b	2.00 or 2.40	1A
▶c	8.89	1C

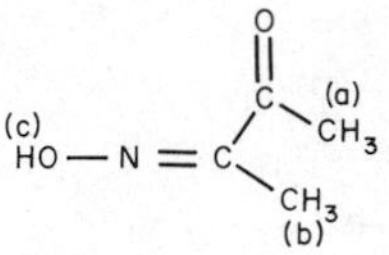

▶ $C_4H_7NO_2$ ▶ 7% $CDCl_3$ ▶ V 72

▶ 13-Ml

	Shift	Peak
a	8.18	1A
b	~8.18	1A
▶c	10.05	1D

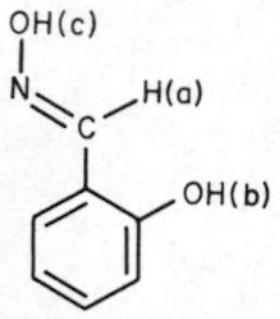

▶ $C_7H_7NO_2$ ▶ 7% $CDCl_3$ ▶ V 156

▶ 13-Ml

(f)H O OH(g) (d)H N CH$_2$(b) (c)CH$_3$O H (e) CH$_3$ (a) CH$_3$ (a)

	Shift	Peak
a	1.42	1A
b	3.04	1A
c	3.91	1A
d	6.88	S
e	6.93	S
f	8.17	2A
▶g	~10.30	1D

▶ $C_{13}H_{15}NO_3$ ▶ 7% $CDCl_3$ ▶ V 611

▶ 13-Vah

OH(c) (d) CH$_3$(b) (a)H$_3$C (f) (e)

	Shift	Peak
a	2.20	1A
b	2.25	1A
▶c	4.88	1A
d	6.55	1A
e	6.70	S
f	6.91	S

▶ $C_8H_{10}O$ ▶ 45.0 mg/0.50 ml $CDCl_3$ ▶ S 82

▶ 13-Nbb

H(e) (c) H H(d) (f) OH (d)H H(e) (c) H H N (b) CH$_2$CH$_2$N CH$_2$CH$_2$ (b) N H (g) (a) (a) (g)

	Shift	Peak
a	3.13	1A
b	3.13	1A
c	7.07	S
d	7.07	S
e	7.53	S
▶f	7.90	S
g	8.48	S

▶ $C_{14}H_{17}N_3O$ ▶ 7% $CDCl_3$ ▶ V 633

▶ 13-Vbe

(b) OH (a) CH$_2$ (b) OH Cl Cl (c)H Cl Cl Cl Cl H(c)

	Shift	Peak
a	4.47	1A
▶b	5.75	1B
c	7.39	1A

▶ $C_{13}H_6Cl_6O_2$ ▶ 7% $CDCl_3$ ▶ V 604

▶ 13-Vaa

(d) CH$_3$ (d)CH$_3$ O (b) CH$_3$ (a) CH$_3$ (a) CH$_3$ (a) CH$_3$ (CH$_2$)$_3$CH–(CH$_2$)$_3$CH–(CH$_2$)$_3$CH–CH$_3$(a) (c) (c) (c) HO (f) CH$_2$ (e) CH$_3$ (d)

	Shift	Peak
a	0.85	2A
b	1.22 ± 0.03	1E
c	1.22 ± 0.03	1E
d	2.12 ± 0.03	1B
e	2.62	3C
▶f	3.96	1D

▶ $C_{29}H_{50}O_2$ ▶ 7% $CDCl_3$ ▶ V 366

▶ 13-Vch

(b) CH$_3$ (f)H H(e) (g)H OH(d) (a)CH$_3$–C–CH$_3$(a) H (c)

	Shift	Peak
a	1.23	2A
b	2.25	1A
c	3.17	5B
▶d	4.75	1A
e	6.55	S
f	6.71	S
g	7.06	S

▶ $C_{10}H_{14}O$ ▶ 7% $CDCl_3$ ▶ V 270

▶ 13-Vaa

(a)CH$_3$ H(d) (b) HO (c) CH$_2$ CH$_3$ (a) H (d)

	Shift	Peak
a	2.12	1A
▶b	4.15	1D
c	4.27	1B
d	6.77	1B

▶ $C_{19}H_{18}O$ ▶ 7% $CDCl_3$ ▶ V 332

▶ 13-Vdh

(c) OH (d) H (a) C(CH$_3$)$_3$ (e)H H(f) CH$_3$ (b)

	Shift	Peak
a	1.40	1A
b	2.27	1A
▶c	4.75	1A
d	6.52	2A
e	6.84	2E or 4A
f	7.05	11

▶ $C_{11}H_{16}O$ ▶ 7% $CDCl_3$ ▶ V 288

▶ 13-Vaa

(a) CH$_3$ (e) H (d) HO (b) CH$_2$ (c) CH$_2$ CH$_3$ (a) H (e)

	Shift	Peak
a	2.20	1A
b	2.90	Q
c	3.30	Q
▶d	4.18	1C
e	6.83	1B

▶ $C_{20}H_{20}O$ ▶ 7% $CDCl_3$ ▶ V 340

▶ 13-Veh

OH(a) Cl NO$_2$

	Shift	Peak
a	6.22	1E

▶ $C_6H_4ClNO_3$ ▶ 44.7 mg/0.50 ml $CDCl_3$ ▶ S 270

▶ 13-Vaa

OH(b) CH$_3$ (a) CH$_3$ (a)

	Shift	Peak
a	2.20	1A
▶b	4.65	1A

▶ $C_8H_{10}O$ ▶ 45.2 mg/0.25 ml $CDCl_3$ ▶ S 36

▶ 13-Vgg

Br (c) H (a) CH$_3$ (c) H Br (b)HO C OH(b) Br H (c) CH$_3$ (a) H (c) Br

	Shift	Peak
a	1.58	1A
▶b	5.66	1D
c	7.27	1A

▶ $C_{15}H_{12}Br_4O_2$ ▶ 7% $CDCl_3$ ▶ V 636

▶ 13-Vhh

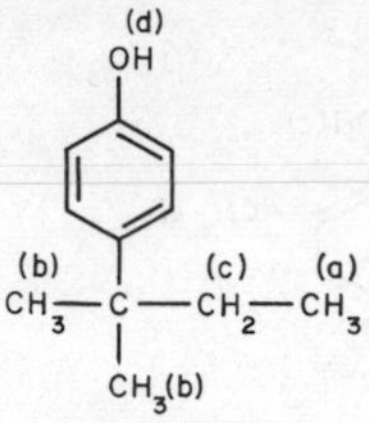

	Shift	Peak
a	0.69	3A
b	1.22	1A
c	1.60	4A
▶d	~4.85	1D

▶ $C_{11}H_{16}O$ ▶ 44.6 mg/0.50 ml $CDCl_3$ ▶ S 256

▶ 13-Vhh

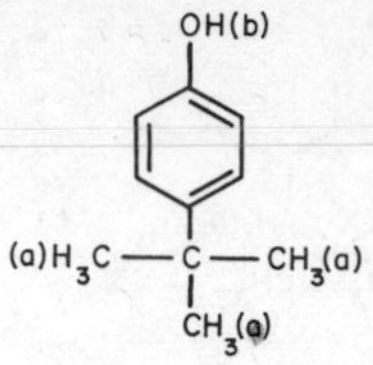

	Shift	Peak
a	1.26	1A
▶b	~5.5	1C

▶ $C_{10}H_{14}O$ ▶ 45.0 mg/0.25 ml $CDCl_3$ ▶ S 47

▶ 13-Vhh

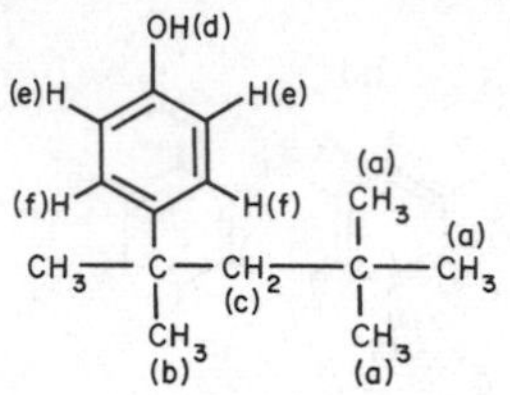

	Shift	Peak
a	0.72	1A
b	1.33	1A
c	1.68	1A
▶d	5.02	1C
e	6.75	Q
f	7.20	Q

▶ $C_{14}H_{22}O$ ▶ 7% $CDCl_3$ ▶ V 315

▶ 13-Vhh

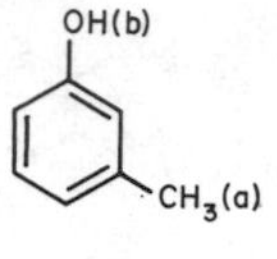

	Shift	Peak
a	2.25	1A
▶b	5.67	1A

▶ C_7H_8O ▶ 7% $CDCl_3$ ▶ V 160

▶ 13-Vhh

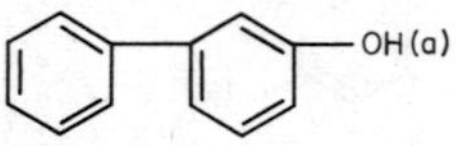

	Shift	Peak
a	5.10	1B

▶ $C_{12}H_{10}O$ ▶ 45.3 mg/0.50 ml $CDCl_3$ ▶ S 262

▶ 13-Vhh

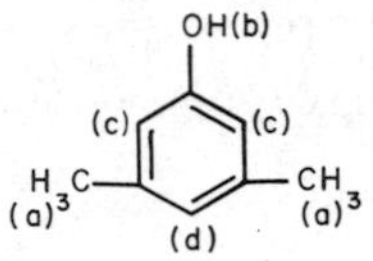

	Shift	Peak
a	2.22	1A
▶b	5.87	1A
c	6.34	1B
d	6.45	1B

▶ $C_8H_{10}O$ ▶ 44.8 mg/0.50 ml $CDCl_3$ ▶ S 83

▶ 13-Vhh

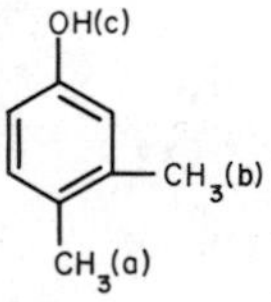

	Shift	Peak
a	2.12	1A
b	2.12	1A
▶c	5.16	1B

▶ $C_8H_{10}O$ ▶ 45.5 mg/0.50 ml $CDCl_3$ ▶ S 193

▶ 13-Vhh

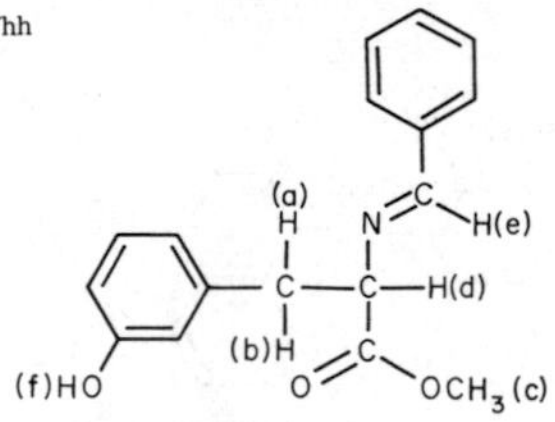

	Shift	Peak
a	3.23	N
b	3.52	N
c	3.72	1A
d	4.28	4A
e	7.89	1A
▶f	8.04	1F

▶ $C_{17}H_{17}NO_3$ ▶ 7% $CDCl_3$ ▶ V 663

▶ 13-Vhh

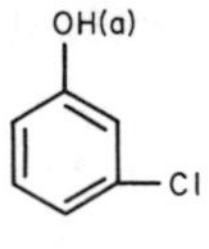

	Shift	Peak
a	5.33	1B

▶ C_6H_5ClO ▶ 45.5 mg/0.50 ml $CDCl_3$ ▶ S 232

▶ 13-Vhj

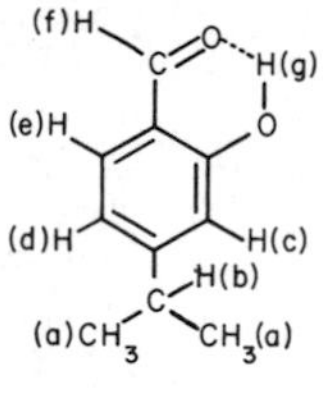

	Shift	Peak
a	1.23	2A
b	2.92	5A
c	6.82	N
d	6.88	N
e	7.42	2A
f	9.79	1A
▶g	11.00	1B

▶ $C_{10}H_{12}O_2$ ▶ 7% $CDCl_3$ ▶ V 259

▶ 13-Vhh

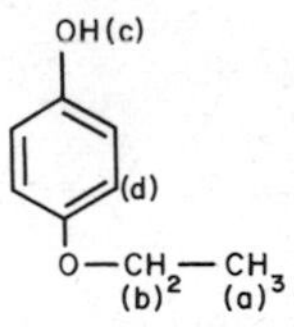

	Shift	Peak
a	1.36	3A
b	3.93	4A
▶c	5.45	1C
d	6.67	1A

▶ $C_8H_{10}O_2$ ▶ 45.0 mg/0.50 ml $CDCl_3$ ▶ S 12

▶ 13-Vhj

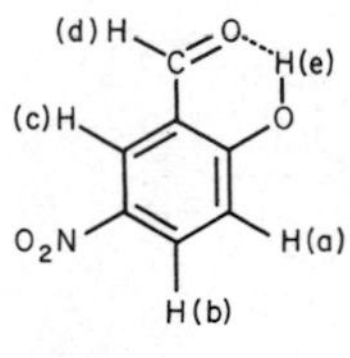

	Shift	Peak
a	7.15	2A
b	8.46	2A
c	8.61	3A
d	10.06	1A
▶e	11.63	1A

▶ $C_7H_5NO_4$ ▶ 7% $CDCl_3$ ▶ V 484

▶ 13-Vhj

	Shift	Peak
a	3.96	1A
b	6.52	1A
c	~ 7.30	1D
d	7.72	Q
e	8.34	Q
▶f	14.22	1A

▶ $C_{15}H_{11}NO_4$ ▶ 7% $CDCl_3$ ▶ V 635

▶ 13-Vho

	Shift	Peak
a	0.96	3B
b	1.37	2A
c	~1.65	S
d	2.58	3B
e	3.42	S
f	3.88	1A
g	5.08	2A
▶h	5.62	1A

▶ $C_{20}H_{24}O_4$ ▶ 7% $CDCl_3$ ▶ V 682

▶ 13-Vhk

	Shift	Peak
a	10.38	1A

▶ $C_{13}H_{10}O_3$ ▶ 45.2 mg/0.50 ml $CDCl_3$ ▶ S 178

▶ 13-Vho

	Shift	Peak
a	1.44	3A
b	4.08	4A
▶c	5.70	1A

▶ $C_8H_{10}O_2$ ▶ 45.3 mg/0.50 ml $CDCl_3$ ▶ S 170

▶ 13-Vhk

	Shift	Peak
a	1.41	3A
b	4.40	4A
c	6.92	O
d	7.37	O
e	7.83	2E
▶f	10.78	1A

▶ $C_9H_{10}O_3$ ▶ 45.6 mg/0.50 ml $CDCl_3$ ▶ S 196

▶ 13-Vho

	Shift	Peak
a	3.93	1A
▶b	6.47	1D
c	7.02	2A
d	7.40	N
e	7.40	N
f	9.78	1A

▶ $C_8H_8O_3$ ▶ 7% $CDCl_3$ ▶ V 197

▶ 13-Vhl

	Shift	Peak
a	8.18	1A
▶b	~8.18	1A
c	10.05	1D

▶ $C_7H_7NO_2$ ▶ 7% $CDCl_3$ ▶ V 156

▶ 13-Vho

	Shift	Peak
a	2.57	1A
b	3.93	1A
▶c	6.62	1C
d	6.97	2A
e	7.57	S
f	7.60	S

▶ $C_9H_{10}O_3$ ▶ 7% $CDCl_3$ ▶ V 234

▶ 13-Vho

	Shift	Peak
a	1.82	2A
b	3.88	1A
▶c	5.48	1B
d	6.03	S
e	6.20	S

▶ $C_{10}H_{12}O_2$ ▶ 45.0 mg/0.5 ml $CDCl_3$ ▶ S 76

▶ 13-Vho

	Shift	Peak
a	1.38	3A
b	3.93	1A
c	4.32	4A
▶d	6.80 or 6.94	1A
e	6.80 or 6.94	1A

▶ $C_{10}H_{12}O_4$ ▶ 45.0 mg/0.50 ml $CDCl_3$ ▶ S 127

▶ 13-Vho

	Shift	Peak
a	3.30	2E
b	3.83	1A
c	5.05	2E
d	5.05	2E
▶e	5.59	1B
f	~5.94	O
g	~6.64	S
h	~6.69	S
i	6.87	S

▶ $C_{10}H_{12}O_2$ ▶ 7% $CDCl_3$ ▶ V 260

▶ 13-Vhp

	Shift	Peak
▶a	5.45	1A
b	6.82	1A

▶ $C_6H_6O_2$ ▶ 7% $CDCl_3$ ▶ V 124

▶ 13-Vov

H(e) (c)H O (d)H O N CH3 (a) (g)HO (b)CH3O H(f) H(f)

	Shift	Peak
a	2.57	1E
b	3.92	1A
c	5.98	2C
d	6.13	2C
e	6.68	1A
f	6.88	1E
▶g	6.88	1E

▶ $C_{19}H_{19}NO_4$ ▶ 7% $CDCl_3$ ▶ V 333

▶ 13-Vvv

(a) OH (b) (b) (c)H H(c) C (d)H O

	Shift	Peak
▶ a	6.11	1B
b	7.52±0.05	1E
c	7.83	1A
d	9.92	1A

▶ $C_{19}H_{14}O_2$ ▶ 7% $CDCl_3$ ▶ V 676

▶ 13-X

(b) OH (c)H N CH3(a) (c)H N CH3(a) OH(b)

	Shift	Peak
a	2.72	1A
▶ b	~7.00	1E
c	7.03	1E

▶ $C_{10}H_{10}N_2O_2$ ▶ 7% $CDCl_3$ ▶ V 553

▶ 13-X

(b) N (c) OH (a)

	Shift	Peak
▶ a	7.25	O
b	8.11	2E
c	8.77	2E

▶ C_9H_7NO ▶ 35.0 mg/0.50 ml $CDCl_3$ ▶ S 24

▶ 13-X

(h) (g)H (c)H OH (e)H H(a) (d)H N H(b) H H (f) (b)

	Shift	Peak
a	7.18	S
b	7.40 ± 0.10	S
c	7.48	S
d	7.60	S
e	7.75	S
f	7.94	S
g	8.17	S
▶ h	~8.67	S

▶ $C_{13}H_9NO$ ▶ 7% $CDCl_3$ ▶ V 605

▶ 13-X

H H OH(c) (a)H N H(b)

	Shift	Peak
a	8.08 or 8.29	3B
b	8.08 or 8.29	1B
▶ c	10.89	1G

▶ C_5H_5NO ▶ 7% $CDCl_3$ ▶ V 428

▶ **14**-AA

OH(b) (c) (c) H3C CH3 (a) (d) (a)

	Shift	Peak
a	2.22	1A
b	5.87	1A
c	6.34	1B
▶ d	6.45	1B

▶ $C_8H_{10}O$ ▶ 44.8 mg/0.50 ml $CDCl_3$ ▶ S 83

▶ 14-AA

CH3 C HC CH(b) CH3 C C CH3(a) C H

	Shift	Peak
a	2.15	1A
▶ b	6.62	1F

▶ C_9H_{12} ▶ run neat ▶ API 179 A

▶ 14-AA

(b) CH3 C (d)HC CH(e) CH3CH2 C C CH3(b) (a) (c) C H (d)

	Shift	Peak
a	1.14	3A
b	2.17	1A
c	2.40	4A
d	6.68	1B
▶ e	6.68	1B

▶ $C_{10}H_{14}$ ▶ liquid ▶ API 12

▶ 14-AA

CH3 C HC CH(b) CH3 C C CH3(a) C H

	Shift	Peak
a	2.23	1A
▶ b	6.69	1F

▶ C_9H_{12} ▶ CCl_4 ▶ API 179 B

▶ 14-AA

CH3 C (b)HC C CH3(a) CH3 C CH(b) C CH3

	Shift	Peak
a	2.16	1A
▶ b	6.74	1B

▶ $C_{10}H_{14}$ ▶ CCl_4 ▶ API 184

▶ 14-AA

(c) NH2 (d) CH3(b) (e) (e) CH3 (a)

	Shift	Peak
a	2.12	1A
b	2.22	1A
c	3.34	1B
d	6.50	2A
▶ e	6.76	2B, 2G

▶ $C_8H_{11}N$ ▶ 44.9 mg/0.50 ml $CDCl_3$ ▶ S 237

▶ 14-AA

Shift	Peak
a 2.25	1A
▶b 6.78	1B

▶ C_9H_{12} ▶ 7% $CDCl_3$ ▶ V 241

▶ 14-AA

Shift	Peak
a 2.26	1A
▶b ~6.87	S
c ~6.87	S
d ~6.93	S

▶ C_8H_{10} ▶ 10% CCl_4 ▶ API 39

▶ 14-AA

Shift	Peak
a 2.17	1A
b 2.23	1A
▶c 6.80	N

▶ C_9H_{12} ▶ CCl_4 ▶ API 178B

▶ 14-AA

Shift	Peak
a 2.18	1A
▶b ~6.93	N

▶ C_8H_{10} ▶ run neat ▶ API 175 A

▶ 14-AA

Shift	Peak
a 2.03	1A
b 2.15	1A
▶c 6.82	N

▶ C_9H_{12} ▶ run neat ▶ API 178 A

▶ 14-AA

Shift	Peak
a 2.03	1A
b 2.37	1A
c 7.02	S
▶d 7.09	S

▶ $C_{16}H_{18}$ ▶ 7% $CDCl_3$ ▶ V 659

▶ 14-AA

Shift	Peak
a 2.20 ± .03	1A
▶b 6.83	1B

▶ $C_{11}H_{16}$ ▶ 7% $CDCl_3$ ▶ V 287

▶ 14-AA

Shift	Peak
a 2.45	1A
b 2.78	1A
▶c 7.33	2C
d 7.74	Q
e 7.99	2C
f 8.23	Q

▶ $C_{16}H_{12}O_2$ ▶ 7% $CDCl_3$ ▶ V 650

▶ 14-AA

Shift	Peak
a 2.29	1A
▶b 6.86	N

▶ C_8H_{10} ▶ CCl_4 ▶ API 175 B

▶ 14-AB

Shift	Peak
a 3.95	1B
▶b 6.28	1B

▶ $C_{22}H_{30}$ ▶ 7% $CDCl_3$ ▶ V 357

▶ 14-AA

Shift	Peak
a 2.19	1A
▶b 6.87	1B

▶ $C_{10}H_{14}$ ▶ 45.9 mg/0.50 ml $CDCl_3$ ▶ S 85

▶ 14-AB

Shift	Peak
a 1.10	3A
b 2.15	2A
c 2.21	4A
▶d 6.67	1E
e 7.08	1E

▶ $C_{10}H_{14}$ ▶ CCl_4 ▶ API 183 A

▶ 14-AB

(b) CH_3 — (d)HC, CH(e) — CH_3CH_2 (a) (c) — CH_3(b) — H (d)

Shift	Peak
a 1.14	3A
b 2.17	1A
c 2.40	4A
▶d 6.68	1B
e 6.68	1B

▶ $C_{10}H_{14}$ ▶ liquid ▶ API 12

▶ 14-AB

(b) CH_3 — (d)HC, (d)HC — CH_3(b) — CH(d) — CH_2CH_3 (c) (a)

Shift	Peak
a 1.18	3A
b 2.20	1A
c 2.60	4A
▶d 6.84	1A

▶ $C_{10}H_{14}$ ▶ CCl_4 ▶ API 182 B

▶ 14-AB

(b) CH_3 — (e)HC, (e)HC — CH_2CH_3 (a) — CH(d) — CH_3 (c)

Shift	Peak
a 1.27	3A
b 2.27	2A
c 2.32	4A
▶d 6.71	1E
e 6.78	1E

▶ $C_{10}H_{14}$ ▶ run neat ▶ API 183 B

▶ 14-AB

(a) CH_3 (c) H (c) H (a) CH_3 — O=C=N— (b) CH_2 —N=C=O — H (c) H (c) H (c) H (c)

Shift	Peak
a 2.26	1A
b 3.82	1A
▶c 6.96	1A

▶ $C_{17}H_{14}N_2O_2$ ▶ 7% $CDCl_3$ ▶ V 661

▶ 14-AB

(a)CH_3 H(d) — (b) HO — (c) CH_2 — CH_3 (a) H (d)

Shift	Peak
a 2.12	1A
b 4.15	1D
c 4.27	1B
▶d 6.77	1B

▶ $C_{19}H_{18}O$ ▶ 7% $CDCl_3$ ▶ V 332

▶ 14-AD

(b) CH_3 — (c)HC, (d)HC — (c) CH — CH_3(a) — C—CH_3 — CH_3 — H (c)

Shift	Peak
a 1.25	1A
b 2.21	1A
▶c ~7.03	S
d 7.11	S

▶ $C_{11}H_{16}$ ▶ CCl_4 ▶ API 185 A

▶ 14-AB

CH_3(c) — (f)HC, CH(e) — (g)HC — $CH_2CH_2CH_3$ (d) (b) (a) — H (h)

Shift	Peak
a .87	3C
b 1.56	6B
c 2.17	1A
d 2.45	3B
▶e 6.82	S
f 6.82	S
g 6.82	S
h 6.82	S

▶ $C_{10}H_{14}$ ▶ liquid run, neat ▶ API 96

▶ 14-AD

(c) OH — (d) H — (a) $C(CH_3)_3$ — (e)H — H(f) — CH_3 (b)

Shift	Peak
a 1.40	1A
b 2.27	1A
c 4.75	1A
d 6.52	2A
e 6.84	2E or 4A
▶f 7.05	1I

▶ $C_{11}H_{16}O$ ▶ 7% $CDCl_3$ ▶ V 288

▶ 14-AB

(b) CH_3 — (d)HC, (d)HC — CH_3(b) — CH(d) — CH_2CH_3 (c) (a)

Shift	Peak
a 1.16	3A
b 2.07	1A
c 2.47	4A
▶d 6.83	2D

▶ $C_{10}H_{14}$ ▶ run neat ▶ API 182 A

▶ 14-AG

(a) CH_3 — (b) (b) — (c) — Br — (c)

Shift	Peak
a 2.32	1A
▶b 7.08	S
c 7.27	S

▶ C_7H_7Br ▶ 45.1 mg/0.50 ml $CDCl_3$ ▶ S 162

▶ 14-AB

(a) CH_3 (e) H — (d) HO — (b) CH_2 — (c) CH_2 — CH_3 (a) H (e)

Shift	Peak
a 2.20	1A
b 2.90	Q
c 3.30	Q
d 4.18	1C
▶e 6.83	1B

▶ $C_{20}H_{20}O$ ▶ 7% $CDCl_3$ ▶ V 340

▶ 14-AH

OH(c) — (d) — CH_3(b) — (a)H_3C — (f) — (e)

Shift	Peak
a 2.20	1A
b 2.25	1A
c 4.88	1A
d 6.55	1A
e 6.70	S
▶f 6.91	S

▶ $C_8H_{10}O$ ▶ 45.0 mg/0.50 ml $CDCl_3$ ▶ S 82

▶ 14-AH

Shift		Peak
a	1.23	2A
b	2.25	1A
c	3.17	5B
d	4.75	1A
e	6.55	S
▶f	6.71	S
g	7.06	S

▶ $C_{10}H_{14}O$ ▶ 7% $CDCl_3$ ▶ V 270

▶ 14-AH

Shift		Peak
a	1.16	3A
b	2.07	1A
c	2.47	4A
▶d	6.83	2D

▶ $C_{10}H_{14}$ ▶ run neat ▶ API 182 A

▶ 14-AH

Shift		Peak
a	2.12	1A
b	2.22	1A
c	3.34	1B
d	6.50	2A
▶e	6.76	2B, 2G

▶ $C_8H_{11}N$ ▶ 44.9 mg/0.50 ml $CDCl_3$ ▶ S 237

▶ 14-AH

Shift		Peak
a	1.18	3A
b	2.20	1A
c	2.60	4A
▶d	6.84	1A

▶ $C_{10}H_{14}$ ▶ CCl_4 ▶ API 182 B

▶ 14-AH

Shift		Peak
a	1.27	3A
b	2.27	2A
c	2.32	4A
d	6.71	1E
▶e	6.78	1E

▶ $C_{10}H_{14}$ ▶ run neat ▶ API 183 B

▶ 14-AH

Shift		Peak
a	1.40	1A
b	2.27	1A
c	4.75	1A
d	6.52	2A
▶e	6.84	2E or 4A
f	7.05	1I

▶ $C_{11}H_{16}O$ ▶ 7% $CDCl_3$ ▶ V 288

▶ 14-AH

Shift		Peak
a	2.17	1A
b	2.23	1A
▶c	6.80	N

▶ C_9H_{12} ▶ CCl_4 ▶ API 178B

▶ 14-AH

Shift		Peak
a	2.29	1A
▶b	6.86	N

▶ C_8H_{10} ▶ CCl_4 ▶ API 175 B

▶ 14-AH

Shift		Peak
a	2.03	1A
b	2.15	1A
▶c	6.82	N

▶ C_9H_{12} ▶ run neat ▶ API 178 A

▶ 14-AH

Shift		Peak
a	2.26	1A
b	~6.87	S
▶c	~6.87	S
d	~6.93	S

▶ C_8H_{10} ▶ 10% CCl_4 ▶ API 39

▶ 14-AH

Shift		Peak
a	.87	3C
b	1.56	6B
c	2.17	1A
d	2.45	3B
e	6.82	S
▶f	6.82	S
g	6.82	S
h	6.82	S

▶ $C_{10}H_{14}$ ▶ liquid run, neat ▶ API 96

▶ 14-AH

Shift		Peak
a	2.28	1A
b	2.40	1A
c	6.10	1I
d	~6.90 or 7.02	N
▶e	~6.90 or 7.02	N
f	7.27	S
g	~7.30	S

▶ $C_{10}H_{11}N$ ▶ 7% $CDCl_3$ ▶ V 255

▶ 14-AH

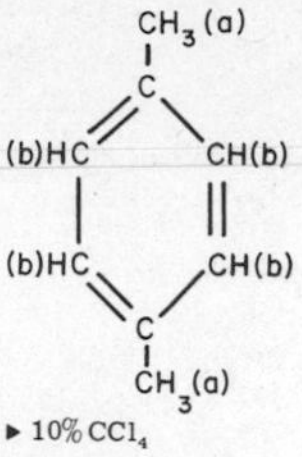

Shift	Peak
a 2.26	1A
▶b 6.93	1A

▶ C_8H_{10} ▶ 10% CCl_4 ▶ API 40

▶ 14-AH

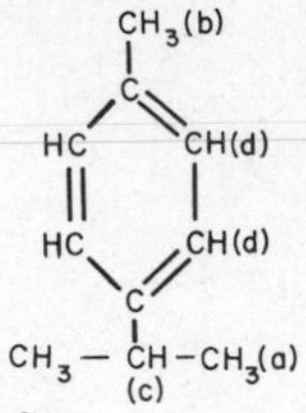

Shift	Peak
a 1.20	2A
b 2.24	1A
c 2.80	5A
▶d 6.99	1A

▶ $C_{10}H_{14}$ ▶ 50% CCl_4 ▶ API 225

▶ 14-AH

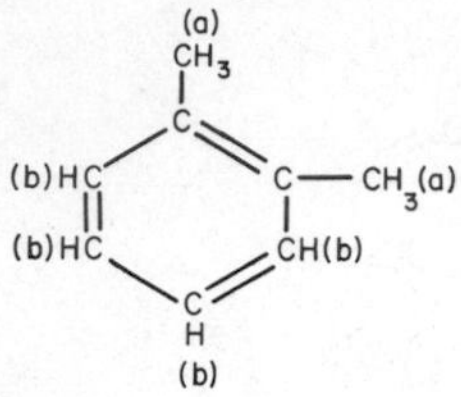

Shift	Peak
a 2.03	1A
▶b 6.93	1A

▶ C_8H_{10} ▶ run neat ▶ API 174 A

▶ 14-AH

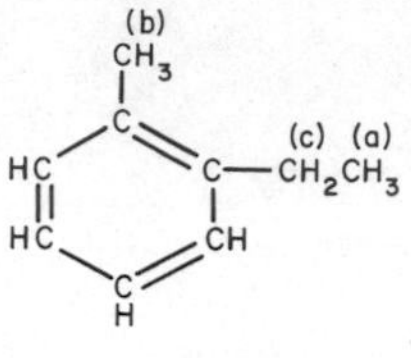

Shift	Peak
a 1.09	3A
b 2.11	1A
c 2.45	4A
▶d 7.00	1A

(d) = all ring H's

▶ C_9H_{12} ▶ run neat ▶ API 177 A

▶ 14-AH

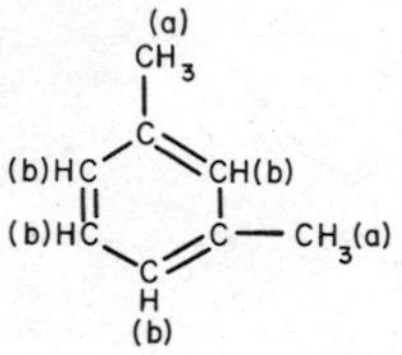

Shift	Peak
a 2.18	1A
▶b ~6.93	N

▶ C_8H_{10} ▶ run neat ▶ API 175 A

▶ 14-AH

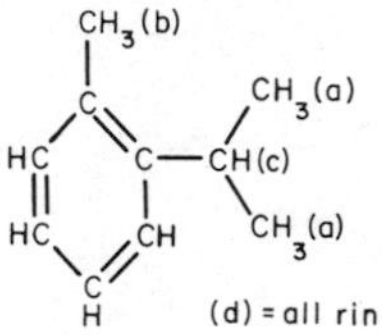

Shift	Peak
a 1.17	2A
b 2.24	1A
c 3.04	5A
▶d 7.00	N

(d) = all ring H's

▶ $C_{10}H_{14}$ ▶ 50% CCl_4 ▶ API 224

▶ 14-AH

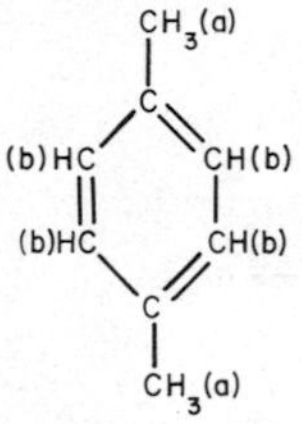

Shift	Peak
a 2.22	1A
▶b 6.95	1A

▶ C_8H_{10} ▶ liquid run, neat ▶ API 82

▶ 14-AH

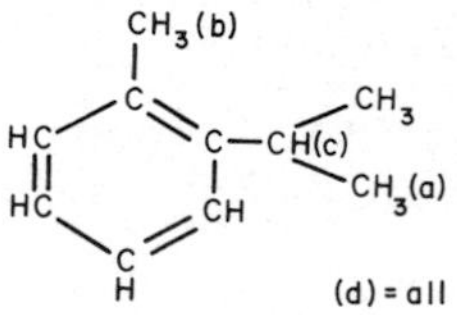

Shift	Peak
a 1.25	2A
b 2.30	1A
c 3.05	5A
▶d 7.01	O

(d) = all ring H's

▶ $C_{10}H_{14}$ ▶ CCl_4 ▶ API 181 B

▶ 14-AH

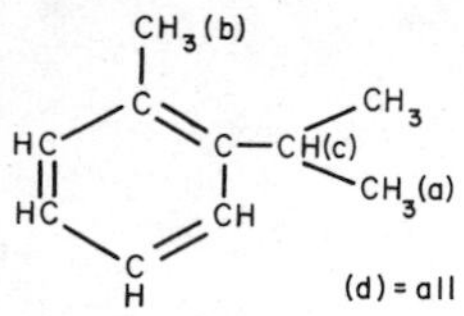

Shift	Peak
a 1.15	2A
b 2.20	1A
c 3.01	5A
▶d 6.95	O

(d) = all ring H's

▶ $C_{10}H_{14}$ ▶ run neat ▶ API 181 A

▶ 14-AH

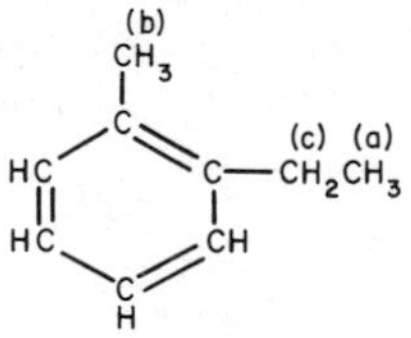

Shift	Peak
a 1.21	3A
b 2.27	1A
c 2.63	4A
▶d 7.02	1A

(d) = all ring H's

▶ C_9H_{12} ▶ CCl_4 ▶ API 177B

▶ 14-AH

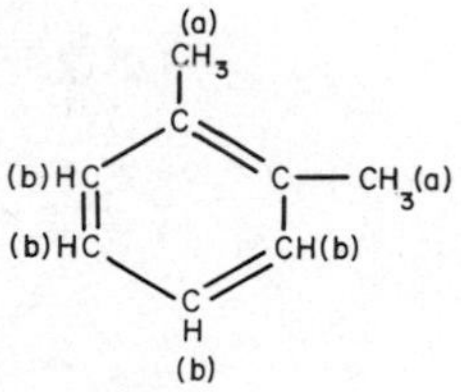

Shift	Peak
a 2.24	1A
▶b 6.98	1A

▶ C_8H_{10} ▶ CCl_4 ▶ API 174 B

▶ 14-AH

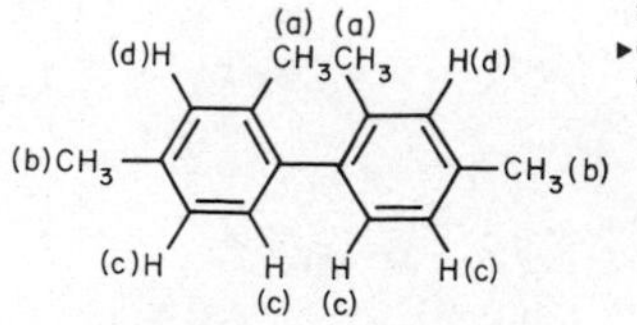

Shift	Peak
a 2.03	1A
b 2.37	1A
▶c 7.02	S
d 7.09	S

▶ $C_{16}H_{18}$ ▶ 7% $CDCl_3$ ▶ V 659

► 14-AH

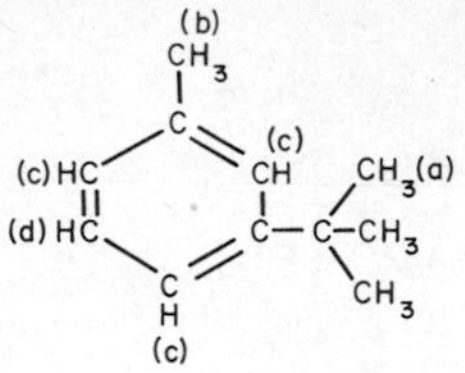

	Shift	Peak
a	1.25	1A
b	2.21	1A
► c	~7.03	S
d	7.11	S

► $C_{11}H_{16}$ ► CCl_4 ► API 185 A

► 14-AH

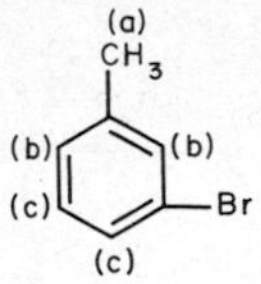

	Shift	Peak
a	2.32	1A
► b	7.08	S
c	7.27	S

► C_7H_7Br ► 45.1 mg/0.50 ml $CDCl_3$ ► S 162

► 14-AH

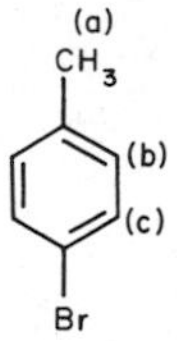

	Shift	Peak
a	2.28	1A
► b	7.05	Q
c	7.33	Q

► C_7H_7Br ► 44.8 mg/0.25 ml $CDCl_3$ ► S 163

► 14-AH

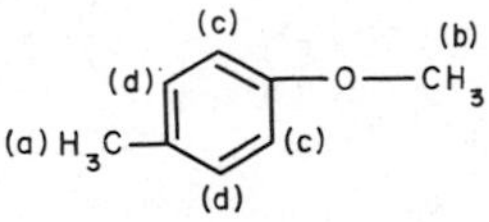

	Shift	Peak
a	2.29	1A
b	3.78	1A
c	6.76	Q
► d	7.08	Q

► $C_8H_{10}O$ ► 45.0 mg/0.50 ml $CDCl_3$ ► S 292

► 14-AH

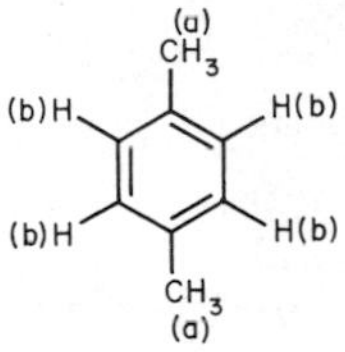

	Shift	Peak
a	2.30	1A
► b	7.05	1A

► C_8H_{10} ► 7% $CDCl_3$ ► V 203

► 14-AH

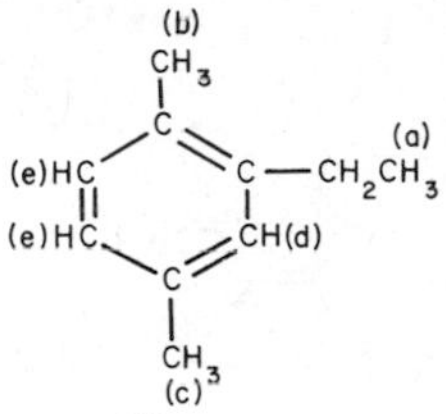

	Shift	Peak
a	1.10	3A
b	2.15	2A
c	2.21	4A
d	6.67	1E
► e	7.08	1E

► $C_{10}H_{14}$ ► CCl_4 ► API 183 A

► 14-AH

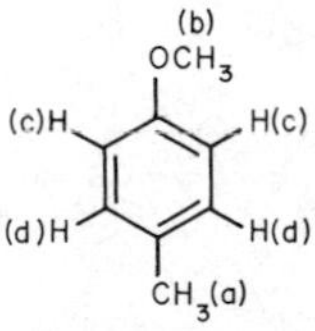

	Shift	Peak
a	2.28	1A
b	3.75	1A
c	6.80	Q
► d	7.05	Q

► $C_8H_{10}O$ ► 7% $CDCl_3$ ► V 205

► 14-AH

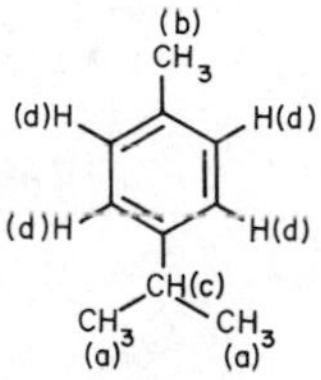

	Shift	Peak
a	1.22	2A
b	2.30	1A
c	2.87	5B
► d	7.08	1A

► $C_{10}H_{14}$ ► 7% $CDCl_3$ ► V 268

► 14-AH

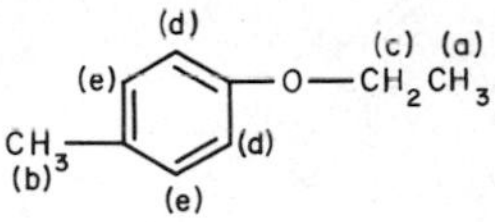

	Shift	Peak
a	1.38	3A
b	2.28	1A
c	3.98	4A
d	6.75	Q
► e	7.05	Q

► $C_9H_{12}O$ ► 45.0 mg/0.50 ml $CDCl_3$ ► S 284

► 14-AH

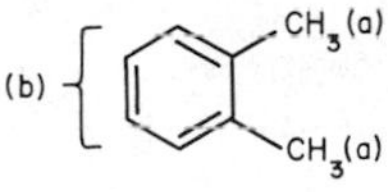

	Shift	Peak
a	2.25	1A
► b	7.10	1A

► C_8H_{10} ► 7% $CDCl_3$ ► V 201

► 14-AH

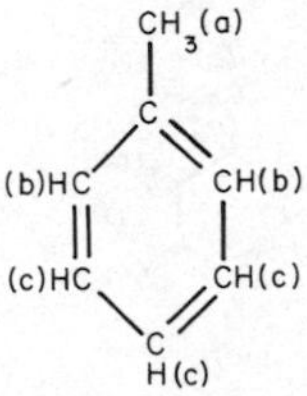

	Shift	Peak
a	2.37	1A
► b	7.07	1A
c	7.07	1A

► C_7H_8 ► liquid run, neat ► API 81

► 14-AH

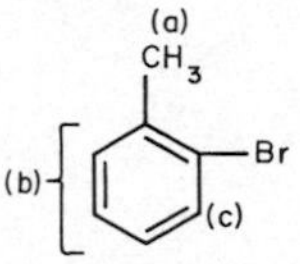

	Shift	Peak
a	2.40	1A
► b	7.12	O
c	7.50	2E

► C_7H_7Br ► 45.4 mg/0.50 ml $CDCl_3$ ► S 161

▶ 14-AH

(d) (e)
H H
(c)H N C CH$_3$(a)
O
(b)CH$_3$ H
(d)
H
(c)

	Shift	Peak
a	2.12	1A
b	2.30	1A
▶c	7.12	Q
d	7.37	Q
e	7.88	1D

▶ $C_9H_{11}NO$ ▶ 7% $CDCl_3$ ▶ V 239

▶ 14-AH

CH$_3$(a)
(b)H H(b)
(c)H H(c)
CF$_3$

	Shift	Peak
a	2.38	1A
▶b	7.25	Q
c	7.48	Q

▶ $C_8H_7F_3$ ▶ 7% $CDCl_3$ ▶ V 190

▶ 14-AH

CH$_3$(a)
(a)CH$_3$—C—CH$_3$(a)
H H(d)
H H(c)
CH$_3$
(b)

	Shift	Peak
a	1.32	1A
b	2.33	1A
▶c	7.12	Q
d	7.28	Q

▶ $C_{11}H_{16}$ ▶ 7% $CDCl_3$ ▶ V 586

▶ 14-AH

(e)H H(f) (f) (e)
(d) (d) H H
(b) CH$_2$—CH$_2$ (b)
CH$_3$ SO$_2$—N N—SO$_2$ CH$_3$
CH$_2$ CH$_2$
(c) CH$_2$ (c)
H H (a) H H
(e) (f) (f) (e)

	Shift	Peak		Shift	Peak
a	1.95	5B	d	3.37	1E
b	2.38	1A	▶e	7.30	Q
c	3.33	3A	f	7.63	Q

▶ $C_{19}H_{24}N_2O_4S_2$
▶ 7% $CDCl_3$
▶ V 336

▶ 14-AH

(c)(c)
H H
(b) (b)
(d)H H—C—C—H H(d)
(d)H H(d)
(d)H CH$_3$ CH$_3$ H(d)
(a) (a)

	Shift	Peak
a	2.29	1A
b	2.61	S
c	2.71	S
▶d	7.15	1A

▶ $C_{16}H_{16}$ ▶ 7% $CDCl_3$ ▶ V 654

▶ 14-AH

O
(b) (d)
(a)CH$_3$ CH$_2$ CH$_2$ CH$_2$ N C N H(h)
(b) (e)H
SO$_2$
(g)H H(g)
(f)H H(f)
CH$_3$(c)

	Shift	Peak
a	0.90	3B
b	~1.40	1F
c	2.46	1A
d	3.23	4C
e	6.53	3B
▶f	7.32	Q
g	7.79	Q
h	~8.60	1D

▶ $C_{12}H_{18}N_2O_3S$ ▶ 7% $CDCl_3$ ▶ V 602

▶ 14-AH

(a)
CH$_3$
(b)

	Shift	Peak
a	2.32	1A
▶b	7.17 ± 0.03	1A

▶ C_7H_8 ▶ 7% $CDCl_3$ ▶ V 157

▶ 14-AH

(d) (e)
H H
O
(b)CH$_3$ S — CH$_2$ — CH$_3$
(c) (a)
H H
(d) (e)

	Shift	Peak
a	1.20	3A
b	2.43	1A
c	2.82	4C
▶d	7.33	Q
e	7.51	Q

▶ $C_9H_{12}OS$ ▶ 7% $CDCl_3$ ▶ V 538

▶ 14-AH

(b)
(c) CH$_2$Br
CH$_3$
(a)

	Shift	Peak
a	2.41	1A
b	4.50	1A
▶c	7.19	1A

▶ C_8H_9Br ▶ 45.3 mg/0.50 ml $CDCl_3$ ▶ S 239

▶ 14-AH

CHO (d)
C
(c)HC CH(c)
(b)HC CH(b)
C
CH$_3$(a)

	Shift	Peak
a	2.40	1A
▶b	7.36	O
c	7.59	O
d	9.87	1A

▶ C_8H_8O ▶ 10% CCl_4 ▶ API 51

▶ 14-AH

(b)
CH$_2$Cl
(c)
CH$_3$(a)

	Shift	Peak
a	2.35	1A
b	4.56	1A
▶c	7.24	1E

▶ C_8H_9Cl ▶ 45.2 mg/0.50 ml $CDCl_3$ ▶ S 68

▶ 14-A(J)

(f) O (b)
H CH$_3$
(d)H C H(c)
(d)H C CH$_3$(a)
H O H
(f) (e)

	Shift	Peak
a	2.45	1A
b	2.78	1A
c	7.33	2C
d	7.74	Q
▶e	7.99	2C
f	8.23	Q

▶ $C_{16}H_{12}O_2$ ▶ 7% $CDCl_3$ ▶ V 650

▶ 14-A(L)

(f) H, (b) CH_3, H(c), (e)H, N, CH_3(a), H (d), H (g)

	Shift	Peak
a	2.28	1A
b	2.40	1A
c	6.10	1I
d	~6.90 or 7.02	N
e	~6.90 or 7.02	N
▶f	7.27	S
g	~7.30	S

▶ $C_{10}H_{11}N$ ▶ 7% $CDCl_3$ ▶ V 255

▶ 14-BH

(f) H, (c) H, (g) H, (a) C=C, CH_2, H(d), (i)H, (e)HO, H(h), OCH_3 (b)

	Shift	Peak
a	3.30	2E
b	3.83	1A
c	5.05	2E
d	5.05	2E
e	5.59	1B
f	~5.94	O
▶g	~6.64	S
h	~6.69	S
i	6.87	S

▶ $C_{10}H_{12}O_2$ ▶ 7% $CDCl_3$ ▶ V 260

▶ 14-A(N)

H(f), (c) CH_3O, H(e), CH_3 (a), N, O, (g)H, H_2C—CH_2, (d), (b)

	Shift	Peak
a	2.33	1A
b	3.13	3A
c	3.83	1A
d	4.28	4A
e	6.85 or 6.98	1A
f	6.85 or 6.98	1A
▶g	7.13	1B

▶ $C_{13}H_{13}NO_2$ ▶ 7% $CDCl_3$ ▶ V 299

▶ 14-BH

(e) O—CH_2—O, (g)H, (h)H, H(f), (a)CH_2, H(b), C=C, (d)H, H(c)

	Shift	Peak
a	3.30	2E
b	5.03	2E
c	5.03	2E
d	~5.88	O
e	5.88	1A
f	6.67 ± .03	1H
g	6.67 ± .03	1H
▶h	6.67 ± .03	1H

▶ $C_{10}H_{10}O_2$ ▶ 7% $CDCl_3$ ▶ V 253

▶ 14-AP

OH(b), (c), (c), H_3C (a), CH_3 (a), (d)

	Shift	Peak
a	2.22	1A
b	5.87	1A
▶c	6.34	1B
d	6.45	1B

▶ $C_8H_{10}O$ ▶ 44.8 mg/0.50 ml $CDCl_3$ ▶ S 83

▶ 14-(B)H

(g)H, (c) CH_2, (b) CH_2, N, CH_3(a), (f) H_2C, O, O, CH_2(d), C=O, CH_2 (e), (i)H, O, (f) CH_2, (h)H, H(h)

	Shift	Peak
a	1.92	1A
b	2.55	1F
c	2.88	1F
d	3.58	1A
e	3.78	1A
f	5.92, 5.96	1A
g	6.65	1A
▶h	6.69	1A
i	6.91	1A

▶ $C_{20}H_{19}NO_5$ ▶ 7% $CDCl_3$ ▶ V 339

▶ 14-AP

OH(c), (d), CH_3(b), (a)H_3C, (f), (e)

	Shift	Peak
a	2.20	1A
b	2.25	1A
c	4.88	1A
▶d	6.55	1A
e	6.70	S
f	6.91	S

▶ $C_8H_{10}O$ ▶ 45.0 mg/0.50 ml $CDCl_3$ ▶ S 82

▶ 14-BH

CH_3(c), C, (f)HC, CH(e), (g)HC, C—$CH_2CH_2CH_3$, (d) (b) (a), C, H (h)

	Shift	Peak
a	.87	3C
b	1.56	6B
c	2.17	1A
d	2.45	3B
e	6.82	S
f	6.82	S
g	6.82	S
▶h	6.82	S

▶ $C_{10}H_{14}$ ▶ liquid run, neat ▶ API 96

▶ 14-AP

(b) CH_3, (f)H, H(e), (g)H, OH(d), (a)CH_3—C—CH_3(a), H (c)

	Shift	Peak
a	1.23	2A
b	2.25	1A
c	3.17	5B
d	4.75	1A
▶e	6.55	S
f	6.71	S
g	7.06	S

▶ $C_{10}H_{14}O$ ▶ 7% $CDCl_3$ ▶ V 270

▶ 14-BH

(b) CH_3, C, (d)HC, C—CH_3(b), (d)HC, CH(d), C, CH_2CH_3 (c) (a)

	Shift	Peak
a	1.16	3A
b	2.07	1A
c	2.47	4A
▶d	6.83	2D

▶ $C_{10}H_{14}$ ▶ run neat ▶ API 182 A

▶ 14-AQ

(a)CH_3, O, (d) H, (c)CH_3, O, CH_3—C—O, (b), H(d)

	Shift	Peak
a	0.95	1A
b	2.28	1A
c	2.33	1A
▶d	6.72	1B

▶ $C_{21}H_{26}O_3$ ▶ 7% $CDCl_3$ ▶ V 350

▶ 14-BH

(b) CH_3, C, (d)HC, C—CH_3(b), (d)HC, CH(d), C, CH_2CH_3 (c) (a)

	Shift	Peak
a	1.18	3A
b	2.20	1A
c	2.60	4A
▶d	6.84	1A

▶ $C_{10}H_{14}$ ▶ CCl_4 ▶ API 182 B

▶ 14-(B)H

Shift	Peak
a 2.57	1E
b 3.92	1A
c 5.98	2C
d 6.13	2C
e 6.68	1A
▶f 6.88	1E
g 6.88	1E

▶ $C_{19}H_{19}NO_4$ ▶ 7% $CDCl_3$ ▶ V 333

▶ 14-BH

Shift	Peak
a 1.10	3A
b 2.5	4A
▶c 7.07	1A

(c)= all ring H's

▶ C_8H_{10} ▶ run neat ▶ API 173 A

▶ 14-BH

Shift	Peak
a 1.09	3A
b 2.11	1A
c 2.45	4A
▶d 7.00	1A

(d) = all ring H's

▶ C_9H_{12} ▶ run neat ▶ API 177 A

▶ 14-(B)H

Shift	Peak
a 1.79	O
b 2.76	O
▶c 7.07	1A

▶ $C_{10}H_{12}$ ▶ 7% $CDCl_3$ ▶ V 557

▶ 14-BH

Shift	Peak
a 1.21	3A
b 2.27	1A
c 2.63	4A
▶d 7.02	1A

(d) = all ring H's

▶ C_9H_{12} ▶ CCl_4 ▶ API 177B

▶ 14-BH

Shift	Peak
a .89	3B
b 1.25	1B
c 2.46	3B
▶d 7.07	1A

(d) = all ring H's

▶ $C_{15}H_{24}$ ▶ 50% CCl_4 ▶ API 358

▶ 14-BH

Shift	Peak
a .89	3B
b 1.36	O
c 2.52	3B
▶d 7.04	1G

(d) = all ring H's

▶ $C_{11}H_{16}$ ▶ 50% CCl_4 ▶ API 355

▶ 14-BH

Shift	Peak
a .88	3B
b 1.25	1B
c 2.50	3B
▶d 7.07	1A

(d) = all ring H's

▶ $C_{16}H_{26}$ ▶ 50% CCl_4 ▶ API 359

▶ 14-(B)H

Shift	Peak
a 2.02	5B
b 2.86	3A
▶c 7.05	1I

▶ C_9H_{10} ▶ 10% CCl_4 ▶ API 188

▶ 14-BH

Shift	Peak
a .86	3B
b 1.28	1B
c 2.54	3B
▶d 7.09	1A

(d) = all ring H's

▶ $C_{13}H_{20}$ ▶ 50% CCl_4 ▶ API 357

▶ 14-BH

Shift	Peak
a 3.97	1A
▶b 7.05 or 7.13	Q
c 7.05 or 7.13	Q
▶d 7.22	1A

▶ $C_{14}H_{13}NO$ ▶ 7% $CDCl_3$ ▶ V 309

▶ 14-BH

Shift	Peak
a .88	3B
b 1.22	1A
c 2.54	3B
▶d 7.09	1G

(d) = all ring H's

▶ $C_{18}H_{30}$ ▶ 50% CCl_4 ▶ API 360

▶ 14-BH

	Shift	Peak
a	1.22	3A
b	2.54	4A
▶ c	7.10	1E

▶ C_8H_{10} ▶ 10% CCl_4 ▶ API 38

▶ 14-BH

	Shift	Peak
a	2.29	1A
b	2.61	S
c	2.71	S
▶ d	7.15	1A

▶ $C_{16}H_{16}$ ▶ 7% $CDCl_3$ ▶ V 654

▶ 14-BH

	Shift	Peak
a	1.10	3A
b	2.47	4A
▶ c	7.10	1A
d	7.10	1A

▶ C_8H_{10} ▶ liquid ▶ API 74

▶ 14-BH

	Shift	Peak
a	1.89	5B
b	2.18	1A
c	2.69	3C
d	3.61	3A
▶ e	7.18	1A

▶ $C_9H_{12}O$ ▶ 44.9 mg/0.50 ml $CDCl_3$ ▶ S 116

▶ 14-BH

	Shift	Peak
a	.92	3C
b	1.58	6B
c	2.57	3C
▶ d	7.10	1E

▶ C_9H_{12} ▶ 10% CCl_4 ▶ API 41

▶ 14-BH

	Shift	Peak
a	3.87	1A
▶ b	7.18	1A

▶ $C_{14}H_{12}$ ▶ 7% $CDCl_3$ ▶ V 307

▶ 14-BH

(d) = all ring H's

	Shift	Peak
a	.89	3B
b	1.30	1F
c	2.54	3B
▶ d	7.10	1A

▶ $C_{12}H_{18}$ ▶ 50% CCl_4 ▶ API 356

▶ 14-BH

	Shift	Peak
a	2.41	1A
b	4.50	1A
▶ c	7.19	1A

▶ C_8H_9Br ▶ 45.3 mg/0.50 ml $CDCl_3$ ▶ S 239

▶ 14-BH

(e) = all ring H's

	Shift	Peak
a	.93	3C
b	1.47	S
c	1.47	S
d	2.57	3B
▶ e	7.11	1A

▶ $C_{10}H_{14}$ ▶ 10% CCl_4 ▶ API 42

▶ 14-(B)H

	Shift	Peak
a	3.65 or 3.80	1A
b	3.80 or 3.65	1A
▶ c	7.20	1A
d	7.83	1A
e	12.69	1A

▶ $C_{16}H_{16}O_5$ ▶ 7% $CDCl_3$ ▶ V 656

▶ 14-BH

(c) = all ring H's

	Shift	Peak
a	1.22	3A
b	2.63	4A
▶ c	7.13	1A

▶ C_8H_{10} ▶ CCl_4 ▶ API 173 B

▶ 14-BH

	Shift	Peak
a	2.74	Q
b	2.97	Q
▶ c	7.22	1A
d	9.81	3A

▶ $C_9H_{10}O$ ▶ 7% $CDCl_3$ ▶ V 529

▶ 14-BH

Shift	Peak
a 2.15	5B
b 2.75	3C
c 3.38	3A
▶d 7.22	1A

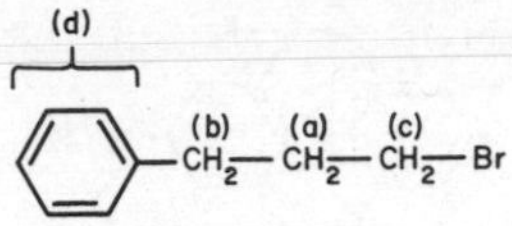

▶ $C_9H_{11}Br$ ▶ 7% $CDCl_3$ ▶ V 237

▶ 14-(B)H

Shift	Peak
a 3.50	1A
▶b 7.25	1A

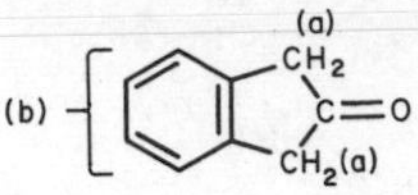

▶ C_9H_8O ▶ 7% $CDCl_3$ ▶ V 523

▶ 14-BH

Shift	Peak
a 1.25	3A
b 2.68	4A
▶c 7.23	1A

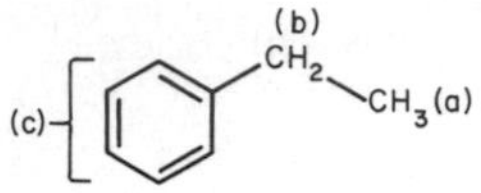

▶ C_8H_{10} ▶ 7% $CDCl_3$ ▶ V 505

▶ 14-BH

Shift	Peak
a 1.90	1A
b 2.80	3B
c 3.48	4A
d 6.50	1D
▶e 7.25	1G

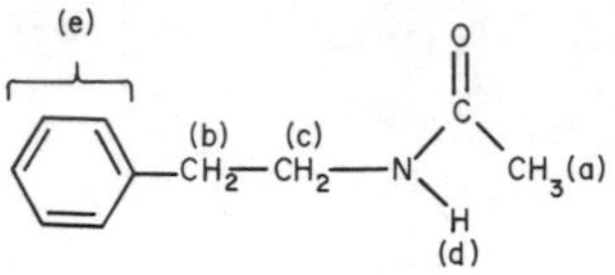

▶ $C_{10}H_{13}NO$ ▶ 7% $CDCl_3$ ▶ V 265

▶ 14-BH

Shift	Peak
a 2.51	Q
b 3.02	Q
c 3.48	1A
▶d 7.23	1A

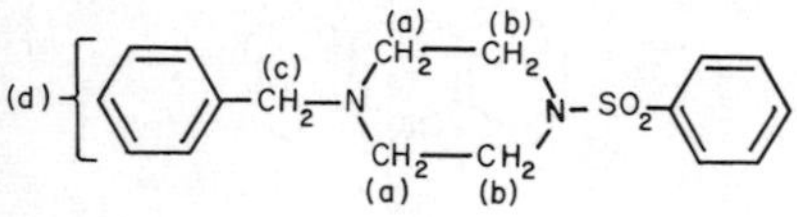

▶ $C_{17}H_{20}N_2O_2S$ ▶ 7% $CDCl_3$ ▶ V 330

▶ 14-BH

Shift	Peak
a 3.58	1A
b 3.78	1A
▶c 7.25	1A

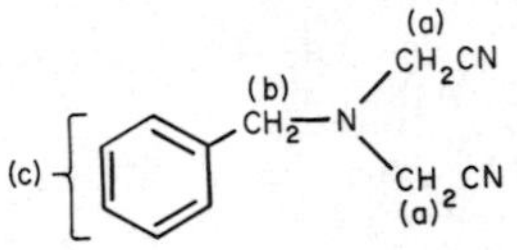

▶ $C_{11}H_{11}N_3$ ▶ 7% $CDCl_3$ ▶ V 285

▶ 14-BH

Shift	Peak
a 2.35	1A
b 4.56	1A
▶c 7.24	1E

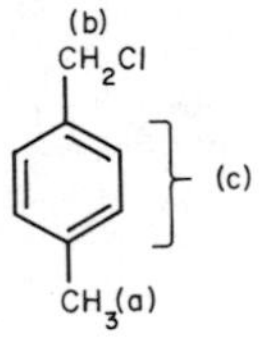

▶ C_8H_9Cl ▶ 45.2 mg/0.50 ml $CDCl_3$ ▶ S 68

▶ 14-BH

Shift	Peak
a 3.10	1A
b 3.62	1A
▶c 7.25 ± .03	1A

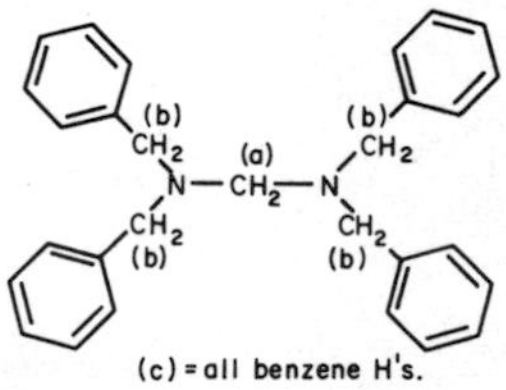

(c) = all benzene H's.

▶ $C_{29}H_{30}N_2$ ▶ 7% $CDCl_3$ ▶ V 365

▶ 14-BH

Shift	Peak
a 2.08	1A
b 2.82	Q
c 3.78	Q
▶d 7.24	1A

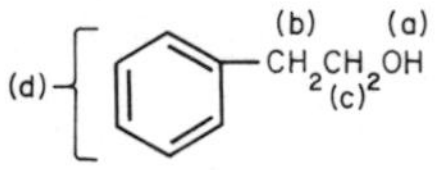

▶ $C_8H_{10}O$ ▶ 45.4 mg/0.50 ml $CDCl_3$ ▶ S 113

▶ 14-BH

Shift	Peak
a 1.57	1A
b 3.07	1A
▶c 7.27	1A

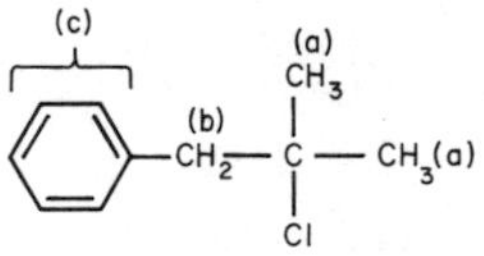

▶ $C_{10}H_{13}Cl$ ▶ 7% $CDCl_3$ ▶ V 263

▶ 14-BH

Shift	Peak
a 1.34	1B
b 2.76	Q
c 2.97	Q
▶d 7.24	1A

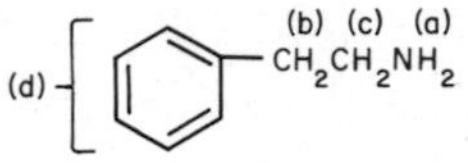

▶ $C_8H_{11}N$ ▶ 45.0 mg/0.50 ml $CDCl_3$ ▶ S 35

▶ 14-BH

Shift	Peak
a 4.18	1A
▶b 7.27	1E
c 7.54	S
d 8.56	3C

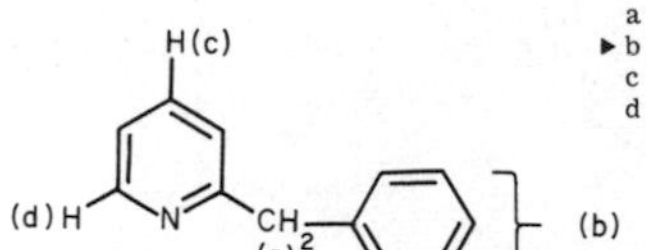

▶ $C_{12}H_{11}N$ ▶ 7% $CDCl_3$ ▶ V 593

▶ 14-BH

	Shift	Peak
a	2.43	1A
b	4.58	1A
▶ c	7.28	1A

(c) (b) (a) $C_6H_5CH_2OH$

▶ C_7H_8O ▶ 7% $CDCl_3$ ▶ V 161

▶ 14-BH

	Shift	Peak
a	3.65	1A
b	4.17	1A
▶ c	7.31 ± .05	1H
d	8.32	1C

(a) CH_2CN; (c) ring; N—H (d); C(=O); (b) CH_2Cl

▶ $C_{10}H_9ClN_2O$ ▶ 7% $CDCl_3$ ▶ V 248

▶ 14-BH

	Shift	Peak
a	1.62	3A
b	3.70	2A
▶ c	7.28	1A

(c) (b) (a) C_6H_5—CH_2—SH

▶ C_7H_8S ▶ 43.6 mg/0.50 ml $CDCl_3$ ▶ S 276

▶ 14-BH

	Shift	Peak
a	4.55	1A
▶ b	7.32	1A

(a) CH_2—Cl; (b) ring

▶ C_7H_7Cl ▶ 45.2 mg/0.50 ml $CDCl_3$ ▶ S 3

▶ 14-BH

	Shift	Peak
a	3.60	1A
b	3.66	1A
▶ c	7.28	1A

(c) ring—(a) CH_2—C(=O)—(b) OCH_3

▶ $C_9H_{10}O_2$ ▶ 44.4 mg/0.50 ml $CDCl_3$ ▶ S 19

▶ 14-BH

	Shift	Peak
a	3.10	1A
b	3.87	1A
▶ c	7.32	1A
d	10.97	1B

(c) ring—(b) CH_2—S—(a) CH_2—C(=O)—OH(d)

▶ $C_9H_{10}O_2S$ ▶ 7% $CDCl_3$ ▶ V 532

▶ 14-BH

	Shift	Peak
a	2.02	1A
b	2.93	3A
c	4.30	3A
▶ d	7.29 ± 0.03	1A

(d) ring—CH_2 (b)—CH_2 (c)—O—C(=O)—CH_3 (a)

▶ $C_{10}H_{12}O_2$ ▶ 7% $CDCl_3$ ▶ V 261

▶ 14-BH

	Shift	Peak
a	3.75	1A
b	5.79	1A
▶ c	7.33	1A

DS, N, N, OD, H (b), (a) CH_2, ring (c)

▶ $C_{11}H_{10}N_2OS$ ▶ D_2O ▶ V 577

▶ 14-BH

	Shift	Peak
a	4.10	1A
▶ b	7.30	1A

(b) ring—(a) CH_2—C(=O)—Cl

▶ C_8H_7ClO ▶ 44.8 mg/0.50 ml $CDCl_3$ ▶ S 218

▶ 14-BH

	Shift	Peak
a	4.52	1A
▶ b	7.33	1A

(b) ring—(a) CH_2—O—(a) CH_2—ring (b)

▶ $C_{14}H_{14}O$ ▶ 45.4 mg/0.50 ml $CDCl_3$ ▶ S 118

▶ 14-BH

	Shift	Peak
a	2.06	1A
b	5.08	1A
▶ c	7.31	1A

(c) ring—CH_2 (b)—O—C(=O)—CH_3(a)

▶ $C_9H_{10}O_2$ ▶ 7% $CDCl_3$ ▶ V 530

▶ 14-BH

	Shift	Peak
a	3.68	1A
▶ b	7.34	1A

(b) ring—(a) CH_2—C≡N

▶ C_8H_7N ▶ 44.5 mg/0.50 ml $CDCl_3$ ▶ S 135

▶ 14-BH

	Shift	Peak
a	2.86	S
b	3.07	S
c	3.59	3C
▶ d	7.35	1A

▶ $C_9H_{11}NO_2$ ▶ D_2O ▶ V 534

▶ 14-(B)O

	Shift	Peak
a	1.53	1A
b	2.83	1A
c	3.67	1A
▶ d	6.45	1A

▶ $C_{20}H_{24}O_2$ ▶ 7% $CDCl_3$ ▶ V 681

▶ 14-BH

	Shift	Peak
a	4.50	1A
▶ b	7.37	1A

▶ C_7H_7Br ▶ 45.2 mg/0.50 ml $CDCl_3$ ▶ S 154

▶ 14-(B)O

	Shift	Peak
a	2.07	1G
b	3.63	2C
c	4.12	2C
d	4.84	1F
e	6.02	2E
▶ f	6.49	1B
g	6.68	2A

▶ $C_{22}H_{25}NO_6$ ▶ 7% $CDCl_3$ ▶ V 690

▶ 14-BH

	Shift	Peak
a	4.46	1A
▶ b	7.50	1A

▶ $C_8H_{11}ClN_2S$ ▶ D_2O ▶ V 507

▶ 14-(B)(O)

	Shift	Peak
a	2.57	1E
b	3.93	1A
c	5.95	2C
d	6.10	2C
▶ e	6.55	1A
f	6.83	1A
g	7.72	1A

▶ $C_{20}H_{21}NO_4$ ▶ 7% $CDCl_3$ ▶ V 342

▶ 14-BH

	Shift	Peak
a	3.92	1A
▶ b	7.57	Q
c	8.25	Q

▶ $C_8H_6N_2O_2$ ▶ 7% $CDCl_3$ ▶ V 495

▶ 14-(B)O

	Shift	Peak
a	1.96	1A
b	2.43	S
c	2.43	S
d	3.67	1A
e	4.62	S
▶ f	6.55	1A
g	6.92	2A
h	7.37	S
i	7.63	1B
j	8.38	2G

▶ $C_{22}H_{25}NO_6$ ▶ 7% $CDCl_3$ ▶ V 689

▶ 14-B(L)

	Shift	Peak
a	2.25 or 2.37	1A
b	2.25 or 2.37	1A
c	2.40	2D
d	3.33	2B
e	5.07	R
f	5.12	R
g	5.85	4D
h	6.22	1I
▶ i	7.30	1I

▶ $C_{16}H_{16}O_4$ ▶ 7% $CDCl_3$ ▶ V 323

▶ 14-(B)(O)

	Shift	Peak
a	1.27	3A
b	3.14	2A
c	4.20	4A
d	4.96	4C
e	5.86	1A
▶ f	6.62 ± 0.04	S

▶ $C_{19}H_{19}NO_5$ ▶ 7% $CDCl_3$ ▶ V 334

▶ 14-(B)(O)

	Shift	Peak
a	2.53	1E
b	3.88*	1A
c	3.88*	1A
d	3.97	S
e	5.47	2A
f	5.88	1A
▶ g	6.38†	1E
h	6.38†	1E
i	6.52	2H
j	7.07	2C

*or 4.05
†or 6.57

▶ $C_{21}H_{21}NO_6$ ▶ 7% $CDCl_3$ ▶ V 347

▶ 14-(B)(O)

	Shift	Peak
a	1.92	1A
b	2.55	1F
c	2.88	1F
d	3.58	1A
e	3.78	1A
f	5.92, 5.96	1A
▶ g	6.65	1A
h	6.69	1A
i	6.91	1A

▶ $C_{20}H_{19}NO_5$ ▶ 7% $CDCl_3$ ▶ V 339

▶ 14-B(O)

H(d) CH2—CH2—N—CH3(a) CH3(a) (c)H2C O O (f)H (b)CH3O H(e) (b)CH3O

	Shift	Peak
a	2.35	1A
b	3.93	1A
c	6.03	1A
▶ d	6.65	1A
e	6.80	1A
f	7.77	1I

▶ $C_{21}H_{25}NO_4$ ▶ 7% $CDCl_3$ ▶ V 349

▶ 14-BO

H(f) (c) (b) (a) CH2—CH2—N CH3 CH3 (a) (e) H2C O O H(i) (j)H H(h) (d) CH3O H(g) CH3O(d)

	Shift	Peak
a	2.38	1A
b	~2.7	Q
c	~3.2	Q
d	4.05 and 4.08	1A
e	6.23	1A
▶ f	7.13 or 7.23	1A
g	7.13 or 7.23	1A
h	7.53 or 7.81	2C
i	7.53 or 7.81	2C
j	8.52	1A

▶ $C_{21}H_{23}NO_4$ ▶ 7% $CDCl_3$ ▶ V 348

▶ 14-B(O)

(e) O—CH2 O (g)H (h)H H(f) (a)CH2 H(b) C=C (d)H H(c)

	Shift	Peak
a	3.30	2E
b	5.03	2E
c	5.03	2E
d	~5.88	O
e	5.88	1A
▶ f	6.67 ± .03	1H
g	6.67 ± .03	1H
h	6.67 ± .03	1H

▶ $C_{10}H_{10}O_2$ ▶ 7% $CDCl_3$ ▶ V 253

▶ 14-(B)P

(a) H3C OH C≡CH(b) H(e) (d)H HO H(c)

	Shift	Peak
a	0.88	1A
b	2.62	1A
▶ c	6.60	1E
d	6.67	2H
e	7.18	2E

▶ $C_{20}H_{24}O_2$ ▶ 7% $CDCl_3$ ▶ V 343

▶ 14-(B)(O)

H(e) (c)H O C (d)H O N CH3 (a) (g)HO (b)CH3O H(f) H(f)

	Shift	Peak
a	2.57	1E
b	3.92	1A
c	5.98	2C
d	6.13	2C
▶ e	6.68	1A
f	6.88	1E
g	6.88	1E

▶ $C_{19}H_{19}NO_4$ ▶ 7% $CDCl_3$ ▶ V 333

▶ 14-BQ

O (a)CH3 (d) (c)CH3 H O C CH3 O (b) H(d)

	Shift	Peak
a	0.95	1A
b	2.28	1A
c	2.33	1A
▶ d	6.72	1B

▶ $C_{21}H_{26}O_3$ ▶ 7% $CDCl_3$ ▶ V 350

▶ 14-BO

(f) (c) (g) H H H (a) C=C CH2 (i)H H(d) (e)HO H(h) OCH3 (b)

	Shift	Peak
a	3.30	2E
b	3.83	1A
c	5.05	2E
d	5.05	2E
e	5.59	1B
f	~5.94	O
g	~6.64	S
▶ h	~6.69	S
i	6.87	S

▶ $C_{10}H_{12}O_2$ ▶ 7% $CDCl_3$ ▶ V 260

▶ 14-CH

(f)H C=O H(g) O (e)H (d)H H(c) H(b) C (a)CH3 CH3(a)

	Shift	Peak
a	1.23	2A
b	2.92	5A
c	6.82	N
▶ d	6.88	N
e	7.42	2A
f	9.79	1A
g	11.00	1B

▶ $C_{10}H_{12}O_2$ ▶ 7% $CDCl_3$ ▶ V 259

▶ 14-(B)O

H(d) CH2—CH2—N—CH3(a) CH3(a) (c)H2C O O (f)H (b)CH3O H(e) (b)CH3O

	Shift	Peak
a	2.35	1A
b	3.93	1A
c	6.03	1A
d	6.65	1A
▶ e	6.80	1A
f	7.77	1I

▶ $C_{21}H_{25}NO_4$ ▶ 7% $CDCl_3$ ▶ V 349

▶ 14-CH

CH3(b) C CH3 HC C—CH(c) HC CH3(a) CH C H

(d) = all ring H's

	Shift	Peak
a	1.15	2A
b	2.20	1A
c	3.01	5A
▶ d	6.95	O

▶ $C_{10}H_{14}$ ▶ run neat ▶ API 181 A

▶ 14-(B)O

H(e) (c)H O C (d)H O N CH3(a) (g)H (b)CH3O H(f) (b)OCH3

	Shift	Peak
a	2.57	1E
b	3.93	1A
c	5.95	2C
d	6.10	2C
e	6.55	1A
▶ f	6.83	1A
g	7.72	1A

▶ $C_{20}H_{21}NO_4$ ▶ 7% $CDCl_3$ ▶ V 342

▶ 14-CH

(a) CH3 (c)H H(d) N C—H(b) (a) CH3 (c)H H(d) 3

	Shift	Peak
a	2.87	1A
b	5.29	1B
c	6.64	Q
▶ d	6.95	Q

▶ $C_{25}H_{31}N_3$ ▶ 7% $CDCl_3$ ▶ V 360

▶ 14-CH

	Shift	Peak
a	1.20	2A
b	2.24	1A
c	2.80	5A
▶ d	6.99	1A

▶ $C_{10}H_{14}$ ▶ 50% CCl_4 ▶ API 225

▶ 14-CH

	Shift	Peak
a	1.22	2A
b	2.30	1A
c	2.87	5B
▶ d	7.08	1A

▶ $C_{10}H_{14}$ ▶ 7% $CDCl_3$ ▶ V 268

▶ 14-CH

	Shift	Peak
a	1.17	2A
b	2.24	1A
c	3.04	5A
▶ d	7.00	N

(d) = all ring H's

▶ $C_{10}H_{14}$ ▶ 50% CCl_4 ▶ API 224

▶ 14-CH

	Shift	Peak
a	1.15	2A
b	2.75	5A
▶ c	7.09	1A

(c) = all ring H's

▶ C_9H_{12} ▶ run neat ▶ API 176

▶ 14-CH

	Shift	Peak
a	1.25	2A
b	2.30	1A
c	3.05	5A
▶ d	7.01	O

(d) = all ring H's

▶ $C_{10}H_{14}$ ▶ CCl_4 ▶ API 181 B

▶ 14-CH

	Shift	Peak
a	.82	3C
b	1.23	2A
c	1.60	5B
d	2.56	4C
▶ e	7.11	1A
f	7.11	1A

▶ $C_{10}H_{14}$ ▶ 10% CCl_4 ▶ API 90

▶ 14-(C)H

	Shift	Peak
a	0.68 ± .05	S
b	0.74 ± .05	S
c	1.80	O
▶ d	7.05	2A

▶ C_9H_{10} ▶ 7% $CDCl_3$ ▶ V 528

▶ 14-(C)H

	Shift	Peak
a	0.88	1A
b	2.62	1A
c	6.60	1E
d	6.67	2H
▶ e	7.18	2E

▶ $C_{20}H_{24}O_2$ ▶ 7% $CDCl_3$ ▶ V 343

▶ 14-CH

	Shift	Peak
a	1.23	2A
b	2.25	1A
c	3.17	5B
d	4.75	1A
e	6.55	S
f	6.71	S
▶ g	7.06	S

▶ $C_{10}H_{14}O$ ▶ 7% $CDCl_3$ ▶ V 270

▶ 14-CH

	Shift	Peak
a	0.80	3A
b	1.22	2A
c	1.59	5B
d	2.53	6A
▶ e	7.20	1A

▶ $C_{10}H_{14}$ ▶ 45.0 mg/0.50 ml $CDCl_3$ ▶ S 31

▶ 14-(C)H

	Shift	Peak
a	2.53	1E
b	3.88*	1A
c	3.88*	1A
d	3.97	S
e	5.47	2A
f	5.88	1A
g	6.38†	1E
h	6.38†	1E
i	6.52	2H
▶ j	7.07	2C

*or 4.05
†or 6.57

▶ $C_{21}H_{21}NO_6$ ▶ 7% $CDCl_3$ ▶ V 347

▶ 14-(C)H

	Shift	Peak
a	4.37	1A
▶ b	7.20	1A

▶ $C_{14}H_{12}O$ ▶ 7% $CDCl_3$ ▶ V 626

▶ 14-CH

(g) (f) H Cl C=C Cl (d)H C B O (e) CH_2 (c) CH_2 (b) CH_2 (a) CH_3 O CH_2 (e) (c) CH_2 CH_2 (b) CH_3 (a)

	Shift	Peak
a	0.88	3C
b	~1.36	S
c	~1.50	S
d	3.58	2H
e	3.84	3B
f	6.24	2A
▶ g	7.20	1A

▶ $C_{17}H_{25}BCl_2O_2$ ▶ 7% $CDCl_3$ ▶ V 667

▶ 14-CH

(h) (f) H CCl_3 C H(d) C B O (g) CH_2 (c) CH_2 (b) CH_2 (a) CH_3 H (e) O (g) CH_2 (c) CH_2 (b) CH_2 CH_3 (a)

	Shift	Peak
a	0.88	3C
b	~1.36	S
c	~1.50	S
d	2.86*	S
e	2.89*	S
f	3.65	S
g	3.88	3C
▶ h	7.26	1A

*±0.04

▶ $C_{17}H_{26}BCl_3O_2$ ▶ 7% $CDCl_3$ ▶ V 668

▶ 14-CH

F (d)H H(d) (e)H H(e) CH(c) CH_3 OH (a) (b)

	Shift	Peak
a	1.38	2A
b	3.20	1B
c	4.75	4A
d	6.98	O
▶ e	7.22	O

▶ C_8H_9FO ▶ 7% $CDCl_3$ ▶ V 200

▶ 14-(C)H

(b)H H(c) (a)H O (d)

	Shift	Peak
a	2.77	4A
b	3.12	4A
c	3.83	4A
▶ d	7.28	1A

▶ C_8H_8O ▶ 7% $CDCl_3$ ▶ V 193

▶ 14-(C)H

(c) (b) (c) (a) (c) (c) (b) (a) (c) (a)

	Shift	Peak
a	~1.4	S
b	~1.8	S
▶ c	7.22	1A

▶ $C_{12}H_{16}$ ▶ 45.4 mg/0.50 ml $CDCl_3$ ▶ S 291

▶ 14-CH

OH(b) (a) (d) CH CH_3 (c)

	Shift	Peak
a	1.42	2A
b	2.38	1A
c	4.78	4A
▶ d	7.30	1A

▶ $C_8H_{10}O$ ▶ 45.2 mg/0.50 ml $CDCl_3$ ▶ S 133

▶ 14-CH

(e)H (g) H (d) H O (b)CH_3O C C H(h) (c) HO H H H H(f) (e) (g) (h) H OCH_3(a) (f)

	Shift	Peak
a	3.75*	1A
b	3.75*	1A
c	4.54	1C
d	5.82	1B
e	6.83	S
f	6.87	S
▶ g	7.24	2E
h	7.89	2E

*or 3.81

▶ $C_{16}H_{16}O_4$ ▶ 7% $CDCl_3$ ▶ V 655

▶ 14-CH

(b) NH_2 (c)H — C — CH_3(a) (d)

	Shift	Peak
a	1.38	2A
b	1.58	1A
c	4.10	4A
▶ d	7.30 ± .03	1A

▶ $C_8H_{11}N$ ▶ 7% $CDCl_3$ ▶ V 207

▶ 14-CH

(a) (a) CH_3 CH_3 CH(b) (c)

	Shift	Peak
a	1.25	2A
b	2.90	7A
▶ c	7.25	1A

▶ C_9H_{12} ▶ 7% $CDCl_3$ ▶ V 240

▶ 14-CH

OH(a) (c) C — H(b) (c)

	Shift	Peak
a	2.29	1B
b	5.80	1G
▶ c	7.32	1E

▶ $C_{13}H_{12}O$ ▶ 7% $CDCl_3$ ▶ V 607

▶ 14-CH

O (b)H — C — C — CH_3(a)

	Shift	Peak
a	2.20	1A
b	5.08	1A
▶ c	7.25 ± .03	1A

(c) = all benzene H's.

▶ $C_{15}H_{14}O$ ▶ 7% $CDCl_3$ ▶ V 318

▶ 14-CH

Cl C — H(a)

	Shift	Peak
a	6.12	1A
▶ b	7.33 ± .05	1I

(b) = all benzene H's.

▶ $C_{13}H_{11}Cl$ ▶ 7% $CDCl_3$ ▶ V 176

▶ 14-(C)H

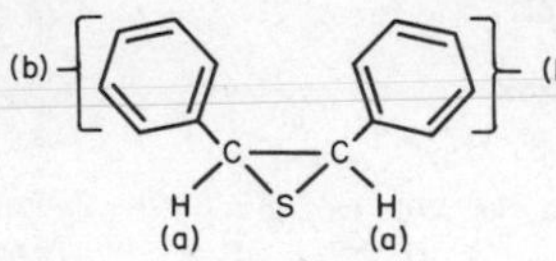

	Shift	Peak
a	3.96	1A
▶ b	7.37	1A

▶ $C_{14}H_{12}S$ ▶ 7% $CDCl_3$ ▶ V 629

▶ 14-(C)(O)

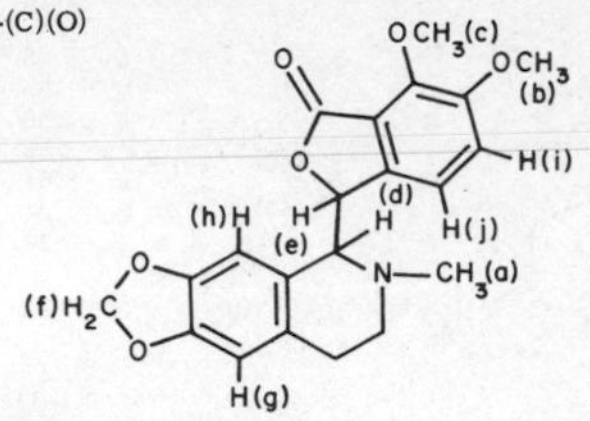

	Shift	Peak
a	2.53	1E
b	3.88*	1A
c	3.88*	1A
d	3.97	S
e	5.47	2A
f	5.88	1A
g	6.38†	1E
▶ h	6.38†	1E
i	6.52	2H
j	7.07	2C

*or 4.05
†or 6.57

▶ $C_{21}H_{21}NO_6$ ▶ 7% $CDCl_3$ ▶ V 347

▶ 14-CH

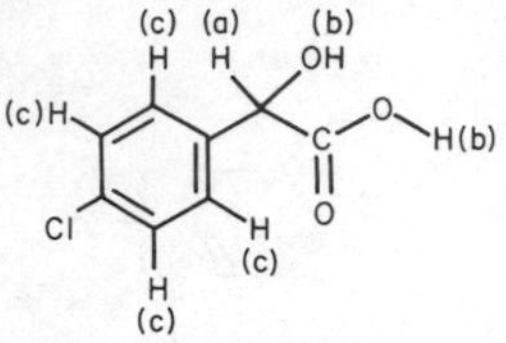

	Shift	Peak
a	5.23	1A
b	6.63	1B
▶ c	7.38	1A

▶ $C_8H_7ClO_3$ ▶ 7% $CDCl_3$ ▶ V 499

▶ 14-CO

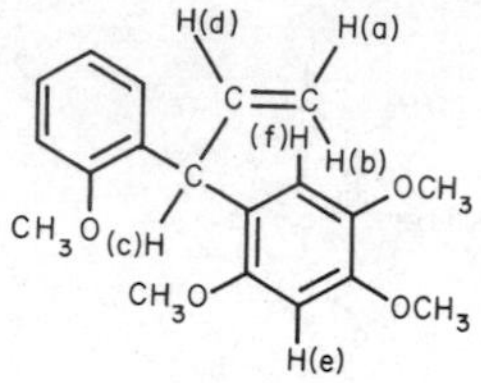

	Shift	Peak
a	4.77	R
b	5.15	R
c	5.47	2E
d	6.27	R
e	6.53	2A
▶ f	6.68	O

▶ $C_{19}H_{22}O_4$ ▶ 7% $CDCl_3$ ▶ V 678

▶ 14-CH

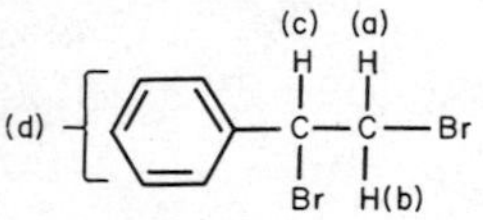

	Shift	Peak
a	4.02	S
b	4.07	S
c	5.15	4A
▶ d	7.38	1A

▶ $C_8H_8Br_2$ ▶ 7% $CDCl_3$ ▶ V 503

▶ 14-CP

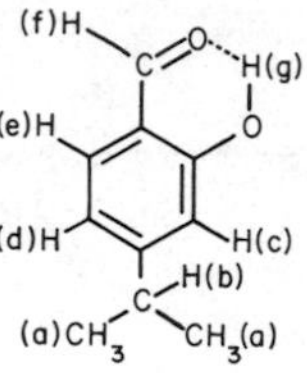

	Shift	Peak
a	1.23	2A
b	2.92	5A
▶ c	6.82	N
d	6.88	N
e	7.42	2A
f	9.79	1A
g	11.00	1B

▶ $C_{10}H_{12}O_2$ ▶ 7% $CDCl_3$ ▶ V 259

▶ 14-(C)H

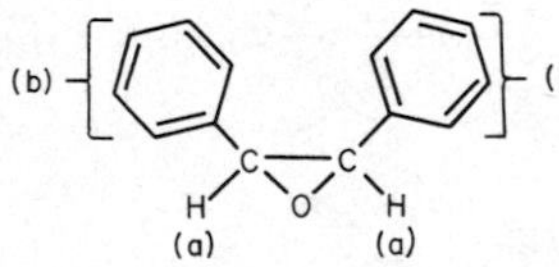

	Shift	Peak
a	3.88	1A
▶ b	7.38	1A

▶ $C_{14}H_{12}O$ ▶ 7% $CDCl_3$ ▶ V 625

▶ 14-DG

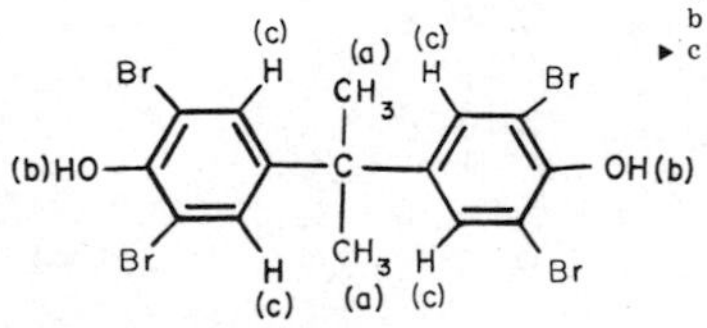

	Shift	Peak
a	1.58	1A
b	5.66	1D
▶ c	7.27	1A

▶ $C_{15}H_{12}Br_4O_2$ ▶ 7% $CDCl_3$ ▶ V 636

▶ 14-(C)H

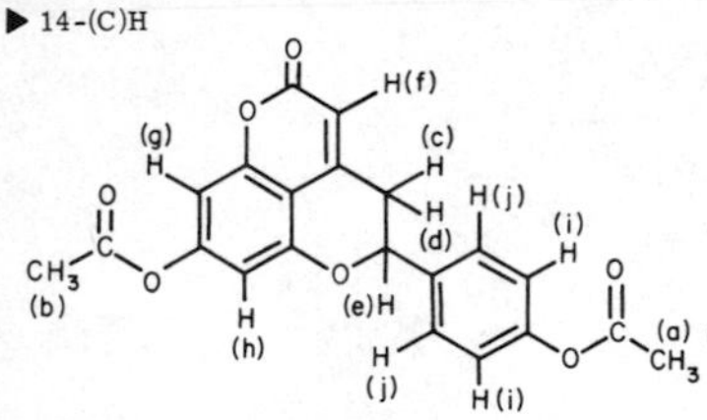

	Shift	Peak
a	2.32	1A
b	2.32	1A
c	3.09	S
d	3.09	A
e	5.18	3B
f	6.06	M
g	6.62*	S
h	6.62*	S
i	7.17	Q
▶ j	7.46	Q

*or 6.71

▶ $C_{21}H_{16}O_7$ ▶ 7% $CDCl_3$ ▶ V 686

▶ 14-DH

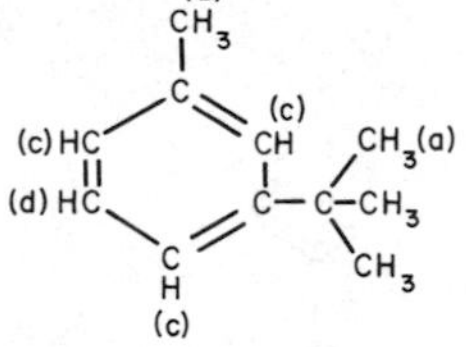

	Shift	Peak
a	1.25	1A
b	2.21	1A
▶ c	~7.03	S
d	7.11	S

▶ $C_{11}H_{16}$ ▶ CCl_4 ▶ API 185 A

▶ 14-CH

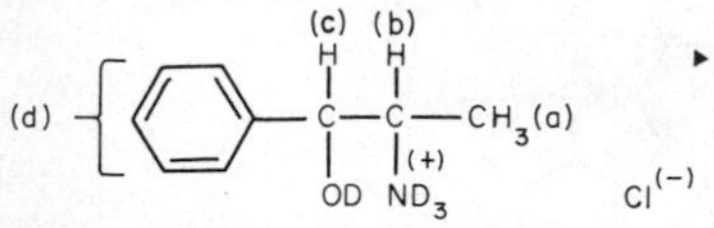

	Shift	Peak
a	1.21	2A
b	3.72	O
c	5.00	2A
▶ d	7.48	1A

▶ $C_9H_{14}ClNO$ ▶ D_2O ▶ V 565

▶ 14-DH

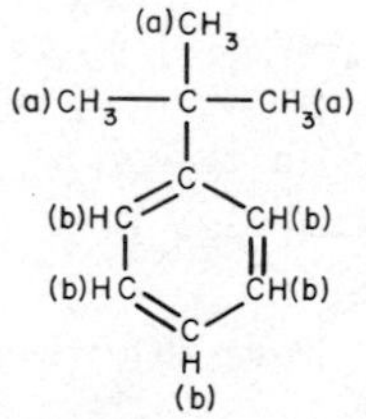

	Shift	Peak
a	1.23	1A
▶ b	7.15	O

▶ $C_{10}H_{14}$ ▶ liquid run ▶ API 20

▶ 14-(D)H

	Shift	Peak
a	0.65 ± .03	N
b	0.81 ± .03	1I
c	1.37	1A
▶ d	7.17	1A

▶ $C_{10}H_{12}$ ▶ 7% $CDCl_3$ ▶ V 556

▶ 14-(D)H

	Shift	Peak
a	1.56	1A
b	2.87	1B
c	4.10	1A
▶ d	7.51 ± 0.05	1G

▶ $C_{12}H_{16}ClN$ ▶ 7% $CDCl_3$ ▶ V 599

▶ 14-DH

	Shift	Peak
a	1.29	1A
▶ b	7.19	1A

▶ $C_{14}H_{22}$ ▶ CCl_4 ▶ API 186

▶ 14-DH

	Shift	Peak
a	7.44	3A
▶ b	7.93	O

▶ $C_7H_5Cl_3$ ▶ 45.1 mg/0.50 ml $CDCl_3$ ▶ S 90

▶ 14-DH

	Shift	Peak
a	0.72	1A
b	1.33	1A
c	1.68	1A
d	5.02	1C
e	6.75	Q
▶ f	7.20	Q

▶ $C_{14}H_{22}O$ ▶ 7% $CDCl_3$ ▶ V 315

▶ 14-(D)O

	Shift	Peak
a	1.42	1A
b	3.04	1A
c	3.91	1A
d	6.88	S
▶ e	6.93	S
f	8.17	2A
g	~10.30	1D

▶ $C_{13}H_{15}NO_3$ ▶ 7% $CDCl_3$ ▶ V 611

▶ 14-DH

	Shift	Peak
a	1.32	1A
b	2.33	1A
c	7.12	Q
▶ d	7.28	Q

▶ $C_{11}H_{16}$ ▶ 7% $CDCl_3$ ▶ V 586

▶ 14-(D)Q

	Shift	Peak
a	1.42	1A
b	1.95	3A
c	2.55	1A
d	2.70	3D
e	2.82	1E
f	2.92	1E
g	4.12	1A
h	5.33	1D
i	6.37	2A
▶ j	6.78	1E
k	6.87	2H

▶ $C_{15}H_{21}N_3O_2$ ▶ 7% $CDCl_3$ ▶ V 319

▶ 14-DH

	Shift	Peak
a	0.92	3A
b	1.80	S
c	2.05	S
d	2.70	2C
e	2.97	2C
▶ f	7.30	1A

▶ $C_{10}H_{12}O$ ▶ 7% $CDCl_3$ ▶ V 558

▶ 14-EE

	Shift	Peak
a	4.47	1A
b	5.75	1B
▶ c	7.39	1A

▶ $C_{13}H_6Cl_6O_2$ ▶ 7% $CDCl_3$ ▶ V 604

▶ 14-DH

	Shift	Peak
a	2.38	1A
b	7.25	Q
▶ c	7.48	Q

▶ $C_8H_7F_3$ ▶ 7% $CDCl_3$ ▶ V 190

▶ 14-EE

	Shift	Peak
a	7.45	2D

▶ $C_6H_3Cl_3$ ▶ 45.7 mg/0.50 ml $CDCl_3$ ▶ S 221

▶ 14-EH

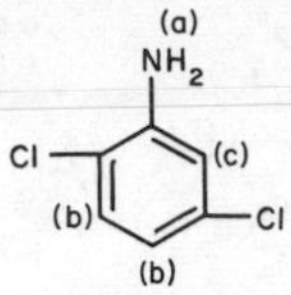

	Shift	Peak
a	3.96	1C
▶ b	6.70	1G
c	7.05	2G

▶ $C_6H_5Cl_2N$ ▶ 45.2 mg/0.50 ml $CDCl_3$ ▶ S 58

▶ 14-EH

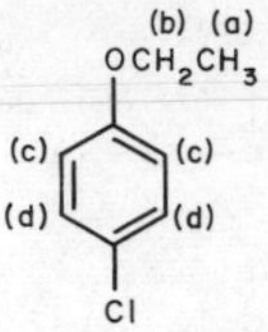

	Shift	Peak
a	0.90	3A
b	3.99	4A
c	6.78	Q
▶ d	7.21	Q

▶ C_8H_9ClO ▶ 45.4 mg/0.50 ml $CDCl_3$ ▶ S 235

▶ 14-EH

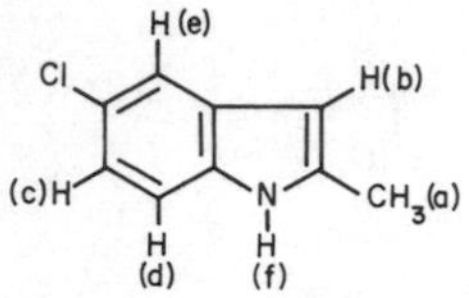

	Shift	Peak
a	2.33	2D
b	6.10	1B
▶ c	7.02	2D
d	7.02	2D
e	7.44	1I
f	~7.65	1D

▶ C_9H_8ClN ▶ 7% $CDCl_3$ ▶ V 228

▶ 14-EH

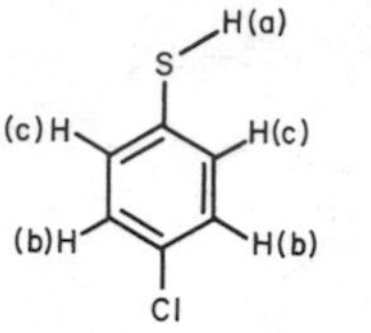

	Shift	Peak
a	3.45	1A
▶ b	7.23	1A
c	7.23	1A

▶ C_6H_5ClS ▶ 7% $CDCl_3$ ▶ V 451

▶ 14-EH

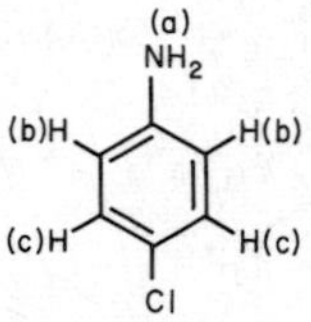

	Shift	Peak
a	3.60	1A
b	6.57	Q
▶ c	7.05	Q

▶ C_6H_6ClN ▶ 7% $CDCl_3$ ▶ V 123

▶ 14-EH

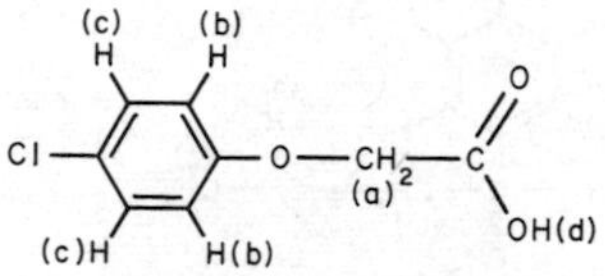

	Shift	Peak
a	4.67	1A
b	6.88	Q
▶ c	7.27	Q
d	8.86	1C

▶ $C_8H_7ClO_3$ ▶ 7% $CDCl_3$ ▶ V 500

▶ 14-EH

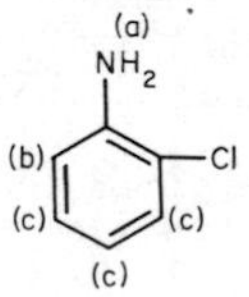

	Shift	Peak
a	3.9	1C
b	6.68	2H
▶ c	7.10	O

▶ C_6H_6ClN ▶ 44.8 mg/0.50 ml $CDCl_3$ ▶ S 59

▶ 14-EH

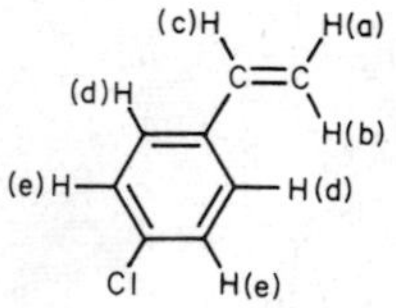

	Shift	Peak
a	5.28	R
b	5.73	R
c	6.69	4D
d	7.32	1A
▶ e	7.32	1A

▶ C_8H_7Cl ▶ 7% $CDCl_3$ ▶ V 498

▶ 14-EH

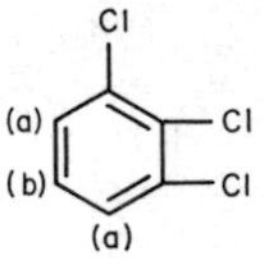

	Shift	Peak
▶ a	7.17	S
b	7.40	S

▶ $C_6H_3Cl_3$ ▶ 44.6 mg/0.50 ml $CDCl_3$ ▶ S 266

▶ 14-EH

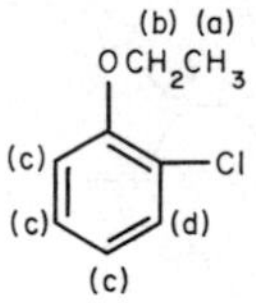

	Shift	Peak
a	0.96	3A
b	4.08	4A
c	6.95	O
▶ d	7.34	N

▶ C_8H_9ClO ▶ 45.2 mg/0.50 ml $CDCl_3$ ▶ S 234

▶ 14-EH

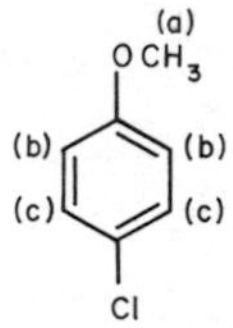

	Shift	Peak
a	3.77	1A
b	6.80	Q
▶ c	7.20	Q

▶ C_7H_7ClO ▶ 44.8 mg/0.50 ml $CDCl_3$ ▶ S 164

▶ 14-EH

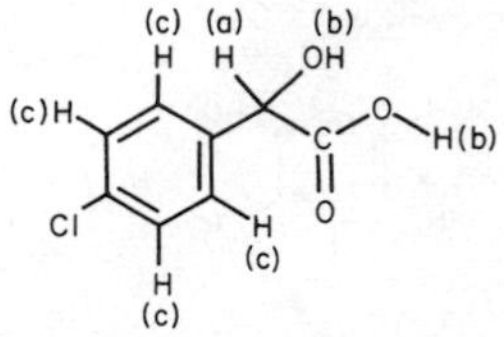

	Shift	Peak
a	5.23	1A
b	6.63	1B
▶ c	7.38	1A

▶ $C_8H_7ClO_3$ ▶ 7% $CDCl_3$ ▶ V 499

14-EH

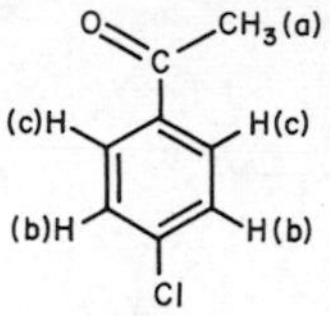

	Shift	Peak
a	2.58	1A
► b	7.45	Q
c	7.90	Q

C_8H_7ClO · 7% $CDCl_3$ · V 188

14-EM

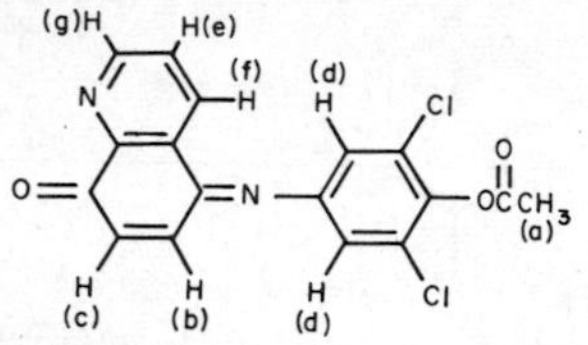

	Shift	Peak
a	2.45	1A
b	6.88*	2H
c	6.88*	2H
► d	6.95	1A
e	7.68	4A
f	8.70	4A
g	8.98	4A

*or 7.30

$C_{17}H_{10}Cl_2N_2O_3$ · 7% $CDCl_3$ · V 327

14-EH

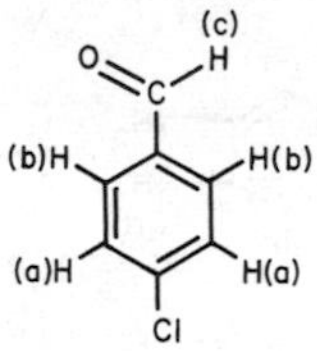

	Shift	Peak
► a	7.50	Q
b	7.75	Q
c	9.97	1A

C_7H_5ClO · 7% $CDCl_3$ · V 146

14-EN

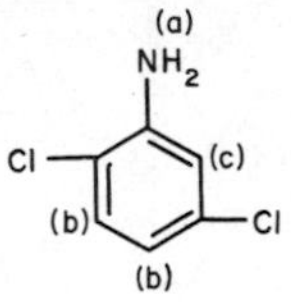

	Shift	Peak
a	3.96	1C
b	6.70	1G
► c	7.05	2G

$C_6H_5Cl_2N$ · 45.2 mg/0.50 ml $CDCl_3$ · S 58

14-EH

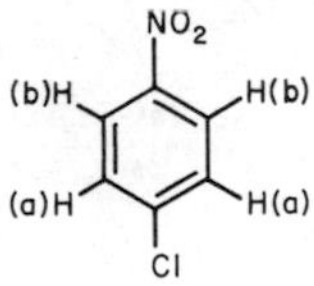

	Shift	Peak
► a	7.52	Q
b	8.17	Q

$C_6H_4ClNO_2$ · 7% $CDCl_3$ · V 122

14-FH

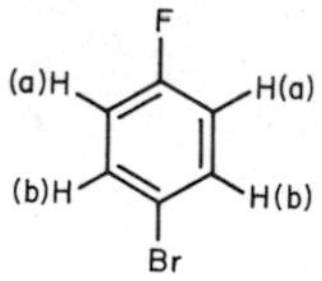

	Shift	Peak
► a	6.00	O
b	7.38	O

C_6H_4BrF · 7% $CDCl_3$ · V 121

14-EH

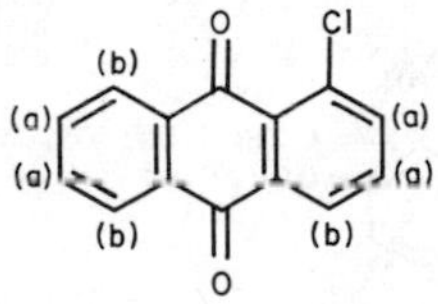

	Shift	Peak
► a	7.72	O
b	8.20	O

$C_{14}H_7ClO_2$ · 44.9 mg/0.50 ml $CDCl_3$ · S 182

14-FH

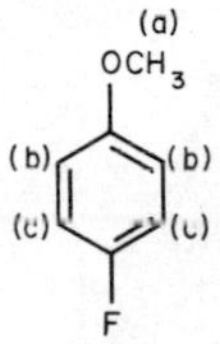

	Shift	Peak
a	3.77	1A
b	6.82	S
► c	6.95	S

C_7H_7FO · 45.2 mg/0.50 ml $CDCl_3$ · S 233

14-EH

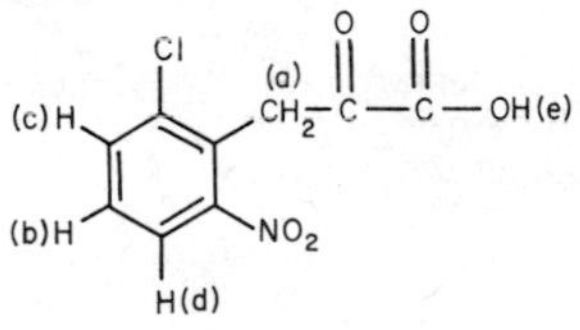

	Shift	Peak
a	4.78	1A
b	7.48	4A
► c	7.75	4A
d	8.00	S
e	9.63	1B

$C_9H_6ClNO_5$ · 7% $CDCl_3$ · V 223

14-FH

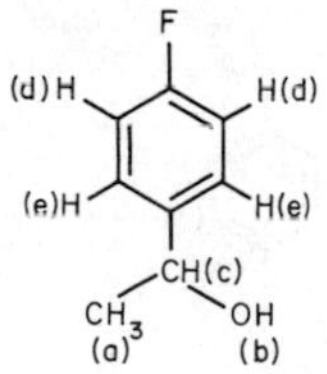

	Shift	Peak
a	1.38	2A
b	3.20	1B
c	4.75	4A
► d	6.98	O
e	7.22	O

C_8H_9FO · 7% $CDCl_3$ · V 200

14-E(L)

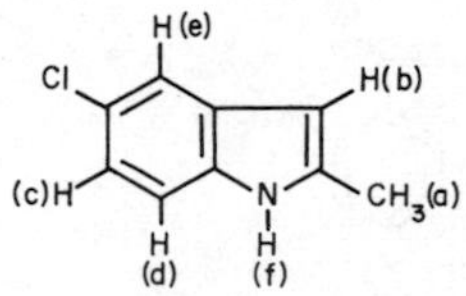

	Shift	Peak
a	2.33	2D
b	6.10	1B
c	7.02	2D
d	7.02	2D
► e	7.44	1I
f	~7.65	1D

C_9H_8ClN · 7% $CDCl_3$ · V 228

14-FH

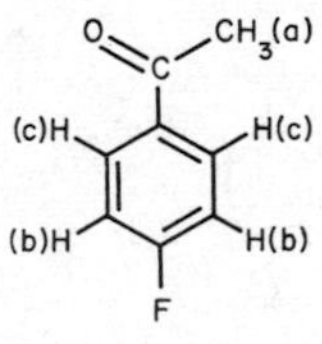

	Shift	Peak
a	2.58	1A
► b	7.12	3C
c	7.97	4C

C_8H_7FO · 7% $CDCl_3$ · V 189

▶ 14-FH

	Shift	Peak
▶ a	7.20	O
b	8.25	O

F; (a) (a); (b) (b); NO_2

▶ $C_6H_4FNO_2$ ▶ 45.5 mg/0.50 ml $CDCl_3$ ▶ S 224

▶ 14-GH

	Shift	Peak
a	2.40	1A
b	7.12	O
▶ c	7.50	2E

(a) CH_3; Br; (b); (c)

▶ C_7H_7Br ▶ 45.4 mg/0.50 ml $CDCl_3$ ▶ S 161

▶ 14-GH

	Shift	Peak
a	2.32	1A
b	7.08	S
▶ c	7.27	S

(a) CH_3; (b) (b); (c) Br; (c)

▶ C_7H_7Br ▶ 45.1 mg/0.50 ml $CDCl_3$ ▶ S 162

▶ 14-HH

	Shift	Peak
a	3.29	1A
▶ b	6.67	1A

(b) (a); (b) NH_2; (b) NH_2; (b) (a)

▶ $C_6H_8N_2$ ▶ 44.6 mg/0.50 ml $CDCl_3$ ▶ S 112

▶ 14-GH

	Shift	Peak
a	1.37	3A
b	3.93	4A
c	6.73	Q
▶ d	7.28	Q

H(c); (b) (a); (d)H; OCH_2CH_3; Br; H(c); H(d)

▶ C_8H_9BrO ▶ 7% $CDCl_3$ ▶ V 198

▶ 14-HH

	Shift	Peak
a	1.40	3A
b	3.72	1B
c	4.03	4A
▶ d	6.72	1A

(b) NH_2; (c) (a); OCH_2CH_3; (d)

▶ $C_8H_{11}NO$ ▶ 45.1 mg/0.50 ml $CDCl_3$ ▶ S 236

▶ 14-GH

	Shift	Peak
a	2.28	1A
b	7.05	Q
▶ c	7.33	Q

(a) CH_3; (b); (c); Br

▶ C_7H_7Br ▶ 44.8 mg/0.25 ml $CDCl_3$ ▶ S 163

▶ 14-HH

	Shift	Peak
a	3.70	1B
b	3.82	1A
▶ c	6.73	1A

(a) NH_2; (b) OCH_3; (c)

▶ C_7H_9NO ▶ 45.2 mg/0.50 ml $CDCl_3$ ▶ S 192

▶ 14-GH

	Shift	Peak
a	3.78	1A
b	6.75	Q
▶ c	7.37	Q

(a) OCH_3; (b) (b); (c) (c); Br

▶ C_7H_7BrO ▶ 45.5 mg/0.50 ml $CDCl_3$ ▶ S 141

▶ 14-HH

	Shift	Peak
a	3.12	1E
b	3.23	3A
c	3.78	3A
d	6.61	S
▶ e	6.78	S
▶ f	7.17	3B

(d); (a) (b) (c) (a); $NHCH_2CH_2OH$; (f); (e) (d); (f)

▶ $C_8H_{11}NO$ ▶ 45.4 mg/0.50 ml $CDCl_3$ ▶ S 295

▶ 14-GH

	Shift	Peak
a	6.90	O
▶ b	7.38	O

F; (a)H H(a); (b)H H(b); Br

▶ C_6H_4BrF ▶ 7% $CDCl_3$ ▶ V 121

▶ 14-HH

	Shift	Peak
a	3.73	1A
b	3.85	1A
▶ c	~6.80	1A

(b) OCH_3; NH_2(a); (c)

▶ C_7H_9NO ▶ 7% $CDCl_3$ ▶ V 172

▶ 14-HH

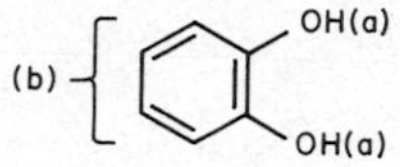

Shift	Peak
a 5.45	1A
▶ b 6.82	1A

▶ $C_6H_6O_2$ ▶ 7% $CDCl_3$ ▶ V 124

▶ 14-HH

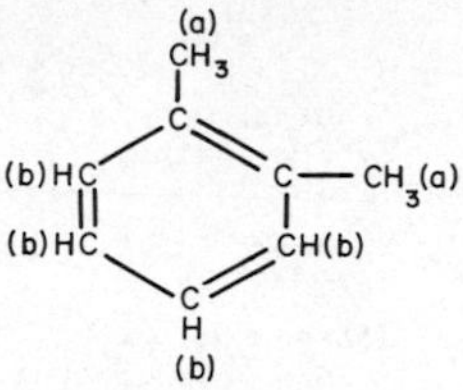

Shift	Peak
a 2.03	1A
▶ b 6.93	1A

▶ C_8H_{10} ▶ run neat ▶ API 174 A

▶ 14-HH

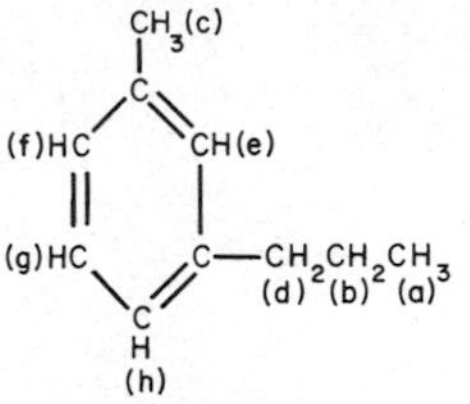

Shift	Peak
a .87	3C
b 1.56	6B
c 2.17	1A
d 2.45	3B
e 6.82	S
f 6.82	S
▶ g 6.82	S
h 6.82	S

▶ $C_{10}H_{14}$ ▶ liquid run, neat ▶ API 96

▶ 14-HH

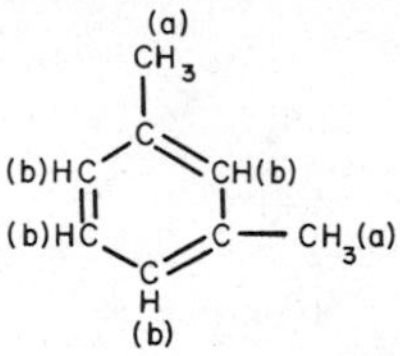

Shift	Peak
a 2.18	1A
▶ b ~6.93	N

▶ C_8H_{10} ▶ run neat ▶ API 175 A

▶ 14-HH

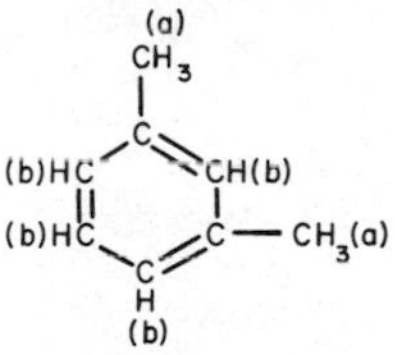

Shift	Peak
a 2.29	1A
▶ b 6.86	N

▶ C_8H_{10} ▶ CCl_4 ▶ API 175 B

▶ 14-HH

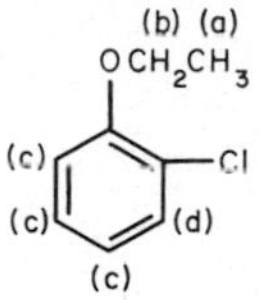

Shift	Peak
a 0.96	3A
b 4.08	4A
▶ c 6.95	O
d 7.34	N

▶ C_8H_9ClO ▶ 45.2 mg/0.50 ml $CDCl_3$ ▶ S 234

▶ 14-HH

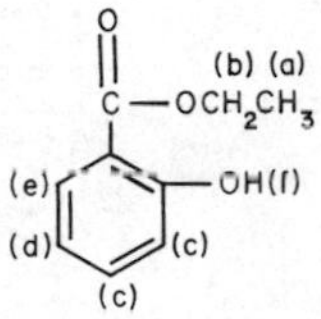

Shift	Peak
a 1.41	3A
b 4.40	4A
▶ c 6.92	O
▶ d 7.37	O
e 7.83	2E
f 10.78	1A

▶ $C_9H_{10}O_3$ ▶ 45.6 mg/0.50 ml $CDCl_3$ ▶ S 196

▶ 14-HH

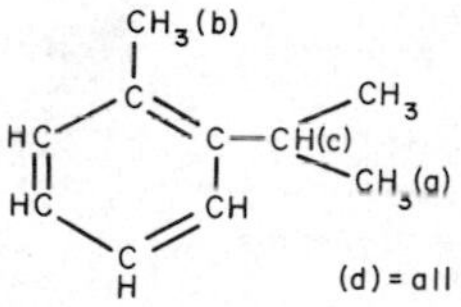

Shift	Peak
a 1.15	2A
b 2.20	1A
c 3.01	5A
▶ d 6.95	O

(d) = all ring H's

▶ $C_{10}H_{14}$ ▶ run neat ▶ API 181 A

▶ 14-HH

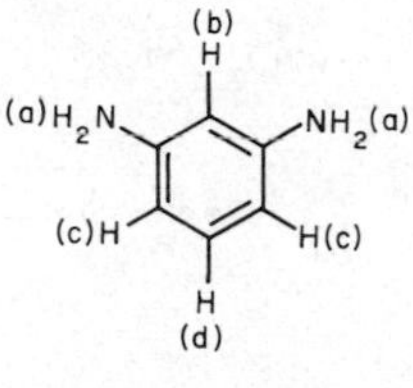

Shift	Peak
a 3.51	1A
b 6.03	1A
c 6.11	2D
▶ d 6.93	3C

▶ $C_6H_8N_2$ ▶ 7% $CDCl_3$ ▶ V 458

▶ 14-HH

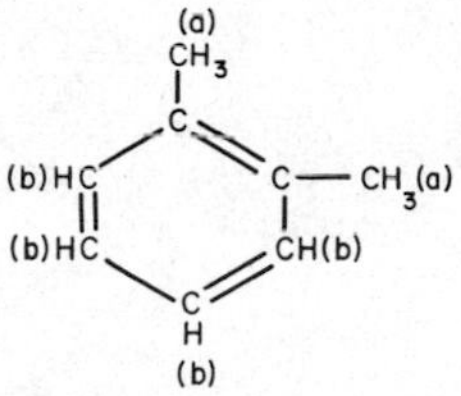

Shift	Peak
a 2.24	1A
▶ b 6.98	1A

▶ C_8H_{10} ▶ CCl_4 ▶ API 174 B

▶ 14-HH

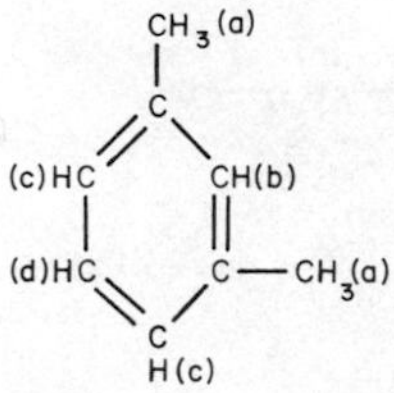

Shift	Peak
a 2.26	1A
b ~6.87	S
c ~6.87	S
▶ d ~6.93	S

▶ C_8H_{10} ▶ 10% CCl_4 ▶ API 39

▶ 14-HH

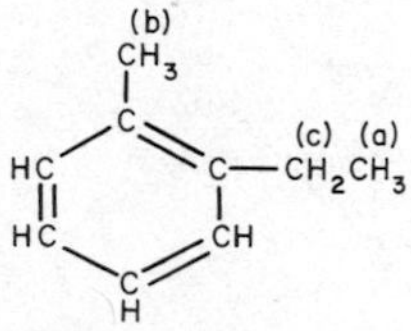

Shift	Peak
a 1.09	3A
b 2.11	1A
c 2.45	4A
▶ d 7.00	1A

(d) = all ring H's

▶ C_9H_{12} ▶ run neat ▶ API 177 A

▶ 14-HH

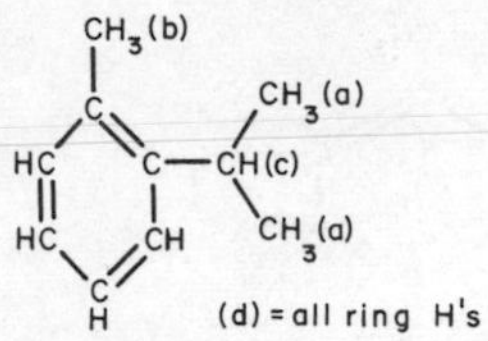

CH_3(b)
CH_3(a)
HC C — CH(c)
HC CH CH_3(a)
C
H
(d) = all ring H's

Shift	Peak
a 1.17	2A
b 2.24	1A
c 3.04	5A
▶ d 7.00	N

▶ $C_{10}H_{14}$ ▶ 50% CCl_4 ▶ API 224

▶ 14-HH

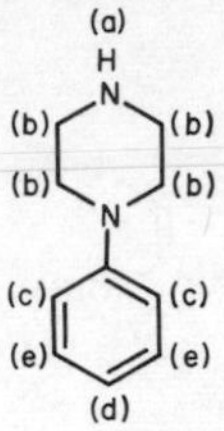

(a)
H
N
(b) (b)
(b) (b)
N
(c) (c)
(e) (e)
(d)

Shift	Peak
a 1.78	1A
b 3.08	1A
c 6.88	2H
▶ d 7.05	S
▶ e 7.18	2H

▶ $C_{10}H_{14}N_2$ ▶ 45.2 mg/0.50 ml $CDCl_3$ ▶ S 300

▶ 14-HH

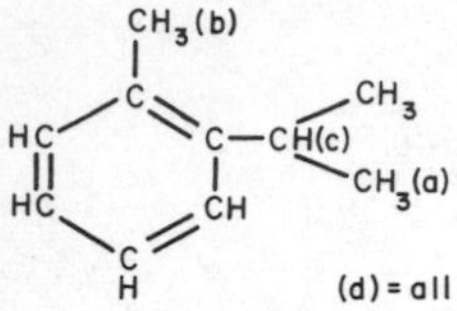

CH_3(b)
CH_3
HC C — CH(c)
HC CH CH_3(a)
C
H
(d) = all ring H's

Shift	Peak
a 1.25	2A
b 2.30	1A
c 3.05	5A
▶ d 7.01	O

▶ $C_{10}H_{14}$ ▶ CCl_4 ▶ API 181 B

▶ 14-HH

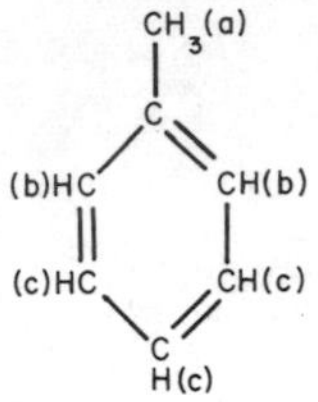

CH_3(a)
C
(b)HC CH(b)
(c)HC CH(c)
C
H(c)

Shift	Peak
a 2.37	1A
b 7.07	1A
▶ c 7.07	1A

▶ C_7H_8 ▶ liquid run, neat ▶ API 81

▶ 14-HH

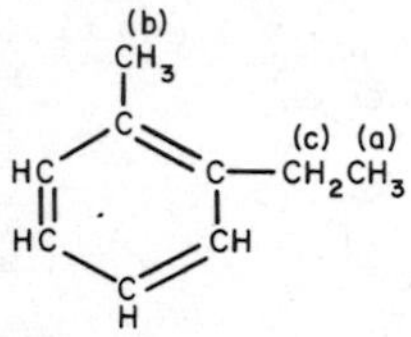

(b)
CH_3
(c) (a)
HC C — CH_2CH_3
HC CH
C
H
(d) = all ring H's

Shift	Peak
a 1.21	3A
b 2.27	1A
c 2.63	4A
▶ d 7.02	1A

▶ C_9H_{12} ▶ CCl_4 ▶ API 177B

▶ 14-HH

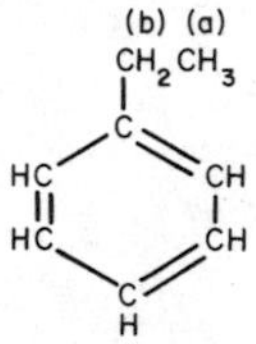

(b) (a)
CH_2CH_3
C
HC CH
HC CH
C
H
(c) = all ring H's

Shift	Peak
a 1.10	3A
b 2.5	4A
▶ c 7.07	1A

▶ C_8H_{10} ▶ run neat ▶ API 173 A

▶ 14-HH

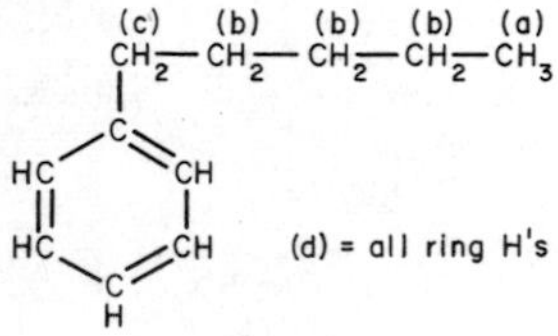

(c) (b) (b) (b) (a)
CH_2—CH_2—CH_2—CH_2—CH_3
C
HC CH
HC CH
C
H
(d) = all ring H's

Shift	Peak
a .89	3B
b 1.36	O
c 2.52	3B
▶ d 7.04	1G

▶ $C_{11}H_{16}$ ▶ 50% CCl_4 ▶ API 355

▶ 14-HH

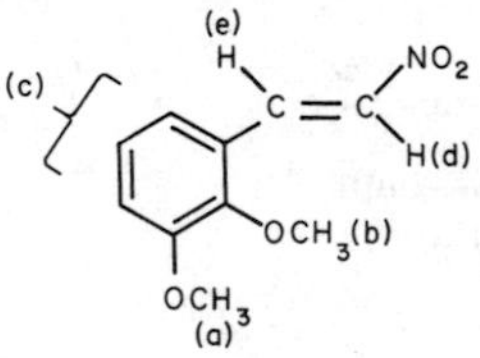

(e)
H NO_2
(c) C = C
H(d)
OCH_3(b)
OCH_3
(a)

Shift	Peak
a 3.88 or 3.92	1A
b 3.88 or 3.92	1A
▶ c 7.07	1A
d 7.73	2A
e 8.20	2A

▶ $C_{10}H_{11}NO_4$ ▶ 7% $CDCl_3$ ▶ V 257

▶ 14-HH

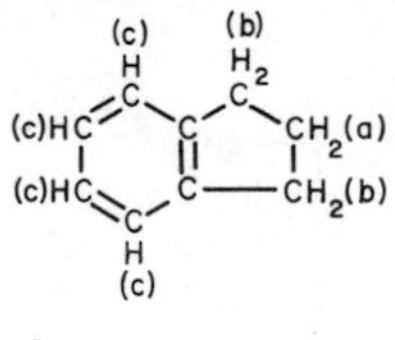

(c) (b)
H H_2
(c)HC C C CH_2(a)
(c)HC C CH_2(b)
H
(c)

Shift	Peak
a 2.02	5B
b 2.86	3A
▶ c 7.05	1I

▶ C_9H_{10} ▶ 10% CCl_4 ▶ API 188

▶ 14-HH

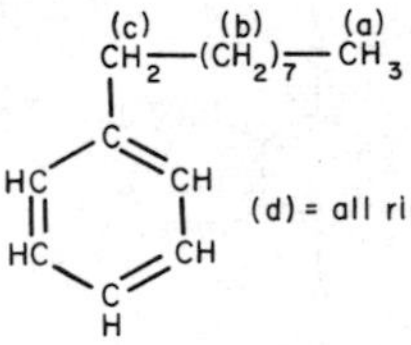

(c) (b) (a)
CH_2—$(CH_2)_7$—CH_3
C
HC CH
HC CH
C
H
(d) = all ring H's

Shift	Peak
a .89	3B
b 1.25	1B
c 2.46	3B
▶ d 7.07	1A

▶ $C_{15}H_{24}$ ▶ 50% CCl_4 ▶ API 358

▶ 14-HH

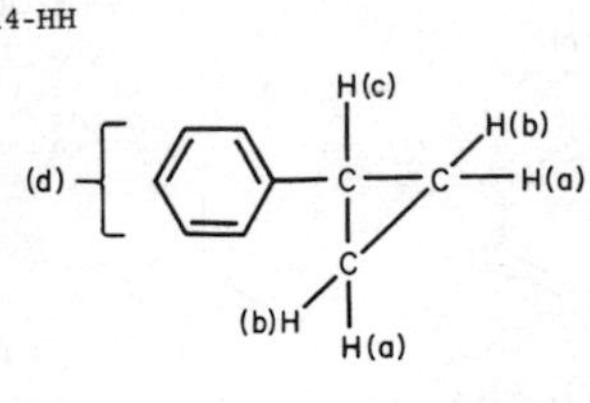

H(c)
H(b)
(d) C C — H(a)
C
(b)H H(a)

Shift	Peak
a 0.68 ± .05	S
b 0.74 ± .05	S
c 1.80	O
▶ d 7.05	2A

▶ C_9H_{10} ▶ 7% $CDCl_3$ ▶ V 528

▶ 14-HH

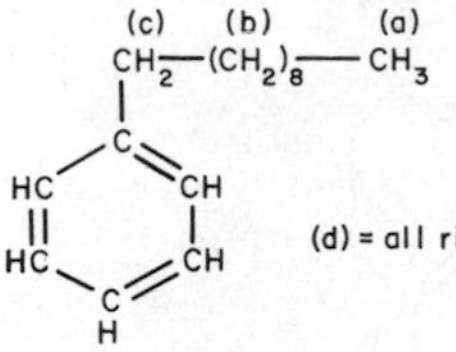

(c) (b) (a)
CH_2—$(CH_2)_8$—CH_3
C
HC CH
HC CH
C
H
(d) = all ring H's

Shift	Peak
a .88	3B
b 1.25	1B
c 2.50	3B
▶ d 7.07	1A

▶ $C_{16}H_{26}$ ▶ 50% CCl_4 ▶ API 359

▶ 14-HH

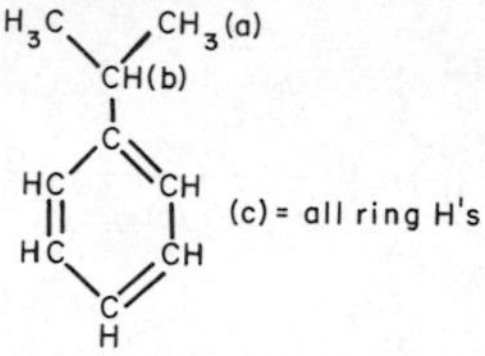

H_3C, CH_3(a), CH(b), C, HC, CH, HC, CH, C, H

(c) = all ring H's

	Shift	Peak
a	1.15	2A
b	2.75	5A
▶ c	7.09	1A

▶ C_9H_{12} ▶ run neat ▶ API 176

▶ 14-HH

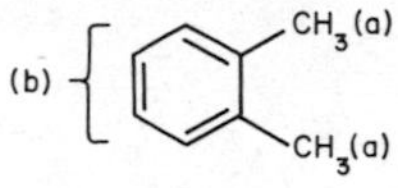

(b) {ring} CH_3(a), CH_3(a)

	Shift	Peak
a	2.25	1A
▶ b	7.10	1A

▶ C_8H_{10} ▶ 7% $CDCl_3$ ▶ V 201

▶ 14-HH

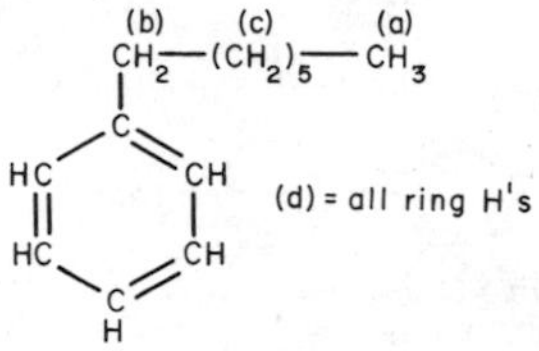

(b) (c) (a)
CH_2—$(CH_2)_5$—CH_3

C, HC, CH, HC, CH, C, H

(d) = all ring H's

	Shift	Peak
a	.86	3B
b	1.28	1B
c	2.54	3B
▶ d	7.09	1A

▶ $C_{13}H_{20}$ ▶ 50% CCl_4 ▶ API 357

▶ 14-HH

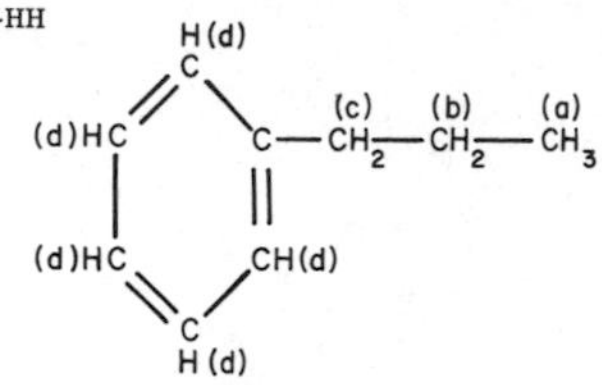

H(d), C, (d)HC, C—CH_2—CH_2—CH_3 (c) (b) (a), (d)HC, CH(d), C, H(d)

	Shift	Peak
a	.92	3C
b	1.58	6B
c	2.57	3C
▶ d	7.10	1E

▶ C_9H_{12} ▶ 10% CCl_4 ▶ API 41

▶ 14-HH

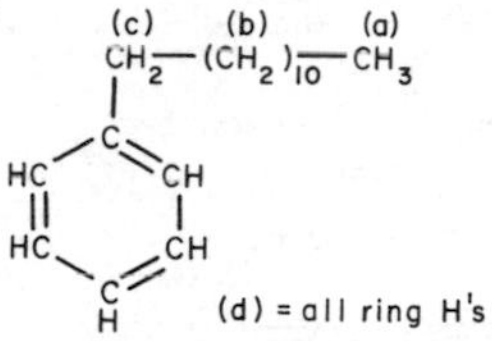

(c) (b) (a)
CH_2—$(CH_2)_{10}$—CH_3

C, HC, CH, HC, CH, C, H

(d) = all ring H's

	Shift	Peak
a	.88	3B
b	1.22	1A
c	2.54	3B
▶ d	7.09	1G

▶ $C_{18}H_{30}$ ▶ 50% CCl_4 ▶ API 360

▶ 14-IIII

(b) (c) (a)
CH_2—$(CH_2)_4$—CH_3

C, HC, CH, HC, CH, C, H

(d) = all ring H's

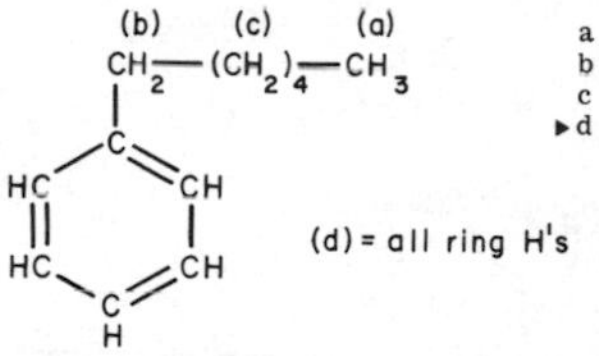

	Shift	Peak
a	.89	3B
b	1.30	1F
c	2.54	3B
▶ d	7.10	1A

▶ $C_{12}H_{18}$ ▶ 50% CCl_4 ▶ API 356

▶ 14-HH

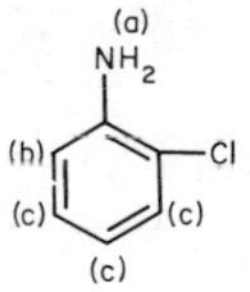

(a) NH_2, Cl, (h), (c), (c), (c)

	Shift	Peak
a	3.9	1C
b	6.68	2H
▶ c	7.10	O

▶ C_6H_6ClN ▶ 44.8 mg/0.50 ml $CDCl_3$ ▶ S 59

▶ 14-HH

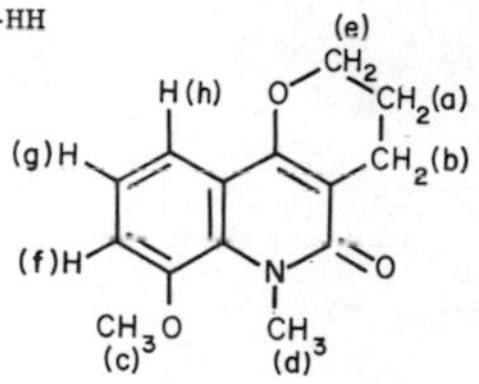

(e) CH_2, O, CH_2(a), CH_2(b), H(h), (g)H, (f)H, N, O, CH_3O (c), CH_3 (d)

	Shift	Peak
a	2.03	5B
b	2.63	3A
c	3.87 or 3.92	1A
d	3.87 or 3.92	1A
e	4.30	3A
f	~7.00	R
▶ g	~7.10	R
h	7.52	4D

▶ $C_{14}H_{15}NO_3$ ▶ 7% $CDCl_3$ ▶ V 312

▶ 14-HH

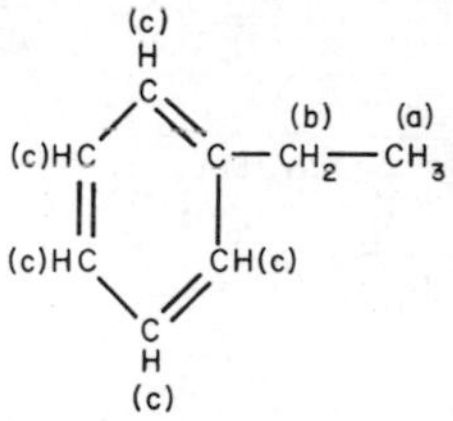

(c) H, C, (c)HC, C—CH_2—CH_3 (b) (a), (c)HC, CH(c), C, H (c)

	Shift	Peak
a	1.22	3A
b	2.54	4A
▶ c	7.10	1E

▶ C_8H_{10} ▶ 10% CCl_4 ▶ API 38

▶ 14-HH

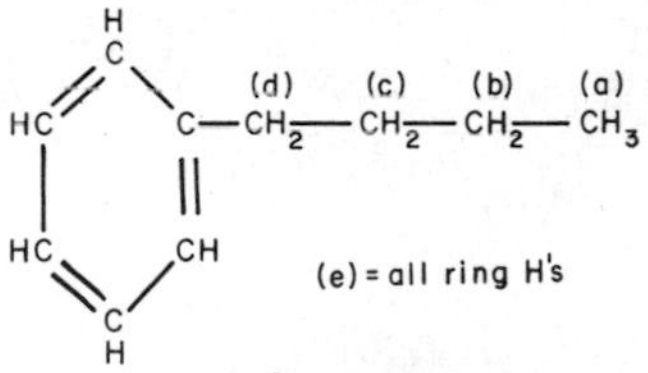

H, C, HC, C—CH_2—CH_2—CH_2—CH_3 (d) (c) (b) (a), HC, CH, C, H

(e) = all ring H's

	Shift	Peak
a	.93	3C
b	1.47	S
c	1.47	S
d	2.57	3B
▶ e	7.11	1A

▶ $C_{10}H_{14}$ ▶ 10% CCl_4 ▶ API 42

▶ 14-HH

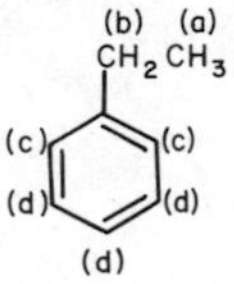

(b) (a) CH_2CH_3, (c), (c), (d), (d), (d)

	Shift	Peak
a	1.10	3A
b	2.47	4A
c	7.10	1A
▶ d	7.10	1A

▶ C_8H_{10} ▶ liquid ▶ API 74

▶ 14-HH

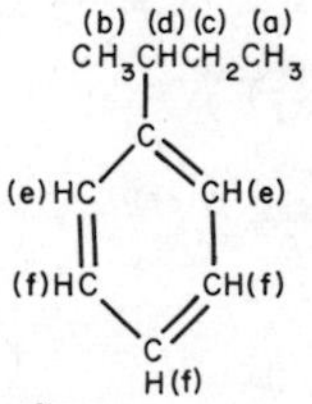

(b) (d) (c) (a)
$CH_3CHCH_2CH_3$

C, (e)HC, CH(e), (f)HC, CH(f), C, H(f)

	Shift	Peak
a	.82	3C
b	1.23	2A
c	1.60	5B
d	2.56	4C
e	7.11	1A
▶ f	7.11	1A

▶ $C_{10}H_{14}$ ▶ 10% CCl_4 ▶ API 90

▶ 14-HH

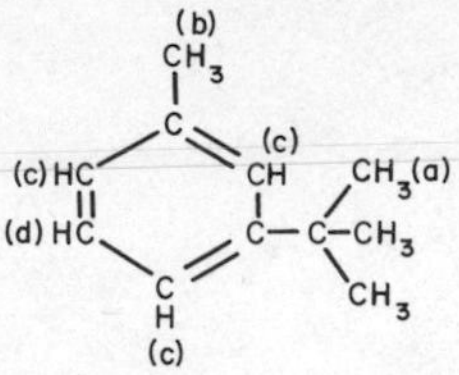

	Shift	Peak
a	1.25	1A
b	2.21	1A
c	~7.03	S
▶ d	7.11	S

▶ $C_{11}H_{16}$ ▶ CCl_4 ▶ API 185 A

▶ 14-HH

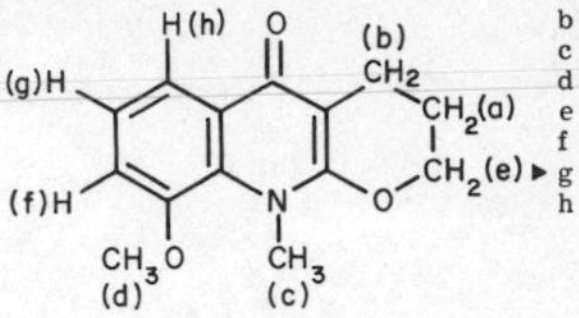

	Shift	Peak
a	2.05	5B
b	2.70	3A
c	3.77 or 3.87	1A
d	3.77 or 3.87	1A
e	4.33	3A
f	~7.00	S
▶ g	~7.15	S
h	8.00	4D

▶ $C_{14}H_{15}NO_3$ ▶ 7% $CDCl_3$ ▶ V 311

▶ 14-HH

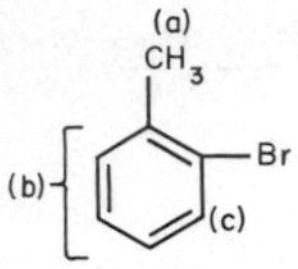

	Shift	Peak
a	2.40	1A
▶ b	7.12	O
c	7.50	2E

▶ C_7H_7Br ▶ 45.4 mg/0.50 ml $CDCl_3$ ▶ S 161

▶ 14-HH

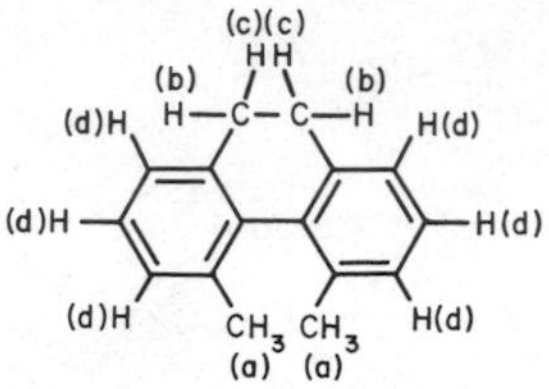

	Shift	Peak
a	2.29	1A
b	2.61	S
c	2.71	S
▶ d	7.15	1A

▶ $C_{16}H_{16}$ ▶ 7% $CDCl_3$ ▶ V 654

▶ 14-HH

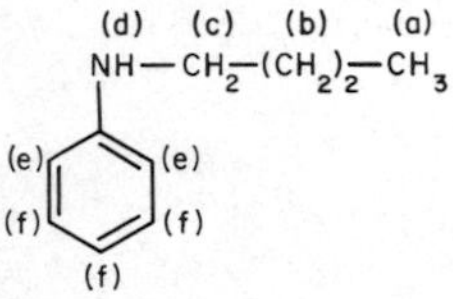

	Shift	Peak
a	0.96	3B
b	1.50	1F
c	3.08	3A
d	3.42	1A
e	6.52	3C
▶ f	7.12	3C

▶ $C_{10}H_{15}N$ ▶ 45.4 mg/0.50 ml $CDCl_3$ ▶ S 280

▶ 14-HH

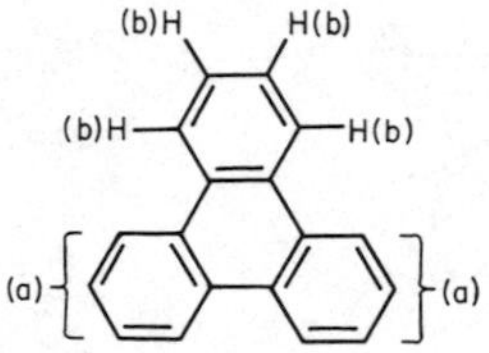

	Shift	Peak
▶ a	7.16	1A
▶ b	7.40	1A

▶ $C_{18}H_{14}$ ▶ 7% $CDCl_3$ ▶ V 670

▶ 14-HH

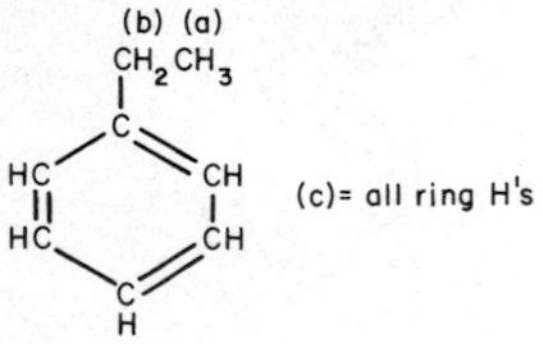

(c)= all ring H's

	Shift	Peak
a	1.22	3A
b	2.63	4A
▶ c	7.13	1A

▶ C_8H_{10} ▶ CCl_4 ▶ API 173 B

▶ 14-HH

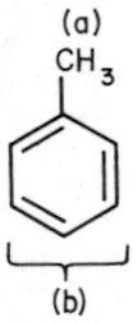

	Shift	Peak
a	2.32	1A
▶ b	7.17 ± 0.03	1A

▶ C_7H_8 ▶ 7% $CDCl_3$ ▶ V 157

▶ 14-HH

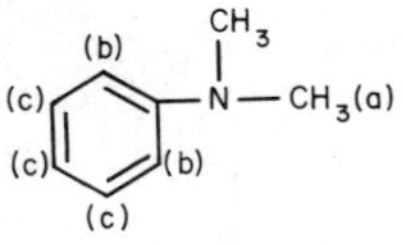

	Shift	Peak
a	2.93	1A
b	~6.67	N
▶ c	~7.15	O

▶ $C_8H_{11}N$ ▶ 45.3 mg/0.50 ml $CDCl_3$ ▶ S 1

▶ 14-HH

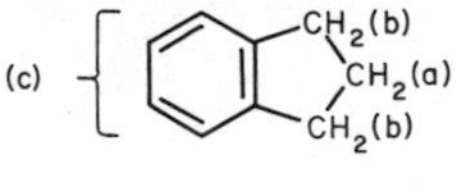

	Shift	Peak
a	2.04	5B
b	2.91	3A
▶ c	7.17	1A

▶ C_9H_{10} ▶ 7% $CDCl_3$ ▶ V 527

▶ 14-HH

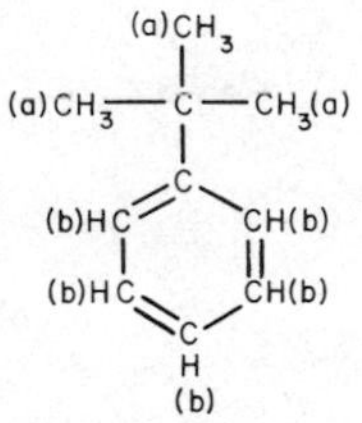

	Shift	Peak
a	1.23	1A
▶ b	7.15	O

▶ $C_{10}H_{14}$ ▶ liquid run ▶ API 20

▶ 14-HH

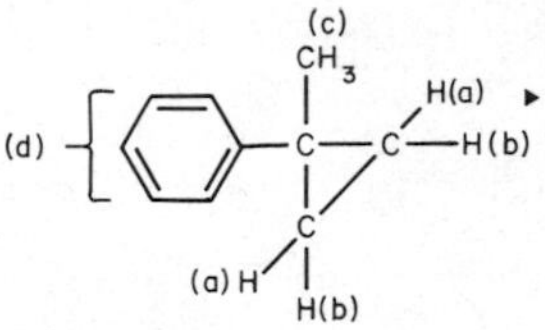

	Shift	Peak
a	0.65 ± .03	N
b	0.81 ± .03	1I
c	1.37	1A
▶ d	7.17	1A

▶ $C_{10}H_{12}$ ▶ 7% $CDCl_3$ ▶ V 556

▶ 14-HH

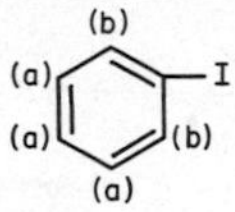

	Shift	Peak
▶ a	7.18	O
b	7.68	N

▶ C_6H_5I ▶ 45.6 mg/0.50 ml $CDCl_3$ ▶ S 172

▶ 14-HH

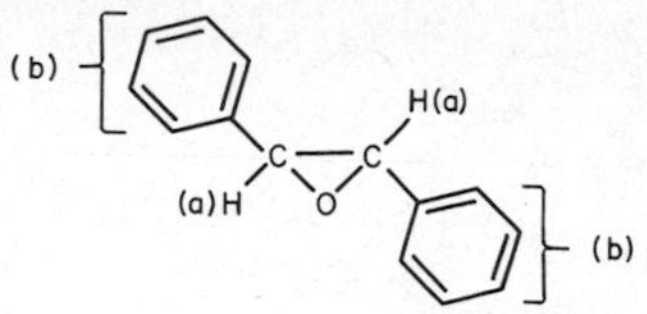

	Shift	Peak
a	4.37	1A
▶ b	7.20	1A

▶ $C_{14}H_{12}O$ ▶ 7% $CDCl_3$ ▶ V 626

▶ 14-HH

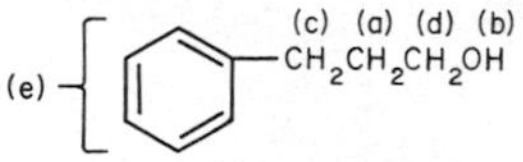

	Shift	Peak
a	1.89	5B
b	2.18	1A
c	2.69	3C
d	3.61	3A
▶ e	7.18	1A

▶ $C_9H_{12}O$ ▶ 44.9 mg/0.50 ml $CDCl_3$ ▶ S 116

▶ 14-HH

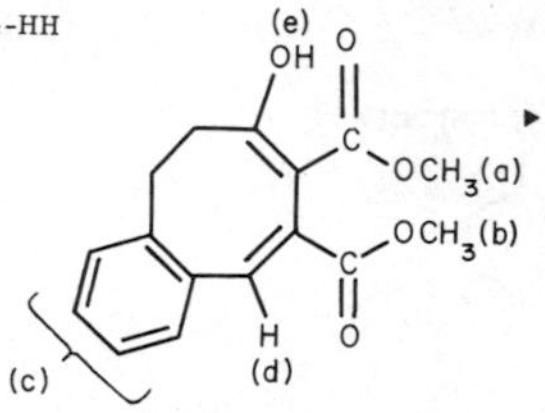

	Shift	Peak
a	3.65 or 3.80	1A
b	3.80 or 3.65	1A
▶ c	7.20	1A
d	7.83	1A
e	12.69	1A

▶ $C_{16}H_{16}O_5$ ▶ 7% $CDCl_3$ ▶ V 656

▶ 14-HH

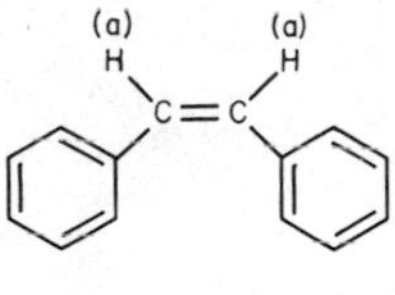

	Shift	Peak
a	6.55	1A
▶ b	7.18 ± .03	1A

(b) = all benzene H's.

▶ $C_{14}H_{12}$ ▶ 7% $CDCl_3$ ▶ V 305

▶ 14-HH

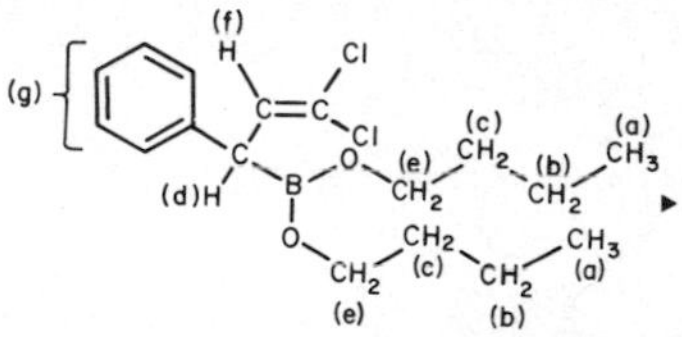

	Shift	Peak
a	0.88	3C
b	~1.36	S
c	~1.50	S
d	3.58	2H
e	3.84	3B
f	6.24	2A
▶ g	7.20	1A

▶ $C_{17}H_{25}BCl_2O_2$ ▶ 7% $CDCl_3$ ▶ V 667

▶ 14-HH

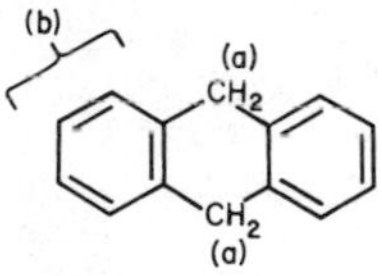

	Shift	Peak
a	3.87	1A
▶ b	7.18	1A

▶ $C_{14}H_{12}$ ▶ 7% $CDCl_3$ ▶ V 307

▶ 14-HH

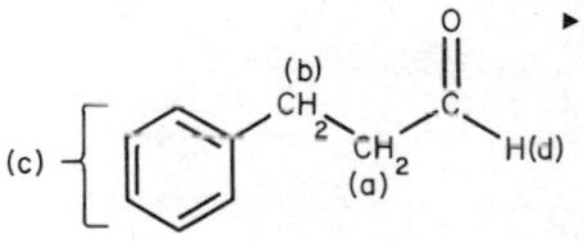

	Shift	Peak
a	2.74	Q
b	2.97	Q
▶ c	7.22	1A
d	9.81	3A

▶ $C_9H_{10}O$ ▶ 7% $CDCl_3$ ▶ V 529

▶ 14-HH

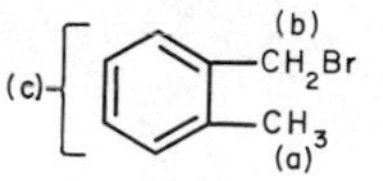

	Shift	Peak
a	2.41	1A
b	4.50	1A
▶ c	7.19	1A

▶ C_8H_9Br ▶ 45.3 mg/0.50 ml $CDCl_3$ ▶ S 239

▶ 14-HH

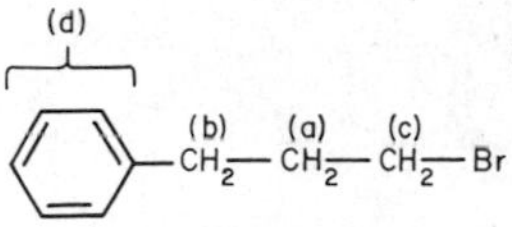

	Shift	Peak
a	2.15	5B
b	2.75	3C
c	3.38	3A
▶ d	7.22	1A

▶ $C_9H_{11}Br$ ▶ 7% $CDCl_3$ ▶ V 237

▶ 14-HH

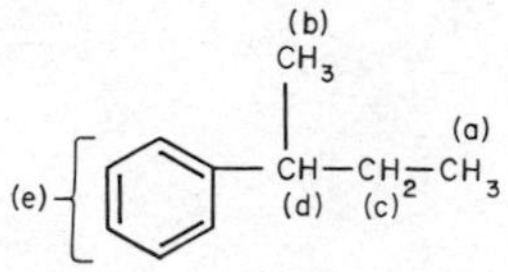

	Shift	Peak
a	0.80	3A
b	1.22	2A
c	1.59	5B
d	2.53	6A
▶ e	7.20	1A

▶ $C_{10}H_{14}$ ▶ 45.0 mg/0.50 ml $CDCl_3$ ▶ S 31

▶ 14-HH

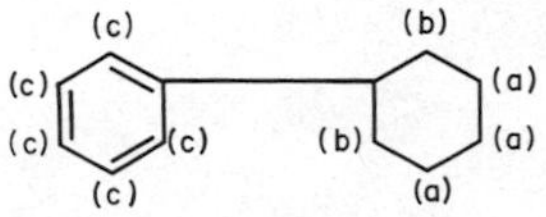

	Shift	Peak
a	~1.4	S
b	~1.8	S
▶ c	7.22	1A

▶ $C_{12}H_{16}$ ▶ 45.4 mg/0.50 ml $CDCl_3$ ▶ S 291

▶ 14-HH

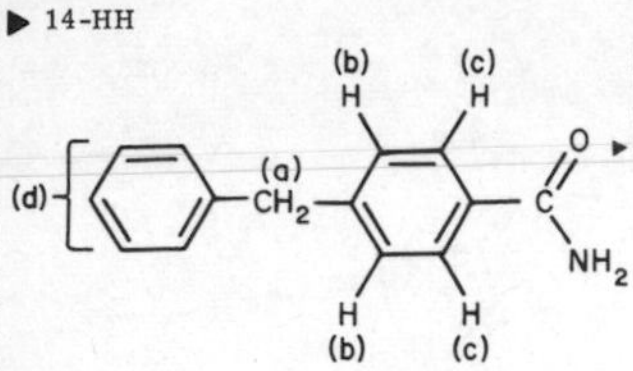

	Shift	Peak
a	3.97	1A
b	7.05 or 7.13	Q
c	7.05 or 7.13	Q
▶ d	7.22	1A

▶ $C_{14}H_{13}NO$ ▶ 7% $CDCl_3$ ▶ V 309

▶ 14-HH

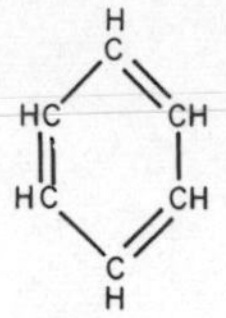

	Shift	Peak
a	7.24	1A

(a) = all ring H's

▶ C_6H_6 ▶ CCl_4 ▶ API 80

▶ 14-HH

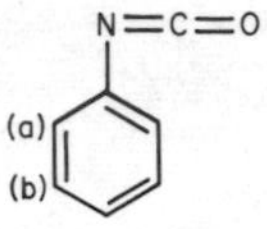

	Shift	Peak
a	7.13	1E
▶ b	7.23	1E

▶ C_7H_5NO ▶ 45.1 mg/0.50 ml $CDCl_3$ ▶ S 80

▶ 14-HH

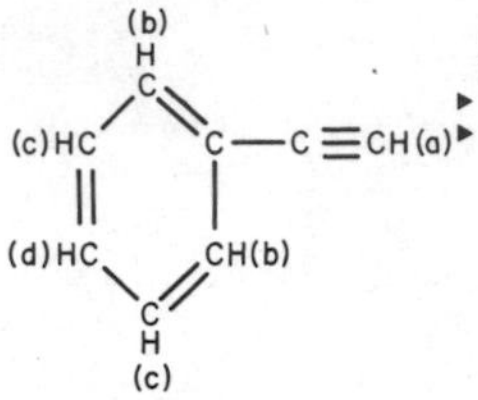

	Shift	Peak
a	3.05	1A
b	7.24 or 7.45	S
▶ c	7.24 or 7.45	S
▶ d	7.32	S

▶ C_8H_6 ▶ $CDCl_3$ ▶ API 136

▶ 14-HH

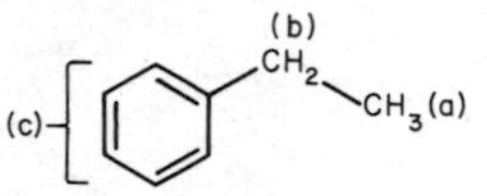

	Shift	Peak
a	1.25	3A
b	2.68	4A
▶ c	7.23	1A

▶ C_8H_{10} ▶ 7% $CDCl_3$ ▶ V 505

▶ 14-HH

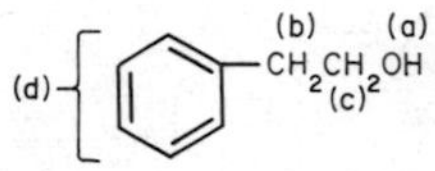

	Shift	Peak
a	2.08	1A
b	2.82	Q
c	3.78	Q
▶ d	7.24	1A

▶ $C_8H_{10}O$ ▶ 45.4 mg/0.50 ml $CDCl_3$ ▶ S 113

▶ 14-HH

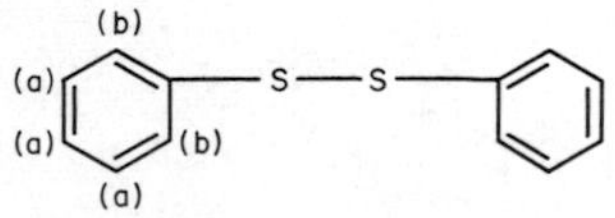

	Shift	Peak
▶ a	7.23	O
b	7.48	N

▶ $C_{12}H_{10}S_2$ ▶ 45.2 mg/0.25 ml $CDCl_3$ ▶ S 286

▶ 14-HH

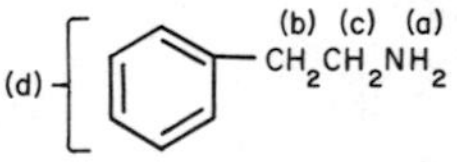

	Shift	Peak
a	1.34	1B
b	2.76	Q
c	2.97	Q
▶ d	7.24	1A

▶ $C_8H_{11}N$ ▶ 45.0 mg/0.50 ml $CDCl_3$ ▶ S 35

▶ 14-HH

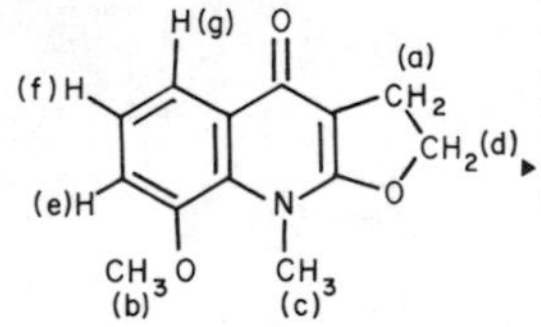

	Shift	Peak
a	3.27	Q
b	3.87 or 3.92	1A
c	3.87 or 3.92	1A
d	4.73	Q
e	~7.05	R
▶ f	~7.23	R
g	8.05	4D

▶ $C_{13}H_{13}NO_3$ ▶ 7% $CDCl_3$ ▶ V 300

▶ 14-HH

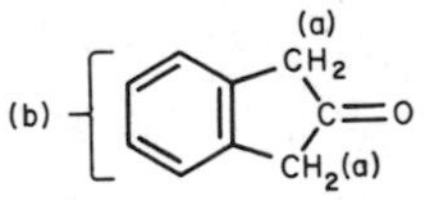

	Shift	Peak
a	3.50	1A
▶ b	7.25	1A

▶ C_9H_8O ▶ 7% $CDCl_3$ ▶ V 523

▶ 14-HH

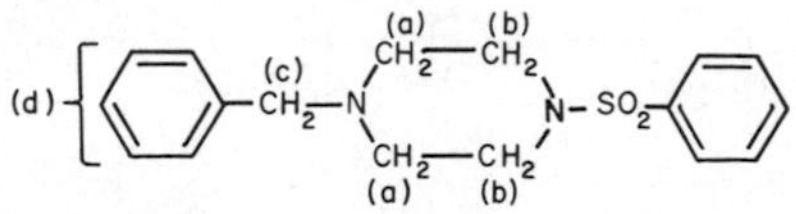

	Shift	Peak
a	2.51	Q
b	3.02	Q
c	3.48	1A
▶ d	7.23	1A

▶ $C_{17}H_{20}N_2O_2S$ ▶ 7% $CDCl_3$ ▶ V 330

▶ 14-HH

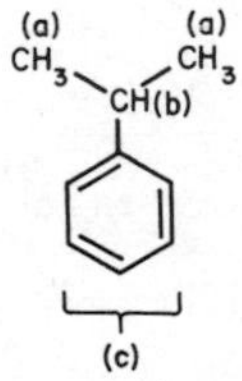

	Shift	Peak
a	1.25	2A
b	2.90	7A
▶ c	7.25	1A

▶ C_9H_{12} ▶ 7% $CDCl_3$ ▶ V 240

▶ 14-HH

(e) C_6H_5—CH_2(b)—CH_2(c)—N(H(d))—C(=O)—CH_3(a)

	Shift	Peak
a	1.90	1A
b	2.80	3B
c	3.48	4A
d	6.50	1D
▶ e	7.25	1G

▶ $C_{10}H_{13}NO$ ▶ 7% $CDCl_3$ ▶ V 265

▶ 14-HH

(c) C_6H_5—CH_2(b)—C(CH_3(a))(Cl)—CH_3(a)

	Shift	Peak
a	1.57	1A
b	3.07	1A
▶ c	7.27	1A

▶ $C_{10}H_{13}Cl$ ▶ 7% $CDCl_3$ ▶ V 263

▶ 14-HH

(c) C_6H_5—CH_2(b)—N(CH_2CN(a))CH_2CN(a)

	Shift	Peak
a	3.58	1A
b	3.78	1A
▶ c	7.25	1A

▶ $C_{11}H_{11}N_3$ ▶ 7% $CDCl_3$ ▶ V 285

▶ 14-HH

H(c), (d) H, N, CH_2(a), (b)

	Shift	Peak
a	4.18	1A
▶ b	7.27	1E
c	7.54	S
d	8.56	3C

▶ $C_{12}H_{11}N$ ▶ 7% $CDCl_3$ ▶ V 593

▶ 14-HH

(b)H—C(C_6H_5)$_2$—C(=O)—CH_3(a)

	Shift	Peak
a	2.20	1A
b	5.08	1A
▶ c	7.25 ± .03	1A

(c) = all benzene H's.

▶ $C_{15}H_{14}O$ ▶ 7% $CDCl_3$ ▶ V 318

▶ 14-HH

(c) C_6H_5—CH_2(b)OH(a)

	Shift	Peak
a	2.43	1A
b	4.58	1A
▶ c	7.28	1A

▶ C_7H_8O ▶ 7% $CDCl_3$ ▶ V 161

▶ 14-HH

($C_6H_5CH_2$(h))($C_6H_5CH_2$(b))N—CH_2(a)—N(CH_2(b)C_6H_5)(CH_2(h)C_6H_5)

	Shift	Peak
a	3.10	1A
b	3.62	1A
▶ c	7.25 ± .03	1A

(c) = all benzene H's.

▶ $C_{29}H_{30}N_2$ ▶ 7% $CDCl_3$ ▶ V 365

▶ 14-HH

(c) C_6H_5—CH_2(b)—SH(a)

	Shift	Peak
a	1.62	3A
b	3.70	2A
▶ c	7.28	1A

▶ C_7H_8S ▶ 43.6 mg/0.50 ml $CDCl_3$ ▶ S 276

▶ 14-HH

(h) C_6H_5, (f) H, CCl_3, H(d), H (e), B, O, (g) CH_2, (c) CH_2, (b) CH_2, (a) CH_3

	Shift	Peak
a	0.88	3C
b	~1.36	S
c	~1.50	S
d	2.86*	S
e	2.89*	S
f	3.05	S
g	3.88	3C
▶ h	7.26	1A

*±0.04

▶ $C_{17}H_{26}BCl_3O_2$ ▶ 7% $CDCl_3$ ▶ V 668

▶ 14-HH

(b)H, (a)H, H(c), O, (d)

	Shift	Peak
a	2.77	4A
b	3.12	4A
c	3.83	4A
▶ d	7.28	1A

▶ C_8H_8O ▶ 7% $CDCl_3$ ▶ V 193

▶ 14-HH

(a) CH_3, (b), (b), (c), (c), Br

	Shift	Peak
a	2.32	1A
b	7.08	S
▶ c	7.27	S

▶ C_7H_7Br ▶ 45.1 mg/0.50 ml $CDCl_3$ ▶ S 162

▶ 14-HH

(c) C_6H_5—CH_2(a)—C(=O)—OCH_3(b)

	Shift	Peak
a	3.60	1A
b	3.66	1A
▶ c	7.28	1A

▶ $C_9H_{10}O_2$ ▶ 44.4 mg/0.50 ml $CDCl_3$ ▶ S 19

▶ 14-HH

Shift	Peak
a 2.02	1A
b 2.93	3A
c 4.30	3A
▶ d 7.29 ± 0.03	1A

▶ $C_{10}H_{12}O_2$ ▶ 7% $CDCl_3$ ▶ V 261

▶ 14-HH

Shift	Peak
a 2.06	1A
b 5.08	1A
▶ c 7.31	1A

▶ $C_9H_{10}O_2$ ▶ 7% $CDCl_3$ ▶ V 530

▶ 14-HH

Shift	Peak
a 4.10	1A
▶ b 7.30	1A

▶ C_8H_7ClO ▶ 44.8 mg/0.50 ml $CDCl_3$ ▶ S 218

▶ 14-HH

Shift	Peak
a 3.65	1A
b 4.17	1A
▶ c 7.31 ± .05	1H
d 8.32	1C

▶ $C_{10}H_9ClN_2O$ ▶ 7% $CDCl_3$ ▶ V 248

▶ 14-HH

Shift	Peak
a 1.42	2A
b 2.38	1A
c 4.78	4A
▶ d 7.30	1A

▶ $C_8H_{10}O$ ▶ 45.2 mg/0.50 ml $CDCl_3$ ▶ S 133

▶ 14-HH

Shift	Peak
▶ a 7.31	S
b 7.58	S

▶ $C_{13}H_8O$ ▶ 45.6 mg/0.25 ml $CDCl_3$ ▶ S 287

▶ 14-HH

Shift	Peak
a 1.38	2A
b 1.58	1A
c 4.10	4A
▶ d 7.30 ± .03	1A

▶ $C_8H_{11}N$ ▶ 7% $CDCl_3$ ▶ V 207

▶ 14-HH

Shift	Peak
▶ a 7.31	2H
b 7.75	Q
c 7.92	Q

▶ $C_{20}H_{14}O_4$ ▶ 45.1 mg/0.50 ml $CDCl_3$ ▶ S 93

▶ 14-HH

Shift	Peak
a 0.92	3A
b 1.80	S
c 2.05	S
d 2.70	2C
e 2.97	2C
▶ f 7.30	1A

▶ $C_{10}H_{12}O$ ▶ 7% $CDCl_3$ ▶ V 558

▶ 14-HH

Shift	Peak
a 4.55	1A
▶ b 7.32	1A

▶ C_7H_7Cl ▶ 45.2 mg/0.50 ml $CDCl_3$ ▶ S 3

▶ 14-HH

Shift	Peak
a 7.30	1A

▶ $C_{12}H_{10}S$ ▶ 44.8 mg/0.50 ml $CDCl_3$ ▶ S 39

▶ 14-HH

Shift	Peak
a 3.10	1A
b 3.87	1A
▶ c 7.32	1A
d 10.97	1B

▶ $C_9H_{10}O_2S$ ▶ 7% $CDCl_3$ ▶ V 532

▶ 14-HH

Shift	Peak
a 2.29	1B
b 5.80	1G
▶c 7.32	1E

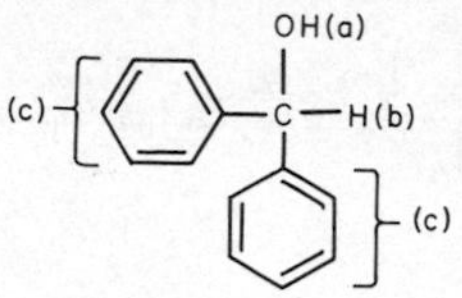

▶ $C_{13}H_{12}O$ ▶ 7% $CDCl_3$ ▶ V 607

▶ 14-HH

Shift	Peak
a 2.86	S
b 3.07	S
c 3.59	3C
▶d 7.35	1A

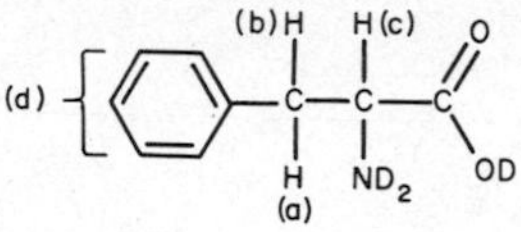

▶ $C_9H_{11}NO_2$ ▶ D_2O ▶ V 534

▶ 14-HH

Shift	Peak
a 3.75	1A
b 5.79	1A
▶c 7.33	1A

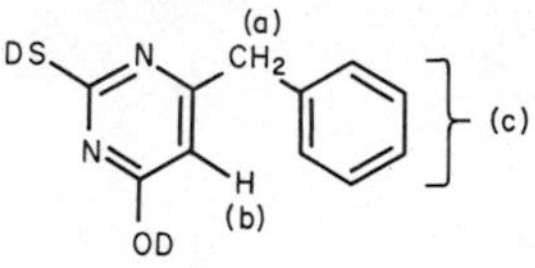

▶ $C_{11}H_{10}N_2OS$ ▶ D_2O ▶ V 577

▶ 14-HH

Shift	Peak
a 4.50	1A
▶b 7.37	1A

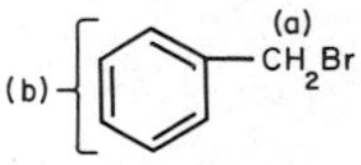

▶ C_7H_7Br ▶ 45.2 mg/0.50 ml $CDCl_3$ ▶ S 154

▶ 14-HH

Shift	Peak
a 6.12	1A
▶b 7.33 ± .05	11

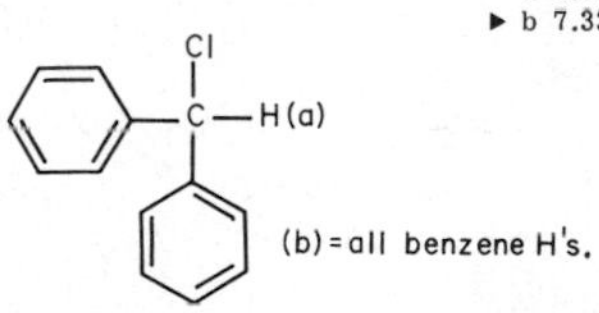

(b) = all benzene H's.

▶ $C_{13}H_{11}Cl$ ▶ 7% $CDCl_3$ ▶ V 176

▶ 14-HH

Shift	Peak
a 3.96	1A
▶b 7.37	1A

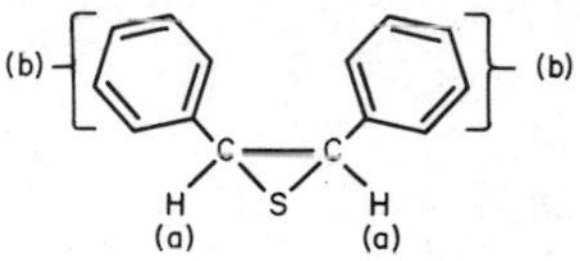

▶ $C_{14}H_{12}S$ ▶ 7% $CDCl_3$ ▶ V 629

▶ 14-HH

Shift	Peak
a 4.52	1A
▶b 7.33	1A

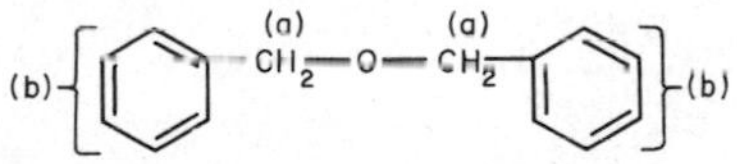

▶ $C_{14}H_{14}O$ ▶ 45.4 mg/0.50 ml $CDCl_3$ ▶ S 118

▶ 14-HH

Shift	Peak
a 4.02	S
b 4.07	S
c 5.15	4A
▶d 7.38	1A

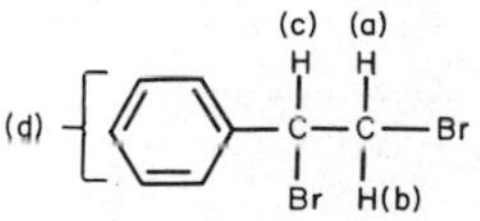

▶ $C_8H_8Br_2$ ▶ 7% $CDCl_3$ ▶ V 503

▶ 14-HH

Shift	Peak
▶a 7.34	4A
b 8.18	2D
c 8.32	2E
d 8.48	2D
e 11.63	1D

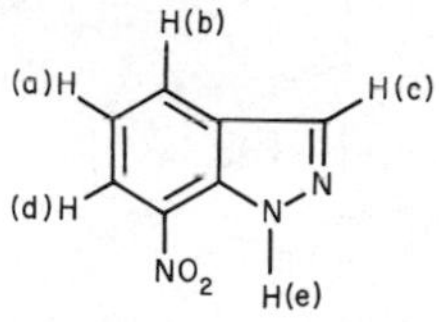

▶ $C_7H_5N_3O_2$ ▶ 7% $CDCl_3$ ▶ V 485

▶ 14-HH

Shift	Peak
a 3.88	1A
▶b 7.38	1A

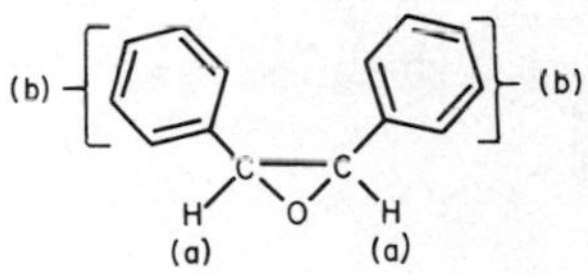

▶ $C_{14}H_{12}O$ ▶ 7% $CDCl_3$ ▶ V 625

▶ 14-HH

Shift	Peak
a 3.68	1A
▶b 7.34	1A

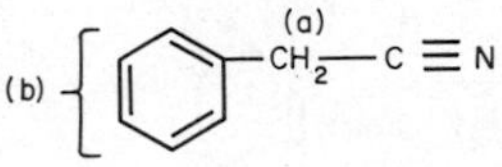

▶ C_8H_7N ▶ 44.5 mg/0.50 ml $CDCl_3$ ▶ S 135

▶ 14-HH

Shift	Peak
a 7.17	S
▶b 7.40	S

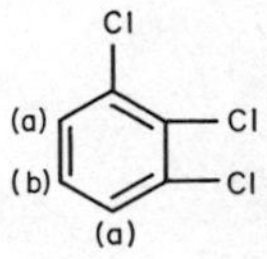

▶ $C_6H_3Cl_3$ ▶ 44.6 mg/0.50 ml $CDCl_3$ ▶ S 266

▶ 14-HH

(b)
(a) —CCl$_3$
(a) (b)
(a)

Shift	Peak
▶ a 7.44	3A
b 7.93	O

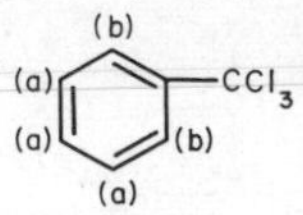

▶ $C_7H_5Cl_3$ ▶ 45.1 mg/0.50 ml $CDCl_3$ ▶ S 90

▶ 14-HH

(e) O (c) (b) (a)
(d) —C—OCH$_2$(CH$_2$)$_2$CH$_3$
(d) (e)
(d)

Shift	Peak
a 0.98	3B
b 1.64	O
c 4.35	3A
▶ d 7.48	N
e 8.06	N

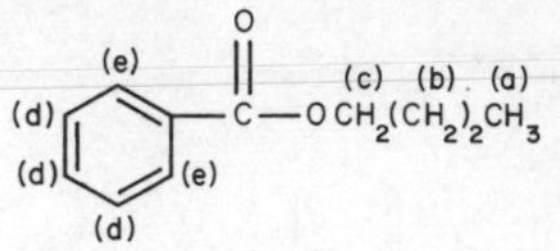

▶ $C_{11}H_{14}O_2$ ▶ 44.8 mg/0.50 ml $CDCl_3$ ▶ S 158

▶ 14-HH

O
(a)H$_3$C CH$_3$(c)
(a)CH$_3$ N
H N
(d) CH$_3$
(b) (e)

Shift	Peak
a 1.28	1A
b 1.60	2A
c 2.57	1A
d 5.87	2A
▶ e 7.46	1A

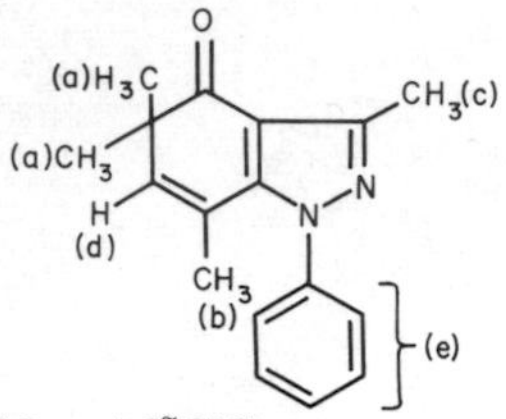

▶ $C_{17}H_{18}N_2O$ ▶ 7% $CDCl_3$ ▶ V 664

▶ 14-HH

O CH$_3$(a)
(c)
—C—O—CH$_2$—CH—CH$_3$
(e) (b)
(d)

Shift	Peak
a 1.02	2A
b 2.05	5B
c 4.12	2A
▶ d 7.5	N
e ~8.05	S

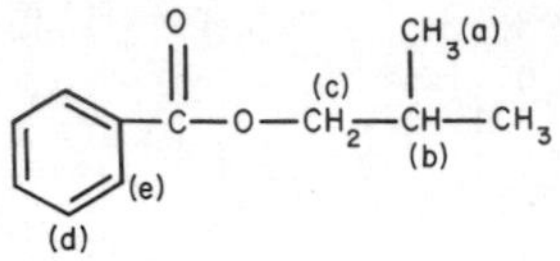

▶ $C_{11}H_{14}O_2$ ▶ 45.5 mg/0.50 ml $CDCl_3$ ▶ S 75

▶ 14-HH

O
(b)
(a) —C—OH(c)
(a) (b)
(a)

Shift	Peak
▶ a 7.48	N
b 8.15	N
c 12.28	1G

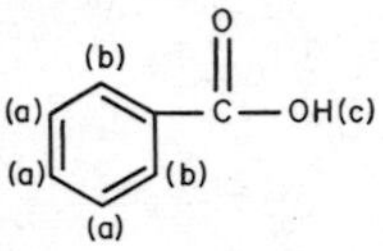

▶ $C_7H_6O_2$ ▶ 45.1 mg/0.50 ml $CDCl_3$ ▶ S 57

▶ 14-HH

(+) ND$_2$
(b) (a)
CH$_2$—S—C Cl$^{(-)}$
ND$_2$

Shift	Peak
a 4.46	1A
▶ b 7.50	1A

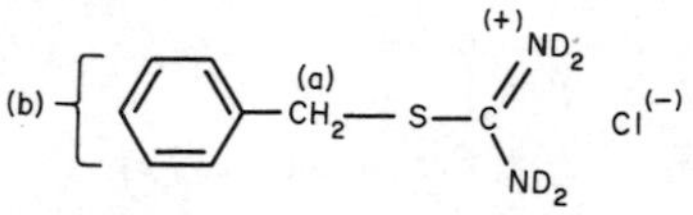

▶ $C_8H_{11}ClN_2S$ ▶ D_2O ▶ V 507

▶ 14-HH

O
(c) (a)
(b) —C—O—CH$_3$
(b) (c)
(b)

Shift	Peak
a 3.92	1A
▶ b 7.48	4A
c ~8.0	O

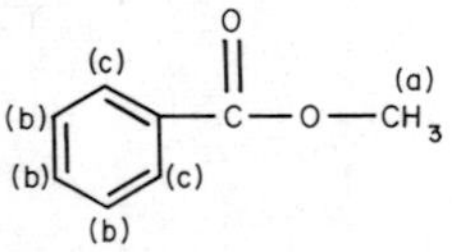

▶ $C_8H_8O_2$ ▶ 45.3 mg/0.50 ml $CDCl_3$ ▶ S 78

▶ 14-HH

O
(d) (b) (a)
(c) —C—CH$_2$—CH$_3$
(c) (d)
(c)

Shift	Peak
a 1.22	3A
b 2.99	4A
▶ c 7.50	N
d 7.98	N

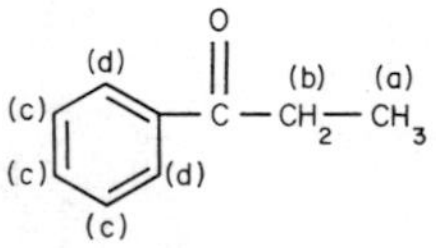

▶ $C_9H_{10}O$ ▶ 44.5 mg/0.50 ml $CDCl_3$ ▶ S 34

▶ 14-HH

Cl O O
(a)
(c)H CH$_2$—C—C—OH(e)
(b)H NO$_2$
H(d)

Shift	Peak
a 4.78	1A
▶ b 7.48	4A
c 7.75	4A
d 8.00	S
e 9.63	1B

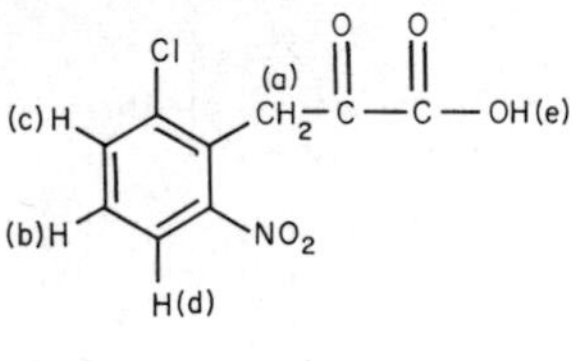

▶ $C_9H_6ClNO_5$ ▶ 7% $CDCl_3$ ▶ V 223

▶ 14-HH

(a)
CH$_3$
CH$_3$(a)
(d) (+) Cl$^{(-)}$
N CH$_3$(b)
CH$_3$(c)

Shift	Peak
a 1.56	1A
b 2.87	1B
c 4.10	1A
▶ d 7.51 ± 0.05	1G

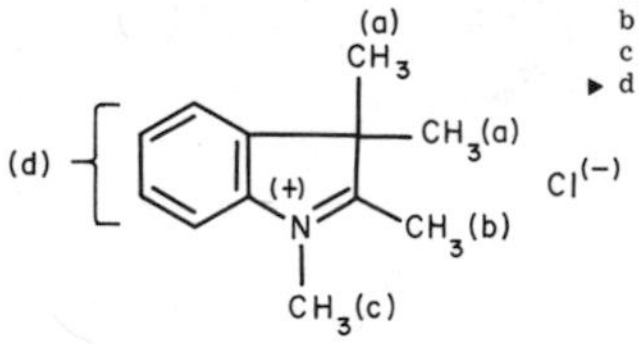

▶ $C_{12}H_{16}ClN$ ▶ 7% $CDCl_3$ ▶ V 599

▶ 14-HH

(c) (b)
H H
(d) —C—C—CH$_3$(a)
OD ND$_3$ (+) Cl$^{(-)}$

Shift	Peak
a 1.21	2A
b 3.72	O
c 5.00	2A
▶ d 7.48	1A

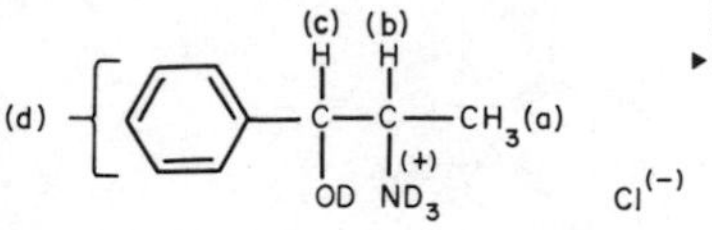

▶ $C_9H_{14}ClNO$ ▶ D_2O ▶ V 565

▶ 14-HH

O
(b)
(a) —C—
(a) (b)
(a)

Shift	Peak
▶ a 7.52	N
b 7.82	O

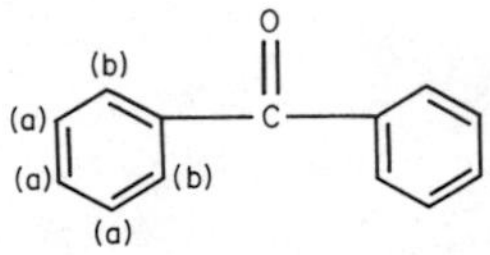

▶ $C_{13}H_{10}O$ ▶ 45.2 mg/0.50 ml $CDCl_3$ ▶ S 153

▶ 14-HH

(a) OH; (b); (b); (c)H; H(c); (d)H; C; O

Shift	Peak
a 6.11	1B
▶ b 7.52±0.05	1E
c 7.83	1A
d 9.92	1A

▶ $C_{19}H_{14}O_2$ ▶ 7% $CDCl_3$ ▶ V 676

▶ 14-HH

(d) H; (e) H; (a) CH_3; N; O; (c)H; (c)H; H(b); H(b); H (d); O; H (e); N; CH_3 (a)

Shift	Peak
a 3.04	2A
b 7.13	1G
▶ c 7.67	Q
d 8.31	Q
e 10.46	1C

▶ $C_{16}H_{14}N_2O_2$ ▶ 7% $CDCl_3$ ▶ V 652

▶ 14-HH

(a); C ≡ N

Shift	Peak
a 7.56	2H

▶ C_7H_5N ▶ 45.4 mg/0.50 ml $CDCl_3$ ▶ S 138

▶ 14-HH

O; Cl; (b); (a); (a); (a); (a); (b); (b); O

Shift	Peak
▶ a 7.72	O
b 8.20	O

▶ $C_{14}H_7ClO_2$ ▶ 44.9 mg/0.50 ml $CDCl_3$ ▶ S 182

▶ 14-HH

O; (c); (a); (b); C—OCH_3; (b); C—OCH_3 (a); (c); O

Shift	Peak
a 3.92	1A
▶ b 7.58	Q
c 7.63	Q

▶ $C_{10}H_{10}O_4$ ▶ 46.0 mg/0.50 ml $CDCl_3$ ▶ S 89

▶ 14-HH

(e)H; O; (c) NH_2; (d)H; OCH_3(a); (d)H; H(b); H (e); O; O; H (f)

Shift	Peak
a 3.96	1A
b 6.52	1A
c ~ 7.30	1D
▶ d 7.72	Q
e 8.34	Q
f 14.22	1A

▶ $C_{15}H_{11}NO_4$ ▶ 7% $CDCl_3$ ▶ V 635

▶ 14-HH

(d) COOH; (c)H; OCH_3(a); H(b)

Shift	Peak
a 4.07	1A
▶ b ~7.60	O
c 8.17	4A
d 11.00	1B

▶ $C_8H_8O_3$ ▶ 7% $CDCl_3$ ▶ V 195

▶ 14-HH

(f) H; O; (b) CH_3; C; (d)H; H(c); (d)H; C; CH_3(a); H (f); O; H (e)

Shift	Peak
a 2.45	1A
b 2.78	1A
c 7.33	2C
▶ d 7.74	Q
e 7.99	2C
f 8.23	Q

▶ $C_{16}H_{12}O_2$ ▶ 7% $CDCl_3$ ▶ V 650

▶ 14-HH

O; O; (b); (a); C—O—C; (a); (b); (a)

Shift	Peak
▶ a 7.60	N
b 8.18	O

▶ $C_{14}H_{10}O_3$ ▶ 45.5 mg/0.25 ml $CDCl_3$ ▶ S 155

▶ 14-HH

O; (c); C—O; (a); (b); (b); C—O; (a); (c); O

Shift	Peak
a 7.31	2H
▶ b 7.75	Q
c 7.92	Q

▶ $C_{20}H_{14}O_4$ ▶ 45.1 mg/0.50 ml $CDCl_3$ ▶ S 93

▶ 14-HH

(b); (a); NO_2; (a); (b); (a)

Shift	Peak
▶ a 7.65	N
b 8.20	O

▶ $C_6H_5NO_2$ ▶ 45.2 mg/0.50 ml $CDCl_3$ ▶ S 4

▶ 14-HH

(c) H; O; (b)H; H(a); (b)H; H(a); H (c); O

Shift	Peak
a 6.97	1A
▶ b 7.77	Q
c 8.07	Q

▶ $C_{10}H_6O_2$ ▶ 7% $CDCl_3$ ▶ V 550

▶ 14-HH

Shift		Peak
▶ a	7.78	Q
b	7.84	Q
c	10.38	1A

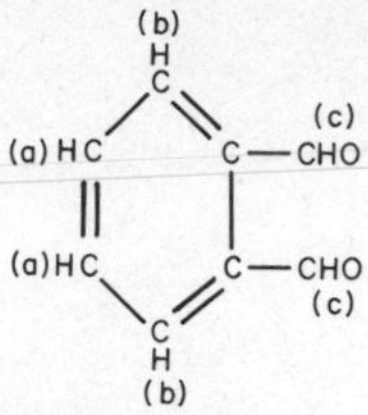

▶ $C_8H_6O_2$ ▶ 10% CCl_4 ▶ API 52

▶ 14-HJ

Shift		Peak
a	3.93	1A
b	6.47	1D
c	7.02	2A
▶ d	7.40	N
e	7.40	N
f	9.78	1A

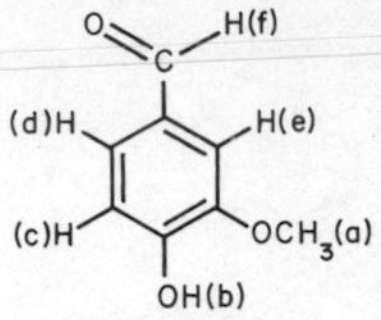

▶ $C_8H_8O_3$ ▶ 7% $CDCl_3$ ▶ V 197

▶ 14-HH

Shift		Peak
a	7.5	N
▶ b	8.0	O

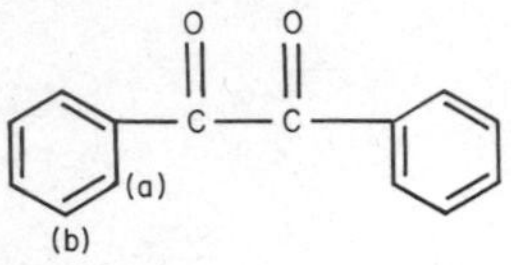

▶ $C_{14}H_{10}O_2$ ▶ 45.0 mg/0.50 ml $CDCl_3$ ▶ S 73

▶ 14-HJ

Shift		Peak
a	1.23	2A
b	2.92	5A
c	6.82	N
d	6.88	N
▶ e	7.42	2A
f	9.79	1A
g	11.00	1B

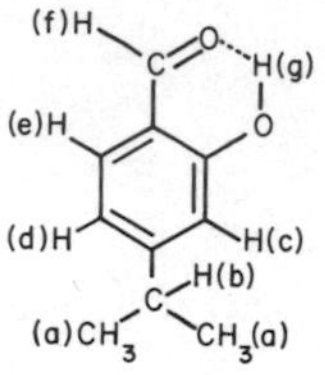

▶ $C_{10}H_{12}O_2$ ▶ 7% $CDCl_3$ ▶ V 259

▶ 14-HI

Shift		Peak
a	7.40	1A

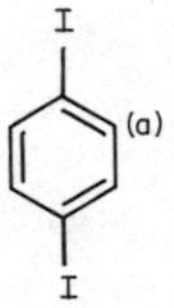

▶ $C_6H_4I_2$ ▶ 44.8 mg/0.50 ml $CDCl_3$ ▶ S 207

▶ 14-HJ

Shift		Peak
a	6.08	1A
b	6.95	2A
c	7.35	1I
▶ d	7.45	2A
e	9.84	1A

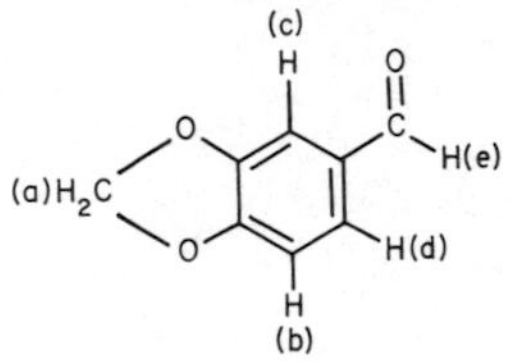

▶ $C_8H_6O_3$ ▶ 7% $CDCl_3$ ▶ V 187

▶ 14-HI

Shift		Peak
a	3.75	1A
b	6.67	Q
▶ c	7.53	Q

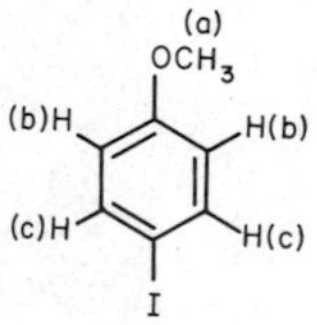

▶ C_7H_7IO ▶ 7% $CDCl_3$ ▶ V 153

▶ 14-HJ

Shift		Peak
a	3.93 or 3.97	1A
b	3.93 or 3.97	1A
c	6.98	2E
d	7.40	S
▶ e	7.45	S
f	9.98	1A

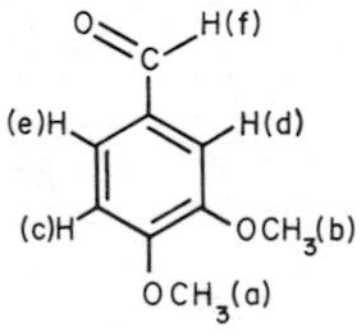

▶ $C_9H_{10}O_3$ ▶ 7% $CDCl_3$ ▶ V 236

▶ 14-HI

Shift		Peak
a	7.18	O
▶ b	7.68	N

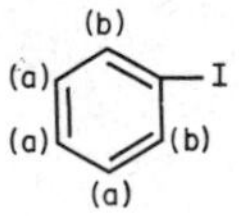

▶ C_6H_5I ▶ 45.6 mg/0.50 ml $CDCl_3$ ▶ S 172

▶ 14-HJ

Shift		Peak
▶ a	7.5	N
b	8.0	O

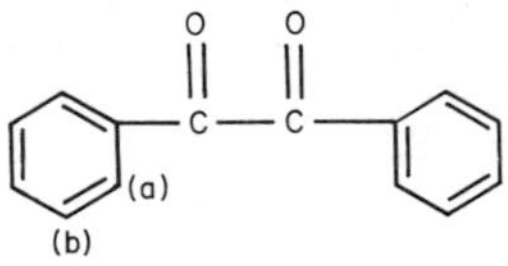

▶ $C_{14}H_{10}O_2$ ▶ 45.0 mg/0.50 ml $CDCl_3$ ▶ S 73

▶ 14-HJ

Shift		Peak
a	3.97	1A
b	7.05 or 7.13	Q
▶ c	7.05 or 7.13	Q
d	7.22	1A

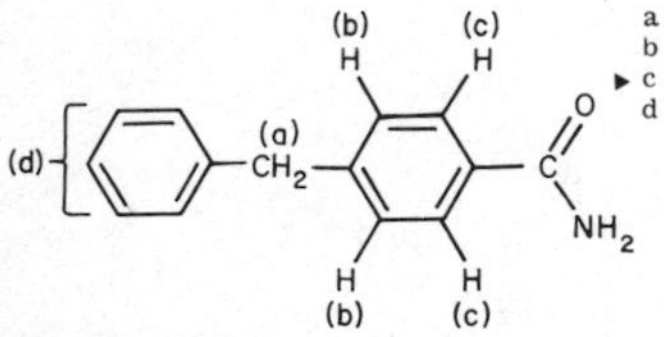

▶ $C_{14}H_{13}NO$ ▶ 7% $CDCl_3$ ▶ V 309

▶ 14-H(J)

Shift		Peak
a	7.31	S
▶ b	7.58	S

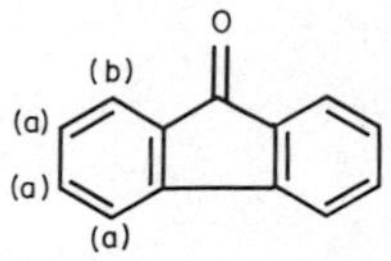

▶ $C_{13}H_8O$ ▶ 45.6 mg/0.25 ml $CDCl_3$ ▶ S 287

▶ 14-HJ

Shift	Peak
a 2.57	1A
b 3.93	1A
c 6.62	1C
d 6.97	2A
e 7.57	S
▶ f 7.60	S

▶ $C_9H_{10}O_3$ ▶ 7% $CDCl_3$ ▶ V 234

▶ 14-HJ

Shift	Peak
a 3.87	1A
b 7.02	Q
▶ c 7.83	Q
d 9.87	1A

▶ $C_8H_8O_2$ ▶ 7% $CDCl_3$ ▶ V 194

▶ 14-HJ

Shift	Peak
a 3.05	1A
b 6.69	Q
▶ c 7.71	Q
d 9.70	1A

▶ $C_9H_{11}NO$ ▶ 7% $CDCl_3$ ▶ V 238

▶ 14-HJ

Shift	Peak
a 7.78	Q
▶ b 7.84	Q
c 10.38	1A

▶ $C_8H_6O_2$ ▶ 10% CCl_4 ▶ API 52

▶ 14-HJ

Shift	Peak
a 7.60	Q
▶ b 7.75	Q
c 9.97	1A

▶ C_7H_5ClO ▶ 7% $CDCl_3$ ▶ V 146

▶ 14-HJ

Shift	Peak
a 3.75*	1A
b 3.75*	1A
c 4.54	1C
d 5.82	1B
e 6.83	S
f 6.87	S
g 7.24	2E
▶ h 7.89	2E

*or 3.81

▶ $C_{16}H_{16}O_4$ ▶ 7% $CDCl_3$ ▶ V 655

▶ 14-HJ

Shift	Peak
a 3.04	1A
b 6.70	Q
▶ c 7.77	Q

▶ $C_{17}H_{20}N_2O$ ▶ 7% $CDCl_3$ ▶ V 329

▶ 14-HJ

Shift	Peak
a 2.58	1A
b 7.45	Q
▶ c 7.90	Q

▶ C_8H_7ClO ▶ 7% $CDCl_3$ ▶ V 188

▶ 14-HJ

Shift	Peak
a 3.35	3B
b 4.10	S
c 5.41	3A
d 7.63	N
▶ e 7.77	2A

▶ $C_{18}H_{20}O_6$ ▶ 7% $CDCl_3$ ▶ V 675

▶ 14-HJ

Shift	Peak
a 2.59	1A
▶ b 7.91	N

▶ C_8H_8O ▶ 7% $CDCl_3$ ▶ V 192

▶ 14-HJ

Shift	Peak
a 7.52	N
▶ b 7.82	O

▶ $C_{13}H_{10}O$ ▶ 45.2 mg/0.50 ml $CDCl_3$ ▶ S 153

▶ 14-H(J)

Shift	Peak
a 4.04	2D
b 7.55	S
▶ c 7.92	S

▶ $C_{16}H_{12}O$ ▶ 7% $CDCl_3$ ▶ V 649

▶ 14-HJ

O
CH(b)
(a)
Cl

	Shift	Peak
▶ a	7.95	S
b	10.52	1A

▶ C_7H_5ClO ▶ 35.7 mg/0.50 ml $CDCl_3$ ▶ S 50

▶ 14-H(J)

H(g) O
(f)H
(a) CH_2
CH_2(d)
(e)H
N
O
CH_3O (b)
CH_3 (c)

	Shift	Peak
a	3.27	Q
b	3.87 or 3.92	1A
c	3.87 or 3.92	1A
d	4.73	Q
e	~7.05	R
f	~7.23	R
▶ g	8.05	4D

▶ $C_{13}H_{13}NO_3$ ▶ 7% $CDCl_3$ ▶ V 300

▶ 14-HJ

O
CH_3(a)
C
(c)H H(c)
(b)H H(b)
F

	Shift	Peak
a	2.58	1A
b	7.12	3C
▶ c	7.97	4C

▶ C_8H_7FO ▶ 7% $CDCl_3$ ▶ V 189

▶ 14-H(J)

(h)H O
(d) H
(d) H
CH_3(a)
CH
(c)
CH_3(b)
N O
H (g)
CH_3O (e)
CH_3 (f)

	Shift	Peak
a	0.97*	2A
b	0.97*	2A
c	2.00	6A
d	~3.20	S
e	3.87†	1A
f	3.87†	1A
g	4.72	4C
▶ h	8.05	4A

*or 1.02
†or 3.90

▶ $C_{16}H_{19}NO_3$ ▶ 7% $CDCl_3$ ▶ V 325

▶ 14-HJ

O
(d) (b) (a)
(c) C — CH_2 — CH_3
(c) (d)
(c)

	Shift	Peak
a	1.22	3A
b	2.99	4A
c	7.50	N
▶ d	7.98	N

▶ $C_9H_{10}O$ ▶ 44.5 mg/0.50 $CDCl_3$ ▶ S 34

▶ 14-H(J)

(c) H O
(b)H H(a)
(b)H H(a)
H O
(c)

	Shift	Peak
a	6.97	1A
b	7.77	Q
▶ c	8.07	Q

▶ $C_{10}H_6O_2$ ▶ 7% $CDCl_3$ ▶ V 550

▶ 14-H(J)

H(h) O (b)
CH_2
(g)H CH_2(a)
(f)H CH_2(e)
N O
CH_3O (d)
CH_3 (c)

	Shift	Peak
a	2.05	5B
b	2.70	3A
c	3.77 or 3.87	1A
d	3.77 or 3.87	1A
e	4.33	3A
f	~7.00	S
g	~7.15	S
▶ h	8.00	4D

▶ $C_{14}H_{15}NO_3$ ▶ 7% $CDCl_3$ ▶ V 311

▶ 14-HJ

(b) (c)
H H
O
C
(a)
CH_2— Br
H H
(b) (c)

	Shift	Peak
a	4.45	1A
b	7.70	Q
▶ c	8.07	Q

▶ $C_{14}H_{11}BrO$ ▶ 7% $CDCl_3$ ▶ V 303

▶ 14-HJ

O
C — Cl
(c) (c)
(b) (b)
O CH_3(a)

	Shift	Peak
a	3.90	1A
b	6.92	Q
▶ c	8.04	Q

▶ $C_8H_7ClO_2$ ▶ 44.6 mg/0.50 ml $CDCl_3$ ▶ S 180

▶ 14-H(J)

(f)H O
OH(g)
(d)H N
CH_2(b)
(c) CH_3O
CH_3 CH_3
H (a) (a)
(e)

	Shift	Peak
a	1.42	1A
b	3.04	1A
c	3.91	1A
d	6.88	S
e	6.93	S
▶ f	8.17	2A
g	~10.30	1D

▶ $C_{13}H_{15}NO_3$ ▶ 7% $CDCl_3$ ▶ V 611

▶ 14-HJ

(a)
H O
C H(b)
C
N
H OH(c)
(a)

	Shift	Peak
▶ a	8.04	O
b	8.13	1E
c	9.06	1B

▶ $C_8H_7NO_2$ ▶ 7% $CDCl_3$ ▶ V 501

▶ 14-H(J)

O Cl
(b)
(a) (a)
(a) (a)
(b) (b)
O

	Shift	Peak
a	7.72	O
▶ b	8.20	O

▶ $C_{14}H_7ClO_2$ ▶ 44.9 mg/0.50 ml $CDCl_3$ ▶ S 182

▶ 14-H(J)

(f) H, O, (b) CH_3, (d)H, C, H(c), (d)H, C, CH_3(a), H (f), O, H (e)

	Shift	Peak
a	2.45	1A
b	2.78	1A
c	7.33	2C
d	7.74	Q
e	7.99	2C
▶ f	8.23	Q

▶ $C_{16}H_{12}O_2$ ▶ 7% $CDCl_3$ ▶ V 650

▶ 14-HK

O, (c), (a), (b), C—OCH_3, (b), C—OCH_3 (a), (c), O

	Shift	Peak
a	3.92	1A
b	7.58	Q
▶ c	7.63	Q

▶ $C_{10}H_{10}O_4$ ▶ 46.0 mg/0.50 ml $CDCl_3$ ▶ S 89

▶ 14-H(J)

(c) H, O, CH_3(a), O, H(b)

	Shift	Peak
a	2.02	2D
b	7.78	2D
▶ c	8.24	2E

▶ $C_{10}H_8O_2$ ▶ 7% $CDCl_3$ ▶ V 552

▶ 14-HK

O, (b) (a), C—OCH_2CH_3, (e), OH(f), (d), (c), (c)

	Shift	Peak
a	1.41	3A
b	4.40	4A
c	6.92	O
d	7.37	O
▶ e	7.83	2E
f	10.78	1A

▶ $C_9H_{10}O_3$ ▶ 45.6 mg/0.50 ml $CDCl_3$ ▶ S 196

▶ 14-H(J)

(d) H, (e) H, (a) CH_3, O, N, (c)H, H(b), (c)H, H(b), H (d), O, H (e), N, CH_3 (a)

	Shift	Peak
a	3.04	2A
b	7.13	1G
c	7.67	Q
▶ d	8.31	Q
e	10.46	1C

▶ $C_{16}H_{14}N_2O_2$ ▶ 7% $CDCl_3$ ▶ V 652

▶ 14-HK

O, (b), (c), (a), C—O—$CH_2CH_2CH_2CH_3$, (f), (f), (e), (e), NH_2(d)

	Shift	Peak
a	0.95	3D
b	~1.60	1F
c	4.25	3D
d	~4.10	1E
e	6.59	Q
▶ f	7.83	Q

▶ $C_{11}H_{15}NO_2$ ▶ 44.6 mg/0.50 ml $CDCl_3$ ▶ S 243

▶ 14-H(J)

(e)H, O, (c) NH_2, (d)H, OCH_3 (a), (d)H, H (b), H (e), O, H, O, (f)

	Shift	Peak
a	3.96	1A
b	6.52	1A
c	~ 7.30	1D
d	7.72	Q
▶ e	8.34	Q
f	14.22	1A

▶ $C_{15}H_{11}NO_4$ ▶ 7% $CDCl_3$ ▶ V 635

▶ 14-HK

(f) H, (g) H, O, (d) H_2N, CO—CH_2—CH_2—N, (e) (c), (b) (a) CH_2—CH_3, CH_2—CH_3 (b) (a), (f)H, H(g)

	Shift	Peak
a	1.05	3A
b	2.62	4E
c	2.82	3D
d	4.13	1B
e	4.33	3D
f	6.63	Q
▶ g	7.83	Q

▶ $C_{13}H_{20}N_2O_2$ ▶ 7% $CDCl_3$ ▶ V 302

▶ 14-H(J)

(b) H, O, (b) H, CH_2 (a)

	Shift	Peak
a	4.32	1A
▶ b	8.36	O

▶ $C_{14}H_{10}O$ ▶ 7% $CDCl_3$ ▶ V 621

▶ 14-HK

O, (a), C—OCH_3, (c), (c), (b), (b), OCH_3(a)

	Shift	Peak
a	3.86	1A
b	6.82	Q
▶ c	7.91	Q

▶ $C_9H_{10}O_3$ ▶ 45.0 mg/0.50 ml $CDCl_3$ ▶ S 124

▶ 14-HJ

CHO (d), C, (c)HC, CH(c), (b)HC, CH(b), C, CH_3(a)

	Shift	Peak
a	2.40	1A
b	7.36	O
c	7.59	O
▶ d	9.87	1A

▶ C_8H_8O ▶ 10% CCl_4 ▶ API 51

▶ 14-H(K)

CH_2(a), O, (b)H, O

	Shift	Peak
a	5.32	1A
▶ b	7.92	S

▶ $C_8H_6O_2$ ▶ 7% $CDCl_3$ ▶ V 496

▶ 14-HK

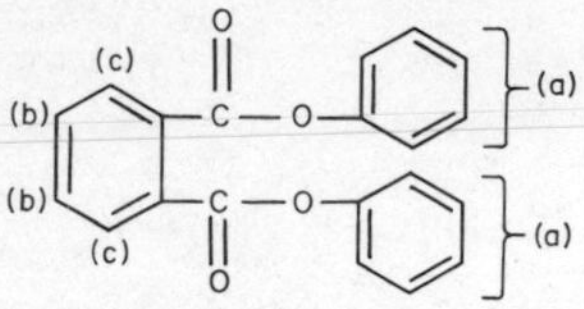

	Shift	Peak
a	7.31	2H
b	7.75	Q
▶ c	7.92	Q

▶ $C_{20}H_{14}O_4$ ▶ 45.1 mg/0.50 ml $CDCl_3$ ▶ S 93

▶ 14-HK

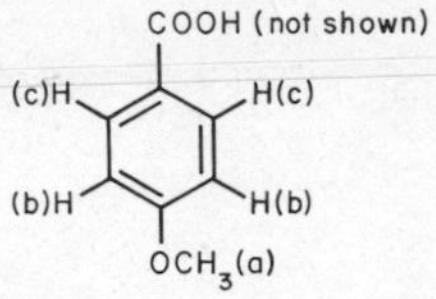

COOH (not shown)

	Shift	Peak
a	3.88	1A
b	6.98	Q
▶ c	8.08	Q

▶ $C_8H_8O_3$ ▶ 7% $CDCl_3$ ▶ V 196

▶ 14-HK

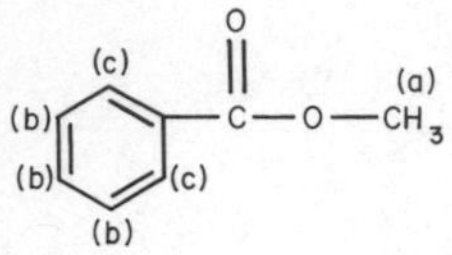

	Shift	Peak
a	3.92	1A
b	7.48	4A
▶ c	~8.0	O

▶ $C_8H_8O_2$ ▶ 45.3 mg/0.50 ml $CDCl_3$ ▶ S 78

▶ 14-HK

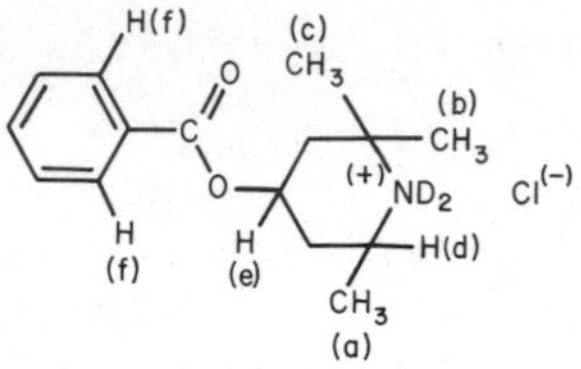

	Shift	Peak
a	1.54	1A
b	1.58	1E
c	1.62	1E
d	3.74	7X
e	5.33	6B
▶ f	8.10	4A

▶ $C_{15}H_{22}ClNO_2$ ▶ D_2O ▶ V 646

▶ 14-HK

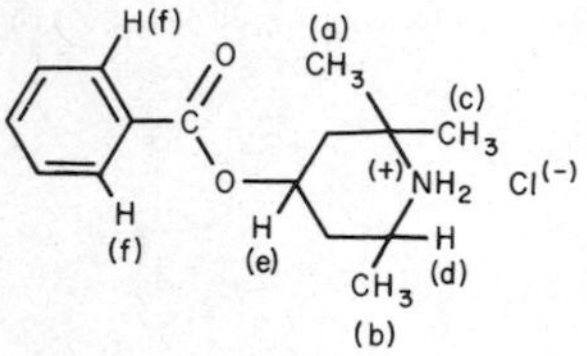

	Shift	Peak
a	1.56	1A
b	1.69	1A
c	1.75	1A
d	3.54	1D
e	5.27	1D
▶ f	8.04	N

▶ $C_{15}H_{22}ClNO_2$ ▶ 7% $CDCl_3$ ▶ V 645

▶ 14-HK

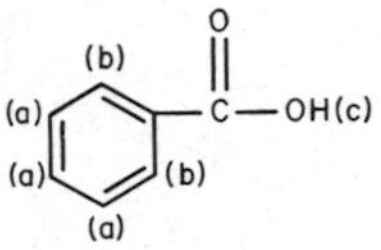

	Shift	Peak
a	7.48	N
▶ b	8.15	N
c	12.28	1G

▶ $C_7H_6O_2$ ▶ 45.1 mg/0.50 ml $CDCl_3$ ▶ S 57

▶ 14-HK

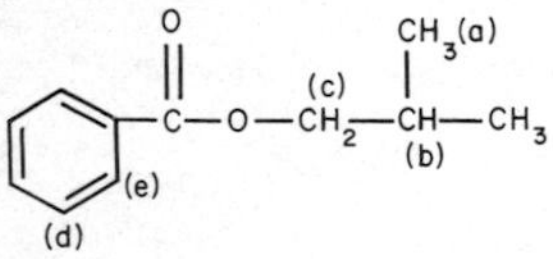

	Shift	Peak
a	1.02	2A
b	2.05	5B
c	4.12	2A
d	7.5	N
▶ e	~8.05	S

▶ $C_{11}H_{14}O_2$ ▶ 45.5 mg/0.50 ml $CDCl_3$ ▶ S 75

▶ 14-HK

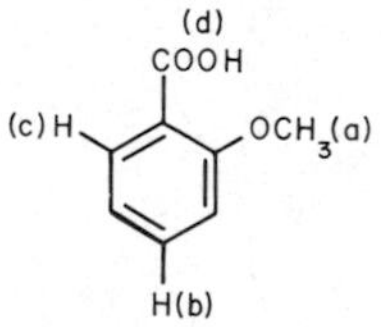

	Shift	Peak
a	4.07	1A
b	~7.60	O
▶ c	8.17	4A
d	11.00	1B

▶ $C_8H_8O_3$ ▶ 7% $CDCl_3$ ▶ V 195

▶ 14-HK

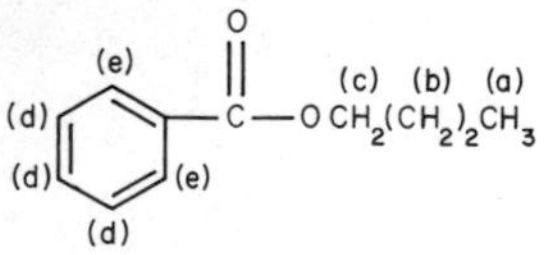

	Shift	Peak
a	0.98	3B
b	1.64	O
c	4.35	3A
d	7.48	N
▶ e	8.06	N

▶ $C_{11}H_{14}O_2$ ▶ 44.8 mg/0.50 ml $CDCl_3$ ▶ S 158

▶ 14-HK

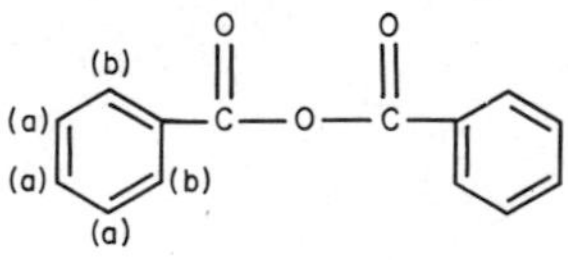

	Shift	Peak
a	7.60	N
▶ b	8.18	O

▶ $C_{14}H_{10}O_3$ ▶ 45.5 mg/0.25 ml $CDCl_3$ ▶ S 155

▶ 14-HK

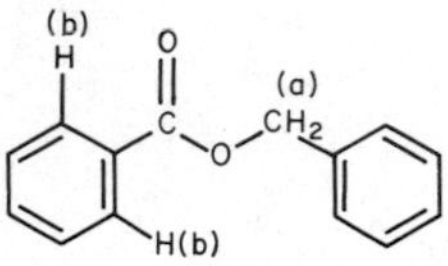

	Shift	Peak
a	5.34	1A
▶ b	8.07	S

▶ $C_{14}H_{12}O_2$ ▶ 7% $CDCl_3$ ▶ V 627

▶ 14-HL

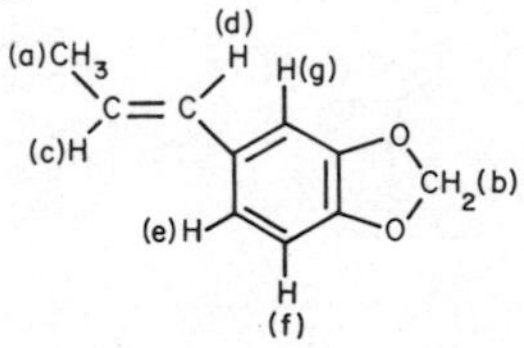

	Shift	Peak
a	1.80	2A
b	5.87	1A
c	~6.08	S
d	6.23	S
▶ e	6.70	S
f	6.70	S
g	6.83	S

▶ $C_{10}H_{10}O_2$ ▶ 7% $CDCl_3$ ▶ V 252

► 14-H(L)

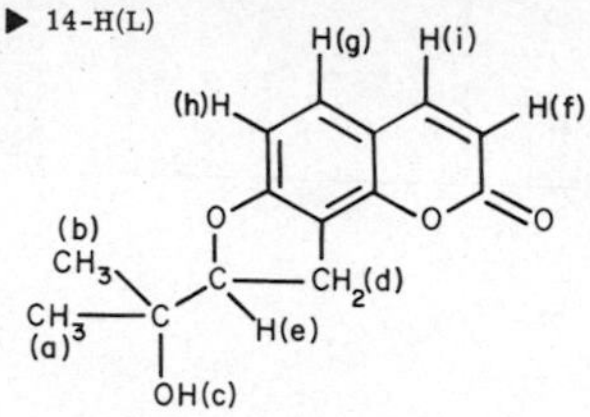

H(g) H(i) (h)H H(f) O O (b) CH_3 C CH_2(d) CH_3 (a) C H(e) OH(c)

	Shift	Peak
a	1.25 or 1.37	1A
b	1.37 or 1.25	1A
c	1.96	1G
d	3.33	2A
e	4.82	3A
f	6.22	2C
► g	6.79	2C
h	7.29	2C
i	7.66	2C

► $C_{14}H_{14}O_4$ ► 7% $CDCl_3$ ► V 310

► 14-HL

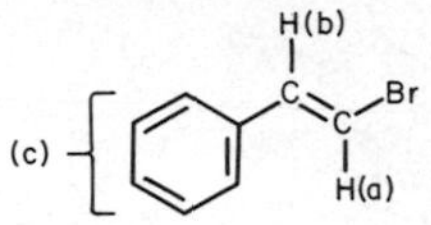

H(b) C C Br (c) H(a)

	Shift	Peak
a	6.75	2C
b	7.10	2C
► c	7.29	1A

► C_8H_7Br ► 7% $CDCl_3$ ► V 497

► 14-HL

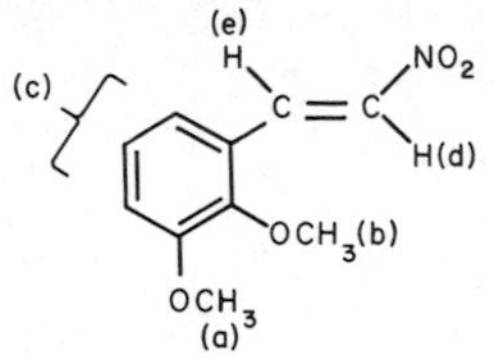

(e) H NO_2 (c) C = C H(d) OCH_3(b) OCH_3 (a)

	Shift	Peak
a	3.88 or 3.92	1A
b	3.88 or 3.92	1A
► c	7.07	1A
d	7.73	2A
e	8.20	2A

► $C_{10}H_{11}NO_4$ ► 7% $CDCl_3$ ► V 257

► 14-HL

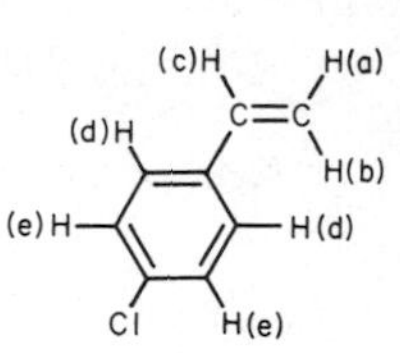

(c)H H(a) C=C (d)H H(b) (e)H H(d) Cl H(e)

	Shift	Peak
a	5.28	R
b	5.73	R
c	6.69	4D
► d	7.32	1A
e	7.32	1A

► C_8H_7Cl ► 7% $CDCl_3$ ► V 498

► 14-HL

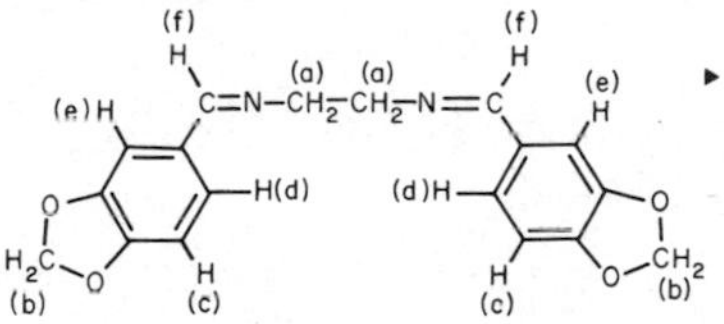

(f) H (a) (a) (f) H C=N–CH_2–CH_2–N=C (e)H (e) H H(d) (d)H O O H_2C O H O CH_2 (b) (c) H (c) (b)

	Shift	Peak
a	3.92	1A
b	5.97	1A
c	6.79	2A
► d	7.07	S
e	7.32	4A
f	8.17	2A

► $C_{18}H_{16}N_2O_4$ ► 7% $CDCl_3$ ► V 673

► 14-H(L)

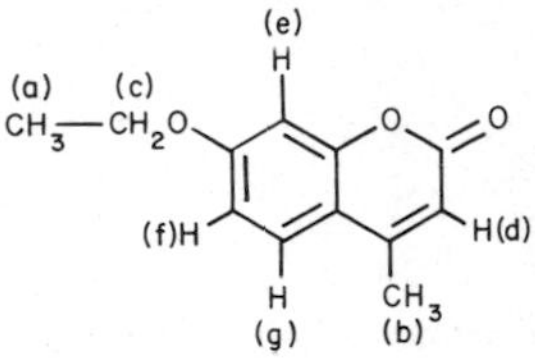

(e) H (a) (c) CH_3–CH_2O O O (f)H H(d) H CH_3 (g) (b)

	Shift	Peak
a	1.45	3A
b	2.37	2D
c	4.08	4A
d	6.08	2E
e	6.73	1G
f	~6.78	2G
► g	7.47	2E

► $C_{12}H_{12}O_3$ ► 7% $CDCl_3$ ► V 294

► 14-HL

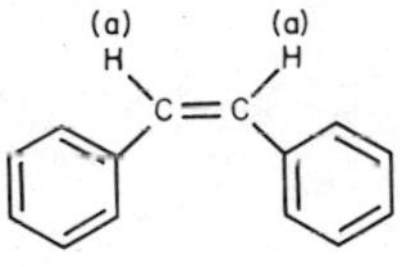

(a) (a) H H C=C

(b) = all benzene H's.

	Shift	Peak
a	6.55	1A
► b	7.18 ± .03	1A

► $C_{14}H_{12}$ ► 7% $CDCl_3$ ► V 305

► 14-H(L)

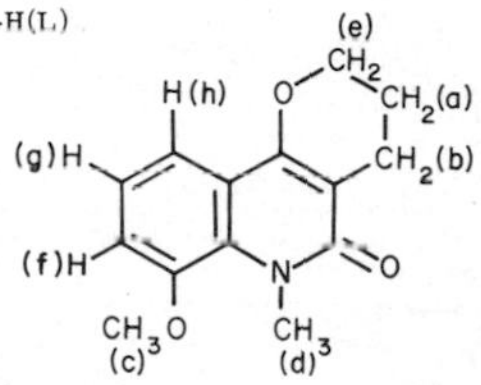

(e) CH_2 H(h) O CH_2(a) (g)H CH_2(b) (f)H N O CH_3O CH_3 (c) (d)

	Shift	Peak
a	2.03	5B
b	2.63	3A
c	3.87 or 3.92	1A
d	3.87 or 3.92	1A
e	4.30	3A
f	~7.00	R
g	~7.10	R
► h	7.52	4D

► $C_{14}H_{15}NO_3$ ► 7% $CDCl_3$ ► V 312

► 14-H(L)

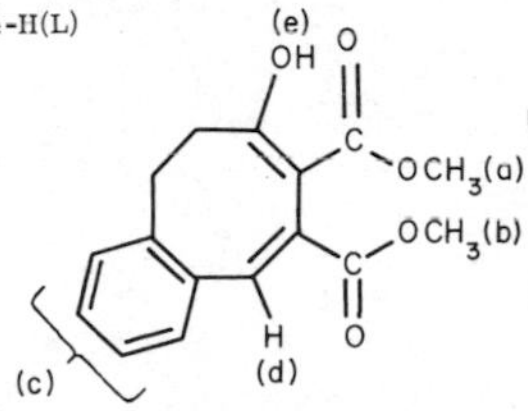

(e) OH O C OCH_3(a) OCH_3(b) C H O (c) (d)

	Shift	Peak
a	3.65 or 3.80	1A
b	3.80 or 3.65	1A
► c	7.20	1A
d	7.83	1A
e	12.69	1A

► $C_{16}H_{16}O_5$ ► 7% $CDCl_3$ ► V 656

► 14-HL

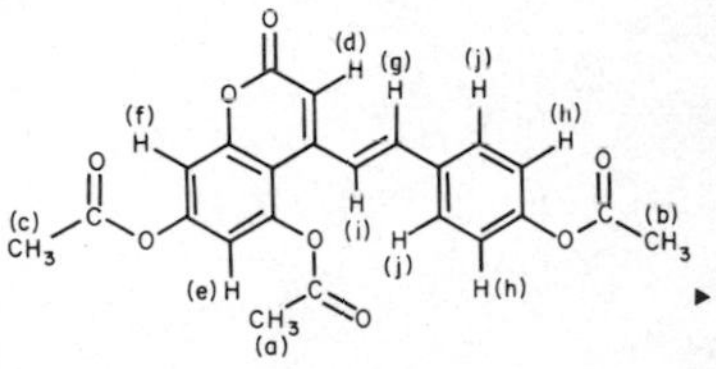

O (d) H (g) H (j) H (f) H (h) H O O O (c) CH_3 C O H (i) H (j) O C (b) CH_3 (e) H H(h) CH_3 C O (a)

	Shift	Peak
a	2.06	1A
b	2.32	1A
c	2.32	1A
d	6.36	1G
e	6.83*	S
f	6.83*	S
g	6.89	S
h	7.15	S
i	7.33	S
► j	7.53	2E

*or 7.08

► $C_{23}H_{18}O_8$ ► 7% $CDCl_3$ ► V 692

► 14-HL

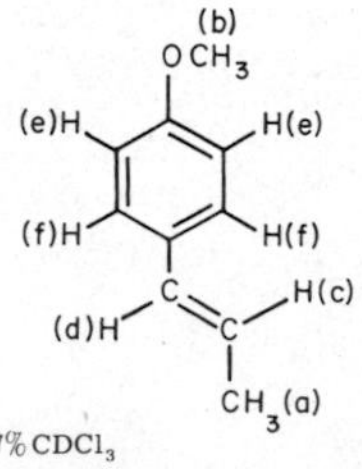

(b) OCH_3 (e)H H(e) (f)H H(f) C H(c) (d)H C CH_3(a)

	Shift	Peak
a	1.83	2A
b	3.75	1A
c	6.08	N
d	6.28	N
e	6.80	Q
► f	7.23	Q

► $C_{10}H_{12}O$ ► 7% $CDCl_3$ ► V 258

► 14-H(L)

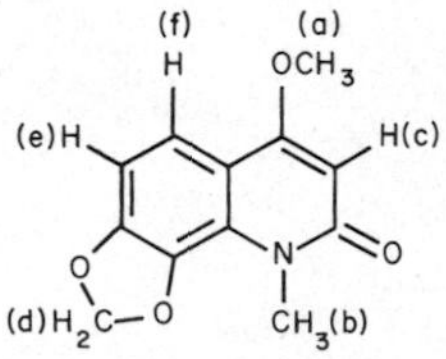

(f) (a) H OCH_3 (e)H H(c) O N O (d)H_2C–O CH_3(b)

	Shift	Peak
a	3.83 or 3.92	1A
b	3.83 or 3.92	1A
c	5.92	1A
d	6.07	1A
e	6.80	2C
► f	7.57	2C

► $C_{12}H_{11}NO_4$ ► 7% $CDCl_3$ ► V 291

▶ 14-HL

	Shift	Peak
a	2.18	1A
b	3.86	1A
c	3.97	1A
d	6.87	Q
▶ e	7.58	Q

▶ $C_{10}H_{13}NO_2$ ▶ 7% $CDCl_3$ ▶ V 561

▶ 14-H(L)

	Shift	Peak
a	2.99	1A
b	3.07	1A
c	7.69	1B
▶ d	8.27	4C
e	10.76	1C

▶ $C_{12}H_{12}N_2O_2$ ▶ 7% $CDCl_3$ ▶ V 594

▶ 14-H(L)

	Shift	Peak
a	1.46*	1A
b	3.58	1A
c	5.62†	2A
d	5.62†	2A
e	6.45‡	2A
f	6.45‡	2A
g	6.62	S
h	7.12	2A
▶ i	7.88§	1A
j	7.88§	1A

*or 1.49 ‡or 6.60
†or 5.70 §or 7.90

▶ $C_{26}H_{24}O_5$ ▶ 7% $CDCl_3$ ▶ V 696

▶ 14-HM

	Shift	Peak
▶ a	7.13	1E
b	7.23	1E

▶ C_7H_5NO ▶ 45.1 mg/0.50 ml $CDCl_3$ ▶ S 80

▶ 14-H(L)

	Shift	Peak
a	1.40	3A
b	3.87	1A
c	4.40	4A
d	6.97	2A
e	7.02	1A
▶ f	7.98	2A
g	~15.33	1D

▶ $C_{12}H_{14}O_4$ ▶ 7% $CDCl_3$ ▶ V 295

▶ 14-H(M)

	Shift	Peak
a	1.56	1A
b	2.87	1B
c	4.10	1A
▶ d	7.51 ± 0.05	1G

▶ $C_{12}H_{16}ClN$ ▶ 7% $CDCl_3$ ▶ V 599

▶ 14-HL

	Shift	Peak
a	3.96	1A
▶ b	8.03	4A
c	8.27	1A

▶ $C_{11}H_9NO_2$ ▶ 7% $CDCl_3$ ▶ V 576

▶ 14-HN

	Shift	Peak
a	3.51	1A
b	6.03	1A
▶ c	6.11	2D
d	6.93	3C

▶ $C_6H_8N_2$ ▶ 7% $CDCl_3$ ▶ V 458

▶ 14-HL

	Shift	Peak
a	2.45	1A
▶ b	8.05	S
c	8.11	S

▶ $C_{11}H_8N_2$ ▶ 7% $CDCl_3$ ▶ V 283

▶ 14-H(N)

	Shift	Peak
a	1.42	1A
b	1.95	3A
c	2.55	1A
d	2.70	3D
e	2.82	1E
f	2.92	1E
g	4.12	1A
h	5.33	1D
▶ i	6.37	2A
j	6.78	1E
k	6.87	2H

▶ $C_{15}H_{21}N_3O_2$ ▶ 7% $CDCl_3$ ▶ V 319

▶ 14-H(L)

	Shift	Peak
a	7.34	4A
▶ b	8.18	2D
c	8.32	2E
d	8.48	2D
e	11.63	1D

▶ $C_7H_5N_3O_2$ ▶ 7% $CDCl_3$ ▶ V 485

▶ 14-HN

	Shift	Peak
a	2.95	1A
b	4.63	1D
▶ c	6.55	Q
d	8.10	Q

▶ $C_7H_8N_2O_2$ ▶ 7% $CDCl_3$ ▶ V 489

▶ 14-HN

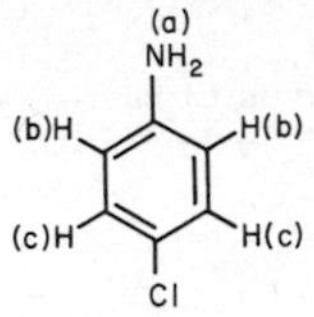

(a) NH_2 (b)H H(b) (c)H H(c) Cl

Shift	Peak
a 3.60	1A
▶ b 6.57	Q
c 7.05	Q

▶ C_6H_6ClN ▶ 7% $CDCl_3$ ▶ V 123

▶ 14-HN

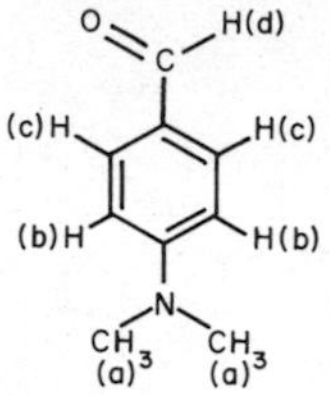

O C H(d) (c)H H(c) (b)H H(b) N CH_3 (a) CH_3 (a)

Shift	Peak
a 3.05	1A
▶ b 6.69	Q
c 7.71	Q
d 9.70	1A

▶ $C_9H_{11}NO$ ▶ 7% $CDCl_3$ ▶ V 238

▶ 14-HN

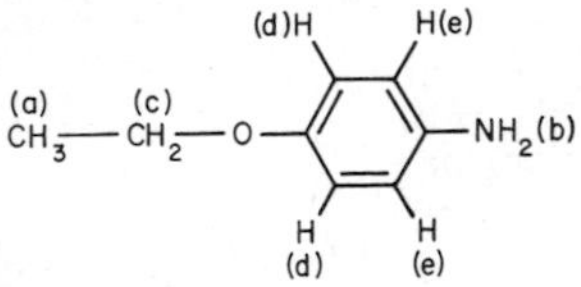

(d)H H(e) (a) CH_3—(c) CH_2—O— —NH_2(b) H (d) H (e)

Shift	Peak
a 1.35	3A
b 3.30	1B
c 3.93	4A
d 6.63 or 6.70	Q
▶ e 6.63 or 6.70	Q

▶ $C_8H_{11}NO$ ▶ 7% $CDCl_3$ ▶ V 208

▶ 14-HN

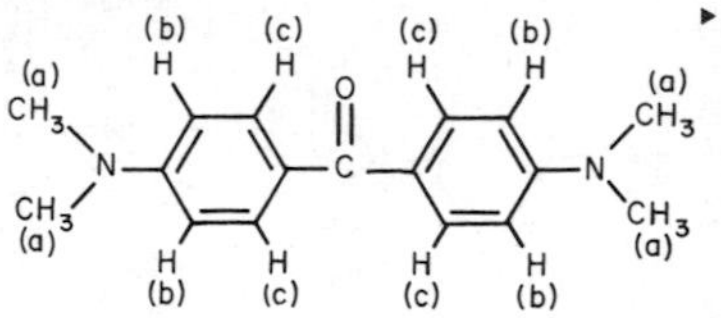

(b) H (c) H (c) H (b) H (a) CH_3 N CH_3 (a) O C (a) CH_3 N CH_3 (a) H (b) H (c) H (c) H (b)

Shift	Peak
a 3.04	1A
▶ b 6.70	Q
c 7.77	Q

▶ $C_{17}H_{20}N_2O$ ▶ 7% $CDCl_3$ ▶ V 329

▶ 14-HN

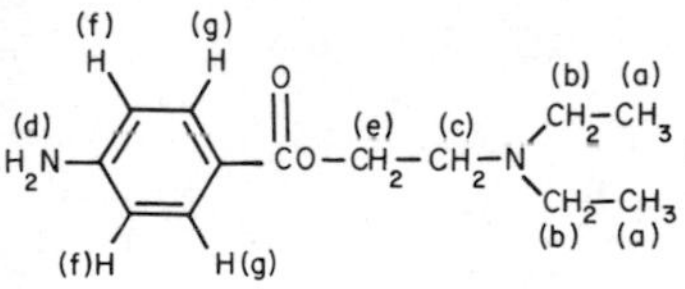

(f) H (g) H O (d) H_2N— —CO—(e) CH_2—(c) CH_2—N (b) CH_2—(a) CH_3 (b) CH_2—(a) CH_3 (f)H H(g)

Shift	Peak
a 1.05	3A
b 2.62	4E
c 2.82	3D
d 4.13	1B
e 4.33	3D
▶ f 6.63	Q
g 7.83	Q

▶ $C_{13}H_{20}N_2O_2$ ▶ 7% $CDCl_3$ ▶ V 302

▶ 14-HN

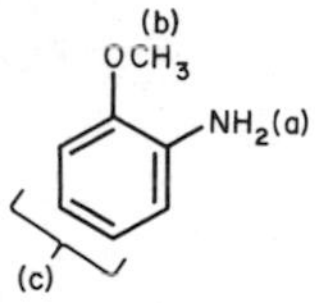

(b) OCH_3 NH_2(a) (c)

Shift	Peak
a 3.73	1A
b 3.85	1A
▶ c ~6.80	1A

▶ C_7H_9NO ▶ 7% $CDCl_3$ ▶ V 172

▶ 14-HN

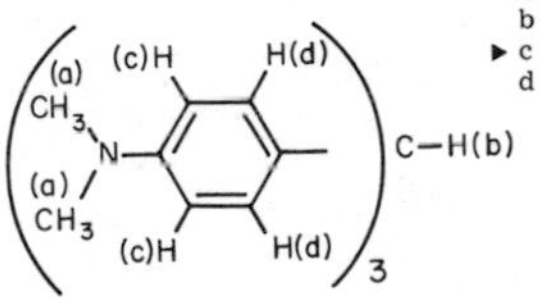

(a) CH_3 N (a) CH_3 (c)H H(d) (c)H H(d) C—H(b) 3

Shift	Peak
a 2.87	1A
b 5.29	1B
▶ c 6.64	Q
d 6.95	Q

▶ $C_{25}H_{31}N_3$ ▶ 7% $CDCl_3$ ▶ V 360

▶ 14-H(N)

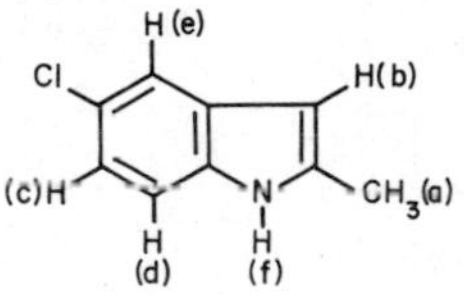

H(e) Cl H(b) (c)H N CH_3(a) H (d) H (f)

Shift	Peak
a 2.33	2D
b 6.10	1B
c 7.02	2D
▶ d 7.02	2D
e 7.44	1I
f ~7.65	1D

▶ C_9H_8ClN ▶ 7% $CDCl_3$ ▶ V 228

▶ 14-HN

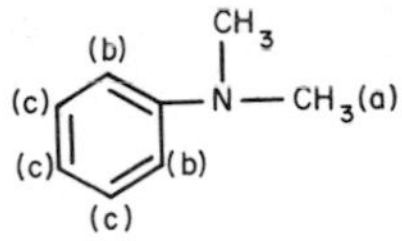

CH_3 (b) (c) N—CH_3(a) (c) (b) (c)

Shift	Peak
a 2.93	1A
▶ b ~6.67	N
c ~7.15	O

▶ $C_8H_{11}N$ ▶ 45.3 mg/0.50 ml $CDCl_3$ ▶ S 1

▶ 14-HN

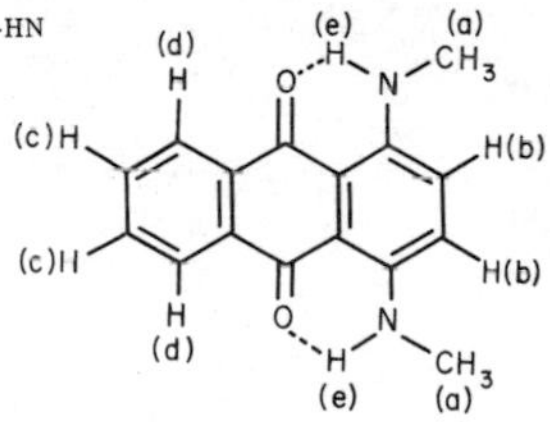

(d) H (e) H (a) O N CH_3 (c)H H(b) (c)H H(b) H (d) O H N CH_3 (e) (a)

Shift	Peak
a 3.04	2A
▶ b 7.13	1G
c 7.67	Q
d 8.31	Q
e 10.46	1C

▶ $C_{16}H_{14}N_2O_2$ ▶ 7% $CDCl_3$ ▶ V 652

▶ 14-HN

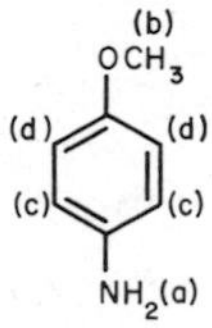

(b) OCH_3 (d) (d) (c) (c) NH_2(a)

Shift	Peak
a 3.40	1A
b 3.73	1A
▶ c ~6.68 or 6.73	Q
d ~6.68 or 6.73	Q

▶ C_7H_9NO ▶ 7% $CDCl_3$ ▶ V 171

▶ 14-HN

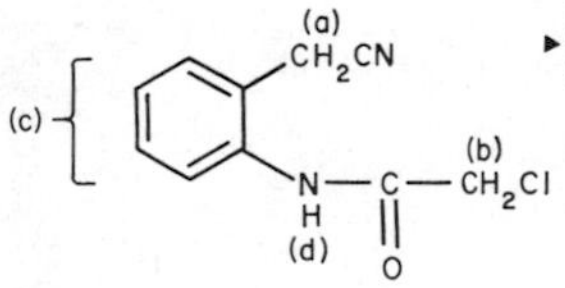

(a) CH_2CN (c) N—C—(b) CH_2Cl H (d) O

Shift	Peak
a 3.65	1A
b 4.17	1A
▶ c 7.31 ± .05	1H
d 8.32	1C

▶ $C_{10}H_9ClN_2O$ ▶ 7% $CDCl_3$ ▶ V 248

▶ 14-HN

(d) (e) H H (c)H N C CH_3(a) O (b)CH_3 H (d) H (c)

	Shift	Peak
a	2.12	1A
b	2.30	1A
c	7.12	Q
▶ d	7.37	Q
e	7.88	1D

▶ $C_9H_{11}NO$ ▶ 7% $CDCl_3$ ▶ V 239

▶ 14-HN

(d) (f) (e) (d) (f) (a) (b) (c) (a) $NH\,CH_2CH_2OH$

	Shift	Peak
a	3.12	1E
b	3.23	3A
c	3.78	3A
▶ d	6.61	S
e	6.78	S
f	7.17	3B

▶ $C_8H_{11}NO$ ▶ 45.4 mg/0.50 ml $CDCl_3$ ▶ S 295

▶ 14-HN

(a) (c) CH_3—CH_2—O (d) (e) H H (f) H N (b) CH_3 C O H H (d) (e)

	Shift	Peak
a	1.38	3A
b	2.12	1A
c	4.00	4A
d	6.83	Q
▶ e	7.41	Q
f	7.91	1C

▶ $C_{10}H_{13}NO_2$ ▶ 7% $CDCl_3$ ▶ V 267

▶ 14-HN

(b) (b) (b) (b) (a) NH_2 NH_2 (a)

	Shift	Peak
a	3.29	1A
▶ b	6.67	1A

▶ $C_6H_8N_2$ ▶ 44.6 mg/0.50 ml $CDCl_3$ ▶ S 112

▶ 14-H(N)

O (a)H_3C (a)CH_3 H (d) CH_3(c) N N CH_3 (b) (e)

	Shift	Peak
a	1.28	1A
b	1.60	2A
c	2.57	1A
d	5.87	2A
▶ e	7.46	1A

▶ $C_{17}H_{18}N_2O$ ▶ 7% $CDCl_3$ ▶ V 664

▶ 14-HN

(a) NH_2 Cl (b) (c) (c) (c)

	Shift	Peak
a	3.9	1C
▶ b	6.68	2H
c	7.10	O

▶ C_6H_6ClN ▶ 44.8 mg/0.50 ml $CDCl_3$ ▶ S 59

▶ 14-HN

(c) NH_2 (d) CH_3(b) (e) (e) CH_3 (a)

	Shift	Peak
a	2.12	1A
b	2.22	1A
c	3.34	1B
▶ d	6.50	2A
e	6.76	2B, 2G

▶ $C_8H_{11}N$ ▶ 44.9 mg/0.50 ml $CDCl_3$ ▶ S 237

▶ 14-HN

(a) NH_2 (c) (c) (d) (d) O—CH_3 (b)

	Shift	Peak
a	3.38	1A
b	3.73	1A
▶ c	~6.68 or 6.72	Q
d	~6.68 or 6.72	Q

▶ C_7H_9NO ▶ 45.2 mg/0.50 ml $CDCl_3$ ▶ S 94

▶ 14-HN

(d) (c) (b) (a) NH—CH_2—$(CH_2)_2$—CH_3 (e) (e) (f) (f) (f)

	Shift	Peak
a	0.96	3B
b	1.50	1F
c	3.08	3A
d	3.42	1A
▶ e	6.52	3C
f	7.12	3C

▶ $C_{10}H_{15}N$ ▶ 45.4 mg/0.50 ml $CDCl_3$ ▶ S 280

▶ 14-HN

(b) NH_2 (c) (a) OCH_2CH_3 (d)

	Shift	Peak
a	1.40	3A
b	3.72	1B
c	4.03	4A
▶ d	6.72	1A

▶ $C_8H_{11}NO$ ▶ 45.1 mg/0.50 ml $CDCl_3$ ▶ S 236

▶ 14-HN

O (b) (c) (a) C—O—$CH_2CH_2CH_2CH_3$ (f) (f) (e) (e) NH_2(d)

	Shift	Peak
a	0.95	3B
b	~1.60	1F
c	4.25	3D
d	~4.10	1E
▶ e	6.59	Q
f	7.83	Q

▶ $C_{11}H_{15}NO_2$ ▶ 44.6 mg/0.50 ml $CDCl_3$ ▶ S 243

▶ 14-HN

(a) NH_2 (b) OCH_3 (c)

	Shift	Peak
a	3.70	1B
b	3.82	1A
▶ c	6.73	1A

▶ C_7H_9NO ▶ 45.2 mg/0.50 ml $CDCl_3$ ▶ S 192

► 14-H(N)

(a) H, N, (b) (b) (b) (b), N, (c) (c) (e) (e) (d)

	Shift	Peak
a	1.78	1A
b	3.08	1A
► c	6.88	2H
d	7.05	S
e	7.18	2H

► $C_{10}H_{14}N_2$ ► 45.2 mg/0.50 ml $CDCl_3$ ► S 300

► 14-HO

(b)CH_3O, H(c), (d)H, H(e), (a) (a) H—H, (d)H, H(e), (b)CH_3O, H(c)

	Shift	Peak
a	2.22	3A
b	3.79	1A
c	4.17	5A
► d	6.50	1A
e	6.83	3A

► $C_{13}H_{14}O_2$ ► 7% $CDCl_3$ ► V 608

► 14-H(N)

(f) H, (b) CH_3, H(c), (e)H, N, CH_3(a), H H, (d) (g)

	Shift	Peak
a	2.28	1A
b	2.40	1A
c	6.10	1I
► d	~6.90 or 7.02	N
e	~6.90 or 7.02	N
f	7.27	S
g	~7.30	S

► $C_{10}H_{11}N$ ► 7% $CDCl_3$ ► V 255

► 14-HO

OCH_3(c), O, OCH_3 (b), O, H(i), (d), (h)H, H, H, H(j), (e), N—CH_3(a), O, (f)H_2C, O, H(g)

	Shift	Peak
a	2.53	1E
b	3.88*	1A
c	3.88*	1A
d	3.97	S
e	5.47	2A
f	5.88	1A
g	6.38†	1E
h	6.38†	1E
► i	6.52	2H
j	7.07	2C

*or 4.05
†or 6.57

► $C_{21}H_{21}NO_6$ ► 7% $CDCl_3$ ► V 347

► 14-HN

(a) CH_3, (b) CH_3—C=N—NH, (f), NO_2, (e), (c), NO_2, (d)

	Shift	Peak
a	2.10 or 2.20	1A
b	2.10 or 2.20	1A
► c	7.85	2A
d	8.18	4A
e	8.99	2A
f	10.04	1C

► $C_9H_{10}N_4O_4$ ► 45.0 mg/0.25 ml $CDCl_3$ ► S 199

► 14 H(O)

H_3C(a), H(h), (j), (b) (e), O, O, H, OCH_3 H, (a)CH_3, (a)H, H(d), CH_3 (a), (f)H (i)H, O, H (i), O, H(g), CH_3(a)

	Shift	Peak
a	1.46*	1A
b	3.58	1A
c	5.62†	2A
d	5.62†	2A
e	6.45‡	2A
f	6.45‡	2A
► g	6.63	S
h	7.12	2A
i	7.88§	1A
j	7.88§	1A

*or 1.49 ‡or 6.60
†or 5.70 §or 7.90

► $C_{26}H_{24}O_5$ ► 7% $CDCl_3$ ► V 696

► 14-HN

(e) H, O_2N, NO_2, (d)H, N—N=C, CH_3(a), CH_3(b), H (c), H (f)

	Shift	Peak
a	2.10 or 2.18	1A
b	2.10 or 2.18	1A
► c	7.90	2A
d	8.23	4A
e	9.04	2A
f	10.95	1B

► $C_9H_{10}N_4O_4$ ► 7% $CDCl_3$ ► V 233

► 14-HO

(d)H H(e), (a) (c) CH_3—CH_2—O, NH_2(b), H H, (d) (e)

	Shift	Peak
a	1.35	3A
b	3.30	1B
c	3.93	4A
► d	6.63 or 6.70	Q
e	6.63 or 6.70	Q

► $C_8H_{11}NO$ ► 7% $CDCl_3$ ► V 208

► 14-HN

O, O, (c) (a), C, CH_2—CH_3, (d)CF_2H, CH_2(b), C, N, (e)H, H(f), N, NO_2, (h)H, NO_2, H(g)

	Shift	Peak
a	1.33	3A
b	3.67	1A
c	4.30	4A
d	6.30	3A*
► e	8.03	2C
f	8.40	4A
g	9.08	2A
h	11.77	1B

*Very large splitting due to fluorine

► $C_{12}H_{12}F_2N_4O_6$ ► 7% $CDCl_3$ ► V 293

► 14-HO

(a) OCH_3, (b)H, H(b), (c)H, H(c), I

	Shift	Peak
a	3.75	1A
► b	6.67	Q
c	7.53	Q

► C_7H_7IO ► 7% $CDCl_3$ ► V 153

► 14-HO

NH_2(c), (e)H, OCH_3(a), (b)CH_3O, H(f), H (d)

	Shift	Peak
a	3.72 or 3.79	1A
b	3.72 or 3.79	1A
c	~3.75	1E
► d	6.25	S
e	6.32	S
► f	6.71	2A

► $C_8H_{11}NO_2$ ► 7% $CDCl_3$ ► V 508

► 14-HO

OH(c), (d), O—CH_2—CH_3, (b) (a)

	Shift	Peak
a	1.36	3A
b	3.93	4A
c	5.45	1C
► d	6.67	1A

► $C_8H_{10}O_2$ ► 45.0 mg/0.50 ml $CDCl_3$ ► S 12

▶ 14-H(O)

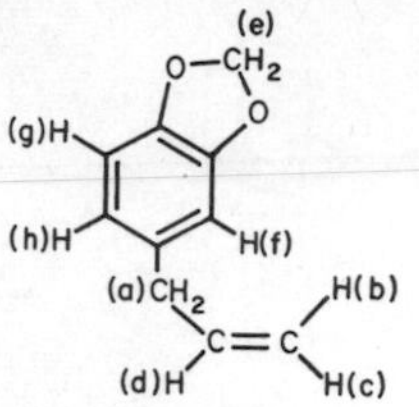

	Shift	Peak
a	3.30	2E
b	5.03	2E
c	5.03	2E
d	~5.88	O
e	5.88	1A
f	6.67 ± .03	1H
▶ g	6.67 ± .03	1H
h	6.67 ± .03	1H

▶ $C_{10}H_{10}O_2$ ▶ 7% $CDCl_3$ ▶ V 253

▶ 14-HO

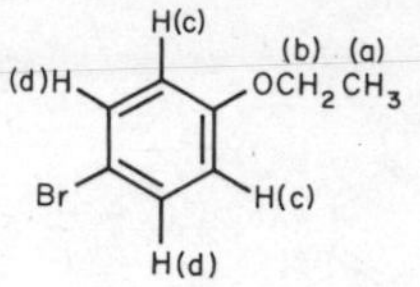

	Shift	Peak
a	1.37	3A
b	3.93	4A
▶ c	6.73	Q
d	7.28	Q

▶ C_8H_9BrO ▶ 7% $CDCl_3$ ▶ V 198

▶ 14-HO

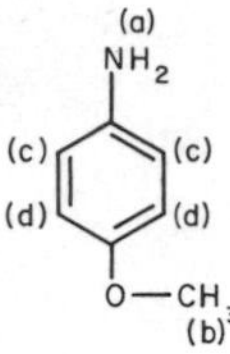

	Shift	Peak
a	3.38	1A
b	3.73	1A
c	~6.68 or 6.72	Q
▶ d	~6.68 or 6.72	Q

▶ C_7H_9NO ▶ 45.2 mg/0.50 ml $CDCl_3$ ▶ S 94

▶ 14-HO

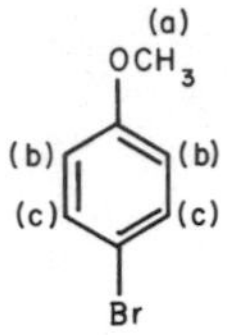

	Shift	Peak
a	3.78	1A
▶ b	6.75	Q
c	7.37	Q

▶ C_7H_7BrO ▶ 45.5 mg/0.50 ml $CDCl_3$ ▶ S 141

▶ 14-HO

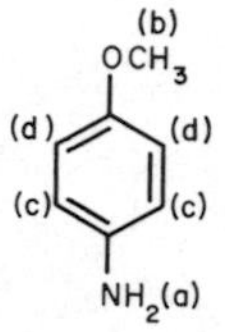

	Shift	Peak
a	3.40	1A
b	3.73	1A
c	~6.68 or 6.73	Q
▶ d	~6.68 or 6.73	Q

▶ C_7H_9NO ▶ 7% $CDCl_3$ ▶ V 171

▶ 14-HO

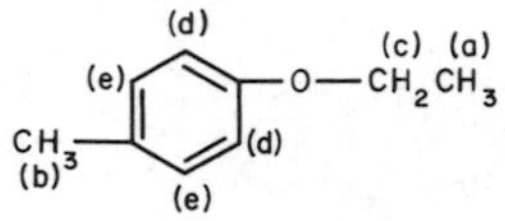

	Shift	Peak
a	1.38	3A
b	2.28	1A
c	3.98	4A
▶ d	6.75	Q
e	7.05	Q

▶ $C_9H_{12}O$ ▶ 45.0 mg/0.50 ml $CDCl_3$ ▶ S 284

▶ 14-H(O)

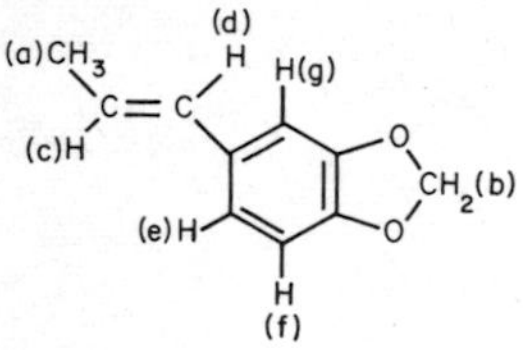

	Shift	Peak
a	1.80	2A
b	5.87	1A
c	~6.08	S
d	6.23	S
e	6.70	S
▶ f	6.70	S
g	6.83	S

▶ $C_{10}H_{10}O_2$ ▶ 7% $CDCl_3$ ▶ V 252

▶ 14-HO

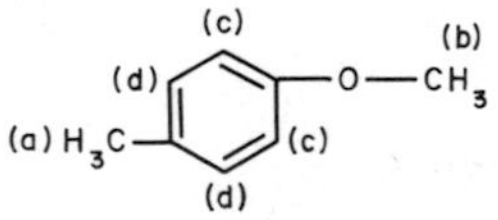

	Shift	Peak
a	2.29	1A
b	3.78	1A
▶ c	6.76	Q
d	7.08	Q

▶ $C_8H_{10}O$ ▶ 45.0 mg/0.50 ml $CDCl_3$ ▶ S 292

▶ 14-HO

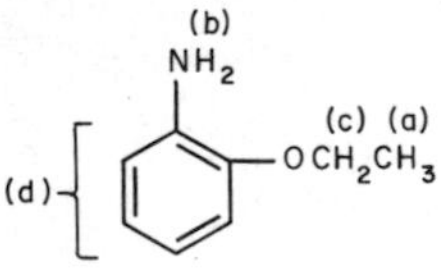

	Shift	Peak
a	1.40	3A
b	3.72	1B
c	4.03	4A
▶ d	6.72	1A

▶ $C_8H_{11}NO$ ▶ 45.1 mg/0.50 ml $CDCl_3$ ▶ S 236

▶ 14-HO

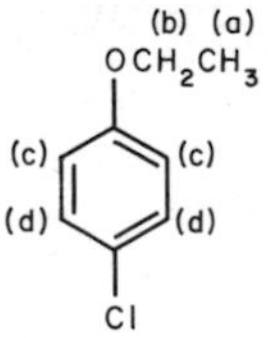

	Shift	Peak
a	0.90	3A
b	3.99	4A
▶ c	6.78	Q
d	7.21	Q

▶ C_8H_9ClO ▶ 45.4 mg/0.50 ml $CDCl_3$ ▶ S 235

▶ 14-HO

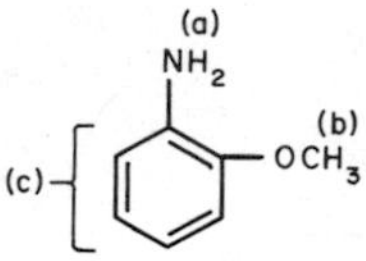

	Shift	Peak
a	3.70	1B
b	3.82	1A
▶ c	6.73	1A

▶ C_7H_9NO ▶ 45.2 mg/0.50 ml $CDCl_3$ ▶ S 192

▶ 14-HO

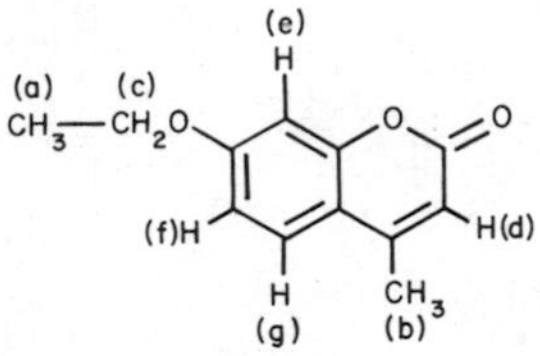

	Shift	Peak
a	1.45	3A
b	2.37	2D
c	4.08	4A
d	6.08	2E
e	6.73	1G
▶ f	~6.78	2G
g	7.47	2E

▶ $C_{12}H_{12}O_3$ ▶ 7% $CDCl_3$ ▶ V 294

14-H(O)

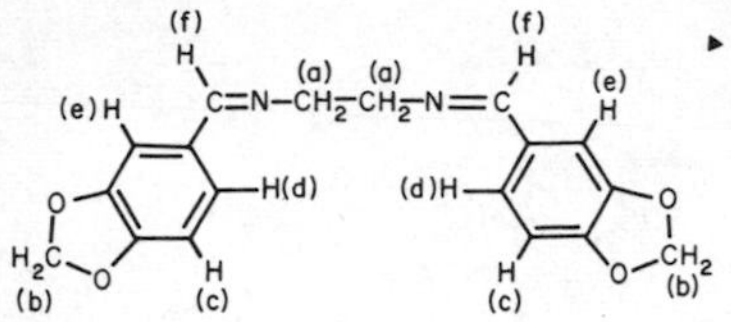

	Shift	Peak
a	3.92	1A
b	5.97	1A
► c	6.79	2A
d	7.07	S
e	7.32	4A
f	8.17	2A

► $C_{18}H_{16}N_2O_4$ ► 7% $CDCl_3$ ► V 673

14-H(O)

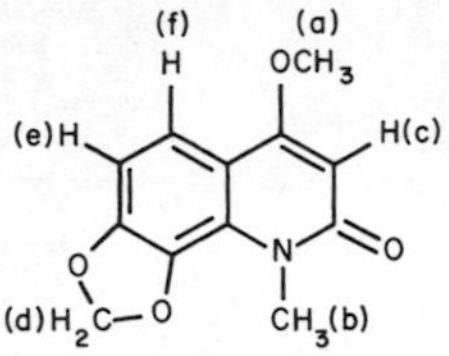

	Shift	Peak
a	3.83 or 3.92	1A
b	3.83 or 3.92	1A
c	5.92	1A
d	6.07	1A
► e	6.80	2C
f	7.57	2C

► $C_{12}H_{11}NO_4$ ► 7% $CDCl_3$ ► V 291

14-HO

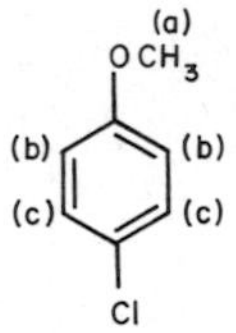

	Shift	Peak
a	3.77	1A
► b	6.80	Q
c	7.20	Q

► C_7H_7ClO ► 44.8 mg/0.50 ml $CDCl_3$ ► S 164

14-HO

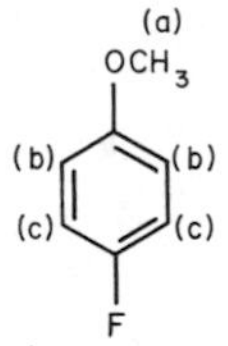

	Shift	Peak
a	3.77	1A
► b	6.82	S
c	6.95	S

► C_7H_7FO ► 45.2 mg/0.50 ml $CDCl_3$ ► S 233

14-HO

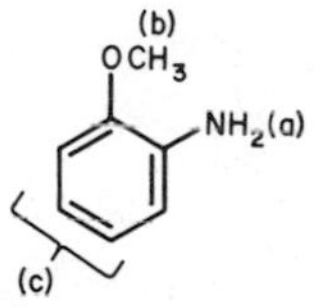

	Shift	Peak
a	3.73	1A
b	3.85	1A
► c	~6.80	1A

► C_7H_9NO ► 7% $CDCl_3$ ► V 172

14-HO

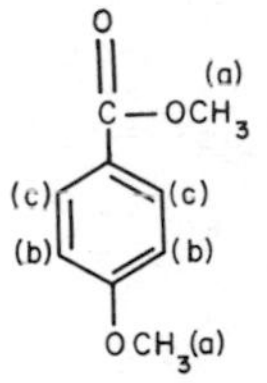

	Shift	Peak
a	3.86	1A
► b	6.82	Q
c	7.91	Q

► $C_9H_{10}O_3$ ► 45.0 mg/0.50 ml $CDCl_3$ ► S 124

14-HO

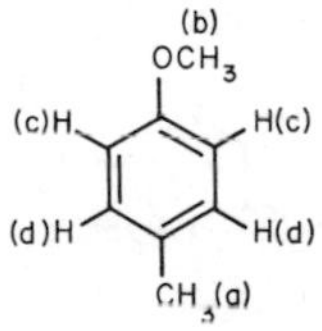

	Shift	Peak
a	2.28	1A
b	3.75	1A
► c	6.80	Q
d	7.05	Q

► $C_8H_{10}O$ ► 7% $CDCl_3$ ► V 205

14-HO

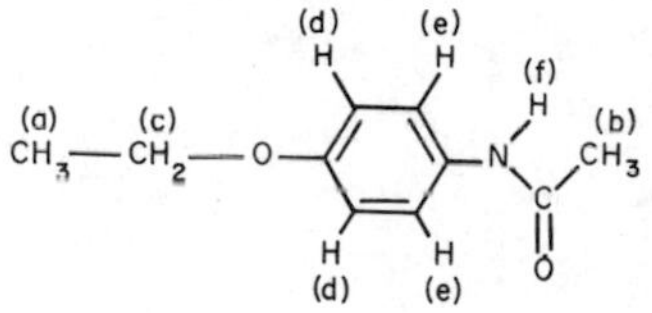

	Shift	Peak
a	1.38	3A
b	2.12	1A
c	4.00	4A
► d	6.83	Q
e	7.41	Q
f	7.91	1C

► $C_{10}H_{13}NO_2$ ► 7% $CDCl_3$ ► V 267

14-HO

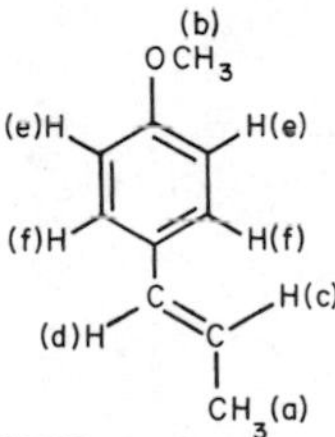

	Shift	Peak
a	1.83	2A
b	3.75	1A
c	6.08	N
d	6.28	N
► e	6.80	Q
f	7.23	Q

► $C_{10}H_{12}O$ ► 7% $CDCl_3$ ► V 258

14-HO

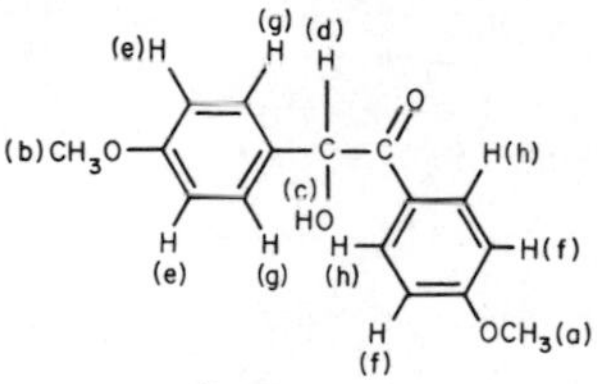

	Shift	Peak
a	3.75*	1A
b	3.75*	1A
c	4.54	1C
d	5.82	1B
► e	6.83	S
► f	6.87	S
g	7.24	2E
h	7.89	2E

*or 3.81

► $C_{16}H_{16}O_4$ ► 7% $CDCl_3$ ► V 655

14-HO

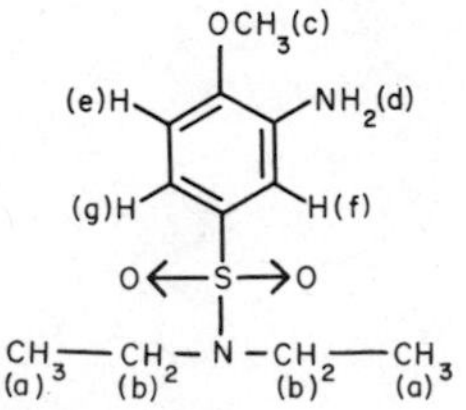

	Shift	Peak
a	1.11	3A
b	3.20	4A
c	3.88	1E
d	3.88	1E
► e	6.80	2E
f	7.13	1A
g	7.17	2A

► $C_{11}H_{18}N_2O_3S$ ► 7% $CDCl_3$ ► V 587

14-HO

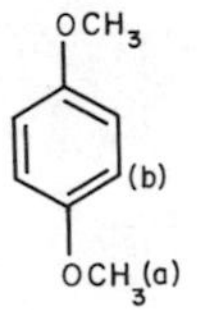

	Shift	Peak
a	3.78	1A
► b	6.85	1A

► $C_8H_{10}O_2$ ► 45.1 mg/0.50 ml $CDCl_3$ ► S 56

▶ 14-HO

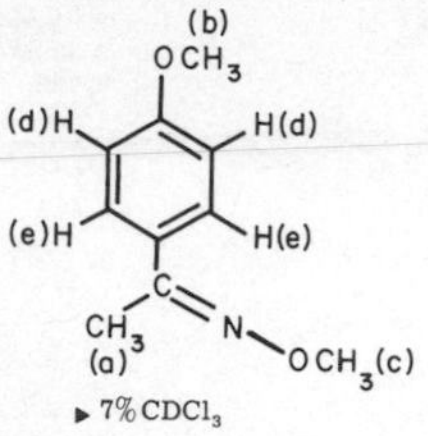

Shift	Peak
a 2.18	1A
b 3.86	1A
c 3.97	1A
▶ d 6.87	Q
e 7.58	Q

▶ $C_{10}H_{13}NO_2$ ▶ 7% $CDCl_3$ ▶ V 561

▶ 14-HO

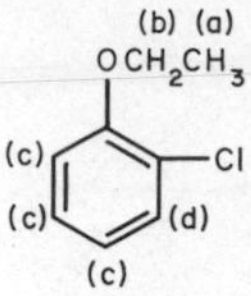

Shift	Peak
a 0.96	3A
b 4.08	4A
▶ c 6.95	O
d 7.34	N

▶ C_8H_9ClO ▶ 45.2 mg/0.50 ml $CDCl_3$ ▶ S 234

▶ 14-HO

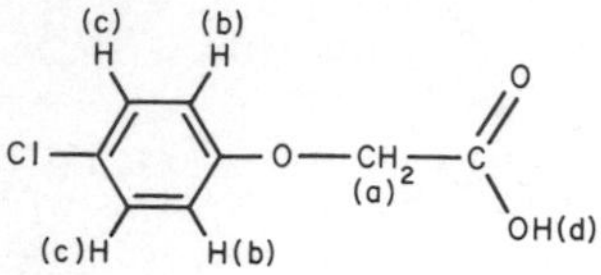

Shift	Peak
a 4.67	1A
▶ b 6.88	Q
c 7.27	Q
d 8.86	1C

▶ $C_8H_7ClO_3$ ▶ 7% $CDCl_3$ ▶ V 500

▶ 14-HO

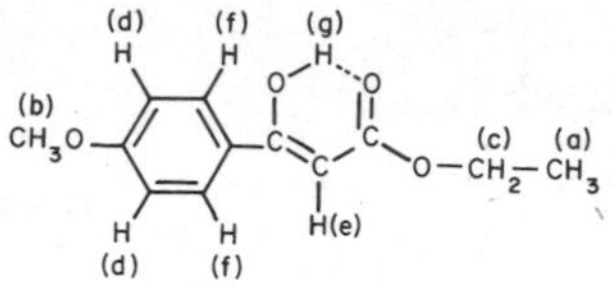

Shift	Peak
a 1.40	3A
b 3.87	1A
c 4.40	4A
▶ d 6.97	2A
e 7.02	1A
f 7.98	2A
g ~15.33	1D

▶ $C_{12}H_{14}O_4$ ▶ 7% $CDCl_3$ ▶ V 295

▶ 14-HO

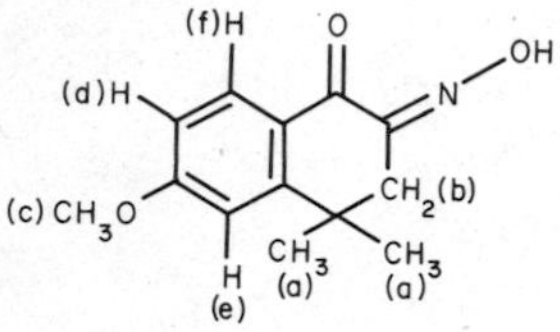

Shift	Peak
a 1.42	1A
b 3.04	1A
c 3.91	1A
▶ d 6.88	S
e 6.93	S
f 8.17	2A
g ~10.30	1D

▶ $C_{13}H_{15}NO_3$ ▶ 7% $CDCl_3$ ▶ V 611

▶ 14-HO

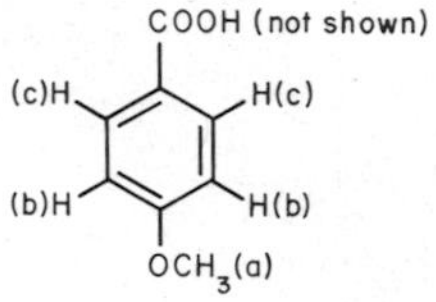

COOH (not shown)

Shift	Peak
a 3.88	1A
▶ b 6.98	Q
c 8.08	Q

▶ $C_8H_8O_3$ ▶ 7% $CDCl_3$ ▶ V 196

▶ 14-HO

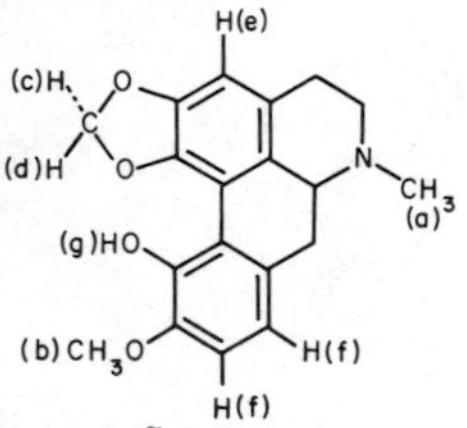

Shift	Peak
a 2.57	1E
b 3.92	1A
c 5.98	2C
d 6.13	2C
e 6.68	1A
▶ f 6.88	1E
g 6.88	1E

▶ $C_{19}H_{19}NO_4$ ▶ 7% $CDCl_3$ ▶ V 333

▶ 14-HO

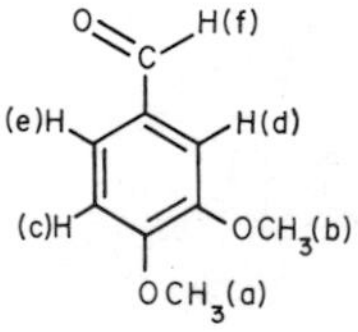

Shift	Peak
a 3.93 or 3.97	1A
b 3.93 or 3.97	1A
▶ c 6.98	2E
d 7.40	S
e 7.45	S
f 9.98	1A

▶ $C_9H_{10}O_3$ ▶ 7% $CDCl_3$ ▶ V 236

▶ 14-HO

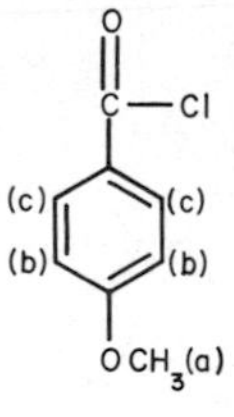

Shift	Peak
a 3.90	1A
▶ b 6.92	Q
c 8.04	Q

▶ $C_8H_7ClO_2$ ▶ 44.6 mg/0.50 ml $CDCl_3$ ▶ S 180

▶ 14-HO

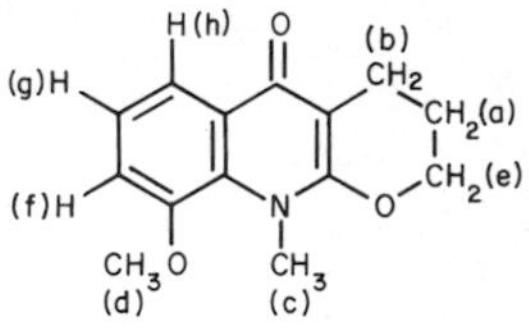

Shift	Peak
a 2.05	5B
b 2.70	3A
c 3.77 or 3.87	1A
d 3.77 or 3.87	1A
e 4.33	3A
▶ f ~7.00	S
g ~7.15	S
h 8.00	4D

▶ $C_{14}H_{15}NO_3$ ▶ 7% $CDCl_3$ ▶ V 311

▶ 14-H(O)

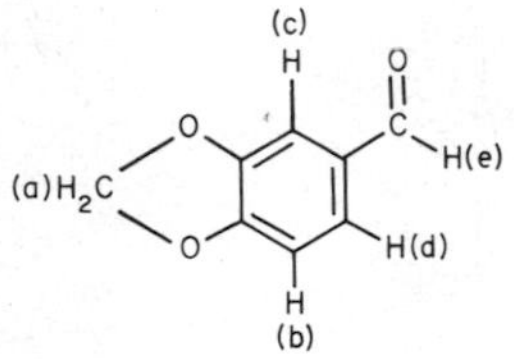

Shift	Peak
a 6.08	1A
▶ b 6.95	2A
c 7.35	1I
d 7.45	2A
e 9.84	1A

▶ $C_8H_6O_3$ ▶ 7% $CDCl_3$ ▶ V 187

▶ 14-HO

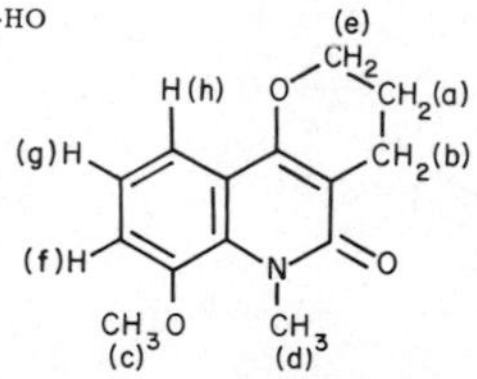

Shift	Peak
a 2.03	5B
b 2.63	3A
c 3.87 or 3.92	1A
d 3.87 or 3.92	1A
e 4.30	3A
▶ f ~7.00	R
g ~7.10	R
h 7.52	4D

▶ $C_{14}H_{15}NO_3$ ▶ 7% $CDCl_3$ ▶ V 312

▶ 14-HO

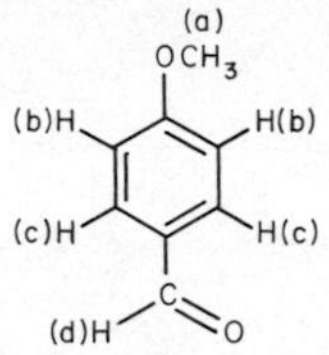

(a) OCH_3; (b)H, H(b); (c)H, H(c); (d)H–C=O

	Shift	Peak
a	3.87	1A
▶ b	7.02	Q
c	7.83	Q
d	9.87	1A

▶ $C_8H_8O_2$ ▶ 7% $CDCl_3$ ▶ V 194

▶ 14-HO

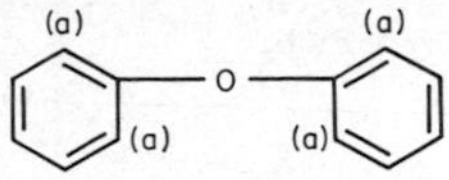

(a), (a), O, (a), (a)

	Shift	Peak
a	7.92	N

▶ $C_{12}H_{10}O$ ▶ 45.1 mg/0.25 ml $CDCl_3$ ▶ S 200

▶ 14-HO

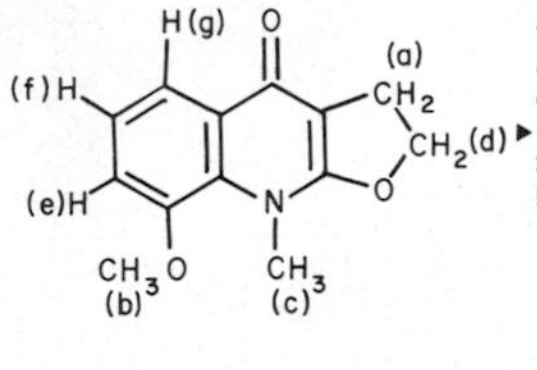

H(g), O, (a) CH_2, CH_2(d), (f)H, (e)H, N, O, CH_3O (b), CH_3 (c)

	Shift	Peak
a	3.27	Q
b	3.87 or 3.92	1A
c	3.87 or 3.92	1A
d	4.73	Q
▶ e	~7.05	R
f	~7.23	R
g	8.05	4D

▶ $C_{13}H_{13}NO_3$ ▶ 7% $CDCl_3$ ▶ V 300

▶ 14-HP

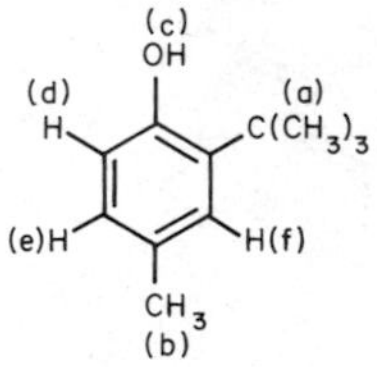

(c) OH, (d) H, (a) $C(CH_3)_3$, (e)H, H(f), CH_3 (b)

	Shift	Peak
a	1.40	1A
b	2.27	1A
c	4.75	1A
▶ d	6.52	2A
e	6.84	2E or 4A
f	7.05	1I

▶ $C_{11}H_{16}O$ ▶ 7% $CDCl_3$ ▶ V 288

▶ 14-HO

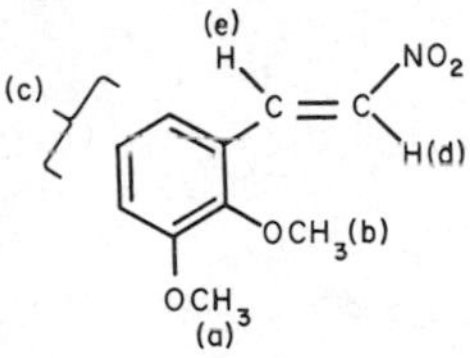

(e) H, C=C, NO_2, H(d), (c), OCH_3(b), OCH_3 (a)

	Shift	Peak
a	3.88 or 3.92	1A
b	3.88 or 3.92	1A
▶ c	7.07	1A
d	7.73	2A
e	8.20	2A

▶ $C_{10}H_{11}NO_4$ ▶ 7% $CDCl_3$ ▶ V 257

▶ 14-HP

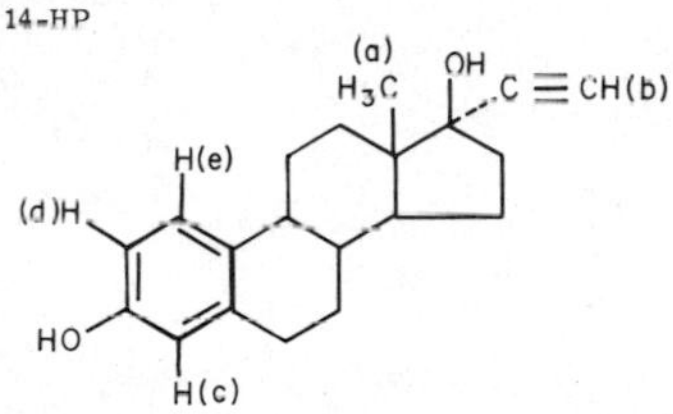

(a) H_3C, OH, C≡CH(b), H(e), (d)H, HO, H(c)

	Shift	Peak
a	0.88	1A
b	2.62	1A
c	6.60	1E
▶ d	6.67	2H
e	7.18	2E

▶ $C_{20}H_{24}O_2$ ▶ 7% $CDCl_3$ ▶ V 343

▶ 14-HO

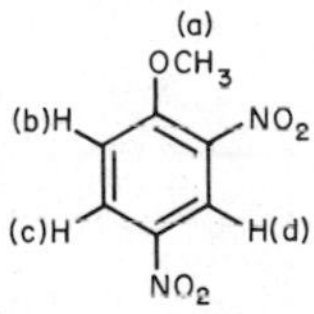

(a) OCH_3, (b)H, NO_2, (c)H, H(d), NO_2

	Shift	Peak
a	4.12	1A
▶ b	7.28	3A
c	8.47	4A
d	8.72	2A

▶ $C_7H_6N_2O_5$ ▶ 7% $CDCl_3$ ▶ V 149

▶ 14-HP

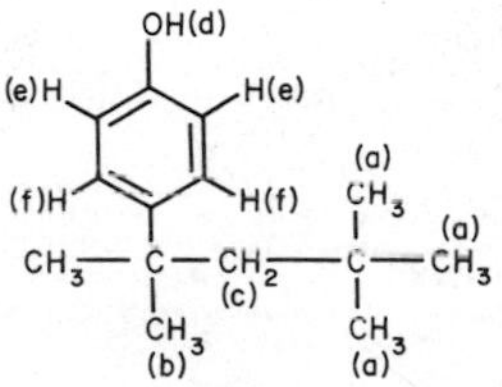

OH(d), (e)H, H(e), (f)H, H(f), CH_3–C–CH_2(c)–C–CH_3 (a), CH_3 (b), CH_3 (a), CH_3 (a)

	Shift	Peak
a	0.72	1A
b	1.33	1A
▶ c	1.68	1A
d	5.02	1C
e	6.75	Q
f	7.20	Q

▶ $C_{14}H_{22}O$ ▶ 7% $CDCl_3$ ▶ V 315

▶ 14-H(O)

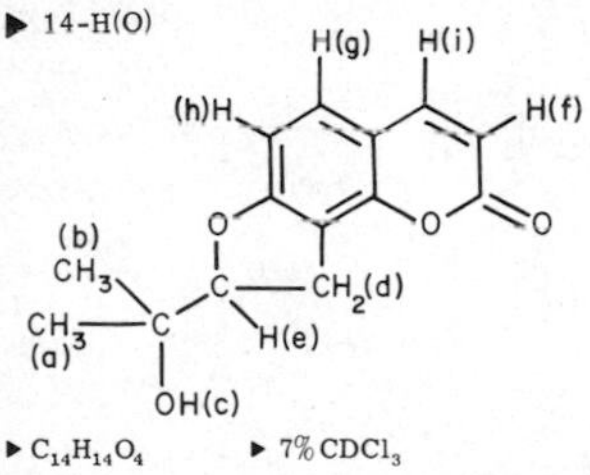

H(g), H(i), (h)H, H(f), O, O, O, (b) CH_3, C, CH_2(d), CH_3 (a), C, H(e), OH(c)

	Shift	Peak
a	1.25 or 1.37	1A
b	1.37 or 1.25	1A
c	1.96	1G
d	3.33	2A
e	4.82	3A
f	6.22	2C
g	6.79	2C
▶ h	7.29	2C
i	7.66	2C

▶ $C_{14}H_{14}O_4$ ▶ 7% $CDCl_3$ ▶ V 310

▶ 14-HP

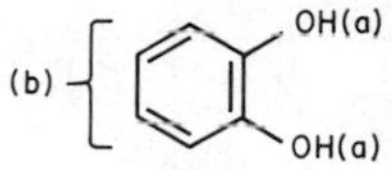

(b), OH(a), OH(a)

	Shift	Peak
a	5.45	1A
▶ b	6.82	1A

▶ $C_6H_6O_2$ ▶ 7% $CDCl_3$ ▶ V 124

▶ 14-HO

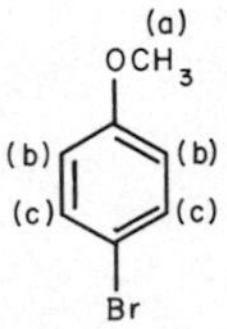

(a) OCH_3, (b), (b), (c), (c), Br

	Shift	Peak
a	3.78	1A
▶ b	6.75	Q
c	7.37	Q

▶ C_7H_7BrO ▶ 45.5 mg/0.50 ml $CDCl_3$ ▶ S 141

▶ 14-HP

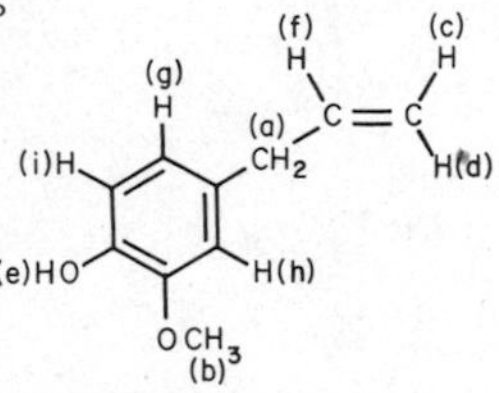

(f) H, (c) H, (g) H, (a) CH_2, C=C, H(d), (i)H, (e)HO, H(h), OCH_3 (b)

	Shift	Peak
a	3.30	2E
b	3.83	1A
c	5.05	2E
d	5.05	2E
e	5.59	1B
f	~5.94	O
g	~6.64	S
h	~6.69	S
▶ i	6.87	S

▶ $C_{10}H_{12}O_2$ ▶ 7% $CDCl_3$ ▶ V 260

▶ 14-HP

	Shift	Peak
a	1.41	3A
b	4.40	4A
▶ c	6.92	O
d	7.37	O
e	7.83	2E
f	10.78	1A

▶ $C_9H_{10}O_3$ ▶ 45.6 mg/0.50 ml $CDCl_3$ ▶ S 196

▶ 14-HP

	Shift	Peak
a	2.57	1A
b	3.93	1A
c	6.62	1C
▶ d	6.97	2A
e	7.57	S
f	7.60	S

▶ $C_9H_{10}O_3$ ▶ 7% $CDCl_3$ ▶ V 234

▶ 14-HP

	Shift	Peak
a	3.93	1A
b	6.47	1D
▶ c	7.02	2A
d	7.40	N
e	7.40	N
f	9.78	1A

▶ $C_8H_8O_3$ ▶ 7% $CDCl_3$ ▶ V 197

▶ 14-HP

	Shift	Peak
▶ a	7.15	2A
b	8.46	2A
c	8.61	3A
d	10.06	1A
e	11.63	1A

▶ $C_7H_5NO_4$ ▶ 7% $CDCl_3$ ▶ V 484

▶ 14-HQ

	Shift	Peak
a	1.42	1A
b	1.95	3A
c	2.55	1A
d	2.70	3D
e	2.82	1E
f	2.92	1E
g	4.12	1A
h	5.33	1D
i	6.37	2A
j	6.78	1E
▶ k	6.87	2H

▶ $C_{15}H_{21}N_3O_2$ ▶ 7% $CDCl_3$ ▶ V 319

▶ 14-HQ

	Shift	Peak
a	2.25	1A
▶ b	7.06	1A

▶ $C_{10}H_{10}O_4$ ▶ 44.8 mg/0.50 ml $CDCl_3$ ▶ S 252

▶ 14-HQ

	Shift	Peak
a	2.06	1A
b	2.32	1A
c	2.32	1A
d	6.36	1G
e	6.83*	S
f	6.83*	S
g	6.89	S
▶ h	7.15	S
i	7.33	S
j	7.53	2E

*or 7.08

▶ $C_{23}H_{18}O_8$ ▶ 7% $CDCl_3$ ▶ V 692

▶ 14-HQ

	Shift	Peak
a	2.32	1A
b	2.32	1A
c	3.09	S
d	3.09	A
e	5.18	3B
f	6.06	M
g	6.62*	S
h	6.62*	S
▶ i	7.17	Q
j	7.46	Q

*or 6.71

▶ $C_{21}H_{16}O_7$ ▶ 7% $CDCl_3$ ▶ V 686

▶ 14-HQ

	Shift	Peak
▶ a	7.31	2H
b	7.75	Q
c	7.92	Q

▶ $C_{20}H_{14}O_4$ ▶ 45.1 mg/0.50 ml $CDCl_3$ ▶ S 93

▶ 14-HR

	Shift	Peak
a	7.75	N

▶ $C_{12}H_9NO_2$ ▶ 44.9 mg/0.50 ml $CDCl_3$ ▶ S 86

▶ 14-HR

	Shift	Peak
a	4.78	1A
b	7.48	4A
c	7.75	4A
▶ d	8.00	S
e	9.63	1B

▶ $C_9H_6ClNO_5$ ▶ 7% $CDCl_3$ ▶ V 223

▶ 14-HR

	Shift	Peak
a	1.19	1A
b	1.30	1A
c	2.74	1A
d	5.84	2A
e	6.93	S
▶ f	8.01 ± .02	S
g	8.01 ± .02	S

▶ $C_{19}H_{18}N_2O_3$ ▶ 7% $CDCl_3$ ▶ V 677

▶ 14-HR

H(c) H(b) (d)H N CH_3(a) O_2N H(c) H(d)

	Shift	Peak
a	2.95	1A
b	4.63	1D
c	6.55	Q
▶ d	8.10	Q

▶ $C_7H_8N_2O_2$ ▶ 7% $CDCl_3$ ▶ V 489

▶ 14-HR

F (a) (a) (b) (b) NO_2

	Shift	Peak
a	7.20	O
▶ b	8.25	O

▶ $C_6H_4FNO_2$ ▶ 45.5 mg/0.50 ml $CDCl_3$ ▶ S 224

▶ 14-HR

NO_2 (b)H H(b) (a)H H(a) Cl

	Shift	Peak
a	7.52	Q
▶ b	8.17	Q

▶ $C_6H_4ClNO_2$ ▶ 7% $CDCl_3$ ▶ V 122

▶ 14-HR

(c)H H(b) (a) NO_2 CH_2—CN (c)H H(b)

	Shift	Peak
a	3.92	1A
b	7.57	Q
▶ c	8.25	Q

▶ $C_8H_6N_2O_2$ ▶ 7% $CDCl_3$ ▶ V 495

▶ 14-HR

(a) CH_3 (b) CH_3—C=N—NH (f) NO_2 (e) (c) NO_2 (d)

	Shift	Peak
a	2.10 or 2.20	1A
b	2.10 or 2.20	1A
c	7.85	2A
▶ d	8.18	4A
e	8.99	2A
f	10.94	1C

▶ $C_9H_{10}N_4O_4$ ▶ 45.0 mg/0.25 ml $CDCl_3$ ▶ S 199

▶ 14 HR

O O (c) (a) C CH_2—CH_3 (d)CF_2H CH_2(b) C N (e)H H(f) N NO_2 (h)H NO_2 H(g)

	Shift	Peak
a	1.33	3A
b	3.67	1A
c	4.30	4A
d	6.30	3A*
e	8.03	2C
▶ f	8.40	4A
g	9.08	2A
h	11.77	1B

*Very large splitting due to fluorine

▶ $C_{12}H_{12}F_2N_4O_6$ ▶ 7% $CDCl_3$ ▶ V 293

▶ 14-HR

H(b) (b)H H(c) H(d) (c)H Fe H (d) H(e) (a) H(e) NO_2

	Shift	Peak
a	4.07	1A
b	4.50	Q
c	4.77	Q
d	7.62	Q
▶ e	8.18	Q

▶ $C_{16}H_{13}FeNO_2$ ▶ 7% $CDCl_3$ ▶ V 321

▶ 14-HR

(d)H C O H(e) (c)H O O_2N H(a) H(b)

	Shift	Peak
a	7.15	2A
▶ b	8.46	2A
c	8.61	3A
d	10.06	1A
e	11.63	1A

▶ $C_7H_5NO_4$ ▶ 7% $CDCl_3$ ▶ V 484

▶ 14-HR

(b) (a) NO_2 (a) (b) (a)

	Shift	Peak
a	7.65	N
▶ b	8.20	O

▶ $C_6H_5NO_2$ ▶ 45.2 mg/0.50 ml $CDCl_3$ ▶ S 4

▶ 14-HR

(a) OCH_3 (b)H NO_2 (c)H H(d) NO_2

	Shift	Peak
a	4.12	1A
b	7.28	3A
▶ c	8.47	4A
d	8.72	2A

▶ $C_7H_6N_2O_5$ ▶ 7% $CDCl_3$ ▶ V 149

▶ 14-HR

(e) H O_2N NO_2 CH_3(a) (d)H N—N=C H CH_3(b) (c) H (f)

	Shift	Peak
a	2.10 or 2.18	1A
b	2.10 or 2.18	1A
c	7.90	2A
▶ d	8.23	4A
e	9.04	2A
f	10.95	1B

▶ $C_9H_{10}N_4O_4$ ▶ 7% $CDCl_3$ ▶ V 233

▶ 14-HR

H(b) (a)H H(c) N (d)H N NO_2 H(e)

	Shift	Peak
a	7.34	4A
b	8.18	2D
c	8.32	2E
▶ d	8.48	2D
e	11.63	1D

▶ $C_7H_5N_3O_2$ ▶ 7% $CDCl_3$ ▶ V 485

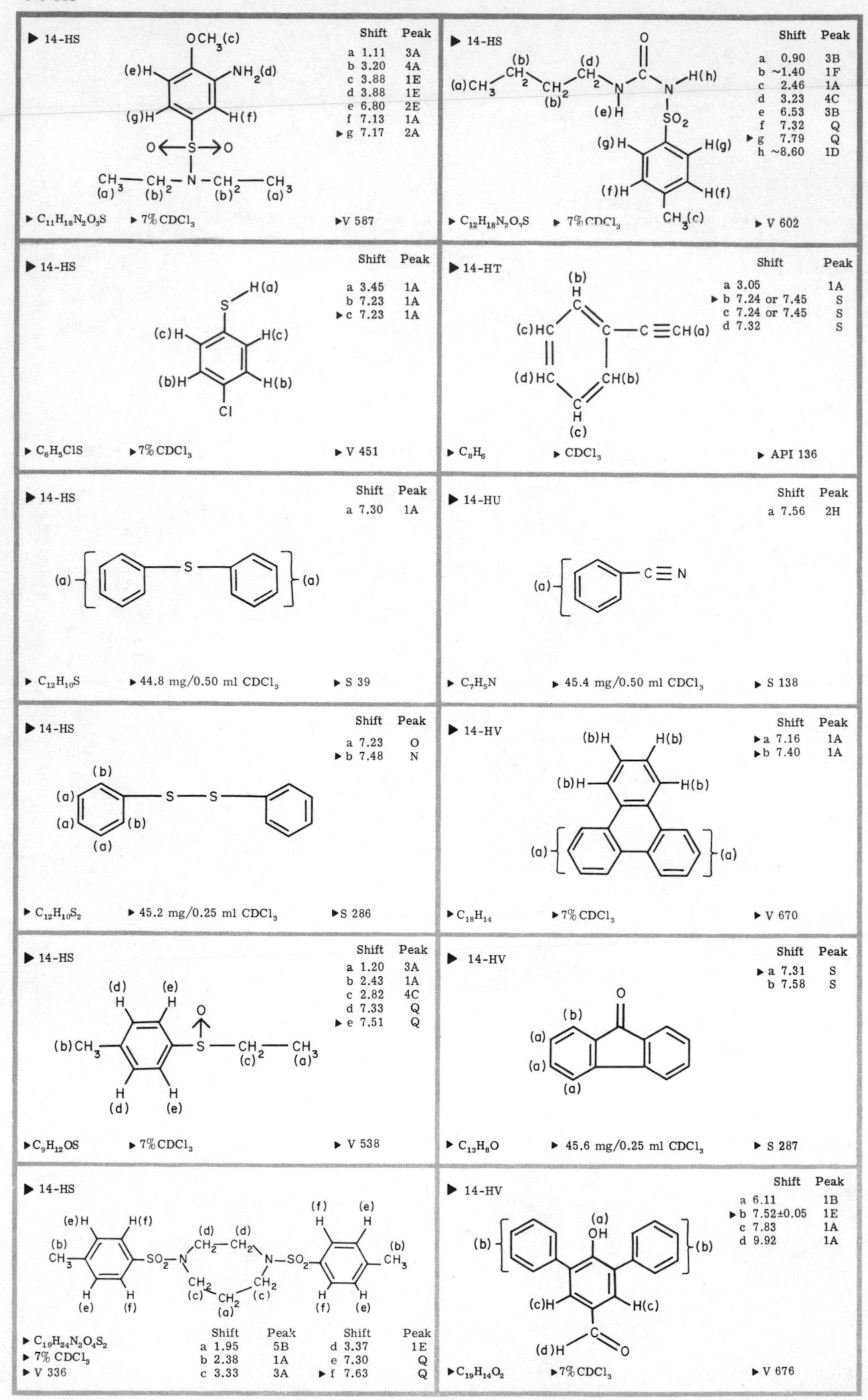

▶ 14-HS
Shift Peak
a 1.11 3A
b 3.20 4A
c 3.88 1E
d 3.88 1E
e 6.80 2E
f 7.13 1A
▶g 7.17 2A
▶ $C_{11}H_{18}N_2O_3S$ ▶ 7% $CDCl_3$ ▶V 587
▶ 14-HS
Shift Peak
a 0.90 3B
b ~1.40 1F
c 2.46 1A
d 3.23 4C
e 6.53 3B
f 7.32 Q
▶g 7.79 Q
h ~8.60 1D
▶ $C_{12}H_{18}N_2O_3S$ ▶ 7% $CDCl_3$ ▶ V 602
▶ 14-HS
Shift Peak
a 3.45 1A
b 7.23 1A
▶c 7.23 1A
▶ C_6H_5ClS ▶ 7% $CDCl_3$ ▶ V 451
▶ 14-HT
Shift Peak
a 3.05 1A
▶ b 7.24 or 7.45 S
c 7.24 or 7.45 S
d 7.32 S
▶ C_8H_6 ▶ $CDCl_3$ ▶ API 136
▶ 14-HS
Shift Peak
a 7.30 1A
▶ $C_{12}H_{10}S$ ▶ 44.8 mg/0.50 ml $CDCl_3$ ▶ S 39
▶ 14-HU
Shift Peak
a 7.56 2H
▶ C_7H_5N ▶ 45.4 mg/0.50 ml $CDCl_3$ ▶ S 138
▶ 14-HS
Shift Peak
a 7.23 O
▶ b 7.48 N
▶ $C_{12}H_{10}S_2$ ▶ 45.2 mg/0.25 ml $CDCl_3$ ▶S 286
▶ 14-HV
Shift Peak
▶a 7.16 1A
▶b 7.40 1A
▶ $C_{18}H_{14}$ ▶ 7% $CDCl_3$ ▶ V 670
▶ 14-HS
Shift Peak
a 1.20 3A
b 2.43 1A
c 2.82 4C
d 7.33 Q
▶ e 7.51 Q
▶ $C_9H_{12}OS$ ▶ 7% $CDCl_3$ ▶ V 538
▶ 14-HV
Shift Peak
▶ a 7.31 S
b 7.58 S
▶ $C_{13}H_8O$ ▶ 45.6 mg/0.25 ml $CDCl_3$ ▶ S 287
▶ 14-HS
▶ $C_{19}H_{24}N_2O_4S_2$
▶ 7% $CDCl_3$
▶ V 336
Shift Peak
a 1.95 5B
b 2.38 1A
c 3.33 3A
Shift Peak
d 3.37 1E
e 7.30 Q
▶ f 7.63 Q
▶ 14-HV
Shift Peak
a 6.11 1B
▶ b 7.52±0.05 1E
c 7.83 1A
d 9.92 1A
▶ $C_{19}H_{14}O_2$ ▶ 7% $CDCl_3$ ▶ V 676

▶ 14-HV

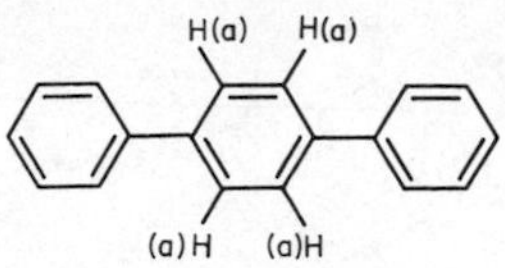

Shift	Peak
a 7.66	1A

▶ $C_{18}H_{14}$ ▶ 7% $CDCl_3$ ▶ V 671

▶ 14-HX

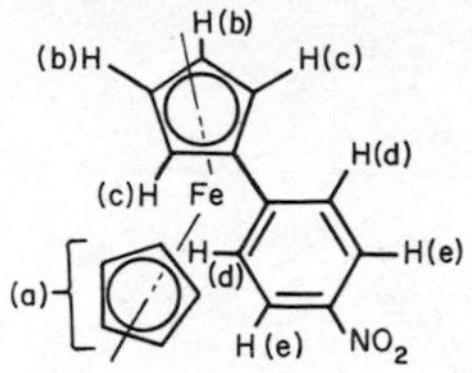

Shift	Peak
a 4.07	1A
b 4.50	Q
c 4.77	Q
▶ d 7.62	Q
e 8.18	Q

▶ $C_{16}H_{13}FeNO_2$ ▶ 7% $CDCl_3$ ▶ V 321

▶ 14-HV

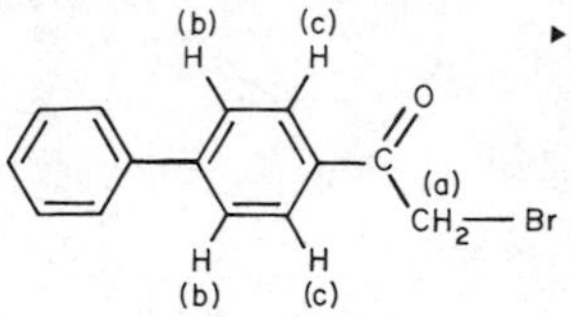

Shift	Peak
a 4.45	1A
▶ b 7.70	Q
c 8.07	Q

▶ $C_{14}H_{11}BrO$ ▶ 7% $CDCl_3$ ▶ V 303

▶ 14-(J)(L)

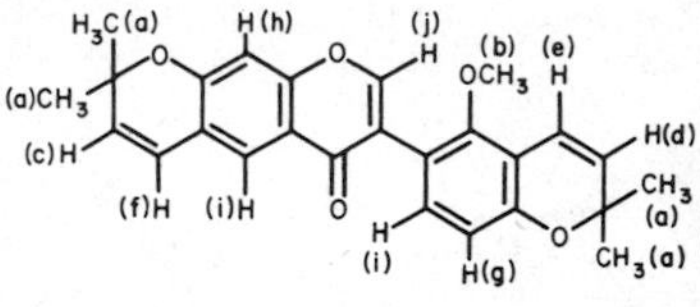

Shift	Peak
a 1.46*	1A
b 3.58	1A
c 5.62†	2A
d 5.62†	2A
e 6.45‡	2A
f 6.45‡	2A
g 6.62	S
h 7.12	2A
▶ i 7.88§	1A
j 7.88§	1A

*or 1.49 ‡or 6.60
†or 5.70 §or 7.90

▶ $C_{26}H_{24}O_5$ ▶ 7% $CDCl_3$ ▶ V 696

▶ 14-HV

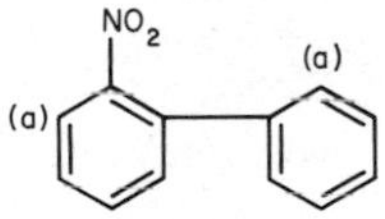

Shift	Peak
a 7.75	N

▶ $C_{12}H_9NO_2$ ▶ 44.9 mg/0.50 ml $CDCl_3$ ▶ S 86

▶ 14-J(O)

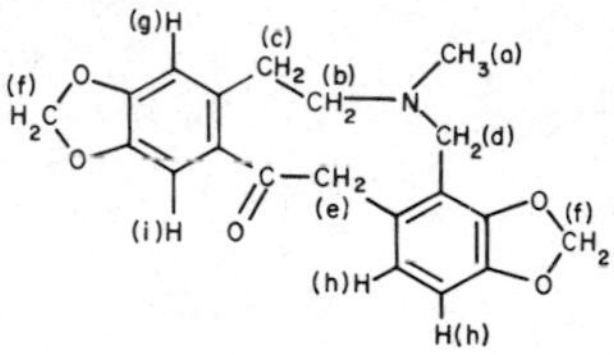

Shift	Peak
a 1.92	1A
b 2.55	1F
c 2.88	1F
d 3.58	1A
e 3.78	1A
f 5.92, 5.96	1A
g 6.05	1A
h 6.69	1A
▶ i 6.91	1A

▶ $C_{20}H_{19}NO_5$ ▶ 7% $CDCl_3$ ▶ V 339

▶ 14-HV

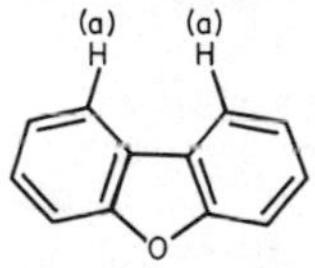

Shift	Peak
a 7.95	S

▶ $C_{12}H_8O$ ▶ 7% $CDCl_3$ ▶ V 589

▶ 14-J(O)

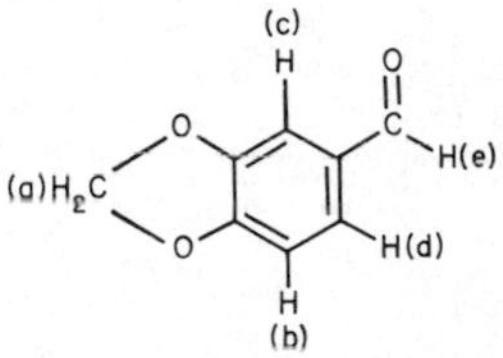

Shift	Peak
a 6.08	1A
b 6.95	2A
▶ c 7.35	1I
d 7.45	2A
e 9.84	1A

▶ $C_8H_6O_3$ ▶ 7% $CDCl_3$ ▶ V 187

▶ 14-HW

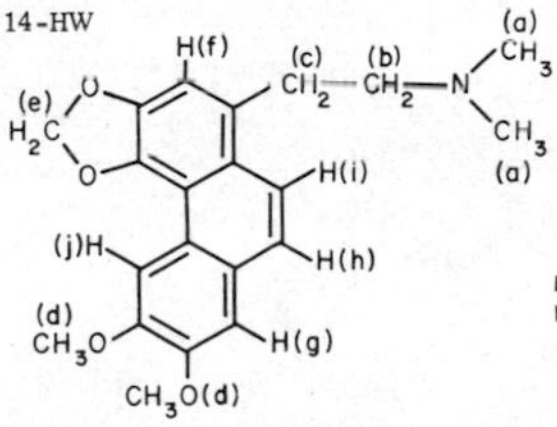

Shift	Peak
a 2.38	1A
b ~2.7	Q
c ~3.2	Q
d 4.05 and 4.08	1A
e 6.23	1A
f 7.13 or 7.23	1A
g 7.13 or 7.23	1A
▶ h 7.53 or 7.81	2C
▶ i 7.53 or 7.81	2C
j 8.52	1A

▶ $C_{21}H_{23}NO_4$ ▶ 7% $CDCl_3$ ▶ V 348

▶ 14-JO

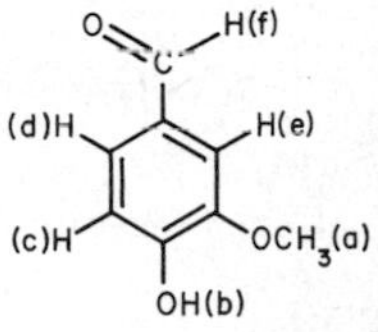

Shift	Peak
a 3.93	1A
b 6.47	1D
c 7.02	2A
d 7.40	N
▶ e 7.40	N
f 9.78	1A

▶ $C_8H_8O_3$ ▶ 7% $CDCl_3$ ▶ V 197

▶ 14-HW

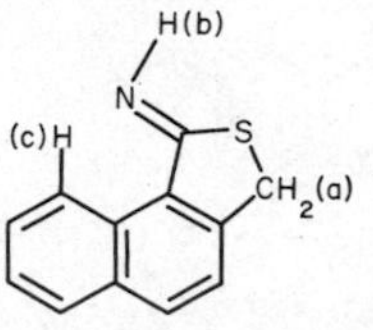

Shift	Peak
a 4.30	1A
b 9.70	1I
▶ c 9.80	2A

▶ $C_{12}H_9NS$ ▶ 7% $CDCl_3$ ▶ V 590

▶ 14-JO

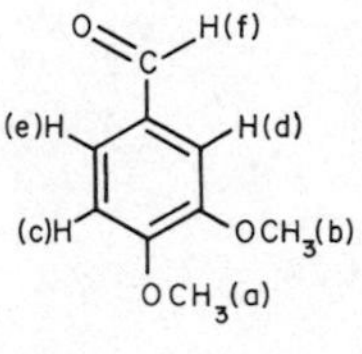

Shift	Peak
a 3.93 or 3.97	1A
b 3.93 or 3.97	1A
c 6.98	2E
▶ d 7.40	S
e 7.45	S
f 9.98	1A

▶ $C_9H_{10}O_3$ ▶ 7% $CDCl_3$ ▶ V 236

14-JO

	Shift	Peak
a	2.57	1A
b	3.93	1A
c	6.62	1C
d	6.97	2A
▶ e	7.57	S
f	7.60	S

▶ $C_9H_{10}O_3$ ▶ 7% $CDCl_3$ ▶ V 234

14-L(O)

	Shift	Peak
a	1.80	2A
b	5.87	1A
c	~6.08	S
d	6.23	S
e	6.70	S
f	6.70	S
▶ g	6.83	S

▶ $C_{10}H_{10}O_2$ ▶ 7% $CDCl_3$ ▶ V 252

14-JO

	Shift	Peak
a	3.35	3B
b	4.10	S
c	5.41	3A
▶ d	7.63	N
e	7.77	2A

▶ $C_{18}H_{20}O_6$ ▶ 7% $CDCl_3$ ▶ V 675

14-(L)O

	Shift	Peak
a	2.33	1A
b	3.13	3A
c	3.83	1A
d	4.28	4A
e	6.85 or 6.98	1A
▶ f	6.85 or 6.98	1A
g	7.13	1B

▶ $C_{13}H_{13}NO_2$ ▶ 7% $CDCl_3$ ▶ V 299

14-JR

	Shift	Peak
a	7.15	2A
b	8.46	2A
▶ c	8.61	3A
d	10.06	1A
e	11.63	1A

▶ $C_7H_5NO_4$ ▶ 7% $CDCl_3$ ▶ V 484

14-L(O)

	Shift	Peak
a	3.92	1A
b	5.97	1A
c	6.79	2A
d	7.07	S
▶ e	7.32	4A
f	8.17	2A

▶ $C_{18}H_{16}N_2O_4$ ▶ 7% $CDCl_3$ ▶ V 673

14-JV

	Shift	Peak
a	6.11	1B
b	7.52±0.05	1E
▶ c	7.83	1A
d	9.92	1A

▶ $C_{19}H_{14}O_2$ ▶ 7% $CDCl_3$ ▶ V 676

14-NN

	Shift	Peak
a	3.51	1A
▶ b	6.03	1A
c	6.11	2D
d	6.93	3C

▶ $C_6H_8N_2$ ▶ 7% $CDCl_3$ ▶ V 458

14-KO

	Shift	Peak
a	1.38	3A
b	3.93	1A
c	4.32	4A
d	6.80 or 6.94	1A
▶ e	6.80 or 6.94	1A

▶ $C_{10}H_{12}O_4$ ▶ 45.0 mg/0.50 ml $CDCl_3$ ▶ S 127

14-NO

	Shift	Peak
a	3.72 or 3.79	1A
b	3.72 or 3.79	1A
c	~3.75	1E
d	6.25	S
▶ e	6.32	S
f	6.71	2A

▶ $C_8H_{11}NO_2$ ▶ 7% $CDCl_3$ ▶ V 508

14-KR

	Shift	Peak
a	1.74	1E
b	2.40	O
c	4.49 or 4.82	1B
d	4.82 or 4.49	1B
e	5.22	R
f	5.24	R
g	6.08	4D
▶ h	~8.01	N

▶ $C_{27}H_{36}N_2O_6$ ▶ 7% $CDCl_3$ ▶ V 697

14-NS

	Shift	Peak
a	1.11	3A
b	3.20	4A
c	3.88	1E
d	3.88	1E
e	6.80	2E
▶ f	7.13	1A
g	7.17	2A

▶ $C_{11}H_{18}N_2O_3S$ ▶ 7% $CDCl_3$ ▶ V 587

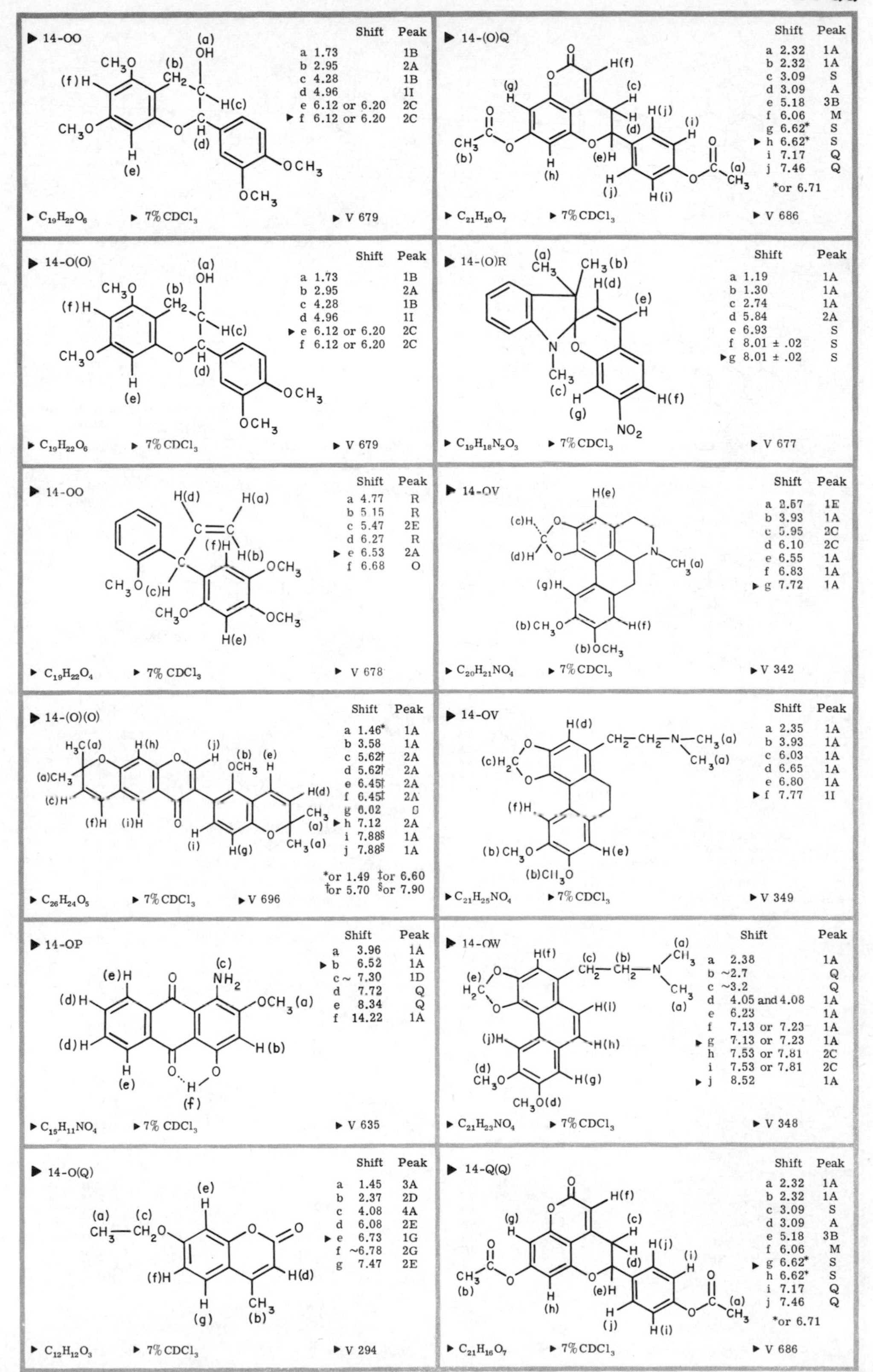

▶ 14-OO

	Shift	Peak
a	1.73	1B
b	2.95	2A
c	4.28	1B
d	4.96	1I
e	6.12 or 6.20	2C
▶ f	6.12 or 6.20	2C

▶ $C_{19}H_{22}O_6$ ▶ 7% $CDCl_3$ ▶ V 679

▶ 14-(O)Q

	Shift	Peak
a	2.32	1A
b	2.32	1A
c	3.09	S
d	3.09	A
e	5.18	3B
f	6.06	M
g	6.62*	S
▶ h	6.62*	S
i	7.17	Q
j	7.46	Q

*or 6.71

▶ $C_{21}H_{16}O_7$ ▶ 7% $CDCl_3$ ▶ V 686

▶ 14-O(O)

	Shift	Peak
a	1.73	1B
b	2.95	2A
c	4.28	1B
d	4.96	1I
▶ e	6.12 or 6.20	2C
f	6.12 or 6.20	2C

▶ $C_{19}H_{22}O_6$ ▶ 7% $CDCl_3$ ▶ V 679

▶ 14-(O)R

	Shift	Peak
a	1.19	1A
b	1.30	1A
c	2.74	1A
d	5.84	2A
e	6.93	S
f	8.01 ± .02	S
▶ g	8.01 ± .02	S

▶ $C_{19}H_{18}N_2O_3$ ▶ 7% $CDCl_3$ ▶ V 677

▶ 14-OO

	Shift	Peak
a	4.77	R
b	5.15	R
c	5.47	2E
d	6.27	R
▶ e	6.53	2A
f	6.68	O

▶ $C_{19}H_{22}O_4$ ▶ 7% $CDCl_3$ ▶ V 678

▶ 14-OV

	Shift	Peak
a	2.67	1E
b	3.93	1A
c	5.95	2C
d	6.10	2C
e	6.55	1A
f	6.83	1A
▶ g	7.72	1A

▶ $C_{20}H_{21}NO_4$ ▶ 7% $CDCl_3$ ▶ V 342

▶ 14-(O)(O)

	Shift	Peak
a	1.46*	1A
b	3.58	1A
c	5.62†	2A
d	5.62†	2A
e	6.45‡	2A
f	6.45‡	2A
g	6.02	O
▶ h	7.12	2A
i	7.88§	1A
j	7.88§	1A

*or 1.49 ‡or 6.60
†or 5.70 §or 7.90

▶ $C_{26}H_{24}O_5$ ▶ 7% $CDCl_3$ ▶ V 696

▶ 14-OV

	Shift	Peak
a	2.35	1A
b	3.93	1A
c	6.03	1A
d	6.65	1A
e	6.80	1A
▶ f	7.77	1I

▶ $C_{21}H_{25}NO_4$ ▶ 7% $CDCl_3$ ▶ V 349

▶ 14-OP

	Shift	Peak
a	3.96	1A
▶ b	6.52	1A
c	~ 7.30	1D
d	7.72	Q
e	8.34	Q
f	14.22	1A

▶ $C_{15}H_{11}NO_4$ ▶ 7% $CDCl_3$ ▶ V 635

▶ 14-OW

	Shift	Peak
a	2.38	1A
b	~2.7	Q
c	~3.2	Q
d	4.05 and 4.08	1A
e	6.23	1A
f	7.13 or 7.23	1A
▶ g	7.13 or 7.23	1A
h	7.53 or 7.81	2C
i	7.53 or 7.81	2C
▶ j	8.52	1A

▶ $C_{21}H_{23}NO_4$ ▶ 7% $CDCl_3$ ▶ V 348

▶ 14-O(Q)

	Shift	Peak
a	1.45	3A
b	2.37	2D
c	4.08	4A
d	6.08	2E
▶ e	6.73	1G
f	~6.78	2G
g	7.47	2E

▶ $C_{12}H_{12}O_3$ ▶ 7% $CDCl_3$ ▶ V 294

▶ 14-Q(Q)

	Shift	Peak
a	2.32	1A
b	2.32	1A
c	3.09	S
d	3.09	A
e	5.18	3B
f	6.06	M
▶ g	6.62*	S
h	6.62*	S
i	7.17	Q
j	7.46	Q

*or 6.71

▶ $C_{21}H_{16}O_7$ ▶ 7% $CDCl_3$ ▶ V 686

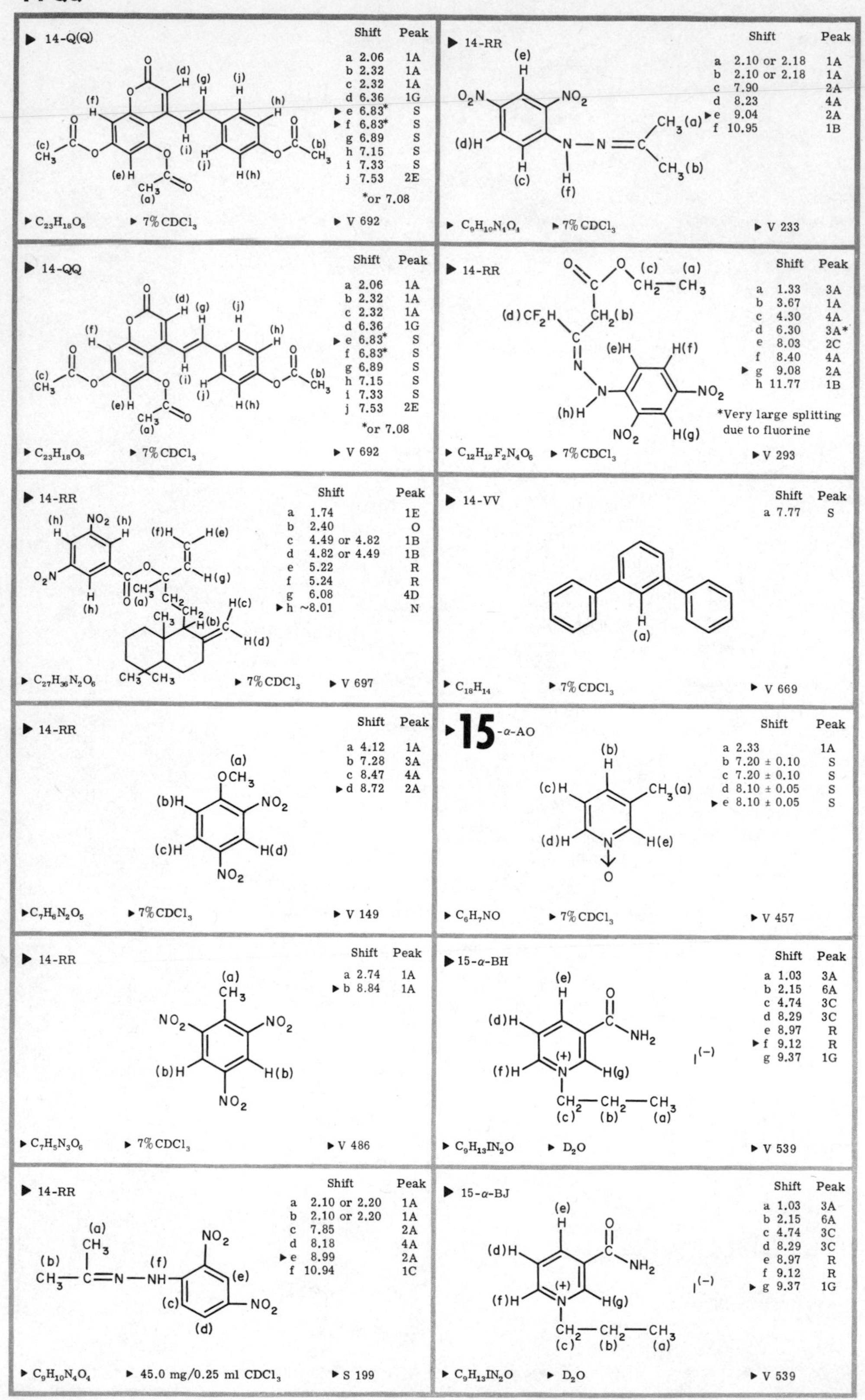

▶ 14-Q(Q)

	Shift	Peak
a	2.06	1A
b	2.32	1A
c	2.32	1A
d	6.36	1G
▶ e	6.83*	S
▶ f	6.83*	S
g	6.89	S
h	7.15	S
i	7.33	S
j	7.53	2E

*or 7.08

▶ $C_{23}H_{18}O_8$ ▶ 7% $CDCl_3$ ▶ V 692

▶ 14-RR

	Shift	Peak
a	2.10 or 2.18	1A
b	2.10 or 2.18	1A
c	7.90	2A
d	8.23	4A
▶ e	9.04	2A
f	10.95	1B

▶ $C_9H_{10}N_4O_4$ ▶ 7% $CDCl_3$ ▶ V 233

▶ 14-QQ

	Shift	Peak
a	2.06	1A
b	2.32	1A
c	2.32	1A
d	6.36	1G
▶ e	6.83*	S
f	6.83*	S
g	6.89	S
h	7.15	S
i	7.33	S
j	7.53	2E

*or 7.08

▶ $C_{23}H_{18}O_8$ ▶ 7% $CDCl_3$ ▶ V 692

▶ 14-RR

	Shift	Peak
a	1.33	3A
b	3.67	1A
c	4.30	4A
d	6.30	3A*
e	8.03	2C
f	8.40	4A
▶ g	9.08	2A
h	11.77	1B

*Very large splitting due to fluorine

▶ $C_{12}H_{12}F_2N_4O_6$ ▶ 7% $CDCl_3$ ▶ V 293

▶ 14-RR

	Shift	Peak
a	1.74	1E
b	2.40	O
c	4.49 or 4.82	1B
d	4.82 or 4.49	1B
e	5.22	R
f	5.24	R
g	6.08	4D
▶ h	~8.01	N

▶ $C_{27}H_{36}N_2O_6$ ▶ 7% $CDCl_3$ ▶ V 697

▶ 14-VV

	Shift	Peak
a	7.77	S

▶ $C_{18}H_{14}$ ▶ 7% $CDCl_3$ ▶ V 669

▶ 14-RR

	Shift	Peak
a	4.12	1A
b	7.28	3A
c	8.47	4A
▶ d	8.72	2A

▶ $C_7H_6N_2O_5$ ▶ 7% $CDCl_3$ ▶ V 149

▶ 15-α-AO

	Shift	Peak
a	2.33	1A
b	7.20 ± 0.10	S
c	7.20 ± 0.10	S
d	8.10 ± 0.05	S
▶ e	8.10 ± 0.05	S

▶ C_6H_7NO ▶ 7% $CDCl_3$ ▶ V 457

▶ 14-RR

	Shift	Peak
a	2.74	1A
▶ b	8.84	1A

▶ $C_7H_5N_3O_6$ ▶ 7% $CDCl_3$ ▶ V 486

▶ 15-α-BH

	Shift	Peak
a	1.03	3A
b	2.15	6A
c	4.74	3C
d	8.29	3C
e	8.97	R
▶ f	9.12	R
g	9.37	1G

▶ $C_9H_{13}IN_2O$ ▶ D_2O ▶ V 539

▶ 14-RR

	Shift	Peak
a	2.10 or 2.20	1A
b	2.10 or 2.20	1A
c	7.85	2A
d	8.18	4A
▶ e	8.99	2A
f	10.94	1C

▶ $C_9H_{10}N_4O_4$ ▶ 45.0 mg/0.25 ml $CDCl_3$ ▶ S 199

▶ 15-α-BJ

	Shift	Peak
a	1.03	3A
b	2.15	6A
c	4.74	3C
d	8.29	3C
e	8.97	R
f	9.12	R
▶ g	9.37	1G

▶ $C_9H_{13}IN_2O$ ▶ D_2O ▶ V 539

▶ 15-α-(C)

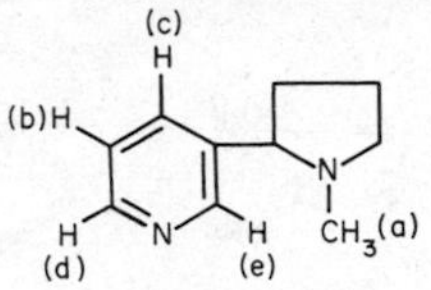

	Shift	Peak
a	2.18	1A
b	7.30	4A
c	7.75	6A
d	8.55	2A
▶ e	8.60	2A

▶ $C_{10}H_{14}N_2$ ▶ 7% $CDCl_3$ ▶ V 269

▶ 15-α-H

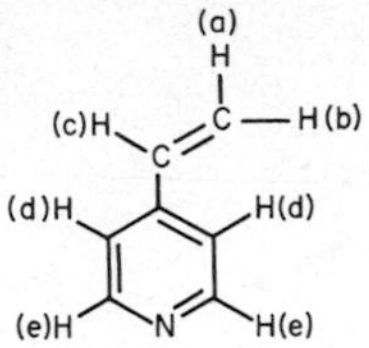

	Shift	Peak
a	5.43	3C
b	5.90	2D
c	6.62	4D
d	7.22	Q
▶ e	8.52	Q

▶ C_7H_7N ▶ 7% $CDCl_3$ ▶ V 155

▶ 15-α-H

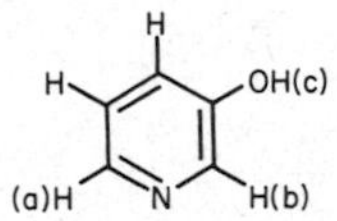

	Shift	Peak
▶ a	8.08 or 8.29	3B
b	8.08 or 8.29	1B
c	10.89	1G

▶ C_5H_5NO ▶ 7% $CDCl_3$ ▶ V 428

▶ 15-α-H

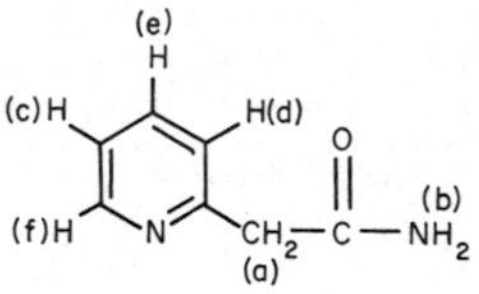

	Shift	Peak
a	3.72	1A
b	6.46	1D
c	~7.17	N
d	~7.25	M
e	7.65	6A
▶ f	8.52	2E

▶ $C_7H_8N_2O$ ▶ 7% $CDCl_3$ ▶ V 159

▶ 15-α-H

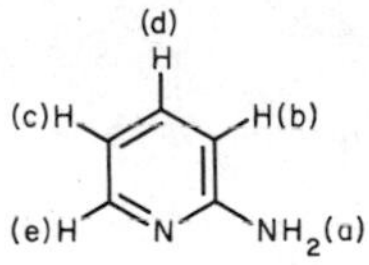

	Shift	Peak
a	4.75	1C
b	6.51	10A
c	6.64	10A
d	7.42	10A
▶ e	8.09	10A

▶ $C_5H_6N_2$ ▶ 7% $CDCl_3$ ▶ V 431

▶ 15-α-H

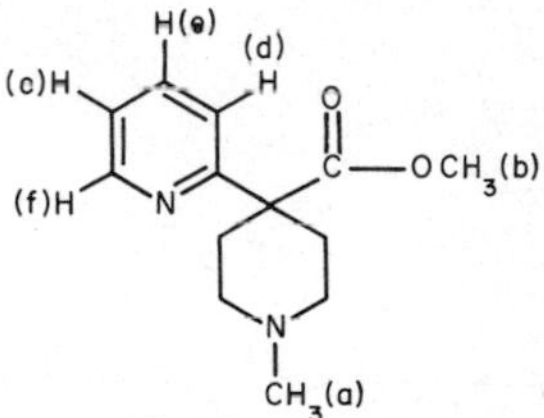

	Shift	Peak
a	2.25	1A
b	3.69	1A
c	7.14	S
d	7.34	S
e	7.64	S
▶ f	8.54	2E

▶ $C_{13}H_{18}N_2O_2$ ▶ 7% $CDCl_3$ ▶ V 301

▶ 15-α-H

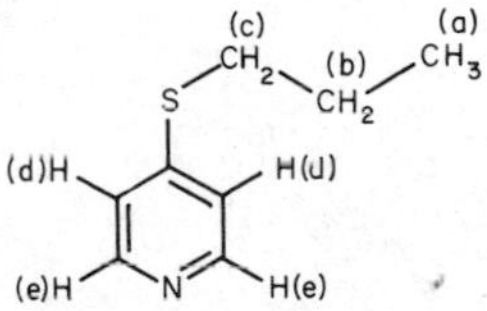

	Shift	Peak
a	1.08	3A
b	1.73	6B
c	2.93	3A
d	7.10	Q
▶ e	8.38	Q

▶ $C_8H_{11}NS$ ▶ 7% $CDCl_3$ ▶ V 509

▶ 15-α-H

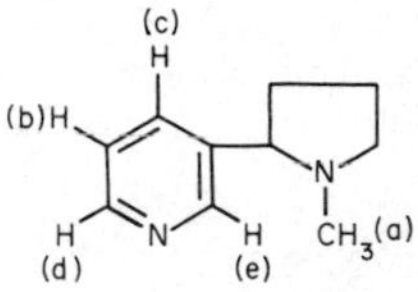

	Shift	Peak
a	2.18	1A
b	7.30	4A
c	7.75	6A
▶ d	8.55	2A
e	8.60	2A

▶ $C_{10}H_{14}N_2$ ▶ 7% $CDCl_3$ ▶ V 269

▶ 15-α-H

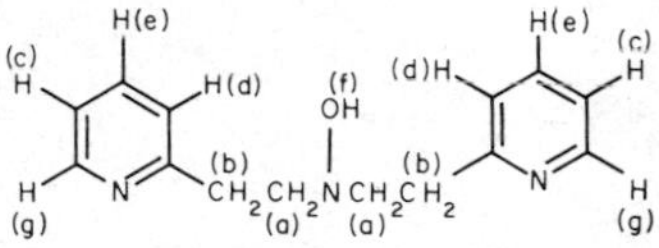

	Shift	Peak
a	3.13	1A
b	3.13	1A
c	7.07	S
d	7.07	S
e	7.53	S
f	7.90	S
▶ g	8.48	S

▶ $C_{14}H_{17}N_3O$ ▶ 7% $CDCl_3$ ▶ V 633

▶ 15-α-H

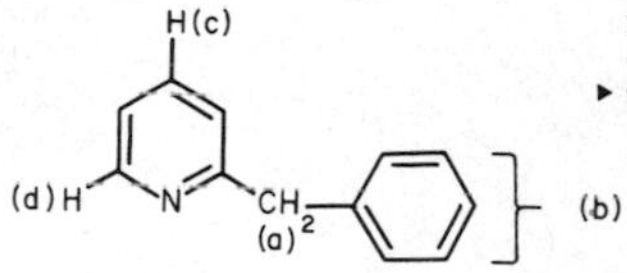

	Shift	Peak
a	4.18	1A
b	7.27	1E
c	7.54	S
▶ d	8.56	3C

▶ $C_{12}H_{11}N$ ▶ 7% $CDCl_3$ ▶ V 593

▶ 15-α-H

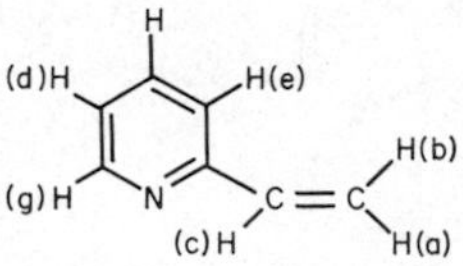

	Shift	Peak
a	5.45	S
b	6.22	S
c	6.75	S
d	7.08	S
e	7.27	S
f	7.55	S
▶ g	8.52	S

▶ C_7H_7N ▶ 7% $CDCl_3$ ▶ V 154

▶ 15-α-H

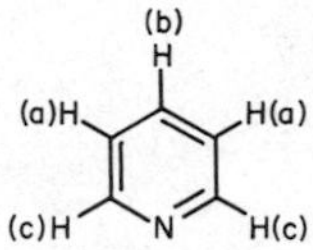

	Shift	Peak
a	~7.00	O
b	~7.60	O
▶ c	8.60	2E

▶ C_5H_5N ▶ 7% $CDCl_3$ ▶ V 96

▶ 15-α-H

(b) H, (a)H, O, ND2, (c)H, N, H(d)

	Shift	Peak
a	7.58	1A
b	8.23	4C
▶ c	8.74	O
d	8.93	O

▶ $C_6H_6N_2O$ ▶ D_2O ▶ V 454

▶ 15-α-P

H, H, OH(c), (a)H, N, H(b)

	Shift	Peak
a	8.08 or 8.29	3B
▶ b	8.08 or 8.29	1B
c	10.89	1G

▶ C_5H_5NO ▶ 7% $CDCl_3$ ▶ V 428

▶ 15-α-H

(c) H, (b)H, O, NH2(a), (d)H, N, H(e)

	Shift	Peak
a	~6.26	1D
b	7.38	4C
c	8.16	O
▶ d	8.75	4A
e	9.04	2E

▶ $C_6H_6N_2O$ ▶ 7% $CDCl_3$ ▶ V 453

▶ 15-β-AH

(c) H, (b)H, H(b), (a) CH3, N, CH3(a)

	Shift	Peak
a	2.52	1A
▶ b	6.95	2A
c	7.45	4A

▶ C_7H_9N ▶ 7% $CDCl_3$ ▶ V 169

▶ 15-α-H

(g)H, H(e), (f) H, (d) H, Cl, N, O, N, O, OCCH3 (a), H (c), H (b), H (d), Cl

	Shift	Peak
a	2.45	1A
b	6.88*	2H
c	6.88*	2H
d	6.95	1A
e	7.68	4A
f	8.70	4A
▶ g	8.98	4A

*or 7.30

▶ $C_{17}H_{10}Cl_2N_2O_3$ ▶ 7% $CDCl_3$ ▶ V 327

▶ 15-β-AH

(c) H, (d) H, H(b), O, (a) CH3, N, C, C, N, CH3(a), O, (b) H, H (d), H (c)

	Shift	Peak
a	2.43	1A
▶ b	7.31	2E
c	7.75	S
d	7.98	S

▶ $C_{14}H_{12}N_2O_2$ ▶ 7% $CDCl_3$ ▶ V 623

▶ 15-α-HO

(b) H, (c)H, CH3(a), (d)H, N, H(e), O

	Shift	Peak
a	2.33	1A
b	7.20 ± 0.10	S
c	7.20 ± 0.10	S
▶ d	8.10 ± 0.05	S
e	8.10 ± 0.05	S

▶ C_6H_7NO ▶ 7% $CDCl_3$ ▶ V 457

▶ 15-β-BH

H(e), H(e), (c) H, H(d), (f) OH, (d)H, (c) H, H (g), N, (b) CH2CH2NCH2CH2 (a) (a) (b), N, H (g)

	Shift	Peak
a	3.13	1A
b	3.13	1A
c	7.07	S
▶ d	7.07	S
e	7.53	S
f	7.90	S
g	8.48	S

▶ $C_{14}H_{17}N_3O$ ▶ 7% $CDCl_3$ ▶ V 633

▶ 15-α-J

(b) H, (a)H, O, ND2, (c)H, N, H(d)

	Shift	Peak
a	7.58	1A
b	8.23	4C
c	8.74	O
▶ d	8.93	O

▶ $C_6H_6N_2O$ ▶ D_2O ▶ V 454

▶ 15-β-BH

(e) H, (c)H, H(d), O, (b), (f)H, N, CH2—C—NH2, (a)

	Shift	Peak
a	3.72	1A
b	6.46	1D
c	~7.17	N
▶ d	~7.25	M
e	7.65	6A
f	8.52	2E

▶ $C_7H_8N_2O$ ▶ 7% $CDCl_3$ ▶ V 159

▶ 15-α-J

(c) H, (b)H, O, NH2(a), (d)H, N, H(e)

	Shift	Peak
a	~6.26	1D
b	7.38	4C
c	8.16	O
d	8.75	4A
▶ e	9.04	2E

▶ $C_6H_6N_2O$ ▶ 7% $CDCl_3$ ▶ V 453

▶ 15-β-(D)H

H(e), (d) H, (c)H, O, (f)H, N, C—OCH3(b), N, CH3(a)

	Shift	Peak
a	2.25	1A
b	3.69	1A
c	7.14	S
▶ d	7.34	S
e	7.64	S
f	8.54	2E

▶ $C_{13}H_{18}N_2O_2$ ▶ 7% $CDCl_3$ ▶ V 301

▶ 15-β-HH

	Shift	Peak
a	4.75	1C
b	6.51	10A
▸ c	6.64	10A
d	7.42	10A
e	8.09	10A

▸ $C_5H_6N_2$ ▸ 7% $CDCl_3$ ▸ V 431

▶ 15-β-HH

	Shift	Peak
a	2.33	1A
b	7.20 ± 0.10	S
▸ c	7.20 ± 0.10	S
d	8.10 ± 0.05	S
e	8.10 ± 0.05	S

▸ C_6H_7NO ▸ 7% $CDCl_3$ ▸ V 457

▶ 15-β-HH

	Shift	Peak
▸ a	~7.00	O
b	~7.60	O
c	8.60	2E

▸ C_5H_5N ▸ 7% $CDCl_3$ ▸ V 96

▶ 15-β-HH

	Shift	Peak
a	2.18	1A
▸ b	7.30	4A
c	7.75	6A
d	8.55	2A
e	8.60	2A

▸ $C_{10}H_{14}N_2$ ▸ 7% $CDCl_3$ ▸ V 269

▶ 15-β-HH

	Shift	Peak
a	3.13	1A
b	3.13	1A
▸ c	7.07	S
d	7.07	S
e	7.53	S
f	7.90	S
g	8.48	S

▸ $C_{14}H_{17}N_3O$ ▸ 7% $CDCl_3$ ▸ V 633

▶ 15 β HH

	Shift	Peak
a	~6.26	1D
▸ b	7.38	4C
c	8.16	O
d	8.75	4A
e	9.04	2E

▸ $C_6H_6N_2O$ ▸ 7% $CDCl_3$ ▸ V 453

▶ 15-β-HH

	Shift	Peak
a	5.45	S
b	6.22	S
c	6.75	S
▸ d	7.08	S
e	7.27	S
f	7.55	S
g	8.52	S

▸ C_7H_7N ▸ 7% $CDCl_3$ ▸ V 154

▶ 15-β-HH

	Shift	Peak
▸ a	7.58	1A
b	8.23	4C
c	8.74	O
d	8.93	O

▸ $C_6H_6N_2O$ ▸ D_2O ▸ V 454

▶ 15-β-HH

	Shift	Peak
a	2.25	1A
b	3.69	1A
▸ c	7.14	S
d	7.34	S
e	7.64	S
f	8.54	2E

▸ $C_{13}H_{18}N_2O_2$ ▸ 7% $CDCl_3$ ▸ V 301

▶ 15-β-HH

	Shift	Peak
a	2.45	1A
b	6.88*	2H
c	6.88*	2H
d	6.95	1A
▸ e	7.68	4A
f	8.70	4A
g	8.98	4A

*or 7.30

▸ $C_{17}H_{10}Cl_2N_2O_3$ ▸ 7% $CDCl_3$ ▸ V 327

▶ 15-β-HH

	Shift	Peak
a	3.72	1A
b	6.46	1D
▸ c	~7.17	N
d	~7.25	M
e	7.65	6A
f	8.52	2E

▸ $C_7H_8N_2O$ ▸ 7% $CDCl_3$ ▸ V 159

▶ 15-β-HH

	Shift	Peak
a	1.03	3A
b	2.15	6A
c	4.74	3C
▸ d	8.29	3C
e	8.97	R
f	9.12	R
g	9.37	1G

▸ $C_9H_{13}IN_2O$ ▸ D_2O ▸ V 539

▶ 15-β-HJ

(c) H, (d) H, H(b), O, (a)CH₃, N, C, C, O, N, CH₃(a), (b)H, H (d), H (c)

	Shift	Peak
a	2.43	1A
b	7.31	2E
c	7.75	S
▶ d	7.98	S

▶ $C_{14}H_{12}N_2O_2$ ▶ 7% $CDCl_3$ ▶ V 623

▶ 15-γ-(C)H

(c) H, (b)H, N, H N H, CH₃(a), (d), (e)

	Shift	Peak
a	2.18	1A
b	7.30	4A
▶ c	7.75	6A
d	8.55	2A
e	8.60	2A

▶ $C_{10}H_{14}N_2$ ▶ 7% $CDCl_3$ ▶ V 269

▶ 15-β-HL

(a) H, (c)H C=C —H(b), (d)H, H(d), (e)H N H(e)

	Shift	Peak
a	5.43	3C
b	5.90	2D
c	6.62	4D
▶ d	7.22	Q
e	8.52	Q

▶ C_7H_7N ▶ 7% $CDCl_3$ ▶ V 155

▶ 15-γ-HH

(d) H, (c)H, H(b), (e)H N NH₂(a)

	Shift	Peak
a	4.75	1C
b	6.51	10A
c	6.64	10A
▶ d	7.42	10A
e	8.09	10A

▶ $C_5H_6N_2$ ▶ 7% $CDCl_3$ ▶ V 431

▶ 15-β-HL

(f) H, (d)H, H(e), H(b), (g)H N C=C, (c)H, H(a)

	Shift	Peak
a	5.45	S
b	6.22	S
c	6.75	S
d	7.08	S
▶ e	7.27	S
f	7.55	S
g	8.52	S

▶ C_7H_7N ▶ 7% $CDCl_3$ ▶ V 154

▶ 15-γ-HH

(c) H, (b)H, H(b), (a)CH₃ N CH₃(a)

	Shift	Peak
a	2.52	1A
b	6.95	2A
▶ c	7.45	4A

▶ C_7H_9N ▶ 7% $CDCl_3$ ▶ V 169

▶ 15-β-HN

(d) H, (c)H, H(b), (e)H N NH₂(a)

	Shift	Peak
a	4.75	1C
▶ b	6.51	10A
c	6.64	10A
d	7.42	10A
e	8.09	10A

▶ $C_5H_6N_2$ ▶ 7% $CDCl_3$ ▶ V 431

▶ 15-γ-HH

H(e), (c) H, H(d), (f) OH, (d)H, H(e), (c) H, H N (b) CH₂ CH₂ N CH₂ CH₂ (b) N H, (a) (a), (g) (g)

	Shift	Peak
a	3.13	1A
b	3.13	1A
c	7.07	S
d	7.07	S
▶ e	7.53	S
f	7.90	S
g	8.48	S

▶ $C_{14}H_{17}N_3O$ ▶ 7% $CDCl_3$ ▶ V 633

▶ 15-β-HS

(c) CH₂, (b) CH₂, (a) CH₃, S, (d)H, H(d), (e)H N H(e)

	Shift	Peak
a	1.08	3A
b	1.73	6B
c	2.93	3A
▶ d	7.10	Q
e	8.38	Q

▶ $C_8H_{11}NS$ ▶ 7% $CDCl_3$ ▶ V 509

▶ 15-γ-HH

H(c), (d) H N CH₂ (a), (b)

	Shift	Peak
a	4.18	1A
b	7.27	1E
▶ c	7.54	S
d	8.56	3C

▶ $C_{12}H_{11}N$ ▶ 7% $CDCl_3$ ▶ V 593

▶ 15-γ-AH

(b) H, (c)H, CH₃(a), (d)H N H(e), O

	Shift	Peak
a	2.33	1A
▶ b	7.20 ± 0.10	S
c	7.20 ± 0.10	S
d	8.10 ± 0.05	S
e	8.10 ± 0.05	S

▶ C_6H_7NO ▶ 7% $CDCl_3$ ▶ V 457

▶ 15-γ-HH

(f) H, (d)H, H(e), H(b), (g)H N C=C, (c)H, H(a)

	Shift	Peak
a	5.45	S
b	6.22	S
c	6.75	S
d	7.08	S
e	7.27	S
▶ f	7.55	S
g	8.52	S

▶ C_7H_7N ▶ 7% $CDCl_3$ ▶ V 154

▶ 15-γ-HH

	Shift	Peak
a	~7.00	O
▶ b	~7.60	O
c	8.60	2E

▶ C_5H_5N ▶ 7% $CDCl_3$ ▶ V 96

▶ 15-γ-HJ

	Shift	Peak
a	1.03	3A
b	2.15	6A
c	4.74	3C
d	8.29	3C
▶ e	8.97	R
f	9.12	R
g	9.37	1G

▶ $C_9H_{13}IN_2O$ ▶ D_2O ▶ V 539

▶ 15-γ-HH

	Shift	Peak
a	2.25	1A
b	3.69	1A
c	7.14	S
d	7.34	S
▶ e	7.64	S
f	8.54	2E

▶ $C_{13}H_{18}N_2O_2$ ▶ 7% $CDCl_3$ ▶ V 301

▶ 15-γ-H(L)

	Shift	Peak
a	2.45	1A
b	6.88*	2H
c	6.88*	2H
d	6.95	1A
e	7.68	4A
▶ f	8.70	4A
g	8.98	4A

*or 7.30

▶ $C_{17}H_{10}Cl_2N_2O_3$ ▶ 7% $CDCl_3$ ▶ V 327

▶ 15-γ-HH

	Shift	Peak
a	3.72	1A
b	6.46	1D
c	~7.17	N
d	~7.25	M
▶ e	7.65	6A
f	8.52	2E

▶ $C_7H_8N_2O$ ▶ 7% $CDCl_3$ ▶ V 159

▶ **16**-α-H

	Shift	Peak
a	4.96	1A
b	6.13	S
c	6.20 ± 0.10	S
d	6.20 ± 0.10	S
e	6.67	3A
▶ f	7.32	3A

▶ C_9H_9NO ▶ 7% $CDCl_3$ ▶ V 525

▶ 15-γ-HH

	Shift	Peak
a	2.43	1A
b	7.31	2E
▶ c	7.75	S
d	7.98	S

▶ $C_{14}H_{12}N_2O_2$ ▶ 7% $CDCl_3$ ▶ V 623

▶ 16-α-H

	Shift	Peak
a	1.90	3A
b	3.73	2A
c	6.17	N
d	6.28	N
▶ e	7.33	M

▶ C_5H_6OS ▶ 7% $CDCl_3$ ▶ V 101

▶ 15-γ-HJ

	Shift	Peak
a	~6.26	1D
b	7.38	4C
▶ c	8.16	O
d	8.75	4A
e	9.04	2E

▶ $C_6H_6N_2O$ ▶ 7% $CDCl_3$ ▶ V 453

▶ 16-α-H

	Shift	Peak
a	1.53	1A
b	3.80	1A
c	6.13	N
d	6.30	N
▶ e	7.33	N

▶ C_5H_7NO ▶ 7% $CDCl_3$ ▶ V 104

▶ 15-γ-HJ

	Shift	Peak
a	7.58	1A
▶ b	8.23	4C
c	8.74	O
d	8.93	O

▶ $C_6H_6N_2O$ ▶ D_2O ▶ V 454

▶ 16-α-H

	Shift	Peak
a	4.02	1C
b	5.81	1A
c	6.37	N
d	6.40	N
e	6.55	N
f	7.27	N
▶ g	7.38	N
▶ h	7.62	N

▶ $C_{10}H_8O_4$ ▶ 7% $CDCl_3$ ▶ V 247

▶ 16-α-H

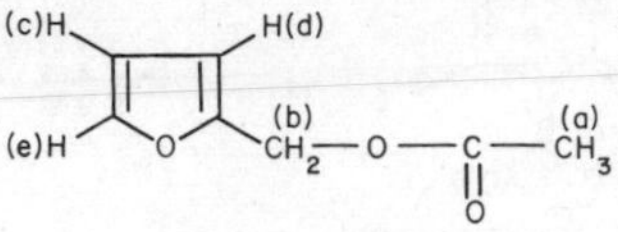

	Shift	Peak
a	2.07	1A
b	5.03	1A
c	6.35	1E
d	6.38	1E
▶ e	7.40	N

▶ $C_7H_8O_3$ ▶ 7% $CDCl_3$ ▶ V 167

▶ 16-α-H

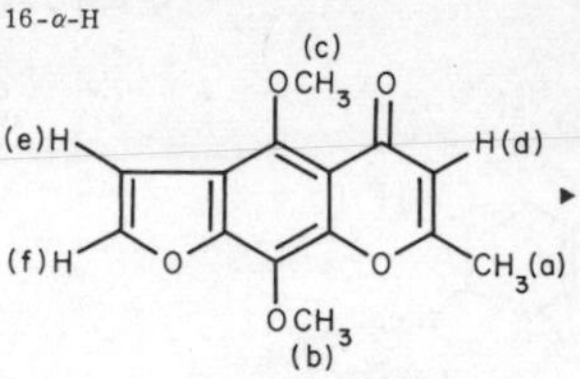

	Shift	Peak
a	2.40	1G
b	4.06 or 4.20	1A
c	4.06 or 4.20	1A
d	6.05	1G
e	7.01	2C
▶ f	7.63	2C

▶ $C_{14}H_{12}O_5$ ▶ 7% $CDCl_3$ ▶ V 628

▶ 16-α-H

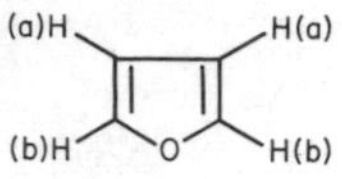

	Shift	Peak
a	6.37	Q
▶ b	7.42	Q

▶ C_4H_4O ▶ 7% $CDCl_3$ ▶ V 50

▶ 16-α-H

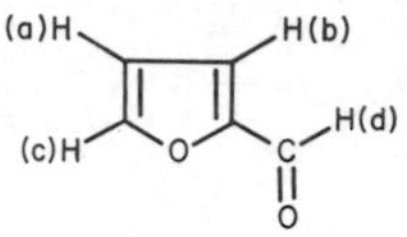

	Shift	Peak
a	6.63	4D
b	7.28	N
▶ c	7.72	M
d	9.67	1A

▶ $C_5H_4O_2$ ▶ 7% $CDCl_3$ ▶ V 95

▶ 16-α-H

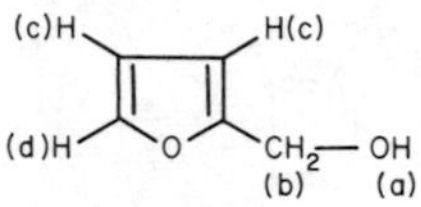

	Shift	Peak
a	2.83	1B
b	4.57	1A
c	6.33 ± 0.03	M
▶ d	7.44	M

▶ $C_5H_6O_2$ ▶ 7% $CDCl_3$ ▶ V 102

▶ 16-β-AH

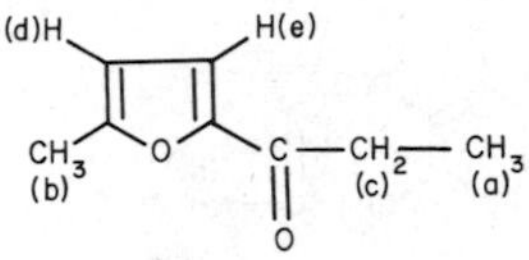

	Shift	Peak
a	1.20	3A
b	2.38	1A
c	2.78	4A
▶ d	6.13	2E
e	7.08	2A

▶ $C_8H_{10}O_2$ ▶ 7% $CDCl_3$ ▶ V 206

▶ 16-α-H

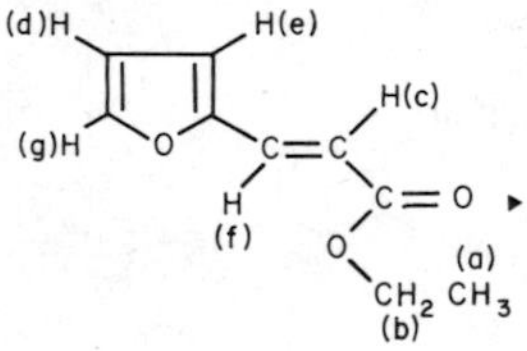

	Shift	Peak
a	1.30	3A
b	4.23	4A
c	6.30	2F
d	6.45	N
e	6.60	S
f	7.40	S
▶ g	7.47	S

▶ $C_9H_{10}O_3$ ▶ 7% $CDCl_3$ ▶ V 235

▶ 16-β-AH

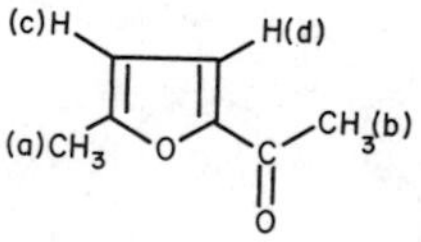

	Shift	Peak
a	2.38	2D
b	2.43	1A
▶ c	6.17	2E
d	7.12	2E

▶ $C_7H_8O_2$ ▶ 7% $CDCl_3$ ▶ V 165

▶ 16-α-H

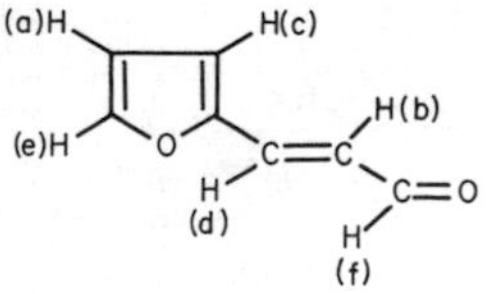

	Shift	Peak
a	6.55	S
b	6.57	S
c	6.77	S
d	7.20	S
▶ e	7.55	S
f	9.57	2A

▶ $C_7H_6O_2$ ▶ 7% $CDCl_3$ ▶ V 152

▶ 16-β-BH

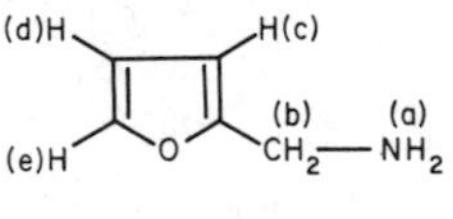

	Shift	Peak
a	1.53	1A
b	3.80	1A
▶ c	6.13	N
d	6.30	N
e	7.33	N

▶ C_5H_7NO ▶ 7% $CDCl_3$ ▶ V 104

▶ 16-α-H

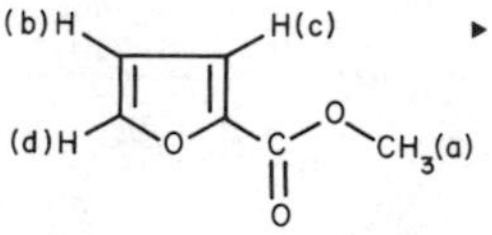

	Shift	Peak
a	3.88	1A
b	6.51	4D
c	7.20	N
▶ d	7.58	N

▶ $C_6H_6O_3$ ▶ 7% $CDCl_3$ ▶ V 125

▶ 16-β-BH

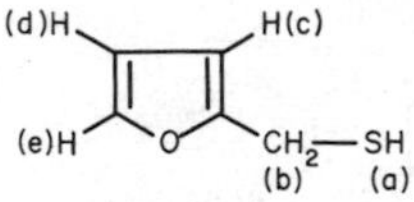

	Shift	Peak
a	1.90	3A
b	3.73	2A
▶ c	6.17	N
d	6.28	N
e	7.33	M

▶ C_5H_6OS ▶ 7% $CDCl_3$ ▶ V 101

▶ 16-β-BH

(c)H H(d) (e)H H(b)
(f)H O (a) CH_2—N
(e)H H(b)

	Shift	Peak
a	4.96	1A
b	6.13	S
c	6.20 ± 0.10	S
▶ d	6.20 ± 0.10	S
e	6.67	3A
f	7.32	3A

▶ C_9H_9NO ▶ 7% $CDCl_3$ ▶ V 525

▶ 16-β-HH

(d)H H(c)
(e)H O CH_2—SH
(b) (a)

	Shift	Peak
a	1.90	3A
b	3.73	2A
c	6.17	N
▶ d	6.28	N
e	7.33	M

▶ C_5H_6OS ▶ 7% $CDCl_3$ ▶ V 101

▶ 16-β-BH

(c)H H(c)
(d)H O CH_2— OH
(b) (a)

	Shift	Peak
a	2.83	1B
b	4.57	1A
▶ c	6.33 ± 0.03	M
d	7.44	M

▶ $C_5H_6O_2$ ▶ 7% $CDCl_3$ ▶ V 102

▶ 16-β-HH

(d)H H(c)
(e)H O (b) CH_2—NH_2 (a)

	Shift	Peak
a	1.53	1A
b	3.80	1A
c	6.13	N
▶ d	6.30	N
e	7.33	N

▶ C_5H_7NO ▶ 7% $CDCl_3$ ▶ V 104

▶ 16-β-BH

(c)H H(d)
(e)H O (b) CH_2— O — C — CH_3 (a)
O

	Shift	Peak
a	2.07	1A
b	5.03	1A
c	6.35	1E
▶ d	6.38	1E
e	7.40	N

▶ $C_7H_8O_3$ ▶ 7% $CDCl_3$ ▶ V 167

▶ 16-β-HH

(c)H H(c)
(d)H O CH_2— OH
(b) (a)

	Shift	Peak
a	2.83	1B
b	4.57	1A
▶ c	6.33 ± 0.03	M
d	7.44	M

▶ $C_5H_6O_2$ ▶ 7% $CDCl_3$ ▶ V 102

▶ 16-β-CH

(c) H (d) H (f) H (e) H
H(b)
C — C
H (g) O O O H (h)
O H (a)

	Shift	Peak
a	4.02	1C
b	5.81	1A
c	6.37	N
▶ d	6.40	N
e	6.55	N
f	7.27	N
g	7.38	N
h	7.62	N

▶ $C_{10}H_8O_4$ ▶ 7% $CDCl_3$ ▶ V 247

▶ 16-β-HH

(c)H H(d)
(e)H O (b) CH_2— O — C — CH_3 (a)
O

	Shift	Peak
a	2.07	1A
b	5.03	1A
▶ c	6.35	1E
d	6.38	1E
e	7.40	N

▶ $C_7H_8O_3$ ▶ 7% $CDCl_3$ ▶ V 167

▶ 16-β-CH

(c)H H(b)
O_2N O C O C CH_3(a)
H(d) O
O=C
CH_3(a)

	Shift	Peak
a	2.20	1A
▶ b	6.77	2C
c	7.32	2C
d	7.75	1A

▶ $C_9H_9NO_6$ ▶ 7% $CDCl_3$ ▶ V 526

▶ 16-β-HH

(a)H H(a)
(b)H O H(b)

	Shift	Peak
▶ a	6.37	Q
b	7.42	Q

▶ C_4H_4O ▶ 7% $CDCl_3$ ▶ V 50

▶ 16-β-HH

(c)H H(d) (e)H H(b)
(f)H O (a) CH_2—N
(e)H H(b)

	Shift	Peak
a	4.96	1A
b	6.13	S
▶ c	6.20 ± 0.10	S
d	6.20 ± 0.10	S
e	6.67	3A
f	7.32	3A

▶ C_9H_9NO ▶ 7% $CDCl_3$ ▶ V 525

▶ 16-β-HH

(c) H (d) H (f) H (e) H
H(b)
C — C
H (g) O O O H (h)
O H (a)

	Shift	Peak
a	4.02	1C
b	5.81	1A
▶ c	6.37	N
d	6.40	N
▶ e	6.55	N
f	7.27	N
g	7.38	N
h	7.62	N

▶ $C_{10}H_8O_4$ ▶ 7% $CDCl_3$ ▶ V 247

▶ 16-β-HH

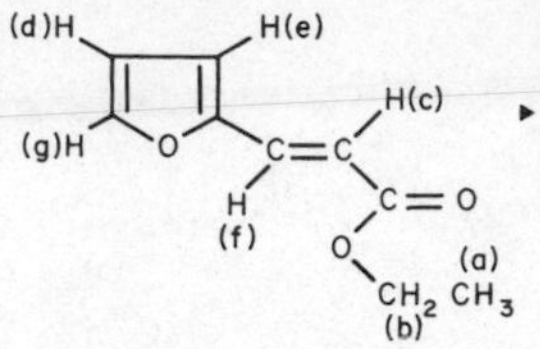

	Shift	Peak
a	1.30	3A
b	4.23	4A
c	6.30	2F
▶ d	6.45	N
e	6.60	S
f	7.40	S
g	7.47	S

▶ $C_9H_{10}O_3$ ▶ 7% $CDCl_3$ ▶ V 235

▶ 16-β-HJ

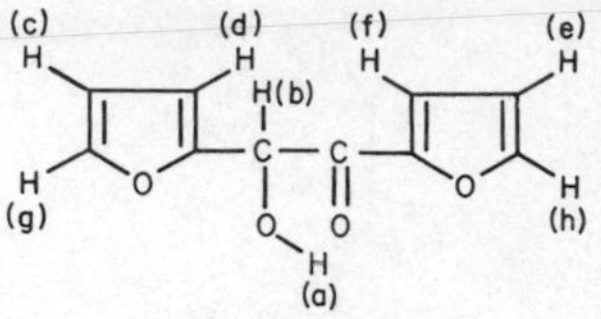

	Shift	Peak
a	4.02	1C
b	5.81	1A
c	6.37	N
d	6.40	N
e	6.55	N
▶ f	7.27	N
g	7.38	N
h	7.62	N

▶ $C_{10}H_8O_4$ ▶ 7% $CDCl_3$ ▶ V 247

▶ 16-β-HH

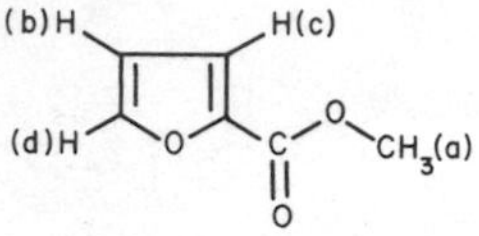

	Shift	Peak
a	3.88	1A
▶ b	6.51	4D
c	7.20	N
d	7.58	N

▶ $C_6H_6O_3$ ▶ 7% $CDCl_3$ ▶ V 125

▶ 16-β-HJ

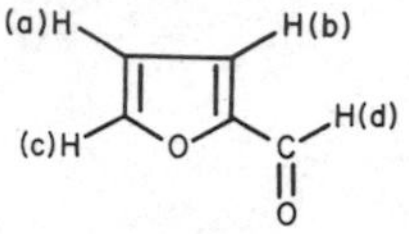

	Shift	Peak
a	6.63	4D
▶ b	7.28	N
c	7.72	M
d	9.67	1A

▶ $C_5H_4O_2$ ▶ 7% $CDCl_3$ ▶ V 95

▶ 16-β-HH

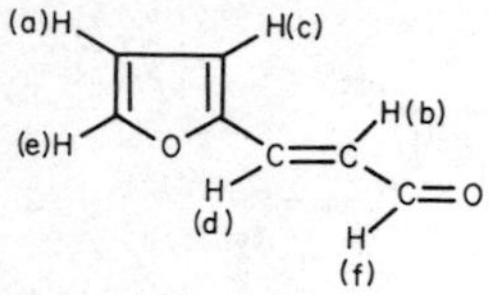

	Shift	Peak
▶ a	6.55	S
b	6.57	S
c	6.77	S
d	7.20	S
e	7.55	S
f	9.57	2A

▶ $C_7H_6O_2$ ▶ 7% $CDCl_3$ ▶ V 152

▶ 16-β-HK

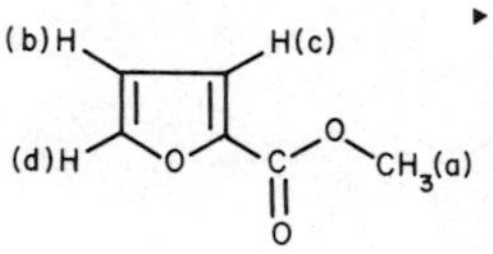

	Shift	Peak
a	3.88	1A
b	6.51	4D
▶ c	7.20	N
d	7.58	N

▶ $C_6H_6O_3$ ▶ 7% $CDCl_3$ ▶ V 125

▶ 16-β-HH

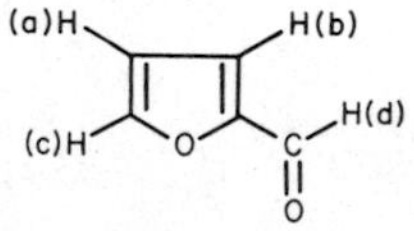

	Shift	Peak
▶ a	6.63	4D
b	7.28	N
c	7.72	M
d	9.67	1A

▶ $C_5H_4O_2$ ▶ 7% $CDCl_3$ ▶ V 95

▶ 16-β-HL

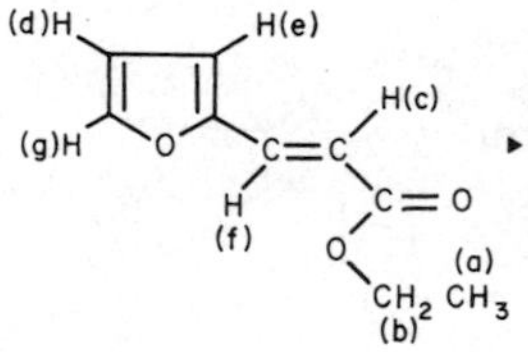

	Shift	Peak
a	1.30	3A
b	4.23	4A
c	6.30	2F
d	6.45	N
▶ e	6.60	S
f	7.40	S
g	7.47	S

▶ $C_9H_{10}O_3$ ▶ 7% $CDCl_3$ ▶ V 235

▶ 16-β-HJ

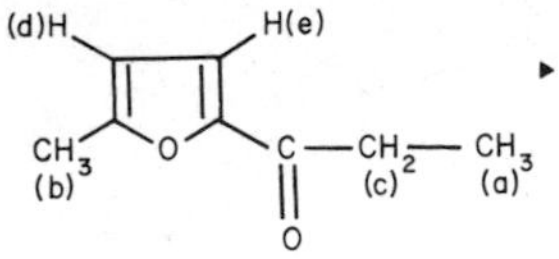

	Shift	Peak
a	1.20	3A
b	2.38	1A
c	2.78	4A
d	6.13	2E
▶ e	7.08	2A

▶ $C_8H_{10}O_2$ ▶ 7% $CDCl_3$ ▶ V 206

▶ 16-β-HL

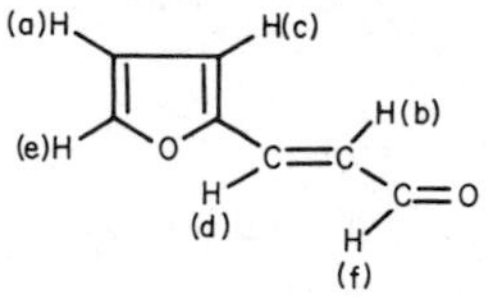

	Shift	Peak
a	6.55	S
b	6.57	S
▶ c	6.77	S
d	7.20	S
e	7.55	S
f	9.57	2A

▶ $C_7H_6O_2$ ▶ 7% $CDCl_3$ ▶ V 152

▶ 16-β-HJ

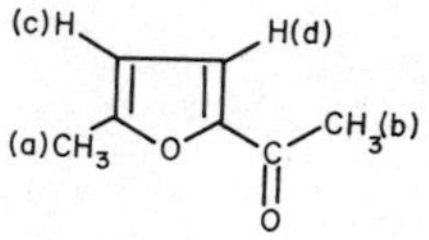

	Shift	Peak
a	2.38	2D
b	2.43	1A
c	6.17	2E
▶ d	7.12	2E

▶ $C_7H_8O_2$ ▶ 7% $CDCl_3$ ▶ V 165

▶ 16-β-HR

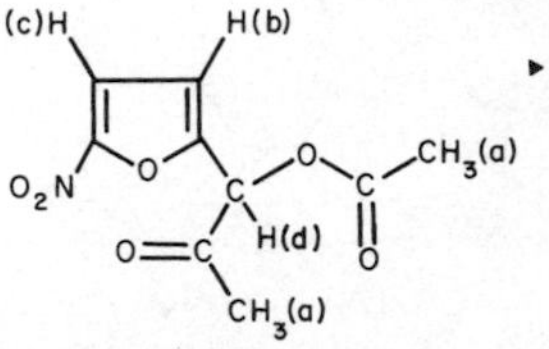

	Shift	Peak
a	2.20	1A
b	6.77	2C
▶ c	7.32	2C
d	7.75	1A

▶ $C_9H_9NO_6$ ▶ 7% $CDCl_3$ ▶ V 526

► 16-β-HW

(c) OCH_3 O
(e)H H(d)
(f)H O O CH_3(a)
OCH_3 (b)

	Shift	Peak
a	2.40	1G
b	4.06 or 4.20	1A
c	4.06 or 4.20	1A
d	6.05	1G
► e	7.01	2C
f	7.63	2C

► $C_{14}H_{12}O_5$ ► 7% $CDCl_3$ ► V 628

► 17-α-H

(d)H H(c)
(e)H N (b) CH_2— (a) CH_3
H (f)

	Shift	Peak
a	1.23	3A
b	2.63	6A
c	5.97	1I
d	6.17	4A
► e	6.68	N
f	~7.85	1A

► C_6H_9N ► 7% $CDCl_3$ ► V 130

►**17**-α-A

(b)CH_3 H(c)
(d)H N CH_3(a)
H (e)

	Shift	Peak
a	2.07 or 2.20	1A
b	2.07 or 2.20	1A
c	5.73	1B
► d	6.37	1B
e	~7.50	1D

► C_6H_9N ► 7% $CDCl_3$ ► V 131

► 17-α-H

(d) H (c) CH_2— (b) CH_2— (a) CH_3
(e) H N H(f)
H (g)

	Shift	Peak
a	0.95	3C
b	1.60	6B
c	2.48	3C
d	6.13	4B
► e	6.70	S
f	6.55	S
g	~7.83	1A

► $C_7H_{11}N$ ► 7% $CDCl_3$ ► V 177

► 17-α-B

(d) H (c) CH_2— (b) CH_2— (a) CH_3
(e) H N H(f)
H (g)

	Shift	Peak
a	0.95	3C
b	1.60	6B
c	2.48	3C
d	6.13	4B
e	6.70	S
► f	6.55	S
g	~7.83	1A

► $C_7H_{11}N$ ► 7% $CDCl_3$ ► V 177

► 17-α-H

(c)H H(d)
(e)H N C CH_3(a)
CH_3 (b) O

	Shift	Peak
a	2.40	1A
b	3.92	1A
c	6.10	4A
d	6.77	3A
► e	6.92	4A

► C_7H_9NO ► 7% $CDCl_3$ ► V 174

► 17-α-H

(d) H (c) $CH_2(CH_2)_3CH_3$ (a)
H (e) N CH_3(b)
H(f)

	Shift	Peak
a	0.90	3B
b	2.15	1A
c	~2.38	S
d	5.98	3A
► e	6.55	3A
f	7.67	1D

► $C_{10}H_{17}N$ ► 7% $CDCl_3$ ► V 278

► 17-α-H

(a)H H(c)
(b)H N C H(d)
H (e) O

	Shift	Peak
a	6.30	N
► b	6.98	N
c	7.17	N
d	9.45	2D
e	~11.08	1D

► C_5H_5NO ► 7% $CDCl_3$ ► V 98

► 17-α-H

(c)H H(d) (e)H H(b)
(f)H O (a) CH_2—N
(e)H H(b)

	Shift	Peak
a	4.96	1A
b	6.13	S
c	6.20 ± 0.10	S
d	6.20 ± 0.10	S
► e	6.67	3A
f	7.32	3A

► C_9H_9NO ► 7% $CDCl_3$ ► V 525

► 17-β-AA

(b)CH_3 H(c)
(d)H N CH_3(a)
H (e)

	Shift	Peak
a	2.07 or 2.20	1A
b	2.07 or 2.20	1A
► c	5.73	1B
d	6.37	1B
e	~7.50	1D

► C_6H_9N ► 7% $CDCl_3$ ► V 131

► 17-α-H

(a)H H(a)
(b)H N H(b)
H (c)

	Shift	Peak
a	6.22	Q
► b	6.68	Q
c	~8.0	1D

► C_4H_5N ► 7% $CDCl_3$ ► V 55

► 17-β-AH

(c)H H(d)
(a) CH_3 N C H(e)
CH_3 (b) O

	Shift	Peak
a	2.27	1A
b	3.87	1A
► c	6.02	2A
d	6.81	2A
e	9.38	1A

► C_7H_9NO ► 7% $CDCl_3$ ► V 170

▶ 17-β-BH

	Shift	Peak
a	1.23	3A
b	2.63	6A
▶c	5.97	1I
d	6.17	4A
e	6.68	N
f	~7.85	1A

▶ C_6H_9N ▶ 7% $CDCl_3$ ▶ V 130

▶ 17-β-HH

	Shift	Peak
▶a	6.22	Q
b	6.68	Q
c	~8.0	1D

▶ C_4H_5N ▶ 7% $CDCl_3$ ▶ V 55

▶ 17-β-BH

	Shift	Peak
a	0.90	3B
b	2.15	1A
c	~2.38	S
▶d	5.98	3A
e	6.55	3A
f	7.67	1D

▶ $C_{10}H_{17}N$ ▶ 7% $CDCl_3$ ▶ V 278

▶ 17-β-HH

	Shift	Peak
▶a	6.30	N
b	6.98	N
c	7.17	N
d	9.45	2D
e	~11.08	1D

▶ C_5H_5NO ▶ 7% $CDCl_3$ ▶ V 98

▶ 17-β-BH

	Shift	Peak
a	0.95	3C
b	1.60	6B
c	2.48	3C
▶d	6.13	4B
e	6.70	S
f	6.55	S
g	~7.83	1A

▶ $C_7H_{11}N$ ▶ 7% $CDCl_3$ ▶ V 177

▶ 17-β-HJ

	Shift	Peak
a	2.40	1A
b	3.92	1A
c	6.10	4A
▶d	6.77	3A
e	6.92	4A

▶ C_7H_9NO ▶ 7% $CDCl_3$ ▶ V 174

▶ 17-β-HH

	Shift	Peak
a	2.40	1A
b	3.92	1A
▶c	6.10	4A
d	6.77	3A
e	6.92	4A

▶ C_7H_9NO ▶ 7% $CDCl_3$ ▶ V 174

▶ 17-β-HJ

	Shift	Peak
a	2.27	1A
b	3.87	1A
c	6.02	2A
▶d	6.81	2A
e	9.38	1A

▶ C_7H_9NO ▶ 7% $CDCl_3$ ▶ V 170

▶ 17-β-HH

	Shift	Peak
a	4.96	1A
▶b	6.13	S
c	6.20 ± 0.10	S
d	6.20 ± 0.10	S
e	6.67	3A
f	7.32	3A

▶ C_9H_9NO ▶ 7% $CDCl_3$ ▶ V 525

▶ 17-β-HJ

	Shift	Peak
a	6.30	N
b	6.98	N
▶c	7.17	N
d	9.45	2D
e	~11.08	1D

▶ C_5H_5NO ▶ 7% $CDCl_3$ ▶ V 98

▶ 17-β-HH

	Shift	Peak
a	1.23	3A
b	2.63	6A
c	5.97	1I
▶d	6.17	4A
e	6.68	N
f	~7.85	1A

▶ C_6H_9N ▶ 7% $CDCl_3$ ▶ V 130

▶ 18-α-H

	Shift	Peak
a	2.48	2D
b	6.72	N
c	6.87	N
▶d	7.03	N

▶ C_5H_6S ▶ 7% $CDCl_3$ ▶ V 103

▶ 18-α-H

(a)H H(a)
(b)H S H(b)

	Shift	Peak
a	7.10	Q
▶ b	7.30	Q

▶ C_4H_4S ▶ 7% $CDCl_3$ ▶ V 52

▶ 18-α-H

(a)H H(c)
(b)H S C H(d)
O

	Shift	Peak
a	7.22	N
▶ b	7.78	M
c	7.78	M
d	9.92	2D

▶ C_5H_4OS ▶ 7% $CDCl_3$ ▶ V 94

▶ 18-α-H

(a)H H(b)
(c)H S I

	Shift	Peak
a	6.77	N
b	7.21	N
▶ c	7.31	N

▶ C_4H_3IS ▶ 7% $CDCl_3$ ▶ V 49

▶ 18-β-AH

(c)H H(b)
(d)H S CH_3 (a)

	Shift	Peak
a	2.48	2D
▶ b	6.72	N
c	6.87	N
d	7.03	N

▶ C_5H_6S ▶ 7% $CDCl_3$ ▶ V 103

▶ 18-α-H

(c)H H(e)
(d)H S C CH_2CH_3 (b) (a)
O

	Shift	Peak
a	1.23	3A
b	2.93	4A
c	7.10	N
▶ d	7.60	N
e	7.70	O

▶ C_7H_8OS ▶ 7% $CDCl_3$ ▶ V 163

▶ 18-β-AH

(c)H H(d)
(b)CH_3 S C CH_3(a)
O

	Shift	Peak
a	2.48	1E
b	2.52	1E
▶ c	6.78	2E
d	7.50	2A

▶ C_7H_8OS ▶ 7% $CDCl_3$ ▶ V 164

▶ 18-α-H

(d) H (f) H
H (e) S C(=O)—CH_2—$(CH_2)_6$—CH_2—C(=O)—OCH_3 (b) (g) (a) (c)

	Shift	Peak
a	2.28	3B
b	2.88	3B
c	3.67	1A
d	7.12	R
▶ e	7.62	R
f	7.70	R
g	1.35	1F

▶ $C_{15}H_{22}O_3S$ ▶ 7% $CDCl_3$ ▶ V 320

▶ 18-β-HH

(a)H H(b)
(c)H S I

	Shift	Peak
▶ a	6.77	N
b	7.21	N
c	7.31	N

▶ C_4H_3IS ▶ 7% $CDCl_3$ ▶ V 49

▶ 18-α-H

(b) (c) O (a)
(c) S C—CH_3

	Shift	Peak
a	2.55	1A
b	7.10	4A
▶ c	7.66	3C

▶ C_6H_6OS ▶ 44.8 mg/0.50 ml $CDCl_3$ ▶ S 123

▶ 18-β-HH

(c)H H(b)
(d)H S CH_3 (a)

	Shift	Peak
a	2.48	2D
b	6.72	N
▶ c	6.87	N
d	7.03	N

▶ C_5H_6S ▶ 7% $CDCl_3$ ▶ V 103

▶ 18-α-H

(b) H (d) H (a) H
(c)H S C=C C CF_3
O H O
(e)

(f) = CH_2 in Keto form.

	Shift	Peak
a	6.42	1A
b	7.18	3C
▶ c	7.73	N
d	7.82	N
e	14.17	1D
f	3.33	1B

▶ $C_8H_5F_3O_2S$ ▶ 7% $CDCl_3$ ▶ V 185

▶ 18-β-HH

(a)H H(a)
(b)H S H(b)

	Shift	Peak
▶ a	7.10	Q
b	7.30	Q

▶ C_4H_4S ▶ 7% $CDCl_3$ ▶ V 52

▶ 18-β-HH

(b) (c) O (a)
(c) S C—CH_3

	Shift	Peak
a	2.55	1A
▶ b	7.10	4A
c	7.66	3C

▶ C_6H_6OS ▶ 44.8 mg/0.50 ml $CDCl_3$ ▶ S 123

▶ 18-β-HJ

(c)H H(d)
(b)CH_3 S C CH_3(a)
O

	Shift	Peak
a	2.48	1E
b	2.52	1E
c	6.78	2E
▶ d	7.50	2A

▶ C_7H_8OS ▶ 7% $CDCl_3$ ▶ V 164

▶ 18-β-HH

(c)H H(e)
(b) (a)
(d)H S C CH_2CH_3
O

	Shift	Peak
a	1.23	3A
b	2.93	4A
▶ c	7.10	N
d	7.60	N
e	7.70	O

▶ C_7H_8OS ▶ 7% $CDCl_3$ ▶ V 163

▶ 18-β-HJ

(b) (c) O (a)
(c) S C—CH_3

	Shift	Peak
a	2.55	1A
b	7.10	4A
▶ c	7.66	3C

▶ C_6H_6OS ▶ 44.8 mg/0.50 ml $CDCl_3$ ▶ S 123

▶ 18-β-HH

(d) H (f) H
(b) (g) (a) (c)
H S C—CH_2—$(CH_2)_6$—CH_2—C—OCH_3
(e) O O

	Shift	Peak
a	2.28	3B
b	2.88	3B
c	3.67	1A
▶ d	7.12	R
e	7.62	R
f	7.70	R
g	1.35	1F

▶ $C_{15}H_{22}O_3S$ ▶ 7% $CDCl_3$ ▶ V 320

▶ 18-β-HJ

(c)H H(e)
(b) (a)
(d)H S C CH_2CH_3
O

	Shift	Peak
a	1.23	3A
b	2.93	4A
c	7.10	N
d	7.60	N
▶ e	7.70	O

▶ C_7H_8OS ▶ 7% $CDCl_3$ ▶ V 163

▶ 18-β-HH

(b) H (d) H (a) H
C C C CF_3
(c)H S O H O
(e)

(f) = CH_2 in Keto form.

	Shift	Peak
a	6.42	1A
▶ b	7.18	3C
c	7.73	N
d	7.82	N
e	14.17	1D
f	3.33	1B

▶ $C_8H_5F_3O_2S$ ▶ 7% $CDCl_3$ ▶ V 185

▶ 18-β-HJ

(d) H (f) H
(b) (g) (a) (c)
H S C—CH_2—$(CH_2)_6$—CH_2—C—OCH_3
(e) O O

	Shift	Peak
a	2.28	3B
b	2.88	3B
c	3.67	1A
d	7.12	R
e	7.62	R
▶ f	7.70	R
g	1.35	1F

▶ $C_{15}H_{22}O_3S$ ▶ 7% $CDCl_3$ ▶ V 320

▶ 18-β-HH

(a)H H(c)
(b)H S C H(d)
O

	Shift	Peak
▶ a	7.22	N
b	7.78	M
c	7.78	M
d	9.92	2D

▶ C_5H_4OS ▶ 7% $CDCl_3$ ▶ V 94

▶ 18-β-HJ

(a)H H(c)
(b)H S C H(d)
O

	Shift	Peak
a	7.22	N
b	7.78	M
▶ c	7.78	M
d	9.92	2D

▶ C_5H_4OS ▶ 7% $CDCl_3$ ▶ V 94

▶ 18-β-HI

(a)H H(b)
(c)H S I

	Shift	Peak
a	6.77	N
▶ b	7.21	N
c	7.31	N

▶ C_4H_3IS ▶ 7% $CDCl_3$ ▶ V 49

▶ 18-β-HL

(b) H (d) H (a) H
C C C CF_3
(c)H S O H O
(e)

(f) = CH_2 in Keto form.

	Shift	Peak
a	6.42	1A
b	7.18	3C
c	7.73	N
▶ d	7.82	N
e	14.17	1D
f	3.33	1B

▶ $C_8H_5F_3O_2S$ ▶ 7% $CDCl_3$ ▶ V 185

▶19(acridine-1)-H

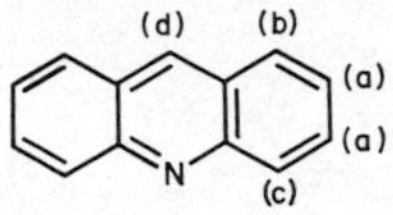

	Shift	Peak
a	7.50	S
▶ b	7.73	S
c	8.18	2A
d	8.48	1A

▶ $C_{13}H_9N$ ▶ 45.0 mg/0.50 ml $CDCl_3$ ▶ S 81

▶ 19(acridyl-4)-H

	Shift	Peak
a	7.18	S
▶ b	7.40 ± 0.10	S
c	7.48	S
d	7.60	S
e	7.75	S
f	7.94	S
g	8.17	S
h	~8.67	S

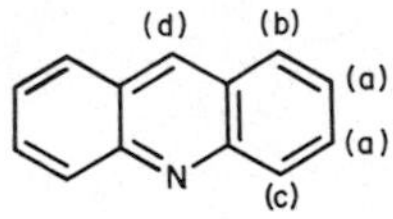

▶ $C_{13}H_9NO$ ▶ 7% $CDCl_3$ ▶ V 605

▶ 19(acridine-2,3)-HH

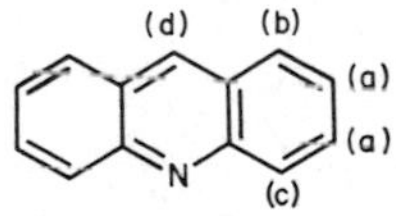

	Shift	Peak
▶ a	7.50	S
b	7.73	S
c	8.18	2A
d	8.48	1A

▶ $C_{13}H_9N$ ▶ 45.0 mg/0.50 ml $CDCl_3$ ▶ S 81

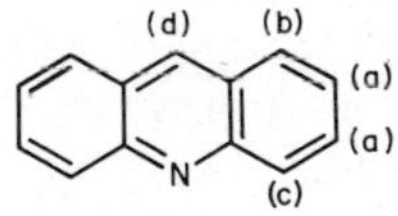

▶ 19(acridyl-5)-H

	Shift	Peak
a	7.18	S
b	7.40 ± 0.10	S
c	7.48	S
d	7.60	S
e	7.75	S
▶ f	7.94	S
g	8.17	S
h	~8.67	S

▶ $C_{13}H_9NO$ ▶ 7% $CDCl_3$ ▶ V 605

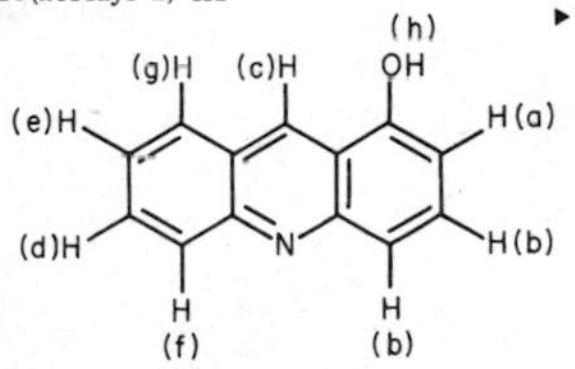

▶ 19(acridine-4)-H

	Shift	Peak
a	7.50	S
b	7.73	S
▶ c	8.18	2A
d	8.48	1A

▶ $C_{13}H_9N$ ▶ 45.0 mg/0.50 ml $CDCl_3$ ▶ S 81

▶ 19(acridyl-6)-HH

	Shift	Peak
a	7.18	S
b	7.40 ± 0.10	S
c	7.48	S
▶ d	7.60	S
e	7.75	S
f	7.94	S
g	8.17	S
h	~8.67	S

▶ $C_{13}H_9NO$ ▶ 7% $CDCl_3$ ▶ V 605

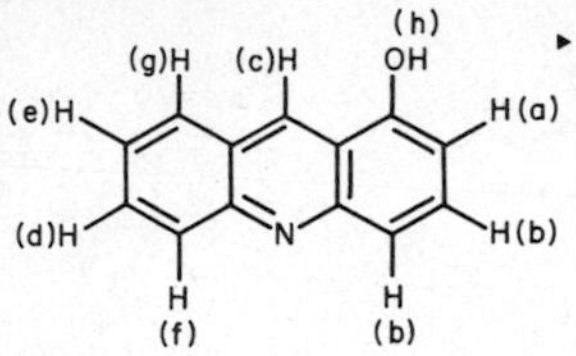

▶ 19(acridine-9)

	Shift	Peak
a	7.50	S
b	7.73	S
c	8.18	2A
▶ d	8.48	1A

▶ $C_{13}H_9N$ ▶ 45.0 mg/0.50 ml $CDCl_3$ ▶ S 81

▶ 19(acridyl-7)-HH

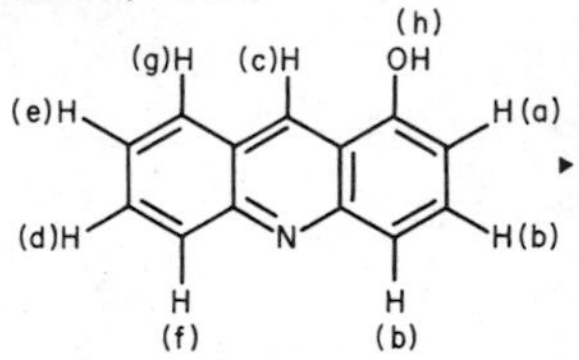

	Shift	Peak
a	7.18	S
b	7.40 ± 0.10	S
c	7.48	S
d	7.60	S
▶ e	7.75	S
f	7.94	S
g	8.17	S
h	~8.67	S

▶ $C_{13}H_9NO$ ▶ 7% $CDCl_3$ ▶ V 605

▶ 19(acridyl-2)-HP

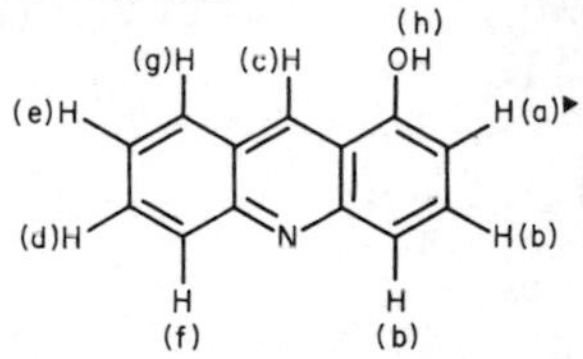

	Shift	Peak
▶ a	7.18	S
b	7.40 ± 0.10	S
c	7.48	S
d	7.60	S
e	7.75	S
f	7.94	S
g	8.17	S
h	~8.67	S

▶ $C_{13}H_9NO$ ▶ 7% $CDCl_3$ ▶ V 605

▶ 19(acridyl-8)-H

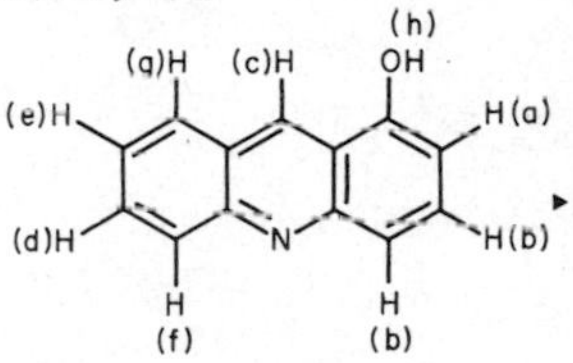

	Shift	Peak
a	7.18	S
b	7.40 ± 0.10	S
c	7.48	S
d	7.60	S
e	7.75	S
f	7.94	S
▶ g	8.17	S
h	~8.67	S

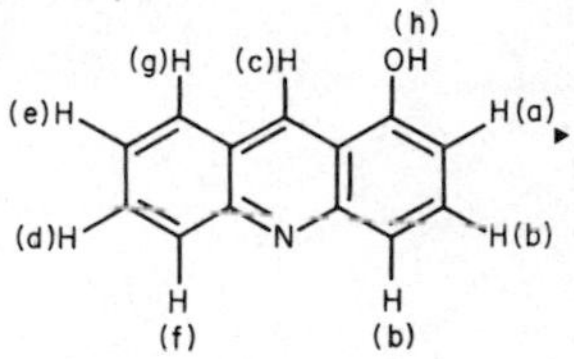

▶ $C_{13}H_9NO$ ▶ 7% $CDCl_3$ ▶ V 605

▶ 19(acridyl-3)-HH

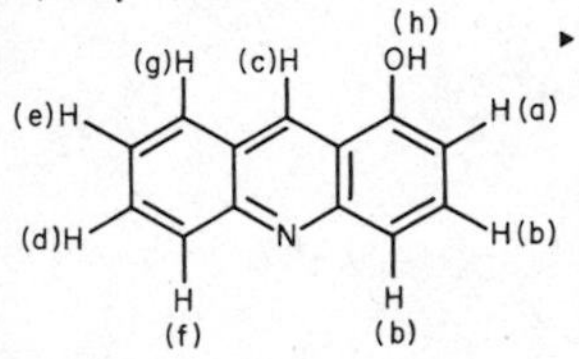

	Shift	Peak
a	7.18	S
▶ b	7.40 ± 0.10	S
c	7.48	S
d	7.60	S
e	7.75	S
f	7.94	S
g	8.17	S
h	~8.67	S

▶ $C_{13}H_9NO$ ▶ 7% $CDCl_3$ ▶ V 605

▶ 19(acridyl-9)

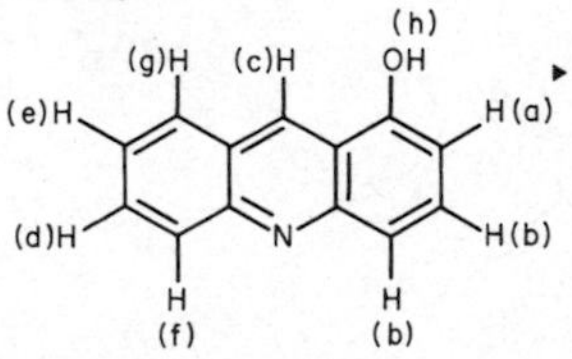

	Shift	Peak
a	7.18	S
b	7.40 ± 0.10	S
▶ c	7.48	S
d	7.60	S
e	7.75	S
f	7.94	S
g	8.17	S
h	~8.67	S

▶ $C_{13}H_9NO$ ▶ 7% $CDCl_3$ ▶ V 605

▶ 19(adeninyl-2)

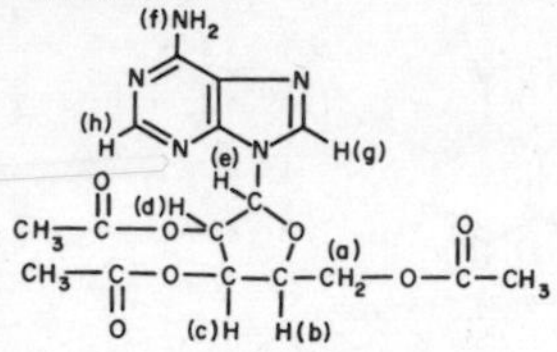

Shift	Peak
a 4.43	1E
b 4.43	1E
c 5.70	1C
d 5.95	3A
e 6.20	2A
f 6.37	1B
g 7.98 or 8.35	1A
▶ h 7.98 or 8.35	1A

▶ $C_{16}H_{19}N_5O_8$ ▶ 7% $CDCl_3$ ▶ V 326

▶ 19(azulene-2)-HH

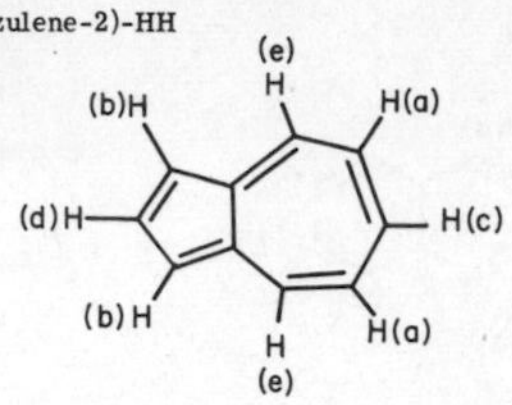

Shift	Peak
a 7.11	S
b 7.39	S
c 7.57	S
▶ d 7.92	3A
e 8.32	2E

▶ $C_{10}H_8$ ▶ 7% $CDCl_3$ ▶ V 551

▶ 19(anthroyl-1,4)-H

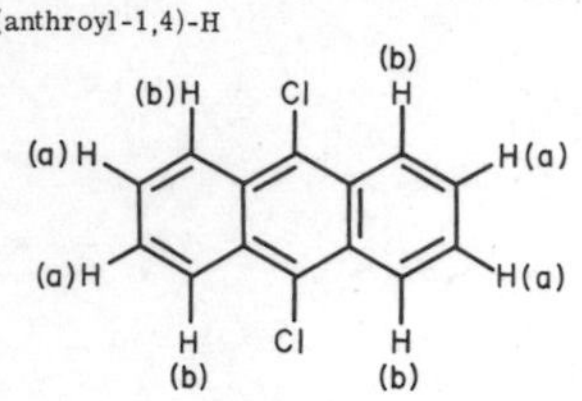

Shift	Peak
a 7.60	Q
▶ b 8.52	Q

▶ $C_{14}H_8Cl_2$ ▶ 7% $CDCl_3$ ▶ V 619

▶ 19(azulene-4,8)-H

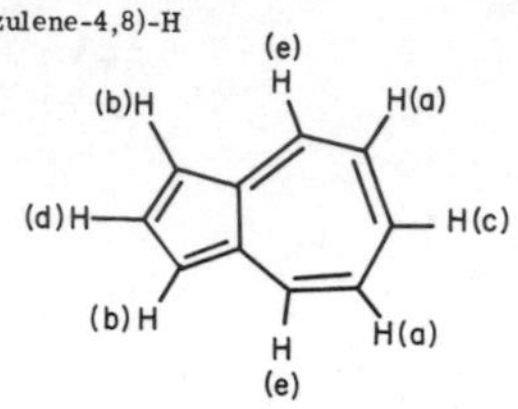

Shift	Peak
a 7.11	S
b 7.39	S
c 7.57	S
d 7.92	3A
▶ e 8.32	2E

▶ $C_{10}H_8$ ▶ 7% $CDCl_3$ ▶ V 551

▶ 19(anthroyl-1,4)-H

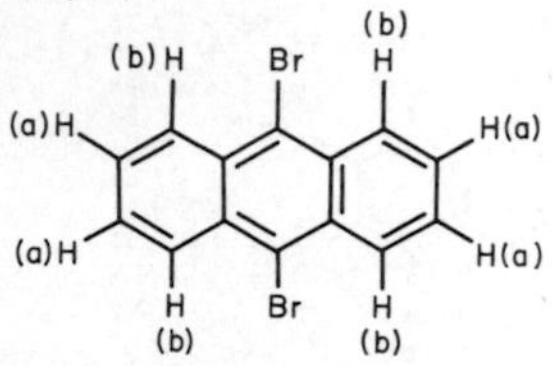

Shift	Peak
a 7.60	Q
▶ b 8.57	Q

▶ $C_{14}H_8Br_2$ ▶ 7% $CDCl_3$ ▶ V 618

▶ 19(azulene-5,7)-HH

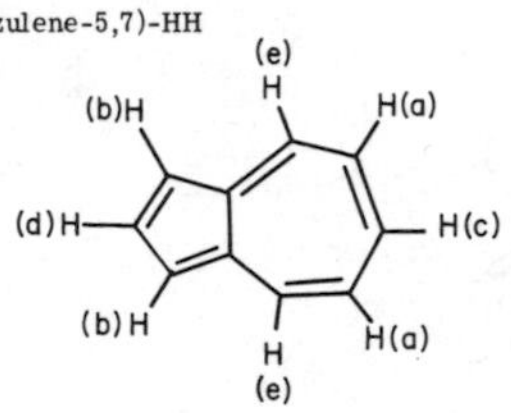

Shift	Peak
▶ a 7.11	S
b 7.39	S
c 7.57	S
d 7.92	3A
e 8.32	2E

▶ $C_{10}H_8$ ▶ 7% $CDCl_3$ ▶ V 551

▶ 19(anthroyl-2,3)-HH

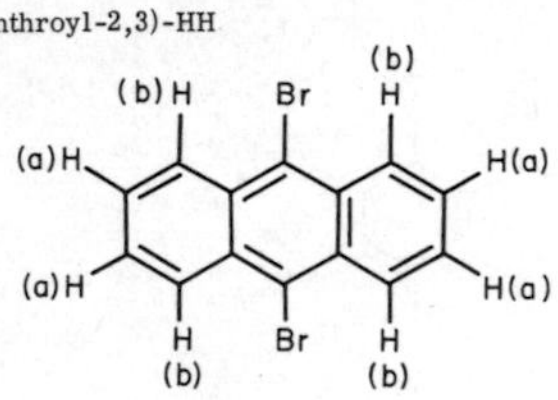

Shift	Peak
▶ a 7.60	Q
b 8.57	Q

▶ $C_{14}H_8Br_2$ ▶ 7% $CDCl_3$ ▶ V 618

▶ 19(azulene-6)-HH

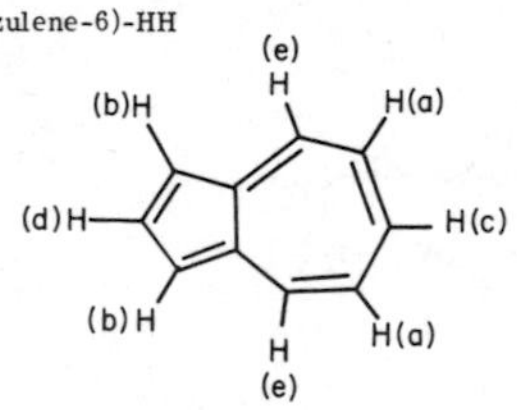

Shift	Peak
a 7.11	S
b 7.39	S
▶ c 7.57	S
d 7.92	3A
e 8.32	2E

▶ $C_{10}H_8$ ▶ 7% $CDCl_3$ ▶ V 551

▶ 19(anthroyl-2,3)-HH

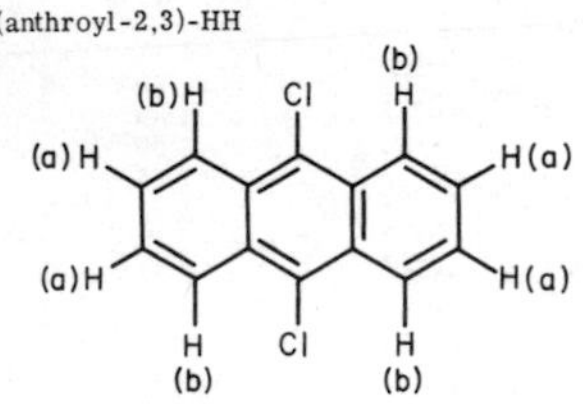

Shift	Peak
▶ a 7.60	Q
b 8.52	Q

▶ $C_{14}H_8Cl_2$ ▶ 7% $CDCl_3$ ▶ V 619

▶ 19(benzothiazole-2)

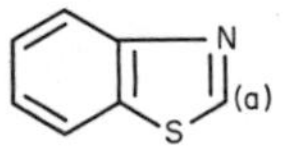

Shift	Peak
a 8.95	1A

▶ C_7H_5NS ▶ 44.8 mg/0.50 ml $CDCl_3$ ▶ S 139

▶ 19(azulene-1,3)-H

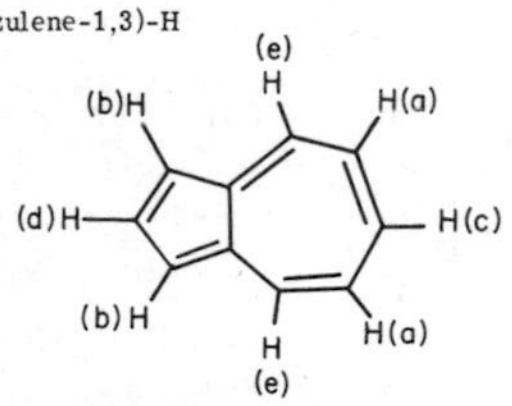

Shift	Peak
a 7.11	S
▶ b 7.39	S
c 7.57	S
d 7.92	3A
e 8.32	2E

▶ $C_{10}H_8$ ▶ 7% $CDCl_3$ ▶ V 551

▶ 19(cyclopentadienyl)-HH

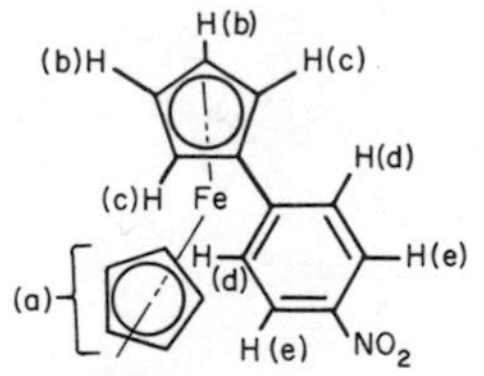

Shift	Peak
▶ a 4.07	1A
b 4.50	Q
c 4.77	Q
d 7.62	Q
e 8.18	Q

▶ $C_{16}H_{13}FeNO_2$ ▶ 7% $CDCl_3$ ▶ V 321

▶ 19(cyclopentadienyl-2,5)-HV

	Shift	Peak
a	4.07	1A
b	4.50	Q
▶ c	4.77	Q
d	7.62	Q
e	8.18	Q

▶ $C_{16}H_{13}FeNO_2$ ▶ 7% $CDCl_3$ ▶ V 321

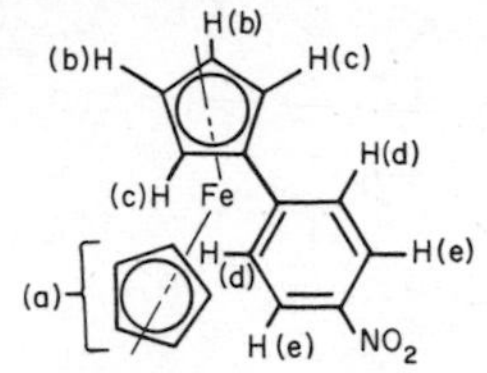

▶ 19(isoquinolinyl-3)-H

	Shift	Peak
a	8.52	2A
▶ b	9.26	1A

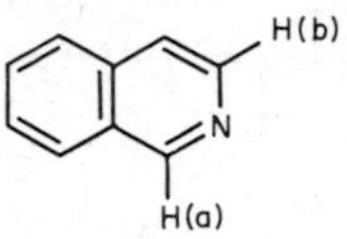

▶ C_9H_7N ▶ 7% $CDCl_3$ ▶ V 520

▶ 19(cyclopentadienyl-3,4)-HH

	Shift	Peak
a	4.07	1A
▶ b	4.50	Q
c	4.77	Q
d	7.62	Q
e	8.18	Q

▶ $C_{16}H_{13}FeNO_2$ ▶ 7% $CDCl_3$ ▶ V 321

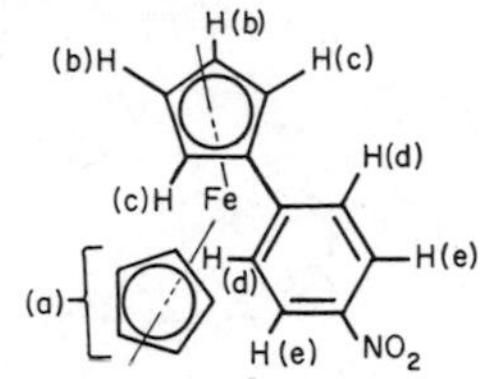

▶ 19(naphthalene-1,4,5,8)-H

	Shift	Peak
▶ a	7.33	Q
b	7.72	Q

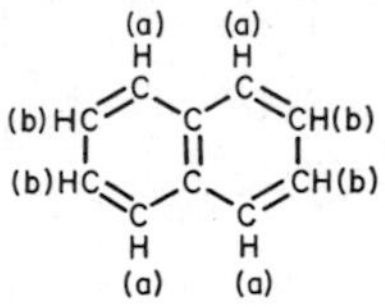

▶ $C_{10}H_8$ ▶ CCl_4 ▶ API 189

▶ 19(dihydropyrenyl-1)-HII

	Shift	Peak
a	−4.23	1A
▶ b	8.13	3B
c	8.63	1A
d	8.66	1A

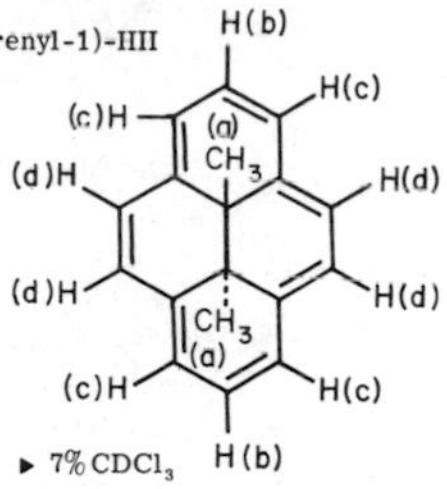

▶ $C_{18}H_{16}$ ▶ 7% $CDCl_3$ ▶ V 672

▶ 19(naphthalene-1,4,5,8)-H

	Shift	Peak
a	7.4	Q
▶ b	7.75	Q

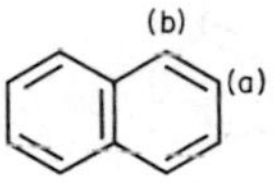

▶ $C_{10}H_8$ ▶ 45.0 mg/0.50 ml $CDCl_3$ ▶ S 62

▶ 19(dihydropyrenyl-2)-H

	Shift	Peak
a	−4.23	1A
b	8.13	3B
▶ c	8.63	1A
d	8.66	1A

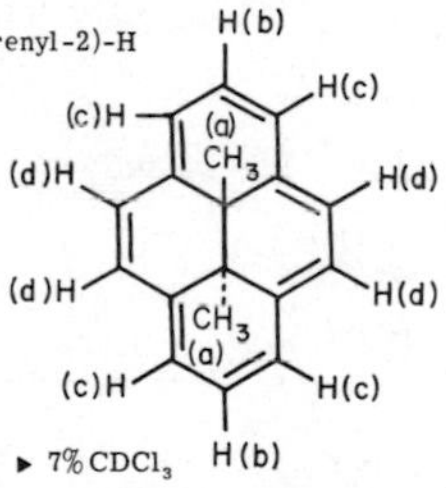

▶ $C_{18}H_{16}$ ▶ 7% $CDCl_3$ ▶ V 672

▶ 19(naphthalene-2,3,6,7)-HH

	Shift	Peak
▶ a	7.4	Q
b	7.75	Q

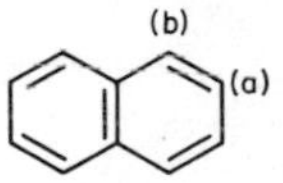

▶ $C_{10}H_8$ ▶ 45.0 mg/0.50 ml $CDCl_3$ ▶ S 62

▶ 19(dihydropyrenyl-4)-H

	Shift	Peak
a	−4.23	1A
b	8.13	3B
c	8.63	1A
▶ d	8.66	1A

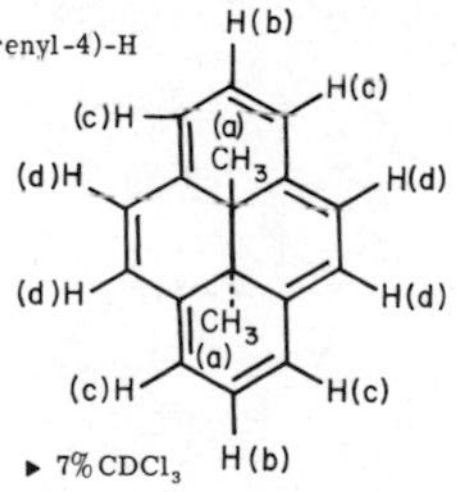

▶ $C_{18}H_{16}$ ▶ 7% $CDCl_3$ ▶ V 672

▶ 19(naphthalene-2,3,6,7)-HH

	Shift	Peak
a	7.33	Q
▶ b	7.72	Q

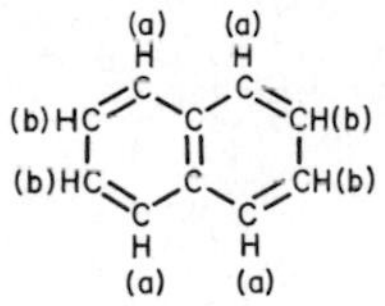

▶ $C_{10}H_8$ ▶ CCl_4 ▶ API 189

▶ 19(isoquinolinyl-1)

	Shift	Peak
▶ a	8.52	2A
b	9.26	1A

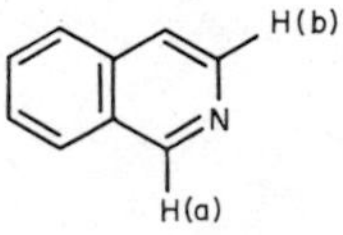

▶ C_9H_7N ▶ 7% $CDCl_3$ ▶ V 520

▶ 19(naphthalene-2)-GH

	Shift	Peak
a	8.14	S

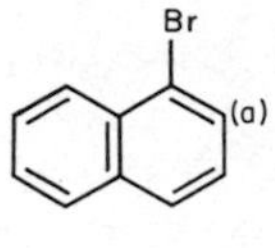

▶ $C_{10}H_7Br$ ▶ 45.8 mg/0.50 ml $CDCl_3$ ▶ S 143

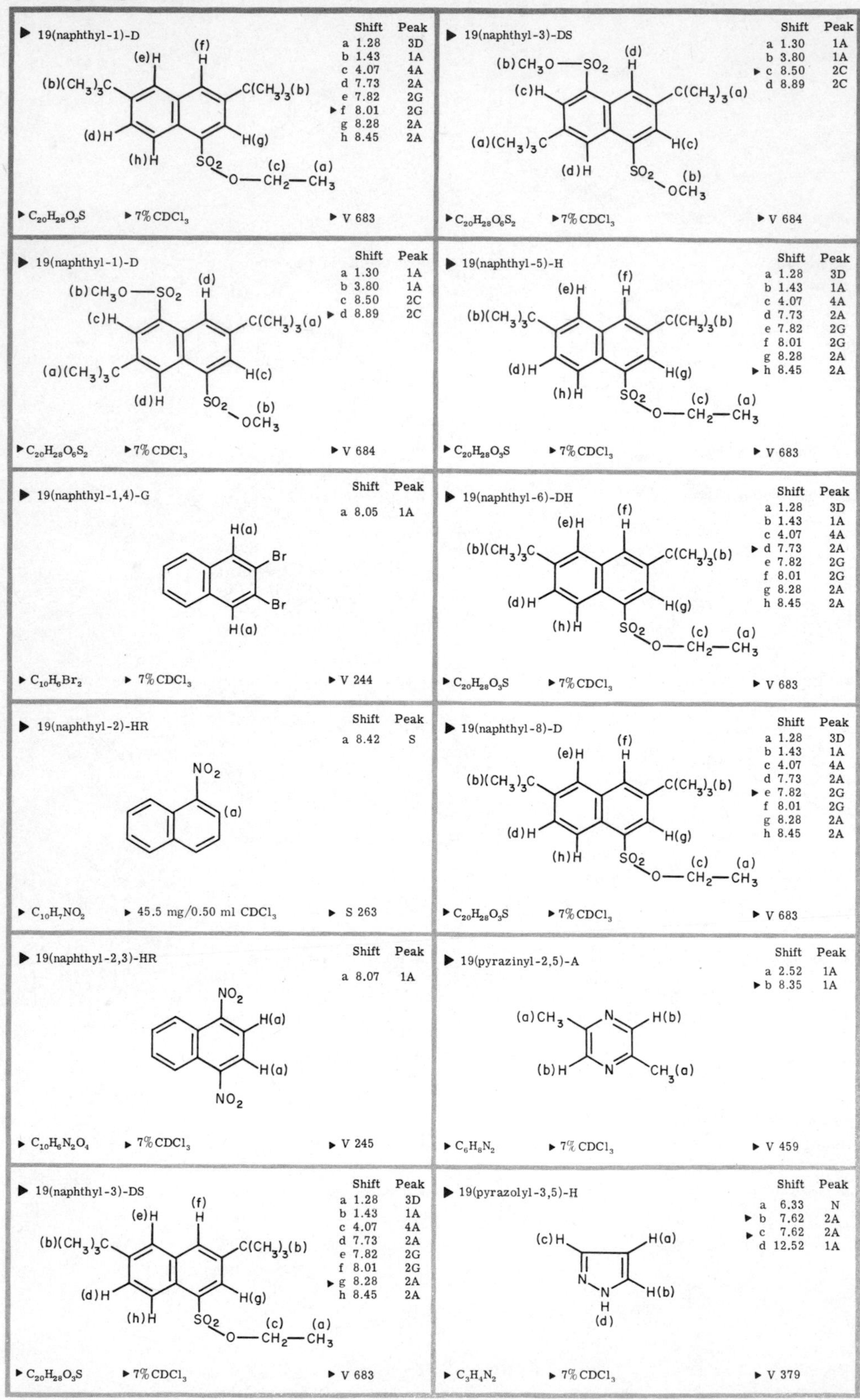
19(naphthyl-1)-D
Shift Peak
a 1.28 3D
b 1.43 1A
c 4.07 4A
d 7.73 2A
e 7.82 2G
f 8.01 2G
g 8.28 2A
h 8.45 2A
$C_{20}H_{28}O_3S$
7% $CDCl_3$
V 683
19(naphthyl-3)-DS
Shift Peak
a 1.30 1A
b 3.80 1A
c 8.50 2C
d 8.89 2C
$C_{20}H_{28}O_6S_2$
7% $CDCl_3$
V 684
19(naphthyl-1)-D
Shift Peak
a 1.30 1A
b 3.80 1A
c 8.50 2C
d 8.89 2C
$C_{20}H_{28}O_6S_2$
7% $CDCl_3$
V 684
19(naphthyl-5)-H
Shift Peak
a 1.28 3D
b 1.43 1A
c 4.07 4A
d 7.73 2A
e 7.82 2G
f 8.01 2G
g 8.28 2A
h 8.45 2A
$C_{20}H_{28}O_3S$
7% $CDCl_3$
V 683
19(naphthyl-1,4)-G
Shift Peak
a 8.05 1A
$C_{10}H_6Br_2$
7% $CDCl_3$
V 244
19(naphthyl-6)-DH
Shift Peak
a 1.28 3D
b 1.43 1A
c 4.07 4A
d 7.73 2A
e 7.82 2G
f 8.01 2G
g 8.28 2A
h 8.45 2A
$C_{20}H_{28}O_3S$
7% $CDCl_3$
V 683
19(naphthyl-2)-HR
Shift Peak
a 8.42 S
$C_{10}H_7NO_2$
45.5 mg/0.50 ml $CDCl_3$
S 263
19(naphthyl-8)-D
Shift Peak
a 1.28 3D
b 1.43 1A
c 4.07 4A
d 7.73 2A
e 7.82 2G
f 8.01 2G
g 8.28 2A
h 8.45 2A
$C_{20}H_{28}O_3S$
7% $CDCl_3$
V 683
19(naphthyl-2,3)-HR
Shift Peak
a 8.07 1A
$C_{10}H_6N_2O_4$
7% $CDCl_3$
V 245
19(pyrazinyl-2,5)-A
Shift Peak
a 2.52 1A
b 8.35 1A
$C_6H_8N_2$
7% $CDCl_3$
V 459
19(naphthyl-3)-DS
Shift Peak
a 1.28 3D
b 1.43 1A
c 4.07 4A
d 7.73 2A
e 7.82 2G
f 8.01 2G
g 8.28 2A
h 8.45 2A
$C_{20}H_{28}O_3S$
7% $CDCl_3$
V 683
19(pyrazolyl-3,5)-H
Shift Peak
a 6.33 N
b 7.62 2A
c 7.62 2A
d 12.52 1A
$C_3H_4N_2$
7% $CDCl_3$
V 379

▶ 19(pyrazolyl-4)-AA

	Shift	Peak
a	2.27	1A
▶ b	5.80	1B
c	11.92	1D

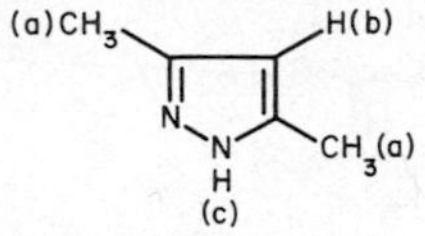

▶ $C_5H_8N_2$ ▶ 7% $CDCl_3$ ▶ V 441

▶ 19(pyridazinyl-4,5)-HH

	Shift	Peak
▶ a	7.50	Q
b	9.21	Q

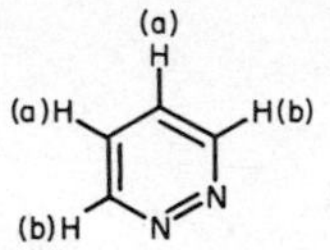

▶ $C_4H_4N_2$ ▶ 7% $CDCl_3$ ▶ V 398

▶ 19(pyrazolyl-4)-HH

	Shift	Peak
▶ a	6.33	N
b	7.62	2A
c	7.62	2A
d	12.52	1A

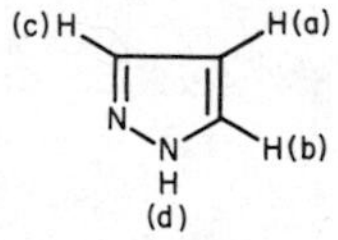

▶ $C_3H_4N_2$ ▶ 7% $CDCl_3$ ▶ V 379

▶ 19(pyrimidyl-2)

	Shift	Peak
▶ a	7.87	1A
b	8.08	1A

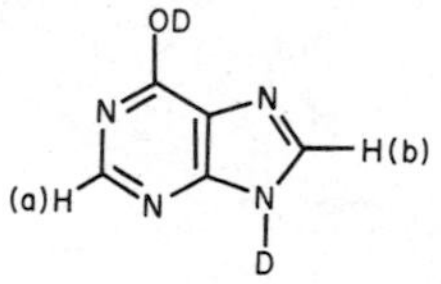

▶ $C_5H_4N_4O$ ▶ D_2O ▶ V 426

▶ 19(pyrene-1)-HH

	Shift	Peak
▶ a	7.95	S
b	7.98	S

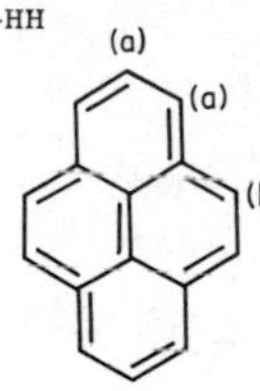

▶ $C_{16}H_{10}$ ▶ 45.0 mg/0.50 ml $CDCl_3$ ▶ S 64

▶ 19(pyrimidyl-2)

	Shift	Peak
▶ a	8.20 or 8.32	1A
b	8.20 or 8.32	1A

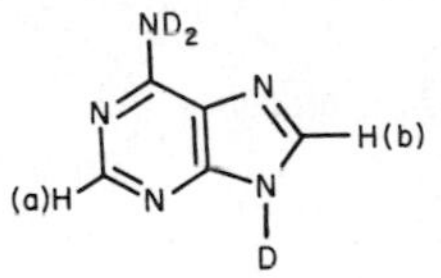

▶ $C_5H_5N_5$ ▶ 7% $CDCl_3$ ▶ V 429

▶ 19(pyrene-2)-H

	Shift	Peak
▶ a	7.95	S
b	7.98	S

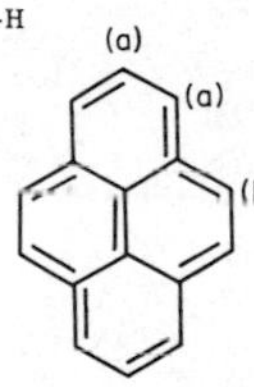

▶ $C_{16}H_{10}$ ▶ 45.0 mg/0.50 ml $CDCl_3$ ▶ S 64

▶ 19(pyrimidyl-4,6)-H

	Shift	Peak
a	5.72	1C
b	6.58	3A
▶ c	8.28	2A

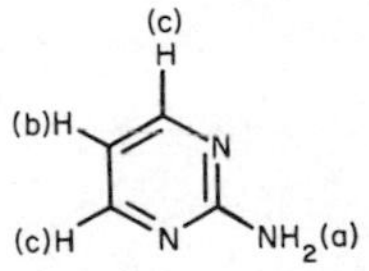

▶ $C_4H_5N_2$ ▶ 7% $CDCl_3$ ▶ V 401

▶ 19(pyrene-4)-H

	Shift	Peak
a	7.95	S
▶ b	7.98	S

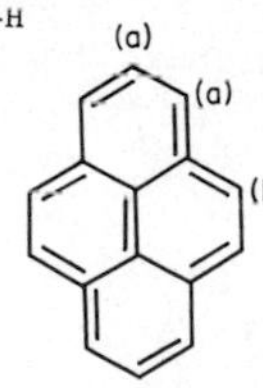

▶ $C_{16}H_{10}$ ▶ 45.0 mg/0.50 ml $CDCl_3$ ▶ S 64

▶ 19(pyrimidyl-5)-AE

	Shift	Peak
a	2.45 or 2.55	1A
b	2.45 or 2.55	1A
▶ c	6.85	1A

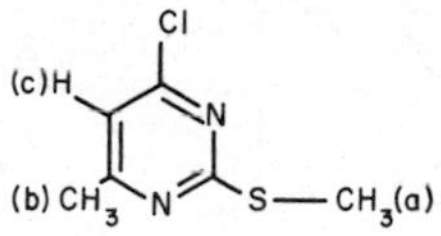

▶ $C_6H_7ClN_2S$ ▶ 7% $CDCl_3$ ▶ V 126

▶ 19(pyridazinyl-3,6)-H

	Shift	Peak
a	7.50	Q
▶ b	9.21	Q

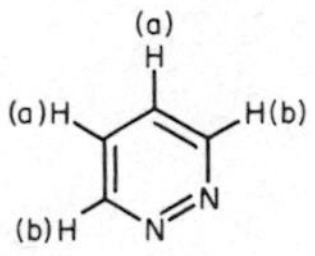

▶ $C_4H_4N_2$ ▶ 7% $CDCl_3$ ▶ V 398

▶ 19(pyrimidyl-5)-BO

	Shift	Peak
a	3.75	1A
▶ b	5.79	1A
c	7.33	1A

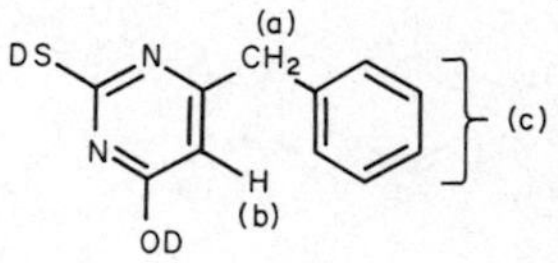

▶ $C_{11}H_{10}N_2OS$ ▶ D_2O ▶ V 577

▶ 19(pyrimidyl-5)-HH

	Shift	Peak
a	5.72	1C
▶ b	6.58	3A
c	8.28	2A

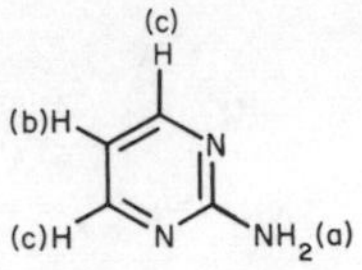

▶ $C_4H_5N_2$ ▶ 7% $CDCl_3$ ▶ V 401

▶ 19(quinoline-4)-H

	Shift	Peak
a	7.52	4A
▶ b	8.25	S
c	8.67 or 8.97	S
d	8.97 or 8.67	S

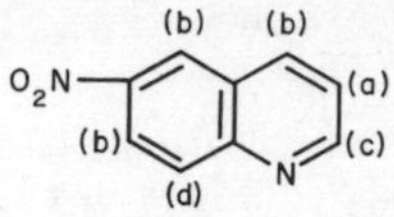

▶ $C_9H_6N_2O_2$ ▶ $CDCl_3$ ▶ S 265

▶ 19(pyrimidyl-5)-HN

	Shift	Peak
▶ a	5.88	2C
b	7.69	2C

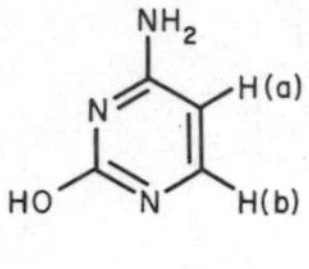

▶ $C_4H_5N_3O$ ▶ D_2O ▶ V 402

▶ 19(quinoline-5)-R

	Shift	Peak
a	7.52	4A
▶ b	8.25	S
c	8.67 or 8.97	S
d	8.97 or 8.67	S

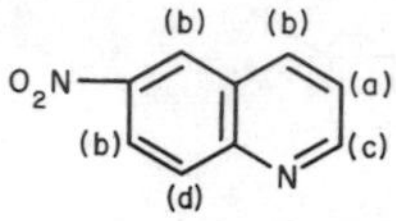

▶ $C_9H_6N_2O_2$ ▶ $CDCl_3$ ▶ S 265

▶ 19(pyrimidyl-6)-A

	Shift	Peak
a	1.83	2D
▶ b	7.42	7B

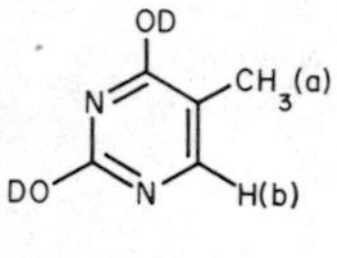

▶ $C_5H_6N_2O_2$ ▶ D_2O ▶ V 432

▶ 19(quinoline-7)-HR

	Shift	Peak
a	7.52	4A
▶ b	8.25	S
c	8.67 or 8.97	S
d	8.97 or 8.67	S

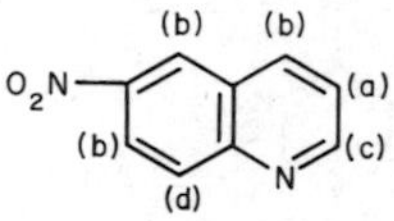

▶ $C_9H_6N_2O_2$ ▶ $CDCl_3$ ▶ S 265

▶ 19(pyrimidyl-6)-H

	Shift	Peak
a	5.88	2C
▶ b	7.69	2C

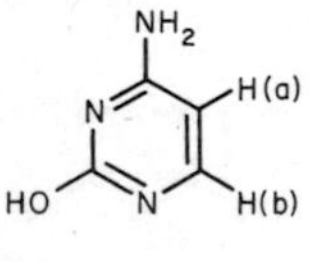

▶ $C_4H_5N_3O$ ▶ D_2O ▶ V 402

▶ 19(quinoline-8)-H

	Shift	Peak
a	7.52	4A
b	8.25	S
c	8.67 or 8.97	S
▶ d	8.97 or 8.67	S

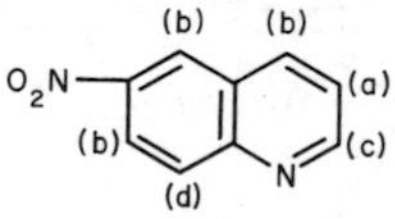

▶ $C_9H_6N_2O_2$ ▶ $CDCl_3$ ▶ S 265

▶ 19(quinoline-2)-H

	Shift	Peak
a	7.52	4A
b	8.25	S
▶ c	8.67 or 8.97	S
d	8.97 or 8.67	S

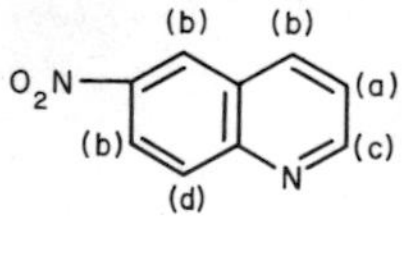

▶ $C_9H_6N_2O_2$ ▶ $CDCl_3$ ▶ S 265

▶ 19(quinolinyl-2)-H

	Shift	Peak
a	3.87	1A
b	7.00	2A
c	7.28	S
d	7.35	S
e	7.97	2E
▶ f	8.71	4A

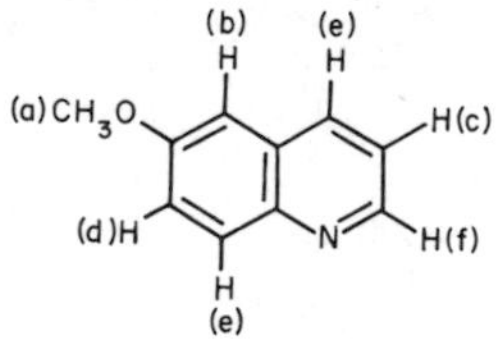

▶ $C_{10}H_9NO$ ▶ 7% $CDCl_3$ ▶ V 249

▶ 19(quinoline-3)-HH

	Shift	Peak
▶ a	7.52	4A
b	8.25	S
c	8.67 or 8.97	S
d	8.97 or 8.67	S

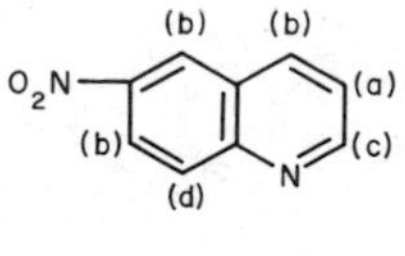

▶ $C_9H_6N_2O_2$ ▶ $CDCl_3$ ▶ S 265

▶ 19(quinolinyl-2)-H

	Shift	Peak
a	7.25	O
b	8.11	2E
▶ c	8.77	2E

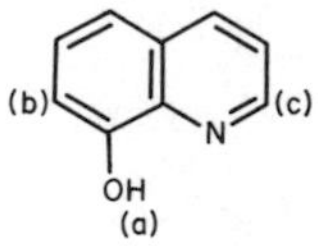

▶ C_9H_7NO ▶ 35.0 mg/0.50 ml $CDCl_3$ ▶ S 24

▶ 19(quinolinyl-3)-AA

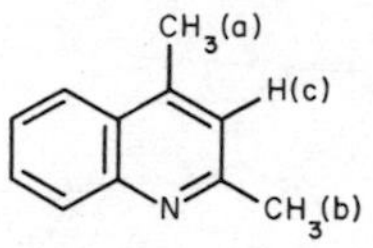

	Shift	Peak
	a 2.66	1A
	b 2.70	1A
▶	c 7.14	1B

▶ $C_{11}H_{11}N$ ▶ 7% $CDCl_3$ ▶ V 578

▶ 19(quinolinyl-4)-H

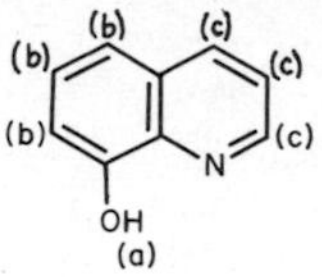

	Shift	Peak
	a 7.25	O
	b 8.11	2E
▶	c 8.77	2E

▶ C_9H_7NO ▶ 35.0 mg/0.50 ml $CDCl_3$ ▶ S 24

▶ 19(quinolinyl-3)-AH

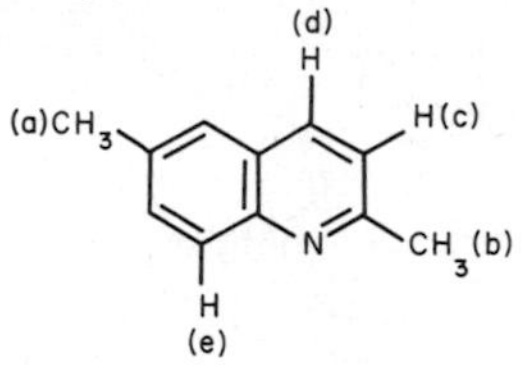

	Shift	Peak
	a 2.50	1A
	b 2.72	1A
▶	c 7.21	2C
	d 7.90	2C
	e 7.90	S

▶ $C_{11}H_{11}N$ ▶ 7% $CDCl_3$ ▶ V 579

▶ 19(quinolinyl-5)-H

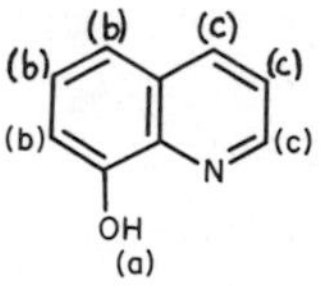

	Shift	Peak
	a 7.25	O
▶	b 8.11	2E
	c 8.77	2E

▶ C_9H_7NO ▶ 35.0 mg/0.50 ml $CDCl_3$ ▶ S 24

▶ 19(quinolinyl-3)-HH

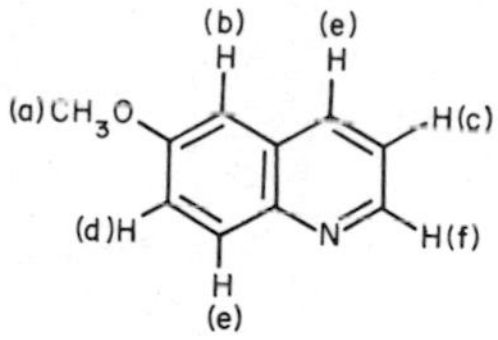

	Shift	Peak
	a 3.87	1A
	b 7.00	2A
▶	c 7.28	S
	d 7.35	S
	e 7.97	2E
	f 8.71	4A

▶ $C_{10}H_9NO$ ▶ 7% $CDCl_3$ ▶ V 249

▶ 19(quinolinyl-5)-O

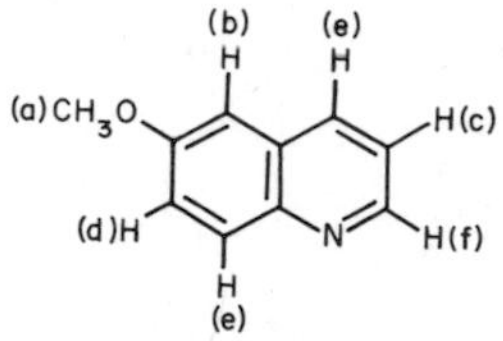

	Shift	Peak
	a 3.87	1A
▶	b 7.00	2A
	c 7.28	S
	d 7.35	S
	e 7.97	2E
	f 8.71	4A

▶ $C_{10}H_9NO$ ▶ 7% $CDCl_3$ ▶ V 249

▶ 19(quinolinyl-3)-HH

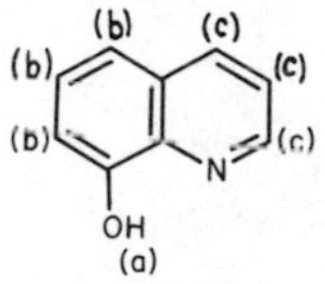

	Shift	Peak
	a 7.25	O
	b 8.11	2E
▶	c 8.77	2E

▶ C_9H_7NO ▶ 35.0 mg/0.50 ml $CDCl_3$ ▶ S 24

▶ 19(quinolinyl-5)-O

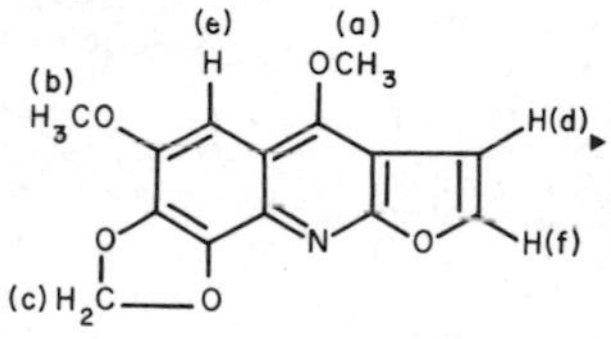

	Shift	Peak
	a 4.28 or 4.40	1A
	b 4.28 or 4.40	1A
	c 6.08	1A
	d 7.05	2C
▶	e 7.28	1H
	f 7.62	2C

▶ $C_{14}H_{11}NO_5$ ▶ 7% $CDCl_3$ ▶ V 304

▶ 19(quinolinyl-4)-H

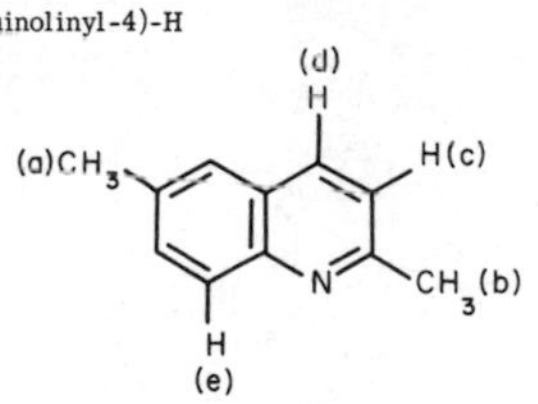

	Shift	Peak
	a 2.50	1A
	b 2.72	1A
	c 7.21	2C
▶	d 7.90	2C
	e 7.90	S

▶ $C_{11}H_{11}N$ ▶ 7% $CDCl_3$ ▶ V 579

▶ 19(quinolinyl-6)-HH

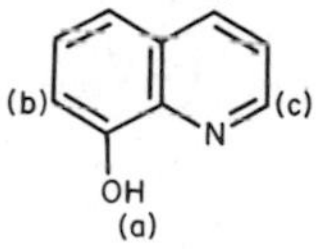

	Shift	Peak
	a 7.25	O
▶	b 8.11	2E
	c 8.77	2E

▶ C_9H_7NO ▶ 35.0 mg/0.50 ml $CDCl_3$ ▶ S 24

▶ 19(quinolinyl-4)-H

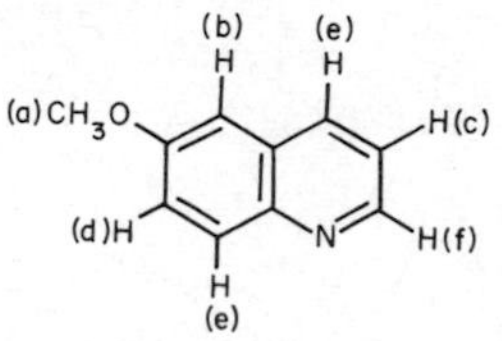

	Shift	Peak
	a 3.87	1A
	b 7.00	2A
	c 7.28	S
	d 7.35	S
▶	e 7.97	2E
	f 8.71	4A

▶ $C_{10}H_9NO$ ▶ 7% $CDCl_3$ ▶ V 249

▶ 19(quinolinyl-7)-HO

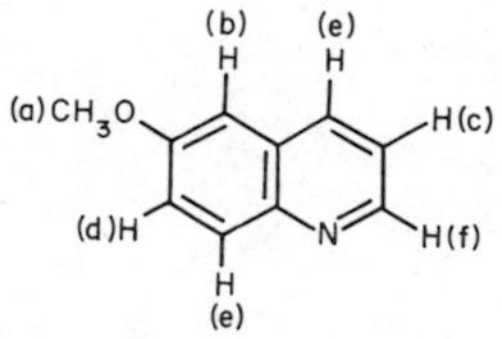

	Shift	Peak
	a 3.87	1A
	b 7.00	2A
	c 7.28	S
▶	d 7.35	S
	e 7.97	2E
	f 8.71	4A

▶ $C_{10}H_9NO$ ▶ 7% $CDCl_3$ ▶ V 249

▶ 19(quinolinyl-7)-HP

	Shift	Peak
a	7.25	O
▶ b	8.11	2E
c	8.77	2E

▶ C_9H_7NO ▶ 35.0 mg/0.50 ml $CDCl_3$ ▶ S 24

▶ 19(thiazolyl-2)

	Shift	Peak
a	7.41	N
b	7.97	2A
▶ c	8.84	2D

▶ C_3H_3NS ▶ 7% $CDCl_3$ ▶ V 378

▶ 19(quinolinyl-8)-H

	Shift	Peak
a	2.50	1A
b	2.72	1A
c	7.21	2C
d	7.90	2C
▶ e	7.90	S

▶ $C_{11}H_{11}N$ ▶ 7% $CDCl_3$ ▶ V 579

▶ 19(thiazolyl-4)-H

	Shift	Peak
a	5.61	O
▶ b	6.48	2C
c	7.08	2C

▶ $C_3H_4N_2S$ ▶ 7% $CDCl_3$ ▶ V 380

▶ 19(quinoxalyl-2,3)-H

	Shift	Peak
a	7.81	O
b	8.10	O
▶ c	8.84	O

▶ $C_8H_6N_2$ ▶ 7% $CDCl_3$ ▶ V 494

▶ 19(thiazolyl-4)-H

	Shift	Peak
a	7.41	N
▶ b	7.97	2A
c	8.84	2D

▶ C_3H_3NS ▶ 7% $CDCl_3$ ▶ V 378

▶ 19(quinoxalyl-5,8)-H

	Shift	Peak
a	7.81	O
▶ b	8.10	O
c	8.84	O

▶ $C_8H_6N_2$ ▶ 7% $CDCl_3$ ▶ V 494

▶ 19(thiazolyl-5)-H

	Shift	Peak
a	5.61	O
b	6.48	2C
▶ c	7.08	2C

▶ $C_3H_4N_2S$ ▶ 7% $CDCl_3$ ▶ V 380

▶ 19(quinoxalyl-6,7)-HH

	Shift	Peak
▶ a	7.81	O
b	8.10	O
c	8.84	O

▶ $C_8H_6N_2$ ▶ 7% $CDCl_3$ ▶ V 494

▶ 19(thiazolyl-5)-H

	Shift	Peak
▶ a	7.41	N
b	7.97	2A
c	8.84	2D

▶ C_3H_3NS ▶ 7% $CDCl_3$ ▶ V 378

▶ 19(quinoxalyl-6,7)-HP

	Shift	Peak
a	2.72	1A
b	~7.00	1E
▶ c	7.03	1E

▶ $C_{10}H_{10}N_2O_2$ ▶ 7% $CDCl_3$ ▶ V 553

▶ 19(tropolonyl-3)-CP

	Shift	Peak
a	1.27	2A
b	1.34	1E
c	1.69	O
d	2.78	O
e	3.28	S
f	5.27	1B
g	5.27	1B
h	7.19*	2F
i	7.19*	2F
▶ j	7.39	1A

*or 7.31

▶ $C_{15}H_{22}O_3$ ▶ 7% $CDCl_3$ ▶ V 647

▶ 19(tropolonyl-6)-BH

	Shift	Peak
a	1.27	2A
b	1.34	1E
c	1.69	O
d	2.78	O
e	3.28	S
f	5.27	1B
g	5.27	1B
▶ h	7.19*	2F
i	7.19*	2F
j	7.39	1A

*or 7.31

▶ $C_{15}H_{22}O_3$ ▶ 7% $CDCl_3$ ▶ V 647

▶ 20-AKv

	Shift	Peak
a	1.42	1A
b	1.95	3A
c	2.55	1A
d	2.70	3D
e	2.82	1E
f	2.92	1E
g	4.12	1A
▶ h	5.33	1D
i	6.37	2A
j	6.78	1E
k	6.87	2H

▶ $C_{15}H_{21}N_3O_2$ ▶ 7% $CDCl_3$ ▶ V 319

▶ 19(tropolonyl-7)-HO

	Shift	Peak
a	1.27	2A
b	1.34	1E
c	1.69	O
d	2.78	O
e	3.28	S
f	5.27	1B
g	5.27	1B
h	7.19*	2F
▶ i	7.19*	2F
j	7.39	1A

*or 7.31

▶ $C_{15}H_{22}O_3$ ▶ 7% $CDCl_3$ ▶ V 647

▶ 20-AVhh

	Shift	Peak
a	2.95	1A
▶ b	4.63	1D
c	6.55	Q
d	8.10	Q

▶ $C_7H_8N_2O_2$ ▶ 7% $CDCl_3$ ▶ V 489

▶ 19(uracil-2)-H

	Shift	Peak
a	5.71	Q
▶ b	7.60	Q

▶ $C_4H_4N_2O_2$ ▶ D_2O ▶ V 399

▶ 20-AVhj

	Shift	Peak
a	3.04	2A
b	7.13	1G
c	7.67	Q
d	8.31	Q
▶ e	10.46	1C

▶ $C_{16}H_{14}N_2O_2$ ▶ 7% $CDCl_3$ ▶ V 652

▶ 19(uracil-3)-HP

	Shift	Peak
▶ a	5.71	Q
b	7.60	Q

▶ $C_4H_4N_2O_2$ ▶ D_2O ▶ V 399

▶ 20-BaBa

	Shift	Peak
a	1.45	3A
b	3.03	4A
▶ c	9.1	1D

▶ $C_4H_{11}N \cdot HCl$ ▶ 44.8 mg/0.50 ml $CDCl_3$ ▶ S 257

▶ 20-AJn

	Shift	Peak
a	2.72	2A
▶ b	5.69	1C

▶ $C_3H_8N_2O$ ▶ 45.2 mg/0.50 ml $CDCl_3$ ▶ S 189

▶ 20-BaBb

	Shift	Peak
a	1.12	3A
b	2.68	4E
c	2.73	3D
▶ d	3.53	1E
e	3.53	1E
f	3.65	3D

▶ $C_4H_{11}NO$ ▶ 7% $CDCl_3$ ▶ V 92

▶ 20-AKb

	Shift	Peak
a	1.23	3A
b	2.78	2A
c	4.14	4A
▶ d	5.16	1D

▶ $C_4H_9NO_2$ ▶ 7% $CDCl_3$ ▶ V 85

▶ (20)-(B)b(B)b

	Shift	Peak
▶ a	0.03	1D
b	1.62	1B

▶ C_2H_5N ▶ 7% $CDCl_3$ ▶ V 372

▶ (20)-(B)b(B)b

	Shift	Peak
a	0.91	2A
▶ b	1.49	S
c	1.53 ± .10	S
d	2.60	3C
e	3.09	2D

▶ $C_6H_{13}N$ ▶ 7% $CDCl_3$ ▶ V 479

▶ (20)-(B)b(B)l

	Shift	Peak
▶ a	1.63	1A
b	2.07	O
c	2.95	3A
d	3.33	N
e	5.72 or 5.77	1E
f	5.72 or 5.77	1E

▶ C_5H_9N ▶ 7% $CDCl_3$ ▶ V 115

▶ (20)-(B)b(B)b

	Shift	Peak
▶ a	1.53	1B
b	1.60 ± 0.05	S
c	1.82 ± 0.05	S
d	2.14	O
e	2.67	4A
f	3.17	6A
g	5.87	1D

▶ $C_6H_{12}N_2O$ ▶ 7% $CDCl_3$ ▶ V 473

▶ 20-BcBc

	Shift	Peak
a	0.91	2A
b	~1.6	S
▶ c	1.77	1B
d	2.42	2A

▶ $C_8H_{19}N$ ▶ 44.7 mg/0.50 ml $CDCl_3$ ▶ S 136

▶ (20)-(B)b(B)b

	Shift	Peak
▶ a	1.78	1A
b	3.08	1A
c	6.88	2H
d	7.05	S
e	7.18	2H

▶ $C_{10}H_{14}N_2$ ▶ 45.2 mg/0.50 ml $CDCl_3$ ▶ S 300

▶ (20)-(B)b(C)ab

	Shift	Peak
a	1.05	2A
▶ b	1.34	S
c	2.57 ± .10	S

▶ $C_6H_{13}N$ ▶ 7% $CDCl_3$ ▶ V 477

▶ (20)-(B)b(B)b

	Shift	Peak
▶ a	1.92	1A
b	2.87	Q
c	3.67	Q

▶ C_4H_9NO ▶ 7% $CDCl_3$ ▶ V 83

▶ 20-BbH

	Shift	Peak
a	0.92	3B
▶ b	1.10	1A
c	2.70	3C

▶ $C_4H_{11}N$ ▶ 7% $CDCl_3$ ▶ V 89

▶ (20)-(B)b(B)b

	Shift	Peak
▶ a	2.12	1A
b	2.27	1A
c	2.37	Q
d	2.88	Q

▶ $C_5H_{12}N_2$ ▶ 7% $CDCl_3$ ▶ V 119

▶ 20-BbH

	Shift	Peak
▶ a	1.34	1B
b	2.76	Q
c	2.97	Q
d	7.24	1A

▶ $C_8H_{11}N$ ▶ 45.0 mg/0.50 ml $CDCl_3$ ▶ S 35

▶ (20)-(B)b(B)c

	Shift	Peak
a	0.82	2A
▶ b	1.47	S
c	1.55 ± 0.10	S
d	2.21	S
e	2.52	S
f	3.00	S
g	3.00	S

▶ $C_6H_{13}N$ ▶ 7% $CDCl_3$ ▶ V 478

▶ 20-BbH

	Shift	Peak
a	0.90	3B
b	1.28	1B
▶ c	1.58	S
d	2.70	1C

▶ $C_6H_{15}N$ ▶ 35.6 mg/0.50 ml $CDCl_3$ ▶ S 46

▶ 20-BbH

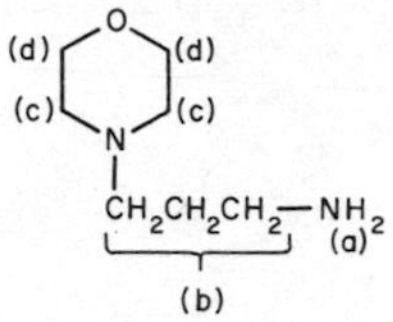

	Shift	Peak
▶ a	1.79	1E
b	2.46	4B
c	2.81	3A
d	3.74	3B

▶ $C_7H_{16}N_2O$ ▶ 33.2 mg/0.50 ml $CDCl_3$ ▶ S 69

▶ 20-BaJa

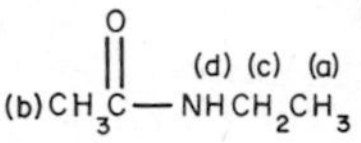

	Shift	Peak
a	1.14	3A
b	2.02	1A
c	3.26	5B
▶ d	8.08	1B

▶ C_4H_9NO ▶ 44.4 mg/0.50 ml $CDCl_3$ ▶ S 146

▶ 20-BbH

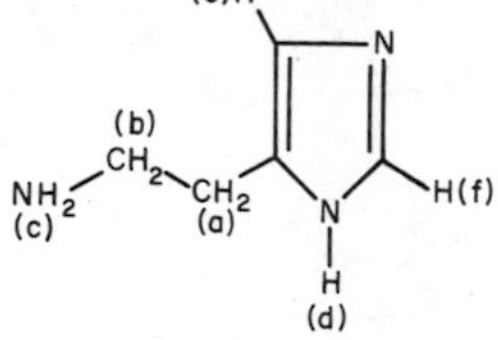

	Shift	Peak
a	2.75	S
b	3.02	S
▶ c	4.96	1E
d	4.96	1E
e	6.82	1A
f	7.55	1A

▶ $C_5H_9N_3$ ▶ 7% $CDCl_3$ ▶ V 443

▶ 20-BbJa

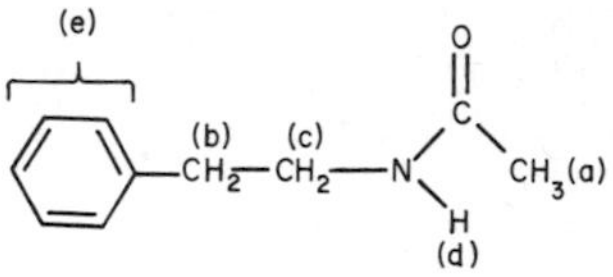

	Shift	Peak
a	1.90	1A
b	2.80	3B
c	3.48	4A
▶ d	6.50	1D
e	7.25	1G

▶ $C_{10}H_{13}NO$ ▶ 7% $CDCl_3$ ▶ V 265

▶ 20-BcH

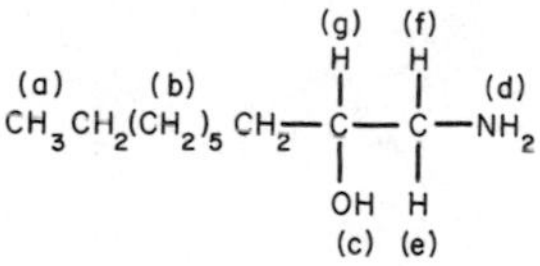

	Shift	Peak
a	0.89	3B
b	1.28	1G
c	1.75	1E
▶ d	1.75	1E
e	2.51	S
f	2.84	S
g	3.50	1D

▶ $C_{10}H_{23}NO$ ▶ 7% $CDCl_3$ ▶ V 575

▶ (20)-(B)b(J)b

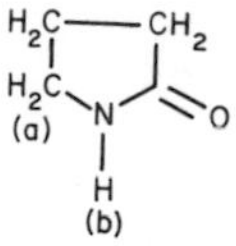

	Shift	Peak
a	3.40	3B
▶ b	7.67	1C

▶ C_4H_7NO ▶ 7% $CDCl_3$ ▶ V 68

▶ 20-BlH

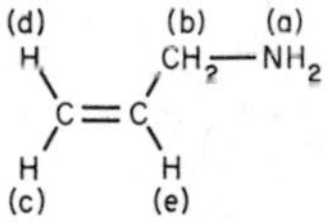

	Shift	Peak
▶ a	1.53	1B
b	3.30	2E
c	5.03	R
d	5.13	R
e	5.92	R

▶ C_3H_7N ▶ 7% $CDCl_3$ ▶ V 38

▶ 20-BbJn

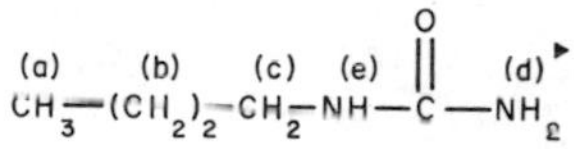

	Shift	Peak
a	0.88	3D
b	~1.2-1.5	N
c	~3.08	4C
d	5.18	1B
▶ e	5.95	3C

▶ $C_5H_{12}N_2O$ ▶ 45.3 mg/0.25 ml $CDCl_3$ ▶ S 282

▶ 20-BlH

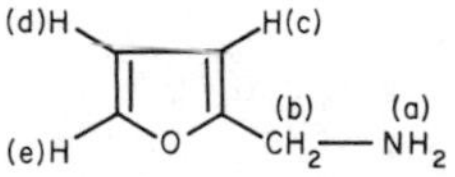

	Shift	Peak
▶ a	1.53	1A
b	3.80	1A
c	6.13	N
d	6.30	N
e	7.33	N

▶ C_5H_7NO ▶ 7% $CDCl_3$ ▶ V 104

▶ 20-BbJn

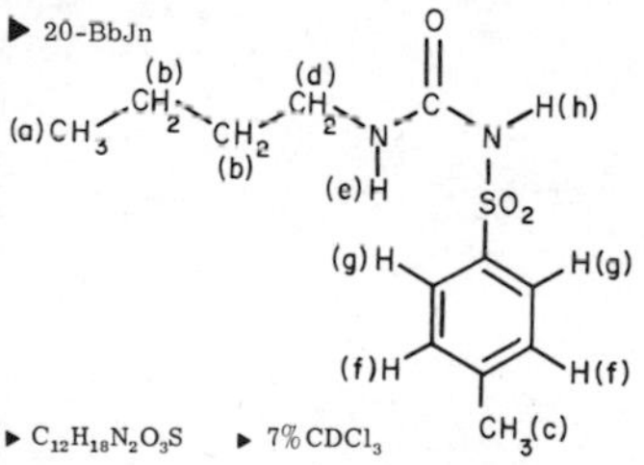

	Shift	Peak
a	0.90	3B
b	~1.40	1F
c	2.46	1A
d	3.23	4C
▶ e	6.53	3B
f	7.32	Q
g	7.79	Q
h	~8.60	1D

▶ $C_{12}H_{18}N_2O_3S$ ▶ 7% $CDCl_3$ ▶ V 602

▶ 20-BvH

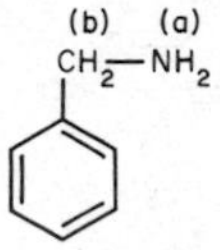

	Shift	Peak
▶ a	1.62	1B
b	3.87	1A

▶ C_7H_9N ▶ 44.9 mg/0.50 ml $CDCl_3$ ▶ S 63

▶ (20)-(B)l(J)l

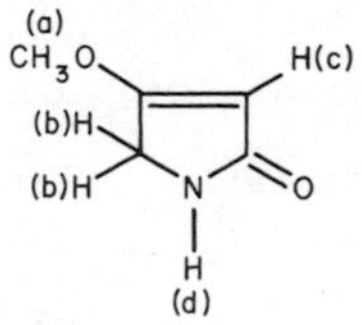

	Shift	Peak
a	3.80	1A
b	3.92	1B
c	5.05	1I or 2D
▶ d	~6.30	1D

▶ $C_5H_7NO_2$ ▶ 7% $CDCl_3$ ▶ V 105

▶ 20-BaLns

	Shift	Peak
a	1.22	3A
b	3.49	5B
▶ c	~6.21	1C

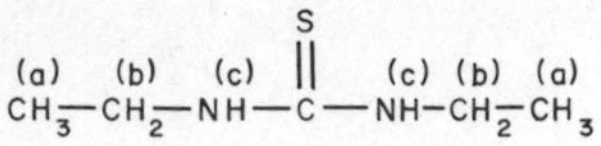

▶ $C_5H_{12}N_2S$ ▶ 45.2 mg/0.50 ml $CDCl_3$ ▶ S 190

▶ 20-CabH

	Shift	Peak
a	.87	3D
b	1.05	2H
c	1.25	1B
▶ d	3.2	1C

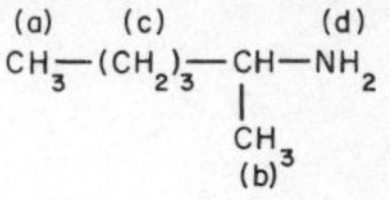

▶ $C_6H_{15}N$ ▶ 34.4 mg/0.50 ml $CDCl_3$ ▶ S 28

▶ (20)-(B)bVbh

	Shift	Peak
a	1.93	5B
b	2.73	3A
c	3.22	3C
▶ d	3.49	1B
e	3.77	1A

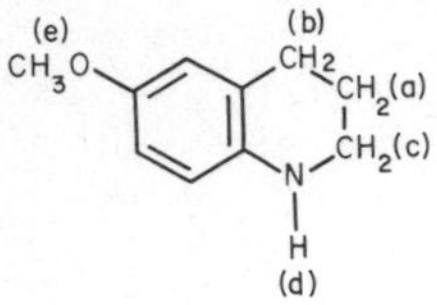

▶ $C_{10}H_{13}NO$ ▶ 7% $CDCl_3$ ▶ V 266

▶ 20-CavH

	Shift	Peak
a	1.38	2A
▶ b	1.58	1A
c	4.10	4A
d	7.30 ± .03	1A

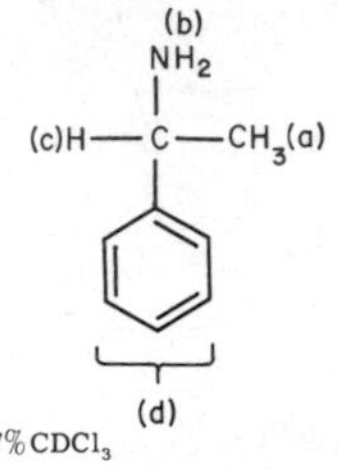

▶ $C_8H_{11}N$ ▶ 7% $CDCl_3$ ▶ V 207

▶ 20-BbVhh

	Shift	Peak
▶ a	3.12	1E
b	3.23	3A
c	3.78	3A
d	6.61	S
e	6.78	S
f	7.17	3B

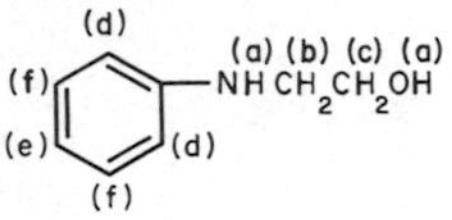

▶ $C_8H_{11}NO$ ▶ 45.4 mg/0.50 ml $CDCl_3$ ▶ S 295

▶ 20-(C)bbH

	Shift	Peak
a	~0.35	O
▶ b	1.83	1A
c	2.30	O

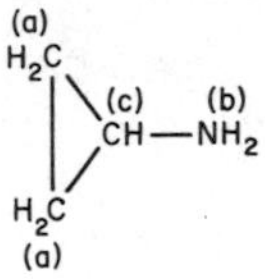

▶ C_3H_7N ▶ 7% $CDCl_3$ ▶ V 37

▶ 20-BbVhh

	Shift	Peak
a	0.96	3B
b	1.50	1F
c	3.08	3A
▶ d	3.42	1A
e	6.52	3C
f	7.12	3C

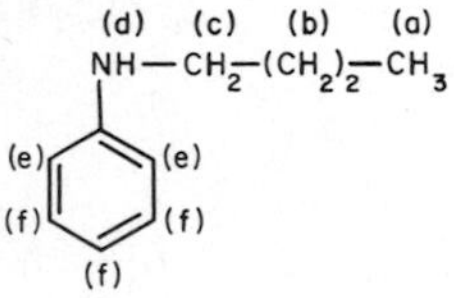

▶ $C_{10}H_{15}N$ ▶ 45.4 mg/0.50 ml $CDCl_3$ ▶ S 280

▶ 20-CaaJl

	Shift	Peak
a	1.18	2A
b	4.17	6B
c	5.59	4A
▶ d	~6.00	S
e	6.14	2A
f	6.24	1A

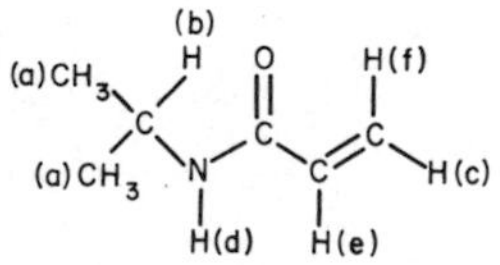

▶ $C_6H_{11}NO$ ▶ 7% $CDCl_3$ ▶ V 468

▶ 20-BvVhh

	Shift	Peak
▶ a	~3.82	1C
b	4.29	1A

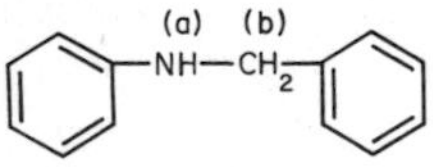

▶ $C_{13}H_{13}N$ ▶ 44.4 mg/0.50 ml $CDCl_3$ ▶ S 148

▶ 20-CbkJv

	Shift	Peak
a	2.10	1A
b	2.25	O
c	2.66	O
d	4.94	4B
▶ e	7.14	3D
f	10.28	1B

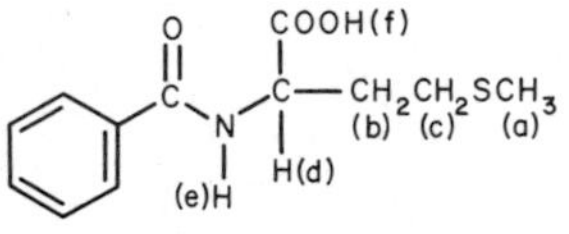

▶ $C_{12}H_{15}NO_3S$ ▶ 7% $CDCl_3$ ▶ V 597

▶ 20-CabH

	Shift	Peak
a	0.90	3D
b	1.05	2F
▶ c	1.25	N
d	2.78	6A

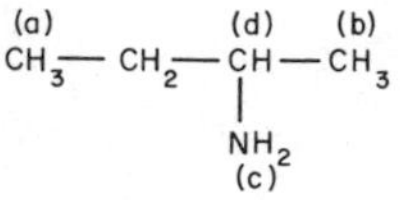

▶ $C_4H_{11}N$ ▶ 7% $CDCl_3$ ▶ V 88

▶ 20-(C)blJa

	Shift	Peak
a	2.07	1G
b	3.63	2C
c	4.12	2C
d	4.84	1F
▶ e	6.02	2E
f	6.49	1B
g	6.68	2A

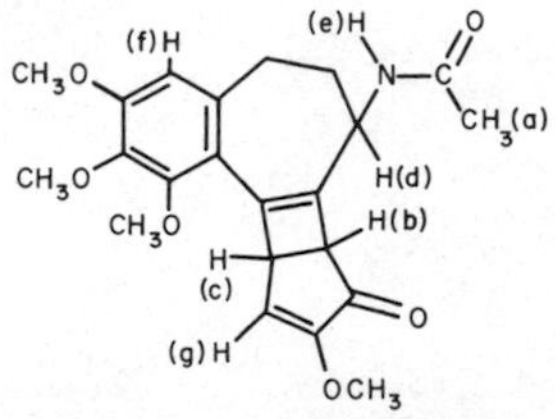

▶ $C_{22}H_{25}NO_6$
▶ 7% $CDCl_3$
▶ V 690

▶ 20-(C)blJa

O, C, CH_3(a), (f) H, (c) H_2C, (b) CH_2, N, H(j), H(e), H(i), CH_3O, CH_3O, CH_3O, H, (g), O, (d), (h)H, OCH_3

	Shift	Peak
a	1.96	1A
b	2.43	S
c	2.43	S
d	3.67	1A
e	4.62	S
f	6.55	1A
g	6.92	2A
h	7.37	S
i	7.63	1B
▶ j	8.38	2G

▶ $C_{22}H_{25}NO_6$ ▶ 7% $CDCl_3$ ▶ V 689

▶ 20-DaapH

(a) CH_3, (b)H_2N—C—OH(c), CH_3 (a)

	Shift	Peak
a	1.12	1A
▶ b	2.17	1B
c	3.30	1A

▶ C_3H_9NO ▶ 7% $CDCl_3$ ▶ V 397

▶ 20-CkuJa

O, (d) H, CN, C, CH_3 (b), N, H(e), C, O, O, CH_2 (c), CH_3(a)

	Shift	Peak
a	1.37	3A
b	2.12	1A
c	4.36	4A
d	5.50	2A
▶ e	6.75	1D

▶ $C_7H_{10}N_2O_3$ ▶ 7% $CDCl_3$ ▶ V 491

▶ 20-HJb

(a) CH_3, (a)CH_3, CH—CH_2—CH_2—C (b) (c), O, N, H(e), H(d)

	Shift	Peak
a	0.90	2E
b	~1.53	3C
c	~2.20	4C
▶ d	6.05 or 6.33	1D
▶ e	6.05 or 6.33	1D

▶ $C_6H_{13}NO$ ▶ 7% $CDCl_3$ ▶ V 142

▶ 20-CqqJa

O, (d) H, O—C—CH_2—CH_3 (c) (a), (b)CH_3, C, N, (e) H, O—C—CH_2—CH_3 (c) (a), O

	Shift	Peak
a	1.30	3A
b	2.08	1A
c	4.28	4A
d	5.19	2A
▶ e	6.60	1D

▶ $C_9H_{15}NO_5$ ▶ 7% $CDCl_3$ ▶ V 547

▶ 20-HJb

(a) (b) CH_3—CH_2—C, O, NH_2 (c)

	Shift	Peak
a	1.13	3A
b	2.23	4C
▶ c	~6.42	1D

▶ C_3H_7NO ▶ 7% $CDCl_3$ ▶ V 40

▶ 20-CabVhh

H(e), (b)CH_3—C—CH_2—CH_3 (c) (a), N, H(d)

	Shift	Peak
a	0.96	3D
b	1.17	2A
c	~1.50	O
▶ d	~3.38	S
e	~3.41	S

▶ $C_{10}H_{15}N$ ▶ 7% $CDCl_3$ ▶ V 568

▶ 20 HJb

(e) H, (c)H, H(d), (f)H, N, CH_2 (a), C, O, NH_2 (b)

	Shift	Peak
a	3.72	1A
▶ b	6.46	1D
c	~7.17	N
d	~7.25	M
e	7.65	6A
f	8.52	2E

▶ $C_7H_8N_2O$ ▶ 7% $CDCl_3$ ▶ V 159

▶ 20-DaaaH

(a) CH_3, (a) CH_3—C—NH_2 (b), CH_3 (a)

	Shift	Peak
a	1.15	1A
▶ b	1.23	1A

▶ $C_4H_{11}N$ ▶ 7% $CDCl_3$ ▶ V 90

▶ 20-HJc

(b) H, (f) H, O, (c) H, H(b), N—H(a), C, (g)H_2N, H(f), H(c) H, H (d), (e) H (e)

	Shift	Peak
a	1.53	1B
b	1.60 ± 0.05	S
c	1.82 ± 0.05	S
d	2.14	O
e	2.67	4A
f	3.17	6A
▶ g	5.87	1D

▶ $C_6H_{12}N_2O$ ▶ 7% $CDCl_3$ ▶ V 473

▶ 20-DaabH

(c) NH_2, (a) CH_3, (b) CH_3—C—CH_2 (d)—C—CH_3 (a), CH_3 (b), CH_3 (a)

	Shift	Peak
a	1.02	1A
b	1.18	1A
▶ c	1.25	1A
d	1.45	1A

▶ $C_8H_{19}N$ ▶ 7% $CDCl_3$ ▶ V 220

▶ 20-HJl

(d) CH_2, O, C, NH_2(f), (e)H, (g)H, N, H(h), CH_2—CH_2—CH_3 (c) (b) (a)

	Shift	Peak
a	0.90	3C
b	1.58	6B
c	3.09	2A
d	3.19	N
e	4.72	5B
▶ f	5.49	1C
g	5.73	4A
h	7.05	2D

▶ $C_9H_{14}N_2O$ ▶ 7% $CDCl_3$ ▶ V 542

▶ 20-HJl

	Shift	Peak
a	1.95	1H
b	5.38	M
c	5.77	M
▶ d	~6.38	1D

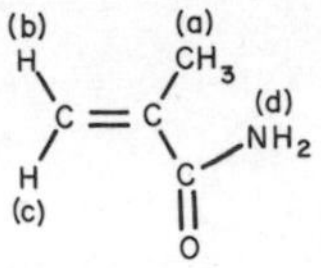

▶ C_4H_7NO ▶ 7% $CDCl_3$ ▶ V 71

▶ 20-H(L)ms

	Shift	Peak
▶ a	5.61	O
b	6.48	2C
c	7.08	2C

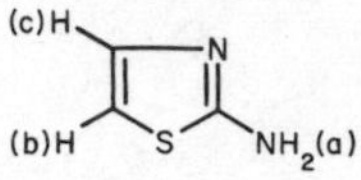

▶ $C_3H_4N_2S$ ▶ 7% $CDCl_3$ ▶ V 380

▶ 20-HJn

	Shift	Peak
a	0.88	3D
b	~1.2-1.5	N
c	~3.08	4C
▶ d	5.18	1B
e	5.95	3C

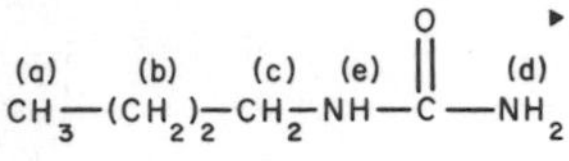

(a) $CH_3-(CH_2)_2-CH_2-NH-C(=O)-NH_2$ (b) (c) (e) (d)

▶ $C_5H_{12}N_2O$ ▶ 45.3 mg/0.25 ml $CDCl_3$ ▶ S 282

▶ 20-H(L)ms

	Shift	Peak
a	6.02	1C

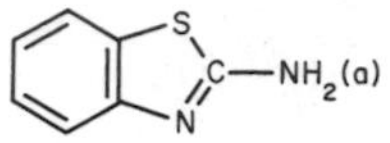

▶ $C_7H_6N_2S$ ▶ 7% $CDCl_3$ ▶ V 150

▶ 20-HJx

	Shift	Peak
▶ a	~6.26	1D
b	7.38	4C
c	8.16	O
d	8.75	4A
e	9.04	2E

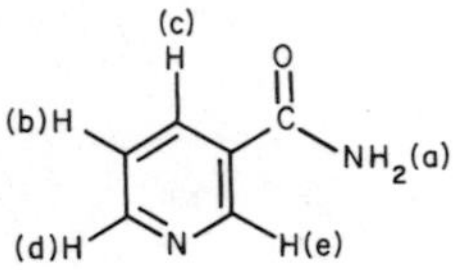

▶ $C_6H_6N_2O$ ▶ 7% $CDCl_3$ ▶ V 453

▶ 20-HVah

	Shift	Peak
a	2.12	1A
b	2.22	1A
▶ c	3.34	1B
d	6.50	2A
e	6.76	2B, 2G

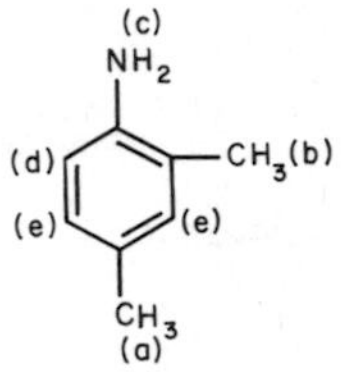

▶ $C_8H_{11}N$ ▶ 44.9 mg/0.50 ml $CDCl_3$ ▶ S 237

▶ 20-HKb

	Shift	Peak
a	1.23	3A
b	4.10	4A
▶ c	~5.17	1D

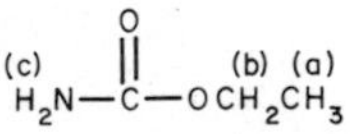

(c) $H_2N-C(=O)-OCH_2CH_3$ (b) (a)

▶ $C_3H_7NO_2$ ▶ 44.4 mg/0.50 ml $CDCl_3$ ▶ S 245

▶ 20-HVah

	Shift	Peak
a	2.13	1A
▶ b	3.43	1B

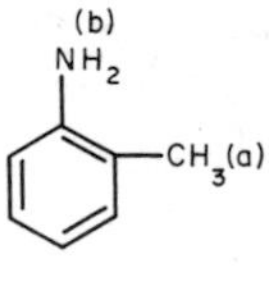

▶ C_7H_9N ▶ 44.3 mg/0.50 ml $CDCl_3$ ▶ S 107

▶ 20-HLal

	Shift	Peak
a	1.92	1A
b	3.62	1A
c	4.53	1B
▶ d	~6.15	1D

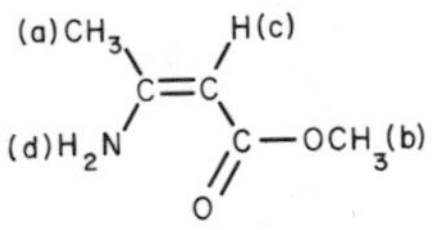

▶ $C_5H_9NO_2$ ▶ 7% $CDCl_3$ ▶ V 442

▶ 20-HVah

	Shift	Peak
a	2.13	1A
▶ b	3.46	1E

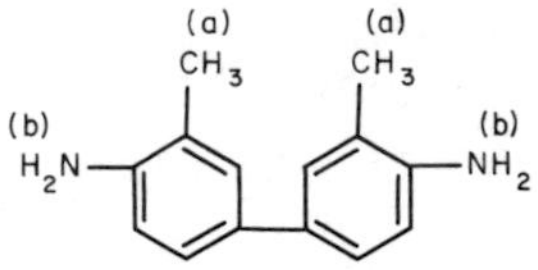

▶ $C_{14}H_{16}N_2$ ▶ 44.7 mg/0.50 ml $CDCl_3$ ▶ S 269

▶ 20-H(L)ls

	Shift	Peak
▶ a	3.89	1C
b	6.19	1A

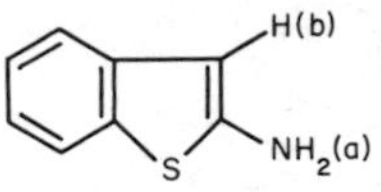

▶ C_8H_7NS ▶ 7% $CDCl_3$ ▶ V 502

▶ 20-HVbh

	Shift	Peak
a	1.23	3A
b	2.46	4A
▶ c	3.47	1A

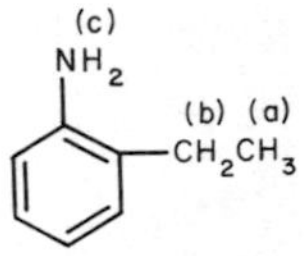

▶ $C_8H_{11}N$ ▶ 45.3 mg/0.10 ml $CDCl_3$ ▶ S 111

▶ 20-HVeh

(a) NH_2 (b) Cl (c) (c) (c)

	Shift	Peak
▶ a	3.9	1C
b	6.68	2H
c	7.10	O

▶ C_6H_6ClN ▶ 44.8 mg/0.50 ml $CDCl_3$ ▶ S 59

▶ 20-HVhh

(b) H (a)H_2N NH_2(a) (c)H H(c) H (d)

	Shift	Peak
▶ a	3.51	1A
b	6.03	1A
c	6.11	2D
d	6.93	3C

▶ $C_6H_8N_2$ ▶ 7% $CDCl_3$ ▶ V 458

▶ 20-HVeh

(a) NH_2 Cl (c) (b) Cl (b)

	Shift	Peak
▶ a	3.96	1C
b	6.70	1G
c	7.05	2G

▶ $C_6H_5Cl_2N$ ▶ 45.2 mg/0.50 ml $CDCl_3$ ▶ S 58

▶ 20-HVhh

(a) NH_2

	Shift	Peak
a	3.56	1A

▶ C_6H_7N ▶ 44.8 mg/0.50 ml $CDCl_3$ ▶ S 191

▶ 20-HVhh

(d)H H(e) (a) (c) CH_3—CH_2—O— —NH_2(b) H (d) H (e)

	Shift	Peak
a	1.35	3A
▶ b	3.30	1B
c	3.93	4A
d	6.63 or 6.70	Q
e	6.63 or 6.70	Q

▶ $C_8H_{11}NO$ ▶ 7% $CDCl_3$ ▶ V 208

▶ 20-HVhh

(a) NH_2 Cl

	Shift	Peak
a	3.58	1B

▶ C_6H_6ClN ▶ 44.9 mg/0.50 ml $CDCl_3$ ▶ S 88

▶ 20-HVhh

(a) NH_2 (c) (c) (d) (d) O—CH_3 (b)

	Shift	Peak
▶ a	3.38	1A
b	3.73	1A
c	~6.68 or 6.72	Q
d	~6.68 or 6.72	Q

▶ C_7H_9NO ▶ 45.2 mg/0.50 ml $CDCl_3$ ▶ S 94

▶ 20-HVhh

(a) NH_2 (b)H H(b) (c)H H(c) Cl

	Shift	Peak
▶ a	3.60	1A
b	6.57	Q
c	7.05	Q

▶ C_6H_6ClN ▶ 7% $CDCl_3$ ▶ V 123

▶ 20-HVhh

(b) OCH_3 (d) (d) (c) (c) NH_2(a)

	Shift	Peak
▶ a	3.40	1A
b	3.73	1A
c	~6.68 or 6.73	Q
d	~6.68 or 6.73	Q

▶ C_7H_9NO ▶ 7% $CDCl_3$ ▶ V 171

▶ 20-HVhh

NH_2(a) Br

	Shift	Peak
a	3.61	1A

▶ C_6H_6BrN ▶ 45.6 mg/0.25 ml $CDCl_3$ ▶ S 278

▶ 20-HVhh

(b) NH_2 CH_3(a)

	Shift	Peak
a	2.25	1A
▶ b	3.46	1A

▶ C_7H_9N ▶ 44.6 mg/0.50 ml $CDCl_3$ ▶ S 108

▶ 20-HVhh

(a) NH_2 Cl Cl

	Shift	Peak
a	3.67	1C

▶ $C_6H_5Cl_2N$ ▶ 45.2 mg/0.50 ml $CDCl_3$ ▶ S 271

▶ 20-HVhh

O
(a)
C — CH_3
NH_2(b)

Shift	Peak
a 2.53	1A
▶ b 3.80	1B

▶ C_8H_9NO ▶ 45.0 mg/0.50 ml $CDCl_3$ ▶ S 119

▶ 20-HVho

(a)
NH_2
(b)
OCH_3
(c)

Shift	Peak
▶ a 3.70	1B
b 3.82	1A
c 6.73	1A

▶ C_7H_9NO ▶ 45.2 mg/0.50 ml $CDCl_3$ ▶ S 192

▶ 20-HVhh

(a)
NH_2
N

Shift	Peak
a 4.02	1C

▶ $C_5H_6N_2$ ▶ 44.7 mg/0.25 ml $CDCl_3$ ▶ S 293

▶ 20-HVho

(b)
NH_2
(c) (a)
OCH_2CH_3
(d)

Shift	Peak
a 1.40	3A
▶ b 3.72	1B
c 4.03	4A
d 6.72	1A

▶ $C_8H_{11}NO$ ▶ 45.1 mg/0.50 ml $CDCl_3$ ▶ S 236

▶ 20-HVhh

O
(b)
(c) (a)
C—O—$CH_2CH_2CH_2CH_3$
(f) (f)
(e) (e)
NH_2(d)

Shift	Peak
a 0.95	3B
b ~1.60	1F
c 4.25	3D
▶ d ~4.10	1E
e 6.59	Q
f 7.83	Q

▶ $C_{11}H_{15}NO_2$ ▶ 44.6 mg/0.50 ml $CDCl_3$ ▶ S 243

▶ 20-HVho

(b)
OCH_3
NH_2(a)
(c)

Shift	Peak
▶ a 3.73	1A
b 3.85	1A
c ~6.80	1A

▶ C_7H_9NO ▶ 7% $CDCl_3$ ▶ V 172

▶ 20-HVhh

(f) (g)
H H
O
(b) (a)
(d) (e) (c) CH_2—CH_3
H_2N— CO—CH_2—CH_2—N
CH_2—CH_3
(f)H H(g) (b) (a)

Shift	Peak
a 1.05	3A
b 2.62	4E
c 2.82	3D
▶ d 4.13	1B
e 4.33	3D
f 6.63	Q
g 7.83	Q

▶ $C_{13}H_{20}N_2O_2$ ▶ 7% $CDCl_3$ ▶ V 302

▶ 20-HVho

NH_2(c)
(e)H OCH_3(a)
(b)CH_3O H(f)
H
(d)

Shift	Peak
a 3.72 or 3.79	1A
b 3.72 or 3.79	1A
▶ c ~3.75	1E
d 6.25	S
e 6.32	S
f 6.71	2A

▶ $C_8H_{11}NO_2$ ▶ 7% $CDCl_3$ ▶ V 508

▶ 20-HVhh

O
(a)
C — CH_3
(b)H_2N—

Shift	Peak
a 2.47	1A
▶ b 4.43	1C

▶ C_8H_9NO ▶ 44.9 mg/0.25 ml $CDCl_3$ ▶ S 242

▶ 20-HVho

OCH_3(c)
(e)H NH_2(d)
(g)H H(f)
O←S→O
CH_3—CH_2—N—CH_2—CH_3
(a) (b) (b) (a)

Shift	Peak
a 1.11	3A
b 3.20	4A
c 3.88	1E
▶ d 3.88	1E
e 6.80	2E
f 7.13	1A
g 7.17	2A

▶ $C_{11}H_{18}N_2O_3S$ ▶ 7% $CDCl_3$ ▶ V 587

▶ 20-HVhn

(b) (a)
(b) NH_2
(b) NH_2
(b) (a)

Shift	Peak
▶ a 3.29	1A
b 6.67	1A

▶ $C_6H_8N_2$ ▶ 44.6 mg/0.50 ml $CDCl_3$ ▶ S 112

▶ 20-HVjo

(e)H O (c)
NH_2
(d)H OCH_3(a)
(d)H H(b)
H O O
(e) H
(f)

Shift	Peak
a 3.96	1A
b 6.52	1A
▶ c ~ 7.30	1D
d 7.72	Q
e 8.34	Q
f 14.22	1A

▶ $C_{15}H_{11}NO_4$ ▶ 7% $CDCl_3$ ▶ V 635

▶ 20-HX

(d) H, (c)H, H(b), (e)H, N, NH_2(a)

	Shift	Peak
▶ a	4.75	1C
b	6.51	10A
c	6.64	10A
d	7.42	10A
e	8.09	10A

▶ $C_5H_6N_2$ ▶ 7% $CDCl_3$ ▶ V 431

▶ 20-JaVah

(c) NH—C(=O)—CH_3 (b); —CH_3(a)

	Shift	Peak
a	2.06 or 2.17	1A
b	2.06 or 2.17	1A
▶ c	~7.5	1C

▶ $C_9H_{11}NO$ ▶ 45.2 mg/0.50 ml $CDCl_3$ ▶ S 175

▶ 20-HX

(c) H, (b)H, N, (c)H, N, NH_2(a)

	Shift	Peak
▶ a	5.72	1C
b	6.58	3A
c	8.28	2A

▶ $C_4H_5N_2$ ▶ 7% $CDCl_3$ ▶ V 401

▶ 20-JaVhh

(d) H, (e) H, (c)H, N, C(=O), CH_3(a), (b)CH_3, H (d), H (c)

	Shift	Peak
a	2.12	1A
b	2.30	1A
c	7.12	Q
d	7.37	Q
▶ e	7.88	1D

▶ $C_9H_{11}NO$ ▶ 7% $CDCl_3$ ▶ V 239

▶ 20-HX

(f)NH_2, N, N, (h) H, N, (e) H, H(g), CH_3—C(=O)—O (d)H, O, (a), CH_3—C(=O)—O—C—C—CH_2—O—C(=O)—CH_3, (c)H, H(b)

	Shift	Peak
a	4.43	1E
b	4.43	1E
c	5.70	1C
d	5.95	3A
e	6.20	2A
▶ f	6.37	1B
g	7.98 or 8.35	1A
h	7.98 or 8.35	1A

▶ $C_{16}H_{19}N_5O_8$ ▶ 7% $CDCl_3$ ▶ V 326

▶ 20-JaVhh

(d) H, (e) H, (f) H, (a) CH_3—(c) CH_2—O—, N, (b) CH_3, C(=O), H (d), H (e)

	Shift	Peak
a	1.38	3A
b	2.12	1A
c	4.00	4A
d	6.83	Q
e	7.41	Q
▶ f	7.91	1C

▶ $C_{10}H_{13}NO_2$ ▶ 7% $CDCl_3$ ▶ V 267

▶ (20)-(J)b(J)b

(e)H—N, C(=O), (d) CH_2, C, CH_2 (c), CH_3(a), CH_2 (d), CH_3 (b)

	Shift	Peak
a	0.91	3A
b	1.06	1A
c	1.43	3B
d	2.43	1A
▶ e	8.47	1D

▶ $C_8H_{13}NO_2$ ▶ 7% $CDCl_3$ ▶ V 514

▶ 20-JaVhh

(b) NH—C(=O)—CH_3 (a)

	Shift	Peak
a	2.16	1A
▶ b	8.14	1F

▶ $C_{12}H_{11}NO$ ▶ 44.8 mg/0.50 ml $CDCl_3$ ▶ S 131

▶ (20)-JI(L)hm

(h)H—N, O, N, (g)H, N, (e) H, H(f), (a)CH_3—C(=O)—O (d)H, O, (a)CH_3—C(=O)—O—C—C—CH_2OC(=O)CH_3 (b) (a), (c)H, H(b)

	Shift	Peak
a	2.10*	1A
b	4.42	1B
c	5.62	S
d	5.88	3A
e	6.17	2A
f	8.03†	1A
g	8.03†	1A
▶ h	~13.1	S

*or 2.15
†or 8.35

▶ $C_{16}H_{18}N_4O_8$ ▶ 7% $CDCl_3$ ▶ V 324

▶ 20 JbVbh

(a) CH_2CN, (c), N—C(=O)—CH_2Cl (b), H (d)

	Shift	Peak
a	3.65	1A
b	4.17	1A
c	7.31 ± .05	1H
▶ d	8.32	1C

▶ $C_{10}H_9ClN_2O$ ▶ 7% $CDCl_3$ ▶ V 248

▶ 20-JnSoov

(a)CH_3—(b) CH_2—(b) CH_2—(d) CH_2—N(H (e))—C(=O)—N—H(h), SO_2, (g) H, H(g), (f)H, H(f), CH_3(c)

	Shift	Peak
a	0.90	3B
b	~1.40	1F
c	2.46	1A
d	3.23	4C
e	6.53	3B
f	7.32	Q
g	7.79	Q
▶ h	~8.60	1D

▶ $C_{12}H_{18}N_2O_3S$ ▶ 7% $CDCl_3$ ▶ V 602

▶ 20-JbVhh

(d) NH—C(=O)—(c) CH_2—(b) CH_2—(a) CH_3

	Shift	Peak
a	0.95	3A
b	1.71	6B
c	2.24	3C
▶ d	~7.8	1C

▶ $C_{10}H_{13}NO$ ▶ 45.3 mg/0.50 ml $CDCl_3$ ▶ S 261

▶ 20-JbVhh

	Shift	Peak
a	2.17	1A
b	3.52	1A
▶ c	9.34	1C

Probably exists about 5% in enol form.

▶ $C_{10}H_{11}NO_2$ ▶ 7% $CDCl_3$ ▶ V 256

▶ (20)-(L)bl(L)hm

	Shift	Peak
a	2.75	S
b	3.02	S
c	4.96	1E
▶ d	4.96	1E
e	6.82	1A
f	7.55	1A

▶ $C_5H_9N_3$ ▶ 7% $CDCl_3$ ▶ V 443

▶ 20-(L)sx

	Shift	Peak
a	4.30	1A
▶ b	9.70	1I
c	9.80	2A

▶ $C_{12}H_9NS$ ▶ 7% $CDCl_3$ ▶ V 590

▶ (20)-(L)hj(L)hl

	Shift	Peak
a	6.30	N
b	6.98	N
c	7.17	N
d	9.45	2D
▶ e	~11.08	1D

▶ C_5H_5NO ▶ 7% $CDCl_3$ ▶ V 98

▶ (20)-(L)al(L)hl

	Shift	Peak
a	2.07 or 2.20	1A
b	2.07 or 2.20	1A
c	5.73	1B
d	6.37	1B
▶ e	~7.50	1D

▶ C_6H_9N ▶ 7% $CDCl_3$ ▶ V 131

▶ (20)-(L)hl(L)hl

	Shift	Peak
a	0.95	3C
b	1.60	6B
c	2.48	3C
d	6.13	4B
e	6.70	S
f	6.55	S
▶ g	~7.83	1A

▶ $C_7H_{11}N$ ▶ 7% $CDCl_3$ ▶ V 177

▶ (20)-(L)al(L)hl

	Shift	Peak
a	0.90	3B
b	2.15	1A
c	~2.38	S
d	5.98	3A
e	6.55	3A
▶ f	7.67	1D

▶ $C_{10}H_{17}N$ ▶ 7% $CDCl_3$ ▶ V 278

▶ (20)-(L)hl(L)hl

	Shift	Peak
a	6.22	Q
b	6.68	Q
▶ c	~8.0	1D

▶ C_4H_5N ▶ 7% $CDCl_3$ ▶ V 55

▶ (20)-(L)al(L)lv

	Shift	Peak
a	0.83	3A
b	2.17	1A
c	2.37	1A
d	3.87	4A
e	5.13	1A
▶ f	6.15	1C

▶ $C_{23}H_{23}NO_3$ ▶ 7% $CDCl_3$ ▶ V 695

▶ (20)-(L)hl(L)hm

	Shift	Peak
a	7.14	1C
b	7.71	1C
▶ c	13.38	1C

▶ $C_3H_4N_2$ ▶ 7% $CDCl_3$ ▶ V 20

▶ (20)-(L)bl(L)hl

	Shift	Peak
a	1.23	3A
b	2.63	6A
c	5.97	1I
d	6.17	4A
e	6.68	N
▶ f	~7.85	1A

▶ C_6H_9N ▶ 7% $CDCl_3$ ▶ V 130

▶ (20)-(L)al(M)l

	Shift	Peak
a	2.27	1A
b	5.80	1B
▶ c	11.92	1D

▶ $C_5H_8N_2$ ▶ 7% $CDCl_3$ ▶ V 441

▶ (20)-(L)hl(M)l

(c)H H(a) N N H(b) H (d)

	Shift	Peak
a	6.33	N
b	7.62	2A
c	7.62	2A
▶ d	12.52	1A

▶ $C_3H_4N_2$ ▶ 7% $CDCl_3$ ▶ V 379

▶ 20-MlVhr

(a) CH_3 (b) CH_3—C=N—NH— (f) NO_2 (e) (c) NO_2 (d)

	Shift	Peak
a	2.10 or 2.20	1A
b	2.10 or 2.20	1A
c	7.85	2A
d	8.18	4A
e	8.99	2A
▶ f	10.94	1C

▶ $C_9H_{10}N_4O_4$ ▶ 45.0 mg/0.25 ml $CDCl_3$ ▶ S 199

▶ (20)-(L)alVhl

(f) H (b) CH_3 H(c) (e)H N CH_3(a) H H H (d) (g)

	Shift	Peak
a	2.28	1A
b	2.40	1A
c	6.10	1I
d	~6.90 or 7.02	N
e	~6.90 or 7.02	N
f	7.27	S
▶ g	~7.30	S

▶ $C_{10}H_{11}N$ ▶ 7% $CDCl_3$ ▶ V 255

▶ 20-MlVhr

H(e) O_2N NO_2 (d)H N—N=C CH_3(a) CH_3(b) H (c) H (f)

	Shift	Peak
a	2.10 or 2.18	1A
b	2.10 or 2.18	1A
c	7.90	2A
d	8.23	4A
e	9.04	2A
▶ f	10.95	1B

▶ $C_9H_{10}N_4O_4$ ▶ 7% $CDCl_3$ ▶ V 233

▶ (20)-(L)alVhl

H(e) Cl H(b) (c)H N CH_3(a) H H (d) (f)

	Shift	Peak
a	2.33	2D
b	6.10	1D
c	7.02	2D
d	7.02	2D
e	7.44	1I
▶ f	~7.65	1D

▶ C_9H_8ClN ▶ 7% $CDCl_3$ ▶ V 228

▶ 20-MlVhr

O O (c) (a) C CH_2—CH_3 (d) CF_2H CH_2(b) C N (e)H H(f) N NO_2 (h)H NO_2 H(g)

	Shift	Peak
a	1.33	3A
b	3.67	1A
c	4.30	4A
d	6.30	3A*
e	8.03	2C
f	8.40	4A
g	9.08	2A
▶ h	11.77	1B

*Very large splitting due to fluorine

▶ $C_{12}H_{12}F_2N_4O_6$ ▶ 7% $CDCl_3$ ▶ V 293

▶ (20)-(L)hlVhl

CH_3(a) N H(b) H(c)

	Shift	Peak
a	2.28	2D
b	6.78	1I
▶ c	~7.40	S

▶ C_9H_9N ▶ 7% $CDCl_3$ ▶ V 231

▶ (20)-(M)lW

H(b) (a)H H(c) N (d)H N N NO_2 H(e)

	Shift	Peak
a	7.34	4A
b	8.18	2D
c	8.32	2E
d	8.48	2D
▶ e	11.63	1D

▶ $C_7H_5N_3O_2$ ▶ 7% $CDCl_3$ ▶ V 485

▶ (20)-(L)hlVhl

(d)H O (a) CH_3 C C N CH_3(b) O N H H (c) (e)

	Shift	Peak
a	2.99	1A
b	3.07	1A
c	7.69	1B
d	8.27	4C
▶ e	10.76	1C

▶ $C_{12}H_{12}N_2O_2$ ▶ 7% $CDCl_3$ ▶ V 594

▶ (20)-VbhVbh

(a) (a) H_2C—CH_2 N H(b)

	Shift	Peak
a	3.07	1A
▶ b	5.90	1C

▶ $C_{14}H_{13}N$ ▶ 7% $CDCl_3$ ▶ V 631

▶ 20-LmnVhh

NH(a) (a) (a) NH—C—NH

	Shift	Peak
a	5.44	1B

▶ $C_{13}H_{13}N_3$ ▶ 45.1 mg/0.50 ml $CDCl_3$ ▶ S 97

▶ 20-VhhVhh

(a) H N

	Shift	Peak
a	5.40	1C

▶ $C_{12}H_{11}N$ ▶ 45.0 mg/0.50 ml $CDCl_3$ ▶ S 11

▶ 21 (phosphorus)-OOX

	Shift	Peak
a	7.10	S

(a) H, O (−3); O — P — P — O; O, O; $Na_3^{(+)}$

▶ $Na_3HP_2O_5$ ▶ D_2O ▶ V 700

▶ 21(sulfur)-Bl

	Shift	Peak
▶ a	1.90	3A
b	3.73	2A
c	6.17	N
d	6.28	N
e	7.33	M

(d)H, H(c), (e)H, O, CH_2—SH, (b) (a)

▶ C_5H_6OS ▶ 7% $CDCl_3$ ▶ V 101

▶ 21(silicon)-AVhhVhh

	Shift	Peak
a	0.62	2A
▶ b	4.97	4A

(a) CH_3, Si, H(b)

▶ $C_{13}H_{14}Si$ ▶ 7% $CDCl_3$ ▶ V 609

▶ 21(sulfur)-Bv

	Shift	Peak
▶ a	1.62	3A
b	3.70	2A
c	7.28	1A

(c) (b) (a); CH_2—SH

▶ C_7H_8S ▶ 43.6 mg/0.50 ml $CDCl_3$ ▶ S 276

▶ 21(silicon)-VhhVhhVhh

	Shift	Peak
a	5.48	1A

Si—H(a)

▶ $C_{18}H_{16}Si$ ▶ 7% $CDCl_3$ ▶ V 331

▶ 21(sulfur)-Caa

	Shift	Peak
a	1.34	2A
▶ b	1.56	2A
c	3.16	6A

(a)CH_3, CH—S—H(b), (c), (a)CH_3

▶ C_3H_8S ▶ 7% $CDCl_3$ ▶ V 396

▶ 21(sulfur)-Bb

	Shift	Peak
a	0.92	3A
▶ b	1.23	S
c	1.53	S
d	2.52	4B

(a) (c) (d) (b)
CH_3—$(CH_2)_2$—CH_2—SH

▶ $C_4H_{10}S$ ▶ 46.0 mg/0.50 ml $CDCl_3$ ▶ S 274

▶ 21(sulfur)-Cak

	Shift	Peak
a	1.56	2A
▶ b	2.26	2A
c	3.58	5B
d	11.03	1B

(b) SH, O; (a)CH_3—C—C; H (c), OH(d)

▶ $C_3H_6O_2S$ ▶ 7% $CDCl_3$ ▶ V 389

▶ 21(sulfur)-Bb

	Shift	Peak
▶ a	1.35	3A
b	1.88	5B
c	2.68	4C

(a) (c) (b) (c) (a)
$HSCH_2CH_2CH_2SH$

▶ $C_3H_8S_2$ ▶ 7% $CDCl_3$ ▶ V 47

▶ 21(sulfur)-Ja

	Shift	Peak
a	2.40	1A
▶ b	4.73	1A

(a) CH_3; O=C; SH (b)

▶ C_2H_4OS ▶ 7% $CDCl_3$ ▶ V 7

▶ 21(sulfur)-Bb

	Shift	Peak
a	0.90	3B
b	1.29	1E
▶ c	2.09	1A
d	2.53	4B

(a) (b) (d) (c)
CH_3—$(CH_2)_5$—CH_2SH

▶ $C_7H_{16}S$ ▶ 46.9 mg/0.50 ml $CDCl_3$ ▶ S 297

▶ 21(sulfur)-Vhh

	Shift	Peak
a	2.30	1A
▶ b	3.27	1A

(a) CH_3; SH(b)

▶ C_7H_8S ▶ 7% $CDCl_3$ ▶ V 168

▶ 21(sulfur)-Vhh

	Shift	Peak
▶ a	3.45	1A
b	7.23	1A
c	7.23	1A

S—H(a)
(c)H H(c)
(b)H H(b)
Cl

▶ C_6H_5ClS ▶ 7% $CDCl_3$ ▶ V 451

INDEX OF MOLECULAR FORMULAS

formula index

C_3H_5NO

1-Ob
2-OaU

C_3H_5NS

1-Bs
2-ASu

C_3H_6BrCl

2-BeBg
2-BbE
2-BbG

$C_3H_6Br_2$

1,2-Dibromopropane

1-Cbg
2-CagG
3-ABgG

1,3-Dibromopropane

2-BgBg
2-BbG

$C_3H_6ClNO_2$

1-Bc
2-ACer
3-BaER

$C_3H_6Cl_2$

1,2-Dichloropropane

1-Cbe
2-CaeE
3-ABeE

1,3-Dichloropropane

2-BeBe
2-BbE

$C_3H_6Cl_2O$

2-CbpE
3-BeBeP
13-Cbb

$C_3H_6N_4$

1-(L)mm
1-(N)mm

C_3H_6O

Propylene oxide

1-(C)bo
(2)-(C)ao(O)c
(3)-A(B)o(O)b

Trimethylene oxide

(2)-(B)o(B)o
(2)-(B)b(O)b

Allyl alcohol

2-LhlP
5-BpHH
5-HBpH
5-HHBp
13-Bl

Acetone-D_5

1-Ja

$C_3H_6O_2S$

Methylvinylsulfone

1-Sloo
5-HHSaoo
5-HSaooH
5-SaooHH

α-Methylthioglycolic acid

1-Cks
3-AKhSh
12-Cas
21(sulfur)-Cak

$C_3H_6O_3S$

(2)-(B)o(B)s
(2)-(B)b(O)s
(2)-(B)b(S)ooo

C_3H_6S

1-Sl
5-HHSa
5-HSaH
5-SaHH

C_3H_7Br

1-Cag
3-AAG

C_3H_7I

2-Iodopropane

1-Cai
3-AAI

1-Iodopropane

1-Bb
2-ABi
2-BaI

C_3H_7N

Cyclopropylamine

(2)-(B)c(C)bn
(3)-(B)b(B)bNhh
20-(C)bbH

Allylamine

2-LhlNhh
5-BnHH
5-HBnH
5-HHBn
20-BlH

C_3H_7NO

N,N-Dimethylformamide

1-Naj
11-Naa

Propionamide

1-Bj
2-AJn
20-HJb

$C_3H_7NO_2$

2-Nitropropane

1-Car
3-AAR

1-Nitropropane

1-Bb
2-ABr
2-BaR

Alanine

1-Ckn
3-AKhNhh

Carbamic acid, ethylester

1-Bq
2-AQn
20-HKb

$C_3H_8N_2O$

1-Nhj
20-AJn

C_3H_8O

n-Propylalcohol

1-Bb
2-ABp
2-BaP
13-Bb

Isopropylalcohol

1-Cap
3-AAP
13-Caa

1-Propanol

1-Bb
2-ABp
2-BaP
13-Bb

Propylalcohol

1-Bb
2-ABp
2-BaP
13-Bb

$C_3H_8O_2$

2-Methoxyethanol

1-Ob
2-BpOa
2-BoP
13-Bb

$C_3H_8O_2$ (cont.)

Propylene glycol

1-Cbp
13-Bc
13-Cab

$C_3H_8O_3$

2-CbpP
3-BpBpP

C_3H_8S

Methylethersulfide

1-Bs
1-Sb
2-ASa

Isopropylmercaptan

1-Cas
3-AASh
21(sulfur)-Caa

$C_3H_8S_2$

2-BsBs
2-BbSh
21(sulfur)-Bb

C_3H_9NO

1-Danp
13-Daan
20-DaapH

C_4HClO_3

9-(K)jE(K)j

$C_4H_2O_3$

(7)-(K)j(K)j

C_4H_3IS

18-α-H
18-β-HI
18-β-HH

$C_4H_4N_2$

19(pyridazinyl-3,6)-H
19(pyridazinyl-4,5)-HH

$C_4H_4N_2O_2$

19(uracil-2)-H
19(uracil-3)-HP

C_4H_4O

16-α-H
16-β-HH

$C_4H_4O_2$

(2)-Lhl(Q)l
(7)-(B)q(K)b
(7)-(K)b(B)q

C_4H_4S

18-α-H
18-β-HH

C_4H_5ClO

1-Ljl
8-AJe
8-JeA

$C_4H_5ClO_2$

3-Chlorocrotonic acid

1-Lel
9-KhAE
12-Lhl

3-Chloroisocrotonic acid

1-Lel
9-KhEA
12-Lhl

C_4H_5N

Pyrrole

17-α-H
17-β-HH
(20)-(L)hl(L)hl

Allyl cyanide

2-LhlU

Methacrylonitrile

1-Llu
8-AU
8-UA

$C_4H_5NO_2$

1-Qb
2-KaU

$C_4H_5N_2$

19(pyrimidyl-4,6)-H
19(pyrimidyl-5)-HH
20-HX

$C_4H_5N_3O$

19(pyrimidyl-5)-HN
19(pyrimidyl-6)-H

C_4H_6

1-Butyne, 10% CCl_4

1-Bt
2-ATh
10-Ba

1-Butyne, liquid

1-Bt
2-ATh
10-Ba

2-Butyne

1-Ta

1,2-Butadiene

1-Lhl
9-ALhh

C_4H_6BrN

2-BgBu
2-BbG
2-BbU

$C_4H_6Br_2O_2$

1-Ccg
3-ACgkG
3-CagGKh
12-Ccg

$C_4H_6Cl_2$

2-ELhl
6-BeBe

$C_4H_6Cl_2O_2$

1-Bq
2-AQc
3-EEKb

C_4H_6O

Crotonaldehyde

1-Lhl
6-AJh
6-JhA
11-Lhl

Cyclobutanone

(2)-(B)j(B)j
(2)-(B)b(J)b

$C_4H_6O_2$

Crotonic acid

1-Lhl
6-AKh
6-KhA
12-Lhl

Methacrylic acid

1-Lkl
8-AKh
8-KhA
12-Lal

Butyrolactone

(2)-(B)b(Q)b

Methylacrylate

1-Ql
5-HHKa
5-HKaH
5-KaHH

Vinylacetate

1-Kl
5-HHQa
5-HQaH
5-QaHH

3,4-Epoxytetrahydrofuran

(2)-(C)co(O)b
3-(B)o(C)bo(O)c

$C_4H_6O_2S$

(2)-(L)hl(S)boo
(7)-(B)s(B)s

Tetramethylenesulfone

(2)-(B)b(B)s
(2)-(B)b(S)boo

$C_4H_6O_4S$

2-CksKh
3-BkKhSh

$C_4H_7BrO_2$

1-Bc
2-ACgk
3-BaGKh
12-Cbg

C_4H_7Cl

1-Lal
9-EAA

$C_4H_7ClO_2$

Ethylchloroacetate

1-Bq
2-AQb
2-EKb

Chloroacetic acid, ethylester

1-Bq
2-AQb
2-EKb

C_4H_7N

Isopropylcyanide

1-Cau
3-AAU

Butyronitrile

1-Bb
2-ABu
2-BaU

C_4H_7NO

2-Pyrrolidone

(2)-(B)b(N)hj
(20)-(B)b(J)b

β-Methoxypropionitrile

1-Ob
2-BuOa
2-BoU

Acetone cyanohydrin

1-Dapu
13-Daau

Methacrylamide

1-Ljl
8-AJn
8-JnA
20-HJl

$C_4H_7NO_2$

1-Jl
1-Ljm
13-Ml

$C_4H_7NO_4$

2-CknKh
3-BkKhNhh

$C_4H_7N_3O$

1-(N)bl
(2)-(J)n(N)al

C_4H_8

1-Butene, degassed liquid

1-Bl
2-ALhl
5-BaHH
5-HBaH
5-HHBa

C_4H_8 (cont.)

1-Butene, liquid

1-Bl
2-ALhl
5-BaHH
5-HBaH
5-HHBa

$C_4H_8Br_2$

1,2-Dibromobutane

1-Bc
2-CbgG
3-BaBgG

Creatinine

1-Dabg
1-DaagG

$C_4H_8Cl_2O$

1-Bo
2-CeoE
3-BeEOb

$C_4H_8N_2O_3$

2-JnNhh
2-KhNhj

C_4H_8O

Methylethylketone

1-Bj
1-Jb
2-AJa

Tetrahydrofuran

(2)-(B)b(B)o
(2) (B)b(O)b

Butyraldehyde

1-Bb
2-ABj
2 BaJh
11-Bb

Crotyl alcohol

1-Lhl
2-LhlP
6-ABp
6-BpA
13-Bl

$C_4H_8O_2$

Ethylacetate

1-Bq
1-Kb
2-AQa

Isopropylformate

1-Caq
3-AAQh
11-Oc

Isobutyric acid

1-Cak
3-AAKh
12-Caa

$C_4H_8O_3$

1-Bo
2-AOb
2-KhOb
12-Bo

C_4H_8S

(2)-(B)b(B)s
(2)-(B)b(S)b

C_4H_9Br

2-Bromobutane

1-Bc
1-Cbg
2-ACag
3-ABaG

2-Bromo-2-methylpropane

1-Daag

$C_4H_9BrO_2$

1-Oc
2-CooG
3-BgOaOa

C_4H_9ClO

2-BbBp
2-BbE
2-BbP
13-Bb

C_4H_9I

1-Iodobutane

1-Bb
2-BbI

2-Iodobutane

1-Bc
1-Cbi
2-ACai
3-ABal

C_4H_9NO

Morpholine

(2)-(B)o(N)bh
(2)-(B)n(O)b
(20)-(B)b(B)b

n-Butyraldoxime

1-Bb
2-ABl
4-BbP
13-Ml

N,N-Dimethylacetamide

1-Jn
1-Naj

N-Ethylacetamide

1-Bn
1-Jn
2-ANhj
20-BaJa

$C_4H_9NO_2$

2-Nitrobutane

1-Bc
1-Cbr
2-ACar
3-ABaR

Ethyl-N-methylcarbamate

1-Bq
1 Nhk
2-AQn
20-AKb

$C_4H_9NO_3$

2-Methyl-2-nitropropan-1-ol

1-Dabr
2-DaarP
13-Bd

2-Methyl-2-nitro-1-propanol

1-Dabr
2-DaarP
13-Bd

C_4H_{10}

n-Butane

1-Bb
2-ABa

Isobutane

1-Caa
3-AAA

$C_4H_{10}O$

1-Butanol

1-Bb
2-ABb
2-BaBp
2-BbP
13-Bb

t-Butyl alcohol

1-Daap
13-Daaa

$C_4H_{10}O_2$

1,3-Butanediol

1-Cpb
2-BpCap
2-BcP
3-ABbP
13-Bb
13-Cab

$C_4H_{10}O_2$ (cont.)

2,3-Butanediol

1-Ccp
3-ACapP
13-Cac

$C_4H_{10}O_3$

2-BpOb
2-BoP
13-Bb

$C_4H_{10}S$

1-Bb
2-ABb
2-BaBs
2-BbSh
21(sulfur)-Bb

$C_4H_{11}N$

sec-Butylamine

1-Bc
1-Cbn
3-ABaNhh
20-CabH

n-Butylamine

1-Bb
2-BbNhh
20-BbH

t-Butylamine

1-Daan
20-DaaaH

Diethylamine, hydrochloride

1-Bn
2-ANbh
20-BaBa

$C_4H_{11}NO$

β-Dimethylamino ethanol

1-Nab
2-BpNaa
2-BnP
13-Bb

2-Ethylaminoethanol

1-Bn
2-ANbh
2-BnP
2-BpNbh
13-Bb
20-BaBb

$C_4H_{12}O_3Si$

1-Oz
1-Z

$C_4H_{12}Sn$

1-Z

$C_5H_2Cl_6O_3$

(3)-Deee(K)c(O)c
(3)-Deee(O)c(Q)c

$C_5H_4N_4O$

(4)-(N)hwW
19(pyrimidyl-2)

C_5H_4OS

11-Y
18-α-H
18-β-HH
18-β-HJ

$C_5H_4O_2$

11-(L)lo
16-α-H
16-β-HH
16-β-HJ

$C_5H_4O_3$

1-(L)kl
(9)-(K)jA(K)j

C_5H_5N

15-α-H
15-β-HH
15-γ-HH

formula index

formula index

formula index

C_7H_8

Methylbenzene

1-Vhh
14-AH
14-HH

Toluene

1-Vhh
14-AH
14-HH

Cycloheptatriene

(2)-(L)hl(L)hl
(7)-(B)l(L)hl
(7)-(L)hl(B)l
(7)-(L)hl(L)hl

Bicycloheptadiene

(2)-(C)ll(C)ll
(3)-(B)c(L)hl(L)hl
(7)-(C)bl(C)bl

$C_7H_8N_2O$

2-Pyridylacetamide

2-JnX
15-α-H
15-β-BH
15-β-HH
15-γ-HH
20-HJb

Methylphenylnitrosamine

1-Nmv

$C_7H_8N_2O_2$

1-Nhv
14-HN
14-HR
20-AVhh

C_7H_8O

m-Cresol

1-Vhh
13-Vhh

Benzyl alcohol

2-PVhh
13-Bv
14-BH
14-HH

Anisole

1-Ov

C_7H_8OS

C-CH_2CH_3

1-Bj
2-AJy
18-α-H
18-β-HH
18-β-HJ

2-Acetyl-5-methylthiophene

1-Jy
1-Yh
18-β-AH
18-β-HJ

$C_7H_8O_2$

2-Acetyl-5-methylfuran

1-Jl
1-(L)lo
16-β-AH
16-β-HJ

2,5-Dimethyl-γ-pyrone

1-(L)jl
(9)-(O)lA(J)l

m-Methoxyphenol

1-Ov

$C_7H_8O_3$

1-Kb
2-(L)loQa
16-α-H
16-β-BH
16-β-HH

C_7H_8S

p-Methylthiophene

1-Vhh
21(sulfur)-Vhh

Thioanisole

1-Sv

α-Toluenethiol

2-ShVhh
14-BH
14-HH
21(sulfur)-Bv

C_7H_9N

2,6-Lutidine

1-X
15-β-AH
15-γ-HH

Benzylamine

2-NhhVhh
20-BvH

o-Toluidine

1-Vhn
20-HVah

m-Toluidine

1-Vhh
20-HVhh

C_7H_9NO

1,5-Dimethylpyrrole-2-aldehyde

1-(L)ln
1-(N)ll
11-(L)ln
17-β-AH
17-β-HJ

p-Anisidine, 7% $CDCl_3$

1-Ov
14-HN
14-HO
20-HVhh

o-Anisidine, 90.4 mg/ml $CDCl_3$

1-Ov
14-HN
14-HO
20-HVhh

o-Anisidine, 7% $CDCl_3$

1-Ov
14-HH
14-HN
14-HO
20-HVho

o-Anisidine, 90.4 mg/ml $CDCl_3$

1-Ov
14-HH
14-HN
14-HO
20-HVho

3,4-Epoxycyclohexylcyanide

(3)-(B)b(B)cU
(3)-(B)c(C)bo(O)c

1-Methyl-2-acetylpyrrole

1-Jl
1-(N)ll
17-α-H
17-β-HH
17-β-HJ

$C_7H_{10}N_2O_3$

1-Bq
1-Jn
2-AQc
3-KbNhjU
20-CkuJa

$C_7H_{10}O_2$

1-(D)akl
1-(L)lq
(9)-(D)aakA(Q)d

$C_7H_{10}O_4$

1-Kc
3-LhlQaQa

$C_7H_{11}N$

1-Bb
2-ABl
2-BaX
17-α-B
17-α-H
17-β-BH
(20)-(L)hl(L)hl

$C_7H_{11}NO$

γ-Cyano-α,α-dimethylbutyraldehyde

1-Dabj
2-BuDaaj
2-BdU
11-Daab

Dicyclopropylketoxime

(3)-(B)b(B)bLcm
13-Ml

C_7H_{12}

Methylenecyclohexane

(8)-(B)b(B)b

1-Methylcyclohexene

1-(L)bl
(2)-(B)b(B)l
(2)-(B)b(L)al
(2)-(B)b(L)hl

$C_7H_{12}O$

1-(C)bb
(2)-(B)j(C)ab
(2)-(B)c(J)b

$C_7H_{12}O_3$

H_3CH_2C

1-Bd
2-A(D)bbb
(2)-(D)bbb(O)c
(3)-(O)b(O)b(O)b

Levulinic acid, ethylester

1-Bq
1-Jb
2-AQb
2-BkJa
2-BjKb

$C_7H_{12}O_4$

Diethylmalonate

1-Bq
2-AQb
2-KbKb

β-Acetoxymethylbutyrate

1-Cbq
1-Kc
1-Qb
2-CaqKa
3-ABkQa

C_7H_{14}

Methylcyclohexane

1-(C)bb

2,3,3-Trimethyl-1-butene, 10% CCl_4

1-Daal
1-Ldl
8-ADaaa
8-DaaaA

2,3,3-Trimethyl-1-butene, 50% CCl_4

1-Daal
1-Ldl
8-ADaaa
8-DaaaA

1,cis-3-Dimethylcyclopentane

1-(C)bb

Ethylcyclopentane

1-Bc

C_7H_{14} (cont.)

1,1-Dimethylcyclopentane

1-(D)abb
(2)-(B)b(B)d
(2)-(B)b(D)aab

3-Ethyl-1-pentene

1-Bc
2-ACbl
3-BaBaLhl
5-CbbHH

3-Ethyl-2-pentene

1-Bl
1-Lhl
2-ALbl
9-ABaBa

2,3-Dimethyl-2-pentene, 50% CCl_4

1-Bl
1-Lal
1-Lbl
2-ALal

2,3-Dimethyl-2-pentene, liquid

1-Bl
1-Lal
1-Lbl
2-ALal

2,4-Dimethyl-1-pentene

1-Cab
1-Lbl
2-CaaLal
3-AABl
8-ABc
8-BcA

2,4-Dimethyl-2-pentene

1-Cal
1-Lal
3-AALhl
9-CaaAA

3,3-Dimethyl-1-pentene

1-Bd
1-Dabl
2-ADaal
5-DaabHH

3,4-Dimethyl-1-pentene

1-Cac
1-Ccl
3-ACaaLhl
5-CacHH
5-HCacH
5-HHCac

3,4-Dimethyl-cis-2-pentene, liquid

1-Cal
1-Lcl
1-Lhl
3-AALal
9-AACaa

3,4-Dimethyl-cis-2-pentene, 50% CCl_4

1-Cal
1-Lcl
1-Lhl
3-AALal
9-AACaa

3,4-Dimethyl-trans-2-pentene, liquid

1-Cal
3-AALal
9-ACaaA

3,4-Dimethyl-trans-2-pentene, 50% CCl_4

1-Cal
1-Lcl
1-Lhl
3-AALal
9-ACaaA

4,4-Dimethyl-1-pentene

1-Daab
2-DaaaLhl
5-BdHH
5-HBdH
5-HHBd

formula index

formula index

formula index

C_9H_{20} (cont.)

2,2,3-Trimethylhexane

1-Bb
1-Cbd
1-Daac
2-BaCad
3-ABbDaaa

2,2,4-Trimethylhexane

1-Bc
1-Cbb
1-Daab
2-CabDaaa
3-ABaBd

2,2,5-Trimethylhexane

1-Cab
1-Daab
2-BdCaa
3-AABb

2,3,3-Trimethylhexane

1-Bb
1-Cad
1-Dabc
2-ABd
2-BaDaac
3-AADaab

2,3,5-Trimethylhexane

1-Cab
1-Cbc
2-CaaCac
3-ABcCaa

3,3-Diethylpentane

1-Bd
2-ADbbb

2,2-Dimethyl-3-ethylpentane

1-Bc
1-Daac
2-ACbd
3-BaBaDaaa

2,4-Dimethyl-3-ethylpentane

1-Bc
1-Cac
2-ACcc
3-AACbc
3-BaCaaCaa

2,2,3,3-Tetramethylpentane

1-Bd
1-Daad
1-Dabd
2-ADaad

2,2,3,4-Tetramethylpentane

1-Cac
1-Ccd
1-Daac
3-AACad

2,3,3,4-Tetramethylpentane

1-Cad
1-Dacc
3-AADaac

2,4,4-Trimethylhexane

1-Bd
1-Cab
1-Dabb
2-ADaab
2-CaaDaab
3-AABd

Nonane, 50% CCl_4

1-Bd
2-ABb
2-BaBb
2-BbBb

Nonane, 90 mg/ml $CDCl_3$

1-Bb
2-ABb
2-BaBb
2-BbBb

$C_9H_{20}O$

1-Nonanol, 20% CCl_4

1-Bb
2-ABb
2-BaBb
2-BbBb
2-BbBp
2-BbP
13-Bb

$C_9H_{20}O$ (cont.)

1-Nonanol, 89.2 mg/ml $CDCl_3$

1-Bb
2-ABb
2-BaBb
2-BbBb
2-BbBp
2-BpP
13-Bb

$C_9H_{20}Si_2$

1-Z
2-TzZ

$C_{10}H_6Br_2$

19(naphthyl-1,2)-G

$C_{10}H_6N_2O_4$

19(naphthyl-2,3)-HR

$C_{10}H_6O_2$

(7)-(J)v(J)v
14-HH
14-H(J)

$C_{10}H_7Br$

19(naphthalene-2)-GH

$C_{10}H_7NO_2$

19(naphthyl-2)-HR

$C_{10}H_8$

Naphthalene, CCl_4

19(naphthalene-1,4,5,8)-H
19(naphthalene-2,3,6,7)-HH

Naphthalene, 90 mg/ml $CDCl_3$

19(naphthalene-1,4,5,8)-H
19(naphthalene-2,3,6,7)-HH

Azulene

19(azulene-1,3)-H
19(azulene-2)-HH
19(azulene-4,8)-H
19(azulene-5,7)-HH
19(azulene-6)-HH

$C_{10}H_8BrNO_2$

2-BnG
2-Bg(N)jj

$C_{10}H_8O_2$

3-Methylcoumarin

1-(L)jl
(9)-VhoA(K)v

3-Methyl-1,4-benzopyrone

1-(L)jl
(9)-(O)vA(J)v
14-H(J)

$C_{10}H_8O_4$

3-Jl(L)loP
13-Cjl
16-α-H
16-β-CH
16-β-HH
16-β-HJ

$C_{10}H_9ClN_2O$

2-EJn
2-UVhn
14-BH
14-HH
14-HN
20-JbVbh

$C_{10}H_9NO$

1-Ox
19(quinolinyl-2)-H
19(quinolinyl-3)-HH
19(quinolinyl-4)-H
19(quinolinyl-5)-OW
19(quinolinyl-7)-HO

$C_{10}H_9NS_2$

2-LhlSl
5-BsHH
5-HBsH
5-HHBs

$C_{10}H_{10}N_2O_2$

1-X
13-X
19(quinoxalyl-6,7)-HP

$C_{10}H_{10}O$

1-Jl
6-JaVhh
6-VhhJa

$C_{10}H_{10}O_2$

Cinnamic acid, methylester

1-Ql
6-KaVhh
6-VhhKa

Isosafrol

1-Lhl
(2)-(O)v(O)v
6-AVhh
6-VhhA
14-HL
14-H(O)
14-L(O)

Safrol

2-LhlVhh
(2)-(O)v(O)v
5-BvHH
5-HBvH
5-HHBv
14-BH
14-B(O)
14-H(O)

$C_{10}H_{10}O_3$

1-Jv
1-Kv

$C_{10}H_{10}O_4$

Phthalic acid, dimethylester

1-Qv
14-HH
14-HK

Hydroquinone, diacetate

1-Kv
14-HQ

O
‖
$C_6H_5CHOCCH_3$
|
O=COH

1-Kc
3-KhQaVhh
12-Cov

$C_{10}H_{11}N$

1-(L)ln
1-Vhh
(9)-VhnA(N)hv
14-AH
14-A(L)
14-H(N)
(20)-(L)alVhl

$C_{10}H_{11}NO$

1-Ljl
8-AJn
8-JnA

$C_{10}H_{11}NO_2$

1-Jb
2-JaJn
20-JbVhh

$C_{10}H_{11}NO_4$

1-Ov
6-RVho
6-VhoR
14-HH
14-HL
14-HO

$C_{10}H_{12}$

1-Methyl-1-phenylcyclopropane

1-(D)bbv
(2)-(B)d(D)abv
14-(D)H
14-HH

Tetralin

(2)-(B)b(B)v
(2)-(B)bVbh
14-(B)H

$C_{10}H_{12}O$

Anethole

1-Lhl
1-Ov
6-AVhh
6-VhhA
14-HL
14-HO

1,2-Epoxy-2-phenylbutane

1-Bd
2-A(D)bov
(2)-(D)bov(O)d
14-DH
14-HH

$C_{10}H_{12}$ (cont.)

Isopropylphenone

1-Caj
3-AAJv

$C_{10}H_{12}O_2$

Phenylacetic acid, ethylester

1-Bq
2-AQb
2-KbVhh

Eugenol

1-Ov
2-LhlVhh
5-BvHH
5-HBvH
5-HHBv
13-Vho
14-BH
14-BO
14-HP

trans-Isoeugenol

1-Lhl
1-Ov
6-AVhh
6-VhhA
13-Vho

4-Isopropyl salicylaldehyde

1-Cav
3-AAVhh
11-Vhp
13-Vhj
14-CH
14-CP
14-HJ

Phenylethylacetate

1-Kb
2-BvQa
2-BqVhh
14-BH
14-HH

γ-Thujaplicin

1-Cal
3-AA(L)ll
(7)-(J)l(L)cl
(7)-(L)cl(J)l
(9)-(L)hlP(J)l
13-(L)jl

$C_{10}H_{12}O_4$

Diallyl maleate

2-LhlQl
5-BqHH
5-HBqH
5-HHBq
7-KbKb

Vanillic acid, ethylester

1-Bq
1-Ov
2-AQv
13-Vho
14-KO

$C_{10}H_{13}Cl$

1-Dabe
2-DaaeVhh
14-BH
14-HH

$C_{10}H_{13}NO$

Calciferol

1-Jn
1-Njv
1-Vhn

$C_6H_5(CH_2)_2NHCOCH_3$

1-Jn
2-BvNhj
2-BnVhh
14-BH
14-HH
20-BbJa

α-Tocopherol

1-Ov
(2)-(B)n(B)v
(2)-(B)b(N)hv
(2)-(B)bVhn
(20)-(B)bVbh

Butyranilide

1-Bb
2-ABj
2-BaJn
20-JbVhh

formula index

$C_{10}H_{22}$

Decane

1-Bb
2-ABb
2-BaBb
2-BbBb

2,2,5,5-Tetramethylhexane

1-Daab
2-BdDaaa

2,2,4,5-Tetramethylhexane

1-Daab

2,7-Dimethyloctane

1-Cab
2-BbBc
2-BbCaa
3-AABb

$C_{10}H_{22}O$

Pentylether

1-Bb
2-ABb
2-BaBb
2-BbBo
2-BbOb

1-Decanol, 20% CCl_4

1-Bb
2-ABb
2-BaBb
2-BbBb
2-BbBp
2-BbP
13-Bb

n-Decanol, 7% $CDCl_3$

1-Bb
2-BbBb
2-BbP
13-Bb

$C_{10}H_{23}NO$

1-Bb
2-BaBb
2-CbpNhh
3-BbBnP
13-Cbb
20-BcH

$C_{11}H_8N_2$

1-Vhl
9-VahUU
14-HL

$C_{11}H_9NO_2$

1-Ql
9-VhhKaU
14-HL

$C_{11}H_{10}$

1-X

$C_{11}H_{10}N_2OS$

2-VhhX
14-BH
14-HH
19(pyrimidyl-5)-BO

$C_{11}H_{10}O$

1-Ox

$C_{11}H_{11}Cl_3O_6$

1-Kc
1-Kl
1-Qc
(3)-(D)ceeKa(L)lq
(3)-(D)cee(L)elQa

$C_{11}H_{11}N$

2,4-Dimethylquinoline

1-X
19(quinolinyl-3)-AA

2,6-Dimethylquinoline

1-X
19(quinolinyl-3)-AH
19(quinolinyl-4)-H
19(quinolinyl-8)-H

$C_{11}H_{11}NO_4S$

1-(L)jl
1-(L)lo

$C_{11}H_{11}N_3$

2-NbbU
2-NbbVhh
14-BH
14-HH

$C_{11}H_{12}N_2O$

Antipyrine, 89.4 mg/ml $CDCl_3$

1-(L)ln
1-(N)ln
(9)-(J)nA(N)an

Antipyrine, 7% $CDCl_3$

1-(L)ln
1-(N)ln
(9)-(J)nA(N)an

$C_{11}H_{12}N_2O_2$

2-Ckn(L)lv
3-BlKhNhh
(9)-(N)hvBcVhn

$C_{11}H_{13}N$

1-Dalv
1-Ldm

$C_{11}H_{14}O_2$

Benzoic acid, isobutylester

1-Cab
2-CaaQv
3-AABq
14-HH
14-HK

Benzoic acid, butylester

1-Bb
2-ABb
2-BaBq
2-BbQv
14-HH
14-HK

$C_{11}H_{14}O_5$

1-Ql
(9)-(B)bKa(L)kl
13-(L)bl

$C_{11}H_{15}NO$

4-Amino-N-methyl-N-phenyl-butanone

1-Jb
1-Nbv
2-BnJa
2-BjNav

Phenyl-2-methylpropylke-toxime

1-Cab
2-CaaLmv
3-AABl
13-Ml

$C_{11}H_{15}NO_2$

1-Bb
2-ABb
2-BaBq
2-BbQv
14-HK
14-HN
20-HVhh

$C_{11}H_{16}$

Pentamethylbenzene

1-Vaa
1-Vah
14-AA

1-Methyl-4-t-butylbenzene

1-Daav
1-Vhh
14-AH
14-DH

1-Methyl-3-tert-butylbenzene

1-Daav
1-Vhh
14-AD
14-AH
14-DH
14-HH

$C_{11}H_{16}$ (cont.)

n-Pentylbenzene

1-Bb
2-ABb
2-BaBb
2-BbBv
2-BbVhh
14-BH
14-HH

$C_{11}H_{16}O$

4-Methyl-2-t-butylphenol

1-Daav
1-Vhh
13-Vdh
14-AD
14-AH
14-HP

p-(1,1-Dimethylpropyl)phenol

1-Bd
1-Dabv
2-ADaav
13-Vhh

$C_{11}H_{17}N$

1-Bn
2-ANbb
2-NbbVhh

$C_{11}H_{18}N_2O_3S$

1-Bn
1-Ov
2-ANbs
14-HO
14-HS
14-NS
20-HVho

$C_{11}H_{20}O_4$

1-(C)bo
(2)-(B)c(O)c
2-(C)oo(C)oo
(3)-A(B)b(O)c
(3)-Bc(O)b(O)c

$C_{11}H_{24}$

2-Methyldecane

1-Bb
1-Cab
2-ABb
2-BaBb
2-BbBb
2-BbBc
2-BbCaa

n-Undecane, 50% CCl_4

1-Bb
2-ABb
2-BaBb
2-BbBb

Undecane, 88.8 mg/ml $CDCl_3$

1-Bb
2-ABb
2-BaBb
2-BbBb

$C_{12}H_8O$

14-HV

$C_{12}H_9NO_2$

14-HR
14-HV

$C_{12}H_9NS$

(2)-(S)lX
14-HW
20-(L)sx

$C_{12}H_{10}$

(2)-(B)xX

$C_{12}H_{10}O$

Phenylether

14-HO

m-Phenylphenol

13-Vhh

1-Acetonaphthone

1-Jv

$C_{12}H_{10}O_2$

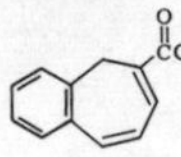

(2)-LjlVhl
(7)-(L)jlVbh
(7)-Vbh(L)hl
(9)-(L)hlKh(B)v
12-(L)bl

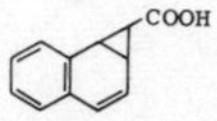

(3)-(C)cl(C)cvKh
(3)-(C)ck(C)cv(L)hl
(3)-(C)ck(C)clVhl
(7)-(C)ccV
(7)-Vch(C)cc
12-(C)cc

$C_{12}H_{10}S$

14-HH
14-HS

$C_{12}H_{10}S_2$

14-HH
14-HS

$C_{12}H_{11}N$

2-Benzylpyridine

2-VhhX
14-BH
14-HH
15-α-H
15-γ-HH

Diphenylamine

20-VhhVhh

$C_{12}H_{11}NO$

1-Jn
20-JaVhh

$C_{12}H_{11}NO_2$

1-Bq
2-AQl
9-VhhUKb

$C_{12}H_{11}NO_4$

1-(N)jv
1-Ol
(2)-(O)v(O)v
(9)-(J)nOaVhn
14-H(L)
14-H(O)

$C_{12}H_{12}$

1,3-Dimethylnaphthalene

1-X

1,5-Dimethylnaphthalene

1-X

1,6-Dimethylnaphthalene

1-X

2,7-Dimethylnaphthalene

1-X

β-Ethylnaphthalene

1-Bv
2-AX

$C_{12}H_{12}F_2N_4O_6$

1-Bq
2-AQb
2-KbLcm
3-FFLbm
14-HN
14-HR
14-RR
20-MlVhr

$C_{12}H_{12}N_2O_2$

1-Naj
(9)-(N)hvJjVhn
14-H(L)
(20)-(L)hlVhl

$C_{13}H_{20}O$ (cont.)

β-Ionone

1-(D)abl
1-Jl
1-(L)bl
(2)-(B)b(L)al
6-Ja(L)dl
6-(L)dlJa

$C_{13}H_{26}O_2$

1-Bb
1-Qb
2-ABb
2-BaBb
2-BbBb
2-BbBk
2-BbKa

$C_{14}H_7ClO_2$

14-EH
14-HH
14-H(J)

$C_{14}H_8Br_2$

19(anthroyl-1,4)-H
19(anthroyl-2,3)-HH

$C_{14}H_8Cl_2$

19(anthroyl-1,4)-H
19(anthroyl-2,3)-HH

$C_{14}H_9Cl_5$

1,1,1-Trichloro-2,2-bis-p-chlorophenylethane

3-DeeeVhhVhh

2,2-bis(p-Chlorophenyl)-1,1,1-trichloroethane

3-DeeeVhhVhh

$C_{14}H_{10}O$

Anthrone, 90.6 mg/ml $CDCl_3$

(2)-VhjVhj

Anthrone, 7% $CDCl_3$

(2)-VhjVhj
14-H(J)

$C_{14}H_{10}O_2$

14-HH
14-HJ

$C_{14}H_{10}O_3$

14-HH
14-HK

$C_{14}H_{11}BrO$

2-GJv
14-HJ
14-HV

$C_{14}H_{11}NO_5$

1-Ol
1-Ox
(2)-(O)x(O)x
(7)-(O)xX
(7)-X(O)x
19(quinolinyl-5)-O

$C_{14}H_{12}$

cis-Stilbene

7-VhhVhh
14-HH
14-HL

trans-Stilbene

6-VhhVhh

9,10-Dihydroanthracene

(2)-VbhVbh
14-BH
14-HH

9,10-Dihydrophenanthracene

(2)-(B)vVhv

$C_{14}H_{12}N_2O_2$

1-X
15-β-AH
15-β-HJ
15-γ-HH

$C_{14}H_{12}O$

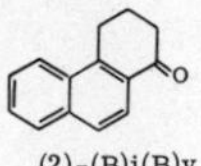

(2)-(B)j(B)v
(2)-(B)b(J)v
(2)-(B)bVjw

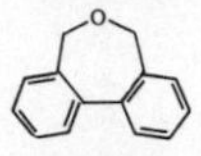

(2)-(O)bVhv

cis-Stilbeneoxide

(3)-(C)ov(O)cVhh
14-(C)H
14-HH

trans-Stilbeneoxide

(3)-(C)o(O)cVhh
14-(C)H
14-HH

$C_{14}H_{12}O_2$

Benzylbenzoate

2-QvVhh
14-HK

Diphenylacetic acid

3-KhVhhVhh
12-Cvv

$C_{14}H_{12}O_5$

1-(L)lo
1-Ox
(9)-(J)vA(O)v
16-α-H
16-β-HW

$C_{14}H_{12}S$

cis-Stilbenesulfide

(3)-(C)sv(S)cVhh
14-(C)H
14-HH

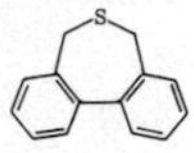

(2)-(S)bVhv

$C_{14}H_{13}N$

(2)-(B)vVhn
(20)-VbhVbh

$C_{14}H_{13}NO$

2-VhhVhh
14-BH
14-HH
14-HJ

$C_{14}H_{14}$

2-BvVhh

$C_{14}H_{14}O$

2-ObVhh
14-BH
14-HH

$C_{14}H_{14}O_4$

1-Dacp
(2)-(C)doVoo
(3)-(B)vDaap(O)v
(7)-(K)vVho
(7)-Vho(K)v
13-Daac
14-H(L)
14-H(O)

$C_{14}H_{15}NO_3$

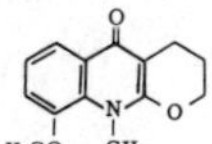

1-(N)lv
1-Ov
(2)-(B)l(B)o
(2)-(B)b(O)l
(2)-(B)b(L)jl
14-HH
14-H(J)
14-HO

$C_{14}H_{15}NO_3$(cont.)

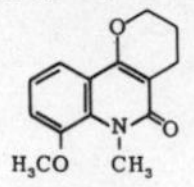

1-(N)jv
1-Ov
(2)-(B)l(B)o
(2)-(B)b(L)jl
(2)-(B)b(O)l
14-HH
14-H(L)
14-HO

$C_{14}H_{16}N_2$

1-Vhn
20-HVah

$C_{14}H_{17}NO_3$

2-CcvR
(3)-(B)bCbv(J)b
3-Br(C)bjVhh

$C_{14}H_{17}NO_4$

1-(N)jv
1-Ox
2-Bp(L)jl
2-BlP
13-Bb

$C_{14}H_{17}N_3O$

2-BxNbp
2-BnX
13-Nbb
15-α-H
15-β-BH
15-β-HH
15-γ-HH

$C_{14}H_{17}N_3O_9$

1-Kb
1-Kc
2-(C)coQa
(3)-(C)cq(C)noQa
(3)-(C)cq(N)jm(O)c
(4)-(J)n(N)cj

$C_{14}H_{20}O_2$

1-Daal
(9)-(J)lDaaa(J)l

$C_{14}H_{22}$

1-Daav
14-DH

$C_{14}H_{22}O$

1-Daab
1-Dabv
2-DaaaDaav
13-Vhh
14-DH
14-HP

$C_{14}H_{23}N$

1-Bb
2-ABb
2-BaBn
2-BbNbv

$C_{14}H_{26}O_4$

1-Bq
2-AQb
2-BbBb
2-BbBk
2-BbKb

$C_{14}H_{28}$

1-Bb
2-ABb
2-BaBb
2-BbBb
2-BbBl
2-BbLhl
5-BbHH
5-HBbH
5-HHBb

$C_{14}H_{28}O_2$

Myristic acid

1-Bb
2-ABb
2-BaBb
2-BbBb
2-BbBk
2-BbKh
12-Bb

$C_{14}H_{28}O_2$ (cont.)

Lauric acid, ethylester

1-Bb
1-Bq
2-ABb
2-AQb
2-BaBb
2-BbBb
2-BbBk
2-BbKb

$C_{14}H_{30}$

1-Bb
2-ABb
2-BaBb
2-BbBb

$C_{14}H_{30}O$

1-Bb
2-BaBb
2-BbBb
2-BbBp
2-BbP

$C_{15}H_{10}Br_2O_2$

9-JvVhhP
13-(L)lv

$C_{15}H_{11}NO_4$

1-Ov
13-Vhj
14-HH
14-H(J)
14-OP
20-HVjo

$C_{15}H_{12}$

1-X

$C_{15}H_{12}Br_4O_2$

1-Davv
13-Vgg
14-DG

$C_{15}H_{12}O$

(2)-(J)bVhv

$C_{15}H_{14}O$

1,1-Diphenylacetone

1-Jc
3-JaVhhVhh
14-CH
14-HH

Dibenzylketone

2-JbVhh

$C_{15}H_{14}O_5$

1-Ql
(2)-(L)lpVhl
(9)-VbhKa(L)kl
13-(L)bl

$C_{15}H_{14}O_6$

1-Lhl
1-Ql
(3)-(C)ll(C)oo(D)clq
(3)-(C)cd(L)hl(L)kl
(3)-(C)cd(O)c(O)l
(3)-(D)clq(L)kl(O)c
(7)-(D)cl(D)ccq
(7)-(D)ccq(C)cl
(9)-A(K)d(C)do
(9)-(O)cKa(C)cl

$C_{15}H_{16}O_6$

1-Bc
1-Ql
2-A(C)ck
(3)-Ba(C)do(K)d
(3)-(C)ll(C)oo(D)clq
(3)-(C)bk(D)clq(O)c
(3)-(C)cd(L)hl(L)kl
(3)-(C)cd(O)c(O)l
(7)-(C)cl(D)ccq
(7)-(D)ccq(C)cl
(9)-(O)cKa(C)cl

$C_{15}H_{18}N_2O_9$

1-Kb
1-Kc
2-(C)coQa
(3)-(C)cq(N)jl(O)c
(7)-(J)n(N)cj
(7)-(N)cj(J)n

formula index

$C_{18}H_{14}$ (cont.)

α-Terphenyl

14-HH
14-HV

p-Terphenyl

14-HV

$C_{18}H_{16}$

1-X
19(dihydropyrenyl-1)-HH
19(dihydropyrenyl-2)-H
19(dihydropyrenyl-4)-H

$C_{18}H_{16}N_2O_4$

2-BmMl
(2)-(O)v(O)v
4-BbVhh
14-HL
14-H(O)
14-L(O)

$C_{18}H_{16}Si$

21(silicon)-VhhVhhVhh

$C_{18}H_{17}D$

1-Lhl
1-Lll
9-ALalVhh

$C_{18}H_{18}O_4$

2-BkKb
2-QbVhh

$C_{18}H_{20}O_6$

2-CjoP
3-BpJvOv
13-Bc
14-HJ
14-JO

$C_{18}H_{30}$

1-Bb
2-ABb
2-BaBb
2-BbBb
2-BbBv
2-BbVhh
14-BH
14-HH

$C_{18}H_{34}O_2$

1-Bb
2-ABb
2-BaBb
2-BbBb
2-BbBk
2-BbBl
2-BbKh
2-BbLhl
7-BbBb
12-Bb

$C_{19}H_{14}O_2$

11-Vhh
13-Vvv
14-HH
14-HV
14-JV

$C_{19}H_{16}$

3-VhhVhhVhh

$C_{19}H_{18}N_2O_3$

1-(D)adv
1-(N)dv
(7)-(D)dnoVho
(7)-Vho(D)dno
14-HR
14-(O)R

$C_{19}H_{18}O$

1-Vhp
2-VhhX
13-Vaa
14-AB

$C_{19}H_{19}NO_4$

1-(N)bc
1-OV
(2)-(O)v(O)v
13-Vov
14-(B)H
14-(B)(O)
14-HO

$C_{19}H_{19}NO_5$

1-Bq
2-AQc
2-CknVhh
(2)-(O)v(O)v
3-BvKbNhj
14-(B)(O)

$C_{19}H_{22}O_2$

1-(D)bcj
1-(D)cll
(7)-(D)acl(J)l

$C_{19}H_{22}O_4$

3-LhlVhoVho
5-CvvHH
5-HCvvH
5-HHCvv
14-CO
14-OO

$C_{19}H_{22}O_6$

(2)-(C)cpVoo
(3)-(B)v(C)ovP
(3)-(C)bp(O)vVhh
13-(C)bc
14-OO
14-O(O)

$C_{19}H_{24}N_2O_4S_2$

1-Vhh
(2)-(B)n(B)b
(2)-(B)b(N)bs
(2)-(B)n(N)bs
14-AH
14-HS

$C_{19}H_{30}O$

1-(D)bbc
(2)-(B)b(D)acc
(2)-(D)abc(J)c

$C_{19}H_{32}O_2$

1-Bl
1-Qb
2-BbBb
2-LhlLhl
7-BaBl
7-BbBl
7-BlBa
7-BlBb
7-BlBl

$C_{19}H_{36}O_2$

1-Bb
1-Qb
2-ABb
2-BaBb
2-BbBb
2-BbBk
2-BbBl
2-BbKa
2-BbLhl
7-BbBb

$C_{20}H_8Cl_{12}$

(3)-(C)cd(C)dv(D)del
(3)-(C)cd(D)delVch

$C_{20}H_{14}O_4$

14-HH
14-HK
14-HQ

$C_{20}H_{19}NO_5$

1-(N)bb
(2)-(B)v(N)ab
(2)-(B)nVhj
(2)-(J)vVbh
(2)-(N)abVbo
(2)-(O)v(O)v
14-(B)H
14-(B)(O)
14-J(O)

$C_{20}H_{20}O$

1-Vhp
2-BxVhh
2-BvX
13-Vaa
14-AB

$C_{20}H_{20}OSi$

1-Bo
2-AOz

$C_{20}H_{21}NO_4$

1-(N)bc
1-Ov
(2)-(O)v(O)v
14-(B)(O)
14-(B)O
14-OV

$C_{20}H_{24}O_2$

17-*α*-Ethynyl estradiol

1-(D)bcd
10-(D)bdp
14-(B)P
14-(C)H
14-HP

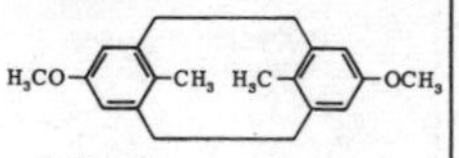

1-Ov
1-Vbb
(2)-(B)vVah
14-(B)O

$C_{20}H_{24}O_4$

1-Bb
1-(C)cv
1-Ov
2-ABv
2-BaVhh
(3)-A(C)ovVho
(3)-(C)av(O)vVhh
13-Vho

$C_{20}H_{24}O_5$

1-(D)alo
1-Ov
2-BvKh
2-BkVoo
(7)-(D)aaoVoo
(7)-Voo(D)aao

$C_{20}H_{28}O_2$

1-(D)bcc
1-Jc
(9)-(J)b(B)b(C)bc

$C_{20}H_{28}O_3S$

1-Bo
1-Daav
2-AOs
19(naphthyl-1)-D
19(naphthyl-3)-DS
19(naphthyl-5)-H
19(naphthyl-6)-DH
19(naphthyl-8)-D

$C_{20}H_{28}O_6S_2$

1-Daav
1-Os
19(naphthyl-1)-D
19(naphthyl-3)-DS

$C_{20}H_{34}O$

1-Dblp
(3)-Bb(D)abc(L)bl
5-DabpHH
5-HDabpH
5-HHDabp
(8)-(B)b(C)bd
(8)-(C)bd(B)b

$C_{20}H_{40}O$

1-Cab
1-Cbb
1-Lbl
2-LhlP
9-BpBbA

$C_{20}H_{40}O_2$

1-Bb
2-ABb
2-BaBb
2-BbBb
2-BbBk
2-BbKh

$C_{21}H_{16}O_7$

1-Kv
(2)-(C)ov(L)lv
(3)-(B)l(O)vVhh
(9)-(K)v(B)cVoq
14-(C)H
14-HQ
14-(O)Q
14-QQ

$C_{21}H_{21}NO_6$

1-(N)bc
1-Ov
(2)-(O)v(O)v
(3)-(C)qv(N)abVbh
(3)-(C)nv(Q)vVhk
14-(B)(O)
14-(C)H
14-(C)(O)
14-HO

$C_{21}H_{22}O_5$

1-Bb
1-(D)ajl
1-Lhl
2-ABb
2-BbBj
2-BbJl
6-A(L)lo
6-(L)loA
(9)-(L)dl(L)hl(L)jl
(9)-(L)llLhl(O)l
(9)-(O)l(J)d(L)ll

$C_{21}H_{23}NO_4$

1-Nab
1-Ox
2-BxNaa
2-BnX
(2)-(O)x(O)x
14-BO
14-HW
14-OW

$C_{21}H_{25}NO_4$

1-Nab
1-Ov
(2)-(O)v(O)v
14-(B)O
14-B(O)
14-OV

$C_{21}H_{26}O_3$

1-(D)bcj
1-Kv
1-Vch
14-AQ
14-BQ

$C_{21}H_{32}O$

1-(D)bcc
1-(D)bcd
10-(D)bdp
13-(D)bdt

$C_{21}H_{32}O_3$

1-(D)bcc
1-Kc
(3)-(B)b(D)abcQa

$C_{21}H_{34}O_3$

1-(D)bcc
1-Qc
13-(C)bb

$C_{21}H_{35}O$

1-(D)bcc
1-(D)bcl
1-Jl
(9)-(B)cJa(D)abc

$C_{22}H_{20}S$

2-LhlSd
5-BsHH
5-HBsH
5-HHBs

$C_{22}H_{22}S_3$

2-ScVhh
3-SbSbSb

$C_{22}H_{24}N_2O_6$

1-Kc
1-Oc
(3)-(C)cm(C)ooQa
(3)-(C)cqOa(O)c
(3)-(O)b(O)cVhh

$C_{22}H_{25}NO_6$

Colchicine

1-Jn
1-Ox
(2)-(B)v(C)ln
(2)-(B)cVhl
(3)-(B)b(L)llNhj
(9)-(J)l(C)bn(L)lv
(9)-(L)hlOa(J)l
(9)-(L)hlVbo(L)cl
14-(B)O
20-(C)blJa

$C_{22}H_{25}NO_6$ (cont.)

β-Lumicolchicine

1-Jn
(3)-(B)b(L)clNhj
(3)-(C)ll(J)l(L)cl
(3)-(C)jl(L)hl(L)lv
(9)-(C)clOa(J)c
14-(B)O
20-(C)blJa

$C_{22}H_{30}$

2-VaaVah
14-AB

$C_{22}H_{31}Cl_2N_7O_5$

1-Nahx
2-CjnVhh
(3)-Bp(C)cn(O)c
3-BvJnNhhh
(3)-(C)bo(C)cpNhj
(3)-(C)cp(N)lx(O)c
(3)-(C)cn(C)noP

$C_{23}H_{18}O_8$

1-Kv
6-(L)lvVhh
6-Vhh(L)lv
(9)-(K)vLhlVqq
14-HL
14-HQ
14-QQ

$C_{23}H_{22}O_2$

1-(D)acc
1-(D)cjl
(3)-(C)dl(D)aac(J)d
(3)-(C)dj(D)aac(L)hl
(3)-(D)ajlVhoVho
(7)-(C)cd(D)acj
(7)-(D)acj(C)cd

$C_{23}H_{22}O_2$ (cont.)

1-(D)abl
1-(L)jl
(2)-(D)aal(J)l
(3)-(L)llVhoVho
(9)-(D)aab(C)vv(L)hl
(9)-(L)clA(J)b

$C_{23}H_{23}NO_3$

1-Bq
1-Jl
1-(L)ln
2-AQl
(3)-(L)jl(L)klVhh
(20)-(L)al(L)lv

$C_{24}H_{18}O_3$

(3)-(C)cj(C)cjJv

$C_{24}H_{40}N_2$

1-(C)cn
1-(D)bcl
1-Nac
1-(N)bc
(9)-(B)c(B)c(D)abc

$C_{25}H_{31}N_3$

1-Nav
3-VhhVhhVhh
14-CH
14-HN

$C_{25}H_{38}O_5$

1-(D)bcc
1-(D)ccc
1-Jc
1-Kc
(3)-(B)b(B)cQa
(3)-(B)b(D)accJa
(3)-(B)c(D)accQa

$C_{26}H_{24}O_5$

1-(D)alo
1-Ov
(7)-(D)aaoVho
(7)-(D)aaoVoo
(7)-Vho(D)aoo
(9)-(O)vVho(J)w
14-H(O)
14-(J)(L)
14-(O)(O)

$C_{27}H_{36}N_2O_6$

1-Dbgl
(3)-Bb(D)abc(L)bl
5-DabqHH
5-HDabqH
5-HHDabq
(8)-(B)b(C)bd
(8)-(C)bd(B)b
14-KR
14-RR

$C_{27}H_{42}O_5$

1-(D)bcc
1-(D)ccj
1-Kc
1-Qb
(3)-(B)b(B)cQa

$C_{27}H_{44}O_2$

1-Cab
1-Cbc
1-(D)bcc
(2)-(C)jo(D)bcc
(2)-(D)bcc(O)c
(3)-(B)d(J)b(O)b

$C_{27}H_{46}O$

1-Cab
1-Cbc
1-(D)bcc
1-(D)bcl
(3)-(B)b(B)lP
(9)-(B)c(B)c(D)abc

$C_{28}H_{22}$

9-LhlVhhVhh

$C_{28}H_{44}$

1-(D)bcc
(3)-(B)b(B)lP
(7)-CacCac
9-Lhl(B)c(L)bb
9-Lhl(C)bd(B)b

$C_{29}H_{30}N_2$

2-NbbNbb
2-NbbVhh
14-BH
14-HH

$C_{29}H_{50}O_2$

1-Cab
1-Cbb
1-(D)bbo
1-Vao
1-Vap
2-BbBc
(2)-(B)dVao
13-Vaa

$C_{45}H_{74}O$

1-Lal
1-Lbl
2-BlLal
2-BlLhl
2-LhlP
9-BbAA
9-BbABb

$C_{57}H_{110}O_6$

1-Bb
2-ABb
2-BbKb
2-BbKc
2-CbqQb
3-BqBqQb

$Na_3HP_2O_5$

21(phosphorus)-OOX

SHIFT INDEX

1.08............1-Bd
1.08............1-Daal
1.09............1-Bv
1.09............1-(D)abb
1.09............1-(D)acc

1.10............1-Bc
1.10............1-Bv
1.10............1-Caj
1.10............1-Daal
1.10............1-Daan
1.10............1-(D)abb
1.10............1-(D)abd
1.10......... 20-BbH
1.11............1-Bl
1.11............1-Bn
1.11............1-Daal
1.12............1-Bk
1.12............1-Bn
1.12............1-Bt
1.12............1-Caj
1.12............1-Cao
1.12............1-Cbp
1.12............1-Dabj
1.12............1-(D)abn
1.12............1-Danp
~1.12............2-CaaDaaa
1.12............3-AABk
1.12............3-AABq
1.13............1-Bc
1.13............1-Bj
1.13............1-Bk
1.13............1-Bn
1.13............1-Ccp
1.13............2-ADaab
~1.13............2-BdCaa
1.13............2-BdDaaa
1.13............2-CaaDaab
1.14............1-Bn
1.14............1-Bv
1.15............1-Bt
1.15............1-Cav
1.15............1-Daan
1.15............2-BdDaaa
~1.15............2-CaaCac
1.16............1-Bv
1.16............2-ADaab
1.17............1-Bl
1.17............1-Cav
1.17............1-(C)bj
1.17............1-Cbn
~1.17............2-BcCaa
1.17............2-CabDaaa
1.17..........(3)-(C)dj(D)aac(L)hl
1.18............1-Bk
1.18............1-Bv
1.18............1-Cak
1.18............1-Can
1.18............1-Cbo
1.18............1-Cbp
1.18............1-Daao
1.18............1-Dabn
1.18............2-ADbbb
1.18............2-BbBd
1.18............2-BbDaaa
1.19............1-(D)adv
1.19............2-ADaaa

~1.2 ....2-ACcc
~1.2-1.5........2-ABb
~1.2-1.5........2-BaBn
1.20............1-Bj
1.20............1-Bo
1.20............1-Bs
1.20............1-Cap
1.20............1-Cav
1.20............1-Daab
1.20............1-(D)abl
1.20............2-ABb
1.20............2-ABc
1.20............2-ABd
1.20............2-ACaa
1.20............2-ACab
1.20............2-ACbc
1.20............2-A(D)bbb
1.20............2-BaBd
1.20............2-BaCaa
1.20............2-BaDaaa
1.20............2-BbDaaa
1.21............1-Bv
1.21............1-Ccn
1.21............2-BaBb
1.21............2-BbBb
1.21............2-BbBp
1.22............1-Bj
1.22............1-Bn
1.22............1-Bo
1.22............1-Bp
1.22............1-Bq
1.22............1-Bv
1.22............1-Caj
1.22............1-Cak
1.22............1-Cav
1.22............1-(C)bo
1.22............1-Cbv
1.22............1-Dabv
1.22±0.03 ....1-(D)bbo
1.22............1-(D)ccj
1.22............2-ABb
1.22............2-ABd
~1.22............2-ACac
1.22............2-BaBb
1.22............2-BbBb
1.22±0.03 ....2-BbBc
1.22............2-BbBv
1.22............2-BaDaac
~1.22............2-CabDaaa
1.23............1-Bj
1.23............1-Bl
1.23............1-Bo
1.23............1-Bq
1.23............1-Bv
1.23............1-Caq
1.23............1-Cav
1.23............1-Cbl
1.23............1-Cbp
1.23............1-Daav
1.23............1-(D)acc
1.23............1-(D)cll
1.23............2-ABb
1.23............2-ACab
1.23............2-BaBc
1.23............2-BbCab
1.23......... 20-DaaaH
1.23......... 21(sulfur)-Bb
1.24............1-Bd
1.24............2-ABb
1.24............2-ABc
1.24............2-ACab
1.24............2-ACal
1.24............2-BaBb
1.24............2-BbBb
1.24............2-BbBk
1.24............2-BbBl
1.24............2-BaCab
1.25............1-Bq
1.25............1-Bv
1.25............1-Cav
~1.25............1-Cbb
1.25............1-(C)bo
1.25............1-Daav
1.25............1-(D)aco
1.25............1-Dacp
1.25............2-ABa
1.25............2-ABb
~1.25............2-ABc
1.25............2-ACal
1.25............2-BaBb
1.25............2-BaBc
1.25............2-BbBb
1.25............2-BbBc
1.25............2-BbBl
1.25............2-BbBv
~1.25............2-BaCab
1.25............2-BbCaa
1.25............2-BbCan
~1.25............3-ABbBb
1.25......... 20-CabH
1.25......... 20-DaabH
1.26............1-Bq
1.26............1-(C)bc
1.26............1-Daav
1.26............1-(D)all
1.26............2-ABb
~1.26............2-ABc
~1.26............2-ACbb
1.26............2-BaBb
1.26............2-BbBb
1.26............2-BbBk
~1.26............2-BaCbb
~1.26............3-ABaBb
1.27............1-Bo
1.27............1-Bq
1.27............1-Bs
1.27............1-Bv
1.27............1-Cal
1.27............1-(C)bo
1.27............1-Daal
1.27............1-(D)acc
1.27............2-ABb
1.27±0.03 ....2-ABb
1.27............2-BaBb
1.27±0.03 ....2-BaBb
1.27............2-BbBb
1.27±0.03 ....2-BbBb
1.27............2-BbBc
1.27............2-BbBl
1.27............2-BbBp
1.27............2-BbCaa
1.27............2-DaaaDaaa
~1.27............3-ABaBb
1.28............1-Bo
1.28............1-Bq
1.28............1-Bv
1.28............1-Daap
1.28............1-(D)ajl
1.28............1-(D)akl
1.28............1-Dalv
1.28............1-Dblp
1.28............2-ABb
1.28............2-ABl
1.28............2-ACac
1.28............2-ADaal
1.28............2-BaBb
1.28............2-BbBb
1.28............2-BbBc
1.28............2-BbBd
1.28............2-BbBj
1.28............2-BbBk
1.28............2-BbBl
1.28............2-BbBn
1.28............2-BbBp
1.28............2-BbCag
1.28............2-BbDaaa
1.28............2-BbVhh
1.29............1-Bq
1.29............1-Caq
1.29............1-(C)bc
1.29............1-Daav
1.29............1-(D)abl
1.29............1-(D)abo
1.29............2-ABb
~1.29............2-ABb
1.29............2-ABc
1.29............2-ACab
1.29............2-ACbb
1.29............2-BaBb
1.29............2-BaBl
1.29............2-BbBb
1.29............2-BbBj
1.29............2-BbBk
1.29............2-BbBl
1.20............2-BbBp
1.29............2-BbBs
1.29............2-BaCal

1.30............1-Bk
1.30............1-Bq
1.30............1-Bv
1.30............1-Daav
1.30............1-(D)adv
~1.30............2-ABb
1.30............2-ABb
1.30............2-ABl
1.30............2-ACab
1.30............2-ACbb
1.30............2-A(C)bb
1.30............2-ACbl
1.30............2-BaBb
1.30............2-BaBc
1.30............2-BbBb
1.30............2-BbBc
1.30............2-BbBj
1.30............2-BbBk
1.30............2-BbCaa
1.30............2-BbCap
1.30............2-BbCbb
1.30............2-BbVhh
1.30..........(2)-(D)aab(D)aab
~1.30............3-AADaaa
~1.30............3-ABaBa
1.30............3-BaBbBp
1.31............1-Bo
1.31............1-(D)abo
1.31............2-ABb
1.31............2-ADaad
1.31............2-BaBb
~1.31............2-BaBl
1.31............2-BaBl
~1.31..........(2)-(B)b(B)b
1.31..........(2)-(B)b(B)b
1.31............2-BbBc
1.31..........(2)-(B)b(B)c
~1.31..........(2)-(B)b(B)d
1.31............2-BbBe
1.31............2-BbBl
1.31............2-BbBp
1.31............2-BbCaa
~1.31............2-(B)b(D)aab
1.32............1-Bq
1.32............1-(C)bo
1.32............1-Cbq
1.32............1-Coo
1.32............1-Daav
1.32............1-Dalp
1.32............2-ABb
1.32............2-ABl
1.32............2-ACal
1.32............2-BaBb
1.32............2-BbBb
1.32..........(2)-(B)b(B)d
1.32............2-BbBk
1.32............2-BbBl
1.32............2-BbBp
1.32............2-BbBq
1.32............2-BlCaa
1.32..........(2)-(B)b(D)bbb
~1.32............3-ABaBd
~1.32............3-ABbDaaa
1.33............1-Bq
1.33............1-Cau
1.33............1-Dabv
1.33............1-(D)adq
1.33............2-ABb
1.33............2-ABc
~1.33............2-ABl
1.33............2-ABl
~1.33............2-BaBb
1.33............2-BaBb
1.33............2-BbBb
1.33............2-BbBk
1.33............2-BbBl
1.33............2-BbBp
1.33............2-BaCap
1.34............1-Cas
1.34............1-Dabp
1.34............1-(D)ado
~1.34............2-ABb
1.34............2-ABl
~1.34............2-BaBl
1.34......... (20)-(B)b(C)ab
1.34......... 20-BbH
1.35............1-Bo
1.35............1-Bq
1.35............1-Daav
1.35............1-(D)alv
1.35............1-(D)bbp
1.35............2-BaBb
1.35............2-BbBb
1.35............2-BbBk
~1.35............2-BbBl
1.35..........(3)-(C)dl(D)aac(J)d
1.35......... 13-Daaa
1.35......... 21(sulfur)-Bb
1.36............1-Bo
~1.36............2-ABb
1.36............2-ABb
1.36............2-BaBb
1.36............2-BbBb
1.36............2-BbBp
1.36............2-BbBv
1.37............1-Bo
1.37............1-Bq
1.37............1-(C)cv
1.37............1-Dacp
1.37............1-(D)bbv
1.37............1-(D)blo
1.37............2-ABb
1.37............2-BaBb
1.37............2-BaBn
1.37............2-BbBb
1.37............2-BbBq
~1.37............3-AABb
1.38............1-Bo
1.38............1-Bq
1.38............1-Cnv
1.38............1-Cpv
1.38............2-ABb
1.38............2-ABl
1.38............2-ADbbb
1.38............2-BaBb
1.38............2-BaBl
1.38............2-BbBj
1.38............3-AABl
1.39............1-Bo

~1.4 ....(2)-(B)b(B)b
~1.4 ....(2)-(B)b(B)c
1.40............1-Bo
1.40............1-Bq
1.40............1-(C)oo
1.40............1-Daav
1.40............1-(D)bbp
1.40............2-ABb
~1.40............2-ABb
1.40............2-BaBb
~1.40............2-BaBn
1.40..........(2)-(B)b(B)c
~1.40............2-BbBj
1.40............2-BbBo
1.40..........(2)-(B)b(C)ac
~1.40............3-AABb
~1.40............3-AABl
1.41............1-Bo
1.41............1-Ckp
1.42............1-Bn
1.42............1-Cpv
1.42............1-(D)abv
1.42............1-(D)alo
1.42............1-(D)bcv
1.42............3-AACaa
1.43............1-Bs
1.43............1-Daav
1.43............2-ABp
1.43............2-A(D)abb
1.43..........(2)-(B)b(B)b
1.43..........(2)-(B)b(B)c
1.43..........(2)-(B)b(C)bc

shift index

shift index

2.52............1-Yh
2.52............1-Z
2.52............2-ANbb
2.52..........(2)-(B)b(N)bh
2.52............2-BbSh
2.52............2-BbVhh
2.52.......... 13-Bd
2.53............1-Jv
2.53............1-(N)bc
2.53............2-ASa
2.53............2-BbSh
2.53..........(2)-(C)jo(D)bcc
2.53............3-ABaVhh
2.53.......... 13-Bb
2.54............2-ABb
2.54............2-AVhh
2.54............2-BaBb
2.54............2-BbBb
2.54............2-BbBv
2.54............2-BkNhh
2.54............2-BbVhh
2.55............1-Jy
2.55............1-(N)bc
2.55............1-Sx
2.55............1-Vhu
2.55............1-X
2.55..........(2)-(B)v(N)ab
2.55............2-DaajNbb
2.55............3-AALhl
2.55..........(3)-A(B)o(J)b
2.55.......... 10-(D)bdp
2.56............2-(B)o(N)bb
2.56............2-Bp(N)bb
2.56..........(2)-(D)aan(L)bm
2.56..........(2)-(D)abb(L)lp
2.56............3-ABaVhh
2.57............1-Jv
2.57............1-(L)lm
2.57............1-(N)bc
2.57............1-X
~2.57..........(2)-(B)b(S)b
2.57............2-BaVhh
2.57............2-BbVhh
2.57..........(2)-(C)ao(J)b
2.57± 0.10 ..(3)-A(B)b(N)bh
2.57.......... 13-Bb
2.58............1-Jv
2.58............1-Kv
2.58............1-Lel
2.58............2-ALmv
2.58............2-AVhh
2.58............2-BbU
2.58............2-BaVhh
2.58.......... 13-Ba
2.59............1-Jv
2.59............1-(L)lm

2.60............2-AVhh
2.60..........(2)-(B)v(J)v
2.60..........(2)-(B)b(N)bh
2.60..........(2)-(B)c(N)bh
2.60..........(2)-(D)acl(J)l
2.60............3-AAKh
2.60..........(3)-(C)ab(D)acj(L)hl
2.61..........(2)-(B)vVhv
2.61..........(2)-(D)bcj(L)al
2.62............1-Sloo
2.62............1-X
2.62............2-ANbb
2.62..........(2)-(B)o(B)s
2.62..........(2)-(B)b(J)n
2.62............2-BkKb
2.62............2-BvKh
2.62............2-BoU
2.62..........(2)-(B)dVao
2.62............3-AALhl
2.62..........(3)-(B)b(B)bLhl
2.62.......... 10-(D)bdp
2.63............2-A(L)ln
2.63............2-ANbb
2.63............2-AVah
2.63............2-AVhh
2.63..........(2)-(B)b(L)jl
2.63............3-AALhl
2.64............1-Naz
2.64............1-Vhk
2.64............2-BkJa
2.64............2-BjKb
2.65± 0.10 ..(2)-(C)cp(C)no
2.65.......... 13-Bb
2.66............1-X
2.66............2-BcSa
2.66..........(3)-(B)bCbv(J)b
2.67............1-(L)ms
2.67..........(2)-(B)c(N)bh
2.68............1-Jl
2.68............1-Jv
2.68............2-ANbh
2.68............2-AVhh
2.68............2-BuKa
2.68± 0.02 ..(2)-(B)c(O)c
2.68............2-BbSh

2.68............2-BkU
2.68..........(2)-(D)abp(K)b
2.68............3-AAKj
2.68.......... 13-Cbb
2.69..........(2)-(B)l(J)n
2.69..........(2)-(B)j(L)ns
2.69............2-BbVhh

~2.7 ....2-BxNaa
2.70............1-Kc
2.70............1-X
2.70............2-BnJa
2.70............2-BkKb
2.70..........(2)-(B)b(L)bl
2.70..........(2)-(B)b(L)jl
2.70............2-BpNbd
2.70............2-BbNhh
2.70..........(2)-(B)d(N)ac
2.70............2-CaqKa
2.70..........(2)-(C)bo(O)c
2.70..........(2)-(D)bov(O)d
2.71..........(2)-(B)vVhv
2.72............1-Nhj
2.72............1-X
2.72..........(2)-(B)o(B)o
2.72............2-BsU
2.72..........(2)-(C)ao(O)c
2.72............3-AAU
~2.72..........(3)-(B)b(B)cU
2.72..........(3)-(C)ck(C)cv(L)hl
2.72.......... 13-Bb
2.72.......... 13-Bd
2.73............2-BpNbh
2.73..........(2)-(B)bVhn
2.73..........(2)-(D)aal(J)l
2.73............2-LhlLhl
2.74............1-(N)dv
2.74............1-Vrr
2.74............2-BvJh
2.74............3-AALal
2.74..........(3)-Ba(C)do(K)d
2.75..........(2)-Bn(L)ln
2.75............2-BbVhh
2.75............3-AAVhh
2.76..........(2)-(B)bVbh
2.76............2-BnVhh
2.76..........(2)-(D)aal(L)ln
2.77............2-AX
2.77..........(2)-(B)b(D)bkk
2.77..........(2)-(B)b(J)v
2.77............2-CaaLmv
2.77..........(2)-(C)ov(O)c
2.78............1-Nhk
2.78............1-Vhj
2.78............2-AJl
2.78............2-Bd(L)ll
2.78............3-ABaNhh
2.78.......... 13-(D)bbt
~2.79............2-BvKh
~2.79............2-BkVhh

2.80............1-(L)ms
2.80............1-X
2.80............2-BnVhh
2.80............2-LhlLhl
2.80............3-AAVhh
2.80± 0.10 ..(3)-A(B)b(O)c
2.80.......... 13-Bb
2.81............1-Nal
2.81............1-Nav
2.81..........(2)-(B)o(N)bb
2.81..........(2)-(D)akn(L)jm
2.82............1-Nhk
2.82............2-ASov
2.82............2-BqNbb
2.82..........(2)-(B)b(S)b
2.82............2-BbSooo
2.82............2-BpVhh
2.82............2-LhlSd
2.82.......... 13-Bt
2.82.......... 13-Cct
2.83............1-(N)bj
2.83..........(2)-(B)bVah
2.83............2-CknKh
2.83............3-AALal
2.83.......... 13-Bd
2.83.......... 13-Bl
2.84............2-ALmv
2.84............2-CbpNhh
2.84............3-AALal
2.86............1-Ov
2.86..........(2)-(B)bVbh
2.86............2-BkVoo
2.86..........(2)-(B)vVhv
2.86............2-CvzDeee
2.86............2-CjkKb
2.86............2-CknVhh
2.86............2-DbkpKb
2.87............1-(L)dm
2.87............1-Nav
2.87..........(2)-(B)o(N)bh
2.87............3-AAVhh

2.87.......... 10-Lal
2.88............1-Naj
2.88............2-BbJy
2.88..........(2)-(B)n(N)bh
2.88............2-BuSb
2.88..........(2)-(B)nVhj
2.88..........(2)-(C)bp(N)ch
2.88..........(2)-(C)ob(O)c
2.89............2-BvVhh
2.89............2-CjkKb
2.89............3-BdVhhZ

2.90............2-BxVhh
2.90............2-CknKh
2.90............3-AA(L)ll
2.90............3-AAVhh
2.91..........(2)-(B)bVbh
2.92............1-(N)cv
2.92............2-AVhr
2.92............2-CksKh
2.92............3-AAVhh
2.92.......... 13-(D)abb
2.93............1-Nav
2.93............2-AJy
2.93............2-BaSx
2.93............2-BqVhh
2.94............1-Naj
2.94............1-Nbv
2.94............2-BbJl
2.95............1-Nhv
2.95..........(2)-(B)l(N)bh
2.95..........(2)-(C)cpVoo
2.96............1-(N)ln
2.96..........(2)-(C)coVch
2.97............1-Nach
2.97............1-Naj
2.97............1-X
2.97............2-BvNhh
2.97............2-BjVhh
2.97..........(2)-(D)bov(O)d
2.98..........(2)-(B)b(J)b
2.98............2-BgU
2.98..........(2)-(C)ck(K)c
~2.98..........(3)-A(B)o(O)b
2.98..........(3)-(B)b(D)accJa
2.99............1-Naj
2.99............1-(N)ln
2.99............2-AJv
2.99............2-ASu

3.00..........(2)-(B)b(N)bh
3.00..........(2)-(B)b(S)boo
3.01............3-AAVah
3.02............1-(N)lv
3.02............2-BlNhh
3.02..........(2)-(B)n(N)bs
3.02............2-CksKh
3.02.......... 13-Cak
3.03............2-ANbh
3.03............2-Bp(L)jl
3.04............1-Nav
3.04............1-Nhv
3.04..........(2)-(D)aav(L)jm
3.04............3-AAVah
3.05............1-Nav
3.05............3-AAVah
3.05.......... 10-Vhh
3.05.......... 13-Bd
3.05.......... 13-Cac
3.06............1-(N)bl
3.06............1-(N)ln
3.07............1-Naj
3.07............2-BiKh
3.07..........(2)-(B)vVhn
3.07............2-CknVhh
3.07............2-DaaeVhh
~3.08............2-BbNhj
3.08............2-BbNhv
3.08..........(2)-(B)n(N)bh
3.08..........(2)-(B)n(N)bv
3.08..........(2)-(B)jVhj
3.08............2-LhlSb
3.08.......... 10-Lhl
3.08.......... 10-Vhh
3.08.......... 13-Bb
3.09............1-Nam
3.09............1-Nan
3.09............2-Ba(N)ll
3.09..........(2)-(B)b(N)bh
3.09..........(2)-(C)ov(L)lv
3.09............2-CjnVhh
3.09..........(2)-(L)lvVhv

~3.10..........(2)-(C)ac(C)cq
3.10............2-Ckn(L)lv
3.10............2-KhSb
3.10............2-NbbNbb
~3.10..........(3)-(C)bq(C)dq(L)kl
3.10.......... 13-Bb
3.12..........(2)-(C)ov(O)c
3.12.......... 13-Bb
3.12.......... 20-BbVhh

3.13..........(2)-(B)n(J)l
3.13............2-BxNbp
3.13............2-BnX
3.13............2-JaLhl
3.13.......... 13-Bt
3.14............2-CknVhh
3.15............2-LhlU
3.15..........(3)-(B)b(B)bKh
3.16............3-AASh
3.17..........(2)-(B)c(N)bh
3.17............2-KbNaa
3.17............3-AAVhp
3.17..........(3)-(C)ck(C)clVhl
3.18............2-ANbb
3.18............2-BaI
3.18............2-BbI
3.18..........(2)-(B)l(J)n
3.18..........(2)-(B)j(L)ns
3.18..........(3)-(B)c(C)bo(O)c
3.19............2-BnKh
3.19..........(2)-(L)hl(L)jl

~3.2 ....2-BnX
3.2 .... 20-CabH
3.20............1-Njv
3.20............2-AI
3.20............2-ANbs
3.20............2-BbI
~3.20..........(2)-(C)co(L)jl
3.20..........(2)-(C)coVch
3.20.......... 13-Cav
3.22............1-(N)lv
3.22............1-Od
3.22..........(2)-(B)b(N)hv
3.23............2-BbNhj
3.23............2-BpNhv
3.23............2-CkmVhh
3.23..........(2)-(J)v(J)v
3.25.......... 13-Daau
3.26............2-ANhj
3.27............1-Ob
3.27..........(2)-(B)o(L)jl
3.27............2-CanE
3.27..........(2)-(S)bVhv
3.27..........(3)-(B)b(B)bJv
3.27.......... 21(sulfur)-Vhh
3.28............1-(N)jl
3.28............2-BbNbv
3.28..........(2)-(B)o(N)bl
3.28..........(2)-(B)b(S)ooo
3.28............3-AA(L)hl
3.28..........(3)-Be(B)o(O)b
3.28.......... 13-(C)ck
3.29............2-CjnVhh
3.29.......... 20-HVhn

3.30............1-(N)jj
3.30............2-BkI
3.30............2-BvX
3.30............2-LhlNhh
3.30............2-LhlVhh
3.30.......... 13-Daan
3.30.......... 20-HVhh
3.31..........(2)-(J)bVhv
3.32..........(2)-(B)c(N)ch
3.32..........(2)-(S)bVhv
3.33............1-Oc
3.33..........(2)-(B)b(N)bs
3.33............2-Ckn(L)lv
3.33..........(2)-(C)doVoo
3.33..........(2)-(L)hl(N)bh
3.33..........(2)-(L)hlVhl
3.33............2-LhlVhq
3.34.......... 20-HVah
3.35............2-CbbP
3.35.......... 13-Bc
3.36..........(2)-(J)b(J)b
3.36..........(2)-(S)bVhv
3.37............1-(N)jj
3.37............1-(N)jl
3.37..........(2)-(B)n(N)bs
3.37..........(2)-(B)Vjw
3.37..........(2)-BxX
3.37............2-CooG
3.37............2-KbKb
3.38............2-ANbv
3.38............2-BbG
3.38............2-BbOb
3.38..........(2)-LjlVhl
3.38..........(2)-(L)lpVhl
3.38............3-CaaDaabP
3.38.......... 13-Bd
~3.38.......... 20-CabVhh
3.38.......... 20-HVhh

3.40............1-Ob
3.40............1-Oc
3.40............2-AG
3.40..........(2)-(B)b(N)aj
3.40..........(2)-(B)b(N)hj
3.40..........(2)-(B)m(N)ll

shift index

shift index

shift index